WILEY

helping teachers and students succeed together

WILEY PLUS

EPONYMS USED
IN THIS TEXT

In the life sciences, an eponym is the name of a structure, drug, or disease that is based on the name of a person. For example, you may be more familiar with the Achilles tendon than you are with its more anatomically descriptive term, the calcaneal tendon. Because eponyms remain in frequent use, this listing correlates common eponyms with their anatomical terms.

EPONYM	ANATOMICAL TERM	EPONYM	ANATOMICAL TERM
Achilles tendon	calcaneal tendon	Kupffer (KOOP-fer) cell	stellate reticuloendothelial cell
Adam's apple	thyroid cartilage		
ampulla of Vater (VA-ter)	hepatopancreatic ampulla	Leydig (LĪ-dig) cell	interstitial endocrinocyte
		loop of Henle (HEN-lē)	loop of the nephron
Bartholin's (BAR-tō-linz) gland	greater vestibular gland	Luschka's (LUSH-kaz) aperture	lateral aperture
Billroth's (BIL-rōtz) cord	splenic cord		
Bowman's (BŌ-manz) capsule	glomerular capsule	Magendie's (ma-JEN-dēz) aperture	median aperture
Bowman's (BŌ-manz) gland	olfactory gland	Meibomian (mi-BŌ-mē-an) gland	tarsal gland
Broca's (BRŌ-kaz) area	motor speech area	Meissner (MĪS-ner) corpuscle	corpuscle of touch
Brunner's (BRUN-erz) gland	duodenal gland	Merkel (MER-kel) disc	tactile disc
bundle of His (HISS)	artrioventricular (AV) bundle	Müllerian (mil-E rē-an) duct	paramesonephric duct
		organ of Corti (KOR-tē)	spiral organ
canal of Schlemm (SHLEM)	scleral venous sinus	Pacinian (pa-SIN-ē-an) corpuscle	lamellated corpuscle
circle of Willis (WIL-is)	cerebral arterial circle	Peyer's (PĪ-erz) patch	aggregated lymphatic follicle
Cooper's (KOO-perz) ligament	suspensory ligament of the breast	plexus of Auerbach (OW-er-bak)	myenteric plexus
		plexus of Meissner (MĪS-ner)	submucosal plexus
Cowper's (KOW-perz) gland	bulbourethral gland	pouch of Douglas	rectouterine pouch
crypt of Lieberkühn (LE-ber-kyūn)	intestinal gland	Purkinje (pur-KIN-jē) fiber	conduction myofiber
duct of Santorini (san′-tō-RĒ-nē)	accessory duct	Rathke's (rath-KĒZ) pouch	hypophyseal pouch
duct of Wirsung (VĒR-sung)	pancreatic duct	Ruffini (roo-FĒ-nē) corpuscle	type II cutaneous mechanoreceptor
Eustachian (yoo-STĀ-kē-an)	auditory tube		
Fallopian (fal-LŌ-pē-an) tube	uterine tube	Sertoli (ser-TŌ-lē) cell	sustentacular cell
		Skene's (SKĒNZ) gland	paraurethral gland
gland of Littré (LĒ-tra)	urethral gland	sphincter of Oddi (OD-dē)	sphincter of the hepatopancreatic ampulla
Golgi (GOL-jē) tendon organ	tendon organ		
Graafian (GRAF-ē-an) follicle	mature ovarian follicle	Volkmann's (FŌLK-manz) canal	perforating canal
Hassall's (HAS-alz) corpuscle	thymic corpuscle	Wernicke's (VER-ni-kēz) area	auditory association area
Haversian (ha-VĒR-shun) canal	central canal	Wharton's (HWAR-tunz) jelly	mucous connective tissue
Haversian (ha-VĒR-shun) system	osteon	Wolffian duct	mesonephric duct
Heimlich (HĪM-lik) maneuver	abdomial thrust maneuver	Wormian (WER-mē-an) bone	sutural bone
islet of Langerhans (LANG-er-hanz)	pancreatic islet		

Experience + Innovation

NINTH EDITION

ESSENTIALS OF ANATOMY AND PHYSIOLOGY

International Student Version

Gerard J. Tortora
Bergen Community College

Bryan Derrickson
Valencia College

John Wiley & Sons, Inc.
WILEY

Essentials of Anatomy and Physiology, Ninth Edition, is designed for courses in human anatomy and physiology or in human biology. It assumes no previous study of the human body. The ninth edition continues to offer a balanced presentation of content under the umbrella of our primary and unifying theme of homeostasis, supported by relevant discussions of disruptions to homeostasis. In addition, years of student feedback have convinced us that readers learn anatomy and physiology more readily when they remain mindful of the relationship between structure and function. As a writing team—an anatomist and a physiologist—our very different specializations offer practical advantages in fine-tuning the balance between anatomy and physiology.

Through years of collaboration with students and instructors alike, we have come to intimately understand not only the material, but also the evolving dynamics of teaching and learning A&P. We believe we bring together experience and innovation, offering a unique solution for A&P, designed to help instructors and students succeed together. From constantly evolving animations and visualizations to design based on optimal learning to lessons firmly grounded in learning outcomes, everything is designed with the goal of helping instructors like you teach in a way that inspires confidence and resilience in students and better learning outcomes.

This edition integrated with *WileyPLUS* builds students' confidence; it takes the guesswork out of studying by providing students with a clear roadmap. Students will take more initiative, so instructors can have greater impact.

On the following pages students will discover the tips and tools needed to make the most of their study time using the integrated text and media. Instructors will gain an overview of the changes to this edition and of the resources available to create dynamic classroom experiences as well as build meaningful assessment opportunities. Both students and instructors will be interested in the outstanding resources available to seamlessly link laboratory activity with lecture presentation and study time.

NOTES TO STUDENTS

Anatomy and Physiology Is a Visual Science

The challenges of learning anatomy and physiology can be complex and time-consuming. This textbook and WileyPLUS for Anatomy and Physiology have been carefully designed to maximize your time studying by simplifying the choices you make in deciding what to study, how to study it, and in assessing your understanding of the content.

Studying the figures in this book is as important as reading the narrative. The tools described here will help you understand the concepts being presented in any figure and assure you get the most out of the visuals.

1 LEGEND. Read this first. It explains what the figure is about.

2 KEY CONCEPT STATEMENT Indicated by a "key" icon, this reveals a basic idea portrayed in the figure and/or described in the related text discussion.

3 ORIENTATION DIAGRAM Added to many figures, this small diagram helps you understand the perspective from which you are viewing a particular piece of anatomical art.

4 FIGURE QUESTION Found at the bottom of each figure and accompanied by a "question mark" icon, each of these serves as a self-check to help you understand the material as you go along.

5 FUNCTIONS BOX Included with selected figures, this provides a brief summary of the functions of the anatomical structure of the system depicted.

6 MP3 DOWNLOADS In each chapter you will find that several illustrations are marked with this icon. This indicates that an audio file which narrates and discusses the important elements of that particular illustration is available. You can access these downloads on the student companion website or within WileyPLUS.

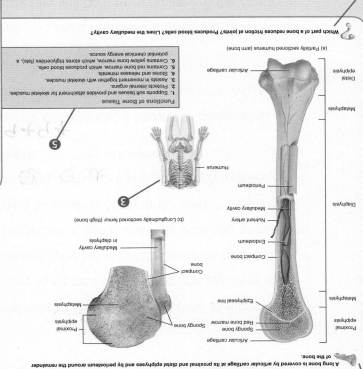

Figure 6.1 Parts of a long bone: epiphysis, metaphysis, and diaphysis. The spongy bone of the epiphysis and metaphyses contains red bone marrow, and the medullary cavity of the diaphysis contains yellow bone marrow in an adult.

A long bone is covered by articular cartilage at its proximal and distal epiphyses and by periosteum around the remainder of the bone.

(a) Partially sectioned humerus (arm bone)

(b) Longitudinally sectioned femur (thigh bone)

Articular cartilage, Distal epiphysis, Metaphysis, Diaphysis, Metaphysis, Proximal epiphysis, Periosteum, Medullary cavity, Nutrient artery, Endosteum, Compact bone, Epiphyseal line, Red bone marrow, Spongy bone, Articular cartilage, Humerus

Medullary cavity in diaphysis, Compact bone, Metaphysis, Proximal epiphysis, Spongy bone

Functions of Bone Tissue
1. Supports soft tissues and provides attachment for skeletal muscles.
2. Protects internal organs.
3. Assists in movement together with skeletal muscles.
4. Stores and releases minerals.
5. Contains red bone marrow, which produces blood cells.
6. Contains yellow bone marrow, which stores triglycerides (fats), a potential chemical energy source.

Which part of a bone reduces friction at joints? Produces blood cells? Lines the medullary cavity?

Figure 9.2 Structure of a typical multipolar neuron. Arrows indicate the direction of information flow: dendrites → cell body → axon → axon terminals → synaptic end bulbs.

The basic parts of a neuron are several dendrites, a cell body, and a single axon.

(a) Parts of a neuron

Neurofibril, Nucleus, Cytoplasm, Rough endoplasmic reticulum, Axon hillock, Mitochondrion, AXON, Axon collateral, CELL BODY, DENDRITES, Nucleus of Schwann cell, Schwann cell: Cytoplasm, Myelin sheath, Neurolemma, Node of Ranvier, Nerve impulse

(b) Motor neuron, LM 400x

Dendrite, Neuroglial cell, Nucleus, Axon, Axon terminal, Synaptic end bulb

What roles do the axon and axon terminals play in the communication of one neuron with another?

Studying physiology requires an understanding of the sequence of processes. Correlation of sequential processes in text and art is achieved through the use of special numbered lists in the narrative that correspond to numbered segments in the accompanying figure. This approach is used extensively throughout the book to lend clarity to the flow of complex processes.

Physiology of Hearing

The events involved in stimulation of hair cells by sound waves are as follows (Figure 12.14):

1 The auricle directs sound waves into the external auditory canal.

2 Sound waves striking the eardrum cause it to vibrate. The distance and speed of its movement depend on the intensity and frequency of the sound waves. More intense (louder) sounds produce larger vibrations. The eardrum vibrates slowly in response to low-frequency (low-pitched) sounds and rapidly in response to high-frequency (high-pitched) sounds.

3 The central area of the eardrum connects to the malleus, which also starts to vibrate. The vibration is transmitted from the malleus to the incus and then to the stapes.

4 As the stapes moves back and forth, it pushes the oval window in and out.

5 The movement of the oval window sets up fluid pressure waves in the perilymph of the cochlea. As the oval window bulges inward, it pushes on the perilymph of the scala vestibuli.

6 The fluid pressure waves are transmitted from the scala vestibuli to the scala tympani and eventually to the membrane covering the round window, causing it to bulge outward into the middle ear. (See **9** in the figure.)

7 As the pressure waves deform the walls of the scala vestibuli and scala tympani, they also push the vestibular membrane back and forth, creating pressure waves in the endolymph inside the cochlear duct.

8 The pressure waves in the endolymph cause the basilar membrane to vibrate, which moves the hair cells of the spiral organ against the tectorial membrane. Bending of their hairs stimulates the hair cells to release neurotransmitter molecules at synapses with sensory neurons that are part of the vestibulocochlear (VIII) nerve (see Figure 12.13b). Then, the sensory neurons generate nerve impulses that conduct along the vestibulocochlear (VIII) nerve.

Figure 12.14 Physiology of hearing shown in the right ear. The numbers correspond to the events listed in the text. The cochlea has been uncoiled in order to visualize more easily the transmission of sound waves and their subsequent distortion of the vestibular and basilar membranes of the cochlear duct.

Sound waves originate from vibrating objects.

Malleus — Incus — Stapes vibrating in oval window — Cochlea — Perilymph — Sound waves — External auditory canal — Tympanic membrane — Round window — Middle ear — Auditory tube — Cochlear duct (contains endolymph) — Vestibular membrane — Tectorial membrane — Spiral organ (organ of Corti) — Basilar membrane — Scala vestibuli — Scala tympani

? What is the function of hair cells?

WILEY PLUS There are many visual resources within **WileyPLUS** in addition to the art from your text. These can help you master the topic you are studying. Resources like animations and anatomy overviews can help you learn the connection between structure and function. Animations and anatomy overviews. Anatomy Drill and Practice lets you test your knowledge of structures with simple-to-use drag and drop labeling exercises, or fill-in-the-blank exercises. You can drill and practice on these activities using illustrations from the text, cadaver photographs, histology micrographs, or lab models.

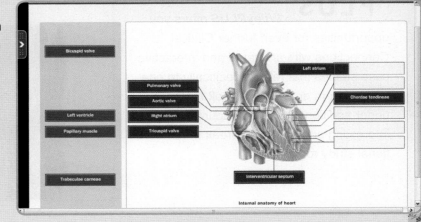

Bicuspid valve — Pulmonary valve — Aortic valve — Left ventricle — Papillary muscle — Trabeculae carneae — Right atrium — Tricuspid valve — Left atrium — Chordae tendineae — Interventricular septum

Internal anatomy of heart

NOTES TO STUDENTS

Clinical Discussions Make your Study Relevant

The relevance of the anatomy and physiology that you are studying is best understood when you make the connection between normal structure and function and what happens when the body doesn't work the way it should. Throughout the chapters of the text you will find **Clinical Connections** that introduce you to interesting clinical perspectives related to the text discussion. In addition, at the end of most of the chapters you will find the **Common Disorders** section, which includes concise discussions of major diseases and disorders. These provide answers to many of your questions about medical problems. The Medical Terminology section that follows the Common Disorders section includes selected terms dealing with both normal and pathological conditions.

CLINICAL CONNECTION | Congenital Adrenal Hyperplasia (CAH)

Congenital adrenal hyperplasia (CAH) is a group of genetic disorders in which one or more enzymes needed for the production of cortisol, or aldosterone, or both are absent. Because the cortisol level is low, secretion of ACTH by the anterior pituitary is high due to lack of negative inhibition. ACTH, in turn, stimulates growth and secretory activ adrenal cortex. As a result, both adrenal glands are enlarged. certain steps leading to synthesis of cortisol are blocked. Thus, molecules build up, and some of these are weak androgens th converted to testosterone. The result is **virilism**, or masculiniza female, virile characteristics include growth of a beard, develop much deeper voice and a masculine distribution of body hair, gro clitoris so it may resemble a penis, atrophy of the breasts, and muscularity that produces a masculine physique. In males, virilis the same characteristics as in females, plus rapid developme male sexual organs and emergence of male sexual desires. •

COMMON DISORDERS

Disorders of the endocrine system often involve either *hyposecretion* (*hypo-* = too little or under), inadequate release of a hormone, or *hypersecretion* (*hyper-* = too much or above), excessive release of a hormone. In other cases, the problem is faulty hormone receptors or an inadequate number of receptors.

Pituitary Gland Disorders

Several disorders of the anterior pituitary involve human growth hormone (hGH). Undersecretion of hGH during the growth years slows bone growth, and the epiphyseal plates close before normal height is reached. This condition is called **pituitary dwarfism**. Other organs of the body also fail to grow, and the body proportions are childlike. A dwarf has a normal-sized head and torso but small limbs; a *midget* has a proportional head, torso, and limbs.

Oversecretion of hGH during childhood results in *giantism* (*gigantism*), an abnormal increase in the length of long bones. The person grows to be very tall, but body proportions are about normal. Figure 13.15a shows identical twins; one brother developed giantism due to a pituitary tumor. Oversecretion of hGH during adulthood is called *acromegaly* (ak′-rō-MEG-a-lē). Although hGH cannot produce further lengthening of the long bones because the epiphyseal plates are already closed, the bones of the hands, feet, cheeks, and jaws thicken and other tissues enlarge (Figure 13.15b).

The most common abnormality of the posterior pituitary is *diabetes insipidus* (dī-a-BĒ-tēs in-SIP-i-dus; *diabetes* = overflow; insipidus = tasteless). This disorder is due to defects in antidiuretic

Figure 13.15 Photographs of people with various endocrine disorders.
Disorders of the endocrine system often involve hyposecretion or hypersecretion of various hormones.

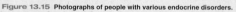

(b) Acromegaly (excess hGH during adulthood)

(c) Goiter (enlargement of thyroid gland)

WileyPLUS offers you opportunities for even further Clinical Connections with animated and interactive case studies that relate specifically to one body system or another. Look for these under additional chapter resources as an interesting and engaging break from traditional study routines.

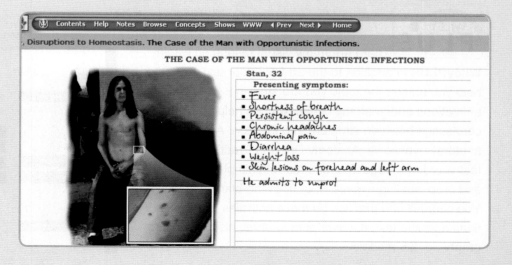

Contents   Help   Notes   Browse   Concepts   Shows   WWW   ◀ Prev   Next ▶   Home

, Disruptions to Homeostasis. The Case of the Man with Opportunistic Infections.

THE CASE OF THE MAN WITH OPPORTUNISTIC INFECTIONS

Stan, 32

Presenting symptoms:
- Fever
- Shortness of breath
- Persistent cough
- Chronic headaches
- Abdominal pain
- Diarrhea
- Weight loss
- Skin lesions on forehead and left arm

He admits to unprot

The First page of Each Chapter Sets the Stage for the Topic That You Will Study.

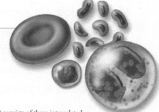

CHAPTER 14

THE CARDIOVASCULAR SYSTEM: BLOOD

The **cardiovascular system** (*cardio-* = heart; *-vascular* = blood vessels) consists of three interrelated components: blood, the heart, and blood vessels. The focus of this chapter is blood; the next two chapters will cover the heart and blood vessels, respectively.

Functionally, the cardiovascular system transports substances to and from body cells. To perform its functions, blood must circulate throughout the body. The heart serves as the pump for circulation, and blood vessels carry blood from the heart to body cells and from body cells back to the heart.

The branch of science concerned with the study of blood, blood-forming tissues, and the disorders associated with them is **hematology** (hēm-a-TOL-ō-jē; *hemo-* or *hemato-* = blood; *-logy* = study of).

Did you know? Red blood cells carry oxygen, which is vital for the energy production processes that fuel all of the body's energy needs. No oxygen, no energy, no life. Athletes use a great deal of energy to meet the increased demands of muscular contractions on the body. Imagine the emotional devastation that endurance athletes experience from a diagnosis of iron-deficiency anemia (low levels of red blood cells or hemoglobin, the oxygen-carrying molecule). While most cases of iron-deficiency anemia can be treated with iron supplements, recovery to normal iron levels can take weeks, or even months, enough time to cause the athlete to miss an entire sports season.

LOOKING BACK TO MOVE AHEAD...
Blood Tissue (Section 4.3)
Positive Feedback System (Section 1.4)
Phagocytosis (Section 3.3)

FOCUS ON WELLNESS
Athletes and Iron-deficiency Anemia

❶ **"Did You Know"** sections help you connect the content of the chapter in question to issues relevant to your everyday experiences.

❷ **Looking Back to Move Ahead** is a listing of concepts, complete with chapter references, that you will need to know to be successful. A look back at what you have already learned can help you prepare for what is to come.

The Focus on Wellness Essays, written by Barbara Brehm-Curtis of Smith College, explore current health topics and controversies. All of the essays have been revised for the new edition and 11 of them are new. They are related to the Focus on Wellness boxes within each chapter. Some of the new topics include muscle dysmorphia, shift work and cancer risk, hypertension prevention, and stimulating metabolic rate.

FOCUS ON WELLNESS

Athletes and Iron-deficiency Anemia

A male athlete in the prime of good health goes to donate blood, but is rejected as a donor because he is diagnosed with iron-deficiency anemia. This has never happened to him before, and he has given blood on several occasions. What could be the cause? He just began an intense conditioning program to prepare for his upcoming sports season. Did the heavy workouts interfere with his red blood cell production?

Sports Anemia

Probably not. This athlete may have a condition called *sports anemia.* When an athlete suddenly increases training volume, one of the first adaptations that occurs is an increase in blood volume. This increase happens within just two or three days of the start of training, and results from an increase in plasma volume. When blood is tested in these early days of training, the blood has fewer red blood cells per unit volume. As a result, both the hematocrit

and the hemoglobin measures appear to be low. But the number of red blood cells has not necessarily declined; the blood is just more dilute, a condition called *hemodilution.*

Sports anemia does not appear to impair performance. It has been proposed that the increased plasma volume and "thinner" blood may even facilitate oxygen delivery to working muscles because the blood can flow more easily. Increased blood flow also improves the ability of the blood to carry heat from the working muscles to the skin, where heat is given off to the environment (except on days that are warmer than body temperature). In any case, with no special treatment red blood cell production accelerates and soon the athlete's blood measures return to normal.

Iron-deficiency Anemia in Athletes

Unlike the athlete described previously, some athletes do develop true iron-

deficiency anemia. Multiple causes usually contribute to this condition, some of them dietary (not consuming enough foods high in iron) and some from participation in physical activity. Women of childbearing age are more prone to iron-deficiency anemia than men, since they lose blood during their menstrual cycles.

Sports may accelerate the loss of red blood cells in a few different ways. Athletes who run long distances may rupture red blood cells when the cells are pounded between the heel bone and the ground, a condition called *foot-strike hemolysis.* In addition, tiny amounts of iron may be lost in sweat. Although small, these losses may add up over time if training is demanding and sweating is heavy.

Think It Over . . . Why do athletes with true iron-deficiency anemia usually see a decline in their aerobic endurance?

NOTES TO STUDENTS

Chapter Resources Help You Focus and Review

Your book has a variety of special features that will make your time studying anatomy and physiology a more rewarding experience. These have been developed based on feedback from students—like you—who have used previous editions of the text. Their effectiveness is even further enhanced within *WileyPLUS for Anatomy and Physiology*.

Chapter Introductions set the stage for the content to come. Each chapter starts with a succinct overview of the particular system's role in maintaining homeostasis in your body, followed by an introduction to the chapter content. The Looking Back to Move Ahead references are hotlinked to the relevant sections in *WileyPLUS*, so you can link back to a particular chapter section as you are studying.

Objectives as you start to read each section of the chapter, be sure to take note of the **Objectives** at the beginning of the section to help you focus on what is important as you read.

Checkpoint Questions at the end of each section, take time to try and answer the **Checkpoint** questions placed there. If you can, then you are ready to move on. If you have trouble answering the questions, you may want to re-read the section before continuing.

Mnemonics are memory aids that can be particularly helpful when learning specific anatomical features. Mnemonics are included throughout the text—some displayed in figures, tables, or Exhibits, and some included within the text discussion. We encourage you not only to use the mnemonics provided, but also to create your own to help you learn the multitude of terms involved in your study of human anatomy.

Chapter Review and Resource Summary is a helpful table at the end each chapter that offers you a concise summary of the important concepts from the chapter and links each section to the media resources available in *WileyPLUS for Anatomy and Physiology*.

Self-Quiz Questions give you an opportunity to evaluate your understanding of the chapter as a whole.

Critical Thinking Questions are word problems that allow you to apply the concepts you have studied in the chapter to specific situations.

Mastering the Language of Anatomy and Physiology

Throughout the text we have included **Pronunciation Guides** and, sometimes, **Word Roots** for many terms that may be new to you. These appear in parentheses immediately following the new words. The Pronunciation Guides are repeated in the Glossary at the back of the book. Look at the words carefully and say them out loud several times. Learning to pronounce a new word will help you remember it and make it a useful part of your medical vocabulary. Take a few minutes to read the Pronunciation Key, found at the beginning of the Glossary at the end of this text, so it will be familiar as you encounter new words.

To provide more assistance in learning the language of anatomy, a full **Glossary** of terms with phonetic pronunciations appears at the end of the book. The basic building blocks of medical terminology—**Combining Forms, Word Roots, Prefixes,** and **Suffixes**—are listed inside the back cover, accompanied by **Eponyms**, traditional terms that include reference to a person's name, along with the current terminology.

WileyPLUS houses help for you in building your new language skills as well. The **Audio Glossary,** which is always available to you, lets you hear all these new, unfamiliar terms pronounced. Throughout the e-text, these terms can be clicked on and heard pronounced as you read. In addition, you can use the helpful **Mastering Vocabulary** program, which creates electronic flashcards of the key terms within each chapter for practice, as well as allowing you to take a self-quiz specifically on the terms introduced in each chapter.

NOTES TO INSTRUCTORS

As active teachers of the course, we recognize both the rewards and challenges in providing a strong foundation for understanding the complexities of the human body. We believe that teaching goes beyond just sharing information. *How* we share information makes all the difference—especially if, as we do, you have an increasingly diverse population of students with varying learning abilities. As we revised this text we focused on those areas that we knew could enhance understanding to provide greater impact in terms of better learning outcomes. Feedback from many of you, as well as from the students we interact with in our own classrooms, guided us in ensuring that the revisions to the text, along with the powerful *WileyPLUS for Anatomy and Physiology,* support the needs and challenges you face day-to-day in your own classrooms.

We focused on several key areas for revision: enhancing the all-important visuals, both drawings and photographs; increasing the use of Exhibits that provide a focused and functional organization of detailed content; adding some new tables and revising many others to increase their effectiveness; updating and adding clinical material that helps students relate what they are learning to the world around them; and making narrative changes aimed at increasing student engagement with—and comprehension of—the material.

The Art of Anatomy and Physiology

Illustrations throughout the text have been refined. Most of the tissue photomicrographs are new, replaced by clearer photomicrographs with high magnification blowouts. The color palette for the skulls in Chapter 6, and for the brain and spinal cord throughout the text, has been adjusted for greater impact. Illustrations in each chapter have been revised and updated to provide greater clarity and more saturated colors. Particular emphasis was placed on revising drawings of bones, joints, muscles, and blood vessels. The new **MyPlate** illustration is included in the chapter on Nutrition and Metabolism.

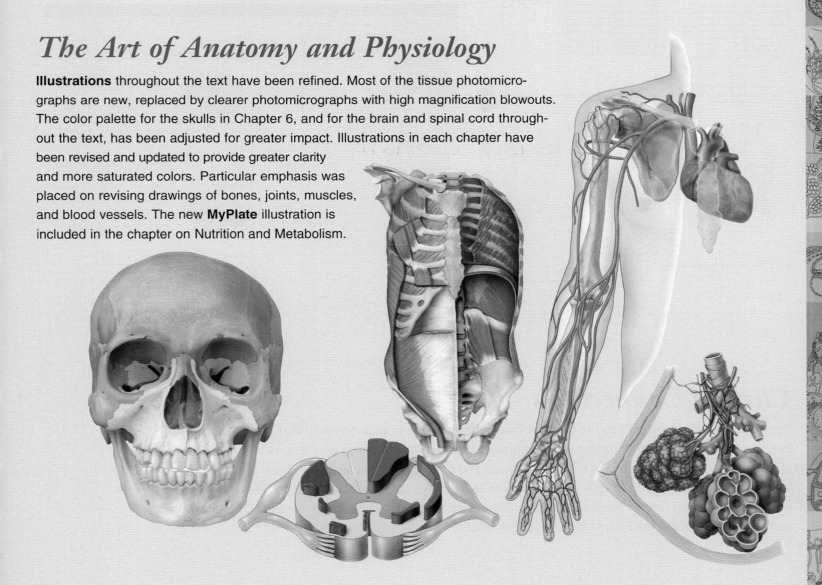

Exhibits and Tables

The use of the pedagogically designed **Exhibits** has been expanded to include tissues, the muscular system, and the axial and appendicular skeletons, providing students with simplified presentations of complex content. Many exhibits also contain Clinical Connections.

The **Tables** have been redesigned to make them easier to read and understand. The text in the tables has been rewritten and updated throughout, and photos have been added to Table 10.1, Summary of Functions of Principal Parts of the Brain.

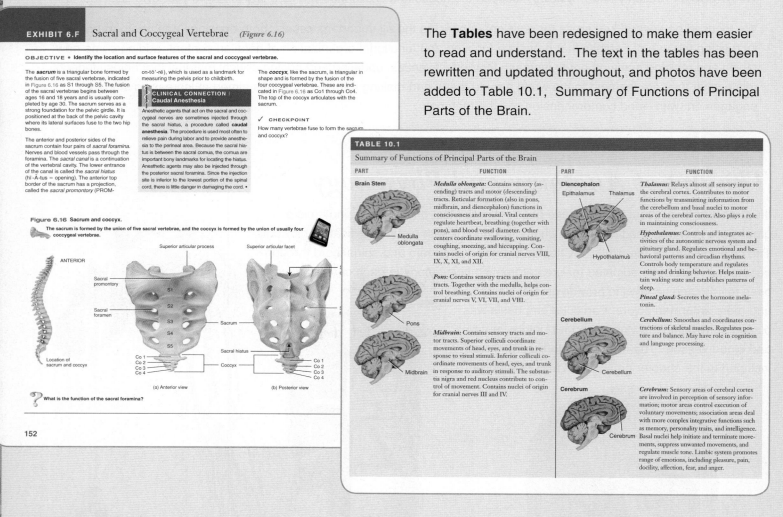

Clinical Connections

Your students are fascinated by the **Clinical Connections** to the normal anatomy and physiology that they are learning. You'll find that the text is liberally peppered with engaging discussions of a wide variety of clinical scenarios, from disease coverage to tests and procedures. As always, we have updated all of the Clinical Connections and Common Disorders to reflect the most current information. We have added several new Clinical Connections and Common Disorders to the text, including Hyperbaric Oxygenation and H1N1 Influenza in the Respiratory System chapter.

RESOURCES FOR INTEGRATING LABORATORY EXPERIENCES

Laboratory Manual for Anatomy and Physiology, 4th edition

Connie Allen and Valerie Harper

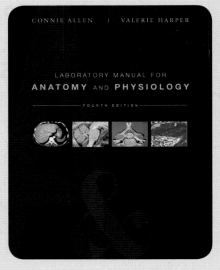

Newly revised, *Laboratory Manual for Anatomy and Physiology with WileyPLUS 5.0* engages your students in active learning and focuses on the most important concepts in A&P. Exercises reflect the multiple ways in which students learn and provide guidance for anatomical exploration and application of critical thinking to analyzing physiological processes. A concise narrative, self-contained exercises that include a wide variety of activities and question types, and two types of lab reports for each exercise keep students focused on the task at hand. Depending on your needs, a *Cat Dissection Manual* or *Fetal Pig Dissection Manual* accompanies the main text. Rich media within *WileyPLUS* further enhance the student experience and include dissection videos, animations, and illustrated drill and practice exercises with illustrations, micrographs, cadaver photos, and popular lab models. Each lab text comes with access to *PowerPhys 2.0*.

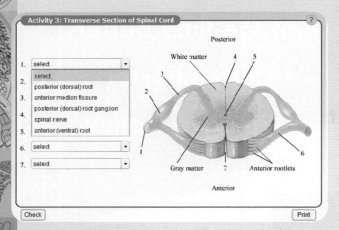

In the Lab, an Essentials Lab Manual Within WileyPLUS

This unique web-based lab manual includes 32 exercises specifically designed to help students apply their knowledge of anatomy and physiology, which can be used in a laboratory classroom or an online or hybrid laboratory section.

Essentials of Anatomy and Physiology Laboratory Manual

Connie Allen and Valerie Harper

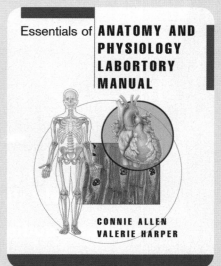

This brief manual has been designed specifically to focus on the needs of a one-semester course and to minimize the expense to the student of an expanded full-color manual for use in a lab. The clear and concise presentation of very hands-on activities and experiments enhances students' ability to both visualize anatomical structures and understand key physiological topics. *The Cat Dissection Laboratory Guide* and *Fetal Pig Dissection Laboratory Guide* are available to package at no additional cost with the main laboratory manual or as stand-alone dissection guides, depending on your dissection needs.

RESOURCES FOR INTEGRATING LABORATORY EXPERIENCES

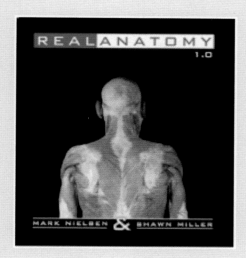

Real Anatomy

Mark Nielsen and Shawn Miller

Real Anatomy is 3-D imaging software that allows you to dissect through multiple layers of a three-dimensional real human body to study and learn the anatomical structures of all body systems.

- Dissect through up to 40 layers of the body and discover the relationships of the structures to the whole.
- Rotate the body, as well as major organs, to view the image from multiple perspectives .
- Use a built-in zoom feature to get a closer look at detail.
- A unique approach to highlighting and labeling structures does not obscure the real anatomy in view.

Interactions: Exploring the Functions of the Human Body 3.0

Thomas Lancraft and Frances Frierson

Interactions 3.0 is the most complete program of interactive animations and activities available for anatomy and physiology. A series of modules encompassing all body systems focuses on a review of anatomy, the examination of physiological processes using animations and interactive exercises, and clinical correlations to enhance student understanding. At the heart of **Interactions** is a focus on core principles—*homeostasis*, *communication*, *energy flow*, *fluid flow*, and *boundaries*—that underscore the key relationships between structure and function as well as interrelationships between systems. It is the reinforcement of these fundamental organizing principles that sets this series apart from others. **Interactions** is available on DVD, web-based, or fully integrated within *WileyPLUS*.

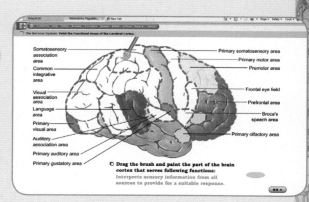

Gerard J. Tortora

is Professor of Biology and former Biology Coordinator at Bergen Community College in Paramus, New Jersey, where he teaches human anatomy and physiology as well as microbiology. He received his bachelor's degree in biology from Fairleigh Dickinson University and his master's degree in science education from Montclair State College. He is a member of many professional organizations, including the Human Anatomy and Physiology Society (HAPS), American Society of Microbiology (ASM), American Association for the Advancement of Science (AAAS), National Education Association (NEA), and Metropolitan Association of College and University Biologists (MACUB).

Above all, Jerry is devoted to his students and their aspirations. In recognition of this commitment, Jerry was the recipient of MACUB's 1992 President's Memorial Award. In 1996, he received a National Institute for Staff and Organizational Development (NISOD) excellence award from the University of Texas and was selected to represent Bergen Community College in a campaign to increase awareness of the contributions of community colleges to higher education.

Jerry is the author of several best-selling science textbooks and laboratory manuals, a calling that often requires an additional 40 hours per week beyond his teaching responsibilities. Nevertheless, he still makes time for four or five weekly aerobic workouts that include biking and running. He also enjoys attending college basketball and professional hockey games and performances at the Metropolitan Opera House.

To my mother, Angelina M. Tortora.
(August 20, 1913–August 14, 2010).
Her love, guidance, faith, support, and example continue to be
the cornerstone of my personal and professional life. **G.J.T.**

Bryan Derrickson

is Professor of Biology at Valencia College in Orlando, Florida, where he teaches human anatomy and physiology as well as general biology and human sexuality. He received his bachelor's degree in biology from Morehouse College and his doctorate in cell biology from Duke University. Bryan's study at Duke was in the Physiology Division within the Department of Cell Biology, so while his degree is in cell biology, his training focused on physiology. At Valencia, he frequently serves on faculty hiring committees. He has served as a member of the Faculty Senate, which is the governing body of the college, and as a member of the Teaching and Learning Academy, which sets the standards for the acquisition of tenure by faculty members. Nationally, he is a member of the Human Anatomy and Physiology Society (HAPS) and the National Association of Biology Teachers (NABT).

Bryan has always wanted to teach. Inspired by several biology professors while in college, he decided to pursue physiology with an eye to teaching at the college level. He is completely dedicated to the success of his students. He particularly enjoys the challenges of his diverse student population, in terms of their age, ethnicity, and academic ability, and he finds being able to reach all of them, despite their differences, a rewarding experience. His students continually recognize Bryan's efforts and care by nominating him for a campus award known as the "Valencia Professor Who Makes Valencia a Better Place to Start." Bryan has received this award three times.

To my family: Rosalind, Hurley, Cherie, and Robb.
Your support and motivation have been invaluable. **B.H.D.**

ACKNOWLEDGMENTS

A talented group of educators has contributed to the high quality of the diverse resource materials that accompany this text. We wish to acknowledge each and thank them for their expertise and work. Thanks to Charles Benton, Madison Area Technical College; Glen Borchert, Heartland Community College; Mona Fam, Passaic County Community College; Melissa Greene, Northwest Mississippi Community College; Brian Kipps, Grand Valley Community College; Deborah Lawson, Maryville University; Gisele Nasr, Brevard Community College and Susan Puglisi, Norwalk Community College. Thanks to Barbara Brehm-Curtis, Smith College, for writing our Focus on Wellness Essays. Thanks to Connie Allen, Edison College and Valerie Harper, Mesa State College, for creating and coordinating the laboratory exercises within *WileyPlus*.

This beautiful textbook would not be possible without the talent and skill of several outstanding medical illustrators. Kevin Sommerville has contributed many illustrations for us over numerous editions. For this edition, many new drawings are the work of his talented hands. We so value the long relationship we have with Kevin. We welcome a new illustrator to our "team," John Gibb. John is responsible for all of the new skeletal art and the outstanding new muscle illustrations. We thank the artists of Imagineering Media Services for all they do to enhance the visuals within this text. Mark Nielsen and Shawn Miller of the University of Utah have our gratitude for excellent dissections in the cadaver photographs they provided as well as the many new histological photomicrographs.

Our hats are off to everyone at Wiley. We enjoy collaborating with this enthusiastic, dedicated, and talented team of publishing professionals. Our thanks to the entire team— Bonnie Roesch, Executive Editor; Karen Trost, Development Editor; Lorraina Raccuia, Project Editor; Christina Picciano, Editorial Assistant; Sandra Dumas, Senior Production Editor; Suzanne Ingrao, Outside Production Manager; Hilary Newman, Photo Manager; Anna Melhorn, Senior Illustration Editor; Madelyn Lesure, Cover and Text Designer; Linda Muriello, Senior Media Editor; Clay Stone, Executive Marketing Manager; Lana Barskaya, Media Project Manager; and Lucy Parkinson, Senior Marketing Assistant.

Finally, we would also like to thank all those who have corresponded with us to offer feedback on the usefulness of this text. Your input helps so much in revising. We particularly want to express our gratitude to the following reviewers who took the time to read and evaluate the draft manuscripts prior to the production of this edition:

Mark Baguma-Nibasheka
Frostburg State University

Michelle Baragona
Northeast Mississippi Community College

Arthur R. Buckley
University of Cincinnati

Rebekah Bunting
Montgomery Community College

Janice Burger
Brevard Community College- Palm Bay

Ruth Ebeling
Biola University

Colin Everhart
Wake Technical Community College

Richard Foreman
Darton College

Dean Furbish
Wake Technical Community College

Roberto Gonzales
Northwest Vista College

Amy Goode
Illinois Central College

Matt Lee
Diablo Valley College

Claire Leonard
William Paterson University

George Spiegel
College of Southern Maryland

Yong Tang
Frontrange Community College

Randall L. Tracy
Worcester State University

We would like to invite all readers and users of the book to continue the tradition of sending comments and suggestions to us so that we can include them in the next edition.

Gerard J. Tortora
Department of Science and Health, S229
Bergen Community College
400 Paramus Road
Paramus, NJ 07652

Bryan Derrickson
Department of Science, PO Box 3028
Valencia Community College
Orlando, Fl 32802
bderrickson@valenciacc.edu

BRIEF CONTENTS

CONTENTS

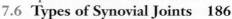

CHAPTER 10

CENTRAL NERVOUS SYSTEM, SPINAL NERVES, AND CRANIAL NERVES 271

CHAPTER 11

AUTONOMIC NERVOUS SYSTEM 301

CHAPTER 12

SOMATIC SENSES AND SPECIAL SENSES 314

CHAPTER 13

THE ENDOCRINE SYSTEM 346

CHAPTER 14

THE CARDIOVASCULAR SYSTEM: BLOOD 377

CHAPTER 15

THE CARDIOVASCULAR SYSTEM: HEART 397

CHAPTER 22

FLUID, ELECTROLYTE, AND ACID–BASE BALANCE 588

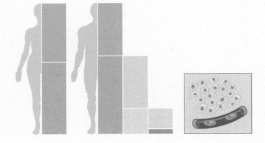

CHAPTER 23

THE REPRODUCTIVE SYSTEMS 602

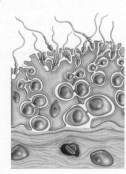

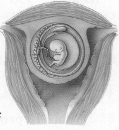

CHAPTER 1

ORGANIZATION OF THE HUMAN BODY

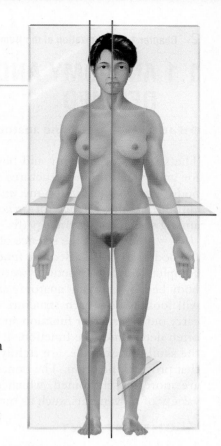

You are beginning a fascinating exploration of the human body in which you'll learn how it is organized and how it functions. First you will be introduced to the scientific disciplines of anatomy and physiology; we'll consider the levels of organization that characterize living things and the properties that all living things share. Then, we will examine how the body is constantly regulating its internal environment. This ceaseless process, called homeostasis, is a major theme in every chapter of this book. We will also discuss how the various individual systems that compose the human body cooperate with one another to maintain the health of the body as a whole. Finally, we will establish a basic vocabulary that allows us to speak about the body in a way that is understood by scientists and health-care professionals alike.

Did you know? *The body's ability to maintain homeostasis (stable conditions) gives it tremendous healing power. As you learn about the many ways your body regulates its physiological processes, you will gain a deeper*

understanding of how your behavior can help or hinder homeostasis and health. Your body's homeostasis is affected by the air you breathe, the food you eat, and even by the thoughts you think. Smoking cigarettes, drinking too much alcohol, and overeating challenge homeostasis and disrupt health. On the other hand, getting enough sleep, engaging in physical activity every day, and consuming plenty of fruits and vegetables supports the body's efforts to stay healthy.

FOCUS ON WELLNESS
Health and Wellness—Homeostasis Is the Basis

1

1.1 ANATOMY AND PHYSIOLOGY DEFINED

OBJECTIVE • **Define anatomy and physiology.**

The sciences of anatomy and physiology are the foundation for understanding the structures and functions of the human body. *Anatomy* (a-NAT-ō-mē; *ana-* = up; *-tomy* = process of cutting) is the science of *structure* and the relationships among structures. *Physiology* (fiz′-ē-OL-ō-jē; *physio-* = nature, *-logy* = study of) is the science of body *functions*, that is, how the body parts work. Because function can never be separated completely from structure, we can understand the human body best by studying anatomy and physiology together. We will look at how each structure of the body is designed to carry out a particular function and how the structure of a part often determines the functions it can perform. The bones of the skull, for example, are tightly joined to form a rigid case that protects the brain. The bones of the fingers, by contrast, are more loosely joined, which enables them to perform a variety of movements, such as turning the pages of this book.

✓ CHECKPOINT

1. What is the basic difference between anatomy and physiology?

2. Give your own example of how the structure of a part of the body is related to its function.

1.2 LEVELS OF ORGANIZATION AND BODY SYSTEMS

OBJECTIVES • **Describe the structural organization of the human body.**

• **Outline the body systems and explain how they relate to one another.**

The structures of the human body are organized into several levels, similar to the way letters of the alphabet, words, sentences, paragraphs, and so on are organized. Listed here, from smallest to largest, are the six levels of organization of the human body: chemical, cellular, tissue, organ, system, and organismal (Figure 1.1).

① The *chemical level* includes *atoms*, the smallest units of matter that participate in chemical reactions, and *molecules*, two or more atoms joined together. Atoms and molecules can be compared to letters of the alphabet. Certain atoms, such as carbon (C), hydrogen (H), oxygen (O), nitrogen (N), phosphorus (P), and others, are essential for maintaining life. Familiar examples of molecules found in the body are DNA (deoxyribonucleic acid), the genetic material passed on from one generation to another; hemo-globin, which carries oxygen in the blood; glucose, commonly known as blood sugar; and vitamins, which are needed for a variety of chemical processes. Chapters 2 and 20 focus on the chemical level of organization.

② Molecules combine to form structures at the next level of organization—the *cellular level*. *Cells* are the basic structural and functional units of an organism. Just as words are the smallest elements of language, cells are the smallest living units in the human body. Among the many types of cells in your body are muscle cells, nerve cells, and blood cells. Figure 1.1 shows a smooth muscle cell, one of three different kinds of muscle cells in your body. As you will see in Chapter 3, cells contain specialized structures called *organelles*, such as the nucleus, mitochondria, and lysosomes, that perform specific functions.

③ The *tissue level* is the next level of structural organization. *Tissues* are groups of cells and the materials surrounding them that work together to perform a particular function. Cells join together to form tissues similar to the way words are put together to form sentences. The four basic types of tissue in your body are *epithelial tissue*, *connective tissue*, *muscular tissue*, and *nervous tissue*. The similarities and differences among the different types of tissues are the focus of Chapter 4. Note in Figure 1.1 that smooth muscle tissue consists of tightly packed smooth muscle cells.

④ At the *organ level*, different kinds of tissues join together to form body structures. *Organs* usually have a recognizable shape, are composed of two or more different types of tissues, and have specific functions. Tissues join together to form organs similar to the way sentences are put together to form paragraphs. Examples of organs are the stomach, heart, liver, lungs, and brain. Figure 1.1 shows several tissues that make up the stomach. The *serous membrane* is a layer around the outside of the stomach that protects it and reduces friction when the stomach moves and rubs against other organs. Underneath the serous membrane are the *smooth muscle tissue layers*, which contract to churn and mix food and push it on to the next digestive organ, the small intestine. The innermost lining of the stomach is an *epithelial tissue layer*, which contributes fluid and chemicals that aid digestion.

⑤ The next level of structural organization in the body is the *system level*. A *system* consists of related organs that have a common function. Organs join together to form systems similar to the way paragraphs are put together to form chapters. The example shown in Figure 1.1 is the digestive system, which breaks down and absorbs molecules in food. In the chapters that follow, we will explore the anatomy and physiology of each of the body systems. Table 1.1 introduces the components and functions of these systems. As you study the body systems, you will discover how they work together to maintain health, protect you from disease, and allow for reproduction of the species.

Figure 1.1 Levels of structural organization in the human body.

The levels of structural organization are the chemical, cellular, tissue, organ, system, and organismal.

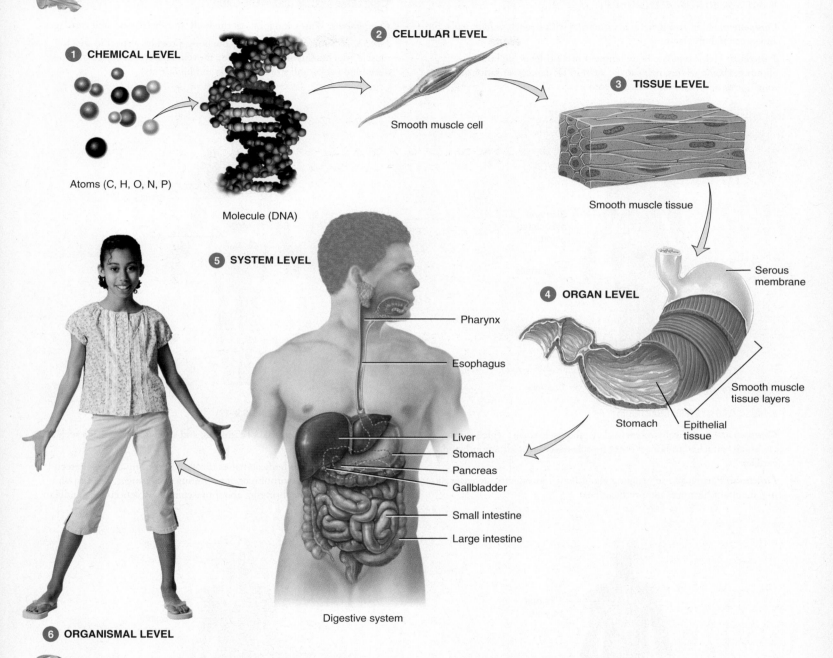

1 CHEMICAL LEVEL

Atoms (C, H, O, N, P)

Molecule (DNA)

2 CELLULAR LEVEL

Smooth muscle cell

3 TISSUE LEVEL

Smooth muscle tissue

4 ORGAN LEVEL

Serous membrane

Smooth muscle tissue layers

Stomach

Epithelial tissue

5 SYSTEM LEVEL

Pharynx

Esophagus

Liver

Stomach

Pancreas

Gallbladder

Small intestine

Large intestine

Digestive system

6 ORGANISMAL LEVEL

Which level of structural organization usually has a recognizable shape and is composed of two or more different types of tissues that have a specific function?

6 The *organismal level* is the largest level of organization. All of the systems of the body combine to make up an *organism* (OR-ga-nizm), that is, one human being. Systems join together to form an organism similar to the way chapters are put together to form a book.

✓ **CHECKPOINT**

3. Define the following terms: atom, molecule, cell, tissue, organ, system, and organism.

4. Referring to Table 1.1, which body systems help eliminate wastes?

TABLE 1.1

Components and Functions of the Eleven Principal Systems of the Human Body

1. INTEGUMENTARY SYSTEM (CHAPTER 5)

Components: Skin and structures associated with it, such as hair, nails, and sweat and oil glands

Functions: Helps regulate body temperature; protects the body; eliminates some wastes; helps make vitamin D; detects sensations such as touch, pressure, pain, warmth, and cold

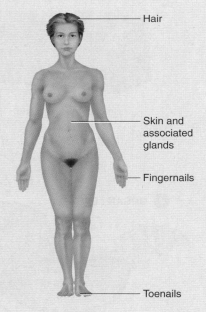

Hair

Skin and associated glands

Fingernails

Toenails

2. SKELETAL SYSTEM (CHAPTERS 6 AND 7)

Components: Bones and joints of the body and their associated cartilages

Functions: Supports and protects the body, provides a specific area for muscle attachment, assists with body movements, stores cells that produce blood cells, and stores minerals and lipids (fats)

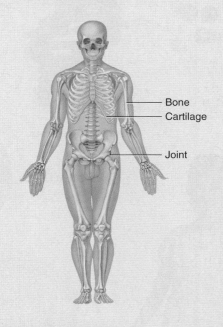

Bone

Cartilage

Joint

3. MUSCULAR SYSTEM (CHAPTER 8)

Components: Specifically refers to skeletal muscle tissue, which is muscle usually attached to bones (other muscle tissues include smooth and cardiac)

Functions: Participates in bringing about body movements such as walking; maintains posture; and produces heat

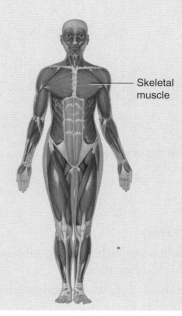

Skeletal muscle

4. NERVOUS SYSTEM (CHAPTERS 9–12)

Components: Brain, spinal cord, nerves, and special sense organs such as the eyes and ears

Functions: Regulates body activities through nerve impulses by detecting changes in the environment, interpreting the changes, and responding to the changes by bringing about muscular contractions or glandular secretions

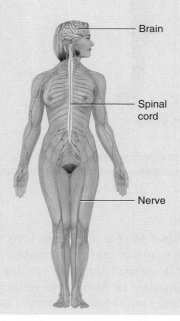

Brain

Spinal cord

Nerve

5. ENDOCRINE SYSTEM (CHAPTER 13)

Components: All glands and tissues that produce chemical regulators of body functions, called hormones

Functions: Regulates body activities through hormones transported by the blood to various target organs

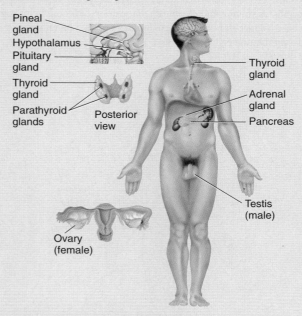

6. CARDIOVASCULAR SYSTEM (CHAPTERS 14–16)

Components: Blood, heart, and blood vessels

Functions: Heart pumps blood through blood vessels; blood carries oxygen and nutrients to cells and carbon dioxide and wastes away from cells, and helps regulate acidity, temperature, and water content of body fluids; blood components help defend against disease and mend damaged blood vessels

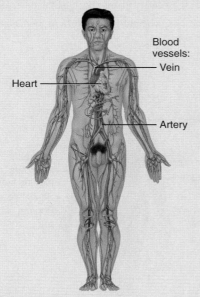

7. LYMPHATIC SYSTEM AND IMMUNITY (CHAPTER 17)

Components: Lymphatic fluid (lymph) and vessels; spleen, thymus, lymph nodes, and tonsils; cells that carry out immune responses (B cells, T cells, and others)

Functions: Returns proteins and fluid to blood; carries lipids from gastrointestinal tract to blood; contains sites of maturation and proliferation of B cells and T cells that protect against disease-causing microbes

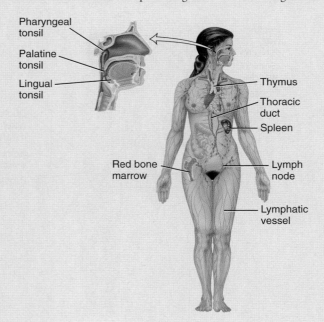

8. RESPIRATORY SYSTEM (CHAPTER 18)

Components: Lungs and air passageways such as the pharynx (throat), larynx (voice box), trachea (windpipe), and bronchial tubes leading into and out of them

Functions: Transfers oxygen from inhaled air to blood and carbon dioxide from blood to exhaled air; helps regulate acidity of body fluids; air flowing out of lungs through vocal cords produces sounds

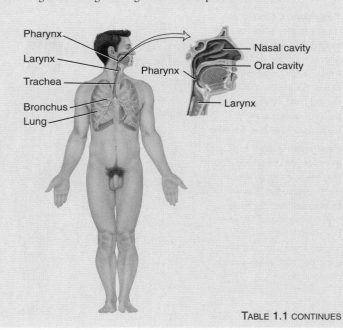

TABLE 1.1 CONTINUES

TABLE 1.1 CONTINUED

Components and Functions of the Eleven Principal Systems of the Human Body

9. DIGESTIVE SYSTEM (CHAPTER 19)

Components: Organs of gastrointestinal tract, including the mouth, pharynx (throat), esophagus, stomach, small and large intestines, rectum, and anus; also includes accessory digestive organs that assist in digestive processes, such as the salivary glands, liver, gallbladder, and pancreas

Functions: Physical and chemical breakdown of food; absorbs nutrients; eliminates solid wastes

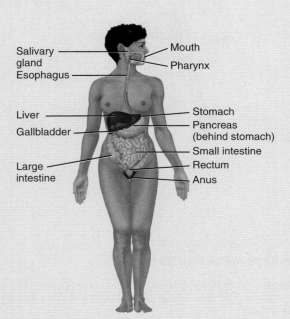

10. URINARY SYSTEM (CHAPTER 21)

Components: Kidneys, ureters, urinary bladder, and urethra

Functions: Produces, stores, and eliminates urine; eliminates wastes and regulates volume and chemical composition of blood; helps regulate acid–base balance of body fluids; maintains body's mineral balance; helps regulate red blood cell production

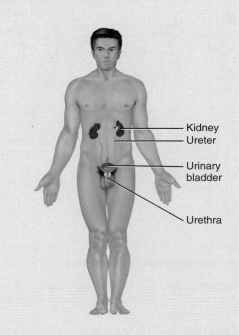

11. REPRODUCTIVE SYSTEMS (CHAPTER 23)

Components: Gonads (testes in males and ovaries in females) and associated organs: uterine (fallopian) tubes, uterus, and vagina in females, and epididymis, ductus (vas) deferens, and penis in males; also, mammary glands in females

Functions: Gonads produce gametes (sperm or oocytes) that unite to form a new organism and release hormones that regulate reproduction and other body processes; associated organs transport and store gametes, mammary glands produce milk

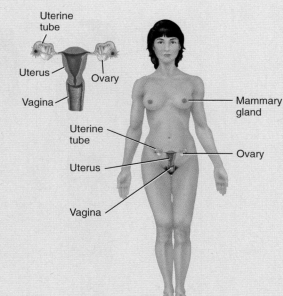

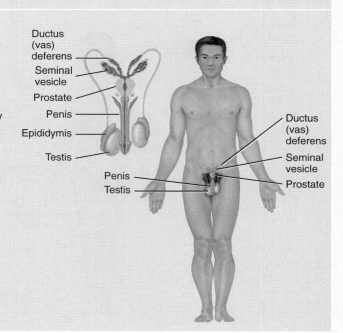

1.3 LIFE PROCESSES

OBJECTIVE • **Define the important life processes of humans.**

All living organisms have certain characteristics that set them apart from nonliving things. The following are six important life processes of humans:

1. *Metabolism* (me-TAB-ō-lizm) is the sum of all the chemical processes that occur in the body. It includes the breakdown of large, complex molecules into smaller, simpler ones and the building up of complex molecules from smaller, simpler ones. For example, proteins in food are split into amino acids. The amino acids are the building blocks that can then be used to build new proteins that make up muscles and bones.

2. *Responsiveness* is the body's ability to detect and respond to changes in its internal (inside the body) or external (outside the body) environment. Different cells in the body detect different sorts of changes and respond in characteristic ways. Nerve cells respond to changes in the environment by generating electrical signals, known as nerve impulses. Muscle cells respond to nerve impulses by contracting, which generates force to move body parts.

3. *Movement* includes motion of the whole body, individual organs, single cells, and even tiny organelles inside cells. For example, the coordinated action of several muscles and bones enables you to move your body from one place to another by walking or running. After you eat a meal that contains fats, your gallbladder (an organ) contracts and squirts bile into the gastrointestinal tract to help in the digestion of fats. When a body tissue is damaged or infected, certain white blood cells move from the blood into the affected tissue to help clean up and repair the area. And inside individual cells, various cell components move from one position to another to carry out their functions.

4. *Growth* is an increase in body size. It may be due to an increase in (1) the size of existing cells, (2) the number of cells, or (3) the amount of material surrounding cells.

5. *Differentiation* (dif'-er-en-shē-Ā-shun) is the process whereby unspecialized cells become specialized cells. Specialized cells differ in structure and function from the unspecialized cells that gave rise to them. For example, specialized red blood cells and several types of white blood cells differentiate from the same unspecialized cells in bone marrow. Similarly, a single fertilized egg cell undergoes tremendous differentiation to develop into a unique individual who is similar to, yet quite different from, either of the parents.

6. *Reproduction* (rē-prō-DUK-shun) refers to either (1) the formation of new cells for growth, repair, or replacement or (2) the production of a new individual.

Although not all of these processes are occurring in cells throughout the body all of the time, when they cease to occur properly cell death may occur. When cell death is extensive and leads to organ failure, the result is death of the organism.

⚕ CLINICAL CONNECTION | Autopsy

An **autopsy** (AW-top-sē = seeing with one's own eyes) is a *postmortem* (after death) examination of the body and dissection of its internal organs to confirm or determine the cause of death. An autopsy can uncover the existence of diseases not detected during life, determine the extent of injuries, and explain how those injuries may have contributed to a person's death. It also may provide more information about a disease, assist in the accumulation of statistical data, and educate health-care students. An autopsy can also reveal conditions that may affect offspring or siblings (such as congenital heart defects). Sometimes an autopsy is legally required, such as during a criminal investigation. It may also be useful in resolving disputes between beneficiaries and insurance companies about the cause of death. •

✓ CHECKPOINT

5. What types of movement can occur in the human body?

1.4 HOMEOSTASIS: MAINTAINING LIMITS

OBJECTIVES • **Define homeostasis and explain its importance.**
• **Describe the components of a feedback system.**
• **Compare the operation of negative and positive feedback systems.**
• **Distinguish between symptoms and signs of a disease.**

The trillions of cells of the human body need relatively stable conditions to function effectively and contribute to the survival of the body as a whole. The maintenance of relatively stable conditions is called *homeostasis* (hō'-mē-ō-STĀ-sis; *homeo-* = sameness; *-stasis* = standing still). Homeostasis ensures that the body's internal environment remains constant despite changes inside and outside the body. A large part of the internal environment consists of the fluid surrounding body cells, called *interstitial fluid* (in'-ter-STISH-al).

Each body system contributes to homeostasis in some way. For instance, in the cardiovascular system, alternating contraction and relaxation of the heart propels blood throughout the body's blood vessels. As blood flows through the blood capillaries, the smallest blood vessels, nutrients and oxygen move into interstitial fluid and wastes move into the blood. Cells, in turn, remove nutrients and oxygen from and release their wastes into interstitial fluid. Homeostasis is *dynamic*; that is, it can change over a narrow range that is compatible with maintaining cellular life processes. For example, the level of glucose in the blood is maintained within a narrow range. It normally does not fall too low between meals or rise too high even after eating a high-glucose meal. The brain needs a steady supply of glucose to keep functioning—a low blood glucose level may lead to unconsciousness or even death. A prolonged high blood glucose level, by contrast, can damage blood vessels and cause excessive loss of water in the urine.

Control of Homeostasis: Feedback Systems

Fortunately, every body structure, from cells to systems, has one or more homeostatic devices that work to keep the internal environment within normal limits. The homeostatic mechanisms of the body are mainly under the control of two systems, the nervous system and the endocrine system. The nervous system detects changes from the balanced state and sends messages in the form of *nerve impulses* to organs that can counteract the change. For example, when body temperature rises, nerve impulses cause sweat glands to release more sweat, which cools the body as it evaporates. The endocrine system corrects changes by secreting molecules called *hormones* into the blood. Hormones affect specific body cells, where they cause responses that restore homeostasis. For example, the hormone insulin reduces blood glucose level when it is too high. Nerve impulses typically cause rapid corrections; hormones usually work more slowly.

Homeostasis is maintained by means of many feedback systems. A ***feedback system*** or *feedback loop* is a cycle of events in which a condition in the body is continually monitored, evaluated, changed, remonitored, reevaluated, and so on. Each monitored condition, such as body temperature, blood pressure, or blood glucose level, is termed a *controlled condition*. Any disruption that causes a change in a controlled condition is called a *stimulus*. Some stimuli come from the external environment, such as intense heat or lack of oxygen. Others originate in the internal environment, such as a blood glucose level that is too low. Homeostatic imbalances may also occur due to psychological stresses in our social environment—the demands of work and school, for example. In most cases, the disruption of homeostasis is mild and temporary, and the responses of body cells quickly restore balance in the internal environment. In other cases, the disruption of homeostasis may be intense and prolonged, as in poisoning, overexposure to temperature extremes, severe infection, or death of a loved one.

Three basic components make up a feedback system: a receptor, a control center, and an effector (Figure 1.2).

1. A ***receptor*** is a body structure that monitors changes in a controlled condition and sends information called the *input* to a control center. Input is in the form of nerve impulses or chemical signals. Nerve endings in the skin that sense temperature are one of the hundreds of different kinds of receptors in the body.

2. A ***control center*** in the body, for example, the brain, sets the range of values within which a controlled condition should be maintained, evaluates the input it receives from receptors, and generates output commands when they are needed. *Output* is information, in the form of nerve impulses or chemical signals, that is relayed from the control center to an effector.

3. An ***effector*** is a body structure that receives output from the control center and produces a *response* that changes the controlled condition. Nearly every organ or tissue in the body can behave as an effector. For example, when

Figure 1.2 Parts of a feedback system. The dashed return arrow to the right symbolizes negative feedback.

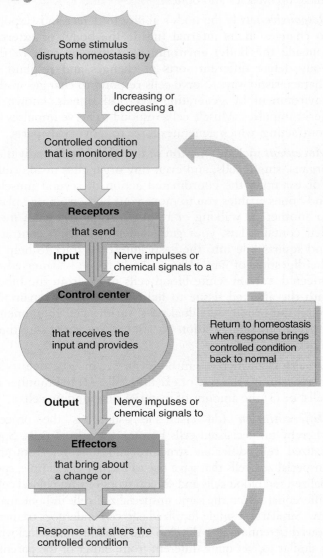

The three basic elements of a feedback system are the receptor, control center, and effector.

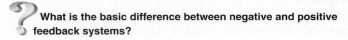

What is the basic difference between negative and positive feedback systems?

your body temperature drops sharply, your brain (control center) sends nerve impulses to your skeletal muscles (effectors) that cause you to shiver, which generates heat and raises your temperature.

Feedback systems can be classified as either negative feedback systems or positive feedback systems.

Negative Feedback Systems

A **negative feedback system** *reverses* a change in a controlled condition. Consider one negative feedback system that helps regulate blood pressure. *Blood pressure (BP)* is the force exerted by blood as it presses against the walls of blood vessels. When the heart beats faster or harder, BP increases. If a stimulus causes BP (controlled condition) to rise, the following sequence of events occurs (Figure 1.3). The higher pressure is detected by *baroreceptors*, pressure-sensitive nerve cells located in the walls of certain blood vessels (the receptors). The baroreceptors send nerve impulses (input) to the brain (control center), which interprets the impulses and responds by sending nerve impulses (output) to the heart (the effector). Heart rate decreases, which causes BP to decrease (response). This sequence of events returns the controlled condition—blood pressure—to normal, and homeostasis is restored. This is a negative feedback system because the activity of the effector produces a result, a drop in BP, that reverses the effect of the stimulus. Negative feedback systems tend to regulate conditions in the body that are held fairly stable over long periods, such as BP, blood glucose level, and body temperature.

Positive Feedback Systems

A **positive feedback system** *strengthens* a change in a controlled condition. Normal positive feedback systems tend to reinforce conditions that don't happen very often, such as childbirth, ovulation, and blood clotting. Because a positive feedback system continually reinforces a change in a controlled condition, it must be shut off by some event outside the system. If the action of a positive feedback system isn't stopped, it can "run away" and produce life-threatening changes in the body.

Homeostasis and Disease

As long as all of the body's controlled conditions remain within certain narrow limits, body cells function efficiently, homeostasis is maintained, and the body stays healthy. Should one or more components of the body lose their ability to contribute to homeostasis, however, the normal balance among all of the body's processes may be disturbed. If the homeostatic imbalance is moderate, a disorder or disease may occur; if it is severe, death may result.

A **disorder** is any abnormality of structure and/or function. **Disease** is a more specific term for an illness characterized by a recognizable set of symptoms and signs. **Symptoms** are *subjective* changes in body functions that are not apparent to an observer,

Figure 1.3 Homeostasis of blood pressure by a negative feedback system. Note that the response is fed back into the system, and the system continues to lower blood pressure until there is a return to normal blood pressure (homeostasis).

If the response reverses a change in a controlled condition, a system is operating by negative feedback.

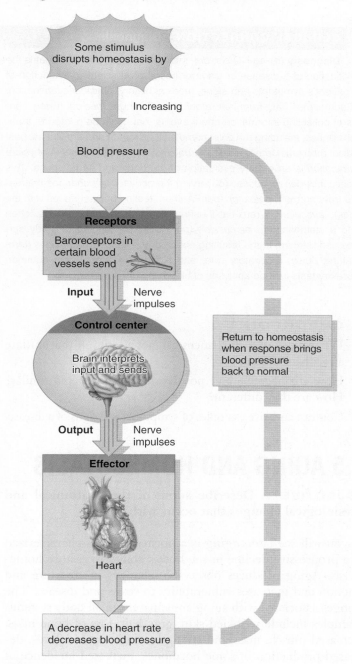

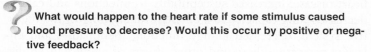
What would happen to the heart rate if some stimulus caused blood pressure to decrease? Would this occur by positive or negative feedback?

for example, headache or nausea. *Signs* are *objective* changes that a clinician can observe and measure, such as bleeding, swelling, vomiting, diarrhea, fever, a rash, or paralysis. Specific diseases alter body structure and function in characteristic ways, usually producing a recognizable set of symptoms and signs.

CLINICAL CONNECTION | Diagnosis

Diagnosis (dī-ag-NŌ-sis; *dia-* through; *-gnosis* = knowledge) is the identification of a disease or disorder based on a scientific evaluation of the patient's symptoms and signs, medical history, physical examination, and sometimes data from laboratory tests. Taking a *medical history* consists of collecting information about events that might be related to a patient's illness, including the chief complaint, history of present illness, past medical problems, family medical problems, and social history. A *physical examination* is an orderly evaluation of the body and its functions. This process includes *inspection* (observing the body for any changes that deviate from normal), *palpation* (pal-PĀ-shun; feeling body surfaces with the hands), *auscultation* (aus-cul-TĀ-shun; listening to body sounds, often using a stethoscope), *percussion* (pur-KUSH-un; tapping on body surfaces and listening to the resulting echo), and *measuring vital signs* (temperature, pulse, respiratory rate, and blood pressure). Some common laboratory tests include analyses of blood and urine. •

✓ CHECKPOINT

6. What types of disturbances can act as stimuli that initiate a feedback system?

7. How are negative and positive feedback systems similar? How are they different?

8. Contrast and give examples of symptoms and signs of a disease.

1.5 AGING AND HOMEOSTASIS

OBJECTIVE • **Describe some of the anatomical and physiological changes that occur with aging.**

As you will see later, *aging* is a normal process characterized by a progressive decline in the body's ability to restore homeostasis. Aging produces observable changes in structure and function and increases vulnerability to stress and disease. The changes associated with aging are apparent in all body systems. Examples include wrinkled skin, gray hair, loss of bone mass, decreased muscle mass and strength, diminished reflexes, decreased production of some hormones, increased incidence of heart disease, increased susceptibility to infections and cancer, decreased lung capacity, less efficient functioning of the digestive system, decreased kidney function, menopause, and enlarged prostate. These and other effects of aging will be discussed in detail in later chapters.

✓ CHECKPOINT

9. What are some of the signs of aging?

1.6 ANATOMICAL TERMS

OBJECTIVES • **Describe the anatomical position.**

• **Identify the major regions of the body and relate the common names to the corresponding anatomical terms for various parts of the body.**

• **Define the directional terms and the anatomical planes and sections used to locate parts of the human body.**

The language of anatomy and physiology is very precise. When describing where the wrist is located, is it correct to say "the wrist is above the fingers"? This description is true if your arms are at your sides. But if you hold your hands up above your head, your fingers would be above your wrists. To prevent this kind of confusion, scientists and health-care professionals refer to one standard anatomical position and use a special vocabulary for relating body parts to one another.

In the study of anatomy, descriptions of any part of the human body assume that the body is in a specific stance called the ***anatomical position*** (an′-a-TOM-i-kal). In the anatomical position, the subject stands erect facing the observer, with the head level and the eyes facing forward. The feet are flat on the floor and directed forward, and the arms are at the sides with the palms turned forward (Figure 1.4). In the anatomical position, the body is upright. Two terms describe a reclining body. If the body is lying face down, it is in the ***prone*** position. If the body is lying face up, it is in the ***supine*** position.

Names of Body Regions

The human body is divided into several major regions that can be identified externally. These are the head, neck, trunk, upper limbs, and lower limbs (Figure 1.4). The ***head*** consists of the skull and face. The *skull* is the part of the head that encloses and protects the brain, and the *face* is the front portion of the head that includes the eyes, nose, mouth, forehead, cheeks, and chin. The ***neck*** supports the head and attaches it to the trunk. The ***trunk*** consists of the chest, abdomen, and pelvis. Each ***upper limb*** is attached to the trunk and consists of the shoulder, armpit, arm (portion of the limb from the shoulder to the elbow), forearm (portion of the limb from the elbow to the wrist), wrist, and hand. Each ***lower limb*** is also attached to the trunk and consists of the buttock, thigh (portion of the limb from the hip to the knee), leg (portion of the limb from the knee to the ankle), ankle, and foot. The *groin* is the area on the front surface of the body, marked by a crease on each side, where the trunk attaches to the thighs.

In Figure 1.4, the corresponding anatomical name for each part of the body appears in parentheses next to the common

Figure 1.4 The anatomical position. The common names and corresponding anatomical terms (in parentheses) indicate specific body regions. For example, the head is the cephalic region.

In the anatomical position, the subject stands erect facing the observer, with the head level and the eyes facing forward. The feet are flat on the floor and directed forward, and the arms are at the sides with the palms facing forward.

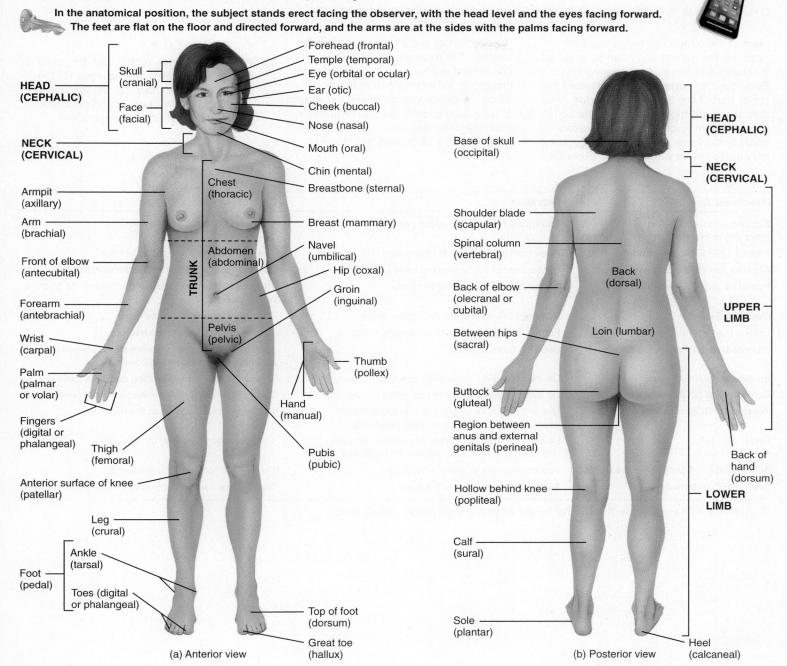

(a) Anterior view

(b) Posterior view

Where is a plantar wart located?

name. For example, if you receive a tetanus shot in your *buttock*, it is a *gluteal* injection. The anatomical name of a body part is based on a Greek or Latin word or "root" for the same part or area. The Latin word for armpit is *axilla* (ak-SIL-a), for example, and thus one of the nerves passing within the armpit is named the axillary nerve. You will learn more about the word roots of anatomical and physiological terms as you read this book.

Directional Terms

To locate various body structures, anatomists use specific *directional terms*, words that describe the position of one body part relative to another. Several directional terms can be grouped in pairs that have opposite meanings, for example, anterior (front) and posterior (back). Study Exhibit 1.1 and Figure 1.5 to determine, among other things, whether your stomach is superior to your lungs.

EXHIBIT 1.1 Directional Terms *(Figure 1.5)*

OBJECTIVE • **Define each directional term used to describe the human body.**

Most of the directional terms used to describe the human body can be grouped into pairs that have opposite meanings. For example, *superior* means toward the upper part of the body, and *inferior* means toward the lower part of the body. It is important to understand that directional terms have *relative* meanings; they only make sense when used to describe the position of one structure relative to another. For example, your knee is superior to your ankle, even though both are located in the inferior half of the body. Study the directional terms and the example of how each is used. As you read each example, refer to Figure 1.5 to see the location of the structures mentioned.

✓ **CHECKPOINT**

Which directional terms can be used to specify the relationships between (1) the elbow and the shoulder, (2) the left and right shoulders, (3) the sternum and the humerus, and (4) the heart and the diaphragm?

Directional Term	Definition	Example of Use
Superior (soo′-PĒR-ē-or) (**cephalic** or **cranial**)	Toward the head, or the upper part of a structure	The heart is superior to the liver.
Inferior (in′-FĒR-ē-or) (**caudal**)	Away from the head, or the lower part of a structure	The stomach is inferior to the lungs.
Anterior (an-TĒR-ē-or) (**ventral**)	Nearer to or at the front of the body	The sternum (breastbone) is anterior to heart.
Posterior (pos-TĒR-ē-or) (**dorsal**)	Nearer to or at the back of the body	The esophagus (food tube) is posterior to the trachea (windpipe).
Medial (MĒ-dē-al)	Nearer to the midline* or midsagittal plane	The ulna is medial to the radius.
Lateral (LAT-er-al)	Farther from the midline or midsagittal plane	The lungs are lateral to the heart.
Intermediate (in′-ter-MĒ-dē-at)	Between two structures	The transverse colon is intermediate between the ascending and descending colons.
Ipsilateral (ip-si-LAT-er-al)	On the same side of the body as another structure	The gallbladder and ascending colon are ipsilateral.
Contralateral (CON-tra-lat-er-al)	On the opposite side of the body from another structure	The ascending and descending colons are contralateral.
Proximal (PROK-si-mal)	Nearer to the attachment of a limb to the trunk; nearer to the point of origin or the beginning	The humerus is proximal to the radius.
Distal (DIS-tal)	Farther from the attachment of a limb to the trunk; farther from the point of origin or the beginning	The phalanges are distal to the carpals.
Superficial (soo′-per-FISH-al) (**external**)	Toward or on the surface of the body	The ribs are superficial to the lungs.
Deep (DĒP) (**internal**)	Away from the surface of the body	The ribs are deep to the skin of the chest and back.

*The **midline** is an imaginary vertical line that divides the body into equal right and left sides.

Figure 1.5 Directional terms.

Directional terms precisely locate various parts of the body in relation to one another.

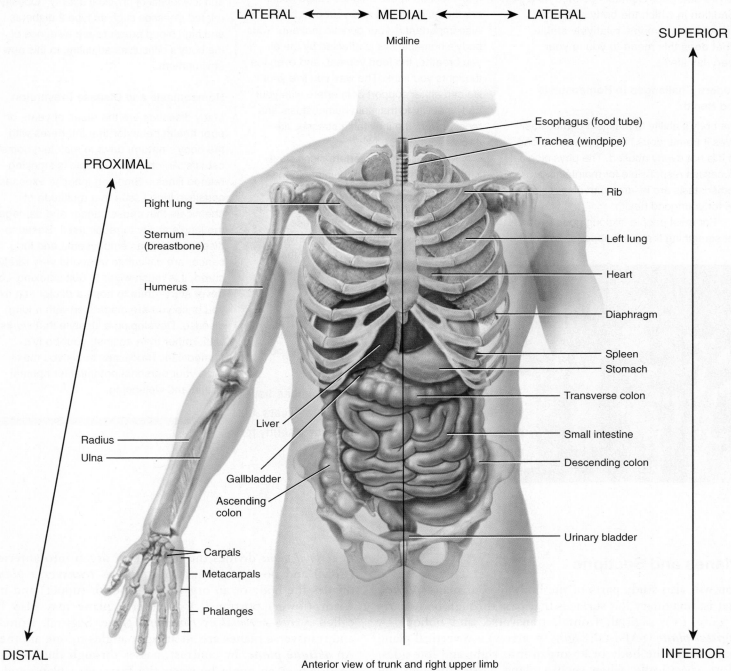

LATERAL ←——→ MEDIAL ←——→ LATERAL

Midline

SUPERIOR

PROXIMAL

Esophagus (food tube)

Trachea (windpipe)

Right lung

Rib

Sternum (breastbone)

Left lung

Humerus

Heart

Diaphragm

Spleen

Stomach

Transverse colon

Liver

Small intestine

Radius

Ulna

Descending colon

Gallbladder

Ascending colon

Urinary bladder

Carpals

Metacarpals

Phalanges

DISTAL

Anterior view of trunk and right upper limb

INFERIOR

Is the radius proximal to the humerus? Is the esophagus anterior to the trachea? Are the ribs superficial to the lungs? Is the urinary bladder medial to the ascending colon? Is the sternum lateral to the descending colon?

Health and Wellness—Homeostasis Is the Basis

You've seen *homeostasis* defined as a condition in which the body's internal environment remains relatively stable. What does this mean to you in your everyday life?

Modern Challenges to Homeostasis and Health

The body's ability to maintain homeostasis gives it tremendous healing power as long as it is not overly abused. The physiological processes responsible for maintaining homeostasis are in large part also responsible for your good health.

For most people, lifelong good health is not something that just happens. Two of the many factors in this balance called health are the environment and your own behavior. Also important is your genetic makeup. Your body's homeostasis is affected by the air you breathe, the food you eat, and even the thoughts you think. The way you live your life can either support or interfere with your body's ability to maintain homeostasis and recover from the inevitable stresses life throws your way.

Modern life can interfere significantly with your body's drive to maintain homeostasis. Homeostatic mechanisms in humans have evolved over hundreds of thousands of years. Until recently, these mechanisms worked well for the environment in which early humans found themselves. Archaeologists who have studied our early ancestors have discovered several differences between the health of these early humans and humans living today, especially in the rates of chronic diseases that plague modern people. For example, obesity was uncommon, yet today obesity rates are very high. Homeostatic processes are not equipped to deal with high intakes of excess calories and sugary, processed foods, and low levels of physical activity. Obesity-related illnesses such as type 2 diabetes and high blood pressure are evidence of the body's difficulties adjusting to this new environment.

Homeostasis and Disease Prevention

Many diseases are the result of years of poor health behavior that interferes with the body's natural drive to maintain homeostasis. An obvious example is smoking-related illness. Smoking tobacco exposes sensitive lung tissue to a multitude of chemicals that cause cancer and damage the lung's ability to repair itself. Because diseases such as emphysema and lung cancer are difficult to treat and very rarely cured, it is much wiser to quit smoking—or never start—than to hope a doctor can fix you once you are diagnosed with a lung disease. Developing a lifestyle that works with, rather than against, your body's homeostatic processes helps you maximize your personal potential for optimal health and well-being.

Think It Over . . . **What health habits have you developed over the past several years to prevent disease or enhance your body's ability to maintain health and homeostasis?**

Planes and Sections

You will also study parts of the body in four major **planes**, that is, imaginary flat surfaces that pass through body parts (Figure 1.6): sagittal, frontal, transverse, and oblique. A **sagittal plane** (SAJ-i-tal; *sagitt-* = arrow) is a vertical plane that divides the body or an organ into right and left sides. More specifically, when such a plane passes through the midline of the body or organ and divides it into *equal* right and left sides, it is called a **midsagittal plane**. If the sagittal plane does not pass through the midline but instead divides the body or an organ into *unequal* right and left sides, it is called a **parasagittal plane** (*para-* = near). A **frontal plane** or *coronal plane* divides the body or an organ into anterior (front) and posterior (back) portions. A **transverse plane** divides the body or an organ into superior (upper) and inferior (lower) portions. A transverse plane may also be called a *cross-sectional* or *horizontal plane*. Sagittal, frontal, and transverse planes are all at right angles to one another. An **oblique plane**, by contrast, passes through the body or an organ at an angle between the transverse plane and a sagittal plane or between the transverse plane and the frontal plane.

When you study a body region, you will often view it in section. A **section** is a cut of the body or an organ made along one of the planes just described. It is important to know the

Figure 1.6 Planes through the human body.

 Frontal, transverse, sagittal, and oblique planes divide the body in specific ways.

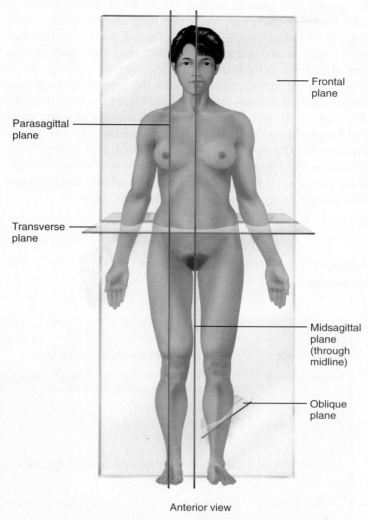

Anterior view

 Which plane divides the heart into anterior and posterior portions?

Figure 1.7 Planes and sections through different parts of the brain. The diagrams (left) show the planes, and the photographs (right) show the resulting sections. (**Note:** The "View" arrows in the diagrams indicate the direction from which each section is viewed. This aid is used throughout the book to indicate viewing perspective.)

Planes divide the body in various ways to produce sections.

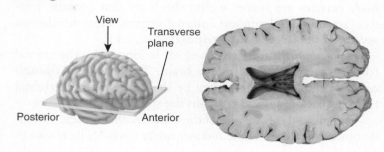

Transverse section

(a)

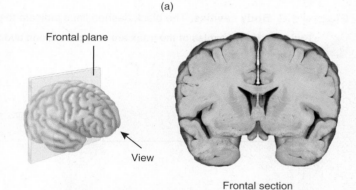

Frontal section

(b)

Midsagittal section

(c)

Which plane divides the brain into equal right and left sides?

plane of the section so you can understand the anatomical relationship of one part to another. Figure 1.7 indicates how three different sections—a *transverse (cross) section*, a *frontal section*, and a *midsagittal section*—provide different views of the brain.

✓ CHECKPOINT

10. Describe the anatomical position and explain why it is used.

11. Locate each region on your own body, and then identify it by its common name and the corresponding anatomical descriptive form.

12. For each directional term listed in Exhibit 1.1, provide your own example.

13. What are the various planes that may be passed through the body? Explain how each divides the body.

1.7 BODY CAVITIES

OBJECTIVES • **Describe the principal body cavities and the organs they contain.**

• **Explain why the abdominopelvic cavity is divided into regions and quadrants.**

Body cavities are spaces within the body that contain, protect, separate, and support internal organs. Here we discuss several of the larger body cavities (Figure 1.8).

The *cranial cavity* (KRĀ-nē-al) is formed by the cranial (skull) bones and contains the brain. The *vertebral (spinal) canal* (VER-te-bral) is formed by the bones of the vertebral column (backbone) and contains the spinal cord.

The major body cavities of the trunk are the thoracic and abdominopelvic cavities. The *thoracic cavity* (thor-AS-ik; *thorac-* =

chest) is the chest cavity. Within the thoracic cavity are three smaller cavities: the *pericardial cavity* (per′-i-KAR-dē-al; *peri-* = around; *-cardial* = heart), a fluid-filled space that surrounds the heart, and two *pleural cavities* (PLOOR-al; *pleur-* = rib or side), each of which surrounds one lung and contains a small amount of fluid (Figure 1.9). The central portion of the thoracic cavity is an anatomical region called the *mediastinum* (mē′-dē-a-STĪ-num; *media-* = middle; *-stinum* = partition). It is between the lungs, extending from the sternum (breastbone) to the vertebral column (backbone), and from the first rib to the diaphragm (Figure 1.9), and contains all thoracic organs except the lungs themselves. Among the structures in the mediastinum are the heart, esophagus, trachea, and several large blood vessels. The *diaphragm* (DĪ-a-fram = partition or wall) is a dome-shaped muscle that powers breathing and separates the thoracic cavity from the abdominopelvic cavity.

Figure 1.8 Body cavities. The black dashed lines indicate the border between the abdominal and pelvic cavities.

🔑 **The major body cavities of the trunk are the thoracic and abdominopelvic cavities.**

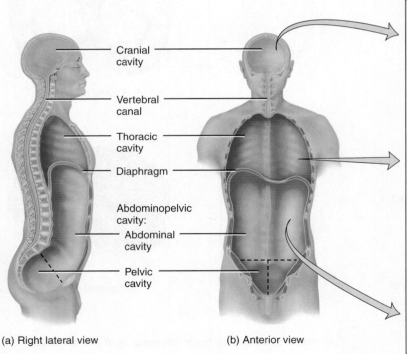

(a) Right lateral view   (b) Anterior view

CAVITY	COMMENTS
Cranial cavity	Formed by cranial bones and contains brain
Vertebral canal	Formed by vertebral column and contains spinal cord and the beginnings of spinal nerves
Thoracic cavity *	Chest cavity; contains pleural and pericardial cavities and mediastinum
Pleural cavity	Each surrounds a lung; the serous membrane of each pleural cavity is the pleura
Pericardial cavity	Surrounds the heart; the serous membrane of the pericardial cavity is the pericardium
Mediastinum	Central portion of thoracic cavity between the lungs; extends from sternum to vertebral column and from first rib to diaphragm; contains heart, thymus, esophagus, trachea, and several large blood vessels
Abdominopelvic cavity	Subdivided into abdominal and pelvic cavities
Abdominal cavity	Contains stomach, spleen, liver, gallbladder, small intestine, and most of large intestine; the serous membrane of the abdominal cavity is the peritoneum
Pelvic cavity	Contains urinary bladder, portions of large intestine, and internal organs of reproduction

* See Figure 1.9 for details of the thoracic cavity.

❓ **In which cavities are the following organs located: urinary bladder, stomach, heart, small intestine, lungs, internal female reproductive organs, thymus, spleen, liver? Use the following symbols for your responses: T = thoracic cavity, A = abdominal cavity, or P = pelvic cavity.**

Figure 1.9 The thoracic cavity. The dashed lines indicate the borders of the mediastinum. **Note:** When transverse sections are viewed inferiorly (from below), the anterior aspect of the body appears on top and the left side of the body appears on the right side of illustration. Notice that the pericardial cavity surrounds the heart, and that the pleural cavities surround the lungs.

 The mediastinum is the anatomical region medial to the lungs that extends from the sternum to the vertebral column and from the first rib to the diaphragm.

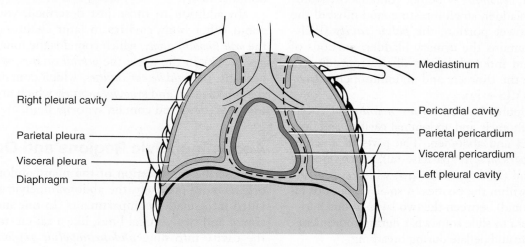

(a) Anterior view of thoracic cavity

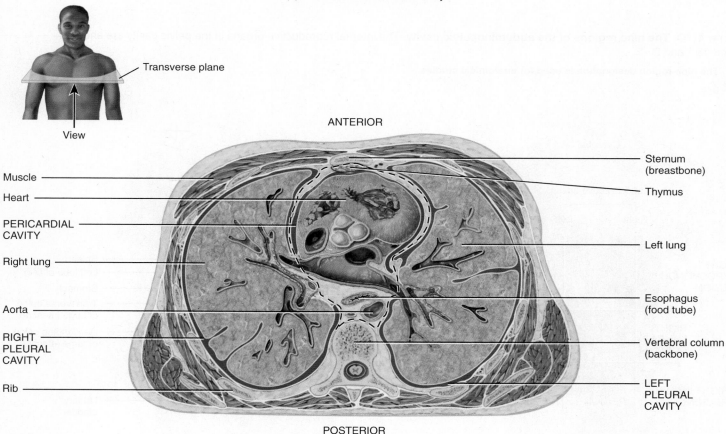

(b) Inferior view of transverse section of thoracic cavity

Which of the following structures are contained in the mediastinum: right lung, heart, esophagus, spinal cord, aorta, left pleural cavity?

The ***abdominopelvic cavity*** (ab-dom′-i-no-PEL-vic) extends from the diaphragm to the groin. As the name suggests, it is divided into two portions, although no wall separates them (see Figure 1.8). The upper portion, the ***abdominal cavity*** (ab-DOM-i-nal; *abdomin-* = belly), contains the stomach, spleen, liver, gallbladder, small intestine, and most of the large intestine. The lower portion, the ***pelvic cavity*** (PEL-vik; *pelv-* = basin), contains the urinary bladder, portions of the large intestine, and internal organs of the reproductive system. Organs inside the thoracic and abdominopelvic cavities are called ***viscera*** (VIS-e-ra).

A thin, slippery, double-layered ***serous membrane*** covers the viscera within the thoracic and abdominal cavities and lines the walls of the thorax and abdomen. The parts of a serous membrane are (1) the *parietal layer* (pa-RĪ-e-tal), which lines the walls of the cavities, and (2) the *visceral layer*, which covers and adheres to the viscera within the cavities. A small amount of lubricating fluid (serous fluid) between the two layers reduces friction, allowing the viscera to slide somewhat during movements, as when the lungs inflate and deflate during breathing.

The serous membrane of the pleural cavities is called the ***pleura*** (PLOO-ra). The serous membrane of the pericardial cavity is the ***pericardium*** (per′-i-KAR-dē-um). The ***peritoneum*** (per-i-tō-NĒ-um) is the serous membrane of the abdominal cavity.

In addition to those just described, you will also learn about other body cavities in later chapters. These include the *oral (mouth) cavity*, which contains the tongue and teeth; the *nasal cavity* in the nose; the *orbital cavities*, which contain the eyeballs; the *middle ear cavities*, which contain small bones in the middle ear; and *synovial cavities*, which are found in freely movable joints and contain synovial fluid.

Abdominopelvic Regions and Quadrants

To describe the location of the many abdominal and pelvic organs more precisely, the abdominopelvic cavity may be divided into smaller compartments. In one method, two horizontal and two vertical lines, like a tic-tac-toe grid, partition the cavity into nine ***abdominopelvic regions*** (Figure 1.10).

Figure 1.10 **The nine regions of the abdominopelvic cavity.** The internal reproductive organs in the pelvic cavity are shown in Figures 23.1 and 23.6.

 The nine-region designation is used for anatomical studies.

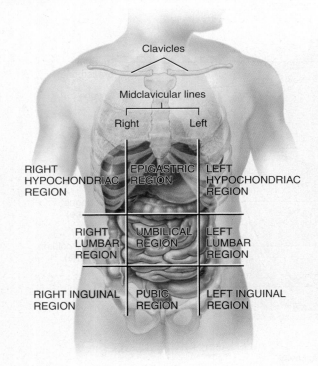

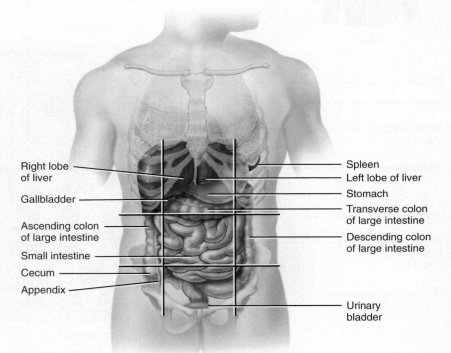

(a) Anterior view showing location of abdominopelvic regions

(b) Anterior superficial view of organs in abdominopelvic regions

In which abdominopelvic region is each of the following found: most of the liver, ascending colon, urinary bladder, appendix?

The names of the nine abdominopelvic regions are the *right hypochondriac* (hī′-pō-KON-drē-ak), *epigastric* (ep-i-GAS-trik), *left hypochondriac, right lumbar, umbilical* (um-BIL-i-kul), *left lumbar, right inguinal (iliac)* (IL-ē-ak), *hypogastric* (hī′-pō-GAS-trik), and *left inguinal (iliac)*. In another method, one horizontal and one vertical line passing through the *umbilicus* (um-BIL-i-kus or um-bi-LĪ-kus; *umbilic-* = navel) or belly button divide the abdominopelvic cavity into **quadrants** (KWOD-rantz; *quad-* = one-fourth) (Figure 1.11). The names of the abdominopelvic quadrants are the *right upper quadrant (RUQ), left upper quadrant (LUQ), right lower quadrant (RLQ),* and *left lower quadrant (LLQ)*. The nine-region division is more widely used for anatomical studies, and quadrants are more commonly used by clinicians to describe the site of an abdominopelvic pain, mass, or other abnormality.

✓ CHECKPOINT

14. What landmarks separate the various body cavities from one another?

15. Locate the nine abdominopelvic regions and the four abdominopelvic quadrants on yourself, and list some of the organs found in each.

• • •

In Chapter 2 we will examine the chemical level of organization. You will learn about the various groups of chemicals, their functions, and how they contribute to homeostasis.

Figure 1.11 Quadrants of the abdominopelvic cavity. The two lines cross at right angles at the umbilicus (navel).

🔑 The quadrant designation is used to locate the site of pain, a mass, or some other abnormality.

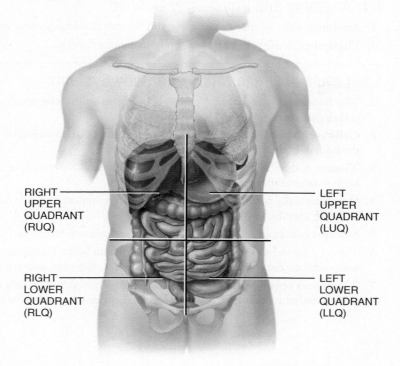

RIGHT UPPER QUADRANT (RUQ)

LEFT UPPER QUADRANT (LUQ)

RIGHT LOWER QUADRANT (RLQ)

LEFT LOWER QUADRANT (LLQ)

Anterior view showing location of abdominopelvic quadrants

 In which abdominopelvic quadrant would the pain from appendicitis (inflammation of the appendix) be felt?

MEDICAL TERMINOLOGY AND CONDITIONS

Most chapters in this text are followed by a glossary of key medical terms that include both normal and pathological conditions. You should familiarize yourself with these terms because they will play an essential role in your medical vocabulary.

Some of these conditions, as well as ones discussed in the text, are referred to as local or systemic. A *local disease* is one that affects one part or a limited area of the body. A *systemic disease* affects the entire body or several parts.

Epidemiology (ep′-i-dē-mē-OL-ō-jē; *epi-* = upon; *-demi* = people) The science that deals with why, when, and where diseases occur and how they are transmitted within a defined human population.

Geriatrics (jer′-ē-AT-riks; *ger-* = old; *-iatrics* = medicine) The science that deals with the medical problems and care of elderly persons.

Pathology (pa-THOL-ō-jē; *patho-* = disease) The science that deals with the nature, causes, and development of abnormal conditions and the structural and functional changes that diseases produce.

Pharmacology (far-ma-KOL-ō-jē; *pharmaco-* = drug) The science that deals with the effects and use of drugs in the treatment of disease.

CHAPTER REVIEW AND RESOURCE SUMMARY

WILEY
PLUS

| **REVIEW** | **RESOURCES** |

1.1 Anatomy and Physiology Defined

1. **Anatomy** is the science of structure and the relationships among structures.
2. **Physiology** is the science of how body structures function.

1.2 Levels of Organization and Body Systems

1. The human body consists of six levels of organization: **chemical, cellular, tissue, organ, system**, and **organismal**.
2. **Cells** are the basic structural and functional units of an organism and the smallest living units in the human body.
3. **Tissues** consist of groups of cells and the materials surrounding them that work together to perform a particular function.
4. **Organs** usually have recognizable shapes, are composed of two or more different types of tissues, and have specific functions.
5. **Systems** consist of related organs that have a common function.
6. Table 1.1 introduces the 11 systems of the human body: integumentary, skeletal, muscular, nervous, endocrine, cardiovascular, lymphatic, respiratory, digestive, urinary, and reproductive.
7. The human organism is a collection of structurally and functionally integrated systems. Body systems work together to maintain health, protect against disease, and allow for reproduction of the species.

RESOURCES

Anatomy Overview—The Integumentary System
Anatomy Overview—The Skeletal System
Anatomy Overview—The Muscular System
Anatomy Overview—The Nervous System
Anatomy Overview—The Endocrine System
Anatomy Overview—The Cardiovascular System
Anatomy Overview—The Lymphatic and Immune Systems
Anatomy Overview—The Respiratory System
Anatomy Overview—The Digestive System
Anatomy Overview—The Urinary System
Anatomy Overview—The Reproductive Systems
Exercise: Concentrate on Systemic Functions
Exercise: Find the System Outsiders

1.3 Life Processes

1. All living organisms have certain characteristics that set them apart from nonliving things.
2. The life processes in humans include **metabolism, responsiveness, movement, growth, differentiation,** and **reproduction**.

Animation—Communication, Regulation and Homeostasis

1.4 Homeostasis: Maintaining Limits

1. **Homeostasis** is a condition in which the internal environment of the body remains stable, within certain limits.
2. A large part of the body's internal environment is interstitial fluid, which surrounds all body cells.
3. Homeostasis is regulated by the nervous and endocrine systems acting together or separately. The nervous system detects body changes and sends nerve impulses to maintain homeostasis. The endocrine system regulates homeostasis by secreting hormones.
4. Disruptions of homeostasis come from external and internal stimuli and from psychological stresses. When disruption of homeostasis is mild and temporary, responses of body cells quickly restore balance in the internal environment. If disruption is extreme, the body's attempts to restore homeostasis may fail.
5. A **feedback system** consists of three parts: (1) **receptors** that monitor changes in a controlled condition and send input to (2) a **control center** that sets the value at which a controlled condition should be maintained, evaluates the input it receives, and generates output commands when they are needed, and (3) **effectors** that receive output from the control center and produce a response (effect) that alters the controlled condition.

Animation—Communication, Regulation and Homeostasis
Animation—Negative Feedback Control of Blood Pressure
Animation—Negative Feedback Control of Body Temperature
Animation—Positive Feedback Control of Labor
Animation—Homeostatic Relationships

REVIEW	**RESOURCES**

6. If a response reverses a change in a controlled condition, the system is called a **negative feedback system**. If a response strengthens a change in a controlled condition, the system is referred to as a **positive feedback system**.

7. One example of negative feedback is the system that regulates blood pressure. If a stimulus causes blood pressure (controlled condition) to rise, baroreceptors (pressure-sensitive nerve cells, the receptors) in blood vessels send impulses (input) to the brain (control center). The brain sends impulses (output) to the heart (effector). As a result, heart rate decreases (response), and blood pressure drops back to normal (restoration of homeostasis).

8. Disruptions of homeostasis—homeostatic imbalances—can lead to disorders, disease, and even death. A **disorder** is any abnormality of structure and/or function. **Disease** is a more specific term for an illness with a definite set of signs and symptoms.

9. **Symptoms** are subjective changes in body functions that are not apparent to an observer. **Signs** are objective changes that can be observed and measured.

10. **Diagnosis** of disease involves identification of symptoms and signs, a medical history, physical examination, and sometimes laboratory tests.

1.5 Aging and Homeostasis

1. **Aging** produces observable changes in structure and function and increases vulnerability to stress and disease.

2. Changes associated with aging occur in all body systems.

1.6 Anatomical Terms

1. Descriptions of any region of the body assume the body is in the **anatomical position**, in which the subject stands erect facing the observer, with the head level and the eyes facing forward, the feet flat on the floor and directed forward, and the arms at the sides, with the palms turned forward.

2. The human body is divided into several major regions: the **head, neck, trunk, upper limbs**, and **lower limbs**.

3. Within body regions, specific body parts have common names and corresponding anatomical names. Examples are chest (thoracic), nose (nasal), and wrist (carpal).

4. **Directional terms** indicate the relationship of one part of the body to another. Exhibit 1.1 summarizes commonly used directional terms.

5. **Planes** are imaginary flat surfaces that divide the body or organs into two parts. A **midsagittal plane** divides the body or an organ into equal right and left sides. A **parasagittal plane** divides the body or an organ into unequal right and left sides. A **frontal plane** divides the body or an organ into anterior and posterior portions. A **transverse plane** divides the body or an organ into superior and inferior portions. An **oblique plane** passes through the body or an organ at an angle between a transverse plane and a sagittal plane, or between a transverse plane and a frontal plane.

6. **Sections** result from cuts through body structures. They are named according to the plane on which the cut is made: transverse, frontal, or sagittal.

Figure 1.4 The anatomical position

Figure 1.6 Directional terms

1.7 Body Cavities

1. Spaces in the body that contain, protect, separate, and support internal organs are called **body cavities**.

2. The **cranial cavity** contains the brain, and the **vertebral canal** contains the spinal cord.

3. The **thoracic cavity** is subdivided into three smaller cavities: a **pericardial cavity**, which contains the heart, and two **pleural cavities**, each of which contains a lung.

4. The central portion of the thoracic cavity is the **mediastinum**. It is located between the lungs and extends from the sternum to the vertebral column and from the neck to the **diaphragm**. It contains all thoracic organs except the lungs.

5. The **abdominopelvic cavity** is separated from the thoracic cavity by the diaphragm and is divided into a superior **abdominal cavity** and an inferior **pelvic cavity**.

Anatomy Overview—
  Serous Membranes

Figure 1.8 Body cavities

Figure 1.9 The thoracic cavity

6. Organs in the thoracic and abdominopelvic cavities are called **viscera**. Viscera of the abdominal cavity include the stomach, spleen, liver, gallbladder, small intestine, and most of the large intestine. Viscera of the pelvic cavity include the urinary bladder, portions of the large intestine, and internal organs of the reproductive system.

7. To describe the location of organs easily, the abdominopelvic cavity may be divided into nine **abdominopelvic regions** by two horizontal and two vertical lines. The names of the nine abdominopelvic regions are right hypochondriac, epigastric, left hypochondriac, right lumbar, umbilical, left lumbar, right inguinal, hypogastric, and left inguinal.

8. The abdominopelvic cavity may also be divided into **quadrants** by passing one horizontal line and one vertical line through the umbilicus (navel). The names of the abdominopelvic quadrants are right upper quadrant (RUQ), left upper quadrant (LUQ), right lower quadrant (RLQ), and left lower quadrant (LLQ).

SELF-QUIZ

1. To properly reconnect the disconnected bones of a human skeleton, you would need to have a good understanding of
   a. physiology.   b. homeostasis.   c. chemistry.
   d. anatomy.   e. feedback systems.

2. Which of the following best illustrates the idea of increasing levels of organizational complexity?
   a. chemical → tissue → cellular → organ → organismal → system
   b. chemical → cellular → tissue → organ → system → organismal
   c. cellular → chemical → tissue → organismal → organ → system
   d. chemical → cellular → tissue → system → organ → organismal
   e. tissue → cellular → chemical → organ → system → organismal

3. Match the following:
   _____ a. transports oxygen, nutrients, and carbon dioxide
   _____ b. breaks down food and absorbs nutrients
   _____ c. functions in body movement, posture, and heat production
   _____ d. regulates body activities through hormones
   _____ e. supports body; helps protects body organs; provides areas for muscle attachment
   _____ f. eliminates wastes and regulates the chemical composition and volume of blood
   _____ g. protects the body, detects sensations, and helps regulate body temperature

   A. urinary system
   B. digestive system
   C. endocrine system
   D. integumentary system
   E. muscular system
   F. skeletal system
   G. cardiovascular system

4. Fill in the missing blanks in the following table.

System	Major Organs	Functions
__a__	__b__	Regulates body activities by nerve impulses
__c__	Lymph vessels, spleen, thymus, tonsils, lymph nodes	__d__
__e__	__f__	Supplies oxygen to cells, eliminates carbon dioxide, helps regulate acidity of body fluids
Reproductive	__g__	__h__

5. Homeostasis is
   a. the sum of all of the chemical processes in the body.
   b. the sign of a disorder or disease.
   c. the combination of growth, repair, and energy release that is basic to life.
   d. the tendency to maintain constant, favorable internal body conditions.
   e. caused by stress.

6. Match the following to the correct life process.
   _____ a. the pupils of your eyes becoming smaller when exposed to strong light
   _____ b. the ability to walk to your car following class
   _____ c. the healing of a broken bone
   _____ d. digesting and absorbing your food from breakfast
   _____ e. the initial development of the nervous system in a fetus
   _____ f. the pediatrician noting the increased head circumference of a four-month-old baby

   A. metabolism
   B. differentiation
   C. movement
   D. growth
   E. responsiveness
   F. reproduction

7. Which of the following body system(s) is(are) the primary controller(s) of homeostasis?
   a. respiratory   b. nervous   c. endocrine
   d. cardiovascular   e. urinary

8. The part of a feedback system that receives the input and generates the output command is the
   a. effector.   **b.** receptor.   **c.** feedback loop.
   d. response.   **e.** control center.

9. Match the following:
   ____ **a.** observable, measurable change   **A.** disease
   ____ **b.** abnormality of function   **B.** symptom
   ____ **c.** a recognizable set of body changes **C.** sign
   ____ **d.** subjective change that isn't   **D.** disorder
              easily observed

10. An itch in your axillary region would cause you to scratch
    a. your armpit.   **b.** the front of your elbow.
    c. your neck.   **d.** the top of your head.
    e. your calf.

11. If you were facing a person who is in the correct anatomical position, you could observe the
    a. crural region.   **b.** lumbar region.
    c. gluteal region.   **d.** popliteal region.
    e. scapular region.

12. Where would you look for the femoral artery?
    a. wrist   **b.** forearm   **c.** face
    d. thigh   **e.** shoulder

13. The right ear is _____ to the right nostril.
    a. intermediate   **b.** inferior   **c.** lateral
    d. distal   **e.** medial

14. Your chin is _____ in relation to your lips.
    a. lateral   **b.** superior   **c.** deep
    d. posterior **e.** inferior

15. Your skull is _____ in relation to your brain.
    a. intermediate   **b.** superior   **c.** deep
    d. superficial   **e.** proximal

16. A magician is about to separate his assistant's body into superior and inferior portions. The plane through which he will pass his magic wand is the
    a. midsagittal.   **b.** frontal.   **c.** transverse.
    d. parasagittal.   **e.** oblique.

17. Which statement is NOT true of body cavities?
    a. The diaphragm separates the thoracic and abdominopelvic cavities.
    b. The organs in the cranial and vertebral cavities are called viscera.
    c. The urinary bladder is in the pelvic cavity.
    d. The abdominal cavity is below the thoracic cavity.
    e. The pelvic cavity terminates below the groin.

18. To expose the heart for open heart surgery, the surgeon would need to cut through the
    a. pericardial cavity.   **b.** pelvic cavity.
    c. diaphragm.   **d.** pleural cavity.
    e. abdominal cavity.

19. To find the urinary bladder, you would look in the
    a. hypochondriac region.
    b. umbilical region.
    c. epigastric region.
    d. inguinal (iliac) region.
    e. hypogastric region.

20. Match the following:
    ____ **a.** contains the urinary bladder   **A.** cranial cavity
              and reproductive organs
    ____ **b.** contains the brain   **B.** abdominal cavity
    ____ **c.** contains the heart   **C.** vertebral cavity
    ____ **d.** region between the lungs,   **D.** pelvic cavity
              from the breastbone to
              the backbone   **E.** pleural cavity
    ____ **e.** separates the thoracic and   **F.** mediastinum
              abdominal cavities
                                                    G. diaphragm
    ____ **f.** contains a lung
                                                    H. pericardial cavity
    ____ **g.** contains the spinal cord
    ____ **h.** contains the stomach and liver

CRITICAL THINKING APPLICATIONS

1. Lily was going for the playground record for the longest upside-down hang from the monkey bars. She didn't make it and may have broken her arm. At the Accident and Emergency Department they would like an x-ray film of Lily's arm in the anatomical position. Use the proper anatomical terms to describe the position of Lily's arm in the x-ray film.

2. You are working in a lab and think you may be observing a new organism. What minimal level of structural organization would you need to be observing? What are some characteristics you would need to observe to ensure that it is a living organism?

3. Guy was trying to impress Jenna with a tale about his last rugby match. "The coach said I suffered a caudal injury to the dorsal sural in my groin." Jenna responded, "I think either you or your coach suffered a cephalic injury." Why wasn't Jenna impressed by Guy's athletic prowess?

4. There's a special fun-house mirror that hides half your body and doubles the image of your other side. In the mirror, you can do amazing feats such as lifting both legs off the ground. Along what plane is the mirror dividing your body? A different mirror in the next room shows your reflection with two heads, four arms, and no legs. Along what plane is this mirror dividing your body?

ANSWERS TO FIGURE QUESTIONS

1.1 Organs have a recognizable shape and consist of two or more different types of tissues that have a specific function.

1.2 The basic difference between negative and positive feedback systems is that in negative feedback systems, the response reverses a change in a controlled condition, and in positive feedback systems, the response strengthens the change in a controlled condition.

1.3 If a stimulus caused blood pressure to decrease, the heart rate would increase due to the operation of this negative feedback system.

1.4 A plantar wart is found on the sole.

1.5 No, the radius is distal to the humerus. No, the esophagus is posterior to the trachea. Yes, the ribs are superficial to the lungs. Yes, the urinary bladder is medial to the ascending colon. No, the sternum is medial to the descending colon.

1.6 The frontal plane divides the heart into anterior and posterior portions.

1.7 The midsagittal plane divides the brain into equal right and left sides.

1.8 Urinary bladder = P, stomach = A, heart = T, small intestine = A, lungs = T, internal female reproductive organs = P, thymus = T, spleen = A, liver = A.

1.9 Some structures in the mediastinum include the heart, esophagus, and aorta.

1.10 The liver is mostly in the epigastric region; the ascending colon is in the right lumbar region; the urinary bladder is in the hypogastric region; the appendix is in the right inguinal region.

1.11 The pain associated with appendicitis would be felt in the right lower quadrant (RLQ).

CHAPTER 2

INTRODUCTORY CHEMISTRY

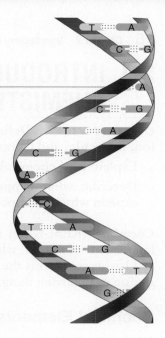

Many common substances we eat and drink—water, sugar, table salt, proteins, starches, fats—play vital roles in keeping us alive. In this chapter, you will learn how these substances function in your body. Because your body is composed of chemicals and all body activities are chemical in nature, it is important to become familiar with the language and basic ideas of chemistry to understand human anatomy and physiology.

Did you know?   *Many people think "chemicals" are somehow artificial, but all living things are composed of chemicals. Many of the "natural" dietary supplements on the market have effects as strong as any pharmaceutical or over-the-counter drug product. Because supplements are made from natural components of plants and animals, they are regulated as food in the U.S. But many of these supplements include real hormones and other strong agents. "Natural" does not necessarily mean "harmless at any dose."*

LOOKING BACK TO MOVE AHEAD...

Levels of Organization and Body Systems (Section 1.2)

FOCUS ON WELLNESS
Dietary Supplements—Handle with Care

2.1 INTRODUCTION TO CHEMISTRY

OBJECTIVES • **Define a chemical element, atom, ion, molecule, and compound.**

• **Explain how chemical bonds form.**

• **Describe what happens in a chemical reaction and explain why it is important to the human body.**

Chemistry (KEM-is-trē) is the science of the structure and interactions of *matter*, which is anything that occupies space and has mass. *Mass* is the amount of matter in any living organism or nonliving thing.

Chemical Elements and Atoms

All forms of matter are made up of a limited number of building blocks called *chemical elements*, substances that cannot be broken down into a simpler form by ordinary chemical means. At present, scientists recognize 112 different elements. Each element is designated by a *chemical symbol*, one or two letters of the element's name in English, Latin, or another language. Examples are H for hydrogen, C for carbon, O for oxygen, N for nitrogen, K for potassium, Na for sodium, Fe for iron, and Ca for calcium.

Twenty-six different elements normally are present in your body. Just four elements, called the *major elements*, constitute about 96 percent of the body's mass: oxygen, carbon, hydrogen, and nitrogen. Eight others, the *lesser elements*, contribute 3.6 percent of the body's mass: calcium (Ca), phosphorus (P), potassium (kalium) (K), sulfur (S), sodium (natrium) (Na), chlorine (Cl), magnesium (Mg), and iron (ferrium) (Fe). An additional 14 elements—the *trace elements*—are present in tiny amounts. Together, they account for the remaining 0.4 percent of the body's mass. Although trace elements are few in number, several have important functions in the body. For example, iodine (I) is needed to make thyroid hormones. The functions of some trace elements are unknown. Table 2.1 lists the main chemical elements of the human body.

Each element is made up of *atoms*, the smallest units of matter that retain the properties and characteristics of the element. A sample of the element carbon, such as pure coal,

---

TABLE 2.1

Main Chemical Elements in the Body

CHEMICAL ELEMENT (SYMBOL)	% OF TOTAL BODY MASS	SIGNIFICANCE
MAJOR ELEMENTS	**about 96%**	
Oxygen (O)	65.0	Part of water and many organic (carbon-containing) molecules; used to generate ATP, a molecule used by cells to temporarily store chemical energy
Carbon (C)	18.5	Forms backbone chains and rings of all organic molecules: carbohydrates, lipids (fats), proteins, and nucleic acids (DNA and RNA)
Hydrogen (H)	9.5	Constituent of water and most organic molecules; ionized form (H^+) makes body fluids more acidic.
Nitrogen (N)	3.2	Component of all proteins and nucleic acids
LESSER ELEMENTS	**about 3.6%**	
Calcium (Ca)	1.5	Contributes to hardness of bones and teeth; ionized form (Ca^{2+}) needed for blood clotting, release of hormones, contraction of muscle, and many other processes
Phosphorus (P)	1.0	Component of nucleic acids and ATP; required for normal bone and tooth structure
Potassium (K)	0.35	Ionized form (K^+) most plentiful cation (positively charged particle) in intracellular fluid; needed to generate action potentials
Sulfur (S)	0.25	Component of some vitamins and many proteins
Sodium (Na)	0.2	Ionized form (Na^+) most plentiful cation in extracellular fluid; essential for maintaining water balance; needed to generate action potentials
Chlorine (Cl)	0.2	Ionized form (Cl^-) most plentiful anion (negatively charged particle) in extracellular fluid; essential for maintaining water balance
Magnesium (Mg)	0.1	Ionized form (Mg^{2+}) needed for action of many enzymes, molecules that increase the rate of chemical reactions in organisms

contains only carbon atoms, and a tank of helium gas contains only helium atoms.

An atom consists of two basic parts: a nucleus and one or more electrons (Figure 2.1). The centrally located *nucleus* contains positively charged *protons* (*p*$^+$) and uncharged (neutral) *neutrons* (*n*^{0}). Because each proton has one positive charge, the nucleus is positively charged. The *electrons* (*e*$^-$) are tiny, negatively charged particles that move about in a large space surrounding the nucleus. They do not follow a fixed path or orbit but instead form a negatively charged "cloud" that surrounds the nucleus (Figure 2.1a). The number of electrons in an atom equals the number of protons. Because each electron carries one negative charge, the negatively charged electrons and the positively charged protons balance each other. As a result, each atom is electrically neutral, meaning its total charge is zero.

The number of protons in the nucleus of an atom is called the atom's *atomic number*. The atoms of each different kind of element have a different number of protons in the nucleus: A hydrogen atom has 1 proton, a carbon atom has 6 protons, a sodium atom has 11 protons, a chlorine atom has

Figure 2.1 Two representations of the structure of an atom. Electrons move about the nucleus, which contains neutrons and protons. (a) In the electron cloud model of an atom, the shading represents the chance of finding an electron in regions outside the nucleus. (b) In the electron shell model, filled circles represent individual electrons, which are grouped into concentric circles according to the shells they occupy. Both models depict a carbon atom, with six protons, six neutrons, and six electrons.

An atom is the smallest unit of matter that retains the properties and characteristics of its element.

- Protons (p$^+$) ⎤
- Neutrons (n^0) ⎦ Nucleus
- Electrons (e$^-$)

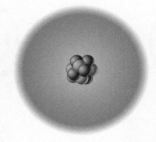

(a) Electron cloud model

(b) Electron shell model

What is the atomic number of carbon?

CHEMICAL ELEMENT (SYMBOL)	% OF TOTAL BODY MASS	SIGNIFICANCE
Iron (Fe)	0.005	Ionized forms (Fe^{2+} and Fe^{3+}) part of hemoglobin (oxygen-carrying protein in red blood cells) and some enzymes
TRACE ELEMENTS	**about 0.4%**	Aluminum (Al), Boron (B), Chromium (Cr), Cobalt (Co), Copper (Cu), Fluorine (F), Iodine (I), Manganese (Mn), Molybdenum (Mo), Selenium (Se), Silicon (Si), Tin (Sn), Vanadium (V), and Zinc (Zn)

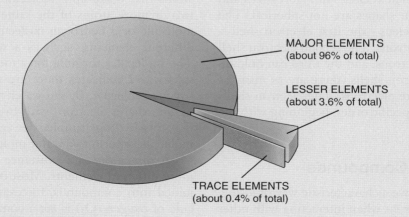

MAJOR ELEMENTS
(about 96% of total)

LESSER ELEMENTS
(about 3.6% of total)

TRACE ELEMENTS
(about 0.4% of total)

Figure 2.2 Atomic structures of several atoms that have important roles in the human body.

The atoms of different elements have different atomic numbers because they have different numbers of protons.

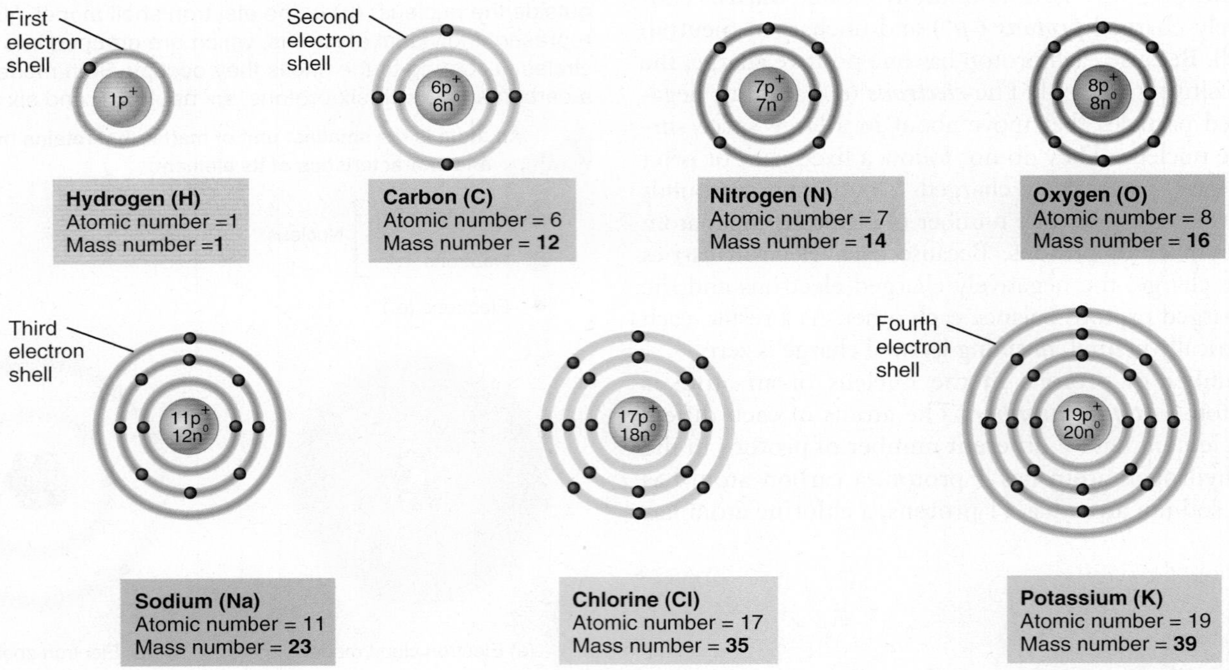

First electron shell
Second electron shell

Hydrogen (H)
Atomic number = 1
Mass number = **1**

Carbon (C)
Atomic number = 6
Mass number = **12**

Nitrogen (N)
Atomic number = 7
Mass number = **14**

Oxygen (O)
Atomic number = 8
Mass number = **16**

Third electron shell

Fourth electron shell

Sodium (Na)
Atomic number = 11
Mass number = **23**

Chlorine (Cl)
Atomic number = 17
Mass number = **35**

Potassium (K)
Atomic number = 19
Mass number = **39**

Atomic number = number of protons in an atom
Mass number = number of protons and neutrons in an atom (boldface indicates most common isotope)

? Which four of these elements are most abundant in living organisms?

17 protons, and so on (Figure 2.2). Thus, each type of atom, or element, has a different atomic number. The total number of protons plus neutrons in an atom is its **mass number**. For instance, an atom of sodium, with 11 protons and 12 neutrons in its nucleus, has a mass number of 23.

Even though their exact positions cannot be predicted, specific groups of electrons are most likely to move about within certain regions around the nucleus. These regions are called **electron shells**, which are depicted as circles in Figures 2.1b and 2.2 even though some of their shapes are not spherical. The electron shell nearest the nucleus—the first electron shell—can hold a maximum of 2 electrons. The second electron shell can hold a maximum of 8 electrons, and the third can hold up to 18 electrons. Higher electron shells (there are as many as seven) can contain many more electrons. The electron shells are filled with electrons in a specific order, beginning with the first shell.

Ions, Molecules, and Compounds

The atoms of each element have a characteristic way of losing, gaining, or sharing their electrons when interacting with other atoms. If an atom either *gives up* or *gains* electrons, it becomes an **ion** (Ī-on), an atom that has a positive or negative charge due to unequal numbers of protons and electrons. An ion of an atom is symbolized by writing its chemical symbol followed by the number of its positive (+) or negative (−) charges. For example, Ca^{2+} stands for a calcium ion that has two positive charges because it has given up two electrons. Refer to Table 2.1 for the important functions of several ions in the body.

In contrast, when two or more atoms *share* electrons, the resulting combination of atoms is called a **molecule** (MOL-e-kūl). A *molecular formula* indicates the number and type of atoms that make up a molecule. A molecule may consist of two or more atoms of the same element, such as an oxygen molecule or a hydrogen molecule, or two or more atoms of different elements, such as a water molecule (Figure 2.3). The molecular formula for a molecule of oxygen is O_2. The subscript 2 indicates there are two atoms of oxygen in the oxygen molecule. In the water molecule, H_2O, one atom of oxygen shares electrons with two atoms of hydrogen. Notice that two hydrogen molecules can combine with one oxygen molecule to form two water molecules (Figure 2.3).

A **compound** is a substance containing atoms of two or more different elements. Most of the atoms in your body are joined into compounds, for example, water (H_2O). A molecule of oxygen (O_2) is *not* a compound because it consists of atoms of only one element.

A **free radical** is an ion or molecule that has an unpaired electron in its outermost shell. (Most of an atom's electrons

Figure 2.3 Molecules.

A molecule may consist of two or more atoms of the same element or two or more atoms of different elements.

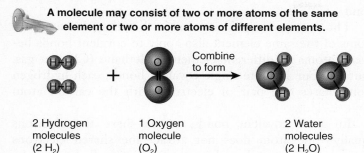

| 2 Hydrogen molecules ($2\ H_2$) | 1 Oxygen molecule (O_2) | 2 Water molecules ($2\ H_2O$) |

 Which of the molecules shown here is a compound?

associate in pairs.) A common example of a free radical is *superoxide*, which is formed by the addition of an electron to an oxygen molecule. Having an unpaired electron makes a free radical unstable and destructive to nearby molecules. Free radicals break apart important body molecules by either giving up their unpaired electron to or taking on an electron from another molecule.

CLINICAL CONNECTION | Free Radicals and Their Effects on Health

In our bodies, several processes can generate free radicals. They may result from exposure to ultraviolet radiation in sunlight or to x-rays. Some reactions that occur during normal metabolic processes produce free radicals. Moreover, certain harmful substances, such as carbon tetrachloride (a solvent used in dry cleaning), give rise to free radicals when they participate in metabolic reactions in the body. Among the many disorders and diseases linked to oxygen-derived free radicals are cancer, the buildup of fatty materials in blood vessels (atherosclerosis), Alzheimer's disease, emphysema, diabetes mellitus, cataracts, macular degeneration, rheumatoid arthritis, and deterioration associated with aging. Consuming more **antioxidants**—substances that inactivate oxygen-derived free radicals—is thought to slow the pace of damage caused by free radicals. Important dietary antioxidants include selenium, zinc, beta-carotene, and vitamins C and E. Red, blue, or purple fruits and vegetables contain high levels of antioxidants. •

✓ CHECKPOINT

1. Compare the meanings of atomic number, mass number, ion, and molecule.

2. What is the significance of the valence (outer) electron shell of an atom?

Chemical Bonds

The forces that bind the atoms of molecules and compounds together, resisting their separation, are ***chemical bonds***. The chance that an atom will form a chemical bond with another atom depends on the number of electrons in its outermost shell, also called the ***valence shell***. An atom with an outer

shell holding eight electrons is *chemically stable*, which means it is unlikely to form chemical bonds with other atoms. Neon, for example, has eight electrons in its outer shell, and for this reason it rarely forms bonds with other atoms.

The atoms of most biologically important elements do not have eight electrons in their outer shells. Given the right conditions, two or more such atoms can interact or bond in ways that produce a chemically stable arrangement of eight electrons in the outer shell of each atom (*octet rule*). Three general types of chemical bonds are ionic bonds, covalent bonds, and hydrogen bonds.

Ionic Bonds

Positively charged ions and negatively charged ions are attracted to one another. This force of attraction between ions of opposite charges is called an **ionic bond**. Consider sodium and chlorine atoms to see how an ionic bond forms (Figure 2.4).

Figure 2.4 Ions and ionic bond formation. The electron that is donated or accepted is colored red.

 An ionic bond is the force of attraction that holds together oppositely charged ions.

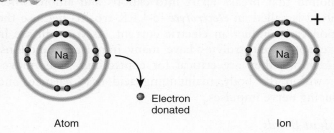

(a) Sodium: 1 valence electron

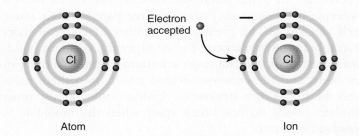

(b) Chlorine: 7 valence electrons

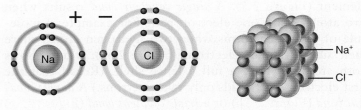

(c) Ionic bond in sodium chloride (NaCl)   (d) Packing of ions in a crystal of sodium chloride

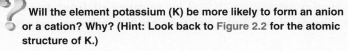 **Will the element potassium (K) be more likely to form an anion or a cation? Why? (Hint: Look back to Figure 2.2 for the atomic structure of K.)**

Sodium has one outer shell electron (Figure 2.4a). If a sodium atom *loses* this electron, it is left with the eight electrons in its second shell. However, the total number of protons (11) now exceeds the number of electrons (10). As a result, the sodium atom becomes a **cation** (KAT-ī-on), a positively charged ion. A sodium ion has a charge of 1+ and is written Na^+. On the other hand, chlorine has seven outer shell electrons (Figure 2.4b), too many to lose. But if chlorine *accepts* one electron from a neighboring atom, it will have eight electrons in its third electron shell. When this happens, the total number of electrons (18) exceeds the number of protons (17), and the chlorine atom becomes an **anion** (AN-ī-on), a negatively charged ion. The ionic form of chlorine is called a chloride ion. It has a charge of 1− and is written Cl^-. When an atom of sodium donates its sole outer shell electron to an atom of chlorine, the resulting positive and negative charges attract each other to form an ionic bond (Figure 2.4c). The resulting ionic compound is sodium chloride, written NaCl.

In the body, ionic bonds are found mainly in teeth and bones, where they give great strength to the tissue. Most other ions in the body are dissolved in body fluids. An ionic compound that breaks apart into cations and anions when dissolved is called an **electrolyte** (e-LEK-trō-līt) because the solution can conduct an electric current. As you will see in later chapters, electrolytes have many important functions. For example, they are critical for controlling water movement within the body, maintaining acid–base balance, and producing nerve impulses.

Covalent Bonds

When a **covalent bond** forms, neither of the combining atoms loses or gains electrons. Instead, the atoms form a molecule by *sharing* one, two, or three pairs of their outer shell electrons. The greater the number of electron pairs shared between two atoms, the stronger the covalent bond. Covalent bonds are the most common chemical bonds in the body, and the compounds that result from them form most of the body's structures. Unlike ionic bonds, most covalent bonds do not break apart when the molecule is dissolved in water.

It is easiest to understand the nature of covalent bonds by considering those that form between atoms of the same element (Figure 2.5). A *single covalent bond* results when two atoms share one electron pair. For example, a molecule of hydrogen forms when two hydrogen atoms share their single valence electrons (Figure 2.5a), which allows both atoms to have a full valence shell. (Recall that the first electron shell holds only two electrons.) A *double covalent bond* (Figure 2.5b) or a *triple covalent bond* (Figure 2.5c) results when two atoms share two or three pairs of electrons. Notice the *structural formulas* for covalently bonded molecules in Figure 2.5. The number of lines between the chemical symbols for two atoms indicates whether the bond is a single (−), double (=), or triple (≡) covalent bond.

The same principles of covalent bonding that apply to atoms of the same element also apply to covalent bonds between atoms of different elements. Methane (CH_4), a gas, contains four separate single covalent bonds; each hydrogen atom shares one pair of electrons with the carbon atom (Figure 2.5d).

In some covalent bonds, atoms share the electrons equally—one atom does not attract the shared electrons more strongly than the other atom. This is called a *nonpolar covalent bond*. The bonds between two identical atoms always are nonpolar covalent bonds (Figure 2.5a–c). Another example of a nonpolar covalent bond is the single covalent bond that forms between carbon and each atom of hydrogen in a methane molecule (Figure 2.5d).

In a *polar covalent bond*, the sharing of electrons between atoms is unequal—one atom attracts the shared electrons more strongly than the other. The partial charges are indicated by a lowercase Greek delta (δ) with a minus or plus sign. For example, when polar covalent bonds form, the resulting molecule has a partial negative charge, written δ^-, near the atom that attracts electrons more strongly. At least one other atom in the molecule then will have a partial positive charge, written δ^+. A very important example of a polar covalent bond in living systems is the bond between oxygen and hydrogen in a molecule of water (Figure 2.5e).

Hydrogen Bonds

The polar covalent bonds that form between hydrogen atoms and other atoms can give rise to a third type of chemical bond, a hydrogen bond. A **hydrogen bond** forms when a hydrogen atom with a partial positive charge (δ^+) attracts the partial negative charge (δ^-) of neighboring electronegative atoms, most often oxygen or nitrogen. Thus, hydrogen bonds result from attraction of oppositely charged parts of molecules rather than from sharing of electrons as in covalent bonds. Hydrogen bonds are weak when compared to ionic and covalent bonds. Thus, they cannot bind atoms into molecules. However, hydrogen bonds do establish important links between molecules, such as water molecules, or between different parts of large molecules, such as proteins and deoxyribonucleic acid (DNA), where they add strength and stability and help determine the molecule's three-dimensional shape (see Figure 2.15).

Chemical Reactions

A **chemical reaction** occurs when new bonds form and/or old bonds break between atoms. Through chemical reactions, body structures are built and body functions are carried out, processes that involve transfers of energy.

Figure 2.5 Covalent bond formation. The red electrons are shared equally in (a)–(d) and unequally in (e). To the right are simpler ways to represent these molecules. In a structural formula, each covalent bond is denoted by a straight line between the chemical symbols for two atoms. In a molecular formula, the number of atoms in each molecule is noted by subscripts.

In a covalent bond, two atoms share one, two, or three pairs of electrons in the outer shell.

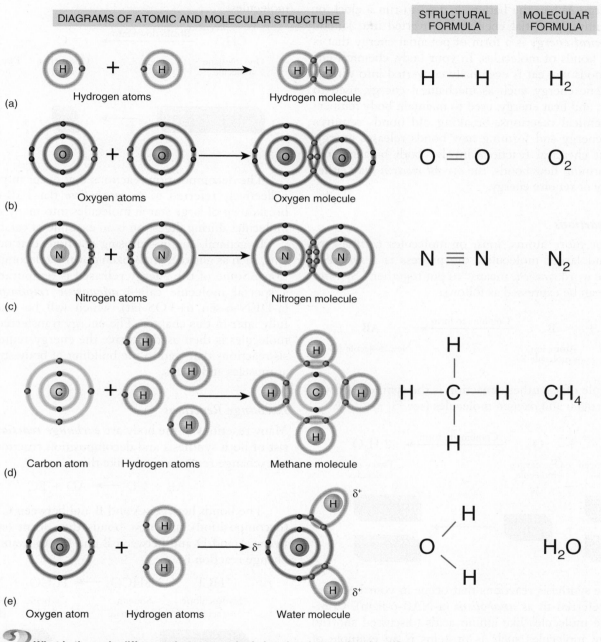

What is the main difference between an ionic bond and a covalent bond?

Forms of Energy and Chemical Reactions

Energy (*en-* = in; *-ergy* = work) is the capacity to do work. The two main forms of energy are **potential energy**, energy stored by matter due to its *position*, and **kinetic energy**, the energy of matter *in motion*. For example, the energy stored in a battery or in a person poised to jump down some steps is potential energy. When the battery is used to run a clock or the person jumps, potential energy is converted into kinetic energy. **Chemical energy** is a form of potential energy that is stored in the bonds of molecules. In your body, chemical energy in the foods you eat is eventually converted into various forms of kinetic energy, such as mechanical energy, used to walk and talk, and heat energy, used to maintain body temperature. In chemical reactions, breaking old bonds requires an input of energy and forming new bonds releases energy. Because most chemical reactions involve both breaking old bonds and forming new bonds, the *overall reaction* may either release energy or require energy.

Synthesis Reactions

When two or more atoms, ions, or molecules combine to form new and larger molecules, the process is a **synthesis reaction**. The word *synthesis* means "to put together." Synthesis reactions can be expressed as follows:

$$A \quad + \quad B \quad \xrightarrow{\text{Combine to form}} \quad AB$$

Atom, ion,   Atom, ion,   New molecule AB
or molecule A   or molecule B

An example of a synthesis reaction is the synthesis of water from hydrogen and oxygen molecules (see Figure 2.3):

$$2\,H_2 \quad + \quad O_2 \quad \xrightarrow{\text{Combine to form}} \quad 2\,H_2O$$

Two hydrogen   One oxygen   Two water
molecules   molecule   molecules

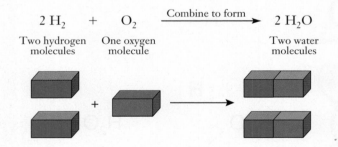

All of the synthesis reactions that occur in your body are collectively referred to as **anabolism** (a-NAB-ō-lizm). Combining simple molecules like amino acids (discussed shortly) to form large molecules such as proteins is an example of anabolism.

Decomposition Reactions

In a **decomposition reaction**, a molecule is split apart. The word *decompose* means to break down into smaller parts. Large molecules are split into smaller molecules, ions, or atoms. A decomposition reaction occurs in this way:

$$AB \quad \xrightarrow{\text{Breaks down into}} \quad A \quad + \quad B$$

Molecule AB   Atom, ion, or   Atom, ion, or
                         molecule A   molecule B

For example, under the proper conditions, a methane molecule can decompose into one carbon atom and two hydrogen molecules:

$$CH_4 \quad \xrightarrow{\text{Breaks down into}} \quad C \quad + \quad 2\,H_2$$

One methane   One carbon   Two hydrogen
molecule   atom   molecules

The decomposition reactions that occur in your body are collectively referred to as **catabolism** (ka-TAB-ō-lizm). The breakdown of large starch molecules into many small glucose molecules during digestion is an example of catabolism.

In general, energy-releasing reactions that occur as nutrients, such as glucose, are broken down via decomposition reactions. Some of the energy released is temporarily stored in a special molecule called **adenosine triphosphate (ATP)** (a-DEN-ō-sēn trī-FOS-fāt), which will be discussed more fully later in this chapter. The energy transferred to the ATP molecules is then used to drive the energy-requiring synthesis reactions that lead to the building of body structures such as muscles and bones.

Exchange Reactions

Many reactions in the body are **exchange reactions**; they consist of both synthesis and decomposition reactions. One type of exchange reaction works like this:

$$AB + CD \longrightarrow AD + BC$$

The bonds between A and B and between C and D break (decomposition), and new bonds then form (synthesis) between A and D and between B and C. An example of an exchange reaction is:

$$HCl \quad + \quad NaHCO_3 \longrightarrow H_2CO_3 + NaCl$$

Hydrochloric   Sodium   Carbonic   Sodium
acid   bicarbonate   acid   chloride

Notice that the atoms or ions in both compounds have "switched partners": The hydrogen ion (H^+) from HCl has combined with the bicarbonate ion (HCO_3^-) from $NaHCO_3$, and the sodium ion (Na^+) from $NaHCO_3$ has combined with the chloride ion (Cl^-) from HCl.

Reversible Reactions

Some chemical reactions proceed in only one direction, as previously indicated by the single arrows. Other chemical reactions may be reversible. *Reversible reactions* can go in either direction under different conditions and are indicated by two half arrows pointing in opposite directions:

$$AB \underset{\text{Combine to form}}{\overset{\text{Breaks down into}}{\rightleftharpoons}} A + B$$

Some reactions are reversible only under special conditions:

$$AB \underset{\text{Heat}}{\overset{\text{Water}}{\rightleftharpoons}} A + B$$

Whatever is written above or below the arrows indicates the condition needed for the reaction to occur. In these reactions, AB breaks down into A and B only when water is added, and A and B react to produce AB only when heat is applied.

The sum of all the chemical reactions in the body is called *metabolism* (me-TAB-ō-lizm; *metabol-* = change). Metabolism and nutrition are discussed in detail in Chapter 20.

✓ CHECKPOINT

3. Distinguish among ionic, covalent, and hydrogen bonds.
4. Explain the difference between anabolism and catabolism. Which involves synthesis reactions?

2.2 CHEMICAL COMPOUNDS AND LIFE PROCESSES

OBJECTIVES ● **Discuss the functions of water and inorganic acids, bases, and salts.**

● **Define pH and explain how the body attempts to keep pH within the limits of homeostasis.**
● **Describe the functions of carbohydrates, lipids, and proteins.**
● **Describe how enzymes function.**
● **Explain the importance of deoxyribonucleic acid (DNA), ribonucleic acid (RNA), and adenosine triphosphate (ATP).**

Chemicals in the body can be divided into two main classes of compounds: inorganic and organic. *Inorganic compounds* usually lack carbon, are structurally simple, and are held together by ionic or covalent bonds. They include water, many salts, acids, and bases. Two inorganic compounds that contain carbon are carbon dioxide (CO_2) and bicarbonate ion (HCO_3^-). *Organic compounds*, by contrast, always contain carbon, usually contain hydrogen, and always have covalent bonds. Examples include carbohydrates, lipids, proteins, nucleic acids, and adenosine triphosphate (ATP). Organic compounds are discussed in detail in Chapters 19 and 20. Large organic molecules called *macromolecules* are formed by covalent bonding of many identical or similar building-block subunits termed *monomers*.

Inorganic Compounds

Water

Water is the most important and most abundant inorganic compound in all living systems, making up 55 to 60 percent of body mass in lean adults. With few exceptions, most of the volume of cells and body fluids is water. Several of its properties explain why water is such a vital compound for life.

1. **Water is an excellent solvent.** A *solvent* is a liquid or gas in which some other material, called a *solute*, has been dissolved. The combination of solvent plus solute is called a *solution*. Water is the solvent that carries nutrients, oxygen, and wastes throughout the body. The versatility of water as a solvent is due to its polar covalent bonds and its "bent" shape (see Figure 2.5e), which allow each water molecule to interact with several neighboring ions or molecules. Solutes that are charged or contain polar covalent bonds are **hydrophilic** (*hydro-* = water; *-philic* = loving), which means they dissolve easily in water. Common examples of hydrophilic solutes are sugar and salt. Molecules that contain mainly nonpolar covalent bonds, by contrast, are **hydrophobic** (*-phobic* = fearing). They are not very water-soluble. Examples of hydrophobic compounds include animal fats and vegetable oils.

2. **Water participates in chemical reactions.** Because water can dissolve so many different substances, it is an ideal medium for chemical reactions. Water also is an active participant in some decomposition and synthesis reactions. During digestion, for example, decomposition reactions break down large nutrient molecules into smaller molecules by the addition of water molecules. This type of reaction is called **hydrolysis** (hī-DROL-i-sis; *-lysis* = to loosen or break apart) (see Figure 2.8). Hydrolysis reactions allow dietary nutrients to be absorbed into the body.

3. **Water absorbs and releases heat very slowly.** In comparison to most other substances, water can absorb or release a relatively large amount of heat with only a slight change in its own temperature. The large amount of water in the body thus moderates the effect of changes in the environmental temperature, thereby helping maintain the homeostasis of body temperature.

4. **Water requires a large amount of heat to change from a liquid to a gas.** When the water in sweat evaporates from the skin's surface, it takes with it large quantities of heat and provides an excellent cooling mechanism.

5. **Water serves as a lubricant.** Water is a major part of saliva, mucus, and other lubricating fluids. Lubrication is especially necessary in the thoracic and abdominal cavities, where internal organs touch and slide over one another. It is also needed at joints, where bones, ligaments, and tendons rub against one another.

Inorganic Acids, Bases, and Salts

Many inorganic compounds can be classified as acids, bases, or salts. An **acid** is a substance that breaks apart or *dissociates* (dis-SŌ-sē-āts′) into one or more *hydrogen ions* (H^+) when it dissolves in water (Figure 2.6a). A **base**, by contrast, usually dissociates into one or more *hydroxide ions* (OH^-) when it dissolves in water (Figure 2.6b). A **salt**, when dissolved in water, dissociates into cations and anions, neither of which is H^+ or OH^- (Figure 2.6c).

Acids and bases react with one another to form salts. For example, the reaction of hydrochloric acid (HCl) and potassium hydroxide (KOH), a base, produces the salt potassium chloride (KCl), along with water (H_2O). This exchange reaction can be written as follows:

$$HCl + KOH \longrightarrow KCl + H_2O$$
$$\text{Acid} \quad \text{Base} \qquad \text{Salt} \quad \text{Water}$$

Figure 2.6 Acids, bases, and salts. (a) When placed in water, hydrochloric acid (HCl) ionizes into H^+ and Cl^-. (b) When the base potassium hydroxide (KOH) is placed in water, it ionizes into OH^- and K^+. (c) When the salt potassium chloride (KCl) is placed in water, it ionizes into positive and negative ions (K^+ and Cl^-), neither of which is H^+ or OH^-.

🔑 Ionization is the separation of inorganic acids, bases, and salts into ions in a solution.

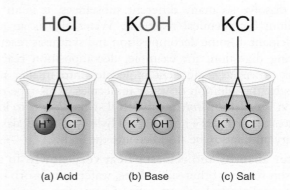

(a) Acid   (b) Base   (c) Salt

❓ The compound $CaCO_3$ (calcium carbonate) dissociates into a calcium ion (Ca^{2+}) and a carbonate ion (CO_3^{2-}). Is it an acid, a base, or a salt? What about H_2SO_4, which dissociates into two H^+ and one SO_4^{2-}?

Acid–Base Balance: The Concept of pH

To ensure homeostasis, body fluids must contain almost balanced quantities of acids and bases. The more hydrogen ions (H^+) dissolved in a solution, the more acidic is the solution; conversely, the more hydroxide ions (OH^-), the more basic (alkaline) is the solution. The chemical reactions that take place in the body are very sensitive to even small changes in the acidity or alkalinity of the body fluids in which they occur. Any departure from the narrow limits of normal H^+ and OH^- concentrations greatly disrupts body functions.

A solution's acidity or alkalinity is expressed on the **pH scale**, which extends from 0 to 14 (Figure 2.7). This scale is based on the number of hydrogen ions in a solution. The midpoint of the pH scale is 7, where the numbers of H^+ and OH^- are equal. A solution with a pH of 7, such as pure water, is neutral—neither acidic nor alkaline. A solution that has more H^+ than OH^- is **acidic** and has a pH below 7. A solution that has more OH^- than H^+ is **basic (alkaline)** and has a pH above 7. A change of one whole number on the pH scale represents a *10-fold* change in the number of H^+. At a pH of 6, there are 10 times more H^+ than at a pH of 7. Put another way, a pH of 6 is 10 times more acidic than a pH of 7, and a pH of 9 is 100 times more alkaline than a pH of 7.

Maintaining pH: Buffer Systems

Although the pH of various body fluids may differ, the normal limits for each are quite narrow. Figure 2.7 also shows the pH values for certain body fluids compared with those of common household substances. Homeostatic mechanisms maintain the pH of blood between 7.35 and 7.45, so that it is slightly more basic than pure water. Even though strong acids and bases may be taken into the body or be formed by body cells, the pH of fluids inside and outside cells remains almost constant. One important reason is the presence of **buffer systems**.

Buffers are chemical compounds that act quickly to temporarily bind H^+, removing the highly reactive, excess H^+ from solution but not from the body. Buffers prevent rapid, drastic changes in the pH of a body fluid by converting strong acids and bases into weak acids and bases. Strong acids release H^+ more readily than weak acids and thus contribute more free hydrogen ions. Similarly, strong bases raise pH more than weak ones.

One example of a buffer system is the **carbonic acid–bicarbonate buffer system**. It is based on the *bicarbonate ion* (HCO_3^-), which can act as a weak base, and *carbonic acid* (H_2CO_3), which can act as a weak acid. HCO_3^- is a significant anion in both intracellular and extracellular fluids. Because the kidneys reabsorb filtered HCO_3^-, this important buffer is not lost in the urine. If there is an excess of H^+, the HCO_3^- can function as a weak base and remove the excess H^+ as follows:

$$H^+ \quad + \quad HCO_3^- \quad \longrightarrow \quad H_2CO_3$$
$$\text{Hydrogen ion} \qquad \text{Bicarbonate ion} \qquad \text{Carbonic acid}$$
$$\text{(weak base)}$$

Figure 2.7 The pH scale. A pH below 7 indicates an acidic solution, or more H^+ than OH^-. The lower the numerical value of the pH, the more acidic the solution because the H^+ concentration becomes progressively greater. A pH above 7 indicates a basic (alkaline) solution; that is, there are more OH^- than H^+. The higher the pH, the more basic the solution.

At pH 7 (neutrality), the concentrations of H^+ and OH^- are equal.

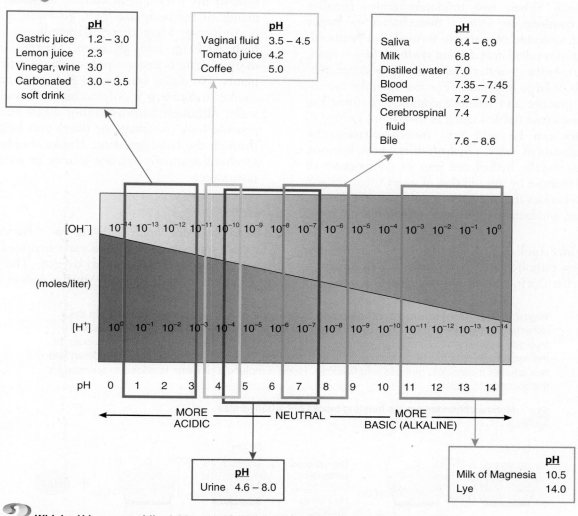

	pH
Gastric juice	1.2 – 3.0
Lemon juice	2.3
Vinegar, wine	3.0
Carbonated soft drink	3.0 – 3.5

	pH
Vaginal fluid	3.5 – 4.5
Tomato juice	4.2
Coffee	5.0

	pH
Saliva	6.4 – 6.9
Milk	6.8
Distilled water	7.0
Blood	7.35 – 7.45
Semen	7.2 – 7.6
Cerebrospinal fluid	7.4
Bile	7.6 – 8.6

	pH
Urine	4.6 – 8.0

	pH
Milk of Magnesia	10.5
Lye	14.0

MORE ACIDIC — NEUTRAL — MORE BASIC (ALKALINE)

Which pH is more acidic, 6.82 or 6.91? Which pH is closer to neutral, 8.41 or 5.59?

Conversely, if there is a shortage of H^+, the H_2CO_3 can function as a weak acid and provide H^+ as follows:

$$H_2CO_3 \longrightarrow H^+ + HCO_3^-$$

Carbonic acid (weak acid)   Hydrogen ion   Bicarbonate ion

You will learn more about buffers in Chapter 22.

Organic Compounds

Carbohydrates

Carbohydrates are organic compounds and include sugars, glycogen, starches, and cellulose. The elements present in carbohydrates are carbon, hydrogen, and oxygen. The ratio of carbon to hydrogen to oxygen atoms is usually 1:2:1. For example, the molecular formula for the small carbohydrate glucose is $C_6H_{12}O_6$. Carbohydrates are divided into three major groups based on their size: monosaccharides, disaccharides, and polysaccharides. Monosaccharides and disaccharides are termed **simple sugars**, and polysaccharides are also known as **complex carbohydrates**.

1. **Monosaccharides** (mon′-ō-SAK-a-rīds; *mono-* = one; *sacchar-* = sugar) are the building blocks of carbohydrates. In your body, the principal function of the monosaccharide glucose is to serve as a source of chemical energy for generating the ATP that fuels metabolic

reactions. Ribose and deoxyribose are monosaccharides used to make ribonucleic acid (RNA) and deoxyribonucleic acid (DNA), which are described later in the chapter.

2. **Disaccharides** (dī-SAK-a-rīds; *di-* = two) are simple sugars that consist of two monosaccharides joined by a covalent bond. When two monosaccharides (smaller molecules) combine to form a disaccharide (a larger molecule), a molecule of water is formed and removed. Such a reaction is called **dehydration synthesis** (*de-* = from, down, or out; *hydra-* = water). Such reactions occur during synthesis of large molecules. For example, the monosaccharides glucose and fructose combine to form the disaccharide sucrose (table sugar), as shown in Figure 2.8. Disaccharides can be split into monosaccharides by adding a molecule of water, a **hydrolysis** reaction. Sucrose, for example, may be hydrolyzed into its components of glucose and fructose by the addition of water (Figure 2.8a). Other disaccharides include maltose (glucose + glucose), or malt sugar, and lactose (glucose + galactose), the sugar in milk.

3. **Polysaccharides** (pol′-ē-SAK-a-rīds; *poly-* = many) are large, complex carbohydrates that contain tens or hundreds of monosaccharides joined through dehydration synthesis

reactions. Like disaccharides, polysaccharides can be broken down into monosaccharides through hydrolysis reactions. The main polysaccharide in the human body is *glycogen*, which is made entirely of glucose units joined together in branching chains (Figure 2.9). Glycogen is stored in cells of the liver and in skeletal muscles. If energy demands of the body are high, glycogen is broken down into glucose; when energy demands are low, glucose is built back up into glycogen. *Starches* are also made of glucose units and are polysaccharides made mostly by plants. We digest starches to glucose as another energy source. *Cellulose* is a polysaccharide found in plant cell walls. Although humans cannot digest cellulose, it does provide bulk (roughage or fiber) that helps move feces through the large intestine. Unlike simple sugars, polysaccharides usually are not soluble in water and do not taste sweet.

Lipids

Like carbohydrates, **lipids** (LIP-ids; *lip-* = fat) contain carbon, hydrogen, and oxygen. Unlike carbohydrates, they do not have a 2:1 ratio of hydrogen to oxygen. The proportion of oxygen atoms in lipids is usually smaller than in carbohydrates,

Figure 2.8 Dehydration synthesis and hydrolysis of a molecule of sucrose. In the dehydration synthesis reaction (read from left to right), two smaller molecules, glucose and fructose, are joined to form a larger molecule of sucrose. Note the loss of a water molecule. In the hydrolysis reaction (read from right to left), the larger sucrose molecule is broken down into two smaller molecules, glucose and fructose. Here, a molecule of water is added to sucrose for the reaction to occur.

Monosaccharides are the building blocks of carbohydrates.

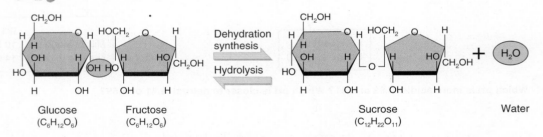

(a) Dehydration synthesis and hydrolysis of sucrose

(b) Lactose

(c) Maltose

How many carbons are there in fructose? in sucrose?

Figure 2.9 Part of a glycogen molecule, the main polysaccharide in the human body.

🔑 Glycogen is made up of glucose units and is the storage form of carbohydrate in the human body.

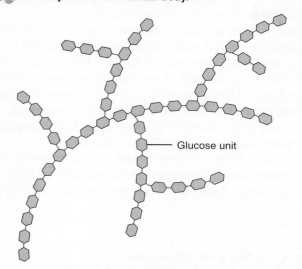

Glucose unit

❓ **Which body cells store glycogen?**

so there are fewer polar covalent bonds. As a result, most lipids are hydrophobic; that is, they are insoluble in water.

The diverse lipid family includes triglycerides (fats and oils), phospholipids (lipids that contain phosphorus), steroids, fatty acids, and fat-soluble vitamins (vitamins A, D, E, and K).

The most plentiful lipids in your body and in your diet are the *triglycerides* (trī-GLI-cer-īdes; *tri-* = three). At room temperature, triglycerides may be either solids (fats) or liquids (oils). They are the body's most highly concentrated form of chemical energy, storing more than twice as much chemical energy per gram as carbohydrates or proteins. Our capacity to store triglycerides in fat tissue, called adipose tissue, is unlimited for all practical purposes. Excess dietary carbohydrates, proteins, fats, and oils all have the same fate: They are deposited in adipose tissue as triglycerides.

A triglyceride consists of two types of building blocks: a single glycerol molecule and three fatty acid molecules. A three-carbon *glycerol* molecule forms the backbone of a triglyceride (Figure 2.10). Three *fatty acids* are attached by dehydration synthesis reactions, one to each carbon of the glycerol backbone. The fatty acid chains of a triglyceride may be saturated, monounsaturated, or polyunsaturated. *Saturated fats* contain only *single covalent bonds* between fatty acid carbon atoms. Because they do not contain any double bonds, each carbon atom is *saturated with hydrogen atoms* (see palmitic acid and stearic acid in Figure 2.10). Triglycerides with mainly saturated fatty acids are solid at room temperature and occur mostly in meats (especially red meats) and nonskim dairy products (whole milk, cheese, and butter). They also occur in a few tropical plants, such as cocoa, palm, and coconut. Diets that contain large amounts of saturated fats are associated with disorders such as heart disease and colorectal cancer. *Monounsaturated fats* (*mono-* = one) contain fatty acids with *one double covalent bond* between two fatty acid carbon atoms and thus are not

Figure 2.10 Triglycerides consist of three fatty acids attached to a glycerol backbone. The fatty acids vary in length and the number and location of double bonds between carbon atoms (C=C). Shown here is a triglyceride molecule that contains two saturated fatty acids and one monounsaturated fatty acid.

🔑 A triglyceride consists of two types of building blocks: a single glycerol molecule and three fatty acid molecules.

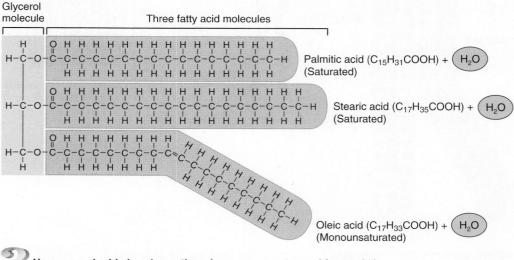

❓ **How many double bonds are there in a monounsaturated fatty acid?**

completely saturated with hydrogen atoms (see oleic acid in Figure 2.10). Olive oil, peanut oil, canola oil, most nuts, and avocados are rich in triglycerides with monounsaturated fatty acids. Monounsaturated fats are thought to decrease the risk of heart disease. *Polyunsaturated fats* (*poly-* = many) contain *more than one double covalent bond* between fatty acid carbon atoms. Corn oil, safflower oil, sunflower oil, soybean oil, and fatty fish (salmon, tuna, and mackerel) contain a high percentage of polyunsaturated fatty acids. Polyunsaturated fats are also believed to decrease the risk of heart disease. However, when products such as margarine and vegetable shortening are made from polyunsaturated fats, compounds called *trans* fatty acids are produced. Trans fatty acids, like saturated fats, increase the risk of cardiovascular disease.

Like triglycerides, *phospholipids* have a glycerol backbone and two fatty acids attached to the first two carbons (Figure 2.11a). Attached to the third carbon is a phosphate group (PO_4^{3-}) that links a small charged group to the glycerol backbone. The nonpolar fatty acids form the hydrophobic "tails" of a phospholipid, and the polar phosphate group and charged group form the hydrophilic "head" (Figure 2.11b). Phospholipids line up tails-to-tails in a double row to make up much of the membrane that surrounds each cell (Figure 2.11c).

The structure of *steroids*, with their four rings of carbon atoms, differs considerably from that of the triglycerides and phospholipids. Cholesterol (Figure 2.12a), which is needed for membrane structure, is the steroid from which other steroids may be synthesized by body cells. For example, cells in the ovaries of females synthesize estradiol (Figure 2.12b), which is one of the *estrogens* (female sex hormones). Estrogens regulate sexual functions. Other steroids include *testosterone* (the main male sex hormone), which also regulates

Figure 2.11 Phospholipids. (a) In the synthesis of phospholipids, two fatty acids attach to the first two carbons of the glycerol backbone. A phosphate group links a small charged group to the third carbon in glycerol. In (b), the circle represents the polar head region, and the two wavy lines represent the two nonpolar tails.

Phospholipids are the main lipids in cell membranes.

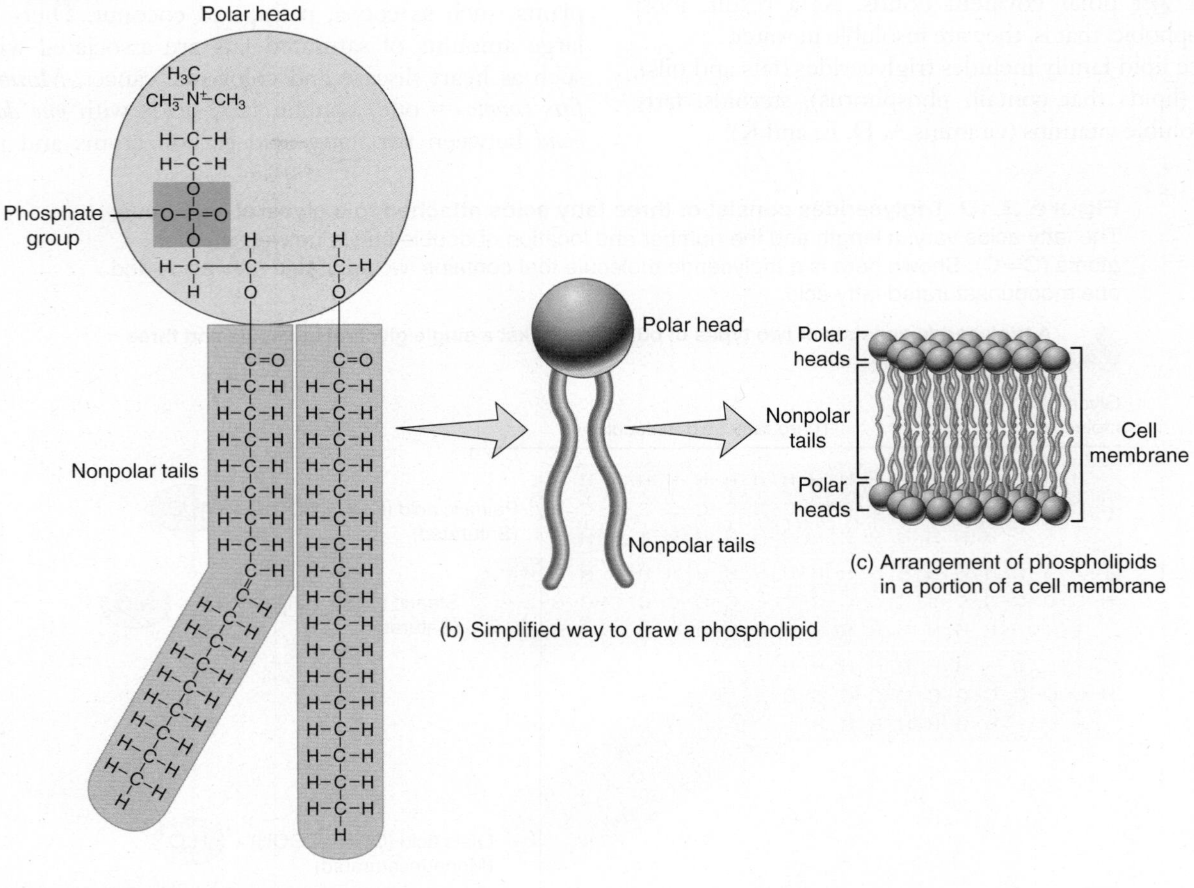

(a) Chemical structure of a phospholipid

(b) Simplified way to draw a phospholipid

(c) Arrangement of phospholipids in a portion of a cell membrane

 How does a phospholipid differ from a triglyceride?

CLINICAL CONNECTION | Fatty Acids in Health and Disease

A group of fatty acids called **essential fatty acids (EFAs)** are essential to human health. However, they cannot be made by the human body and must be obtained from foods or supplements. Among the more important EFAs are *omega-3 fatty acids, omega-6 fatty acids*, and cis-*fatty acids*.

Omega-3 and omega-6 fatty acids are polyunsaturated fatty acids that may have a protective effect against heart disease and stroke by lowering total cholesterol, raising HDL (high-density lipoproteins or "good cholesterol") and lowering LDL (low-density lipoproteins or "bad cholesterol"). In addition, they decrease bone loss; reduce symptoms of arthritis due to inflammation; promote wound healing; improve certain skin disorders (psoriasis, eczema, and acne); and improve mental functions. Primary sources of omega-3 fatty acids include flaxseed, fatty fish, oils that have large amounts of polyunsaturated fats, fish oils, and walnuts. Primary sources of omega-6 fatty acids include most processed foods (cereals, breads, white rice), eggs, baked goods, oils with large amounts of polyunsaturated fats, and meats (especially organ meats, such as liver).

Cis-fatty acids are nutritionally beneficial monounsaturated fatty acids that are used by the body to produce hormone-like regulators and cell membranes. However, when *cis*-fatty acids are heated, pressurized, and combined with a catalyst (usually nickel) in a process called *hydrogenation*, they are changed to unhealthy *trans* fatty acids. Hydrogenation is used by manufacturers to make vegetable oils solid at room temperature and less likely to turn rancid. Hydrogenated or *trans* fatty acids are common in commercially baked goods (crackers, cakes, and cookies), salty snack foods, some margarines, and fried foods (donuts and french fries). If a product label contains the words "hydrogenated" or "partially hydrogenated," then the product contains *trans* fatty acids. Among the adverse effects of *trans* fatty acids are an increase in total cholesterol, a decrease in HDL, an increase in LDL, and an increase in triglycerides. These effects, which can increase the risk of heart disease and other cardiovascular diseases, are similar to those caused by saturated fats. •

Figure 2.12 **Steroids.** All steroids have four rings of carbon atoms. The individual rings are designated by the letters A, B, C, and D.

 Cholesterol is the starting material for synthesis of other steroids in the body.

(a) Cholesterol

(b) Estradiol (an estrogen or female sex hormone)

 Which dietary lipids are thought to contribute to atherosclerosis?

sexual functions; cortisol, which is necessary for maintaining normal blood sugar levels; bile salts, which are needed for lipid digestion and absorption; and vitamin D, which is related to bone growth.

Proteins

Proteins are large molecules that contain carbon, hydrogen, oxygen, and nitrogen; some proteins also contain sulfur. Much more complex in structure than carbohydrates or lipids, proteins have many roles in the body and are largely responsible for the structure of body cells. For example, proteins termed enzymes speed up particular chemical reactions, other proteins are responsible for contraction of muscles, proteins called antibodies help defend the body against invading microbes, and some hormones are proteins.

Amino acids (a-MĒ-nō) are the building blocks of proteins. All amino acids have an *amino group* ($—NH_2$) at one end and a *carboxyl group* ($—COOH$) at the other end. Each of the 20 different amino acids has a different *side chain* (R group) (Figure 2.13a). The covalent bonds that join amino acids together to form more complex molecules are called *peptide bonds* (Figure 2.13b).

The union of two or more amino acids produces a **peptide** (PEP-tīd). When two amino acids combine, the molecule is called a **dipeptide** (Figure 2.13b). Adding another amino acid to a dipeptide produces a **tripeptide**. A **polypeptide** contains a large number of amino acids. Proteins are polypeptides that contain as few as 50 or as many as 2000 amino acids. Because each variation in the number and sequence of amino acids produces a different protein, a great variety of proteins is possible. The situation is similar to using an alphabet of 20 letters to form words. Each letter would be equivalent to an amino acid, and each word would be a different protein.

An alteration in the sequence of amino acids can have serious consequences. For example, a single substitution of an amino acid in hemoglobin, a blood protein, can result in a deformed molecule that produces **sickle-cell disease** (see Common Disorders section in Chapter 14).

A protein may consist of only one polypeptide or several intertwined polypeptides. A given type of protein has a unique three-dimensional shape because of the ways that each individual polypeptide twists and folds as associated polypeptides come together. If a protein encounters a hostile environment in which temperature, pH, or ion concentration is significantly altered, it may unravel and lose its characteristic shape. This process is called **denaturation** (dē-nā′-chur-Ā-shun). Denatured proteins are no longer functional. A common example of denaturation is seen in frying an egg. In a raw egg, the egg-white protein (albumin) is soluble and the egg white appears as a clear, viscous fluid. When heat is applied to the egg, however, the albumin denatures; it changes shape, becomes insoluble, and turns white.

Figure 2.13 Amino acids. (a) In keeping with their name, amino acids have an amino group (shaded blue) and a carboxyl (acid) group (shaded red). The side chain (R group) is shaded gold and is different in each type of amino acid. (b) When two amino acids are chemically united by dehydration synthesis (read from left to right), the resulting covalent bond between them is called a peptide bond. The peptide bond is formed at the point where water is lost. Here, the amino acids glycine and alanine are joined to form the dipeptide glycylalanine. Breaking a peptide bond occurs by hydrolysis (read from right to left).

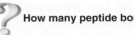 **Amino acids are the building blocks of proteins.**

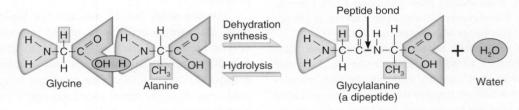

(a) Structure of an amino acid

(b) Protein formation

How many peptide bonds would there be in a tripeptide?

Enzymes

As we have seen, chemical reactions occur when chemical bonds are made or broken as atoms, ions, or molecules collide with one another. At normal body temperature, such collisions occur too infrequently to maintain life. *Enzymes* (EN-zīms) are the living cell's solution to this problem, because they speed up chemical reactions by increasing the frequency of collisions and by properly orienting the colliding molecules. Substances such as enzymes that can speed up chemical reactions without themselves being altered are called *catalysts* (KAT-a-lists). In living cells, most enzymes are proteins. The names of enzymes usually end in *-ase*. All enzymes can be grouped according to the types of chemical reactions they catalyze. For example, *oxidases* add oxygen, *kinases* add phosphate, *dehydrogenases* remove hydrogen, *anhydrases* remove water, *ATPases* split ATP, *proteases* break down proteins, and *lipases* break down lipids.

Enzymes catalyze selected reactions with great efficiency and with many built-in controls. Three important properties of enzymes are their specificity, efficiency, and control.

1. **Specificity.** Enzymes are highly specific. Each particular enzyme catalyzes a particular chemical reaction that involves specific *substrates*, the molecules on which the en-

zyme acts, and that gives rise to specific *products*, the molecules produced by the reaction. In some cases, the enzyme fits the substrate like a key fits in a lock. In other cases, the enzyme changes its shape to fit snugly around the substrate once the substrate and enzyme come together. Each of the more than 1000 known enzymes in your body has a characteristic three-dimensional shape with a specific surface configuration that allows it to fit specific substrates.

2. **Efficiency.** Under optimal conditions, enzymes can catalyze reactions at rates that are millions to billions of times more rapid than those of similar reactions occurring without enzymes. A single enzyme molecule can convert substrate molecules to product molecules at rates as high as 600,000 per second.

3. **Control.** Enzymes are subject to a variety of cellular controls. Their rate of synthesis and their concentration at any given time are under the control of a cell's genes. Substances within the cell may either enhance or inhibit activity of a given enzyme. Many enzymes exist in both active and inactive forms within the cell. The rate at which the inactive form becomes active or vice versa is determined by the chemical environment inside the cell. Many enzymes require a nonprotein

substance, known as a **cofactor** or **coenzyme**, to operate properly. Ions of iron, zinc, magnesium, or calcium are cofactors; niacin or riboflavin, derivatives of B vitamins, act as coenzymes.

Figure 2.14 illustrates the actions of an enzyme.

1 The substrates attach to the **active site** of the enzyme molecule, the specific part of the enzyme that catalyzes the reaction, forming a temporary compound called the **enzyme–substrate complex**. In this reaction, the substrates are the disaccharide sucrose and a molecule of water.

2 The substrate molecules are transformed by the rearrangement of existing atoms, the breakdown of the substrate molecule, or the combination of several substrate molecules into products of the reaction. Here the products are two monosaccharides: glucose and fructose.

3 After the reaction is completed and the reaction products move away from the enzyme, the unchanged enzyme is free to attach to another substrate molecule.

Figure 2.14 How an enzyme works.

An enzyme speeds up a chemical reaction without being altered or consumed.

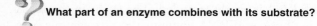

Substrates
Sucrose
Water

H_2O

Enzyme
Sucrase

Active site
of enzyme

1 Enzyme and substrate come together at active site of enzyme, forming an enzyme–substrate complex

Products
Glucose
Fructose

3 When reaction is complete, enzyme is unchanged and free to catalyze same reaction again on a new substrate

2 Enzyme catalyzes reaction and transforms substrate into products

What part of an enzyme combines with its substrate?

CLINICAL CONNECTION | Lactose Intolerance

Enzyme deficiencies may lead to certain disorders. For example, some people do not produce enough lactase, an enzyme that breaks down the disaccharide lactose into the monosaccharides glucose and galactose. This deficiency causes a condition called **lactose intolerance**, in which undigested lactose retains fluid in the feces, and bacterial fermentation of lactose results in the production of gases. Symptoms of lactose intolerance include diarrhea, gas, bloating, and abdominal cramps after consumption of milk and other dairy products. The severity of symptoms varies from relatively minor to sufficiently serious to require medical attention. Persons with lactose intolerance can take dietary enzyme supplements to aid in the digestion of lactose. •

Nucleic Acids: Deoxyribonucleic Acid (DNA) and Ribonucleic Acid (RNA)

Nucleic acids (noo-KLĒ-ic), so named because they were first discovered in the nuclei of cells, are huge organic molecules that contain carbon, hydrogen, oxygen, nitrogen, and phosphorus. The two kinds of nucleic acids are **deoxyribonucleic acid (DNA)** (dē-ok′-sē-rī′-bō-noo-KLĒ-ik) and **ribonucleic acid (RNA)**.

A nucleic acid molecule is composed of repeating building blocks called **nucleotides**. Each nucleotide of DNA consists of three parts (Figure 2.15a):

■ One of four different **nitrogenous bases**, ring-shaped molecules that contain atoms of C, H, O, and N.

■ A five-carbon monosaccharide called *deoxyribose*.

■ A *phosphate group* (PO_4^{3-}).

In DNA, the four bases are adenine (A), thymine (T), cytosine (C), and guanine (G). Figure 2.15b shows the following structural characteristics of the DNA molecule:

1. The molecule consists of two strands, with crossbars. The strands twist about each other in the form of a **double helix** so that the shape resembles a twisted rope ladder.

2. The uprights (strands) of the DNA ladder consist of alternating phosphate groups and the deoxyribose portions of the nucleotides.

3. The rungs of the ladder contain paired nitrogenous bases, which are held together by hydrogen bonds. Adenine always pairs with thymine, and cytosine always pairs with guanine.

About 1000 rungs of DNA comprise a **gene**, a portion of a DNA strand that performs a specific function, for example, providing instructions to synthesize the hormone insulin. Humans have about 30,000 genes. Genes determine which traits we inherit, and they control all the activities that take place in our cells throughout a lifetime. Any

Figure 2.15 DNA molecule. (a) A nucleotide consists of a nitrogenous base, a five-carbon sugar, and a phosphate group. (b) The paired nitrogenous bases project toward the center of the double helix. The structure is stabilized by hydrogen bonds (dotted lines) between each base pair. There are two hydrogen bonds between adenine and thymine and three between cytosine and guanine.

Nucleotides are the building blocks of nucleic acids.

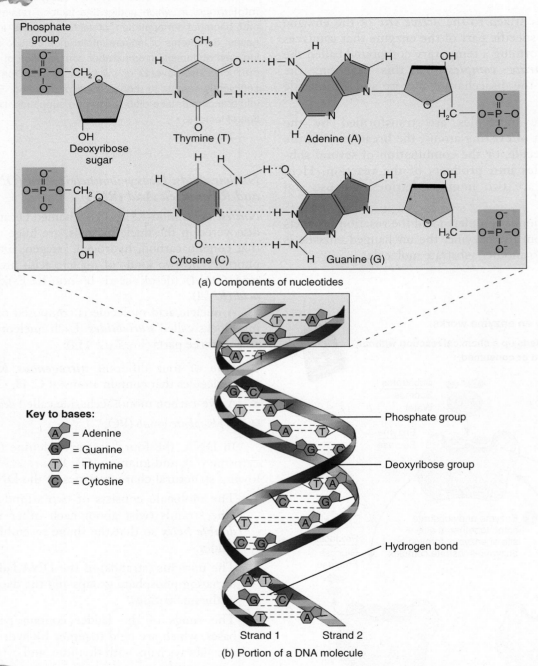

(a) Components of nucleotides

Key to bases:

Ⓐ = Adenine
Ⓖ = Guanine
Ⓣ = Thymine
Ⓒ = Cytosine

Phosphate group

Deoxyribose group

Hydrogen bond

Strand 1   Strand 2

(b) Portion of a DNA molecule

Which nitrogenous base is not present in RNA? Which nitrogenous base is not present in DNA?

change that occurs in the sequence of nitrogenous bases of a gene is called a *mutation*. Some mutations can result in the death of a cell, cause cancer, or produce genetic defects in future generations.

RNA, the second kind of nucleic acid, is copied from DNA but differs from DNA in several respects. DNA is double-stranded, RNA is single-stranded. The sugar in the RNA nucleotide is ribose, and RNA contains the nitrogenous base

Dietary Supplements—Handle with Care

Sales of dietary supplements are booming. Preparation of echinacea and ginseng stand next to bottles of vitamin C and calcium in medicine cabinets across North America. But although these products look like drugs, it is important for users to realize they are more loosely regulated. While some supplements are useful for specific problems, others are a waste of money, and several can even be harmful to your health.

Does Natural Mean Safe?

Dietary supplements are manufactured from ingredients that could appear in the diet: components of plants and animals. Keep in mind, however, that these ingredients may be concentrated and changed in other ways, so that in the end they bear little resemblance to chemicals that might appear in your breakfast, lunch, or dinner.

Dietary supplements bear a closer resemblance to drugs than food. People often assume that just because they are "natural" they are harmless. But even vitamins and minerals can be harmful if you consume too much of them. Natural chemicals are still chemicals. They participate in chemical reactions in your body. They have chemical effects in the same way that manufactured drugs do. Dietary supplements can't be effective and harmless at the same time, because anything that has a physiological effect can be harmful at some dose. All drugs become toxic if you take too much of them.

Handle with Care

If you want to use dietary supplements, you must also use your head. Because regulation of these supplements is currently fairly relaxed in most countries, you can't believe everything the manufacturer says on the label or in advertising literature. If a product sounds too good to be true, beware!

If you decide to use supplements, look for scientific information about the product, especially research findings in scientific journals. Keep track of what you take, since you may take several products that all have the same ingredient. Avoid supplements if you are pregnant or breastfeeding. And check with your health care provider if you are taking any medications or have health conditions that might be affected by the supplement.

> *Think It Over . . .* **Your Aunt Clara tells you she is taking a weight-loss product. "It's natural, so it's safe," she says. Since it's not working very well, she is now taking double the recommended dose. Based on what you have just read, what would you say to her?**

---

uracil (U) rather than thymine. Cells contain three different kinds of RNA: messenger RNA, ribosomal RNA, and transfer RNA. Each has a specific role to perform in carrying out the instructions encoded in DNA for protein synthesis, as will be described in Chapter 3.

Adenosine Triphosphate

Adenosine triphosphate (ATP) (a-DEN-ō-sēn trī-FOS-fāt) is the "energy currency" of living organisms. As you learned earlier in the chapter, ATP transfers energy from energy-releasing reactions to energy-requiring reactions that maintain cellular activities. Among these cellular activities are contraction of muscles, movement of chromosomes during cell division, movement of structures within cells, transport of substances across cell membranes, and synthesis of larger molecules from smaller ones.

Structurally, ATP consists of three phosphate groups attached to adenosine, which is composed of adenine and ribose (Figure 2.16). The energy-transferring reaction occurs via hydrolysis: Removal of the last phosphate group (PO_4^{3-}), symbolized by (P) in the following discussion, by addition of a water molecule liberates energy and leaves a molecule called **adenosine diphosphate (ADP)**. The enzyme that catalyzes the hydrolysis of ATP is called *ATPase*. This reaction may be represented as follows:

$$\text{ATP} + \text{H}_2\text{O} \xrightarrow{ATPase} \text{P} + \text{E}$$

Adenosine triphosphate   Water   Phosphate group   Energy

The energy released by the breakdown of ATP into ADP is constantly being used by the cell. As the supply of ATP at any given time is limited, a mechanism exists to replenish it: The enzyme *ATP synthase* promotes the addition of a phosphate group to ADP. The reaction may be represented as follows:

$$\text{ADP} + \text{P} + \text{E} \xrightarrow{ATP\ synthase} \text{ATP} + \text{H}_2\text{O}$$

Adenosine diphosphate   Phosphate group   Energy   Adenosine triphosphate   Water

Figure 2.16 Structure of ATP and ADP. The two phosphate bonds that can be used to transfer energy are indicated by "squiggles" (~). Most often energy transfer involves hydrolysis of the terminal phosphate bond of ATP.

🔑 **ATP transfers chemical energy to power cellular activities.**

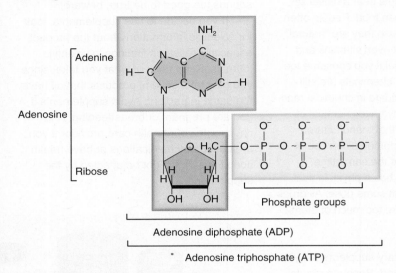

Adenosine diphosphate (ADP)

Adenosine triphosphate (ATP)

❓ **What are some cellular activities that depend on energy supplied by ATP?**

As you can see from this reaction, energy is required to produce ATP. The energy needed to attach a phosphate group to ADP is supplied mainly by the breakdown of glucose in a process called cellular respiration, which you will learn more about in Chapter 20.

✓ CHECKPOINT

5. How do inorganic compounds differ from organic compounds?
6. What functions does water perform in the body?
7. What is a buffer?
8. Distinguish among saturated, monounsaturated, and polyunsaturated fats.
9. What are the important properties of enzymes?
10. How do DNA and RNA differ?
11. Why is ATP important?

• • •

In Chapter 1, you learned that the human body is characterized by various levels of organization and that the chemical level consists of atoms and molecules. Now that you have an understanding of the chemicals in the body, you will see in the next chapter how they are organized to form the structures of cells and perform the activities of cells that contribute to homeostasis.

CHAPTER REVIEW AND RESOURCE SUMMARY

REVIEW

2.1 Introduction to Chemistry

1. Chemistry is the science of the structure and interactions of matter, which is anything that occupies space and has mass. Matter is made up of **chemical elements**. The elements oxygen (O), carbon (C), hydrogen (H), and nitrogen (N) make up 96% of the body's mass.

2. Each element is made up of units called **atoms**, which consist of a **nucleus** that contains **protons** and **neutrons**, and **electrons** that move about the nucleus in **electron shells**. The number of electrons is equal to the number of protons in an atom. The **atomic number**, the number of protons, distinguishes the atoms of one element from those of another element. The combined total of protons and neutrons in an atom is its **mass number**.

3. An atom that *gives up or gains* electrons becomes an **ion**—an atom that has a positive or negative charge due to unequal numbers of protons and electrons.

4. A **molecule** is a substance that consists of two or more chemically combined atoms. The molecular formula indicates the number and type of atoms that make up a molecule.

5. A **compound** is a substance that can be broken down into two or more different elements by ordinary chemical means.

6. A **free radical** is a destructive ion or molecule that has an unpaired electron in its outermost shell.

7. **Chemical bonds** hold the atoms of a molecule together. Electrons in the **valence shell** (outermost shell) are the parts of an atom that participate in chemical reactions (are involved in forming and breaking bonds).

8. When outer shell electrons are transferred from one atom to another, the transfer forms ions with unlike charges that attract the ions to each other and form **ionic bonds**. Positively charged ions are

RESOURCES

Anatomy Overview—Common Biomolecules: Water
Anatomy Overview—Common Biomolecules: Blood Gases
Anatomy Overview—Common Biomolecules: Electrolytes
Animation—Atomic Structure and Basis of Bonds
Animation—Chemical Bonding
Animation—Types of Reactions and Equilibrium
Exercise—Bond Boulevard
Exercise—Reaction Race
Concepts and Connections—Chemical Bonds

REVIEW	RESOURCES

called **cations**; negatively charged ions are called **anions**. In a **covalent bond**, pairs of outer shell electrons are shared between two atoms. **Hydrogen bonds** are weak bonds between hydrogen and certain other atoms. They establish important links between water molecules and between different parts of a large molecule such as proteins and deoxyribonucleic acid (DNA), where they add strength and stability and help determine the molecule's three-dimensional shape.

9. **Energy** is the capacity to do work. **Potential energy** is energy stored by matter due to its position. **Kinetic energy** is the energy of matter in motion. **Chemical energy** is a form of potential energy stored in the bonds of molecules. In chemical reactions, breaking old bonds requires energy and forming new bonds releases energy.

10. In a **synthesis reaction** (anabolic reaction), two or more atoms, ions, or molecules combine to form a new and larger molecule. In a **decomposition reaction** (catabolic reaction), a molecule is split apart into smaller molecules, ions, or atoms.

11. When nutrients such as glucose are broken down via decomposition reactions, some of the energy released is temporarily stored in **adenosine triphosphate (ATP)** and then later used to drive energy-requiring synthesis reactions that build body structures, such as muscles and bones.

12. **Exchange reactions** are combination synthesis and decomposition reactions. **Reversible reactions** can proceed in both directions under different conditions.

2.2 Chemical Compounds and Life Processes

1. **Inorganic compounds** usually are structurally simple and lack carbon. **Organic compounds** always contain carbon, usually contain hydrogen, and always have covalent bonds.

2. **Water** is the most abundant substance in the body. It is an excellent **solvent**, participates in chemical reactions, absorbs and releases heat slowly, requires a large amount of heat to change from a liquid to a gas, and serves as a lubricant.

3. Inorganic **acids**, **bases**, and **salts** dissociate into ions in water. An acid ionizes into hydrogen ions (H^+); a base usually ionizes into hydroxide ions (OH^2). A salt ionizes into neither H^+ nor OH^- ions.

4. The pH of body fluids must remain fairly constant for the body to maintain homeostasis. On the **pH scale**, 7 represents neutrality. Values below 7 indicate **acidic** solutions, and values above 7 indicate **basic (alkaline)** solutions.

5. **Buffer systems** help maintain pH by converting strong acids or bases into weak acids or bases.

6. **Carbohydrates** include sugars, glycogen, and starches. They may be **monosaccharides**, **disaccharides**, or **polysaccharides**. Carbohydrates provide most of the chemical energy needed to generate ATP. Carbohydrates, and other large, organic molecules, are synthesized via **dehydration synthesis** reactions, in which a molecule of water is lost. In the reverse process, called **hydrolysis**, large molecules are broken down into smaller ones upon the addition of water.

7. **Lipids** are a diverse group of compounds that include **triglycerides** (fats and oils), **phospholipids**, and **steroids**. Triglycerides protect, insulate, provide energy, and are stored in adipose tissue. Phospholipids are important membrane components. Steroids are synthesized from cholesterol.

8. **Proteins** are constructed from **amino acids**. They give structure to the body, regulate processes, provide protection, help muscles to contract, transport substances, and serve as enzymes.

9. **Enzymes** are molecules, usually proteins, that speed up chemical reactions and are subject to a variety of cellular controls.

10. **Deoxyribonucleic acid (DNA)** and **ribonucleic acid (RNA)** are nucleic acids consisting of repeating units called **nucleotides**. A nucleotide consists of **nitrogenous bases**, five-carbon sugars, and phosphate groups. DNA is a double helix and is the primary chemical in genes. RNA differs in structure and chemical composition from DNA; its main function is to carry out the instructions encoded in DNA for protein synthesis.

11. **Adenosine triphosphate (ATP)** is the principal energy-transferring molecule in living systems. When it transfers energy, ATP is decomposed by hydrolysis to **adenosine diphosphate (ADP)** and Ⓟ. ATP is synthesized from ADP and Ⓟ using primarily the energy supplied by the breakdown of glucose.

Animation—Polarity and
   Solubility of Molecules
Animation—Water and
   Fluid Flow
Animation—Acids and
   Bases
Animation—Enzyme Function
   and ATP

Figure 2.13 Amino acids

Figure 2.14 How an
enzyme works

Exercise—Destination:
   Acid/Base Balance

SELF-QUIZ

1. The nucleus of an atom contains
   a. electrons.　　b. protons and electrons.　　c. neutrons.
   d. protons and neutrons.　　e. electrons and neutrons.

2. Ionic bonds
   a. involve sharing electrons between atoms.
   b. are strong, stable bonds.
   c. are created by atoms giving away and taking electrons.
   d. are the type of bonding found in most of the body's structures.
   e. are an attraction between water molecules.

3. Combining atoms of two or more different elements results in a(n)
   a. compound.　　b. ion.　　c. free radical.
   d. electrolyte.　　e. superoxide.

4. Matter that cannot be broken down into simpler substances by ordinary chemical reactions is known as
   a. a molecule.　　b. an antioxidant.　　c. a compound.
   d. a buffer.　　e. a chemical element.

5. Chlorine (Cl) has an atomic number of 17. An atom of chlorine may become a chloride ion (Cl^-) by
   a. losing one electron.　　b. losing one neutron.
   c. gaining one proton.　　d. gaining one electron.
   e. gaining two electrons.

6. Which of the following is NOT true?
   a. A substance that separates in water to form some cation other than H^+ and some anion other than OH^- is known as a salt.
   b. A solution that has a pH of 9.4 is acidic.
   c. A solution with a pH of 5 is 100 times more acidic than distilled water, which has a pH of 7.
   d. Buffers help to make the body's pH more stable.
   e. Salts are formed by the interaction between acids and bases.

7. Which of the following organic compounds are NOT paired with their correct subunits (building blocks)?
   a. glycogen/glucose　　b. proteins/monosaccharides
   c. DNA/nucleotides　　d. lipids/glycerol and fatty acids
   e. ATP/ADP and P

8. The type of reaction by which a disaccharide is formed from two monosaccharides is known as a
   a. decomposition reaction.　　b. hydrolysis reaction.
   c. dehydration synthesis reaction.　　d. reversible reaction.
   e. catabolic reaction.

9. Which of the following contains the genetic code in human cells?
   a. DNA　　b. enzymes　　c. RNA
   d. glucose　　e. ATP

10. What is the principal energy-transferring molecule in the body?
    a. ADP　　b. RNA　　c. DNA
    d. ATP　　e. coenzyme

11. Which of the following statements about water is NOT true?
    a. It is involved in many chemical reactions in the body.
    b. It is an important solvent in the human body.
    c. It helps lubricate a variety of structures in the body.
    d. It can absorb a large amount of heat without drastically changing its temperature.
    e. It requires very little heat to change from a liquid to a gas.

12. The difference in H^+ concentration between solutions with a pH of 3 and a pH of 5 is that the solution with the pH of 3 has _____ H^+.
    a. 2 times more　　b. 5 times more　　c. 10 times more
    d. 100 times more　　e. 200 times less

13. Which of the following is NOT a true statement about enzyme activity?
    a. Enzymes form a temporary complex with their substrates.
    b. Enzymes are not permanently altered by the chemical reactions they catalyze.
    c. All proteins are enzymes.
    d. Enzymes are considered organic catalysts.
    e. Enzymes are subject to cellular control.

14. For each item in the following list, place an R if it applies to RNA or a D if it refers to DNA; use R and D if it applies to both RNA and DNA.
    ____ a. composed of nucleotides
    ____ b. forms a double helix
    ____ c. contains thymine
    ____ d. contains the sugar ribose
    ____ e. contains the nitrogenous base uracil
    ____ f. is the hereditary material of cells
    ____ g. contains the sugar deoxyribose
    ____ h. is single-stranded
    ____ i. contains adenosine
    ____ j. contains phosphate groups

15. Classify the following substances as being either inorganic (I) or organic (O).
    ____ a. water
    ____ b. DNA
    ____ c. salt
    ____ d. disaccharide
    ____ e. protein
    ____ f. acid
    ____ g. phospholipid

16. Which type of lipid is the least desirable for maintaining good health?
    a. essential fatty acid　　b. phospholipid
    c. polyunsaturated fat　　d. saturated fat
    e. monounsaturated fat

17. If an enzyme is exposed to an extremely high temperature, it will
    a. divide.　　b. release energy.　　c. become an electrolyte.
    d. form hydrogen bonds.　　e. denature.

18. In what form are lipids stored in the adipose (fat) tissue of the body?
    a. triglycerides   **b.** glycogen   **c.** cholesterol
    d. polypeptides   **e.** disaccharides

19. Approximately 96% of your body's mass is composed of which of the following elements? Place an X beside each correct answer.

    ____ calcium   ____ iron   ____ nitrogen
    ____ phosphorus   ____ sodium   ____ chlorine
    ____ carbon   ____ oxygen   ____ sulfur
    ____ hydrogen   ____ potassium   ____ magnesium

20. Match the following:
    ____ **a.** inorganic compound   **A.** glycogen
    ____ **b.** monosaccharide   **B.** enzyme
    ____ **c.** polysaccharide   **C.** glucose
    ____ **d.** component of triglycerides   **D.** water
    ____ **e.** lipase   **E.** glycerol

CRITICAL THINKING APPLICATIONS

1. While having a tea party, your three-year-old cousin Olivia added milk, lemon juice, and lots of sugar to her tea. The tea now has strange white lumps floating in it. What caused the milk to curdle?

2. You are determined to change to more healthy eating habits and buy a piece of salmon for dinner. You can't decide whether to cook it using margarine made of pure corn oil or the liquid corn oil in your cabinet. Which would be a better choice and why?

3. Alfie was trying out the new Super Genius Home Chemistry Kit that he got for his birthday. He decided to check the pH of his secret formula: lemon juice and diet cola. The pH was 2.5. Next he added tomato juice. Now he has a really disgusting mixture with a pH of 3.5. "Wow! That's twice as strong!" Does Alfie have the makings of a "Super Genius"? Explain.

4. During chemistry lab, Gemma places sucrose (table sugar) in a glass beaker, adds water, and stirs. As the table sugar disappears, she loudly proclaims that she has chemically broken down the sucrose into fructose and glucose. Is Gemma's chemical analysis correct?

ANSWERS TO FIGURE QUESTIONS

2.1 The atomic number of carbon is 6.

2.2 The four most plentiful elements in living organisms are oxygen, carbon, hydrogen, and nitrogen.

2.3 Water is a compound because it contains atoms of two different elements (hydrogen and oxygen).

2.4 K is an electron donor; when it ionizes, it becomes a cation, K^+, because losing one electron from the fourth electron shell leaves eight electrons in the third shell.

2.5 An ionic bond involves the *loss* and *gain* of electrons; a covalent bond involves the *sharing* of pairs of electrons.

2.6 $CaCO_3$ is a salt, and H_2SO_4 is an acid.

2.7 A pH of 6.82 is more acidic than a pH of 6.91. Both pH = 8.41 and pH = 5.59 are 1.41 pH units from neutral (pH = 7).

2.8 There are 6 carbons in fructose, 12 in sucrose.

2.9 Glycogen is stored in liver and skeletal muscle cells.

2.10 A monounsaturated fatty acid has one double bond.

2.11 A triglyceride has three fatty acid molecules attached to a glycerol backbone, and a phospholipid has two fatty acid tails and a phosphate group attached to a glycerol backbone.

2.12 The dietary lipids thought to contribute to atherosclerosis are cholesterol and saturated fats.

2.13 A tripeptide would have two peptide bonds, each linking two amino acids.

2.14 The enzyme's active site combines with the substrate.

2.15 Thymine is present in DNA but not in RNA, and uracil is present in RNA but not in DNA.

2.16 A few cellular activities that depend on energy supplied by ATP are muscular contractions, movement of chromosomes, transport of substances across cell membranes, and synthesis reactions.

CHAPTER 3
CELLS

About 200 different types of cells compose your body. Each *cell* is a living structural and functional unit that is enclosed by a membrane. All cells arise from existing cells by the process of *cell division*, in which one cell divides into two new cells. In your body, different types of cells fulfill unique roles that support homeostasis and contribute to the many functional capabilities of the human organism. *Cell biology* is the study of cellular structure and function. As you study the various parts of a cell and their relationships to each other, you will learn that cell structure and function are intimately related.

Did you know? *Why is it so important to eat a variety of fruits and vegetables? Because your parents wouldn't let you have dessert unless you did? Another good reason to eat plenty of fruits and vegetables is that these foods contain important compounds, known as phytochemicals (literally, "plant chemicals"), which help to keep cells healthy. Some phytochemicals block other chemicals that can cause damage to cells. Others enhance your body's production of enzymes that render potentially cancer-causing substances harmless. Collectively, the actions of phytochemicals promote healthy cellular function, and prevent the types of cellular damage associated with cancer, aging, and heart disease.*

FOCUS ON WELLNESS

Phytochemicals—Protecting Cellular Function

LOOKING BACK TO MOVE AHEAD...

Levels of Organization and Body Systems (Section 1.2)

Ions, Molecules, and Compounds (Section 2.1)

Carbohydrates (Section 2.2)

Lipids (Section 2.2)

Proteins (Section 2.2)

Deoxyribonucleic Acid (DNA) and Ribonucleic Acid (RNA) (Section 2.2)

3.1 A GENERALIZED VIEW OF THE CELL

OBJECTIVE • **Name and describe the three main parts of a cell.**

Figure 3.1 is a generalized view of a cell that shows the main cellular components. Though some body cells lack some cellular structures shown in this diagram, many body cells include most of these components. For ease of study, we can divide a cell into three main parts: the plasma membrane, cytoplasm, and nucleus.

1. The *plasma membrane* forms a cell's flexible outer surface, separating the cell's internal environment (inside the cell) from its external environment (outside the cell). It regulates the flow of materials into and out of a cell to maintain the appropriate environment for normal cellular activities. The plasma membrane also plays a key role in communication among cells and between cells and their external environment.

2. The *cytoplasm* (SĪ-tō-plazm; *-plasm* = formed or molded) consists of all the cellular contents between the plasma membrane and the nucleus. Cytoplasm can be divided into two components: cytosol and organelles. *Cytosol* (SĪ-tō-sol) is the liquid portion of cytoplasm that consists mostly of water plus dissolved solutes and suspended particles. It is also called *intracellular fluid*. Within the cytosol are several different types of **organelles** (or-ga-NELZ = little organs), each of which has a characteristic structure and specific functions.

Figure 3.1 Generalized view of a body cell.

🗝 **The cell is the basic, living, structural and functional unit of the body.**

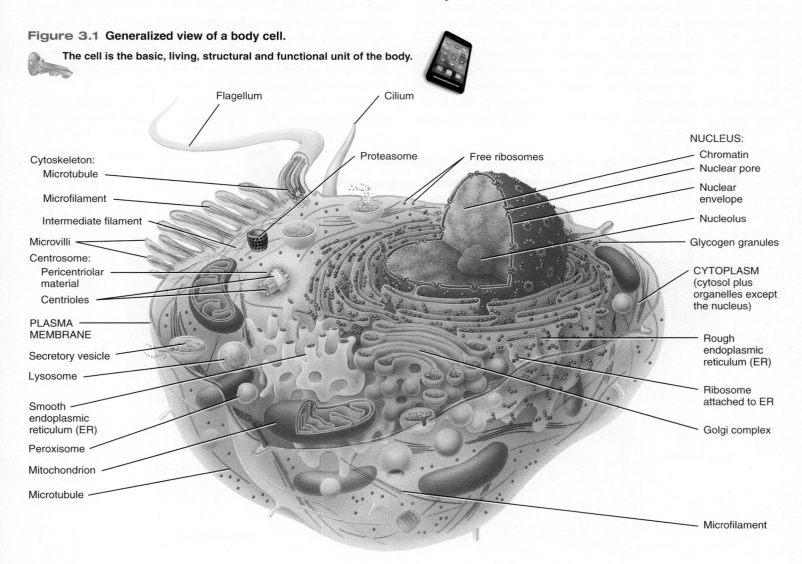

Flagellum — Cilium

Cytoskeleton:
  Microtubule
Microfilament
Intermediate filament
Microvilli
Centrosome:
  Pericentriolar material
  Centrioles
PLASMA MEMBRANE
Secretory vesicle
Lysosome
Smooth endoplasmic reticulum (ER)
Peroxisome
Mitochondrion
Microtubule

Proteasome — Free ribosomes

NUCLEUS:
  Chromatin
  Nuclear pore
  Nuclear envelope
  Nucleolus
Glycogen granules
CYTOPLASM (cytosol plus organelles except the nucleus)
Rough endoplasmic reticulum (ER)
Ribosome attached to ER
Golgi complex
Microfilament

Sectional view

 What are the three principal parts of a cell?

3. The **nucleus** (NOO-klē-us = nut kernel) is the largest organelle of a cell. The nucleus acts as the control center for a cell because it contains the genes, which control cellular structure and most cellular activities.

✓ CHECKPOINT

1. What are the general functions of the three main parts of a cell?

3.2 THE PLASMA MEMBRANE

OBJECTIVE • **Describe the structure and functions of the plasma membrane.**

The plasma membrane is a flexible yet sturdy barrier that consists mostly of lipids and proteins. The basic framework of the plasma membrane is the **lipid bilayer**, two back-to-back layers made up of three types of lipid molecules: **phospholipids** (lipids that contain phosphorus), **cholesterol**, and **glycolipids** (lipids attached to carbohydrates; *glyco* = carbohydrate) (Figure 3.2). The proteins in a membrane are of two types— integral and peripheral (Figure 3.2). **Integral proteins** extend into or through the lipid bilayer. **Peripheral proteins** are loosely attached to the exterior or interior surface of the membrane. Some peripheral proteins, called **glycoproteins**, are proteins attached to carbohydrates.

The plasma membrane allows some substances to move into and out of the cell but restricts the passage of other substances. This property of membranes is called **selective permeability** (per′-mē-a-BIL-i-tē). The lipid bilayer part of the membrane is permeable to water and to nonpolar (lipid-soluble) molecules, such as fatty acids, fat-soluble vitamins, steroids, oxygen, and carbon dioxide. The lipid bilayer is *not* permeable to ions and large, uncharged polar molecules such as glucose and amino acids. These small and medium-sized water-soluble materials may cross the membrane with the assistance of integral proteins. Some integral proteins form **ion channels** through which specific ions, such as potassium ions (K^+), can move into and out of cells (see Figure 3.5). Other membrane proteins act as **carriers (transporters)**, which change shape as they move a substance from one side of the membrane to the other (see Figure 3.6). Large molecules such as proteins are unable to pass through the plasma membrane except by transport within vesicles (discussed later in this chapter).

Most functions of the plasma membrane depend on the types of proteins that are present. Integral proteins called

Figure 3.2 Chemistry and structure of the plasma membrane.

The plasma membrane consists mostly of phospholipids, arranged in a bilayer, and proteins, most of which are glycoproteins.

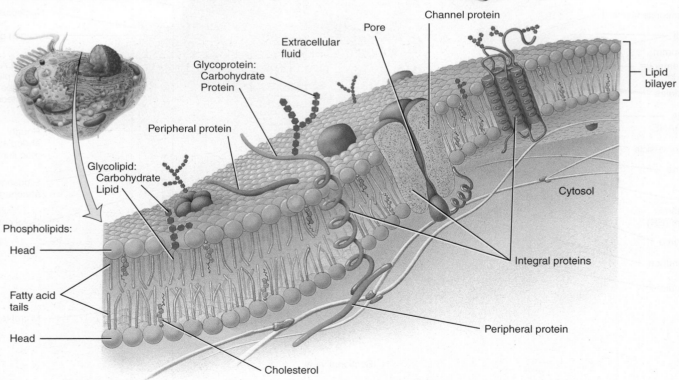

Name several functions carried out by membrane proteins.

receptors recognize and bind a specific molecule that governs some cellular function, for example, a hormone such as insulin. Some integral proteins act as *enzymes*, speeding up specific chemical reactions. Membrane glycoproteins and glycolipids often are *cell identity markers*. They enable a cell to recognize other cells of its own kind during tissue formation, or to recognize and respond to potentially dangerous foreign cells.

✓ CHECKPOINT

2. What molecules make up the plasma membrane and what are their functions?

3. What is meant by selective permeability?

3.3 TRANSPORT ACROSS THE PLASMA MEMBRANE

OBJECTIVE • Describe the processes that transport substances across the plasma membrane.

Movement of materials across its plasma membrane is essential to the life of a cell. Certain substances must move into the cell to support metabolic reactions. Other materials must be moved out because they have been produced by the cell for export or are cellular waste products. Before discussing how materials move into and out of a cell, we need to understand what exactly is being moved as well as the form it needs to take to make its journey.

About two-thirds of the fluid in your body is contained inside body cells and is called *intracellular fluid* (**ICF**) (*intra-* = within). ICF, as indicated earlier, is actually the cytosol of a cell. Fluid outside body cells is called *extracellular fluid* (**ECF**) (*extra-* = outside). The ECF in the microscopic spaces between the cells of tissues is *interstitial fluid* (in′-ter-STISH-al; *inter-* = between). The ECF in blood vessels is called *plasma* (PLAZ-ma), and that in lymphatic vessels is called *lymph*. The ECF within and around the brain and spinal cord is called *cerebrospinal fluid* (**CSF**) (se-rē′-brō-SPĪ-nal).

Materials dissolved in body fluids include gases, nutrients, ions, and other substances needed to maintain life. Any material dissolved in a fluid is called a *solute*, and the fluid in which it is dissolved is the *solvent*. Body fluids are dilute solutions in which a variety of solutes are dissolved in a very familiar solvent, water. The amount of a solute in a solution is its *concentration*. A *concentration gradient* is a difference in concentration between two different areas, for example, the ICF and ECF. Solutes moving from a high-concentration area (where there are more of them) to a low-concentration area (where there are fewer of them) are said to move *down* or *with* the concentration gradient. Solutes moving from a low-concentration area to a high-concentration area are said to move *up* or *against* the concentration gradient.

Substances move across cellular membranes by passive processes and active processes. *Passive processes*, in which a

substance moves down its concentration gradient through the membrane, using only its own energy of motion (kinetic energy), include simple diffusion and osmosis. In *active processes*, cellular energy, usually in the form of ATP, is used to "push" the substance through the membrane "uphill" against its concentration gradient. An example is active transport. Another way that some substances may enter and leave cells is an active process in which tiny membrane sacs referred to as *vesicles* are used (see Figure 3.10).

Passive Processes

Diffusion: The Principle

Diffusion (di-FŪ-zhun; *diffus-* = spreading) is a passive process in which a substance moves due to its kinetic energy. If a particular substance is present in high concentration in one area and in low concentration in another area, more particles of the substance diffuse from the region of high concentration to the region of low concentration than in the opposite direction. The diffusion of more molecules in one direction than the other is called *net* diffusion. Substances undergoing net diffusion move from a high to a low concentration, or *down their concentration gradient*. After some time, *equilibrium* (ē′-kwi-LIB-rē-um) is reached: The substance becomes evenly distributed throughout the solution and the concentration gradient disappears.

Placing a crystal of dye in a water-filled container provides an example of diffusion (Figure 3.3). At the beginning, the color is most intense just next to the crystal because the crystal is dissolving and the dye concentration is greatest there. At

Figure 3.3 Principle of diffusion. A crystal of dye placed in a cylinder of water dissolves (a), and there is net diffusion from the region of higher dye concentration to regions of lower dye concentration (b). At equilibrium (c), dye concentration is uniform throughout the solution.

 At equilibrium, net diffusion stops but random movements continue.

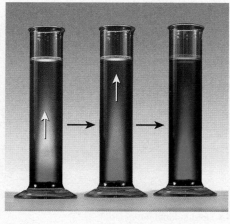

| Beginning (a) | Intermediate (b) | Equilibrium (c) |

 How does simple diffusion differ from facilitated diffusion?

increasing distances, the color is lighter and lighter because the dye concentration is lower and lower. The dye molecules undergo net diffusion, down their concentration gradient, until they are evenly mixed in the water. At equilibrium the solution has a uniform color. In the example of dye diffusion, no membrane was involved. Substances may also diffuse across a membrane, if the membrane is permeable to them.

Now that you have a basic understanding of the nature of diffusion, we will consider two types of diffusion: simple diffusion and facilitated diffusion.

SIMPLE DIFFUSION   In *simple diffusion*, substances diffuse across a membrane through the lipid bilayer (Figure 3.4). Lipid-soluble substances that move across membranes by simple diffusion through the lipid bilayer include oxygen, carbon dioxide, and nitrogen gases; fatty acids; steroids; and fat-soluble vitamins (A, D, E, and K). Polar molecules such as water and urea also move through the lipid bilayer. Simple diffusion through the lipid bilayer is important in the exchange of oxygen and carbon dioxide between blood and body cells and between blood and air within the lungs during breathing. It also is the transport method for absorption of lipid-soluble nutrients and release of some wastes from body cells.

FACILITATED DIFFUSION   Some substances that cannot move through the lipid bilayer by simple diffusion do cross the plasma membrane by a passive process called *facilitated diffusion*. In this process, an integral membrane protein assists a specific substance to move across the membrane. The membrane protein can be either a membrane channel or a carrier.

In facilitated diffusion involving *ion channels*, ions move down their concentration gradients across the lipid bilayer. Most membrane channels are ion channels, which allow a specific type of ion to move across the membrane through the channel's pore. In typical plasma membranes, the most common ion channels are selective for K⁺ (potassium ions) or Cl⁻ (chloride ions); fewer channels are available for Na⁺ (sodium ions) or Ca²⁺ (calcium ions). Many ion channels are gated; that is, a portion of the channel protein acts as a "gate," moving in one direction to open the pore and in another direction to close it (Figure 3.5). When the gates are open, ions diffuse into or out of cells, down their concentration gradient. Gated channels are important for the production of electrical signals by body cells.

In facilitated diffusion involving a *carrier*, the substance binds to a specific carrier on one side of the membrane and is released on the other side after the carrier undergoes a change in shape.

Substances that move across plasma membranes by facilitated diffusion involving carriers include glucose, fructose, galactose, and some vitamins. Glucose enters many body cells by facilitated diffusion as follows (Figure 3.6):

1 Glucose binds to a glucose carrier protein on the outside surface of the membrane.

2 As the carrier undergoes a change in shape, glucose passes through the membrane.

3 The carrier releases glucose on the other side of the membrane.

The selective permeability of the plasma membrane is often regulated to achieve homeostasis. For example, the hormone insulin promotes the insertion of more glucose carriers

Figure 3.4 Simple diffusion. Lipid-soluble molecules diffuse through the lipid bilayer.

🔑 In simple diffusion there is a net (greater) movement of substances from a region of their higher concentration to a region of their lower concentration.

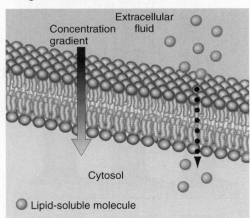

❓ What are some examples of substances that diffuse through the lipid bilayer?

Figure 3.5 Facilitated diffusion of potassium ions (K⁺) through a gated K⁺ channel. A gated channel is one in which a portion of the channel protein acts as a gate to open or close the channel's pore to the passage of ions.

🔑 Ion channels are integral membrane proteins that allow specific small, inorganic ions to pass across the membrane.

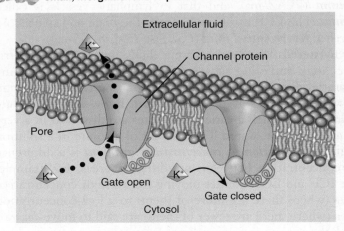

Details of the K⁺ channel

❓ Is the concentration of K⁺ in body cells higher in the cytosol or in the extracellular fluid?

Figure 3.6 Facilitated diffusion of glucose across a plasma membrane using a carrier. The carrier protein binds to glucose in the extracellular fluid and releases it into the cytosol.

Facilitated diffusion across a membrane involving a carrier is an important mechanism for transporting sugars such as glucose, fructose, and galactose into cells.

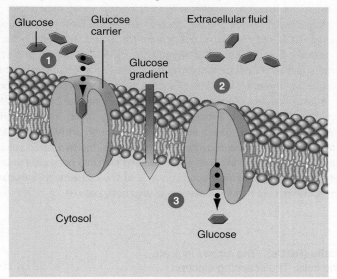

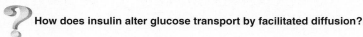

How does insulin alter glucose transport by facilitated diffusion?

Figure 3.7 Principle of osmosis. (a) At the start of the experiment, a cellophane sac containing a 20 percent sucrose solution is immersed in a beaker of pure (100 percent) water. The sac is a selectively permeable membrane that permits water but not sucrose to pass through it. Osmosis begins (arrows) as water moves down its concentration gradient into the sac. (b) As the volume of the sucrose solution increases, the solution moves up the glass tubing. The added fluid in the tube exerts a pressure that drives some water molecules back into the beaker. At equilibrium, osmosis has stopped because the number of water molecules entering and the number leaving the cellophane sac are equal.

Osmosis is the net movement of water molecules through a selectively permeable membrane.

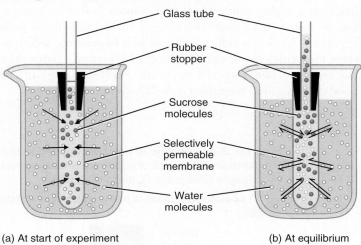

(a) At start of experiment   (b) At equilibrium

Will the fluid level in the tube continue to rise until the sucrose concentrations are the same in the beaker and in the sac?

into the plasma membranes of certain cells. Thus, the effect of insulin is to increase entry of glucose into body cells by means of facilitated diffusion.

Osmosis

Osmosis (oz-MŌ-sis) is a passive process in which there is a net movement of water through a selectively permeable membrane. Water moves by osmosis from an area of *higher water concentration* to an area of *lower water concentration* (or from an area of *lower solute concentration* to an area of *higher solute concentration*). Water molecules pass through plasma membranes in two places: through the lipid bilayer and through integral membrane proteins that function as water channels.

The device in Figure 3.7 demonstrates osmosis. A sac made of cellophane, a selectively permeable membrane that permits water but not sucrose (sugar) molecules to pass, is filled with a solution that is 20 percent sucrose and 80 percent water. The upper part of the cellophane sac is wrapped tightly about a stopper through which a glass tube is fitted. The sac is then placed into a beaker containing pure (100 percent) water (Figure 3.7a). Notice that the cellophane now separates two fluids having different water concentrations. As a result, water begins to move by osmosis from the region where its concentration is higher (100 percent water in the beaker) through the cellophane to where its concentration is lower (80 percent water inside the sac). Because the cellophane is not permeable to sucrose, however, all the su-

crose molecules remain inside the sac. As water moves into the sac, the volume of the sucrose solution increases and the fluid rises into the glass tube (Figure 3.7b). As the fluid rises in the tube, its water pressure forces some water molecules from the sac back into the beaker. At equilibrium, just as many water molecules are moving into the beaker due to the water pressure as are moving into the sac due to osmosis.

A solution containing solute particles that cannot pass through a membrane exerts a pressure on the membrane, called **osmotic pressure**. The osmotic pressure of a solution depends on the concentration of its solute particles—the higher the solute concentration, the higher the solution's osmotic pressure. Because the osmotic pressure of cytosol and interstitial fluid is the same, cell volume remains constant. Cells neither shrink due to water loss by osmosis nor swell due to water gain by osmosis.

Any solution in which cells maintain their normal shape and volume is called an **isotonic solution** (ī′-sō-TON-ik; *iso-* = same; *tonic-* = tension). This is a solution in which the concentrations of solutes are the *same* on both sides. For example, a 0.9% NaCl (sodium chloride, or table salt) solution, called a *normal saline solution*, is isotonic for red blood cells. When red blood cells are bathed in 0.9% NaCl, water molecules enter and exit the cells at the same rate, allowing the red blood cells to maintain their normal shape and volume (Figure 3.8a).

If red blood cells are placed in a ***hypotonic solution*** (hī′-pō-TON-ik; *hypo-* = less than), a solution that has a *lower* concentration of solutes (higher concentration of water) than the cytosol inside the red blood cells (Figure 3.8b), water molecules enter the cells by osmosis faster than they leave. This situation causes the red blood cells to swell and eventually to burst. Rupture of red blood cells is called ***hemolysis*** (hē-MOL-i-sis).

A ***hypertonic solution*** (hī′-per-TON-ik; *hyper-* = greater than) has a *higher* concentration of solutes (lower concentration of water) than does the cytosol inside red blood cells (Figure 3.8c). When cells are placed in a hypertonic solution, water molecules move out of the cells by osmosis faster than they enter, causing the cells to shrink. Such shrinkage of red blood cells is called ***crenation*** (kre-NĀ-shun).

CLINICAL CONNECTION | Medical Uses of Isotonic, Hypertonic, and Hypotonic Solutions

RBCs and other body cells may be damaged or destroyed if exposed to hypertonic or hypotonic solutions. For this reason, most *intravenous (IV) solutions,* liquids infused into the blood of a vein, are **isotonic solutions**. Examples are isotonic saline (0.9 percent NaCl) and D5W, which stands for dextrose 5 percent in water. Sometimes infusion of **hypertonic solutions** is useful to treat patients who have *cerebral edema,* excess interstitial fluid in the brain. Infusion of such a solution relieves fluid overload by causing osmosis of water from interstitial fluid into the blood. The kidneys then excrete the excess water from the blood into the urine. **Hypotonic solutions**, given either orally or through an IV, can be used to treat people who are dehydrated. The water in the hypotonic solution moves from the blood into interstitial fluid and then into body cells to rehydrate them. Water and most sports drinks that you consume to "rehydrate" after a workout are hypotonic relative to your body cells. •

Figure 3.8 Principle of osmosis applied to red blood cells (RBCs). The arrows indicate the direction and degree of water movement into and out of the cells. The scanning electron micrographs have magnifications of 15,000X.

An isotonic solution is one in which cells maintain their normal shape and volume.

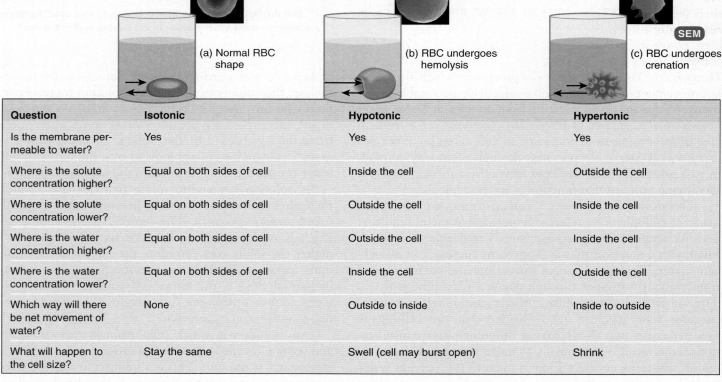

Isotonic solution — (a) Normal RBC shape

Hypotonic solution — (b) RBC undergoes hemolysis

Hypertonic solution — (c) RBC undergoes crenation

SEM

Question	Isotonic	Hypotonic	Hypertonic
Is the membrane permeable to water?	Yes	Yes	Yes
Where is the solute concentration higher?	Equal on both sides of cell	Inside the cell	Outside the cell
Where is the solute concentration lower?	Equal on both sides of cell	Outside the cell	Inside the cell
Where is the water concentration higher?	Equal on both sides of cell	Outside the cell	Inside the cell
Where is the water concentration lower?	Equal on both sides of cell	Inside the cell	Outside the cell
Which way will there be net movement of water?	None	Outside to inside	Inside to outside
What will happen to the cell size?	Stay the same	Swell (cell may burst open)	Shrink

? **Will a 2 percent solution of NaCl cause hemolysis or crenation of RBCs?**

Active Processes

Active Transport

Active transport is an active process in which cellular energy is used to transport substances across the membrane against a concentration gradient (from an area of low concentration to an area of high concentration).

Energy derived from splitting ATP changes the shape of a carrier protein, called a *pump*, which moves a substance across a cellular membrane against its concentration gradient. A typical body cell expends about 40 percent of its ATP on active transport. Drugs that turn off ATP production, such as the poison cyanide, are lethal because they shut down active transport in cells throughout the body. Substances transported across the plasma membrane by active transport are mainly ions, primarily Na^+, K^+, H^+, Ca^{2+}, I^-, and Cl^-.

The most important active transport pump expels sodium ions (Na^+) from cells and brings in potassium ions (K^+). The pump protein also acts as an enzyme to split ATP. Because of the ions it moves, this pump is called the *sodium–potassium* (Na^+–K^+) *pump*. All cells have thousands of sodium–potassium pumps in their plasma membranes. These pumps maintain a low concentration of sodium ions in the cytosol by pumping Na^+ into the extracellular fluid against the Na^+ concentration gradient. At the same time, the pump moves potassium ions into cells against the K^+ concentration gradient. Because K^+ and Na^+ slowly leak back across the plasma membrane down their gradients, the sodium–potassium pumps must operate continually to maintain a low concentration of Na^+ and a high concentration of K^+ in the cytosol. These differing concentrations are crucial for osmotic balance of the two

fluids and also for the ability of some cells to generate electrical signals such as action potentials.

Figure 3.9 shows how the sodium–potassium pump operates.

1 Three sodium ions (Na^+) in the cytosol bind to the pump protein.

2 Na^+ binding triggers the splitting of ATP into ADP plus a phosphate group (P), which also becomes attached to the pump protein. This chemical reaction changes the shape of the pump protein, expelling the three Na^+ into the extracellular fluid. The changed shape of the pump protein then favors binding of two potassium ions (K^+) in the extracellular fluid to the pump protein.

3 The binding of K^+ causes the pump protein to release the phosphate group, which causes the pump protein to return to its original shape.

4 As the pump protein returns to its original shape, it releases the two K^+ into the cytosol. At this point, the pump is ready again to bind Na^+, and the cycle repeats.

Transport in Vesicles

A *vesicle* (VES-i-kul) is a small round sac formed by budding off from an existing membrane. Vesicles transport substances from one structure to another within cells, take in substances from extracellular fluid, and release substances into extracellular fluid. Movement of vesicles requires energy supplied by ATP and is therefore an active process. The two main types of transport in vesicles between a cell and the extracellular fluid that surrounds it are (1) *endocytosis* (en′-dō-sī-TŌ-sis; *endo-* = within), in which materials move *into* a cell in a vesicle formed from the plasma membrane, and (2) *exocytosis* (eks-ō-sī-TŌ-sis;

Figure 3.9 Operation of the sodium–potassium pump. Sodium ions (Na^+) are expelled from the cell, and potassium ions (K^+) are imported into the cell. The pump does not work unless Na^+ and ATP are present in the cytosol and K^+ is present in the extracellular fluid.

The sodium–potassium pump maintains a low intracellular concentration of Na^+.

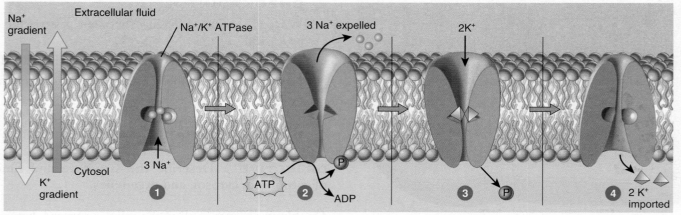

 What is the role of ATP in the operation of this pump?

exo- = out), in which materials move *out of* a cell by the fusion of a vesicle formed inside a cell with the plasma membrane.

ENDOCYTOSIS Substances brought into the cell by endocytosis are surrounded by a piece of the plasma membrane, which buds off inside the cell to form a vesicle containing the ingested substances. The two types of endocytosis we will consider are phagocytosis and bulk-phase endocytosis.

1. **Phagocytosis.** In *phagocytosis* (fag′-ō-sī-TŌ-sis; *phago- =* to eat), large solid particles, such as whole bacteria or viruses or aged or dead cells, are taken in by the cell (Figure 3.10). Phagocytosis begins as the particle binds to a plasma membrane receptor, causing the cell to extend projections of its plasma membrane and cytoplasm, called *pseudopods* (SOO-dō-pods; *pseudo- =* false; *-pods =* feet). Two or more pseudopods surround the particle, and portions of their membranes fuse to form a vesicle called a *phagosome* that enters the cytoplasm. The vesicle fuses with one or more lysosomes, and lysosomal enzymes break down the ingested material. In most cases, any undigested materials remain indefinitely in a vesicle called a *residual body.*

Phagocytosis occurs only in *phagocytes*, cells that are specialized to engulf and destroy bacteria and other foreign substances. Phagocytes include certain types of white blood cells and macrophages, which are present in most body tissues. The process of phagocytosis is a vital defense mechanism that helps protect the body from disease.

2. **Bulk-phase endocytosis.** In *bulk-phase endocytosis (pinocytosis)*, cells take up tiny droplets of extracellular fluid. The process occurs in most body cells and takes in any and all solutes dissolved in the extracellular fluid. During bulk-phase endocytosis the plasma membrane folds inward and forms a vesicle containing a droplet of extracellular fluid. The vesicle detaches or "pinches off" from the plasma membrane and enters the cytosol. Within the cell, the vesicle fuses with a lysosome, where enzymes degrade the engulfed solutes. The resulting smaller molecules, such as amino acids and fatty acids, leave the lysosome to be used elsewhere in the cell.

EXOCYTOSIS In contrast with endocytosis, which brings materials into a cell, exocytosis results in *secretion*, the liberation of materials from a cell. All cells carry out exocytosis, but it is especially important in two types of cells: (1) secretory cells that liberate digestive enzymes, hormones, mucus, or other secretions; (2) nerve cells that release substances called *neurotransmitters* via exocytosis (see Figure 9.7). During exocytosis, membrane-enclosed vesicles called *secretory vesicles* form inside the cell, fuse with the plasma membrane, and release their contents into the extracellular fluid.

Segments of the plasma membrane lost through endocytosis are recovered or recycled by exocytosis. The balance between endocytosis and exocytosis keeps the surface area of a cell's plasma membrane relatively constant.

Table 3.1 summarizes the processes by which materials move into and out of cells.

✓ CHECKPOINT

4. What is the key difference between passive and active processes?

5. How does simple diffusion compare to facilitated diffusion?

6. In what ways are endocytosis and exocytosis similar and different?

3.4 CYTOPLASM

OBJECTIVE • **Describe the structure and functions of cytoplasm, cytosol, and organelles.**

Cytoplasm consists of all of the cellular contents between the plasma membrane and the nucleus and includes both cytosol and organelles.

Figure 3.10 Phagocytosis.

🔑 **Phagocytosis is a vital defense mechanism that helps protect the body from disease.**

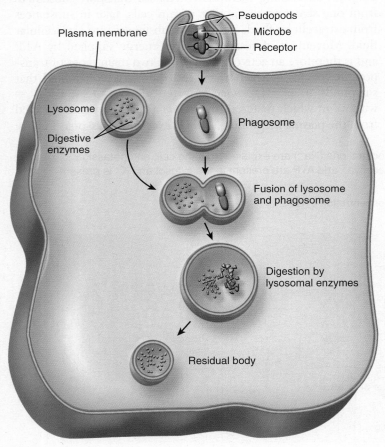

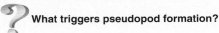 **What triggers pseudopod formation?**

TABLE 3.1

Transport of Materials into and out of Cells

TRANSPORT PROCESS	DESCRIPTION	SUBSTANCES TRANSPORTED
Passive Processes	Movement of substances down a concentration gradient until equilibrium is reached; do not require cellular energy in the form of ATP	
	Movement of a substance by kinetic energy down a concentration gradient until equilibrium is reached	
Diffusion		
Simple diffusion	Passive movement of a substance through the lipid bilayer of the plasma membrane	Lipid-soluble molecules: oxygen, carbon dioxide, and nitrogen gases; fatty acids, steroids, and fat-soluble vitamins (A, D, E, K) Polar molecules: water and urea
Facilitated diffusion	Passive movement of a substance down its concentration gradient aided by ion channels and carriers	K^+, Cl^-, Na^+, Ca^{2+}, glucose, fructose, galactose, and some vitamins
Osmosis	Movement of water molecules across a selectively permeable membrane from an area of higher water concentration to an area of lower water concentration	Water
Active Processes	Movement of substances against a concentration gradient; requires cellular energy in the form of ATP	
Active Transport	Transport in which cell expends energy to move a substance across the membrane against its concentration gradient aided by membrane proteins that act as pumps; these integral membrane proteins use energy supplied by ATP	Na^+, K^+, Ca^{2+}, H^+, I^-, Cl^-, and other ions
Transport In Vesicles	Movement of substances into or out of a cell in vesicles that bud from the plasma membrane; requires energy supplied by ATP	
Endocytosis	Movement of substances into a cell in vesicles	
Phagocytosis	"Cell eating"; movement of a solid particle into a cell after pseudopods engulf it	Bacteria, viruses, and aged or dead cells
Bulk-phase endocytosis	"Cell drinking"; movement of extracellular fluid into a cell by infolding of plasma membrane	Solutes in extracellular fluid
Exocytosis	Movement of substances out of a cell in secretory vesicles that fuse with the plasma membrane and release their contents into the extracellular fluid	Neurotransmitters, hormones, and digestive enzymes

Cytosol

The **cytosol** (*intracellular fluid*) is the liquid portion of the cytoplasm that surrounds organelles and accounts for about 55 percent of the total cell volume. Although cytosol varies in composition and consistency from one part of a cell to another, typically it is 75 percent to 90 percent water plus various dissolved solutes and suspended particles. Among these are various ions, glucose, amino acids, fatty acids, proteins, lipids, ATP, and waste products. Some cells also contain *lipid droplets* that contain triglycerides and *glycogen granules*, clusters of glycogen molecules (see Figure 3.1). The cytosol is the site of many of the chemical reactions that maintain cell structures and allow cellular growth.

Extending throughout the cytosol, the **cytoskeleton** is a network of three different types of protein filaments: microfilaments, intermediate filaments, and microtubules.

The thinnest elements of the cytoskeleton are the **microfilaments** (mī-krō-FIL-a-ments), which are concentrated at the periphery of a cell and contribute to the cell's strength and shape (Figure 3.11a). Microfilaments have two general functions: providing mechanical support and helping generate movements. They also anchor the cytoskeleton to integral proteins in the plasma membrane and provide support for microscopic, fingerlike projections of the plasma membrane called **microvilli** (mī'-krō-VIL-ē; *micro-* = small; *-villi* = tufts of hair; singular is *microvillus*). Because they greatly increase the surface area of the cell, microvilli are abundant on cells involved in absorption, such as the cells that line the small intestine. Some microfilaments extend beyond the plasma membrane and help cells attach to one another or to extracellular materials.

With respect to movement, microfilaments are involved in muscle contraction, cell division, and cell locomotion. Microfilament-assisted movements include the migration of embryonic cells during development, the invasion of tissues by white blood cells to fight infection, and the migration of skin cells during wound healing.

As their name suggests, **intermediate filaments** are thicker than microfilaments but thinner than microtubules

Figure 3.11 Cytoskeleton.

Extending throughout the cytosol, the cytoskeleton is a network of three kinds of protein filaments: microfilaments, intermediate filaments, and microtubules.

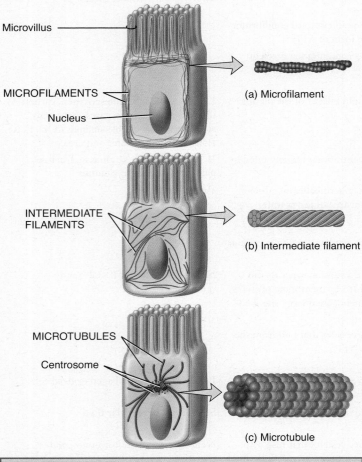

Microvillus

MICROFILAMENTS

Nucleus

(a) Microfilament

INTERMEDIATE FILAMENTS

(b) Intermediate filament

MICROTUBULES

Centrosome

(c) Microtubule

Functions of the Cytoskeleton

1. Serves as a scaffold that helps to determine a cell's shape and to organize the cellular contents.
2. Aids movement of organelles within the cell, of chromosomes during cell division, and of whole cells such as phagocytes.

? Which cytoskeletal components help form the structure of centrioles, cilia, and flagella?

(Figure 3.11b). They are found in parts of cells subject to tension (such as stretching), help hold organelles such as the nucleus in place, and help attach cells to one another.

The largest of the cytoskeletal components, ***microtubules*** (mī-krō-TOO-būls) are long, hollow tubes (Figure 3.11c). Microtubules help determine cell shape and function in both the movement of organelles, such as secretory vesicles, within a cell and the migration of chromosomes during cell division. They also are responsible for movements of cilia and flagella.

Organelles

Organelles are specialized structures inside cells that have characteristic shapes and specific functions. Each type of or-

ganelle is a functional compartment where specific processes take place, and each has its own unique set of enzymes.

Centrosome

The ***centrosome***, located near the nucleus, has two components—a pair of centrioles and pericentriolar material (Figure 3.12). The two *centrioles* are cylindrical structures, each of which is composed of nine clusters of three microtubules (a triplet) arranged in a circular pattern. Surrounding the centrioles is the *pericentriolar material* (per'-ē-sen'-trē-Ō-lar), containing hundreds of ring-shaped proteins called *tubulins*. The tubulins are the organizing centers for growth of the mitotic spindle, which plays a critical role in cell division, and for microtubule formation in nondividing cells.

Cilia and Flagella

Microtubules are the main structural and functional components of cilia and flagella, both of which are motile projections of the cell surface. ***Cilia*** (SIL-ē-a; singular is *cilium* = eyelash) are numerous, short, hairlike projections that extend from the surface of the cell (see Figure 3.1). In the human body, cilia propel fluids across the surfaces of cells that are firmly anchored in place. The coordinated movement of many cilia on the surface of a cell causes a steady movement of fluid along the cell's

Figure 3.12 Centrosome.

The pericentriolar material of a centrosome organizes the mitotic spindle during cell division.

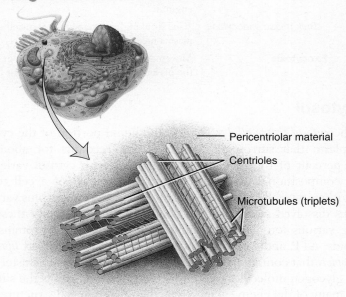

Pericentriolar material

Centrioles

Microtubules (triplets)

Functions of Centrosomes

The pericentriolar material of the centrosome contains tubulins that build microtubules in nondividing cells and form the mitotic spindle during cell division.

 What are the components of the centrosome?

surface. Many cells of the respiratory tract, for example, have hundreds of cilia that help sweep foreign particles trapped in mucus away from the lungs. Their movement is paralyzed by nicotine in cigarette smoke. For this reason, smokers cough often to remove foreign particles from their airways. Cells that line the uterine (fallopian) tubes also have cilia that sweep oocytes (egg cells) toward the uterus.

Flagella (fla-JEL-a; singular is *flagellum* = whip) are similar in structure to cilia but are much longer (see Figure 3.1). Flagella usually move an entire cell. The only example of a flagellum in the human body is a sperm cell's tail, which propels the sperm toward its possible union with an oocyte.

Ribosomes

Ribosomes (RĪ-bō-sōms; *-somes* = bodies) are the sites of protein synthesis. Ribosomes are named for their high content of *ribo*nucleic acid (RNA). Besides ribosomal RNA (rRNA), these tiny organelles contain ribosomal proteins. Structurally, a ribosome consists of two subunits, large and small, one about half the size of the other (Figure 3.13). The large and small subunits are made in the nucleolus of the nucleus. Later, they exit the nucleus and are assembled in the cytoplasm, where they form a functional ribosome.

Figure 3.13 Ribosomes.

Ribosomes, the sites of protein synthesis, consist of a large subunit and a small subunit.

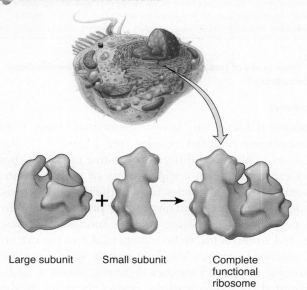

Large subunit   Small subunit   Complete functional ribosome

Details of ribosomal subunits

Functions of Ribosomes

1. Ribosomes associated with endoplasmic reticulum synthesize proteins destined for insertion in the plasma membrane or secretion from the cell.
2. Free ribosomes synthesize proteins used in the cytosol.

 Where are ribosomal subunits synthesized and assembled?

Some ribosomes are attached to the outer surface of the nuclear membrane and to an extensively folded membrane called the endoplasmic reticulum. These ribosomes synthesize proteins destined for specific organelles, for insertion in the plasma membrane, or for export from the cell. Other ribosomes are called free ribosomes because they are not attached to other cytoplasmic structures. Free ribosomes synthesize proteins used in the cytosol. Ribosomes are also located within mitochondria, where they synthesize mitochondrial proteins.

Endoplasmic Reticulum

The *endoplasmic reticulum* **(ER)** (en′-dō-PLAS-mik re-TIK-ū-lum; *-plasmic* = cytoplasm; *reticulum* = network) is a network of folded membranes in the form of flattened sacs or tubules (Figure 3.14). The ER extends throughout the cytoplasm and is so extensive that it constitutes more than half of the membranous surfaces within the cytoplasm of most cells.

Figure 3.14 Endoplasmic reticulum (ER).

The ER is a network of folded membranes that extend throughout the cytoplasm and connect to the nuclear envelope.

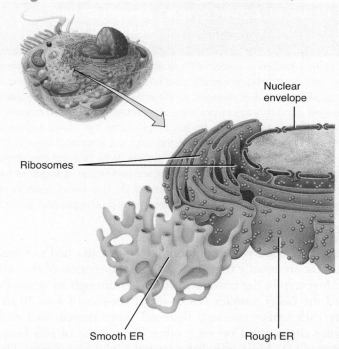

Nuclear envelope

Ribosomes

Smooth ER   Rough ER

Functions of Endoplasmic Reticulum

1. Rough ER synthesizes glycoproteins and phospholipids that are transferred into cellular organelles, inserted into the plasma membrane, or secreted during exocytosis.
2. Smooth ER synthesizes fatty acids and steroids, such as estrogens and testosterone; inactivates or detoxifies drugs and other potentially harmful substances; removes the phosphate group from glucose-6-phosphate; and stores and releases calcium ions that trigger contraction in muscle cells.

 How do rough ER and smooth ER differ structurally and functionally?

Cells contain two distinct forms of ER that differ in structure and function. ***Rough ER*** extends from the nuclear envelope (membrane around the nucleus) and appears "rough" because its outer surface is studded with ribosomes. Proteins synthesized by ribosomes attached to rough ER enter the spaces within the ER for processing and sorting. These molecules (glycoproteins and phospholipids) may be incorporated into organelle membranes or the plasma membrane. Thus, rough ER is a factory for synthesizing secretory proteins and membrane molecules.

Smooth ER extends from the rough ER to form a network of membranous tubules (Figure 3.14). As you may already have guessed, smooth ER appears "smooth" because it lacks ribosomes. Smooth ER is where fatty acids and steroids such as estrogens and testosterone are synthesized. In liver cells, enzymes of the smooth ER also help release glucose into the bloodstream and inactivate or detoxify a variety of drugs and potentially harmful substances, including alcohol, pesticides, and *carcinogens* (cancer-causing agents). In muscle cells, calcium ions needed for muscle contraction are stored and released from a form of smooth ER called sarcoplasmic reticulum.

CLINICAL CONNECTION | Smooth ER and Increased Drug Tolerance

One of the functions of smooth ER, as noted earlier, is to detoxify certain drugs. Individuals who repeatedly take such drugs, such as the sedative phenobarbital, develop changes in the smooth ER in their liver cells. Prolonged administration of phenobarbital results in increased tolerance to the drug; the same dose no longer produces the same degree of sedation. With repeated exposure to the drug, the amount of smooth ER and its enzymes increases to protect the cell from its toxic effects. As the amount of smooth ER increases, higher and higher dosages of the drug are needed to achieve the original effect. This could result in an increased possibility of overdose and increased drug dependence. •

Golgi Complex

After proteins are synthesized on a ribosome attached to rough ER, most are usually transported to another region of the cell. The first step in the transport pathway is through an organelle called the ***Golgi complex*** (GOL-jē). It consists of 3 to 20 ***cisterns*** (SIS-terns = cavities), flattened membranous sacs with bulging edges, piled on each other like a stack of pita bread (Figure 3.15). Most cells have several Golgi complexes. The Golgi complex is more extensive in cells that secrete proteins.

The main function of the Golgi complex is to modify and package proteins. Proteins synthesized by ribosomes on rough ER enter the Golgi complex and are modified to form glycoproteins and lipoproteins. Then, they are sorted and packaged into vesicles. Some of the processed proteins are discharged from the cell by exocytosis. Certain cells of the pancreas release the hormone insulin this way. Other processed proteins become part of the plasma membrane as existing parts of the membrane are lost. Still other processed proteins become incorporated into organelles called lysosomes.

Figure 3.15 Golgi complex.

Most proteins synthesized by ribosomes attached to rough ER pass through the Golgi complex for processing.

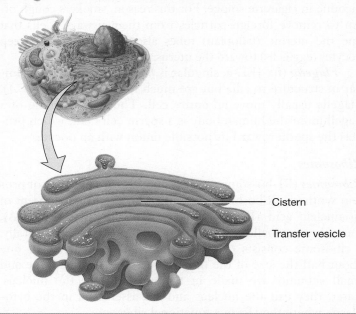

Cistern

Transfer vesicle

Functions of Golgi Complex

1. Modifies, sorts, packages, and transports proteins received from the rough ER.
2. Forms secretory vesicles that discharge processed proteins via exocytosis into extracellular fluid; forms membrane vesicles that ferry new molecules to the plasma membrane; forms transport vesicles that carry molecules to other organelles, such as lysosomes.

 What types of body cells are likely to have extensive Golgi complexes?

Lysosomes

Lysosomes (LĪ-sō-sōms; *lyso-* = dissolving; *-somes* = bodies) are membrane-enclosed vesicles (see Figure 3.1) that may contain as many as 60 different digestive enzymes; these enzymes can break down a wide variety of molecules once the lysosome fuses with vesicles formed during endocytosis. The lysosomal membrane contains carrier proteins that allow the final products of digestion, such as monosaccharides, fatty acids, and amino acids, to be transported into the cytosol.

Lysosomal enzymes also help recycle worn-out structures. A lysosome can engulf another organelle, digest it, and return the digested components to the cytosol for reuse. In this way, old organelles are continually replaced. The process by which worn-out organelles are digested is called ***autophagy*** (aw-TOF-a-jē; *auto-* = self; *-phagy* = eating). During autophagy, the organelle to be digested is enclosed by a membrane derived from the ER to create a vesicle that then fuses with a lysosome. In this way, a human liver cell, for example, recycles about half its contents every week. Lysosomal enzymes may also destroy the entire cell, a process known as ***autolysis*** (aw-TOL-i-sis). Autolysis occurs in some pathological conditions and also is responsible for the tissue deterioration that occurs just after death.

Peroxisomes

Another group of organelles similar in structure to lysosomes, but smaller, are called *peroxisomes* (pe-ROK-si-sōms; *peroxi-* = peroxide; see Figure 3.1). Peroxisomes contain several *oxidases*, which are enzymes that can oxidize (remove hydrogen atoms from) various organic substances. For example, amino acids and fatty acids are oxidized in peroxisomes as part of normal metabolism. In addition, enzymes in peroxisomes oxidize toxic substances. Thus, peroxisomes are very abundant in the liver, where detoxification of alcohol and other damaging substances takes place. A byproduct of the oxidation reactions is hydrogen peroxide (H_2O_2), a potentially toxic compound, and associated free radicals, such as superoxide. However, peroxisomes also contain an enzyme called *catalase* that decomposes the H_2O_2. Because the generation and degradation of H_2O_2 occurs within the same organelle, peroxisomes protect other parts of the cell from the toxic effects of H_2O_2. Peroxisomes also have enzymes that destroy superoxide.

Proteasomes

Although lysosomes degrade proteins delivered to them in vesicles, proteins in the cytosol also require disposal at certain times in the life of a cell. Continuous destruction of unneeded, damaged, or faulty proteins is the function of tiny barrel-shaped structures called *proteasomes* (PRŌ-tē-a-sōmes = protein bodies). A typical body cell contains many thousands of proteasomes, in both the cytosol and the nucleus. Proteasomes were so named because they contain myriad *proteases*, enzymes that cut proteins into small peptides. Once the enzymes of a proteasome have chopped up a protein into smaller chunks, other enzymes then break down the peptides into amino acids, which can be recycled into new proteins.

Mitochondria

Because they are the site of most ATP production, the "powerhouses" of a cell are its *mitochondria* (mī-tō-KON-drē-a; *mito-* = thread; *-chondria* = granules; singular is *mitochondrion*). A cell may have as few as one hundred or as many as several thousand mitochondria, depending on how active the cell is. For example, active cells such as those found in muscles, the liver, and kidneys use ATP at a high rate and have large numbers of mitochondria. A mitochondrion consists of two membranes, each of which is similar in structure to the plasma membrane (Figure 3.16). The *outer mitochondrial membrane* is smooth, but the *inner mitochondrial membrane* is arranged in a series of folds called mitochondrial *cristae* (KRIS-tē; singular is *crista* = ridge). The large central fluid-filled cavity of a mitochondrion, enclosed by the inner membrane and cristae, is the *mitochondrial matrix*. The elaborate folds of the cristae provide an enormous surface area for a series of chemical reactions that provide most of a cell's ATP. Enzymes that catalyze

Figure 3.16 Mitochondrion.

🔑 **Within mitochondria, chemical reactions generate most of a cell's ATP.**

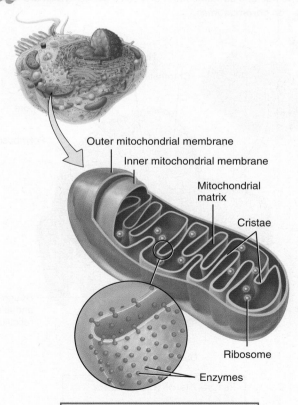

Outer mitochondrial membrane

Inner mitochondrial membrane

Mitochondrial matrix

Cristae

Ribosome

Enzymes

Function of Mitochondria

Generates ATP through reactions of aerobic cellular respiration.

? How do the cristae of a mitochondrion contribute to its ATP-producing function?

these reactions are located in the matrix and on the cristae. Mitochondria also contain a small number of genes and a few ribosomes, enabling them to synthesize some proteins.

✓ **CHECKPOINT**

7. What does cytoplasm have that cytosol does not?

8. What is an organelle?

9. Describe the structure and function of ribosomes, the Golgi complex, and mitochondria.

3.5 NUCLEUS

OBJECTIVE • Describe the structure and functions of the nucleus.

The *nucleus* is a spherical or oval structure that usually is the most prominent feature of a cell (Figure 3.17). Most body cells have a single nucleus, although some, such as mature

Figure 3.17 Nucleus.

The nucleus contains most of a cell's genes, which are located on chromosomes.

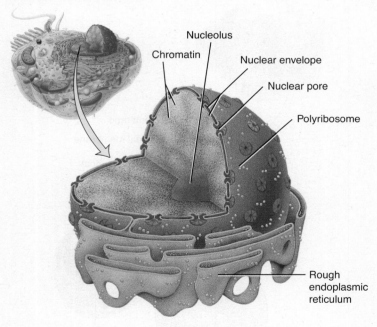

Details of the nucleus

Functions of the Nucleus

1. Controls cellular structure.
2. Directs cellular activities.
3. Produces ribosomes in nucleoli.

? What are the functions of nuclear genes?

red blood cells, have none. In contrast, skeletal muscle cells and a few other types of cells have several nuclei. A double membrane called the *nuclear envelope* separates the nucleus from the cytoplasm. Both layers of the nuclear envelope are lipid bilayers similar to the plasma membrane. The outer membrane of the nuclear envelope is continuous with the rough endoplasmic reticulum and resembles it in structure. Many openings called *nuclear pores* pierce the nuclear envelope. Nuclear pores control the movement of substances between the nucleus and the cytoplasm.

Inside the nucleus are one or more spherical bodies called *nucleoli* (noo′-KLĒ-ō-lī; singular is *nucleolus*). These clusters of protein, DNA, and RNA are the sites of assembly of ribosomes, which exit the nucleus through nuclear pores and participate in protein synthesis in the cytoplasm. Cells that synthesize large amounts of protein, such as muscle and liver cells, have prominent nucleoli.

Also within the nucleus are most of the cell's hereditary units, called *genes*, which control cellular structure and direct most cellular activities. The nuclear genes are arranged along *chromosomes* (*chromo-* = colored) (see Figure 3.21). Human somatic (body) cells have 46 chromosomes, 23 inherited from each parent. In a cell that is not dividing, the 46 chromosomes appear as a diffuse, granular mass, which is called *chromatin* (KRŌ-ma-tin) (Figure 3.17). The total genetic information carried in a cell or organism is called its *genome* (JĒ-nōm).

CLINICAL CONNECTION | Genomics

In the last decade of the twentieth century, the genomes of humans, mice, fruit flies, and more than 50 microbes were sequenced. As a result, research in the field of **genomics**, the study of the relationships between the genome and the biological functions of an organism, has flourished. The Human Genome Project began in 1990 as an effort to sequence all of the nearly 3.2 billion nucleotides of our genome and was completed in April 2003. Scientists now know that the total number of genes in the human genome is about 30,000. Information regarding the human genome and how it is affected by the environment seeks to identify and discover the functions of the specific genes that play a role in genetic diseases. Genomic medicine also aims to design new drugs and to provide screening tests to enable physicians to provide more effective counseling and treatment for disorders with significant genetic components such as hypertension (high blood pressure), obesity, diabetes, and cancer. •

The main parts of a cell and their functions are summarized in Table 3.2.

✓ **CHECKPOINT**

10. Why is the nucleus so important in the life of a cell?

TABLE 3.2

Cell Parts and Their Functions

PART	DESCRIPTION	FUNCTION(S)
PLASMA MEMBRANE	Composed of a lipid bilayer consisting of phospholipids, cholesterol, and glycolipids with various proteins inserted; surrounds cytoplasm	Protects cellular contents; makes contact with other cells; contains channels, transporters, receptors, enzymes, and cell-identity markers; mediates the entry and exit of substances
CYTOPLASM	Cellular contents between the plasma membrane and nucleus, including cytosol and organelles	Site of all intracellular activities except those occurring in the nucleus
Cytosol	Composed of water, solutes, suspended particles, lipid droplets, and glycogen granules	Liquid in which many of the cell's chemical reactions occur
	The cytoskeleton is a network in the cytoplasm composed of three protein filaments: microfilaments, intermediate filaments, and microtubules	Maintains shape and general organization of cellular contents; responsible for cell movements
Organelles	Specialized cellular structures with characteristic shapes and specific functions	Each organelle has one or more specific functions
Centrosome	Paired centrioles plus pericentriolar material	Pericentriolar material is organizing center for microtubules and mitotic spindle
Cilia and flagella	Motile cell surface projections with inner core of microtubules	Cilia move fluids over a cell's surface; a flagellum moves an entire cell
Ribosome	Composed of two subunits containing ribosomal RNA and proteins; may be free in cytosol or attached to rough ER	Protein synthesis
Endoplasmic reticulum (ER)	Membranous network of folded membranes; rough ER is studded with ribosomes and is attached to the nuclear membrane; smooth ER lacks ribosomes	Rough ER is the site of synthesis of glycoproteins and phospholipids; smooth ER is the site of fatty acid and steroid synthesis; smooth ER also releases glucose into the bloodstream, inactivates or detoxifies drugs and potentially harmful substances, and stores and releases calcium ions for muscle contraction
Golgi complex	A stack of 3–20 flattened membranous sacs called cisterns	Accepts proteins from rough ER; forms glycoproteins and lipoproteins; stores, packages, and exports proteins
Lysosome	Vesicle formed from Golgi complex; contains digestive enzymes	Fuses with and digests contents of vesicles; digests worn-out organelles (autophagy), entire cells (autolysis), and extracellular materials
Peroxisome	Vesicle containing oxidative enzymes	Detoxifies harmful substances, such as hydrogen peroxide and associated free radicals
Proteasome	Tiny barrel-shaped structure that contains proteases, enzymes that cut proteins	Degrades unneeded, damaged, or faulty proteins by cutting them into small peptides
Mitochondrion	Consists of outer and inner membranes, cristae, and matrix	Site of reactions that produce most of a cell's ATP
NUCLEUS	Consists of nuclear envelope with pores, nucleoli, and chromatin (or chromosomes)	Contains genes, which control cellular structure and direct most cellular activities

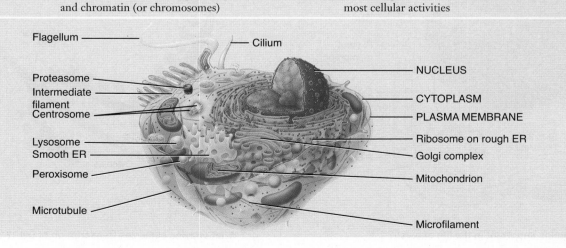

Flagellum — Cilium

Proteasome
Intermediate filament
Centrosome

Lysosome
Smooth ER

Peroxisome

Microtubule

NUCLEUS
CYTOPLASM
PLASMA MEMBRANE
Ribosome on rough ER
Golgi complex
Mitochondrion

Microfilament

3.6 GENE ACTION: PROTEIN SYNTHESIS

OBJECTIVE • **Outline the sequence of events involved in protein synthesis.**

Although cells synthesize many chemicals to maintain homeostasis, much of the cellular machinery is devoted to protein production. Cells constantly synthesize large numbers of diverse proteins. The proteins in turn determine the physical and chemical characteristics of cells and, on a larger scale, of organisms.

The DNA contained in genes provides the instructions for making proteins. To synthesize a protein, the information contained in a specific region of DNA is first *transcribed* (copied) to produce a specific molecule of RNA (ribonucleic acid). The RNA then attaches to a ribosome, where the information contained in the RNA is *translated* into a corresponding specific sequence of amino acids to form a new protein molecule (Figure 3.18).

Information is stored in DNA in four types of nucleotides, the repeating units of nucleic acids (see Figure 2.15). Each sequence of three DNA nucleotides is transcribed as a complementary (corresponding) sequence of three RNA nucleotides. Such a sequence of three successive DNA nucleotides is called a ***base triplet***. The three successive RNA nucleotides are called a ***codon***. When translated, a given codon specifies a particular amino acid.

Figure 3.18 **Overview of transcription and translation.**

Transcription occurs in the nucleus; translation takes place in the cytoplasm.

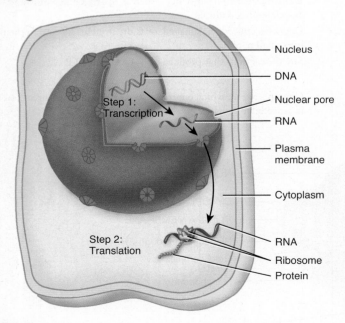

Why are proteins important in the life of a cell?

Transcription

During ***transcription***, which occurs in the nucleus, the genetic information in DNA base triplets is copied into a complementary sequence of codons in a strand of RNA. Transcription of DNA is catalyzed by the enzyme *RNA polymerase*, which must be instructed where to start the transcription process and where to end it. The segment of DNA where RNA polymerase attaches to it is a special sequence of nucleotides called a ***promoter***, located near the beginning of a gene (Figure 3.19a). Three kinds of RNA are made from DNA:

■ **Messenger *RNA (mRNA)*** directs synthesis of a protein.

Figure 3.19 **Transcription.**

During transcription, the genetic information in DNA is copied to RNA.

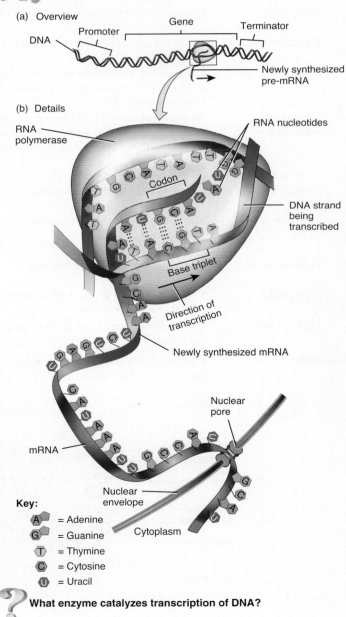

What enzyme catalyzes transcription of DNA?

- **Ribosomal RNA (rRNA)** joins with ribosomal proteins to make ribosomes.
- **Transfer RNA (tRNA)** binds to an amino acid and holds it in place on a ribosome until it is incorporated into a protein during translation. Each of the more than 20 different types of tRNA binds to only one of the 20 different amino acids.

During transcription, nucleotides pair in a complementary manner: The nitrogenous base cytosine (C) in DNA dictates the complementary nitrogenous base guanine (G) in the new RNA strand, a G in DNA dictates a C in RNA, a thymine (T) in DNA dictates an adenine (A) in RNA, and an A in DNA dictates a uracil (U) in RNA (Figure 3.19b). As an example, if a segment of DNA had the base sequence ATGCAT, the newly transcribed RNA strand would have the complementary base sequence UACGUA.

Transcription of DNA ends at another special nucleotide sequence on DNA called a **terminator**, which specifies the end of the gene (see Figure 3.19a). Upon reaching the terminator, RNA polymerase detaches from the transcribed RNA molecule and the DNA strand. Once synthesized, mRNA, rRNA (in ribosomes), and tRNA leave the nucleus of the cell by passing through a nuclear pore. In the cytoplasm, they participate in the next step in protein synthesis, translation.

Translation

Translation is the process in which mRNA associates with ribosomes and directs synthesis of a protein by converting the sequence of nucleotides in mRNA into a specific sequence of amino acids. Translation occurs in the following way (Figure 3.20):

1 An mRNA molecule binds to the small ribosomal subunit, and a special tRNA, called *initiator tRNA*, binds to the start codon (AUG) on mRNA, where translation begins.

2 The large ribosomal subunit attaches to the small subunit, creating a functional ribosome. The initiator tRNA fits into position on the ribosome. One end of a tRNA carries a specific amino acid, and the opposite end consists of a triplet of nucleotides called an **anticodon**. By pairing between complementary nitrogenous bases, the tRNA anticodon attaches to the mRNA codon. For example, if the mRNA codon is AUG, then a tRNA with the anticodon UAC would attach to it.

3 The anticodon of another tRNA with its amino acid attaches to the complementary mRNA codon next to the initiator tRNA.

4 A peptide bond is formed between the amino acids carried by the initiator tRNA and the tRNA next to it.

5 After the peptide bond forms, the empty tRNA detaches from the ribosome, and the ribosome shifts the mRNA strand by one codon. As the tRNA bearing the newly forming protein shifts, another tRNA with its amino acid binds to a newly exposed codon. Steps **3** through **5** repeat again and again as the protein lengthens.

6 Protein synthesis ends when the ribosome reaches a stop codon, at which time the completed protein detaches from the final tRNA. When the tRNA vacates the ribosome, the ribosome splits into its large and small subunits.

Protein synthesis progresses at a rate of about 15 amino acids per second. As the ribosome moves along the mRNA and before it completes synthesis of the whole protein, another ribosome may attach behind it and begin translation of the same mRNA strand. In this way, several ribosomes may be attached to the same mRNA. Such a group of ribosomes is called a **polyribosome**. The simultaneous movement of several ribosomes along the same mRNA strand permits a large amount of protein to be produced from each mRNA.

✓ CHECKPOINT

11. Define protein synthesis.

12. Distinguish between transcription and translation.

3.7 SOMATIC CELL DIVISION

OBJECTIVE • **Discuss the stages, events, and significance of somatic cell division.**

As body cells become damaged, diseased, or worn out, they are replaced by **cell division**, the process whereby cells reproduce themselves. The two types of cell division are reproductive cell division and somatic cell division. **Reproductive cell division** or **meiosis** is the process that produces gametes—sperm and oocytes—the cells needed to form the next generation of sexually reproducing organisms. Meiosis is described in Chapter 23; here we will focus on somatic cell division.

All body cells, except the gametes, are called **somatic cells** (sō-MAT-ik; *soma* = body). In **somatic cell division**, a cell divides into two identical cells. An important part of somatic cell division is replication (duplication) of the DNA sequences that make up genes and chromosomes so that the same genetic material can be passed on to the newly formed cells. After somatic cell division, each newly formed cell has the same number of chromosomes as the original cell. Somatic cell division replaces dead or injured cells and adds new ones for tissue growth. For example, skin cells are continually replaced by somatic cell divisions.

Figure 3.20 Protein elongation and termination of protein synthesis during translation.

During protein synthesis the ribosomal subunits join, but they separate when the process is complete.

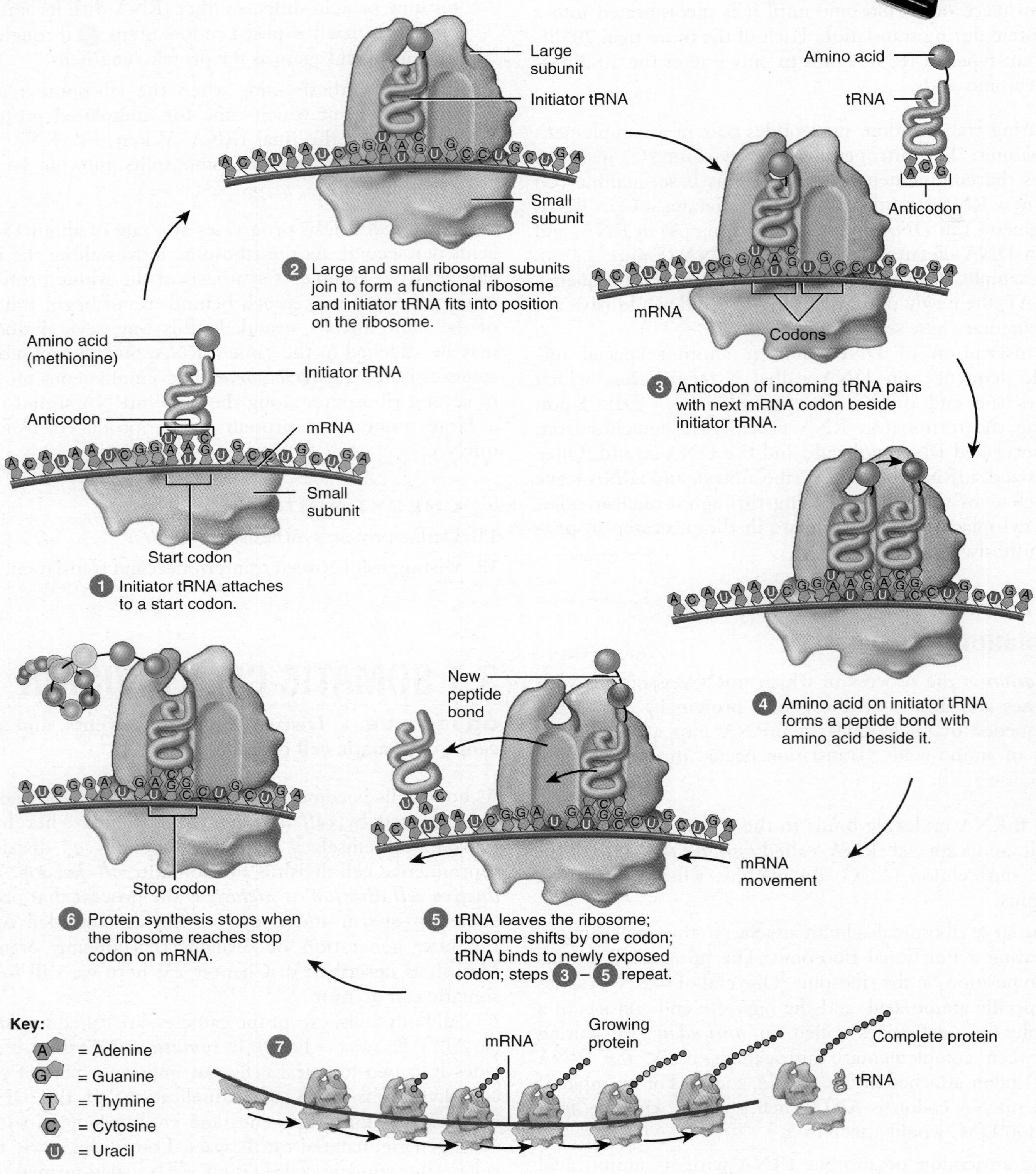

1 Initiator tRNA attaches to a start codon.

2 Large and small ribosomal subunits join to form a functional ribosome and initiator tRNA fits into position on the ribosome.

3 Anticodon of incoming tRNA pairs with next mRNA codon beside initiator tRNA.

4 Amino acid on initiator tRNA forms a peptide bond with amino acid beside it.

5 tRNA leaves the ribosome; ribosome shifts by one codon; tRNA binds to newly exposed codon; steps **3** – **5** repeat.

6 Protein synthesis stops when the ribosome reaches stop codon on mRNA.

Key:

- (A) = Adenine
- (G) = Guanine
- (T) = Thymine
- (C) = Cytosine
- (U) = Uracil

What is the function of a stop codon?

The **cell cycle** is the name for the sequence of changes that a cell undergoes from the time it forms until it duplicates its contents and divides into two cells. In somatic cells, the cell cycle consists of two major periods: interphase, when a cell is not dividing, and the mitotic phase, when a cell is dividing.

Interphase

During **interphase** the cell replicates its DNA. It also manufactures additional organelles and cytosolic components, such as centrosomes, in anticipation of cell division. Interphase is a state of high metabolic activity, and during this time the cell does most of its growing.

A microscopic view of a cell during interphase shows a clearly defined nuclear envelope, a nucleolus, and a tangled mass of chromatin (Figure 3.21a). Once a cell completes its replication of DNA and other activities of interphase, the mitotic phase begins.

Mitotic Phase

The **mitotic phase** (mī-TOT-ik) of the cell cycle consists of *mitosis*, division of the nucleus, followed by *cytokinesis*, division of the cytoplasm into two cells. The events that take place during mitosis and cytokinesis are plainly visible under a microscope because chromatin condenses into chromosomes.

Nuclear Division: Mitosis

During **mitosis** (mī-TŌ-sis; *mitos* = thread), the duplicated chromosomes become exactly segregated, one set into each of two separate nuclei. For convenience, biologists divide the process into four stages: prophase, metaphase, anaphase, and telophase. However, mitosis is a continuous process, with one stage merging imperceptibly into the next.

PROPHASE   During early **prophase** (PRŌ-fāz), the chromatin fibers condense and shorten into chromosomes that are visible under the light microscope (Figure 3.21b). The condensation process may prevent entangling of the long DNA strands as they move during mitosis. Recall that DNA replication took place during interphase. Thus, each prophase chromosome consists of a pair of identical, double-stranded **chromatids**. A constricted region of the chromosome, called a **centromere** (SEN-trō-mer), holds the chromatid pair together.

Later in prophase, the pericentriolar material of the two centrosomes starts to form the **mitotic spindle**, a football-shaped assembly of microtubules (Figure 3.21b). Lengthening of the microtubules between centrosomes pushes the centrosomes to opposite poles (ends) of the cell. Finally, the spindle extends from pole to pole. Then the nucleolus and nuclear envelope break down.

METAPHASE   During **metaphase** (MET-a-fāz), the centromeres of the chromatid pairs are aligned along the microtubules of the mitotic spindle at the exact center of the mitotic spindle (Figure 3.21c). This midpoint region is called the **metaphase plate**.

ANAPHASE   During **anaphase** (AN-a-fāz) the centromeres split, separating the two members of each chromatid pair, which move to opposite poles of the cell (Figure 3.21d). Once separated, the chromatids are called chromosomes. As the chromosomes are pulled by the microtubules of the mitotic spindle during anaphase, they appear V-shaped because the centromeres lead the way and seem to drag the trailing arms of the chromosomes toward the pole.

TELOPHASE   The final stage of mitosis, **telophase** (TEL-ō-fāz), begins after chromosomal movement stops (Figure 3.21e). The identical sets of chromosomes, now at opposite poles of the cell, uncoil and revert to the threadlike chromatin form. A new nuclear envelope forms around each chromatin mass, nucleoli appear, and eventually the mitotic spindle breaks up.

Cytoplasmic Division: Cytokinesis

Division of a cell's cytoplasm and organelles is called **cytokinesis** (sī′-tō-ki-NĒ-sis; *-kinesis* = motion). This process usually begins late in anaphase with formation of a **cleavage furrow**, a slight indentation of the plasma membrane, that extends around the center of the cell (Figure 3.21d, e). Microfilaments in the cleavage furrow pull the plasma membrane progressively inward, constricting the center of the cell like a belt around a waist, and ultimately pinching it in two. After cytokinesis there are two new and separate cells, each with equal portions of cytoplasm and organelles and identical sets of chromosomes. When cytokinesis is complete, interphase begins (Figure 3.21f).

⚕ **CLINICAL CONNECTION | Chemotherapy**

One of the distinguishing features of cancer cells is uncontrolled division. The mass of cells resulting from this division is called a neoplasm or tumor. One of the ways to treat cancer is by **chemotherapy**, the use of anticancer drugs. Some of these drugs stop cell division by inhibiting the formation of the mitotic spindle. Unfortunately, these types of anticancer drugs also kill all types of rapidly dividing cells in the body, causing side effects, as will be described in the Common Disorders section. •

✓ CHECKPOINT

13. Distinguish between somatic and reproductive cell division. Why is each important?

14. What are the major events of each stage of the mitotic phase?

Figure 3.21 Cell division: mitosis and cytokinesis. Begin the sequence at (a) at the top of the figure and read clockwise until you complete the process.

In somatic cell division, a single cell divides to produce two identical cells.

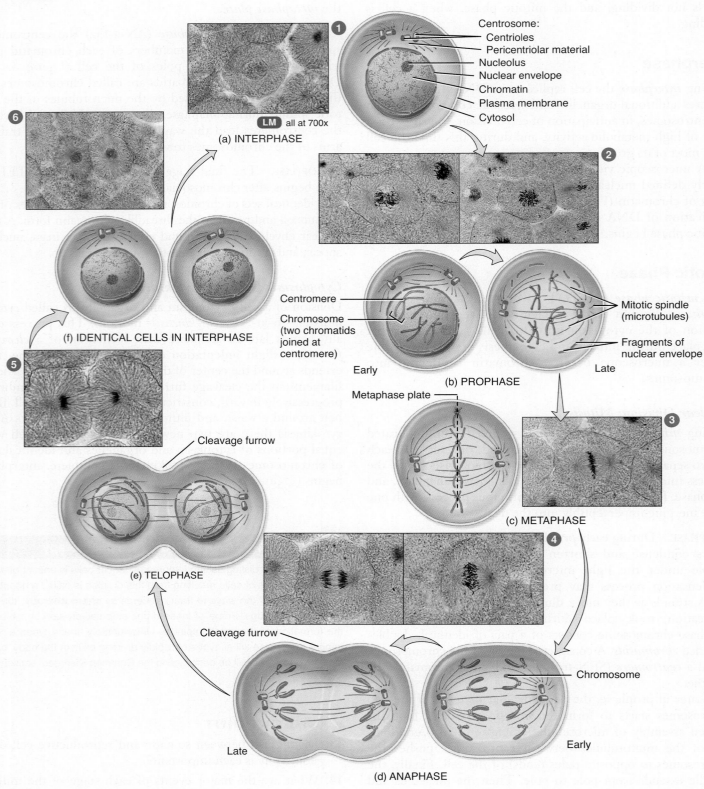

LM all at 700x

(a) INTERPHASE

Centrosome:
- Centrioles
- Pericentriolar material
- Nucleolus
- Nuclear envelope
- Chromatin
- Plasma membrane
- Cytosol

(f) IDENTICAL CELLS IN INTERPHASE

Centromere

Chromosome (two chromatids joined at centromere)

Mitotic spindle (microtubules)

Fragments of nuclear envelope

Early

Late

(b) PROPHASE

Metaphase plate

(c) METAPHASE

Cleavage furrow

(e) TELOPHASE

Chromosome

Cleavage furrow

Late

Early

(d) ANAPHASE

During which phase of mitosis does cytokinesis begin?

3.8 CELLULAR DIVERSITY

OBJECTIVE • **Describe how cells differ in size and shape.**

The body of an average human adult is composed of nearly 100 trillion cells. Cells vary considerably in size.

The sizes of cells are measured in units called *micrometers*. One micrometer (μm) is equal to 1 one-millionth of a meter, or 10^{-6} m (1/25,000 of an inch). High-powered microscopes are needed to see the smallest cells of the body. The largest cell, a single oocyte, has a diameter of about 140 μm and is barely visible to the unaided eye. A red blood cell has a diameter of 8 μm. To better visualize this, an average hair from the top of your head is approximately 100 μm in diameter.

The shapes of cells also vary considerably (Figure 3.22). They may be round, oval, flat, cube-shaped, column-shaped, elongated, star-shaped, cylindrical, or disc-shaped. A cell's shape is related to its function in the body. For example, a sperm cell has a long whiplike tail (flagellum) that it uses for

Figure 3.22 Diverse shapes and sizes of human cells. The relative difference in size between the smallest and largest cells is actually much greater than shown here.

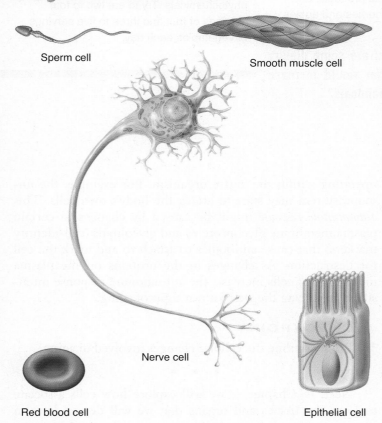

The nearly 100 trillion cells in an average adult can be classified into about 200 different cell types.

Sperm cell

Smooth muscle cell

Nerve cell

Red blood cell

Epithelial cell

 Why are sperm the only body cells that need to have a flagellum?

locomotion. The disc shape of a red blood cell gives it a large surface area that enhances its ability to pass oxygen to other cells. The long, spindle shape of a relaxed smooth muscle cell shortens as it contracts. This change in shape allows groups of smooth muscle cells to narrow or widen the passage for blood flowing through blood vessels. In this way, they regulate blood flow through various tissues. Recall that some cells contain microvilli, which greatly increase their surface area. Microvilli are common in the epithelial cells that line the small intestine, where the large surface area speeds the absorption of digested food. Nerve cells have long extensions that permit them to conduct nerve impulses over great distances. As you will see in the following chapters, cellular diversity also permits organization of cells into more complex tissues and organs.

✓ CHECKPOINT

15. How is cell shape related to function? Give examples.

3.9 AGING AND CELLS

OBJECTIVE • **Describe the cellular changes that occur with aging.**

Aging is a normal process accompanied by a progressive alteration of the body's homeostatic adaptive responses. It produces observable changes in structure and function and increases vulnerability to environmental stress and disease. The specialized branch of medicine that deals with the medical problems and care of elderly persons is *geriatrics* (jer′-ē-AT-riks; *ger-* = old age; *-iatrics* = medicine). *Gerontology* (jer′-on-TOL-ō-jē) is the scientific study of the process and problems associated with aging.

Although many millions of new cells normally are produced each minute, several kinds of cells in the body—including skeletal muscle cells and nerve cells—do not divide. Experiments have shown that many other cell types have only a limited capability to divide. Normal cells grown outside the body divide only a certain number of times and then stop. These observations suggest that cessation of mitosis is a normal, genetically programmed event. According to this view, "aging genes" are part of the genetic blueprint at birth. These genes have an important function in normal cells, but their activities slow over time. They bring about aging by slowing down or halting processes vital to life.

Another aspect of aging involves *telomeres* (TĒ-lō-merz), specific DNA sequences found only at the tips of each chromosome. These pieces of DNA protect the tips of chromosomes from erosion and from sticking to one another. However, in most normal body cells each cycle of cell division shortens the telomeres. Eventually, after many cycles of cell division, the telomeres can be completely gone, and even some of the functional chromosomal material may be lost. These observations suggest that erosion of DNA from the tips of our chromosomes contributes greatly to the aging and

Phytochemicals—Protecting Cellular Function

Many studies over the years have shown that people who consume plenty of plant foods, including vegetables, beans, fruits, and grains, have a lower risk of cancer and heart disease than their meat-and-potato-eating peers. Scientists are just beginning to uncover the biochemical explanations for these associations. Their investigations have led to the discovery of compounds in plants that appear to promote healthy cellular function, and to prevent the types of cellular damage associated with cancer, aging, and heart disease. Collectively, these compounds are called *phytochemicals*, literally "plant chemicals."

A Radical Notion?

Phytochemicals appear to protect cells and interrupt cancerous tumor growth in a number of interesting ways. Some phytochemicals block chemicals that can cause oxidative damage to cells. You will learn about the process of oxidation in Chapter 20 when you read about metabolism. Oxidative damage commonly occurs in cells when by-products of metabolism, known as oxygen free radicals, "steal" electrons from other molecules. This electron theft causes chain reactions of electron transfers that can damage cell membranes, the membranes of cellular organelles, and even the cells' genetic material.

Some phytochemicals act as antioxidants, donating electrons to free radical molecules, thus protecting cellular structures. Antioxidants include polyphenols, which are found in green tea; lycopenes, which are found in tomato products; and anthocyanins, found in berries and apples.

Disabling the Opponent

Many substances entering the body are potentially carcinogenic, depending on their interaction with certain enzymes in the liver. Some phytochemicals, such as the allyl sulfides in garlic and onions, enhance the production of enzymes that may render potentially carcinogenic substances harmless. The sulforaphane in broccoli, cauliflower, and other cruciferous vegetables performs a similar function.

Promoting Health

Some phytochemicals protect against cancer by blocking the action of substances called promoters. Promoters encourage the aggressive cellular division of cells that have undergone cancer-causing genetic changes. For example, estrogens are hormones that promote the division of cancerous cells in the breast. Isoflavonoids, found in soy products, weaken the action of estrogens in breast tissue.

Variety is the key to consuming more phytochemicals. Try to eat two to four servings of fruit and three to five servings of vegetables each day.

Think It Over . . . **What are some dietary changes you could make that would increase your intake of helpful phytochemicals?**

---

death of cells. Individuals who experience high levels of stress have significantly shorter telomere length.

Glucose, the most abundant sugar in the body, plays a role in the aging process. It is haphazardly added to proteins inside and outside cells, forming irreversible cross-links between adjacent protein molecules. With advancing age, more cross-links form, which contributes to the stiffening and loss of elasticity that occur in aging tissues.

Free radicals produce oxidative damage in lipids, proteins, or nucleic acids. Some effects are wrinkled skin, stiff joints, and hardened arteries. Naturally occurring enzymes in peroxisomes and in the cytosol normally dispose of free radicals. Certain dietary substances, such as vitamin E, vitamin C, beta carotene, zinc, and selenium, are antioxidants that inhibit free radical formation.

Some theories of aging explain the process at the cellular level, while others concentrate on regulatory mechanisms operating within the entire organism. For example, the immune system may start to attack the body's own cells. This *autoimmune response* might be caused by changes in certain plasma membrane glycoproteins and glycolipids (cell-identity markers) that cause antibodies to attach to and mark the cell for destruction. As changes in the proteins on the plasma membrane of cells increase, the autoimmune response intensifies, producing the well-known signs of aging.

✓ CHECKPOINT

16. Briefly outline the cellular changes involved in aging.

• • •

Next, in Chapter 4, we will explore how cells associate to form the tissues and organs that we will discuss later in the text.

COMMON DISORDERS

Cancer

Cancer is a group of diseases characterized by uncontrolled or abnormal cell proliferation. When cells in a part of the body divide without control, the excess tissue that develops is called a *tumor* or *neoplasm* (NĒ-ō-plazm; *neo-* = new). The study of tumors is called *oncology* (on-KOL-ō-jē; *onco-* = swelling or mass). Tumors may be cancerous and often fatal, or they may be harmless. A cancerous neoplasm is called a *malignant tumor* or *malignancy*. One property of most malignant tumors is their ability to undergo *metastasis* (me-TAS-ta-sis), the spread of cancerous cells to other parts of the body. A *benign tumor* is a neoplasm that does not metastasize. An example is a wart. Most benign tumors may be surgically removed if this interferes with normal body function or they become disfiguring. Some can be inoperable and perhaps fatal.

Growth and Spread of Cancer

Cells of malignant tumors duplicate rapidly and continuously. Cells of the body that have a high rate of cell division are more at risk for developing cancer. As malignant cells invade surrounding tissues, they often trigger *angiogenesis* (an'-jē-ō-JEN-e-sis), the growth of new networks of blood vessels. Proteins that stimulate angiogenesis in tumors are called *tumor angiogenesis factors (TAFs)*. The formation of new blood vessels can occur either by overproduction of TAFs or by the lack of naturally occurring angiogenesis inhibitors. As the cancer grows, it begins to compete with normal tissues for space and nutrients. Eventually, the normal tissue decreases in size and dies. Some malignant cells may detach from the initial (primary) tumor and invade a body cavity or enter the blood or lymph, then circulate to and invade other body tissues, establishing secondary tumors. Malignant cells resist the antitumor defenses of the body. The pain associated with cancer develops when the tumor presses on nerves or blocks a passageway in an organ so that secretions build up pressure, or as a result of dying tissue or organs.

Causes of Cancer

Several factors may trigger a normal cell to lose control and become cancerous. One cause is environmental agents: substances in the air we breathe, the water we drink, and the food we eat. A chemical agent or radiation that produces cancer is called a *carcinogen* (car-SIN-ō-jen). Carcinogens induce *mutations* (mū-TĀ-shuns), permanent changes in the DNA base sequence of a gene. The World Health Organization estimates that carcinogens are associated with 60–90% of all human cancers. Examples of carcinogens are hydrocarbons found in cigarette tar, radon gas from the earth, and ultraviolet (UV) radiation in sunlight.

Intensive research efforts are now directed toward studying cancer-causing genes, or *oncogenes* (ON-kō-jēnz). When inappropriately activated, these genes have the ability to transform a normal cell into a cancerous cell. Most oncogenes derive from normal genes called *proto-oncogenes* that regulate growth and development. The proto-oncogene undergoes some change that either causes it to be expressed inappropriately or make its products in ex-cessive amounts or at the wrong time. Some oncogenes cause excessive production of growth factors, chemicals that stimulate cell growth. Others may trigger changes in a cell-surface receptor, causing it to send signals as though it were being activated by a growth factor. As a result, the growth pattern of the cell becomes abnormal.

Some cancers have a viral origin. Viruses are tiny packages of nucleic acids, either RNA or DNA, that can reproduce only while inside the cells they infect. Some viruses, termed *oncogenic viruses*, cause cancer by stimulating abnormal proliferation of cells; for instance, the *human papillomavirus (HPV)* causes virtually all cervical cancers in women. The virus produces a protein that causes proteasomes to destroy a protein that normally suppresses unregulated cell division. In the absence of this suppressor protein, cells proliferate without control.

Recent studies suggest that certain cancers may be linked to a cell having abnormal numbers of chromosomes. As a result, the cell could potentially have extra copies of oncogenes or too few copies of tumor-suppressor genes, which in either case could lead to uncontrolled cell proliferation. There is also some evidence suggesting that cancer may be caused by normal stem cells that develop into cancerous stem cells capable of forming malignant tumors.

Later in the book, we will discuss the process of inflammation, which is a defensive response to tissue damage. It appears that inflammation contributes to various steps in the development of cancer. Some evidence suggests that chronic inflammation stimulates the proliferation of mutated cells and enhances their survival, promotes angiogenesis, and contributes to invasion and metastasis of cancer cells. There is a clear relationship between certain chronic inflammatory conditions and the transformation of inflamed tissue into a malignant tissue. For example, chronic gastritis (inflammation of the stomach lining) and peptic ulcers may be a causative factor in 60–90% of stomach cancers. Chronic hepatitis (inflammation of the liver) and cirrhosis of the liver are believed to be responsible for about 80% of liver cancers. Colorectal cancer is 10 times more likely to occur in patients with chronic inflammatory diseases of the colon, such as ulcerative colitis and Crohn's disease. And the relationship between asbestosis and silicosis, two chronic lung inflammatory conditions, and lung cancer has long been recognized. Chronic inflammation is also an underlying contributor to rheumatoid arthritis, Alzheimer's disease, depression, schizophrenia, cardiovascular disease, and diabetes.

Carcinogenesis: A Multistep Process

Carcinogenesis (kar'-si-nō-JEN-e-sis), the process by which cancer develops, is a multistep process in which as many as 10 distinct mutations may have to accumulate in a cell before it becomes cancerous. In colon cancer, the tumor begins as an area of increased cell proliferation that results from one mutation. This growth then progresses to abnormal, but noncancerous, growths called adenomas. After several more mutations, a carcinoma develops. The fact that so many mutations are needed for a cancer to develop indicates that cell growth is normally controlled with many sets of checks and balances.

Treatment of Cancer

Many cancers are removed surgically. However, when cancer is widely distributed throughout the body or exists in organs such as the brain whose functioning would be greatly harmed by surgery, chemotherapy and radiation therapy may be used instead. Sometimes surgery, chemotherapy, and radiation therapy are used in combination. Chemotherapy involves administering drugs that cause the death of cancerous cells. Radiation therapy breaks chromosomes, thus blocking cell division. Because cancerous cells divide rapidly, they are more vulnerable to the destructive effects of chemotherapy and radiation therapy than are normal cells. Unfortunately for the patients, hair follicle cells, red bone marrow cells, and cells lining the gastrointestinal tract also are rapidly dividing. Hence, the side effects of chemotherapy and radiation therapy include hair loss due to death of hair follicle cells, vomiting and nausea due to death of cells lining the stomach and intestines, and susceptibility to infection due to slowed production of white blood cells in red bone marrow.

Treating cancer is difficult because it is not a single disease and because the cells in a single tumor population rarely behave all in the same way. Although most cancers are thought to derive from a single abnormal cell, by the time a tumor reaches a clinically detectable size, it may contain a diverse population of abnormal cells. For example, some cancerous cells metastasize readily, and others do not. Some are sensitive to chemotherapy drugs and some are drug-resistant. Because of differences in drug resistance, a single chemotherapeutic agent may destroy susceptible cells but permit resistant cells to proliferate.

Another potential treatment for cancer that is currently under development is **virotherapy**, the use of viruses to kill cancer cells. The viruses employed in this strategy are designed so that they specifically target cancer cells without affecting the healthy cells of the body. For example, proteins (such as antibodies) that specifically bind to receptors found only in cancer cells are attached to viruses. Once inside the body, the viruses bind to cancer cells and then infect them. The cancer cells are eventually killed once the viruses cause cellular lysis.

Researchers are also investigating the role of **metastasis regulatory genes** that control the ability of cancer cells to undergo metastasis. Scientists hope to develop therapeutic drugs that can manipulate these genes and, therefore, block metastasis of cancer cells.

MEDICAL TERMINOLOGY AND CONDITIONS

Anaplasia (an′-a-PLĀ-zē-a; *an-* = not; *-plasia* = to shape) The loss of tissue differentiation and function that is characteristic of most malignancies.

Apoptosis (ap′-op-TŌ-sis; a falling off, like dead leaves from a tree) An orderly, genetically programmed cell death in which "cell-suicide" genes become activated. Enzymes produced by these genes disrupt the cytoskeleton and nucleus; the cell shrinks and pulls away from neighboring cells; the DNA within the nucleus fragments; and the cytoplasm shrinks, although the plasma membrane remains intact. Phagocytes in the vicinity then ingest the dying cell. Apoptosis removes unneeded cells during development before birth and continues after birth both to regulate the number of cells in a tissue and to eliminate potentially dangerous cells such as cancer cells.

Atrophy (AT-rō-fē; *a-* = without; *-trophy* = nourishment) A decrease in the size of cells with subsequent decrease in the size of the affected tissue or organ; wasting away.

Biopsy (BĪ-op-sē; *bio-* = life; *-opsy* = viewing) The removal and microscopic examination of tissue from the living body for diagnosis.

Dysplasia (dis-PLĀ-zē-a; *dys-* = abnormal) Alteration in the size, shape, and organization of cells due to chronic irritation or inflammation; may progress to a neoplasm (tumor formation, usually malignant) or revert to normal if the irritation is removed.

Hyperplasia (hī′-per-PLĀ-zē-a; *hyper-* = over) Increase in the number of cells of a tissue due to an increase in the frequency of cell division.

Hypertrophy (hī-PER-trō-fē) Increase in the size of cells in a tissue without cell division.

Metaplasia (met′-a-PLĀ-zē-a; *meta-* = change) The transformation of one type of cell into another.

Necrosis (ne-KRŌ-sis = death) A pathological type of cell death, resulting from tissue injury, in which many adjacent cells swell, burst, and spill their cytoplasm into the interstitial fluid; the cellular debris usually stimulates an inflammatory response, which does not occur in apoptosis.

Progeny (PROJ-e-nē; *pro-* = forward; *-geny* = production) Offspring or descendants.

Progeria (prō-JER-ē-a) A disease characterized by normal development in the first year of life followed by rapid aging. It is caused by a genetic defect in which telomeres are considerably shorter than normal. Symptoms include dry and wrinkled skin, total baldness, and birdlike facial features. Death usually occurs around age 13.

Proteomics (prō′-tē-Ō-miks; *proteo-* = protein) The study of the proteome (all of an organism's proteins) in order to identify all of the proteins produced; it involves determining how the proteins interact and ascertaining the three-dimensional structure of proteins so that drugs can be designed to alter protein activity to help in the treatment and diagnosis of disease.

Tumor marker A substance introduced into circulation by tumor cells that indicates the presence of a tumor, as well as the specific type. Tumor markers may be used to screen, diagnose, make a prognosis, evaluate a response to treatment, and monitor for recurrence of cancer.

Werner syndrome A rare, inherited disease that causes a rapid acceleration of aging, usually while the person is only in his or her twenties. It is characterized by wrinkling of the skin, graying of the hair and baldness, cataracts, muscular atrophy, and a tendency to develop diabetes mellitus, cancer, and cardiovascular disease. Most afflicted individuals die before age 50. Recently, the gene that causes Werner syndrome has been identified. Researchers hope to use the information to gain insight into the mechanisms of aging, as well as to help those suffering from the disorder.

CHAPTER REVIEW AND RESOURCE SUMMARY

REVIEW	RESOURCES

Introduction

1. A **cell** is the basic, living, structural and functional unit of the body.
2. **Cell biology** is the study of cell structure and function.

3.1 A Generalized View of the Cell

1. Figure 3.1 shows a generalized view of a cell that is a composite of many different cells in the body.
2. The principal parts of a cell are the **plasma membrane**; the **cytoplasm**, which consists of **cytosol** and **organelles**; and the **nucleus**.

Anatomy Overview—
Cell Structure and Function
Concepts and
Connections—Human Cell

Figure 3.1 Generalized view of a body cell

3.2 The Plasma Membrane

1. The plasma membrane surrounds and contains the cytoplasm of a cell; it is composed of lipids and proteins.
2. The **lipid bilayer** consists of two back-to-back layers of **phospholipids, cholesterol,** and **glycolipids**.
3. **Integral proteins** extend into or through the lipid bilayer; **peripheral proteins** associate with the inner or outer surface of the membrane.
4. The membrane's **selective permeability** permits some substances to pass across it more easily than others. The lipid bilayer is permeable to water and to most lipid-soluble molecules. Small- and medium-sized water-soluble materials may cross the membrane with the assistance of integral proteins.
5. Membrane proteins have several functions. **Ion channels** and **carriers (transporters)** are integral proteins that help specific solutes across the membrane; **receptors** serve as cellular recognition sites; some membrane proteins are **enzymes**, and others are **cell identity markers**.

Anatomy Overview—
Plasma Membrane
Animation—Membrane
Function

Figure 3.2 Chemistry and structure of the plasma membrane

Exercise—Paint a Cell
Membrane
Concepts and Connections—
Membrane Functions

3.3 Transport Across the Plasma Membrane

1. Fluid inside body cells is called **intracellular fluid (ICF)**; fluid outside body cells is **extracellular fluid (ECF)**. The ECF in the microscopic spaces between the cells of tissues is **interstitial fluid**. The ECF in blood vessels is **plasma**, and that in lymphatic vessels is **lymph**.
2. Any material dissolved in a fluid is called a **solute**, and the fluid that dissolves materials is the **solvent**. Body fluids are dilute solutions in which a variety of solutes are dissolved in the solvent water.
3. The selective permeability of the plasma membrane supports the existence of **concentration gradients**, differences in the concentration of chemicals between one side of the membrane and the other.
4. Materials move through cell membranes by **passive processes** or by **active processes**. In passive processes, a substance moves down its concentration gradient across the membrane. In active processes, cellular energy is used to drive the substance "uphill" against its concentration gradient.
5. In transport in vesicles, tiny vesicles either detach from the plasma membrane while bringing materials into the cell or merge with the plasma membrane to release materials from the cell.
6. **Diffusion** is the movement of substances due to their kinetic energy. In net diffusion, substances move from an area of higher concentration to an area of lower concentration until **equilibrium** is reached. At equilibrium the concentration is the same throughout the solution. In **simple diffusion**, lipid-soluble substances move through the lipid bilayer. In **facilitated diffusion**, substances cross the membrane with the assistance of ion channels and carriers.

Animation—Membrane
Functions
Animation—Transport across
Plasma Membrane
Exercise—Membrane Function
Matchup
Exercise—Create a Transport
Condition
Exercise—Osmosis—Move It!
Concepts and Connection—
Membrane Transport
Processes

REVIEW	RESOURCES

7. **Osmosis** is the movement of water molecules through a selectively permeable membrane from an area of higher water concentration to an area of lower water concentration. In an **isotonic solution**, red blood cells maintain their normal shape; in a **hypotonic solution**, they gain water and undergo **hemolysis**; in a **hypertonic solution**, they lose water and undergo **crenation**.

8. With the expenditure of cellular energy, usually in the form of ATP, solutes can cross the membrane against their concentration gradient by means of **active transport**. Actively transported solutes include several ions such as Na$^+$, K$^+$, H$^+$, Ca^{2+}, I$^-$, and Cl$^-$; amino acids; and monosaccharides. The most important active transport pump is the **sodium–potassium pump**, which expels Na$^+$ from cells and brings in K$^+$.

9. Transport in **vesicles** includes both **endocytosis** (**phagocytosis** and **bulk-phase endocytosis [pinocytosis]**) and **exocytosis**. Phagocytosis is the ingestion of solid particles. It is an important process used by some white blood cells to destroy bacteria that enter the body. Bulk-phase endocytosis is the ingestion of extracellular fluid. Exocytosis involves movement of secretory or waste products out of a cell by fusion of vesicles within the cell with the plasma membrane.

3.4 Cytoplasm

1. **Cytoplasm** includes all the cellular contents between the plasma membrane and nucleus; it consists of **cytosol** and **organelles**. The liquid portion of cytoplasm is cytosol, composed mostly of water, plus ions, glucose, amino acids, fatty acids, proteins, lipids, ATP, and waste products; the cytosol is the site of many chemical reactions required for a cell's existence. Organelles are specialized cellular structures with characteristic shapes and specific functions.

2. The **cytoskeleton** is a network of several kinds of protein filaments that extend throughout the cytoplasm; they provide a structural framework for the cell and generate movements. Components of the cytoskeleton include **microfilaments**, **intermediate filaments**, and **microtubules**.

3. The **centrosome** is an organelle that consists of two centrioles and pericentriolar material. The centrosome serves as a center for organizing microtubules in interphase cells and the mitotic spindle during cell division.

4. **Cilia** and **flagella** are motile projections of the cell surface. Cilia move fluid along the cell surface; a flagellum moves an entire cell.

5. **Ribosomes**, composed of ribosomal RNA and ribosomal proteins, consist of two subunits and are the sites of protein synthesis.

6. **Endoplasmic reticulum (ER)** is a network of membranes that extends from the nuclear envelope throughout the cytoplasm. **Rough ER** is studded with ribosomes. Proteins synthesized on the ribosomes enter the ER for processing and sorting. The ER is also where glycoproteins and phospholipids form. **Smooth ER** lacks ribosomes. It is the site where fatty acids and steroids are synthesized. Smooth ER also participates in releasing glucose from the liver into the bloodstream, inactivating or detoxifying drugs and other potentially harmful substances, and storing and releasing calcium ions that trigger contraction in muscle cells.

7. The **Golgi complex** consists of flattened sacs called **cisterns** that receive proteins synthesized in the rough ER. Within the Golgi cisterns the proteins are modified, sorted, and packaged into vesicles for transport to different destinations. Some processed proteins leave the cell in secretory vesicles, some are incorporated into the plasma membrane, and some enter lysosomes.

8. **Lysosomes** are membrane-enclosed vesicles that contain digestive enzymes. They function in digestion of worn-out organelles (**autophagy**) and even in digestion of their own cell (**autolysis**).

9. **Peroxisomes** are similar to lysosomes but smaller. They oxidize various organic substances such as amino acids, fatty acids, and toxic substances and, in the process, produce hydrogen peroxide and associated free radicals, such as superoxide. The hydrogen peroxide is degraded by an enzyme in peroxisomes called catalase.

10. **Proteasomes** contain proteases that continually degrade unneeded, damaged, or faulty proteins.

11. **Mitochondria** consist of a smooth outer membrane, an inner membrane containing folds called mitochondrial **cristae**, and a fluid-filled cavity called the **matrix**. They are called "powerhouses" of the cell because they produce most of a cell's ATP.

Anatomy Overview—Cell Structure and Function
Exercise—Concentrate on Cellular Functions
Exercise—Target Practice
Concepts and Connections— Human Cell
Concepts and Connections— Membranous Organelles

REVIEW	**RESOURCES**

3.5 Nucleus

1. The **nucleus** consists of a double **nuclear envelope**; **nuclear pores**, which control the movement of substances between the nucleus and cytoplasm; **nucleoli**, which produce ribosomes; and **genes** arranged on **chromosomes**.

2. Most body cells have a single nucleus; some (red blood cells) have none, and others (skeletal muscle cells) have several.

3. Genes control cellular structure and most cellular functions.

Anatomy Overview—The Nucleus

3.6 Gene Action: Protein Synthesis

1. Most of the cellular machinery is devoted to protein synthesis.

2. Cells make proteins by transcribing and translating the genetic information encoded in the sequence of four types of nitrogenous bases in DNA.

3. In **transcription**, genetic information encoded in the DNA base sequence (**base triplet**) is copied into a complementary sequence of bases in a strand of **messenger RNA (mRNA)** called a **codon**. Transcription begins on DNA in a region called a **promoter**.

4. **Translation** is the process in which mRNA associates with ribosomes and directs synthesis of a protein, converting the nucleotide sequence in mRNA into a specific sequence of amino acids in the protein.

5. In translation, mRNA binds to a ribosome, specific amino acids attach to **transfer RNA (tRNA)**, and **anticodons** of tRNA bind to codons of mRNA, bringing specific amino acids into position on a growing protein. Translation begins at the start codon and terminates at the stop codon.

Animation—Protein Synthesis

Figure 3.20 Protein elongation and termination of protein synthesis during translation

Exercise—Protein-Producing Processes
Exercise—Synthesize and Transport a Protein

3.7 Somatic Cell Division

1. **Cell division** is the process by which cells reproduce themselves. Cell division that results in an increase in the number of body cells is called **somatic cell division**; it involves a nuclear division called **mitosis** plus division of the cytoplasm, called **cytokinesis**. Cell division that results in the production of sperm and oocytes is called **reproductive cell division**.

2. The **cell cycle** is an orderly sequence of events in somatic cell division in which a cell duplicates its contents and divides in two. It consists of **interphase** and a **mitotic phase**.

3. During interphase, the DNA molecules, or chromosomes, replicate themselves so that identical chromosomes can be passed on to the next generation of cells. A cell that is between divisions and is carrying on every life process except division is said to be in interphase.

4. Mitosis is the replication and distribution of two sets of chromosomes into separate and equal nuclei; it consists of **prophase**, **metaphase**, **anaphase**, and **telophase**.

5. During cytokinesis, which usually begins late in anaphase and ends in telophase, a **cleavage furrow** forms and progresses inward, cutting through the cell to form two separate identical cells, each with equal portions of cytoplasm, organelles, and chromosomes.

Animation—The Cell Cycle and Division Processes
Exercise—Mitosis Matchup
Exercise—Cell Cycle Quiz

3.8 Cellular Diversity

1. The different types of cells in the body vary considerably in size and shape.

2. The sizes of cells are measured in micrometers. One micrometer (μm) equals 10^{-6} m (1/25,000 of an inch). Cells in the body range from 8 μm to 140 μm in size.

3. A cell's shape is related to its function.

Concepts and Connections—Human Cell

3.9 Aging and Cells

1. **Aging** is a normal process accompanied by progressive alteration of the body's homeostatic adaptive responses.

2. Many theories of aging have been proposed, including genetically programmed cessation of cell division, shortening of telomeres, addition of glucose to proteins, buildup of free radicals, and an intensified autoimmune response.

SELF-QUIZ

1. If the extracellular fluid contains a greater concentration of solutes than the cytosol of the cell, the extracellular fluid is said to be
   a. isotonic.   b. hypertonic.   c. in equilibrium.
   d. normal.   e. hypotonic.

2. The proteins found in the plasma membrane
   a. are primarily glycoproteins.
   b. allow the passage of many substances into and out of the cell.
   c. allow cells to recognize other cells.
   d. help anchor cells to each other.
   e. have all of the above functions.

3. To enter many body cells, glucose must bind to a specific membrane carrier protein, which assists glucose to cross the membrane without using ATP. This type of movement is known as
   a. facilitated diffusion.   b. simple diffusion
   c. vesicular transport.   d. osmosis.   e. active transport.

4. A red blood cell placed in a hypotonic solution undergoes
   a. hemolysis.   b. crenation.   c. equilibrium.
   d. a decrease in osmotic pressure.   e. shrinkage.

5. Many white blood cells engulf bacteria through
   a. osmosis.   b. bulk-phase endocytosis.
   c. facilitated diffusion.   d. exocytosis.   e. phagocytosis.

6. Which organelles are involved in the breakdown of hydrogen peroxide in the liver?
   a. Golgi complexes   b. proteasomes   c. lysosomes
   d. peroxisomes   e. phagosomes

7. Which of the following processes requires ATP?
   a. simple diffusion   b. active transport   c. osmosis
   d. facilitated diffusion using ion channels
   e. facilitated diffusion using carriers

8. Nicotine in cigarette smoke interferes with the ability of cells to rid the breathing passageways of debris. Which organelles are "paralyzed" by nicotine?
   a. flagella   b. ribosomes   c. microfilaments
   d. cilia   e. lysosomes

9. Many proteins found in the plasma membrane are formed by the _____ and packaged by the _____.
   a. ribosomes, Golgi complex
   b. smooth endoplasmic reticulum, Golgi complex
   c. Golgi complex, lysosomes
   d. mitochondria, Golgi complex
   e. nucleus, smooth endoplasmic reticulum

10. Match the following:
   ____ a. cellular movement   A. centrosome
   ____ b. selective permeability   B. cytoskeleton
   ____ c. protein synthesis   C. Golgi complex
   ____ d. lipid synthesis, detoxification   D. lysosomes
   ____ e. packages proteins   E. mitochondria

   ____ f. ATP production   F. plasma membrane
   ____ g. digest bacteria and worn-out   G. ribosomes
         organelles   H. smooth ER
   ____ h. forms mitotic spindle

11. If the smooth endoplasmic reticulum were destroyed, a cell would not be able to
   a. form lysosomes.   b. synthesize certain proteins.
   c. generate energy.   d. phagocytize bacteria.
   e. synthesize fatty acids and steroids.

12. Water moves into and out of red blood cells through
   a. endocytosis.   b. phagocytosis.   c. osmosis.
   d. active transport.   e. facilitated diffusion.

13. A cell undergoing mitosis goes through the following stages in which sequence?
   a. interphase, metaphase, prophase, cytokinesis
   b. interphase, prophase, cytokinesis, telophase
   c. anaphase, metaphase, prophase, telophase
   d. anaphase, metaphase, prophase, cytokinesis
   e. prophase, metaphase, anaphase, telophase

14. Transcription involves
   a. transferring information from mRNA to tRNA.
   b. codons binding with anticodons.
   c. joining amino acids by peptide bonds.
   d. copying information contained in DNA to mRNA.
   e. synthesizing the protein on the ribosome.

15. If a DNA strand has a nitrogenous base sequence TACGA, then the sequence of bases on the corresponding mRNA would be
   a. ATGCT.   b. AUGCU.   c. GUACU.
   d. CTGAT.   e. AUCUG.

16. Place the following events of protein synthesis in the proper order.
   1. DNA uncoils and mRNA is transcribed.
   2. tRNA with an attached amino acid pairs with mRNA.
   3. mRNA passes from the nucleus into the cytoplasm and attaches to a ribosome.
   4. Protein is formed.
   5. Two amino acids are linked by a peptide bond.
   a. 1, 2, 3, 4, 5   b. 1, 3, 2, 5, 4   c. 1, 2, 3, 5, 4
   d. 1, 5, 3, 2, 4   e. 2, 1, 3, 4, 5

17. Match the following descriptions with the phases shown.
   ____ a. nuclear envelope (membrane) and   A. prophase
         nucleoli reappear   B. cytokinesis
   ____ b. centromeres of the chromatid pairs line   C. telophase
         up in the center of the mitotic spindle   D. anaphase
   ____ c. DNA replicates   E. metaphase
   ____ d. cleavage furrow splits cell into two   F. interphase
         identical cells
   ____ e. chromosomes move toward opposite
         poles of cell
   ____ f. chromatids are attached at
         centromeres; mitotic spindle forms

18. In which phase is a cell highly active and growing?

    a. anaphase   **b.** prophase   **c.** metaphase

    d. telophase   **e.** interphase

19. Movement of materials into and out of the nucleus is controlled by the

    a. plasma membrane.   **b.** nuclear pores.

    c. mitochondria.   **d.** cilia.   **e.** nucleoli.

20. Which of the following statements concerning cancer is NOT true?

    a. A benign tumor is noncancerous.

    b. When a cancerous growth presses on nerves, it can cause pain.

    c. Angiogenesis is the spread of cancerous cells to other parts of the body.

    d. Ultraviolet radiation and radon gas are carcinogens.

    e. Cancer is uncontrolled mitosis in abnormal cells.

CRITICAL THINKING APPLICATIONS

1. One function of bone is to store minerals, especially calcium. The bone tissue must be dissolved to release the calcium for use by the body's systems. Which organelle would be involved in breaking down bone tissue?

2. In your dream, you're on a deserted tropical island. The sun's hot, you're very thirsty, and you're surrounded by water. You want to take a long, cool drink of seawater, but something you learned in A&P stops you from drinking and saves your life! Why shouldn't you drink seawater?

3. Mucin is a glycoprotein present in saliva. When mixed with water, mucin becomes the slippery substance known as mucus. Trace the route taken by mucin through the cells of the salivary glands, starting with the organelle where it is synthesized and ending with its release from the cells.

4. Your friend Steve works a highly stressful job as an air traffic controller. His diet during his shift consists mainly of chocolate and fizzy drinks. He has been sick more often and jokingly exclaims to you that his job is "aging him prematurely." Your response to Steve is that this may not be far from the truth. Why?

ANSWERS TO FIGURE QUESTIONS

3.1 The three main parts of a cell are the plasma membrane, cytoplasm, and nucleus.

3.2 Some integral proteins function as channels or carriers to move substances across membranes. Other integral proteins function as receptors. Membrane glycolipids and glycoproteins are involved in cellular recognition.

3.3 In simple diffusion, substances cross a membrane through the lipid bilayer; in facilitated diffusion, ion channels or carriers are involved.

3.4 Oxygen, carbon dioxide, fatty acids, fat-soluble vitamins, and steroids can cross the plasma membrane by simple diffusion through the lipid bilayer.

3.5 The concentration of K^+ is higher in the cytosol of body cells than in extracellular fluids.

3.6 Insulin promotes insertion of glucose carriers in the plasma membrane, which increases cellular glucose uptake by facilitated diffusion.

3.7 No, the water concentrations can never be the same because the beaker always contains pure (100%) water and the sac contains a solution that is less than 100% water.

3.8 A 2% solution of NaCl will cause crenation of RBCs because it is hypertonic.

3.9 ATP adds a phosphate group to the pump protein, which changes the pump's three-dimensional shape.

3.10 The trigger that causes pseudopod extension is binding of a particle to a membrane receptor.

3.11 Clusters of microtubules form the structure of centrioles, cilia, and flagella.

3.12 The components of the centrosome are two centrioles and the pericentriolar material.

3.13 Large and small ribosomal subunits are synthesized in a nucleolus in the nucleus and then join together in the cytoplasm.

3.14 Rough ER has attached ribosomes where proteins that will be used in organelles or plasma membranes or exported from the cell are synthesized; smooth ER lacks ribosomes and is associated with lipid synthesis and other metabolic reactions.

3.15 Cells that secrete proteins into extracellular fluid have extensive Golgi complexes.

3.16 Mitochondrial cristae provide a large surface area for chemical reactions and contain enzymes needed for ATP production.

3.17 Nuclear genes control cellular structure and direct most cellular activities.

3.18 Proteins determine the physical and chemical characteristics of cells.

3.19 RNA polymerase catalyzes transcription of DNA.

3.20 When a ribosome encounters a stop codon in mRNA, the completed protein detaches from the final tRNA.

3.21 Cytokinesis usually begins late in anaphase.

3.22 Sperm, which use flagella for locomotion, are the only body cells required to move considerable distances.

CHAPTER 4

TISSUES

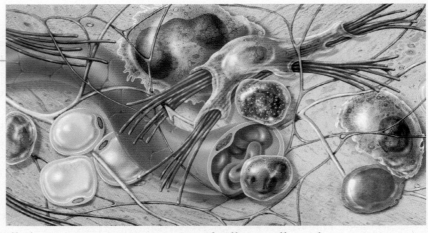

As you learned in the previous chapter, cells are highly organized living units, but they typically do not function alone. Instead, cells work together in groups called tissues. A *tissue* is a group of cells, usually with a common embryonic origin, that function together to carry out specialized activities. *Histology* (hiss-TOL-ō-jē; *hist-* = tissue; *-logy* = study of) is the science that deals with the study of tissues. A *pathologist* (pa-THOL-ō-gist; *patho-* = disease) is a physician who examines cells and tissues to help other physicians make accurate diagnoses. One of the principal functions of a pathologist is to examine tissues for any changes that might indicate disease.

Did you know? People who have never taken biology courses have been known to remark, "I don't want to lift weights, because if I build muscular tissue, I am afraid it will turn to fat when I stop lifting." Of course, muscular tissue is specialized for its contractile function, and does not "turn into" fat tissue, which is specialized for energy storage. As people age, skeletal muscles do atrophy, or shrink, over the years. Adipose tissue tends to grow. While some of these changes in body composition appear to be an inevitable part of the aging process, scientists believe much of the loss of muscular tissue can be prevented with exercise training.

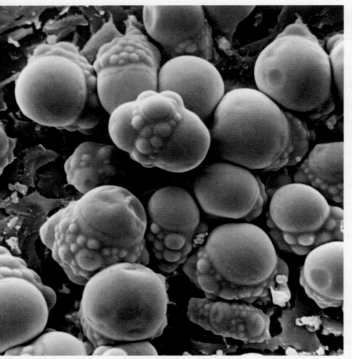

FOCUS ON WELLNESS
Fat Talk

LOOKING BACK TO MOVE AHEAD...

4.1 TYPES OF TISSUES

OBJECTIVE • **Name four basic types of tissue that make up the human body and state the characteristics of each.**

Body tissues are classified into four basic types based on their structure and functions:

1. *Epithelial tissue* (ep′-i-THĒ-lē-al) covers body surfaces; lines body cavities, hollow organs, and ducts (tubes); and forms glands.

2. *Connective tissue* protects and supports the body and its organs, binds organs together, stores energy reserves as fat, and provides immunity.

3. *Muscular tissue* generates the physical force needed to make body structures move.

4. *Nervous tissue* detects changes inside and outside the body and initiates and transmits nerve impulses (action potentials) that coordinate body activities to help maintain homeostasis.

Epithelial tissue and most types of connective tissue are discussed in detail in this chapter. The structure and functions of bone tissue and blood (connective tissues), muscular tissue, and nervous tissue are examined in detail in later chapters.

Most epithelial cells and some muscle and nerve cells are tightly joined into functional units by points of contact between their plasma membranes called **cell junctions**. Some cell junctions fuse cells together so tightly that they prevent substances from passing between the cells. This fusion is very important for tissues that line the stomach, intestines, and urinary bladder because it prevents the contents of these organs from leaking out. Other cell junctions hold cells together so that they don't separate while performing their functions. Still other cell junctions form channels that allow ions and molecules to pass between cells. This permits cells in a tissue to communicate with each other and it also enables nerve or muscle impulses to spread rapidly among cells.

✓ CHECKPOINT

1. Define a tissue. What are the four basic types of body tissues?

2. Why are cell junctions important?

4.2 EPITHELIAL TISSUE

OBJECTIVES • **Discuss the general features of epithelial tissue.**

• **Describe the structure, location, and function of the various types of epithelial tissue.**

Epithelial tissue, or more simply **epithelium** (plural is *epithelia*), may be divided into two types: (1) *covering and lining*

epithelium and (2) *glandular epithelium*. As its name suggests, covering and lining epithelium forms the outer covering of the skin and the outer covering of some internal organs. It also lines body cavities; blood vessels; ducts; and the interiors of the respiratory, digestive, urinary, and reproductive systems. It makes up, along with nervous tissue, the parts of the sense organs for hearing, vision, and touch. Glandular epithelium makes up the secreting portion of glands, such as sweat glands.

General Features of Epithelial Tissue

As you will see shortly, there are many different types of epithelia, each with characteristic structure and functions. However, all of the different types of epithelial tissue also have features in common. General features of epithelial tissue include the following:

■ Epithelial tissue consists largely or entirely of closely packed cells with little extracellular material between them, and the cells are arranged in continuous sheets, in either single or multiple layers.

■ The cells of epithelial tissue have an *apical (free) surface*, which is exposed to a body cavity, lining of an internal organ, or the exterior of the body; *lateral surfaces*, which face adjacent cells on either side; and a *basal surface*, which is attached to a basement membrane. In discussing epithelial tissues with multiple layers, the term *apical layer* refers to the most superficial layer of cells; the term *basal layer* refers to the deepest layer of cells. The **basement membrane** is a thin extracellular structure composed mostly of protein fibers. It is located between the epithelial tissue and the underlying connective tissue layer and helps bind and support the epithelial tissue (see Figure 4.1).

■ Epithelial tissue is **avascular** (*a-* = without; *vascular* = blood vessels); that is, they lack blood vessels. The vessels that supply nutrients to and remove wastes from epithelial tissue are located in adjacent connective tissues. The exchange of materials between epithelial and connective tissue occurs by diffusion.

■ Epithelial tissue has a nerve supply.

■ Because epithelial tissue is subject to a certain amount of wear and tear and injury, it has a high capacity for renewal by cell division.

Classification of Epithelial Tissues

Covering and lining epithelium, which covers or lines various parts of the body, is classified according to the arrangement of cells into layers and the shape of the cells (Figure 4.1):

1. **Arrangement of cells in layers.** The cells of covering and lining epithelia are arranged in one or more layers depending on the functions the epithelium performs:

Figure 4.1 Cell shapes and arrangement of layers for covering and lining epithelium.

Cell shapes and arrangement of layers are the bases for classifying covering and lining epithelium.

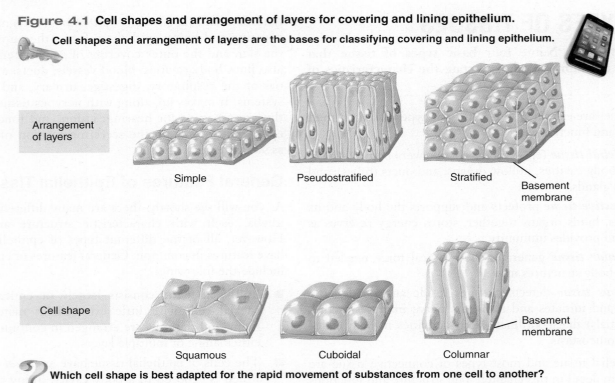

Arrangement of layers

Simple   Pseudostratified   Stratified
                                      Basement membrane

Cell shape

Squamous   Cuboidal   Columnar
                                Basement membrane

Which cell shape is best adapted for the rapid movement of substances from one cell to another?

a. *Simple epithelium* is a single layer of cells that functions in diffusion, osmosis, filtration, secretion, and absorption. *Secretion* (se-KRĒ-shun) is the production and release of substances such as mucus, sweat, or enzymes. *Absorption* (ab-SORP-shun) is the intake of fluids or other substances such as digested food from the intestinal tract.

b. *Pseudostratified epithelium* (*pseudo-* = false) appears to have multiple layers of cells because the cell nuclei lie at different levels and not all cells reach the apical surface. Cells that do extend to the apical surface may contain cilia; others (goblet cells) secrete mucus. Pseudostratified epithelium is actually a simple epithelium because all of its cells rest on the basement membrane.

c. *Stratified epithelium* (*stratum* = layer) consists of two or more layers of cells that protect underlying tissues in locations where there is considerable wear and tear.

2. **Cell shapes.**

a. *Squamous* cells (SKWĀ-mus = flat) are thin, and this allows for the rapid passage of substances through them.

b. *Cuboidal* cells are as tall as they are wide and are shaped like cubes or hexagons. They may have microvilli at their apical surface and function in either secretion or absorption.

c. *Columnar* cells are much taller than they are wide, like columns, and protect underlying tissues. Their apical surfaces may have cilia or microvilli, and they often are specialized for secretion and absorption.

d. *Transitional* cells change shape, from flat to cuboidal and back, as organs such as the urinary bladder

stretch (distend) to a larger size and then collapse to a smaller size.

Combining the two characteristics (arrangements of layers and cell shapes), the types of covering and lining epithelia are as follows:

I.   Simple epithelium

   A. Simple squamous epithelium

   B. Simple cuboidal epithelium

   C. Simple columnar epithelium (nonciliated and ciliated)

   D. Pseudostratified columnar epithelium (nonciliated and ciliated)

II. Stratified epithelium*

   A. Stratified squamous epithelium (keratinized and nonkeratinized)

   B. Stratified cuboidal epithelium

   C. Stratified columnar epithelium

   D. Transitional epithelium

Each of these covering and lining epithelia is included in Table 4.1. Each table entry consists of a photomicrograph, a corresponding diagram, and an inset that identifies a major location of the tissue in the body. Descriptions, locations, and functions of the tissues accompany each illustration.

———

*This classification is based on the shape of the cells in the *apical* layer.

TABLE 4.1

Epithelial Tissues: Covering and Lining Epithelium

A. Simple squamous epithelium

Description: Single layer of flat cells that resembles a tiled floor when viewed from apical surface; centrally located nucleus that is flattened and oval or spherical in shape.

Location: Lines heart, blood vessels, lymphatic vessels, air sacs of lungs, glomerular (Bowman's) capsule of kidneys, and inner surface of the tympanic membrane (eardrum); forms epithelial layer of serous membranes (mesothelium), such as the peritoneum. The simple squamous epithelium that lines the heart, blood vessels, and lymphatic vessels is known as ***endothelium*** (en'-dō-THĒ-lē-um; *endo-* = within; *-thelium* = covering); the type that forms the epithelial layer of serous membranes, such as the peritoneum, pleura, or pericardium, is called ***mesothelium*** (*meso-* = middle). See Figure 4.3b.

Function: Filtration, diffusion, osmosis, and secretion in serous membranes.

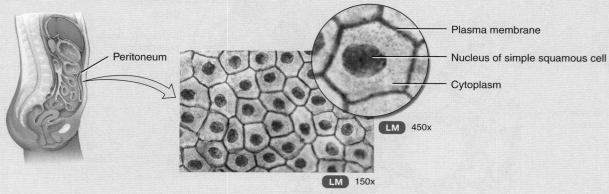

Peritoneum

Plasma membrane

Nucleus of simple squamous cell

Cytoplasm

LM 450x

LM 150x

Surface view of simple squamous epithelium of mesothelial lining of peritoneum

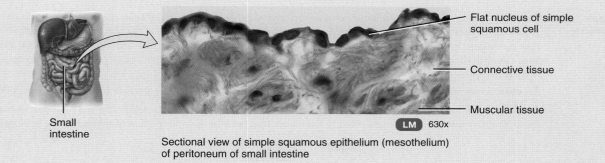

Small intestine

Flat nucleus of simple squamous cell

Connective tissue

Muscular tissue

LM 630x

Sectional view of simple squamous epithelium (mesothelium) of peritoneum of small intestine

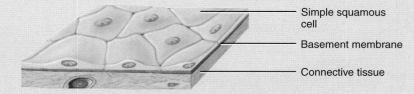

Simple squamous cell

Basement membrane

Connective tissue

Simple squamous epithelium

TABLE 4.1 CONTINUES

TABLE 4.1 CONTINUED

Epithelial Tissues: Covering and Lining Epithelium

B. Simple cuboidal epithelium

Description: Single layer of cube-shaped cells; round, centrally located nucleus. Cuboidal shape is obvious when tissue is sectioned and viewed from the side.

Location: Lines kidney tubules and smaller ducts of many glands, makes up the secreting portion of some glands such as the thyroid gland, covers surface of ovary, lines anterior surface of capsule of the lens of the eye, and forms the pigmented epithelium at the posterior surface of the eye.

Function: Secretion and absorption.

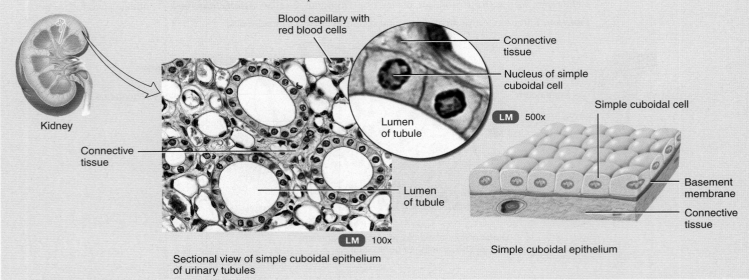

Sectional view of simple cuboidal epithelium of urinary tubules

Simple cuboidal epithelium

C. Nonciliated simple columnar epithelium

Description: Single layer of nonciliated column-like cells with nuclei near bases of cells; contains cells with microvilli and goblet cells. *Microvilli,* microscopic fingerlike projections, increase the surface area of the plasma membrane (see Figure 3.1), thus increasing the rate of absorption by the cell. *Goblet cells* are modified columnar cells that secrete mucus, a slightly sticky fluid, at their apical surfaces. Before release, mucus accumulates in upper portion of cell, causing it to bulge and making the whole cell resemble a goblet or wine glass.

Location: Lines most of the gastrointestinal tract (from the stomach to the anus), ducts of many glands, and gallbladder.

Function: Secretion and absorption. Secreted mucus lubricates linings of digestive, respiratory, and reproductive tracts, and most of urinary tract; helps trap dust entering respiratory tract; and prevents destruction of stomach lining by stomach acid.

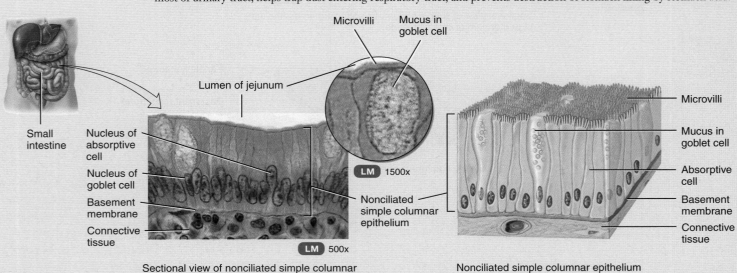

Sectional view of nonciliated simple columnar epithelium of lining of jejunum of small intestine

Nonciliated simple columnar epithelium

D. Ciliated simple columnar epithelium

Description: Single layer of ciliated column-like cells with nuclei near bases; contains goblet cells in some locations.

Location: Lines a few portions of upper respiratory tract, uterine (fallopian) tubes, uterus, some paranasal sinuses, and central canal of spinal cord.

Function: Mucus secreted by goblet cells forms a film over respiratory surface that traps inhaled foreign particles. Cilia wave in unison and move mucus and any trapped foreign particles toward throat, where it can be coughed up and swallowed or spit out. Cilia also help move oocytes expelled by the ovaries through uterine tubes into uterus.

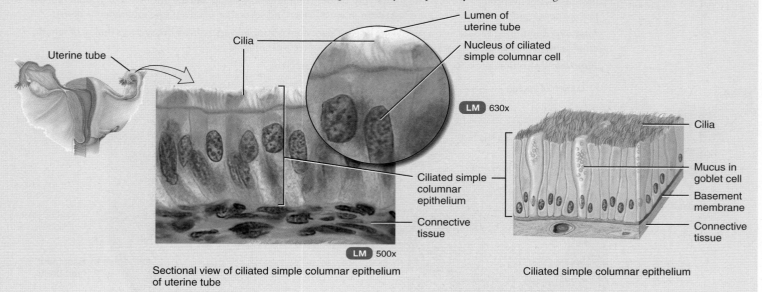

Sectional view of ciliated simple columnar epithelium of uterine tube

Ciliated simple columnar epithelium

E. Pseudostratified columnar epithelium

Description: Not a true stratified tissue; nuclei of cells are at different levels; all cells are attached to basement membrane, but not all reach the apical surface.

Location: Pseudostratified ciliated columnar epithelium lines the airways of most of upper respiratory tract; pseudostratified nonciliated columnar epithelium lines larger ducts of many glands, epididymis, and part of male urethra.

Function: Ciliated variety secretes mucus that traps foreign particles, and cilia sweep away mucus for elimination from body; nonciliated variety functions in absorption and protection. Appears to have several layers because the nuclei of the cells are at various depths. All cells are attached to basement membrane in a single layer, but some cells do not extend to apical surface. When viewed from side, these features give false impression of a multilayered tissue—thus the name *pseudo*stratified epithelium (*pseudo-* = false). *Pseudostratified ciliated columnar epithelium* contains cells that extend to surface and secrete mucus (goblet cells) or bear cilia. *Pseudostratified nonciliated columnar epithelium* contains cells without cilia and lacks goblet cells.

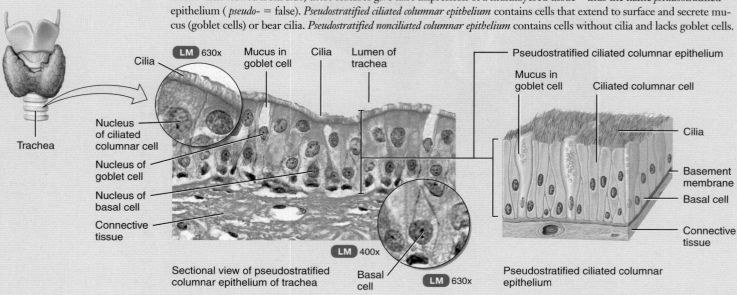

Sectional view of pseudostratified columnar epithelium of trachea

Pseudostratified ciliated columnar epithelium

TABLE 4.1 CONTINUES

TABLE 4.1 CONTINUED

Epithelial Tissues: Covering and Lining Epithelium

F. Stratified squamous epithelium

Description: Two or more layers of cells; cells in apical layer and several layers deep to it are squamous; those in the deep layers vary in shape from cuboidal to columnar. Basal (deepest) cells continually undergo cell division. As new cells grow, cells of basal layer are pushed upward toward surface. As they move farther from the deeper layers and from their blood supply in the underlying connective tissue, they become dehydrated, shrunken, and harder. At apical layer, cells lose their cell junctions and are sloughed off, but are replaced as new cells continually emerge from basal cells. *Keratinized stratified squamous epithelium* develops a tough layer of keratin in apical layer and several layers deep to it. (*Keratin* is a tough protein that helps protect the skin and underlying tissues from microbes, heat, and chemicals.) *Nonkeratinized stratified squamous epithelium* does not contain keratin in apical layer and several layers deep to it and remains moist.

Location: Keratinized variety forms superficial layer of skin; nonkeratinized variety lines wet surfaces (lining of mouth, esophagus, part of epiglottis, part of pharynx, and vagina), and covers tongue.

Function: Protection; provides first line of defense against microbes.

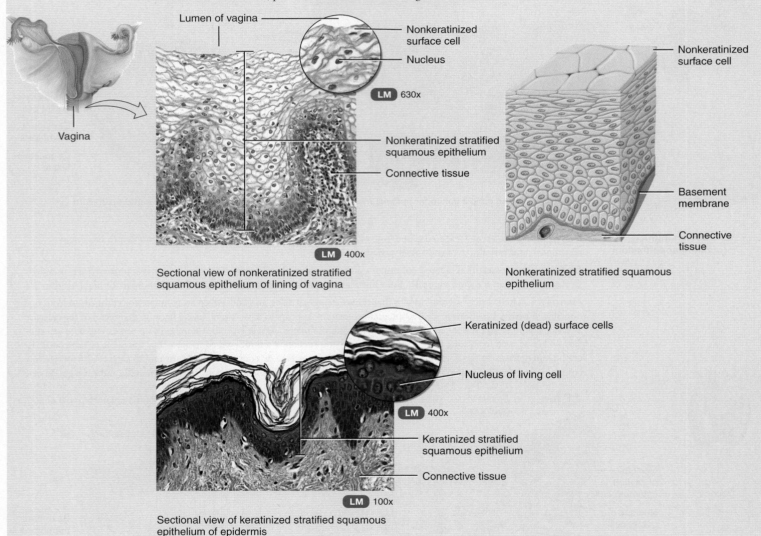

Sectional view of nonkeratinized stratified squamous epithelium of lining of vagina

LM 630x
LM 400x

Nonkeratinized stratified squamous epithelium

LM 400x
LM 100x

Sectional view of keratinized stratified squamous epithelium of epidermis

G. Stratified cuboidal epithelium

Description: Two or more layers of cells; cells in the apical layer are cube-shaped; fairly rare type.

Location: Ducts of adult sweat glands and esophageal glands and part of male urethra.

Function: Protection and limited secretion and absorption.

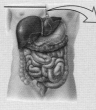

Esophagus

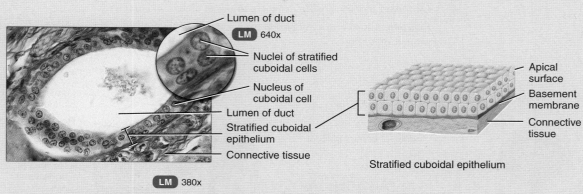

Lumen of duct

LM 640x

Nuclei of stratified cuboidal cells

Nucleus of cuboidal cell

Lumen of duct

Stratified cuboidal epithelium

Connective tissue

LM 380x

Sectional view of stratified cuboidal epithelium of the duct of an esophageal gland

Apical surface

Basement membrane

Connective tissue

Stratified cuboidal epithelium

H. Stratified columnar epithelium

Description: Basal layers usually consist of shortened, irregularly shaped cells; only apical layer has columnar cells; uncommon.

Location: Lines part of urethra, large excretory ducts of some glands such as esophageal glands, small areas in anal mucous membrane, and part of conjunctiva of eye.

Function: Protection and secretion.

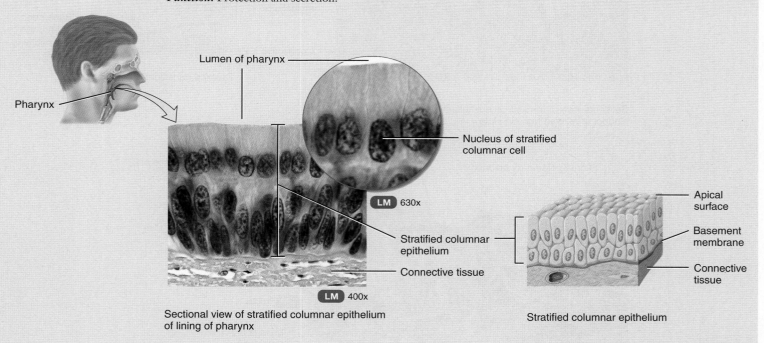

Pharynx

Lumen of pharynx

Nucleus of stratified columnar cell

LM 630x

Stratified columnar epithelium

Connective tissue

LM 400x

Sectional view of stratified columnar epithelium of lining of pharynx

Apical surface

Basement membrane

Connective tissue

Stratified columnar epithelium

TABLE 4.1 CONTINUES

TABLE 4.1 CONTINUED

Epithelial Tissues: Covering and Lining Epithelium

I. Transitional epithelium

Description: Variable in appearance (transitional). In relaxed or unstretched state, looks similar to stratified cuboidal epithelium, except apical layer cells tend to be large and rounded. As tissue is stretched, cells become flatter, giving the appearance of stratified squamous epithelium. Multiple layers and elasticity make it ideal for lining hollow structures (urinary bladder) subject to expansion from within.

Location: Lines urinary bladder and portions of ureters and urethra.

Function: Allows urinary organs to stretch and maintain protective lining while holding variable amounts of fluid without rupturing.

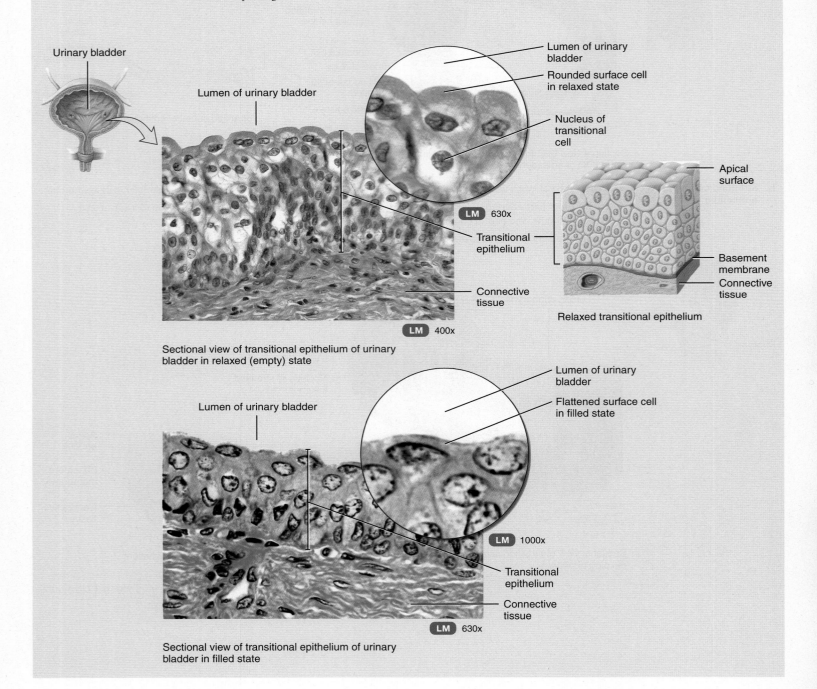

Urinary bladder

Lumen of urinary bladder

Lumen of urinary bladder

Rounded surface cell in relaxed state

Nucleus of transitional cell

LM 630x

Transitional epithelium

Connective tissue

Apical surface

Basement membrane

Connective tissue

Relaxed transitional epithelium

LM 400x

Sectional view of transitional epithelium of urinary bladder in relaxed (empty) state

Lumen of urinary bladder

Lumen of urinary bladder

Flattened surface cell in filled state

LM 1000x

Transitional epithelium

Connective tissue

LM 630x

Sectional view of transitional epithelium of urinary bladder in filled state

Glandular Epithelium

The function of glandular epithelium is secretion, which is accomplished by glandular cells that often lie in clusters deep to the covering and lining epithelium. A **gland** may consist of one cell or a group of highly specialized epithelial cells that secrete substances into ducts (tubes), onto a surface, or into the blood. All glands of the body are classified as either endocrine or exocrine.

The secretions of **endocrine glands** (EN-dō-krin; *endo-* = within; *-crine* = secretion) (Table 4.2A) enter the interstitial

TABLE 4.2

Epithelial Tissues: Glandular Epithelium

A. Endocrine glands

Description: Secretory products (hormones) diffuse into blood after passing through interstitial fluid.

Location: Examples include pituitary gland at base of brain, pineal gland in brain, thyroid and parathyroid glands near larynx (voice box), adrenal glands superior to kidneys, pancreas near stomach, ovaries in pelvic cavity, testes in scrotum, and thymus in thoracic cavity.

Function: Produce hormones that regulate various body activities.

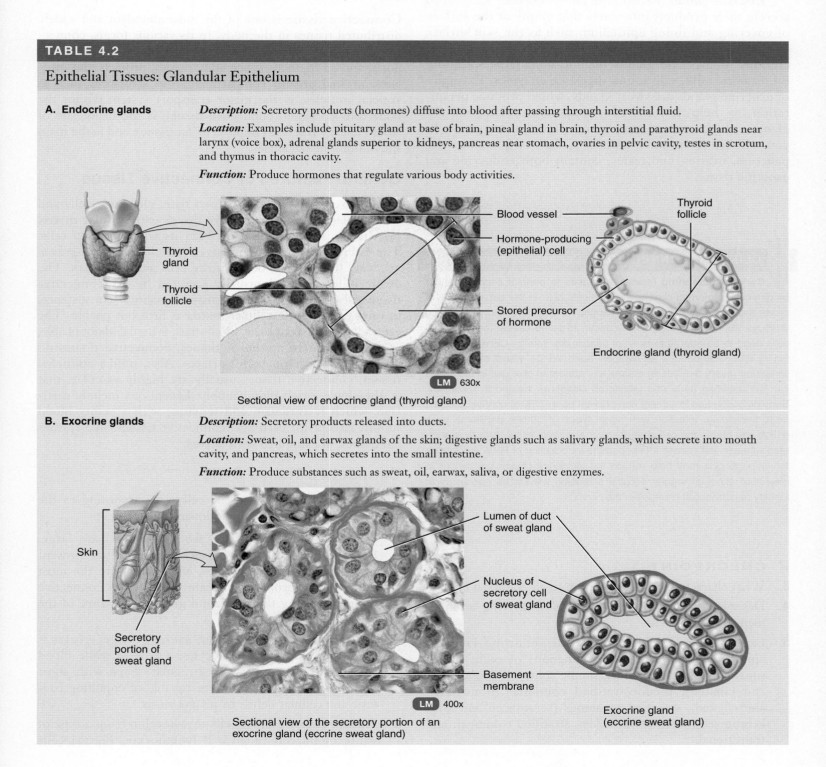

Sectional view of endocrine gland (thyroid gland)

LM 630x

Endocrine gland (thyroid gland)

B. Exocrine glands

Description: Secretory products released into ducts.

Location: Sweat, oil, and earwax glands of the skin; digestive glands such as salivary glands, which secrete into mouth cavity, and pancreas, which secretes into the small intestine.

Function: Produce substances such as sweat, oil, earwax, saliva, or digestive enzymes.

Sectional view of the secretory portion of an exocrine gland (eccrine sweat gland)

LM 400x

Exocrine gland (eccrine sweat gland)

fluid and then diffuse into the bloodstream without flowing through a duct. These secretions, called *hormones*, regulate many metabolic and physiological activities to maintain homeostasis. The pituitary, thyroid, and adrenal glands are examples of endocrine glands. Endocrine glands will be described in detail in Chapter 13.

Exocrine glands (EK-sō-krin; *exo-* = outside; Table 4.2B) secrete their products into ducts that empty at the surface of covering and lining epithelium such as the skin surface or the lumen (interior space) of a hollow organ. The secretions of exocrine glands include mucus, perspiration, oil, earwax, milk, saliva, and digestive enzymes. Examples of exocrine glands are sweat glands, which produce perspiration to help lower body temperature, and salivary glands, which secrete mucus and digestive enzymes. As you will see later, some glands of the body, such as the pancreas, ovaries, and testes, contain both endocrine and exocrine tissue.

CLINICAL CONNECTION | Cervical Screening

Cervical Screening (smear test) aims to prevent cervical cancer by detecting abnormal changes to epithelial cells in the cervix. In 5% of cases the test may show some changes in these cells however, most changes will not lead to cervical cancer. In some cases, the abnormal cells need to be treated to prevent cancer developing. Cervical screening is recommended for all women aged 25–64 years old. Regular screening (every 3–5 years) means any abnormal changes in the cervix can be identified early and monitored closely or treated if necessary. Early detection and treatment can prevent up to 75% of cervical cancers from developing. Infection with some types of human papilloma virus (HPV) can cause the cell changes and abnormal tissue growth associated with cervical cancer. It is hoped the introduction of HPV vaccination for teenage girls may help reduce the incidence of cervical cancer in the future although regular cervical screening will still be necessary to identify abnormal cell changes in the cervix. •

✓ CHECKPOINT

3. What characteristics are common to all epithelial tissues?

4. Describe the various cell shapes and layering arrangements of epithelium.

5. Explain how the structure of the following kinds of epithelium is related to the functions of each: simple squamous, simple cuboidal, simple columnar (nonciliated and ciliated), pseudostratified columnar (nonciliated and ciliated), stratified squamous (keratinized and nonkeratinized), stratified cuboidal, stratified columnar, and transitional.

4.3 CONNECTIVE TISSUE

OBJECTIVES • **Discuss the general features of connective tissue.**

• **Describe the structure, location, and function of the various types of connective tissue.**

Connective tissue is one of the most abundant and widely distributed tissues in the body. In its various forms, connective tissue has a variety of functions. It binds together, supports, and strengthens other body tissues; protects and insulates internal organs; compartmentalizes structures such as skeletal muscles; is the major transport system within the body (blood, a fluid connective tissue); is the major site of stored energy reserves (adipose, or fat tissue); and is the main site of immune responses.

General Features of Connective Tissue

Connective tissue consists of two basic elements: cells and extracellular matrix. A connective tissue's **extracellular matrix** (MĀ-triks) is the material between its widely spaced cells. The extracellular matrix consists of protein fibers and *ground substance*, the material between the cells and the fibers. The extracellular matrix is usually secreted by the connective tissue cells and determines the tissue's qualities. For instance, in cartilage, the extracellular matrix is firm but pliable. The extracellular matrix of bone, by contrast, is hard and not pliable.

In contrast to epithelial tissues, connective tissues do not usually occur on body surfaces. Also, unlike epithelial tissues, connective tissues usually are highly vascular; that is, they have a rich blood supply. Exceptions include cartilage, which is avascular, and tendons, with a scanty blood supply. Except for cartilage, connective tissues, like epithelial tissues, are supplied with nerves.

Connective Tissue Cells

The types of connective tissue cells vary according to the type of tissue and include the following (Figure 4.2):

■ *Fibroblasts* (FĪ-brō-blasts; *fibro-* = fibers) are large, flat cells with branching processes. They are present in several connective tissues, and usually are the most numerous. Fibroblasts migrate through the connective tissue, secreting the fibers and ground substance of the extracellular matrix.

■ *Macrophages* (MAK-rō-fā-jez; *macro-* = large; *-phages* = eaters) develop from monocytes, a type of white blood cell. Macrophages have an irregular shape with short branching projections and are capable of engulfing bacteria and cellular debris by phagocytosis.

■ *Plasma cells* are small cells that develop from a type of white blood cell called a B lymphocyte. Plasma cells

Figure 4.2 Representative cells and fibers present in connective tissues.

 Fibroblasts are usually the most numerous connective tissue cells.

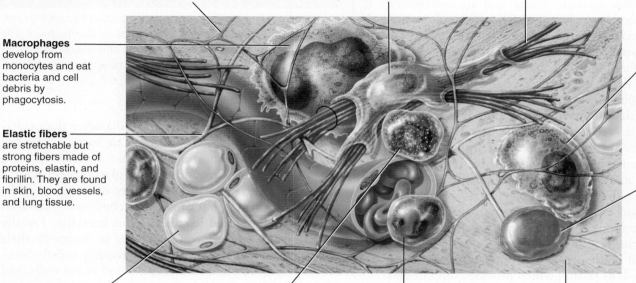

Reticular fibers
are made of collagen and glycoproteins. They provide support in blood vessel walls and form branching networks around various cells (fat, smooth muscle, nerve).

Fibroblasts
are large flat cells that move through connective tissues and secrete fibers and ground substance.

Collagen fibers
are strong, flexible bundles of the protein collagen, the most abundant protein in your body.

Macrophages
develop from monocytes and eat bacteria and cell debris by phagocytosis.

Mast cells
are abundant along blood vessels. They produce histamine, which dilates small blood vessels during inflammation and kill bacteria.

Elastic fibers
are stretchable but strong fibers made of proteins, elastin, and fibrillin. They are found in skin, blood vessels, and lung tissue.

Plasma cells
are small cells that develop from B lymphocytes. They secrete antibodies that attack and neutralize foreign substances.

Adipocytes
or fat cells store fats. They are found below the skin and around organs (heart, kidney).

Eosinophils
are white blood cells that migrate to sites of parasitic infection and allergic responses.

Neutrophils
are white blood cells that migrate to sites of infection that destroy microbes by phagocytosis.

Ground substance
is the material between cells and fibers. It is made of water and organic molecules (hyaluronic acid, chondroitin sulfate, glucosamine). It supports cells and fibers, binds them together, and provides a medium for exchanging substances between blood and cells.

? **What is the function of fibroblasts?**

secrete antibodies, proteins that attack or neutralize foreign substances in the body. Thus, plasma cells are an important part of the body's immune response.

- *Mast cells* are abundant alongside the blood vessels that supply connective tissue. They produce histamine, a chemical that dilates small blood vessels as part of the inflammatory response, the body's reaction to injury or infection. Mast cells can also kill bacteria.

- *Adipocytes* (A-di-pō-sīts), also called fat cells or adipose cells, are connective tissue cells that store triglycerides (fats). They are found below the skin and around organs such as the heart and kidneys.

White blood cells are not normally found in significant numbers in connective tissues. However, in response to certain conditions, white blood cells can leave blood and enter connective tissues. For example, *neutrophils* gather at sites of infection and *eosinophils* migrate to sites of parasitic invasion and allergic responses.

Connective Tissue Extracellular Matrix

Each type of connective tissue has unique properties, based on the specific extracellular materials between the cells. The extracellular matrix consists of a fluid, gel, or solid ground substance plus protein fibers.

Ground Substance

Ground substance, the component of a connective tissue between the cells and fibers, supports cells, binds them together, and provides a medium through which substances are exchanged between the blood and cells. The ground substance plays an active role in how tissues develop, migrate, proliferate, and change shape, and in how they carry out their metabolic functions.

Ground substance contains water and an assortment of large organic molecules, many of which are complex combinations of polysaccharides and proteins. For example, the polysaccharide *hyaluronic acid* (hī'-a-loo-RON-ik) is a viscous, slippery substance that binds cells together, lubricates joints, and helps maintain the shape of the eyeballs. It also appears to play a role in helping phagocytes migrate through connective tissue during development and wound repair. White blood cells, sperm cells, and some bacteria produce *hyaluronidase,* an enzyme that breaks apart hyaluronic acid and causes the ground substance of connective tissue to become watery. The ability to produce hyaluronidase enables white blood cells to move through connective tissues to reach sites of infection and enables sperm cells to penetrate the ovum during fertilization. It also accounts for how bacteria spread through connective tissues.

Another ground substance is the polysaccharide *chondroitin sulfate* (kon-DROY-tin), which provides support and adhesiveness in connective tissues in bone, cartilage, skin, and blood vessels. *Glucosamine* is a protein–polysaccharide molecule.

CLINICAL CONNECTION | Chondroitin Sulfate, Glucosamine, and Joint Disease

In recent years, **chondroitin sulfate** and **glucosamine** (a proteoglycan) have been used as nutritional supplements either alone or in combination to promote and maintain the structure and function of joint cartilage, to provide pain relief from osteoarthritis, and to reduce joint inflammation. Although these supplements have benefited some individuals with moderate to severe osteoarthritis, the benefit is minimal in lesser cases. More research is needed to determine how they act and why they help some people and not others. •

Fibers

Fibers in the extracellular matrix strengthen and support connective tissues. Three types of fibers are embedded in the extracellular matrix between the cells: collagen fibers, elastic fibers, and reticular fibers.

Collagen fibers (KOL-a-jen; *colla* = glue) are very strong and resist pulling forces, but they are not stiff, which promotes tissue flexibility. These fibers often occur in bundles lying parallel to one another (Figure 4.2). The bundle arrangement affords great strength. Chemically, collagen fibers consist of the protein *collagen.* This is the most abundant protein in your body, representing about 25 percent of total protein. Collagen fibers are found in most types of connective tissues, especially bone, cartilage, tendons, and ligaments.

CLINICAL CONNECTION | Sprain

Despite their strength, ligaments may be stressed beyond their normal capacity. This results in **sprain**, a stretched or torn ligament. The ankle joint is most frequently sprained. Because of their poor blood supply, the healing of even partially torn ligaments is a very slow process; completely torn ligaments require surgical repair. •

Elastic fibers, which are smaller in diameter than collagen fibers, branch and join together to form a network within a tissue. An elastic fiber consists of molecules of a protein called *elastin* surrounded by a glycoprotein named *fibrillin,* which is essential to the stability of an elastic fiber. Elastic fibers are strong but can be stretched up to one-and-a-half times their relaxed length without breaking. Equally important, elastic fibers have the ability to return to their original shape after being stretched, a property called *elasticity.* Elastic fibers are plentiful in skin, blood vessel walls, and lung tissue.

CLINICAL CONNECTION | Marfan Syndrome

Marfan syndrome (MAR-fan) is an inherited disorder caused by a defective fibrillin gene. The result is abnormal development of elastic fibers. Tissues rich in elastic fibers are malformed or weakened. Structures affected most seriously are the covering layer of bones (periosteum), the ligament that suspends the lens of the eye, and the walls of the large arteries. People with Marfan syndrome tend to be tall and have disproportionately long arms, legs, fingers, and toes. A common symptom is blurred vision caused by displacement of the lens of the eye. The most life-threatening complication of Marfan syndrome is weakening of the aorta (the main artery that emerges from the heart), which can suddenly burst. •

Reticular fibers (*reticul-* = net), consisting of *collagen* and a coating of glycoprotein, provide support in the walls of blood vessels and form branching networks around fat cells, nerve fibers, and skeletal and smooth muscle cells. Produced by fibroblasts, they are much thinner than collagen fibers. Like collagen fibers, reticular fibers provide support and strength and also form the *stroma* (STRŌ-ma = bed or covering) or supporting framework of many soft organs, such as the spleen and lymph nodes. These fibers also help form the basement membrane.

Classification of Connective Tissues

Because of the diversity of cells and extracellular matrix and the differences in their relative proportions, the classification of connective tissues is not always clear-cut. We offer the following scheme:

I. Loose connective tissue
 A. Areolar connective tissue
 B. Adipose tissue
 C. Reticular connective tissue

II. Dense connective tissue
 D. Dense regular connective tissue
 E. Dense irregular connective tissue
 F. Elastic connective tissue

III. Cartilage
 G. Hyaline cartilage
 H. Fibrocartilage
 I. Elastic cartilage

IV. Bone tissue

V. Liquid connective tissue (blood tissue and lymph)

Loose Connective Tissue

The fibers in *loose connective tissue* are loosely arranged among the many cells. The types of loose connective tissue are areolar connective tissue, adipose tissue, and reticular connective tissue (Table 4.3).

CLINICAL CONNECTION | Liposuction

A surgical procedure called **liposuction** (*lip-* = fat) or **suction lipectomy** (*-ectomy* = to cut out) involves suctioning small amounts of adipose tissue from various areas of the body. After an incision is made in the skin, the fat is removed through a hollow tube, called a cannula, with the assistance of a powerful vacuum pressure unit that suctions out the fat. The technique can be used as a body contouring procedure in regions such as the thighs, buttocks, arms, breasts, and abdomen, and to transfer fat to another area of the body. Postsurgical complications that may develop include fat that may enter blood vessels broken during the procedure and obstruct blood flow, infection, loss of feeling in the area, fluid depletion, injury to internal structures, and severe postoperative pain. •

TABLE 4.3

Connective Tissues: Loose Connective Tissue

A. Areolar connective tissue (a-RĒ-ō-lar; areol- = a small space)

Description: One of the most widely distributed connective tissues; consists of fibers (collagen, elastic, and reticular) arranged randomly and several kinds of cells (fibroblasts, macrophages, plasma cells, adipocytes, mast cells, and a few white blood cells) embedded in a semifluid ground substance. Combined with adipose tissue, areolar connective tissue forms the *subcutaneous layer*, the layer of tissue that attaches the skin to underlying tissues and organs.

Location: In and around nearly every body structure (thus called "packing material" of the body); subcutaneous layer deep to skin; superficial region of dermis of skin; connective tissue layer of mucous membranes; and around blood vessels, nerves, and body organs.

Function: Strength, elasticity, and support.

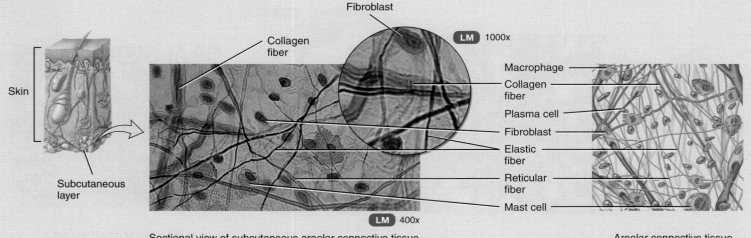

Sectional view of subcutaneous areolar connective tissue

Areolar connective tissue

TABLE 4.3 CONTINUES

TABLE 4.3 CONTINUED

Connective Tissues: Loose Connective Tissue

B. Adipose tissue

Description: Has cells called ***adipocytes*** (*adipo-* = fat) specialized for storage of triglycerides (fats). Because cell fills up with a single, large triglyceride droplet, cytoplasm and nucleus are pushed to periphery of the cell. As amount of adipose tissue increases with weight gain, new blood vessels form. Thus, an obese person has many more blood vessels than a lean person, a situation that can cause high blood pressure, since the heart has to work harder.

Location: Wherever areolar connective tissue is located; subcutaneous layer deep to skin, around heart and kidneys, yellow bone marrow, and padding around joints and behind eyeball in eye socket.

Function: Reduces heat loss through skin; serves as an energy reserve; supports and protects organs.

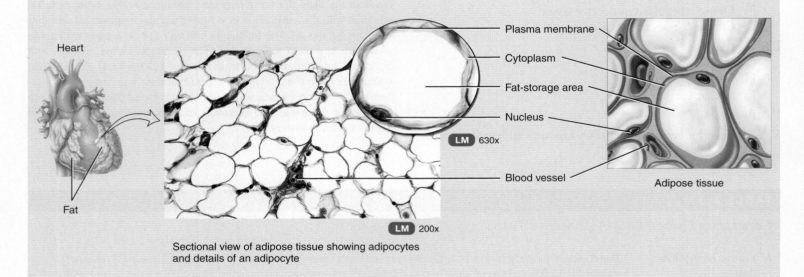

Sectional view of adipose tissue showing adipocytes and details of an adipocyte

Adipose tissue

C. Reticular connective tissue

Description: Fine interlacing network of *reticular fibers* (thin form of collagen fiber) and reticular cells.

Location: Stroma (supporting framework) of liver, spleen, lymph nodes; red bone marrow, which gives rise to blood cells; part of the basement membrane; and around blood vessels and muscles.

Function: Forms stroma of organs; binds together smooth muscle tissue cells; filters and removes worn-out blood cells in spleen and microbes in lymph nodes.

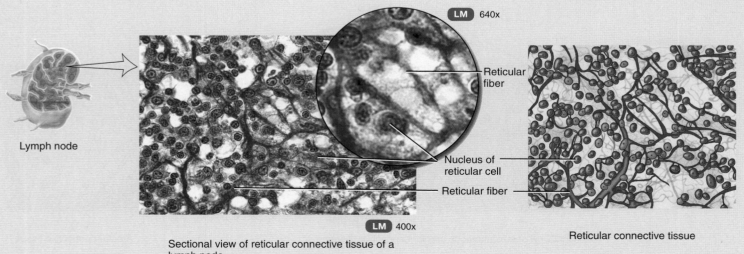

Sectional view of reticular connective tissue of a lymph node

Reticular connective tissue

Dense Connective Tissue

Dense connective tissue contains more numerous, thicker, and denser fibers (more closely packed), but fewer cells than loose connective tissue. There are three types: dense regular connective tissue, dense irregular connective tissue, and elastic connective tissue (Table 4.4).

TABLE 4.4

Connective Tissues: Dense Connective Tissue

A. Dense regular connective tissue

Description: Extracellular matrix looks shiny white; consists mainly of collagen fibers regularly arranged in bundles; fibroblasts present in rows between bundles. Collagen fibers are not living cells, but protein structures secreted by fibroblasts, so damaged tendons and ligaments heal slowly.

Location: Forms *tendons* (attach muscle to bone), most *ligaments* (attach bone to bone), and *aponeuroses* (sheetlike tendons that attach muscle to muscle or muscle to bone).

Function: Provides strong attachment between various structures. Tissue structure resists pulling (tension) along long axis of fibers.

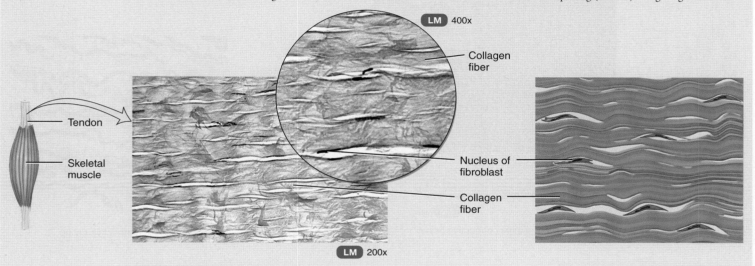

Sectional view of dense regular connective tissue of a tendon

Dense regular connective tissue

B. Dense irregular connective tissue

Description: Consists predominantly of collagen fibers randomly arranged and a few fibroblasts.

Location: Often occurs in sheets, such as *fasciae* (tissue beneath skin and around muscles and other organs), deeper region of dermis of skin, periosteum of bone, perichondrium of cartilage, joint capsules, membrane capsules around various organs (kidneys, liver, testes, lymph nodes), pericardium of the heart; also in heart valves.

Function: Provides tensile (pulling) strength in many directions.

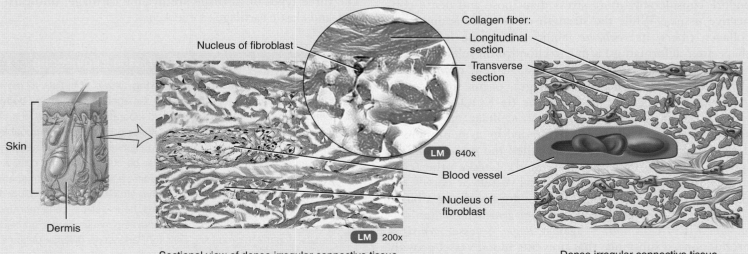

Sectional view of dense irregular connective tissue of reticular region of dermis

Dense irregular connective tissue

TABLE 4.4 CONTINUES

TABLE 4.4 CONTINUED

Connective Tissues: Dense Connective Tissue

C. Elastic connective tissue

Description: Consists predominantly of elastic fibers; fibroblasts are present in spaces between fibers; unstained tissue is yellowish.

Location: Lung tissue, walls of elastic arteries, trachea, bronchial tubes, true vocal cords, suspensory ligaments of penis, some ligaments between vertebrae.

Function: Allows stretching of various organs; is strong and can recoil to original shape after being stretched. Elasticity is important to normal functioning of lung tissue, which recoils as you exhale, and elastic arteries, whose recoil between heartbeats helps maintain blood flow.

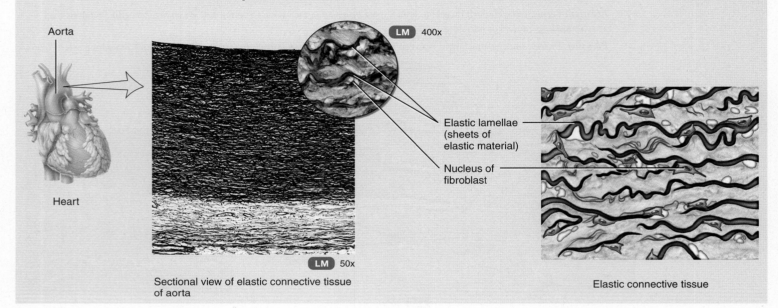

Aorta

Heart

LM 400x

Elastic lamellae (sheets of elastic material)

Nucleus of fibroblast

LM 50x

Sectional view of elastic connective tissue of aorta

Elastic connective tissue

Cartilage

Cartilage consists of a dense network of collagen fibers or elastic fibers firmly embedded in chondroitin sulfate, a rubbery component of the ground substance. Cartilage can endure considerably more stress than loose and dense connective tissues. While the strength of cartilage is due to its collagen fibers, its *resilience* (ability to assume its original shape after deformation) is due to chondroitin sulfate.

The cells of mature cartilage, called **chondrocytes** (KON-drō-sīts; *chondro-* = cartilage), occur singly or in groups within spaces called **lacunae** (la-KOO-nē = little lakes; singular is *lacuna*) in the extracellular matrix. The surface of most cartilage is surrounded by a membrane of dense irregular connective tissue called the **perichondrium** (per′-i-KON-drē-um; *peri-* = around). Unlike other connective tissues, cartilage has no blood vessels or nerves, except in the perichondrium. Cartilage does not have a blood supply because it secretes an *antiangiogenesis factor* (an′-tē-an′-gē-o-JEN-e-sis; *anti* = against; *angio* = vessel; *genesis* = production), a substance that prevents blood vessel growth. Because of this property, antiangiogenesis factor is being

studied as a possible cancer treatment to stop cancer cells from promoting new blood vessel growth that supports their rapid rate of cell division and expansion. Since cartilage has no blood supply, it heals poorly following an injury. The three types of cartilage are hyaline cartilage, fibrocartilage, and elastic cartilage (Table 4.5).

CLINICAL CONNECTION | Tissue Engineering

The technology of **tissue engineering** allows scientists to grow new tissues in the laboratory to replace damaged tissues in the body. Tissue engineers have already developed laboratory-grown versions of skin and cartilage. In the procedure, scaffolding beds of biodegradable synthetic materials or collagen are used as substrates that permit body cells such as skin cells or cartilage cells to be cultured. As the cells divide and assemble, the scaffolding degrades, and the new, permanent tissue is then implanted in the patient. Other structures being developed by tissue engineers include bones, tendons, heart valves, bone marrow, and intestines. Work is also underway to develop insulin-producing cells for diabetics, dopamine-producing cells for Parkinson's disease patients, and even entire livers and kidneys. •

TABLE 4.5

Connective Tissues: Cartilage

A. Hyaline cartilage
(*hyalinos* = glassy)

Description: Contains a resilient gel as ground substance and appears in the body as a bluish-white, shiny substance (can stain pink or purple when prepared for microscopic examination). Fine collagen fibers are not visible with ordinary staining techniques; prominent chondrocytes are found in lacunae surrounded by perichondrium (exceptions: articular cartilage in joints and the epiphyseal plates, where bones lengthen as a person grows); most abundant type of cartilage in the body.

Location: Ends of long bones, anterior ends of ribs, nose, parts of larynx, trachea, bronchi, bronchial tubes, and embryonic and fetal skeleton.

Function: Provides smooth surfaces for movement at joints, as well as flexibility and support; weakest type of cartilage (can be fractured).

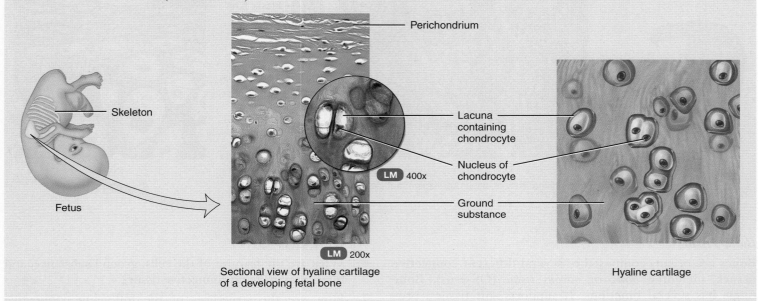

Sectional view of hyaline cartilage of a developing fetal bone

Hyaline cartilage

B. Fibrocartilage

Description: Consists of chondrocytes scattered among clearly visible thick bundles of collagen fibers within extracellular matrix; lacks perichondrium.

Location: *Pubic symphysis* (point where hip bones join anteriorly), *intervertebral discs* (discs between vertebrae), *menisci* (cartilage pads) of knee, and portions of tendons that insert into cartilage.

Function: Support and joining structures together. Strength and rigidity make it the strongest type of cartilage.

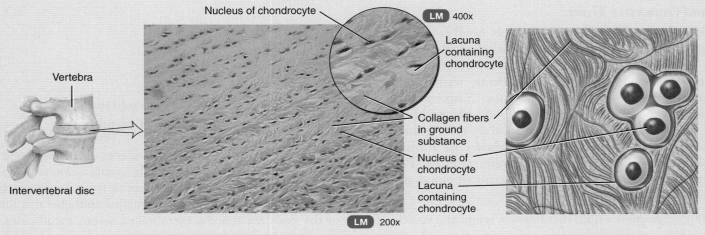

Sectional view of fibrocartilage of intervertebral disc

Fibrocartilage

TABLE 4.5 CONTINUES

TABLE 4.5 CONTINUED

Connective Tissues: Cartilage

C. Elastic cartilage

Description: Consists of chondrocytes located in a threadlike network of elastic fibers within the extracellular matrix; perichondrium present.

Location: Lid on top of larynx (epiglottis), part of external ear (auricle), and auditory (eustachian) tubes.

Function: Provides strength and elasticity; maintains shape of certain structures.

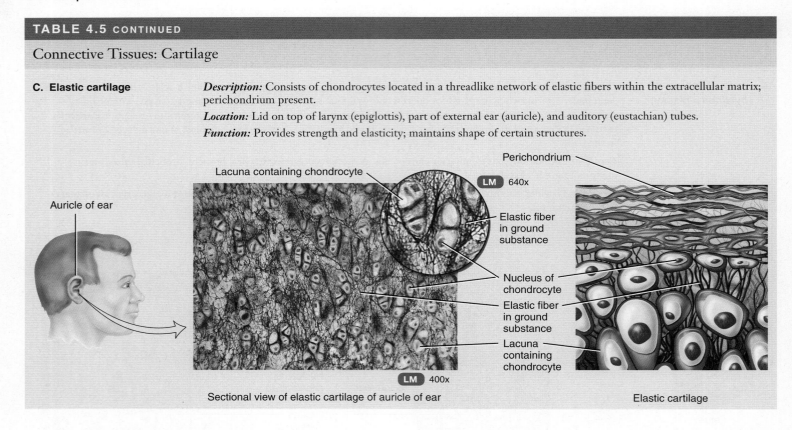

Sectional view of elastic cartilage of auricle of ear

Elastic cartilage

Bone Tissue

Bones are organs composed of several different connective tissues, including **bone** or *osseous tissue* (OS-ē-us). Bone tissue has several functions. It supports soft tissues, protects delicate structures, and works with skeletal muscles to generate movement. Bone stores calcium and phosphorus; stores red bone marrow, which produces blood cells; and houses yellow bone marrow, a storage site for triglycerides. The details of bone tissue are presented in Chapter 6.

Liquid Connective Tissue

BLOOD TISSUE   **Blood tissue** (or simply *blood*) is a connective tissue with a liquid extracellular matrix called **blood plasma**, a pale yellow fluid that consists mostly of water with a wide variety of dissolved substances: nutrients, wastes, enzymes, hormones, respiratory gases, and ions. Suspended in the plasma are red blood cells, white blood cells, and platelets. **Red blood cells** transport oxygen to body cells and remove carbon dioxide from them. **White blood cells** are involved in phagocytosis, immunity, and allergic reactions. **Platelets** participate in blood clotting. The details of blood are considered in Chapter 14.

LYMPH   **Lymph** is a fluid that flows in lymphatic vessels. It is a connective tissue that consists of several types of cells in a clear extracellular matrix similar to blood plasma but with much less protein. The details of lymph are considered in Chapter 17.

✓ CHECKPOINT

6. What are the features of the cells, ground substance, and fibers that make up connective tissue?

7. How are the structures of the following connective tissues related to their functions: areolar connective tissue, adipose tissue, reticular connective tissue, dense regular connective tissue, dense irregular connective tissue, elastic connective tissue, hyaline cartilage, fibrocartilage, elastic cartilage, bone tissue, blood tissue, and lymph?

4.4 MEMBRANES

OBJECTIVES • **Define a membrane.**

• **Describe the classification of membranes.**

Membranes (Figure 4.3) are flat sheets of pliable tissue that cover or line a part of the body. The combination of an epithelial layer and an underlying connective tissue layer constitutes an **epithelial membrane**. The principal epithelial membranes of the body are mucous membranes, serous membranes, and the cutaneous membrane, or skin. (Skin is discussed in detail in Chapter 5 and will not be discussed here.) Another kind of membrane, a synovial membrane, lines joints and contains connective tissue but no epithelium.

Figure 4.3 Membranes.

A membrane is a flat sheet of pliable tissues that covers or lines a part of the body.

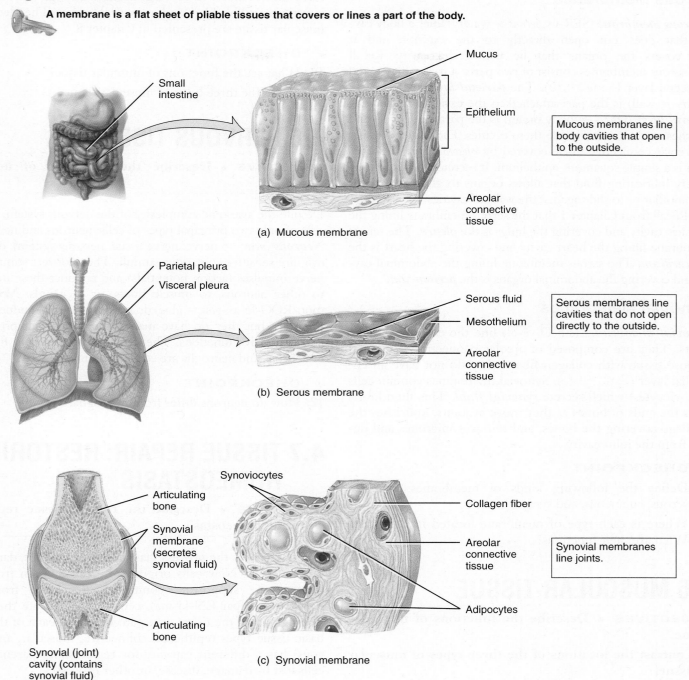

Small
intestine

Mucus

Epithelium

Areolar
connective
tissue

Mucous membranes line
body cavities that open
to the outside.

(a) Mucous membrane

Parietal pleura

Visceral pleura

Serous fluid

Mesothelium

Areolar
connective
tissue

Serous membranes line
cavities that do not open
directly to the outside.

(b) Serous membrane

Synoviocytes

Articulating
bone

Synovial
membrane
(secretes
synovial fluid)

Articulating
bone

Synovial (joint)
cavity (contains
synovial fluid)

Collagen fiber

Areolar
connective
tissue

Adipocytes

Synovial membranes
line joints.

(c) Synovial membrane

What is an epithelial membrane?

Mucous Membranes

A *mucous membrane* or *mucosa* (mū-KŌ-sa) lines a body cavity that opens directly to the exterior. Mucous membranes line the entire digestive, respiratory, and reproductive systems and much of the urinary system. The epithelial layer of a mucous membrane secretes mucus, which prevents the cavities from drying out (Figure 4.3a). It also traps particles in the respiratory passageways, lubricates and absorbs food as it moves through the gastrointestinal tract, and secretes digestive enzymes. The connective tissue layer (areolar connective tissue) helps bind the epithelium to the underlying structures. It also provides the epithelium with oxygen and nutrients and removes wastes via its blood vessels.

Serous Membranes

A *serous membrane* (SĒR-us; *serous* = watery) lines a body cavity that does not open directly to the exterior, and it also covers the organs that lie within the cavity. Recall that serous membranes consist of two parts: a parietal layer and a visceral layer (Figure 4.3b). The *parietal layer* (pa-RĪ-e-tal; *pariet-* = wall) is the part attached to the cavity wall, and the *visceral layer* (*viscer-* = body organ) is the part that covers and attaches to the organs inside these cavities. Each layer consists of areolar connective tissue covered by *mesothelium*. Mesothelium is a simple squamous epithelium. It secretes *serous fluid*, a watery lubricating fluid that allows organs to glide easily over one another or to slide against the walls of cavities.

Recall from Chapter 1 that the serous membrane lining the thoracic cavity and covering the lungs is the *pleura*. The serous membrane lining the heart cavity and covering the heart is the *pericardium*. The serous membrane lining the abdominal cavity and covering the abdominal organs is the *peritoneum*.

Synovial Membranes

Synovial membranes (sin-Ō-vē-al) line the cavities of some joints. They are composed of areolar connective tissue and adipose tissue with collagen fibers; they do not have an epithelial layer (Figure 4.3c). Synovial membranes contain cells (synoviocytes) which secrete *synovial fluid*. This fluid lubricates the ends of bones as they move at joints, nourishes the cartilage covering the bones, and removes microbes and debris from the joint cavity.

✓ CHECKPOINT

8. Define the following kinds of membranes: mucous, serous, cutaneous, and synovial.
9. Where is each type of membrane located in the body? What are their functions?

4.5 MUSCULAR TISSUE

OBJECTIVES • **Describe the functions of muscular tissue.**
• **Contrast the locations of the three types of muscular tissue.**

Muscular tissue consists of elongated cells called *muscle fibers* that are highly specialized to generate force. As a result of this characteristic, muscular tissue produces motion, maintains posture, and generates heat. It also offers protection. Based on its location and certain structural and functional characteristics, muscular tissue is classified into three types: skeletal, cardiac, and smooth. *Skeletal muscle tissue* is named for its location— it is usually attached to the bones of the skeleton. *Cardiac muscle tissue* forms the bulk of the wall of the heart. *Smooth muscle tissue* is located in the walls of hollow internal structures such as blood vessels, airways to the lungs, the stomach, intestines, gallbladder, and urinary bladder. The details of muscular tissue are presented in Chapter 8.

✓ CHECKPOINT

10. What are the functions of muscular tissue?
11. Name the three types of muscular tissue.

4.6 NERVOUS TISSUE

OBJECTIVE • **Describe the functions of nervous tissue.**

Despite the awesome complexity of the nervous system, it consists of only two principal types of cells: neurons and neuroglia. *Neurons* (*neur-* = nerve, nerve tissue, nervous system), or *nerve cells*, are sensitive to various stimuli. They convert stimuli into nerve impulses (action potentials) and conduct these impulses to other neurons, to muscle fibers, or to glands. *Neuroglia* (noo-RŌG-lē-a; *-glia* = glue) do not generate or conduct nerve impulses, but they do have many other important protective and supportive functions. The detailed structure and function of neurons and neuroglia are considered in Chapter 9.

✓ CHECKPOINT

12. How do neurons differ from neuroglia?

4.7 TISSUE REPAIR: RESTORING HOMEOSTASIS

OBJECTIVE • **Describe the role of tissue repair in restoring homeostasis.**

Tissue repair is the process that replaces worn-out, damaged, or dead cells. New cells originate by cell division from the *stroma*, the supporting connective tissue, or from the *parenchyma* (par-EN-ki-ma), cells that constitute the functioning part of the tissue or organ. In adults, each of the four basic tissue types (epithelial, connective, muscular, and nervous) has a different capacity for replenishing parenchymal cells lost by damage, disease, or other processes.

Epithelial cells, which endure considerable wear and tear (and even injury) in some locations, have a continuous capacity for renewal. In some cases, immature, undifferentiated cells called *stem cells* divide to replace lost or damaged cells. For example, stem cells reside in protected locations in the epithelia of the skin and gastrointestinal tract to replenish cells sloughed from the apical layer.

Some connective tissues also have a continuous capacity for renewal. One example is bone, which has an ample blood supply. Other connective tissues such as cartilage can replenish cells less readily in part because of a poor blood supply.

Fat Talk

Adipose tissue is composed of adipocytes, specialized cells whose primary function is the storage of triglycerides. Triglycerides (fats) are made by the liver from excess calories. The liver is very clever in this regard: it can make triglycerides from any kind of excess calorie hanging around: protein, carbohydrate, alcohol, or fat. Adipocytes have a nucleus and the organelles found in most cells, but up to 85 percent of their volume is composed of stored triglycerides. Triglyceride storage ensures an adequate energy supply, even during fairly lean times. Great idea.

In addition to a fleet of adipocytes, adipose tissue also contains other structural elements such as blood vessels and connective tissue. Scientists used to regard adipose tissue as fairly inert storage depots that took in or released triglycerides depending on the energy balance in the body. Excess calorie consumption was thought to lead to increased fat storage, while a calorie deficit would signal the adipocytes to release triglycerides for the body to use as fuel.

Fat cells still do these things, but scientists are beginning to unravel some of the adipose tissue activity involved in fat storage and metabolism. Of special interest has been the discovery that adipose tissue can talk to the rest of the body, to provide information about energy storage levels.

The role of adipose tissue in the endocrine system

Researchers have identified a number of chemical messengers that allow adipose tissue to help regulate fat storage, and allow it to communicate with other organs and systems in the body. Some of these messengers act as hormones, sending signals to other parts of the body. Leptin, for example, is a messenger produced by adipose tissue. Leptin concentration in the blood is thought to inform the brain about triglyceride storage levels. When the brain understands that levels are getting low, it can turn on the feeling of hunger, and the drive to look for some food. Thus, by releasing hormones, adipose tissue acts as a component of the endocrine system.

Leptin signaling: Primarily for low energy alerts

Pharmaceutical manufacturers hoped that increasing leptin levels in obese people would reduce the drive to eat. However, while low leptin levels seem to catch the brain's attention, high leptin levels, signaling the presence of nice fat adipocyte cells, do not seem to have much effect. People keep eating. It appears that the brain is not worried about having too much to eat.

Think It Over . . . Why might evolution favor a mechanism designed to ensure adequate calorie intake over one that limits food consumption? (Clue: think 40,000 years ago.)

Muscular tissue has a relatively poor capacity for renewal of lost cells. Cardiac muscle fibers can be produced from stem cells under special conditions (see Section 24.1). Skeletal muscle tissue does not divide rapidly enough to replace extensively damaged muscle fibers. Smooth muscle fibers can proliferate to some extent, but they do so much more slowly than the cells of epithelial or connective tissues.

Nervous tissue has the poorest capacity for renewal. Although experiments have revealed the presence of some stem cells in the brain, they normally do not undergo mitosis to replace damaged neurons.

If parenchymal cells accomplish the repair, *tissue regeneration* is possible, and a near-perfect reconstruction of the injured tissue may occur. However, if fibroblasts of the stroma are active in the repair, the replacement tissue will be a new connective tissue. The fibroblasts synthesize collagen and other extracellular matrix materials that aggregate to form scar tissue, a process known as *fibrosis*. Because scar tissue is not specialized to perform the functions of the parenchymal tissue, the original function of the tissue or organ is impaired.

CLINICAL CONNECTION | Adhesions

Scar tissue can form **adhesions** (ad-HĒ-shuns), abnormal joining of tissues. Adhesions commonly form in the abdomen around a site of previous inflammation such as an inflamed appendix, and they can develop after surgery. Although adhesions do not always cause problems, they can decrease tissue flexibility, cause obstruction (such as in the intestine), and make a subsequent operation more difficult. The surgical removal of adhesions may be necessary. •

✓ CHECKPOINT

13. How are stromal and parenchymal repair of a tissue different?

4.8 AGING AND TISSUES

OBJECTIVE • Describe the effects of aging on tissues.

Generally, tissues heal faster and leave less obvious scars in the young than in the aged. In fact, surgery performed on fetuses leaves no scars. The younger body is generally in a better nutritional state, its tissues have a better blood supply, and its cells have a higher metabolic rate. Thus, cells can synthesize needed materials and divide more quickly. The extracellular components of tissues also change with age. Glucose, the most abundant sugar in the body, plays a role in the aging process. Glucose is haphazardly added to proteins inside and outside cells, forming irreversible cross-links between adjacent protein molecules. With advancing age, more cross-links form, which contributes to the stiffening and loss of elasticity that occur in aging tissues. Collagen fibers, responsible for the strength of tendons, increase in number and change in quality with aging. Elastin, another extracellular component, is responsible for the elasticity of blood vessels and skin. It thickens, fragments, and acquires a greater affinity for calcium with age—changes that may also be associated with the development of atherosclerosis, the deposition of fatty materials in arterial walls.

✓ CHECKPOINT

14. What common changes occur in epithelial and connective tissues with aging?

• • •

Now that you have an understanding of tissues, we will look at the organization of tissues into organs and organs into systems. In the next chapter we will consider how the skin and other organs function as components of the integumentary system.

COMMON DISORDERS

Sjögren's Syndrome

Sjögren's syndrome (SHŌ-grenz) is a common autoimmune disorder that causes inflammation and destruction of exocrine glands, especially the lacrimal (tear) glands and salivary glands. Signs include dryness of the eyes, mouth, nose, ears, skin, and vagina, and salivary gland enlargement. Systemic effects include fatigue, arthritis, difficulty in swallowing, pancreatitis (inflammation of the pancreas), pleuritis (inflammation of the pleurae of the lungs), and muscle and joint pain. The disorder affects females more than males by a ratio of 9 to 1. About 20 percent of older adults experience some signs of Sjögren's. Treatment is supportive and includes using artificial tears to moisten the eyes, sipping fluids, chewing sugarless gum, using a saliva substitute to moisten the mouth, and applying moisturizing creams for the skin. If symptoms or complications are severe, medications may be used. These include cyclosporine eyedrops, pilocarpine to increase saliva production, immunosuppressants, nonsteroidal anti-inflammatory drugs, and corticosteroids.

Systemic Lupus Erythematosus

Systemic lupus erythematosus (er-i-thē-ma-TŌ-sus), *SLE*, or simply *lupus*, is a chronic inflammatory disease of connective tissue occurring mostly in nonwhite women during their childbearing years. It is an autoimmune disease that can cause tissue damage in every body system. The disease, which can range from a mild condition in most patients to a rapidly fatal disease, is marked by periods of exacerbation and remission. Although the cause of SLE is unknown, genetic, environmental, and hormonal factors all have been implicated. The genetic component is suggested by studies of twins and family history. Environmental factors include viruses, bacteria, chemicals, drugs, exposure to excessive sunlight, and emotional stress. Sex hormones, such as estrogens, may also trigger SLE.

Signs and symptoms of SLE include painful joints, low-grade fever, fatigue, mouth ulcers, weight loss, enlarged lymph nodes and spleen, sensitivity to sunlight, rapid loss of large amounts of scalp hair, and anorexia. A distinguishing feature of lupus is an eruption across the bridge of the nose and cheeks called a "butterfly rash." Other skin lesions may occur, including blistering and ulceration. The erosive nature of some SLE skin lesions was thought to resemble the damage inflicted by the bite of a wolf—thus, the name *lupus* (= wolf). The most serious complications of the disease involve inflammation of the kidneys, liver, spleen, lungs, heart, brain, and gastrointestinal tract. Because there is no cure for SLE, treatment is supportive, including anti-inflammatory drugs, such as aspirin, and immunosuppressive drugs.

MEDICAL TERMINOLOGY AND CONDITIONS

Tissue rejection   An immune response of the body directed at foreign proteins in a transplanted tissue or organ; immunosuppressive drugs, such as cyclosporine, have largely overcome tissue rejection in heart-, kidney-, and liver-transplant patients.

Tissue transplantation   The replacement of a diseased or injured tissue or organ; the most successful transplants involve use of a person's own tissues or those from an identical twin.

Xenotransplantation (zen'-ō-trans'-plan-TĀ-shun; *xeno-* = strange, foreign)   The replacement of a diseased or injured tissue or organ with cells or tissues from an animal. Only a few cases of successful xenotransplantation exist to date.

CHAPTER REVIEW AND RESOURCE SUMMARY

WILEY
PLUS

REVIEW	RESOURCES

4.1 Types of Tissues

1. A tissue is a group of similar cells that usually has a similar embryological origin and is specialized for a particular function.

2. The various tissues of the body are classified into four basic types: **epithelial tissue**, **connective tissue**, **muscular tissue**, and **nervous tissue**.

Anatomy Overview—Tissues

4.2 Epithelial Tissue

1. The general types of epithelial tissue (**epithelium**) include covering and lining epithelium and glandular epithelium. Epithelium has the following general characteristics: It consists mostly of cells with little extracellular material, is arranged in sheets, is attached to connective tissue by a **basement membrane**, is **avascular** (no blood vessels), has a nerve supply, and can replace itself.

2. Epithelial layers can be simple (one layer) or stratified (several layers). The cell shapes may be squamous (flat), cuboidal (cubelike), columnar (rectangular), or transitional (variable).

3. **Simple squamous epithelium** consists of a single layer of flat cells (Table 4.1A). It is found in parts of the body where filtration or diffusion are priority processes. One type, **endothelium**, lines the heart and blood vessels. Another type, **mesothelium**, forms the serous membranes that line the thoracic and abdominal cavities and cover the organs within them.

4. **Simple cuboidal epithelium** consists of a single layer of cube-shaped cells that function in secretion and absorption (Table 4.1B). It is found covering the ovaries, in the kidneys and eyes, and lining some glandular ducts.

5. **Nonciliated simple columnar epithelium** consists of a single layer of nonciliated rectangular cells (Table 4.1C). It lines most of the gastrointestinal tract. Specialized cells containing **microvilli** perform absorption. **Goblet cells** secrete mucus.

6. **Ciliated simple columnar epithelium** consists of a single layer of ciliated rectangular cells (Table 4.1D). It is found in a few portions of the upper respiratory tract, where it moves foreign particles trapped in mucus out of the respiratory tract.

7. **Pseudostratified columnar epithelium** has only one layer but gives the appearance of many (Table 4.1E). The ciliated variety moves mucus in the respiratory tract. The nonciliated variety functions in absorption and protection.

8. **Stratified squamous epithelium** consists of several layers of cells; cells in the apical layer and several layers deep to it are flat (Table 4.1F). It is protective. A nonkeratinized variety lines the mouth; a keratinized variety forms the epidermis, the most superficial layer of the skin.

9. **Stratified cuboidal epithelium** consists of several layers of cells; cells in the apical layer are cube-shaped (Table 4.1G). It is found in adult sweat glands and a portion of the male urethra. It protects and provides limited secretion and absorption.

10. **Stratified columnar epithelium** consists of several layers of cells; cells in the apical layer are column-shaped (Table 4.1H). It is found in a portion of the male urethra and large excretory ducts of some glands. It functions in protection and secretion.

11. **Transitional epithelium** consists of several layers of cells whose appearance varies with the degree of stretching (Table 4.1I). It lines the urinary bladder.

12. A **gland** is a single cell or a group of epithelial cells adapted for secretion. **Endocrine glands** secrete hormones into interstitial fluid and then the blood (Table 4.2A). **Exocrine glands** (mucous, sweat, oil, and digestive glands) secrete into ducts or directly onto a free surface (Table 4.2B).

Figure 4.1 Cell shapes and arrangement of layers for covering and lining epithelium

Anatomy Overview—Epithelial Tissues

4.3 Connective Tissue

1. **Connective tissue**, one of the most abundant body tissues, consists of cells and an **extracellular matrix** of ground substance and fibers; it has abundant matrix with relatively few cells. It does not usually occur on free surfaces, has a nerve supply (except for cartilage), and is highly vascular (except for cartilage, tendons, and ligaments).

Figure 4.2 Representative cells and fibers present in connective tissues

REVIEW

2. Cells in connective tissue include **fibroblasts** (secrete matrix), **macrophages** (perform phagocytosis), **plasma cells** (secrete antibodies), **mast cells** (produce histamine), and **adipocytes** (store fat).

3. The **ground substance** and **fibers** make up the extracellular matrix. The ground substance supports and binds cells together, provides a medium for the exchange of materials, and is active in influencing cell functions.

4. The fibers in the extracellular matrix provide strength and support and are of three types: (a) **collagen fibers** (composed of collagen) are found in large amounts in bone, tendons, and ligaments; (b) **elastic fibers** (composed of elastin, fibrillin, and other glycoproteins) are found in skin, blood vessel walls, and lungs; and (c) **reticular fibers** (composed of collagen and glycoprotein) are found around fat cells, nerve fibers, and skeletal and smooth muscle cells.

5. Connective tissue is subdivided into loose connective tissue, dense connective tissue, cartilage, bone, and liquid connective tissue (blood tissue and lymph).

6. **Loose connective tissue** includes areolar connective tissue, adipose tissue, and reticular connective tissue. **Areolar connective tissue** consists of the three types of fibers, several cells, and a semifluid ground substance (Table 4.3A). It is found in the subcutaneous layer; in mucous membranes; and around blood vessels, nerves, and body organs. **Adipose tissue** consists of **adipocytes**, which store triglycerides (Table 4.3B). It is found in the subcutaneous layer, around organs, and in the yellow bone marrow. **Reticular connective tissue** consists of reticular fibers and reticular cells and is found in the liver, spleen, and lymph nodes (Table 4.3C).

7. **Dense connective tissue** includes dense regular connective tissue, dense irregular connective tissue, and elastic connective tissue. **Dense regular connective tissue** consists of parallel bundles of collagen fibers and fibroblasts (Table 4.4A). It forms tendons, most ligaments, and aponeuroses. **Dense irregular connective tissue** consists of usually randomly arranged collagen fibers and a few fibroblasts (Table 4.4B). It is found in fasciae, the dermis of skin, and membrane capsules around organs. **Elastic connective tissue** consists of branching elastic fibers and fibroblasts (Table 4.4C). It is found in the walls of large arteries, lungs, trachea, and bronchial tubes.

8. **Cartilage** contains **chondrocytes** and has a rubbery matrix (chondroitin sulfate) containing collagen and elastic fibers. **Hyaline cartilage** is found in the embryonic skeleton, at the ends of bones, in the nose, and in respiratory structures (Table 4.5A). It is flexible, allows movement, and provides support. **Fibrocartilage** is found in the pubic symphysis, intervertebral discs, and menisci (cartilage pads) of the knee joint (Table 4.5B). **Elastic cartilage** maintains the shape of organs such as the epiglottis of the larynx, auditory (eustachian) tubes, and external ear (Table 4.5C).

9. **Bone** or osseous tissue supports, protects, helps provide movement, stores minerals, and houses blood-forming tissue.

10. **Blood tissue** is liquid connective tissue that consists of **blood plasma** in which red blood cells, white blood cells, and platelets are suspended. Its cells transport oxygen and carbon dioxide, carry on phagocytosis, participate in allergic reactions, provide immunity, and bring about blood clotting. **Lymph**, the extracellular fluid that flows in lymphatic vessels, is also a liquid connective tissue. It is a clear fluid similar to blood plasma but with less protein.

Anatomy Overview—
Connective Tissues

4.4 Membranes

1. An **epithelial membrane** consists of an epithelial layer overlying a connective tissue layer. Examples are mucous membranes, serous membranes, and synovial membranes.

2. **Mucous membranes** line cavities that open to the exterior, such as the gastrointestinal tract.

3. **Serous membranes** line closed cavities (pleura, pericardium, peritoneum) and cover the organs in the cavities. These membranes consist of a **parietal layer** and a **visceral layer**.

Anatomy Overview—Epithelial
Membranes

REVIEW	RESOURCES
4. **Synovial membranes** line joint cavities, bursae, and tendon sheaths. They consist of areolar connective tissue and do not have an epithelial layer.	

4.5 Muscular Tissue

1. **Muscular tissue** consists of cells (called muscle fibers) that are specialized for contraction. It provides motion, maintenance of posture, heat production, and protection.
2. **Skeletal muscle tissue** is attached to bones, **cardiac muscle tissue** forms most of the heart wall, and **smooth muscle tissue** is found in the walls of hollow internal structures (blood vessels and viscera).

Anatomy Overview—Muscle Tissue
Animation—Intercellular Junctions
Exercise—Getting It Together
Anatomy Overview—Nervous Tissue
Exercise—Tissue identification
Exercise—Concentrate on Tissue Function
Exercise—Name that Tissue
Concepts and Connections— Human Tissue

4.6 Nervous Tissue

1. The nervous system is composed of **neurons** (nerve cells) and **neuroglia** (protective and supporting cells).
2. Neurons are sensitive to stimuli, convert stimuli into nerve impulses, and conduct nerve impulses.

4.7 Tissue Repair: Restoring Homeostasis

1. **Tissue repair** is the replacement of worn-out, damaged, or dead cells by healthy ones.
2. **Stem cells** may divide to replace lost or damaged cells. The formation of scar tissue is called **fibrosis**.

4.8 Aging and Tissues

1. Tissues heal faster and leave less obvious scars in the young than in the aged; surgery performed on fetuses leaves no scars.
2. The extracellular components of tissues, such as collagen and elastic fibers, also change with age.

SELF-QUIZ

1. Epithelial tissue
   a. conducts nerve impulses.
   b. stores fat.
   c. covers and lines the body and its parts.
   d. moves the body.
   e. stores minerals.
2. Epithelial tissue can be classified according to
   a. its location.
   b. its function.
   c. the composition of the extracellular matrix.
   d. the shape and arrangement of its cells.
   e. whether it is under voluntary or involuntary control.
3. Associated with the inner lining of some joints between bones are
   a. mucous membranes.   b. glands.
   c. serous fluids.   d. serous membranes.
   e. synovial membranes.

4. The presence of _____ in a tissue allows stretching and ability to return to the original shape.
   a. ground substance   b. collagen fibers
   c. reticular fibers   d. elastic fibers
   e. cartilage
5. Which of the following is true concerning connective tissue?
   a. Except for cartilage, connective tissue has a rich blood supply.
   b. Connective tissue is classified according to cell shape and arrangement.
   c. The cells of connective tissue are generally closely joined.
   d. Loose connective tissue consists of many fibers arranged in a regular pattern.
   e. The fibers in connective tissue are composed of chondroitin sulfate.

6. Match the following tissue types with their descriptions.

   _____ **a.** fat storage
   _____ **b.** protects the skin from damage
   _____ **c.** forms the stroma (framework) of many organs
   _____ **d.** composes the intervertebral discs
   _____ **e.** stores red bone marrow, protects, supports
   _____ **f.** found in the walls of hollow organs
   _____ **g.** found in lungs, involved in diffusion
   _____ **h.** found in kidney tubules, involved in absorption

   A. simple cuboidal epithelium
   B. simple squamous epithelium
   C. adipose
   D. fibrocartilage
   E. reticular connective
   F. smooth muscle
   G. keratinized stratified squamous epithelium
   H. bone

7. If you were going to design a hollow organ that needed to expand and have stretchability, which of the following epithelial and connective tissues might you use?
   a. transitional epithelium and elastic connective tissue
   b. stratified columnar epithelium and adipose tissue
   c. simple columnar epithelium and dense regular connective tissue
   d. simple squamous epithelium and hyaline cartilage
   e. transitional epithelium and reticular connective tissue

8. Which of the following statements is NOT true concerning epithelial tissue?
   a. The cells of epithelial tissue are closely packed.
   b. The basal layer of cells rests on a basement membrane.
   c. Epithelial tissue has a nerve supply.
   d. Epithelial tissue undergoes rapid rates of cell division.
   e. Epithelial tissue is well supplied with blood vessels.

9. In the repair of tissue, scar tissue appears when
   a. parenchymal cells dominate the repair process.
   b. stem cells replace the damaged tissue cells.
   c. fibroblasts secrete extracellular materials and collagen.
   d. the process of tissue repair is complete.
   e. the stroma is not involved in the repair process.

10. A connective tissue with a liquid extracellular matrix is
    a. elastic cartilage.   **b.** blood.   **c.** areolar.
    d. reticular.   **e.** osseous.

11. The interior of your nose is lined with
    a. a mucous membrane.
    b. smooth muscle tissue.
    c. a synovial membrane.
    d. keratinized stratified squamous epithelium.
    e. a serous membrane.

12. The four main types of tissue are
    a. epithelial, membranous, blood, and nervous.
    b. blood, connective, muscular, and nervous.
    c. connective, epithelial, muscular, and nervous.
    d. glandular, muscular, connective, and nervous.
    e. epithelial, connective, muscular, and membranous.

13. Which of the following materials would NOT be found in the extracellular matrix of connective tissue?
    a. collagen fibers   **b.** elastic fibers   **c.** keratin
    d. reticular fibers   **e.** hyaluronic acid

14. Which connective tissue cells secrete antibodies?
    a. mast cells   **b.** adipocytes   **c.** macrophages
    d. plasma cells   **e.** chondrocytes

15. Modified columnar epithelial cells that secrete mucus are _____ cells.
    a. ciliated   **b.** keratinized   **c.** mast
    d. fibroblast   **e.** goblet

16. Which tissue forms the bulk of the heart wall?
    a. skeletal muscle   **b.** nervous   **c.** bone
    d. cardiac muscle   **e.** smooth muscle

17. Stratified squamous epithelium functions in
    a. protection and secretion   **b.** contraction
    c. absorption   **d.** stretching
    e. transport

18. What tissue type is found in tendons?
    a. dense irregular connective tissue
    b. elastic connective tissue
    c. dense regular connective tissue
    d. pseudostratified epithelium
    e. areolar tissue

19. In what tissue type would you find stores of calcium and phosphorus?
    a. bone
    b. hyaline cartilage
    c. fibrocartilage
    d. dense irregular connective tissue
    e. elastic cartilage

20. Which of the following statements is true concerning glandular tissue?
    a. Endocrine glands are composed of connective tissue; exocrine glands are composed of modified epithelium.
    b. Endocrine gland secretions diffuse into the bloodstream; exocrine gland secretions enter ducts.
    c. A sweat gland is an example of an endocrine gland.
    d. Endocrine glands contain ducts; exocrine glands do not.
    e. Exocrine glands produce substances known as hormones.

CRITICAL THINKING APPLICATIONS

1. Your young nephew can't wait to get his eyebrow pierced like his big brother. In the meantime, he's walking around with sewing needles stuck through his fingertips. There is no visible bleeding. What type of tissue has he pierced? (Be specific.) How do you know?

2. Collagen is the new "miracle" cosmetic. It's advertised to give you shiny hair and glowing skin, and can be injected to reduce wrinkles. What is collagen? If you wanted to launch your own line of cosmetics, what tissue or structure would supply you with abundant collagen?

3. Your lab partner Sophie put a tissue slide labeled uterine tube under the microscope. She focused the slide and exclaimed, "Look! It's all hairy." Explain to Sophie what the "hair" really is.

4. Three-year-old Alex jumped off the couch and injured his right tibia (lower leg bone). The Accident and Emergency nurse said he "broke his cartilage." Alex's mother feels relieved that he didn't break his bone. Two years later, Alex seems to be limping and his mother notices that Alex's right leg appears shorter than his left. What happened to Alex?

ANSWERS TO FIGURE QUESTIONS

4.1 Substances would move most rapidly through squamous cells because they are so thin.

4.2 Fibroblasts secrete the fibers and ground substance of the extracellular matrix.

4.3 An epithelial membrane is a membrane that consists of an epithelial layer and an underlying layer of connective tissue.

## CHAPTER 5

THE INTEGUMENTARY SYSTEM

Of all the body's organs, none is more easily inspected or more exposed to infection, disease, and injury than the skin. Because of its visibility, skin reflects our emotions and some aspects of normal physiology, as evidenced by frowning, blushing, and sweating. Changes in skin color or condition may indicate homeostatic imbalances in the body. For example, a skin rash such as occurs in chickenpox reveals a systemic infection, but a yellowing of the skin is an indication of jaundice, usually due to disease of the liver, an internal organ. Other disorders may be limited to the skin, such as warts, age spots, or pimples. The skin's location makes it vulnerable to damage from trauma, sunlight, microbes, or pollutants in the environment. Major damage to the skin, as occurs in third-degree burns, can be life threatening due to the loss of the protective skin functions.

Many interrelated factors may affect both the appearance and health of the skin, including nutrition, hygiene, circulation, age, immunity, genetic traits, psychological state, and drugs. So important is the skin to body image that people spend much time and money to restore it to a more youthful appearance. **Dermatology** (der′-ma-TOL-ō-jē; *dermato-* = skin; *-logy* = study of) is the branch of medicine that specializes in diagnosing and treating skin disorders.

Did you know?   Protecting skin from the damaging ultraviolet (UV) rays of the sun or tanning bed helps prevent premature aging and cancers of the skin. Pale skin ages most quickly, showing lines and discoloration at the earliest ages. Pale skin is also most vulnerable to skin cancers. People with darker skin can relax a little in the sun, although it is still important for everyone to monitor any changes in the appearance of the skin. While people with very dark skin, such as those of African descent, have a lower likelihood of developing skin cancer, they tend to have cancers found at a more advanced stage. Skin cancers caught early are more treatable and less likely to be fatal.

FOCUS ON WELLNESS
Skin Care for Outdoor Exercisers

LOOKING BACK TO MOVE AHEAD...

5.1 SKIN

OBJECTIVES • **Describe the structure and functions of the skin.**

• **Explain the basis for different skin colors.**

Recall from Chapter 1 that a system consists of a group of organs working together to perform specific activities. The *integumentary system* (in-teg-ū-MEN-tar-ē; *in* = inward; *tegere* = to cover) is composed of the skin, hair, oil and sweat glands, nails, and sensory receptors. The *skin* or *cutaneous membrane* covers the external surface of the body. It is the largest organ of the body in surface area and weight. In adults, the skin covers an area of about 2 square meters (22 square feet) and weighs 4.5–5kg (10–11 lb), about 7 percent of total body weight.

Structure of Skin

Structurally, the skin consists of two main parts (Figure 5.1). The superficial, thinner portion, which is composed of

Figure 5.1 Components of the integumentary system. The skin consists of a thin, superficial epidermis and a deep, thicker dermis. Deep to the skin is the subcutaneous layer, which attaches the dermis to underlying organs and tissues.

The integumentary system includes the skin and its accessory structures—hair, nails, and glands—along with associated muscles and nerves.

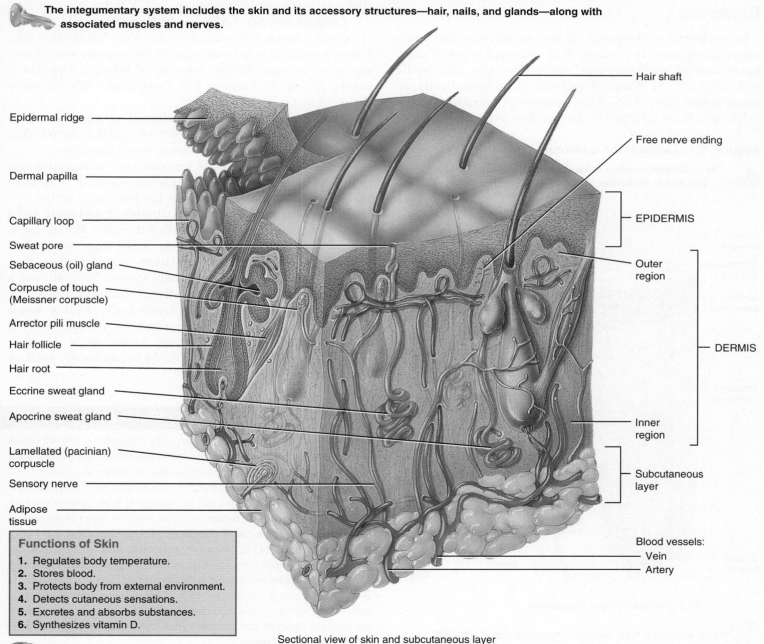

Epidermal ridge

Dermal papilla

Capillary loop

Sweat pore

Sebaceous (oil) gland

Corpuscle of touch (Meissner corpuscle)

Arrector pili muscle

Hair follicle

Hair root

Eccrine sweat gland

Apocrine sweat gland

Lamellated (pacinian) corpuscle

Sensory nerve

Adipose tissue

Hair shaft

Free nerve ending

EPIDERMIS

Outer region

Inner region

DERMIS

Subcutaneous layer

Blood vessels:
Vein
Artery

Sectional view of skin and subcutaneous layer

Functions of Skin

1. Regulates body temperature.
2. Stores blood.
3. Protects body from external environment.
4. Detects cutaneous sensations.
5. Excretes and absorbs substances.
6. Synthesizes vitamin D.

 What types of tissues make up the epidermis and the dermis?

epithelial tissue, is the **epidermis** (ep'-i-DERM-is; *epi-* = above). The deeper, thicker *connective tissue* portion is the **dermis.**

Deep to the dermis, but not part of the skin, is the **subcutaneous (subQ) layer.** Also called the **hypodermis** (*hypo-* = below), this layer consists of areolar and adipose tissues. Fibers that extend from the dermis anchor the skin to the subcutaneous layer, which, in turn, attaches to underlying tissues and organs. The subcutaneous layer serves as a storage depot for fat and contains large blood vessels that supply the skin. This region (and sometimes the dermis) also contains nerve endings called **lamellated (pacinian) corpuscles** (pa-SIN-ē-an) that are sensitive to pressure (Figure 5.1).

Epidermis

The epidermis is composed of keratinized stratified squamous epithelium. It contains four principal types of cells: keratinocytes, melanocytes, Langerhans cells, and tactile cells (Figure 5.2). About 90 percent of epidermal cells are **keratinocytes** (ker-a-TIN-ō-sīts; *keratino-* = hornlike; *-cytes* = cells), which are arranged in four or five layers and produce

the protein **keratin** (KER-a-tin). Recall from Chapter 4 that keratin is a tough, fibrous protein that helps protect the skin and underlying tissues from abrasions, heat, microbes, and chemicals. Keratinocytes also produce lamellar granules, which release a water-repellent sealant.

About 8 percent of the epidermal cells are **melanocytes** (MEL-a-nō-sīts; *melano-* = black), which produce the pigment melanin. Their long, slender projections extend between the keratinocytes and transfer melanin granules to them. **Melanin** (MEL-a-nin) is a yellow-red or brown-black pigment that contributes to skin color and absorbs damaging ultraviolet (UV) light. Although keratinocytes gain some protection from melanin granules, melanocytes themselves are particularly susceptible to damage by UV light.

Langerhans cells (LANG-er-hans) participate in immune responses mounted against microbes that invade the skin. Langerhans cells help other cells of the immune system recognize an antigen (foreign microbe or substance) so that it can be destroyed (Chapter 17). Langerhans cells are easily damaged by UV light.

Tactile cells contact the flattened process of a sensory neuron (nerve cell), a structure called a **tactile disc.** Tactile cells and tactile discs detect touch sensations.

Several distinct layers of keratinocytes in various stages of development form the epidermis (see Figure 5.2). In most regions of the body the epidermis has four strata or layers—stratum basale, stratum spinosum, stratum granulosum, and a thin stratum corneum. This is called *thin skin*. Where exposure to friction is greatest, such as in the fingertips, palms, and soles, the epidermis has five layers—stratum basale, stratum spinosum, stratum granulosum, stratum lucidum, and a thick stratum corneum. This is referred to as *thick skin.*

The deepest layer of the epidermis is the **stratum basale** (ba-SA-lē; *basal-* = base), composed of a single row of cuboidal or columnar keratinocytes. Some cells in this layer are *stem cells* that undergo cell division to continually produce new keratinocytes.

Figure 5.2 Layers of the epidermis.

The epidermis consists of keratinized stratified squamous epithelium.

Stratum corneum — Dead keratinocytes — Superficial

Stratum lucidum

Stratum granulosum — Lamellar granules

Keratinocyte

Stratum spinosum — Langerhans cell

Tactile cell
Tactile disc
Sensory neuron

Melanocyte

Stratum basale — Dermis — Deep

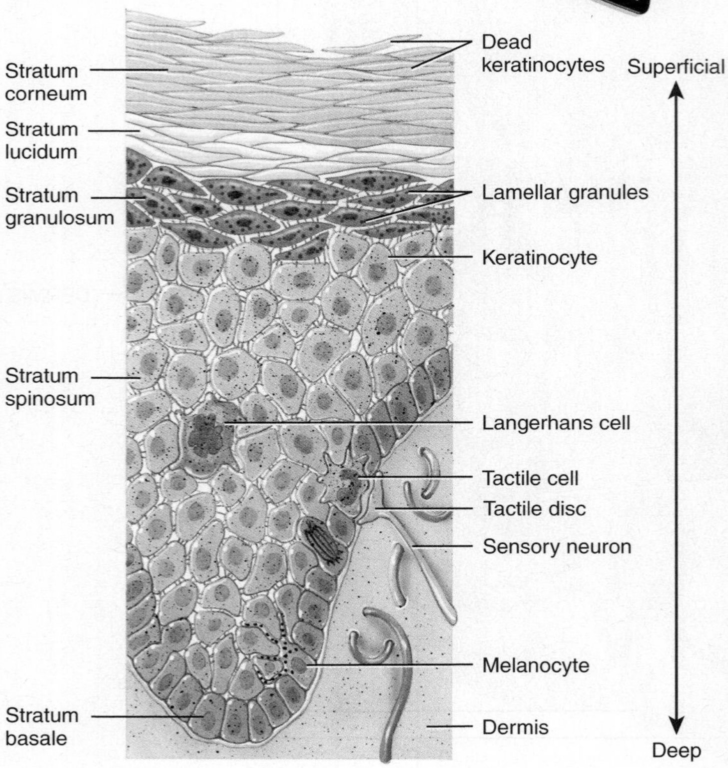

Four principal cell types in epidermis

Which epidermal layer includes stem cells that continually undergo cell division?

CLINICAL CONNECTION | Skin Grafts

New skin cannot regenerate if an injury destroys the stratum basale and its stem cells. Skin wounds of this magnitude require skin grafts in order to heal. A **skin graft** is the transfer of a patch of healthy skin taken from a donor site to cover a wound. To avoid tissue rejection, the transplanted skin is usually taken from the same individual (*autograft*) or an identical twin (*isograft*). If skin damage is so extensive that an autograft would cause harm, a self-donation procedure called *autologous skin transplantation* (aw-TOL-ō-gus) may be used. In this procedure, performed most often for severely burned patients, small amounts of an individual's epidermis are removed, and the keratinocytes are cultured in the laboratory to produce thin sheets of skin. The new skin is transplanted back to the patient so that it covers the burn wound and generates a permanent skin. Also available as skin grafts are products (Apligraft and Transite) grown in the laboratory from the foreskins of circumcised infants. •

Superficial to the stratum basale is the ***stratum spinosum*** (spi-NŌ-sum; *spinos-* = thornlike), where 8 to 10 layers of many-sided keratinocytes fit closely together. This layer provides strength and flexibility to the skin. Cells in the more superficial portions of this layer become somewhat flattened.

At about the middle of the epidermis, the ***stratum granulosum*** (gran-ū-LŌ-sum; *granulos-* = little grains) consists of three to five layers of flattened keratinocytes that are undergoing apoptosis, genetically programmed cell death in which the nucleus fragments before the cells die. The nuclei and other organelles of these cells begin to degenerate. A distinctive feature of cells in this layer is the presence of keratin. Also present in the keratinocytes are membrane-enclosed ***lamellar granules***, which release a lipid-rich secretion that acts as a water-repellent sealant, retarding loss of body fluids and entry of foreign materials.

The ***stratum lucidum*** (LOO-si-dum; *lucid-* = clear) is present only in the thick skin of areas such as the fingertips, palms, and soles. It consists of three to five layers of flattened clear, dead keratinocytes that contain large amounts of keratin.

The ***stratum corneum*** (COR-nē-um; *corne-* = horn or horny) consists of 25 to 30 layers of flattened dead keratinocytes. These cells are continuously shed and replaced by cells from the deeper strata. The interior of the cells contains mostly keratin. Its multiple layers of dead cells help to protect deeper layers from injury and microbial invasion. Constant exposure of skin to friction stimulates the formation of a *callus*, an abnormal thickening of the stratum corneum.

Newly formed cells in the stratum basale are slowly pushed to the surface. As the cells move from one epidermal layer to the next, they accumulate more and more keratin, a process called ***keratinization*** (ker'-a-tin-i-ZĀ-shun). Eventually the keratinized cells slough off and are replaced by underlying cells that, in turn, become keratinized. The whole process by which cells form in the stratum basale, rise to the surface, become keratinized, and slough off, takes about four weeks in an average epidermis of 0.1 mm (0.004 in.) thickness. An excessive amount of keratinized cells shed from the skin of the scalp is called *dandruff*.

Dermis

The second, deeper part of the skin, the dermis, is composed mainly of connective tissue containing collagen and elastic fibers. The superficial part of the dermis makes up about one-fifth of the thickness of the total layer (see Figure 5.1). It consists of areolar connective tissue containing fine elastic fibers. Its surface area is greatly increased by small, fingerlike projections called ***dermal papillae*** (pa-PIL-ē = nipples). These nipple-shaped structures project into the undersurface of the epidermis. Some contain *capillary loops* (blood capillaries). Other dermal papillae also contain tactile receptors called *corpuscles of touch* or *Meissner corpuscles*, nerve endings that are sensitive to touch. Also present in the dermal papillae are *free nerve endings* that are associated with sensations of warmth, coolness, pain, tickling, and itching.

The deeper part of the dermis, which is attached to the subcutaneous layer, consists of dense irregular connective tissue containing bundles of collagen and some coarse elastic fibers. Adipose cells, hair follicles, nerves, oil glands, and sweat glands are found between the fibers.

The combination of collagen and elastic fibers in the deeper part of the dermis provides the skin with strength, *extensibility* (ability to stretch), and *elasticity* (ability to return to original shape after stretching). The extensibility of skin can readily be seen in pregnancy and obesity. Extreme stretching, however, may produce small tears in the dermis, causing *striae* (STRĪ-ē = streaks), or stretch marks, that are visible as red or silvery white streaks on the skin surface.

Skin Color

Melanin, hemoglobin, and carotene are three pigments that impart a wide variety of colors to skin. The amount of ***melanin*** causes the skin's color to vary from pale yellow to reddish-brown to black. Melanocytes are most plentiful in the epidermis of the penis, nipples of the breasts, the area just around the nipples (areolae), face, and limbs. They are also present in mucous membranes. Because the *number* of melanocytes is about the same in all people, differences in skin color are due mainly to the *amount of pigment* the melanocytes produce and transfer to keratinocytes. In some people, melanin accumulates in patches called ***freckles***. As a person grows older, ***age (liver) spots*** may develop. These flat blemishes look like freckles and range in color from light brown to black. Like freckles, age spots are accumulations of melanin. A round, flat, or raised area that represents a benign localized overgrowth of melanocytes and usually develops in childhood or adolescence is called a ***nevus*** (NĒ-vus), or a ***mole***.

Exposure to UV light stimulates melanin production. Both the amount and darkness of melanin increase, which gives the skin a tanned appearance and further protects the body against UV radiation. Thus, within limits, melanin serves a protective function. Nevertheless, repeatedly exposing the skin to UV light causes skin cancer. A tan is lost when the melanin-containing keratinocytes are shed from the stratum corneum. ***Albinism*** (AL-bin-izm; *albin-* = white) is the inherited inability of an individual to produce melanin. Most ***albinos*** (al-BĪ-nōs), people affected by albinism, do not have melanin in their hair, eyes, and skin. In another condition, called ***vitiligo*** (vit-i-LĪ-gō), the partial or complete loss of melanocytes from patches of skin produces irregular white spots. The loss of melanocytes may be related to an immune system malfunction in which antibodies attack the melanocytes.

Dark-skinned individuals have large amounts of melanin in the epidermis. Consequently, the epidermis has a dark pigmentation and skin color ranges from yellow to red to tan to black. Light-skinned individuals have little melanin in the epidermis. Thus, the epidermis appears translucent and skin color ranges from pink to red depending on the amount and oxygen content of the blood moving through capillaries in the dermis. The red color is due to **hemoglobin**, the oxygen-carrying pigment in red blood cells.

Carotene (KAR-ō-tēn; *carot* = carrot) is a yellow-orange pigment that gives egg yolk and carrots their color. This precursor of vitamin A, which is used to synthesize pigments needed for vision, accumulates in the stratum corneum and fatty areas of the dermis and subcutaneous layer in response to excessive dietary intake. In fact, so much carotene may be deposited in the skin after eating large amounts of carotene-rich foods that the skin color actually turns orange, which is especially apparent in light-skinned individuals. Decreasing carotene intake eliminates the problem.

CLINICAL CONNECTION | Skin and Mucous Membrane Color as a Diagnostic Clue

The color of skin and mucous membranes can provide clues for diagnosing certain conditions. When blood is not picking up an adequate amount of oxygen from the lungs, as in someone who has stopped breathing, the mucous membranes, nail beds, and skin appear bluish or **cyanotic** (sī-a-NOT-ik; *cyan-* = blue). **Jaundice** (JON-dis; *jaund-* = yellow) is due to a buildup of the yellow pigment bilirubin in the skin. This condition gives a yellowish appearance to the skin and the whites of the eyes, and usually indicates liver disease. **Erythema** (er-i-THĒ-ma; *eryth-* = red), redness of the skin, is caused by engorgement of capillaries in the dermis with blood due to skin injury, exposure to heat, infection, inflammation, or allergic reactions. **Pallor** (PAL-or), or paleness of the skin, may occur in conditions such as shock and anemia. All skin color changes are observed most readily in people with lighter-colored skin and may be more difficult to discern in people with darker skin. However, examination of the nail beds and gums can provide some information about circulation in individuals with darker skin. •

Tattooing and Body Piercing

Tattooing is a permanent coloration of the skin in which a foreign pigment is deposited with a needle into the dermis. It is believed that the practice originated in ancient Egypt between 4000 and 2000 B.C. Today, tattooing is performed in one form or another by nearly all peoples of the world, and it is estimated that about one in five U.S. college students has one or more. Tattoos are created by injecting ink with a needle that punctures the epidermis, moves between 50 and 3000 times per minute, and deposits the ink in the dermis. Since the dermis is stable (unlike the epidermis, which is shed about every four weeks), tattoos are permanent. However, they can fade over time due to exposure to sunlight, improper healing, picking scabs, and flushing away of ink particles by the lymphatic system. Tattoos can be removed by lasers, which use concentrated beams of light. In the procedure, which requires a series of

treatments, the tattoo inks and pigments selectively absorb the high-intensity laser light without destroying normal surrounding skin tissue. The laser causes the tattoo to dissolve into small ink particles that are eventually removed by the immune system. Laser removal of tattoos involves a considerable investment in time and money and can be quite painful.

Body piercing, the insertion of jewelry through an artificial opening, is also an ancient practice employed by Egyptian pharaohs and Roman soldiers, and a current tradition among many Americans. Today it is estimated that about one in three U.S. college students has had a body piercing. For most piercing locations, the piercer cleans the skin with an antiseptic, retracts the skin with forceps, and pushes a needle through the skin. Then the jewelry is connected to the needle and pushed through the skin. Total healing can take up to a year. Among the sites that are pierced are the ears, nose, eyebrows, lips, tongue, nipples, navel, and genitals. Potential complications of body piercing are infections, allergic reactions, and anatomical damage (such as nerve damage or cartilage deformation). In addition, body piercing jewelry may interfere with certain medical procedures such as masks used for resuscitation, airway management procedures, urinary catheterization, radiographs, and delivery of a baby.

✓ CHECKPOINT

1. What structures are included in the integumentary system?
2. What are the main differences between the epidermis and dermis of the skin?
3. What are the three pigments found in the skin, and how do they contribute to skin color?
4. What is a tattoo? What are some potential problems associated with body piercing?

5.2 ACCESSORY STRUCTURES OF THE SKIN

OBJECTIVE • Describe the structure and functions of hair, skin glands, and nails.

Accessory structures of the skin that develop from the epidermis of an embryo—hair, glands, and nails—perform vital functions. For example, hair and nails protect the body and sweat glands help regulate body temperature.

Hair

Hairs, or *pili* (PI-lī), are present on most skin surfaces except the palms, palmar surfaces of the fingers, soles, and plantar surfaces of the toes. In adults, hair usually is most heavily distributed across the scalp, over the brows of the eyes, and around the external genitalia. Genetic and hormonal influences largely determine the thickness and pattern of distribution of hair. Hair on the head guards the scalp from injury and the sun's rays; eyebrows and eyelashes protect the eyes from foreign particles; and hair in the nostrils protects against inhaling insects and foreign particles.

Each hair is a thread of fused, dead, keratinized epidermal cells that consists of a shaft and a root (Figure 5.3). The **shaft** is the superficial portion that projects above the surface of the skin. The **root** is the portion below the surface that penetrates into the dermis and sometimes into the subcutaneous layer. Surrounding the root is the **hair follicle**, which is composed of two layers of epidermal cells, *external* and *internal root sheaths*, surrounded by a *connective tissue sheath*. Surrounding each hair follicle are nerve endings, called *hair root plexuses*, that are sensitive to touch. If a hair shaft is moved, its hair root plexus responds.

The base of each follicle is enlarged into an onion-shaped structure, the *bulb*. In the bulb is a nipple-shaped indentation, the *papilla of the hair*, that contains many blood vessels and provides nourishment for the growing hair. The bulb also contains a region of cells called the *hair matrix*, which produces new hairs by cell division when older hairs are shed.

Figure 5.3 Hair.

Hairs are growths of dead, keratinized epidermal cells.

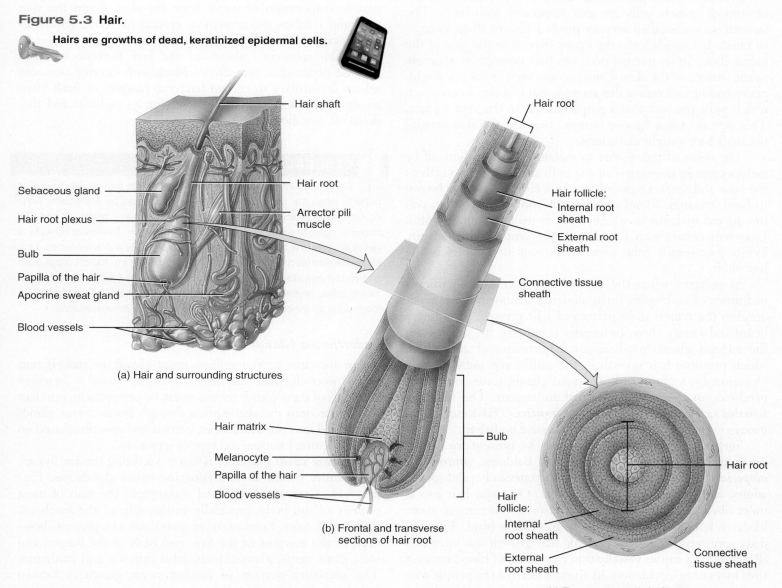

(a) Hair and surrounding structures

(b) Frontal and transverse sections of hair root

(c) Transverse section of hair root

Which part of a hair produces a new hair by cell division?

Chemotherapy is the treatment of disease, usually cancer, by means of chemical substances or drugs. Chemotherapeutic agents interrupt the life cycle of rapidly dividing cancer cells. Unfortunately, the drugs also affect other rapidly dividing cells in the body, such as the hair matrix cells. It is for this reason that individuals undergoing chemotherapy experience hair loss. Since about 15 percent of the hair matrix cells of scalp hairs are in the resting stage, these cells are not affected by chemotherapy. Once chemotherapy is stopped, the hair matrix cells replace lost hair follicles and hair growth resumes. •

Sebaceous (oil) glands (discussed shortly) and a bundle of smooth muscle cells are also associated with hairs. The smooth muscle is called **arrector pili** (a-REK-tor PI-lē; *arrect =* to raise). It extends from the upper dermis to the side of the hair follicle. In its normal position, hair emerges at an angle to the surface of the skin. Under stress, such as cold or fright, nerve endings stimulate the arrector pili muscles to contract, which pulls the hair shafts perpendicular to the skin surface. This action causes "goose bumps" because the skin around the shaft forms slight elevations.

The color of hair is due to melanin. It is synthesized by melanocytes in the matrix of the bulb and passes into cells of the root and shaft. Dark-colored hair contains mostly brown to black melanin. Blond and red hair contain variants of yellow to red melanin in which there is iron and more sulfur. Gray hair occurs with a decline in the synthesis of melanin. White hair results from accumulation of air bubbles in the hair shaft.

At puberty, when the testes begin secreting significant quantities of androgens (masculinizing sex hormones), males develop the typical male pattern of hair growth, including a beard and a hairy chest. In females at puberty, the ovaries and the adrenal glands produce small quantities of androgens, which promote hair growth in the axillae and pubic region. Occasionally, a tumor of the adrenal glands, testes, or ovaries produces an excessive amount of androgens. The result in females or prepubertal males is **hirsutism** (HER-soo-tizm; *hirsut- =* shaggy), a condition of excessive body hair.

Surprisingly, androgens also must be present for occurrence of the most common form of baldness, **androgenic alopecia** or **male-pattern baldness**. In genetically predisposed adults, androgens inhibit hair growth. On men, hair loss is most obvious at the temples and crown. Women are more likely to have thinning of hair on top of the head. The first drug approved for enhancing scalp hair growth was minoxidil (Rogaine®). It causes vasodilation (widening of blood vessels), thus increasing circulation. In about a third of the people who try it, minoxidil improves hair growth, causing scalp follicles to enlarge and lengthening the growth cycle. For many, however, the hair growth is meager. Minoxidil does not help people who already are bald.

Glands

Recall from Chapter 4 that glands are single or groups of epithelial cells that secrete a substance. The glands associated with the skin include sebaceous, sudoriferous, and ceruminous glands.

Sebaceous Glands

Sebaceous glands (se-BĀ-shus; *sebace- =* greasy) or **oil glands**, with few exceptions, are connected to hair follicles (Figure 5.3a). The secreting portions of the glands lie in the dermis and open into the hair follicles or directly onto a skin surface. There are no sebaceous glands in the palms and soles.

Sebaceous glands secrete an oily substance called **sebum** (SĒ-bum). Sebum keeps hair from drying out, prevents excessive evaporation of water from the skin, keeps the skin soft, and inhibits the growth of certain bacteria. Sebaceous gland activity increases during adolescence.

When sebaceous glands of the face become enlarged because of accumulated sebum, **blackheads** develop. Because sebum is nutritive to certain bacteria, **pimples** or **boils** often result. The color of blackheads is due to melanin and oxidized oil, not dirt.

Acne is an inflammation of sebaceous glands that usually begins at puberty, when the sebaceous glands are stimulated by androgens. Acne occurs predominantly in sebaceous follicles that have been colonized by bacteria, some of which thrive in the lipid-rich sebum. Treatment consists of gently washing the affected areas once or twice daily with a mild soap, topical antibiotics (such as clindamycin and erythromycin), topical drugs such as benzoyl peroxide or tretinoin, and oral antibiotics (such as tetracycline, minocycline, erythromycin, and isotretinoin). Contrary to popular belief, foods such as chocolate or fried foods do not cause or worsen acne. •

Sudoriferous Glands

There are three to four million **sweat glands**, or **sudoriferous glands** (soo'-dor-IF-er-us; *sudor- =* sweat; *-ferous =* bearing). The cells of these glands release sweat, or perspiration, into hair follicles or onto the skin surface through pores. Sweat glands are divided into two main types, eccrine and apocrine, based on their structure, location, and type of secretion.

Eccrine sweat glands (EK-rin = secreting outwardly) are much more common than apocrine sweat glands (see Figure 5.1). They are distributed throughout the skin of most regions of the body, especially in the skin of the forehead, palms, and soles. Eccrine sweat glands are not present, however, in the margins of the lips, nail beds of the fingers and toes, glans penis, glans clitoris, labia minora, and eardrums. The secretory portion of eccrine sweat glands is located mostly in the deep dermis (sometimes in the upper subcutaneous layer). The excretory duct projects through the dermis and epidermis and ends as a pore at the surface of the epidermis (see Figure 5.1).

The sweat produced by eccrine sweat glands (about 600 mL per day) consists of water, ions (mostly Na⁺ and Cl⁻), urea, uric acid, ammonia, amino acids, glucose, and lactic acid. The main function of eccrine sweat glands is to help regulate body temperature through evaporation. As sweat evaporates, large quantities of heat energy leave the body surface. Eccrine sweat glands also release sweat in response to an emotional stress such as fear or embarrassment. This type of sweating is referred to as **emotional sweating** or a *cold sweat*. In contrast to the regulation of body temperature through sweating, emotional sweating first occurs on the palms, soles, and axillae and then spreads to other areas of the body. As you will soon learn, apocrine sweat glands are also active during emotional sweating.

Apocrine sweat glands (AP-ō-krin; *apo-* = separated from) are also simple, coiled tubular glands (see Figure 5.1). They are found mainly in the skin of the axilla (armpit), groin, areolae (pigmented areas around the nipples) of the breasts, and bearded regions of the face in adult males. The secretory portion of these sweat glands is located mostly in the subcutaneous layer, and the excretory duct opens into hair follicles (see Figure 5.1).

Compared to eccrine sweat, apocrine sweat is slightly viscous and appears milky or yellowish in color. Apocrine sweat contains the same components as eccrine sweat plus lipids and proteins. Sweat secreted from apocrine sweat glands is odorless. However, when apocrine sweat interacts with bacteria on the surface of the skin, the bacteria metabolize its components, causing apocrine sweat to have a musky odor that is often referred to as *body odor*. Eccrine sweat glands start to function soon after birth, but apocrine sweat glands do not begin to function until puberty.

Apocrine sweat glands, along with eccrine sweat glands, are active during emotional sweating. In addition, apocrine sweat glands secrete sweat during sexual activities. In contract to eccrine sweat glands, apocrine sweat glands do not play a role in the regulation of body temperature.

Ceruminous Glands

Ceruminous glands (se-ROO-mi-nus; *cer-* = wax) are present in the external auditory canal (outer ear canal). The combined secretion of the ceruminous and sebaceous glands is a yellowish secretion called **cerumen** (se-ROO-men) or earwax. Cerumen, together with hairs in the external auditory canal, provides a sticky barrier that impedes the entrance of foreign bodies and insects. Cerumen also waterproofs the canal and prevents bacteria and fungi from entering cells.

Nails

Nails are plates of tightly packed, hard, dead, keratinized cells of the epidermis. Each nail (Figure 5.4) consists of a nail body, a free edge, and a nail root. The **nail body** is the portion of the nail that is visible; the **free edge** is the part of

Figure 5.4 Nails. Shown is a fingernail.

Nail cells arise by transformation of superficial cells of the nail matrix into nail cells.

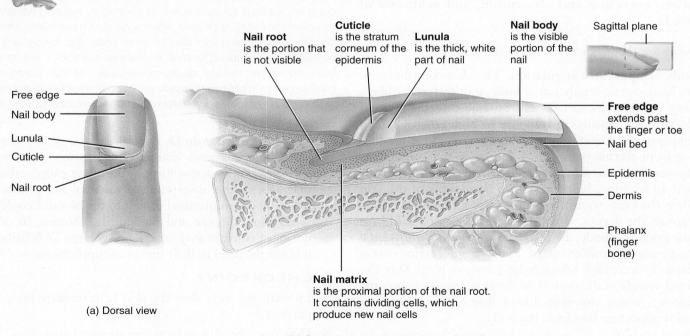

(a) Dorsal view

(b) Sagittal section showing internal detail

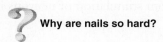 Why are nails so hard?

the body that extends past the end of the finger or toe; the **nail root** is the portion that is not visible. Most of the nail body is pink because of the underlying blood capillaries. The whitish semilunar area near the nail root is called the **lunula** (LOO-nyū-la = little moon). It appears whitish because the vascular tissue underneath does not show through due to the thickened stratum basale in the area. The proximal portion of the epithelium deep to the nail root is called the **nail matrix**. It is in this region that the superficial cells divide by mitosis to produce new nail cells. The average growth of fingernails is about 1 mm (0.04 inch) per week. The **cuticle** consists of stratum corneum.

Functionally, nails help us grasp and manipulate small objects, provide protection to the ends of the fingers and toes, and allow us to scratch various parts of the body.

✓ CHECKPOINT

5. Describe the structure of a hair. What causes "goose bumps"?

6. Contrast the locations and functions of sebaceous (oil) glands and sudoriferous (sweat) glands.

7. Describe the parts of a nail.

5.3 FUNCTIONS OF THE SKIN

OBJECTIVE • **Describe how the skin contributes to the regulation of body temperature, protection, sensation, excretion and absorption, and synthesis of vitamin D.**

Following are the major functions of the skin:

1. Body temperature regulation. The skin contributes to the homeostatic regulation of body temperature by liberating sweat at its surface and by adjusting the flow of blood in the dermis. In response to high environmental temperature or heat produced by exercise, sweat production from eccrine sweat glands increases; the evaporation of sweat from the skin surface helps lower body temperature. In addition, blood vessels in the dermis of the skin dilate (become wider); consequently, more blood flows through the dermis, which increases the amount of heat loss from the body. In response to low environmental temperature, production of sweat from eccrine sweat glands is decreased, which helps conserve heat. Also, the blood vessels in the dermis of the skin constrict (become narrow), which decreases blood flow through the skin and reduces heat loss from the body.

2. Protection. Keratin in the skin protects underlying tissues from microbes, abrasion, heat, and chemicals, and the tightly interlocked keratinocytes resist invasion by microbes. Lipids released by lamellar granules inhibit evaporation of water from the skin surface, thus protecting the body from dehydration. Oily sebum prevents hairs from drying out and contains bactericidal chemicals that kill surface bacteria. The acidic pH of perspiration retards the growth of some microbes. Melanin provides some protection against the damaging effects of UV light. Hair and nails also have protective functions.

3. Cutaneous sensations. *Cutaneous sensations* are those that arise in the skin. These include tactile sensations—touch, pressure, vibration, and tickling—as well as thermal sensations such as warmth and coolness. Another cutaneous sensation, pain, usually is an indication of impending or actual tissue damage. Chapter 12 provides more details on the topic of cutaneous sensations.

4. Excretion and absorption. The skin normally has a small role in *excretion*, the elimination of substances from the body, and *absorption*, the passage of materials from the external environment into body cells.

CLINICAL CONNECTION | **Transdermal Drug Administration**

Most drugs are either absorbed into the body through the digestive system or injected into subcutaneous tissue or muscle. An alternative route, **transdermal (transcutaneous) drug administration**, enables a drug contained within an adhesive skin patch to pass across the epidermis and into the blood vessels of the dermis. The drug is released continuously at a controlled rate over one to several days. A growing number of drugs are available for transdermal administration, including nitroglycerin, for prevention of angina pectoris, which is chest pain associated with heart disease (nitroglycerine can also be given under the tongue and intravenously); scopolamine, for motion sickness; estradiol, used for estrogen-replacement therapy during menopause; ethinyl estradiol and norelgestromin in contraceptive patches; nicotine, used to help people stop smoking; and fentanyl, used to relieve severe pain in cancer patients. •

5. Synthesis of vitamin D. Exposure of the skin to ultraviolet radiation activates vitamin D. Ultimately, vitamin D is converted to its active form, a hormone called calcitriol, that aids in the absorption of calcium and phosphorus from the gastrointestinal tract into the blood. People who avoid sun exposure and individuals who live in colder, northern climates may experience vitamin D deficiency if it is not included in their diet or as supplements.

✓ CHECKPOINT

8. In what two ways does the skin help regulate body temperature?

9. How does the skin serve as a protective barrier?

10. What sensations arise from stimulation of neurons in the skin?

Skin Care for Outdoor Exercisers

Physical activity is good for your skin. During exercise, the body shunts blood to the skin to help release excess heat produced by the contracting muscles. This increased blood flow provides the skin with nutrients and gets rid of wastes.

Fun in the Sun

From your skin's point of view, the main problem with outdoor exercise is that sun exposure over the years can lead to wrinkles, age spots, and cancers of the skin. To prevent these, do what you can to minimize sun exposure. The most effective skin protection is some form of sun block. Tightly woven clothing (hold it up to a light and see how much shines through) helps keep the sun's rays from reaching the skin, and wide-brimmed hats provide some protection. Zinc oxide blocks the sun and is good for noses and lips when long-term exposure is unavoidable.

When a sun block is not practical, a sunscreen should be used. These do not shield the skin completely, but they do reduce the damaging effects of the ultraviolet rays. Evidence suggests that the skin can repair some damage when sunscreens are consistently applied. But researchers warn that sunscreens can provide a false sense of security. Because they prevent burning, sunscreens may lull you into thinking the sun is not causing harm, while damage may still be occurring.

Unless you burn very easily, a little sun exposure is fine. UV rays stimulate precursor molecules in the skin to start the process of becoming vitamin D, an important vitamin that strengthens bones and the immune system.

Barriers to Skin Protection

Chemists have yet to invent a sunscreen that is fun to wear. Many exercisers can't take the grease, especially, as an avid bicyclist put it, "as it mingles with sweat and dead bugs." Advice for heavy sweaters is to exercise in the early or late part of the day, take as shady a route as possible, wear a hat and protective clothing, and use as much sunscreen as you can tolerate.

Swimmers should note that "waterproof" sunscreen stays on for only about 30 minutes in the water and should be reapplied after that time.

Dry Skin Care

Although not life threatening, dry skin can be very uncomfortable. Frequent showers and water exposure can strip the skin of its natural protective oils. The only solution is frequent moisturizing. Use of a good moisturizing cream immediately after drying off will counteract the drying effect of a "wash-and-wear" lifestyle.

Think It Over . . . **Imagine you have a friend who is training for a marathon and must exercise outdoors for an hour or more on most days. He has fair skin and a family history of skin cancer. What advice would you give him for minimizing sun exposure while continuing his training?**

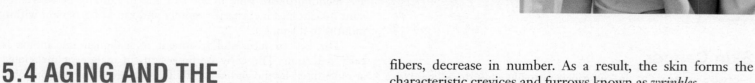

5.4 AGING AND THE INTEGUMENTARY SYSTEM

OBJECTIVE • **Describe the effects of aging on the integumentary system.**

The pronounced effects of skin aging do not become noticeable until people reach their late forties. Most of the age-related changes occur in the proteins in the dermis. Collagen fibers in the dermis begin to decrease in number, stiffen, break apart, and disorganize into a shapeless, matted tangle. Elastic fibers lose some of their elasticity, thicken into clumps, and fray, an effect that is greatly accelerated in the skin of smokers. Fibroblasts, which produce both collagen and elastic fibers, decrease in number. As a result, the skin forms the characteristic crevices and furrows known as *wrinkles*.

With further aging, Langerhans cells dwindle in number and macrophages become less-efficient phagocytes, thus decreasing the skin's immune responsiveness. Moreover, decreased size of sebaceous glands leads to dry and broken skin that is more susceptible to infection. Production of sweat diminishes, which probably contributes to the increased incidence of heat stroke in the elderly. There is a decrease in the number of functioning melanocytes, resulting in gray hair and atypical skin pigmentation. Hair loss increases with aging as hair follicles stop producing hairs. About 25 percent of males begin to show signs of hair loss by age 30 and about two-thirds have significant hair loss by age 60. Both males and females develop pattern baldness. An increase in the size of some melanocytes produces pigmented blotching (age spots). Walls of blood vessels in the dermis become thicker and less permeable, and subcutaneous adipose tissue is lost. Aged skin (especially the dermis) is thinner than young skin, and the migration of cells from the basal layer to the epidermal surface slows considerably. With the onset of old age, skin heals poorly and becomes more susceptible to pathological conditions such as skin cancer and pressure sores. *Rosacea* (ro-ZĀ-shē-a = rosy) is a skin condition that affects mostly light-skinned adults between the ages of 30 and 60. It is characterized by redness, tiny pimples, and noticeable blood vessels, usually in the central area of the face.

Growth of nails and hair slows during the second and third decades of life. The nails also may become more brittle with age, often due to dehydration or repeated use of cuticle remover or nail polish.

Several cosmetic anti-aging treatments are available to diminish the effects of aging or sun-damaged skin, including *topical products* that bleach the skin to tone down blotches and blemishes (hydroquinone) or decrease fine wrinkles and roughness (retinoic acid); *microdermabrasion* (mī-krō-DER-ma-brā'-zhun; *mikros-* = small; *derm* = skin; *-abrasio* = to wear away), the use of tiny crystals under pressure to remove and vacuum the skin's surface cells to improve skin texture and reduce blemishes; *chemical peel*, the application of a mild acid (such as glycolic acid) to the skin to remove surface cells to improve skin texture and reduce blemishes; *laser resurfacing*, the use of a laser to clear up blood vessels near the skin surface, even out blotches and blemishes, and decrease fine wrinkles; *dermal fillers*, injections of collagen from cows, hyaluronic acid, or calcium hydroxylapatite that plumps up the skin to smooth out wrinkles and fill in furrows, such as those around the nose and mouth and between the eyebrows; *fat transplantation*, in which fat from one part of the body is injected into another location such as around the eyes; *botulinum toxin* or *Botox*®, a diluted version of a toxin that is injected into the skin to paralyze skeletal muscles that cause the skin to wrinkle; *radio frequency nonsurgical facelift*, the use of radio frequency emissions to tighten the deeper layers of the skin to the jowls, neck, and sagging eyebrows and eyelids; and *facelift*, *browlift*, or *necklift*, invasive surgery in which loose skin and fat are removed surgically and the underlying connective tissue and muscle are tightened.

✓ CHECKPOINT

11. Which portion of the skin is involved in most age-related changes? Give several examples.

• • •

To appreciate the many ways that skin contributes to homeostasis of other body systems, examine Focus on Homeostasis: The Integumentary System. This focus box is the first of 10, found at the end of selected chapters, that explain how the body system under consideration contributes to homeostasis of all of the other body systems. The Focus on Homeostasis feature will help you understand how individual body systems interact to contribute to the homeostasis of the entire body. Next, in Chapter 6, we will explore how bone tissue is formed and how bones are assembled into the skeletal system, which protects many of our internal organs.

COMMON DISORDERS

Skin Cancer

Excessive exposure to the sun causes virtually all of the one million cases of *skin cancer* diagnosed annually in the United States. There are three common forms of skin cancer. *Basal cell carcinomas* account for about 78% of all skin cancers. The tumors arise from cells in the stratum basale of the epidermis and rarely metastasize. *Squamous cell carcinomas*, which account for about 20% of all skin cancers, arise from the stratum spinosum of the epidermis, and they have a variable tendency to metastasize. Basal and squamous cell carcinomas are together known as *nonmelanoma skin cancer.*

Malignant melanomas arise from melanocytes and account for about 2% of all skin cancers. They are the most prevalent life threatening cancer in young women. The estimated lifetime risk of developing melanoma is now 1 in 75, double the risk only 15 years ago. In part, this increase is due to depletion of the ozone layer, which absorbs some UV light high in the atmosphere. But the main reason for the increase is that more people are spending more time in the sun and in tanning beds. Malignant melanomas metastasize rapidly and can kill a person within months of diagnosis.

The key to successful treatment of malignant melanoma is early detection. The early warning signs of malignant melanoma are identified by the acronym ABCDE (Figure 5.5). *A* is for *asymmetry*; malignant melanomas tend to lack symmetry. *B* is for *border*; malignant melanomas have irregular—notched, indented, scalloped, or indistinct—borders. *C* is for *color*; malignant melanomas have uneven coloration and may contain several colors. *D* is for *diameter*; ordinary moles typically are smaller than 6 mm (0.25 in.), about the size of a pencil eraser. *E* is for *evolving*; malignant melanomas change in size, shape, and color. Once a malignant melanoma has the characteristics of A, B, and C, it is usually larger than 6 mm.

Among the risk factors for skin cancer are the following:

1. *Skin type.* Individuals with light-colored skin who never tan but always burn are at high risk.

2. *Sun exposure.* People who live in areas with many days of sunlight per year and at high altitudes (where ultraviolet light is more intense) have a higher risk of developing skin cancer. Likewise,

Focus on Homeostasis

BODY SYSTEM	CONTRIBUTION OF THE INTEGUMENTARY SYSTEM

THE INTEGUMENTARY SYSTEM

For all body systems
The skin and hair provide barriers that protect all internal organs from damaging agents in the external environment; sweat glands and skin blood vessels help regulate body temperature, needed for proper functioning of other body systems.

Skeletal system
The skin helps activate vitamin D, needed for proper absorption of dietary calcium and phosphorus to build and maintain bones.

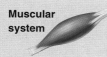
Muscular system
Through activation of vitamin D, the skin helps provide calcium ions, needed for muscle contraction; the skin also rids the body of heat produced by muscular activity.

Nervous system
Nerve endings in the skin and subcutaneous tissue provide input to the brain for touch, pressure, thermal, and pain sensations.

Endocrine system
Keratinocytes in the skin help activate vitamin D, initiating its conversion to calcitriol, a hormone that aids absorption of dietary calcium and phosphorus.

Cardiovascular system
Local chemical changes in the dermis cause widening and narrowing of skin blood vessels, which help adjust blood flow to the skin.

Lymphatic system and immunity
The skin is the "first line of defense" in immunity, providing mechanical barriers and chemical secretions that discourage penetration and growth of microbes; Langerhans cells in the epidermis participate in immune responses by recognizing foreign antigens for destruction by immune cells.

Respiratory system
Hairs in the nose filter dust particles from inhaled air; stimulation of pain nerve endings in the skin may alter breathing rate.

Digestive system
The skin helps activate vitamin D to become the hormone calcitriol, which promotes absorption of dietary calcium and phosphorus in the small intestine.

Urinary system
Kidney cells receive partially activated vitamin D hormone from the skin and convert it to calcitriol; some waste products are excreted from the body in sweat, contributing to excretion by the urinary system.

Reproductive systems
Nerve endings in the skin and subcutaneous tissue respond to erotic stimuli, thereby contributing to sexual pleasure; suckling of a baby stimulates nerve endings in the skin, leading to milk ejection; mammary glands (modified sweat glands) produce milk; the skin stretches during pregnancy as the fetus enlarges.

117

Figure 5.5 Comparison of a normal nevus (mole) and a malignant melanoma.

 Excessive exposure to the sun is the cause of most skin cancers.

 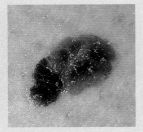

(a) Normal nevus (mole)   (b) Malignant melanoma

Which type of skin cancer is the most common type?

people who engage in outdoor occupations and those who have suffered three or more severe sunburns have a higher risk.

3. *Family history.* Skin cancer rates are higher in some families than in others.

4. *Age.* Older people are more prone to skin cancer owing to longer total exposure to sunlight.

5. *Immunological status.* Individuals who are immunosuppressed have a higher incidence of skin cancer.

Sun Damage

Although basking in the warmth of the sun may feel good, it's not a healthy practice. There are two forms of ultraviolet radiation that affect the health of the skin. Longer-wavelength ultraviolet A (UVA) rays make up nearly 95 percent of the ultraviolet radiation that reaches the earth. UVA rays are not absorbed by the ozone layer. They penetrate the furthest into the skin, where they are absorbed by melanocytes and thus are involved in sun tanning. UVA rays also depress the immune system. Shorter-wavelength ultraviolet B (UVB) rays are partially absorbed by the ozone layer and do not penetrate the skin as deeply as UVA rays. UVB rays cause sunburn and are responsible for most of the tissue damage (production of oxygen free radicals that disrupt collagen and elastic fibers) that results in wrinkling and aging of the skin and cataract formation. Both UVA and UVB rays are thought to cause skin cancer. Long-term overexposure to sunlight results in dilated blood vessels, age spots, freckles, and changes in skin texture.

Exposure to ultraviolet radiation (either natural sunlight or the artificial light of a tanning booth) may also produce *photosensitivity*, a heightened reaction of the skin after consumption of certain medications or contact with certain substances. Photosensitivity is characterized by redness, itching, blistering, peeling, hives, and even shock. Among the medications or substances that may cause a photosensitivity reaction are certain antibiotics (tetracycline), non-steroidal antiinflammatory drugs (ibuprofen or naproxen), certain herbal supplements (Saint-John's Wort), some birth control pills, some high blood pressure medications, some antihistamines, and certain artificial sweeteners, perfumes, aftershaves, lotions, detergents, and medicated cosmetics.

Self-tanning lotions (sunless tanners), topically applied substances, contain a color additive (dihydroxyacetone) that produces a tanned appearance by interacting with proteins in the skin.

Sunscreens are topically applied preparations that contain various chemical agents (such as benzophenone or one of its derivatives) that absorb UVB rays, but let most of the UVA rays pass through.

Sunblocks are topically applied preparations that contain substances such as zinc oxide that reflect and scatter both UVB and UVA rays.

Both sunscreens and sunblocks are graded according to a *sun protection factor (SPF)* rating, which measures the level of protection they supposedly provide against UV rays. The higher the rating, presumably the greater the degree of protection. As a precautionary measure, individuals who plan to spend a significant amount of time in the sun should use a sunscreen or a sunblock with an SPF of 15 or higher. Although sunscreens protect against sunburn, there is considerable debate as to whether they actually protect against skin cancer. In fact, some studies suggest that sunscreens increase the incidence of skin cancer because of the false sense of security they provide.

Burns

A *burn* is tissue damage caused by excessive heat, electricity, radioactivity, or corrosive chemicals that denature (destroy) the proteins in the skin cells. Burns destroy some of the skin's important contributions to homeostasis—protection against microbial invasion and dehydration, and regulation of body temperature.

Burns are graded according to their severity. A *first-degree burn* involves only the epidermis (Figure 5.6a). It is characterized by mild pain and *erythema* (redness) but no blisters. Skin functions remain intact. Immediate flushing with cold water may lessen the pain and damage caused by a first-degree burn. Generally, healing of a first-degree burn will occur in 3 to 6 days and may be accompanied by flaking or peeling. One example of a first-degree burn is mild sunburn.

A *second-degree burn* destroys the epidermis and part of the dermis (Figure 5.6b). Some skin functions are lost. In a second-degree burn, redness, blister formation, edema, and pain result. In a blister the epidermis separates from the dermis due to the accumulation of tissue fluid between them. Associated structures, such as hair follicles, sebaceous glands, and sweat glands, usually are not injured. If there is no infection, second-degree burns heal without skin grafting in about 3 to 4 weeks, but scarring may result. First- and second-degree burns are collectively referred to as *partial-thickness burns*.

A *third-degree burn* or *full-thickness burn* destroys the epidermis, dermis, and subcutaneous layer (Figure 5.6c). Most skin functions are lost. Such burns vary in appearance from marble-white to mahogany colored to charred, dry wounds. There is marked edema, and the burned region is numb because sensory nerve endings have been destroyed. Regeneration occurs slowly, and much granulation tissue forms before being covered by epithelium. Skin grafting may be required to promote healing and to minimize scarring.

The injury to the skin tissues directly in contact with the damaging agent is the *local effect* of a burn. Generally, however, the *systemic effects* of a major burn are a greater threat to life. The systemic effects of a burn may include (1) a large loss of water, plasma, and plasma proteins, which cause shock; (2) bacterial

Figure 5.6 Burns.

 A burn is tissue damage caused by agents that destroy the proteins in skin cells.

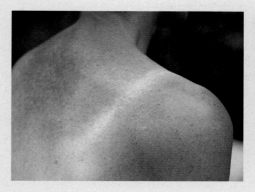 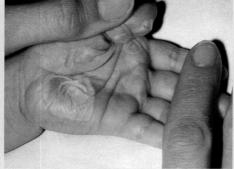 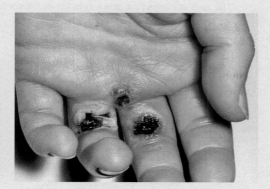

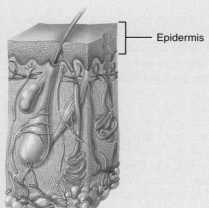 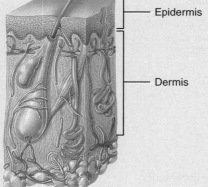 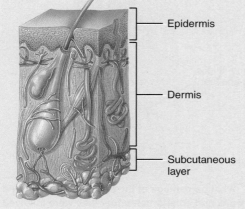

Epidermis

Epidermis

Dermis

Epidermis

Dermis

Subcutaneous layer

(a) First-degree burn (sunburn)

(b) Second-degree burn (note the blisters in the photograph above)

(c) Third-degree burn

- Mild pain
- Redness (no blisters)
- Skin functions normally
- Treatment: flush with cold water to relieve pain
- Heals within 3-6 days
- Example: sunburn

- Pain
- Redness
- Blisters (epidermis separates from underlying layers and fluid fills void)
- Edema
- Hair follicles and glands are not injured
- Some skin function is lost
- If there is no infection and no grafting is required, heals within 3–4 weeks

- Severe pain (burned region is numb due to nerve damage)
- Marked edema
- Marble-white to black color
- Most skin functions are lost
- Tissue damage
- Susceptible to infection
- Slow healing
- May require skin graft to promote healing and minimize scarring

 What factors determine the seriousness of a burn?

infection; (3) reduced circulation of blood; (4) decreased production of urine; and (5) diminished immune responses.

The seriousness of a burn is determined by its depth and extent of area involved, as well as the person's age and general health. According to the American Burn Association's classification of burn injury, a major burn includes third-degree burns over 10 percent of body surface area; or second-degree burns over 25 percent of body surface area; or any third-degree burns on the face, hands, feet, or *perineum* (per-i-NĒ-um), which includes the anal and urogenital regions. When the burn area exceeds 70 percent, more than half the

victims die. A quick means for estimating the surface area affected by a burn in an adult is the rule of nines (Figure 5.7):

1. Count 9 percent if both the anterior and posterior surfaces of the head and neck are affected.

2. Count 9 percent for both the anterior and posterior surfaces of each upper limb (total of 18 percent for both upper limbs).

3. Count 4 times 9 or 36 percent for both the anterior and posterior surfaces of the trunk, including the buttocks.

Figure 5.7 Rule-of-nines method for determining the extent of a burn. The percentages are the approximate proportions of the body surface area.

🔑 The rule of nines is a quick rule for estimating the surface area affected by a burn in an adult.

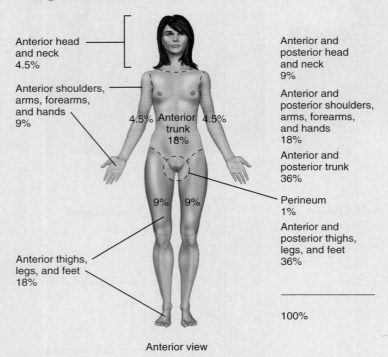

Anterior head and neck 4.5%

Anterior shoulders, arms, forearms, and hands 9%

Anterior and posterior head and neck 9%

Anterior and posterior shoulders, arms, forearms, and hands 18%

4.5% Anterior trunk 18% 4.5%

Anterior and posterior trunk 36%

Perineum 1%

Anterior and posterior thighs, legs, and feet 36%

9% 9%

Anterior thighs, legs, and feet 18%

100%

Anterior view

❓ What percentage of the body would be burned if only the anterior trunk and anterior left upper limb were involved?

4. Count 9 percent for the anterior and 9 percent for the posterior surfaces of each lower limb as far up as the buttocks (total of 36 percent for both lower limbs).

5. Count 1 percent for the perineum.

Many people who have been burned in fires also inhale smoke. If the smoke is unusually hot or dense or if inhalation is prolonged, serious problems can develop. The hot smoke can damage the trachea (windpipe), causing its lining to swell. As the swelling narrows the trachea, airflow into the lungs is obstructed. Further, small airways inside the lungs can also narrow, producing wheezing or shortness of breath. A person who has inhaled smoke is given oxygen through a face mask, and a tube may be inserted into the trachea to assist breathing.

Pressure Ulcers

Pressure ulcers, also known as *decubitus ulcers* (dē-KŪ-bi-tus) or *bedsores*, are caused by a constant deficiency of blood flow to tissues. Typically the affected tissue overlies a bony projection that has been subjected to prolonged pressure against an object such as a bed, cast, or splint. If the pressure is relieved in a few hours, redness occurs but no lasting tissue damage results. Blistering of the affected area may indicate superficial damage; a reddish-blue discoloration may indicate deep tissue damage. Prolonged pressure causes tissue ulceration. Small breaks in the epidermis become infected, and the sensitive subcutaneous layer and deeper tissues are damaged. Eventually, the tissue dies. Pressure ulcers occur most often in bedridden patients. With proper care, pressure ulcers are preventable, but they can develop very quickly in patients who are very old or very ill.

MEDICAL TERMINOLOGY AND CONDITIONS

Abrasion (a-BRĀ-shun; *ab-* = away; *-raison* = scraped) A portion of the epidermis that has been scraped away.

Athlete's foot (ATH-lēts) A superficial fungus infection of the skin of the foot.

Blister A collection of serous fluid within the epidermis or between the epidermis and dermis, due to short-term but severe friction.

Cold sore A lesion, usually in the oral mucous membrane, caused by type 1 herpes simplex virus (HSV) transmitted by oral or respiratory routes. The virus remains dormant until triggered by factors such as ultraviolet light, hormonal changes, and emotional stress. Also called a *fever blister*.

Contact dermatitis (der′-ma-TĪ-tis; *dermat-* = skin; *-itis* = inflammation) Inflammation of the skin characterized by redness, itching, and swelling and caused by exposure of the skin to chemicals that bring about an allergic reaction, such as poison ivy toxin.

Corn (KORN) A painful thickening of the stratum corneum of the epidermis found principally over toe joints and between the toes, often caused by friction or pressure. Corns may be hard or soft, depending on their location. Hard corns are usually found over toe joints, and soft corns are usually found between the fourth and fifth toes.

Frostbite Local destruction of skin and subcutaneous tissue on exposed surfaces as a result of extreme cold. In mild cases, the skin is blue and swollen and there is slight pain. In severe cases there is considerable swelling, some bleeding, no pain, and blistering. If untreated, gangrene may develop. Frostbite is treated by rapid rewarming.

Hemangioma (hē-man′-jē-Ō-ma; *hem-* = blood; *-angi* = blood vessel; *-oma* = tumor) Localized tumor of the skin and subcutaneous layer that results from an abnormal increase in blood vessels. One type is a *port-wine stain*, a flat pink, red, or purple lesion present at birth, usually at the nape of the neck.

Hives (HĪVZ) Skin condition marked by reddened elevated patches that are often itchy. Most commonly caused by infections, physical trauma, medications, emotional stress, food additives, and certain food allergies.

Impetigo (im′-pe-TĪ-gō) Superficial skin infection caused by *Staphylococcus* bacteria; most common in children.

Intradermal (in-tra-DER-mal; *intra-* = within) Within the skin. Also called *intracutaneous*.

Keloid (KĒ-loid; *kelis* = tumor) An elevated, irregular darkened area of excess scar tissue caused by collagen formation during healing. It extends beyond the original injury and is tender and frequently painful. It occurs in the dermis and underlying subcutaneous tissue, usually after trauma, surgery, a burn, or severe acne; more common in people of African descent.

Keratosis (ker′-a-TŌ-sis; *kera-* = horn) Formation of a hardened growth of epidermal tissue, such as a *solar keratosis*, a premalignant lesion of the sun-exposed skin of the face and hands.

Laceration (las-er-Ā-shun; *lacer-* = torn) An irregular tear of the skin.

Psoriasis (sō-RĪ-a-sis, *psora* = itch) A common, chronic skin disorder in which keratinocytes divide and move more quickly than normal from the stratum basale to the stratum corneum and form flaky scales, most often on the knees, elbows, and scalp.

Pruritus (proo-RĪ-tus; *pruri-* = to itch) Itching, one of the most common dermatological disorders. It may be caused by skin disorders (infections), systemic disorders (cancer, kidney failure), psychogenic factors (emotional stress), or allergic reactions.

Topical Refers to a medication applied to the skin surface rather than ingested or injected.

Wart Mass produced by uncontrolled growth of epithelial skin cells, caused by a papilloma virus. Most warts are noncancerous.

CHAPTER REVIEW AND RESOURCE SUMMARY

REVIEW	RESOURCES

5.1 Skin

1. The skin and hairs and other structures such as nails constitute the **integumentary system**.

2. The principal parts of the skin are the superficial **epidermis** and deeper **dermis**. The dermis overlies and attaches to the **subcutaneous (subQ) layer**.

3. Epidermal cells include **keratinocytes**, **melanocytes**, **Langerhans cells**, and **Tactile cells**. The epidermal layers, from deepest to most superficial, are the **stratum basale** (undergoes cell division and produces all other layers), **stratum spinosum** (provides strength and flexibility), **stratum granulosum** (contains keratin and lamellar granules), **stratum lucidum** (present only in palms and soles), and **stratum corneum** (sloughs off dead skin).

4. The dermis consists of two regions. The superficial region is areolar connective tissue containing blood vessels, nerves, hair follicles, **dermal papillae**, and corpuscles of touch (Meissner corpuscles). The deeper region is dense, irregularly arranged connective tissue containing adipose tissue, hair follicles, nerves, oil glands, and sweat glands.

5. Skin color is due to the pigments **melanin, carotene**, and **hemoglobin**.

6. In **tattooing**, a pigment is deposited with a needle in the dermis. **Body piercing** is the insertion of jewelry through an artificial opening.

Anatomy Overview—The Integumentary System
Anatomy Overview— The Skin

Figure 5.2 Layers of the epidermis

5.2 Accessory Structures of the Skin

1. **Accessory structures of the skin** develop from the epidermis of an embryo and include hair, skin glands (sebaceous, sudoriferous, and ceruminous), and nails.

2. **Hairs** are threads of fused, dead keratinized cells that function in protection. They consist of a **shaft** above the surface, a **root** that penetrates the dermis and subcutaneous layer, and a **hair follicle**.

3. Associated with hairs are bundles of smooth muscle called **arrector pili** and **sebaceous glands** or **oil glands**. Sebaceous glands are usually connected to hair follicles; they are absent in the palms and soles. Sebaceous glands produce **sebum**, which moistens hairs and waterproofs the skin.

4. There are two types of **sweat glands** or **sudoriferous glands**: eccrine and apocrine. **Eccrine sweat glands** have an extensive distribution; their ducts terminate at pores at the surface of the epidermis, and their main function is to help regulate body temperature. **Apocrine sweat glands** are limited in distribution, and their ducts open into hair follicles. They begin functioning at puberty and are stimulated during emotional stress and sexual excitement.

5. **Ceruminous glands** are modified sudoriferous glands that secrete **cerumen**. They are found in the external auditory canal.

6. **Nails** are hard, dead, keratinized epidermal cells covering the terminal portions of the fingers and toes. The principal parts of a nail are the **nail body, free edge, nail root, lunula, cuticle**, and **nail matrix**. Cell division of the matrix cells produces new nails.

Anatomy Overview—Hair
Anatomy Overview—Nails

Figure 5.3 Hair

REVIEW	RESOURCES

5.3 Functions of the Skin

1. Skin functions include body temperature regulation, protection, sensation, excretion and absorption, and synthesis of vitamin D.
2. The skin participates in body temperature regulation by liberating sweat at its surface and by adjusting the flow of blood in the dermis.
3. The skin provides physical, chemical, and biological barriers that help protect the body.
4. **Cutaneous sensations** include tactile sensations, thermal sensations, and pain.

Anatomy Overview—The Integument and Disease Resistance

5.4 Aging and the Integumentary System

1. Most effects of aging occur when an individual reaches the late forties.
2. Among the effects of aging are wrinkling, loss of subcutaneous fat, atrophy of sebaceous glands, and decreases in the number of melanocytes and Langerhans cells.

SELF-QUIZ

1. The hair structure that responds to touch is the
   a. shaft. b. root plexus. c. root sheath.
   d. bulb. e. follicle.

2. Skin coloration
   a. is due to melanin found in the subcutaneous layer.
   b. is related to the large differences in the number of melanocytes between individuals.
   c. is most often determined by the amount of carotene in the skin.
   d. changes when exposed to ultraviolet radiation of the sun.
   e. tends not to change with age.

3. In which portion of the skin will you find dermal papillae?
   a. superficial region of the dermis
   b. epidermis
   c. hypodermis
   d. stratum spinosum
   e. deeper region of the dermis

4. If you pricked your fingertip with a needle, the first layer of epidermis that it would penetrate is the
   a. stratum basale. b. stratum spinosum.
   c. stratum granulosum. d. stratum lucidum.
   e. stratum corneum.

5. Match the following
   ____ a. localized overgrowth of melanocytes   A. vitiligo
   ____ b. accumulation of melanin   B. albinism
   ____ c. inherited inability to produce melanin   C. nevus
   ____ d. shedding of excess keratinized cells   D. freckle
   ____ e. white patches from loss of melanocytes   E. dandruff

6. The red or pink tones seen in some skin are due to
   a. hemoglobin in the blood moving through capillaries in the dermis.
   b. the presence of carotene.
   c. the lack of oxygen.

   d. the destruction of melanocytes.
   e. an increased production of melanin.

7. When you have your hair cut, scissors are cutting through the hair
   a. follicle. b. root. c. shaft.
   d. papilla. e. bulb.

8. Which of the following is NOT true concerning eccrine sweat glands?
   a. They are most numerous on the palms and the soles.
   b. They help regulate body temperature.
   c. They produce a viscous secretion.
   d. They function throughout life.
   e. They terminate at pores on the skin's surface.

9. Which tissue is the main type found in the deeper region of the dermis?
   a. dense irregular connective
   b. stratified squamous epithelium
   c. smooth muscle
   d. nervous
   e. cartilage

10. Which of the following is NOT a function of skin?
   a. calcium production
   b. vitamin D synthesis
   c. protection
   d. immunity
   e. temperature regulation

11. Which of the following is NOT true concerning hair?
   a. Hair is mainly composed of keratin.
   b. Hirsutism is another name for male-pattern baldness.
   c. Hair color is due to melanin.
   d. Sebaceous glands are associated with hair.
   e. Contraction of the arrector pili muscles makes hair stand erect.

12. Sebaceous glands

    a. secrete an oily substance.

    b. are located on the palms and soles.

    c. are responsible for breaking out in a "cold sweat."

    d. are involved in body temperature regulation.

    e. are found in the external auditory meatus.

13. As keratinocytes in the stratum basale are pushed toward the skin's surface, they

    a. begin to divide more rapidly.

    b. become more elastic.

    c. begin to die.   **d.** lose their melanin.

    e. begin to assume a columnar shape.

14. To activate vitamin D, the skin cells need to be exposed to

    a. calcium and phosphorus.   **b.** ultraviolet light.

    c. heat.   **d.** pressure.   **e.** keratin.

15. To prevent an unwanted hair from growing back, you must destroy the

    a. shaft.   **b.** lunula.   **c.** root sheath.

    d. hair matrix.   **e.** papilla.

16. Aging can result in

    a. an increase in collagen and elastic fibers in the skin.

    b. a steady increase in the activity of sudoriferous glands.

    c. a greater immune response from Langerhans cells.

    d. more efficient activity by macrophages.

    e. a decline in the activity of sebaceous glands.

17. The portion of the nail that is responsible for nail growth is the

    a. cuticle.   **b.** nail matrix.   **c.** lunula.

    d. nail body.   **e.** nail root.

18. Match the following:

    _____ **a.** Langerhans cell

    _____ **b.** Merkel cell

    _____ **c.** keratin

    _____ **d.** melanin

    _____ **e.** lamellated (pacinian) corpuscle

    _____ **f.** cerumen

    _____ **g.** carotene

    _____ **h.** striae

    _____ **i.** corpuscle of touch (Meissner corpuscle)

    A. earwax

    B. silvery white streaks

    C. yellow-orange precursor of vitamin A

    D. function in immune responses

    E. protective protein of skin, hair

    F. touch receptor found in epidermis

    G. yellow to black pigment

    H. nerve endings sensitive to pressure

    I. touch receptor found in dermal papillae

19. Which of the following is NOT an accessory structure of the skin?

    a. dermal papillae   **b.** sudoriferous glands

    c. sebaceous glands   **d.** ceruminous glands

    e. nails

20. What is the response by effectors when the body temperature is elevated?

    a. Blood vessels in the dermis constrict.

    b. Sweat glands increase production of sweat.

    c. Skeletal muscles begin to contract involuntarily.

    d. The body's metabolic rate increases.

    e. The ceruminous glands increase production.

CRITICAL THINKING APPLICATIONS

1. Three-year-old Ewan was having his first haircut. As the barber started to snip his hair, Ewan cried, "Stop! You're killing it!" He then pulled his own hair, yelling, "Ouch! See! It's alive!" Is Ewan right about his hair?

2. Ewan's twin sister Ella scraped her knee at the playground. She told her mother that she wanted "new skin that doesn't leak." Her mother promised that new skin would soon appear under the bandage. How does new skin grow?

3. Georgia is seven months pregnant with her first child. She can't believe how big her "belly" has gotten, but is disturbed by the white streaks appearing on her abdomen. What region of the skin and what structures are responsible for the stretching to accommodate the pregnancy? What is causing the white streaks?

4. Fifteen-year-old Frank has a bad case of "blackheads." According to his mother, Frank's skin problems are from too much late-night TV, fast food, and sweets. Explain the real cause of blackheads to Frank's mother.

ANSWERS TO FIGURE QUESTIONS

5.1 The epidermis is made up of epithelial tissue, and the dermis is composed of connective tissue.

5.2 The stratum basale is the layer of the epidermis that contains stem cells that continually undergo cell division.

5.3 The hair matrix produces a new hair by cell division.

5.4 Nails are hard because they are composed of tightly packed, hard, dead keratinized epidermal cells.

5.5 Basal cell carcinoma is the most common type of skin cancer.

5.6 The seriousness of a burn is determined by the depth and extent of the area involved, and the individual's age and general health.

5.7 About 22.5 percent of the body would be involved (4.5 percent [arm] + 18 percent [anterior trunk]).

CHAPTER 6
THE SKELETAL SYSTEM

Despite its simple appearance, bone is a complex and dynamic living tissue that is remodeled continuously—new bone is built while old bone is broken down. Each individual bone is an organ composed of several different tissues working together: bone, cartilage, dense connective tissues, epithelium, blood-forming tissue, adipose tissue, and nervous tissue. The entire framework of bones and their cartilages constitute the *skeletal system*. The study of bone structure and the treatment of bone disorders is termed *osteology* (os′-tē-OL-ō-jē; *osteo-* = bone; *-logy* = study of).

Did you know?   *Over the years, bone strength may decline to a point where bones fracture easily. In many cases, this can lead to continuing disability and a reduced quality of life. Osteoporosis is diagnosed when bone mineral density reaches a critically low level. While many medications are available to help slow bone loss, all have side effects, and result in only minor gains in bone mineral density. Several lifestyle factors, including regular physical activity and good nutrition, can help slow the rate of bone loss in midlife, and possibly in old age as well.*

FOCUS ON WELLNESS
Making an Impact on Bone Strength

LOOKING BACK TO MOVE AHEAD...

Connective Tissue Extracellular Matrix (Section 4.3)

Cartilage (Section 4.3)

Bone Tissue (Section 4.3)

Collagen Fibers (Section 4.3)

Dense Irregular Connective Tissue (Section 4.3)

6.1 FUNCTIONS OF BONE AND THE SKELETAL SYSTEM

OBJECTIVE • **Discuss the six functions of bone and the skeletal system.**

Bone tissue and the skeletal system perform several basic functions:

1. **Support.** The skeleton provides a structural framework for the body by supporting soft tissues and providing points of attachment for the tendons of most skeletal muscles.

2. **Protection.** The skeleton protects many internal organs from injury. For example, cranial bones protect the brain, vertebrae (backbones) protect the spinal cord, and the rib cage protects the heart and lungs.

3. **Assistance in movement.** Because most skeletal muscles attach to bones, when muscles contract, they pull on bones. Together bones and muscles produce movement. This function is discussed in detail in Chapter 8.

4. **Mineral homeostasis.** Bone tissue stores several minerals, especially calcium and phosphorus. On demand, bone releases minerals into the blood to maintain critical mineral balances (homeostasis) and to distribute the minerals to other parts of the body.

5. **Blood cell production.** Within certain bones a connective tissue called *red bone marrow* produces red blood cells, white blood cells, and platelets, a process called *hemopoiesis* (hēm-ō-poy-Ē-sis; *hemo-* = blood; *poiesis* = making). Red bone marrow consists of developing blood cells, adipocytes, fibroblasts, and macrophages. It is present in developing bones of the fetus and in some adult bones, such as the pelvis, ribs, sternum (breastbone), vertebrae (backbones), skull, and ends of the arm bones and thigh bones.

6. **Triglyceride storage.** *Yellow bone marrow* consists mainly of adipose cells, which store triglycerides. The stored triglycerides are a potential chemical energy reserve. Yellow bone marrow also contains a few blood cells. In the newborn, all bone marrow is red and is involved in hemopoiesis. With increasing age, much of the bone marrow changes from red to yellow.

✓ **CHECKPOINT**

1. What types of tissues make up the skeletal system?
2. How do red and yellow bone marrow differ in composition, location, and function?

6.2 TYPES OF BONES

OBJECTIVE • **Classify bones on the basis of their shape and location.**

Almost all the bones of the body may be classified into four main types based on their shape: long, short, flat, or irregular. *Long bones* have greater length than width and consist of a shaft and a variable number of ends. They are usually somewhat curved for strength. Long bones include those in the thigh (femur), leg (tibia and fibula), arm (humerus), forearm (ulna and radius), and fingers and toes (phalanges).

Short bones are somewhat cube-shaped and nearly equal in length and width. Examples of short bones include most wrist and ankle bones.

Flat bones are generally thin, afford considerable protection, and provide extensive surfaces for muscle attachment. Bones classified as flat bones include the cranial bones, which protect the brain; the sternum (breastbone) and ribs, which protect organs in the thorax; and the scapulae (shoulder blades).

Irregular bones have complex shapes and cannot be grouped into any of the previous categories. Such bones include the vertebrae and some facial bones.

✓ **CHECKPOINT**

3. Give several examples of long, short, flat, and irregular bones.

6.3 STRUCTURE OF BONE

OBJECTIVES • **Describe the parts of a long bone.**
• **Describe the histological features of bone tissue.**

We will now explore the structure of bone at both the macroscopic and microscopic levels.

Macroscopic Structure of Bone

The structure of a bone may be analyzed by considering the parts of a long bone, for instance, the humerus (the arm bone), as shown in Figure 6.1. A typical long bone consists of the following seven parts:

1. The *diaphysis* (dī-AF-i-sis = growing between) is the bone's shaft or body—the long, cylindrical, main portion of the bone.

2. The *epiphyses* (e-PIF-i-sēz = growing over; singular is *epiphysis*) are the distal and proximal ends of the bone.

Figure 6.1 Parts of a long bone: epiphysis, metaphysis, and diaphysis. The spongy bone of the epiphysis and metaphyses contains red bone marrow, and the medullary cavity of the diaphysis contains yellow bone marrow in an adult.

🔑 A long bone is covered by articular cartilage at its proximal and distal epiphyses and by periosteum around the remainder of the bone.

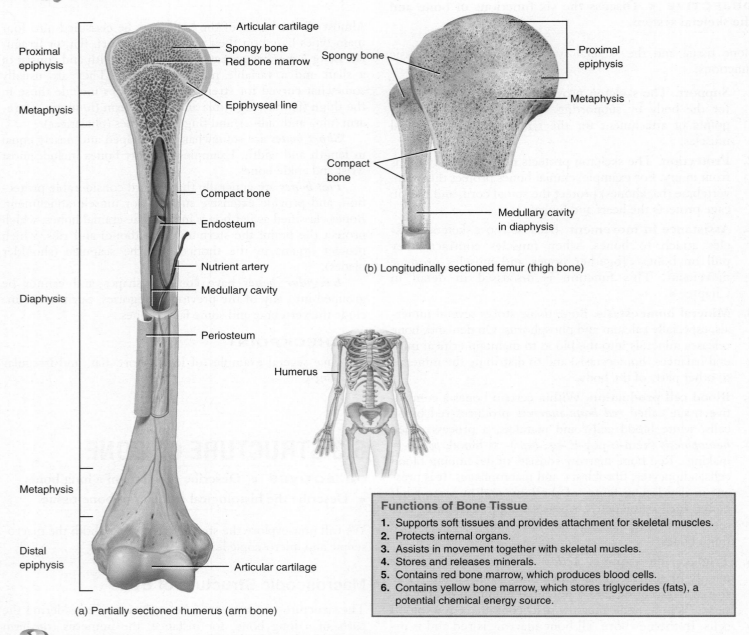

(a) Partially sectioned humerus (arm bone)

(b) Longitudinally sectioned femur (thigh bone)

Functions of Bone Tissue
1. Supports soft tissues and provides attachment for skeletal muscles.
2. Protects internal organs.
3. Assists in movement together with skeletal muscles.
4. Stores and releases minerals.
5. Contains red bone marrow, which produces blood cells.
6. Contains yellow bone marrow, which stores triglycerides (fats), a potential chemical energy source.

❓ Which part of a bone reduces friction at joints? Produces blood cells? Lines the medullary cavity?

3. The ***metaphyses*** (me-TAF-i-sēz; *meta-* = between; singular is *metaphysis*) are the regions in a *mature bone* where the diaphysis joins the epiphyses. In a *growing bone*, each metaphysis contains an *epiphyseal (growth) plate* (ep′-i-FIZ-ē-al), a layer of hyaline cartilage that allows the diaphysis of the bone to grow in length (described later in the chapter). When bone growth in length stops, the cartilage in the epiphyseal plate is replaced by bone and the resulting bony structure is known as the *epiphyseal line.*

4. The ***articular cartilage*** is a thin layer of hyaline cartilage covering the part of the epiphysis where the bone forms an articulation (joint) with another bone. Articular cartilage reduces friction and absorbs shock at freely movable

joints. Because articular cartilage lacks a perichondrium, repair of damage is limited.

5. The ***periosteum*** (per′-ē-OS-tē-um; *peri-* = around) is a tough sheath of dense irregular connective tissue and its associated blood vessels that surrounds the bone surface wherever it is not covered by articular cartilage. The periosteum contains bone-forming cells that enable bone to grow in diameter or thickness, but not in length. It also protects the bone, assists in fracture repair, helps nourish bone tissue, and serves as an attachment point for ligaments and tendons.

6. The ***medullary cavity*** (MED-ū-lar′-ē; *medulla-* = marrow, pith) or *marrow cavity* is a hollow, cylindrical space within the diaphysis that contains fatty yellow bone marrow in adults.

7. The ***endosteum*** (end-OS-tē-um; *endo-* = within) is a thin membrane that lines the medullary cavity. It contains a single layer of bone-forming cells.

Microscopic Structure of Bone

Like other connective tissues, ***bone***, or ***osseous tissue*** (OS-ē-us), contains abundant extracellular matrix that surrounds widely separated cells. The extracellular matrix is about 25 percent water, 25 percent collagen fibers, and 50 percent crystallized mineral salts. As these mineral salts are deposited in the framework formed by the collagen fibers of the extracellular matrix, they crystallize and the tissue hardens. This process of ***calcification*** is initiated by osteoblasts, the bone-building cells.

Although a bone's *hardness* depends on the crystallized inorganic mineral salts, a bone's *flexibility* depends on its collagen fibers. Like reinforcing metal rods in concrete, collagen fibers and other organic molecules provide *tensile strength*, which is resistance to being stretched or torn apart.

Four major types of cells are present in bone tissue: osteogenic cells, osteoblasts, osteocytes, and osteoclasts (Figure 6.2a).

1. ***Osteogenic cells*** (os-tē-ō-JEN-ik; *-genic* = producing) are unspecialized stem cells derived from ***mesenchyme***, the tissue from which almost all connective tissues are formed. They are the only bone cells to undergo cell division; the resulting cells develop into osteoblasts. Osteogenic cells are found along the inner portion of the periosteum, in the endosteum, and in the canals within bone that contain blood vessels.

2. ***Osteoblasts*** (OS-tē-ō-blasts′; *-blasts* = buds or sprouts) are bone-building cells. They synthesize and secrete collagen fibers and other organic components needed to build the extracellular matrix of bone tissue. As osteoblasts surround themselves with extracellular matrix, they become trapped in their secretions and become osteocytes. (Note: *Blasts* in bone or any other connective tissue secrete extracellular matrix.)

3. ***Osteocytes*** (OS-tē-ō-sīts′; *-cytes* = cells), mature bone cells, are the main cells in bone tissue and maintain its daily metabolism, such as the exchange of nutrients and wastes with the blood. Like osteoblasts, osteocytes do not undergo cell division. (Note: *Cytes* in bone or any other tissue maintain the tissue.)

4. ***Osteoclasts*** (OS-tē-ō-clasts′; *-clast* = break) are huge cells derived from the fusion of as many as 50 monocytes (a type of white blood cell) and are concentrated in the endosteum. They release powerful lysosomal enzymes and acids that digest the protein and mineral components of the bone extracellular matrix. This breakdown of bone extracellular matrix, termed *resorption*, is part of the normal development, growth, maintenance, and repair of bone. (Note: *Clasts* in bone break down extracellular matrix.)

Bone is not completely solid but has many small spaces between its cells and extracellular matrix components. Some spaces are channels for blood vessels that supply bone cells with nutrients. Other spaces are storage areas for red bone marrow. Depending on the size and distribution of the spaces, the regions of a bone may be categorized as compact or spongy (see Figure 6.1). Overall, about 80 percent of the skeleton is compact bone and 20 percent is spongy bone.

Compact Bone Tissue

Compact bone tissue contains few spaces and is arranged in repeating structural units called ***osteons*** or ***haversian systems*** (Figure 6.2c). Each osteon consists of a central (haversian) canal with its concentrically arranged lamellae. A ***central*** or ***haversian canal*** (ha-VER-shun) is a channel that contains blood vessels, nerves, and lymphatic vessels. The central canals run longitudinally through the bone. Around the canals are ***concentric lamellae*** (la-MEL-ē)—rings of hard, calcified extracellular matrix that resemble the growth rings of a tree. The tube-like osteons form a series of cylinders that run parallel to each other in long bones along the long axis of the bone. Between the lamellae are small spaces called ***lacunae*** (la-KOO-nē = little lakes; singular is *lacuna*), which contain osteocytes. Radiating in all directions from the lacunae are tiny ***canaliculi*** (kan′-a-LIK-ū-lī = small channels), which are filled with extracellular fluid. Inside the canaliculi are slender fingerlike processes of osteocytes (see inset at right of Figure 6.2c). The canaliculi connect lacunae with one another and with the central canals. Thus, an intricate, miniature canal system throughout the bone provides many routes for nutrients and oxygen to reach the osteocytes and for wastes to diffuse away. This is very important because diffusion through the lamellae is extremely slow.

Blood vessels, lymphatic vessels, and nerves from the periosteum penetrate the compact bone through transverse ***perforating (Volkmann) canals***. The vessels and nerves of the perforating canals connect with those of the medullary cavity, periosteum, and central (haversian) canals.

Figure 6.2 Histology of bone.

Osteocytes lie in lacunae arranged in concentric circles around a central (haversian) canal in compact bone, and in lacunae arranged irregularly in the trabeculae of spongy bone.

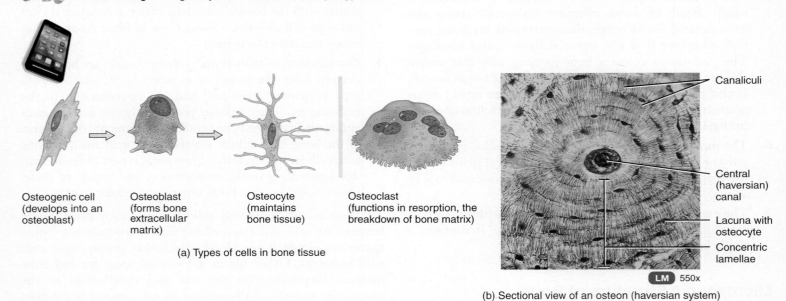

Osteogenic cell (develops into an osteoblast)

Osteoblast (forms bone extracellular matrix)

Osteocyte (maintains bone tissue)

Osteoclast (functions in resorption, the breakdown of bone matrix)

(a) Types of cells in bone tissue

Canaliculi

Central (haversian) canal

Lacuna with osteocyte

Concentric lamellae

LM 550x

(b) Sectional view of an osteon (haversian system)

Compact bone
Spongy bone
Periosteum
Medullary cavity

Medullary cavity

Trabeculae

Concentric lamellae
Blood vessels
Lymphatic vessel
Lacuna
Canaliculi
Osteon

Osteocyte

Periosteum

Central canal

Perforating canal

Spongy bone

Compact bone

(c) Osteons (haversian systems) in compact bone and trabeculae in spongy bone

 As people age, some central (haversian) canals may become blocked. What effect would this have on the osteocytes?

Compact bone tissue is the strongest type of bone tissue. It is found beneath the periosteum of all bones and makes up the bulk of the diaphyses of long bones. Compact bone tissue provides protection and support and resists the stresses produced by weight and movement.

Spongy Bone Tissue

In contrast to compact bone tissue, **spongy bone tissue** does not contain osteons. As shown in Figure 6.2c, it consists of units called *trabeculae* (tra-BEK-ū-lē = little beams; singular is *trabecula*), irregular latticeworks of thin columns of bone. The macroscopic spaces between the trabeculae of some bones are filled with red bone marrow. Within each trabecula are concentric lamellae, osteocytes that lie in lacunae, and canaliculi radiating from the lacunae.

Spongy bone tissue makes up most of the bone tissue of short, flat, and irregularly shaped bones. It also forms most of the epiphyses of long bones and a narrow rim around the medullary cavity of the diaphysis of long bones.

Spongy bone tissue is different from compact bone tissue in two respects. First, spongy bone tissue is light, which reduces the overall weight of a bone so that it moves more readily when pulled by a skeletal muscle. Second, the trabeculae of spongy bone tissue support and protect the red bone marrow. The spongy bone tissue in the hip bones, ribs, breastbone, backbones, and the ends of long bones is the only site where red bone marrow is found and, thus, the site of blood cell production in adults.

CLINICAL CONNECTION | Bone Scan

A **bone scan** is a diagnostic procedure that takes advantage of the fact that bone is living tissue. A small amount of a radioactive tracer compound that is readily absorbed by bone is injected intravenously. The degree of uptake of the tracer is related to the amount of blood flow to the bone. A scanning device (gamma camera) measures the radiation emitted from the bones, and the information is translated into a photograph that can be read like an x-ray on a monitor. Normal bone tissue is identified by a consistent gray color throughout because of its uniform uptake of the radioactive tracer. Darker or lighter areas may indicate bone abnormalities. Darker areas called "hot spots" are areas of increased metabolism that absorb more of the radioactive tracer due to increased blood flow. Hot spots may indicate bone cancer, abnormal healing of fractures, or abnormal bone growth. Lighter areas called "cold spots" are areas of decreased metabolism that absorb less of the radioactive tracer due to decreased blood flow. Cold spots may indicate problems such as degenerative bone disease, decalcified bone, fractures, bone infections, Paget's disease, and rheumatoid arthritis. A bone scan detects abnormalities 3 to 6 months sooner than standard x-ray procedures and exposes the patient to less radiation. A bone scan is the standard test for bone screening, particularly important in screening for osteoporosis in females. •

✓ CHECKPOINT

4. Diagram the parts of a long bone, and list the functions of each part.

5. What are the four types of cells in bone tissue?

6. How are spongy and compact bone tissue different in terms of their microscopic appearance, location, and function?

6.4 BONE FORMATION

OBJECTIVES • **Explain the importance of bone formation during different phases of a person's lifetime.**

• **Describe the factors that affect bone growth during a person's lifetime.**

The process by which bone forms is called *ossification* (os'-i-fi-KĀ-shun; *ossi-* = bone; *-fication* = making). Bone formation occurs in four principal situations: (1) the initial formation of bones in an embryo and fetus, (2) the growth of bones during infancy, childhood, and adolescence until their adult sizes are reached, (3) the remodeling of bone (replacement of old bone tissue by new bone tissue throughout life); and (4) the repair of fractures (breaks in bones) throughout life.

Initial Bone Formation in an Embryo and Fetus

We will first consider the initial formation of bone in an embryo and fetus. The embryonic "skeleton" is at first composed of mesenchyme shaped like bones; these are the sites where ossification occurs. These "bones" provide the template for subsequent ossification, which begins during the sixth week of embryonic development and follows one of two patterns.

The two patterns of bone formation, which both involve the replacement of a preexisting connective tissue with bone, do not lead to differences in the structure of mature bones, but are simply different methods of bone development. In the first type of ossification, called *intramembranous ossification* (in'-tra-MEM-bra-nus; *intra-* = within; *membrana-* = membrane), bone forms directly within mesenchyme arranged in sheetlike layers that resemble membranes. In the second type, *endochondral ossification* (en'-dō-KON-dral; *endo-* = within; *-chondral* = cartilage), bone forms within hyaline cartilage that develops from mesenchyme.

Intramembranous Ossification

Intramembranous ossification is the simpler of the two methods of bone formation. The flat bones of the skull, most of the facial bones, mandible (lower jawbone), and part of the clavicle (collar bone) are formed in this way. Also, the "soft spots" that help the fetal skull pass through the birth canal later harden as they undergo intramembranous ossification, which occurs as follows (Figure 6.3):

① **Development of the ossification center.** At the site where bone will develop, called the *ossification center*, cells of the mesenchyme cluster together and differentiate, first

Figure 6.3 Intramembranous ossification. Illustrations ① and ② show a smaller field of vision at higher magnification than illustrations ③ and ④.

Intramembranous ossification involves the formation of bone within mesenchyme arranged in sheetlike layers that resemble membranes.

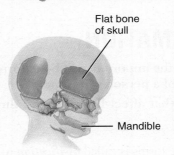

Flat bone of skull

Mandible

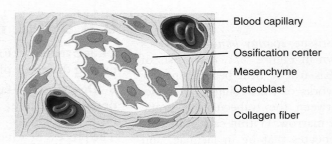

Blood capillary

Ossification center

Mesenchyme

Osteoblast

Collagen fiber

① Development of ossification center: osteoblasts secrete organic extracellular matrix

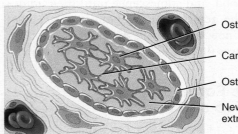

Osteocyte in lacuna

Canaliculus

Osteoblast

Newly calcified bone extracellular matrix

② Calcification: calcium and other mineral salts are deposited and extracellular matrix calcifies (hardens)

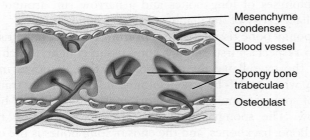

Mesenchyme condenses

Blood vessel

Spongy bone trabeculae

Osteoblast

③ Formation of trabeculae: extracellular matrix develops into trabeculae that fuse to form spongy bone

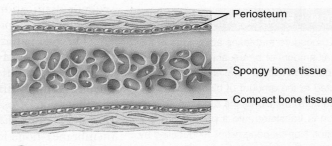

Periosteum

Spongy bone tissue

Compact bone tissue

④ Development of the periosteum: mesenchyme at the periphery of the bone develops into the periosteum

Which bones of the body develop by intramembranous ossification?

into osteogenic cells and then into osteoblasts. Osteoblasts secrete the organic extracellular matrix of bone.

② **Calcification.** Next, the secretion of extracellular matrix stops and the cells, now called osteocytes, lie in lacunae and extend their narrow cytoplasmic processes into canaliculi that radiate in all directions. Within a few days, calcium and other mineral salts are deposited and the extracellular matrix hardens or calcifies (calcification).

③ **Formation of trabeculae.** As the bone extracellular matrix forms, it develops into trabeculae that fuse with one another to form spongy bone. Blood vessels grow into the spaces between the trabeculae. Connective tissue that is associated with the blood vessels in the trabeculae differentiates into red bone marrow.

④ **Development of the periosteum.** In conjunction with the formation of trabeculae, mesenchyme condenses at the periphery and develops into the periosteum. Eventually, a thin layer of compact bone replaces the surface layers of the spongy bone, but spongy bone remains in the center.

Endochondral Ossification

The replacement of cartilage by bone is called *endochondral ossification*. Most bones of the body are formed in this way, but as shown in Figure 6.4, this type of ossification is best observed in a long bone and proceeds as follows:

① **Development of the cartilage model.** At the site where the bone is going to form, the cells in mesenchyme crowd together in the shape of the future bone and then develop into chondroblasts. The chondroblasts secrete cartilage extracellular matrix, producing a *cartilage model*

Figure 6.4 Endochondral ossification.

During endochondral ossification, bone gradually replaces a cartilage model.

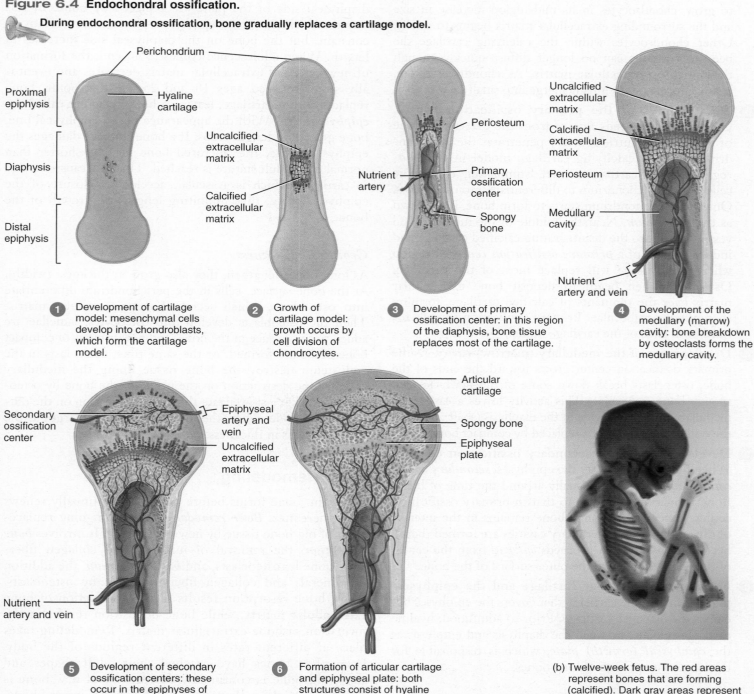

① Development of cartilage model: mesenchymal cells develop into chondroblasts, which form the cartilage model.

② Growth of cartilage model: growth occurs by cell division of chondrocytes.

③ Development of primary ossification center: in this region of the diaphysis, bone tissue replaces most of the cartilage.

④ Development of the medullary (marrow) cavity: bone breakdown by osteoclasts forms the medullary cavity.

⑤ Development of secondary ossification centers: these occur in the epiphyses of the bone.

⑥ Formation of articular cartilage and epiphyseal plate: both structures consist of hyaline cartilage.

(b) Twelve-week fetus. The red areas represent bones that are forming (calcified). Dark gray areas represent cartilage (uncalcified).

(a) Sequence of events

 Which structure signals that bone growth in length has stopped?

consisting of hyaline cartilage. A membrane called the *perichondrium* (per'-i-KON-drē-um) develops around the cartilage model.

2 **Growth of the cartilage model.** Once chondroblasts become deeply buried in cartilage extracellular matrix, they are called chondrocytes. As the cartilage model continues to grow, chondrocytes in its mid-region increase in size and the surrounding extracellular matrix begins to calcify. Other chondrocytes within the calcifying cartilage die because nutrients can no longer diffuse quickly enough through the extracellular matrix. As chondrocytes die, lacunae form and eventually merge into small cavities.

3 **Development of the primary ossification center.** Ossification proceeds *inward* from the external surface of the bone. A nutrient artery penetrates the perichondrium and the calcifying cartilage model in the mid-region of the cartilage model, stimulating osteogenic cells in the perichondrium to differentiate into osteoblasts. Once the perichondrium starts to form bone, it is known as the *periosteum*. Near the middle of the model, blood vessels grow into the disintegrating calcified cartilage and induce growth of a *primary ossification center*, a region where bone tissue will replace most of the cartilage. Osteoblasts then begin to deposit bone extracellular matrix over the remnants of calcified cartilage, forming spongy bone trabeculae. Primary ossification spreads toward both ends of the cartilage model.

4 **Development of the medullary (marrow) cavity.** As the primary ossification center grows toward the ends of the bone, osteoclasts break down some of the newly formed spongy bone trabeculae. This activity leaves a cavity, the medullary (marrow) cavity, in the diaphysis (shaft). Most of the wall of the diaphysis is replaced by compact bone.

5 **Development of the secondary ossification centers.** When blood vessels enter the epiphyses, *secondary ossification centers* develop, usually around the time of birth. Bone formation is similar to that in primary ossification centers except that spongy bone remains in the interior of the epiphyses (no medullary cavities are formed there). Secondary ossification proceeds *outward* from the center of the epiphysis toward the outer surface of the bone.

6 **Formation of articular cartilage and the epiphyseal plate.** The hyaline cartilage that covers the epiphyses becomes the articular cartilage. Prior to adulthood, hyaline cartilage remains between the diaphysis and epiphysis as the *epiphyseal (growth) plate*, which is responsible for the lengthwise growth of long bones.

Bone Growth in Length and Thickness

During infancy, childood, and adolescence, long bones grow in length and thickness.

Growth in Length

Bone growth in length is related to the activity of the epiphyseal plate. Within the epiphyseal plate is a group of young chondrocytes that are constantly dividing. As a bone grows in length, new chondrocytes are formed on the epiphyseal side of the plate, while old chondrocytes on the diaphyseal side of the plate are replaced by bone. In this way the thickness of the epiphyseal plate remains relatively constant, but the bone on the diaphyseal side increases in length. When adolescence comes to an end, the formation of new cells and extracellular matrix decreases and eventually stops between ages 18 and 25. At this point, bone replaces all the cartilage, leaving a bony structure called the *epiphyseal line*. With the appearance of the epiphyseal line, bone growth in length stops. If a bone fracture damages the epiphyseal plate, the fractured bone may be shorter than normal once adult stature is reached. This is because damage to cartilage, which is avascular, accelerates closure of the epiphyseal plate, thus inhibiting lengthwise growth of the bone.

Growth in Thickness

As long bones lengthen, they also grow in thickness (width). At the bone surface, cells in the perichondrium differentiate into osteoblasts, which secrete bone extracellular matrix. Then the osteoblasts develop into osteocytes, lamellae are added to the surface of the bone, and new osteons of compact bone tissue are formed. At the same time, osteoclasts in the endosteum destroy the bone tissue lining the medullary cavity. Bone destruction on the inside of the bone by osteoclasts occurs at a slower rate than bone formation on the outside of the bone. Thus, the medullary cavity enlarges as the bone increases in thickness.

Bone Remodeling

Like skin, bone forms before birth but continually renews itself thereafter. *Bone remodeling* is the ongoing replacement of old bone tissue by new bone tissue. It involves *bone resorption*, the removal of minerals and collagen fibers from bone by osteoclasts, and *bone deposition*, the addition of minerals and collagen fibers to bone by osteoblasts. Thus, bone resorption results in the destruction of bone extracellular matrix, while bone deposition results in the formation of bone extracellular matrix. Remodeling takes place at different rates in different regions of the body. Even after bones have reached their adult shapes and sizes, old bone is continually destroyed and new bone is formed in its place. Remodeling also removes injured bone, replacing it with new bone tissue. Remodeling may be triggered by factors such as exercise, lifestyle, and changes in diet.

Orthodontics (or-thō-DON-tiks) is the branch of dentistry concerned with the prevention and correction of poorly aligned teeth. The movement of teeth by braces places a stress on the bone that forms the sockets that anchor the teeth. In response to this artificial stress, osteoclasts and osteblasts remodel the sockets so that the teeth align properly. •

A delicate balance exists between the actions of osteoclasts and osteoblasts. Should too much new tissue be formed, the bones become abnormally thick and heavy. If too much mineral material is deposited in the bone, the surplus may form thick bumps, called *spurs*, on the bone that interfere with movement at joints. Excessive loss of calcium or tissue weakens the bones, and they may break, as occurs in osteoporosis, or they may become too flexible, as in rickets and osteomalacia. (For more on these disorders, see the Common Disorders section at the end of the chapter.) Abnormal acceleration of the remodeling process results in a condition called Paget's disease, in which the newly formed bone, especially that of the pelvis, limbs, lower vertebrae, and skull, becomes hard and brittle and fractures easily.

Fractures

A *fracture* (FRAK-choor) is any break in a bone. Types of fractures include the following:

- **Partial:** an incomplete break across the bone, such as a crack.

- **Complete:** a complete break across the bone; that is, the bone is broken into two or more pieces.

- **Closed (simple):** the fractured bone does not break through the skin.

- **Open (compound):** the broken ends of the bone protrude through the skin.

Repair of a fracture involves several steps. First, phagocytes begin to remove any dead bone tissue. Then, chondroblasts form fibrocartilage at the fracture site that bridges the broken ends of the bone. Next, the fibrocartilage is converted to spongy bone tissue by osteoblasts. Finally, bone remodeling occurs, in which dead portions of bone are absorbed by osteoclasts and spongy bone is converted to compact bone.

Although bone has a generous blood supply, healing sometimes takes months. The calcium and phosphorus needed to strengthen and harden new bone are deposited only gradually, and bone cells generally grow and reproduce slowly. The temporary disruption in their blood supply also helps explain the slowness of healing of severely fractured bones.

Factors Affecting Bone Growth and Remodeling

Bone growth in the young, bone remodeling in the adult, and the repair of fractured bone depend on several factors (see Table 6.1). These include (1) adequate minerals, most importantly calcium, phosphorus, and magnesium; (2) vitamins A, C, and D; (3) several hormones; and (4) weight-bearing exercise (exercise that places stress on bones). Before puberty, the main hormones that stimulate bone growth are human growth hormone (hGH), which is produced by the anterior lobe of the pituitary gland, and insulinlike growth factors (IGFs), which are produced locally by bone and also by the liver in response to hGH stimulation. Oversecretion of hGH produces *giantism*, in which a person becomes much taller and heavier than normal, and undersecretion of hGH produces *dwarfism* (short stature). Thyroid hormones (from the thyroid gland) and insulin (from the pancreas) also stimulate normal bone growth. At puberty, *estrogens* (sex hormones produced by the ovaries) and *androgens* (sex hormones produced by the testes in males and the adrenal glands in both sexes) start to be released in larger quantities. These hormones are responsible for the sudden growth spurt that occurs during the teenage years. Estrogens also promote changes in the skeleton that are typical of females, such as widening of the pelvis.

Bone's Role in Calcium Homeostasis

Bone is the major reservoir of calcium, storing 99 percent of the total amount of calcium present in the body. Calcium (Ca^{2+}) becomes available to other tissues when bone is broken down during remodeling. However, even small changes in blood calcium levels can be deadly—the heart may stop (cardiac arrest) if the level is too high or breathing may cease (respiratory arrest) if the level is too low. In addition, most functions of nerve cells depend on just the right level of Ca^{2+}, many enzymes require Ca^{2+} as a cofactor, and blood clotting requires Ca^{2+}. The role of bone in calcium homeostasis is to "buffer" the blood calcium level, releasing Ca^{2+} to the blood when the blood calcium level falls (using osteoclasts) and depositing Ca^{2+} back in bone when the blood level rises (using osteoblasts).

The most important hormone that regulates Ca^{2+} exchange between bone and blood is **parathyroid hormone (PTH)**, secreted by the parathyroid glands (see Figure 13.10). PTH secretion operates via a negative feedback system (Figure 6.5). If some stimulus causes blood Ca^{2+} level to decrease, parathyroid gland cells (receptors) detect this change and increase their production of a molecule known as cyclic adenosine monophosphate (cyclic AMP). The gene for PTH within the nucleus of a parathyroid gland cell, which acts as the control center, detects the increased production of cyclic AMP (the input). As a result, PTH synthesis speeds up,

Figure 6.5 Negative feedback system for the regulation of blood calcium (Ca²⁺) level.

Release of calcium from bone extracellular matrix and retention of calcium by the kidneys are the two main ways that blood calcium level can be increased.

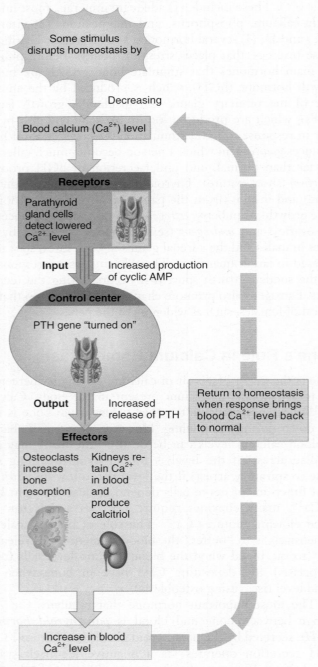

What body functions depend on proper levels of Ca²⁺?

and more PTH (the output) is released into the blood. The presence of higher levels of PTH increases the number and activity of osteoclasts (effectors), which step up the pace of bone resorption. The resulting release of Ca²⁺ from bone into blood returns the blood Ca²⁺ level to normal.

PTH also decreases loss of Ca²⁺ in the urine, so more is retained in the blood, and it stimulates formation of calcitriol, a hormone that promotes absorption of calcium from the gastrointestinal tract. Both of these effects also help elevate the blood Ca²⁺ level.

As you will learn in Chapter 13, another hormone involved in calcium homeostasis is *calcitonin (CT)* (kal-si-TŌ-nin). This hormone is produced by the thyroid gland and decreases blood Ca²⁺ level by inhibiting the action of osteoclasts, thus decreasing bone resorption.

✓ CHECKPOINT

7. Distinguish between intramembranous and endochondral ossification.
8. Explain how bones grow in length and thickness.
9. What is bone remodeling? Why is it important?
10. Define a fracture and explain how fracture repair occurs.
11. What factors affect bone growth?
12. What are some of the important functions of calcium in the body?

6.5 EXERCISE AND BONE TISSUE

OBJECTIVE • Describe how exercise and mechanical stress affect bone tissue.

Within limits, bone tissue has the ability to alter its strength in response to mechanical stress. When placed under stress, bone tissue becomes stronger through increased deposition of mineral salts and production of collagen fibers. Without mechanical stress, bone does not remodel normally because resorption outpaces bone formation. The absence of mechanical stress weakens bone through decreased numbers of collagen fibers and ***demineralization*** (dē-mi′-ne-ra-li-ZĀ-shun), loss of bone minerals.

The main mechanical stresses on bone are those that result from the pull of skeletal muscles and the pull of gravity. If a person is bedridden or has a fractured bone in a cast, the strength of the unstressed bones diminishes. Astronauts subjected to the weightlessness of space also lose bone mass. In both cases, the bone loss can be dramatic, as much as 1 percent per week. Bones of athletes, which are repetitively and highly stressed, become notably thicker than those of nonathletes. Weight-bearing activities, such as walking or moderate weightlifting, help build and retain bone mass. Adolescents and young adults should engage in regular weight-bearing exercise prior to the closure of the epiphyseal plates to help build total mass before its inevitable reduction with aging. However, the benefits of exercise do not end in young adulthood. Even elderly people can strengthen their bones by engaging in weight-bearing exercise.

Table 6.1 summarizes all of the factors that influence bone metabolism: growth, remodeling, and repair of fractured bones.

TABLE 6.1

Summary of Factors That Influence Bone Metabolism

FACTOR	COMMENT
Minerals	
Calcium and phosphorus	Make bone extracellular matrix hard
Magnesium	Helps form bone extracellular matrix
Fluoride	Helps strengthen bone extracellular matrix
Manganese	Activates enzymes involved in synthesis of bone extracellular matrix
Vitamins	
Vitamin A	Needed for activity of osteoblasts during remodeling of bone; deficiency stunts bone growth; toxic in high doses
Vitamin C	Needed for synthesis of collagen, the main bone protein; deficiency leads to decreased collagen production, which slows down bone growth and delays repair of broken bones
Vitamin D	Active form (calcitriol) is produced by kidneys; helps build bone by increasing absorption of calcium from gastrointestinal tract into blood; deficiency causes faulty calcification and slows down bone growth; may reduce risk of osteoporosis but is toxic if taken in high doses. People who use sunscreens, have minimal exposure to ultraviolet rays, or do not take vitamin D supplements may not have sufficient vitamin D to absorb calcium. This interferes with calcium metabolism.
Vitamins K and B_{12}	Needed for synthesis of bone proteins; deficiency leads to abnormal protein production in bone extracellular matrix and decreased bone density
Hormones	
Human growth hormone (hGH)	Secreted by anterior lobe of pituitary gland; promotes general growth of all body tissues, including bone, mainly by stimulating production of insulinlike growth factors
Insulinlike growth factors (lGFs)	Secreted by liver, bones, and other tissues upon stimulation by human growth hormone; promotes normal bone growth by stimulating osteoblasts and by increasing synthesis of proteins needed to build new bone
Thyroid hormones (thyroxine and triiodothyronine)	Secreted by thyroid gland; promote normal bone growth by stimulating osteoblasts
Insulin	Secreted by pancreas; promotes normal bone growth by increasing synthesis of bone proteins
Sex hormones (estrogens and testosterone)	Secreted by ovaries in women (estrogens) and by testes in men (testosterone); stimulate osteoblasts and promote the sudden "growth spurt" that occurs during the teenage years; shut down growth at epiphyseal plates around age 18–21, causing lengthwise growth of bone to end; contribute to bone remodeling during adulthood by slowing bone resorption by osteoclasts and promoting bone deposition by osteoblasts
Parathyroid hormone (PTH)	Secreted by parathyroid glands; promotes bone resorption by osteoclasts, enhances recovery of calcium ions from urine; promotes formation of active form of vitamin D (calcitriol)
Calcitonin (CT)	Secreted by thyroid gland; inhibits bone resorption by osteoclasts
Exercise	Weight-bearing activities stimulate osteoblasts and, consequently, help build thicker, stronger bones and retard loss of bone mass that occurs as people age.
Aging	As the level of sex hormones diminishes during middle age to older adulthood, especially in women after menopause, bone resorption by osteoclasts outpaces bone deposition by osteoblasts, which leads to a decrease in bone mass and an increased risk of osteoporosis.

✓ **CHECKPOINT**

13. What types of mechanical stress may be used to strengthen bone tissue?

6.6 DIVISIONS OF THE SKELETAL SYSTEM

OBJECTIVE • **Group the bones of the body into axial and appendicular divisions.**

Because the skeletal system forms the framework of the body, a familiarity with the names, shapes, and positions of individual bones will help you locate other organs. For example, the radial artery, the site where the pulse is usually taken, is named for its closeness to the radius, the lateral bone of the forearm. The ulnar nerve is named for its closeness to the ulna, the medial bone of the forearm. The frontal lobe of the brain lies deep to the frontal (forehead) bone. The tibialis anterior muscle lies along the anterior surface of the tibia (shin bone).

The adult human skeleton consists of 206 bones grouped in two principal divisions: 80 in the *axial skeleton* and 126 in the *appendicular skeleton* (Table 6.2 and Figure 6.6). The axial skeleton consists of the bones that lie around the longitudinal *axis* of the human body, an imaginary line that runs through the body's center of gravity from the head to the space between

TABLE 6.2

The Bones of the Adult Skeletal System

DIVISION OF THE SKELETON	STRUCTURE	NUMBER OF BONES
Axial Skeleton		
	Skull	
	Cranium	8
	Face	14
	Hyoid	1
	Auditory ossicles	6
	Vertebral column	26
	Thorax	
	Sternum	1
	Ribs	24
		Subtotal = 80
Appendicular Skeleton	**Pectoral (shoulder) girdles**	
	Clavicle	2
	Scapula	2
	Upper limbs	
	Humerus	2
	Ulna	2
	Radius	2
	Carpals	16
	Metacarpals	10
	Phalanges	28
	Pelvic (hip) girdle	
	Hip or pelvic bone	2
	Lower limbs	
	Femur	2
	Patella	2
	Fibula	2
	Tibia	2
	Tarsals	14
	Metatarsals	10
	Phalanges	28
		Subtotal = 126
	Total in an adult skeleton = 206	

Figure 6.6 Divisions of the skeletal system. The axial skeleton is indicated in blue. (Note the position of the hyoid bone in Figure 6.7c.)

🔑 The adult human skeleton consists of 206 bones grouped into axial and appendicular divisions.

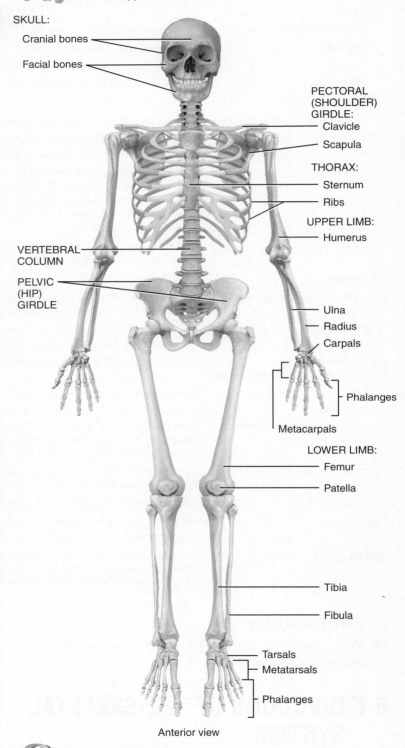

Anterior view

❓ Identify each of the following bones as part of the axial skeleton or the appendicular skeleton: skull, clavicle, vertebral column, shoulder girdle, humerus, pelvic girdle, and femur.

the feet: the bones of the skull, auditory ossicles (ear bones), hyoid bone, ribs, sternum, and vertebrae. The appendicular skeleton contains the bones of the upper and lower limbs or *appendages* plus the bone groups called *girdles* that connect the limbs to the axial skeleton. The skeletons of infants and children have more than 206 bones because some of their bones, such as the hip bones and vertebrae, fuse later in life.

✓ CHECKPOINT

14. How are the limbs connected to the axial skeleton?

6.7 SKULL AND HYOID BONE

OBJECTIVE • Name the cranial and facial bones and indicate their locations and major structural features.

The *skull*, which contains 22 bones, rests on top of the vertebral column. It includes two sets of bones: cranial bones and facial bones. The eight *cranial bones*, collectively called the *cranium*, form the cranial cavity that encloses and protects the brain. They are the frontal bone, two parietal bones, two temporal bones, occipital bone, sphenoid bone, and ethmoid bone. Fourteen *facial bones* form the face: two nasal bones, two maxillae, two zygomatic bones, the mandible, two lacrimal bones, two palatine bones, two inferior nasal conchae, and the vomer.

Together, the cranial and facial bones protect and support the delicate special sense organs for vision, taste, smell, hearing, and equilibrium (balance). Exhibits 6.A and 6.B provide more details about the cranial bones and facial bones, respectively.

CLINICAL CONNECTION | Cleft Palate and Cleft Lip

Usually the left and right maxillary bones unite during weeks 10 to 12 of embryonic development. Failure to do so can result in one type of **cleft palate**. The condition may also involve incomplete fusion of the palatine bones (see Figure 6.8). Another form of this condition, called **cleft lip**, involves a split in the upper lip. Cleft lip and cleft palate often occur together. Depending on the extent and position of the cleft, speech and swallowing may be affected. Facial and oral surgeons recommend closure of cleft lip during the first few weeks following birth, and surgical results are excellent. Repair of cleft palate typically is completed between 12 and 18 months of age, ideally before the child begins to talk. Because the palate is important for pronouncing consonants, speech therapy may be required, and orthodontic therapy may be needed to align the teeth. Again, results are usually excellent. Supplementation with folic acid (one of the B vitamins) during pregnancy decreases the incidence of cleft palate and cleft lip. •

Unique Features of the Skull

Now that you are familiar with the names of the skull bones, we will take a closer look at three unique features of the skull: sutures, paranasal sinuses, and fontanels.

Sutures

A *suture* (SOO-chur = seam) is an immovable joint in most cases in an adult that holds skull bones together. Of the many sutures that are found in the skull, we will identify only four prominent ones (see Figure 6.7):

1. The *coronal suture* (kō-RŌ-nal; *coron-* = crown) unites the frontal bone and two parietal bones.
2. The *sagittal suture* (SAJ-i-tal; *sagitt-* = arrow) unites the two parietal bones.
3. The *lambdoid suture* (LAM-doyd; so named because its shape resembles the Greek letter lambda, Λ) unites the parietal bones to the occipital bone.
4. The *squamous sutures* (SKWĀ-mus; *squam-* = flat) unite the parietal bones to the temporal bones.

(Text continues on page 143)

EXHIBIT 6.A Cranial Bones *(Figures 6.7–6.10)*

OBJECTIVE • Describe the locations and functions of each of the eight cranial bones.

The cranial bones have functions besides protection of the brain. Their inner surfaces attach to membranes (meninges) that stabilize the positions of the brain, blood vessels, and nerves. Their outer surfaces provide large areas of attachment for muscles that move various parts of the head.

The *frontal bone* forms the forehead (the anterior part of the cranium), the roofs of the *or-bits* (eye sockets; Figure 6.7a, b), and most of the anterior (front) part of the cranial floor. The *frontal sinuses* lie deep within the frontal bone (Figure 6.7c). These mucous membrane-lined cavities act as sound chambers that give the voice resonance. Other functions of the sinuses are given later in the chapter.

The two *parietal bones* (pa-RĪ-e-tal; *pariet-* = wall) form most of the sides and roof of the cranial cavity (Figure 6.7d).

The two *temporal bones* (*tempor-* = temples) form the inferior (lower) sides of the cranium and part of the cranial floor. In the lateral view of the skull (Figure 6.7b), note that the temporal and zygomatic bones join to form the *zygomatic arch*. The *mandibular fossa* (depression) forms a joint with a projection on the mandible (lower jawbone) called the condylar process to form the *temporomandibular joint (TMJ)*. The

Figure 6.7 Skull. Although the hyoid bone is not part of the skull, it is included in (c) for reference.

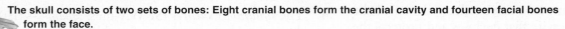

The skull consists of two sets of bones: Eight cranial bones form the cranial cavity and fourteen facial bones form the face.

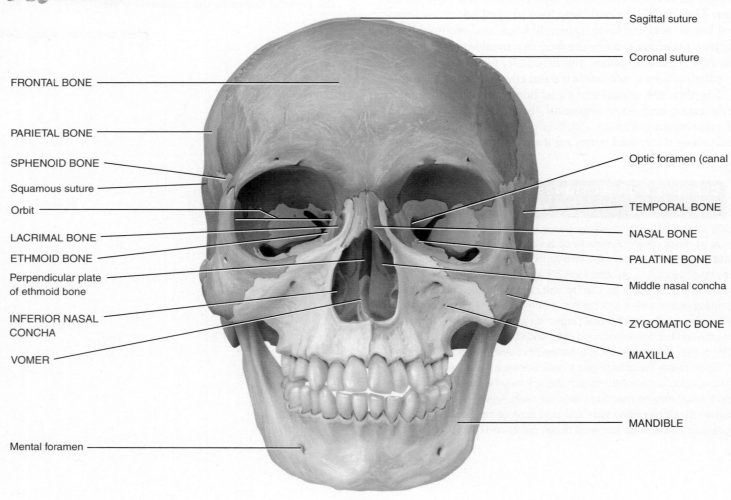

FRONTAL BONE

PARIETAL BONE

SPHENOID BONE

Squamous suture

Orbit

LACRIMAL BONE

ETHMOID BONE

Perpendicular plate of ethmoid bone

INFERIOR NASAL CONCHA

VOMER

Mental foramen

Sagittal suture

Coronal suture

Optic foramen (canal

TEMPORAL BONE

NASAL BONE

PALATINE BONE

Middle nasal concha

ZYGOMATIC BONE

MAXILLA

MANDIBLE

(a) Anterior view

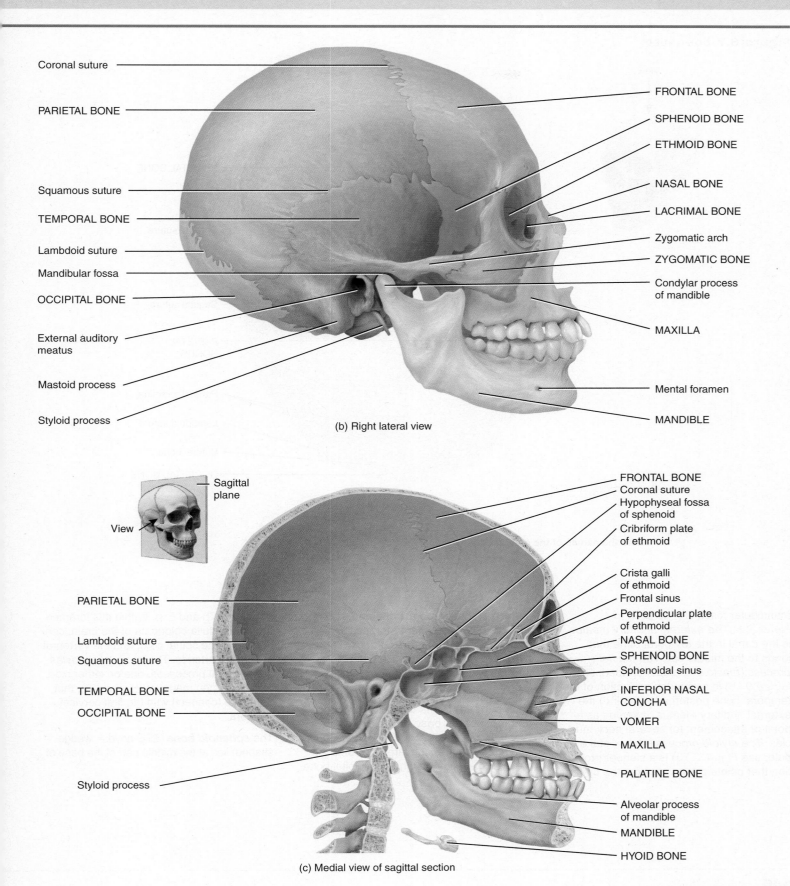

Coronal suture

PARIETAL BONE

Squamous suture

TEMPORAL BONE

Lambdoid suture

Mandibular fossa

OCCIPITAL BONE

External auditory meatus

Mastoid process

Styloid process

FRONTAL BONE

SPHENOID BONE

ETHMOID BONE

NASAL BONE

LACRIMAL BONE

Zygomatic arch

ZYGOMATIC BONE

Condylar process of mandible

MAXILLA

Mental foramen

MANDIBLE

(b) Right lateral view

Sagittal plane

View

PARIETAL BONE

Lambdoid suture

Squamous suture

TEMPORAL BONE

OCCIPITAL BONE

Styloid process

FRONTAL BONE
Coronal suture
Hypophyseal fossa of sphenoid
Cribriform plate of ethmoid

Crista galli of ethmoid
Frontal sinus
Perpendicular plate of ethmoid
NASAL BONE
SPHENOID BONE
Sphenoidal sinus
INFERIOR NASAL CONCHA
VOMER
MAXILLA
PALATINE BONE
Alveolar process of mandible
MANDIBLE
HYOID BONE

(c) Medial view of sagittal section

EXHIBIT 6.A CONTINUES

EXHIBIT 6.A Cranial Bones *(Figures 6.7–6.10)* CONTINUED

Figure 6.7 CONTINUED

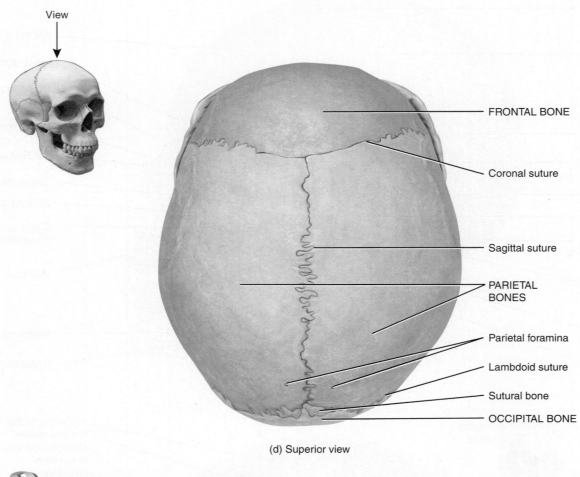

View

FRONTAL BONE

Coronal suture

Sagittal suture

PARIETAL BONES

Parietal foramina

Lambdoid suture

Sutural bone

OCCIPITAL BONE

(d) Superior view

? What are the names of the cranial bones?

mandibular fossa can be seen in Figure 6.8. The *external auditory meatus* is the canal in the temporal bone that leads to the middle ear. The *mastoid process* (*mastoid* = breast-shaped; see Figure 6.7b) is a rounded projection of the temporal bone posterior to (behind) the external auditory meatus. It serves as a point of attachment for several neck muscles. The *styloid process* (*styl-* = stake or pole; see Figure 6.7b) is a slender projection that points downward from the under-

surface of the temporal bone and serves as a point of attachment for muscles and ligaments of the tongue and neck. The *carotid foramen* (Figure 6.8) is a hole through which the carotid artery passes.

The ***occipital bone*** (ok-SIP-i-tal; *occipit-* = back of head) forms the posterior part and most of the base of the cranium (Figures 6.7b, c and 6.8). The *foramen magnum* (*magnum* = large), the largest foramen in the skull, passes through the occipital bone

(Figures 6.7b and 6.8). Within this foramen are the medulla oblongata of the brain, connecting to the spinal cord, and the vertebral and spinal arteries. The *occipital condyles* are two oval processes, one on either side of the foramen magnum (Figure 6.8), that articulate (connect) with the first cervical vertebra.

The ***sphenoid bone*** (SFĒ-noyd = wedge-shaped) lies at the middle part of the base of

Figure 6.8 Inferior view of the skull.

The occipital bone forms most of the posterior and inferior portion of the cranium.

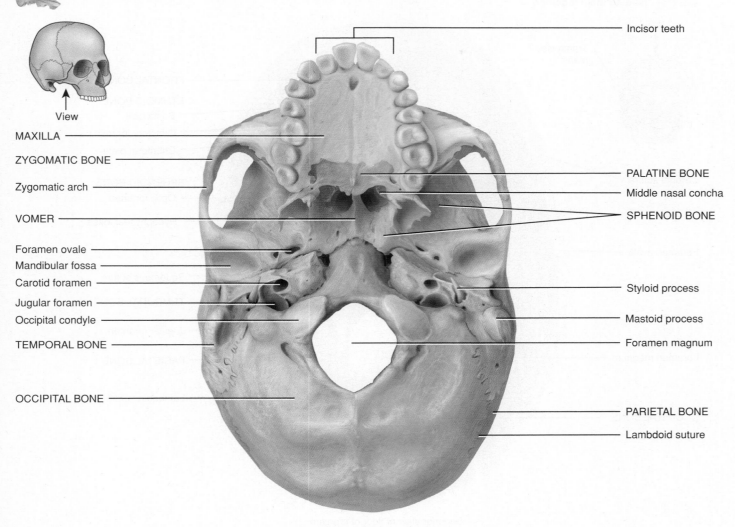

View

MAXILLA

ZYGOMATIC BONE

Zygomatic arch

VOMER

Foramen ovale

Mandibular fossa

Carotid foramen

Jugular foramen

Occipital condyle

TEMPORAL BONE

OCCIPITAL BONE

Incisor teeth

PALATINE BONE

Middle nasal concha

SPHENOID BONE

Styloid process

Mastoid process

Foramen magnum

PARIETAL BONE

Lambdoid suture

Inferior view, mandible removed

 What is the largest foramen in the skull?

the skull (Figures 6.7, 6.8, and 6.9). This bone is called the keystone of the cranial floor because it articulates with all other cranial bones, holding them together. The shape of the sphenoid bone resembles a bat with outstretched wings. The cubelike central portion of the sphenoid bone contains the *sphenoidal sinuses*, which drain into the nasal cavity (see Figures 6.7c and 6.11). On the superior surface of the sphenoid is a depression called the **hypophyseal fossa**

(hī -pō-FIZ-ē-al), which contains the pituitary gland. Two nerves pass through foramina in the sphenoid bone: the mandibular nerve through the *foramen ovale* and the optic nerve through the *optic foramen (canal)*.

The **ethmoid bone** (ETH-moid = sievelike) is spongelike in appearance and is located in the anterior part of the cranial floor between the orbits (Figure 6.10). It

forms part of the anterior portion of the cranial floor, the medial wall of the orbits, the superior portions of the *nasal septum*, a partition that divides the nasal cavity into right and left sides, and most of the side walls of the nasal cavity. The ethmoid bone contains 3 to 18 air spaces, or "cells," that give this bone a sievelike appearance. The ethmoidal cells together form the *ethmoidal sinuses* (Figure 6.10b).

EXHIBIT 6.A CONTINUES

EXHIBIT 6.A Cranial Bones *(Figures 6.7–6.10)* **CONTINUED**

Figure 6.9 Sphenoid bone.

The sphenoid bone is called the keystone of the cranial floor because it articulates with all other cranial bones, holding them together.

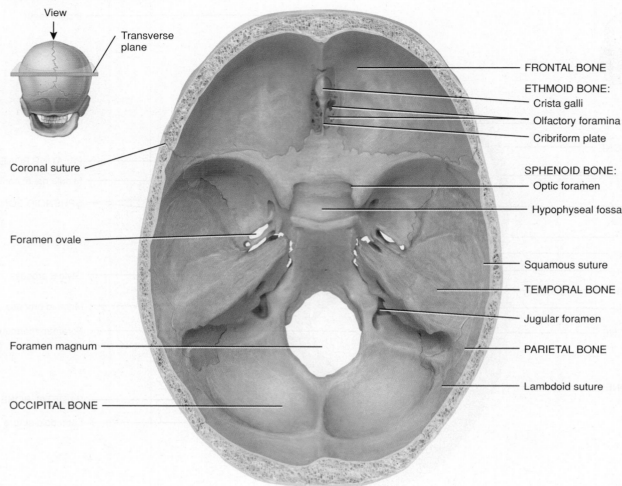

Superior view of floor of cranium

Starting at the crista galli of the ethmoid bone and going in a clockwise direction, what are the names of the bones that articulate with the sphenoid bone?

The *perpendicular plate* forms the upper portion of the nasal septum. The *cribriform plate* (KRIB-ri-form) forms the roof of the nasal cavity (Figure 6.10). It contains the *olfactory foramina* (*olfact-* = to smell), holes through which fibers of the olfactory nerves pass (see Figure 6.9). Projecting upward from the cribriform plate is a triangular process called the *crista galli* (= cock's comb), which serves as a point of attachment for the membranes (meninges) that cover the brain (Figure 6.10).

Also part of the ethmoid bone are two thin, scroll-shaped bones on either side

of the nasal septum. These are called the *superior nasal concha* (KONG-ka; *conch-* = shell; plural is *conchae* [KONG-kē]) and the *middle nasal concha*. A third pair of conchae, the inferior nasal conchae, are separate bones (discussed shortly). The conchae greatly increase the vascular and mucous membrane surface area in the nasal cavity, which warms and moistens (humidifies) inhaled air before it passes into the lungs. The conchae also cause inhaled air to swirl, and the result is that many inhaled particles become trapped in the mucus that lines the nasal

cavity. This action of the conchae helps cleanse inhaled air before it passes into the rest of the respiratory passageways. The superior nasal conchae are near the olfactory foramina of the cribriform plate where the sensory receptors for olfaction (smell) terminate in the mucous membrane of the superior nasal conchae. Thus, they increase the surface area for the sense of smell.

✓ **CHECKPOINT**

Why is the sphenoid bone referred to as the keystone of the cranial floor?

Figure 6.10 Ethmoid bone.

 The ethmoid bone is the major supporting structure of the nasal cavity.

Sagittal plane

View

ETHMOID BONE:
- Crista galli
- Cribriform plate
- Olfactory foramen
- Superior nasal concha
- Middle nasal concha
- Sphenoidal sinus
- Inferior nasal concha
- Palatine bone
- Maxilla

(a) Sagittal section

POSTERIOR

- Ethmoidal cells of ethmoidal sinus
- Lateral mass
- Cribriform plate
- Crista galli
- Olfactory foramina
- Perpendicular plate

ANTERIOR

(b) Superior view

- Crista galli
- Left orbit
- Perpendicular plate
- Superior nasal concha
- Middle nasal concha

(c) Anterior view of position of ethmoid bone in skull

? What part of the ethmoid bone forms the top part of the nasal septum?

Paranasal Sinuses

Paired cavities, the *paranasal sinuses* (*para-* = beside), are located in certain skull bones near the nasal cavity (Figure 6.11). The paranasal sinuses are lined with mucous membranes that are continuous with the lining of the nasal cavity. Skull bones containing paranasal sinuses include the frontal bone (*frontal sinus*), sphenoid bone (*sphenoid sinus*), ethmoid bone (*ethmoidal sinuses*), and maxillae (*maxillary sinuses*). Besides producing mucus, the paranasal sinuses serve as resonating (echo) chambers, producing the unique sounds of each of our speaking and singing voices, and lighten the weight of the skull.

CLINICAL CONNECTION | Sinusitis

Sinusitis (sīn-ū-SĪ-tis) is an inflammation of the mucous membrane of one or more paranasal sinuses. It may be caused by a microbial infection (virus, bacteria, or fungus), allergic reactions, nasal polyps, or a severely deviated nasal septum. If the inflammation or an obstruction blocks the drainage of mucus into the nasal cavity, fluid pressure builds up in the paranasal sinuses, and a sinus headache may develop. Other symptoms may include nasal congestion, inability to smell, fever, and cough. Treatment options include decongestant sprays or drops, oral decongestants, nasal corticosteroids, antibiotics, analgesics to relieve pain, warm compresses, and surgery. •

143

EXHIBIT 6.B   Facial Bones

OBJECTIVE • **Describe the structural features of each of the facial bones.**

Besides forming the framework of the face, the facial bones protect and provide support for the entrances to the digestive and respiratory systems. The facial bones also provide attachment for some muscles that are involved in producing various facial expressions.

The shape of the face changes dramatically during the first two years after birth. The brain and cranial bones expand, the teeth form and erupt (emerge), and the paranasal sinuses increase in size. Growth of the face ceases at about 16 years of age.

The paired *nasal bones* form part of the bridge of the nose (see Figure 6.7a). The rest of the supporting tissue of the nose consists of cartilage.

The paired *maxillae* (mak-SIL-ē = jawbones; singular is *maxilla*) unite to form the upper jawbone and articulate with every bone of the face except the mandible (lower jawbone) (see Figure 6.7a, b). Each maxilla contains a *maxillary sinus* that empties into the nasal cavity (see Figure 6.11). The *alveolar process* (al-VĒ-ō-lar; *alveol-* = small cavity) of the maxilla is an arch that contains the *alveoli* (sockets) for the maxillary (upper) teeth. The maxilla forms the anterior three-quarters of the *hard palate*, which forms the roof of the mouth.

The two L-shaped *palatine bones* (PAL-a-tīn; *palat-* = roof of mouth) are fused and form the posterior portion of the hard palate, part of the floor and lateral wall of the nasal cavity, and a small portion of the floors of the orbits (see Figure 6.8). In cleft palate, the palatine bones may also be incompletely fused.

The *mandible* (*mand-* = to chew), or lower jawbone, is the largest, strongest

facial bone (see Figure 6.7b). It is the only movable skull bone. Recall from our discussion of the temporal bone that the mandible has a *condylar process* (KON-di-lar). This process articulates with the mandibular fossa of the temporal bone to form the temporomandibular joint. The mandible, like the maxilla, has an *alveolar process* containing the *alveoli* (sockets) for the mandibular (lower) teeth (see Figure 6.7c). The *mental foramen* (*ment-* = chin) is a hole in the mandible that can be used by dentists to reach the mental nerve when injecting anesthetics (see Figure 6.7a).

CLINICAL CONNECTION |
Temporomandibular Joint Syndrome

One problem associated with the temporomandibular joint is **temporomandibular joint (TMJ) syndrome**. It is characterized by dull pain around the ear, tenderness of the jaw muscles, a clicking or popping noise when opening or closing the mouth, limited or abnormal opening of the mouth, headache, tooth sensitivity, and abnormal wearing of the teeth. TMJ syndrome can be caused by improperly aligned teeth, grinding or clenching the teeth, trauma to the head and neck, or arthritis. Treatments include application of moist heat or ice, limiting the diet to soft foods, administration of pain relievers such as aspirin, muscle retraining, use of a splint or bite plate to reduce clenching and teeth grinding (especially when worn at night), adjustment or reshaping of the teeth (orthodontic treatment), and surgery. •

The two *zygomatic bones* (*zygo-* = like a yoke), commonly called cheekbones, form the prominences of the cheeks and part of the lateral wall and floor of each orbit (see Figure 6.7a). They articulate with the frontal, maxilla, sphenoid, and temporal bones.

The paired *lacrimal bones* (LAK-ri-mal; *lacrim-* = teardrop), the smallest bones of the face, are thin and roughly resemble a fingernail in size and shape. The lacrimal bones can be seen in the anterior and lateral views of the skull in Figure 6.7a, b.

The two *inferior nasal conchae* are scroll-like bones that project into the nasal cavity below the superior and middle nasal conchae of the ethmoid bone (see Figures 6.7a, c and 6.10). They serve the same function as the other nasal conchae: the filtration of air before it passes into the lungs.

The *vomer* (VŌ-mer = plowshare) is a roughly triangular bone on the floor of the nasal cavity that articulates inferiorly with both the maxillae and palatine bones along the midline of the skull. The vomer, clearly seen in the anterior view of the skull in Figure 6.7a and the inferior view in Figure 6.8, is one of the components of the nasal septum. The nasal septum is formed by the vomer, septal cartilage, and the perpendicular plate of the ethmoid bone (see Figure 6.7a). The anterior border of the vomer articulates with the septal cartilage (hyaline cartilage) to form the more anterior portion of the septum. The upper border of the vomer articulates with the perpendicular plate of the ethmoid bone to form the remainder of the nasal septum.

✓ **CHECKPOINT**

Which is the largest, strongest facial bone?

Fontanels

Recall that the skeleton of a newly formed embryo consists of cartilage or membranous layers of mesenchyme shaped like bones. Gradually, ossification occurs—bone replaces the cartilage or mesenchyme. Mesenchyme-filled spaces called *fontanels* (fonta-NELZ = little fountains) or "soft spots" are found between cranial bones at birth. They include the anterior fontanel, the posterior fontanel, the anterolateral fontanels, and the posterolateral fontanels. These areas of unossified mesenchyme will eventually be replaced with bone by intramembranous ossification and will become sutures. Functionally, the fontanels enable the fetal skull to be compressed as it passes through the birth canal and permit rapid growth of the brain during infancy. Several fontanels are shown and described in Table 6.3.

Figure 6.11 Paranasal sinuses.

 Paranasal sinuses are mucous membrane-lined spaces in the frontal, sphenoid, ethmoid, and maxillary bones that connect to the nasal cavity.

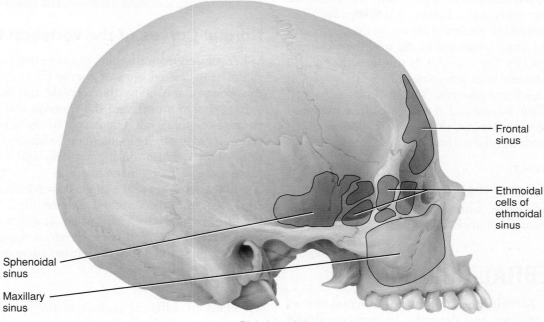

Frontal sinus

Ethmoidal cells of ethmoidal sinus

Sphenoidal sinus

Maxillary sinus

Right lateral view

What are two main functions of the paranasal sinuses?

TABLE 6.3		
Fontanels		
FONTANEL	**LOCATION**	**DESCRIPTION**
Anterior	Between the two parietal bones and the frontal bone	Roughly diamond-shaped, the largest of the fontanels; usually closes 18–24 months after birth
Posterior	Between the two parietal bones and the occipital bone	Diamond-shaped, considerably smaller than the anterior fontanel; generally closes about 2 months after birth
Anterolateral	One on each side of the skull between the frontal, parietal, temporal, and sphenoid bones	Small and irregular in shape; normally close about 3 months after birth
Posterolateral	One on each side of the skull between the parietal, occipital, and temporal bones	Irregularly shaped; begin to close 1 or 2 months after birth, but closure is generally not complete until 12 months

Anterior

Posterior

Anterolateral

Posterolateral

Hyoid Bone

The single **hyoid bone** (HĪ-oyd = U-shaped) is a unique component of the axial skeleton because it does not articulate with or attach to any other bone. Rather, it is suspended from the styloid processes of the temporal bones by ligaments and muscles. The hyoid bone is located in the neck between the mandible and larynx (see Figure 6.7c). It supports the tongue and provides attachment sites for some tongue muscles and for muscles of the neck and pharynx. The hyoid bone, as well as the cartilage of the larynx and trachea, is often fractured during strangulation. As a result, they are carefully examined in an autopsy when strangulation is suspected.

✓ CHECKPOINT

15. Describe the general features of the skull.
16. Define the following: suture, foramen, nasal septum, paranasal sinus, and fontanel.

6.8 VERTEBRAL COLUMN

OBJECTIVE • Identify the regions and normal curves of the vertebral column and describe its structural and functional features.

The **vertebral column**, also called the *spine*, *spinal column*, or *backbone*, is composed of a series of bones called **vertebrae** (VER-te-brē; singular is *vertebra*). The vertebral column functions as a strong, flexible rod that can rotate and move forward, backward, and sideways. It encloses and protects the spinal cord, supports the head, and serves as a point of attachment for the ribs, pelvic girdle, and the muscles of the back.

Regions of the Vertebral Column

The total number of vertebrae during early development is 33. Then, several vertebrae in the sacral and coccygeal regions fuse. As a result, the adult vertebral column typically contains 26 vertebrae (Figure 6.12). These are distributed as follows:

- **7 cervical vertebrae** (*cervic-* = neck) in the neck region.
- **12 thoracic vertebrae** (*thorax* = chest) posterior to the thoracic cavity.
- **5 lumbar vertebrae** (*lumb-* = loin) support the lower back.
- **1 sacrum** (SĀ-krum = sacred bone) consists of five fused *sacral vertebrae*.
- **1 coccyx** (KOK-siks = cuckoo, because the shape resembles the bill of a cuckoo bird) usually consists of four fused *coccygeal vertebrae* (kok-SIJ-ē-al).

The cervical, thoracic, and lumbar vertebrae are movable, but the sacrum and coccyx are immovable. Between adjacent vertebrae from the second cervical vertebra to the sacrum are

intervertebral discs (in'-ter-VER-te-bral; *inter-* = between). Each disc has an outer ring of fibrocartilage and a soft, pulpy, highly elastic interior. The discs form strong joints, permit various movements of the vertebral column, and absorb vertical shock.

Normal Curves of the Vertebral Column

When viewed from the side, the vertebral column shows four slight bends called **normal curves** (Figure 6.12). Relative to the front of the body, the **cervical** and **lumbar curves** are convex (bulging out), and the **thoracic** and **sacral curves** are concave (cupping in). The curves of the vertebral column increase its strength, help maintain balance in the upright position, absorb shocks during walking and running, and help protect the vertebrae from breaks.

In the fetus, there is a single concave curve throughout the entire length of the vertebral column (Figure 6.12b). At about the third month after birth, when an infant begins to hold its head erect, the cervical curve develops. Later, when the child sits up, stands, and walks, the lumbar curve develops.

Vertebrae

Vertebrae in different regions of the spinal column vary in size, shape, and detail, but they are similar enough that we can discuss the structure and functions of a typical vertebra (see Figure 6.14).

- The **body**, the thick, disc-shaped front portion, is the weight-bearing part of a vertebra.
- The **vertebral arch** extends backwards from the body of the vertebra. It is formed by two short, thick processes, the *pedicles* (PED-i-kuls = little feet), which project backward from the body to unite with the laminae. The *laminae* (LAM-i-nē = thin layers) are the flat parts of the arch and end in a single sharp, slender projection called a *spinous process*. The hole between the vertebral arch and body contains the spinal cord and is known as the *vertebral foramen*. Together, the vertebral foramina of all vertebrae form the *vertebral cavity*. When the vertebrae are stacked on top of one another, there is an opening between adjoining vertebrae on both sides of the column. Each opening, called an *intervertebral foramen*, permits the passage of a single spinal nerve.
- Seven **processes** arise from the vertebral arch. At the point where a lamina and pedicle join, a *transverse process* extends laterally on each side. A single *spinous process (spine)* projects from the junction of the laminae. These three processes serve as points of attachment for muscles. The remaining four processes form joints with other vertebrae above or below. The two *superior articular processes* of a vertebra articulate with the vertebra immediately above them. The two *inferior articular processes* of a vertebra articulate with the vertebra immediately below them. The smooth articulating surfaces of the articular processes are

Figure 6.12 Vertebral column.

The adult vertebral column typically contains 26 vertebrae.

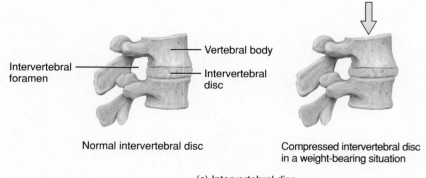

POSTERIOR   ANTERIOR

Cervical curve (formed by 7 cervical vertebrae)

Thoracic curve (formed by 12 thoracic vertebrae)

Intervertebral disc

Lumbar curve (formed by 5 lumbar vertebrae)

Intervertebral foramen

Sacrum

Sacral curve (formed by 5 fused sacral vertebrae)

Coccyx

(a) Right lateral view showing four normal curves

Single curve in fetus   Four curves in adult

(b) Fetal and adult curves

Functions of the Vertebral Column
1. Permits movement.
2. Encloses and protects the spinal cord.
3. Serves as a point of attachment for the ribs and muscles of the back.

Intervertebral foramen

Vertebral body

Intervertebral disc

Normal intervertebral disc

Compressed intervertebral disc in a weight-bearing situation

(c) Intervertebral disc

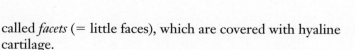 **Which curves are concave (relative to the front of the body)?**

called *facets* (= little faces), which are covered with hyaline cartilage.

Vertebrae in each region are numbered in sequence from top to bottom. Exhibits 6.C–6.F provide details about the vertebrae in the different regions of the vertebral column.

✓ CHECKPOINT

17. What are the functions of the vertebral column?

18. What are the main distinguishing characteristics of the bones of the various regions of the vertebral column?

EXHIBIT 6.C   Cervical Vertebrae   *(Figure 6.13)*

OBJECTIVE • Identify the location and surface features of the cervical vertebrae.

The seven **cervical vertebrae** are termed C1 through C7 (Figure 6.13). The spinous processes of the second through sixth cervical vertebrae are often *bifid*, or split into two parts (Figure 6.13c). All cervical vertebrae have three foramina: one vertebral foramen and two transverse foramina. Each cervical transverse process contains a *transverse foramen* through which blood vessels and nerves pass.

The first two cervical vertebrae differ considerably from the others. The first cervical vertebra (C1), the **atlas**, supports the head and is named for the mythological Atlas who supported the world on his shoulders. The atlas lacks a body and a

spinous process. The upper surface contains *superior articular facets* that articulate with the occipital bone of the skull. This articulation permits you to nod your head to indicate "yes." The inferior surface contains *inferior articular facets* that articulate with the second cervical vertebra.

The second cervical vertebra (C2), the **axis**, does have a body and a spinous process. A tooth-shaped process called the *dens* (= tooth) projects up through the vertebral foramen of the atlas. The dens is a pivot on which the atlas and head move, as in side-to-side movement of the head to signify "no."

The third through sixth cervical vertebrae (C3 through C6), represented by the vertebra in Figure 6.13c, correspond to the structural pattern of the typical cervical vertebra described previously. The seventh cervical vertebra (C7), called the *vertebra prominens*, is somewhat different. It is marked by a single, large spinous process that can be seen and felt at the base of the neck.

✓ CHECKPOINT

How are the atlas and axis different from the other cervical vertebrae?

Figure 6.13 Cervical vertebrae.

The cervical vertebrae are found in the neck region.

POSTERIOR

Posterior arch

Groove for vertebral artery and first cervical spinal nerve

Vertebral foramen

Transverse foramen

Superior articular facet

ANTERIOR

(a) Superior view of atlas (C1)

Location of cervical vertebrae

Atlas

Axis

Typical cervical vertebra

ANTERIOR

POSTERIOR

Spinous process

Lamina

Vertebral foramen

Transverse foramen

Transverse process

Superior articular facet

Dens

ANTERIOR

(b) Superior view of axis (C2)

POSTERIOR

Bifid spinous process

Lamina

Vertebral foramen

Transverse process

Transverse foramen

Superior articular facet

Pedicle

Vertebral body

ANTERIOR

(c) Superior view of a typical cervical vertebra

Which bones permit the movement of the head to signify "no"?

EXHIBIT 6.D   Thoracic Vertebrae   *(Figure 6.14)*

OBJECTIVE • **Identify the location and surface features of the thoracic vertebrae.**

Thoracic vertebrae (T1 through T12) are considerably larger and stronger than cervical vertebrae. Distinguishing features of the thoracic vertebrae are their facets for articulating with the ribs (Figure 6.14). Movements of the thoracic region are limited by the attachment of the ribs to the sternum.

✓ **CHECKPOINT**

Describe several distinguishing features of thoracic vertebrae.

Figure 6.14 Structure of a typical vertebra, as illustrated by a thoracic vertebra. (Note the facets for the ribs, which only thoracic vertebrae have.) In (b), only one spinal nerve has been included, and it has been extended beyond the intervertebral foramen for clarity.

 A vertebra consists of a body, a vertebral arch, and several processes.

ANTERIOR

Location of thoracic vertebrae

POSTERIOR

POSTERIOR

- Spinous process
- Transverse process
- Vertebral arch:
  - Lamina
  - Pedicle
- Facet for rib
- Vertebral body

Facet for rib

Facet of superior articular process

Spinal cord

Vertebral foramen

ANTERIOR

(a) Superior view

Spinous process

ANTERIOR

- Spinal cord
- Facet of superior articular process
- Pedicle
- Spinal nerve
- Intervertebral disc
- Intervertebral foramen
- Facets for rib
- Vertebral body
- Inferior articular process

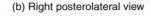

(b) Right posterolateral view

What are the functions of the vertebral and intervertebral foramina?

EXHIBIT 6.E | Lumbar Vertebrae *(Figure 6.15)*

OBJECTIVE • Identify the location and surface features of the lumber vertebrae.

The *lumbar vertebrae* (L1 through L5) are the largest and strongest of the unfused bones in the vertebral column (Figure 6.15). Their various projections are short and thick, and the spinous processes are well adapted for the attachment of the large back muscles.

✓ **CHECKPOINT**

What are the distinguishing features of the lumbar vertebrae?

Figure 6.15 Lumbar vertebrae.

Lumbar vertebrae are found in the lower back.

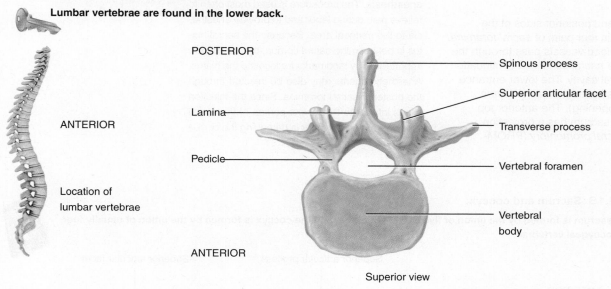

POSTERIOR

ANTERIOR

Location of lumbar vertebrae

Lamina

Pedicle

ANTERIOR

Spinous process

Superior articular facet

Transverse process

Vertebral foramen

Vertebral body

Superior view

? Why are the lumbar vertebrae the largest and strongest in the vertebral column?

OBJECTIVE • Identify the location and surface features of the sacral and coccygeal vertebrae.

The *sacrum* is a triangular bone formed by the fusion of five sacral vertebrae, indicated in Figure 6.16 as S1 through S5. The fusion of the sacral vertebrae begins between ages 16 and 18 years and is usually completed by age 30. The sacrum serves as a strong foundation for the pelvic girdle. It is positioned at the back of the pelvic cavity where its lateral surfaces fuse to the two hip bones.

The anterior and posterior sides of the sacrum contain four pairs of *sacral foramina.* Nerves and blood vessels pass through the foramina. The *sacral canal* is a continuation of the vertebral cavity. The lower entrance of the canal is called the *sacral hiatus* (hī -Ā-tus = opening). The anterior top border of the sacrum has a projection, called the *sacral promontory* (PROM-

on-tō′-rē), which is used as a landmark for measuring the pelvis prior to childbirth.

CLINICAL CONNECTION | Caudal Anesthesia

Anesthetic agents that act on the sacral and coccygeal nerves are sometimes injected through the sacral hiatus, a procedure called **caudal anesthesia**. The procedure is used most often to relieve pain during labor and to provide anesthesia to the perineal area. Because the sacral hiatus is between the sacral cornua, the cornua are important bony landmarks for locating the hiatus. Anesthetic agents may also be injected through the posterior sacral foramina. Since the injection site is inferior to the lowest portion of the spinal cord, there is little danger in damaging the cord. •

The *coccyx*, like the sacrum, is triangular in shape and is formed by the fusion of the four coccygeal vertebrae. These are indicated in Figure 6.16 as Co1 through Co4. The top of the coccyx articulates with the sacrum.

✓ **CHECKPOINT**

How many vertebrae fuse to form the sacrum and coccyx?

Figure 6.16 Sacrum and coccyx.

The sacrum is formed by the union of five sacral vertebrae, and the coccyx is formed by the union of usually four coccygeal vertebrae.

ANTERIOR

Location of sacrum and coccyx

Sacral promontory

Sacral foramen

Superior articular process

S1
S2
S3
S4
S5

Co 1
Co 2
Co 3
Co 4

(a) Anterior view

Superior articular facet

Sacral canal

Sacral foramen

Sacrum

Sacral hiatus

Coccyx

Co 1
Co 2
Co 3
Co 4

(b) Posterior view

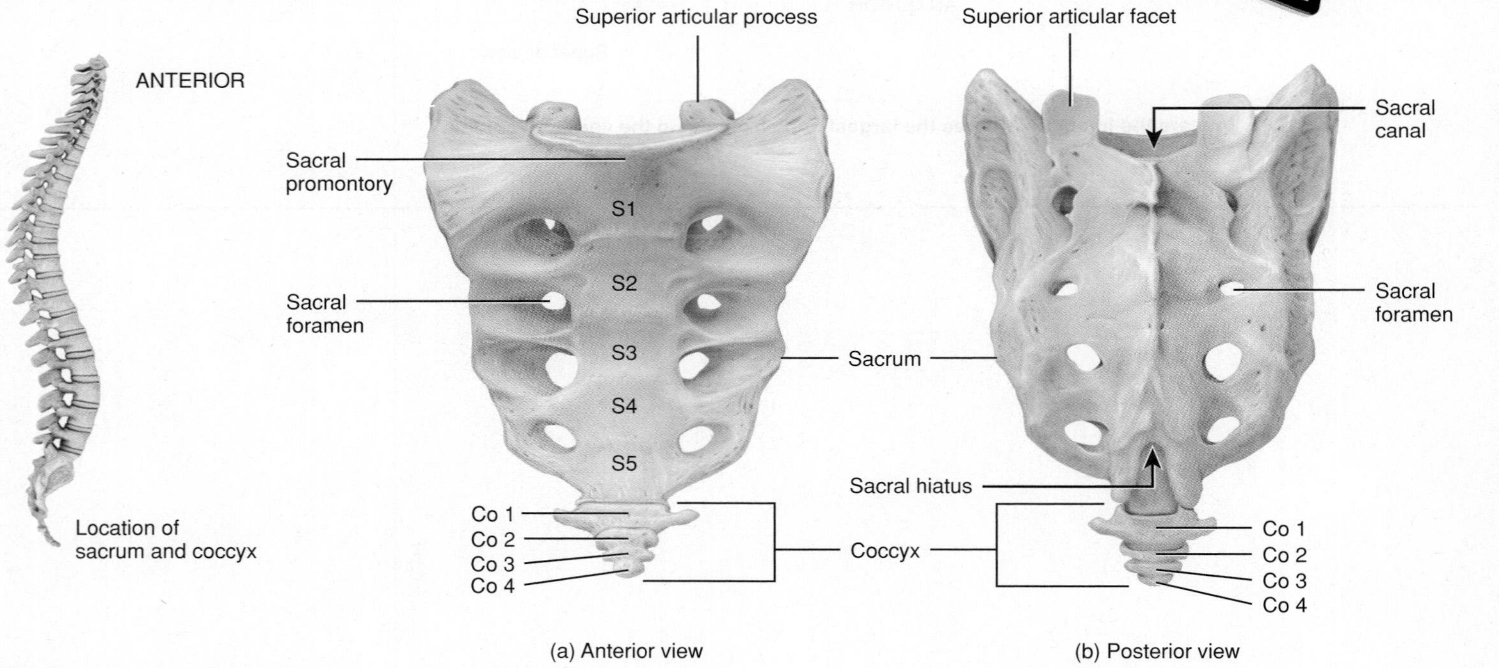

What is the function of the sacral foramina?

6.9 THORAX

OBJECTIVE • Identify the bones of the thorax and their principal markings.

The term ***thorax*** refers to the entire chest. The skeletal portion of the thorax, the ***thoracic cage***, is a bony cage formed by the sternum, costal cartilages, ribs, and the bodies of the thoracic vertebrae (Figure 6.17). The thoracic cage encloses and protects the organs in the thoracic cavity and upper abdominal cavity. It also provides support for the bones of the shoulder girdle and upper limbs.

Sternum

The ***sternum***, or breastbone, is a flat, narrow bone located in the center of the anterior thoracic wall and consists of three parts that usually fuse by age 25 (Figure 6.17). The upper part is the *manubrium* (ma-NOO-brē-um = handle-like); the middle and largest part is the *body*; and the lowest, smallest part is the *xiphoid process* (ZĪ-foyd = sword-shaped).

The manubrium articulates with the clavicles, first rib, and part of the second rib. The body of the sternum articulates directly or indirectly with part of the second rib and the third through tenth ribs. The xiphoid process consists of hyaline cartilage during infancy and childhood and does not ossify completely until about age 40. It has no ribs attached to it but provides attachment for some abdominal muscles. If the hands of a rescuer are incorrectly positioned during cardiopulmonary resuscitation (CPR), there is danger of fracturing the xiphoid process and driving it into internal organs.

Ribs

Twelve pairs of ***ribs*** make up the sides of the thoracic cavity (Figure 6.17). The ribs increase in length from the first through seventh ribs, then decrease in length to the twelfth rib. Each rib articulates posteriorly with its corresponding thoracic vertebra.

The first through seventh pairs of ribs have a direct anterior attachment to the sternum by a strip of hyaline cartilage called *costal cartilage* (*cost-* = rib). These ribs are called *true ribs*. The remaining five pairs of ribs are termed *false ribs* because their costal cartilages either attach indirectly to the sternum or do not attach to the sternum at all. The cartilages of the eighth, ninth, and tenth pairs of ribs attach to each other and then to the cartilages of the seventh pair of ribs. The eleventh and twelfth false ribs are also known as *floating ribs* because the costal cartilage at their anterior ends does not attach to the sternum at all. Floating ribs attach only posteriorly to the thoracic vertebrae. Spaces between ribs, called *intercostal spaces*, are occupied by intercostal muscles, blood vessels, and nerves.

Figure 6.17 Skeleton of the thorax.

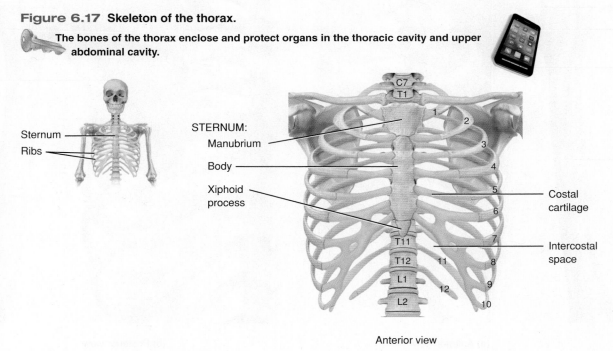

The bones of the thorax enclose and protect organs in the thoracic cavity and upper abdominal cavity.

Anterior view

 Which ribs are true ribs? False ribs? Floating ribs?

CLINICAL CONNECTION | Rib Fractures

Rib fractures are the most common chest injuries, and they usually result from direct blows, most often from impact with a steering wheel, falls, and crushing injuries to the chest. In some cases, fractured ribs may puncture the heart, great vessels of the heart, lungs, trachea, bronchi, esophagus, spleen, liver, and kidneys. Rib fractures are usually quite painful. Rib fractures are no longer bound with bandages because of the pneumonia that would result from lack of proper lung ventilation. •

✓ CHECKPOINT

19. What are the functions of the bones of the thorax?

20. What are the parts of the sternum?

6.10 PECTORAL (SHOULDER) GIRDLE

OBJECTIVE • **Identify the bones of the pectoral (shoulder) girdle and their principal markings.**

The *pectoral girdles* (PEK-tō-ral) or *shoulder girdles* attach the bones of the upper limbs to the axial skeleton (Figure 6.18). The right and left pectoral girdles each consist of two bones: a clavicle and a scapula. The clavicle, the anterior component, articulates with the sternum, and the scapula, the posterior component, articulates with the clavicle and the humerus. The pectoral girdles do not articulate with the

Figure 6.18 Right pectoral (shoulder) girdle.

 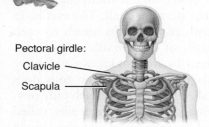

The pectoral girdle attaches the bones of the upper limb to the axial skeleton.

Pectoral girdle:
Clavicle
Scapula

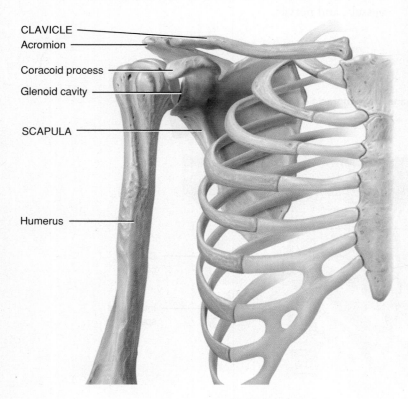

CLAVICLE
Acromion
Coracoid process
Glenoid cavity
SCAPULA
Humerus

(a) Anterior view

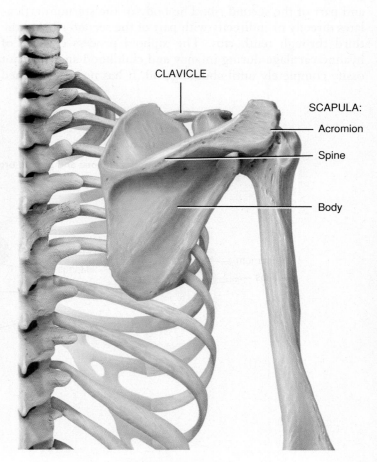

CLAVICLE

SCAPULA:
Acromion
Spine
Body

(b) Posterior view

 Which bones make up a pectoral girdle?

vertebral column. The joints of the shoulder girdles are freely movable and thus allow movements in many directions.

Clavicle

Each *clavicle* (KLAV-i-kul = key) or collarbone is a long, slender S-shaped bone that is positioned horizontally above the first rib. The medial end of the clavicle articulates with the sternum, and the lateral end articulates with the acromion of the scapula (Figure 6.18). Because of its position, the clavicle transmits mechanical force from the upper limb to the trunk. If the force transmitted to the clavicle is excessive, as when you fall on your outstretched arm, a *fractured clavicle* may result.

Scapula

Each *scapula* (SCAP-ū-la), or *shoulder blade*, is a large, flat, triangular bone situated in the posterior part of the thorax (Figure 6.18). A sharp ridge, the *spine*, runs diagonally across the posterior surface of the flattened, triangular *body* of the scapula. The lateral end of the spine, the *acromion* (a-KRŌ-mē-on; *acrom-* = topmost), is easily felt as the high point of the shoulder and is the site of articulation with the clavicle. Inferior to the acromion is a depression called the *glenoid cavity*. This cavity articulates with

the head of the humerus (arm bone) to form the shoulder joint. Also present on the scapula is a projection called the *coracoid process* (KOR-a-koyd = like a crow's beak) to which muscles attach.

✓ CHECKPOINT

21. What bones make up the pectoral girdle? What is the function of the pectoral girdle?

6.11 UPPER LIMB

OBJECTIVE • **Identify the bones of the upper limb and their principal markings.**

Each *upper limb* consists of 30 bones. Each upper limb includes a humerus in the arm; ulna and radius in the forearm; and 8 carpals (wrist bones), 5 metacarpals (palm bones), and 14 phalanges (finger bones) in the hand (see Figure 6.6). Exhibits 6.G–6.I describe the bones of the upper limb in more detail.

✓ CHECKPOINT

22. What bones form the upper limb, from proximal to distal?

CLINICAL CONNECTION | Fractured Clavicle

The clavicle transmits mechanical force from the free upper limb to the trunk. If the force transmitted to the clavicle is excessive, as when you fall on your outstretched arm, a **fractured clavicle** may result. A fractured clavicle may also result from a blow to the superior part of the anterior thorax, for example, as a result of an impact following an automobile accident. In fact, the clavicle is one of the most frequently broken bones in the body. Because the junction of the clavicle's two curves is its weakest point, the clavicular midregion is the site most frequently fractured. Even in the absence of fracture, compression of the clavicle as a result of automobile accidents involving the use of shoulder-harness seatbelts often causes damage to the brachial plexus (the network of nerves that enter the upper limb), which lies between the clavicle and the second rib. A fractured clavicle is usually treated with a figure-of-eight sling to keep the arm from moving outward. •

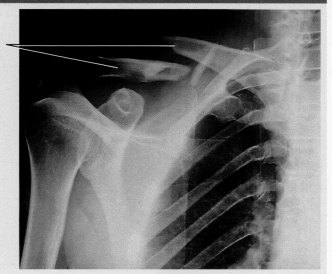

Fractured clavicle

EXHIBIT 6.G · Humerus *(Figure 6.19)*

OBJECTIVE • Identify the location and surface landmarks of the humerus.

The **humerus** (HŪ-mer-us), or arm bone, is the longest and largest bone of the upper limb (Figure 6.19). At the shoulder it articulates with the scapula, and at the elbow it articulates with both the ulna and radius. The proximal end of the humerus consists of a *head* that articulates with the glenoid cavity of the scapula. It also has an *anatomical neck*, the former site of the epiphyseal (growth) plate, which is a groove just distal to the head. The *surgical neck* is below the anatomical neck and is so named because fractures often occur here. The *body* of the humerus contains a roughened, V-shaped area called the *deltoid tuberosity* where the deltoid muscle attaches. At the distal end of the humerus, the *capitulum* (ka-PIT-ū-lum = small head) is a rounded knob that articulates with the head of the radius. The *radial fossa* is a depression that receives the head of the radius when the forearm is flexed (bent). The *trochlea* (TRŌK-lē-a) is a spool-shaped surface that articulates with the ulna. The *coronoid fossa* (KOR-ō-noyd = crown-shaped) is a depression that receives part of the ulna when the forearm is flexed. The *olecranon fossa* (ō-LEK-ra-non) is a depression on the back of the bone that receives the olecranon of the ulna when the forearm is extended (straightened).

✓ **CHECKPOINT**

Distinguish between the anatomical neck and surgical neck of the humerus.

Figure 6.19 Right humerus in relation to the scapula, ulna, and radius.

The humerus is the longest and largest bone of the upper limb.

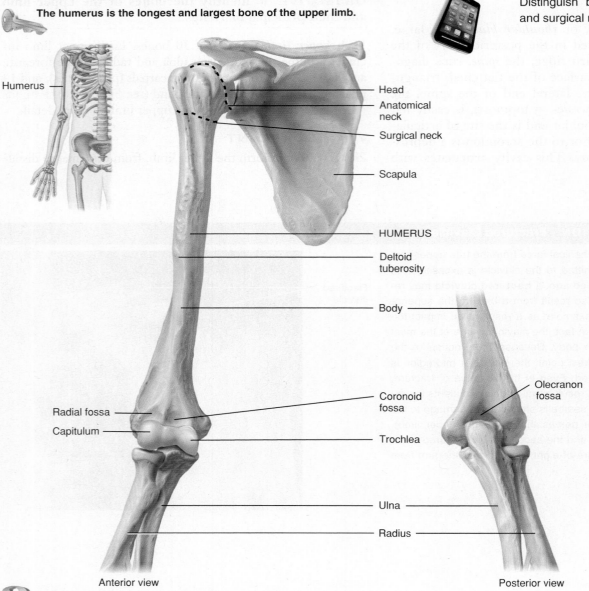

Humerus

Head
Anatomical neck
Surgical neck

Scapula

HUMERUS

Deltoid tuberosity

Body

Radial fossa
Capitulum

Coronoid fossa

Trochlea

Olecranon fossa

Ulna

Radius

Anterior view

Posterior view

With which part of the scapula does the humerus articulate?

EXHIBIT 6.H Ulna and Radius *(Figure 6.20)*

OBJECTIVE • **Identify the location and surface landmarks of the ulna and radius.**

The **ulna** is on the medial aspect (little-finger side) of the forearm and is longer than the radius (Figure 6.20). At the proximal end of the ulna is the *olecranon*, which forms the prominence of the elbow. The *coronoid process*, together with the olecranon, receives the trochlea of the humerus. The trochlea of the humerus also fits into the *trochlear notch*, a large curved area between

the olecranon and the coronoid process. The *radial notch* is a depression for the head of the radius. A *styloid process* is at the distal end of the ulna.

The **radius** is located on the lateral aspect (thumb side) of the forearm. The proximal end of the radius has a disc-shaped *head* that articulates with the capitulum of the humerus and radial notch of the ulna. It has

a raised, roughened area called the *radial tuberosity* that provides a point of attachment for the biceps brachii muscle. The distal end of the radius articulates with three carpal bones of the wrist. Also at the distal end is a *styloid process*. Fracture of the distal end of the radius is the most common fracture in adults older than 50 years.

✓ **CHECKPOINT**

What structure serves as a point of attachment for the biceps brachii muscle?

Figure 6.20 Right ulna and radius in relation to the humerus and carpals.

In the forearm, the longer ulna is on the medial side, and the radius is on the lateral side.

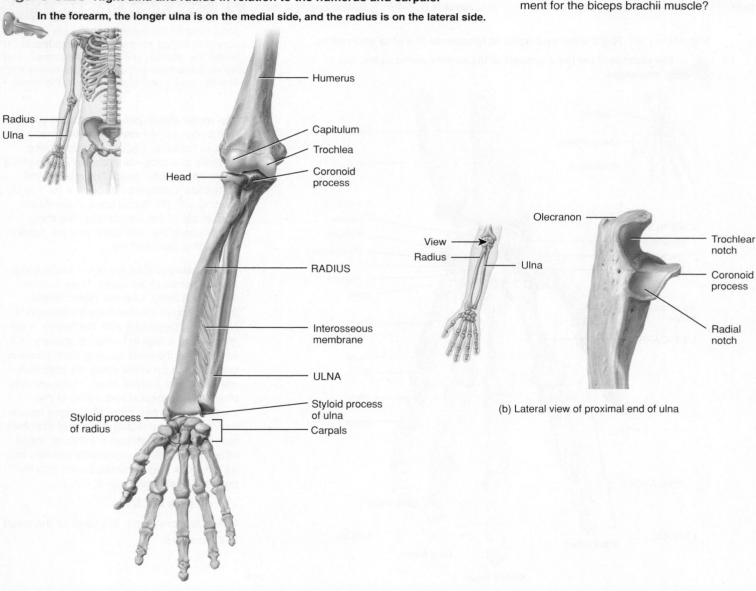

(a) Anterior view

(b) Lateral view of proximal end of ulna

? **What part of the ulna is called the elbow?**

EXHIBIT 6.I

Carpals, Metacarpals, and Phalanges *(Figure 6.21)*

OBJECTIVE • **Identify the location and surface landmarks of the bones of the hand.**

The **carpus *(wrist)*** of the hand contains eight small bones, the **carpals**, held together by ligaments (Figure 6.21). The carpals are arranged in two transverse rows, with four bones in each row, and they are named for their shapes. In the anatomical position, the carpals in the top row, from the lateral to medial position, are the **scaphoid** (SKAF-oid = boatlike), **lunate** (LOO-nāt = moon-shaped), **triquetrum** (trī-KWĒ-trum = three-cornered), and **pisiform** (PĪ-si-form = pea-shaped). In about

70 percent of carpal fractures, only the scaphoid is broken because of the force transmitted through it to the radius. The carpals in the bottom row, from the lateral to medial position, are the **trapezium** (tra-PĒ-zē-um = four-sided figure with no two sides parallel), **trapezoid** (TRAP-e-zoid = four-sided figure with two sides parallel), **capitate** (KAP-i-tāt = head-shaped; the largest carpal bone, whose rounded projection, the head, articulates with the lunate), and **hamate** (HAM-āt = hooked; named for

a large hook-shaped projection on its anterior surface). Together, the concavity formed by the pisiform and hamate (on the ulnar side) and the scaphoid and trapezium (on the radial side) constitute a space called the **carpal tunnel**. Through it pass the long flexor tendons of the digits and thumb and the median nerve.

CLINICAL CONNECTION | Carpal Tunnel Syndrome

Narrowing of the carpal tunnel gives rise to a condition called **carpal tunnel syndrome**, in which the median nerve is compressed. The nerve compression causes pain, numbness, tingling, and muscle weakness in the hand. •

The **metacarpus *(palm)*** of the hand contains five bones called **metacarpals** (*meta-* = after or beyond). Each metacarpal bone consists of a proximal *base*, an intermediate *body*, and a distal *head*. The metacarpal bones are numbered I through V (or 1 to 5), starting with the lateral bone in the thumb. The heads of the metacarpals are commonly called the "knuckles" and are readily visible in a clenched fist.

The **phalanges** (fa-LAN-jēz = battle lines) are the bones of the digits. They number 14 in each hand. Like the metacarpals, the phalanges are numbered I through V (or 1 to 5), beginning with the thumb. A single bone of a digit is termed a *phalanx* (FĀ-lanks). Like the metacarpals, each phalanx consists of a proximal *base*, an intermediate *body*, and a distal *head*. There are two phalanges (proximal and distal) in the thumb (*pollex*) and three phalanges (proximal, middle, and distal) in each of the other four digits. In order from the thumb, these other four digits are commonly referred to as the index finger, middle finger, ring finger, and little finger (Figure 6.21).

✓ CHECKPOINT

Which is more distal, the base or the head of the carpals?

Figure 6.21 Right wrist and hand in relation to the ulna and radius.

The skeleton of the hand consists of the carpals, metacarpals, and phalanges.

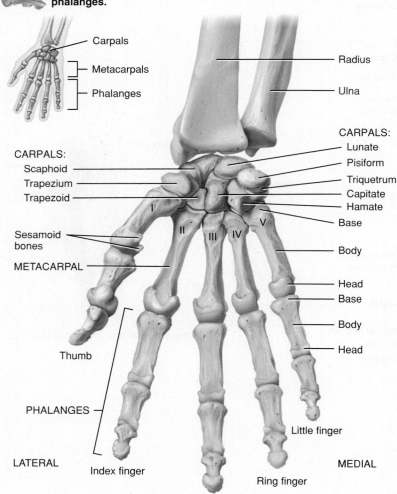

Anterior view

 What part of which bones are commonly called the knuckles?

6.12 PELVIC (HIP) GIRDLE

OBJECTIVE • Identify the bones of the pelvic (hip) girdle and their principal markings.

The *pelvic (hip) girdle* consists of the two *hip bones*, also called *coxal bones* (Figure 6.22). The pelvic girdle provides a

strong, stable support for the vertebral column, protects the pelvic viscera, and attaches the lower limbs to the axial skeleton. The hip bones are united to each other in front at a joint called the *pubic symphysis* (PŪ-bik SIM-fi-sis); posteriorly they unite with the sacrum at the sacroiliac joint.

Together with the sacrum and coccyx, the two hip bones of the pelvic girdle form a basinlike structure called the *pelvis*

Figure 6.22 Female pelvic (hip) girdle.

🔑 **The hip bones are united in front at the pubic symphysis and in back at the sacrum.**

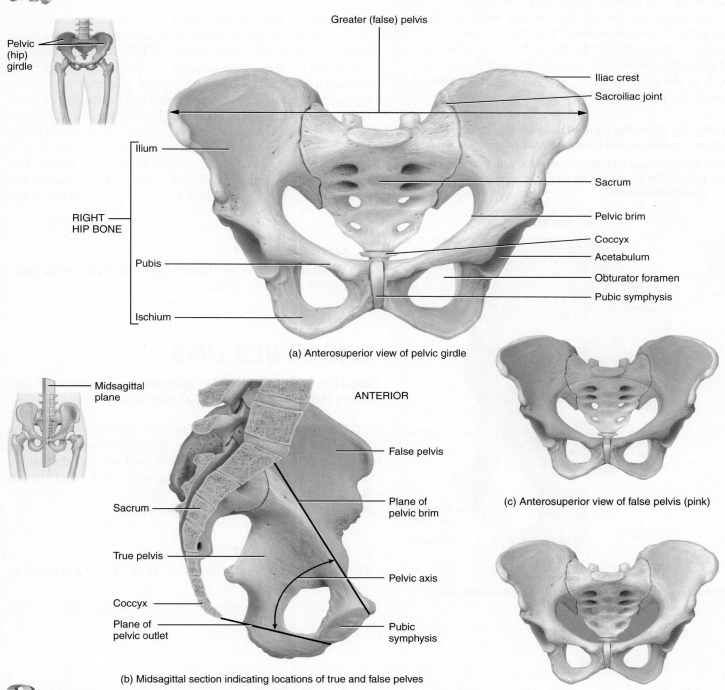

(a) Anterosuperior view of pelvic girdle

(b) Midsagittal section indicating locations of true and false pelves

(c) Anterosuperior view of false pelvis (pink)

(d) Anterosuperior view of true pelvis (blue)

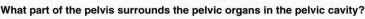

❓ **What part of the pelvis surrounds the pelvic organs in the pelvic cavity?**

(plural is *pelvises* or *pelves*). In turn, the bony pelvis is divided into upper and lower portions by a boundary called the *pelvic brim* (Figure 6.22). The part of the pelvis above the pelvic brim is called the *false (greater) pelvis*. The false pelvis is actually part of the abdomen and does not contain any pelvic organs, except for the urinary bladder, when it is full, and the uterus during pregnancy. The part of the pelvis below the pelvic brim is called the *true (lesser) pelvis*. The true pelvis surrounds the pelvic cavity (see Figure 1.8). The upper opening of the true pelvis is called the *pelvic inlet*, and the lower opening of the true pelvis is called the *pelvic outlet*. The *pelvic axis* is an imaginary curved line passing through the true pelvis; it joins the central points of the planes of the pelvic inlet and outlet. During childbirth, the pelvic axis is the course taken by the baby's head as it descends through the pelvis.

Each of the two hip bones of a newborn is composed of three parts: the ilium, the pubis, and the ischium (Figure 6.23).

Figure 6.23 Right hip bone. The lines of fusion of the ilium, ischium, and pubis are not always visible in an adult hip bone.

The two hip bones form the pelvic girdle, which attaches the lower limbs to the axial skeleton and supports the vertebral column and viscera.

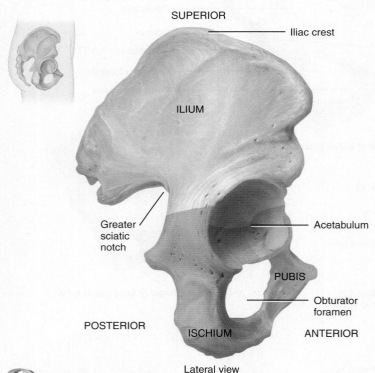

SUPERIOR

Iliac crest

ILIUM

Greater
sciatic
notch

Acetabulum

PUBIS

Obturator
foramen

POSTERIOR

ISCHIUM

ANTERIOR

Lateral view

Which bone fits into the socket formed by the acetabulum?

The *ilium* (= flank) is the largest of the three subdivisions of the hip bone. Its upper border is the *iliac crest*. On the lower surface is the *greater sciatic notch* (sī-AT-ik) through which the sciatic nerve, the longest nerve in the body, passes. The *ischium* (IS-kē-um = lip) is the lower, posterior part of the hipbone. The *pubis* (PŪ-bis = pubic hair) is the lower, anterior part of the hipbone. By age 23 years, the three separate bones have fused into one. The deep fossa (depression) where the three bones meet is the *acetabulum* (as-e-TAB-ū-lum = vinegar cup). It is the socket for the head of the femur. The ischium joins with the pubis, and together they surround the *obturator foramen* (OB-too-rā-ter), the largest foramen in the skeleton.

✓ CHECKPOINT

23. What bones make up the pelvic girdle? What is the function of the pelvic girdle?

6.13 LOWER LIMB

OBJECTIVE • List the skeletal components of the lower limb and their principal markings.

Each *lower limb* is composed of 30 bones: the femur in the thigh; the patella (kneecap); the tibia and fibula in the leg (the part of the lower limb between the knee and the ankle); and 7 tarsals (ankle bones), 5 metatarsals, and 14 phalanges (toes) in the foot (see Figure 6.6). Exhibits 6.J–6.L describe the bones of the lower limb in more detail.

✓ CHECKPOINT

24. What bones form the lower limb, from proximal to distal?

25. What are the functions of the arches of the foot?

EXHIBIT 6.J Femur and Patella *(Figure 6.24)*

OBJECTIVE • Identify the location and surface features of the femur and patella.

FEMUR

The ***femur*** (thigh bone) is the longest, heaviest, and strongest bone in the body (Figure 6.24). Its proximal end articulates with the hip bone, and its distal end articulates with the tibia and patella. The body of the femur bends medially, and as a result, the knee joints are brought nearer to the midline of the body. The bend is greater in females because the female pelvis is broader.

The *head* of the femur articulates with the acetabulum of the hip bone to form the *hip joint*. The *neck* of the femur is a constricted region below the head. A fairly common fracture in the elderly occurs at the neck of the femur, which becomes so weak that it fails to support the weight of the body. Although it is actually the femur that is fractured, this condition is commonly known as a broken hip. The *greater trochanter* (trō-KAN-ter) is a projection felt and seen in front of the hollow on the side of the hip. It is where some of the thigh and buttock muscles attach and serves as a landmark for intramuscular injections in the thigh.

The distal end of the femur expands into the *medial condyle* and *lateral condyle* projections that articulate with the tibia. The *patellar surface* is located on the anterior surface of the femur between the condyles.

PATELLA

The ***patella*** (= little dish), or kneecap, is a small, triangular bone in front of the joint between the femur and tibia, commonly known as the knee joint (Figure 6.24). The patella develops in the tendon of the quadriceps femoris muscle. Its functions are to increase the leverage of the tendon, maintain the position of the tendon when the knee is flexed, and protect the knee joint. During normal flexion and extension of the knee, the patella tracks (glides) up and down in the groove between the two femoral condyles.

**CLINICAL CONNECTION |
Patellofemoral Stress
Syndrome**

Patellofemoral stress syndrome ("runner's knee") is one of the most common problems runners experience. During normal flexion and extension of the knee, the patella tracks (glides) superiorly and inferiorly in the groove between the femoral condyles. In patellofemoral stress syndrome, normal tracking does not occur; instead, the patella tracks laterally as well as superiorly and inferiorly, and the increased pressure on the joint causes aching or tenderness around or under the patella. The pain typically occurs after a person has been sitting for awhile, especially after exercise. It is worsened by squatting or walking down stairs. One cause of runner's knee is constantly walking, running, or jogging on the same side of the road. Because roads slope down on the sides, the knee that is closer to the center of the road endures greater mechanical stress because it does not fully extend during a stride. Other predisposing factors include running on hills, running long distances, and an anatomical deformity called knock-knee. •

✓ **CHECKPOINT**

What is the clinical importance of the greater trochanter?

EXHIBIT 6.J CONTINUES

EXHIBIT 6.J   Femur and Patella   *(Figure 6.24)*   CONTINUED

Figure 6.24 Right femur in relation to the hip bone, patella, tibia, and fibula.

 The head of the femur articulates with the acetabulum of the hip bone to form the hip joint.

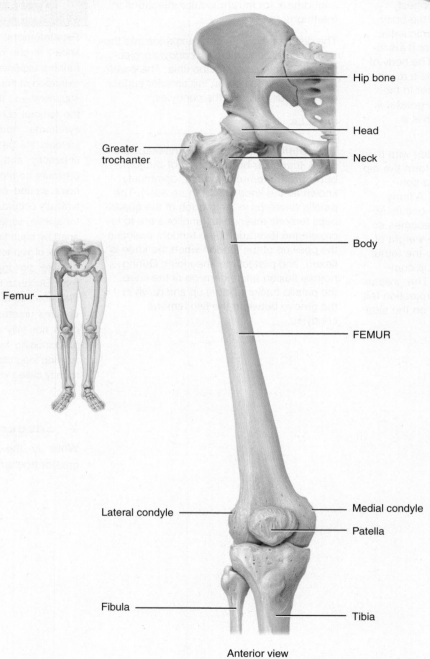

Hip bone

Head

Greater trochanter

Neck

Body

Femur

FEMUR

Lateral condyle

Medial condyle

Patella

Fibula

Tibia

Anterior view

 With which bones does the distal end of the femur articulate?

EXHIBIT 6.K Tibia and Fibula *(Figure 6.25)*

OBJECTIVE • **Identify the location and surface features of the tibia and fibula.**

The ***tibia***, or shin bone, is the larger, medial, weight-bearing bone of the leg (Figure 6.25). The tibia articulates at its proximal end with the femur and fibula, and at its distal end with the fibula and talus of the ankle. The proximal end of the tibia expands into a *lateral condyle* and a *medial condyle*, projections that articulate with the condyles of the femur to form the *knee joint*. The *tibial tuberosity* is on the anterior surface below the condyles and is a point of attachment for the patellar ligament. The medial surface of the distal end of the tibia forms the *medial malleolus* (ma-LĒ-ō-lus = little hammer), which articulates with the talus of the ankle and forms the prominence that can be felt on the medial surface of your ankle.

CLINICAL CONNECTION |
Shin Splints

Shin splints is the name given to soreness or pain along the tibia. Probably caused by inflammation of the periosteum brought about by repeated tugging of the attached muscles and tendons, it is often the result of walking or running up and down hills. •

The ***fibula*** is parallel and lateral to the tibia (Figure 6.25) and is considerably smaller than the tibia. The *head* of the fibula articulates with the lateral condyle of the tibia below the knee joint. The distal end has a projection called the *lateral malleolus* that articulates with the talus of the ankle. This forms the prominence on the lateral surface of the ankle. As shown in Figure 6.25, the fibula also articulates with the tibia at the *fibular notch*.

✓ **CHECKPOINT**

Which structures form the medial and lateral prominences of the ankle?

EXHIBIT 6.K CONTINUES

EXHIBIT 6.K Tibia and Fibula *(Figure 6.25)* **CONTINUED**

Figure 6.25 Right tibia and fibula in relation to the femur, patella, and talus.

 The tibia articulates with the femur and fibula proximally and with the fibula and talus distally, while the fibula articulates proximally with the tibia below the knee joint and distally with the talus.

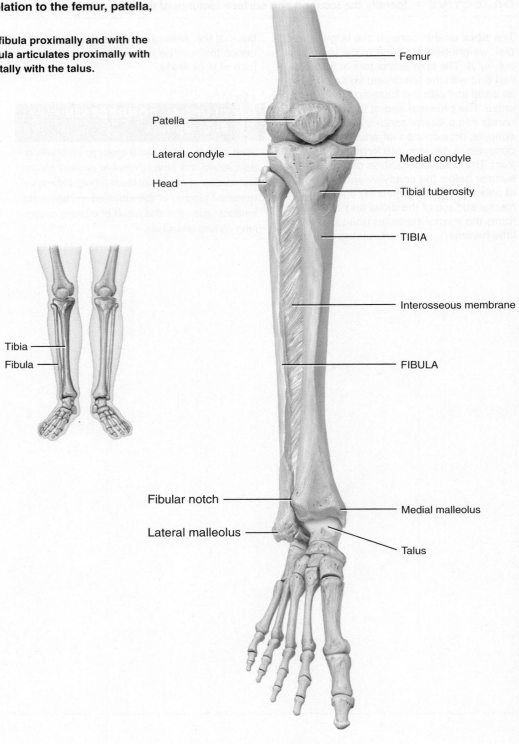

Femur

Patella

Lateral condyle

Head

Medial condyle

Tibial tuberosity

TIBIA

Interosseous membrane

FIBULA

Tibia

Fibula

Fibular notch

Lateral malleolus

Medial malleolus

Talus

Anterior view

 Which leg bone bears the weight of the body?

EXHIBIT 6.L | Tarsals, Metatarsals, and Phalanges (*Figures 6.26 and 6.27*)

OBJECTIVE • **Identify the location and surface features of the bones of the foot.**

The **tarsus (ankle)** of the foot contains seven bones, the **tarsals**, held together by ligaments (Figure 6.26). Of these, the **talus** (TĀ-lus = ankle bone) and **calcaneus** (kal-KĀ-nē-us = heel bone) are located on the posterior part of the foot. The anterior part of the ankle contains the **cuboid** (KŪ-boyd), **navicular** (na-VIK-ū-lar), and three

cuneiform bones (KŪ-nē-i-form) called the *first, second*, and *third cuneiforms*. The talus is the only bone of the foot that articulates with the fibula and tibia. It articulates medially with the medial malleolus of the tibia and laterally with the lateral malleolus of the fibula. During walking, the talus initially bears the entire weight of the body. About half the

weight is then transmitted to the calcaneus. The remainder is transmitted to the other tarsal bones. The calcaneus is the largest and strongest of the tarsals.

Five bones called **metatarsals** and numbered I to V (or 1 to 5) from the medial to lateral position form the skeleton of the

Figure 6.26 Right foot.

The skeleton of the foot consists of the tarsals, metatarsals, and phalanges.

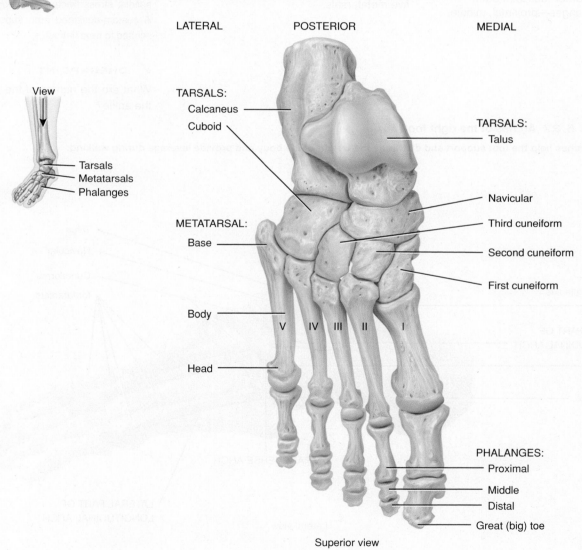

LATERAL   POSTERIOR   MEDIAL

View

TARSALS:
- Calcaneus
- Cuboid

Tarsals
Metatarsals
Phalanges

METATARSAL:
- Base
- Body
- Head

V   IV   III   II   I

TARSALS:
- Talus

Navicular
Third cuneiform
Second cuneiform
First cuneiform

PHALANGES:
- Proximal
- Middle
- Distal
- Great (big) toe

Superior view

 Which tarsal bone articulates with the tibia and fibula?

EXHIBIT 6.L CONTINUES

metatarsus. Like the metacarpals of the palm, each metatarsal consists of a proximal *base*, an intermediate *body*, and a distal *head*. The first metatarsal, which is connected to the big toe, is thicker than the others because it bears more weight.

The ***phalanges*** of the foot resemble those of the hand both in number and arrangement. Each also consists of a proximal *base*, an intermediate *body*, and a distal *head*. The great or big toe (*hallux*) has two large, heavy phalanges—proximal and distal. The other four toes each have three phalanges—proximal, middle, and distal.

The bones of the foot are arranged in two ***arches*** (Figure 6.27). These arches enable the foot to support the weight of the body, provide an ideal distribution of body weight over the hard and soft tissues of the foot, and provide leverage while walking. The arches are not rigid—they yield as weight is applied and spring back when the weight is lifted, thus helping to absorb shocks. The *longitudinal arch* extends from the front to the back of the foot and has two parts, medial and lateral. The *transverse arch* is formed by the navicular, three cuneiforms, and the bases of the five metatarsals.

CLINICAL CONNECTION |
Flatfoot

The bones composing the arches of the foot are held in position by ligaments and tendons. If these ligaments and tendons are weakened, the height of the medial longitudinal arch may decrease or "fall." The result is **flatfoot**, the causes of which include excessive weight, postural abnormalities, weakened supporting tissues, and genetic predisposition. Fallen arches may lead to inflammation of the deep fascia of the sole (plantar fasciitis), Achilles tendinitis, shin splints, stress fractures, bunions, and calluses. A custom-designed arch support often is prescribed to treat flatfoot. •

✓ **CHECKPOINT**

What are the names of the seven bones of the ankle?

Figure 6.27 **Arches of the right foot.**

Arches help the foot support and distribute the weight of the body and provide leverage during walking.

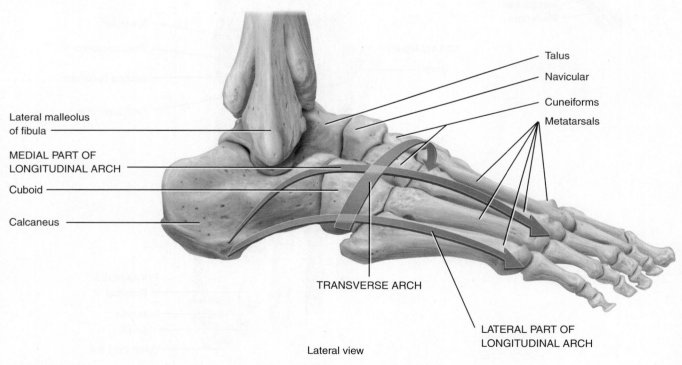

Lateral malleolus of fibula

MEDIAL PART OF LONGITUDINAL ARCH

Cuboid

Calcaneus

Talus

Navicular

Cuneiforms

Metatarsals

TRANSVERSE ARCH

LATERAL PART OF LONGITUDINAL ARCH

Lateral view

What structural aspect of the arches allows them to absorb shocks?

6.14 COMPARISON OF FEMALE AND MALE SKELETONS

OBJECTIVE • **Identify the principal structural differences between female and male skeletons.**

The bones of a male are generally larger and heavier than those of a female. The articular ends are thicker in relation to the shafts. In addition, because certain muscles of the male are larger than those of the female, the points of muscle attachment—tuberosities, lines, and ridges—are larger in the male skeleton.

Many significant structural differences between the skeletons of females and males are related to pregnancy and childbirth. Because the female's pelvis is wider and shallower than the male's, there is more space in the true pelvis of the female, especially in the pelvic inlet and pelvic outlet, which accommodate the passage of the infant's head at birth. Several of the significant differences between the female and male pelves are shown in Table 6.4.

✓ **CHECKPOINT**
26. What features of the female skeleton differ from the male skeleton to allow for pregnancy and childbirth?

6.15 AGING AND THE SKELETAL SYSTEM

OBJECTIVE • **Describe the effects of aging on the skeletal system.**

From birth through adolescence, more bone is produced than is lost during bone remodeling. In young adults, the rates of bone production and loss are about the same. As the levels of sex hormones diminish during middle age, especially in women after menopause, a decrease in bone mass occurs because bone destruction outpaces bone formation. Because women's bones generally are smaller than men's bones to begin with, loss of bone mass in old age typically causes greater

TABLE 6.4

Comparison of the Pelvis in Females and Males

POINT OF COMPARISON	FEMALE	MALE
General structure	Light and thin	Heavy and thick
False (greater) pelvis	Shallow	Deep
Pelvic inlet	Larger and more oval	Smaller and heart-shaped
Acetabulum	Small and faces anteriorly	Large and faces laterally
Obturator foramen	Oval	Round
Pubic arch	Greater than 90° angle	Less than 90° angle

Anterior views

Making an Impact on Bone Strength

Just as muscle responds to appropriate exercise training by becoming stronger, so does bone tissue. Researchers have evaluated the impact that certain exercise variables have on bone strength. Bone responds to activities that cause physical deformation of bone cells, rather than a high cardiovascular workload. In general, bones respond best to weight-bearing exercise that is dynamic and applies relatively high force to bone tissue. What does this mean?

Weight-bearing Exercise

Weight-bearing means weight is placed on the bones. Walking, running, and jumping place weight on the bones. So do running

sports such as soccer, racquet sports, and basketball. In these activities, the weight is supplied by the body's own mass, affected by gravity. Adding weight, as in strength training, also applies weight to the bones. Training with free weights generally applies more force to the spine than using machines that put you in a seated or lying down position. Supported activities such as cycling and swimming are less effective than other activities for building bone strength, but are still much better than no exercise at all.

Dynamic Idea

Dynamic, for bones, means that the forces placed on bone are moving and changing, rather than stationary and continuous. Racquet sports and basketball are examples of activities that apply dynamic forces. They involve bursts of running, stopping, turning,

and, for basketball, jumping. The lines of force on the bones are constantly changing. Hitting the ball in racquet sports, volleyball, baseball, and softball provide dynamic forces to the arm that vibrate into the rest of the body.

The opposite of dynamic is continuous exercise, activities that do not offer much variation in force. The bones seem to get used to continuous, repetitive motions like cross-country skiing and distance running. While these are great weight-bearing activities, they do not stimulate the bones as much as dynamic activities. Research also suggests that two shorter workouts, with six to eight hours of rest between sessions, provide more bone stimulation than a single, longer period of continuous activity.

Impact Intensity

For bone adaptation, high-intensity exercise means maximizing the mechanical force placed on the bone. Running and jumping activities are good bone builders because they place more force on bone tissue than walking. When heavier resistance is applied in strength training, bone response is greater.

> *Think It Over . . .* Adding relatively high-intensity, high-impact activity to your exercise program can cause injury. Describe how you could begin slowly and build gradually to prevent injury from these new bone-building activities.

problems in women. These factors contribute to a higher incidence of osteoporosis in women.

Aging has two main effects on the skeletal system: Bones become more brittle and lose mass. Bone brittleness results from a decrease in the rate of protein synthesis and a decrease in the production of human growth hormone, which diminishes the production of the collagen fibers that give bone its strength and flexibility. As a result, inorganic minerals gradually constitute a greater proportion of the bone extracellular matrix. Loss of bone mass results from demineralization and usually begins after age 30 in females, accelerates greatly around age 45 as levels of estrogens decrease, and continues until as much as 30 percent of the calcium in bones is lost by age 70. Once bone loss begins in females, about 8 percent of bone mass is lost every 10 years. In males, calcium loss from bone typically does not begin

until after age 60, and about 3 percent of bone mass is lost every 10 years. The loss of calcium from bones is one of the problems in osteoporosis (described in the Common Disorders section). Loss of bone mass also leads to bone deformity, pain, stiffness, some loss of height, and loss of teeth.

✓ CHECKPOINT

27. How does aging affect the composition of bone and bone mass?

• • •

To appreciate the many ways that the skeletal system contributes to homeostasis of other body systems, examine Focus on Homeostasis: The Skeletal System. Next, in Chapter 7, we will see how joints both hold the skeleton together and permit it to participate in movements.

BODY SYSTEM	CONTRIBUTION OF THE SKELETAL SYSTEM
For all body systems	Bones provide support and protection for internal organs; bones store and release calcium, which is needed for proper functioning of most body tissues.
Integumentary system	Bones provide strong support for overlying muscles and skin; joints provide flexibility while skin compensates for the change in joint angle.
Muscular system	Bones provide attachment points for skeletal muscles and leverage for the muscles to bring about body movements; contraction of skeletal muscle requires calcium ions.
Nervous system	The skull and vertebrae protect the brain and spinal cord; a normal blood level of calcium is needed for normal functioning of neurons and neuroglia.
Endocrine system	Bones store and release calcium, which is regulated by parathyroid hormone (PTH) and calcitonin (CT).
Cardiovascular system	Red bone marrow carries out hemopoiesis (blood cell formation); rhythmic beating of the heart requires calcium ions.
Lymphatic system and immunity	Red bone marrow produces white blood cells involved in immune responses.
Respiratory system	The axial skeleton of the thorax protects the lungs; rib movements assist breathing; some muscles used for breathing attach to bones by means of tendons.
Digestive system	Teeth masticate (chew) food; the rib cage protects the esophagus, stomach, and liver; the pelvis protects portions of the intestines.
Urinary system	Ribs partially protect the kidneys, and the pelvis protects the urinary bladder and urethra.
Reproductive systems	The pelvis protects the ovaries, uterine (fallopian) tubes, and uterus in females and part of the ductus (vas) deferens and accessory glands in males; bones are an important source of calcium needed for milk synthesis during lactation.

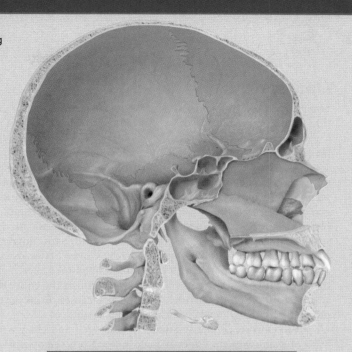

THE SKELETAL SYSTEM

COMMON DISORDERS

Osteoporosis

Osteoporosis (os'-tē-ō-pō-RŌ-sis; *por-* = passageway; *-osis* = condition), literally a condition of porous bones, affects 10 million people a year in the United States (Figure 6.28). In addition, 18 million people have low bone mass (*osteopenia*), which puts them at risk for osteoporosis. The basic problem is that bone resorption (breakdown) outpaces bone deposition (formation). In large part this is due to depletion of calcium from the body—more calcium is lost in urine, feces, and sweat than is absorbed from the diet. Bone mass becomes so depleted that bones fracture, often spontaneously, under the mechanical stresses of everyday living. For example, a hip fracture might result from simply sitting down too quickly. In the United States, osteoporosis causes more than one and a half million fractures a year, mainly in the hips, wrists, and vertebrae. Osteoporosis afflicts the entire skeletal system. In addition to fractures, osteoporosis causes shrinkage of vertebrae, height loss, hunched backs, and bone pain.

Osteoporosis primarily affects middle-aged and elderly people, 80% of them women. Older women suffer from osteoporosis more often than men for two reasons: (1) Women's bones are less massive than men's bones, and (2) production of estrogens in women declines dramatically at menopause, while production of the main androgen, testosterone, in older men wanes gradually and only slightly. Estrogens and testosterone stimulate osteoblast activity and synthesis of bone matrix. Besides gender, risk factors for developing osteoporosis include a family history of the disease, European or Asian ancestry, thin or small body build, an inactive lifestyle, cigarette smoking, a diet low in calcium and vitamin D, more than two alcoholic drinks a day, and the use of certain medications.

Figure 6.28 Comparison of spongy bone tissue from (a) a normal young adult and (b) a person with osteoporosis. Notice the weakened trabeculae in (b). Compact bone tissue is similarly affected by osteoporosis.

 In osteoporosis, bone resorption outpaces bone formation, so bone mass decreases.

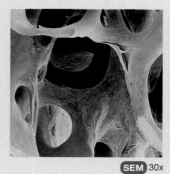 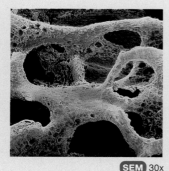

(a) Normal bone   (b) Osteoporotic bone

 If you wanted to develop a drug to lessen the effects of osteoporosis, would you look for a chemical that inhibits the activity of osteoblasts or that of osteoclasts?

Osteoporosis is diagnosed by taking a family history and undergoing a *bone mineral density* (BMD) test. Performed like x-rays, BMD tests measure bone density. They can also be used to confirm a diagnosis of osteoporosis, determine the rate of bone loss, and monitor the effects of treatment.

Treatment options for osteoporosis are varied. With regard to nutrition, a diet high in calcium is important to reduce the risk of fractures. Vitamin D is necessary for the body to utilize calcium. In terms of exercise, regular performance of weight-bearing exercises has been shown to maintain and build bone mass. These exercises include walking, jogging, hiking, climbing stairs, playing tennis, and dancing. Resistance exercises, such as weight lifting, also build bone strength and muscle mass.

Medications used to treat osteoporosis are generally of two types: (1) **antireabsorptive drugs** slow down the progression of bone loss and (2) **bone-building drugs** promote increasing bone mass. Among the antireabsorptive drugs are (1) *bisphosphonates*, which inhibit osteoclasts (Fosamax®, Actonel®, Boniva®, and calcitonin); (2) *selective estrogen receptor modulators*, which mimic the effects of estrogens without unwanted side effects (Raloxifene®, Evista®); and (3) estrogen replacement therapy (ERT), which replaces estrogens lost during and after menopause (Premarin®), and hormone replacement therapy (HRT), which replaces estrogens and progesterone lost during and after menopause (Prempro®). ERT helps maintain and increase bone mass after menopause. Women on ERT have a slightly increased risk of stroke and blood clots. HRT also helps maintain and increase bone mass. Women on HRT have an increased risk of heart disease, breast cancer, stroke, blood clots, and dementia.

Among the bone-building drugs is parathyroid hormone (PTH), which stimulates osteoblasts to produce new bone (Forteo®). Others are under development.

Rickets and Osteomalacia

Rickets and *osteomalacia* (os-tē-ō-ma-LĀ-shē-a; *-malacia* = softness) are two forms of the same disease that result from inadequate calcification of the extracellular bone matrix, usually caused by a vitamin D deficiency. Rickets is a disease of children in which the growing bones become "soft" or rubbery and are easily deformed. Because new bone formed at the epiphyseal (growth) plates fails to ossify, bowed legs and deformities of the skull, rib cage, and pelvis are common. Osteomalacia is the adult counterpart of rickets, sometimes called *adult rickets*. New bone formed during remodeling fails to calcify, and the person experiences varying degrees of pain and tenderness in bones, especially the hip and legs. Bone fractures also result from minor trauma. Prevention and treatment for rickets and osteomalacia consist of the administration of adequate doses of vitamin D.

Deviated Nasal Septum

A **deviated nasal septum** is one that does not run along the midline of the nasal cavity. It *deviates* (bends) to one side. A blow to the nose can easily damage, or break, this delicate septum of bone and displace and damage the cartilage. Often, when a broken nasal septum heals, the bones and cartilage deviate to one side or the other. This deviated septum can block airflow into the constricted side of the nose, making it difficult to breathe through that half of the nasal cavity. The deviation usually occurs at the junction of the vomer bone with the septal cartilage. Septal deviations may also occur due to developmental abnormality. If the deviation is severe, it may block the nasal passageway entirely. Even a partial blockage may lead to infection. If inflammation occurs, it may cause nasal congestion, blockage of the paranasal sinus openings, chronic sinusitis, headache, and nosebleeds. The condition usually can be corrected or improved surgically.

Herniated Disc

If the ligaments of the intervertebral discs become injured or weakened, the resulting pressure may be great enough to rupture the surrounding fibrocartilage. When this occurs, the material inside may herniate (protrude). This condition is called a *herniated (slipped) disc*. It occurs most often in the lumbar region because that part of the vertebral column bears much of the weight of the body and is the region of the most bending.

Spina Bifida

Spina bifida (SPĪ-na BIF-i-da) is a congenital defect of the vertebral column in which laminae fail to unite at the midline. In serious cases, protrusion of the membranes (meninges) around the spinal cord or the spinal cord itself may produce partial or complete paralysis, partial or complete loss of urinary bladder control, and the absence of reflexes. Because an increased risk of spina bifida is associated with a low level of folic acid (one of the B vitamins) early in pregnancy, all women who might become pregnant are encouraged to take folic acid supplements.

Hip Fracture

Although any region of the hip girdle may fracture, the term *hip fracture* most commonly applies to a break in the bones associated with the hip joint—the head, neck, or trochanteric regions of the femur, or the bones that form the acetabulum. In the United States, 300,000 to 500,000 people sustain hip fractures each year. The incidence of hip fractures is increasing, in part due to longer life spans. Decreases in bone mass due to osteoporosis and an increased tendency to fall predispose elderly people to hip fractures.

Hip fractures often require surgical treatment, the goal of which is to repair and stabilize the fracture, increase mobility, and decrease pain. Sometimes the repair is accomplished by using surgical pins, screws, nails, and plates to secure the head of the femur. In severe hip fractures, the femoral head or the acetabulum of the hip bone may be replaced by prostheses (artificial devices). The procedure of replacing either the femoral head or the acetabulum is *hemiarthroplasty* (hem-ē-AR-thrō-plas-tē; *hemi-* = one half; *-arthro-* = joint; *-plasty* = molding). Replacement of both the femoral head and acetabulum is *total hip arthroplasty*. The acetabular prosthesis is made of plastic, and the femoral prosthesis is metal; both are designed to withstand a high degree of stress. The prostheses are attached to healthy portions of bone with acrylic cement and screws.

MEDICAL TERMINOLOGY AND CONDITIONS

Bunion (BUN-yun) A deformity of the great toe that typically is caused by wearing tightly fitting shoes. The condition produces inflammation of bursae (fluid-filled sacs at the joint), bone spurs, and calluses.

Chiropractic (kī-rō-PRAK-tik; *cheir-* = hand; *-praktikos* = efficient) A holistic health-care discipline that focuses on nerves, muscles, and bones. A *chiropractor* is a health-care professional who is concerned with the diagnoses, treatment, and prevention of mechanical disorders of the musculoskeletal system and the effects of these disorders on the nervous system and health in general. Treatment involves using the hands to apply specific force to adjust joints of the body (manual adjustment), especially the vertebral column. Chiropractors may also use massage, heat therapy, ultrasound, electric stimulation, and acupuncture. Chiropractors often provide information about diet, exercise, changes in lifestyle, and stress management. Chiropractors do not prescribe drugs or perform surgery.

Clawfoot A condition in which the medial part of the longitudinal arch is abnormally elevated. It is often caused by muscle deformities, such as may result from diabetes.

Kyphosis (kī-FŌ-sis; *kypho-* = bent; *-osis* = condition) An exaggeration of the thoracic curve of the vertebral column. In the elderly, degeneration of the intervertebral discs leads to kyphosis; it may also be caused by osteoporosis, rickets, and poor posture.

Lordosis (lor-DŌ-sis; *lord-* = bent backward) An exaggeration of the lumbar curve of the vertebral column, also called *hollow back*. It may result from increased weight of the abdomen as in pregnancy or extreme obesity, poor posture, rickets, or tuberculosis of the spine.

Osteoarthritis (os′-tē-ō-ar-THRĪ-tis; *arthr* = joint) The degeneration of articular cartilage such that the bony ends touch; the resulting friction of bone against bone worsens the condition. Usually associated with the elderly.

Osteogenic sarcoma (os′-tē-Ō-JEN-ik sar-KŌ-ma; *sarcoma* = connective tissue tumor) Bone cancer that primarily affects osteoblasts and occurs most often in teenagers during their growth spurt; the most common sites are the metaphyses of the thigh bone (femur), shin bone (tibia), and arm bone (humerus). Metastases occur most often in lungs; treatment consists of multidrug chemotherapy and removal of the malignant growth, or amputation of the limb.

Osteomyelitis (os′-tē-ō-mī-e-LĪ-tis) An infection of bone characterized by high fever, sweating, chills, pain, nausea, pus formation, edema, and warmth over the affected bone and rigid overlying muscles. Bacteria, usually *Staphylococcus aureus*, often cause it. The bacteria may reach the bone from outside the body (through open fractures, penetrating wounds, or orthopedic surgical procedures); from other sites of infection in the body (abscessed teeth, burn infections, urinary tract infections, or upper respiratory infections) via the blood; and from adjacent soft tissue infections (as occurs in diabetes mellitus).

Osteopenia (os′-tē-ō-PĒ-nē-a; *penia* = poverty) Reduced bone mass due to a decrease in the rate of bone synthesis to a level insufficient to compensate for normal bone resorption; any decrease in bone mass below normal. An example is osteoporosis.

Scoliosis (skō′-lē-Ō-sis; *scolio-* = crooked) A sideways bending of the vertebral column, usually in the thoracic region. It may result from congenitally (present at birth) malformed vertebrae, chronic sciatica, paralysis of muscles on one side of the vertebral column, poor posture, or one leg being shorter than the other.

Whiplash injury Injury to the neck region due to severe hyperextension (backward tilting) of the head followed by severe hyperflexion (forward tilting) of the head, usually associated with a rear-end automobile collision. Symptoms are related to stretching and tearing of ligaments and muscles, vertebral fractures, and herniated vertebral discs.

CHAPTER REVIEW AND RESOURCE SUMMARY

WILEY
PLUS

REVIEW	RESOURCES
6.1 Functions of Bone and the Skeletal System 1. The skeletal system consists of all bones attached at joints and cartilage between joints. 2. The functions of the skeletal system include support, protection, movement, mineral homeostasis, housing blood-forming tissue, and storage of energy.	
6.2 Types of Bones 1. On the basis of shape, bones are classified as **long**, **short**, **flat**, or **irregular**.	Real Anatomy Viewpoint— Bones by Type
6.3 Structure of Bone 1. Parts of a long bone include the **diaphysis** (shaft), **epiphyses** (ends), **metaphyses**, **articular cartilage**, **periosteum**, **medullary cavity**, and **endosteum**. The diaphysis is covered by periosteum. 2. Bone tissue consists of widely separated cells surrounded by large amounts of extracellular matrix. The four principal types of cells are **osteogenic cells**, **osteoblasts** (bone-building cells), **osteocytes** (maintain daily activities of bone), and **osteoclasts** (bone-destroying cells). The extracellular matrix contains collagen fibers (organic) and mineral salts that consist mainly of calcium phosphate (inorganic). 3. **Compact bone tissue** consists of **osteons (haversian systems)** with little space between them. Compact bone composes most of the bone tissue of the diaphysis. Functionally, compact bone protects, supports, and resists stress. 4. **Spongy bone tissue** consists of **trabeculae** surrounding many red bone marrow–filled spaces. It forms most of the structure of short, flat, and irregular bones and the epiphyses of long bones. Functionally, spongy bone stores red bone marrow and provides some support.	Anatomy Overview— Bone Structure and Tissues Animation—Bone Dynamics and Tissue **Figure 6.2a** Types of cells in bone tissue **Figure 6.2c** Osteons in compact bone and trabeculae in spongy bone Exercise—Growing Long Bone
6.4 Bone Formation 1. Bone forms by a process called **ossification**. Bone formation in an embryo or fetus occurs by intramembranous ossification and endochondral ossification, which involve the replacement of preexisting connective tissue with bone. 2. **Intramembranous ossification** occurs within mesenchyme arranged in sheetlike layers that resemble membranes. 3. **Endochondral ossification** occurs within a **cartilage model** derived from mesenchyme. The **primary ossification center** of a long bone is in the diaphysis. Cartilage degenerates, leaving cavities that merge to form the medullary (marrow) cavity. Osteoblasts lay down bone. Next, ossification occurs in the epiphyses, where bone replaces cartilage, except for articular cartilage and the **epiphyseal plate**. 4. Because of the activity of the epiphyseal plate, the diaphysis of a bone increases in length. 5. Bone grows in diameter as a result of the addition of new bone tissue around the outer surface of the bone. 6. Old bone is constantly destroyed by osteoclasts, while new bone is constructed by osteoblasts. This process is called **bone remodeling**.	**Figure 6.3** Intramembranous ossification Animation—Bone Formation Animation—Bone Elongation and Widening Animation—Bone Remodeling Animation—Regulation of Bone Growth

REVIEW	RESOURCES

7. A **fracture** is any break in a bone. Fracture repair involves bone remodeling.

8. Normal bone growth depends on minerals (calcium, phosphorus, magnesium), vitamins (A, C, D), and hormones (human growth hormone, insulinlike growth factors, insulin, thyroid hormones, sex hormones, and parathyroid hormone).

9. Bones store and release calcium and phosphate, controlled mainly by **parathyroid hormone (PTH)**. PTH raises blood calcium level. Calcitonin (CT) lowers blood calcium level.

Animation—Regulation of Blood Calcium

Concepts and Connections—Blood Calcium Regulation

6.5 Exercise and Bone Tissue

1. Mechanical stress increases bone strength by increasing deposition of mineral salts and production of collagen fibers.

2. Removal of mechanical stress weakens bone through **demineralization** and collagen fiber reduction.

3. Table 6.1 summarizes the factors that influence bone metabolism.

Animation: Bone Processes— Response to Stress in Adult Bones

6.6 Divisions of the Skeletal System

1. The **axial skeleton** consists of bones arranged along the longitudinal axis of the body. The parts of the axial skeleton are the skull, hyoid bone, auditory ossicles, vertebral column, sternum, and ribs.

2. The **appendicular skeleton** consists of the bones of the girdles and the upper and lower limbs. The parts of the appendicular skeleton are the pectoral (shoulder) girdles, bones of the upper limbs, pelvic (hip) girdle, and bones of the lower limbs.

Anatomy Overview— The Skeletal System

6.7 Skull and Hyoid Bone

1. The **skull** consists of cranial bones and facial bones.

2. The eight **cranial bones** include the frontal (1), parietal (2), temporal (2), occipital (1), sphenoid (1), and ethmoid (1) (Exhibit 6.A).

3. The 14 **facial bones** are the nasal (2), maxillae (2), zygomatic (2), mandible (1), lacrimal (2), palatine (2), inferior nasal conchae (2), and vomer (1) (Exhibit 6.B).

4. **Sutures** are immovable joints between bones of the skull. Examples are the coronal, sagittal, lambdoid, and squamous sutures.

5. **Paranasal sinuses** are cavities in bones of the skull that communicate with the nasal cavity. They are lined by mucous membranes. Cranial bones containing paranasal sinuses are the frontal, sphenoid, ethmoid, and maxillae.

6. Fontanels are mesenchyme-filled spaces between the cranial bones of fetuses and infants. The major fontanels are the anterior, posterior, anterolaterals, and posterolaterals.

7. The **hyoid bone**, a U-shaped bone that does not articulate with any other bone, supports the tongue and provides attachment for some of its muscles as well as some neck muscles.

Anatomy Overview: Skull Anatomy Overview: Facial Bones

6.8 Vertebral Column

1. The bones of the adult **vertebral column** are the **cervical vertebrae** (7), **thoracic vertebrae** (12), **lumbar vertebrae** (5), the **sacrum** (5, fused), and the **coccyx** (4, fused). (See Exhibits 6.C–6.F.)

2. The vertebral column contains **normal curves** that give strength, support, and balance.

3. The vertebrae are similar in structure, each consisting of a **body**, **vertebral arch**, and seven **processes**. Vertebrae in the different regions of the column vary in size, shape, and detail.

Anatomy Overview—Vertebral Column

Figure 6.12 Vertebral column

Figure 6.16 Sacrum and coccyx

6.9 Thorax

1. The **thorax** consists of the **sternum**, **ribs**, costal cartilages, and thoracic vertebrae. The ribs are classified as true (pairs 1–7) and false (pairs 8–12).

2. The **thoracic cage** protects vital organs in the chest area.

Anatomy Overview— Thorax

Figure 6.17 Skeletal of the thorax

REVIEW	RESOURCES

6.10 Pectoral (Shoulder) Girdle

1. Each **pectoral (shoulder) girdle** consists of a **clavicle** and **scapula**.
2. Each pectoral girdle attaches an upper limb to the trunk.

Anatomy Overview—
Pectoral Girdle

Figure 6.18 Right pectoral (shoulder) girdle

6.11 Upper Limb

1. There are 30 bones in each **upper limb**.
2. The upper limb bones include the **humerus, ulna, radius, carpals, metacarpals**, and **phalanges**. (See Exhibits 6.G–6.I.)

Anatomy Overview—
Upper Limb

Figure 6.19 Right humerus in relation to the scapula, ulna, and radius

6.12 Pelvic (Hip) Girdle

1. The **pelvic (hip) girdle** consists of two **hip bones**.
2. It attaches the lower limbs to the trunk at the sacrum.
3. Each hip bone consists of three fused components: **ilium, pubis**, and **ischium**.

Anatomy Overview—
Pelvic Girdle

6.13 Lower Limb

1. There are 30 bones in each **lower limb**.
2. The lower limb bones include the **femur, patella, tibia, fibula, tarsals, metatarsals**, and **phalanges**. (See Exhibits 6.J–6.L.)
3. The bones of the foot are arranged in two **arches**, the longitudinal arch and the transverse arch, to provide support and leverage.

Anatomy Overview—
Lower Limb

Figure 6.24 Right femur in relation to the hip bone, patella, tibia, and fibula

6.14 Comparison of Female and Male Skeletons

1. Male bones are generally larger and heavier than female bones and have more prominent markings for muscle attachment.
2. The female pelvis is adapted for pregnancy and childbirth. Differences in pelvic structure are listed in Table 6.4.

6.15 Aging and the Skeletal System

1. The main effect of aging is a loss of calcium from bones, which may result in osteoporosis.
2. Another effect of aging is a decreased production of extracellular matrix proteins (mostly collagen fibers), which makes bones more brittle and thus more susceptible to fracture.

Homeostatic Imbalances: The Case of the Brittle Woman with Back Pain

SELF-QUIZ

1. Match the following cell types to their functions:
   - _____ **a.** chondroblasts
   - _____ **b.** osteoclasts
   - _____ **c.** chondrocytes
   - _____ **d.** osteocytes
   - _____ **e.** osteoblasts

   A. mature bone cells
   B. cells that form bone
   C. secrete cartilage matrix
   D. mature cartilage cells
   E. involved in bone resorption

2. If hands are not properly positioned when performing cardiopulmonary resuscitation (CPR), there is a risk of fracturing which bony part?
   a. lateral end of the clavicle
   b. head of the humerus
   c. manubrium of sternum
   d. xiphoid process of the sternum
   e. spine of the scapula

3. The ribs articulate with the
   a. thoracic vertebrae.
   b. sacrum.
   c. cervical vertebrae.
   d. lumbar vertebrae.
   e. atlas and axis.

4. Match the following:

_____ **a.** run lengthwise through bone; contain blood vessels and nerves

_____ **b.** connect central canals with lacunae

_____ **c.** concentric rings of matrix

_____ **d.** connect nutrient arteries and nerves from the periosteum to the central canals

_____ **e.** spaces that contain osteocytes

A. lamellae

B. lacunae

C. perforating (Volkmann's) canal

D. canaliculi

E. central (haversian) canal

5. The presence of an epiphyseal line in a long bone indicates that the bone

a. is undergoing resorption.

b. has stopped growing in length.

c. is growing in diameter.

d. is still capable of growing in length.

e. is broken.

6. The hyoid bone is unique because it

a. is the smallest bone in the skull.

b. can malform causing a cleft palate.

c. forms the paranasal sinuses.

d. is often broken when an individual falls forward.

e. does not articulate with any other bone.

7. The bones that form the pectoral girdle are the

a. clavicle and scapula.

b. scapula and sternum.

c. humerus and scapula.

d. clavicle and humerus.

e. coxal bones.

8. The main hormone that regulates the Ca^{2+} balance between bone and blood is

a. parathyroid hormone.

b. thyroid hormones.

c. testosterone.

d. insulinlike growth factors.

e. human growth hormone.

9. Which of the following does NOT describe spongy bone?

a. tends to be lightweight

b. found primarily in the diaphyses of long bones

c. composed of units called trabeculae

d. site of blood cell production

e. contain large spaces between trabeculae

10. In which of the following individuals might you expect to find the smallest bone mass?

a. 20-year-old male weightlifter

b. 45-year-old female weightlifter

c. 45-year-old male astronaut

d. 80-year-old bedridden female

e. 65-year-old bedridden male

11. Place the following steps of endochondral ossification in the correct order:

1. Hyaline cartilage remains on the articular surfaces and epiphyseal plates.

2. Chondroblasts produce a growing hyaline cartilage model surrounded by the perichondrium.

3. Nutrient artery penetrates perichondrium and osteogenic cells differentiate into osteoblasts, which begin to produce compact bone.

4. Secondary ossification centers form at epiphyses.

5. Primary ossification center and medullary cavity form.

a. 2, 3, 4, 5, 1   **b.** 2, 3, 5, 4, 1   **c.** 5, 2, 1, 3, 4

d. 3, 2, 5, 4, 1   **e.** 5, 3, 2, 1, 4

12. Match each bone to its shape:

_____ **a.** humerus

_____ **b.** carpus

_____ **c.** vertebra

_____ **d.** sternum

A. flat

B. irregular

C. long

D. short

13. Where long bones form joints, the epiphyses are covered with

a. yellow bone marrow.

b. osteoclasts.

c. periosteum.

d. endosteum.

e. hyaline cartilage.

14. What substance in bone contributes to its tensile strength?

a. red bone marrow

b. collagen

c. yellow bone marrow

d. mineral salts

e. loose fibrous connective tissue

15. The skeletal system is responsible for

a. protecting internal organs from injury.

b. assisting in movement.

c. providing a supporting framework for the body.

d. hemopoiesis.

e. all of the above.

16. For each of the following bones, place an AX in the blank if it belongs to the axial skeleton and an AP in the blank if it is part of the appendicular skeleton.

_____ **a.** lacrimal

_____ **b.** clavicle

_____ **c.** radius

_____ **d.** mandible

_____ **e.** patella

_____ **f.** carpals

_____ **g.** scapula

_____ **h.** sternum

_____ **i.** phalanges

_____ **j.** tarsals

_____ **k.** ethmoid

_____ **l.** metatarsals

_____ **m.** temporal

_____ **n.** metacarpals

_____ **o.** vomer

_____ **p.** fibula

_____ **q.** palatine

_____ **r.** hyoid

_____ **s.** tibia

_____ **t.** sphenoid

_____ **u.** vertebrae

_____ **v.** coxal

_____ **w.** maxilla

_____ **x.** frontal

_____ **y.** inferior nasal concha

_____ **z.** humerus

_____ **aa.** ulna

_____ **bb.** femur

_____ **cc.** ribs

_____ **dd.** occipital

CRITICAL THINKING APPLICATIONS

1. Iain was riding his motorcycle home late one night when he took a turn too fast. In the resulting crash, Iain crushed his left leg, fracturing both leg bones; snapped the pointy distal end of his lateral forearm bone; and broke the most lateral and proximal bone in his wrist. Name the bones that Iain broke.

2. You are starting a class in forensic anatomy. The instructor gives you and your lab partner two complete sets of bones of adult humans. Your assignment is to determine which set of bones is a male and which is a female. What characteristics will you use to determine the gender of the skeletons?

3. Your great-grandmother is a tiny, stooped woman with a big sense of humor. Her favorite movie line is from *The Wizard of Oz*, when the wicked witch says, "I'm melting." "That's me," she laughs, "melting away, getting shorter every year." What is happening to your great-grandmother?

4. During the volleyball game, Keith jumped, twisted, spiked, scored, and screamed! He couldn't put any weight on his left leg. X-rays revealed a fracture of the proximal tibia. In layman's terms, what is the location of Keith's fracture? What are the body's requirements for bone healing?

ANSWERS TO FIGURE QUESTIONS

6.1 The articular cartilage reduces friction at joints; red bone marrow produces blood cells; and the endosteum lines the medullary cavity.

6.2 Because the central (haversian) canals are the main blood supply to the osteocytes, their blockage would lead to death of osteocytes.

6.3 The flat bones of the skull, mandible, and part of the clavicle develop by intramembranous ossification.

6.4 The epiphyseal lines are indications of growth zones that have ceased to function.

6.5 Heartbeat, respiration, nerve cell functioning, enzyme functioning, and blood clotting are all processes that depend on proper levels of calcium.

6.6 Axial skeleton: skull and vertebral column. Appendicular skeleton: clavicle, shoulder girdle, humerus, pelvic girdle, and femur.

6.7 The cranial bones are the frontal, parietal, occipital, sphenoid, ethmoid, and temporal bones.

6.8 The foramen magnum is the largest foramen in the skull.

6.9 Crista galli of ethmoid bone, frontal, parietal, temporal, occipital, temporal, parietal, frontal, and crista galli of ethmoid bone articulate in clockwise order with the sphenoid bone.

6.10 The perpendicular plate of the ethmoid bone forms the top part of the nasal septum.

6.11 The paranasal sinuses produce mucus and serve as resonating chambers for vocalization.

6.12 The thoracic and sacral curves are concave.

6.13 The atlas and axis permit movement of the head to signify "no."

6.14 The vertebral foramina enclose the spinal cord, and the intervertebral foramina provide spaces for spinal nerves to exit the vertebral column.

6.15 The lumbar vertebrae support more weight than the thoracic and cervical vertebrae.

6.16 The sacral foramina are passageways for nerves and blood vessels.

6.17 The true ribs are pairs 1 through 7; the false ribs are pairs 8 through 12; and the floating ribs are pairs 11 and 12.

6.18 A pectoral girdle consists of a clavicle and a scapula.

6.19 The glenoid cavity of the scapula articulates with the humerus.

6.20 The elbow part of the ulna is the olecranon.

6.21 The knuckles are the heads of the metacarpals.

6.22 The true pelvis surrounds the pelvic organs in the pelvic cavity.

6.23 The femur fits into the acetabulum.

6.24 The distal end of the femur articulates with the tibia and the patella.

6.25 The tibia is the weight-bearing bone of the leg.

6.26 The talus articulates with the tibia and fibula.

6.27 The arches are not rigid, yielding when weight is applied and springing back when weight is lifted to allow them to absorb the shock of walking and running.

6.28 A drug that inhibits the activity of osteoclasts might lessen the effects of osteoporosis.

CHAPTER 7

JOINTS

Bones are too rigid to bend without being damaged. Fortunately, flexible connective tissues form joints that hold bones together while in most cases permitting some degree of movement. If you have ever damaged these areas, you know how difficult it is to walk with a cast over your knee or to turn a doorknob with a splint on your finger. A *joint* (also called an *articulation*) is a point of contact between bones, between cartilage and bones, or between teeth and bones. When we say one bone articulates with another bone, we mean that the two bones form a joint. *Arthrology* (ar-THROL-ō-jē; *arthr-* = joint; *-logy* = study of) is the scientific study of joints. Many joints of the body permit movement. The study of motion of the human body is called *kinesiology* (ki-nē′-sē-OL-ō-jē; *kinesi-* = movement).

Did you know?   Ligaments attach bones to each other, and help to stabilize joints. One of the ligaments most vulnerable to injury is the anterior cruciate ligament (ACL) of the knee. This small ligament can be torn when the knee receives a forceful impact, such as during a bicycling or skiing accident, or a collision during sports. But injury to the ACL can also occur without contact. The ACL sustains the greatest impact when people are stopping or changing direction quickly, common moves in court and field sports such as basketball and soccer.

FOCUS ON WELLNESS
Gender Differences in Anterior Cruciate Ligament Injury

LOOKING BACK TO MOVE AHEAD...

Collagen Fibers (Section 4.3)

Dense Regular Connective Tissue (Section 4.3)

Cartilage (Section 4.3)

Synovial Membranes (Section 4.4)

Divisions of the Skeletal System (Section 6.6)

7.1 CLASSIFICATION OF JOINTS

OBJECTIVES • **Describe how the structure of a joint determines its function.**

• **Describe the structural and functional classes of joints.**

A joint's structure determines its combination of strength and flexibility. At one end of the spectrum are joints that permit no movement and are thus very strong, but inflexible. In contrast, other joints afford fairly free movement and are thus flexible but not as strong. In general, the closer the fit at the point of contact, the stronger the joint. At tightly fitted joints, movement is obviously more restricted. The looser the fit, the greater the movement. However, loosely fitted joints are prone to displacement of the articulating bones from their normal positions (dislocation). Movement at joints is also determined by (1) the shape of the articulating bones, (2) the flexibility (tension or tautness) of the ligaments that bind the bones together, and (3) the tension of associated muscles and tendons. Joint flexibility may also be affected by hormones. For example, toward the end of pregnancy, a hormone called relaxin increases the flexibility of the fibrocartilage of the pubic symphysis and loosens the ligaments between the sacrum and hip bone. These changes enlarge the pelvic outlet, which assists in delivery of the baby.

Joints are classified structurally, based on their anatomical characteristics, and functionally, based on the type of movement they permit.

The structural classification of joints is based on two criteria: (1) the presence or absence of a space between the articulating bones, called a synovial cavity, and (2) the type of connective tissue that holds the bones together. Structurally, joints are classified as one of the following types:

■ **Fibrous joints** (FĪ-brus). There is no synovial cavity and the bones are held together by dense irregular connective tissue that is rich in collagen fibers.

■ **Cartilaginous joints** (kar-ti-LAJ-i-nus). There is no synovial cavity and the bones are held together by cartilage.

■ **Synovial joints** (si-NŌ-vē-al). The bones forming the joint have a synovial cavity and are united by the dense irregular connective tissue of an articular capsule, and often by accessory ligaments.

The functional classification of joints relates to the degree of movement they permit. Functionally, joints are classified as one of the following types:

■ **Synarthrosis** (sin′-ar-THRŌ-sis; *syn-* = together). An immovable joint. The plural is *synarthroses*.

■ **Amphiarthrosis** (am′-fē-ar-THRŌ-sis; *amphi-* = on both sides). A slightly movable joint. The plural is *amphiarthroses*.

■ **Diarthrosis** (dī′-ar-THRŌ-sis = movable joint). A freely movable joint. The plural is *diarthroses*. All diarthroses are synovial joints. They have a variety of shapes and permit several different types of movements.

The following sections present the joints of the body according to their structural classifications. As we examine the structure of each type of joint, we will also describe its functions.

✓ CHECKPOINT

1. What factors determine movement at joints?

2. How are joints classified on the basis of structure and function?

7.2 FIBROUS JOINTS

OBJECTIVE • **Describe the structure and functions of the three types of fibrous joints.**

Fibrous joints permit little or no movement. The three types of fibrous joints are (1) sutures, (2) syndesmoses, and (3) interosseous membranes.

1. A **suture** (SOO-cher; *sutur-* = seam) is a fibrous joint composed of a thin layer of dense irregular connective tissue. Sutures unite the bones of the skull. An example is the coronal suture between the frontal and parietal bones (Figure 7.1a). The irregular, interlocking edges of sutures give them added strength and decrease their chance of fracturing. A suture is classified as an amphiarthrosis (slightly movable) in infants and children and as a synarthrosis (immovable) in older individuals.

2. A **syndesmosis** (sin′-dez-MŌ-sis; *syndesmo-* = band or ligament) is a fibrous joint in which there is a greater distance between the articulating surfaces and more dense irregular connective tissue than in a suture. The dense irregular connective tissue is typically arranged as a bundle (ligament) and the joint permits limited movement. One example of a syndesmosis is the distal tibiofibular joint, where the anterior tibiofibular ligament connects the tibia and fibula (Figure 7.1b, left). It permits slight movement so it is classified as an amphiarthrosis. Another example of a syndesmosis is called a **gomphosis** (gom-FŌ-sis; *gompho-* = bolt or nail) or *dentoalveolar joint*, in which a cone-shaped peg fits into a socket. The only examples of gomphoses in the human body are the articulations between the roots of the teeth and their sockets (alveoli) in the maxillae and mandible (Figure 7.1b, right). The dense irregular connective tissue between a tooth and its socket is the thin periodontal ligament (membrane). A gomphosis permits no movement so it is classified as a synarthrosis. Inflammation and degeneration of the gums, periodontal ligament, and bone is called *periodontal disease*.

3. The final category of fibrous joint is the **interosseous membrane**, a substantial sheet of dense irregular connective tissue that binds neighboring long bones and permits slight movement (amphiarthrosis). There are two principal interosseous membrane joints in the human body. One

Figure 7.1 Fibrous joints.

At a fibrous joint, the bones are held together by dense irregular connective tissue.

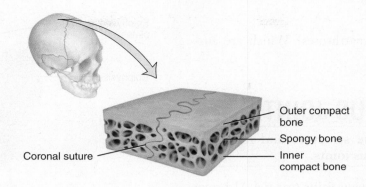

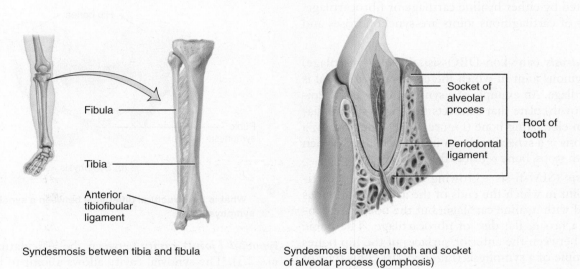

(a) Suture between skull bones

Coronal suture

Outer compact bone

Spongy bone

Inner compact bone

Fibula

Tibia

Anterior tibiofibular ligament

Syndesmosis between tibia and fibula

Socket of alveolar process

Root of tooth

Periodontal ligament

Syndesmosis between tooth and socket of alveolar process (gomphosis)

(b) Syndesmosis

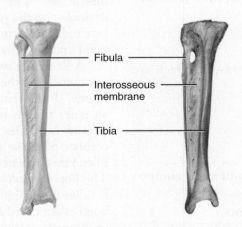

Fibula

Interosseous membrane

Tibia

(c) Interosseous membrane between tibia and fibula

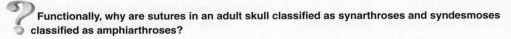

Functionally, why are sutures in an adult skull classified as synarthroses and syndesmoses classified as amphiarthroses?

occurs between the radius and ulna in the forearm (see Figure 6.20) and the other occurs between the tibia and fibula in the leg (Figure 7.1c).

✓ **CHECKPOINT**

3. Which fibrous joints are synarthroses? Which are amphiarthroses?

7.3 CARTILAGINOUS JOINTS

OBJECTIVE • **Describe the structure and functions of the two types of cartilaginous joints.**

Like a fibrous joint, a *cartilaginous joint* (car-ti-LAJ-i-nus) allows little or no movement. Here the articulating bones are tightly connected by either hyaline cartilage or fibrocartilage. The two types of cartilaginous joints are synchondroses and symphyses.

1. A *synchondrosis* (sin′-kon-DRŌ-sis; *chondro-* = cartilage) is a cartilaginous joint in which the connecting material is hyaline cartilage. An example of a synchondrosis is the epiphyseal (growth) plate that connects the epiphysis and diaphysis of an elongating bone (Figure 7.2a). Functionally, a synchondrosis is a synarthrosis, an immovable joint. When bone growth stops, bone replaces the hyaline cartilage.

2. A *symphysis* (SIM-fi-sis = growing together) is a cartilaginous joint in which the ends of the articulating bones are covered with hyaline cartilage, but the bones are connected by a broad, flat disc of fibrocartilage. The pubic symphysis between the anterior surfaces of the hip bones is one example of a symphysis (Figure 7.2b). This type of joint is also found at the intervertebral joints between bodies of vertebrae. Functionally, a symphysis is an amphiarthrosis, a slightly movable joint.

✓ **CHECKPOINT**

4. Which cartilaginous joints are synarthroses? Which are amphiarthroses?

7.4 SYNOVIAL JOINTS

OBJECTIVES • **Explain the function of each component of a synovial joint.**
• **Describe the structure of synovial joints.**

Structure of Synovial Joints

Synovial joints (si-NŌ-vē-al) have certain characteristics that distinguish them from other joints. The unique characteristic of a synovial joint is the presence of a space called a

Figure 7.2 Cartilaginous joints.

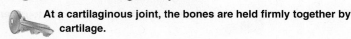

At a cartilaginous joint, the bones are held firmly together by cartilage.

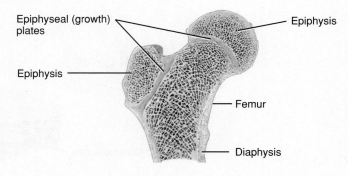

(a) Synchondrosis

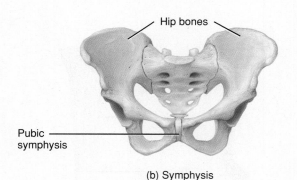

(b) Symphysis

? What is the structural difference between a synchondrosis and a symphysis?

synovial (joint) cavity between the articulating bones (Figure 7.3). The synovial cavity allows a joint to be freely movable. Hence, all synovial joints are classified functionally as diarthroses. The bones at a synovial joint are covered by *articular cartilage*, which is hyaline cartilage. Articular cartilage reduces friction between bones in the joint during movement and helps to absorb shock.

A sleevelike *articular (joint) capsule* surrounds a synovial joint, encloses the synovial cavity, and unites the articulating bones. The articular capsule is composed of two layers, an outer fibrous membrane and an inner synovial membrane (Figure 7.3). The outer layer, the *fibrous membrane*, usually consists of dense irregular connective tissue (mostly collagen fibers) that attaches to the periosteum of the articulating bones. The fibers of some fibrous membranes are arranged in parallel bundles that are highly adapted for resisting strains. Such fiber bundles are called *ligaments* (*liga-* = bound or tied) and are one of the main mechanical factors that hold bones close together in a synovial joint. The inner layer of the articular capsule, the *synovial membrane*, is composed of areolar connective tissue with elastic fibers. At many synovial joints the synovial membrane includes accumulations of adipose tissue, called *articular fat pads* (see Figure 7.11c).

Figure 7.3 Structure of a typical synovial joint. Note the two layers of the articular capsule: the fibrous membrane and the synovial membrane. Synovial fluid lubricates the synovial cavity, which is located between the synovial membrane and the articular cartilage.

 The distinguishing feature of a synovial joint is the synovial cavity between the articulating bones.

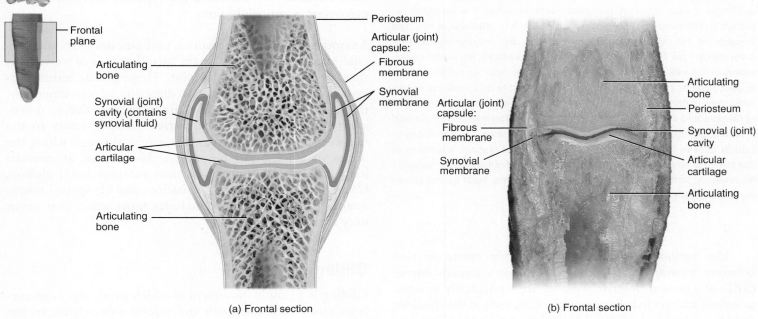

(a) Frontal section

(b) Frontal section

What is the functional classification of synovial joints?

A *"double-jointed"* person does not really have extra joints. Individuals who are "double-jointed" have greater flexibility in their articular capsules and ligaments; the resulting increase in range of motion allows them to entertain fellow partygoers with activities such as touching their thumbs to their wrists and putting their ankles or elbows behind their necks. Unfortunately, such flexible joints are less structurally stable and are more easily dislocated.

The synovial membrane secretes *synovial fluid* (*ov-* = egg), which forms a thin film over the surfaces within the articular capsule. This viscous, clear or pale yellow fluid was named for its similarity in appearance and consistency to uncooked egg white (albumin) and consists of hyaluronic acid. Its several functions include reducing friction by lubricating the joint, and supplying nutrients to and removing metabolic wastes from the chondrocytes within articular cartilage. When a synovial joint is immobile for a time, the fluid is quite viscous (gel-like), but as joint movement increases, the fluid becomes less viscous. One of the benefits of a warm-up before exercise is that it stimulates the production and secretion of synovial fluid. More fluid means less stress on the joint during exercise.

We are all familiar with the cracking sounds heard as certain joints move, or the popping sounds that arise when

people *crack their knuckles*. According to one theory, when the synovial cavity expands, the pressure of the synovial fluid decreases, creating a partial vacuum. The suction draws carbon dioxide and oxygen out of blood vessels in the synovial membrane, forming bubbles in the fluid. When the bubbles burst, as when the fingers are flexed (bent), the cracking or popping sound is heard.

Many synovial joints also contain *accessory ligaments* that lie outside and inside the articular capsule. Examples of accessory ligaments outside the articular capsule are the fibular (lateral) and tibial (medial) collateral ligaments of the knee joint (see Figure 7.11d). Examples of accessory ligaments inside the articular capsule are the anterior and posterior cruciate ligaments of the knee joint (see Figure 7.11d).

Inside some synovial joints, such as the knee, are pads of fibrocartilage that lie between the articular surfaces of the bones and are attached to the fibrous capsule. These pads are called *articular discs* or *menisci* (me-NIS-sī; singular is *meniscus*). Figure 7.11d–e depicts the lateral and medial menisci in the knee joint. By modifying the shape of the joint surfaces of the articulating bones, articular discs allow two bones of different shapes to fit more tightly. Articular discs also help to maintain the stability of the joint and direct the flow of synovial fluid to the areas of greatest friction.

CLINICAL CONNECTION | Torn Cartilage and Arthroscopy

The tearing of articular discs (menisci) in the knee, commonly called **torn cartilage**, occurs often among athletes. Such damaged cartilage will begin to wear and may precipitate arthritis unless it is surgically removed (*meniscectomy*). Surgical repair of the torn cartilage is required because of the avascular nature of cartilage and may be assisted by **arthroscopy** (ar-THROS-kō-pē; -*scopy* = observation), the visual examination of the interior of a joint, usually the knee, with an *arthroscope*, a lighted, pencil-thin instrument. Arthroscopy is used to determine the nature and extent of damage following knee injury and to monitor the progression of disease and the effects of therapy. In addition, the insertion of surgical instruments through the arthroscope or other incisions enables a physician to remove torn cartilage and repair damaged cruciate ligaments in the knee; to remodel poorly formed cartilage; to obtain tissue samples for analysis; and to perform surgery on other joints, such as the shoulder, elbow, ankle, and wrist. •

The various movements of the body create friction between moving parts. Saclike structures called *bursae* (BER-sē = purses; singular is *bursa*) are strategically situated to reduce friction in some synovial joints, such as the shoulder and knee joints (see Figure 7.11c). Bursae are not strictly part of synovial joints, but do resemble joint capsules because their walls consist of connective tissue lined by a synovial membrane. They are also filled with a fluid similar to synovial fluid. Bursae are located between the skin and bone in places where skin rubs over bone. They are also found between tendons and bones, muscles and bones, and ligaments and bones. The fluid-filled bursal sacs cushion the movement of one body part over another.

CLINICAL CONNECTION | Bursitis

An acute or chronic inflammation of a bursa, for example in the shoulder and knee, is called **bursitis** (bur-SĪ-tis). The condition may be caused by trauma, by an acute or chronic infection (including syphilis and tuberculosis), or by rheumatoid arthritis (described in the Common Disorders section). Repeated, excessive exertion of a joint often results in bursitis, with local inflammation and the accumulation of fluid. Symptoms include pain, swelling, tenderness, and limited movement. Treatment may include oral anti-inflammatory agents and injections of cortisol-like steroids. •

✓ **CHECKPOINT**

5. How does the structure of synovial joints classify them as diarthroses?

6. What are the functions of articular cartilage, the articular capsule, synovial fluid, articular discs, and bursae?

7.5 TYPES OF MOVEMENTS AT SYNOVIAL JOINTS

OBJECTIVE • **Describe the types of movements that can occur at synovial joints.**

Anatomists, physical therapists, and kinesiologists use specific terminology to designate specific types of movement that can occur at a synovial joint. These precise terms indicate the form of motion, the direction of movement, or the relationship of one body part to another during movement. The term *range of motion (ROM)* refers to the range, measured in degrees in a circle, through which the bones of a joint can be moved. Movements at synovial joints are grouped into four main categories: (1) gliding, (2) angular movements, (3) rotation, and (4) special movements. The last category includes movements that occur only at certain joints.

Gliding

Gliding is a simple movement in which nearly flat bone surfaces move back-and-forth and side-to-side relative to one another (Figure 7.4). This can be illustrated between the clavicle and acromion of the scapula by placing your upper limb at your side, raising it above your head, and lowering it again. Gliding movements are limited in range due to the loose-fitting structure of the articular capsule and associated ligaments and bones.

Figure 7.4 Gliding movements at synovial joints.

Gliding movements consist of side-to-side and back-and-forth motions.

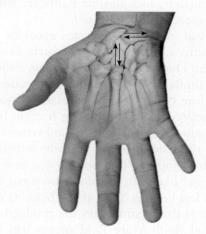

Gliding between intercarpals (arrows)

 What are two examples of joints that permit gliding movements?

Angular Movements

In *angular movements*, there is an increase or a decrease in the angle between articulating bones. The principal angular movements are flexion, extension, hyperextension, abduction, adduction, and circumduction and are discussed with respect to the body in the anatomical position. In *flexion* (FLEK-shun = to bend), there is a decrease in the angle between articulating bones; in *extension* (eks-TEN-shun = to stretch out), there is an increase in the angle between articulating bones, often to restore a part of the body to the anatomical position after it has been flexed. Flexion and extension usually occur along the sagittal plane (Figure 7.5). Examples of flexion include bending the head toward the chest (Figure 7.5a); moving the humerus forward at the shoulder joint as in swinging the arms forward while walking (Figure 7.5b); moving the forearm toward the arm (Figure 7.5c); moving the palm toward the forearm (Figure 7.5d); moving the femur forward, as in walking

(Figure 7.5e); and bending the knee (Figure 7.5f). Extension is simply the reverse of these movements.

Continuation of extension beyond the anatomical position is called *hyperextension* (*hyper-* = beyond or excessive). Examples of hyperextension include bending the head backward (Figure 7.5a); moving the humerus backward, as in swinging the arms backward while walking (Figure 7.5b); moving the palm backward at the wrist joint (Figure 7.5d); and moving the femur backward, as in walking (Figure 7.5e). Hyperextension of other joints, such as the elbow, interphalangeal joints (fingers and toes), and knee joints, is usually prevented by the arrangement of ligaments and bones.

Abduction (ab-DUK-shun; *ab-* = away; *-duct* = to lead) is the movement of a bone away from the midline, and *adduction* (ad-DUK-shun; *ad-* = toward) is the movement of a bone toward the midline. Abduction and adduction usually occur along the frontal plane. Examples of abduction include lateral movement of the humerus upward (Figure 7.6a), lateral

Figure 7.5 Angular movements at synovial joints: flexion, extension, and hyperextension.

In angular movements, there is an increase or decrease in the angle between articulating bones.

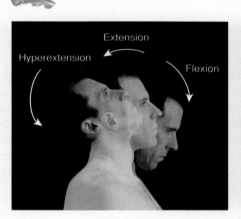

(a) Joints between atlas and occipital bone and between cervical vertebrae

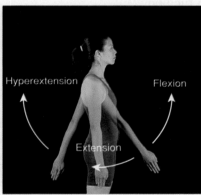

(b) Shoulder joint

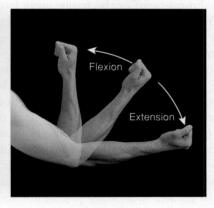

(c) Elbow joint

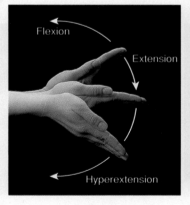

(d) Wrist joint

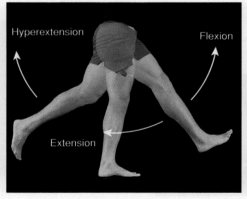

(e) Hip joint

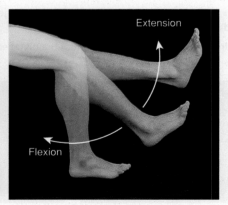

(f) Knee joint

What prevents hyperextension at some synovial joints?

Figure 7.6 Angular movements at synovial joints: abduction and adduction.

Abduction and adduction usually occur along the frontal plane.

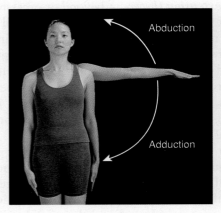

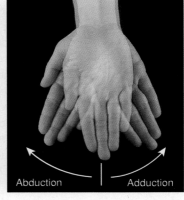

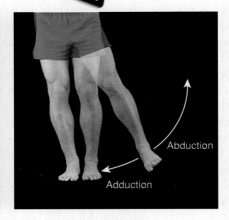

(a) Shoulder joint

(b) Wrist joint

(c) Hip joint

One way to remember what adduction means is use of the phrase "adding your limb to your trunk." Why is this an effective learning device?

movement of the palm away from the body (Figure 7.6b), and lateral movement of the femur away from the body (Figure 7.6c). Movement in the opposite direction (medially) in each case produces adduction (Figure 7.6a–c).

Circumduction (ser-kum-DUK-shun; *circ-* = circle) is movement of the distal end of a part of the body in a circle (Figure 7.7). It is not an isolated movement by itself but rather a continuous sequence of flexion, abduction, extension, and adduction. Therefore, circumduction does not occur along a separate plane of movement. Examples of joints that allow circumduction include the humerus at the shoulder joint (making a circle with your arm) and the femur at the hip joint (making a circle with your leg). Circumduction is more limited at the hip due to greater tension on the ligaments and muscles and the depth of the acetabulum in the hip joint.

Figure 7.7 Angular movements at synovial joints: circumduction.

Circumduction is the movement of the distal end of a body part in a circle.

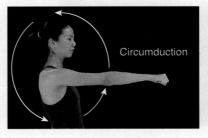

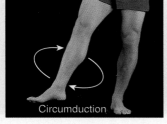

(a) Shoulder joint

(b) Hip joint

List two joints where circumduction occurs.

Rotation

In ***rotation*** (rō-TĀ-shun; *rota-* = to revolve) a bone revolves around its own longitudinal axis. An example is turning the head from side to side, as in signifying "no" (Figure 7.8a). In the limbs, rotation is defined relative to the midline. If the anterior surface of a bone of the limb is turned toward the midline, the movement is called *medial (internal) rotation*. You can medially rotate the humerus at the shoulder joint as follows: Starting in the anatomical position, flex your elbow and then draw your palm across the chest (Figure 7.8b). If the anterior surface of the bone of a limb is turned away from the midline, the movement is called *lateral (external) rotation* (see Figure 7.8b).

Figure 7.8 Rotation at synovial joints.

In rotation, a bone revolves around its own longitudinal axis.

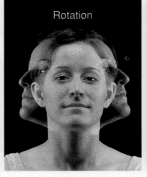

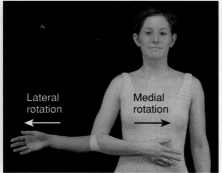

(a) Atlanto-axial joint

(b) Shoulder joint

How do medial and lateral rotation differ?

Special Movements

The ***special movements***, which occur only at certain joints, include elevation, depression, protraction, retraction, inversion, eversion, dorsiflexion, plantar flexion, supination, and pronation (Figure 7.9).

- ***Elevation*** (el'-e-VĀ-shun = to lift up) is the upward movement of a part of the body, such as closing the mouth to elevate the mandible (Figure 7.9a) or shrugging the shoulders to elevate the scapula.

- ***Depression*** (dē-PRESH-un = to press down) is the downward movement of a part of the body, such as opening the mouth to depress the mandible (Figure 7.9b) or

returning shrugged shoulders to the anatomical position to depress the scapula.

- ***Protraction*** (prō-TRAK-shun = to draw forth) is the movement of a part of the body forward. You can protract your mandible by thrusting it outward (Figure 7.9c) or protract your clavicles by crossing your arms.

- ***Retraction*** (rē-TRAK-shun = to draw back) is the movement of a protracted part of the body back to the anatomical position (Figure 7.9d).

- ***Inversion*** (in-VER-zhun = to turn inward) is movement of the soles medially so that they face each other (Figure 7.9e).

Figure 7.9 Special movements at synovial joints.

 Special movements occur only at certain synovial joints.

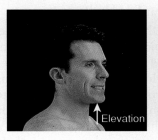   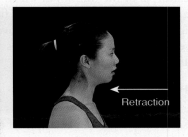

(a) Temporomandibular joint (b) (c) Temporomandibular joint (d)

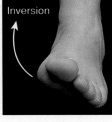 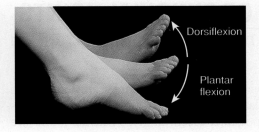

(e) Intertarsal joints (f) (g) Ankle joint

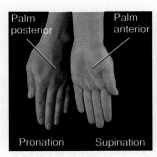 

(h) Radioulnar joint (i) Carpometacarpal joint

? **What movement of the shoulder girdle is involved in bringing the arms forward until the elbows touch?**

- *Eversion* (ē-VER-zhun = to turn outward) is movement of the soles laterally so that they face away from each other (Figure 7.9f).

- *Dorsiflexion* (dor′-si-FLEK-shun) is bending of the foot in the direction of the dorsum (superior surface), as when you stand on your heels (Figure 7.9g).

- *Plantar flexion* involves bending of the foot in the direction of the plantar surface (Figure 7.9g), as when standing on your toes.

- *Supination* (soo′-pi-NĀ-shun) is movement of the forearm so that the palm is turned forward (Figure 7.9h). Supination of the palms is one of the defining features of the anatomical position (see Figure 1.4).

- *Pronation* (prō-NĀ-shun) is movement of the forearm so that the palm is turned backward (Figure 7.9h).

- *Opposition* (op-ō-ZISH-un) is the movement of the thumb at the carpometacarpal joint (between the trapezium and metacarpal of the thumb) in which the thumb moves across the palm to touch the tips of the fingers on the same hand (Figure 7.9i). This is the distinctive digital movement that gives humans and other primates the ability to grasp and manipulate objects very precisely.

✓ CHECKPOINT

7. Define each of the movements at synovial joints just described and give an example of each.

7.6 TYPES OF SYNOVIAL JOINTS

OBJECTIVE • **Describe the six subtypes of synovial joints.**

Although all synovial joints have a similar structure, the shapes of the articulating surfaces vary and thus various types of movement are possible. Accordingly, synovial joints are divided into six subtypes: plane, hinge, pivot, condyloid, saddle, and ball-and-socket joints.

1. The articulating surfaces of bones in a *plane (planar) joint* are flat or slightly curved (Figure 7.10a). Plane joints primarily permit back-and-forth and side-to-side movements between the flat surfaces of bones, but they may also rotate against one another. Many plane joints are *biaxial* because they permit movement around two axes. An *axis* is a straight line around which a rotating (revolving) bone moves. If plane joints rotate in addition to gliding, then they are *triaxial (multiaxial)*, permitting movement in three axes. Some examples of plane joints are the intercarpal (between carpal bones at the wrist), intertarsal (between tarsal bones at the ankle), sternoclavicular (between the sternum and the clavicle), and

acromioclavicular (between the acromion of the scapula and the clavicle) joints.

2. In *hinge joints*, the convex surface of one bone fits into the concave surface of another bone (Figure 7.10b). As the name implies, hinge joints produce an angular, opening-and-closing motion like that of a hinged door. Hinge joints permit only flexion and extension. Hinge joints are *uniaxial (monaxial)* because they typically allow motion around a single axis. Examples of hinge joints are the knee, elbow, ankle, and interphalangeal joints (between the phalanges of the fingers and toes).

3. In *pivot joints*, the rounded or pointed surface of one bone articulates with a ring formed partly by another bone and partly by a ligament (Figure 7.10c). A pivot joint is *uniaxial* because it allows rotation only around its own longitudinal axis. Examples of pivot joints are the atlantoaxial joint, in which the atlas rotates around the axis and permits you to turn your head from side to side as in signifying "no," and the radioulnar joints that allow you to move your palms forward and backward.

4. In *condyloid joints* (KON-di-loyd = knucklelike), the convex oval-shaped projection of one bone fits into the concave oval-shaped depression of another bone (Figure 7.10d). A condyloid joint is *biaxial* because the movement it permits is around two axes (flexion–extension and abduction–adduction) plus limited circumduction (remember that circumduction is not an isolated movement). Examples are the wrist and metacarpophalangeal joints (between the metacarpals and phalanges) of the second through fifth digits.

5. In *saddle joints*, the articular surface of one bone is saddle-shaped, and the articular surface of the other bone fits into the saddle like a rider sitting on a horse (Figure 7.10e). The movements at a saddle joint are the same as those at a condyloid joint: *biaxial* (flexion–extension and abduction–adduction) plus limited circumduction. An example of a saddle joint is the carpometacarpal joint between the trapezium of the carpus and metacarpal of the thumb.

6. In *ball-and-socket joints*, the ball-like surface of one bone fits into a cuplike depression of another bone (Figure 7.10f). Ball-and-socket joints are *triaxial (multiaxial)* and permit movements around three axes (flexion–extension, abduction–adduction, and rotation); the only examples in the human body are the shoulder and hip joints.

To give you an idea of the complexity of a synovial joint, in Exhibit 7.A we will examine some of the structural features of the knee joint, a modified hinge joint which is the largest and most complex joint in the body.

✓ CHECKPOINT

8. Where in the body can each subtype of synovial joint be found?

Figure 7.10 Types of synovial joints. For each type, a drawing of the actual joint and a simplified diagram are shown.

 Synovial joints are classified into types on the basis of the shapes of the articulating bone surfaces.

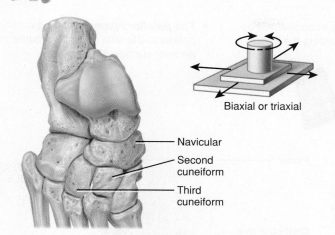

(a) Plane joint between navicular and second and third cuneiforms of tarsus in foot

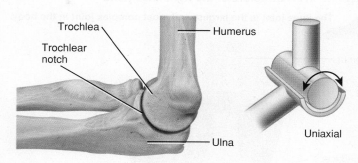

(b) Hinge joint between trochlea of humerus and trochlear notch of ulna at the elbow

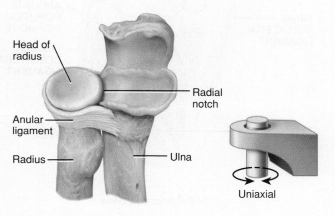

(c) Pivot joint between head of radius and radial notch of ulna

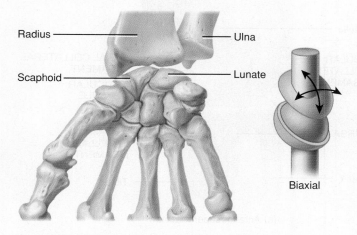

(d) Condyloid joint between radius and scaphoid and lunate bones of carpus (wrist)

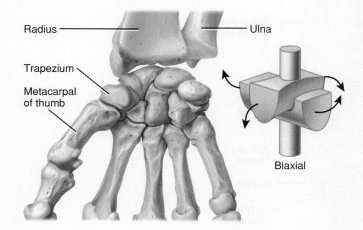

(e) Saddle joint between trapezium of carpus (wrist) and metacarpal of thumb

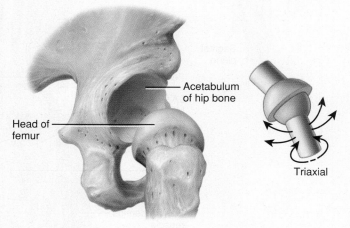

(f) Ball-and-socket joint between head of femur and acetabulum of hip bone

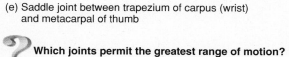 **Which joints permit the greatest range of motion?**

EXHIBIT 7.A — The Knee Joint *(Figures 7.11 and 7.12)*

OBJECTIVE • Describe the principal structures and functions of the knee joint.

Among the main structures of the knee joint are the following (Figure 7.11).

- The articular capsule is strengthened by muscle tendons surrounding the joint.

- The **patellar ligament** extends from the patella to the tibia and strengthens the anterior surface of the joint.

Figure 7.11 Structure of the right knee joint.

The knee joint is the largest and most complex joint in the body.

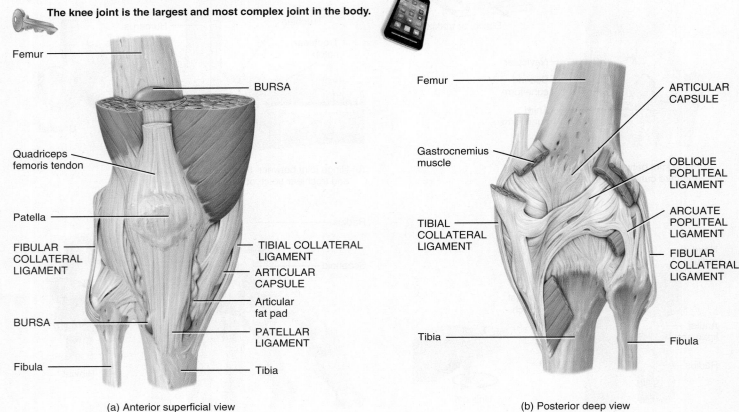

(a) Anterior superficial view

(b) Posterior deep view

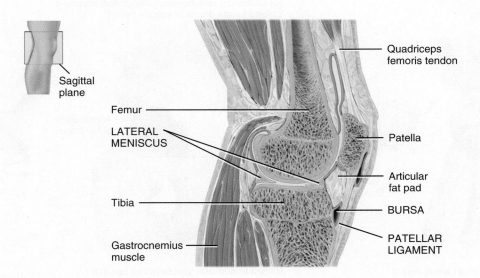

(c) Sagittal section

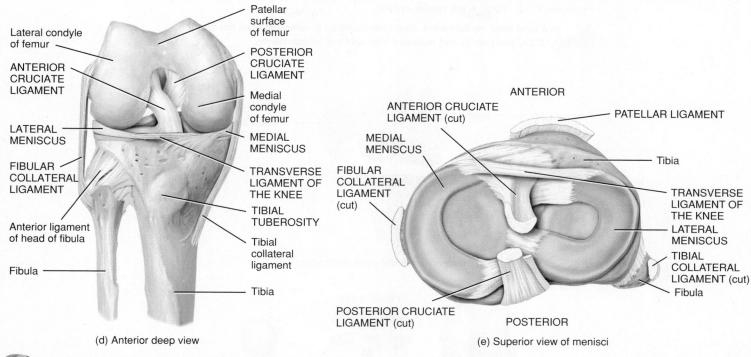

(d) Anterior deep view

(e) Superior view of menisci

 What structures are damaged in the knee injury called torn cartilage?

- The **oblique popliteal ligament** (pop-LIT-ē-al) strengthens the posterior surface of the joint.

- The **arcuate popliteal ligament** strengthens the lower lateral part of the posterior surface of the joint.

- The **tibial (medial) collateral ligament** strengthens the medial aspect of the joint.

- The **fibular (lateral) collateral ligament** strengthens the lateral aspect of the joint.

- The **anterior cruciate ligament (ACL)** extends posteriorly and laterally from the tibia to the femur. The ACL limits hyperextension of the knee and prevents anterior sliding of the tibia on the femur. The ACL is stretched or torn in about 70 percent of all serious knee injuries.

- The **posterior cruciate ligament (PCL)** extends anteriorly and medially from the tibia to the femur. The ACL prevents the posterior sliding of the tibia on the femur.

- The **menisci**, fibrocartilage discs between the tibial and femoral condyles, help compensate for the irregular shapes of the articulating bones and circulate

synovial fluid. The two menisci of the knee joint are the **medial meniscus**, a semicircular piece of fibrocartilage on the medial aspect of the knee, and the **lateral meniscus**, a nearly circular piece of fibrocartilage on the lateral aspect of the knee. The menisci are connected to each other by the **transverse ligament of the knee**.

- The **bursae**, saclike structures filled with fluid, help reduce friction.

Joints that have been severely damaged by diseases such as arthritis, or by injury, may be replaced surgically with artifical joints in a procedure referred to as **arthroplasty** (AR-thrō-plas′-tē; *arthr-* = joint; *-plasty* = plastic repair of). Although most joints in the body can undergo arthroplasty, the ones most commonly replaced are the hips, knees, and shoulders. During the procedure, the ends of the damaged bones are removed and the metal, ceramic, or plastic components are fixed in place. The goals of arthroplasty are to relieve pain and increase range of motion.

Knee replacements are actually a resurfacing of cartilage and may be partial or total. In a **total knee replacement (TKR)**, the damaged cartilage is removed from the distal end of the femur, proximal end of the tibia, and the back surface of the patella (if the back surface of the patella is not badly damaged, it may be left intact) (Figure 7.12). The femur is reshaped and fitted with a metal femoral component and cemented in place. The tibia is reshaped and fitted with a plastic tibial component that is cemented in place. If the back surface of the patella is badly damaged, it is replaced with a plastic patellar component.

In a **partial knee replacement (PKR)**, only one side of the knee joint is replaced. Once the damaged cartilage is removed from the distal end of the femur, the femur is reshaped and a metal femoral component is cemented in place. Then the damaged cartilage from the proximal end of the tibia is removed, along with the meniscus. The tibia is reshaped and fitted with a plastic tibial component that is cemented into place.

EXHIBIT 7.A CONTINUES

EXHIBIT 7.A The Knee Joint *(Figures 7.11 and 7.12)* **CONTINUED**

Figure 7.12 Total knee replacement.

In a total knee replacement, damaged cartilage is removed from the femur, tibia, and patella and replaced with artificial components.

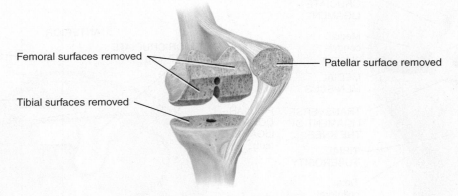

Femoral surfaces removed

Patellar surface removed

Tibial surfaces removed

(a) Preparation for total knee replacement

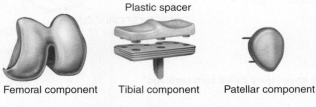

Plastic spacer

Femoral component   Tibial component   Patellar component

(b) Components of artificial knee joint prior to implantation

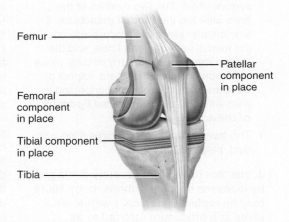

Femur

Patellar component in place

Femoral component in place

Tibial component in place

Tibia

(c) Implanted components of a total knee replacement

What are the goals of arthroplasty?

Researchers are continually seeking to improve the strength of the cement and devise ways to stimulate bone growth around the implanted area. Potential complications of arthroplasty include infection, blood clots, loosening or dislocation of the replacement components, and nerve injury.

With increasing sensitivity of metal detectors at airports and other public areas, it is possible that metal joint replacements may activate metal detectors.

✓ **CHECKPOINT**

Which ligaments strengthen the posterior aspect of the knee joint?

Gender Differences in Anterior Cruciate Ligament Injury

One of the most common knee injuries is damage to the anterior cruciate ligament (ACL). The ACL may be torn or even completely severed. Only about 30 percent of all ACL injuries are caused by contact, such as a hit to the knee. The majority result from the athlete's own movements: landing "wrong" after a jump, or twisting the knee as the athlete pivots on the field. The rate of noncontact ACL injury in females is quite a bit higher than the rate in men. Why the difference? And what can be done?

Raging Hormones?

Estrogen relaxes soft tissues such as ligaments somewhat, so it is possible that female hormones put women at a disadvantage in the ACL injury category. However, studies comparing injury rates to phases of the menstrual cycle have not revealed a clear relationship between injury and higher estrogen levels.

Anatomic Angles

Women generally have a wider pelvis for their size than men. This means that the femur comes in toward the knee at a slightly sharper angle. This angle changes the force placed on the knee, especially when the knee is under increased pressure from higher speeds and twisting motions.

Strength and Coordination

The muscles of the leg help to stabilize the knee joint. Weak muscles make the knee more vulnerable to injury. Most people have relatively stronger quadriceps muscles (the big muscle group on the front of the thigh) than the hamstrings muscle group (on back of the thigh). Some studies have found that the quadriceps:hamstrings strength ratio tends to be more out of balance in women than men. Weaker hamstring muscles mean less knee stability against the force of the strong quadriceps.

Differences in muscle recruitment patterns may also explain the higher ACL injury rate in women. Some studies have suggested that men may contract their hamstrings earlier when landing from a jump. These unconscious movement patterns may be the result of boys learning sports from earlier ages, receiving better coaching at early ages, or simply being more active as children.

Playing Posture

Some researchers have observed that women tend to play field and court sports with a more upright playing posture. Less flexion of the knee makes the ACL more vulnerable to changes in direction and twisting motions.

Think It Over . . . Think about the moves you have seen players execute in sports like basketball, soccer, football, hockey, and lacrosse. Which moves might be most likely to cause noncontact ACL injuries?

7.7 AGING AND JOINTS

OBJECTIVE • **Explain the effects of aging on joints.**

Aging usually results in decreased production of synovial fluid in joints. In addition, the articular cartilage becomes thinner with age, and ligaments shorten and lose some of their flexibility. The effects of aging on joints are influenced by genetic factors and by wear and tear, and vary considerably from one person to another. Although degenerative changes in joints may begin as early as age 20, most changes do not occur until much later. By age 80, almost everyone develops some type of degeneration in the knees, elbows, hips, and shoulders. It is also common for elderly individuals to develop degenerative changes in the vertebral column, resulting in a hunched-over posture and pressure on nerve roots. One type of arthritis, called osteoarthritis, is at least partially age-related. Nearly everyone over age 70 has evidence of some osteoarthritic changes. (For more on osteoarthritis, see the Common Disorders section.) Stretching and aerobic exercises that attempt to maintain full range of motion are helpful in minimizing the effects of aging. They help to maintain the effective functioning of ligaments, tendons, muscles, synovial fluid, and articular cartilage.

✓ CHECKPOINT

9. Which joints show evidence of degeneration in nearly all individuals as aging progresses?

• • •

Now that you have a basic understanding of bones and joints, in the next chapters we will examine the structure and functions of muscular tissue and muscles. In this way you will understand how bones, joints, and muscles work together to produce various movements.

COMMON DISORDERS

Common Joint Injuries

Rotator cuff injury is a strain or tear in the rotator cuff muscles (see Figure 8.19) and is a common injury among baseball pitchers and volleyball players, racket sports players, swimmers, and violinists, due to shoulder movements that involve vigorous circumduction. It also occurs as a result of wear and tear, aging, trauma, poor posture, improper lifting, and repetitive motions in certain jobs, such as placing items on a shelf above your head. Most often, there is tearing of the supraspinatus muscle tendon of the rotator cuff. This tendon is especially predisposed to wear and tear because of its location between the head of the humerus and acromion of the scapula, which compresses the tendon during shoulder movements. Poor posture and poor body mechanics also increase compression of the supraspinatus muscle tendon.

A *separated shoulder* is an injury of the acromioclavicular joint, the joint formed by the acromion of the scapula and the acromial end of the clavicle. It most often happens with forceful trauma, as may happen when the shoulder strikes the ground in a fall.

Tennis elbow most commonly refers to pain at or near the lateral epicondyle of the humerus, usually caused by an improperly executed backhand. The extensor muscles strain or sprain, resulting in pain. *Little-league elbow* is an inflammation of the medial epicondyle and typically develops because of a heavy pitching schedule or throwing many curve balls, especially in youngsters. In this injury, the elbow may enlarge, fragment, or separate.

A *dislocation of the radial head* is the most common upper limb dislocation in children. In this injury, the head of the radius slides past or ruptures the ligament that forms a collar around the head of the radius at the proximal radioulnar joint. Dislocation is most apt to occur when a strong pull is applied to the forearm while it is extended and supinated, for instance while swinging a child around with outstretched arms.

The knee joint is the joint most vulnerable to damage because it is a mobile, weight-bearing joint and its stability depends almost entirely on its associated ligaments and muscles. Further, there is no correspondence of the articulating bones. A *swollen knee* may occur immediately or hours after an injury. The initial swelling is due to escape of blood from damaged blood vessels adjacent to areas involving rupture of the anterior cruciate ligament, damage to synovial membranes, torn menisci, fractures, or collateral ligament sprains. Delayed swelling is due to excessive production of synovial fluid, a condition commonly referred to as "water on the knee." A common type of knee injury in football is *rupture of the tibial collateral ligaments*, often associated with tearing of the anterior cruciate ligament and medial meniscus (torn cartilage). Usually, a hard blow to the lateral side of the knee while the foot is fixed on the ground causes the damage. A *dislocated knee* refers to the displacement of the tibia relative to the femur. The most common type is dislocation anteriorly, resulting from hyperextension of the knee. A frequent consequence of a dislocated knee is damage to the popliteal artery.

Rheumatism and Arthritis

Rheumatism (ROO-ma-tizm) is any painful disorder of the supporting structures of the body—bones, ligaments, tendons, or muscles—that is not caused by infection or injury. *Arthritis* is a form of rheumatism in which the joints are swollen, stiff, and painful. It afflicts about 45 million people in the United States, and is the leading cause of physical disability among adults over age 65.

Rheumatoid arthritis (RA) is an autoimmune disease in which the immune system of the body attacks its own tissues—in this case, its own cartilage and joint linings. The primary symptom of RA is inflammation of the synovial membrane. RA is characterized by inflammation of the joint, which causes redness, warmth, swelling, pain, and loss of function.

Osteoarthritis (os′-tē-ō-ar-THRĪ-tis) is a degenerative joint disease in which joint cartilage is gradually lost. It results from a combination of aging, irritation of the joints, muscle weakness, and wear and abrasion. Commonly known as "wear and tear" arthritis, osteoarthritis is the most common type of arthritis. A major distinction between osteoarthritis and rheumatoid arthritis is that osteoarthritis strikes the larger joints (knees, hips) first, and rheumatoid arthritis first strikes smaller joints such as those in the fingers. A relatively new treatment for osteoarthritis of some joints is called *viscosupplementation*, in which hyaluronic acid is injected into a joint to improve lubrication. Results are usually as good as those involving corticosteroids.

Sprain and Strain

A *sprain* is the forcible wrenching or twisting of a joint that stretches or tears its ligaments but does not dislocate the bones. It occurs when the ligaments are stressed beyond their normal capacity. Severe sprains may be so painful that the joint cannot be moved. There is considerable swelling, which results from chemicals released by the damaged cells and hemorrhage of ruptured blood vessels. The lateral ankle joint is most often sprained; the wrist is another area that is frequently sprained. A *strain* is a stretched or partially torn muscle or muscle and tendon. It often occurs when a muscle contracts suddenly and powerfully—such as the leg muscles of sprinters when they spring from the blocks.

Initially sprains should be treated with *PRICE:* protection, rest, ice, compression, and elevation. PRICE therapy may be used on muscle strains, joint inflammation, suspected fractures, and bruises. The five components of PRICE therapy are as follows:

- *Protection* means protecting the injury from further damage; for example, stop the activity and use padding and protection, and use splints, a sling, or crutches if necessary.
- *Rest* the injured area to avoid further damage to the tissues. Avoid exercise or other activities that cause pain or swelling to the injured area. Rest is needed for repair. Exercising before an injury has healed may increase the probability of re-injury.
- *Ice* the injured area as soon as possible. Applying ice slows blood flow to the area, reduces swelling, and relieves pain. Ice works effectively when applied for 20 minutes, off for 40 minutes, back on for 20 minutes, and so on.

- *Compression* by wrap or bandage helps to reduce swelling. Care must be taken to compress the injured area but not to block blood flow.

- *Elevation* of the injured area above the level of the heart, when possible, will reduce potential swelling.

MEDICAL TERMINOLOGY AND CONDITIONS

Arthralgia (ar-THRAL-jē-a; *arthr-* = joint; *-algia* = pain) Pain in a joint.

Bursectomy (bur-SEK-tō-mē; *-ectomy* = to cut out) Removal of a bursa.

Chondritis (kon-DRĪ-tis; *chondro-* = cartilage) Inflammation of cartilage.

Dislocation (*dis-* = apart) or **luxation** (luks-A-shun; *lux-* = dislocation) The displacement of a bone from a joint with tearing of ligaments, tendons, and articular capsules. A partial or incomplete dislocation is called a **subluxation**.

Synovitis (sin'-ō-VĪ-tis) Inflammation of a synovial membrane in a joint.

CHAPTER REVIEW AND RESOURCE SUMMARY

WILEY PLUS

REVIEW	RESOURCES
7.1 Classification of Joints **1.** A joint (articulation) is a point of contact between two bones, cartilage and bone, or teeth and bone. **2.** A joint's structure determines its combination of strength and flexibility. **3.** Structural classification is based on the presence or absence of a synovial cavity and the type of connecting tissue. Structurally, joints are classified as **fibrous**, **cartilaginous**, or **synovial**. **4.** Functional classification of joints is based on the degree of movement permitted. A joint may be a **synarthrosis** (immovable), **amphiarthrosis** (slightly movable), or **diarthrosis** (freely movable).	Anatomy Overview—Joints
7.2 Fibrous Joints **1.** In **fibrous joints** there is no joint cavity and the bones are held together by dense irregular connective tissue. **2.** A fibrous joint may be a slightly movable or immovable **suture** (found between skull bones), a slightly movable **syndesmosis** (such as the distal joint between the tibia and fibula), an immovable **gomphosis** (such as the root of a tooth in an alveolus of the mandible and maxilla), or a slightly movable **interosseous membrane** (found between the radius and ulna and tibia and fibula).	Anatomy Overview—Fibrous Joints
7.3 Cartilaginous Joints **1.** There is no joint cavity and the bones are held together by cartilage in **cartilaginous joints**. **2.** These joints can be an immovable **synchondrosis** united by hyaline cartilage (epiphyseal plates) or a slightly movable **symphysis** united by fibrocartilage (pubic symphysis).	Anatomy Overview—Cartilaginous Joints Concepts and Connections—Joint Classification
7.4 Synovial Joints **1.** A **synovial joint** contains a **synovial (joint) cavity**. All synovial joints are diarthroses. **2.** Other characteristics of a synovial joint are the presence of articular cartilage and an **articular (joint) capsule**, made up of a **fibrous membrane** and a **synovial membrane**. **3.** The synovial membrane secretes **synovial fluid**, which forms a thin, viscous film over the surfaces within the articular capsule.	Anatomy Overview—Synovial Joints **Figure 7.3** Structure of a typical synovial joint

REVIEW	RESOURCES

4. Many synovial joints also contain **accessory ligaments** and **articular discs**.

5. **Bursae** are saclike structures, similar in structure to joint capsules, that reduce friction in joints such as the shoulder and knee joints.

7.5 Types of Movements at Synovial Joints

1. In a **gliding** movement, the nearly flat surfaces of bones move back-and-forth and side-to-side.

2. In **angular movements**, there is a change in the angle between bones. Examples are **flexion–extension**, **hyperextension**, **abduction–adduction**, and **circumduction**.

3. In **rotation**, a bone moves around its own longitudinal axis.

4. **Special movements** occur at specific synovial joints in the body. Examples are as follows: **elevation–depression**, **protraction–retraction**, **inversion–eversion**, **dorsiflexion–plantar flexion**, and **supination–pronation**.

7.6 Types of Synovial Joints

1. Types of synovial joints are plane, hinge, pivot, condyloid, saddle, and ball-and-socket.

2. In **plane (planar) joints** the articulating surfaces are flat, and the bones glide back-and-forth and side-to-side (many are biaxial); they may also permit rotation (triaxial). Examples of plane joints are the joints between carpals and the joints between tarsals.

3. In **hinge joints**, the convex surface of one bone fits into the concave surface of another, and the motion is angular around one axis (uniaxial); examples are the elbow, knee (a modified hinge joint), and ankle joints.

4. In **pivot joints**, a round or pointed surface of one bone fits into a ring formed by another bone and a ligament, and movement is rotational (uniaxial); examples are the atlanto-axial and radioulnar joints.

5. In **condyloid joints**, an oval projection of one bone fits into an oval cavity of another, and motion is angular around two axes (biaxial); examples include the wrist joint and metacarpophalangeal joints of the second through fifth digits.

6. In **saddle joints**, the articular surface of one bone is shaped like a saddle and the other bone fits into the "saddle" like a sitting rider; motion is angular around two axes (biaxial). An example is the carpometacarpal joint between the trapezium and the metacarpal of the thumb.

7. In **ball-and-socket joints**, the ball-shaped surface of one bone fits into the cuplike depression of another; motion is around three axes (triaxial). Examples include the shoulder and hip joints.

8. The knee joint is a diarthrosis that illustrates the complexity of this type of joint (Exhibit 7.A). It contains an articular capsule, several ligaments within and around the outside of the joint, **menisci**, and bursae. **Arthroplasty** refers to the surgical replacement of severely damaged natural joints with artificial joints.

7.7 Aging and Joints

1. With aging, a decrease in synovial fluid, thinning of articular cartilage, and decreased flexibility of ligaments occur.

2. Most individuals experience some degeneration in the knees, elbows, hips, and shoulders due to the aging process.

SELF-QUIZ

1. A joint that has a _____ fit offers a great amount of movement and is _____ likely to become dislocated.
   a. tight, less
   b. tight, more
   c. loose, less
   d. loose, more
   e. flexible, less

2. An example of a fibrous joint in which the bones are immovable in an adult is a
   a. suture.
   b. syndesmosis.
   c. synovial joint.
   d. symphysis.
   e. synchondrosis.

3. Pulling out a tooth would disarticulate which type of joint?
   a. symphysis
   b. synovial
   c. gomphosis
   d. cartilaginous
   e. suture

4. Which of the following is NOT a function of synovial fluid?
   a. It acts as a lubricant.
   b. It helps strengthen the joint.
   c. It reduces friction.
   d. It provides nutrients to the tissues around the joints.
   e. It removes metabolic wastes.

5. Articular cartilage and bursae would most likely be found in which of the following?
   a. a gomphosis
   b. a suture
   c. the pubic symphysis
   d. the knee
   e. a synchondrosis

6. At a synovial joint, the articulating bones are held together and strengthened by the
   a. bursae.
   b. articular cartilage.
   c. synovial fluid.
   d. synovial membrane.
   e. articular capsule.

7. The joints between the vertebrae and the joint between the hip bones are examples of which joint type?
   a. synovial
   b. symphysis
   c. fibrous
   d. synchondrosis
   e. suture

8. Match the following:
   _____ a. the joint between the atlas and axis
   _____ b. allows gliding movements
   _____ c. the joint between the carpal and metacarpal of the thumb
   _____ d. hip joint
   _____ e. knee joint

   A. plane joint
   B. hinge joint
   C. ball-and-socket joint
   D. pivot joint
   E. saddle joint

9. Which of the following diarthrotic joints allows for the greatest degree of movement?
   a. ball-and-socket
   b. hinge
   c. condyloid
   d. pivot
   e. saddle

10. As your leg moves in the most anterior position to kick a ball, the femur will be _____ and the knee _____.
    a. extended; flexed
    b. extended; extended
    c. extended; hyperextended
    d. flexed; extended
    e. flexed; flexed

11. Shaking your head to indicate "yes" involves _____; moving your head to indicate "no" involves _____.
    a. rotation; hyperextension
    b. elevation and depression; rotation
    c. flexion and extension; rotation
    d. flexion; circumduction
    e. extension; flexion

12. In the anatomical position, the palms are
    a. supinated.
    b. flexed.
    c. inverted.
    d. pronated.
    e. protracted.

13. A fluid-filled sac found between skin and bone that helps reduce friction between the skin and bone is a
    a. meniscus.
    b. bursa.
    c. ligament.
    d. articular capsule.
    e. synovial membrane.

14. Moving the mandible for chewing involves
    a. protraction and retraction.
    b. inversion and eversion.
    c. depression and extension.
    d. protraction and elevation.
    e. depression and elevation.

15. Match the following:
    _____ a. movement of a bone around its own axis
    _____ b. movement away from the midline of the body
    _____ c. turning the palm so it faces forward
    _____ d. downward movement of a body part
    _____ e. movement toward the midline of the body
    _____ f. movement of the mandible or shoulder backward
    _____ g. turning the palm so it faces backward
    _____ h. upward movement of a body part
    _____ i. movement of the distal end of a body part in a circle
    _____ j. movement beyond the plane of extension

    A. rotation
    B. supination
    C. depression
    D. adduction
    E. retraction
    F. pronation
    G. abduction
    H. hyperextension
    I. circumduction
    J. elevation

CRITICAL THINKING APPLICATIONS

1. After your second A & P exam, you dropped to one knee, tipped your head back, raised one arm over your head, clenched your fist, pumped your arm up and down, and yelled "Yes!" Use the proper terms to describe the movements undertaken by the various joints.

2. Aunt Bella's hip has been bothering her for years, and now she can hardly walk. Her doctor suggested a hip replacement. "It's one of those synonymous joints," Aunt Bella explained. What type of joint is the hip joint? What types of movements can it perform?

3. Remember Keith, the volleyball player from Chapter 6? His cast finally came off today. The orthopedist tested his knee's range of motion and declared that the ACL appeared to be intact. What is the ACL? How does the ACL contribute to the knee joint's stability?

4. Your Uncle Bob's in good health but has had more difficulty walking the past year. He doesn't complain, but simply states "It's the curse of being 82! I just need to buy me some new legs!" Why do you suspect Bob is having trouble walking? Are "new legs" an option?

ANSWERS TO FIGURE QUESTIONS

7.1 Sutures in an adult skull are synarthroses because they are immovable; syndesmoses are classified as amphiarthroses because they are slightly movable.

7.2 Hyaline cartilage holds a synchondrosis together, and fibrocartilage holds a symphysis together.

7.3 Synovial joints are diarthroses, freely movable joints.

7.4 Gliding movements occur at intercarpal joints and intertarsal joints.

7.5 The arrangement of ligaments and bones prevents hyperextension at some synovial joints.

7.6 When you adduct your arm or leg, you bring it closer to the midline of the body, thus "adding" it to the trunk.

7.7 Circumduction can occur at the shoulder joint and at the hip joint.

7.8 The anterior surface of a bone or limb rotates toward the midline in medial rotation, and away from the midline in lateral rotation.

7.9 Bringing the arms forward until the elbows touch is an example of protraction.

7.10 Ball-and-socket joints permit the greatest range of motion.

7.11 In torn cartilage injuries of the knee, the menisci are damaged.

7.12 The goals of arthroplasty are to relieve pain and increase range of motion.

CHAPTER 8
THE MUSCULAR SYSTEM

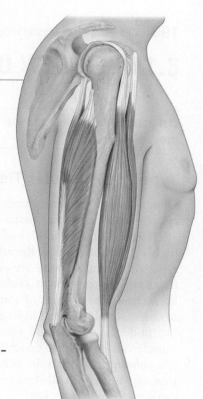

Movements such as throwing a ball, biking, and walking require an interaction between bones and muscles. To understand how muscles produce different movements, you will learn where the muscles attach on individual bones and the types of joints acted on by the contracting muscles. The bones, muscles, and joints together form an integrated system called the ***musculoskeletal system***. The scientific study of muscles is known as ***myology*** (mī-OL-ō-jē; *my-* = muscle; *-logy* = study of). The branch of medical science concerned with the prevention or correction of disorders of the musculoskeletal system is called ***ortho-pedics*** (or'-thō-PĒ-diks; *ortho-* = correct; *pedi* = child).

Did you know?   *People used to think body image dissatisfaction occurred primarily in girls and women. This dissatisfaction often manifests itself as a drive for impossible thinness, which can then lead to the development of abnormal eating and exercise behaviors. But at the other end of the size spectrum is another body image problem: muscle dysmorphia, which occurs more commonly in boys and men. People with muscle dysmorphia want to look like superheroes, and perceive their own bodies as woefully inadequate. These people go to extreme measures to build more muscle mass.*

FOCUS ON WELLNESS
Superhero Dreams—Muscle Dysmorphia

LOOKING BACK TO MOVE AHEAD...

Muscular Tissue (Section 4.5)

Adenosine Triphosphate (Section 2.2)

Divisions of the Skeletal System (Section 6.6)

Joints (Section 7.1)

Types of Movements at Synovial Joints (Section 7.5)

8.1 OVERVIEW OF MUSCULAR TISSUE

OBJECTIVE • **Describe the types and functions of muscular tissue.**

Types of Muscular Tissue

Depending on the percentage of body fat, gender, and exercise regimen, muscular tissue constitutes about 40 percent to 50 percent of the total body weight and is composed of highly specialized cells. Recall from Chapter 4 that the three types of muscular tissue are skeletal, cardiac, and smooth. As its name suggests, most *skeletal muscle tissue* is attached to bones and moves parts of the skeleton. It is *striated*; that is, *striations*, or alternating light and dark protein bands, are visible under a microscope (see Figure 8.2). Because skeletal muscle can be made to contract and relax by conscious control, it is *voluntary*. Due to the presence of a small number of cells that can undergo cell division, skeletal muscle has a limited capacity for regeneration.

Cardiac muscle tissue, found only in the heart, forms the bulk of the heart wall. The heart pumps blood through blood vessels to all parts of the body. Like skeletal muscle tissue, cardiac muscle tissue is *striated*. However, unlike skeletal muscle tissue, it is *involuntary*: Its contractions are not under conscious control. Cardiac muscle can regenerate under certain conditions. For example, in response to damage to heart cells, it appears that stem cells can migrate from the blood into the heart and develop into functional cardiac muscle cells to repair the damage.

Smooth muscle tissue is located in the walls of hollow internal structures, such as blood vessels, airways, the stomach, and the intestines. It participates in internal processes such as digestion and the regulation of blood pressure. Smooth muscle is *nonstriated* (lacks striations) and *involuntary* (not under conscious control). Although smooth muscle tissue has considerable capacity to regenerate when compared with other muscle tissues, this capacity is limited when compared to other types of tissues, for example, epithelial tissue.

Functions of Muscular Tissue

Through sustained contraction or alternating contraction and relaxation, muscular tissue has four key functions: producing body movements, stabilizing body positions, storing and moving substances within the body, and producing heat.

1. **Producing body movements.** Body movements such as walking, running, writing, or nodding the head rely on the integrated functioning of skeletal muscles, bones, and joints.
2. **Stabilizing body positions.** Skeletal muscle contractions stabilize joints and help maintain body positions, such as standing or sitting. Postural muscles contract continuously when a person is awake; for example, sustained contractions of your neck muscles hold your head upright.
3. **Storing and moving substances within the body.** Storage is accomplished by sustained contractions of ring-like bands of smooth muscle called *sphincters*, which prevent outflow of the contents of a hollow organ. Temporary storage of food in the stomach or urine in the urinary bladder is possible because smooth muscle sphincters close off the outlets of these organs. Cardiac muscle contractions of the heart pump blood through the blood vessels of the body. Contraction and relaxation of smooth muscle in the walls of blood vessels help adjust blood vessel diameter and thus regulate blood flow. Smooth muscle contractions also move food and other substances through the gastrointestinal tract, push gametes (sperm and oocytes) through the reproductive systems, and propel urine through the urinary system. Skeletal muscle contractions aid the return of blood in veins to the heart.
4. **Producing heat.** As muscular tissue contracts, it produces heat. Much of the heat released by muscles is used to maintain normal body temperature. Involuntary contractions of skeletal muscle, known as shivering, can help warm the body by greatly increasing the rate of heat production.

✓ CHECKPOINT

1. What features distinguish the three types of muscular tissue?
2. What are the general functions of muscular tissue?

8.2 SKELETAL MUSCLE TISSUE

OBJECTIVES • **Explain the relationships of connective tissue components, blood vessels, and nerves to skeletal muscles.**

• **Describe the histology of a skeletal muscle fiber.**

Each skeletal muscle is a separate organ composed of hundreds to thousands of cells, which are called *muscle fibers* because of their elongated shapes. Connective tissues surround muscle fibers and whole muscles, and blood vessels and nerves penetrate muscles (Figure 8.1).

Connective Tissue Components

Connective tissue surrounds and protects muscular tissue. The *subcutaneous layer* or *hypodermis*, which separates muscle from skin, is composed of areolar connective tissue and adipose tissue. It provides a pathway for nerves, blood vessels, and lymphatic vessels to enter and exit muscles. The adipose

Figure 8.1 Organization of skeletal muscle and its connective tissue coverings.

A skeletal muscle consists of individual muscle fibers (cells) bundled into fascicles and surrounded by three connective tissue layers.

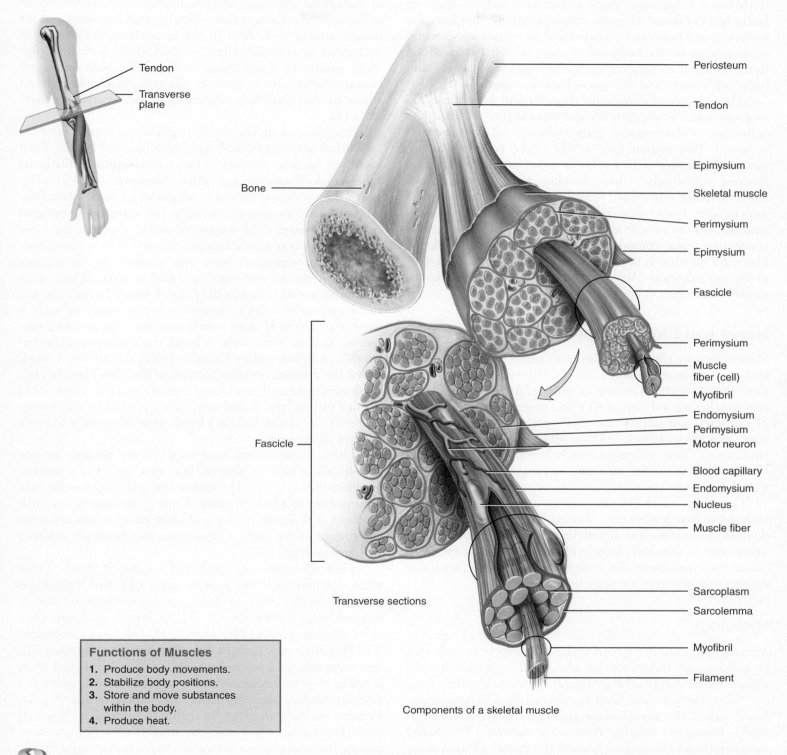

Tendon

Transverse plane

Bone

Fascicle

Transverse sections

Periosteum

Tendon

Epimysium

Skeletal muscle

Perimysium

Epimysium

Fascicle

Perimysium

Muscle fiber (cell)

Myofibril

Endomysium

Perimysium

Motor neuron

Blood capillary

Endomysium

Nucleus

Muscle fiber

Sarcoplasm

Sarcolemma

Myofibril

Filament

Components of a skeletal muscle

Functions of Muscles

1. Produce body movements.
2. Stabilize body positions.
3. Store and move substances within the body.
4. Produce heat.

Starting with the connective tissue that surrounds an individual muscle fiber (cell) and working toward the outside, list the connective tissue layers in order.

tissue of the subcutaneous layer stores most of the body's triglycerides, serves as an insulating layer that reduces heat loss, and protects muscles from physical trauma. *Fascia* (FASH-ē-a = bandage; plural is *fasciae*) is a dense sheet or broad band of dense irregular connective tissue that lines the body wall and limbs and supports and surrounds muscles and other organs of the body (see Figure 8.20c). Fascia allows free movement of muscles, carries nerves, blood vessels, and lymphatic vessels, and fills spaces between muscles.

Three layers of connective tissue extend from the fascia to protect and strengthen skeletal muscle (Figure 8.1). The entire muscle is wrapped in *epimysium* (ep′-i-MĪZ-ē-um; *epi-* = upon*)*. *Perimysium* (per′-i-MĪZ-ē-um; *peri-* = around) surrounds bundles of 10 to 100 or more muscle fibers called *fascicles* (FAS-i-kuls = little bundle). Finally, *endomysium* (en′-dō-MĪZ-ē-um; *endo-* = within) wraps each individual muscle fiber. Epimysium, perimysium, and endomysium extend beyond the muscle as a *tendon*—a cord of dense regular connective tissue composed of parallel bundles of collagen fibers. Its function is to attach a muscle to a bone. An example is the calcaneal (Achilles) tendon of the gastrocnemius muscle (see Figure 8.24a).

Nerve and Blood Supply

Skeletal muscles are well supplied with nerves and blood vessels (Figure 8.1), both of which are directly related to contraction, the chief characteristic of muscle. Muscle contraction also requires a good deal of ATP and therefore large amounts of nutrients and oxygen for ATP synthesis. Moreover, the waste products of these ATP-producing reactions must be eliminated. Thus, prolonged muscle action depends on a rich blood supply to deliver nutrients and oxygen and to remove wastes.

Generally, an artery and one or two veins accompany each nerve that penetrates a skeletal muscle. Within the endomysium, microscopic blood vessels called capillaries are distributed so that each muscle fiber is in close contact with one or more capillaries. Each skeletal muscle fiber also makes contact with the terminal portion of a neuron.

Histology

Microscopic examination of a skeletal muscle reveals that it consists of thousands of elongated, cylindrical cells called *muscle fibers* arranged parallel to one another (Figure 8.2a). Each muscle fiber is covered by a plasma membrane called the *sarcolemma* (sar′-kō-LEM-ma; *sarco-* = flesh; *-lemma* = sheath). *Transverse tubules (T tubules)* tunnel in from the surface toward the center of each muscle fiber. Multiple nuclei lie at the periphery of the fiber, under the sarcolemma. The muscle fiber's cytoplasm, called *sarcoplasm* (SAR-kō-plazm), contains many mitochondria

that produce large amounts of ATP during muscle contraction. Extending throughout the sarcoplasm is *sarcoplasmic reticulum* (sar′-kō-PLAZ-mik re-TIK-ū-lum), a network of fluid-filled membrane-enclosed tubules (similar to smooth endoplasmic reticulum) that stores calcium ions required for muscle contraction. Also in the sarcoplasm are numerous molecules of *myoglobin* (mī′-ō-GLŌ-bin), a reddish pigment similar to hemoglobin in blood. In addition to the characteristic color it lends to skeletal muscle, myoglobin stores oxygen until it is needed by mitochondria to generate ATP.

Extending along the entire length of the muscle fiber are cylindrical structures called *myofibrils* (mī′-ō-FĪ-brils). Each myofibril, in turn, consists of two types of protein filaments called *thin filaments* and *thick filaments* (Figure 8.2b), which do not extend the entire length of a muscle fiber. Filaments overlap in specific patterns and form compartments called *sarcomeres* (SAR-kō-mērs; *-meres* = parts), the basic functional units of striated muscle fibers (Figure 8.2b, c). Sarcomeres are separated from one another by zig-zagging zones of dense protein material called *Z discs*. Within each sarcomere a darker area, called the *A band*, extends the entire length of the thick filaments. At the center of each A band is a narrow *H zone*, which contains only the thick filaments. At both ends of the A band, thick and thin filaments overlap. A lighter-colored area to either side of the A band, called the *I band*, contains the rest of the thin filaments but no thick filaments. Each I band extends into two sarcomeres, divided in half by a Z disc (see Figure 8.2c). The alternating darker A bands and lighter I bands give the muscle fiber its striated appearance.

Thick filaments are composed of the protein *myosin* (MĪ-ō-sin), which is shaped like two golf clubs twisted together (Figure 8.3a). The *myosin tails* (golf club handles) are arranged parallel to each other, forming the shaft of the thick filament. The heads of the golf clubs project outward from the surface of the shaft. These projecting heads are referred to as *myosin heads*.

Thin filaments are anchored to the Z discs. Their main component is the protein *actin* (AK-tin). Individual actin molecules join to form an actin filament that is twisted into a helix (Figure 8.3b). Each actin molecule contains a *myosin-binding site*, where a myosin head can attach. The thin filaments contain two other proteins, *tropomyosin* and *troponin*. In a relaxed muscle, myosin is blocked from binding to actin because strands of tropomyosin cover the myosin-binding sites on actin. The tropomyosin strands, in turn, are held in place by troponin molecules. You will soon learn that when calcium ions (Ca^{2+}) bind to troponin, it undergoes a change in shape; this change moves tropomyosin away from myosin-binding sites on actin and muscle contraction subsequently begins as myosin binds to actin.

Figure 8.2 Organization of skeletal muscle from macroscopic (gross) to microscopic (molecular) levels.

The structural organization of a skeletal muscle from macroscopic to microscopic is as follows: skeletal muscle, fascicle (bundle of muscle fibers), muscle fiber, myofibril, and thin and thick filaments.

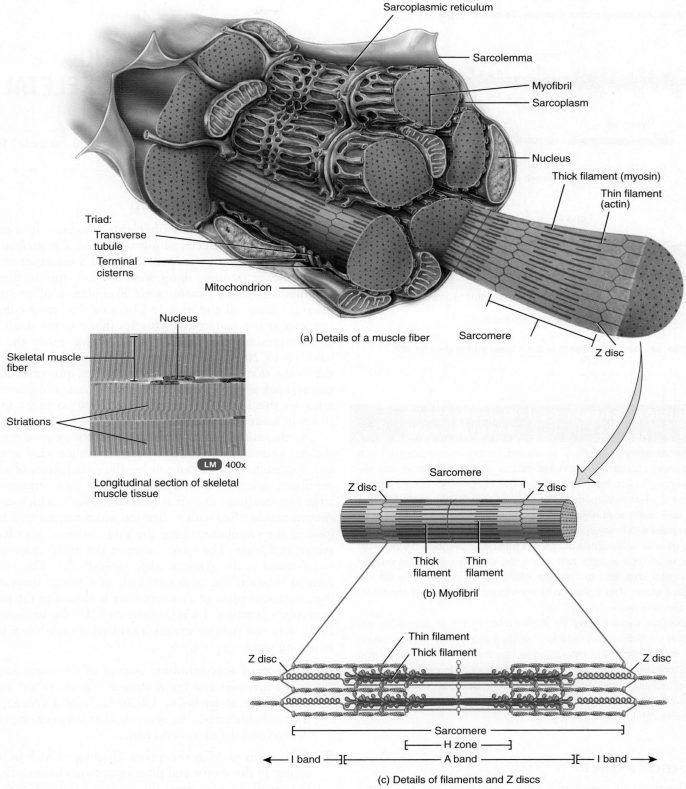

(a) Details of a muscle fiber

LM 400x

Longitudinal section of skeletal muscle tissue

(b) Myofibril

(c) Details of filaments and Z discs

Which filaments are part of the A band and I band?

Figure 8.3 Detailed structure of filaments. (a) About 300 myosin molecules compose a thick filament. The myosin tails all point toward the center of the sarcomere. (b) Thin filaments contain actin, troponin, and tropomyosin.

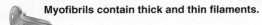

Myofibrils contain thick and thin filaments.

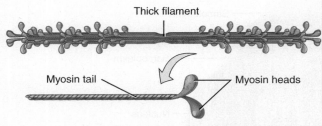

Thick filament

Myosin tail — Myosin heads

(a) One thick filament and a myosin molecule

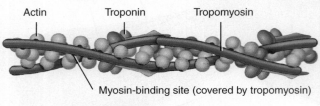

Actin   Troponin   Tropomyosin

Myosin-binding site (covered by tropomyosin)

(b) Portion of a thin filament

 What proteins are present in the A band and in the I band?

CLINICAL CONNECTION | Muscular Atrophy and Hypertrophy

Muscular atrophy (A-trō-fē; *a-* = without, *-trophy* = nourishment) is a wasting away of muscles. Individual muscle fibers decrease in size because of progressive loss of myofibrils. The atrophy that occurs if muscles are not used is termed *disuse atrophy*. Bedridden individuals and people with casts experience disuse atrophy because the number of nerve impulses to inactive muscle is greatly reduced. If the nerve supply to a muscle is disrupted or cut, the muscle undergoes *denervation atrophy*. In about 6 months to 2 years, the muscle will be one-quarter of its original size, and the muscle fibers will be replaced by fibrous connective tissue. The transition to connective tissue, when complete, cannot be reversed.

Muscular hypertrophy (hī-PER-trō-fē; *hyper-* = above or excessive) is an increase in muscle fiber diameter owing to the production of more myofibrils, mitochondria, sarcoplasmic reticulum, and other cytoplasmic structures. It results from very forceful, repetitive muscular activity, such as strength training. Because hypertrophied muscles contain more myofibrils, they are capable of contractions that are more forceful. •

✓ **CHECKPOINT**

3. What type of connective tissue coverings are associated with skeletal muscle?

4. Why is a rich blood supply important for muscle contraction?

5. What is a sarcomere? What does a sarcomere contain?

8.3 CONTRACTION AND RELAXATION OF SKELETAL MUSCLE

OBJECTIVE • **Explain how skeletal muscle fibers contract and relax.**

Neuromuscular Junction

Before a skeletal muscle fiber can contract, it must be stimulated by an electrical signal called a ***muscle action potential*** delivered by its neuron called a ***motor neuron***. A single motor neuron along with all the muscle fibers it stimulates is called a ***motor unit***. Stimulation of one motor neuron causes all the muscle fibers in that motor unit to contract at the same time. Muscles that control small, precise movements, such as the muscles that move the eyes, have 10 to 20 muscles fibers per motor unit. Muscles of the body that are responsible for large, powerful movements, such as the biceps brachii in the arm and gastrocnemius in the leg, have as many as 2000 to 3000 muscle fibers in some motor units.

As the *axon* (long process) of a motor neuron enters a skeletal muscle, it divides into branches called *axon terminals* that approach—but do not touch—the sarcolemma of a muscle fiber (Figure 8.4a, b). The ends of the axon terminals enlarge into swellings known as *synaptic end bulbs*, which contain *synaptic vesicles* filled with a chemical *neurotransmitter*. The region of the sarcolemma near the axon terminal is called the ***motor end plate***. The space between the axon terminal and sarcolemma is the ***synaptic cleft*** (sin-AP-tik). The synapse formed between the axon terminals of a motor neuron and the motor end plate of a muscle fiber is known as the ***neuromuscular junction (NMJ)*** (noo-rō-MUS-kū-lar). At the NMJ, a motor neuron excites a skeletal muscle fiber in the following way (Figure 8.4c):

1 **Release of acetylcholine**. Arrival of the nerve impulse at the synaptic end bulbs triggers release of the neurotransmitter ***acetylcholine*** (***ACh***) (as′-e-til-KŌ-lēn). ACh then diffuses across the synaptic cleft between the motor neuron and the motor end plate.

2 **Activation of ACh receptors**. Binding of ACh to its receptor in the motor end plate opens ion channels that allow small cations, especially sodium ions (Na^+), to flow across the membrane.

Figure 8.4 Neuromuscular junction (NMJ).

A neuromuscular junction includes the axon terminal of a motor neuron plus the motor end plate of a muscle fiber.

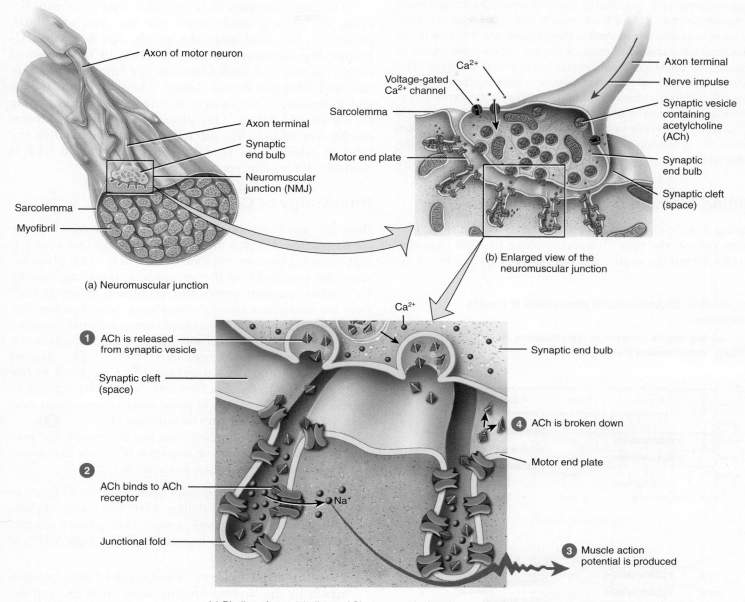

Axon of motor neuron

Axon terminal

Synaptic end bulb

Neuromuscular junction (NMJ)

Sarcolemma

Myofibril

(a) Neuromuscular junction

Ca²⁺

Voltage-gated Ca²⁺ channel

Sarcolemma

Motor end plate

Axon terminal

Nerve impulse

Synaptic vesicle containing acetylcholine (ACh)

Synaptic end bulb

Synaptic cleft (space)

(b) Enlarged view of the neuromuscular junction

Ca²⁺

1 ACh is released from synaptic vesicle

Synaptic cleft (space)

2 ACh binds to ACh receptor

Junctional fold

Synaptic end bulb

4 ACh is broken down

Motor end plate

Na⁺

3 Muscle action potential is produced

(c) Binding of acetylcholine to ACh receptors in the motor end plate

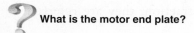 **What is the motor end plate?**

3 **Generation of muscle action potential.** The inflow of Na⁺ (down its concentration gradient) generates a muscle action potential. The muscle action potential then travels along the sarcolemma and through the T tubules. Each nerve impulse normally elicits one muscle action potential. If another nerve impulse releases more acetyl-

choline, then steps **2** and **3** repeat. See Section 9.3 for the details of nerve impulse generation.

4 **Breakdown of ACh.** The effect of ACh lasts only briefly because the neurotransmitter is rapidly broken down in the synaptic cleft by an enzyme called ***acetylcholinesterase (AChE)*** (as′-e-til-kō′-lin-ES-ter-ās).

Sliding-filament Mechanism

During muscle contraction, myosin heads of the thick filaments pull on the thin filaments, causing the thin filaments to slide toward the center of a sarcomere (Figure 8.5a, b). As

Figure 8.5 Sliding-filament mechanism of muscle contraction.

During muscle contraction, thin filaments move inward toward the H zone.

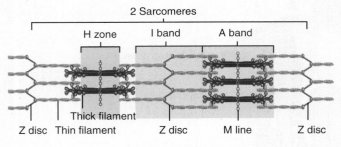

(a) Relaxed muscle

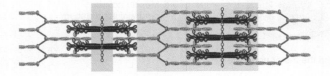

(b) Partially contracted muscle

(c) Maximally contracted muscle

What happens to the I bands as muscle contracts? Do the lengths of the thick and thin filaments change during contraction?

the thin filaments slide, the I bands and H zones become narrower (Figure 8.5b) and eventually disappear altogether when the muscle is maximally contracted (Figure 8.5c).

The thin filaments slide past the thick filaments because the myosin heads move like the oars of a boat, pulling on the actin molecules of the thin filaments. Although the sarcomere shortens because of the increased overlap of thin and thick filaments, the lengths of the thin and thick filaments do not change. The sliding of filaments and shortening of sarcomeres in turn cause the shortening of the muscle fibers. This process, the ***sliding-filament mechanism*** of muscle contraction, occurs only when the level of calcium ions (Ca^{2+}) is high enough and ATP is available, for reasons you will see shortly.

Physiology of Contraction

Both Ca^{2+} and energy, in the form of ATP, are needed for muscle contraction. When a muscle fiber is relaxed (not contracting), there is a low concentration of Ca^{2+} in the sarcoplasm because the membrane of the sarcoplasmic reticulum contains Ca^{2+} active transport pumps that continually transport Ca^{2+} from the sarcoplasm into the sarcoplasmic reticulum (see Figure 8.7 **7**). However, when a muscle action potential (impulse) travels along the sarcolemma and into the transverse tubule system, Ca^{2+} release channels open (see Figure 8.7 **4**), allowing Ca^{2+} to escape into the sarcoplasm. The Ca^{2+} binds to troponin molecules in the thin filaments, causing the troponin to change shape. This change in shape moves tropomyosin away from the myosin-binding sites on actin (see Figure 8.7 **5**).

Once the myosin-binding sites are uncovered, the ***contraction cycle***—the repeating sequence of events that causes the filaments to slide—begins, as shown in Figure 8.6:

1 **Splitting ATP.** The myosin heads contain ATPase, an enzyme that splits ATP into ADP (adenosine diphosphate) and P (a phosphate group). This splitting reaction transfers energy to the myosin head, although ADP and P remain attached to it.

2 **Forming cross-bridges.** The energized myosin heads attach to the myosin-binding sites on actin, and release the phosphate groups. When myosin heads attach to actin during contraction, they are referred to as ***cross-bridges***.

3 **Power stroke.** After the cross-bridges form, the ***power stroke*** occurs. During the power stroke, the cross-bridge rotates or swivels and releases the ADP. The force produced as hundreds of cross-bridges swivel slides the thin filament past the thick filament toward the center of the sarcomere.

4 **Binding ATP and detaching.** At the end of the power stroke, the cross-bridges remain firmly attached to actin. When they bind another molecule of ATP, the myosin heads detach from actin.

Figure 8.6 The contraction cycle. Sarcomeres shorten through repeated cycles in which the myosin heads (cross-bridges) attach to actin, rotate, and detach.

🔑 **During the power stroke of contraction, cross-bridges rotate and move the thin filaments past the thick filaments toward the center of the sarcomere.**

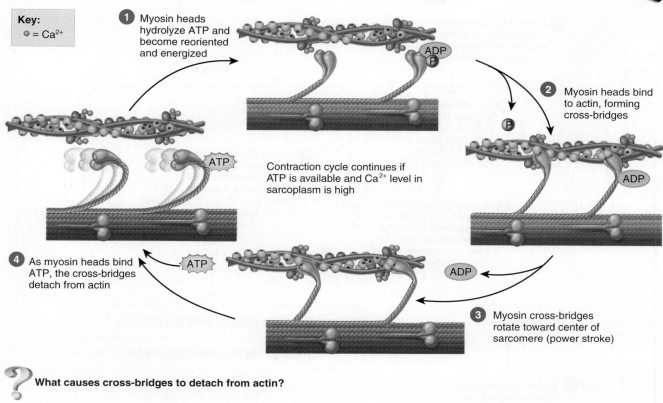

Key:
• = Ca^{2+}

① Myosin heads hydrolyze ATP and become reoriented and energized

ADP
P

② Myosin heads bind to actin, forming cross-bridges

P

ADP

ATP

Contraction cycle continues if ATP is available and Ca^{2+} level in sarcoplasm is high

④ As myosin heads bind ATP, the cross-bridges detach from actin

ATP

ADP

③ Myosin cross-bridges rotate toward center of sarcomere (power stroke)

❓ **What causes cross-bridges to detach from actin?**

As the myosin ATPase again splits ATP, the myosin head is reoriented and energized, ready to combine with another myosin-binding site farther along the thin filament. The contraction cycle repeats as long as ATP and Ca^{2+} are available in the sarcoplasm. At any one instant, some of the myosin heads are attached to actin, forming cross-bridges and generating force, and other myosin heads are detached from actin and getting ready to bind again. During a maximal contraction, the sarcomere can shorten by as much as half its resting length.

CLINICAL CONNECTION | Rigor mortis

After death, cellular membranes become leaky. Calcium ions leak out of the sarcoplasmic reticulum into the cytosol and allow myosin heads to bind to actin. ATP synthesis ceases shortly after breathing stops, however, so the cross-bridges cannot detach from actin. The resulting condition, in which muscles are in a state of rigidity (cannot contract or stretch), is called **rigor mortis** (rigidity of death). Rigor mortis begins 3 to 4 hours after death and lasts about 24 hours; then it disappears as proteolytic enzymes from lysosomes digest the cross-bridges. •

Relaxation

Two changes permit a muscle fiber to relax after it has contracted. First, the neurotransmitter acetylcholine is rapidly broken down by the enzyme acetylcholinesterase (AChE). When nerve action potentials cease, release of ACh stops, and AChE rapidly breaks down the ACh already present in the synaptic cleft. This ends the generation of muscle action potentials, and the Ca^{2+} release channels in the sarcoplasmic reticulum membrane close.

Second, calcium ions are rapidly transported from the sarcoplasm into the sarcoplasmic reticulum. As the level of Ca^{2+} in the sarcoplasm falls, tropomyosin slides back over the myosin-binding sites on actin. Once the myosin-binding sites are covered, the thin filaments slip back to their relaxed positions. Figure 8.7 summarizes the events of contraction and relaxation in a muscle fiber.

Muscle Tone

Even when a whole muscle is not contracting, a small number of its motor units are involuntarily activated to produce

Figure 8.7 Summary of the events of contraction and relaxation in a skeletal muscle fiber.

Acetylcholine released at the neuromuscular junction triggers a muscle action potential, which leads to muscle contraction.

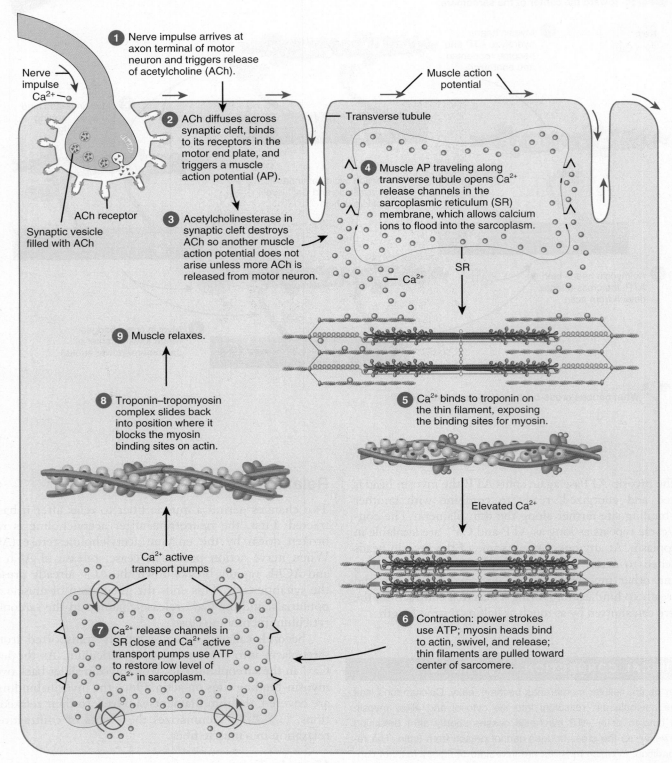

1 Nerve impulse arrives at axon terminal of motor neuron and triggers release of acetylcholine (ACh).

Nerve impulse
Ca^{2+}

ACh receptor

Synaptic vesicle filled with ACh

2 ACh diffuses across synaptic cleft, binds to its receptors in the motor end plate, and triggers a muscle action potential (AP).

3 Acetylcholinesterase in synaptic cleft destroys ACh so another muscle action potential does not arise unless more ACh is released from motor neuron.

Muscle action potential

Transverse tubule

4 Muscle AP traveling along transverse tubule opens Ca^{2+} release channels in the sarcoplasmic reticulum (SR) membrane, which allows calcium ions to flood into the sarcoplasm.

Ca^{2+}

SR

9 Muscle relaxes.

8 Troponin–tropomyosin complex slides back into position where it blocks the myosin binding sites on actin.

5 Ca^{2+} binds to troponin on the thin filament, exposing the binding sites for myosin.

Elevated Ca^{2+}

Ca^{2+} active transport pumps

7 Ca^{2+} release channels in SR close and Ca^{2+} active transport pumps use ATP to restore low level of Ca^{2+} in sarcoplasm.

6 Contraction: power strokes use ATP; myosin heads bind to actin, swivel, and release; thin filaments are pulled toward center of sarcomere.

? The power stroke occurs during which numbered step in this figure?

a sustained contraction of their muscle fibers. This process results in *muscle tone* (*tonos* = tension). To sustain muscle tone, small groups of motor units are alternately active and inactive in a constantly shifting pattern. Muscle tone keeps skeletal muscles firm, but it does not result in a contraction strong enough to produce movement. For example, the tone of muscles in the back of the neck keeps the head upright and prevents it from slumping forward on the chest. Recall that skeletal muscle contracts only after it is activated by acetylcholine released by nerve impulses in its motor neurons. Hence, muscle tone is established by neurons in the brain and spinal cord that excite the muscle's motor neurons. When the motor neurons serving a skeletal muscle are damaged or cut, the muscle becomes *flaccid* (FLAS-id = flabby), a state of limpness in which muscle tone is lost.

✓ CHECKPOINT

6. Explain how a skeletal muscle contracts and relaxes.

7. What is the importance of the neuromuscular junction?

8.4 METABOLISM OF SKELETAL MUSCLE TISSUE

OBJECTIVES • **Describe the sources of ATP and oxygen for muscle contraction.**

• **Define muscle fatigue and list its possible causes.**

Energy for Contraction

Unlike most cells of the body, skeletal muscle fibers often switch between virtual inactivity, when they are relaxed and using only a modest amount of ATP, and great activity, when they are contracting and using ATP at a rapid pace. However, the ATP present inside muscle fibers is enough to power contraction for only a few seconds. If strenuous exercise is to continue, additional ATP must be synthesized. Muscle fibers have three sources for ATP production: (1) creatine phosphate, (2) anaerobic cellular respiration, and (3) aerobic cellular respiration.

While at rest, muscle fibers produce more ATP than they need. Some of the excess ATP is used to make *creatine phosphate*, an energy-rich molecule that is unique to muscle fibers (Figure 8.8a). One of ATP's high-energy phosphate groups is transferred to creatine, forming creatine phosphate and ADP (adenosine diphosphate). *Creatine* is a small, amino acid-like molecule that is synthesized in the liver, kidneys, and pancreas and derived from certain foods

(milk, red meat, fish), then transported to muscle fibers. While muscle is contracting, the high-energy phosphate group can be transferred from creatine phosphate back to ADP, quickly forming new ATP molecules. Together, creatine phosphate and ATP provide enough energy for muscles to contract maximally for about 15 seconds. This energy is sufficient for short bursts of intense activity, for example, running a 100-meter dash.

CLINICAL CONNECTION | Creatine Supplementation

Creatine is both synthesized in the body (in the liver, kidneys, and pancreas) and derived from foods such as milk, red meat, and some fish. Adults need to synthesize and ingest a total of about 2 grams of creatine daily to make up for the urinary loss of creatinine, the breakdown product of creatine. Some studies have demonstrated improved performance during explosive movements, such as sprinting. Other studies, however, have failed to find a performance-enhancing effect of creatine supplementation. Moreover, ingesting extra creatine decreases the body's own synthesis of creatine, and it is not known whether natural synthesis recovers after long-term creatine supplementation. In addition, creatine supplementation can cause dehydration and may cause kidney dysfunction. Further research is needed to determine both the long-term safety and the value of **creatine supplementation**. •

When muscle activity continues past the 15-second mark, the supply of creatine phosphate is depleted. The next source of ATP is *glycolysis*, a series of cytosolic reactions that produces 2 ATPs by breaking down a glucose molecule to pyruvic acid. Glucose passes easily from the blood into contracting muscle fibers and also is produced within muscle fibers by breakdown of glycogen (Figure 8.8b). When oxygen levels are low as a result of vigorous muscle activity, most of the pyruvic acid is converted to lactic acid, a process called *anaerobic cellular respiration* because it occurs without using oxygen. Anaerobic cellular respiration can provide enough energy for about 30 to 40 seconds of maximal muscle activity. Together, conversion of creatine phosphate and glycolysis can provide enough ATP to run a 400-meter race.

Muscle activity that lasts longer than half a minute depends increasingly on *aerobic cellular respiration*, a series of oxygen-requiring reactions that produce ATP in mitochondria. Muscle fibers have two sources of oxygen: (1) oxygen that diffuses into them from the blood and (2) oxygen released by myoglobin in the sarcoplasm. *Myoglobin* is an oxygen-binding protein found only in muscle fibers. It binds oxygen when oxygen is plentiful and releases oxygen when it is scarce. If enough oxygen is present, pyruvic acid enters the mitochondria, where it is completely oxidized in reactions that generate ATP, carbon dioxide,

Figure 8.8 Production of ATP for muscle contraction. (a) Creatine phosphate, formed from ATP while the muscle is relaxed, transfers a high-energy phosphate group to ADP, forming ATP, during muscle contraction. (b) Breakdown of muscle glycogen into glucose and production of pyruvic acid from glucose via glycolysis produce both ATP and lactic acid. Because no oxygen is needed, this is an anaerobic pathway. (c) Within mitochondria, pyruvic acid, fatty acids, and amino acids are used to produce ATP via aerobic cellular respiration, an oxygen-requiring set of reactions.

During a long-term event such as a marathon race, most ATP is produced aerobically.

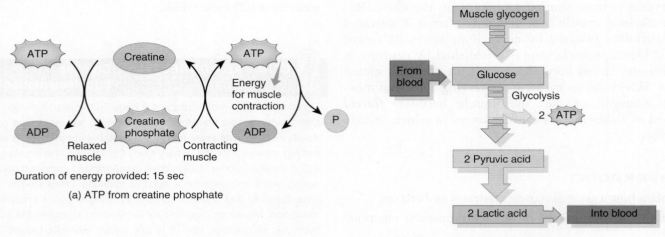

Duration of energy provided: 15 sec

(a) ATP from creatine phosphate

Duration of energy provided: 30–40 sec

(b) ATP from anaerobic glycolysis

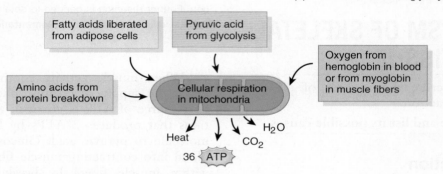

Duration of energy provided: Minutes to hours

(c) ATP from aerobic cellular respiration

 Where inside a skeletal muscle fiber are the events shown here occurring?

water, and heat (Figure 8.8c). In comparison with anaerobic cellular respiration, aerobic cellular respiration yields much more ATP, about 36 molecules of ATP from each glucose molecule. In activities that last more than 10 minutes, aerobic cellular respiration provides most of the needed ATP.

Muscle Fatigue

The inability of a muscle to contract forcefully after prolonged activity is called *muscle fatigue* (fa-TĒG). One important factor in muscle fatigue is lowered release of calcium ions from the sarcoplasmic reticulum, resulting in a decline of Ca^{2+} level in the sarcoplasm. Other factors that contribute to muscle fatigue include depletion of creatine phosphate, insufficient oxygen, depletion of glycogen and other nutrients, buildup of lactic acid and ADP, and failure of nerve impulses in the motor neuron to release enough acetylcholine.

Oxygen Consumption After Exercise

During prolonged periods of muscle contraction, increases in breathing and blood flow enhance oxygen delivery to muscular tissue. After muscle contraction has stopped, heavy breathing continues for a period of time, and oxygen consumption remains above the resting level. The term **oxygen debt** refers to the added oxygen, over and above the oxygen consumed at rest, that is taken into the body after exercise. This extra oxygen is used to "pay back" or restore metabolic conditions to the resting level in three ways: (1) to convert lactic acid back into glycogen stores in the liver, (2) to resynthesize creatine phosphate and ATP, and (3) to replace the oxygen removed from myoglobin.

The metabolic changes that occur *during exercise*, however, account for only some of the extra oxygen used *after exercise*. Only a small amount of resynthesis of glycogen occurs from lactic acid. Instead, glycogen stores are replenished much later from dietary carbohydrates. Much of the lactic acid that remains after exercise is converted back to pyruvic acid and used for ATP production via aerobic cellular respiration. Ongoing changes after exercise also boost oxygen use. First, the elevated body temperature after strenuous exercise increases the rate of chemical reactions throughout the body. Faster reactions use ATP more rapidly, and more oxygen is needed to produce ATP. Second, the heart and muscles used in breathing are still working harder than they were at rest, and thus they consume more ATP. Third, tissue repair processes are occurring at an increased pace. For these reasons, **recovery oxygen uptake** is a better term than oxygen debt for the elevated use of oxygen after exercise.

✓ **CHECKPOINT**

8. What are the sources of ATP for muscle fibers?

9. What factors contribute to muscle fatigue?

10. Why is the term *recovery oxygen uptake* more accurate than *oxygen debt*?

8.5 CONTROL OF MUSCLE TENSION

OBJECTIVES • **Explain the three phases of a twitch contraction.**

• **Describe how the frequency of stimulation and motor unit recruitment affect muscle tension.**

• **Compare the three types of skeletal muscle fibers.**

The contraction that results from a single muscle action potential, a muscle twitch, has significantly smaller force than the maximum force or tension the fiber is capable of producing. The total tension that a *single* muscle fiber can produce depends mainly on the rate at which nerve impulses arrive at

its neuromuscular junction. The number of impulses per second is the *frequency of stimulation*. When considering the contraction of a *whole* muscle, the total tension it can produce depends on the number of muscle fibers that are contracting in unison.

Twitch Contraction

A **twitch contraction** is a brief contraction of all of the muscle fibers in a motor unit in response to a single action potential in its motor neuron. Figure 8.9 shows a recording of a muscle contraction, called a **myogram** (MĪ-ō-gram). Note that a brief delay, called the *latent period*, occurs between application of the stimulus (time zero on the graph) and the beginning of contraction. During the latent period, the muscle action potential sweeps over the sarcolemma and calcium ions are released from the sarcoplasmic reticulum. During the second phase, the *contraction period* (upward tracing), repetitive power strokes are occurring, generating tension or force of contraction. In the third phase, the *relaxation period* (downward tracing), power strokes cease because the level of Ca²⁺ in the sarcoplasm is decreasing to the resting level. (Recall that calcium ions are actively transported back into the sarcoplasmic reticulum.)

Frequency of Stimulation

If a second stimulus arrives before a muscle fiber has completely relaxed, the second contraction will be stronger than the first because the second contraction begins when

Figure 8.9 **Myogram of a twitch contraction.** The arrow indicates the time at which the stimulus occurred.

 A myogram is a record of a muscle contraction.

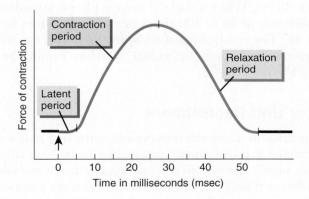

? **During which period do sarcomeres shorten?**

Figure 8.10 **Myograms showing the effects of different frequencies of stimulation.** (a) Single twitch. (b) When a second stimulus occurs before the muscle has relaxed, wave summation occurs, and the second contraction is stronger than the first. (The dashed line indicates the force of contraction expected in a single twitch.) (c) In unfused tetanus, the curve looks jagged due to partial relaxation of the muscle between stimuli. (d) In fused tetanus, the contraction force is steady and sustained.

🔑 **Due to wave summation, the tension produced during a sustained contraction is greater than during a single twitch.**

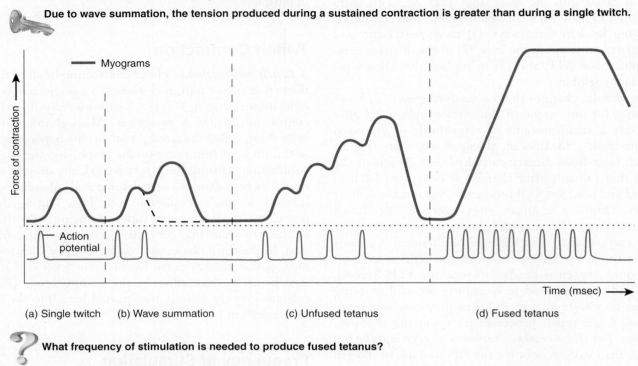

(a) Single twitch   (b) Wave summation   (c) Unfused tetanus   (d) Fused tetanus

❓ **What frequency of stimulation is needed to produce fused tetanus?**

the fiber is at a higher level of tension (Figure 8.10a, b). This phenomenon, in which stimuli arriving one after the other before a muscle fiber has completely relaxed causes larger contractions, is called *wave summation*. When a skeletal muscle fiber is stimulated at a rate of 20 to 30 times per second, it can only partially relax between stimuli. The result is a sustained but wavering contraction called *unfused (incomplete) tetanus* (*tetan-* = rigid, tense; Figure 8.10c). When a skeletal muscle fiber is stimulated at a higher rate of 80 to 100 times per second, it does not relax at all. The result is *fused (complete) tetanus*, a sustained contraction in which individual twitches cannot be detected (Figure 8.10d).

Motor Unit Recruitment

The process in which the number of contracting motor units is increased is called *motor unit recruitment*. Normally, the various motor neurons to a whole muscle fire *asynchronously* (at different times): While some motor units are contracting, others are relaxed. This pattern of motor unit activity delays muscle fatigue by allowing alternately contracting motor units to relieve one another, so that the contraction can be sustained for long periods.

Recruitment is one factor responsible for producing smooth movements rather than a series of jerky movements. Precise movements are brought about by small changes in muscle contraction. Typically, the muscles that produce precise movements are composed of small motor units. In this way, when a motor unit is recruited or turned off, only slight changes occur in muscle tension. On the other hand, large motor units are active where large amounts of tension are needed and precision is less important.

Types of Skeletal Muscle Fibers

Skeletal muscles contain three types of muscle fibers, which are present in varying proportions in different muscles of the body: The fiber types are (1) slow oxidative fibers, (2) fast oxidative–glycolytic fibers, and (3) fast glycolytic fibers.

Slow oxidative (SO) fibers or *red fibers* are small in diameter and appear dark red because they contain a large amount of myoglobin. Because they have many large mitochondria, SO fibers generate ATP mainly by aerobic cellular respiration, which is why they are called oxidative fibers. These fibers are said to be "slow" because the contraction cycle proceeds at a slower pace than in "fast" fibers. SO fibers are very resistant to fatigue and are capable of prolonged, sustained contractions.

Fast oxidative–glycolytic (FOG) fibers are intermediate in diameter between the other two types. Like slow oxidative fibers, they contain a large amount of myoglobin, and thus appear dark red. FOG fibers can generate considerable ATP by aerobic cellular respiration, which gives them a moderately high resistance to fatigue. Because their glycogen content is high, they also generate ATP by anaerobic glycolysis. These fibers are "fast" because they contract and relax more quickly than SO fibers.

Fast glycolytic (FG) fibers or *white fibers* are largest in diameter, contain the most myofibrils, and generate the most powerful and most rapid contractions. They have a low myoglobin content and few mitochondria. FG fibers contain large amounts of glycogen and generate ATP mainly by anaerobic glycolysis. They are used for intense movements of short duration, but they fatigue quickly. Strength-training programs that engage a person in activities requiring great strength for short times produce increases in the size, strength, and glycogen content of FG fibers.

Most skeletal muscles are a mixture of all three types of skeletal muscle fibers, about half of which are SO fibers. The proportions vary somewhat, depending on the action of the muscle, the person's training program, and genetic factors. For example, the continually active postural muscles of the neck, back, and legs have a high proportion of SO fibers. Muscles of the shoulders and arms, in contrast, are not constantly active but are used intermittently and briefly to produce large amounts of tension, such as in lifting and throwing. These muscles have a high proportion of FG fibers. Leg muscles, which not only support the body but are also used for walking and running, have large numbers of both SO and FOG fibers.

Even though most skeletal muscles are a mixture of all three types of skeletal muscle fibers, the skeletal muscle fibers of any given motor unit are all of the same type. The different motor units in a muscle are recruited in a specific order, depending on need. For example, if weak contractions suffice to perform a task, only SO motor units are activated. If more force is needed, the motor units of FOG fibers are also recruited. Finally, if maximal force is required, motor units of FG fibers are also called into action.

✓ **CHECKPOINT**

11. Define the following terms: myogram, twitch contraction, wave summation, unfused tetanus, and fused tetanus.

12. Why is motor unit recruitment important?

13. What characteristics distinguish the three types of skeletal muscle fibers?

8.6 EXERCISE AND SKELETAL MUSCLE TISSUE

OBJECTIVE • Describe the effects of exercise on skeletal muscle tissue.

The relative ratio of fast glycolytic (FG) and slow oxidative (SO) fibers in each muscle is genetically determined and helps account for individual differences in physical performance. For example, people with a higher proportion of FG fibers often excel in activities that require periods of intense activity, such as weight lifting or sprinting. People with higher percentages of SO fibers are better at activities that require endurance, such as long-distance running.

Although the total number of skeletal muscle fibers usually does not increase, the characteristics of those present can change to some extent. Various types of exercises can induce changes in the fibers in a skeletal muscle. Endurance-type (aerobic) exercises, such as running or swimming, cause a gradual transformation of some FG fibers into fast oxidative–glycolytic (FOG) fibers. The transformed muscle fibers show slight increases in diameter, number of mitochondria, blood supply, and strength. Endurance exercises also result in cardiovascular and respiratory changes that cause skeletal muscles to receive better supplies of oxygen and nutrients but do not increase muscle mass. By contrast, exercises that require great strength for short periods produce an increase in the size and strength of FG fibers. The increase in size is due to increased synthesis of thick and thin filaments. The overall result is muscle enlargement (hypertrophy), as evidenced by the bulging muscles of bodybuilders.

✓ **CHECKPOINT**

14. Explain how the characteristics of skeletal muscle fibers may change with exercise.

8.7 CARDIAC MUSCLE TISSUE

OBJECTIVE • Describe the structure and function of cardiac muscle tissue.

Most of the heart consists of cardiac muscle tissue. Like skeletal muscle, cardiac muscle is also *striated*, but its action is *involuntary*: Its alternating cycles of contraction and relaxation are not consciously controlled. Cardiac muscle fibers often are branched; are shorter in length and larger in diameter than skeletal muscle fibers; and have a single, centrally located nucleus (see Figure 15.2b). Cardiac muscle fibers interconnect with one another by irregular transverse thickenings of the sarcolemma called *intercalated discs* (in-TER-ka-lāt-ed = to insert between). The intercalated discs hold the fibers together and contain *gap junctions*, which allow muscle action potentials to spread quickly from one cardiac muscle fiber to another. Cardiac muscle tissue has an endomysium and perimysium, but lacks an epimysium.

A major difference between skeletal muscle and cardiac muscle is the source of stimulation. We have seen that skeletal muscle tissue contracts only when stimulated by acetylcholine released by a nerve impulse in a motor neuron. In contrast, the heart beats because some of the cardiac muscle fibers act as a pacemaker to initiate each cardiac contraction. The built-in or intrinsic rhythm of heart contractions is called *autorhythmicity* (aw'-tō-rith-MIS-i-tē). Several hormones and neurotransmitters can increase or decrease heart rate by speeding or slowing the heart's pacemaker.

Under normal resting conditions, cardiac muscle tissue contracts and relaxes an average of about 75 times a minute. Thus, cardiac muscle tissue requires a constant supply of oxygen and nutrients. The mitochondria in cardiac muscle fibers are larger and more numerous than in skeletal muscle fibers and produce most of the needed ATP via aerobic cellular respiration. In addition, cardiac muscle fibers can use lactic acid, released by skeletal muscle fibers during exercise, to make ATP.

✓ CHECKPOINT

15. What are the major structural and functional differences between cardiac and skeletal muscle tissue?

8.8 SMOOTH MUSCLE TISSUE

OBJECTIVE • **Describe the structure and function of smooth muscle tissue.**

Smooth muscle tissue is found in many internal organs and blood vessels. Like cardiac muscle, smooth muscle is *involuntary*. Smooth muscle fibers are considerably smaller in length and diameter than skeletal muscle fibers and are tapered at both ends. Within each fiber is a single, oval, centrally located nucleus (Figure 8.11). In addition to thick and thin filaments, smooth muscle fibers also contain *intermediate filaments*. Because the various filaments have

no regular pattern of overlap, smooth muscle fibers lack alternating dark and light bands and thus appear *nonstriated*, or smooth.

In smooth muscle fibers, the thin filaments attach to structures called *dense bodies*, which are functionally similar to Z discs in striated muscle fibers. Some dense bodies are dispersed throughout the sarcoplasm; others are attached to the sarcolemma. Bundles of intermediate filaments also attach to dense bodies and stretch from one dense body to another. During contraction, the sliding filament mechanism involving thick and thin filaments generates tension that is transmitted to intermediate filaments. These, in turn, pull on the dense bodies attached to the sarcolemma, causing a lengthwise shortening of the muscle fiber.

There are two kinds of smooth muscle tissue, visceral and multiunit. The more common type is *visceral (single-unit) muscle tissue*. It is found in sheets that wrap around to form part of the walls of small arteries and veins and hollow viscera such as the stomach, intestines, uterus, and urinary bladder. The fibers in visceral muscle tissue are tightly bound together in a continuous network. Like cardiac muscle, visceral smooth muscle is autorhythmic. Because the fibers connect to one another by gap junctions, muscle action potentials spread throughout the network. When a neurotransmitter, hormone, or autorhythmic signal stimulates one fiber, the muscle action potential spreads to neighboring fibers, which then contract in unison, as a single unit.

The second kind of smooth muscle tissue, *multiunit smooth muscle tissue*, consists of individual fibers, each with its own motor nerve endings. Unlike stimulation of a single visceral muscle fiber, which causes contraction of many adjacent fibers, stimulation of a single multiunit smooth muscle fiber causes contraction of that fiber only. Multiunit smooth muscle tissue is found in the walls of large arteries, in large airways to the lungs, in the arrector pili muscles attached to hair follicles, and in the internal eye muscles.

Compared with contraction in a skeletal muscle fiber, contraction in a smooth muscle fiber starts more slowly and lasts much longer. Calcium ions enter smooth muscle fibers slowly and also move slowly out of the muscle fiber when excitation declines, which delays relaxation. The prolonged presence of Ca^{2+} in the cytosol provides for *smooth muscle tone*, a state of continuous partial contraction. Smooth muscle tissue can thus sustain long-term tone, which is important in the walls of blood vessels and in the walls of organs that maintain pressure on their contents. Finally, smooth muscle can both shorten and stretch to a greater extent than other muscle types. Stretchiness permits smooth muscle in the wall of hollow organs such as the uterus, stomach, intestines, and urinary bladder to

Figure 8.11 Histology of smooth muscle tissue. A smooth muscle fiber is shown in the relaxed state (left) and the contracted state (right).

 Smooth muscle lacks striations—it looks "smooth"—because the thick and thin filaments and intermediate filaments are irregularly arranged.

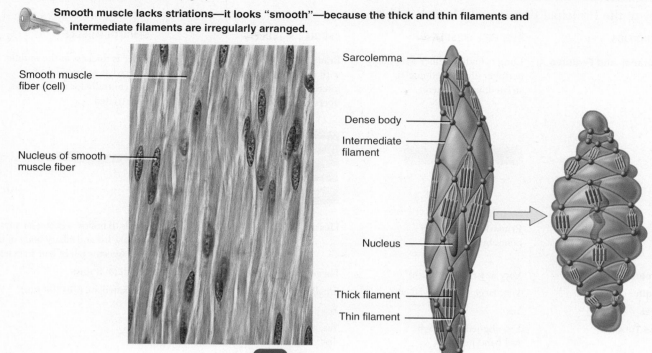

Longitudinal section of smooth muscle tissue

LM 500x

Relaxed   Contracted

? Which type of smooth muscle is found in the walls of hollow organs?

---

expand as their contents enlarge, while still retaining the ability to contract.

Most smooth muscle fibers contract or relax in response to nerve impulses from the autonomic (involuntary) nervous system. In addition, many smooth muscle fibers contract or relax in response to stretching; hormones; or local factors such as changes in pH, oxygen and carbon dioxide levels, temperature, and ion concentrations. For example, the hormone epinephrine, released by the adrenal medulla, causes relaxation of smooth muscle in the airways and in some blood vessel walls.

Table 8.1 presents a summary of the major characteristics of the three types of muscular tissue.

✓ CHECKPOINT

16. How do visceral and multiunit smooth muscle differ?

17. What are the major structural and functional differences between smooth and skeletal muscle tissue?

8.9 AGING AND MUSCULAR TISSUE

OBJECTIVE • **Explain the effects of aging on skeletal muscle.**

Beginning at about 30 years of age, humans undergo a slow, progressive loss of skeletal muscle mass that is replaced largely by fibrous connective tissue and adipose tissue. In part, this decline is due to decreased levels of physical activity. Accompanying the loss of muscle mass is a decrease in maximal strength, a slowing of muscle reflexes, and a loss of flexibility. In some muscles, a selective loss of muscle fibers of a given type may occur. With aging, the relative number of slow oxidative fibers appears to increase. This could be due either to atrophy of the other fiber types or their conversion into slow oxidative fibers. Whether this is an effect of aging itself or mainly reflects

TABLE 8.1

Summary of the Principal Features of Muscular Tissue

CHARACTERISTICS	SKELETAL MUSCLE	CARDIAC MUSCLE	SMOOTH MUSCLE
Cell Appearance and Features	Long cylindrical fiber with many peripherally located nuclei; striated; unbranched	Branched cylindrical fiber, usually with one centrally located nucleus; intercalated discs join neighboring fibers; striated	Fiber is thickest in the middle, tapered at each end, has one centrally located nucleus; not striated
Location	Primarily attached to bones by tendons	Heart	Walls of hollow viscera, airways, blood vessels, iris and ciliary body of the eye, arrector pili of hair follicles
Fiber Diameter	Very large (10–100 μm)*	Large (10–20 μm)	Small (3–8 μm)
Fiber Length	Very large (100 μm–30 cm)	Small (50–100 μm)	Intermediate (30–200 μm)
Sarcomeres	Yes	Yes	No
Transverse Tubules	Yes, aligned with each A–I band junction	Yes, aligned with each Z disc	No
Speed of Contraction	Fast	Moderate	Slow
Nervous Control	Voluntary	Involuntary	Involuntary
Capacity for Regeneration	Limited	Limited	Considerable compared with other muscle tissues, but limited compared with tissues such as epithelium

*1 micrometer (μm) = 1/25,000 of an inch.

the more limited physical activity of older people is still an unresolved question. Nevertheless, aerobic activities and strength training programs are effective in older people and can slow or even reverse the age-associated decline in muscular performance.

✓ CHECKPOINT

18. Why does muscle strength decrease with aging?

8.10 HOW SKELETAL MUSCLES PRODUCE MOVEMENT

OBJECTIVE • Describe how skeletal muscles cooperate to produce movement.

Now that you have a basic understanding of the structure and functions of muscular tissue, we will examine how skeletal muscles cooperate to produce various body movements.

Origin and Insertion

Based on the description of muscular tissue, we can define a *skeletal muscle* as an organ composed of several types of tissues. These include skeletal muscle tissue, vascular tissue (blood vessels and blood), nervous tissue (motor neurons), and several types of connective tissues.

Skeletal muscles are not attached directly to bones; they produce movements by pulling on tendons, which, in turn, pull on bones. Most skeletal muscles cross at least one joint and are attached to the articulating bones that form the joint (Figure 8.12). When the muscle contracts, it draws one bone toward the other. The two bones do not move equally. One is held nearly in its original position; the attachment of a muscle (by means of a tendon) to the stationary bone is called the *origin*. The other end of the muscle is attached by means of a tendon to the movable bone at a point called the *insertion*. The fleshy portion of the muscle between the tendons of the origin and insertion is called the *belly*. A good analogy is a spring on a door. The part of the spring

Figure 8.12 Relationship of skeletal muscles to bones.
Skeletal muscles produce movements by pulling on tendons attached to bones.

 In the limbs, the origin of a muscle is proximal and the insertion is distal.

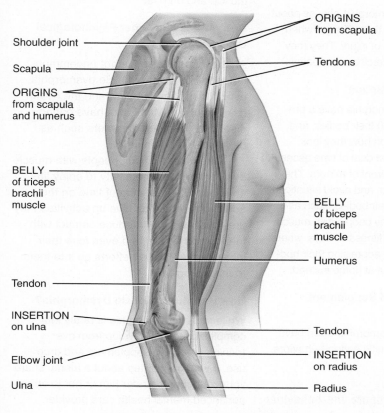

- Shoulder joint
- Scapula
- ORIGINS from scapula and humerus
- BELLY of triceps brachii muscle
- Tendon
- INSERTION on ulna
- Elbow joint
- Ulna
- ORIGINS from scapula
- Tendons
- BELLY of biceps brachii muscle
- Humerus
- Tendon
- INSERTION on radius
- Radius

Origin and insertion of a skeletal muscle

 Which muscle produces the desired action?

attached to the door represents the insertion, the part attached to the frame is the origin, and the coils of the spring are the belly.

CLINICAL CONNECTION | Tenosynovitis

Tenosynovitis (ten′-ō-sin-ō-VĪ-tis) is an inflammation of the tendons, tendon sheaths, and synovial membranes surrounding certain joints. The tendons most often affected are at the wrists, shoulders, elbows (resulting in *tennis elbow*), finger joints (resulting in *trigger finger*), ankles, and feet. The affected sheaths sometimes become visibly swollen because of fluid accumulation. Tenderness and pain are frequently associated with movement of the body part. The condition often follows trauma, strain, or excessive exercise. Tenosynovitis of the top of the foot may be caused by tying shoelaces too tightly. Gymnasts are prone to developing the condition as a result of chronic, repetitive, and maximum hyperextension at the wrists. Other repetitive movements involved in activities such as typing, haircutting, carpentry, and assembly-line work can also result in tenosynovitis. •

Group Actions

Most movements occur because several skeletal muscles are acting in groups rather than individually. Also, most skeletal muscles are arranged in opposing pairs at joints, that is, flexors–extensors, abductors–adductors, and so on. A muscle that causes a desired action is referred to as the *prime mover* or *agonist* (= leader). Often, another muscle, called the *antagonist* (*ant-* = against), relaxes while the prime mover contracts. The antagonist has an effect opposite to that of the prime mover; that is, the antagonist stretches and yields to the movement of the prime mover. When you bend (flex) your elbow, the biceps brachii is the prime mover. While the biceps brachii is contracting, the triceps brachii, the antagonist, is relaxing (see Figure 8.20). Do not assume, however, that the biceps brachii is always the prime mover and the triceps brachii is always the antagonist. For example, when straightening (extending) the elbow, the triceps brachii serves as the prime mover and the biceps brachii functions as the antagonist. If the prime mover and antagonist contracted together with equal force, there would be no movement.

Most movements also involve muscles called *synergists* (SIN-er-gists; *syn-* = together; *erg-* = work), which help the prime mover function more efficiently by reducing unnecessary movement. Some muscles in a group also act as *fixators*, stabilizing the origin of the prime mover so that the prime mover can act more efficiently. Under different conditions and depending on the movement, many muscles act at various times as prime movers, antagonists, synergists, or fixators.

✓ CHECKPOINT

19. Distinguish between the origin and insertion of a skeletal muscle.

20. Explain why most body movements occur because several skeletal muscles act in groups rather than individually.

8.11 PRINCIPAL SKELETAL MUSCLES

OBJECTIVES • **List and describe the ways that skeletal muscles are named.**

• **Describe the location of skeletal muscles in various regions of the body and identify their functions.**

The names of most of the nearly 700 skeletal muscles are based on specific characteristics. Learning the terms used to indicate specific characteristics will help you remember the names of the muscles (Table 8.2).

Superhero Dreams—Muscle Dysmorphia

The term "muscle dysmorphia" describes people abnormally preoccupied with the appearance of their muscles. People with muscle dysmorphia worry excessively that their muscles are not large enough and go to great lengths to increase muscle size. Muscle dysmorphia is generally diagnosed when a preoccupation with muscularity interferes with daily life or good health. People with muscle dysmorphia may have several of the following characteristics.

Obsession with Exercise

People with muscle dysmorphia have a devotion to daily resistance exercise, even when injured. They are afraid to take even one day off, for fear their muscles will begin to atrophy.

People with muscle dysmorphia obsess about exercise, and can't take time off even when they develop symptoms of injury. They may feel unable to control exercise participation.

Body Image Dissatisfaction

Those with muscle dysmorphia have a profound dissatisfaction with their bodies, and may focus excessively on how they look. They often spend a great deal of time assessing their appearance in front of mirrors. They may wear baggy clothes, and avoid situations that require exposing their bodies, such as going to the beach. Many people with muscle dysmorphia even avoid fitness centers, where they worry about others appraising their bodies, choosing to work out at home instead.

Elaborate Dietary and Supplement Use Patterns

People with muscle dysmorphia often have extremely rigid or ritualistic eating behaviors.

They may consume special diets to lose fat and build muscle. They often consume dietary supplements that claim to build muscle and burn fat.

Drug Abuse and Other Psychological Disorders

Anabolic steroid use is not uncommon among people with muscle dysmorphia, even though they understand the health risks involved. Many also have eating disorders, or mood disorders such as depression and anxiety.

Like other addicts, people with muscle dysmorphia lose the ability to enjoy life, and spend a great deal of time on their addiction. They may give up activities they previously enjoyed, reduce contact with friends and family, and even lose their jobs when all of their efforts go into their training programs.

How Harmful is Muscle Dysmorphia?

While muscle dysmorphia is rarely fatal, complications can develop from overtraining, and from supplement and drug use. If you are worried about a friend, share your concern, and refer him or her to an experienced mental health care provider.

Think It Over . . . **Compare the behaviors associated with muscle dysmorphia to the behaviors associated with other addictions, such as alcohol or drug abuse. What are some similarities? Differences?**

Exhibits 8.A through 8.M list the principal skeletal muscles of the body with their origins, insertions, and actions. (By no means have all the muscles of the body been included.) For each exhibit, an overview section provides a general orientation to the muscles and their functions or unique characteristics. To make it easier for you to learn to say the names of skeletal muscles and understand how they are named, we have provided phonetic pronunciations and word roots that indicate how the muscles are named (refer also to Table 8.2). Once you have mastered the naming of the muscles, their actions will have more meaning and be easier to remember.

The muscles are divided into groups according to the part of the body on which they act. Figure 8.13 shows an-

terior and posterior views of the muscular system. As you study groups of muscles in the following exhibits, refer to Figure 8.13 to see how each group is related to all of the others.

• • •

To appreciate the many ways that the muscular system contributes to homeostasis of other body systems, examine Focus on Homeostasis: The Muscular System at the end of the chapter following the exhibits. Next, in Chapter 9, we will see how the nervous system is organized, how neurons generate nerve impulses that activate muscle tissues as well as other neurons, and how synapses function.

TABLE 8.2

Characteristics Used to Name Skeletal Muscles

NAME	MEANING	EXAMPLE	FIGURE
Direction: Orientation of muscle fibers relative to the body's midline			
Rectus	Parallel to midline	Rectus abdominis	8.16b
Transverse	Perpendicular to midline	Transverse abdominis	8.16b
Oblique	Diagonal to midline	External oblique	8.16a
Size: Relative size of the muscle			
Maximus	Largest	Gluteus maximus	8.23b
Minimus	Smallest	Gluteus minimus	8.23c
Longus	Longest	Adductor longus	8.23a
Latissimus	Widest	Latissimus dorsi	8.13b
Longissimus	Longest	Longissimus muscles	8.22
Magnus	Large	Adductor magnus	8.23b
Major	Larger	Pectoralis major	8.13a
Minor	Smaller	Pectoralis minor	8.19a
Vastus	Great	Vastus lateralis	8.23a
Shape: Relative shape of the muscle			
Deltoid	Triangular	Deltoid	8.13b
Trapezius	Trapezoid	Trapezius	8.13b
Serratus	Saw-toothed	Serratus anterior	8.18a
Rhomboid	Diamond-shaped	Rhomboid major	8.19b
Orbicularis	Circular	Orbicularis oculi	8.14
Pectinate	Comblike	Pectineus	8.23a
Piriformis	Pear-shaped	Piriformis	8.23c
Platys	Flat	Platysma	8.13a
Quadratus	Square	Quadratus lumborum	8.17b
Gracilis	Slender	Gracilis	8.23a
Action: Principal action of the muscle			
Flexor	Decreases joint angle	Flexor carpi radialis	8.21a
Extensor	Increases joint angle	Extensor carpi ulnaris	8.21b
Abductor	Moves bone away from midline	Abductor pollicis longus	8.13b
Adductor	Moves bone closer to midline	Adductor longus	8.23a
Levator	Produces superior movement	Levator scapulae	8.18
Depressor	Produces inferior movement	Depressor labii inferioris	8.14
Supinator	Turns palm anteriorly	Supinator	
Pronator	Turns palm posteriorly	Pronator teres	8.21a
Sphincter	Decreases size of opening	External anal sphincter	19.15b
Tensor	Makes a body part rigid	Tensor fasciae latae	8.23a
Number of Origins: Number of tendons of origin			
Biceps	Two origins	Biceps brachii	8.20a
Triceps	Three origins	Triceps brachii	8.20b
Quadriceps	Four origins	Quadriceps femoris	8.23a

Location: Structure near which a muscle is found

Example: Temporalis, a muscle near the temporal bone (Figure 8.14).

Origin and Insertion: Sites where muscle originates and inserts

Example: Brachioradialis, originating on the humerus and inserting on the radius (Figure 8.21a).

Figure 8.13 Principal superficial skeletal muscles.

Most movements require contraction of several skeletal muscles acting in groups rather than individually.

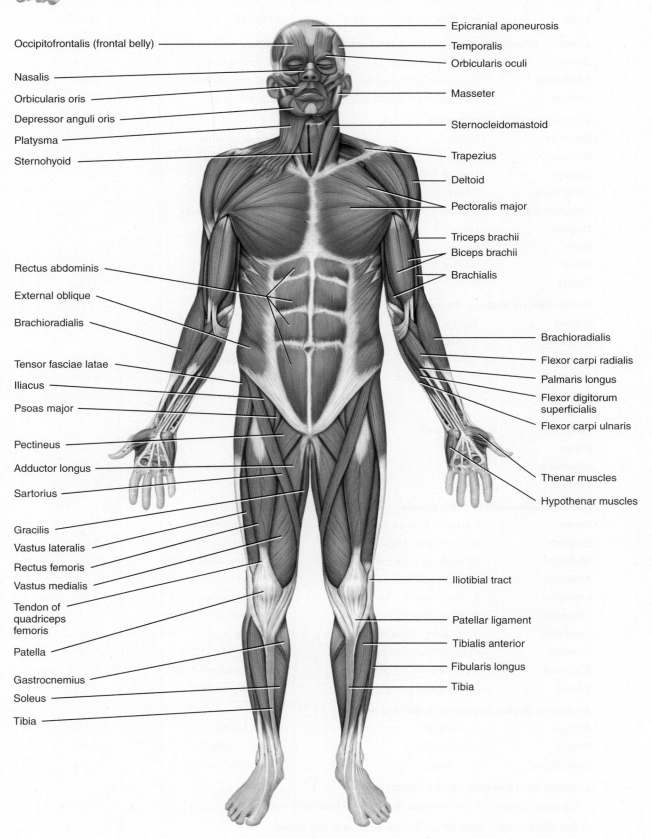

Occipitofrontalis (frontal belly)

Nasalis

Orbicularis oris

Depressor anguli oris

Platysma

Sternohyoid

Epicranial aponeurosis

Temporalis

Orbicularis oculi

Masseter

Sternocleidomastoid

Trapezius

Deltoid

Pectoralis major

Triceps brachii

Biceps brachii

Brachialis

Rectus abdominis

External oblique

Brachioradialis

Tensor fasciae latae

Iliacus

Psoas major

Pectineus

Adductor longus

Sartorius

Gracilis

Vastus lateralis

Rectus femoris

Vastus medialis

Tendon of quadriceps femoris

Patella

Gastrocnemius

Soleus

Tibia

Brachioradialis

Flexor carpi radialis

Palmaris longus

Flexor digitorum superficialis

Flexor carpi ulnaris

Thenar muscles

Hypothenar muscles

Iliotibial tract

Patellar ligament

Tibialis anterior

Fibularis longus

Tibia

(a) Anterior view

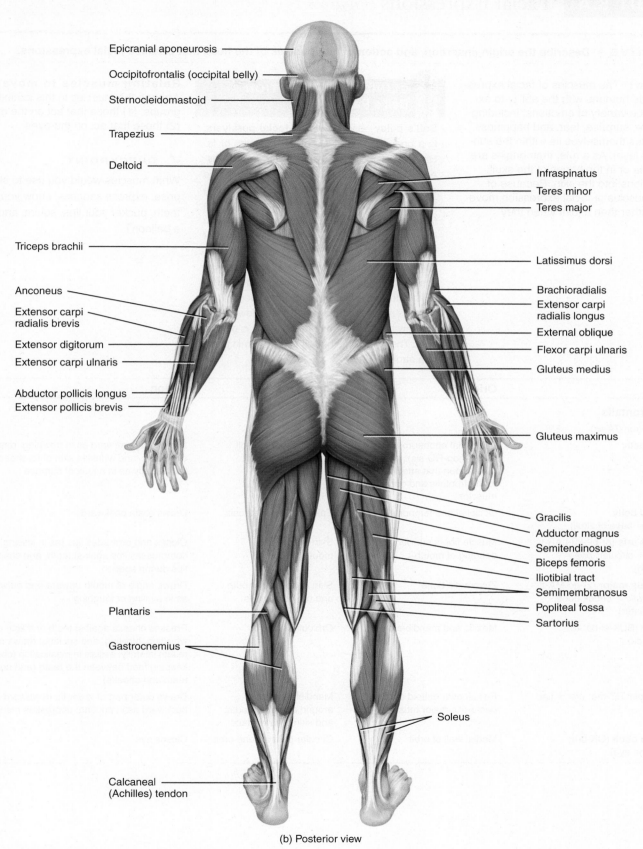

Epicranial aponeurosis

Occipitofrontalis (occipital belly)

Sternocleidomastoid

Trapezius

Deltoid

Triceps brachii

Anconeus

Extensor carpi radialis brevis

Extensor digitorum

Extensor carpi ulnaris

Abductor pollicis longus

Extensor pollicis brevis

Plantaris

Gastrocnemius

Calcaneal (Achilles) tendon

Infraspinatus

Teres minor

Teres major

Latissimus dorsi

Brachioradialis

Extensor carpi radialis longus

External oblique

Flexor carpi ulnaris

Gluteus medius

Gluteus maximus

Gracilis

Adductor magnus

Semitendinosus

Biceps femoris

Iliotibial tract

Semimembranosus

Popliteal fossa

Sartorius

Soleus

(b) Posterior view

Which is an example of a muscle named for the following characteristics: direction of fibers, shape, action, size, origin and insertion, location, and number of origins?

OBJECTIVE • Describe the origin, insertion, and action of the muscles of the head that produce facial expressions.

Overview: The muscles of facial expression provide humans with the ability to express a wide variety of emotions, including displeasure, surprise, fear, and happiness. The muscles themselves lie within the subcutaneous layer. As a rule, their origins are in the fascia or in the bones of the skull, with insertions into the skin. Because of this, the muscles of facial expression move the skin rather than a joint when they contract.

CLINICAL CONNECTION |
Bell's Palsy

Bell's palsy, also known as **facial paralysis**, is a one-sided paralysis of the muscles of facial expression as a result of damage or disease of the facial (VII) nerve. Although the cause is unknown, a relationship between the herpes simplex virus and inflammation of the facial nerve has been suggested. In severe cases, the paralysis causes the entire side of the face to droop, and the person cannot wrinkle the forehead, close the eye, or pucker the lips on the affected side. Drooling and difficulty in swallowing also occur. Eighty percent of patients recover completely within a few weeks to a few months. For others, paralysis is permanent. The symptoms of Bell's palsy mimic those of a stroke. •

Relating muscles to movements: Arrange the muscles in this exhibit into two groups: (1) those that act on the mouth and (2) those that act on the eyes.

✓ **CHECKPOINT**

What muscles would you use to show surprise, express sadness, show your upper teeth, pucker your lips, squint, and blow up a balloon?

Muscle	Origin	Insertion	Action
Occipitofrontalis (ok-sip-i-tō-frun-TĀ-lis)			
Frontal belly	Epicranial aponeurosis (ap'-ō-noo-RŌ-sis) (flat tendon that attaches to the frontalis and occipitalis muscles)	Skin superior to orbit	Draws scalp forward as in frowning, raises eyebrows, and wrinkles skin of forehead horizontally as in a look of surprise
Occipital belly (*occipit-* = base of skull)	Occipital and temporal	Epicranial aponeurosis	Draws scalp backward
Orbicularis oris (or-bi'-kū-LAR-is OR-is; *orb* = circular; *oris* = of the mouth)	Muscle fibers surrounding opening of mouth	Skin at corner of mouth	Closes and protrudes lips (as in kissing), compresses lips against teeth, and shapes lips during speech
Zygomaticus major (zī-gō-MA-ti-kus; *zygomatic* = cheek bone; *major* = greater)	Zygomatic bone	Skin at angle of mouth and orbicularis oris	Draws angle of mouth upward and outward, as in smiling or laughing
Buccinator (BUK-si-nā'-tor; *bucc* = cheek)	Maxilla and mandible	Orbicularis oris	Presses cheeks against teeth and lips, as in whistling, blowing, and sucking; draws corner of mouth laterally; assists in mastication (chewing) by keeping food between the teeth (and not between teeth and cheeks)
Platysma (pla-TIZ-ma; *plat* = flat, broad)	Fascia over deltoid and pectoralis major muscles	Mandible, muscles around angle of mouth, and skin of lower face	Draws outer part of lower lip downward and backward as in pouting; depresses mandible
Orbicularis oculi (OK-ū-lī; *oculi* = of the eye)	Medial wall of orbit	Circular path around orbit	Closes eye

Figure 8.14 Muscles of the head that produce facial expressions. In this and subsequent figures in the chapter, the muscles indicated in all uppercase letters are the ones specifically referred to in the corresponding exhibit or another exhibit in this chapter.

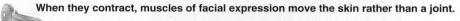

When they contract, muscles of facial expression move the skin rather than a joint.

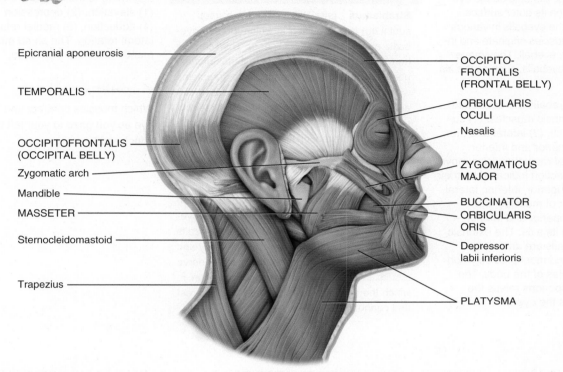

Epicranial aponeurosis

OCCIPITO-FRONTALIS (FRONTAL BELLY)

TEMPORALIS

ORBICULARIS OCULI

Nasalis

OCCIPITOFRONTALIS (OCCIPITAL BELLY)

Zygomatic arch

ZYGOMATICUS MAJOR

Mandible

MASSETER

BUCCINATOR

ORBICULARIS ORIS

Sternocleidomastoid

Depressor labii inferioris

Trapezius

PLATYSMA

Right lateral superficial view

 Which muscles of facial expression cause smiling, pouting, and squinting?

EXHIBIT 8.B

Muscles That Move the Mandible and Assist in Mastication and Speech *(See Figure 8.14)*

OBJECTIVE • Describe the origin, insertion, and action of the muscles that move the mandible and assist in mastication and speech.

Overview: Muscles that move the mandible (lower jaw) are also known as muscles of **mastication** (mas'-ti-KĀ-shun = to chew) because they are involved in biting and chewing. These muscles also assist in speech.

Relating muscles to movements: Arrange the muscles in this exhibit and the previous exhibit according to their actions on the mandible: (1) elevation, (2) depression, and (3) retraction. The same muscle may be mentioned more than once.

✓ **CHECKPOINT**

What would happen if you lost tone in the masseter and temporalis muscles?

Muscle	Origin	Insertion	Action
Masseter (MA-se-ter; *maseter* = chewer) (See Figure 8.14.)	Maxilla and zygomatic arch	Mandible	Elevates mandible as in closing mouth
Temporalis (tem'-por-Ā-lis; *tempor* = temples) (See Figure 8.14.)	Temporal bone	Mandible	Elevates and retracts (draws back) mandible

OBJECTIVE • Describe the origin, insertion, and action of the extrinsic muscles of the eyeballs.

Overview: Two types of muscles are associated with the eyeball, extrinsic and intrinsic. *Extrinsic muscles* originate outside the eyeball and are inserted on its outer surface (sclera). They move the eyeballs in various directions. *Intrinsic muscles* originate and insert entirely within the eyeball. They move structures within the eyeballs, such as the iris and the lens.

Movements of the eyeballs are controlled by three pairs of extrinsic muscles: (1) superior and inferior recti, (2) lateral and medial recti, and (3) superior and inferior obliques. Two pairs of rectus muscles move the eyeball in the direction indicated by their respective names: superior, inferior, lateral, and medial. One pair of muscles, the oblique muscles—superior and inferior—rotate the eyeball on its axis. The extrinsic muscles of the eyeballs are among the fastest contracting and most precisely controlled skeletal muscles of the body. The levator palpebrae superioris raises the upper eyelids (opens the eyes).

CLINICAL CONNECTION | Strabismus

Strabismus (stra-BIZ-mus; *strabismos* = squinting) is a condition in which the two eyeballs are not properly aligned. This can be hereditary or it can be due to birth injuries, poor attachments of the muscles, problems with the brain's control center, or localized disease. Strabismus can be constant or intermittent. In strabismus, each eye sends an image to a different area of the brain and because the brain usually ignores the messages sent by one of the eyes, the ignored eye becomes weaker, hence "lazy eye" or *amblyopia*, develops. *External strabismus* results when a lesion in the oculomotor (III) nerve causes the eyeball to move laterally when at rest, and results in an inability to move the eyeball medially and inferiorly. A lesion in the abducens (VI) nerve results in *internal strabismus*, a condition in which the eyeball moves medially when at rest and cannot move laterally. •

Relating muscles to movements: Arrange the muscles in this exhibit according to their actions on the eyeballs: (1) elevation, (2) depression, (3) abduction, (4) adduction, (5) medial rotation, and (6) lateral rotation. The same muscle may be mentioned more than once.

✓ **CHECKPOINT**

Which muscles contract and relax in each eye as you gaze to your left without moving your head?

Muscle	Origin	Insertion	Action
Superior rectus (REK-tus; *superior* = above; *rect-* = straight; here, muscle fascicles that are parallel to long axis of eyeball)	Tendinous ring attached to bony orbit around optic foramen	Superior and central part of eyeball	Moves eyeball upward (elevation) and medially (adduction), and rotates it medially
Inferior rectus (*inferior* = below)	Same as above	Inferior and central part of eyeball	Moves eyeball downward (depression) and medially (adduction), and rotates it medially
Lateral rectus	Same as above	Lateral side of eyeball	Moves eyeball laterally (abduction)
Medial rectus	Same as above	Medial side of eyeball	Moves eyeball medially (adduction)
Superior oblique (ō-BLĒK; *oblique* = slanting; here, muscle fascicles run diagonally to long axis of eyeball)	Same as above	Eyeball between superior and lateral recti; moves through a ring of fibrocartilaginous tissue called the trochlea (*trochlea* = pulley)	Moves eyeball downward (depression) and laterally (abduction), and rotates it medially
Inferior oblique	Maxilla	Eyeball between inferior and lateral recti	Moves eyeball upward (elevation) and laterally (abduction), and rotates it laterally
Levator palpebrae superioris (le-VĀ-tor PAL-pe-brē soo-per′-ē-OR-is; *palpebrae* = eyelids)	Roof of orbit	Skin of upper eyelid	Elevates upper eyelid (opens eye)

Figure 8.15 Muscles that move the eyeballs (extrinsic muscles) and upper eyelids.

 The extrinsic muscles of the eyeball are among the fastest contracting and most precisely controlled skeletal muscles in the body.

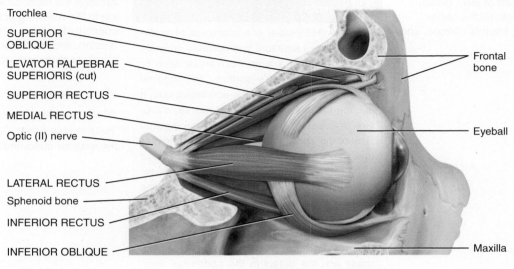

Trochlea
SUPERIOR OBLIQUE
LEVATOR PALPEBRAE SUPERIORIS (cut)
SUPERIOR RECTUS
MEDIAL RECTUS
Optic (II) nerve
LATERAL RECTUS
Sphenoid bone
INFERIOR RECTUS
INFERIOR OBLIQUE

Frontal bone
Eyeball
Maxilla

(a) Right lateral view of right eyeball

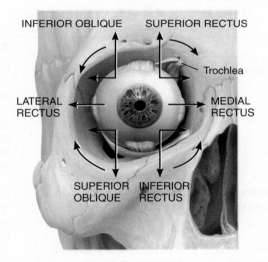

INFERIOR OBLIQUE   SUPERIOR RECTUS
Trochlea
LATERAL RECTUS   MEDIAL RECTUS
SUPERIOR OBLIQUE   INFERIOR RECTUS

(b) Movements of right eyeball in response to contraction of extrinsic muscles

 Which muscle passes through the trochlea?

OBJECTIVE • Describe the origin, insertion, and action of the muscles of the abdomen that protect the abdominal organs and move the vertebral column.

Overview: The anterior and lateral abdominal wall is composed of skin; fascia; and four pairs of muscles: rectus abdominis, external oblique, internal oblique, and transverse abdominis.

CLINICAL CONNECTION | Hernia

A **hernia** (HER-nē-a) is a protrusion of an organ through a structure that normally contains it, which creates a lump that can be seen or felt through the skin's surface. The inguinal region is a weak area in the abdominal wall. It is often the site of an **inguinal hernia**, a rupture or separation of a portion of the inguinal area of the abdominal wall resulting in the protrusion of a part of the small intestine. Hernia is much more common in males than in females because the inguinal canals in males are larger to accommodate the spermatic cord and ilioinguinal nerve. Treatment of hernias most often involves surgery. The organ that protrudes is "tucked" back into the abdominal cavity and the defect in the abdominal muscles is repaired. In addition, a mesh is often applied to reinforce the area of weakness. •

Relating muscles to movements: Arrange the muscles in this exhibit according to the following actions on the vertebral column: (1) flexion, (2) lateral flexion, (3) extension, and (4) rotation. The same muscle may be mentioned more than once.

✓ **CHECKPOINT**

Which muscles do you contract when you "suck in your tummy," thereby compressing the anterior abdominal wall?

Muscle	Origin	Insertion	Action
Rectus abdominis (REK-tus ab-DOM-in-is; *rect-* = straight, fibers parallel to midline; *abdomin-* = abdomen)	Pubis and pubic symphysis	Cartilage of fifth to seventh ribs and xiphoid process of sternum	Flexes vertebral column, and compresses abdomen to aid in defecation, urination, forced expiration, and childbirth
External oblique (ō-BLĒK; *external* = closer to surface; *oblique* = slanting; here, fibers that are diagonal to midline)	Ribs 5–12	Ilium and linea alba (a tough connective tissue band that runs from the xiphoid process of the sternum to the pubic symphysis)	Contraction of both external obliques compresses abdomen and flexes vertebral column; contraction of one side alone bends vertebral column laterally and rotates it
Internal oblique (*internal* = farther from surface)	Ilium, inguinal ligament, and thoracolumbar fascia	Cartilage of last three or four ribs and linea alba	Contraction of both internal obliques compresses abdomen and flexes vertebral column; contraction of one side alone bends vertebral column laterally and rotates it
Transverse abdominis (*transverse* = fibers that are perpendicular to midline)	Ilium, inguinal ligament, lumbar fascia, and cartilages of last six ribs	Xiphoid process of sternum, linea alba, and pubis	Compresses abdomen

Figure 8.16 Muscles of the abdomen that protect the abdominal organs and move the vertebral column. Shown here are the muscles in a male.

🗝 **The inguinal ligament separates the thigh from the body wall.**

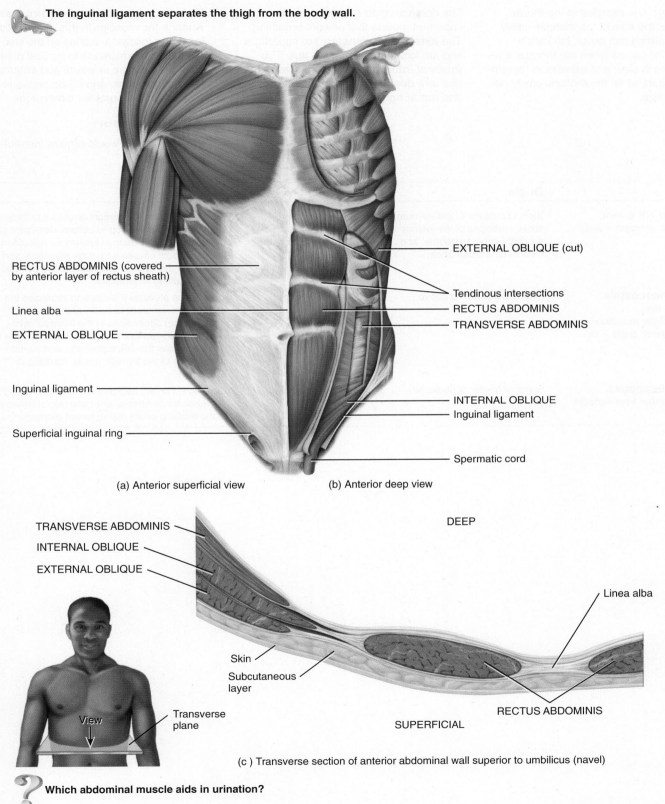

RECTUS ABDOMINIS (covered by anterior layer of rectus sheath)

Linea alba

EXTERNAL OBLIQUE

Inguinal ligament

Superficial inguinal ring

EXTERNAL OBLIQUE (cut)

Tendinous intersections

RECTUS ABDOMINIS

TRANSVERSE ABDOMINIS

INTERNAL OBLIQUE

Inguinal ligament

Spermatic cord

(a) Anterior superficial view

(b) Anterior deep view

TRANSVERSE ABDOMINIS

INTERNAL OBLIQUE

EXTERNAL OBLIQUE

DEEP

Linea alba

Skin

Subcutaneous layer

RECTUS ABDOMINIS

View

Transverse plane

SUPERFICIAL

(c) Transverse section of anterior abdominal wall superior to umbilicus (navel)

❓ **Which abdominal muscle aids in urination?**

EXHIBIT 8.E Muscles of the Thorax That Assist in Breathing *(Figure 8.17)*

OBJECTIVE • Describe the origin, insertion, and action of the muscles of the thorax that assist in breathing.

Overview: The muscles of the thorax (chest) alter the size of the thoracic cavity so that breathing can occur. Inhalation (breathing in) occurs when the thoracic cavity increases in size, and exhalation (breathing out) occurs when the thoracic cavity decreases in size.

The dome-shaped *diaphragm* is the most important muscle that powers breathing. The *external intercostals* are superficial and are located between the ribs. The *internal intercostals*, also between the ribs, are deep to the external intercostals and run at right angles to them.

Relating muscles to movements: Arrange the muscles in this exhibit according to the following actions on the size of the thorax: (1) increase in vertical dimension, (2) increase in lateral and anteroposterior dimensions, and (3) decrease in lateral and anteroposterior dimensions.

✓ **CHECKPOINT**

What situations would require forceful breathing?

Muscle	Origin	Insertion	Action
Diaphragm (DĪ-a-fram; *dia* = across; *-phragm* = wall)	Xiphoid process of the sternum, costal cartilages of the inferior six ribs, lumbar vertebrae, and their intervertebral discs	Central tendon (strong aponeurosis near the center of the diaphragm)	Contraction of the diaphragm causes it to flatten and increases the vertical (top-to-bottom) dimension of the thoracic cavity, resulting in inhalation; relaxation of the diaphragm causes it to move superiorly and decreases the vertical dimension of the thoracic cavity, resulting in exhalation
External intercostals (in'-ter-KOS-tals; *external* = closer to surface; *inter-* = between; *costa* = rib)	Inferior border of rib above	Superior border of rib below	Contraction elevates the ribs and increases the anteroposterior (front-to-back) and lateral (side-to-side) dimensions of the thoracic cavity, resulting in inhalation; relaxation depresses the ribs and decreases the anteroposterior and lateral dimensions of the thoracic cavity, resulting in exhalation
Internal intercostals (*internal* = farther from surface)	Superior border of rib below	Inferior border of rib above	Contraction draws adjacent ribs together to further decrease the anteroposterior and lateral dimensions of the thoracic cavity during forced exhalation

Figure 8.17 Muscles of the thorax that assist in breathing.

The muscles used in breathing alter the size of the thoracic cavity.

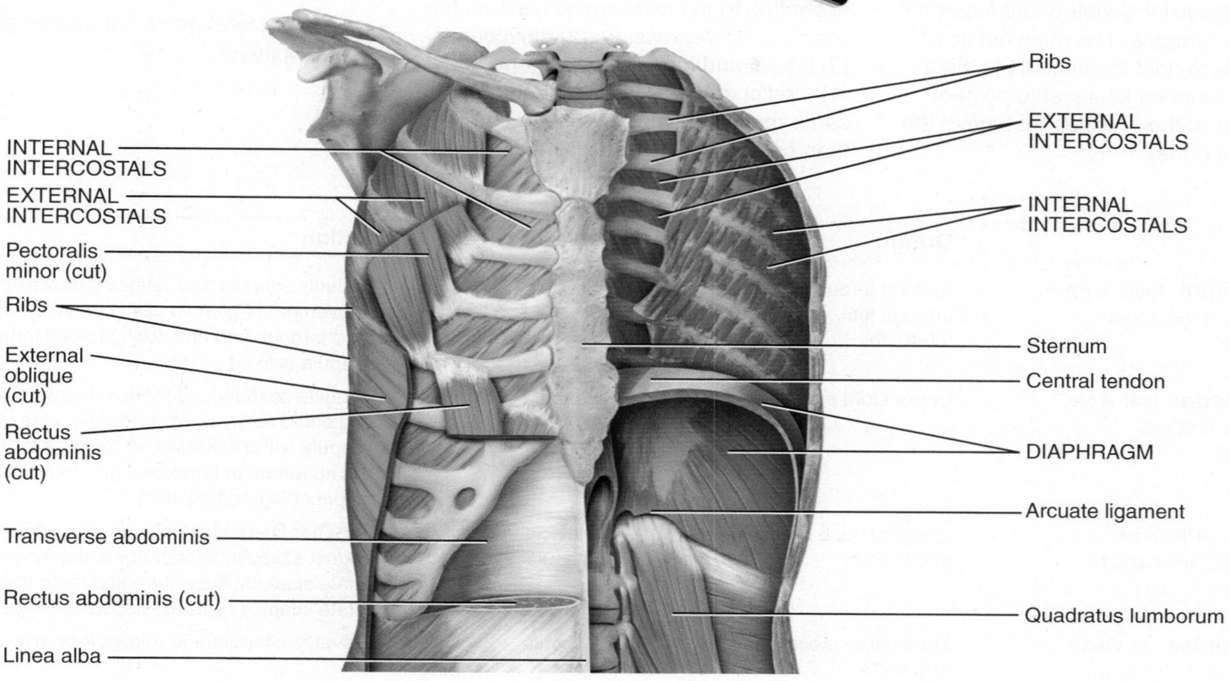

INTERNAL
INTERCOSTALS

EXTERNAL
INTERCOSTALS

Pectoralis
minor (cut)

Ribs

External
oblique
(cut)

Rectus
abdominis
(cut)

Transverse abdominis

Rectus abdominis (cut)

Linea alba

Ribs

EXTERNAL
INTERCOSTALS

INTERNAL
INTERCOSTALS

Sternum

Central tendon

DIAPHRAGM

Arcuate ligament

Quadratus lumborum

(a) Anterior superficial view   (b) Anterior deep view

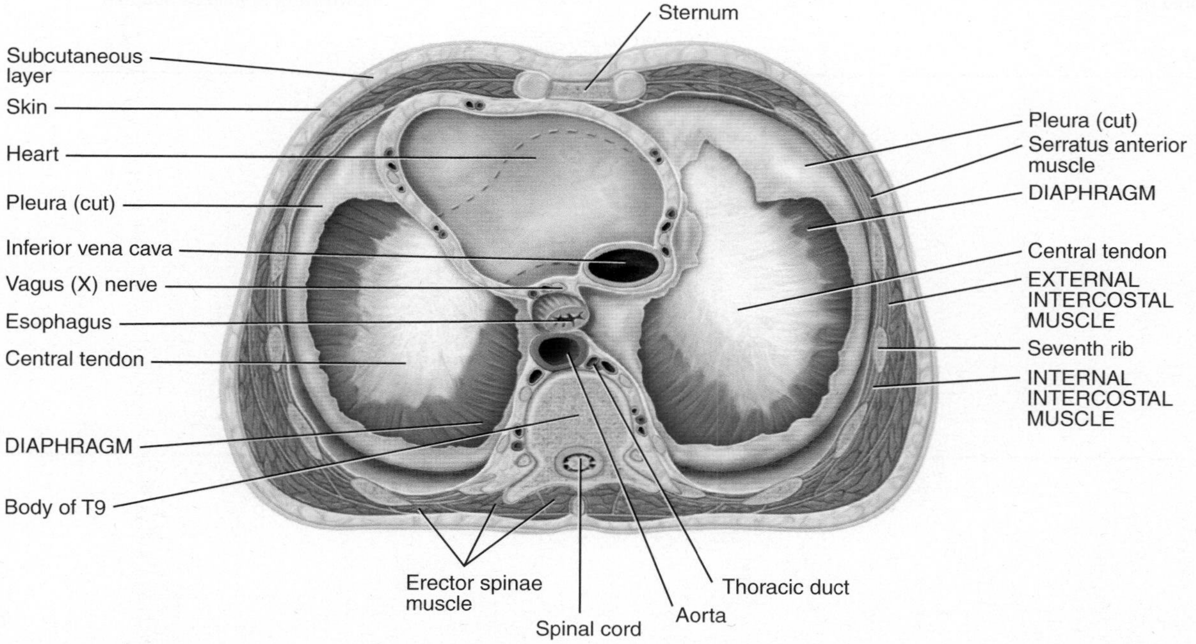

Sternum

Subcutaneous
layer

Skin

Heart

Pleura (cut)

Inferior vena cava

Vagus (X) nerve

Esophagus

Central tendon

DIAPHRAGM

Body of T9

Pleura (cut)

Serratus anterior
muscle

DIAPHRAGM

Central tendon

EXTERNAL
INTERCOSTAL
MUSCLE

Seventh rib

INTERNAL
INTERCOSTAL
MUSCLE

Erector spinae
muscle

Spinal cord

Thoracic duct

Aorta

(c) Superior view of diaphragm

Which muscles contract during a normal quiet inhalation?

227

EXHIBIT 8.F

Muscles of the Thorax That Move the Pectoral Girdle (Figure 8.18)

OBJECTIVE • **Describe the origin, insertion, and action of the muscles that move the pectoral girdle.**

Overview: Muscles that move the pectoral (shoulder) girdle (clavicle and scapula) originate on the axial skeleton and insert on the clavicle or scapula. The main action of the muscles is to hold the scapula in place so that it can function as a stable point of origin for most of the muscles that move the humerus (arm bone).

Relating muscles to movements: Arrange the muscles in this exhibit according to the following actions on the scapula: (1) depression, (2) elevation, (3) lateral and forward movement, and (4) medial and backward movement. The same muscle may be mentioned more than once.

✓ **CHECKPOINT**

Which muscle in this exhibit not only moves the pectoral girdle but also assists in forced inhalation?

Muscle	Origin	Insertion	Action
Pectoralis minor (pek′-tor-Ā-lis; *pect-* = breast, chest, thorax; *minor* = lesser)	Second through fifth, third through fifth, or second through fourth ribs	Scapula	Abducts scapula and rotates it downward (movement of glenoid cavity upward); elevates third through fifth ribs during forced inhalation when scapula is fixed
Serratus anterior (ser-Ā-tus; *serratus* = saw-toothed; *anterior* = front)	Upper eight or nine ribs	Scapula	Abducts scapula and rotates it upward (movement of glenoid cavity downward); elevates ribs when scapula is fixed; known as "boxer's muscle" because it is important in horizontal arm movements such as punching and pushing
Trapezius (tra-PĒ-zē-us; *trapezi-* = trapezoid-shaped) (See also Figure 8.13b.)	Occipital bone and spines of C7–T12	Clavicle and scapula	Superior fibers elevate scapula; middle fibers adduct scapula; inferior fibers depress and upward rotate scapula; superior and inferior fibers together rotate scapula upward; stabilizes scapula
Levator scapulae (le-VĀ-tor SKA-pū-lē; *levator* = to raise; *scapulae* = of the scapula)	Transverse processes of C1–C4	Scapula	Elevates scapula and rotates it downward
Rhomboid major (ROM-boyd; *rhomboid* = rhomboid or diamond-shaped) (See Figure 8.19b.)	Spines of T2–T5	Scapula	Elevates and adducts scapula and rotates it downward; stabilizes scapula

Figure 8.18 Muscles of the thorax that move the pectoral girdle.

Muscles that move the pectoral girdle originate on the axial skeleton and insert on the clavicle or scapula.

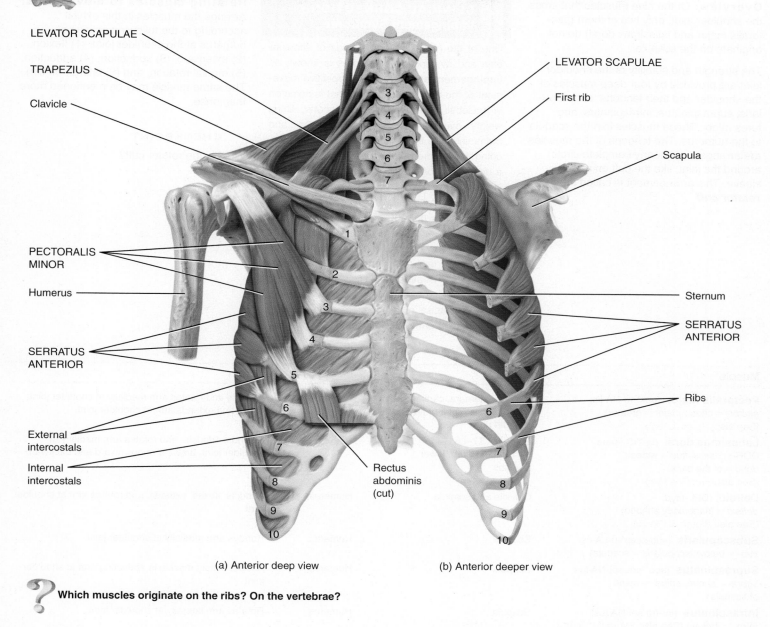

LEVATOR SCAPULAE

TRAPEZIUS

Clavicle

PECTORALIS MINOR

Humerus

SERRATUS ANTERIOR

External intercostals

Internal intercostals

LEVATOR SCAPULAE

First rib

Scapula

Sternum

SERRATUS ANTERIOR

Ribs

Rectus abdominis (cut)

(a) Anterior deep view

(b) Anterior deeper view

Which muscles originate on the ribs? On the vertebrae?

OBJECTIVE • Describe the origin, insertion, and action of the muscles of the thorax that move the humerus.

Overview: Of the nine muscles that cross the shoulder joint, only two of them (pectoralis major and latissimus dorsi) do not originate on the scapula.

The strength and stability of the shoulder joint are provided by four deep muscles of the shoulder and their tendons: subscapularis, supraspinatus, infraspinatus, and teres minor. These muscles join the scapula to the humerus. The tendons of the muscles are arranged in a nearly complete circle around the joint, like the cuff on a shirt sleeve. This arrangement is called the **rotator cuff**.

CLINICAL CONNECTION | Impingement Syndrome

One of the most common causes of shoulder pain and dysfunction in athletes is known as **impingement syndrome**. The repetitive movement of the arm over the head that is common in baseball, overhead racquet sports, lifting weights over the head, spiking a volleyball, and swimming puts these athletes at risk for developing this syndrome. It may also be caused by a direct blow or stretch injury. Continual pinching of the supraspinatus tendon as a result of overhead motions causes it to become inflamed and results in pain. If movement is continued despite the pain, the tendon may degenerate near the attachment to the humerus and ultimately may tear away from the bone (rotator cuff injury). Treatment consists of resting the injured tendons, strengthening the shoulder through exercise, massage therapy, and surgery if the injury is particularly severe. •

Relating muscles to movements: Arrange the muscles in this exhibit according to the following actions on the humerus at the shoulder joint: (1) flexion, (2) extension, (3) abduction, (4) adduction, (5) medial rotation, and (6) lateral rotation. The same muscle may be mentioned more than once.

✓ **CHECKPOINT**

What is the rotator cuff?

Muscle	Origin	Insertion	Action
Pectoralis major (pek′-tō-RĀ-lis; *pector-* = chest; *major* = greater) (See also Figure 8.13a.)	Clavicle, sternum, cartilages of second to sixth ribs or first to seventh ribs	Humerus	Adducts and rotates arm medially at shoulder joint; flexes and extends arm at shoulder joint
Latissimus dorsi (la-TIS-i-mus DOR-sī; *latissimus* = widest; *dorsi* = of the back) (See also Figure 8.13b.)	Spines of T7–L5, sacrum and ilium, lower four ribs	Humerus	Extends, adducts, and rotates arm medially at shoulder joint; draws arm downward and backward
Deltoid (DEL-toyd; *deltoid* = triangularly shaped) (See also Figure 8.13a, b.)	Clavicle and scapula	Humerus	Abducts, flexes, extends, and rotates arm at shoulder joint
Subscapularis (sub-scap′-ū-LĀ-ris; *sub-* = below; *scapularis* = scapula)	Scapula	Humerus	Rotates arm medially at shoulder joint
Supraspinatus (soo′-pra-spi-NĀ-tus; *supra-* = above; *spina-* = spine of scapula)	Scapula	Humerus	Assists deltoid muscle in abducting arm at shoulder joint
Infraspinatus (in′-fra-spi-NĀ-tus; *infra-* = below) (See also Figure 8.13b.)	Scapula	Humerus	Rotates arm laterally at shoulder joint
Teres major (TE-rēz) (*teres* = long and round)	Scapula	Humerus	Extends arm at shoulder joint; assists in adduction and rotation of arm medially at shoulder joint
Teres minor	Scapula	Humerus	Rotates arm laterally and extends arm at shoulder joint
Coracobrachialis (kor′-a-kō-brā-kē-Ā-lis; *coraco-* = coracoid process; *brachi-* = arm)	Scapula	Humerus	Flexes and adducts arm at shoulder joint

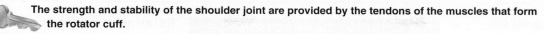

Figure 8.19 Muscles of the thorax and shoulder that move the humerus.

The strength and stability of the shoulder joint are provided by the tendons of the muscles that form the rotator cuff.

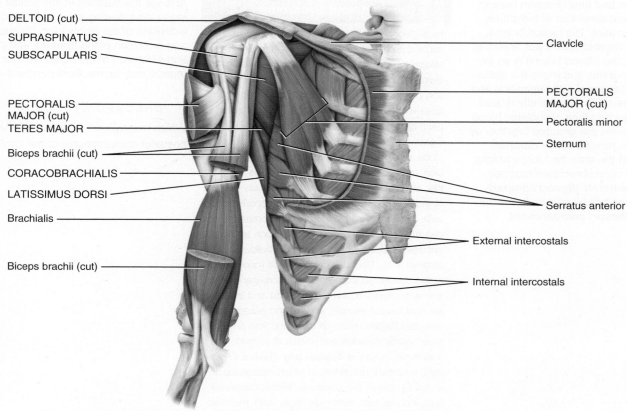

DELTOID (cut)
SUPRASPINATUS
SUBSCAPULARIS

PECTORALIS
MAJOR (cut)
TERES MAJOR

Biceps brachii (cut)
CORACOBRACHIALIS
LATISSIMUS DORSI
Brachialis

Biceps brachii (cut)

Clavicle

PECTORALIS
MAJOR (cut)
Pectoralis minor
Sternum

Serratus anterior

External intercostals

Internal intercostals

(a) Anterior deep view (the intact pectoralis major muscle is shown in Figure 8.16a)

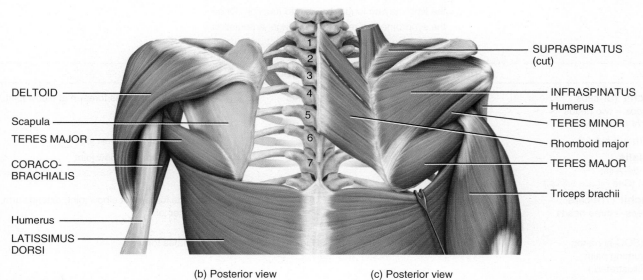

DELTOID
Scapula
TERES MAJOR
CORACO-
BRACHIALIS

Humerus
LATISSIMUS
DORSI

SUPRASPINATUS
(cut)

INFRASPINATUS
Humerus
TERES MINOR
Rhomboid major
TERES MAJOR

Triceps brachii

(b) Posterior view   (c) Posterior view

Of the nine muscles that cross the shoulder joint, which two muscles do not originate on the scapula?

EXHIBIT 8.H

Muscles of the Arm That Move the Radius and Ulna *(Figure 8.20)*

OBJECTIVE • Describe the origin, insertion, and action of the muscles that move the radius and ulna.

Overview: Most of the muscles that move the radius and ulna (forearm bones) cause flexion and extension of the elbow, which is a hinge joint. The biceps brachii, brachialis, and brachioradialis are flexors of the elbow joint; the triceps brachii is an extensor. Other muscles that move the radius and ulna are concerned with supination and pronation. In the limbs, functionally related skeletal muscles and their associated blood vessels and nerves are grouped together by deep fascia into regions called *compartments*. Thus, in the arm, the biceps brachii, brachialis, and coracobrachialis muscles constitute the *anterior (flexor) compartment*; the triceps brachii muscle forms the *posterior (extensor) compartment*.

CLINICAL CONNECTION | Compartment Syndrome

In a disorder called **compartment syndrome**, some external or internal pressure constricts the structures within a compartment, resulting in damaged blood vessels and subsequent reduction of the blood supply (ischemia) to the structures within the compartment. Symptoms include pain, burning, pressure, pale skin, and paralysis. Common causes of compartment syndrome include crushing and penetrating injuries, contusion (damage to subcutaneous tissues without the skin being broken), muscle strain (overstretching of a muscle), or an improperly fitted cast. The pressure increase in the compartment can have serious consequences, such as hemorrhage, tissue injury, and edema (buildup of interstitial fluid). Because deep fasciae (connective tissue coverings) that enclose the compartments are very strong, accumulated blood and interstitial fluid cannot escape, and the increased pressure can literally choke off the blood flow and deprive nearby muscles and nerves of oxygen. One treatment option is **fasciotomy** (fash-ē-OT-ō-mē), a surgical procedure in which muscle fascia is cut to relieve the pressure. Without intervention, nerves can suffer damage, and muscles can develop scar tissue that results in permanent shortening of the muscles, a condition called *contracture*. If left untreated, tissues may die and the limb may no longer be able to function. Once the syndrome has reached this stage, amputation may be the only treatment option. •

Relating muscles to movements: Arrange the muscles in this exhibit according to the following actions: (1) flexion and extension of the elbow joint; (2) supination and pronation of the forearm; and (3) flexion and extension of the humerus. The same muscle may be mentioned more than once.

✓ **CHECKPOINT**

Which muscles are in the anterior and posterior compartments of the arm?

Muscle	Origin	Insertion	Action
Biceps brachii (BĪ-ceps BRĀ-kē-ī; *biceps* = two heads of origin; *brachi* = arm)	Scapula	Radius	Flexes and supinates forearm at elbow joint; flexes arm at shoulder joint
Brachialis (brā'-kē-Ā-lis)	Humerus	Ulna	Flexes forearm at elbow joint
Brachioradialis (brā'-kē-ō-rā'-dē-Ā-lis; *radi-* = radius) (See Figure 8.21a.)	Humerus	Radius	Flexes forearm at elbow joint
Triceps brachii (TRĪ-ceps BRĀ-kē-ī; *triceps* = three heads of origin)	Scapula and humerus	Ulna	Extends forearm at elbow joint; extends arm at shoulder joint
Supinator (SOO-pi-nā-tor; *supination* = turning palm forward) (Not illustrated.)	Humerus and ulna	Radius	Supinates forearm
Pronator teres (PRŌ-nā-tor TE-rēz; *pronation* = turning palm backward) (See Figure 8.21a.)	Humerus and ulna	Radius	Pronates forearm

Figure 8.20 Muscles of the arm that move the radius and ulna.

The anterior arm muscles flex the forearm, but the posterior arm muscles extend it.

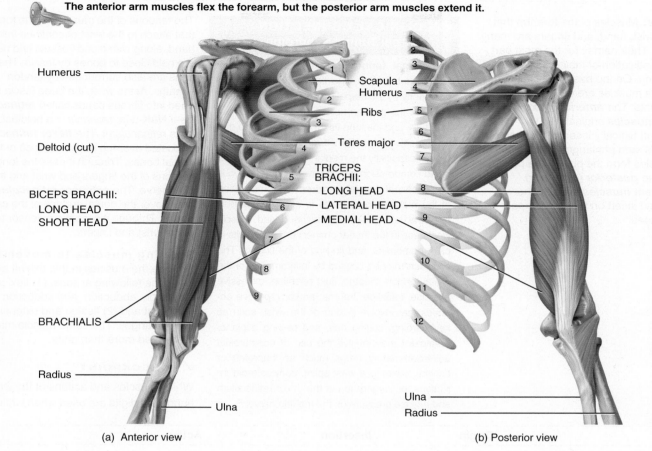

(a) Anterior view

(b) Posterior view

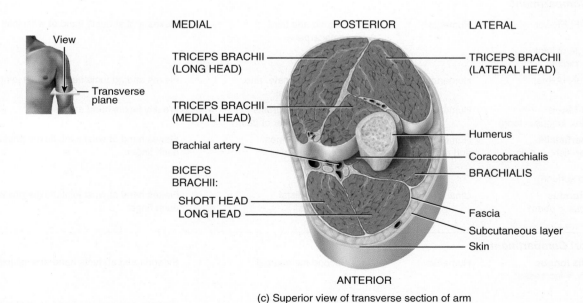

(c) Superior view of transverse section of arm

? **What is a compartment?**

233

OBJECTIVE • Describe the origin, insertion, and action of the muscles that move the wrist, hand, and fingers.

Overview: Muscles of the forearm that move the wrist, hand, and fingers are many and varied. Their names for the most part give some indication of their origin, insertion, or action. On the basis of location and function, the muscles are divided into two compartments. The *anterior (flexor) compartment muscles* originate on the humerus and typically insert on the carpals, metacarpals, and phalanges. The bellies of these muscles form the bulk of the proximal forearm. The *posterior (extensor) compartment muscles* arise on the humerus and insert on the metacarpals and phalanges.

CLINICAL CONNECTION | Carpal Tunnel Syndrome

The **carpal tunnel** is a narrow passageway formed anteriorly by the flexor retinaculum and posteriorly by the carpal bones. Through this tunnel pass the median nerve, the most superficial structure, and the long flexor tendons for the digits (Figure 8.21c). Structures within the carpal tunnel, especially the median nerve, are vulnerable to compression, and the resulting condition is called **carpal tunnel syndrome**. Compression of the median nerve leads to sensory changes over the lateral side of the hand and muscle weakness in the thenar eminence. This results in pain, numbness, and tingling of the fingers. The condition may be caused by inflammation of the digital tendon sheaths, fluid retention, excessive exercise, infection, trauma, and/or repetitive activities that involve flexion of the wrist, such as keyboarding, cutting hair, and playing a piano. Treatment may involve the use of nonsteroidal anti-inflammatory drugs (such as ibuprofen or aspirin), wearing a wrist splint, corticosteroid injections, or surgery to cut the flexor retinaculum and release pressure on the median nerve. •

The tendons of the muscles of the forearm that attach to the wrist or continue into the hand, along with blood vessels and nerves, are held close to bones by fascia. The tendons are also surrounded by tendon sheaths. At the wrist, the deep fascia is thickened into fibrous bands called *retinacula* (re-ti-NAK-ū-la; *retinacul* = a holdfast; singular is *retinaculum*). The *flexor retinaculum* is located over the palmar surface of the carpal bones. Through it pass the long flexor tendons of the fingers and wrist and the median nerve. The *extensor retinaculum* is located over the dorsal surface of the carpal bones. Through it pass the extensor tendons of the wrist and fingers.

Relating muscles to movements: Arrange the muscles in this exhibit according to the following actions: (1) flexion, extension, abduction, and adduction of the wrist joint and (2) flexion and extension of the phalanges. The same muscle may be mentioned more than once.

✓ **CHECKPOINT**

Which muscles and actions of the wrist, hand, and digits are used when writing?

Muscle	Origin	Insertion	Action
Anterior (Flexor) Compartment			
Flexor carpi radialis (FLEK-sor KAR-pē rā′-dē-Ā-lis; *flexor* = decreases angle at joint; *carpus* = wrist; *radi-* = radius)	Humerus	Second and third metacarpals	Flexes and abducts hand at wrist joint
Flexor carpi ulnaris (ul-NAR-is; *ulnar-* = ulna)	Humerus and ulna	Pisiform, hamate, and fifth metacarpal	Flexes and adducts hand at wrist joint
Palmaris longus (pal-MA-ris LON-gus; *palma* = palm; *longus* = long)	Humerus	Palmar aponeurosis (fascia in center of palm)	Weakly flexes hand at wrist joint
Flexor digitorum superficialis (di′-ji-TOR-um soo′-per- fish′-ē-Ā-lis; *digit* = finger or toe; *superficialis* = closer to surface)	Humerus, ulna, and radius	Middle phalanges of each finger*	Flexes hand at wrist joint; flexes phalanges of each finger
Flexor digitorum profundus (pro-FUN-dus; *profundus* = deep) (Not illustrated.)	Ulna	Bases of distal phalanges	Flexes hand at wrist joint; flexes phalanges of each finger
Posterior (Extensor) Compartment			
Extensor carpi radialis longus (eks-TEN-sor; *extensor* = increases angle at joint)	Humerus	Second metacarpal	Extends and abducts hand at wrist joint
Extensor carpi ulnaris	Humerus and ulna	Fifth metacarpal	Extends and adducts hand at wrist joint
Extensor digitorum	Humerus	Second through fifth phalanges of each finger	Extends hand at wrist joint; extends phalanges of each finger

*Reminder: The thumb or pollex is the first digit and has two phalanges: proximal and distal. The remaining digits, the fingers, are numbered II–V (2–5), and each has three phalanges: proximal, middle, and distal.

Figure 8.21 Muscles of the forearm that move the wrist, hand, and fingers.

🔑 The anterior compartment muscles function as flexors, and the posterior compartment muscles function as extensors.

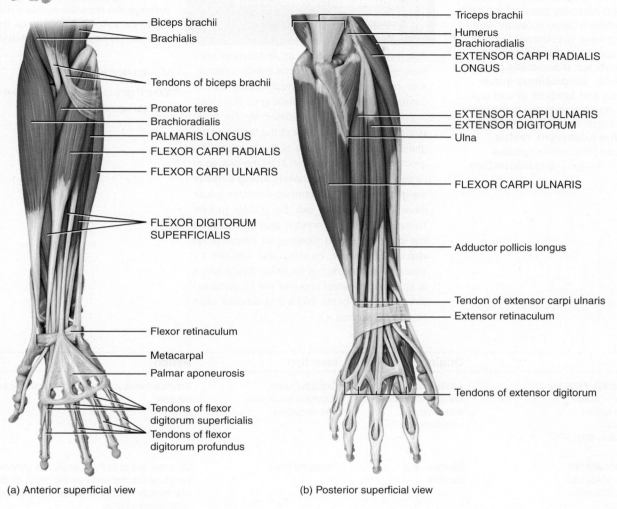

(a) Anterior superficial view

- Biceps brachii
- Brachialis
- Tendons of biceps brachii
- Pronator teres
- Brachioradialis
- PALMARIS LONGUS
- FLEXOR CARPI RADIALIS
- FLEXOR CARPI ULNARIS
- FLEXOR DIGITORUM SUPERFICIALIS
- Flexor retinaculum
- Metacarpal
- Palmar aponeurosis
- Tendons of flexor digitorum superficialis
- Tendons of flexor digitorum profundus

(b) Posterior superficial view

- Triceps brachii
- Humerus
- Brachioradialis
- EXTENSOR CARPI RADIALIS LONGUS
- EXTENSOR CARPI ULNARIS
- EXTENSOR DIGITORUM
- Ulna
- FLEXOR CARPI ULNARIS
- Adductor pollicis longus
- Tendon of extensor carpi ulnaris
- Extensor retinaculum
- Tendons of extensor digitorum

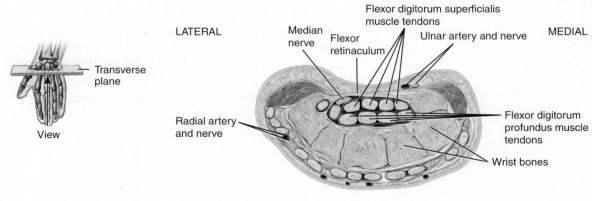

LATERAL

MEDIAL

- Transverse plane
- View
- Radial artery and nerve
- Median nerve
- Flexor retinaculum
- Flexor digitorum superficialis muscle tendons
- Ulnar artery and nerve
- Flexor digitorum profundus muscle tendons
- Wrist bones

(c) Inferior view of transverse section

? Which nerve is associated with the flexor retinaculum?

OBJECTIVE • **Describe the origin, insertion, and action of the muscles that move the vertebral column.**

Overview: The *erector spinae muscles* form the largest muscular mass of the back, forming a prominent bulge on either side of the vertebral column (backbone) (Figure 8.22). These consist of three groups of overlapping muscles: *iliocostalis group* (il′-ē-ō-kos-TĀ-lis), *longissimus group* (lon′-JI-si-mus), and *spinalis group* (spi-NĀ-lis). Other muscles that move the vertebral column include the *sternocleidomastoid, quadratus lumborum, rectus abdominis* (see Exhibit 8.D), *psoas major* (see Exhibit 8.K), and *iliacus* (see Exhibit 8.K).

CLINICAL CONNECTION |
Back Injuries and Heavy Lifting

Full flexion at the waist, as in touching your toes, overstretches the erector spinae muscles, and muscles that are overstretched cannot contract effectively. Straightening up from such a position is therefore initiated by the hamstring muscles on the back of the thigh and the gluteus maximus muscles of the buttocks. The erector spinae muscles join in as the degree of flexion decreases. **Improperly lifting a heavy weight**, however, can strain the erector spinae muscles. The result can be painful muscle spasms, tearing of tendons and ligaments of the lower back, and rupturing of intervertebral discs. The lumbar muscles are adapted for maintaining posture, not for lifting. This is why it is important to kneel and use the powerful extensor muscles of the thighs and buttocks while lifting a heavy load. •

Relating muscles to movements: Arrange the muscles in this exhibit according to the following actions on the vertebral column: (1) flexion and (2) extension.

✓ **CHECKPOINT**

Which groups of muscles make up the erector spinae?

Muscle	Origin	Insertion	Action
Erector spinae (e-REK-tor SPI-nē; *erector* = raise; *spinae* = of the spine) (iliocostalis group, longissimus group, and spinalis group)	All ribs plus cervical, thoracic, and lumbar vertebrae	Occipital bone, temporal bone, ribs, and vertebrae	Extends head; extends and laterally flexes vertebral column
Sternocleidomastoid (ster′-nō-klī-dō-MAS-toid; *sternum* = breastbone; *cleido-* = clavicle; *mastoid* = mastoid process of temporal bone) (See Figure 8.13b.)	Sternum and clavicle	Temporal bone	Contractions of both muscles flex cervical part of the vertebral column and flex the head; contraction of one muscle rotates head toward side opposite contracting muscle.
Quadratus lumborum (kwod-RĀ-tus lum-BOR-um; *quadratus* = four-sided; *lumbo* = lumbar region) (See Figure 8.17b.)	Ilium	Twelfth rib and upper four lumbar vertebrae	Contractions of both muscles extend lumbar part of the vertebral column; contraction of one muscle flexes lumbar part of vertebral column.

Figure 8.22 Muscles of the neck and back that move the vertebral column.

 The erector spinae muscles extend the vertebral column.

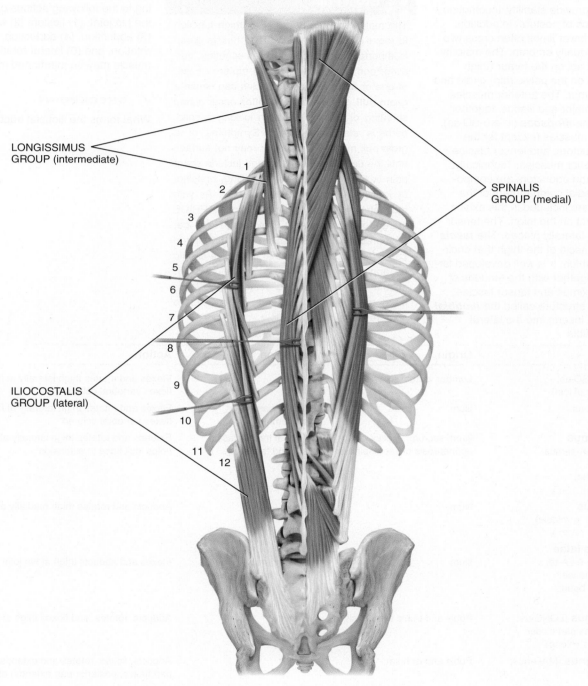

LONGISSIMUS
GROUP (intermediate)

SPINALIS
GROUP (medial)

ILIOCOSTALIS
GROUP (lateral)

1
2
3
4
5
6
7
8
9
10
11
12

Posterior view of erector spinae muscles

Which muscles constitute the erector spinae?

EXHIBIT 8.K

Muscles of the Gluteal Region That Move the Femur *(Figure 8.23)*

OBJECTIVE • Describe the origin, insertion, and action of the muscles that move the femur.

Overview: Muscles of the lower limbs are larger and more powerful than those of the upper limbs to provide stability, locomotion, and maintenance of posture. In addition, muscles of the lower limbs often cross two joints and act equally on both. The majority of muscles that act on the femur (thigh bone) originate on the pelvic (hip) girdle and insert on the femur. The anterior muscles are the psoas major and iliacus, together referred to as the *iliopsoas* (il'-ē-ō-SŌ-as). The remaining muscles (except for the pectineus, adductors, and tensor fasciae latae) are posterior muscles. Technically, the pectineus and adductors are components of the medial compartment of the thigh, but they are included in this exhibit because they act on the thigh. The tensor fasciae latae is laterally placed. The *fascia lata* is a deep fascia of the thigh that encircles the entire thigh. It is well developed laterally, where together with the tendons of the gluteus maximus and tensor fasciae latae it forms a structure called the *iliotibial tract*. The tract inserts into the lateral condyle of the tibia.

> **CLINICAL CONNECTION | Groin Pull**
>
> The major muscles of the inner thigh function to move the legs medially. This muscle group is important in activities such as sprinting, hurdling, and horseback riding. A rupture or tear of one or more of these muscles can cause a **groin pull**. Groin pulls most often occur during sprinting or twisting, or from kicking a solid, perhaps stationary object. Symptoms of a groin pull may be sudden, or may not surface until the day after the injury, and include sharp pain in the inguinal region, swelling, bruising, or inability to contract the muscles. As with most strain injuries, treatment involves PRICE therapy, which stands for *P*rotection, *R*est, *I*ce, *C*ompression, and *E*levation. After the injured part is protected from further injury, ice should be applied immediately, and the injured part should be elevated and rested. An elastic bandage should be applied, if possible, to compress the injured tissue. •

Relating muscles to movements: Arrange the muscles in this exhibit according to the following actions on the thigh at the hip joint: (1) flexion, (2) extension, (3) abduction, (4) adduction, (5) medial rotation, and (6) lateral rotation. The same muscle may be mentioned more than once.

✓ **CHECKPOINT**

What forms the iliotibial tract?

Muscle	Origin	Insertion	Action
Psoas major (SŌ-as; *psoa* = a muscle of loin)	Lumbar vertebrae	Femur	Flexes and rotates thigh laterally at hip joint; flexes vertebral column
Iliacus (il'-ē-AK-us; *iliac* = ilium)	Ilium	With psoas major into femur	Flexes and rotates thigh laterally at hip joint; flexes vertebral column
Gluteus maximus (GLOO-tē-us MAK-si-mus; *glute-* = buttock; *maximus* = largest) (See also Figure 8.13b.)	Ilium, sacrum, coccyx, and aponeurosis of sacrospinalis	Iliotibial tract of fascia lata and femur	Extends and rotates thigh laterally at hip joint; helps lock knee in extension
Gluteus medius (ME-dē-us; *medi-* = middle) (See also Figure 8.13b.)	Ilium	Femur	Abducts and rotates thigh medially at hip joint
Tensor fasciae latae (TEN-sor FA-shē-ē LĀ-tē; *tensor* = makes tense; *fasciae-* = of the band; *lat-* = wide)	Ilium	Tibia by means of the iliotibial tract	Flexes and abducts thigh at hip joint
Adductor longus (LONG-us; *adductor* = moves part closer to midline; *longus* = long)	Pubis and pubic symphysis	Femur	Adducts, rotates, and flexes thigh at hip joint
Adductor magnus (MAG-nus; *magnus* = large)	Pubis and ischium	Femur	Adducts, flexes, rotates and extends thigh (anterior part flexes, posterior part extends) at hip joint
Piriformis (pir-i-FOR-mis; *piri-* = pear; *form-* = shape)	Sacrum	Femur	Rotates thigh laterally and abducts it at hip joint
Pectineus (pek-TIN-ē-us; *pectin-* = comb-shaped)	Pubis	Femur	Flexes and adducts thigh at hip joint

Figure 8.23 Muscles of the gluteal region that move the femur and muscles of the thigh that move the femur and tibia and fibula.

🔑 Most muscles that move the femur originate on the pelvic (hip) girdle and insert on the femur.

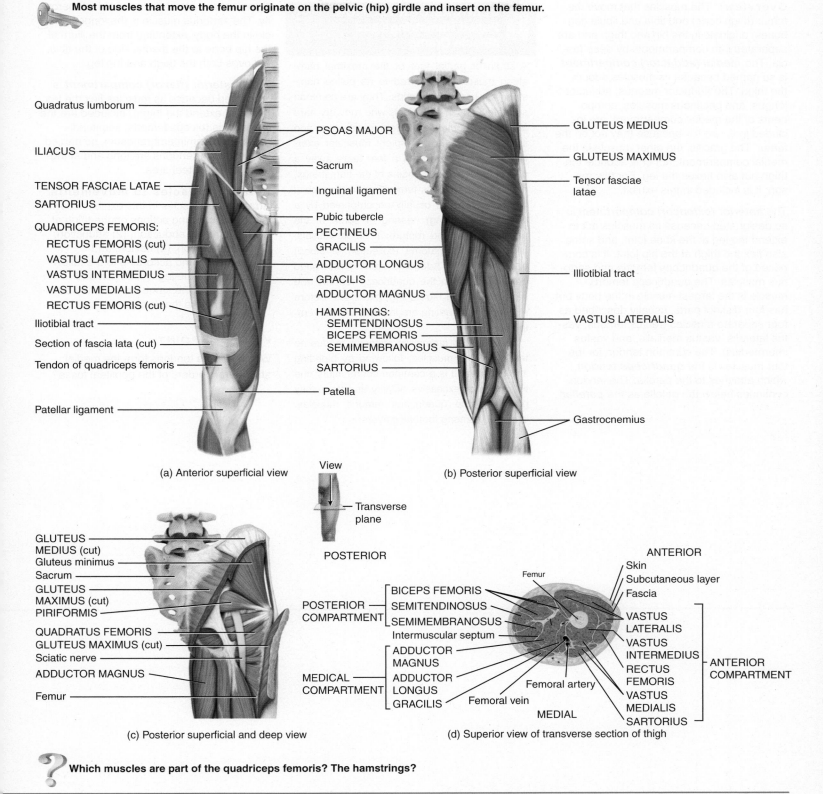

(a) Anterior superficial view

Quadratus lumborum

ILIACUS

TENSOR FASCIAE LATAE

SARTORIUS

QUADRICEPS FEMORIS:
   RECTUS FEMORIS (cut)
   VASTUS LATERALIS
   VASTUS INTERMEDIUS
   VASTUS MEDIALIS
   RECTUS FEMORIS (cut)

Iliotibial tract

Section of fascia lata (cut)

Tendon of quadriceps femoris

Patellar ligament

PSOAS MAJOR

Sacrum

Inguinal ligament

Pubic tubercle
PECTINEUS
GRACILIS
ADDUCTOR LONGUS
GRACILIS
ADDUCTOR MAGNUS

HAMSTRINGS:
   SEMITENDINOSUS
   BICEPS FEMORIS
   SEMIMEMBRANOSUS

SARTORIUS

Patella

(b) Posterior superficial view

GLUTEUS MEDIUS

GLUTEUS MAXIMUS

Tensor fasciae latae

Illiotibial tract

VASTUS LATERALIS

Gastrocnemius

View

Transverse plane

POSTERIOR

(c) Posterior superficial and deep view

GLUTEUS MEDIUS (cut)
Gluteus minimus
Sacrum
GLUTEUS MAXIMUS (cut)
PIRIFORMIS
QUADRATUS FEMORIS
GLUTEUS MAXIMUS (cut)
Sciatic nerve
ADDUCTOR MAGNUS
Femur

(d) Superior view of transverse section of thigh

ANTERIOR

Femur

POSTERIOR COMPARTMENT
   BICEPS FEMORIS
   SEMITENDINOSUS
   SEMIMEMBRANOSUS
Intermuscular septum

MEDICAL COMPARTMENT
   ADDUCTOR MAGNUS
   ADDUCTOR LONGUS
   GRACILIS

Femoral artery
Femoral vein

Skin
Subcutaneous layer
Fascia

VASTUS LATERALIS
VASTUS INTERMEDIUS
RECTUS FEMORIS
VASTUS MEDIALIS
SARTORIUS

ANTERIOR COMPARTMENT

MEDIAL

❓ **Which muscles are part of the quadriceps femoris? The hamstrings?**

OBJECTIVE • Describe the origin, insertion, and action of the muscles that move the femur, tibia, and fibula.

Overview: The muscles that move the femur (thigh bone) and tibia and fibula (leg bones) originate in the hip and thigh and are separated into compartments by deep fascia. The *medial (adductor) compartment* is so named because its muscles adduct the thigh. The adductor magnus, adductor longus, and pectineus muscles, components of the medial compartment, are included in Exhibit 8.K because they act on the femur. The gracilis, the other muscle in the medial compartment, not only adducts the thigh but also flexes the leg. For this reason, it is included in this exhibit.

The *anterior (extensor) compartment* is so designated because its muscles act to extend the leg at the knee joint, and some also flex the thigh at the hip joint. It is composed of the quadriceps femoris and sartorius muscles. The quadriceps femoris muscle is the largest muscle in the body but has four distinct parts, usually described as four separate muscles (rectus femoris, vastus lateralis, vastus medialis, and vastus intermedius). The common tendon for the four muscles is the *quadriceps tendon*, which attaches to the patella. The tendon continues below the patella as the *patellar*

**CLINICAL CONNECTION |
Pulled Hamstrings and
Muscle Cramp**

A strain or partial tear of the proximal hamstring muscles is referred to as pulled hamstrings or hamstring strains. They are common sports injuries in individuals who run very hard and/or are required to perform quick starts and stops. Sometimes the violent muscular exertion required to perform a feat tears away a part of the tendinous origins of the hamstrings, especially the biceps femoris, from the ischial tuberosity. This is usually accompanied by a contusion (bruising), tearing of some of the muscle fibers, and rupture of blood vessels, producing a hematoma (collection of blood) and sharp pain. Adequate training with good balance between the quadriceps femoris and hamstrings and stretching exercises before running or competing are important in preventing this injury.

A muscle cramp or stiffness is due to tearing of the muscles, following by bleeding into the area. It is a common sports injury due to trauma or excessive activity and frequently occurs in the quadriceps femoris muscles, especially among football players. •

ligament and attaches to the tibial tuberosity. The sartorius muscle is the longest muscle in the body, extending from the ilium of the hip bone to the medial side of the tibia. It moves both the thigh and the leg.

The *posterior (flexor) compartment* is so named because its muscles flex the leg (but also extend the thigh). Included are the hamstrings (biceps femoris, semitendinosus, and semimembranosus), so named because their tendons are long and stringlike in the popliteal area.

Relating muscles to movements: Arrange the muscles in this exhibit according to the following actions on the thigh at the hip joint: (1) abduction, (2) adduction, (3) lateral rotation, (4) flexion, and (5) extension; and according to the following actions on the leg: (1) flexion and (2) extension. The same muscle may be mentioned more than once.

✓ **CHECKPOINT**

Which muscle tendons form the medial and lateral borders of the popliteal fossa?

Muscle	Origin	Insertion	Action
Medial (Adductor) Compartment			
Adductor magnus (MAG-nus)			
Adductor longus (LONG-us)	See Exhibit 8.K.		
Pectineus (pek-TIN-ē-us)			
Gracilis (GRAS-i-lis; *gracilis* = slender)	Pubis	Tibia	Adducts and medially rotates thigh at hip joint; flexes leg at knee joint
Medial (Extensor) Compartment			
Quadriceps femoris (KWOD-ri-seps FEM-or-is; *quadriceps* = four heads of origin; *femoris* = femur)			
Rectus femoris (REK-tus *rectus* = straight; here, fascicles run parallel to midline)	Ilium	Patella by means of quadriceps tendon and then tibial tuberosity by means of patellar ligament	All four heads extend leg at knee joint; rectus femoris muscle alone also flexes thigh at hip joint
Vastus lateralis (VAS-tus lat′-er-Ā-lis; *vast-* = large; *lateralis* = lateral)	Femur		
Vastus medialis (mē′-dē-Ā-lis; *medialis* = medial)	Femur		
Vastus intermedius (in′-ter-MĒ-dē-us; *intermedius* = middle)	Femur		
Sartorius (sar-TOR-ē-us; *sartor-* = tailor; refers to cross-legged position of tailors) Longest muscle in the body	Ilium	Tibia	Weakly flexes leg at knee joint; flexes, abducts, and laterally rotates thigh at hip joint, thus crossing leg
Posterior (Flexor) Compartment			
Hamstrings			
Biceps femoris (BĪ-ceps FEM-or-is; *biceps* = two heads of origin)	Ischium and femur	Fibula and tibia	Flexes leg at knee joint; extends thigh at hip joint
Semitendinosus (sem′-ē-TEN-di-nō′-sus; *semi-* = half; *tendo-* = tendon)	Ischium	Tibia	Flexes leg at knee joint; extends thigh at hip joint
Semimembranosus (sem′-ē-MEM-bra-nō′-sus; *membran-* = membrane)	Ischium	Tibia	Flexes leg at knee joint; extends thigh at hip joint

EXHIBIT 8.M

Muscles of the Leg That Move the Foot and Toes *(Figure 8.24)*

OBJECTIVE • Describe the origin, insertion, and action of the muscles of the leg that move the foot and toes.

Overview: Muscles that move the foot and toes are located in the leg. The muscles of the leg, like those of the thigh, are divided into three compartments by deep fascia. The **anterior compartment** consists of muscles that dorsiflex the foot. In a situation like that at the wrist, the tendons of the muscles of the anterior compartment are held firmly to the ankle bones by thickenings of deep fascia called the **superior extensor retinaculum** and **inferior extensor retinaculum**. The **lateral compartment** contains muscles that plantar flex and evert the foot. The **posterior compartment** consists of superficial and deep muscles. The superficial muscles (gastrocnemius and soleus) share a common tendon of insertion, the calcaneal (Achilles) tendon, the strongest tendon of the body.

CLINICAL CONNECTION |
Shin Splint Syndrome

Shin splint syndrome, or simply **shin splints**, refers to pain or soreness along the medial, distal two-thirds of the tibia. It may be caused by tendinitis of the tibialis anterior or toe flexors, inflammation of the periosteum around the tibia, or stress fractures of the tibia. The tendinitis usually occurs when poorly conditioned runners run on hard or banked surfaces with poorly supportive running shoes or from walking or running up and down hills. The condition may also occur as a result of vigorous activity of the legs following a period of relative inactivity. The muscles in the anterior compartment (mainly the tibialis anterior) can be strengthened to balance the stronger posterior compartment muscles. •

Relating muscles to movements:
Arrange the muscles in this exhibit according to the following actions on the foot: (1) dorsiflexion, (2) plantar flexion, (3) inversion, and (4) eversion; and according to the following actions on the toes: (1) flexion and (2) extension. The same muscle may be mentioned more than once.

✓ **CHECKPOINT**

What is the function of the superior and inferior extensor retinaculum?

Figure 8.24 Muscles of the leg that move the foot and toes.

The superficial muscles of the posterior compartment share a common tendon of insertion, the calcaneal (Achilles) tendon, that inserts into the calcaneal bone of the ankle.

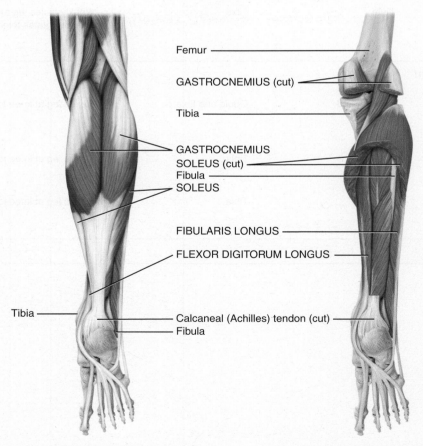

Femur
GASTROCNEMIUS (cut)
Tibia
GASTROCNEMIUS
SOLEUS (cut)
Fibula
SOLEUS
FIBULARIS LONGUS
FLEXOR DIGITORUM LONGUS
Tibia
Calcaneal (Achilles) tendon (cut)
Fibula

(a) Posterior superficial view

(b) Posterior deep view

Muscle	Origin	Insertion	Action
Anterior Compartment			
Tibialis anterior (tib´-ē-Ā-lis; *tibialis* = tibia; *anterior* = front)	Tibia	First metatarsal and first cuneiform	Dorsiflexes and inverts (supinates) foot
Extensor digitorum longus (eks-TEN-sor di´-ji-TOR-um LON-gus; *extensor* = increases angle at joint; *digitorum* = finger or toe; *longus* = long)	Tibia and fibula	Middle and distal phalanges of each toe (except the great toe)	Dorsiflexes and everts foot; extends toes
Lateral Compartment			
Fibularis (Peroneus) longus (fib-ū-LAR-is LON-gus)	Fibula and tibia	First metatarsal and first cuneiform	Plantar flexes and everts (pronates) foot
Posterior Compartment			
Gastrocnemius (gas´-trok-NĒ-mē-us; *gastro-* = belly; *-cnem* = leg)	Femur	Calcaneus by means of calcaneal (Achilles) tendon	Plantar flexes foot; flexes leg at knee joint
Soleus (SŌ-lē-us; *soleus* = a type of flatfish)	Fibula and tibia	Calcaneus by means of calcaneal (Achilles) tendon	Plantar flexes foot
Tibialis posterior (*posterior* = back)	Tibia and fibula	Second, third, and fourth metatarsals; navicular; all three cuneiforms; and cuboid	Plantar flexes and inverts foot
Flexor digitorum longus (FLEK-sor; *flexor* = decreases angle at joint)	Tibia	Distal phalanges of each toe (except great toe)	Plantar flexes foot; flexes toes

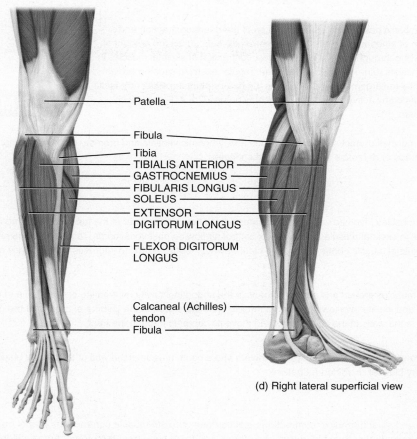

Patella
Fibula
Tibia
TIBIALIS ANTERIOR
GASTROCNEMIUS
FIBULARIS LONGUS
SOLEUS
EXTENSOR DIGITORUM LONGUS
FLEXOR DIGITORUM LONGUS
Calcaneal (Achilles) tendon
Fibula

(d) Right lateral superficial view

(c) Anterior superficial view

 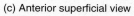 **Which muscle is primarily affected in shin splint syndrome?**

BODY SYSTEM	CONTRIBUTION OF THE MUSCULAR SYSTEM

For all body systems

The muscular system and muscular tissues produce body movements, stabilize body positions, move substances within the body, and produce heat that helps maintain normal body temperature.

Integumentary system

Pull of skeletal muscles on attachments to skin of face causes facial expressions; muscular exercise increases skin blood flow.

Skeletal system

Skeletal muscle causes movement of body parts by pulling on attachments to bones; skeletal muscle provides stability for bones and joints.

Nervous system

Smooth, cardiac, and skeletal muscles carry out commands for the nervous system; shivering—involuntary contraction of skeletal muscles that is regulated by the brain—generates heat to raise body temperature.

Endocrine system

Regular activity of skeletal muscles (exercise) improves the action of some hormones, such as insulin; muscles protect some endocrine glands.

Cardiovascular system

Cardiac muscle powers the pumping action of the heart; contraction and relaxation of smooth muscle in blood vessel walls help adjust the amount of blood flowing through various body tissues; contraction of skeletal muscles in the legs assists return of blood to the heart; regular exercise causes cardiac hypertrophy (enlargement) and increases the heart's pumping efficiency; lactic acid produced by active skeletal muscles may be used for ATP production by the heart.

Lymphatic system and immunity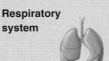

Skeletal muscles protect some lymph nodes and lymphatic vessels and promote the flow of lymph inside lymphatic vessels; exercise may increase or decrease some immune responses.

Respiratory system

Skeletal muscles involved with breathing cause air to flow into and out of the lungs; smooth muscle fibers adjust the size of airways; vibrations in skeletal muscles of the larynx control air flowing past vocal cords, regulating voice production; coughing and sneezing, due to skeletal muscle contractions, help clear airways; regular exercise improves the efficiency of breathing.

Digestive system

Skeletal muscles protect and support organs in the abdominal cavity; alternating contraction and relaxation of skeletal muscles power chewing and initiate swallowing; smooth muscle sphincters control the volume of organs of the gastrointestinal (GI) tract; smooth muscles in the walls of the GI tract mix and move its contents through the tract.

Urinary system

Skeletal muscle and smooth muscle sphincters and smooth muscle in the wall of the urinary bladder control whether urine is stored in the urinary bladder or voided (urination).

Reproductive systems

Skeletal and smooth muscle contractions eject semen; smooth muscle contractions propel oocytes through uterine tubes, help regulate flow of menstrual blood from the uterus, and force baby from the uterus during childbirth; during intercourse, skeletal muscle contractions are associated with orgasm and pleasurable sensations in both sexes.

THE MUSCULAR SYSTEM

COMMON DISORDERS

Skeletal muscle function may be abnormal due to disease or damage of any of the components of a motor unit: somatic motor neurons, neuromuscular junctions, or muscle fibers. The term *neuromuscular disease* encompasses problems at all three sites; the term *myopathy* (mī-OP-a-thē; *-pathy* = disease) signifies a disease or disorder of the skeletal muscle tissue itself.

Myasthenia Gravis

Myasthenia gravis (mī-as-THĒ-nē-a GRAV-is; *mys* = muscle; *-asthenia* = weakness) is an autoimmune disease that causes chronic, progressive damage of the neuromuscular junction. In people with myasthenia gravis, the immune system inappropriately produces antibodies that bind to and block some ACh receptors, thereby decreasing the number of functional ACh receptors at the motor end plates of skeletal muscles (see Figure 8.4). Because 75% of patients with myasthenia gravis have hyperplasia or tumors of the thymus, it is possible that thymic abnormalities cause the disorder. As the disease progresses, more ACh receptors are lost. Thus, muscles become increasingly weaker, fatigue more easily, and may eventually cease to function.

Myasthenia gravis occurs in about 1 in 10,000 people and is more common in women, who typically are ages 20 to 40 at onset, than in men, who usually are ages 50 to 60 at onset. The muscles of the face and neck are most often affected. Initial symptoms include weakness of the eye muscles, which may produce double vision, and difficulty in swallowing. Later, the person has difficulty chewing and talking. Eventually the muscles of the limbs may become involved. Death may result from paralysis of the respiratory muscles, but often the disorder does not progress to this stage.

Muscular Dystrophy

The term *muscular dystrophy* (DIS-trō-fē; *dys-* = difficult; *-trophy* = nourishment) refers to a group of inherited muscle-destroying diseases that cause progressive degeneration of skeletal muscle fibers. The most common form of muscular dystrophy is *DMD—Duchenne muscular dystrophy* (doo-SHĀN). Because the mutated gene is on the X chromosome, which males have only one of, DMD strikes boys almost exclusively. (Sex-linked inheritance is described in Chapter 24.) Worldwide, about 1 in every 3500 male babies—21,000 in all—are born with DMD each year. The disorder usually becomes apparent between the ages of 2 and 5, when parents notice the child falls often and has difficulty running, jumping, and hopping. By age 12 most boys with DMD are unable to walk. Respiratory or cardiac failure usually causes death between the ages of 20 and 30.

In DMD, the gene that codes for the protein dystrophin is mutated and little or no dystrophin is present (dystrophin provides structural reinforcement for the skeletal muscle fiber sarcolemma). Without the reinforcing effect of dystrophin, the sarcolemma easily tears during muscle contraction. Because their plasma membranes are damaged, muscle fibers slowly rupture and die.

Fibromyalgia

Fibromyalgia (fī-brō-mī-AL-jē-a; *algia* = painful condition) is a painful, nonarticular rheumatic disorder that usually appears between the ages of 25 and 50. An estimated 3 million people in the United States suffer from fibromyalgia, which is 15 times more common in women than in men. The disorder affects the fibrous connective tissue components of muscles, tendons, and ligaments. A striking sign is pain that results from gentle pressure at specific "tender points." Even without pressure, there is pain, tenderness, and stiffness of muscles, tendons, and surrounding soft tissues. Besides muscle pain, those with fibromyalgia report severe fatigue, poor sleep, headaches, depression, and inability to carry out their daily activities. Treatment includes therapy, medication for pain, and a low-dose antidepressant to help improve sleep.

Abnormal Contractions of Skeletal Muscle

One kind of abnormal muscular contraction is a *spasm*, a sudden involuntary contraction of a single muscle in a large group of muscles. A painful spasmodic contraction is known as a *cramp*. A *tic* is a spasmodic twitching made involuntarily by muscles that are ordinarily under voluntary control. Twitching of the eyelid and facial muscles are examples of tics. A *tremor* is a rhythmic, involuntary, purposeless contraction that produces a quivering or shaking movement. A *fasciculation* (fa-sik-ū-LĀ-shun) is an involuntary, brief twitch of an entire motor unit that is visible under the skin; it occurs irregularly and is not associated with movement of the affected muscle. Fasciculations may be seen in multiple sclerosis (see Chapter 9 Common Disorders Section) or in amyotrophic lateral sclerosis (Lou Gehrig's disease). A *fibrillation* is a spontaneous contraction of a single muscle fiber that is not visible under the skin but can be recorded by electromyography. Fibrillations may signal destruction of motor neurons.

Running Injuries

Many individuals who jog or run sustain some type of *running injury*. Although such injuries may be minor, some can be quite serious. Untreated or inappropriately treated minor injuries may become chronic. Among runners, common sites of injury include the ankle, knee, calcaneal (Achilles) tendon, hip, groin, foot, and back. Of these, the knee often is the most severely injured area.

Running injuries are frequently related to faulty training techniques. This may involve improper or lack of sufficient warm-up routines, running too much, or running too soon after an injury. Or it might involve extended running on hard and/or uneven surfaces.

Poorly constructed or worn-out running shoes can also contribute to injury, as can any biomechanical problem (such as a fallen arch) aggravated by running.

Most sports injuries should be treated initially with PRICE therapy, which stands for *Protection, Rest, Ice, Compression,* and *Elevation.* After the injured part is protected from further damage, ice should be applied immediately, and the injured part should be elevated and protected. Then apply an elastic bandage, if possible, to compress the injured tissue. Continue using PRICE for 2 to 3 days, and resist the temptation to apply heat, which may worsen the swelling. Follow-up treatment may include alternating moist heat and ice massage to enhance blood flow in the injured area. Sometimes it is helpful to take nonsteroidal anti-inflammatory drugs (NSAIDs) or to have local injections of corticosteroids. During the recovery period, it is important to keep active, using an alternative fitness program that does not worsen the original injury. This activity should be determined in consultation with a physician. Finally, careful exercise is needed to rehabilitate the injured area itself. Massage therapy may also be used to prevent or treat many sports injuries.

Effects of Anabolic Streroids

The use of ***anabolic steroids*** by athletes has received widespread attention. These steroid hormones, similar to testosterone, are taken to increase muscle size and strength. The large doses needed to produce an effect, however, have damaging, sometimes even devastating side effects, including liver cancer, kidney damage, increased risk of heart disease, stunted growth, wide mood swings, and increased irritability and aggression. Additionally, females who take anabolic steroids may experience atrophy of the breasts and uterus, menstrual irregularities, sterility, facial hair growth, and deepening of the voice. Males may experience diminished testosterone secretion, atrophy of the testes, and baldness.

MEDICAL TERMINOLOGY AND CONDITIONS

Electromyography or **EMG** (e-lek'-trō-mī-OG-ra-fē; *electro-* = electricity; *myo-* = muscle; *-graphy* = to write) The recording and study of electrical changes that occur in muscular tissue.

Hypertonia (*hyper-* = above) Increased muscle tone, characterized by increased muscle stiffness and sometimes associated with a change in normal reflexes.

Hypotonia (*hypo-* = below) Decreased or lost muscle tone.

Muscle strain Tearing of a muscle because of forceful impact, accompanied by bleeding and severe pain. Also known as a *charley horse* or pulled muscle. It often occurs in contact sports and typically affects the quadriceps femoris muscle on the anterior surface of the thigh.

Myalgia (mī-AL-jē-a; *-algia* = painful condition) Pain in or associated with muscles.

Myoma (mī-Ō-ma; *-oma* = tumor) A tumor consisting of muscular tissue.

Myomalacia (mī'-ō-ma-LĀ-shē-a; *-malacia* = soft) Pathological softening of muscle tissue.

Myositis (mī'-ō-SĪ-tis; *-itis* = inflammation of) Inflammation of muscle fibers (cells).

Myotonia (mī'-ō-TŌ-nē-a; *-tonia* = tension) Increased muscular excitability and contractility, with decreased power of relaxation; tonic spasm of the muscle.

CHAPTER REVIEW AND RESOURCE SUMMARY

REVIEW	RESOURCES
### 8.1 Overview of Muscular Tissue	Anatomy Overview—The Muscular System
1. The three types of muscular tissue are skeletal muscle, cardiac muscle, and smooth muscle.	Anatomy Overview—Muscle Tissue
2. **Skeletal muscle tissue** is mostly attached to bones. It is **striated** and **voluntary**.	
3. **Cardiac muscle tissue** forms most of the wall of the heart. It is striated and **involuntary**.	
4. **Smooth muscle tissue** is located in viscera. It is **nonstriated** and involuntary.	
5. Through contraction and relaxation, muscular tissue has five key functions: producing body movements, stabilizing body positions, regulating organ volume, moving substances within the body, and producing heat.	
### 8.2 Skeletal Muscle Tissue	Anatomy Overview—Cross-section of Skeletal Muscle
1. Connective tissue coverings associated with skeletal muscle include the **epimysium**, covering an entire muscle; **perimysium**, covering **fascicles**; and **endomysium**, covering individual **muscle fibers**. **Tendons** are extensions of connective tissue beyond muscle fibers that attach the muscle to bone.	Anatomy Overview—Muscle Cell Structure

REVIEW	RESOURCES

2. Skeletal muscles are well supplied with nerves and blood vessels, which provide nutrients and oxygen for contraction.

3. Skeletal muscle consists of muscle fibers (cells) covered by a **sarcolemma** that features tunnel-like extensions, the **transverse tubules**. The fibers contain **sarcoplasm**, multiple nuclei, many mitochondria, **myoglobin**, and **sarcoplasmic reticulum**.

4. Each fiber also contains **myofibrils** that contain **thin filaments** and **thick filaments**. The filaments are arranged in functional units called **sarcomeres**.

5. Thick filaments consist of **myosin**; thin filaments are composed of **actin**, **tropomyosin**, and **troponin**.

8.3 Contraction and Relaxation of Skeletal Muscle

1. Muscle contraction occurs when myosin heads attach to and "walk" along the thin filaments at both ends of a sarcomere, progressively pulling the thin filaments toward the center of a sarcomere. As the thin filaments slide inward, the **Z discs** come closer together, and the sarcomere shortens.

2. **The neuromuscular junction (NMJ)** is the synapse between a **motor neuron** and a skeletal muscle fiber. The NMJ includes the axon terminals and synaptic end bulbs of a motor neuron plus the adjacent **motor end plate** of the muscle fiber sarcolemma.

3. A motor neuron and all of the muscle fibers it stimulates form a **motor unit**. A single motor unit may include as few as 10 or as many as 2000 muscle fibers.

4. When a nerve impulse reaches the synaptic end bulbs of a somatic motor neuron, it triggers the release of **acetylcholine (ACh)** from synaptic vesicles. ACh diffuses across the synaptic cleft and binds to ACh receptors, initiating a muscle action potential. **Acetylcholinesterase** then quickly destroys ACh.

5. The **sliding-filament mechanism** of muscle contraction is the sliding of filaments and shortening of sarcomeres that cause the shortening of muscle fibers.

6. An increase in the level of Ca^{2+} in the sarcoplasm, caused by the muscle action potential, starts the contraction cycle; a decrease in the level of Ca^{2+} turns off the contraction cycle.

7. The **contraction cycle** is the repeating sequence of events that causes sliding of the filaments: (1) myosin ATPase splits ATP and becomes energized, (2) the myosin head attaches to actin, forming a **cross-bridge**, (3) the cross-bridge generates force as it swivels or rotates toward the center of the sarcomere (**power stroke**), and (4) binding of ATP to myosin detaches myosin from actin. The myosin head again splits ATP, returns to its original position, and binds to a new site on actin as the cycle continues.

8. Ca^{2+} active transport pumps continually remove Ca^{2+} from the sarcoplasm into the sarcoplasmic reticulum (SR). When the level of Ca^{2+} in the sarcoplasm decreases, tropomyosin slides back over and covers the myosin-binding sites, and the muscle fiber relaxes.

9. Continual involuntary activation of a small number of motor units produces **muscle tone**, which is essential for maintaining posture.

8.4 Metabolism of Skeletal Muscle Tissue

1. Muscle fibers have three sources for ATP production: creatine phosphate, anaerobic cellular respiration, and aerobic cellular respiration.

2. The transfer of a high-energy phosphate group from **creatine phosphate** to ADP forms new ATP molecules. Together, creatine phosphate and ATP provide enough energy for muscles to contract maximally for about 15 seconds.

3. Glucose is converted to pyruvic acid in the reactions of glycolysis, which yield two ATPs without using oxygen. These reactions, referred to as **anaerobic cellular respiration**, can provide enough ATP for about 30 to 40 seconds of maximal muscle activity.

4. Muscular activity that lasts longer than half a minute depends on **aerobic cellular respiration**, mitochondrial reactions that require oxygen to produce ATP. Aerobic cellular respiration yields about 36 molecules of ATP from each glucose molecule.

5. The inability of a muscle to contract forcefully after prolonged activity is **muscle fatigue**.

6. Elevated oxygen use after exercise is called **recovery oxygen uptake**.

RESOURCES

Animation—Neuromuscular Junction
Animation—Muscle Cell Structure
Animation—Contraction of a Sarcomere

Figure 8.6 The contraction cycle

Exercise—Contraction Connections
Concepts and Connections—Events at the Neuromuscular Junction
Concepts and Connections—Excitation of Skeletal Muscle
Concepts and Connections—Skeletal Muscle Contraction Cycle

Animation—Muscle Metabolism
Exercise—Fueling Contraction
Concepts and Connections—Muscle Metabolism

REVIEW	RESOURCES

8.5 Control of Muscle Tension

1. A **twitch contraction** is a brief contraction of all of the muscle fibers in a motor unit in response to a single action potential.
2. A record of a contraction is called a **myogram**. It consists of a latent period, a contraction period, and a relaxation period.
3. **Wave summation** is the increased strength of a contraction that occurs when a second stimulus arrives before the muscle has completely relaxed after a previous stimulus.
4. Repeated stimuli can produce **unfused (incomplete) tetanus**, a sustained muscle contraction with partial relaxation between stimuli; more rapidly repeating stimuli will produce **fused (complete) tetanus**, a sustained contraction without partial relaxation between stimuli.
5. **Motor unit recruitment** is the process of increasing the number of active motor units.
6. On the basis of their structure and function, skeletal muscle fibers are classified as **slow oxidative (SO) fibers**, **fast oxidative-glycolytic (FOG) fibers**, and **fast glycolytic (FG) fibers**.
7. Most skeletal muscles contain a mixture of all three fiber types; their proportions vary with the typical action of the muscle.
8. The motor units of a muscle are recruited in the following order: first SO fibers, then FOG fibers, and finally FG fibers.

Animation—Control of Muscle Tension
Exercise—Increase Muscle Tension

8.6 Exercise and Skeletal Muscle Tissue

1. Various types of exercises can induce changes in the fibers in a skeletal muscle. Endurance-type (aerobic) exercises cause a gradual transformation of some fast glycolytic (FG) fibers into fast oxidative–glycolytic (FOG) fibers.
2. Exercises that require great strength for short periods produce an increase in the size and strength of fast glycolytic (FG) fibers. The increase in size is due to increased synthesis of thick and thin filaments.

Homeostatic Imbalance—The Case of Muscles and More

8.7 Cardiac Muscle Tissue

1. Cardiac muscle tissue, which is striated and involuntary, is found only in the heart.
2. Each cardiac muscle fiber usually contains a single centrally located nucleus and exhibits branching.
3. Cardiac muscle fibers are connected by means of **intercalated discs**, which hold the muscle fibers together and allow muscle action potentials to quickly spread from one cardiac muscle fiber to another.
4. Cardiac muscle tissue contracts when stimulated by its own autorhythmic fibers. Due to its continuous, rhythmic activity (**autorhythmicity**), cardiac muscle depends greatly on aerobic cellular respiration to generate ATP.

Anatomy Overview—Cardiac Muscle

8.8 Smooth Muscle Tissue

1. Smooth muscle tissue is nonstriated and involuntary.
2. In addition to thin and thick filaments, smooth muscle fibers contain **intermediate filaments** and **dense bodies**.
3. **Visceral (single-unit) smooth muscle tissue** is found in the walls of hollow viscera and of small blood vessels. Many visceral fibers form a network that contracts in unison.
4. **Multiunit smooth muscle tissue** is found in large blood vessels, large airways to the lungs, arrector pili muscles, and the eye. The fibers contract independently rather than in unison.
5. The duration of contraction and relaxation is longer in smooth muscle than in skeletal muscle. **Smooth muscle tone** is a state of continuous partial contraction of smooth muscle tissue.
6. Smooth muscle fibers can be stretched considerably and still retain the ability to contract.
7. Smooth muscle fibers contract in response to nerve impulses, stretching, hormones, and local factors.
8. The characteristics of the three types of muscular tissue are summarized in Table 8.1.

Anatomy Overview—Smooth Muscle

REVIEW	RESOURCES
## 8.9 Aging and Muscular Tissue	
1. Beginning at about 30 years of age, there is a slow, progressive loss of skeletal muscle, which is replaced by fibrous connective tissue and fat.	Homeostatic Imbalance—The Case of the Weakened Teenager
2. Aging also results in a decrease in muscle strength, slower muscle reflexes, and loss of flexibility.	
## 8.10 How Skeletal Muscles Produce Movement	
1. **Skeletal muscles** produce movement by pulling on tendons attached to bones.	Animation—Contraction and Movement
2. The attachment to the stationary bone is the **origin**. The attachment to the movable bone is the **insertion**.	
3. The **prime mover (agonist)** produces the desired action. The **antagonist** produces an opposite action. The **synergist** assists the prime mover by reducing unnecessary movement. The **fixator** stabilizes the origin of the prime mover so that it can act more efficiently.	
## 8.11 Principal Skeletal Muscles	
1. The principal skeletal muscles of the body are grouped according to region, as shown in Exhibits 8.A through 8.M.	**Figure 8.17** Muscles of the thorax that assist in breathing
2. In studying muscle groups, refer to Figure 8.13 to see how each group is related to all others.	**Figure 8.23d** Muscles of the gluteal region that move the femur and muscles of the thigh that move the femur and tibia and fibula
3. The names of most skeletal muscles indicate specific characteristics.	
4. The major descriptive categories are direction of fibers, location, size, number of origins, shape, origin and insertion, and action (see Table 8.2).	

SELF-QUIZ

1. The sarcolemma of muscle cells is equivalent to the _____ of other cells.
   a. cytoplasm   b. nucleus
   c. plasma membrane   d. endoplasmic reticulum
   e. mitochondria

2. The connective tissue component that surrounds fascicles is the
   a. endomysium.   b. sarcoplasmic reticulum.
   c. fascia.   d. epimysium.   e. perimysium.

3. Which of the following statements about skeletal muscle tissue is NOT true?
   a. Skeletal muscle requires a large blood supply.
   b. Skeletal muscle fibers have many mitochondria.
   c. The arrangement of thick and thin filaments produces the striations in skeletal muscle tissue.
   d. Skeletal muscle fibers contain gap junctions that help conduct action potentials from one fiber to another.
   e. A skeletal muscle fiber has many nuclei.

4. Match the following:
   ____ a. network of tubules that stores calcium
   ____ b. pigment that stores oxygen
   ____ c. composed of myosin
   ____ d. composed of actin, tropomyosin, and troponin
   ____ e. tunnel-like extensions of sarcolemma

   A. thick filaments
   B. transverse tubules
   C. sarcoplasmic reticulum
   D. myoglobin
   E. thin filaments

5. The reddish color of slow oxidative and fast oxidative-glycolytic skeletal muscle fibers is due to
   a. the mitochondria.   b. the large amount of myoglobin.
   c. the presence of large numbers of myofibrils.
   d. blood.   e. genetics.

6. You begin an intensive weightlifting plan because you want to enter a weightlifting contest. During the activity of weightlifting, your skeletal muscles will obtain energy (ATP) primarily through
   a. anaerobic glycolysis.
   b. the complete breakdown of pyruvic acid in the mitochondria.
   c. creatine phosphate.   d. faster breathing.
   e. aerobic cellular respiration.

7. Muscle contractions are smooth, coordinated movements because of
   a. motor unit recruitment.
   b. wave summation.
   c. tetanic contractions.
   d. the presence of fast oxidative–glycolytic fibers.
   e. latent periods.

8. For each of the following descriptions, indicate whether it refers to skeletal muscle, cardiac muscle, or smooth muscle. Use the abbreviations SK for skeletal, CA for cardiac, and SM for smooth. The same response may be used more than once.
   ____ a. involuntary
   ____ b. multinucleated
   ____ c. striated

_____ **d.** contain intercalated discs

_____ **e.** elongated, cylindrical cells

_____ **f.** voluntary

_____ **g.** cells that taper at both ends

_____ **h.** nonstriated

_____ **i.** muscle fibers contract individually

_____ **j.** autorhythmic

9. When ATP in the sarcoplasm is exhausted, the muscle must rely on _____ to quickly produce more ATP from ADP for contraction.

   a. acetylcholine   **b.** creatine phosphate   **c.** lactic acid

   d. pyruvic acid   **e.** acetylcholinesterase

10. A motor unit consists of

    a. a transverse tubule and its associated sarcomeres.

    b. a motor neuron and all of the muscle fibers it stimulates.

    c. a muscle and all of its motor neurons.

    d. all of the filaments encased within a sarcomere.

    e. the motor end plate and the transverse tubules.

11. Thick filaments

    a. include actin, troponin, and tropomyosin.

    b. compose the I band.

    c. stretch the entire length of a sarcomere.

    d. have binding sites for Ca^{2+}.

    e. have myosin heads used for the power stroke.

12. The chemical that prevents the continuous stimulation of a muscle fiber is

    a. Ca^{2+}.   **b.** acetylcholinesterase.   **c.** ATP.

    d. acetylcholine.   **e.** actin.

13. Which of the following is NOT associated with muscle fatigue?

    a. depletion of creatine phosphate

    b. lack of oxygen

    c. decrease in Ca^{2+} levels in the sarcoplasm

    d. decrease in lactic acid levels

    e. lack of glycogen

14. Which of the following binds to Ca^{2+}?

    a. actin   **b.** troponin   **c.** tropomyosin

    d. myosin   **e.** acetylcholine

15. Skeletal muscles are named using several characteristics. Which characteristic is NOT used to name skeletal muscles?

    a. direction of fibers   **b.** size   **c.** speed of contraction

    d. location   **e.** shape

16. Arrange the following in the correct order for skeletal muscle fiber contraction.

    1. Sarcoplasmic reticulum releases Ca^{2+}.

    2. Ca^{2+} combines with troponin, uncovering myosin-binding sites.

    3. Acetylcholine is released from the axon terminal and binds to receptors in the motor end plate.

    4. Action potential travels along the sarcolemma and into transverse tubules.

5. Energized myosin heads (crossbridges) attach to actin.

6. Thin filaments slide toward the center of the sarcomere.

   a. 3, 4, 1, 2, 5, 6   **b.** 4, 3, 2, 1, 5, 6   **c.** 1, 2, 3, 4, 5, 6

   d. 4, 1, 3, 5, 2, 6   **e.** 3, 1, 4, 5, 2, 6

17. What would happen if ATP were suddenly unavailable after the sarcomere had begun to shorten?

    a. Nothing. The contraction would proceed normally.

    b. The myosin heads would be unable to detach from actin.

    c. Troponin would bind with the myosin heads.

    d. Actin and myosin filaments would separate completely and be unable to recombine.

    e. The myosin heads would detach completely from actin and bind to the troponin–tropomyosin complex.

18. Match the following:

    _____ **a.** extend from the thick filaments

    _____ **b.** contain myosin-binding sites

    _____ **c.** dense area of protein that separates sarcomeres

    _____ **d.** contain acetylcholine

    _____ **e.** striated zone of the sarcomere composed of thick and thin filaments

    _____ **f.** space between axon terminal and the sarcolemma

    _____ **g.** striated zone of the sarcomere composed of thin filaments only

    _____ **h.** region of sarcolemma near the adjoining axon terminal

    A. I band

    B. synaptic vesicles

    C. myosin heads

    D. Z discs

    E. motor end plate

    F. actin molecules

    G. A band

    H. synaptic cleft

19. Match the following:

    _____ **a.** extends and laterally rotates thigh at the hip joint

    _____ **b.** adducts, rotates, and flexes thigh at the hip joint

    _____ **c.** compresses abdomen and flexes vertebral column

    _____ **d.** flexes leg at knee joint and extends thigh at hip joint

    _____ **e.** flexes and abducts wrist joint

    _____ **f.** extends phalanges

    _____ **g.** flexes, adducts, and rotates arm medially at shoulder joint

    _____ **h.** extends leg at the knee and flexes thigh at hip joint

    _____ **i.** plantar flexes foot at ankle joint and flexes leg at knee joint

    _____ **j.** dorsiflexes and inverts foot

    _____ **k.** abducts, flexes, extends, and rotates arm at shoulder joint

    A. trapezius

    B. biceps brachii

    C. tibialis anterior

    D. adductor longus

    E. gluteus maximus

    F. quadriceps femoris group

    G. rectus abdominis

    H. hamstring group

    I. frontal belly of occipitofrontalis

    J. gastrocnemius

    K. deltoid

    L. masseter

    M. extensor digitorum

    N. latissimus dorsi

    O. pectoralis major

    P. flexor carpi radialis

_____ **l.** dependent on contracting fibers; can elevate clavicle, depress or elevate scapula

_____ **m.** elevates mandible; closes mouth

_____ **n.** wrinkles skin of forehead horizontally as in a look of surprise

_____ **o.** extends, adducts, and rotates arm medially at shoulder joint; draws arm downward and backward

_____ **p.** flexes and supinates forearm at elbow joint

20. Match the following:

_____ **a.** works with prime mover to reduce unnecessary movement

_____ **b.** muscle in a group that contracts to produce desired movement

_____ **c.** stationary end of a muscle

_____ **d.** muscle that has an action opposite to that of another muscle

_____ **e.** helps stabilize the origin of the prime mover

_____ **f.** the end of a muscle attached to the movable bone

A. insertion
B. origin
C. synergist
D. antagonist
E. prime mover
F. fixator

CRITICAL THINKING APPLICATIONS

1. The newspaper reported several cases of botulism poisoning following a celebration meal at a company's sales conference. The cause appeared to be three-bean salad "flavored" with the bacterium _Clostridium botulinum_. What would be the result of botulism poisoning on muscle function?

2. Sarah's nephew was squealing with laughter. She was entertaining him by sticking her thumb in her pursed lips, raising her eyebrows, pumping her arm up and down, and puffing her cheeks in and out. Name the muscles Sarah was using to maneuver her face.

3. When his cast finally came off after six long weeks, Keith thought he'd be all set to rejoin his volleyball team, but now his left thigh is only half the size of his right. Explain what happened to his thigh and what he needs to do to get back in the game.

4. While watching the Olympic track-and-field trials on television, your friend asked you why the sprinters have such large leg muscles compared to the marathon runners. How could you explain this observation?

ANSWERS TO FIGURE QUESTIONS

8.1 In order from the inside toward the outside, the connective tissue layers are endomysium, perimysium, and epimysium.

8.2 The A band is composed of thick filaments in its center and overlapping thick and thin filaments at each end; the I band is composed of thin filaments.

8.3 A band: myosin, actin, troponin, and tropomyosin. I band: actin, troponin, and tropomyosin.

8.4 The motor end plate is the region of the sarcolemma near the axon terminal.

8.5 The I bands disappear. The lengths of the thick and thin filaments do not change.

8.6 Binding of ATP to the myosin heads detaches them from actin.

8.7 The power stroke occurs during step ⑥.

8.8 Glycolysis, exchange of phosphate between creatine phosphate and ADP, and glycogen breakdown occur in the cytosol. Oxidation of pyruvic acid, amino acids, and fatty acids (aerobic cellular respiration) occurs in the mitochondria.

8.9 Sarcomeres shorten during the contraction period.

8.10 Fused tetanus occurs when the frequency of stimulation reaches 80 to 100 stimuli per second.

8.11 The walls of hollow organs contain visceral (single-unit) smooth muscle.

8.12 The prime mover or agonist produces the desired action.

8.13 The following are some possible responses (there are other correct answers): direction of fibers—external oblique; shape— deltoid; action—extensor digitorum; size—gluteus maximus; origin and insertion—sternocleidomastoid; location—tibialis anterior; number of origins—biceps brachii.

8.14 Smiling—zygomaticus major; pouting—platysma; squinting—orbicularis oculi.

8.15 The superior oblique passes through the trochlea.

8.16 The rectus abdominis aids in urination.

8.17 The diaphragm and external intercostals contract during a normal quiet inhalation.

8.18 The pectoralis minor and serratus anterior have origins on the ribs; the trapezius, levator scapulae, and rhomboid major have origins on the vertebrae.

8.19 The pectoralis major and latissimus dorsi are muscles that cross the shoulder joint but do not originate on the scapula.

8.20 A compartment is a group of functionally related skeletal muscles in a limb, along with their blood vessels and nerves.

8.21 The median nerve is associated with the flexor retinaculum.

8.22 The iliocostalis, longissimus, and spinalis groups constitute the erector spinae.

8.23 Quadriceps femoris—rectus femoris, vastus lateralis, vastus medialis, and vastus intermedius; hamstrings—biceps femoris, semitendinosus, and semimembranosus.

8.24 Shinsplint syndrome affects the tibialis anterior.

CHAPTER 9

NERVOUS TISSUE

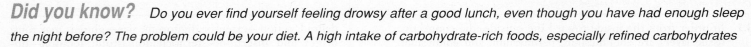

Together, all nervous tissues in the body comprise the ***nervous system***. Among the 11 body systems, the nervous system and the endocrine system play the most important roles in maintaining homeostasis. The nervous system, the subject of this and the next three chapters, can respond rapidly to help adjust body processes using nerve impulses. The endocrine system typically operates more slowly and exerts its influence on homeostasis by releasing hormones that the blood delivers to cells throughout the body. Besides helping maintain homeostasis, the nervous system is responsible for our perceptions, behaviors, and memories. It also initiates all voluntary movements. The branch of medical science that deals with the normal functioning and disorders of the nervous system is called ***neurology*** (noo-ROL-ō-jē; *neuro-* = nerve or nervous system; *-logy* = study of).

Did you know? Do you ever find yourself feeling drowsy after a good lunch, even though you have had enough sleep the night before? The problem could be your diet. A high intake of carbohydrate-rich foods, especially refined carbohydrates such as sugary desserts, can make some people sleepy. Large, heavy meals also lead to feelings of drowsiness.

Sleepiness after large, starchy meals can be due to an increase in the levels of a neurotransmitter in the brain called serotonin. While serotonin can make some people feel relaxed, too much can make others sleepy.

FOCUS ON WELLNESS
Neurotransmitters—Why Food Could Affect Your Mood

LOOKING BACK TO MOVE AHEAD...

Ion Channels (Section 3.3)

Sodium–Potassium Pump (Section 3.3)

Nervous Tissue (Section 4.6)

Sensory Nerve Endings and Sensory Receptors in the Skin (Section 5.1)

Release of Acetylcholine at the Neuromuscular Junction (Section 8.3)

9.1 OVERVIEW OF THE NERVOUS SYSTEM

OBJECTIVES • Describe the organization of the nervous system.

• Explain the three basic functions of the nervous system.

Organization of the Nervous System

The nervous system is an intricate network of billions of neurons and even more neuroglia. This system is organized into two main subdivisions: the central nervous system and the peripheral nervous system.

Central Nervous System

The ***central nervous system (CNS)*** consists of the brain and spinal cord (Figure 9.1a). The ***brain*** is the part of the CNS that is located in the skull. The ***spinal cord*** connects to the brain and is encircled by the bones of the vertebral column. The CNS processes many different kinds of incoming sen-sory information. It is also the source of thoughts, emotions, and memories. Most nerve impulses that stimulate muscles to contract and glands to secrete originate in the CNS.

Peripheral Nervous System

The ***peripheral nervous system (PNS)*** includes all nervous tissue outside the CNS (Figure 9.1a). Components of the PNS include nerves, ganglia, enteric plexuses, and sensory receptors. A ***nerve*** is a bundle of hundreds to thousands of axons plus associated connective tissue and blood vessels that lies outside the brain and spinal cord. Twelve pairs of ***cranial nerves*** emerge from the brain and thirty-one pairs of ***spinal nerves*** emerge from the spinal cord. Each nerve follows a de-fined path and serves a specific region of the body. ***Ganglia*** (GANG-lē-a = swelling or knot; singular is *ganglion*) are small masses of nervous tissue, consisting primarily of neuron cell bodies, that are located outside the brain and spinal cord. Ganglia are closely associated with cranial and spinal nerves. ***Enteric plexuses*** (PLEK-sus-ez) are extensive networks of neurons located in the walls of organs of the gastrointestinal tract. The neurons of these plexuses help regulate the digestive

Figure 9.1 Organization of the nervous system. (a) Subdivisions of the nervous system. (b) Nervous system organizational chart; blue boxes represent sensory components of the peripheral nervous system, red boxes represent motor components of the PNS, and green boxes represent effectors (muscles and glands).

The nervous system includes the brain, cranial nerves, spinal cord, spinal nerves, ganglia, enteric plexuses, and sensory receptors.

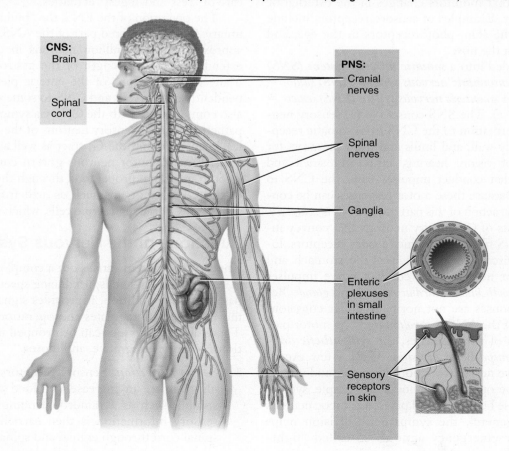

CNS:
Brain
Spinal cord

PNS:
Cranial nerves
Spinal nerves
Ganglia
Enteric plexuses in small intestine
Sensory receptors in skin

FIGURE 9.1 CONTINUES

Figure 9.1 CONTINUED

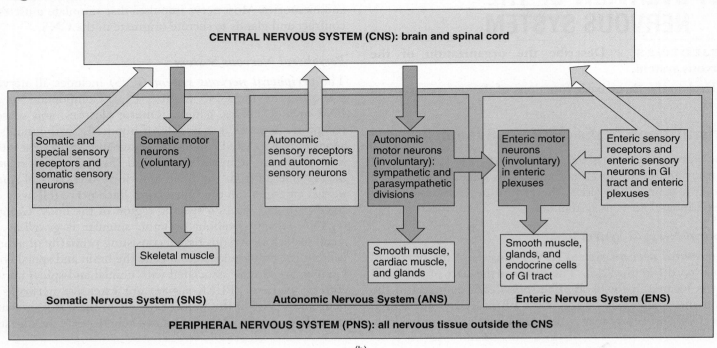

(b)

? What is the total number of cranial and spinal nerves in your body?

---

system. The term *sensory receptor* refers to a structure of the nervous system that monitors changes in the external or internal environment. Examples of sensory receptors include touch receptors in the skin, photoreceptors in the eye, and olfactory receptors in the nose.

The PNS is divided into a *somatic nervous system (SNS)* (*somat-* = body), an *autonomic nervous system (ANS)* (*auto-* = self; *-nomic* = law), and an *enteric nervous system (ENS)* (*enter-* = intestines) (Figure 9.1b). The SNS consists of (1) sensory neurons that convey information to the CNS from somatic receptors in the head, body wall, and limbs and from receptors for the special senses of vision, hearing, taste, and smell and (2) motor neurons that conduct impulses from the CNS to *skeletal muscles* only. Because these motor responses can be consciously controlled, the action of this part of the PNS is *voluntary*.

The ANS consists of (1) sensory neurons that convey information to the CNS from autonomic sensory receptors, located primarily in visceral organs such as the stomach and lungs, and (2) motor neurons that conduct nerve impulses from the CNS to *smooth muscle, cardiac muscle,* and *glands*. Because its motor responses are not normally under conscious control, the action of the ANS is *involuntary*. The motor part of the ANS consists of two branches, the *sympathetic division* and the *parasympathetic division*. With a few exceptions, effectors receive nerves from both divisions, and usually the two divisions have opposing actions. For example, sympathetic neurons increase heart rate, and parasympathetic neurons slow it down. In general, the sympathetic division helps support exercise or emergency actions, so-called "fight-or-flight" responses, and the parasympathetic division takes care of "rest-and-digest" activities.

The operation of the ENS, the "brain of the gut," is involuntary. Once considered part of the ANS, the ENS consists of approximately 100 million neurons in enteric plexuses that extend most of the length of the gastrointestinal (GI) tract. Many of the neurons of the enteric plexuses function independently of the ANS and CNS to some extent, although they also communicate with the CNS via sympathetic and parasympathetic neurons. Sensory neurons of the ENS monitor chemical changes within the GI tract as well as the stretching of its walls. Enteric motor neurons govern contraction of GI tract smooth muscle to propel food through the GI tract, secretions of the GI tract organs such as acid from the stomach, and activity of GI tract endocrine cells, which secrete hormones.

Functions of the Nervous System

The nervous system carries out a complex array of tasks, such as sensing various smells, producing speech, and remembering past events; in addition, it provides signals that control body movements, and regulates the operation of internal organs. These diverse activities can be grouped into three basic functions: sensory, integrative, and motor.

1. *Sensory function.* Sensory receptors *detect* internal stimuli, such as an increase in blood acidity, and external stimuli, such as a raindrop landing on your arm. This sensory information is then carried into the brain and spinal cord through cranial and spinal nerves.

2. *Integrative function.* The nervous system *integrates* (processes) sensory information by analyzing and storing some of it and by making decisions for appropriate responses—an activity called *integration*.

3. *Motor function.* Once sensory information is integrated, the nervous system may elicit an appropriate motor response by activating *effectors* (muscles and glands) through cranial and spinal nerves. Stimulation of the effectors causes muscles to contract and glands to secrete.

✓ CHECKPOINT

1. What is the purpose of a sensory receptor? An effector?
2. What are the components and functions of the SNS, ANS, and ENS?
3. Which subdivisions of the PNS control voluntary actions? Involuntary actions?

9.2 HISTOLOGY OF NERVOUS TISSUE

OBJECTIVES • **Contrast the histological characteristics and the functions of neurons and neuroglia.**
• **Distinguish between gray matter and white matter.**

Nervous tissue consists of two types of cells: neurons and neuroglia. Neurons provide most of the unique functions of the nervous system, such as sensing, thinking, remembering, controlling muscle activity, and regulating glandular secretions. Neuroglia support, nourish, and protect the neurons and maintain homeostasis in the interstitial fluid that bathes them.

Neurons

Like muscle cells, *neurons (nerve cells)* possess *electrical excitability*, the ability to respond to a stimulus and convert it into an action potential. A *stimulus* is any change in the environment that is strong enough to initiate an action potential. An *action potential* or *impulse* is an electrical signal that propagates (travels) along the surface of the membrane of a neuron or a muscle fiber.

Parts of a Neuron

Most neurons have three parts: (1) a cell body, (2) dendrites, and (3) an axon (Figure 9.2). The *cell body*, or *soma*, contains a nucleus surrounded by cytoplasm that includes typical organelles such as rough endoplasmic reticulum, lysosomes, mitochondria, and a Golgi complex. Most cellular molecules needed for a neuron's operation are synthesized in the cell body.

Two kinds of processes (extensions) emerge from the cell body of most neurons: multiple dendrites and a single axon. The cell body and the *dendrites* (= little trees) are the receiving or input parts of a neuron. Usually, dendrites are short, tapering, and highly branched, forming a tree-shaped array of processes that emerge from the cell body. The second type of

process, the *axon*, conducts nerve impulses toward another neuron, a muscle cell, or a gland cell. An axon is a long, cylindrical projection that often joins the cell body at a cone-shaped elevation called the *axon hillock* (= small hill). Nerve impulses usually arise at the axon hillock and then travel along the axon. Some axons have side branches called *axon collaterals*. The axon and axon collaterals end by dividing into many fine processes called *axon terminals*.

The site where two neurons or a neuron and an effector cell can communicate is termed a *synapse* (SIN-aps). The tips of most axon terminals swell into *synaptic end bulbs*. These bulb-shaped structures contain *synaptic vesicles*, tiny sacs that store chemicals called *neurotransmitters*. The neurotransmitter molecules released from synaptic vesicles are the means of communication at a chemical synapse.

Classification of Neurons

Both structural and functional features are used to classify the various neurons in the body.

STRUCTURAL CLASSIFICATION Structurally, neurons are classified according to the number of processes extending from the cell body (Figure 9.3):

■ *Multipolar neurons* usually have several dendrites and one axon (Figure 9.3a). Most neurons in the brain and spinal cord are of this type.

■ *Bipolar neurons* have one main dendrite and one axon (Figure 9.3b). They are found in the retina of the eye, in the inner ear, and in the olfactory (*olfact* = to smell) area of the brain.

■ *Unipolar neurons* have dendrites and one axon that are fused together to form a continuous process that emerges from the cell body (Figure 9.3c). These neurons begin in the embryo as bipolar neurons. During development, the dendrites and axon fuse together and become a single process. The dendrites of most unipolar neurons function as *sensory receptors* that detect a sensory stimulus such as touch, pressure, pain, or thermal stimuli. Nerve impulses in a unipolar neuron arise at the junction of the dendrites and axon. The impulses then propagate toward the synaptic end bulbs. The cell bodies of most unipolar neurons are located in the ganglia of spinal and cranial nerves.

FUNCTIONAL CLASSIFICATION Functionally, neurons are classified according to the direction in which the nerve impulse (action potential) is conveyed with respect to the CNS.

■ *Sensory* or *afferent neurons* (AF-er-ent NOO-ronz; *af-* = toward; *-ferrent* = carried) either contain sensory receptors at their distal ends (dendrites) or are located just after sensory receptors that are separate cells. Once an appropriate stimulus activates a sensory receptor, the sensory neuron forms an action potential in its axon and the action potential is conveyed *into* the CNS through cranial or spinal nerves. Most sensory neurons are unipolar in structure.

Figure 9.2 Structure of a typical multipolar neuron. Arrows indicate the direction of information flow: dendrites → cell body → axon → axon terminals → synaptic end bulbs.

 The basic parts of a neuron are several dendrites, a cell body, and a single axon.

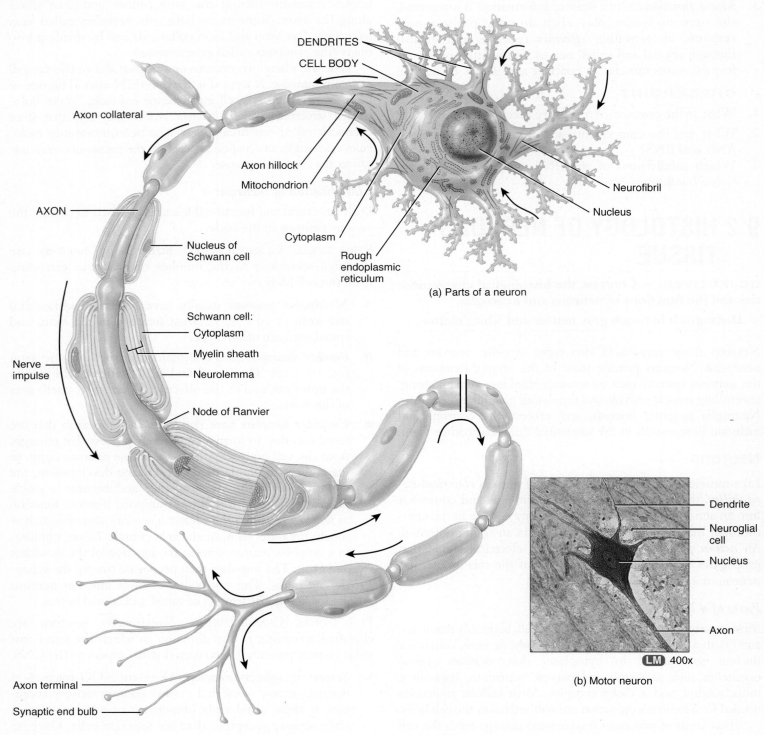

(a) Parts of a neuron

(b) Motor neuron

LM 400x

 What roles do the axon and axon terminals play in the communication of one neuron with another?

Figure 9.3 Structural classification of neurons. Breaks indicate that axons are longer than shown.

A multipolar neuron has many processes extending from the cell body, a bipolar neuron has two, and a unipolar neuron has one.

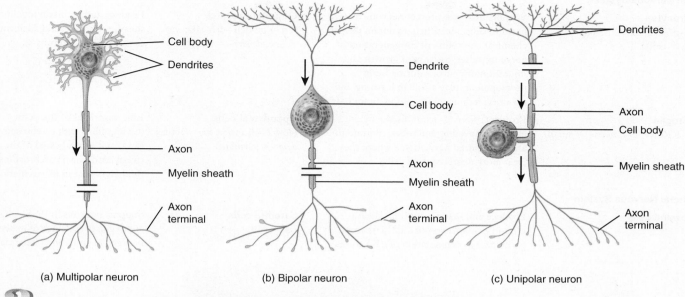

(a) Multipolar neuron

(b) Bipolar neuron

(c) Unipolar neuron

Which type of neuron shown in this figure is the most abundant type of neuron in the CNS?

- *Motor* or *efferent neurons* (EF-e-rent; *ef-* = away from) convey action potentials *away* from the CNS to *effectors* (muscles and glands) in the periphery (PNS) through cranial and spinal nerves. Most motor neurons are multipolar in structure.

- *Interneurons* or *association neurons* are located within the CNS between sensory and motor neurons. Interneurons integrate (process) incoming sensory information from sensory neurons and then elicit a motor response by activating the appropriate motor neurons. Most interneurons are multipolar in structure.

Neuroglia

Neuroglia (noo-RŌG-lē-a; *-glia* = glue) or *glia* make up about half the volume of the CNS. Their name derives from the idea of early histologists that they were the "glue" that held nervous tissue together. We now know that neuroglia are not merely passive bystanders but rather actively participate in the activities of nervous tissue. Generally, neuroglia are smaller than neurons, and they are 5 to 25 times more numerous. In contrast to neurons, glia do not generate or conduct nerve impulses, and they can multiply and divide in the mature nervous system. In cases of injury or disease, neuroglia multiply to fill in the spaces formerly occupied by neurons. Brain tumors derived from glia, called *gliomas*, tend to be highly malignant and to grow rapidly. Of the six types of neuroglia, four—astrocytes, oligodendrocytes, microglia, and ependymal cells—are found only in the CNS. The remaining two types—Schwann cells and satellite cells—are present in the PNS. Table 9.1 shows the appearance of neuroglia and lists their functions.

Myelination

The axons of most neurons are surrounded by a *myelin sheath*, a many-layered covering composed of lipid and protein (see Figure 9.2). Like insulation covering an electrical wire, the myelin sheath insulates the axon of a neuron and increases the speed of nerve impulse conduction. Recall that Schwann cells in the PNS and oligodendrocytes in the CNS produce myelin sheaths by wrapping themselves around and around axons. Eventually, as many as 100 layers cover the axon, much as multiple layers of paper cover the cardboard tube in a roll of toilet paper. Gaps in the myelin sheath, called *nodes of Ranvier* (RON-vē-ā), appear at intervals along the axon (see Figure 9.2). Axons with a myelin sheath are said to be *myelinated*, and those without it are said to be *unmyelinated*.

The amount of myelin increases from birth to maturity, and its presence greatly increases the speed of nerve impulse conduction. By the time a baby starts to talk, most myelin sheaths are partially formed, but myelination continues into the teenage years. An infant's responses to stimuli are neither as rapid nor as coordinated as those of an older child or an adult, in part because myelination is still in progress during infancy. Certain diseases, such as multiple sclerosis (see Common Disorders at the end of this chapter) and Tay-Sachs disease (see Section 3.4), destroy myelin sheaths.

TABLE 9.1

Neuroglia in the CNS and PNS

TYPE OF NEUROGLIAL CELL	FUNCTIONS	TYPE OF NEUROGLIAL CELL	FUNCTIONS
Central Nervous System			
Astrocytes (AS-trō-sītz; *astro-* = star; *-cyte* = cell)	Support neurons; protect neurons from harmful substances; help maintain proper chemical environment for generation of nerve impulses; assist with growth and migration of neurons during brain development; play a role in learning and memory; help form the blood–brain barrier	**Oligodendrocytes** (OL-i-gō-den′-drō- sītz; *oligo-* = few; *dendro-* = tree)	Produce and maintain myelin sheath around several adjacent axons of CNS neurons
Microglia (mī-KROG-lē-a; *micro-* = small)	Protect CNS cells from disease by engulfing invading microbes; migrate to areas of injured nerve tissue where they clear away debris of dead cells	**Ependymal cells** (ep-EN-di-mal; *epen-* = above; *dym-* = garment)	Line ventricles of the brain (cavities filled with cerebrospinal fluid) and central canal of the spinal cord; form cerebrospinal fluid and assist in its circulation
Peripheral Nervous System			
Schwann cells	Produce and maintain myelin sheath around a single axon of a PNS neuron; participate in regeneration of PNS axons	**Satellite cells** (SAT-i-līt)	Support neurons in PNS ganglia and regulate exchange of materials between neurons and interstitial fluid

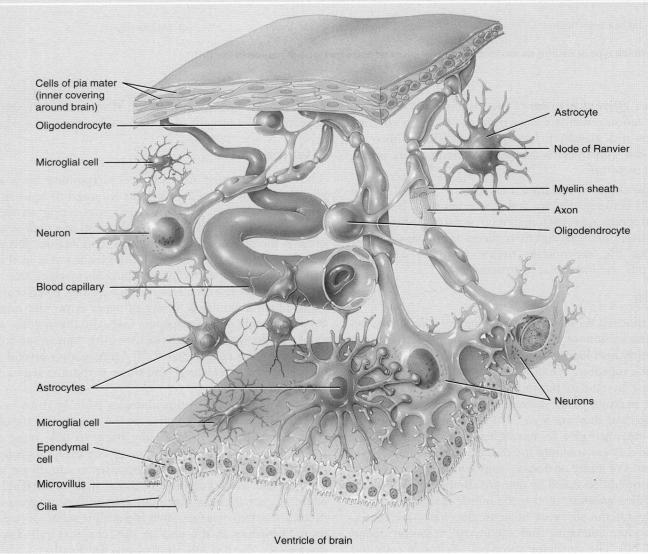

Cells of pia mater (inner covering around brain)

Oligodendrocyte

Microglial cell

Neuron

Blood capillary

Astrocytes

Microglial cell

Ependymal cell

Microvillus

Cilia

Astrocyte

Node of Ranvier

Myelin sheath

Axon

Oligodendrocyte

Neurons

Ventricle of brain

Collections of Nervous Tissue

The components of nervous tissue are grouped together in a variety of ways. Neuronal cell bodies are often grouped together in clusters. The axons of neurons are usually grouped together in bundles. In addition, widespread regions of nervous tissue are grouped together as either gray matter or white matter.

Clusters of Neuronal Cell Bodies

A **ganglion** (plural is *ganglia*) refers to a cluster of neuronal cell bodies located in the *PNS*. As mentioned earlier, ganglia are closely associated with cranial and spinal nerves. By contrast, a **nucleus** is a cluster of neuronal cell bodies located in the *CNS*.

Bundles of Axons

A **nerve** is a bundle of axons that is located in the *PNS*. Cranial nerves connect the brain to the periphery; spinal nerves connect the spinal cord to the periphery. A **tract** is a bundle of axons that is located in the *CNS*. Tracts interconnect neurons in the spinal cord and brain.

Gray and White Matter

In a freshly dissected section of the brain or spinal cord, some regions look white and glistening, and others appear gray. **White matter** is composed primarily of myelinated axons. The whitish color of myelin gives white matter its name. The **gray matter** of the nervous system contains neuronal cell bodies, dendrites, unmyelinated axons, axon terminals, and neuroglia. It appears grayish, rather than white, because the cellular organelles impart a gray color and there is little or no myelin in these areas. Blood vessels are present in both white and gray matter. In the spinal cord, the outer white matter surrounds an inner core of gray matter that, depending on how imaginative you are, is shaped like a butterfly or the letter H (see Figure 10.1). In the brain, a thin shell of gray matter (cortex) covers the surface of the largest portions of the brain, the cerebrum and cerebellum (see Figures 10.10 and 10.11).

CLINICAL CONNECTION | Neuron Regeneration

Human neurons have very limited powers of **regeneration**, the capability to replicate or repair themselves. In the PNS, axons and dendrites may undergo repair if the cell body is intact and if the Schwann cells are functional. The Schwann cells on either side of an injured site multiply by mitosis, grow toward each other, and may form a **regeneration tube** across the injured area. The tube guides axonal regrowth from the proximal area across the injured area into the distal area previously occupied by the original axon. Regrowth is slow, in part, because many needed materials must be transported from their sites of synthesis in the cell body several inches or feet down the axon to the growth region. New axons cannot grow if the gap becomes filled with scar tissue. In the CNS, a cut axon is usually not repaired even when the cell body remains intact. The inhibition of neuron regeneration in the brain and spinal cord seems to result from two factors: (1) inhibitory influences from neuroglia, particularly oligodendrocytes, and (2) absence of growth-stimulating cues that were present during fetal development. •

4. Give several examples of the structural and functional classifications of neurons.
5. What is the myelin sheath and why is it important?

9.3 ACTION POTENTIALS

OBJECTIVE • **Describe how a nerve impulse is generated and conducted.**

Neurons communicate with one another by means of nerve action potentials (nerve impulses). Recall from Chapter 8 that a muscle fiber (cell) contracts in response to a muscle action potential. The generation of action potentials in both muscle cells and neurons depends on two basic features of the plasma membrane: the existence of a resting membrane potential and the presence of specific types of ion channels. Body cells exhibit a **membrane potential**, a difference in the amount of electrical charge on the inside of the plasma membrane as compared to the outside. The membrane potential is like voltage stored in a battery. A cell that has a membrane potential is said to be **polarized**. When muscle cells and neurons are "at rest" (not conducting action potentials), the voltage across the plasma membrane is termed the **resting membrane potential**.

If you connect the positive and negative terminals of a battery with a piece of metal (look in the battery compartment of your wireless video game controller), an *electrical current* carried by electrons flows from the battery, allowing you to control the action in your favorite game. In living tissues, the flow of *ions* (rather than electrons) constitutes electrical currents. The main sites where ions can flow across the membrane are through the pores of various types of ion channels.

Ion Channels

When they are open, ion channels allow specific ions to diffuse across the plasma membrane from where the ions are more concentrated to where they are less concentrated. Similarly, positively charged ions will move toward a negatively charged area, and negatively charged ions will move toward a positively charged area. As ions diffuse across a plasma membrane to equalize differences in charge or concentration, the result is a flow of current that can change the membrane potential.

Two types of ion channels are leakage channels and gated channels. **Leak channels** allow a small but steady stream of ions to leak across the membrane. Because plasma membranes typically have many more potassium ion (K^+) leak channels than sodium ion (Na^+) leak channels, the membrane's permeability to K^+ is much higher than its permeability to Na^+. **Gated channels**, in contrast, open and close on command (see Figure 3.5). **Voltage-gated channels**—channels that open in response to a change in membrane potential—are used to generate and conduct action potentials.

Resting Membrane Potential

In a resting neuron, the outside surface of the plasma membrane has a positive charge and the inside surface has a negative charge. The separation of positive and negative electrical charges is a form of potential energy, which can be measured in volts. For example, two 1.5-volt batteries can power a portable CD player. Voltages produced by cells typically are much smaller and are measured in millivolts (1 millivolt [1 mV] = 1/1000 volt). In neurons, the resting membrane potential is about −70 mV. The minus sign indicates that the inside of the membrane is negative relative to the outside.

The resting membrane potential arises from the unequal distributions of various ions in cytosol and extracellular fluid (Figure 9.4). Extracellular fluid is rich in sodium ions (Na^+) and chloride ions (Cl^-). Inside cells, the main positively charged ions in the cytosol are potassium ions (K^+), and the two dominant negatively charged ions are phosphates attached to organic molecules (such as the three phosphates in ATP [adenosine triphosphate]), and amino acids in proteins. Because the concentration of K^+ is higher in cytosol and because plasma membranes have many K^+ leak channels, potassium ions diffuse down their concentration gradient—out of cells into the extracellular fluid. As more and more positive potassium ions exit, the inside of the membrane becomes increasingly negative, and the outside of the membrane becomes increasingly positive. Another factor contributes to the negativity inside: Most negatively charged ions inside the

cell are not free to leave. They cannot follow the K^+ out of the cell because they are attached either to large proteins or to other large molecules.

Membrane permeability to Na^+ is very low because there are only a few sodium leak channels. Nevertheless, sodium ions do slowly diffuse inward, down their concentration gradient. Left unchecked, such inward leakage of Na^+ would eventually destroy the resting membrane potential. The small inward Na^+ leak and outward K^+ leak are offset by the sodium–potassium pumps (see Figure 3.9). These pumps help maintain the resting membrane potential by pumping out Na^+ as fast as it leaks in. At the same time, the sodium–potassium pumps bring in K^+.

Generation of Action Potentials

An *action potential (AP)* or *impulse* is a sequence of rapidly occurring events that decrease and reverse the membrane potential and then eventually restore it to the resting state. If a stimulus causes the membrane to depolarize to a critical level, called *threshold* (typically, about −55 mV), then an action potential arises (Figure 9.5). An action potential has two main phases: a depolarizing phase (depolarization) and a repolarizing phase (repolarization). During the *depolarizing phase*, the negative membrane potential becomes less negative, reaches zero, and then becomes positive. Then, during the *repolarizing phase*, the membrane polarization is restored to its resting state of −70 mV. Following the repolarizing phase there may be an *after-hyperpolarizing phase*, also called *hyperpolarization*, during which the membrane potential temporarily becomes more negative than the resting level. In neurons, the depolarizing and repolarizing phases of an action potential typically last about one millisecond (1/1000 sec).

Figure 9.4 The distribution of ions that produces the resting membrane potential.

 The resting membrane potential is due to a small buildup of negatively charged ions, mainly organic phosphates (PO_4^{3-}) and proteins, in the cytosol just inside the membrane and an equal buildup of positively charged ions, mainly sodium ions (Na^+), in the interstitial fluid just outside the membrane.

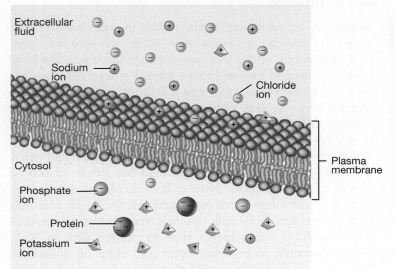

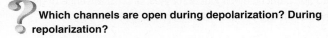 **What is a typical value for the resting membrane potential of a neuron?**

Figure 9.5 Action potential (AP). When a stimulus depolarizes the membrane to threshold, an action potential is generated.

An action potential consists of depolarizing and repolarizing phases.

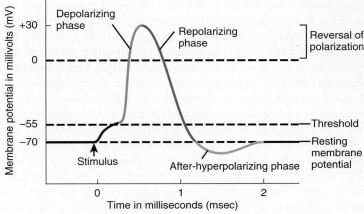

Which channels are open during depolarization? During repolarization?

During an action potential, depolarization to threshold briefly opens two types of voltage-gated ion channels. In neurons, these channels are present mainly in the plasma membrane of the axon and axon terminals. First, a threshold depolarization opens voltage-gated Na$^+$ channels. As these channels open, sodium ions rush into the cell, causing the depolarizing phase. The inflow of Na$^+$ causes the membrane potential to pass 0 mV and finally reach +30 mV (Figure 9.5). Second, the threshold depolarization also opens voltage-gated K$^+$ channels. The voltage-gated K$^+$ channels open more slowly, so their opening occurs at about the same time the voltage-gated Na$^+$ channels are automatically closing. As the K$^+$ channels open, potassium ions flow out of the cell, producing the repolarizing phase.

While the voltage-gated K$^+$ channels are open, outflow of K$^+$ may be large enough to cause an after-hyperpolarizing phase of the action potential (Figure 9.5). During hyperpolarization, the membrane potential becomes even *more negative* than the resting level. Finally, as K$^+$ channels close, the membrane potential returns to the resting level of −70 mV.

Action potentials arise according to the ***all-or-none principle***. As long as a stimulus is strong enough to cause depolarization to threshold, the voltage-gated Na$^+$ and K$^+$ channels open, and an action potential occurs. A much stronger stimulus cannot cause a larger action potential because the size of an action potential is always the same. A weak stimulus that fails to cause a threshold-level depolarization does not elicit an action potential. For a brief time after an action potential begins, a muscle fiber or neuron cannot generate another action potential. This time is called the *refractory period*.

Conduction of Nerve Impulses

To communicate information from one part of the body to another, nerve impulses must travel from where they arise, usually at the axon hillock, along the axon to the axon terminals (Figure 9.6). This mode of conduction is called ***propagation***, and it depends on positive feedback. Depolarization to threshold at the axon hillock opens voltage-gated Na$^+$ channels. The resulting

Figure 9.6 Conduction of a nerve impulse after it arises at the axon hillock. Dotted lines indicate ionic current flow.
(a) In continuous conduction along an unmyelinated axon, ionic currents flow across each adjacent portion of the plasma membrane.
(b) In saltatory conduction along a myelinated axon, the nerve impulse at the first node generates ionic currents in the cytosol and interstitial fluid that open voltage-gated Na$^+$ channels at the second node, and so on at each subsequent node.

Unmyelinated axons exhibit continuous conduction, and myelinated axons exhibit saltatory conduction.

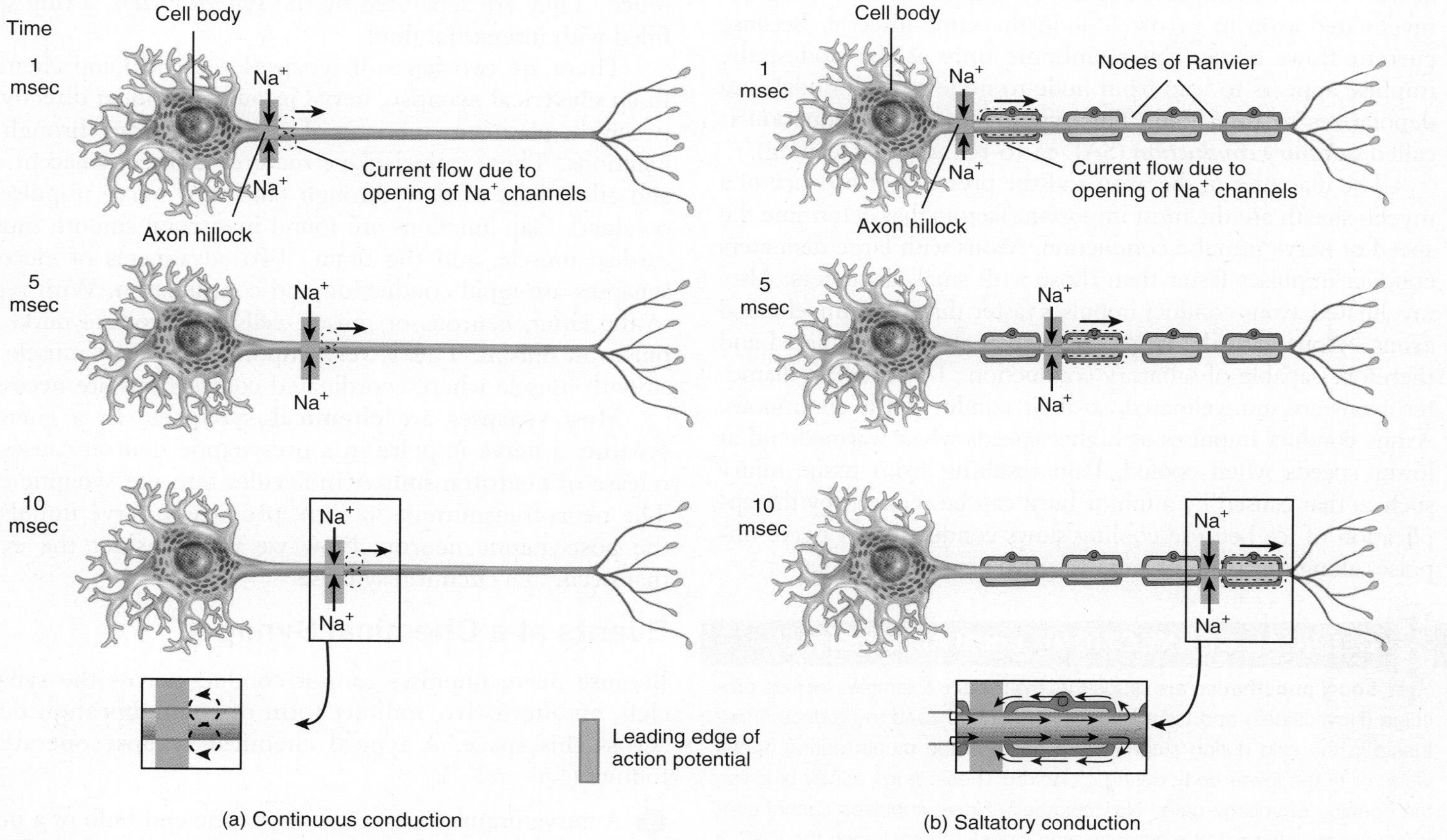

(a) Continuous conduction

(b) Saltatory conduction

What factors influence the speed of nerve impulse conduction?

inflow of sodium ions depolarizes the adjacent membrane to threshold, which opens even more voltage-gated Na$^+$ channels, a positive feedback effect. Thus, a nerve impulse self-conducts along the axon plasma membrane. This situation is similar to pushing on the first domino in a long row: When the push on the first domino is strong enough, that domino falls against the second domino, and eventually the entire row topples.

The type of action potential conduction that occurs in unmyelinated axons (and in muscle fibers) is called *continuous conduction*. In this case, each adjacent segment of the plasma membrane depolarizes to threshold and generates an action potential that depolarizes the next patch of the membrane (Figure 9.6a). Note that the impulse has traveled only a relatively short distance after 10 milliseconds (10 msec).

In myelinated axons, conduction is somewhat different. The voltage-gated Na$^+$ and K$^+$ channels are located primarily at the nodes of Ranvier, the gaps in the myelin sheath. When a nerve impulse conducts along a myelinated axon, current carried by Na$^+$ and K$^+$ flows through the interstitial fluid surrounding the myelin sheath and through the cytosol from one node to the next (Figure 9.6b). The nerve impulse at the first node generates ionic currents that open voltage-gated Na$^+$ channels at the second node and trigger a nerve impulse there. Then the nerve impulse from the second node generates an ionic current that opens voltage-gated Na$^+$ channels at the third node, and so on. Each node depolarizes and then repolarizes. Note the impulse has traveled much farther along the myelinated axon in Figure 9.6b in the same interval. Because current flows across the membrane only at the nodes, the impulse appears to leap from node to node as each nodal area depolarizes to threshold. This type of impulse conduction is called *saltatory conduction* (SAL-ta-tō-rē; *saltat-* = leaping).

The diameter of the axon and the presence or absence of a myelin sheath are the most important factors that determine the speed of nerve impulse conduction. Axons with large diameters conduct impulses faster than those with small diameters. Also, myelinated axons conduct impulses faster than do unmyelinated axons. Axons with the largest diameters are all myelinated and therefore capable of saltatory conduction. The smallest diameter axons are unmyelinated, so their conduction is continuous. Axons conduct impulses at higher speeds when warmed and at lower speeds when cooled. Pain resulting from tissue injury such as that caused by a minor burn can be reduced by the application of ice because cooling slows conduction of nerve impulses along the axons of pain-sensitive neurons.

CLINICAL CONNECTION | Local Anesthetics

Local anesthetics are drugs that block pain. Examples include procaine (Novocaine®) and Lidocaine, which may be used to produce anesthesia in the skin during suturing of a gash, in the mouth during dental work, or in the lower body during childbirth. These drugs act by blocking the opening of voltage-gated Na$^+$ channels. Nerve impulses cannot conduct past the obstructed region, so pain signals do not reach the CNS. •

✓ CHECKPOINT

6. What are the meanings of the terms resting membrane potential, depolarization, repolarization, nerve impulse, and refractory period?

7. How is saltatory conduction different from continuous conduction?

9.4 SYNAPTIC TRANSMISSION

OBJECTIVE • **Explain the events of chemical synaptic transmission and the types of neurotransmitters used.**

Now that you know how action potentials arise and conduct along the axon of an individual neuron, we will explore how neurons communicate with one another. At synapses, neurons communicate with other neurons or with effectors by a series of events known as *synaptic transmission*. In Chapter 8 we examined the events occurring at the neuromuscular junction, the synapse between a somatic motor neuron and a skeletal muscle fiber (see Figure 8.4). Synapses between neurons operate in a similar way. The neuron sending the signal is called the *presynaptic neuron* (*pre-* = before), and the neuron receiving the message is called the *postsynaptic neuron* (*post-* = after). Although the presynaptic and postsynaptic neurons are in close proximity at a synapse, their plasma membranes do not touch. They are separated by the *synaptic cleft*, a tiny space filled with interstitial fluid.

There are two types of synapses: electrical and chemical. In an **electrical synapse**, nerve impulses conduct directly between the plasma membranes of adjacent neurons through gap junctions. These tunnel-like structures connect adjacent cells and allow ions to flow through (and thus nerve impulses to conduct). Gap junctions are found in visceral smooth muscle, cardiac muscle, and the brain. Two advantages of electrical synapses are rapid conduction and coordination. With regard to the latter, neurons or muscle cells can produce nerve impulses in unison. This is very important in heart muscle and smooth muscle where coordinated contractions are necessary.

Most synapses are **chemical synapses**. In a chemical synapse, a nerve impulse in a presynaptic neuron causes the release of neurotransmitter molecules into the synaptic cleft. The neurotransmitters, in turn, produce a nerve impulse in the postsynaptic neuron. Now we will consider the events that occur at a chemical synapse.

Events at a Chemical Synapse

Because nerve impulses cannot conduct across the synaptic cleft, an alternative, indirect form of communication occurs across this space. A typical chemical synapse operates as follows (Figure 9.7):

1 A nerve impulse arrives at a synaptic end bulb of a presynaptic axon.

Figure 9.7 Signal transmission at a chemical synapse.

🔑 At a chemical synapse, a presynaptic neuron converts an electrical signal (nerve impulse) into a chemical signal (neurotransmitter release). The postsynaptic neuron then converts the chemical signal back into an electrical signal (nerve impulse).

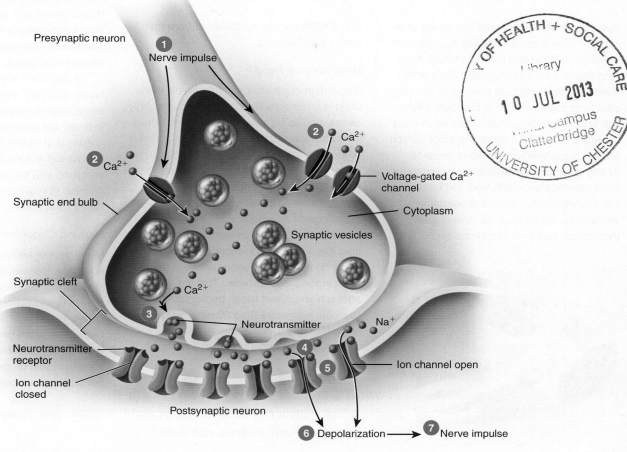

❓ Why may electrical synapses work in two directions, but chemical synapses can transmit a signal in only one direction?

2 The depolarizing phase of the nerve impulse opens *voltage-gated Ca^{2+} channels*, which are present in the membrane of synaptic end bulbs. Because calcium ions are more concentrated in the interstitial fluid, Ca^{2+} flows into the synaptic end bulb through the opened channels.

3 An increase in the concentration of Ca^{2+} inside the synaptic end bulb triggers exocytosis of some of the synaptic vesicles, which releases thousands of neurotransmitter molecules into the synaptic cleft.

4 The neurotransmitter molecules diffuse across the synaptic cleft and bind to **neurotransmitter receptors** in the postsynaptic neuron's plasma membrane.

5 Binding of neurotransmitter molecules opens ion channels, which allows certain ions to flow across the membrane.

6 As ions flow through the opened channels, the voltage across the membrane changes. Depending on which ions the channels admit, the voltage change may be a depolarization or a hyperpolarization.

7 If a depolarization occurs in the postsynaptic neuron and reaches threshold, then it triggers one or more nerve impulses.

At chemical synapses, only *one-way information transfer* can occur—from a presynaptic neuron to a postsynaptic neuron or to an effector, such as a muscle fiber or a gland cell. For example, synaptic transmission at a neuromuscular junction (NMJ) proceeds from a somatic motor neuron to a skeletal muscle fiber (but not in the opposite direction). Only synaptic end bulbs of presynaptic neurons can release neurotransmitters, and only the postsynaptic neuron's membrane has

Neurotransmitters—Why Food Could Affect Your Mood

Everyone who has enjoyed the sense of relaxation after a good meal has experienced the effect of food can have on mood. Neurons manufacture neurotransmitters from chemicals that originate from food, so you could say that the story of the food–mood link begins with digestion. Many neurotransmitters are derived from amino acids, which are the basic building blocks of proteins. Amino acids are made available when your body digests the protein from the food you eat. For example, the neurotransmitter serotonin is made from the amino acid tryptophan, and both dopamine and norepinephrine are synthesized from the amino acid tyrosine.

Mind-altering Food?

Regulation of neurotransmitter levels in the brain is quite complex and depends not only on the availability of amino acid (and other) precursors, but also on competition of these precursors for entry into the brain. Serotonin, is an important neurotransmitter that appears to have an important effect on mood. Elevated levels of serotonin can lead to feelings of relaxation and sleepiness.

Although serotonin is manufactured from the amino acid tryptophan, high protein foods do not lead to higher levels of tryptophan in the blood or brain. This is because, after a high-protein meal, tryptophan has to compete with more than 20 other amino acids for entry into the central nervous system, so its concentration in the brain remains relatively low. On the other hand, consumption of carbohydrate-rich foods, such as bread, pasta, potatoes, or sweets, is associated with an increase in the synthesis and release of serotonin in the brain. The result: Carbohydrates help us feel relaxed and sleepy.

If you don't want to fall asleep after lunch, and stay alert, it may be wise to avoid large amounts of carbohydrates. Instead, eat proteins and vegetables, and maybe skip dessert. The higher levels of protein will reduce serotonin production, and keep you focused and alert. A little caffeine (if you can tolerate it) might not be a bad idea, either.

High on Chocolate

Many people believe that chocolate can be addictive. Research does suggest that the ingestion of chocolate can stimulates the brain's opioid production, which in turn raises the level of dopamine in the brain. Dopamine is the neurotransmitter that gives feelings of pleasure, and reinforces addictions to opioid drugs like morphine and heroin (and to foods like chocolate).

Think It Over . . . **Why do you think that consuming a high-protein diet for several days could lead to cravings for carbohydrate-rich foods?**

the correct receptor proteins to recognize and bind that neurotransmitter. As a result, nerve impulses move along their pathways in one direction.

When a postsynaptic neuron depolarizes, the effect is excitatory: If threshold is reached, one or more nerve impulses occur. By contrast, hyperpolarization has an inhibitory effect on the postsynaptic neuron: As the membrane potential moves farther away from threshold, nerve impulses are less likely to arise. A typical neuron in the CNS receives input from 1000 to 10,000 synapses. Some of this input is excitatory and some is inhibitory. The sum of all the excitatory and inhibitory effects at any given time determines whether one or more impulses will occur in the postsynaptic neuron.

A neurotransmitter affects the postsynaptic neuron, muscle fiber, or gland cell as long as it remains bound to its receptors. Thus, removal of the neurotransmitter is essential for normal synaptic function. Neurotransmitter is removed in three ways.

(1) Some of the released neurotransmitter molecules diffuse away from the synaptic cleft. Once a neurotransmitter molecule is out of reach of its receptors, it can no longer exert an effect. (2) Some neurotransmitters are destroyed by enzymes. (3) Many neurotransmitters are actively transported back into the neuron that released them (reuptake). Others are transported into neighboring neuroglia (uptake).

CLINICAL CONNECTION | Selective Serotonin Reuptake Inhibitor (SSRI)

Several therapeutically important drugs selectively block reuptake of specific neurotransmitters. For example, the drug fluoxetine (Prozac®) is a **selective serotonin reuptake inhibitor (SSRI)**. By blocking reuptake of serotonin, Prozac prolongs the activity of this neurotransmitter at synapses in the brain. SSRIs provide relief for those suffering from some forms of depression. •

Neurotransmitters

About 100 substances are either known or suspected neurotransmitters. Most neurotransmitters are synthesized and loaded into synaptic vesicles in the synaptic end bulbs, close to their site of release. One of the best-studied neurotransmitters is *acetylcholine (ACh)*, which is released by many PNS neurons and by some CNS neurons. ACh is an excitatory neurotransmitter at some synapses, such as the neuromuscular junction. It is also known to be an inhibitory neurotransmitter at other synapses. For example, parasympathetic neurons slow heart rate by releasing ACh at inhibitory synapses.

Several amino acids are neurotransmitters in the CNS. *Glutamate* and *aspartate* have powerful excitatory effects. Two other amino acids, *gamma aminobutyric acid (GABA)* (GAM-ma am-i-nō-bū-TIR-ik) and *glycine*, are important inhibitory neurotransmitters. Antianxiety drugs such as diazepam (Valium®) enhance the action of GABA.

Some neurotransmitters are modified amino acids. These include norepinephrine, dopamine, and serotonin. *Norepinephrine (NE)* plays roles in arousal (awakening from deep sleep), dreaming, and regulating mood. Brain neurons containing the neurotransmitter *dopamine (DA)* are active during emotional responses, addictive behaviors, and pleasurable experiences. In addition, dopamine-releasing neurons help regulate skeletal muscle tone and some aspects of movement due to contraction of skeletal muscles. One form of schizophrenia is due to accumulation of excess dopamine. *Serotonin* is thought to be involved in sensory perception, temperature regulation, control of mood, appetite, and the onset of sleep.

Neurotransmitters consisting of amino acids linked by peptide bonds are called *neuropeptides* (noor-ō-PEP-tīds). The *endorphins* (en-DOR-fins) are neuropeptides that are the body's natural painkillers. Acupuncture may produce analgesia (loss of pain sensation) by increasing the release of endorphins. They have also been linked to improved memory and learning and to feelings of pleasure or euphoria.

An important newcomer to the ranks of recognized neurotransmitters is the simple gas *nitric oxide (NO)*, which is different from all previously known neurotransmitters because it is not synthesized in advance and packaged into synaptic vesicles. Rather, it is formed on demand, diffuses out of cells that produce it and into neighboring cells, and acts immediately. Some research suggests that NO plays a role in learning and memory.

Carbon monoxide (CO), like NO, is not produced in advance and packaged into synaptic vesicles. It too is formed as needed and diffuses out of cells that produce it into adjacent cells. CO is an excitatory neurotransmitter produced in the brain and in response to some neuromuscular and neuroglandular functions. CO might protect against excess neuronal activity and might be related to dilation of blood vessels, memory, olfaction (sense of smell), vision, thermoregulation, insulin release, and anti-inflammatory activity.

CLINICAL CONNECTION | Modifying the Effects of Neurotransmitters

Substances naturally present in the body as well as drugs and toxins can **modify the effects of neurotransmitters** in several ways. Cocaine produces euphoria—intensely pleasurable feelings—by blocking reuptake of dopamine. This action allows dopamine to linger longer in synaptic clefts, producing excessive stimulation of certain brain regions. Isoproterenol (Isuprel®) can be used to dilate the airways during an asthma attack because it binds to and activates receptors for norepinephrine. Zyprexa®, a drug prescribed for schizophrenia, is effective because it binds to and blocks receptors for serotonin and dopamine. •

✓ CHECKPOINT

8. How do electrical and chemical synapses differ?

9. How are neurotransmitters removed after they are released from synaptic vesicles?

COMMON DISORDERS

Multiple Sclerosis

Multiple sclerosis (MS) is a disease that causes progressive destruction of myelin sheaths of neurons in the CNS. It afflicts about 2 million people worldwide and affects females twice as often as males. The condition's name describes the anatomical pathology: In *multiple* regions, the myelin sheaths deteriorate to *scleroses*, which are hardened scars or plaques. The destruction of myelin sheaths slows and then short-circuits conduction of nerve impulses.

The most common form of the condition is relapsing–remitting MS, which usually appears in early adulthood. The first symptoms may include a feeling of heaviness or weakness in the muscles, abnormal sensations, or double vision. An attack is followed by a period of remission during which the symptoms temporarily disappear. One attack follows another over the years. The result is a progressive loss of function interspersed with remission periods, during which symptoms abate.

MS is an autoimmune disease—the body's own immune system spearheads the attack. Although the trigger of MS is unknown, both genetic susceptibility and exposure to some environmental factor (perhaps a herpes virus) appear to contribute. Many patients with relapsing–remitting MS are treated with injections of beta interferon. This treatment lengthens the time between relapses, decreases the severity of relapses, and slows formation of new lesions in some cases. Unfortunately, not all MS patients can tolerate beta interferon, and therapy becomes less effective as the disease progresses.

Epilepsy

Epilepsy is a disorder characterized by short, recurrent, periodic attacks of motor, sensory, or psychological malfunction, although it almost never affects intelligence. The attacks, called *epileptic seizures*, afflict about 1% of the world's population. They are initiated by abnormal, synchronous electrical discharges from millions of neurons in the brain. As a result, lights, noise, or smells may be sensed when the eyes, ears, and nose have not been stimulated. In addition, the skeletal muscles of a person having a seizure may contract involuntarily. *Partial seizures* begin in a small area called a *focus* on one side of the brain and produce milder symptoms; *generalized seizures* involve larger areas on both sides of the brain and loss of consciousness.

Epilepsy has many causes, including brain damage at birth (the most common cause); metabolic disturbances such as insufficient glucose or oxygen in the blood; infections; toxins; loss of blood or low blood pressure; head injuries; and tumors and abscesses of the brain. However, most epileptic seizures have no demonstrable cause.

Epileptic seizures often can be eliminated or alleviated by antiepileptic drugs, such as phenytoin, carbamazepine, and valproate sodium. An implantable device that stimulates the vagus (X) nerve also has produced dramatic results in reducing seizures in some patients whose epilepsy was not well-controlled by drugs.

MEDICAL TERMINOLOGY AND CONDITIONS

Demyelination (dē-mī-e-li-NĀ-shun) Loss or destruction of myelin sheaths around axons in the CNS or PNS.

Guillain-Barré Syndrome (GBS) (gē-an ba-RĀ) A demyelinating disorder in which macrophages remove myelin from PNS axons. It is a common cause of sudden paralysis and may result from the immune system's response to a bacterial infection. Most patients recover completely or partially, but about 15% remain paralyzed.

Neuroblastoma (noor-ō-blas-TŌ-ma) A malignant tumor that consists of immature nerve cells (neuroblasts); occurs most commonly in the abdomen and most frequently in the adrenal glands. Although rare, it is the most common tumor in infants.

Neuropathy (noo-ROP-a-thē; *neuro-* = a nerve; *-pathy* = disease) Any disorder that affects the nervous system but particularly a disorder of a cranial or spinal nerve. An example is *facial neuropathy* (Bell's palsy), a disorder of the facial (VII) nerve.

Rabies (RĀ-bēz; *rabi-* = mad, raving) A fatal disease caused by a virus that reaches the CNS via fast axonal transport. It is usually transmitted by the bite of an infected dog or other meat-eating animal. The symptoms are excitement, aggressiveness, and madness, followed by paralysis and death.

CHAPTER REVIEW AND RESOURCE SUMMARY

WILEY **PLUS**

REVIEW	**RESOURCES**
### 9.1 Organization of the Nervous System	

REVIEW

9.1 Organization of the Nervous System

1. The **central nervous system (CNS)** consists of the **brain** and **spinal cord**. The **peripheral nervous system (PNS)** consists of all nervous tissue outside the CNS.

2. Components of the PNS include the **somatic nervous system (SNS)**, **autonomic nervous system (ANS)**, and **enteric nervous system (ENS)**.

3. The SNS consists of sensory neurons that conduct impulses from somatic and special sense receptors to the CNS and motor neurons from the CNS to skeletal muscles.

4. The ANS contains sensory neurons from visceral organs and motor neurons that convey impulses from the CNS to smooth muscle tissue, cardiac muscle tissue, and glands.

5. The ENS consists of neurons in enteric plexuses in the gastrointestinal (GI) tract that function somewhat independently of the ANS and CNS. The ENS monitors sensory changes in and controls operation of the GI tract.

6. Three basic functions of the nervous system are detecting stimuli (sensory function); analyzing, integrating, and storing sensory information (integrative function); and responding to integrative decisions (motor function).

RESOURCES

Anatomy Overview—Nervous System

Animation—Introduction to Structure and Function of the Nervous System

REVIEW	**RESOURCES**

9.2 Histology of Nervous Tissue

1. Nervous tissue consists of two types of cells: neurons and neuroglia. Neurons are cells specialized for nerve impulse conduction and provide most of the unique functions of the nervous system, such as sensing, thinking, remembering, controlling muscle activity, and regulating glandular secretions. Neuroglia support, nourish, and protect the neurons and maintain homeostasis in the interstitial fluid that bathes neurons.

2. Most **neurons** have three parts. The **dendrites** are the main receiving or input region. Integration occurs in the **cell body**. The output part typically is a single **axon**, which conducts nerve impulses toward another neuron, a muscle fiber, or a gland cell.

3. On the basis of their structure, neurons are classified as **multipolar**, **bipolar**, or **unipolar**.

4. Neurons are functionally classified as **sensory (afferent) neurons**, **motor (efferent) neurons**, and **interneurons**. Sensory neurons carry sensory information into the CNS. Motor neurons carry information out of the CNS to **effectors** (muscles and glands). Interneurons are located within the CNS between sensory and motor neurons.

5. **Neuroglia** support, nurture, and protect neurons and maintain the interstitial fluid that bathes them. Neuroglia in the CNS include **astrocytes**, **oligodendrocytes**, **microglia**, and **ependymal cells**. Neuroglia in the PNS include **Schwann cells** and **satellite cells**.

6. Two types of neuroglia produce **myelin sheaths**: Oligodendrocytes myelinate axons in the CNS, and Schwann cells myelinate axons in the PNS.

7. **White matter** consists of aggregates of myelinated axons: **gray matter** contains cell bodies, dendrites, and axon terminals of neurons, unmyelinated axons, and neuroglia.

8. In the spinal cord, gray matter forms an H-shaped inner core that is surrounded by white matter. In the brain, a thin, superficial shell of gray matter covers the cerebrum and cerebellum.

Anatomy Overview—Nervous Tissue
Animation—Neuron Structure and Function

Figure 9.2 Structure of a typical multipolar neuron

Exercise—Paint a Neuron

9.3 Action Potentials

1. Neurons communicate with one another using nerve action potentials, also called nerve impulses.

2. Generation of action potentials depends on the existence of a **membrane potential** and the presence of **voltage-gated channels** for Na^+ and K^+.

3. A typical value for the **resting membrane potential** (difference in electrical charge across the plasma membrane) is -70 mV. A cell that exhibits a membrane potential is **polarized**.

4. The resting membrane potential arises due to an unequal distribution of ions on either side of the plasma membrane and a higher membrane permeability to K^+ than to Na^+. The level of K^+ is higher inside and the level of Na^+ is higher outside, a situation that is maintained by sodium–potassium pumps.

5. During an **action potential**, voltage-gated Na^+ and K^+ channels open in sequence. Opening of voltage-gated Na^+ channels results in **depolarization**, the loss and then reversal of membrane polarization (from -70 mV to $+30$ mV). Then, opening of voltage-gated K^+ channels allows **repolarization**, recovery of the membrane potential to the resting level.

6. According to the **all-or-none principle**, if a stimulus is strong enough to generate an action potential, the impulse generated is of a constant size.

7. During the refractory period, another action potential cannot be generated.

8. Nerve impulse conduction that occurs as a step-by-step process along an unmyelinated axon is called **continuous conduction**. In **saltatory conduction**, a nerve impulse "leaps" from one node of Ranvier to the next along a myelinated axon.

9. Axons with larger diameters conduct impulses faster than those with smaller diameters; myelinated axons conduct impulses faster than unmyelinated axons.

Animation—Membrane Potentials
Animation—Propagation of Nerve Impulses
Animation—Action Potentials
Animation: Factors That Affect Conduction Rates—Myelination
Exercise—Ruling the Gated Channels
Exercise—Keep the Resting Potential
Exercise—Nervous Race
Concepts and Connections—Membrane Potential

REVIEW | **RESOURCES**

9.4 Synaptic Transmission

1. Neurons communicate with other neurons and with effectors at synapses in a series of events known as **synaptic transmission**.

2. At a synapse, a neurotransmitter is released from a **presynaptic neuron** into the synaptic cleft and then binds to receptors on the plasma membrane of the **postsynaptic neuron**.

3. An excitatory neurotransmitter depolarizes the postsynaptic neuron's membrane, brings the membrane potential closer to threshold, and increases the chance that one or more action potentials will arise. An inhibitory neurotransmitter hyperpolarizes the membrane of the postsynaptic neuron, thereby inhibiting action potential generation.

4. Neurotransmitter is removed in three ways: diffusion, enzymatic destruction, and reuptake by neurons or neuroglia.

5. Important neurotransmitters include **acetylcholine**, **glutamate**, **aspartate**, **gamma aminobutyric acid (GABA)**, **glycine**, **norepinephrine**, **dopamine**, **serotonin**, **neuropeptides**, and **nitric oxide**.

Anatomy Overview—
Neurotransmitters
Animation: Events at the Synapse

Figure 9.7 Signal transmission at a chemical synapse

Concepts and Connections—Synaptic Summation

SELF-QUIZ

1. Which of the following are incorrectly matched?
   a. central nervous system: composed of the brain and spinal cord
   b. somatic nervous system: includes motor neurons to skeletal muscles
   c. sympathetic nervous system: includes motor neurons to smooth and cardiac muscles
   d. peripheral nervous system: includes cranial and spinal nerves
   e. autonomic nervous system: includes parasympathetic, sympathetic, and enteric divisions

2. Match the following:
   ____ a. consists of one fused dendrite and axon
   ____ b. found primarily in the sensory organs such as the eye and ear
   ____ c. most common neuron in the CNS
   ____ d. contains several dendrites and one axon
   ____ e. dendrite functions as a sensory receptor for touch, temperature and pain
   ____ f. consists of one dendrite and one axon

   A. multipolar neuron
   B. bipolar neuron
   C. unipolar neuron

3. The release of neurotransmitters from the synaptic vesicles is dependent on the presence of ____ in the synaptic end bulb.
   a. glucose   b. calcium   c. proteins
   d. sodium   e. potassium

4. The type of cell that produces myelin sheaths around axons in the CNS is the
   a. astrocyte.   b. ependymal cell.   c. Schwann cell.
   d. oligodendrocyte.   e. microglia.

5. White matter in the CNS is a
   a. tract.   b. nucleus.   c. nerve.
   d. ganglion.   e. group of cell bodies.

6. Which of the following is NOT true concerning the repair of nervous tissue?
   a. If the cell body is not damaged, neurons in the CNS may be able to repair themselves.
   b. In the CNS, inhibition by neuroglia influences regeneration.
   c. Injury to the PNS is permanent if scar tissue forms in the axonal gap.
   d. Active Schwann cells contribute to the repair process in the PNS.
   e. In the PNS, a regeneration tube guides neuronal repair.

7. In a resting neuron
   a. there is a high concentration of K^+ outside the cell.
   b. negatively-charged ions move freely through the plasma membrane.
   c. the sodium–potassium pumps help maintain the low concentration of Na^+ inside the cell.
   d. the outside surface of the plasma membrane has a negative charge.
   e. the plasma membrane is highly permeable to Na^+.

8. In the depolarization phase of a nerve impulse, voltage-gated ion channels open, resulting in a(n)
   a. rush of Na^+ into the neuron.
   b. rush of Na^+ out of the neuron.
   c. equal exchange of Na^+ and K^+.
   d. rush of K^+ out of the neuron.
   e. pumping of K^+ into the neuron.

9. If a stimulus is strong enough to generate an action potential, the impulse generated is of a constant size. A stronger stimulus cannot generate a larger impulse. This is known as

a. the principle of polarization–depolarization.

b. saltatory conduction. **c.** the all-or-none principle.

d. continuous conduction. **e.** the refractory period.

10. Place the following events in the correct order of occurrence:

   1. Voltage-gated Na^+ channels open and permit Na^+ to rush inside the neuron.

   2. The Na^+/K^+ pump restores the ions to their original sites.

   3. A stimulus of threshold strength is applied to the neuron.

   4. The membrane polarization changes from negative (–55 mV) to positive (+30 mV).

   5. Voltage-gated K^+ channels open, and K^+ flows out of the neurons.

      a. 4, 1, 2, 3, 5   **b.** 4, 3, 1, 2, 5   **c.** 3, 1, 4, 2, 5

      d. 5, 3, 1, 4, 2   **e.** 3, 1, 4, 5, 2

11. Saltatory conduction occurs

   a. in unmyelinated axons.

   b. at the nodes of Ranvier.

   c. in the smallest diameter axons.

   d. in skeletal muscle fibers.

   e. in cardiac muscle fibers.

12. The speed of nerve impulse conduction is increased by

   a. cold.   **b.** a very strong stimulus.

   c. small diameter of the axon.   **d.** myelination.

   e. astrocytes.

13. For a signal to be transmitted by means of a chemical synapse from a presynaptic neuron to a postsynaptic neuron,

   a. the presynaptic neuron must be touching the postsynaptic neuron.

   b. the postsynaptic neuron must contain neurotransmitter receptors.

   c. there must be gap junctions present between the two neurons.

   d. the postsynaptic neuron needs to release neurotransmitters from its synaptic vesicles.

   e. the neurons must be myelinated.

14. What would happen at the postsynaptic neuron if the total inhibitory effects of the neurotransmitters were greater than the total excitatory effects?

   a. A nerve impulse would be generated.

   b. It would be easier to generate a nerve impulse when the next stimulus was received.

   c. The nerve impulse would be rerouted to another neuron.

   d. No nerve impulse would be generated.

   e. The neurotransmitter would be broken down more quickly.

15. Match the following neurotransmitters with their descriptions:

   ____ **a.** inhibitory amino acid in the CNS

   ____ **b.** a gaseous neurotransmitter that is not packaged into synaptic vesicles

   ____ **c.** excitatory amino acid in the CNS

   ____ **d.** body's natural painkillers

   ____ **e.** helps regulate mood and sleep

   ____ **f.** neurotransmitter that activates skeletal muscle fibers

   A. serotonin
   B. acetylcholine
   C. endorphins
   D. GABA
   E. nitric oxide
   F. glutamate

16. Match the following:

   ____ **a.** the portion of a neuron containing the nucleus

   ____ **b.** rounded structure at the distal end of an axon terminal

   ____ **c.** highly branched, input part of a neuron

   ____ **d.** sac in which neurotransmitter is stored

   ____ **e.** neuron located entirely within the CNS

   ____ **f.** long, cylindrical process that conducts impulses toward another neuron

   ____ **g.** produces myelin sheath in PNS

   ____ **h.** unmyelinated gap in the myelin sheath

   ____ **i.** substance that increases the speed of nerve impulse conduction

   ____ **j.** neuron that conveys information from a receptor to the CNS

   ____ **k.** neuron that conveys information from the CNS to an effector

   ____ **l.** bundle of many axons in the PNS

   ____ **m.** bundle of many axons in the CNS

   ____ **n.** group of cell bodies in the PNS

   ____ **o.** group of cell bodies in the CNS

   ____ **p.** substance used for communication at chemical synapses

   A. synaptic end bulb
   B. motor neuron
   C. sensory neuron
   D. dendrite
   E. interneuron
   F. nucleus
   G. myelin sheath
   H. Schwann cell
   I. cell body
   J. node of Ranvier
   K. ganglion
   L. nerve
   M. neurotransmitter
   N. tract
   O. synaptic vesicle
   P. axon

CRITICAL THINKING APPLICATIONS

1. The buzzing of the alarm clock awoke Rachel. She stretched, yawned, and started to salivate as she smelled the brewing coffee. List the divisions of the nervous system that are involved in each of these activities.

2. Prior to surgery, James was given a curare-like drug that temporarily "paralyzed" his muscles so that he could be more easily intubated and would not move during surgery. What is the neurotransmitter involved and how do you think the drug prevents skeletal muscle contraction?

3. Angela really looks forward to the great feeling she has after going for a nice long run on the weekends. By the end of her run, she doesn't even feel the pain in her sore feet. Angela read in a magazine that some kind of natural brain chemical was responsible for the "runner's high" that she feels. Are there such chemicals in Angela's brain?

4. The pediatrician was trying to educate the anxious new parents of a six-month-old baby. "No, don't worry about him not walking yet. The myelination of the baby's nervous system is not finished yet." Explain what the pediatrician means by this reassurance.

ANSWERS TO FIGURE QUESTIONS

9.1 The total number of cranial and spinal nerves in your body is $(12 \times 2) + (31 \times 2) = 86$.

9.2 The axon conducts nerve impulses and transmits the message to another neuron or effector cell by releasing a neurotransmitter at its axon terminals.

9.3 Most neurons in the CNS are multipolar neurons.

9.4 A typical value for the resting membrane potential in a neuron is -70 mV.

9.5 Voltage-gated Na^+ channels are open during the depolarizing phase, and voltage-gated K^+ channels are open during the repolarizing phase of an action potential.

9.6 The diameter of an axon, presence or absence of a myelin sheath, and temperature influence the speed of nerve impulse conduction.

9.7 In electrical synapses (gap junctions), ions may flow equally well in either direction, so either neuron may be the presynaptic one. At a chemical synapse, a presynaptic neuron releases neurotransmitter and the postsynaptic neuron has receptors that bind this chemical. Thus, the signal can proceed in only one direction.

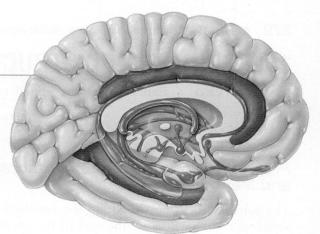

CHAPTER 10

CENTRAL NERVOUS SYSTEM, SPINAL NERVES, AND CRANIAL NERVES

Now that you understand how the nervous system functions on the cellular level, in this chapter we will explore the structure and functions of the *central nervous system (CNS)*, which consists of the brain and spinal cord. We will also examine spinal nerves and cranial nerves, which are part of the *peripheral nervous system (PNS)* (see Figure 9.1).

The spinal cord and its associated spinal nerves contain neural pathways that control some of your most rapid reactions to environmental changes. If you pick up something hot, the grasping muscles may relax and you may drop the hot object even before you are consciously aware of the extreme heat or pain. This is an example of a spinal cord reflex—a quick, automatic response to certain kinds of stimuli that involves neurons only in the spinal nerves and spinal cord. The white matter of the spinal cord contains a dozen major sensory and motor pathways, which function as the "highways" along which sensory input travels to the brain and motor output travels from the brain to skeletal muscles and other effectors. Recall that the spinal cord is continuous with the brain and that together they make up the CNS.

The brain is the control center for registering sensations, correlating them with one another and with stored information, making decisions, and taking actions. It also is the center for the intellect, emotions, behavior, and memory. But the brain encompasses yet a larger domain: It directs our behavior toward others. With ideas that excite, artistry that dazzles, or rhetoric that mesmerizes, one person's thoughts and actions may influence and shape the lives of many others.

Did you know? People around the world have enjoyed caffeine since the beginning of history. Caffeine occurs naturally in coffee beans and tea leaves, and in many other plants. Caffeine is also added to energy drinks, soft drinks, and over-the-counter drugs. Caffeine acts directly on the CNS, reducing feelings of fatigue and increasing alertness and anxiety. Scientists believe the effects of caffeine on the CNS result from the way caffeine changes the action of certain neurotransmitters in the brain. Caffeine readily crosses the blood–brain barrier, and binds with certain postsynaptic receptors, thus altering the communication of neurons in the brain.

FOCUS ON WELLNESS
The Many Faces of Caffeine

LOOKING BACK TO MOVE AHEAD...

Skull and Hyoid Bone (Section 6.7)

Vertebral Column (Section 6.8)

Structures of the Nervous System (Section 9.1)

Structure of a Neuron (Section 9.2)

Gray and White Matter (Section 9.2)

10.1 SPINAL CORD STRUCTURE

OBJECTIVES • **Explain how the spinal cord is protected.**

• **Describe the structure of the spinal cord.**

Protection and Coverings: Vertebral Canal and Meninges

The spinal cord is located within the vertebral canal of the vertebral column. Because the wall of the vertebral canal is essentially a ring of bone, the cord is well protected. The vertebral ligaments, meninges, and cerebrospinal fluid provide additional protection.

The *meninges* (me-NIN-jēz) are three layers of connective tissue coverings that extend around the spinal cord and brain. The meninges that protect the spinal cord, the *spinal meninges* (Figure 10.1), are continuous with those that protect the brain, the *cranial meninges* (see Figure 10.7). The outermost of the three layers of the meninges is called the *dura mater* (DOO-ra MĀ-ter = tough mother). Its tough, dense irregular connective tissue helps protect the delicate structures of the CNS. The tube of spinal dura mater extends to the second sacral vertebra, well beyond the spinal cord, which ends at about the level of the second lumbar vertebra. The spinal cord is also protected by a cushion of fat and connective tissue located in the *epidural space*, a space between the dura mater and vertebral column.

Figure 10.1 Spinal meninges.

🔑 **Meninges are connective tissue coverings that surround the brain and spinal cord.**

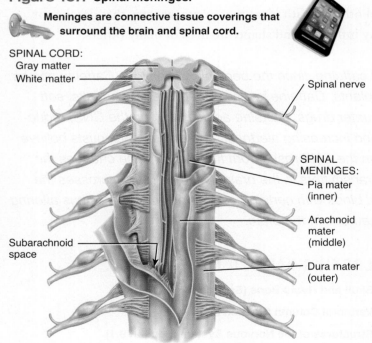

SPINAL CORD:
Gray matter
White matter

Spinal nerve

SPINAL MENINGES:
Pia mater (inner)

Arachnoid mater (middle)

Subarachnoid space

Dura mater (outer)

Anterior view and transverse section through spinal cord

❓ **In which meningeal space does cerebrospinal fluid circulate?**

The middle layer of the meninges is called the *arachnoid mater* (a-RAK-noyd; *arachn-* = spider; *-oid* = similar to) because the arrangement of its collagen and elastic fibers resembles a spider's web. The inner layer, the *pia mater* (PĒ-a MĀ-ter; *pia* = delicate), is a transparent layer of collagen and elastic fibers that adheres to the surface of the spinal cord and brain. It contains numerous blood vessels. Between the arachnoid mater and the pia mater is the *subarachnoid space*, where cerebrospinal fluid circulates.

CLINICAL CONNECTION | Spinal Tap

In a **spinal tap (lumbar puncture)**, a local anesthetic is given, and a long needle is inserted into the subarachnoid space. In adults, a spinal tap is normally performed between the third and fourth or fourth and fifth lumbar vertebrae. Because this region is inferior to the lowest portion of the spinal cord, it provides relatively safe access. The procedure is used to withdraw cerebrospinal fluid (CSF) for diagnostic purposes; to introduce antibiotics, contrast media for myelography, or anesthetics; to administer chemotherapy; to measure CSF pressure; and/or to evaluate the effects of treatment for diseases such as meningitis. •

Gross Anatomy of the Spinal Cord

The length of the adult *spinal cord* ranges from 42 to 45 cm (16 to 18 in.). It extends from the lowest part of the brain, the medulla oblongata, to the upper border of the second lumbar vertebra in the vertebral column (Figure 10.2). Because the spinal cord is shorter than the vertebral column, nerves that arise from the lumbar, sacral, and coccygeal regions of the spinal cord do not leave the vertebral column at the same level they exit the cord. The roots of these spinal nerves angle down the vertebral canal like wisps of flowing hair. They are appropriately named the *cauda equina* (KAW-da ē-KWĪ-na), meaning horse's tail. The spinal cord has two conspicuous enlargements: The *cervical enlargement* contains nerves that supply the upper limbs, and the *lumbar enlargement* contains nerves supplying the lower limbs.

Two grooves, the deep *anterior median fissure* and the shallow *posterior median sulcus*, divide the spinal cord into right and left halves (see Figure 10.3). In the spinal cord, white matter surrounds a centrally located H-shaped mass of gray matter. In the center of the gray matter is the *central canal*, a small space that extends the length of the cord and contains cerebrospinal fluid.

Spinal nerves are the paths of communication between the spinal cord and specific regions of the body. The spinal cord appears to be segmented because 31 pairs of spinal nerves emerge from it at regular intervals (Figure 10.2). Two bundles of axons, called *roots*, connect each spinal nerve to a segment of the cord (see Figure 10.3). The *posterior (dorsal) root* contains only sensory axons, which conduct nerve impulses from sensory receptors in the skin, muscles, and internal organs into the central nervous system. Each posterior root has a swelling, the *posterior (dorsal) root ganglion*, which contains the cell bodies of

Figure 10.2 Spinal cord and spinal nerves. Selected nerves are labeled on the left side of the figure. Together, the lumbar and sacral plexuses are called the lumbosacral plexus.

The spinal cord extends from the base of the skull to the superior border of the second lumbar vertebra.

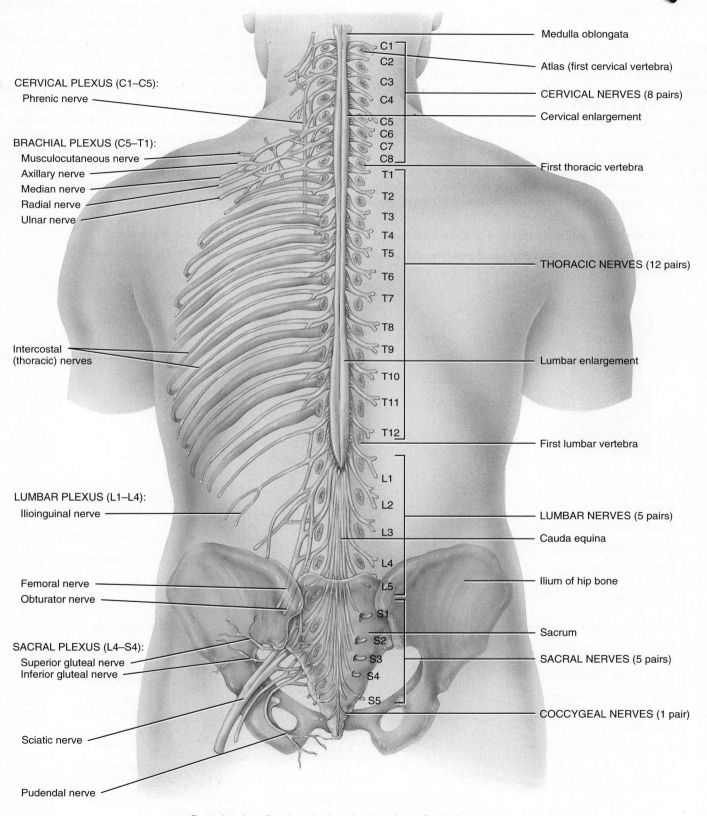

CERVICAL PLEXUS (C1–C5):
  Phrenic nerve

BRACHIAL PLEXUS (C5–T1):
  Musculocutaneous nerve
  Axillary nerve
  Median nerve
  Radial nerve
  Ulnar nerve

Intercostal (thoracic) nerves

LUMBAR PLEXUS (L1–L4):
  Ilioinguinal nerve

Femoral nerve
Obturator nerve

SACRAL PLEXUS (L4–S4):
  Superior gluteal nerve
  Inferior gluteal nerve

Sciatic nerve

Pudendal nerve

C1
C2
C3
C4
C5
C6
C7
C8
T1
T2
T3
T4
T5
T6
T7
T8
T9
T10
T11
T12
L1
L2
L3
L4
L5
S1
S2
S3
S4
S5

Medulla oblongata
Atlas (first cervical vertebra)
CERVICAL NERVES (8 pairs)
Cervical enlargement
First thoracic vertebra
THORACIC NERVES (12 pairs)
Lumbar enlargement
First lumbar vertebra
LUMBAR NERVES (5 pairs)
Cauda equina
Ilium of hip bone
Sacrum
SACRAL NERVES (5 pairs)
COCCYGEAL NERVES (1 pair)

Posterior view of entire spinal cord and portions of spinal nerves

 Are spinal nerves part of the CNS or the PNS?

273

sensory neurons. The ***anterior (ventral) root*** contains axons of motor neurons, which conduct nerve impulses from the CNS to effectors (muscles and glands).

Internal Structure of the Spinal Cord

The gray matter of the spinal cord contains neuronal cell bodies, dendrites, unmyelinated axons, axon terminals, and neuroglia. On each side of the spinal cord, the gray matter is subdivided into regions called ***horns***, named relative to their location: anterior, lateral, and posterior (Figure 10.3). The ***posterior (dorsal) gray horns*** contain cell bodies and axons of interneurons as well as axons of incoming sensory neurons. Recall that cell bodies of sensory neurons are located in the posterior (dorsal) root ganglion of a spinal nerve. The ***anterior (ventral) gray horns***

contain cell bodies of somatic motor neurons that provide nerve impulses for contraction of skeletal muscles. Between the posterior and anterior gray horns are the ***lateral gray horns***, which are present only in thoracic and upper lumbar segments of the spinal cord. The lateral gray horns contain cell bodies of autonomic motor neurons that regulate the activity of cardiac muscle, smooth muscle, and glands.

The white matter of the spinal cord consists primarily of myelinated axons of neurons and is organized into regions called anterior, lateral, and posterior ***white columns***. Each column contains one or more ***tracts***, which are distinct bundles of axons having a common origin or destination and carrying similar information. ***Sensory (ascending) tracts*** consist of axons that conduct nerve impulses toward the brain. Tracts consisting of axons that carry nerve impulses down the spinal

Figure 10.3 Internal structure of the spinal cord. Columns of white matter surround the gray matter.

🔑 **The spinal cord conducts nerve impulses along tracts and serves as an integrating center for spinal reflexes.**

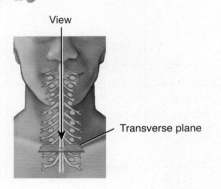

View

Transverse plane

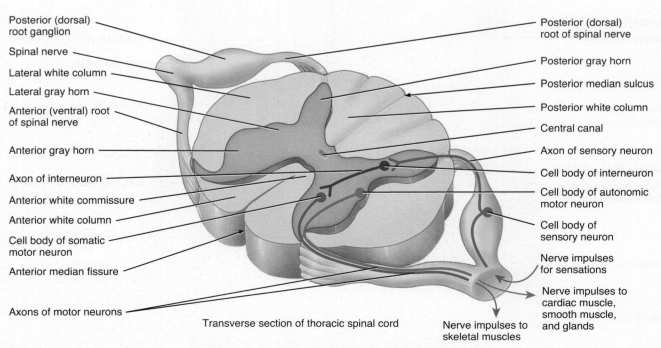

Posterior (dorsal) root ganglion

Spinal nerve

Lateral white column

Lateral gray horn

Anterior (ventral) root of spinal nerve

Anterior gray horn

Axon of interneuron

Anterior white commissure

Anterior white column

Cell body of somatic motor neuron

Anterior median fissure

Axons of motor neurons

Posterior (dorsal) root of spinal nerve

Posterior gray horn

Posterior median sulcus

Posterior white column

Central canal

Axon of sensory neuron

Cell body of interneuron

Cell body of autonomic motor neuron

Cell body of sensory neuron

Nerve impulses for sensations

Nerve impulses to cardiac muscle, smooth muscle, and glands

Nerve impulses to skeletal muscles

Transverse section of thoracic spinal cord

❓ **What is the difference between a *horn* and a *column* in the spinal cord?**

cord are called *motor (descending) tracts*. Sensory and motor tracts of the spinal cord are continuous with sensory and motor tracts in the brain. Often, the name of a tract indicates its position in the white matter, where it begins and ends, and the direction of nerve impulse conduction. For example, the anterior corticospinal tract is located in the *anterior* white column; it begins in the *cerebral cortex* (a region of the brain) and ends in the *spinal cord* (see Figure 10.15).

✓ CHECKPOINT

1. How is the spinal cord protected?
2. What body regions are served by nerves from the cervical and lumbar enlargements?
3. Distinguish between a horn and a column in the spinal cord.

10.2 SPINAL NERVES

OBJECTIVE • Describe the composition, coverings, and distribution of spinal nerves.

Spinal nerves and the nerves that branch from them are part of the peripheral nervous system (PNS). They connect the CNS to sensory receptors, muscles, and glands in all parts of the body. The 31 pairs of spinal nerves are named and numbered according to the region and level of the vertebral column from which they emerge (see Figure 10.2). There are 8 pairs of cervical nerves, 12 pairs of thoracic nerves, 5 pairs of lumbar nerves, 5 pairs of sacral nerves, and 1 pair of coccygeal nerves. The first cervical pair emerges above the atlas. All other spinal nerves leave the vertebral column by passing through the *intervertebral foramina*, the holes between vertebrae.

As noted earlier, a typical spinal nerve has two connections to the cord: a posterior root and an anterior root (see Figure 10.3). The posterior and anterior roots unite to form a spinal nerve at the intervertebral foramen. Because the posterior root contains sensory axons and the anterior root contains motor axons, a spinal nerve is classified as a *mixed nerve*. The posterior root contains a posterior root ganglion in which cell bodies of sensory neurons are located.

Spinal Nerve Coverings

Each spinal nerve (and cranial nerve) contains layers of protective connective tissue coverings (Figure 10.4). Individual axons, whether myelinated or unmyelinated, are wrapped in *endoneurium* (en′-dō-NOO-rē-um; *endo-* = within or inner). Groups of axons with their endoneurium are arranged in bundles, called *fascicles*, each of which is wrapped in *perineurium* (per′-i-NOO-rē-um; *peri-* = around). The superficial covering over the entire nerve is the *epineurium* (ep′-i-NOO-rē-um; *epi-* = over). The dura mater of the spinal meninges fuses with the epineurium as a spinal nerve passes through the intervertebral foramen. Note the presence of many blood vessels, which nourish nerves, within the perineurium and epineurium.

Distribution of Spinal Nerves

Plexuses

A short distance after passing through its intervertebral foramen, a spinal nerve divides into several branches. Many of the spinal nerve branches do not extend directly to the body structures they supply. Instead, they form networks on either side of the body by joining with axons from adjacent nerves. Such a network is called a *plexus* (= braid or network). Emerging from the plexuses are nerves bearing names that are often descriptive of the general regions they supply or the course they take. Each of the nerves, in turn, may have several branches named for the specific structures they supply.

The major plexuses are the cervical plexus, brachial plexus, lumbar plexus, and sacral plexus (see Figure 10.2). The *cervical plexus* supplies the skin and muscles of the posterior head, neck, upper part of the shoulders, and the diaphragm. The phrenic nerves, which stimulate the diaphragm to contract, arise from the cervical plexus. Damage to the spinal cord above the origin of the phrenic nerves may cause respiratory failure. The *brachial plexus* constitutes the nerve supply for the upper limbs and several neck and shoulder muscles. Among the nerves that arise from the brachial plexus are the musculocutaneous, axillary, median, radial, and ulnar nerves. The *lumbar plexus* supplies the abdominal wall, external genitals, and part of the lower limbs. Arising from this plexus are the ilioinguinal, femoral, and obturator nerves. The *sacral plexus* supplies the

Figure 10.4 Composition and connective tissue coverings of a spinal nerve.

Three layers of connective tissue wrappings protect axons: endoneurium surrounds individual axons, perineurium surrounds bundles of axons, and epineurium surrounds an entire nerve.

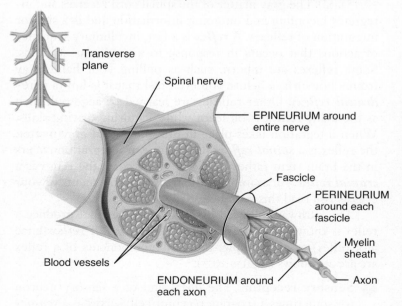

Transverse section showing the coverings of a spinal nerve

Why are all spinal nerves classified as mixed nerves?

buttocks, perineum, and lower limbs. Among the nerves that arise from this plexus are the gluteal, sciatic, and pudendal nerves. The sciatic nerve is the longest nerve in the body.

Intercostal Nerves

Spinal nerves T2 to T11 do not form plexuses. They are known as **intercostal nerves** and extend directly to the structures they supply, including the muscles between ribs, abdominal muscles, and skin of the chest and back (see Figure 10.2).

✓ CHECKPOINT

4. How do spinal nerves connect to the spinal cord?

5. Which regions of the body are supplied by plexuses, and which are served by intercostal nerves?

10.3 SPINAL CORD FUNCTIONS

OBJECTIVES • **Describe the functions of the spinal cord.**
• **Outline the components of a reflex arc.**

The spinal cord white matter and gray matter have two major functions in maintaining homeostasis. (1) The white matter of the spinal cord consists of tracts that serve as highways for nerve impulse conduction. Along these highways, sensory impulses travel toward the brain and motor impulses travel from the brain toward skeletal muscles and other effector tissues. The route that nerve impulses follow from a neuron in one part of the body to other neurons elsewhere in the body is called a **pathway**. After describing the functions of various regions of the brain, we will depict some important pathways that connect the spinal cord and brain (see Figures 10.14 and 10.15). (2) The gray matter of the spinal cord receives and integrates incoming and outgoing information and is a site for integration of reflexes. A **reflex** is a fast, involuntary sequence of actions that occurs in response to a particular stimulus. Some reflexes are inborn, such as pulling your hand away from a hot surface before you even feel that it is hot (a **withdrawal reflex**). Other reflexes are learned or acquired, such as the many reflexes you learn while acquiring driving skills. When integration takes place in the spinal cord gray matter, the reflex is a **spinal reflex**. By contrast, if integration occurs in the brain stem rather than the spinal cord, the reflex is a **cranial reflex**. An example is the tracking movements of your eyes as you read this sentence.

The pathway followed by nerve impulses that produce a reflex is known as a **reflex arc**. Using the **patellar reflex** (knee jerk reflex) as an example, the basic components of a reflex arc are as follows (Figure 10.5):

1 **Sensory receptor.** The distal end of a sensory neuron (or sometimes a separate receptor cell) serves as a **sensory receptor**. Sensory receptors respond to a specific type of stimulus by generating one or more nerve impulses. In the patellar reflex, sensory receptors known as *muscle spindles* detect slight stretching of the quadriceps femoris (anterior thigh) muscle when the patellar (knee cap) ligament is tapped with a reflex hammer.

2 **Sensory neuron.** The nerve impulses conduct from the sensory receptor along the axon of a **sensory neuron** to its axon terminals, which are located in the CNS gray matter. Axon branches of the sensory neuron also relay nerve impulses to the brain, allowing conscious awareness that the reflex has occurred.

3 **Integrating center.** One or more regions of gray matter in the CNS act as an **integrating center**. In the simplest type of reflex, such as the patellar reflex, the integrating center is a single synapse between a sensory neuron and a motor neuron. In other types of reflexes, the integrating center includes one or more interneurons.

4 **Motor neuron.** Impulses triggered by the integrating center pass out of the spinal cord (or brain stem, in the case of a cranial reflex) along a **motor neuron** to the part of the body that will respond. In the patellar reflex, the axon of the motor neuron extends to the quadriceps femoris muscle.

5 **Effector.** The part of the body that responds to the motor nerve impulse, such as a muscle or gland, is the **effector**. Its action is a reflex. If the effector is skeletal muscle, the reflex is a **somatic reflex**. If the effector is smooth muscle, cardiac muscle, or a gland, the reflex is an **autonomic (visceral) reflex**. For example, the acts of swallowing, urinating, and defecating all involve autonomic reflexes. The patellar reflex is a somatic reflex because its effector is the quadriceps femoris muscle, which contracts and thereby relieves the stretching that initiated the reflex. In sum, the patellar reflex causes extension of the knee by contraction of the quadriceps femoris muscle in response to tapping the patellar ligament.

CLINICAL CONNECTION | Reflexes and Diagnosis

Damage or disease anywhere along a reflex arc can cause the reflex to be absent or abnormal. For example, **absence of the patellar reflex** could indicate damage of the sensory or motor neurons, or a spinal cord injury, in the lumbar region. Somatic reflexes generally can be tested simply by tapping or stroking the body surface. Most autonomic reflexes, by contrast, are not practical diagnostic tools because it is difficult to stimulate visceral receptors, which are deep inside the body. An exception is the pupillary light reflex, in which the pupils of both eyes decrease in diameter when either eye is exposed to light. Because the reflex arc includes synapses in lower parts of the brain, the **absence of a normal pupillary light reflex** may indicate brain damage or injury. •

✓ CHECKPOINT

6. What is the significance of the white matter tracts of the spinal cord?

7. How are somatic and autonomic reflexes similar and different?

Figure 10.5 Patellar reflex, showing general components of a reflex arc. The arrows show the direction of nerve impulse conduction.

 A reflex is a fast, involuntary sequence of actions that occurs in response to a particular stimulus.

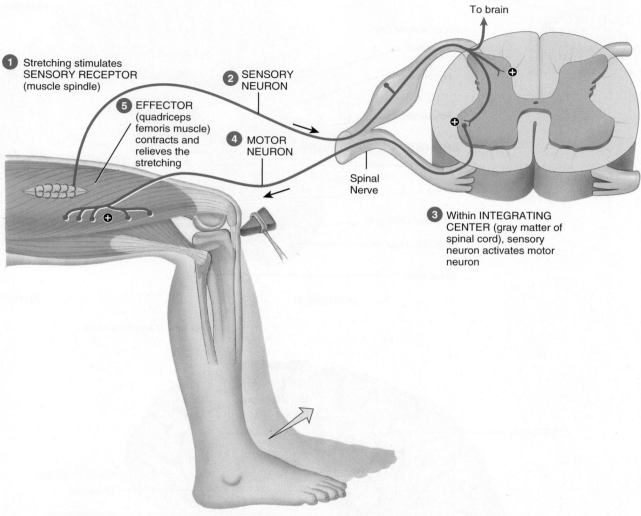

To brain

① Stretching stimulates SENSORY RECEPTOR (muscle spindle)

② SENSORY NEURON

⑤ EFFECTOR (quadriceps femoris muscle) contracts and relieves the stretching

④ MOTOR NEURON

Spinal Nerve

③ Within INTEGRATING CENTER (gray matter of spinal cord), sensory neuron activates motor neuron

Which root of a spinal nerve contains axons of sensory neurons? Which root contains axons of motor neurons?

10.4 BRAIN

OBJECTIVES • **Discuss how the brain is protected and supplied with blood.**

• **Name the major parts of the brain and explain the function of each part.**

• **Describe three somatic sensory and somatic motor pathways.**

Next we will consider the major parts of the brain, how the brain is protected, and how it is related to the spinal cord and cranial nerves.

Major Parts and Protective Coverings

The ***brain*** is one of the largest organs of the body, consisting of about 100 billion neurons and 10–50 trillion neuroglia with a mass of about 1300 g (almost 3 lb). The four major parts are the brain stem, diencephalon, cerebrum, and cerebellum (Figure 10.6). The ***brain stem*** is continuous with the spinal cord and consists of the medulla oblongata, pons, and midbrain. Above the brain stem is the ***diencephalon*** (dī′-en-SEF-a-lon; *di-* = through; *-encephalon* = brain), consisting mostly of the thalamus, hypothalamus, and pineal gland. Supported on the diencephalon and brain stem and forming the bulk of the brain is the ***cerebrum*** (se-RĒ-brum = brain).

Figure 10.6 Brain. The pituitary gland is discussed together with the endocrine system in Chapter 13.

 The four major parts of the brain are the brain stem, cerebellum, diencephalon, and cerebrum.

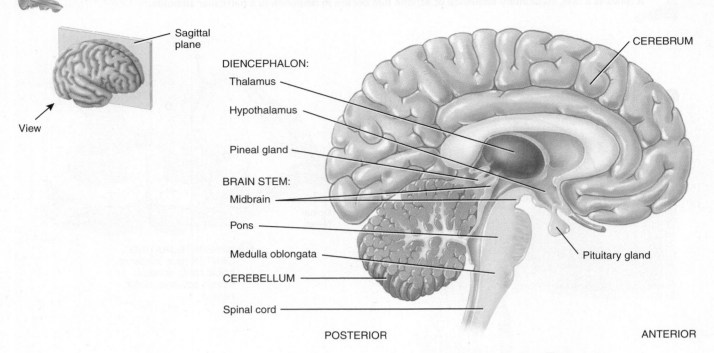

Sagittal plane

View

DIENCEPHALON:
Thalamus
Hypothalamus
Pineal gland

BRAIN STEM:
Midbrain
Pons
Medulla oblongata
CEREBELLUM
Spinal cord

CEREBRUM
Pituitary gland

POSTERIOR

ANTERIOR

(a) Medial view of sagittal section

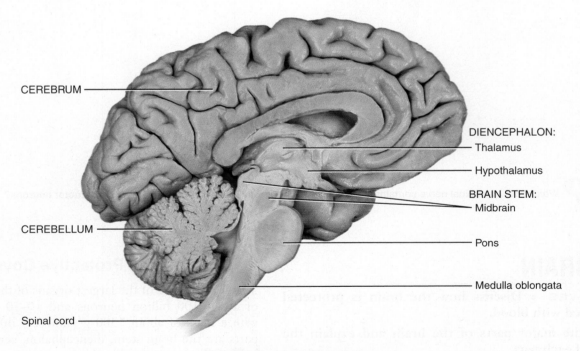

CEREBRUM

CEREBELLUM

Spinal cord

DIENCEPHALON:
Thalamus
Hypothalamus
BRAIN STEM:
Midbrain
Pons
Medulla oblongata

(b) Medial view of sagittal section

? Which part of the brain attaches to the spinal cord?

The surface of the cerebrum is composed of a thin layer of gray matter, the *cerebral cortex* (*cortex* = rind or bark), beneath which lies the cerebral white matter. Posterior to the brain stem is the *cerebellum* (ser'-e-BEL-um = little brain).

As you learned earlier in the chapter, the brain is protected by the cranium and cranial meninges. The *cranial meninges* have the same names as the spinal meninges: the outermost *dura mater*, middle *arachnoid mater*, and innermost *pia mater* (Figure 10.7).

Brain Blood Supply and the Blood–Brain Barrier

Although the brain constitutes only about 2 percent of total body weight, it requires about 20 percent of the body's oxygen supply. If blood flow to the brain stops, even briefly, unconsciousness may result. Brain neurons that are totally deprived of oxygen for four or more minutes may be permanently injured. Blood supplying the brain also contains glucose, the main source of energy for brain cells. Because virtually no glucose is stored in the brain, the supply of glucose also must be continuous. If blood entering the brain has a low level of glucose, mental confusion, dizziness, convulsions, and loss of consciousness may occur.

The existence of a *blood–brain barrier (BBB)* protects brain cells from harmful substances and pathogens by preventing passage of many substances from blood into brain tissue. This barrier consists basically of very tightly sealed blood capillaries (microscopic blood vessels) in the brain assisted by astrocytes. However, lipid-soluble substances such as oxygen, carbon dioxide, alcohol, and most anesthetic agents, easily cross the blood–brain barrier. Trauma, certain toxins, and inflammation can cause a breakdown of the blood–brain barrier.

Cerebrospinal Fluid

The spinal cord and brain are further protected against chemical and physical injury by *cerebrospinal fluid (CSF)*. CSF is a clear, colorless liquid that carries oxygen, glucose, and other needed chemicals from the blood to neurons and neuroglia and removes wastes and toxic substances produced by brain and spinal cord cells. CSF circulates through the subarachnoid space (between the arachnoid mater and pia mater), around the brain and spinal cord, and through cavities in the brain known as *ventricles* (VEN-tri-kuls = little cavities). There are four ventricles: two *lateral ventricles*, one *third ventricle*, and one *fourth ventricle* (Figure 10.7). Openings connect them with one another, with the central canal of the spinal cord, and with the subarachnoid space.

The sites of CSF production are the *choroid plexuses* (KŌ-royd = membranelike), which are specialized networks of capillaries in the walls of the ventricles (Figure 10.7).

Covering the choroid plexus capillaries are ependymal cells, which form cerebrospinal fluid from blood plasma by filtration and secretion. From the fourth ventricle, CSF flows into the central canal of the spinal cord and into the subarachnoid space around the surface of the brain and spinal cord. CSF is gradually reabsorbed into the blood through *arachnoid villi*, which are fingerlike extensions of the arachnoid mater. The CSF drains primarily into a vein called the *superior sagittal sinus* (Figure 10.7). Normally, the volume of CSF remains constant at 80 to 150 mL (3 to 5 oz) because it is absorbed as rapidly as it is formed.

CLINICAL CONNECTION | Hydrocephalus

Abnormalities in the brain—tumors, inflammation, or developmental malformations—can interfere with the drainage of CSF from the ventricles into the subarachnoid space. When *excess* CSF accumulates in the ventricles, the CSF pressure rises. Elevated CSF pressure causes a condition called **hydrocephalus** (hī'-drō-SEF-a-lus; *hydro-* = water; *cephal-* = head). In a baby whose fontanels have not yet closed, the head bulges due to the increased pressure. If the condition persists, the fluid buildup compresses and damages the delicate nervous tissue. Hydrocephalus is relieved by draining the excess CSF. A neurosurgeon may implant a drain line, called a shunt, into the lateral ventricle to divert CSF into the superior vena cava or abdominal cavity, where it can be absorbed by the blood. In adults, hydrocephalus may occur after head injury, meningitis, or subarachnoid hemorrhage. This condition can quickly become life-threatening and requires immediate intervention; since the adult skull bones have already fused, nervous tissue damage occurs quickly. •

Brain Stem

The brain stem is the part of the brain between the spinal cord and the diencephalon. It consists of three regions: (1) the medulla oblongata, (2) pons, and (3) midbrain. Extending through the brain stem is the reticular formation, a region where gray and white matter are intermingled.

Medulla Oblongata

The *medulla oblongata* (me-DOOL-la ob'-long-GA-ta), or simply *medulla*, is a continuation of the spinal cord (see Figure 10.6). It forms the inferior part of the brain stem (see Figure 10.8). Within the medulla's white matter are all sensory (ascending) and motor (descending) tracts extending between the spinal cord and other parts of the brain.

The medulla also contains several *nuclei*, which are masses of gray matter where neurons form synapses with one another. Two major nuclei are the *cardiovascular center*, which regulates the rate and force of the heartbeat and the diameter of blood vessels (see Figure 15.9), and the *medullary rhythmicity area*, which adjusts the basic rhythm of breathing (see Figure 18.12). Nuclei associated with sensations of touch, pressure, vibration, and conscious proprioception (awareness of the position of body

Figure 10.7 Meninges and ventricles of the brain.

 Cerebrospinal fluid (CSF) protects the brain and spinal cord and delivers nutrients from the blood to the brain and spinal cord; CSF also removes wastes from the brain and spinal cord to the blood.

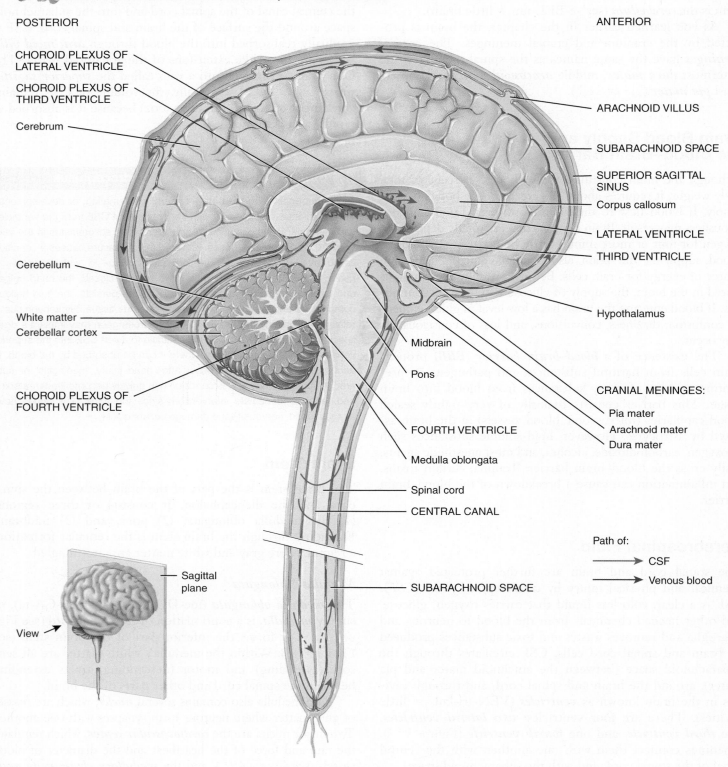

POSTERIOR

CHOROID PLEXUS OF LATERAL VENTRICLE

CHOROID PLEXUS OF THIRD VENTRICLE

Cerebrum

Cerebellum

White matter

Cerebellar cortex

CHOROID PLEXUS OF FOURTH VENTRICLE

ANTERIOR

ARACHNOID VILLUS

SUBARACHNOID SPACE

SUPERIOR SAGITTAL SINUS

Corpus callosum

LATERAL VENTRICLE

THIRD VENTRICLE

Hypothalamus

Midbrain

Pons

CRANIAL MENINGES:

Pia mater

Arachnoid mater

Dura mater

FOURTH VENTRICLE

Medulla oblongata

Spinal cord

CENTRAL CANAL

SUBARACHNOID SPACE

Sagittal plane

View

Path of:

→ CSF

→ Venous blood

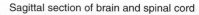

Sagittal section of brain and spinal cord

Where is CSF formed and absorbed?

parts) are located in the posterior part of the medulla. Many ascending sensory axons form synapses in these nuclei (see Figure 10.14a). Other nuclei in the medulla control reflexes for swallowing, vomiting, coughing, hiccupping, and sneezing. Finally, the medulla contains nuclei associated with five pairs of cranial nerves (Figure 10.8): vestibulocochlear (VIII) nerves, glossopharyngeal (IX) nerves, vagus (X) nerves, accessory (XI) nerves (cranial portion), and hypoglossal (XII) nerves.

CLINICAL CONNECTION | Injury of the Medulla

Given the many vital activities controlled by the medulla, it is not surprising that a hard blow to the back of the head or upper neck can be fatal. **Damage to the medullary rhythmicity area** is particularly serious and can rapidly lead to death. Symptoms of nonfatal injury to the medulla may include paralysis and loss of sensation on the opposite side of the body, and irregularities in breathing or heart rhythm. •

Figure 10.8 Inferior aspect of the brain, showing the brain stem and cranial nerves.

The brain stem consists of the medulla oblongata, pons, and midbrain.

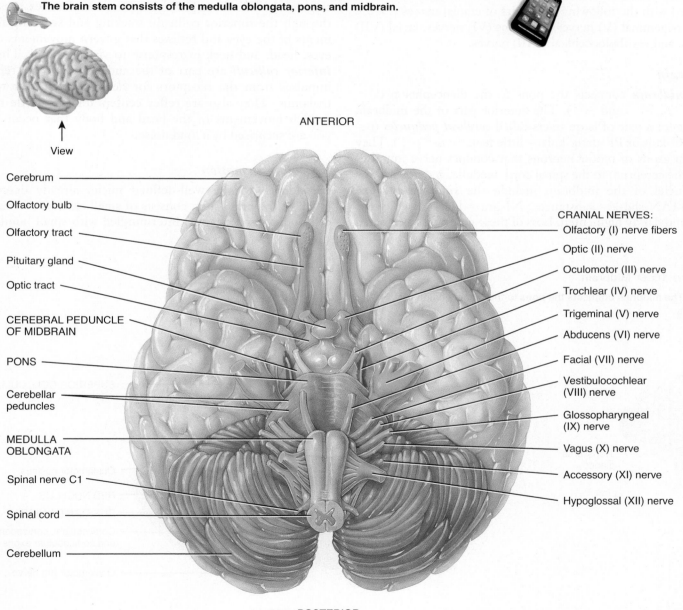

ANTERIOR

Cerebrum

Olfactory bulb

Olfactory tract

Pituitary gland

Optic tract

CEREBRAL PEDUNCLE OF MIDBRAIN

PONS

Cerebellar peduncles

MEDULLA OBLONGATA

Spinal nerve C1

Spinal cord

Cerebellum

CRANIAL NERVES:

Olfactory (I) nerve fibers

Optic (II) nerve

Oculomotor (III) nerve

Trochlear (IV) nerve

Trigeminal (V) nerve

Abducens (VI) nerve

Facial (VII) nerve

Vestibulocochlear (VIII) nerve

Glossopharyngeal (IX) nerve

Vagus (X) nerve

Accessory (XI) nerve

Hypoglossal (XII) nerve

View

POSTERIOR

Inferior aspect of brain

 Which part of the brain stem contains the cerebral peduncles?

Pons

The *pons* (= bridge) is above the medulla and anterior to the cerebellum (Figures 10.6, 10.7, and 10.8). Like the medulla, the pons consists of both nuclei and tracts. As its name implies, the pons is a bridge that connects parts of the brain with one another. These connections are bundles of axons. Some axons of the pons connect the right and left sides of the cerebellum. Others are part of ascending sensory tracts and descending motor tracts. Several nuclei in the pons are the sites where signals for voluntary movements that originate in the cerebral cortex are relayed into the cerebellum. Other nuclei in the pons help control breathing. The pons also contains nuclei associated with the following four pairs of cranial nerves (Figure 10.8): trigeminal (V) nerves, abducens (VI) nerves, facial (VII) nerves, and vestibulocochlear (VIII) nerves.

Midbrain

The *midbrain* connects the pons to the diencephalon (Figures 10.6, 10.7, and 10.8). The anterior part of the midbrain consists of a pair of large tracts called *cerebral peduncles* (pe-DUNG-kuls or PĒ-dung-kuls = little feet; Figure 10.9). They contain axons of motor neurons that conduct nerve impulses from the cerebrum to the spinal cord, medulla, and pons.

Nuclei of the midbrain include the *substantia nigra* (sub-STAN-shē-a = substance; NĪ-gra = black), which is large and darkly pigmented. Loss of these neurons is associated with Parkinson's disease (see Common Disorders section). Also present are the right and left *red nuclei*, which look reddish due to their rich blood supply and an iron-containing pigment in their neuronal cell bodies. Axons from the cerebellum and cerebral cortex form synapses in the red nuclei, which function with the cerebellum to coordinate muscular movements. Other nuclei in the midbrain are associated with two pairs of cranial nerves (see Figure 10.8): oculomotor (III) nerves and trochlear (IV) nerves.

The midbrain also contains nuclei that appear as four rounded bumps on the posterior surface. The two superior bumps are the *superior colliculi* (ko-LIK-ū-lī = little hills; singular is *colliculus*) (Figure 10.9). Several reflex arcs pass through the superior colliculi: tracking and scanning movements of the eyes and reflexes that govern movements of the eyes, head, and neck in response to visual stimuli. The two *inferior colliculi* are part of the auditory pathway, relaying impulses from the receptors for hearing in the ear to the thalamus. They also are reflex centers for the startle reflex, sudden movements of the head and body that occur when you are surprised by a loud noise.

Reticular Formation

In addition to the well-defined nuclei already described, much of the brain stem consists of small clusters of neuronal cell bodies (gray matter) intermingled with small bundles of

Figure 10.9 Midbrain.

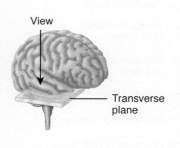

The midbrain connects the pons to the diencephalon.

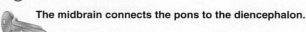

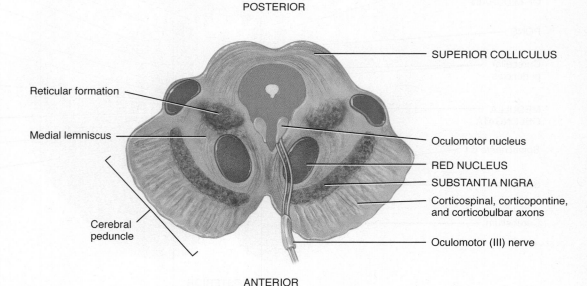

Transverse section of midbrain

 What functions are carried out by the superior colliculi?

myelinated axons (white matter). This region is known as the *reticular formation* (*ret-* = net) due to its netlike arrangement of white matter and gray matter. Neurons within the reticular formation have both ascending (sensory) and descending (motor) functions.

The ascending part of the reticular formation is called the *reticular activating system (RAS)*, which consists of sensory axons that project to the cerebral cortex. When the RAS is stimulated, many nerve impulses pass upward to widespread areas of the cerebral cortex. The result is a state of wakefulness called *consciousness*. The RAS helps maintain consciousness and is active during awakening from sleep. Inactivation of the RAS produces *sleep*, a state of partial unconsciousness from which an individual can be aroused. The reticular formation's main descending function is to help regulate muscle tone, which is the slight degree of contraction in normal resting muscles.

✓ CHECKPOINT

8. What is the significance of the blood–brain barrier?

9. What structures are the sites of CSF production, and where are they located?

10. Where are the medulla, pons, and midbrain located relative to one another?

11. What functions are governed by nuclei in the brain stem?

12. What are two important functions of the reticular formation?

Diencephalon

Major regions of the diencephalon include the thalamus, hypothalamus, and pineal gland (see Figure 10.6).

Thalamus

The *thalamus* (THAL-a-mus = inner chamber) consists of paired oval masses of gray matter organized into nuclei with interspersed tracts of white matter (Figure 10.10). The thalamus is the major relay station for most sensory impulses that reach the cerebral cortex from the spinal cord and brain stem. In addition, the thalamus contributes to motor functions by transmitting information from the cerebellum and basal nuclei to motor areas of the cerebral cortex. The thalamus also relays nerve impulses between different areas of the cerebrum and plays a role in the maintenance of consciousness.

Hypothalamus

The *hypothalamus* (*hypo-* = under) is the small portion of the diencephalon that lies below the thalamus and above the pituitary gland (see Figures 10.6 and 10.10). Although its size is small, the hypothalamus controls many important body activities, most of them related to homeostasis. The chief functions of the hypothalamus are as follows:

1. **Control of the ANS**. The hypothalamus controls and integrates activities of the autonomic nervous system, which regulates contraction of smooth and cardiac muscle and the secretions of many glands. Through the ANS,

Figure 10.10 Diencephalon: thalamus and hypothalamus. Also shown are the basal nuclei—the caudate nucleus, putamen, and globus pallidus.

The thalamus is the principal relay station for sensory impulses that reach the cerebral cortex from other parts of the brain and the spinal cord.

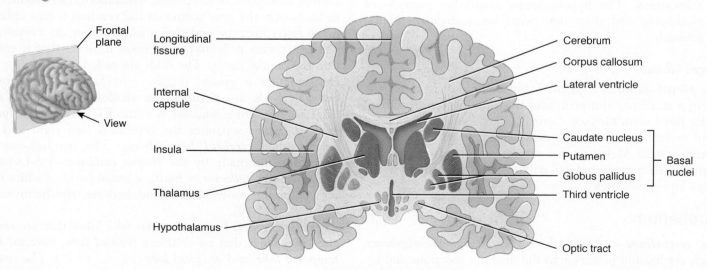

Anterior view of frontal section

In which major part of the brain are the basal nuclei located, and what kind of tissue composes them?

the hypothalamus helps to regulate activities such as heart rate, movement of food through the gastrointestinal tract, and contraction of the urinary bladder.

2. **Control of the pituitary gland and production of hormones**. The hypothalamus controls the release of several hormones from the pituitary gland and thus serves as a primary connection between the nervous system and endocrine system. The hypothalamus also produces two hormones (antidiuretic hormone and oxytocin) that are stored in the pituitary gland prior to their release.

3. **Regulation of emotional and behavioral patterns**. Together with the limbic system (described shortly), the hypothalamus regulates feelings of rage, aggression, pain, and pleasure, and the behavioral patterns related to sexual arousal.

4. **Regulation of eating and drinking**. The hypothalamus regulates food intake. It contains a **feeding center**, which promotes eating, and a **satiety center**, which causes a sensation of fullness and cessation of eating. The hypothalamus also contains a **thirst center**. When certain cells in the hypothalamus are stimulated by rising osmotic pressure of the extracellular fluid, they cause the sensation of thirst. The intake of water by drinking restores the osmotic pressure to normal, removing the stimulation and relieving the thirst.

5. **Control of body temperature**. If the temperature of blood flowing through the hypothalamus is above normal, the hypothalamus directs the autonomic nervous system to stimulate activities that promote heat loss. If, however, blood temperature is below normal, the hypothalamus generates impulses that promote heat production and retention.

6. **Regulation of circadian rhythms and states of consciousness**. The hypothalamus establishes patterns of awakening and sleep that occur on a circadian (daily) schedule.

Pineal Gland

The **pineal gland** (PĪN-ē-al = pinecone-like) is about the size of a small pea and protrudes from the posterior midline of the third ventricle (see Figure 10.6a). Because the pineal gland secretes the hormone *melatonin*, it is part of the endocrine system. Melatonin is a very powerful antioxidant that promotes sleepiness and contributes to the setting of the body's biological clock.

Cerebellum

The **cerebellum** consists of two **cerebellar hemispheres**, which are located posterior to the medulla and pons and inferior to the cerebrum (see Figure 10.6). The surface of the cerebellum, called the **cerebellar cortex**, consists of gray matter. Beneath the cortex is **white matter** (*arbor vitae*) that resembles the branches of a tree (see Figure 10.7). Deep

within the white matter are masses of gray matter, the **cerebellar nuclei**. The cerebellum attaches to the brain stem by bundles of axons called **cerebellar peduncles** (see Figure 10.8).

The cerebellum compares intended movements programmed by the cerebral cortex with what is actually happening. It constantly receives sensory impulses from muscles, tendons, joints, equilibrium receptors, and visual receptors. The cerebellum helps to smooth and coordinate complex sequences of skeletal muscle contractions. It regulates posture and balance and is essential for all skilled motor activities, from catching a baseball to dancing.

CLINICAL CONNECTION | Ataxia

Damage to the cerebellum through trauma or disease disrupts muscle coordination, a condition called **ataxia** (a-TAK-sē-a; *a-* = without; *-taxia* = order). Blindfolded people with ataxia cannot touch the tip of their nose with a finger because they cannot coordinate movement with their sense of where a body part is located. Another sign of ataxia is a changed speech pattern due to uncoordinated speech muscles. Cerebellar damage may also result in staggering or abnormal walking movements. People who consume too much alcohol show signs of ataxia because alcohol inhibits activity of the cerebellum. Alcohol overdose also suppresses the medullary rhythmicity area and may result in death. •

Cerebrum

The cerebrum consists of the cerebral cortex (an outer rim of gray matter), an internal region of cerebral white matter, and gray matter nuclei deep within the white matter (Figure 10.10). The cerebrum provides us with the ability to read, write, and speak; to make calculations and compose music; to remember the past and plan for the future; and to create. During embryonic development, when there is a rapid increase in brain size, the gray matter of the cerebral cortex enlarges much faster than the underlying white matter. As a result, the cerebral cortex rolls and folds upon itself so that it can fit into the cranial cavity. The folds are called **gyri** (JĪ-rī = circles; singular is *gyrus*) (Figure 10.11). The deep grooves between folds are **fissures**; the shallow grooves are **sulci** (SUL-sī = groove; singular is *sulcus*, SUL-kus). The **longitudinal fissure** separates the cerebrum into right and left halves called **cerebral hemispheres**. The hemispheres are connected internally by the **corpus callosum** (kal-LŌ-sum; *corpus* = body; *callosum* = hard), a broad band of white matter containing axons that extend between the hemispheres (see Figure 10.10).

Each cerebral hemisphere has four lobes that are named after the bones that cover them: **frontal lobe**, **parietal lobe**, **temporal lobe**, and **occipital lobe** (Figure 10.11). The **central sulcus** separates the frontal and parietal lobes. A major gyrus, the **precentral gyrus**, is located immediately anterior to the central sulcus. The precentral gyrus contains the primary motor area of the cerebral cortex. The **postcentral gyrus**,

Figure 10.11 Cerebrum. The inset in (a) indicates the differences among a gyrus, a sulcus, and a fissure. Because the insula cannot be seen externally, it has been projected to the surface in (b).

 The cerebrum provides us with the ability to read, write, and speak; make calculations and compose music; remember the past and make future plans; and create.

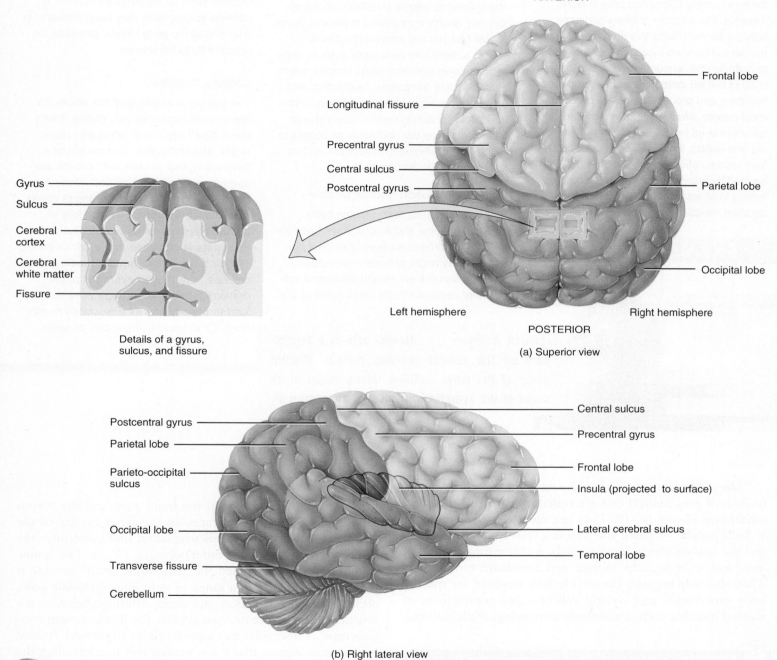

Gyrus

Sulcus

Cerebral cortex

Cerebral white matter

Fissure

Details of a gyrus, sulcus, and fissure

ANTERIOR

Frontal lobe

Longitudinal fissure

Precentral gyrus

Central sulcus

Postcentral gyrus

Parietal lobe

Occipital lobe

Left hemisphere

Right hemisphere

POSTERIOR

(a) Superior view

Postcentral gyrus

Parietal lobe

Parieto-occipital sulcus

Occipital lobe

Transverse fissure

Cerebellum

Central sulcus

Precentral gyrus

Frontal lobe

Insula (projected to surface)

Lateral cerebral sulcus

Temporal lobe

(b) Right lateral view

 What structure separates the right and left cerebral hemispheres?

located immediately posterior to the central sulcus, contains the primary somatosensory area of the cerebral cortex, which is discussed shortly. A fifth part of the cerebrum, the *insula*, cannot be seen at the surface of the brain because it lies within the lateral cerebral sulcus, deep to the parietal, frontal, and temporal lobes (see Figure 10.10).

The *cerebral white matter* consists of myelinated and unmyelinated axons that transmit impulses between gyri in the same hemisphere, from the gyri in one cerebral hemisphere to the corresponding gyri in the opposite cerebral hemisphere via the corpus callosum, and from the cerebrum to other parts of the brain and spinal cord.

The Many Faces of Caffeine

Found naturally in over 60 plants, caffeine is the most widely consumed drug in North America. The majority of users consume caffeine for the effects this drug exerts on the central nervous system. Moderate doses of caffeine (the amount found in two to three cups of coffee) increase alertness and reaction time, and produce a positive mood in most people. Students and writers sing caffeine's praises for its assistance with studying and writing. Caffeine boosts athletic performance so effectively that high levels are banned by organizations that oversee drug testing in athletes. Caffeine improves endurance, reaction time, and coordination.

Individual Results May Vary

The immediate effects of caffeine consumption vary greatly from person to person. Some people find that any amount of caffeine causes undesirable symptoms, such as anxiety, increased blood pressure, irregular heartbeat, digestive complaints, headaches, and difficulty sleeping. Before getting swept into the caffeine cheering section, users should be aware of their own individual responses to caffeine and caffeinated beverages, and adjust consumption accordingly.

Caffeine + Alcohol: Double Trouble

Beverage products that combine hefty doses of caffeine and alcohol have been implicated in numerous cases of alcohol poisoning. Why might caffeine make alcohol consumption riskier? Health educators suspect that caffeine effects mask some of the symptoms of excessive alcohol consumption, making users feel less inebriated than they actually are, and leading to poor choices such as the decision to drive. A caffeine energy buzz may keep drinkers up later and at the party longer, providing an opportunity to drink more.

Caffeine Calories

The variety of coffee and tea drinks on the market can tempt any palate. Don't like coffee? Try it with whipped cream, sugar, and chocolate. Or how about a bottomless cup of chai, with cream, cardamom, nutmeg, and ginger? Consumers must be aware that many of these drinks, especially the large sizes with lots of cream, have hundreds of calories. Since most people consume these drinks on top of their three or four square meals a day, calories from these beverages push drinkers into the fat storage zone that can lead to obesity and its associated health risks. Or at least a larger pair of jeans.

Think It Over . . . **Alcohol acts as a depressant on the central nervous system. Explain some of the ways caffeine intake might mask some of the symptoms of alcohol consumption.**

Deep within each cerebral hemisphere are three nuclei (masses of gray matter) that are collectively termed the **basal nuclei** (see Figure 10.10). They are the *globus pallidus* (*globus* = ball; *pallidus* = pale), the *putamen* (pū-TA-men = shell), and the *caudate nucleus* (*caud-* = tail). A major function of the basal nuclei is to help initiate and terminate movements. They also help regulate the muscle tone required for specific body movements and control subconscious contractions of skeletal muscles, such as automatic arm swings while walking.

CLINICAL CONNECTION | Damage to the Basal Nuclei

Damage to the basal nuclei results in uncontrollable shaking (tremor), muscular rigidity (stiffness), and involuntary muscle movements. Movement disruptions also are a hallmark of Parkinson's disease (see Common Disorders section). In this disorder, neurons that extend from the substantia nigra to the putamen and caudate nucleus degenerate, causing the disruptions. •

Limbic System

Encircling the upper part of the brain stem and the corpus callosum is a ring of structures on the inner border of the cerebrum and floor of the diencephalon that constitutes the **limbic system** (*limbic* = border) (Figure 10.12). The limbic system is sometimes called the "emotional brain" because it plays a primary role in a range of emotions, including pain, pleasure, docility, affection, and anger. Although behavior is a function of the entire nervous system, the limbic system controls most of its involuntary aspects related to survival. Animal experiments suggest that it has a major role in controlling the overall pattern of behavior. Together with parts of the cerebrum, the limbic system also functions in memory; damage to the limbic system causes memory impairment.

Functional Areas of the Cerebral Cortex

Specific types of sensory, motor, and integrative signals are processed in certain regions of the cerebral cortex. Generally, **sensory areas** receive sensory information and

Figure 10.12 The limbic system. The components of the limbic system are shaded olive green.

The limbic system governs emotional aspects of behavior.

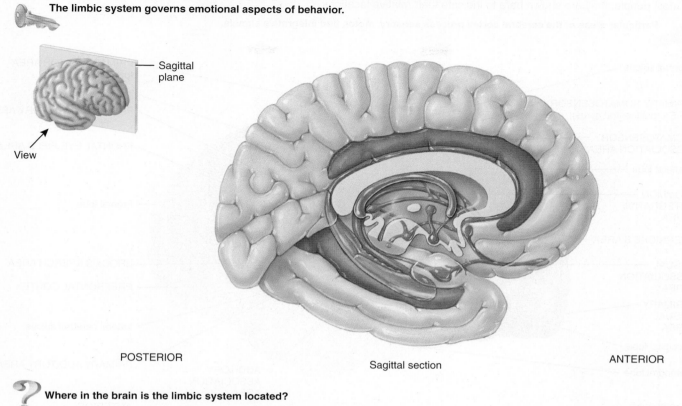

POSTERIOR

Sagittal section

ANTERIOR

Where in the brain is the limbic system located?

are involved in ***perception***, the conscious awareness of a sensation; ***motor areas*** initiate movements; and ***association areas*** deal with more complex integrative functions such as memory, emotions, reasoning, will, judgment, personality traits, and intelligence.

SENSORY AREAS Sensory input to the cerebral cortex flows mainly to the posterior half of the cerebral hemispheres, to regions behind the central sulci. In the cerebral cortex, primary sensory areas receive sensory information that has been relayed from peripheral sensory receptors through lower regions of the brain.

The *primary somatosensory area* (sō′-mat-ō-SEN-sō-rē) is posterior to the central sulcus of each cerebral hemisphere in the postcentral gyrus of the parietal lobe (Figure 10.13). It receives nerve impulses for touch, proprioception (joint and muscle position), pain, itching, tickle, and temperature and is involved in the perception of these sensations. The primary somatosensory area allows you to pinpoint where sensations originate, so that you know exactly where on your body to swat that mosquito. The *primary visual area*, located in the occipital lobe, receives visual information and is involved in visual perception. The *primary auditory area*, located in the temporal lobe, receives information for sound and is involved in auditory perception. The *primary gustatory area*, located at the base of the postcentral gyrus, receives impulses for taste and is involved in gustatory perception. The *primary olfactory area*, located on the medial aspect of the temporal lobe (and thus not visible in Figure 10.13), receives impulses for smell and is involved in olfactory perception.

MOTOR AREAS Motor output from the cerebral cortex flows mainly from the anterior part of each hemisphere. Among the most important motor areas are the primary motor area and Broca's speech area (Figure 10.13). The *primary motor area* is located in the precentral gyrus of the frontal lobe in each hemisphere. Each region in the primary motor area controls voluntary contractions of specific muscles on the opposite side of the body. *Broca's speech area* (BRŌ-kaz) is located in the frontal lobe close to the lateral cerebral sulcus. Speaking and understanding language are complex activities that involve several sensory, association, and motor areas of the cortex. In 97 percent of the population, these language areas are localized in the *left* hemisphere. Neural connections between Broca's speech area, the premotor area, and primary motor area activate muscles needed for speaking and breathing muscles.

ASSOCIATION AREAS The association areas of the cerebrum consist of large areas of the occipital, parietal, and temporal lobes and of the frontal lobes anterior to the motor areas.

Figure 10.13 Functional areas of the cerebrum. Broca's speech area and Wernicke's area are in the left cerebral hemisphere of most people; they are shown here to indicate their relative locations.

 Particular areas of the cerebral cortex process sensory, motor, and integrative signals.

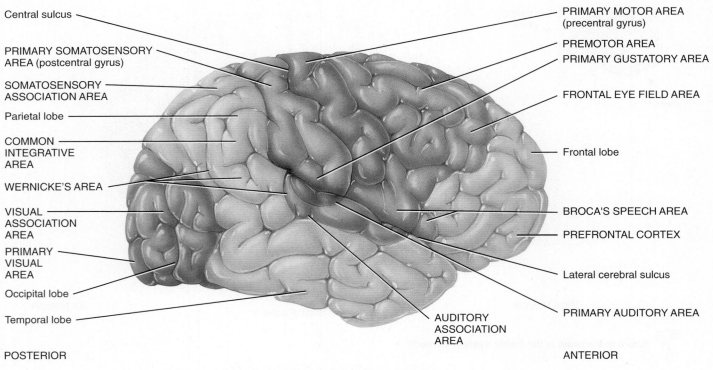

Lateral view of right cerebral hemisphere

? **Which part of the cerebrum localizes exactly where somatic sensations occur?**

Tracts connect association areas to one another. The *somatosensory association area*, just posterior to the primary somatosensory area, integrates and interprets somatic sensations such as the exact shape and texture of an object. Another role of the somatosensory association area is the storage of memories of past sensory experiences, enabling you to compare current sensations with previous experiences. For example, the somatosensory association area allows you to recognize objects such as a pencil and a paper clip simply by touching them. The *visual association area*, located in the occipital lobe, relates present and past visual experiences and is essential for recognizing and evaluating what is seen. The *auditory association area*, located below the primary auditory area in the temporal cortex, allows you to recognize a particular sound as speech, music, or noise.

Wernicke's area (VER-ni-kēz), a broad region in the *left* temporal and parietal lobes, interprets the meaning of speech by recognizing spoken words. It is active as you translate words into thoughts. The regions in the *right* hemisphere that correspond to Broca's and Wernicke's areas in the left hemisphere also contribute to verbal communication by adding emotional content, for instance, anger or joy, to spoken words. The *common integrative area* receives and interprets nerve impulses from the somatosensory, visual, and auditory association areas, and from

the primary gustatory area, primary olfactory area, the thalamus, and parts of the brain stem. The *premotor area*, immediately anterior to the primary motor area, generates nerve impulses that cause a specific group of muscles to contract in a specific sequence, for example, to write a word. The *frontal eye field area* in the frontal cortex controls voluntary scanning movements of the eyes, such as those that occur while you are reading this sentence. The *prefrontal cortex* in the anterior portion of the frontal lobe is concerned with the makeup of a person's personality, intellect, complex learning abilities, recall of information, initiative, judgment, foresight, reasoning, conscience, intuition, mood, planning for the future, and development of abstract ideas. A person with bilateral damage to the prefrontal cortices typically becomes rude, inconsiderate, incapable of accepting advice, moody, inattentive, less creative, unable to plan for the future, and incapable of anticipating the consequences of rash or reckless words or behavior.

Somatic Sensory and Somatic Motor Pathways

Somatic sensory information from the body ascends to the primary somatosensory area via two main somatic sensory pathways: (1) the posterior column–medial lemniscus pathway

Injury to language areas of the cerebral cortex results in **aphasia** (a-FĀ-zē-a; *a-* = without; *-phasia* = speech), an inability to use or comprehend words. Damage to Broca's speech area results in *nonfluent aphasia*, an inability to properly form words. People with nonfluent aphasia know what they wish to say but cannot properly speak the words. Damage to Wernicke's area, the common integrative area, or auditory association area results in *fluent aphasia*, characterized by faulty understanding of spoken or written words. A person experiencing this type of aphasia may produce strings of words that have no meaning ("word salad"). For example, someone with fluent aphasia might say, "I rang car porch dinner light river pencil." •

and (2) the anterolateral (spinothalamic) pathway. By contrast, nerve impulses that cause contraction of skeletal muscles descend along many pathways that originate mainly in the primary motor area of the brain and in the brain stem.

Somatic sensory pathways relay information from somatic sensory receptors to the primary somatosensory area in the cerebral cortex. The pathways consist of thousands of sets of three neurons (Figure 10.14). Nerve impulses for touch, pressure, vibration, and conscious proprioception (awareness of the position of body parts) ascend to the cerebral cortex along the ***posterior column–medial lemniscus pathway*** (Figure 10.14a). The name of the pathway comes from the

Figure 10.14 Somatic sensory pathways. Circles represent cell bodies and dendrites, lines represent axons, and Y-shaped forks represent axon terminals. Arrows indicate the direction of nerve impulse conduction. (a) In the posterior column–medial lemniscus pathway, the first-order neuron in the pathway ascends to the medulla oblongata via the posterior column (white matter located on the posterior side of the spinal cord). In the medulla, it synapses with a second-order neuron, which then extends through the medial lemniscus to the thalamus on the opposite side. The third-order neuron extends from the thalamus to the cerebral cortex. (b) In the anterolateral pathway, the first-order neuron synapses with a second-order neuron in the spinal cord gray matter. The second-order neuron extends to the thalamus on the opposite side, and the third-order neuron extends from the thalamus to the cerebral cortex.

Nerve impulses for somatic sensations conduct to the primary somatosensory area in the postcentral gyrus of the cerebral cortex.

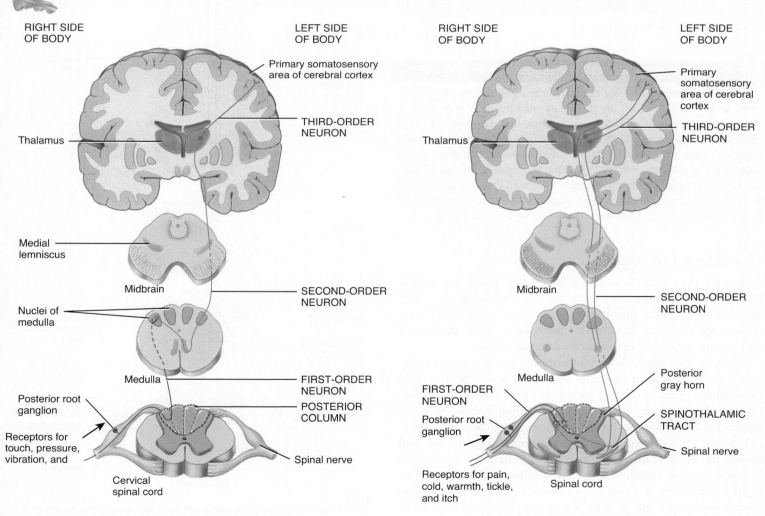

(a) Posterior column–medial lemniscus pathway

(b) Anterolateral (spinothalamic) pathway

Which somatic sensations could be lost due to damage of the spinothalamic tracts?

names of two white matter tracts that convey the impulses: the posterior column of the spinal cord and the medial lemniscus of the brain stem. The *anterolateral* or *spinothalamic pathway* (spī-nō-tha-LAM-ik) begins as a white matter tract of the spinal cord known as the **spinothalamic tract** (Figure 10.14b). This tract relays impulses for pain, temperature, itch, and tickle.

Neurons in the brain and spinal cord coordinate all voluntary and involuntary movements of the body. Ultimately, all *somatic motor pathways* that control movement converge on neurons known as *lower motor neurons*. The axons of lower motor neurons extend out of the brain stem to stimulate skeletal muscles in the head and out of the spinal cord to stimulate skeletal muscles in the limbs and trunk.

Lower motor neurons receive input from *upper motor neurons* (Figure 10.15). Upper motor neurons are essential for the execution of voluntary movements of the body. Two major tracts that conduct nerve impulses from upper motor neurons in the cerebral cortex are the *lateral corticospinal tract* and *anterior corticospinal tract*. Notice that axons of upper motor neurons from one cerebral hemisphere cross over and synapse with lower motor neurons in the other side of the spinal cord (Figure 10.15).

CLINICAL CONNECTION | Epidural haematoma

Epidural haematoma is a traumatic brain injury that develops when blood accumulates between the dura mater and the skull. It may occur after a blunt trauma to the head and is often associated with a "lucid interval" following the trauma (lasting minutes or hours) when the patient's condition appears to improve. The condition then rapidly deteriorates, potentially becoming life-threatening, as the accumulation of blood raises the intracranial pressure, compressing brain tissue. The condition is fatal in 15–20% of cases. Any significant head trauma should receive emergency medical attention, even when the patient remains conscious. •

Hemispheric Lateralization

Although the brain is quite symmetrical, there are subtle anatomical differences between the two hemispheres. They are also functionally different in some ways, with each hemisphere specializing in certain functions. This functional asymmetry is termed *hemispheric lateralization*.

As you have seen, the left hemisphere receives sensory signals from and controls the right side of the body, and the right hemisphere receives sensory signals from and controls the left side of the body. In addition, the left hemisphere is more important for spoken and written language, numerical and scientific skills, ability to use and understand sign language, and reasoning in most people. Patients with damage in the left hemisphere, for example, often have difficulty speaking. The right hemisphere is more important for musical and artistic awareness; spatial and pattern perception; recognition of faces and emotional content of language; and for generating mental images of sight, sound, touch, taste, and smell.

Figure 10.15 Somatic motor pathways. Shown here are the two most direct pathways through which signals initiated by the primary motor area in one hemisphere control skeletal muscles on the opposite side of the body. Circles represent cell bodies and dendrites, lines represent axons, and Y-shaped forks represent axon terminals.

Lower motor neurons stimulate skeletal muscles to produce movements.

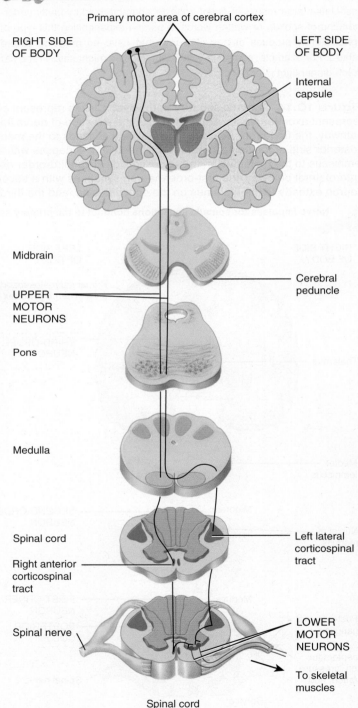

What two spinal cord tracts conduct impulses along axons of upper motor neurons?

Memory

Without memory, we would repeat mistakes and be unable to learn. Similarly, we would not be able to repeat our successes or accomplishments, except by chance. *Memory* is the process by which information acquired through learning is stored and retrieved. For an experience to become part of memory, it must produce structural and functional changes in the brain. The parts of the brain known to be involved with memory include the association areas of the frontal, parietal, occipital, and temporal lobes; parts of the limbic system; and the diencephalon. Memories for motor skills, such as how to serve a tennis ball, are stored in the basal ganglia and cerebellum as well as in the cerebral cortex.

Electroencephalogram (EEG)

At any instant, brain neurons are generating millions of nerve impulses. Taken together, these electrical signals are called *brain waves*. Brain waves generated by neurons close to the brain surface, mainly neurons in the cerebral cortex, can be detected by metal electrodes placed on the forehead and scalp. A record of such waves is called an *electroencephalogram (EEG)* (e-lek′-trō-en-SEF-a-lō-gram). Electroencephalograms are useful for studying normal brain functions, such as changes that occur during sleep. Neurologists also use them to diagnose a variety of brain disorders, such as epilepsy, tumors, metabolic abnormalities, sites of trauma, and degenerative diseases.

Table 10.1 summarizes the principal parts of the brain and their functions.

TABLE 10.1

Summary of Functions of Principal Parts of the Brain

PART	FUNCTION	PART	FUNCTION
Brain Stem 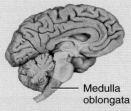	***Medulla oblongata:*** Contains sensory (ascending) tracts and motor (descending) tracts. Reticular formation (also in pons, midbrain, and diencephalon) functions in consciousness and arousal. Vital centers regulate heartbeat, breathing (together with pons), and blood vessel diameter. Other centers coordinate swallowing, vomiting, coughing, sneezing, and hiccupping. Contains nuclei of origin for cranial nerves VIII, IX, X, XI, and XII.	**Diencephalon** 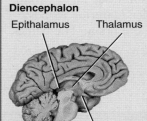	***Thalamus:*** Relays almost all sensory input to the cerebral cortex. Contributes to motor functions by transmitting information from the cerebellum and basal nuclei to motor areas of the cerebral cortex. Also plays a role in maintaining consciousness.
	Pons: Contains sensory tracts and motor tracts. Together with the medulla, helps control breathing. Contains nuclei of origin for cranial nerves V, VI, VII, and VIII.		***Hypothalamus:*** Controls and integrates activities of the autonomic nervous system and pituitary gland. Regulates emotional and behavioral patterns and circadian rhythms. Controls body temperature and regulates eating and drinking behavior. Helps maintain waking state and establishes patterns of sleep.
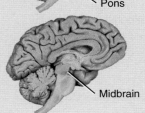	***Midbrain:*** Contains sensory tracts and motor tracts. Superior colliculi coordinate movements of head, eyes, and trunk in response to visual stimuli. Inferior colliculi coordinate movements of head, eyes, and trunk in response to auditory stimuli. The substantia nigra and red nucleus contribute to control of movement. Contains nuclei of origin for cranial nerves III and IV.		***Pineal gland:*** Secretes the hormone melatonin.
		Cerebellum 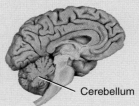	***Cerebellum:*** Smoothes and coordinates contractions of skeletal muscles. Regulates posture and balance. May have role in cognition and language processing.
		Cerebrum 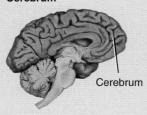	***Cerebrum:*** Sensory areas of cerebral cortex are involved in perception of sensory information; motor areas control execution of voluntary movements; association areas deal with more complex integrative functions such as memory, personality traits, and intelligence. Basal nuclei help initiate and terminate movements, suppress unwanted movements, and regulate muscle tone. Limbic system promotes range of emotions, including pleasure, pain, docility, affection, fear, and anger.

✓ **CHECKPOINT**

13. Why is the hypothalamus considered part of both the nervous system and the endocrine system?

14. What are the functions of the cerebellum and basal nuclei?

15. Where are the primary somatosensory area and primary motor area located in the brain? What are their functions?

16. What areas of the cerebral cortex are needed for normal language abilities?

17. Compare and contrast the posterior column–medial lemniscus pathway and the spinothalamic pathway.

10.5 CRANIAL NERVES

OBJECTIVE • **Identify the 12 pairs of cranial nerves by name and number and give the functions of each.**

The 12 pairs of *cranial nerves*, like spinal nerves, are part of the peripheral nervous system. The cranial nerves are designated with roman numerals and with names (see Figure 10.8). The roman numerals indicate the order (anterior to posterior) in which the nerves emerge from the brain. The names indicate the distribution or function.

Cranial nerves emerge from the nose (cranial nerve I), the eyes (cranial nerve II), the inner ear (cranial nerve VIII), the brain stem (cranial nerves III–XII), and the spinal cord (cranial nerve XI). Three cranial nerves (I, II, and VIII) contain only axons of sensory neurons and thus are called *sensory nerves*. Five cranial nerves (III, IV, VI, XI, and XII) contain only axons of motor neurons as they leave the brain stem and are called *motor nerves*. The other four cranial nerves (V, VII, IX, and X) are *mixed nerves* because they contain axons of both sensory and motor neurons. The cell bodies of sensory neurons are located in ganglia outside the brain. The cell bodies of motor neurons lie in nuclei within the brain. Cranial nerves III, VII, IX, and X include both somatic and autonomic motor axons. The somatic axons innervate skeletal muscles; the autonomic axons, which are part of the parasympathetic division, innervate glands, smooth muscle, and cardiac muscle.

Table 10.2 lists the cranial nerves, along with their components (sensory, motor, or mixed) and functions.

✓ **CHECKPOINT**

18. What is the difference between a mixed cranial nerve and a sensory cranial nerve?

TABLE 10.2

Summary of Cranial Nerves (see also Figure 10.8)

NUMBER	NAME*	COMPONENTS	FUNCTION
I	**Olfactory nerve** (ol-FAK-tō-rē; *olfact-* = to smell)	*Sensory:* Axons in the lining of the nose	Smell
II	**Optic nerve** (OP-tik; *opti-* = eye, vision)	*Sensory:* Axons from the retina of the eye	Vision
III	**Oculomotor nerve** (ok′-ū-lō-MŌ-tor; *oculo-* = eye; *-motor* = mover)	*Motor:* Axons of somatic motor neurons that stimulate muscles of upper eyelid and four muscles that move the eyeballs (superior rectus, medial rectus, inferior rectus, and inferior oblique) plus axons of parasympathetic neurons that pass to two sets of smooth muscles—the ciliary muscle of the eyeball and the sphincter muscle of the iris	Movement of upper eyelid and eyeball; alters shape of lens for near vision and constricts pupil
IV	**Trochlear nerve** (TRŌK-lē-ar; *trochle-* = a pulley)	*Motor:* Axons of somatic motor neurons that stimulate the superior oblique muscles	Movement of the eyeball
V	**Trigeminal nerve** (trī-JEM-i-nal = triple, for its three branches)	*Sensory part:* Consists of three branches: the *ophthalmic nerve* contains axons from the scalp and forehead skin; the *maxillary nerve* contains axons from the lower eyelid, nose, upper teeth, upper lip, and pharynx; and the *mandibular nerve* contains axons from the tongue, lower teeth, and the lower side of the face	Touch, pain, and temperature sensations and muscle sense (proprioception)
		Motor part: Axons of somatic motor neurons that stimulate muscles used in chewing	Chewing

NUMBER	NAME*	COMPONENTS	FUNCTION
VI	**Abducens nerve** (ab-DOO-senz; *ab-* = away; *-ducens* = to lead)	*Motor:* Axons of somatic motor neurons that stimulate the lateral rectus muscles	Movement of eyeball
VII	**Facial nerve** (FĀ-shal = face)	*Sensory part:* Axons from taste buds on tongue and axons from proprioceptors in muscles of face and scalp	Taste; muscle sense (proprioception); touch, pain, and temperature sensations
		Motor part: Axons of somatic motor neurons that stimulate facial, scalp, and neck muscles plus parasympathetic axons that stimulate lacrimal (tear) glands and salivary glands	Facial expressions; secretion of tears and saliva
VIII	**Vestibulocochlear nerve** (vest-tib-ū-lō-KOK-lē-ar; *vestibulo-* = small cavity; *-cochlear* = a spiral, snail-like)	*Vestibular branch, sensory:* Axons from semicircular canals, saccule, and utricle (organs of equilibrium)	Equilibrium
		Cochlear branch, sensory: Axons from spiral organ (organ of hearing)	Hearing
IX	**Glossopharyngeal nerve** (glos'-ō-fa-RIN-jē-al; *glosso-* = tongue; *-pharyngeal* = throat)	*Sensory part:* Axons from taste buds and somatic sensory receptors on part of tongue, from proprioceptors in some swallowing muscles, and from stretch receptors in carotid sinus and chemoreceptors in carotid body	Taste and somatic sensations (touch, pain, and temperature) from tongue; muscle sense (proprioception) in some swallowing muscles; monitoring blood pressure; monitoring oxygen and carbon dioxide in blood for regulation of breathing
		Motor part: Axons of somatic motor neurons that stimulate swallowing muscles of throat plus parasympathetic axons that stimulate a salivary gland	Swallowing, speech, secretion of saliva
X	**Vagus nerve** (VĀ-gus; *vagus* = vagrant or wandering)	*Sensory part:* Axons from taste buds in pharynx (throat) and epiglottis; proprioceptors in muscles of neck and throat, from stretch receptors and chemoreceptors in carotid sinus and carotid body, from chemoreceptors in aortic body, and from visceral sensory receptors in most organs of the thoracic and abdominal cavities	Taste and somatic sensations (touch, pain, temperature) from pharynx and epiglottis; monitoring of blood pressure; monitoring of oxygen and carbon dioxide in blood for regulation of breathing; sensations from visceral organs in thorax and abdomen
		Motor part: Axons of somatic motor neurons that stimulate skeletal muscles of the throat and neck plus parasympathetic axons that supply smooth muscle in the airways, esophagus, stomach, small intestine, most of the large intestine, and gallbladder; cardiac muscle in the heart; and glands of the gastrointestinal tract	Swallowing, coughing, and voice production; smooth muscle contraction and relaxation in organs of the gastrointestinal tract; slowing of the heart rate; secretion of digestive fluids
XI	**Accessory nerve** (ak-SES-ō-re = assisting)	*Motor:* Axons of somatic motor neurons that stimulate the sternocleidomastoid and trapezius muscles of the throat and neck	Movements of head and shoulders
XII	**Hypoglossal nerve** (hī'-pō-GLOS-al; *hypo-* = below; *-glossal* = tongue)	*Motor:* Axons of somatic motor neurons that stimulate muscles of the tongue	Movement of tongue during speech and swallowing

*A mnemonic device that can be used to remember the names of the nerves is: "**O**h, **o**h, **o**h, **t**o **t**ouch **a**nd **f**eel **v**ery **g**reen **v**egetables-**AH**!" Each boldfaced letter corresponds to the first letter of a pair of cranial nerves.

10.6 AGING AND THE NERVOUS SYSTEM

OBJECTIVE • **Describe the effects of aging on the nervous system.**

The brain grows rapidly during the first few years of life. Growth is due mainly to an increase in the size of neurons already present, the proliferation and growth of neuroglia, the development of dendritic branches and synaptic contacts, and continuing myelination of axons. From early adulthood onward, brain mass declines. By the time a person reaches age 80, the brain weighs about 7 percent less than it did in young adulthood. Although the number of neurons present does not decrease very much, the number of synaptic contacts declines. Associated with the decrease in brain mass is a decreased capacity for sending nerve impulses to and from the brain. As a result, processing of information diminishes. Conduction velocity decreases, voluntary motor movements slow down, and reflex times increase.

✓ CHECKPOINT

19. How is brain mass related to age?

COMMON DISORDERS

Spinal Cord Injury

Most spinal cord injuries are due to trauma as a result of factors such as automobile accidents, falls, contact sports, diving, or acts of violence. The effects of the injury depend on the extent of direct trauma to the spinal cord or compression of the cord by fractured or displaced vertebrae or blood clots. Although any segment of the spinal cord may be involved, most common sites of injury are in the cervical, lower thoracic, and upper lumbar regions. Depending on the location and extent of spinal cord damage, paralysis may occur. *Monoplegia* (*mono-* = one; *-plegia* = blow or strike) is paralysis of *one limb only*. *Diplegia* (*di-* = two) is paralysis of the same part on *both* sides of the body. It usually affects the lower limbs more severely than the upper limbs. *Paraplegia* (*para-* = beyond) is paralysis of *both lower limbs*. *Hemiplegia* (*hemi-* = half) is paralysis of the upper limb, trunk, and lower limb on *one* side of the body, and *quadriplegia* (*quad-* = four) is paralysis of *all four limbs*.

Shingles

Shingles is an acute infection of the peripheral nervous system caused by reactivation of the varicella-zoster virus (VZV), the virus that also causes chickenpox. After a person recovers from chickenpox, the virus retreats to a posterior root ganglion. If the virus is reactivated, it may leave the ganglion and travel down sensory axons to the skin. The result is pain, discoloration of the skin, and a characteristic line of skin blisters. The line of blisters marks the distribution of the particular sensory nerve belonging to the infected posterior root ganglion.

Amyotrophic Lateral Sclerosis

Amyotrophic lateral sclerosis (ALS) (ā'-mī-ō'-TROF-ik; *a-* = without; *myo-* = muscle; *trophic* = nourishment) is a progressive degenerative disease that attacks motor areas of the cerebral cortex, axons of upper motor neurons, and lower motor neuron cell bodies. ALS is commonly known as *Lou Gehrig's disease* after the New York Yankees baseball player who died of it at age 37 in 1941. ALS causes progressive muscle weakness and atrophy. ALS often begins in sections of the spinal cord that serve the hands and arms but rapidly spreads to involve the whole body and face, without affecting intellect or sensations. Death typically occurs in 2 to 5 years. ALS may be caused by the buildup in the synaptic cleft of the neurotransmitter glutamate released by motor neurons. The excess glutamate causes motor neurons to malfunction and eventually die. The drug riluzole, which is used to treat ALS, reduces damage to motor neurons by decreasing the release of glutamate. Other factors implicated in the development of ALS include damage to motor neurons by free radicals, autoimmune responses, viral infection, deficiency of nerve growth factor, apoptosis (programmed cell death), environmental toxins, and trauma.

Cerebrovascular Accident

The most common brain disorder is a *cerebrovascular accident (CVA)*, also called a *stroke* or *brain attack*. CVAs affect 500,000 people a year in the United States and represent the third leading cause of death, behind heart attacks and cancer. A CVA is characterized by abrupt onset of persisting symptoms, such as paralysis or loss of sensation, that arise from destruction of brain tissue. Common causes of CVAs are hemorrhage from a blood vessel in the pia mater or brain, blood clots, and formation of cholesterol-containing atherosclerotic plaques that block brain blood flow. The risk factors implicated in CVAs are high blood pressure, high blood cholesterol, heart disease, narrowed carotid arteries, transient ischemic attacks (discussed next), diabetes, smoking, obesity, and excessive alcohol intake.

Transient Ischemic Attack

A *transient ischemic attack (TIA)* is a temporary cerebral dysfunction caused by impaired blood flow to part of the brain. Symptoms include dizziness, weakness, numbness, or paralysis in a limb or in one side of the body; drooping of one side of the face; headache; slurred speech or difficulty understanding speech; and a partial loss of vision or double vision. Sometimes nausea or vomiting also occurs. The onset of symptoms is sudden and reaches maximum intensity almost immediately. A TIA usually persists for 5 to 10 minutes and only rarely lasts as long as 24 hours. It leaves no persistent neurological deficits. The causes of TIAs include blood clots, atherosclerosis, and certain blood disorders.

Poliomyelitis

Poliomyelitis, or simply *polio*, is caused by a virus called *poliovirus*. The onset of the disease is marked by fever, severe headache, a stiff neck and back, deep muscle pain and weakness, and loss of certain somatic reflexes. In its most serious form, the virus produces paralysis by destroying cell bodies of motor neurons, specifically those in the anterior horns of the spinal cord and in the nuclei of the cranial nerves. Polio can cause death from respiratory or heart failure if the virus invades neurons in vital centers that control breathing and heart functions in the brain stem. Even though polio vaccines have virtually eradicated polio in the United States, outbreaks of polio continue throughout the world. Due to international travel, polio could easily be reintroduced into North America if individuals are not vaccinated appropriately.

Several decades after suffering a severe attack of polio and following their recovery from it, some individuals develop a condition called *post-polio syndrome*. This neurological disorder is characterized by progressive muscle weakness, extreme fatigue, loss of function, and pain, especially in muscles and joints. Post-polio syndrome seems to involve a slow degeneration of motor neurons that innervate muscle fibers. Triggering factors appear to be a fall, a minor accident, surgery, or prolonged bed rest. Possible causes include overuse of surviving motor neurons over time, smaller motor neurons because of the initial infection by the virus, reactivation of dormant polio viruses, immune-mediated responses, hormone deficiencies, and environmental toxins. Treatment consists of muscle-strengthening exercises, administration of drugs to enhance the action of acetylcholine in stimulating muscle contraction, and administration of nerve growth factors to stimulate both nerve and muscle growth.

Parkinson's Disease

Parkinson's disease (PD) is a progressive disorder of the CNS that typically affects its victims around age 60. Neurons that extend from the substantia nigra to the putamen and caudate nucleus, where they release the neurotransmitter dopamine (DA), degenerate in PD. The cause of PD is unknown, but toxic environmental chemicals, such as pesticides, herbicides, and carbon monoxide, are suspected contributing agents. Only 5 percent of PD patients have a family history of the disease.

In PD patients, involuntary skeletal muscle contractions often interfere with voluntary movement. For instance, the muscles of the upper limb may alternately contract and relax, causing the hand to shake. This shaking, called *tremor*, is the most common symptom of PD. Also, muscle tone may increase greatly, causing rigidity of the involved body part. Rigidity of the facial muscles gives the face a masklike appearance. The expression is characterized by a wide-eyed, unblinking stare and a slightly open mouth with uncontrolled drooling.

Motor performance is also impaired by *bradykinesia* (brady- = slow), slowness of movements. Activities such as shaving, cutting food, and buttoning a shirt take longer and become increasingly more difficult as the disease progresses. Muscular movements also exhibit *hypokinesia* (hypo- = under), decreasing range of motion. For example, words are written smaller, letters are poorly formed, and eventually handwriting becomes illegible. Often, walking is impaired; steps become shorter and shuffling, and arm swing diminishes. Even speech may be affected.

Alzheimer's Disease

Alzheimer's disease (AD) (ALTZ-hī-merz) is a disabling senile dementia, the loss of reasoning and ability to care for oneself, that afflicts about 11 percent of the population over age 65. In the United States, AD afflicts about 4 million people and claims over 100,000 lives a year. The cause of most AD cases is still unknown, but evidence suggests it is due to a combination of genetic factors, environmental or lifestyle factors, and the aging process. Mutations in three different genes (coding for presenilin-1, presenilin-2, and amyloid precursor protein) lead to early-onset forms of AD in afflicted families but account for less than 1 percent of all cases. An environmental risk factor for developing AD is a history of head injury. A similar dementia occurs in boxers, probably caused by repeated blows to the head.

Individuals with AD initially have trouble remembering recent events. They then become confused and forgetful, often repeating questions or getting lost while traveling to previously familiar places. Disorientation increases; memories of past events disappear; and episodes of paranoia, hallucination, or violent changes in mood may occur. As their minds continue to deteriorate, AD patients lose their ability to read, write, talk, eat, or walk. At autopsy, brains of AD victims show three distinct structural abnormalities: (1) loss of neurons that liberate acetylcholine from a brain region called the nucleus basalis, located below the globus pallidus; (2) beta-amyloid plaques, clusters of abnormal proteins deposited outside neurons; and (3) neurofibrillary tangles, abnormal bundles of protein filaments inside neurons in affected brain regions. A person with AD usually dies of some complication that afflicts bedridden patients, such as pneumonia.

MEDICAL TERMINOLOGY AND CONDITIONS

Analgesia (an'-al-JĒ-zē-a; *an-* = without; *-algesia* = painful condition) Pain relief.

Anesthesia (an'-es-THĒ-zē-a; *-esthesia* = feeling) Loss of sensation.

Consciousness (KON-shus-nes) A state of wakefulness in which an individual is fully alert, aware, and oriented, partly as a result of feedback between the cerebral cortex and reticular activating system.

Dementia (de-MEN-shē-a; *de-* = away from; *-mentia* = mind) Permanent or progressive general loss of intellectual abilities, including impairment of memory, judgment, and abstract thinking, and changes in personality.

Encephalitis (en'-sef-a-LĪ-tis) An acute inflammation of the brain caused by either a direct attack by any of several viruses or an allergic reaction to any of the many viruses that are normally harmless to the central nervous system. If the virus affects the spinal cord as well, the condition is called *encephalomyelitis*.

Meningitis (men-in-JĪ-tis) Inflammation of the meninges.

Nerve block   Loss of sensation due to injection of a local anesthetic; an example is local dental anesthesia.

Neuralgia (noo-RAL-jē-a; *neur-* = nerve; *-algia* = pain)   Attacks of pain along the entire length or a branch of a peripheral sensory nerve.

Neuritis (*neur-* = nerve; *-itis* = inflammation)   Inflammation of one or several nerves, resulting from irritation caused by bone fractures, contusions, or penetrating injuries. Additional causes include infections; vitamin deficiency (usually thiamine); and poisons such as carbon monoxide, carbon tetrachloride, heavy metals, and some drugs.

Reye (RĪ) **syndrome**   Occurs after a viral infection, particularly chickenpox or influenza, most often in children or teens who have taken aspirin; characterized by vomiting and brain dysfunction (disorientation, lethargy, and personality changes) that may progress to coma and death.

Sciatica (sī-AT-i-ka)   A type of neuritis characterized by severe pain along the path of the sciatic nerve or its branches; may be caused by a slipped disc, pelvic injury, osteoarthritis of the backbone, or pressure from an expanding uterus during pregnancy.

CHAPTER REVIEW AND RESOURCE SUMMARY

REVIEW	RESOURCES
10.1 Spinal Cord Structure	
1. The spinal cord is protected by the vertebral column, **meninges**, and cerebrospinal fluid. The meninges are three connective tissue coverings of the spinal cord and brain: **dura mater**, **arachnoid mater**, and **pia mater**.	Anatomy Overview–The Spinal Cord
2. Removal of cerebrospinal fluid from the **subarachnoid space** is called a **spinal tap**. The procedure is used to remove CSF and to introduce antibiotics, anesthetics, and chemotherapy.	Figure 10.1 Spinal meninges
3. The **spinal cord** extends from the lowest part of the brain, the medulla oblongata, to the upper border of the second lumbar vertebra in the vertebral column. It contains **cervical** and **lumbar enlargements** that serve as points of origin for nerves to the limbs.	Figure 10.2 Spinal cord and spinal nerves
4. The roots of the nerves arising from the lumbar, sacral, and coccygeal regions of the cord are called the **cauda equina**. **Spinal nerves** are attached to the spinal cord by means of a **posterior root** and an **anterior root**.	
5. All spinal nerves are mixed nerves containing sensory and motor axons.	
6. The gray matter in the spinal cord is divided into horns and the white matter is divided into columns. Parts of the spinal cord observed in cross section are the central canal; **anterior, posterior,** and **lateral gray horns**; anterior, posterior, and lateral **white columns**; and **sensory (ascending)** and **motor** (descending) tracts.	
10.2 Spinal Nerves	
1. The 31 pairs of spinal nerves are named and numbered according to the region and level of the spinal cord from which they emerge.	Anatomy Overview– Spinal Nerves
2. There are 8 pairs of cervical nerves, 12 pairs of thoracic nerves, 5 pairs of lumbar nerves, 5 pairs of sacral nerves, and 1 pair of coccygeal nerves.	
3. Branches of spinal nerves, except for T2 to T11, form networks of nerves called plexuses. Nerves T2 to T11 do not form plexuses and are called intercostal nerves.	
4. The major plexuses are the cervical, brachial, lumbar, and sacral plexuses.	
10.3 Spinal Cord Functions	
1. The spinal cord white matter and gray matter have two major functions in maintaining homeostasis. The white matter serves as a highway for nerve impulse conduction. The gray matter receives and integrates incoming and outgoing information and is a site for integration of reflexes.	Animation–Reflexes

REVIEW	RESOURCES

2. A **reflex** is a fast, involuntary sequence of actions that occurs in response to a particular stimulus. The basic components of a **reflex arc** are a **receptor**, a **sensory neuron**, an **integrating center**, a **motor neuron**, and an **effector**.

10.4 Brain

1. The major parts of the **brain** are the **brain stem, diencephalon, cerebellum,** and **cerebrum** (see Table 10.1). The brain is well supplied with oxygen and nutrients. Any interruption of the oxygen supply to the brain can weaken, permanently damage, or kill brain cells. Glucose deficiency may produce dizziness, convulsions, and unconsciousness.

2. The **blood–brain barrier (BBB)** limits the passage of certain material from the blood into the brain. The brain is also protected by cranial bones, meninges, and cerebrospinal fluid. The **cranial meninges** are continuous with the spinal meninges and are named **dura mater, arachnoid mater,** and **pia mater.** Cerebrospinal fluid (CSF) is formed in the **choroid plexuses** and circulates continually through the subarachnoid space, **ventricles,** and central canal. CSF protects by serving as a shock absorber, delivers nutrients from the blood, and removes wastes.

3. The brain stem consists of the **medulla oblongata, pons,** and **midbrain,** along with clusters of neuronal cells bodies referred to as the **reticular formation.** The medulla oblongata (medulla) is continuous with the upper part of the spinal cord and contains regions for regulating heart rate, diameter of blood vessels, breathing, swallowing, coughing, vomiting, sneezing, and hiccupping. Cranial nerves VIII–XII originate at the medulla. The pons links parts of the brain with one another; it relays impulses for voluntary skeletal movements from the cerebral cortex to the cerebellum, and it contains two regions that control breathing. Cranial nerves V–VII and part of VIII originate at the pons. The midbrain, located between the pons and diencephalon, conveys motor impulses from the cerebrum to the cerebellum and spinal cord, sends sensory impulses from the spinal cord to the thalamus, and mediates auditory and visual reflexes. It also contains **nuclei** associated with cranial nerves III and IV. The reticular formation is a netlike arrangement of gray and white matter extending throughout the brain stem that alerts the cerebral cortex to incoming sensory signals and helps regulate muscle tone.

4. The diencephalon consists of the **thalamus, hypothalamus,** and **pineal gland.** The thalamus contains nuclei that serve as relay stations for sensory impulses to the cerebral cortex, and contributes to motor functions by transmitting information from the cerebellum and basal ganglia to motor areas of the cerebral cortex. The hypothalamus, located inferior to the thalamus, controls the autonomic nervous system, secretes hormones, functions in rage and aggression, governs body temperature, regulates food and fluid intake, and establishes circadian rhythms. The pineal gland secretes melatonin, which is involved in the setting of the body's biological clock.

5. The cerebellum, which occupies the inferior and posterior aspects of the cranial cavity, attaches to the brain stem by **cerebellar peduncles.** It coordinates movements and helps maintain normal muscle tone, posture, and balance.

6. The cerebrum is the largest part of the brain. Its cortex contains **gyri** (convolutions), **fissures,** and **sulci.** The cerebral lobes are **frontal, parietal, temporal,** and **occipital.** The **cerebral white matter** is deep to the cerebral cortex and consists of myelinated and unmyelinated axons extending to other CNS regions. The **basal nuclei** are several groups of nuclei in each cerebral hemisphere that help control automatic movements of skeletal muscles and help regulate muscle tone.

7. The **limbic system,** which encircles the upper part of the brain stem and the **corpus callosum,** functions in emotional aspects of behavior and memory.

8. The **sensory areas** of the cerebral cortex receive and perceive sensory information. The **motor areas** govern muscular movement. The **association areas** are concerned with emotional and intellectual processes. **Somatic sensory pathways** from receptors to the cerebral cortex involve sets of three neurons. The **posterior column–medial lemniscus pathway** relays nerve impulses for touch, pressure, vibration, and conscious proprioception. The **spinothalamic pathway** relays impulses for pain, temperature, itch, and tickle. All **somatic motor pathways** that control movement converge on **lower motor neurons.** Input to lower motor neurons comes from local interneurons, **upper motor neurons,** basal nuclei neurons, and cerebellar neurons.

Anatomy Overview–The Brain

Animation–Somatic Sensory and Motor Pathways

Figure 10.8 Inferior aspect of the brain, showing the brain stem and cranial nerves

Figure 10.11 Cerebrum

REVIEW	RESOURCES

9. Subtle anatomical differences exist between the two cerebral hemispheres, and each has some unique functions. This functional asymmetry is called **hemispheric lateralization**. **Memory**, the ability to store and recall thoughts, involves persistent changes in the brain. Brain waves generated by the cerebral cortex are recorded as an **electroencephalogram (EEG)**, which may be used to diagnose epilepsy, infections, and tumors.

10.5 Cranial Nerves

1. Twelve pairs of **cranial nerves** emerge from the brain.
2. Like spinal nerves, cranial nerves are part of the PNS. See Table 10.2 for the names, components, and functions of each of the cranial nerves.

Anatomy Overview–Cranial Nerves

10.6 Aging and the Nervous System

1. The brain grows rapidly during the first few years of life.
2. Age-related effects involve loss of brain mass and decreased capacity for sending nerve impulses.

Homeostatic Imbalances–The Case of the Forgetful Father

SELF-QUIZ

1. Which sequence best represents a reflex arc from the stimulus to the response?
   1. effector
   2. integrating center
   3. motor neuron
   4. receptor
   5. sensory neuron
      a. 3, 1, 4, 5, 2   **b.** 1, 5, 2, 3, 4   **c.** 4, 3, 2, 5, 1
      d. 5, 2, 3, 4, 1   **e.** 4, 5, 2, 3, 1

2. Which of the following would carry sensory nerve impulses?
   a. spinothalamic tract   **b.** anterior root
   c. lateral corticospinal tract   **d.** anterior gray horns
   e. upper motor neurons

3. The basal nuclei
   a. release melatonin to regulate individuals' biological clocks.
   b. help regulate muscle tone needed for specific movements.
   c. are sometimes known as the "emotional brain."
   d. are masses of gray matter in the PNS.
   e. are the sites where motor impulses cross over to the opposite side of the spinal cord.

4. Carpal tunnel syndrome is due to damage to a nerve in the
   a. lumbar plexus.
   b. cervical plexus.
   c. brachial plexus.
   d. cauda equina.
   e. sacral plexus.

5. A needle used in a spinal tap would penetrate (in order):
   1. arachnoid mater   2. dura mater   3. epidural space
   4. subarachnoid space
   a. 1, 2, 3, 4   **b.** 2, 3, 1, 4   **c.** 3, 1, 4, 2
   d. 3, 2, 1, 4   **e.** 4, 1, 2, 3

6. Axons of the lower motor neurons leave the CNS through the
   a. cerebellum.
   b. brain stem and spinal cord.
   c. cerebellum and midbrain.
   d. corticospinal tracts.
   e. cerebral hemispheres.

7. Which of the following statements about the blood supply to the brain is NOT true?
   a. The brain needs a constant supply of glucose.
   b. The structure of the brain capillaries allows selective passage of certain materials from the blood into the brain.
   c. The glucose brought to the brain can be stored for future use.
   d. Depriving brain neurons of oxygen for four minutes or more may cause permanent injury.
   e. The brain requires about 20% of the body's oxygen supply.

8. After a car accident, Joe exhibits severe dizziness, difficulty in walking, and slurred speech. He may have damaged his
   a. cerebellum.   **b.** pons.
   c. reticular activating system.   **d.** fifth cranial nerve.
   e. midbrain.

9. Which of the following is NOT a function of cerebrospinal fluid?
   a. protection   **b.** circulation
   c. conduction of nerve impulses
   d. nutrition   **e.** waste removal

10. Which part of the brain contains the centers that control the heart rate and breathing rhythm?
   a. medulla   **b.** midbrain   **c.** cerebellum
   d. thalamus   **e.** pons

11. The part of the brain that serves as a primary link between the nervous and endocrine systems is the
    a. thalamus.   **b.** hypothalamus.   **c.** pons.
    d. brain stem.   **e.** cerebellum.

12. Which of the following is NOT a function of the hypothalamus?
    a. regulates food intake
    b. controls body temperature
    c. regulates feelings of rage and aggression
    d. helps establish sleep patterns
    e. coordinates the startle reflex

13. The part(s) of the brain concerned with memory, reasoning, judgment, and intelligence is (are) the
    a. sensory areas.   **b.** limbic system.   **c.** motor areas.
    d. cerebellum.   **e.** association areas.

14. A broad band of white matter that connects the two cerebral hemispheres is the
    a. corpus callosum.   **b.** gyrus.   **c.** insula.
    d. ascending tract.   **e.** upper motor neurons.

15. The ringing of your alarm clock in the morning wakes you up by stimulating your
    a. thalamus.   **b.** reticular activating system.
    c. Broca's area.   **d.** basal nuclei.   **e.** spinal cord.

16. Match the following functions to the primary lobe in which they are located:
    ____ **a.** contains primary visual area that allows interpretation of shape and color
    ____ **b.** receives impulses for smell
    ____ **c.** contains primary motor area that controls muscle movement
    ____ **d.** receives sensory impulses for touch, pain, and temperature

    A. frontal lobe
    B. parietal lobe
    C. occipital lobe
    D. temporal lobe

17. When entering a restaurant, you are bombarded with many different sensory stimuli. The part of the brain that combines all of those sensory inputs so that you can respond appropriately is the
    a. somatosensory association area.   **d.** Wernicke's area.
    b. common integrative area.   **e.** hypothalamus.
    c. premotor area.

18. Which cranial nerves contain only sensory fibers?
    a. olfactory, optic, and glossopharyngeal
    b. optic and oculomotor
    c. optic and trochlear
    d. optic and olfactory
    e. vagus and facial

19. Which cranial nerve(s) help(s) control swallowing?
    a. glossopharyngeal   **b.** vagus   **c.** facial
    d. hypoglossal   **e.** trochlear

20. Match the following:
    ____ **a.** organization of white matter in the spinal cord
    ____ **b.** absorb cerebrospinal fluid
    ____ **c.** extension of nerves beyond the end of the spinal cord
    ____ **d.** folds of the cerebral cortex
    ____ **e.** contains the sensory axons of a spinal nerve
    ____ **f.** contains the motor axons of a spinal nerve
    ____ **g.** separates the cerebrum into right and left halves
    ____ **h.** divides spinal cord into right and left sides
    ____ **i.** brain cavities where CSF circulates
    ____ **j.** shallow grooves in the cerebrum
    ____ **k.** contains CSF in the spinal cord

    A. longitudinal fissure
    B. sulci
    C. ventricles
    D. anterior median fissure
    E. central canal
    F. posterior (dorsal) root
    G. columns
    H. arachnoid villi
    I. anterior (ventral) root
    J. gyri
    K. cauda equina

CRITICAL THINKING APPLICATIONS

1. After a few days of using his new crutches, Keith's arms and hands felt tingly and numb. The physical therapist said Keith had a case of "crutch palsy" from improper use of his crutches. Keith had been leaning his armpits on the crutches while hobbling along. What caused the numbness in his arms and hands?

2. A few days after a minor car accident, Alexa suffers from vision problems and is feeling pressure in the back of her head. After a series of diagnostic procedures, her physician indicates that Alexa immediately needs to have "water drained from her brain." Explain to Alexa what the surgeon plans to do and why she might have "water on her brain."

3. An elderly relative suffered a stroke and now has difficulty with the movement of her right upper limb. She is also working with a therapist due to some speech problems. What areas of the brain were damaged by the stroke?

4. Joanne ran into the room when she heard her husband's cry. Jack was bouncing on his left foot while holding his right foot in his hand. A pin was sticking out of the bottom of his foot. Explain Jack's response to stepping on the pin.

ANSWERS TO FIGURE QUESTIONS

10.1 CSF circulates in the subarachnoid space.

10.2 Spinal nerves are part of the PNS (peripheral nervous system).

10.3 A horn is an area of gray matter, and a column is a region of white matter in the spinal cord.

10.4 All spinal nerves are mixed (have sensory and motor components) because the posterior root containing sensory axons and the anterior root containing motor axons unite to form the spinal nerve.

10.5 Axons of sensory neurons are part of the posterior root, and axons of motor neurons are part of the anterior root.

10.6 The medulla oblongata of the brain attaches to the spinal cord.

10.7 CSF is formed in the choroid plexuses and is reabsorbed through arachnoid villi into blood in the superior sagittal sinus.

10.8 The midbrain contains the cerebral peduncles.

10.9 The superior colliculi govern eye movements for tracking moving images and scanning stationary images and are responsible for reflexes that govern movements of the eyes, head, and neck in response to visual stimuli.

10.10 The basal nuclei are located in the cerebrum and are composed of gray matter.

10.11 The longitudinal fissure separates the right and left cerebral hemispheres.

10.12 The limbic system is located on the inner border of the cerebrum and floor of the diencephalon.

10.13 The primary somatosensory area localizes somatic sensations.

10.14 Damage to the spinothalamic tracts could produce loss of pain, thermal, tickle, and itch sensations.

10.15 In the spinal cord, the lateral and anterior corticospinal tracts conduct impulses along axons of upper motor neurons.

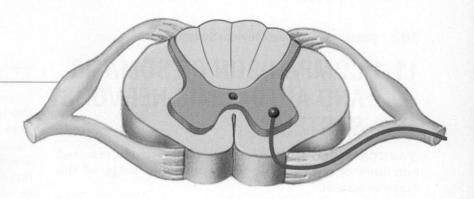

CHAPTER 11

AUTONOMIC NERVOUS SYSTEM

It is the end of the semester, you have studied diligently for your anatomy and physiology final, and now it is time to take the exam. As you enter the crowded lecture hall and take a seat, you sense tension in the room as the other students nervously chatter about last-minute details they think will be important to know for the test. Suddenly you feel your heart race with excitement—or is that apprehension? You notice that your mouth becomes somewhat dry, and you break out in a cold sweat. You also notice that your breathing is a little bit faster and deeper. As you wait for the professor to pass out the test, these symptoms become more and more pronounced. Finally the test arrives at your desk. As you slowly flip through the exam to get a feel for the questions being asked, you recognize that you can answer them all with confidence. What a relief! Your symptoms begin to disappear as you focus on transferring your knowledge from your brain to the paper. Most of the effects just decribed fall under the control of the ***autonomic nervous system (ANS)***, that part of the nervous system that regulates smooth muscle, cardiac muscle, and certain glands. Recall that together the ANS and somatic nervous system compose the peripheral nervous system; see Figure 9.1. The ANS was originally named *autonomic* (*auto-* = self; *-nomic* = law) because it was thought to function in a self-governing manner. Although the ANS usually does operate without conscious control from the cerebral cortex, it is regulated by other brain regions, mainly the hypothalamus and brain stem. In this chapter, we compare the structural and functional features of the somatic and autonomic nervous systems. Then we discuss the anatomy of the motor portion of the ANS and compare the organization and actions of its two major branches, the sympathetic and parasympathetic divisions.

Did you know?   *The "fight-or-flight" response of the sympathetic nervous system is very helpful when you encounter a snarling dog or need to escape from a burning building. But when the emergency is over, your parasympathetic nervous system needs time to help your body relax and recover. What happens when stress builds up, and no recovery occurs? When your days are filled with negative stress and an overactivated sympathetic nervous system, stress-related health problems may develop. Chronic, unrelenting, overwhelming stress interferes with the body's ability to maintain homeostasis and health. Learning relaxation and stress reduction skills can reduce the harmful effects of stress on the body. Mind–body exercise forms such as yoga, tai chi, Pilates, and martial arts provide physical conditioning along with relaxation and stress reduction.*

FOCUS ON WELLNESS

Mind–Body Exercise—An Antidote to Stress

LOOKING BACK TO MOVE AHEAD...

Structures of the Nervous System (Section 9.1)

Sensory and Motor Components of the ANS and ANS Effectors (Section 9.1)

301

11.1 COMPARISON OF SOMATIC AND AUTONOMIC NERVOUS SYSTEMS

OBJECTIVE • **Compare the main structures and functions of the somatic and autonomic parts of the nervous system.**

As you learned in Chapter 10, the somatic nervous system includes both sensory and motor neurons. The sensory neurons convey input from receptors for the special senses (vision, hearing, taste, smell, and equilibrium, described in Chapter 12) and from receptors for somatic senses (pain, temperature, touch, and proprioceptive sensations). All of these sensations normally are consciously perceived. In turn, somatic motor neurons synapse with skeletal muscle—the effector tissue of the somatic nervous system—and produce conscious, voluntary movements. When a somatic motor neuron stimulates a skeletal muscle, the muscle contracts. If somatic motor neurons cease to stimulate a muscle, the result is a paralyzed, limp muscle that has no muscle tone. In addition, even though we are generally not conscious of breathing, the muscles that generate breathing movements are skeletal muscles controlled by somatic motor neurons. If the respiratory motor neurons become inactive, breathing stops.

The main input to the ANS comes from *autonomic sensory neurons*. These neurons are associated with sensory receptors that monitor internal conditions, such as blood CO_2 level or the degree of stretching in the walls of internal organs or blood vessels. When the viscera are functioning properly, these sensory signals usually are not consciously perceived.

Autonomic motor neurons regulate ongoing activities in their effector tissues, which are cardiac muscle, smooth muscle, and glands, by both excitation and inhibition. Unlike skeletal muscle, these tissues often function to some extent even if their nerve supply is damaged. The heart continues to beat, for instance, when it is removed for transplantation into another person. Examples of autonomic responses are changes in the diameter of the pupil, dilation and constriction of blood vessels, and changes in the rate and force of the heartbeat. Because most autonomic responses cannot be consciously altered or suppressed to any great degree, they are the basis for polygraph ("lie detector") tests. However, practitioners of yoga or other techniques of meditation and those who employ biofeedback methods may learn how to modulate ANS activities. For example, they may be able to voluntarily decrease their heart rate or blood pressure.

Recall from Chapter 8 that the axon of a single somatic motor neuron extends from the central nervous system (CNS) all the way to the skeletal muscle fibers in its motor unit (Figure 11.1a). By contrast, most autonomic motor pathways consist of *two* motor neurons (Figure 11.1b). The first neuron has its cell body in the CNS; its axon extends from the CNS as part of a cranial or spinal nerve to an *autonomic ganglion*. (Recall that a *ganglion* is a collection of neuronal cell bodies in the peripheral nervous system or PNS.) The cell body of the second neuron is also in that same autonomic ganglion; its axon extends directly from the ganglion to the effector (smooth muscle, cardiac muscle, or a gland). Alternatively, in some autonomic pathways, the first motor neuron extends to the adrenal medullae (inner portion of the adrenal glands) rather than an autonomic ganglion. In addition, all somatic motor neurons release only acetylcholine (ACh) as their neurotransmitter, but autonomic motor neurons release either ACh or norepinephrine (NE).

The output (motor) part of the ANS has two main branches: the *sympathetic division* and the *parasympathetic division*. Most organs have *dual innervation*; that is, they receive impulses from both sympathetic and parasympathetic neurons. In general, nerve impulses from one division stimulate the organ to increase its activity (excitation), and impulses from the other division decrease the organ's activity (inhibition). For example, an increased rate of nerve impulses from the sympathetic division increases heart rate, and an increased rate of nerve impulses from the parasympathetic division decreases heart rate. Table 11.1 summarizes the similarities and differences between the somatic and autonomic nervous systems.

✓ CHECKPOINT

1. Why is the autonomic nervous system so named?
2. What are the main input and output components of the autonomic nervous system?

11.2 STRUCTURE OF THE AUTONOMIC NERVOUS SYSTEM

OBJECTIVES • **Identify the structural features of the autonomic nervous system.**
• **Compare the organization of the autonomic pathways in the sympathetic and parasympathetic divisions.**

Anatomical Components

The first of the two motor neurons in any autonomic motor pathway is called a *preganglionic neuron* (Figure 11.1b). Its cell body is in the brain or spinal cord, and its axon exits the CNS as part of a cranial or spinal nerve. The axon of a preganglionic neuron usually extends to an autonomic ganglion, where it synapses with a *postganglionic neuron*, the second neuron in the autonomic motor pathway (Figure 11.1b). Notice that the postganglionic neuron lies entirely outside the CNS. Its cell body and dendrites are located in an autonomic ganglion, where it forms synapses with one or more preganglionic axons. The axon of a postganglionic neuron terminates

Figure 11.1 Comparison of somatic and autonomic motor neuron pathways to their effector tissues.

Stimulation by the autonomic motor neurons can either excite or inhibit smooth muscle, cardiac muscle, and glands. Stimulation by somatic motor neurons always causes contraction of skeletal muscle.

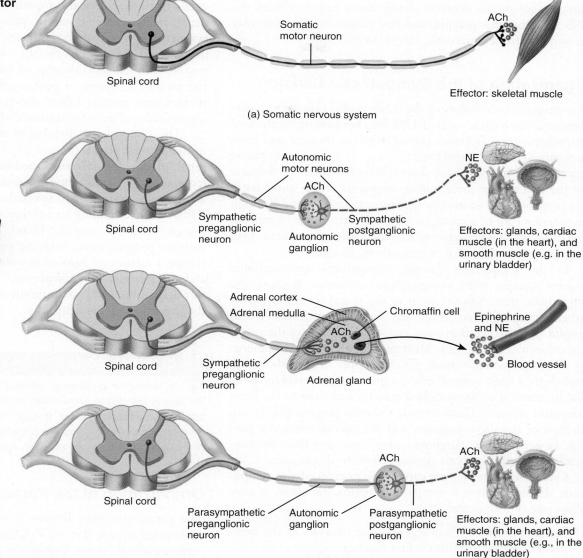

? **What does "dual innervation" mean?**

TABLE 11.1		
Comparison of Somatic and Autonomic Nervous Systems		
PROPERTY	**SOMATIC**	**AUTONOMIC**
Effectors	Skeletal muscles	Cardiac muscle, smooth muscle, and glands
Type of control	Mainly voluntary	Mainly involuntary
Neural pathway	One motor neuron extends from CNS and synapses directly with a skeletal muscle fiber	One motor neuron extends from the CNS and synapses with another motor neuron in a ganglion; the second motor neuron synapses with an autonomic effector
Neurotransmitter	Acetylcholine	Acetylcholine or norepinephrine
Action of neurotransmitter on effector	Always excitatory (causing contraction of skeletal muscle)	May be excitatory (causing contraction of smooth muscle, increased heart rate, increased force of heart contraction, or increased secretions from glands) or inhibitory (causing relaxation of smooth muscle, decreased heart rate, or decreased secretions from glands)

in an effector (smooth muscle, cardiac muscle, or a gland). Thus, preganglionic neurons convey nerve impulses from the CNS to autonomic ganglia, and postganglionic neurons relay the impulses from autonomic ganglia to effectors.

Organization of the Sympathetic Division

The sympathetic division of the ANS is also called the *thoracolumbar division* (thōr′-a-kō-LUM-bar) because the outflow of sympathetic nerve impulses comes from the thoracic and lumbar segments of the spinal cord (Figure 11.2). The sympathetic preganglionic neurons have their cell bodies in the 12 thoracic and the first two or three lumbar segments of the spinal cord. The preganglionic axons emerge from the spinal cord through the anterior root of a spinal nerve along with axons of somatic motor neurons. After exiting the cord, the sympathetic preganglionic axons extend to a sympathetic ganglion.

In the sympathetic ganglia, sympathetic preganglionic neurons synapse with postganglionic neurons. Because the sympathetic trunk ganglia are near the spinal cord, most sympathetic preganglionic axons are short. **Sympathetic trunk ganglia** lie in two vertical rows, one on either side of the vertebral column (Figure 11.2). Most postganglionic axons emerging from sympathetic trunk ganglia supply organs above the diaphragm. Other sympathetic ganglia, the **prevertebral ganglia**, lie anterior to the vertebral column and close to the large abdominal arteries. These include the *celiac ganglion* (SĒ-lē-ak), the *superior mesenteric ganglion*, and the *inferior mesenteric ganglion*. In general, postganglionic axons emerging from the prevertebral ganglia innervate organs below the diaphragm.

Once the axon of a preganglionic neuron of the sympathetic division enters a sympathetic trunk ganglion, it may follow one of four paths:

1. It may synapse with postganglionic neurons in the sympathetic trunk ganglion it first reaches.
2. It may ascend or descend to a higher or lower sympathetic trunk ganglion before synapsing with postganglionic neurons.
3. It may continue, without synapsing, through the sympathetic trunk ganglion to end at a prevertebral ganglion and synapse with postganglionic neurons there.
4. It may extend to and terminate in the adrenal medulla.

A single sympathetic preganglionic axon has many branches and may synapse with 20 or more postganglionic neurons. Thus, nerve impulses that arise in a single preganglionic neuron may activate many different postganglionic neurons that in turn synapse with several autonomic effectors. This pattern helps explain why sympathetic responses can affect organs throughout the body almost simultaneously.

Most postganglionic axons leaving the cervical sympathetic trunk ganglia serve the head. They are distributed to sweat glands, smooth muscles of the eye, blood vessels of the face, nasal mucosa, and salivary glands. A few postganglionic axons from the cervical sympathetic trunk ganglia supply the heart. In the thoracic region, postganglionic axons from the sympathetic trunk serve the heart, lungs, and bronchi. Some axons from thoracic levels also supply sweat glands, blood vessels, and smooth muscles of hair follicles in the skin. In the abdomen, axons of postganglionic neurons leaving the prevertebral ganglia follow the course of various arteries to abdominal and pelvic autonomic effectors.

The sympathetic division of the ANS also includes part of the adrenal glands (Figure 11.2). The inner part of the adrenal gland, the **adrenal medulla** (me-DUL-a), develops from the same embryonic tissue as the sympathetic ganglia, and its cells are similar to sympathetic postganglionic neurons. Rather than extending to another organ, however, these cells release hormones into the blood. Upon stimulation by sympathetic preganglionic neurons, cells of the adrenal medulla release a mixture of hormones—about 80 percent **epinephrine** and 20 percent **norepinephrine**. These hormones circulate throughout the body and intensify responses elicited by sympathetic postganglionic neurons.

CLINICAL CONNECTION | Horner's Syndrome

In **Horner's syndrome**, sympathetic stimulation of one side of the face is lost due to an inherited mutation, an injury, or a disease that affects sympathetic outflow through the superior cervical ganglion. Symptoms occur in the head on the affected side and include drooping of the upper eyelid, constricted pupil, and lack of sweating. •

Organization of the Parasympathetic Division

The parasympathetic division (Figure 11.3) is also called the *craniosacral division* (krā′-nē-ō-SĀ-kral) because the outflow of parasympathetic nerve impulses comes from cranial nerve nuclei and sacral segments of the spinal cord. The cell bodies of parasympathetic preganglionic neurons are located in the nuclei of four cranial nerves (III, VII, IX, and X) in the brain stem and in the second through fourth sacral segments of the spinal cord (S2, S3, and S4) (Figure 11.3). Parasympathetic preganglionic axons emerge from the CNS as part of a cranial nerve or as part of the anterior root of a spinal nerve. Axons of the vagus (X) nerve carry nearly 80 percent of the total parasympathetic outflow. In the thorax, axons of the vagus nerve extend to ganglia in the heart and the airways of the lungs. In the abdomen, axons of the vagus nerve extend to ganglia in the liver, stomach, pancreas, small intestine, and part of the large intestine. Parasympathetic preganglionic axons exit the sacral spinal cord in the anterior roots of the second through fourth sacral nerves. The axons then extend to ganglia in the walls of the colon, ureters, urinary bladder, and reproductive organs.

Preganglionic axons of the parasympathetic division synapse with postganglionic neurons in **terminal ganglia**, which are

Figure 11.2 Structure of the sympathetic division of the autonomic nervous system. Although some innervated structures are diagrammed only for one side of the body, the sympathetic division actually innervates tissues and organs on both sides.

Cell bodies of sympathetic preganglionic neurons are located in the gray matter in the 12 thoracic and first two or three lumbar segments of the spinal cord.

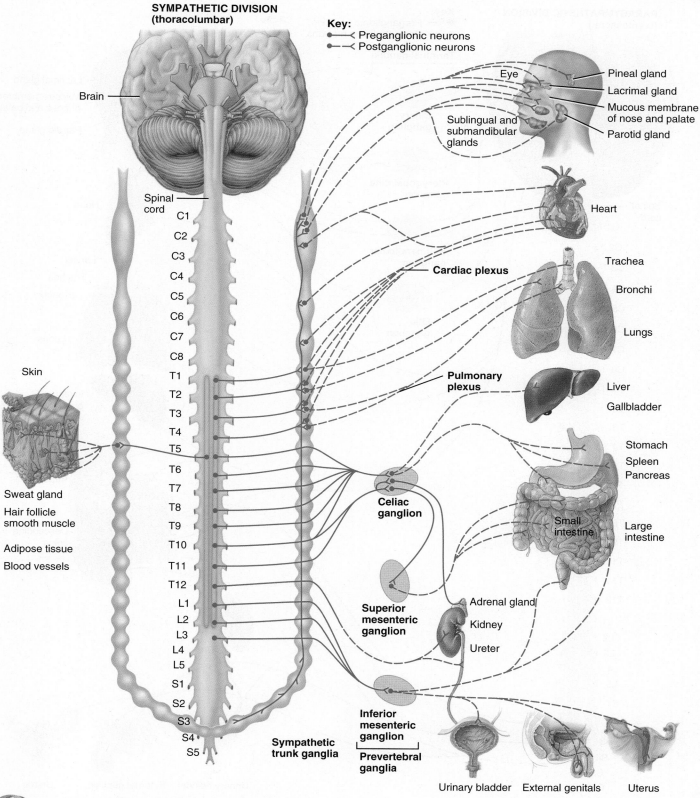

SYMPATHETIC DIVISION
(thoracolumbar)

Key:
● —— < Preganglionic neurons
● -- -- < Postganglionic neurons

Eye
Pineal gland
Lacrimal gland
Mucous membrane of nose and palate
Sublingual and submandibular glands
Parotid gland

Brain
Spinal cord

C1
C2
C3
C4
C5
C6
C7
C8
T1
T2
T3
T4
T5
T6
T7
T8
T9
T10
T11
T12
L1
L2
L3
L4
L5
S1
S2
S3
S4
S5

Skin
Sweat gland
Hair follicle smooth muscle
Adipose tissue
Blood vessels

Heart

Cardiac plexus

Trachea
Bronchi
Lungs

Pulmonary plexus

Liver
Gallbladder

Stomach
Spleen
Pancreas

Celiac ganglion

Small intestine
Large intestine

Superior mesenteric ganglion

Adrenal gland
Kidney
Ureter

Inferior mesenteric ganglion

Sympathetic trunk ganglia

Prevertebral ganglia

Urinary bladder   External genitals   Uterus

 Which neurons synapse in a sympathetic trunk ganglion?

305

Figure 11.3 Structure of the parasympathetic division of the autonomic nervous system. Although some innervated structures are diagrammed on one side of the body, the parasympathetic division actually innervates organs on both sides.

Cell bodies of parasympathetic preganglionic neurons are located in brain stem nuclei and in the gray matter in the second through fourth sacral segments of the spinal cord.

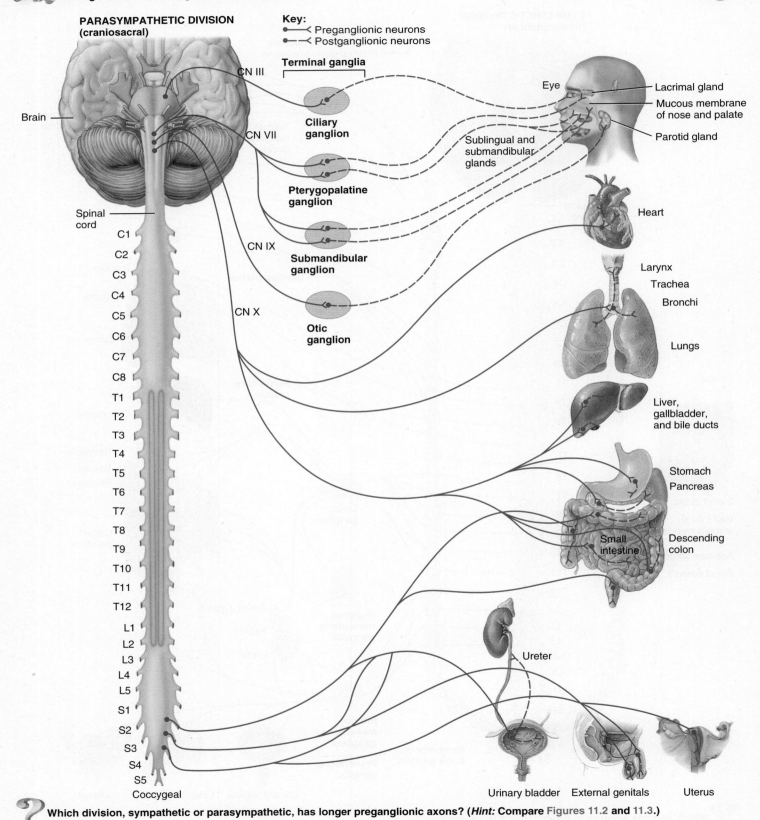

Which division, sympathetic or parasympathetic, has longer preganglionic axons? (*Hint:* Compare Figures 11.2 and 11.3.)

located close to or actually within the wall of the innervated organ. Terminal ganglia in the head receive preganglionic axons from the oculomotor (III), facial (VII), or glossopharyngeal (IX) cranial nerves and supply structures in the head (Figure 11.3). Axons in the vagus (X) nerve extend to many terminal ganglia in the thorax and abdomen. Because the axons of parasympathetic preganglionic neurons extend from the brain stem or sacral spinal cord to a terminal ganglion in an innervated organ, they are longer than most of the axons of sympathetic preganglionic neurons (compare Figures 11.2 and 11.3).

In contrast to the preganglionic axons, most parasympathetic postganglionic axons are very short because the terminal ganglia lie in the walls of their autonomic effectors. In the ganglion, the preganglionic neuron usually synapses with only four or five postganglionic neurons, all of which supply the same effector. Thus, parasympathetic responses are localized to a single effector.

CLINICAL CONNECTION | Megacolon

A **megacolon** (*mega-* = big) is an abnormally large colon. In congenital megacolon, parasympathetic nerves to the distal segment of the colon do not develop properly. Loss of motor function in the segment causes massive dilation of the normal proximal colon. The condition results in extreme constipation, abdominal distension, and occasionally, vomiting. Surgical removal of the affected segment of the colon corrects the disorder. •

✓ CHECKPOINT

3. Describe the locations of sympathetic trunk ganglia, prevertebral ganglia, and terminal ganglia. Which types of autonomic neurons synapse in each type of ganglion?

4. How can the sympathetic division produce simultaneous effects throughout the body, when parasympathetic effects typically are localized to specific organs?

11.3 FUNCTIONS OF THE AUTONOMIC NERVOUS SYSTEM

OBJECTIVE • **Describe the functions of the sympathetic and parasympathetic divisions of the autonomic nervous system.**

ANS Neurotransmitters

As you learned in Chapter 9, neurotransmitters are chemical substances released by neurons at synapses. Autonomic neurons release neurotransmitters at synapses between neurons

(preganglionic to postganglionic) and at synapses with autonomic effectors (smooth muscle, cardiac muscle, and glands). Some ANS neurons release acetylcholine; others release norepinephrine.

ANS neurons that release *acetylcholine* include (1) all sympathetic and parasympathetic preganglionic neurons, (2) all parasympathetic postganglionic neurons, and (3) a few sympathetic postganglionic neurons. Because acetylcholine is quickly inactivated by the enzyme *acetylcholinesterase (AChE)*, parasympathetic effects are short-lived and localized.

Most sympathetic postganglionic neurons release the neurotransmitter *norepinephrine (NE)*. Because norepinephrine is inactivated much more slowly than acetylcholine and because the adrenal medulla also releases epinephrine and norepinephrine into the bloodstream, the effects of activation of the sympathetic division are longer lasting and more widespread than those of the parasympathetic division. For instance, your heart continues to pound for several minutes after a near miss at a busy intersection due to the long lasting effects of the sympathetic division.

Activities of the ANS

As noted earlier, most body organs receive instructions from both divisions of the ANS, which typically work in opposition to one another. The balance between sympathetic and parasympathetic activity or "tone" is regulated by the hypothalamus. Typically, the hypothalamus turns up sympathetic tone at the same time it turns down parasympathetic tone, and vice versa. A few structures receive only sympathetic innervation—sweat glands, arrector pili muscles attached to hair follicles in the skin, the kidneys, the spleen, most blood vessels, and the adrenal medullae (see Figure 11.2). In these structures there is no opposition from the parasympathetic division. Still, an increase in sympathetic tone has one effect, and a decrease in sympathetic tone produces the opposite effect.

Sympathetic Activities

During physical or emotional stress, high sympathetic tone favors body functions that can support vigorous physical activity and rapid production of ATP. At the same time, the sympathetic division reduces body functions that favor the storage of energy. Besides physical exertion, a variety of emotions—such as fear, embarrassment, or rage—stimulate the sympathetic division. Visualizing body changes that occur during "E situations" (exercise, emergency, excitement, embarrassment) will help you remember most of the sympathetic responses. Activation of the sympathetic division and release of hormones by the adrenal medullae result in a series of physiological responses collectively called the *fight-or-flight response*, in which the following events occur:

1. The pupils of the eyes dilate.

2. Heart rate, force of heart contraction, and blood pressure increase.

Mind–Body Exercise—An Antidote to Stress

When we think of exercise, we usually think of toning up our muscles and maybe our hearts. But when some people think of exercise, they focus on toning up neural input from the parasympathetic division of the autonomic nervous system. As you learned in this chapter, activation of the parasympathetic division helps restore homeostasis in many systems and is associated with feelings of relaxation.

Mind–Body Harmony

Mind–body exercise refers to exercise systems such as tai chi, Pilates, hatha yoga, and many forms of the martial arts, all of which couple muscular activity with an internally directed focus. These exercise systems exercise the mind as well as the body, and usually include an awareness of breathing, energy, muscle tension, and other physical sensations.

Practitioners often refer to this internal awareness as "mindful," meaning that the exerciser is open to physical and emotional sensations with an understanding, nonjudgmental attitude. A mindful attitude is typical of many kinds of meditation and relaxation practices. For example, when practicing a yoga pose, you would think something like "Deep, steady breathing; relax into the pose; shoulders pulling back, neck lengthening,"

Think It Over . . . How could you make walking more of a mind–body activity?

rather than "That girl next to me sure is flexible; I'm really a failure at this stuff." Of course, in real life such external thoughts do sneak in, but we can redirect our attention back to a more neutral, nonjudgmental style.

Mind–Body Benefits

People practicing mind–body activities reap benefits from both the physical and mental activity. Hatha yoga, tai chi, and the martial arts increase muscular strength and flexibility, posture, balance, and coordination, and if performed vigorously, they can even improve cardiovascular health and endurance to some extent. In addition, the stress relief provided by the activities extends into both the physical and psychological realms. Feelings of mental relaxation and emotional well-being translate into better resting blood pressure, a healthier immune system, and more relaxed muscles. Less stress can also mean an improvement in health habits. Those who practice mind-body exercise often improve their eating habits and reduce harmful behaviors such as cigarette smoking.

3. The airways dilate, allowing faster movement of air into and out of the lungs.

4. The blood vessels that supply nonessential organs such as the kidneys and gastrointestinal tract constrict, which reduces blood flow through these tissues. The result is a slowing of urine formation and digestive activities, which are not essential during exercise.

5. Blood vessels that supply organs involved in exercise or fighting off danger—skeletal muscles, cardiac muscle, liver, and adipose tissue—dilate, which allows greater blood flow through these tissues.

6. Liver cells break down glycogen to glucose, and adipose cells break down triglycerides to fatty acids and glycerol, providing molecules that can be used by body cells for ATP production.

7. Release of glucose by the liver increases blood glucose level.

8. Processes that are not essential for meeting the stressful situation are inhibited. For example, muscular movements of the gastrointestinal tract and digestive secretions decrease or even stop.

Parasympathetic Activities

In contrast to the "fight-or-flight" activities of the sympathetic division, the parasympathetic division enhances "rest-and-digest" activities. Parasympathetic responses support body functions that conserve and restore body energy during times of rest and recovery. In the quiet intervals between periods of exercise, parasympathetic impulses to the digestive glands and the smooth muscle of the gastrointestinal tract predominate over sympathetic impulses. This allows energy-supplying food to be digested and absorbed. At the same time, parasympathetic responses reduce body functions that support physical activity.

The acronym *SLUDD* can be helpful in remembering five parasympathetic responses. It stands for salivation (S), lacrimation (L), urination (U), digestion (D), and defecation (D). Mainly the parasympathetic division stimulates all of these activities. Besides the increasing SLUDD responses, other important parasympathetic responses are "three decreases": decreased heart rate, decreased diameter of airways, and decreased diameter (constriction) of the pupils.

Table 11.2 lists the responses of glands, cardiac muscle, and smooth muscle to stimulation by the sympathetic and parasympathetic divisions of the ANS.

TABLE 11.2

Functions of the Autonomic Nervous System

EFFECTOR	EFFECT OF SYMPATHETIC STIMULATION	EFFECT OF PARASYMPATHETIC STIMULATION
Glands		
Sweat	Increased sweating	No known effect
Lacrimal (tear)	Slight secretion of tears	Secretion of tears
Adrenal medulla	Secretion of epinephrine and norepinephrine	No known effect
Pancreas	Inhibition of secretion of digestive enzymes and insulin (hormone that lowers blood glucose level); secretion of glucagon (hormone that raises blood glucose level)	Secretion of digestive enzymes and insulin
Posterior pituitary	Secretion of antidiuretic hormone (ADH)	No known effect
Liver*	Breakdown of glycogen into glucose, synthesis of new glucose, and release of glucose into the blood; decreased bile secretion	Promotes synthesis of glycogen; increases bile secretion
Adipose tissue*	Breakdown of triglycerides and release of fatty acids into blood	No known effect
Cardiac Muscle		
Heart	Increased heart rate and increased force of atrial and ventricular contraction	Decreased heart rate and decreased force of atrial contraction
Smooth Muscle		
Radial muscle of iris of eye	Dilation of pupil	No known effect
Circular muscle of iris of eye	No known effect	Constriction of pupil
Ciliary muscle of eye	Relaxation to adjust shape of lens for distant vision	Contraction to adjust shape of lens for close vision
Gallbladder and ducts	Storage of bile in the gallbladder	Release of bile into the small intestine
Stomach and intestines	Decreased motility (movement); contraction of sphincters	Increased motility; relaxation of sphincters
Lungs (smooth muscle of bronchi)	Widening of the airways (bronchodilation)	Narrowing of the airways (bronchoconstriction)
Urinary bladder	Relaxation of muscular wall; contraction of internal sphincter	Contraction of muscular wall; relaxation of internal sphincter
Spleen	Contraction and discharge of stored blood into general circulation	No known effect
Smooth muscle of hair follicles	Contraction that results in erection of hairs, producing "goose bumps"	No known effect
Uterus	Inhibits contraction in nonpregnant women; stimulates contraction in pregnant women	Minimal effect
Sex organs	In men, causes ejaculation of semen	Vasodilation; erection of clitoris (women) and penis (men)
Salivary glands (arterioles)	Decreases secretion of saliva	Stimulates secretion of saliva
Gastric glands and intestinal glands (arterioles)	Inhibits secretion	Promotes secretion
Kidney (arterioles)	Decreases production of urine	No known effect
Skeletal muscle (arterioles)	Vasodilation in most, which increases blood flow	No known effect
Heart (coronary arterioles)	Vasodilation in most, which increases blood flow	Slight vasoconstriction, which decreases blood flow

*Listed with glands because they release substances into the blood.

CLINICAL CONNECTION | Bell's Palsy

Bell's Palsy causes temporary weakness or paralysis of muscles on one side of the face, the severity of which varies. The exact cause is unknown but the herpes virus has been implicated. It is more common in immune-compromised individuals, including pregnant women and diabetics. Around 70% of patients make a complete recovery. •

✓ **CHECKPOINT**

5. What are some examples of the opposite effects of the sympathetic and parasympathetic divisions of the autonomic nervous system?

6. What happens during the fight-or-flight response?

7. Why is the parasympathetic division of the ANS considered the rest-and-digest division?

• • •

Now that we have discussed the structure and function of the nervous system, we will next consider in Chapter 12 how sensory information is relayed to the nervous system and how the nervous system responds to it.

COMMON DISORDERS

Autonomic Dysreflexia

Autonomic dysreflexia is an exaggerated response of the sympathetic division of the ANS that occurs in about 85% of individuals with spinal cord injury at or above the level of T6. The condition occurs due to interruption of the control of ANS neurons by higher centers. When certain sensory impulses, such as those resulting from stretching of a full urinary bladder, are unable to ascend the spinal cord, mass stimulation of the sympathetic nerves below the level of injury occurs. Among the effects of increased sympathetic activity is severe vasoconstriction, which elevates blood pressure. In response, the cardiovascular center in the medulla oblongata (1) increases parasympathetic output via the vagus nerve, which decreases heart rate, and (2) decreases sympathetic output, which causes dilation of blood vessels above the level of the injury.

Autonomic dysreflexia is characterized by a pounding headache; severe high blood pressure (hypertension); flushed, warm skin with profuse sweating above the injury level; pale, cold, and dry skin below the injury level; and anxiety. It is an emergency condition that requires immediate intervention. If untreated, autonomic dysreflexia can cause seizures, stroke, or heart attack.

Raynaud Phenomenon

In *Raynaud phenomenon* (rā-NŌ), the fingers and toes become ischemic (lack blood) after exposure to cold or with emotional stress. The condition is due to excessive sympathetic stimulation of smooth muscle in the arterioles of the fingers and toes. When the arterioles constrict in response to sympathetic stimulation, blood flow is greatly diminished. Symptoms are colorful—red, white, and blue. Fingers and toes may look white due to blockage of blood flow or look blue (cyanotic) due to deoxygenated blood in capillaries. With rewarming after cold exposure, the arterioles may dilate, causing the fingers and toes to look red. The disorder is most common in young women and occurs more often in cold climates.

CHAPTER REVIEW AND RESOURCE SUMMARY

REVIEW

RESOURCES

11.1 Comparison of Somatic and Autonomic Nervous Systems

1. The part of the nervous system that regulates smooth muscle, cardiac muscle, and certain glands is the **autonomic nervous system (ANS)**. The ANS usually operates without conscious control from the cerebral cortex, but other brain regions, mainly the hypothalamus and brain stem, regulate it.

2. The axons of somatic motor neurons extend from the CNS and synapse directly with an effector (skeletal muscle). Autonomic motor pathways consist of two motor neurons. The axon of the first motor neuron extends from the CNS and synapses in an **autonomic ganglion** with the second motor neuron; the second neuron synapses with an effector (smooth muscle, cardiac muscle, or a gland).

3. The output (motor) portion of the ANS has two divisions: the **sympathetic division** and the **parasympathetic division**. Most body organs receive **dual innervation**; usually one ANS division causes excitation and the other causes inhibition.

Anatomy Overview—Nervous System
Animation—Reflex Arcs
Animation—Reflexes

Figure 11.1 Comparison of somatic and autonomic motor neuron pathways to their effector tissues

REVIEW	RESOURCES

4. Somatic motor neurons release acetylcholine (ACh), and autonomic motor neurons release either acetylcholine or norepinephrine (NE).

5. Somatic nervous system effectors are skeletal muscles; ANS effectors include cardiac muscle, smooth muscle, and glands.

6. Table 11.1 compares the somatic and autonomic nervous systems.

11.2 Structure of the Autonomic Nervous System

1. The sympathetic division of the ANS is also called the thoracolumbar division because the outflow of sympathetic nerve impulses comes from the thoracic and lumbar segments of the spinal cord. Cell bodies of sympathetic **preganglionic neurons** are in the 12 thoracic and the first two lumbar segments of the spinal cord.

2. Sympathetic ganglia are classified as **sympathetic trunk ganglia** (lateral to the vertebral column) or **prevertebral ganglia** (anterior to the vertebral column).

3. A single sympathetic preganglionic axon may synapse with 20 or more **postganglionic neurons**. Sympathetic responses can affect organs throughout the body almost simultaneously.

4. The parasympathetic division is also called the craniosacral division because the outflow of parasympathetic nerve impulses comes from cranial nerve nuclei and sacral segments of the spinal cord. The cell bodies of parasympathetic preganglionic neurons are located in the nuclei of cranial nerves III, VII, IX, and X in the brain stem and in three sacral segments of the spinal cord (S2, S3, and S4).

5. Parasympathetic ganglia are called **terminal ganglia** and are located near or within autonomic effectors. Parasympathetic terminal ganglia are close to or in the walls of their autonomic effectors, so most parasympathetic postganglionic axons are very short. In the ganglion, the preganglionic neuron usually synapses with only four or five postganglionic neurons, all of which supply the same effector. Thus, parasympathetic responses are localized to a single effector.

Anatomy Overview—Nervous System: Organization of the ANS

Animation—ANS: Motor Pathways

Figure 11.2 Structure of the sympathetic division of the autonomic nervous system

Figure 11.3 Structure of the parasympathetic division of the autonomic nervous system

Anatomy Overview—Assemble an Arc

Exercise—Assemble the Structures of the ANS

11.3 Functions of the Autonomic Nervous System

1. Some ANS neurons release acetylcholine, and others release norepinephrine; the result is excitation in some cases and inhibition in others.

2. ANS neurons that release acetylcholine include (1) all sympathetic and parasympathetic preganglionic neurons, (2) all parasympathetic postganglionic neurons, and (3) a few sympathetic postganglionic neurons.

3. Most sympathetic postganglionic neurons release the neurotransmitter norepinephrine (NE). The effects of NE are longer-lasting and more widespread than those of acetylcholine.

4. Activation of the sympathetic division causes widespread responses collectively referred to as the **fight-or-flight response**. Activation of the parasympathetic division produces more restricted responses that typically are concerned with rest-and-digest activities.

5. Table 11.2 summarizes the main functions of the sympathetic and parasympathetic divisions of the ANS.

Anatomy Overview— Neurotransmitters

Animation—The ANS: Type of Neurotransmitters and Neurons

Animation—Physiological Effects of the ANS

Animation—The Alarm Reaction

Exercise—Sort ANS Functions

Exercise—What Is Your (ANS) Status?

SELF-QUIZ

1. The sympathetic division of the ANS is also called the
   a. somatic division.
   b. thoracolumbar division.
   c. peripheral nervous system.
   d. voluntary nervous system.
   e. craniosacral division.

2. The output portion of the ANS includes
   a. two motor neurons and one ganglion.
   b. one motor neuron and two ganglia.
   c. two motor neurons and two ganglia.

   d. one motor and one sensory neuron, and no ganglia.
   e. one motor and one sensory neuron, and one ganglion.

3. Which statement is NOT true?
   a. Most sympathetic postganglionic neurons release norepinephrine.
   b. Parasympathetic preganglionic neurons release acetylcholine.
   c. Sympathetic effects are more localized and short-lived than parasympathetic effects.
   d. The effects from norepinephrine tend to be long-lasting.
   e. Branches of a single postganglionic neuron in the sympathetic division extend to many organs.

4. Which of the following pairs is mismatched?
   a. acetylcholine:parasympathetic nervous system
   b. fight-or-flight:sympathetic nervous system
   c. conserves body energy:parasympathetic nervous system
   d. rest-and-digest:parasympathetic nervous system
   e. norepinephrine:parasympathetic nervous system

5. Which of the following statements is NOT true concerning the autonomic nervous system?
   a. Most autonomic responses cannot be consciously controlled.
   b. In general, if the sympathetic division increases the activity in a specific organ, then the parasympathetic division decreases the activity of that organ.
   c. Sensory receptors monitor internal body conditions.
   d. Sensory neurons include pre- and postganglionic neurons.
   e. Most visceral effectors receive dual innervation.

6. Which part of the central nervous system regulates autonomic tone?
   a. hypothalamus   b. cerebellum
   c. spinal cord   d. limbic system
   e. thalamus

7. Place the following structures in the correct order as they relate to an autonomic nervous system response from receipt of the stimulus to response:
   1. visceral effector
   2. centers in the CNS
   3. autonomic ganglion
   4. receptor and autonomic sensory neuron
   5. preganglionic neuron
   6. postganglionic neuron
      a. 4, 5, 2, 3, 6, 1   b. 5, 6, 2, 3, 1, 4   c. 1, 6, 3, 5, 2, 4
      d. 4, 2, 5, 3, 6, 1   e. 2, 4, 5, 6, 3, 1

8. Which of the following activities would NOT be monitored by autonomic sensory neurons?
   a. carbon dioxide levels in the blood
   b. hearing and equilibrium
   c. blood pressure
   d. stretching of the walls of visceral organs
   e. nausea from damaged viscera

9. The autonomic ganglia that are located near the innervated organs are
   a. trunk ganglia.
   b. prevertebral ganglia.
   c. posterior root ganglia.
   d. terminal ganglia.
   e. basal ganglia.

10. Which of these statements about the parasympathetic division of the autonomic nervous system is NOT true? The parasympathetic division
    a. arises from the cranial nerves in the brain stem and sacral spinal cord segments.

    b. is concerned with conserving and restoring energy.
    c. uses acetylcholine as its neurotransmitter.
    d. has ganglia near or within visceral effectors.
    e. initiates responses in a preganglionic neuron that synapses with 20 or more postganglionic neurons.

11. Which nerve carries most of the parasympathetic output from the brain?
    a. spinal   b. vagus   c. oculomotor   d. facial
    e. glossopharyngeal

12. Which of the following would NOT be affected by the autonomic nervous system?
    a. heart   b. intestines   c. urinary bladder
    d. skeletal muscle   e. reproductive organs

13. Which of the following neurons release norepinephrine?
    a. somatic motor neurons
    b. sympathetic postganglionic neurons
    c. sympathetic preganglionic neurons
    d. parasympathetic postganglionic neurons
    e. parasympathetic preganglionic neurons

14. Match the following:
    _____ a. cluster of cell bodies outside the CNS
    _____ b. cell body located in ganglion; unmyelinated axon extends to effector
    _____ c. cell body lies inside the CNS; myelinated axon extends to ganglion
    _____ d. their postganglionic axons innervate organs below the diaphragm
    _____ e. their postganglionic axons supply organs above the diaphragm
    _____ f. contain the cell bodies and dendrites of parasympathetic postganglionic neurons

    A. sympathetic trunk ganglia
    B. prevertebral ganglia
    C. ganglion
    D. terminal ganglia
    E. preganglionic neuron
    F. postganglionic neuron

15. For each of the following, place a P if it refers to increased activity of the parasympathetic division or an S if it refers to increased activity of the sympathetic division.
    _____ a. dilates pupils
    _____ b. decreases heart rate
    _____ c. causes airway constriction
    _____ d. stimulates breakdown of triglycerides
    _____ e. inhibits secretion from digestive organs
    _____ f. stimulates the digestive tract
    _____ g. occurs during exercise
    _____ h. causes release of glucose from the liver
    _____ i. dilates blood vessels to cardiac muscle

CRITICAL THINKING APPLICATIONS

1. It's Sunday and you've just eaten a huge roast dinner. Now you're going to watch a film on TV, if you can make it to the couch! Which division of the nervous system will be handling your body's post-dinner activities? Give examples of some organs and the effects on their functions.

2. It is your turn to present a report to your fellow students. You break into a sweat, heart pounding, with a mouth so dry you can barely speak. You notice lingering effects on your body even after you've returned to your seat. Describe what kind of reaction is occurring in your body.

3. Claire was watching a scary horror movie when she heard a door slam and a dog barking. The hair on her arms rose and she was covered with goose bumps. Trace the pathway taken by the impulses from her CNS to her arms.

4. In the novel *The Hitchhiker's Guide to the Galaxy*, the character Zaphod Beeblebrox has two heads and therefore two brains. Is this what is meant by dual innervation? Explain.

ANSWERS TO FIGURE QUESTIONS

11.1 Dual innervation means that an organ receives impulses from both the sympathetic and parasympathetic divisions of the ANS.

11.2 In the sympathetic trunk ganglia, sympathetic preganglionic axons form synapses with cell bodies and dendrites of sympathetic postganglionic neurons.

11.3 Most parasympathetic preganglionic axons are longer than most sympathetic preganglionic axons because parasympathetic ganglia are located in the walls of visceral organs, while most sympathetic ganglia are close to the spinal cord in the sympathetic trunk.

CHAPTER 12
SOMATIC SENSES AND SPECIAL SENSES

Imagine a camping trip to a beautiful rocky coastline cradling a sandy patch of beach. As you rouse from your night of slumber on the packed sand, you slowly stretch your stiffened joints and cautiously climb out of your sleeping bag to greet the crisp morning air. You rub the sleep from your eyes and see the fog rolling in off the white crests of the slapping waves. As you walk toward the ocean you take a deep breath, smell the salty scent of the tide, and feel individual grains of sand between your wriggling toes. Suddenly you stop to rub your exposed arms vigorously, as the cool, crisp air sends a chill through your still-sleepy body. You see and hear noisy gulls gliding through the air overhead, and listen to a distant boat sounding its horn. As you walk toward the water line, where the sounds of the water are playing their tune against the rocks, you glance down into the tide pools left behind by the receding waves, and notice a colorful array of intertidal life—sea stars, mussels, anemones, and scurrying crabs. Bending down to take a closer look, your face is splashed by an incoming wave, giving you a taste of the salty sea. You think for a minute about the beauty you have *sensed* in the past few minutes. Your mind is flooded with what you have *seen*, what you have *felt*, what you have *smelled*, what you have *heard*, and what you have *tasted*.

Did you know?   *People's perceptions of their situations affect the way their brains interpret pain sensations. You have probably heard stories of soldiers losing limbs in battle who felt no pain until much later, or athletes who continued playing with severe injuries. On the other hand, fear and anxiety can intensify sensations of pain. Childbirth education programs teach pregnant women breathing and other relaxation techniques to reduce their experiences of pain. Knowing what to expect during childbirth reduces fear, which in turn also reduces pain.*

LOOKING BACK TO MOVE AHEAD...

Sensory Nerve Endings and Sensory Receptors in the Skin (Section 5.1)

Somatic Sensory Pathways (Section 10.4)

FOCUS ON WELLNESS
Coping with Chronic Pain—Sensation Modulation

12.1 OVERVIEW OF SENSATIONS

OBJECTIVE • Define a sensation and describe the conditions needed for a sensation to occur.

Most of us are aware of sensory input to the central nervous system (CNS) from structures associated with smell, taste, vision, hearing, and balance. These five senses are known as the **special senses**. The other senses are termed **general senses** and include both somatic senses and visceral senses. **Somatic senses** (*somat-* = of the body) include tactile sensations (touch, pressure, and vibration); thermal sensations (warm and cold); pain sensations; and proprioceptive sensations (joint and muscle position sense and movements of the limbs and head). **Visceral senses** provide information about conditions within internal organs.

Definition of Sensation

Sensation is the conscious or subconscious awareness of changes in the external or internal environment. For a sensation to occur, four conditions must be satisfied:

1. A *stimulus*, or change in the environment, capable of activating certain sensory neurons, must occur. A stimulus that activates a sensory receptor may be in the form of light, heat, pressure, mechanical energy, or chemical energy.

2. A *sensory receptor* must convert the stimulus to an electrical signal, which ultimately produces one or more nerve impulses if it is large enough.

3. The nerve impulses must be *conducted* along a neural pathway from the sensory receptor to the brain.

4. A region of the brain must receive and *integrate* the nerve impulses into a sensation.

Characteristics of Sensations

As you learned in Chapter 10, **perception** is the conscious awareness and interpretation of sensations and is primarily a function of the cerebral cortex. You seem to see with your eyes, hear with your ears, and feel pain in an injured part of your body. This is because sensory nerve impulses from each part of the body arrive in a specific region of the cerebral cortex, which interprets the sensation as coming from the stimulated sensory receptors. A given sensory neuron carries information for one type of sensation only. Neurons relaying impulses for touch, for example, do not also conduct impulses for pain. The specialization of sensory neurons enables nerve impulses from the eyes to be perceived as sight and those from the ears to be perceived as sounds.

A characteristic of most sensory receptors is **adaptation**, a decrease in the strength of a sensation during a prolonged stimulus. Adaptation is caused in part by a decrease in the responsiveness of sensory receptors. As a result of adaptation, the perception of a sensation may fade or disappear even though the stimulus persists. For example, when you first step into a hot shower, the water may feel very hot, but soon the sensation decreases to one of comfortable warmth even though the stimulus (the high temperature of the water) does not change. Receptors vary in how quickly they adapt. **Rapidly adapting receptors**, also called *phasic receptors*, adapt very quickly. They are specialized for signaling *changes* in a stimulus. Receptors associated with pressure, touch, and smell are rapidly adapting. By contrast, **slowly adapting receptors**, also known as *tonic receptors*, adapt slowly and continue to trigger nerve impulses as long as the stimulus persists. Slowly adapting receptors monitor stimuli associated with pain, body position, and chemical composition of the blood.

Types of Sensory Receptors

Both structural and functional characteristics of sensory receptors can be used to group them into different classes (Table 12.1). Structurally, the simplest are *free nerve endings*,

TABLE 12.1

Classification of Sensory Receptors

BASIS OF CLASSIFICATION	DESCRIPTION
Structure	
Free nerve endings	Bare dendrites associated with pain, thermal, tickle, itch, and some touch sensations
Encapsulated nerve endings	Dendrites enclosed in a connective tissue capsule for pressure, vibration, and some touch sensations
Separate cells	Receptor cell that synapses with first-order neuron; located in the retina of the eye (photoreceptors), inner ear (hair cells), and taste buds of the tongue (gustatory receptor cells)
Function	
Mechanoreceptors	Detect mechanical pressure; provide sensations of touch, pressure, vibration, proprioception, and hearing and equilibrium; also monitor stretching of blood vessels and internal organs
Thermoreceptors	Detect changes in temperature
Nociceptors	Respond to painful stimuli resulting from physical or chemical damage to tissue
Photoreceptors	Detect light that strikes the retina of the eye
Chemoreceptors	Detect chemicals in mouth (taste), nose (smell), and body fluids
Osmoreceptors	Sense the osmotic pressure of body fluids

which are bare dendrites that lack any structural specializations at their ends that can be seen under a light microscope (Figure 12.1). Receptors for pain, temperature, tickle, itch, and some touch sensations are free nerve endings. Receptors for other somatic and visceral sensations, such as some touch, pressure, and vibration, have *encapsulated nerve endings.* Their dendrites are enclosed in a connective tissue capsule with a distinctive microscopic structure. Still other sensory receptors consist of specialized, *separate cells* that synapse with sensory neurons, for example, hair cells in the inner ear.

Another way to group sensory receptors is functionally—according to the type of stimulus they detect. Most stimuli are in the form of mechanical energy, such as sound waves or pressure changes; electromagnetic energy, such as light or heat; or chemical energy, such as in a molecule of glucose.

■ *Mechanoreceptors* are sensitive to mechanical stimuli such as the deformation, stretching, or bending of cells.

Mechanoreceptors provide sensations of touch, pressure, vibration, proprioception, and hearing and equilibrium. They also monitor the stretching of blood vessels and internal organs.

■ *Thermoreceptors* detect changes in temperature.

■ *Nociceptors* respond to painful stimuli resulting from physical or chemical damage to tissue.

■ *Photoreceptors* detect light that strikes the retina of the eye.

■ *Chemoreceptors* detect chemicals in the mouth (taste), nose (smell), and body fluids.

■ *Osmoreceptors* detect the osmotic pressure of body fluids.

✓ CHECKPOINT

1. Which senses are "special senses"?
2. How is a sensation different from a perception?

Figure 12.1 Structure and location of sensory receptors in the skin and subcutaneous layer.

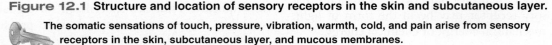
The somatic sensations of touch, pressure, vibration, warmth, cold, and pain arise from sensory receptors in the skin, subcutaneous layer, and mucous membranes.

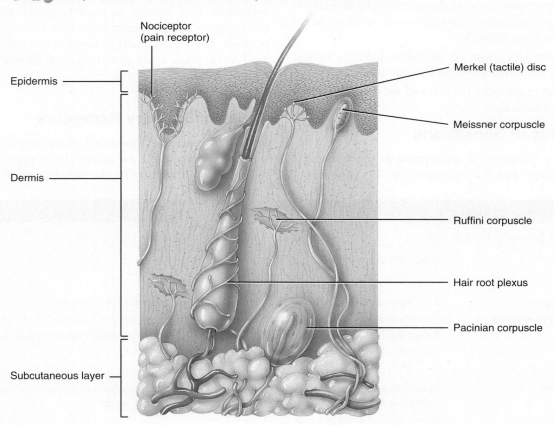

 Which receptors are especially abundant in the fingertips, palms, and soles?

12.2 SOMATIC SENSES

OBJECTIVES • Describe the location and function of the receptors for tactile, thermal, and pain sensations.

• Identify the receptors for proprioception and describe their functions.

Somatic sensations arise from stimulation of sensory receptors in the skin, mucous membranes, muscles, tendons, and joints. The sensory receptors for somatic sensations are distributed unevenly. Some parts of the body surface are densely populated with receptors, and other parts contain only a few. The areas with the largest numbers of sensory receptors are the tip of the tongue, the lips, and the fingertips.

Tactile Sensations

The *tactile sensations* (TAK-tĭl; *tact-* = touch) include touch, pressure, vibration, itch, and tickle. Although we perceive differences among these sensations, they arise by activation of some of the same types of receptors. Several types of encapsulated mechanoreceptors detect sensations of touch, pressure, and vibration. Other tactile sensations, such as itch and tickle sensations, are detected by free nerve endings. Tactile receptors in the skin or subcutaneous layer include Meissner corpuscles, hair root plexuses, Merkel discs, Ruffini corpuscles, pacinian corpuscles, and free nerve endings (see Figure 12.1).

Touch

Sensations of *touch* generally result from stimulation of tactile receptors in the skin or subcutaneous layer. There are two types of rapidly adapting touch receptors. *Meissner corpuscles* (MĪS-ner) or *corpuscles of touch* are touch receptors that are located in the dermal papillae of hairless skin. Each corpuscle is an egg-shaped mass of dendrites enclosed by a capsule of connective tissue. They are abundant in the fingertips, hands, eyelids, tip of the tongue, lips, nipples, soles, clitoris, and tip of the penis. *Hair root plexuses* are found in hairy skin; they consist of free nerve endings wrapped around hair follicles. Hair root plexuses detect movements on the skin surface that disturb hairs. For example, an insect landing on a hair causes movement of the hair shaft that stimulates the free nerve endings.

There are also two types of slowly adapting touch receptors. *Merkel discs*, also known as *tactile discs* or *type I cutaneous mechanoreceptors*, are saucer-shaped, flattened free nerve endings that make contact with Merkel cells of the stratum basale (see Figure 5.2a). These touch receptors are plentiful in the fingertips, hands, lips, and external genitalia. *Ruffini corpuscles* or *type II cutaneous mechanoreceptors* are elongated, encapsulated receptors located deep in the dermis, and in ligaments and tendons. Present in the hands and abundant on the soles, they are most sensitive to stretching that occurs as digits or limbs are moved.

Pressure

Pressure, a sustained sensation that is felt over a larger area and occurs in deeper tissues than touch, occurs with deformation of deeper tissues. Receptors that contribute to sensations of pressure include Meissner corpuscles, Merkel discs, and pacinian corpuscles. A *pacinian* (pa-SIN-ē-an), or *lamellated*, *corpuscle* is a large oval structure composed of a multilayered connective tissue capsule that encloses a dendrite. Like Meissner corpuscles, pacinian corpuscles adapt rapidly. They are widely distributed in the body: in the dermis and subcutaneous layer; in tissues that underlie mucous and serous membranes; around joints, tendons, and muscles; in the periosteum; and in the mammary glands, external genitalia, and certain viscera, such as the pancreas and urinary bladder.

Vibration

Sensations of *vibration* result from rapidly repetitive sensory signals from tactile receptors. The receptors for vibration sensations are Meissner corpuscles and pacinian corpuscles. Meissner corpuscles can detect lower-frequency vibrations, and pacinian corpuscles detect higher-frequency vibrations.

Itch and Tickle

The *itch* sensation results from stimulation of free nerve endings by certain chemicals, such as bradykinin, often because of a local inflammatory response. Free nerve endings are thought to mediate the *tickle* sensation. This intriguing sensation typically arises only when someone else touches you, not when you touch yourself. The solution to this puzzle seems to lie in the impulses that conduct to and from the cerebellum when you are moving your fingers and touching yourself that don't occur when someone else is tickling you.

CLINICAL CONNECTION | Phantom Limb Sensation

Patients who have had a limb amputated may still experience sensations such as itching, pressure, tingling, or pain as if the limb were still there. This phenomenon is called **phantom limb sensation**. One explanation for phantom limb sensations is that the cerebral cortex interprets impulses arising in the proximal portions of sensory neurons that previously carried impulses from the limb as coming from the nonexistent (phantom) limb. Another explanation for phantom limb sensations is that neurons in the brain that previously received sensory impulses from the missing limb are still active, giving rise to false sensory perceptions. •

Thermal Sensations

Thermoreceptors are free nerve endings. Two distinct *thermal sensations*—coldness and warmth—are mediated by different receptors. Temperatures between 10° and 40°C (50–105°F) activate *cold receptors*, which are located in the epidermis. *Warm receptors* are located in the dermis and are activated by temperatures between 32° and 48°C (90–118°F). Cold and warm receptors both adapt rapidly at the onset of a stimulus but continue to generate nerve impulses more slowly throughout a prolonged stimulus. Temperatures below 10°C and above 48°C stimulate mainly nociceptors, rather than thermoreceptors, producing painful sensations.

Pain Sensations

The sensory receptors for pain, called *nociceptors* (nō′-sē-SEP-tors; *noci-* = harmful), are free nerve endings (see Figure 12.1). Nociceptors are found in practically every tissue of the body except the brain, and they respond to several types of stimuli. Excessive stimulation of sensory receptors, excessive stretching of a structure, prolonged muscular contractions, inadequate blood flow to an organ, or the presence of certain chemical substances can all produce the sensation of pain. Pain may persist even after a pain-producing stimulus is removed because pain-causing chemicals linger and because nociceptors exhibit very little adaptation.

There are two types of pain: fast and slow. The perception of *fast pain* occurs very rapidly, usually within 0.1 second after a stimulus is applied. This type of pain is also known as acute, sharp, or pricking pain. The pain felt from a needle puncture or knife cut to the skin are examples of fast pain. Fast pain is not felt in deeper tissues of the body. The perception of *slow pain* begins a second or more after a stimulus is applied. It then gradually increases in intensity over a period of several seconds or minutes. This type of pain, which may be excruciating, is also referred to as chronic, burning, aching, or throbbing pain. Slow pain can occur both in the skin and in deeper tissues or internal organs. An example is the pain associated with a toothache.

Fast pain is very precisely localized to the stimulated area. For example, if someone pricks you with a pin, you know exactly which part of your body was stimulated. Somatic slow pain is well localized but more diffuse (involves large areas); it usually appears to come from a larger area of the skin. In many instances of visceral pain, the pain is felt in or just deep to the skin that overlies the stimulated organ, or in a surface area far from the stimulated organ. This phenomenon is called *referred pain* (Figure 12.2). In general, the visceral organ involved and the area in which the pain is referred are served by the same segment of the spinal cord. For example, sensory neurons from the heart, the skin over the heart, and the skin along the medial aspect of the left arm enter spinal cord segments T1 to T5. Thus, the pain of a heart attack typically is felt in the skin over the heart and along the left arm.

Figure 12.2 Distribution of referred pain. The colored parts of the diagrams indicate skin areas to which visceral pain is referred.

🔑 **Nociceptors are present in almost every tissue of the body.**

(a) Anterior view

(b) Posterior view

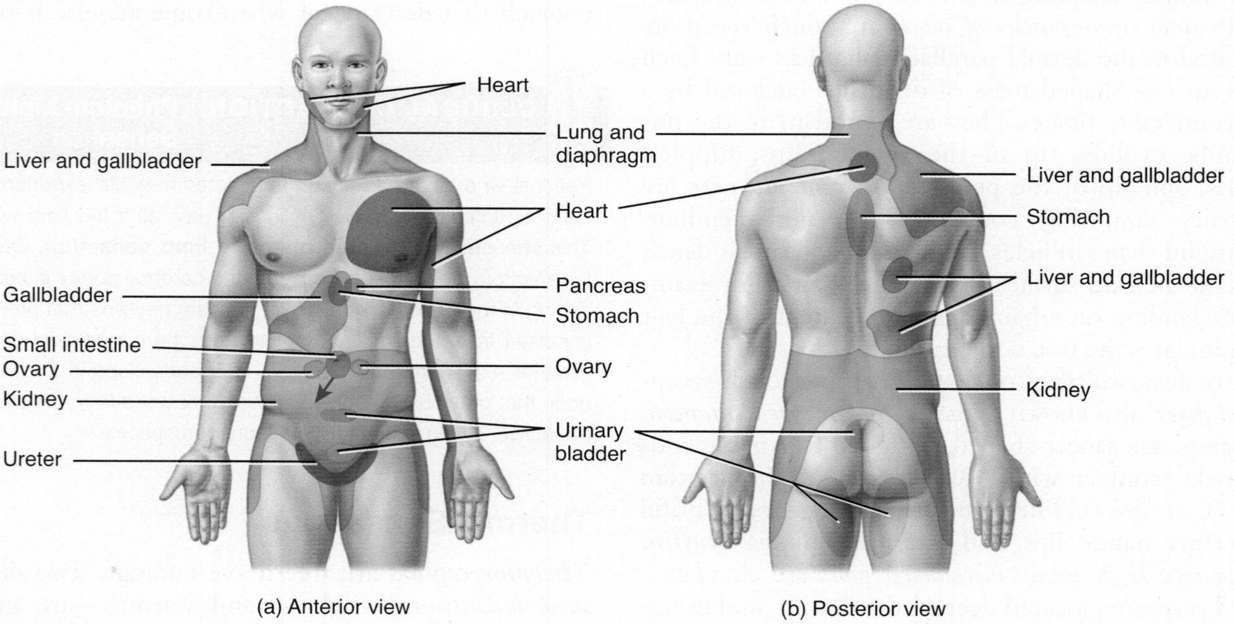

❓ **Which visceral organ has the broadest area for referred pain?**

Coping with Chronic Pain—Sensation Modulation

Pain is a useful sensation when it alerts us to an injury that needs attention. We pull our finger away from a hot stove, we take off shoes that are too tight, and we rest an ankle that has been sprained. The injury heals, and we quit thinking about it.

Pain that persists for longer than it "should" given the physical trauma endured, and despite appropriate treatment, is called *chronic pain*. The most common forms of chronic pain are low back pain and headache. Cancer, arthritis, fibromyalgia, and many other disorders are associated with chronic pain. People experiencing chronic pain often experience chronic frustration as they are sent from one specialist to another in search of a diagnosis.

The goal of pain management programs, developed to help people with chronic pain, is to decrease pain as much as possible, and then help patients learn to live with whatever pain remains. Because no single treatment works for everyone, pain management programs typically offer a wide variety of treatments, from surgery and nerve blocks to acupuncture and exercise therapy. Following are some of the therapies that complement medical and surgical treatment for the management of chronic pain.

Education and Counseling

Pain used to be regarded as a purely physical response to physical injury. Psychological factors are now understood to serve as important mediators in the perception of pain. The brain can learn to amplify or to suppress signals from the nerves. Feelings such as fear and anxiety may strengthen pain perceptions. Laughter, joy, and engagement in rewarding activities can decrease feelings of pain. Pain may be used to avoid certain situations, or to gain attention. Depression and associated symptoms such as sleep disturbances can contribute to chronic pain. Education and counseling techniques can help people with chronic pain confront issues that may be worsening pain, and think about pain in ways that reduce its intensity.

Relaxation and Exercise

Relaxation and meditation techniques may reduce pain by decreasing anxiety and giving people a sense of personal control. Some of these techniques include deep breathing, visualization of positive images, and muscular relaxation. Others encourage people to become more aware of thoughts and situations that increase or decrease pain or provide a mental distraction from the sensations of pain.

Exercise stimulates the production of endorphins, chemicals produced by the body to relieve pain. It also improves strength and functional capacity, increases self-confidence, and can serve as a distraction from pain.

Think It Over . . . **Why might long periods of inactivity make pain worse?**

CLINICAL CONNECTION | Analgesia

Some pain sensations occur out of proportion to minor damage or persist chronically for no obvious reason. In such cases, **analgesia** (*an-* = without; *-algesia* = pain) or pain relief is needed. Analgesic drugs such as aspirin and ibuprofen (for example, Advil®) block formation of some chemicals that stimulate nociceptors. Local anesthetics, such as Novocaine®, provide short-term pain relief by blocking conduction of nerve impulses. Morphine and other opiate drugs alter the quality of pain perception in the brain; pain is still sensed, but it is no longer perceived as so unpleasant. •

Proprioceptive Sensations

Proprioceptive sensations (prō-prē-ō-SEP-tive; *proprio-* = one's own) allow us to know where our head and limbs are located and how they are moving even if we are not looking at them, so that we can walk, type, or dress without using our eyes. *Kinesthesia* (kin′-es-THĒ-zē-a; *kin-* = motion; *-esthesia* = perception) is the perception of body movements. Proprioceptive sensations arise in receptors termed *proprioceptors*. Proprioceptors are located in skeletal muscles (muscle spindles), in tendons (tendon organs), in and around synovial joints (joint kinesthetic receptors), and in the inner ear (hair cells). Those proprioceptors embedded in muscles, tendons, and synovial joints inform us of the degree to which muscles are contracted, the amount of tension on tendons, and the positions of joints. Hair cells of the inner ear monitor the orientation of the head relative to the ground and head position during movements. Proprioceptive sensations also allow us to estimate the weight of objects and determine the muscular effort necessary to perform a task. For example, as you pick up an object you quickly realize how heavy it is, and you then exert the correct amount of effort needed to lift it.

Nerve impulses for conscious proprioception pass along sensory tracts in the spinal cord and brain stem and are relayed to the primary somatosensory area (postcentral gyrus) in the parietal lobe of the cerebral cortex (see Figure 10.13). Proprioceptive impulses also pass to the cerebellum, where they contribute to the cerebellum's role in coordinating skilled movements. Because proprioceptors adapt slowly and only slightly, the brain continually receives nerve impulses related to the position of different body parts and makes adjustments to ensure coordination.

✓ CHECKPOINT

3. Why is it beneficial to your well-being that nociceptors and proprioceptors exhibit very little adaptation?

4. Which somatic sensory receptors detect touch sensations?

5. What is referred pain, and how is it useful in diagnosing internal disorders?

12.3 SPECIAL SENSES

OBJECTIVE ● **Define the special senses.**

Receptors for the special senses—smell, taste, sight, hearing, and equilibrium—are housed in complex sensory organs such as the eyes and ears. Like the general senses, the special senses allow us to detect changes in our environment. *Ophthalmology* (of′-thal-MOL-ō-jē; *ophthalmo-* = eye; *-logy* = study of) is the science that deals with the eye and its disorders. The other special senses are, in large part, the concern of *otorhinolaryngology* (ō′-tō-rī′-nō-lar′-in-GOL-ō-jē; *oto-* = ear; *rhino-* = nose; *laryngo-* = larynx), the science that deals with the ears, nose, and throat and their disorders.

12.4 OLFACTION: SENSE OF SMELL

OBJECTIVE ● **Describe the receptors for olfaction and the olfactory pathway to the brain.**

The nose contains 10–100 million receptors for the sense of smell, or *olfaction* (ol-FAK-shun; *olfact-* = smell). Because some nerve impulses for smell and taste propagate to the limbic system, certain odors and tastes can evoke strong emotional responses or a flood of memories.

Structure of the Olfactory Epithelium

The olfactory epithelium occupies the upper portion of the nasal cavity (Figure 12.3a) and consists of three types of cells: olfactory receptors, supporting cells, and basal stem cells (Figure 12.3b). *Olfactory receptors* are the first-order neu-

rons of the olfactory pathway. Several cilia called *olfactory hairs* project from a knob-shaped tip on each olfactory receptor. The olfactory hairs are the parts of the olfactory receptor that respond to inhaled chemicals. Chemicals that have an odor and can therefore stimulate the olfactory hairs are called *odorants*. The axons of olfactory receptors extend from the olfactory epithelium to the olfactory bulb. *Supporting cells* are columnar epithelial cells of the mucous membrane lining the nose. They provide physical support, nourishment, and electrical insulation for the olfactory receptors, and they help detoxify chemicals that come in contact with the olfactory epithelium. *Basal cells* are stem cells located between the bases of the supporting cells and continually undergo cell division to produce new olfactory receptors, which live for only a month or so before being replaced. This process is remarkable because olfactory receptors are neurons, and in general, mature neurons are not replaced. *Olfactory glands* produce mucus that moistens the surface of the olfactory epithelium and serves as a solvent for inhaled odorants.

Stimulation of Olfactory Receptors

Many attempts have been made to distinguish among and classify "primary" sensations of smell. Genetic evidence now suggests the existence of hundreds of primary odors. Our ability to recognize about 10,000 different odors probably depends on patterns of activity in the brain that arise from activation of many different combinations of olfactory receptors. Olfactory receptors react to odorant molecules by producing an electrical signal that triggers one or more nerve impulses. Adaptation (decreasing sensitivity) to odors occurs rapidly. Olfactory receptors adapt by about 50 percent in the first second or so after stimulation and very slowly thereafter.

The Olfactory Pathway

On each side of the nose, about 40 bundles of the slender, unmyelinated axons of olfactory receptors extend through about 20 holes in the cribriform plate of the ethmoid bone (Figure 12.3b). These bundles of axons collectively form the right and left *olfactory (I) nerves*. The olfactory nerves terminate in the brain in paired masses of gray matter called the *olfactory bulbs*, which are located below the frontal lobes of the cerebrum. Within the olfactory bulbs, the axon terminals of olfactory receptors—the first-order neurons—form synapses with the dendrites and cell bodies of second-order neurons in the olfactory pathway.

The axons of the neurons extending from the olfactory bulb form the *olfactory tract*. Some of the axons of the olfactory tract project to the *primary olfactory area* in the temporal lobe of the cerebral cortex (see Figure 10.13), where conscious awareness of smell begins. Other axons of the olfactory tract project to the limbic system and hypothalamus; these connections account for emotional and memory-evoked responses to odors. Examples include sexual excitement upon smelling a certain perfume or nausea upon smelling a food that once made you violently ill.

Figure 12.3 Olfactory epithelium and olfactory receptors. (a) Location of olfactory epithelium in the nasal cavity. (b) Anatomy of olfactory receptors, whose axons extend through the cribriform plate to the olfactory bulb.

🔑 **The olfactory epithelium consists of olfactory receptors, supporting cells, and basal cells.**

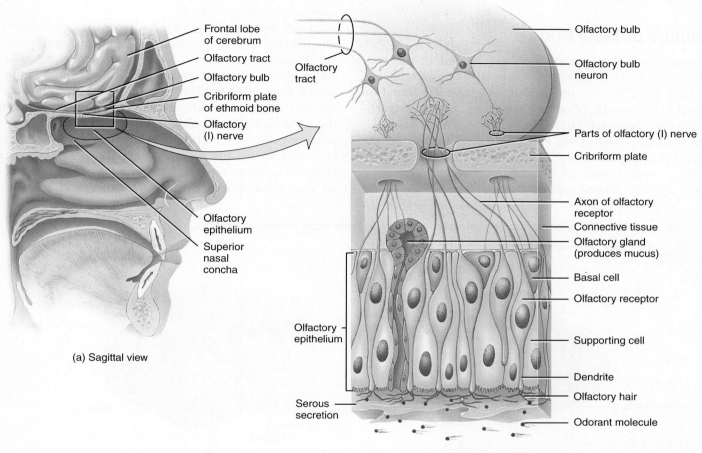

(a) Sagittal view

(b) Enlarged aspect of olfactory receptors

❓ What is the function of basal stem cells?

✓ CHECKPOINT

6. What functions are carried out by the three types of cells in the olfactory epithelium?

7. Define the following terms: olfactory nerve, olfactory bulb, and olfactory tract.

12.5 GUSTATION: SENSE OF TASTE

OBJECTIVE • Describe the receptors for gustation and the gustatory pathway to the brain.

Taste or *gustation* (gus-TĀ-shun; *gust-* = taste) is much simpler than olfaction because only five primary tastes can be distinguished: *sour, sweet, bitter, salty,* and *umami* (ū-MAM-ē). The umami taste is described as "meaty" or "savory." All other flavors, such as chocolate, pepper, and coffee, are combinations of the five primary tastes, plus the accompanying olfactory and tactile (touch) sensations. Odors from food can pass upward from the mouth into the nasal cavity, where they stimulate olfactory receptors. Because olfaction is much more sensitive than taste, a given concentration of a food substance may stimulate the olfactory system thousands of times more strongly than it stimulates the gustatory system. When you have a cold or are suffering from allergies and cannot taste your food, it is actually olfaction that is blocked, not taste.

Structure of Taste Buds

The receptors for taste sensations are located in the *taste buds* (Figure 12.4). Most of the nearly 10,000 taste buds of a young adult are on the tongue, but some are also found on the roof of the mouth, pharynx (throat), and epiglottis (cartilage lid over the voice box). The number of taste buds declines with age. Taste buds are found in elevations on the tongue called *papillae* (pa-PIL-ē; singular is *papilla*), which provide a rough texture to the upper surface of the tongue (Figure 12.4a,b). *Vallate papillae* (VAL-āt = wall-like) form an inverted V-shaped row at the back of the tongue. *Fungiform papillae* (FUN-ji-form = mushroomlike) are mushroom-shaped elevations scattered over the entire surface of the tongue. In addition, the entire surface of the tongue has *filiform papillae* (FIL-i-form = threadlike), which contain touch receptors but no taste buds.

Each *taste bud* is an oval body consisting of three types of epithelial cells: supporting cells, gustatory receptor cells, and basal cells (Figure 12.4c). The *supporting cells* surround about 50 *gustatory receptor cells*. A single, long *gustatory hair* projects from each gustatory receptor cell to the external surface through the *taste pore*, an opening in the taste bud. *Basal cells* are stem cells that produce supporting cells, which then develop into gustatory receptor cells that have a life span of about 10 days. The gustatory receptor cells are separate receptor cells. They do not have an axon (like olfactory receptors) but rather synapse with

Figure 12.4 The relationship of gustatory receptors in taste buds to tongue papillae.

Gustatory (taste) receptor cells are located in taste buds.

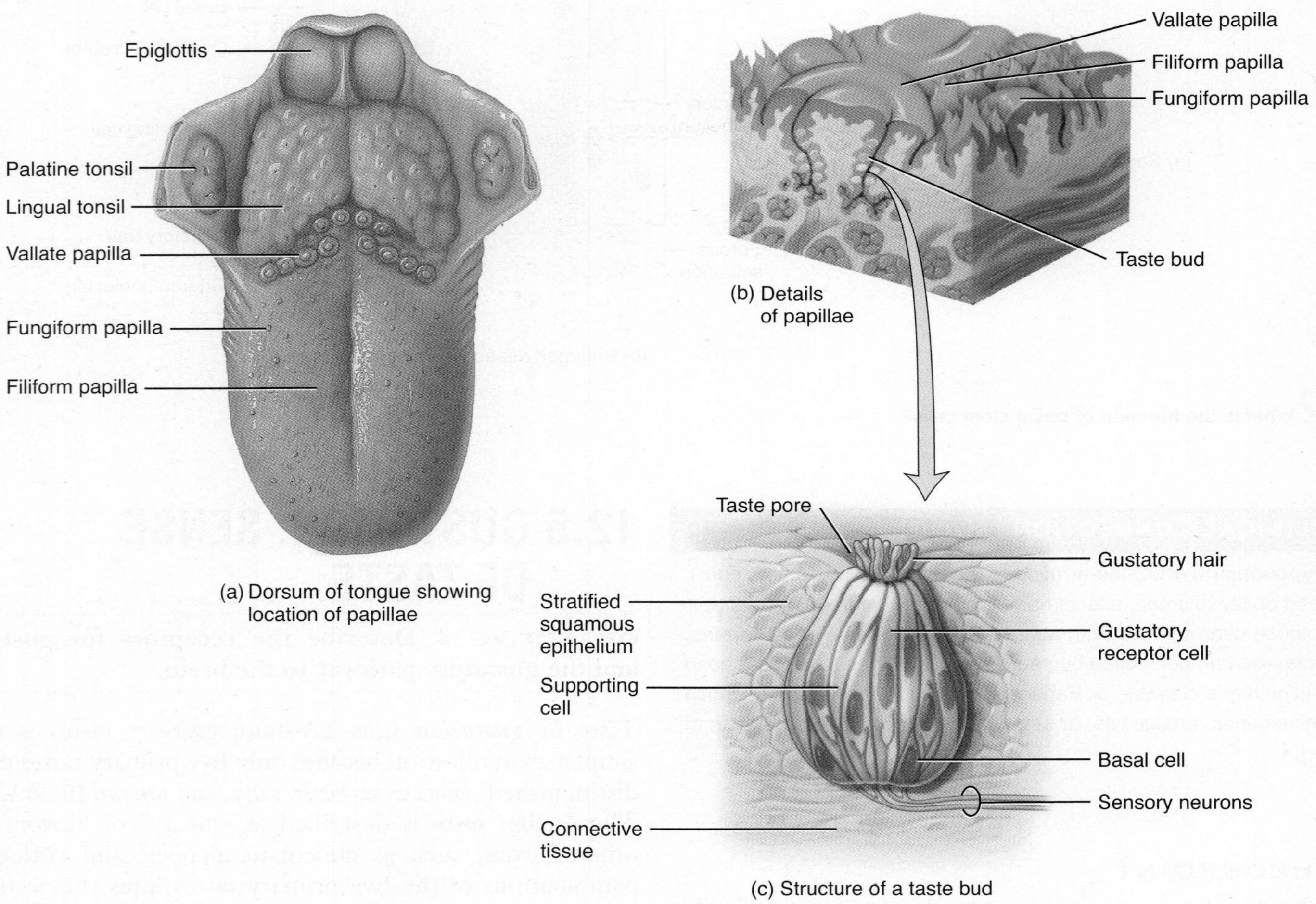

(a) Dorsum of tongue showing location of papillae

(b) Details of papillae

(c) Structure of a taste bud

In order, from the tongue to the brain, what structures form the gustatory pathway?

dendrites of the first-order sensory neurons of the gustatory pathway.

Stimulation of Gustatory Receptors

Chemicals that stimulate gustatory receptor cells are known as *tastants*. Once a tastant is dissolved in saliva, it can enter taste pores and make contact with the plasma membrane of the gustatory hairs. The result is an electrical signal that stimulates release of neurotransmitter molecules from the gustatory receptor cell. Nerve impulses are triggered when these neurotransmitter molecules bind to their receptors on the dendrites of the first-order sensory neuron. The dendrites branch profusely and contact many gustatory receptors in several taste buds. Individual gustatory receptor cells may respond to more than one of the five primary tastes. Complete adaptation (loss of sensitivity) to a specific taste can occur in 1 to 5 minutes of continuous stimulation.

If all tastants cause release of neurotransmitter from many gustatory receptor cells, why do foods taste different? The answer to this question is thought to lie in the patterns of nerve impulses in groups of first-order taste neurons that synapse with the gustatory receptor cells. Different tastes arise from activation of different groups of taste neurons. In addition, although each individual gustatory receptor cell responds to more than one of the five primary tastes, it may respond more strongly to some tastants than to others.

The Gustatory Pathway

Three cranial nerves contain axons of first-order gustatory neurons that innervate the taste buds. The facial (VII) nerve and glossopharyngeal (IX) nerve serve the tongue; the vagus (X) nerve serves the throat and epiglottis. From taste buds, impulses propagate along these cranial nerves to the medulla oblongata. From the medulla, some axons carrying taste signals project to the limbic system and the hypothalamus, and others project to the thalamus. Taste signals that project from the thalamus to the *primary gustatory area* in the parietal lobe of the cerebral cortex (see Figure 10.13) give rise to the conscious perception of taste.

CLINICAL CONNECTION | Taste Aversion

Probably because of taste projections to the hypothalamus and limbic system, there is a strong link between taste and pleasant or unpleasant emotions. Sweet foods evoke reactions of pleasure while bitter ones cause expressions of disgust, even in newborn babies. This phenomenon is the basis for **taste aversion**, in which people and animals quickly learn to avoid a food if it upsets the digestive system. Because the drugs and radiation treatments used to combat cancer often cause nausea and gastrointestinal upset regardless of what foods are consumed, cancer patients may lose their appetite because they develop taste aversions for most foods. •

✓ CHECKPOINT

8. How do olfactory receptors and gustatory receptor cells differ in structure and function?

9. Compare the olfactory and gustatory pathways.

12.6 VISION

OBJECTIVES • Describe the accessory structures of the eye, the layers of the eyeball, the lens, the interior of the eyeball, image formation, and binocular vision.

• Describe the receptors for vision and the visual pathway to the brain.

Vision, the act of seeing, is extremely important to human survival. More than half the sensory receptors in the human body are located in the eyes, and a large part of the cerebral cortex is devoted to processing visual information. In this section of the chapter, we examine the accessory structures of the eye, the eyeball itself, the formation of visual images, the physiology of vision, and the visual pathway from the eye to the brain.

Accessory Structures of the Eye

The *accessory structures* of the eye are the eyebrows, eyelashes, eyelids, extrinsic muscles that move the eyeballs, and lacrimal (tear-producing) apparatus. The *eyebrows* and *eyelashes* help protect the eyeballs from foreign objects, perspiration, and direct rays of the sun (Figure 12.5). The upper and lower *eyelids* shade the eyes during sleep, protect the eyes from excessive light and foreign objects, and spread lubricating secretions over the eyeballs (by blinking). Six extrinsic eye muscles cooperate to move each eyeball right, left, up, down, and diagonally: the *superior rectus, inferior rectus, lateral rectus, medial rectus, superior oblique,* and *inferior oblique*. Neurons in the brain stem and cerebellum coordinate and synchronize the movements of the eyes.

The *lacrimal apparatus* (*lacrima* = tear) is a group of glands, ducts, canals, and sacs that produce and drain *lacrimal fluid* or *tears* (Figure 12.5). The right and left *lacrimal glands* are each about the size and shape of an almond. They secrete tears through the *lacrimal ducts* onto the surface of the upper eyelid. Tears then pass over the surface of the eyeball toward the nose into two *lacrimal canals* and a *nasolacrimal duct*, which allow the tears to drain into the nasal cavity.

Tears are a watery solution containing salts, some mucus, and a bacteria-killing enzyme called *lysozyme*. Tears clean, lubricate, and moisten the portion of the eyeball exposed to the air to prevent it from drying. Normally, tears are cleared away by evaporation or by passing into the nasal cavity as fast

Figure 12.5 Accessory structures of the eye.

Accessory structures of the eye are the eyebrows, eyelashes, eyelids, extrinsic eye muscles, and the lacrimal apparatus.

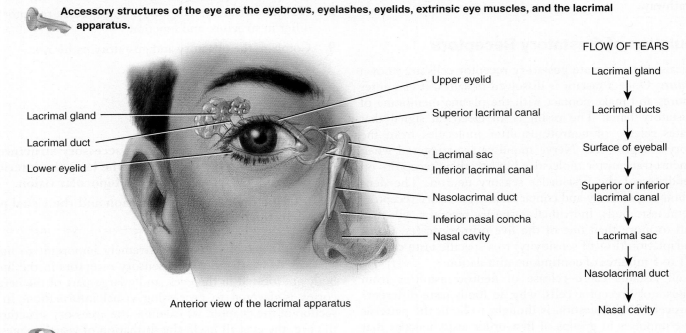

Anterior view of the lacrimal apparatus

FLOW OF TEARS

Lacrimal gland
↓
Lacrimal ducts
↓
Surface of eyeball
↓
Superior or inferior lacrimal canal
↓
Lacrimal sac
↓
Nasolacrimal duct
↓
Nasal cavity

What are the functions of tears?

as they are produced. If, however, an irritating substance makes contact with the eye, the lacrimal glands are stimulated to oversecrete and tears accumulate. This protective mechanism dilutes and washes away the irritant. Only humans express emotions, both happiness and sadness, by *crying*. In response to parasympathetic stimulation, the lacrimal glands produce excessive tears that may spill over the edges of the eyelids and even fill the nasal cavity with fluid. This is how crying produces a runny nose.

Layers of the Eyeball

The adult *eyeball* measures about 2.5 cm (1 inch) in diameter and is divided into three layers: fibrous tunic, vascular tunic, and retina (Figure 12.6).

Fibrous Tunic

The *fibrous tunic* is the outer coat of the eyeball. It consists of an anterior cornea and a posterior sclera. The *cornea* (KOR-nē-a) is a transparent fibrous coat that covers the colored iris. Because it is curved, the cornea helps focus light rays onto the retina. The *sclera* (SKLER-a = hard), the "white" of the eye, is a coat of dense connective tissue that covers all of the entire eyeball except the cornea. The sclera gives shape to the eyeball, makes it more rigid, and protects its inner parts. An epithelial layer called the *conjunctiva* (kon-junk-TĪ-va) covers the sclera but not the cornea and lines the inner surface of the eyelids.

Vascular Tunic

The *vascular tunic* is the middle layer of the eyeball and is composed of the choroid, ciliary body, and iris. The *choroid* (KŌ-royd) is a thin membrane that lines most of the internal surface of the sclera. It contains many blood vessels that help nourish the retina. The choroid also contains melanocytes that produce the pigment melanin, which causes this layer to appear dark brown in color. Melanin in the choroid absorbs stray light rays, which prevents reflection and scattering of light within the eyeball. As a result, the image cast on the retina by the cornea and lens remains sharp and clear.

At the front of the eye, the choroid becomes the *ciliary body* (SIL-ē-ar'-ē). The ciliary body consists of the *ciliary processes*, folds on the inner surface of the ciliary body whose capillaries secrete a fluid called aqueous humor, and the *ciliary muscle*, a smooth muscle that alters the shape of the lens for viewing objects up close or at a distance. The *lens*, a transparent structure that focuses light rays onto the retina, is constructed of many layers of elastic protein fibers. *Zonular fibers* attach the lens to the ciliary muscle and hold the lens in position.

The *iris* (= colored circle) is the colored part of the eyeball. It includes both circular and radial smooth muscle fibers. The hole in the center of the iris, through which light enters the eyeball, is the *pupil*. The smooth muscle of the iris regulates the amount of light passing through the lens. When the eye is stimulated by bright light, the parasympathetic

Figure 12.6 Structure of the eyeball.

The wall of the eyeball consists of three layers: the fibrous tunic, the vascular tunic, and the retina.

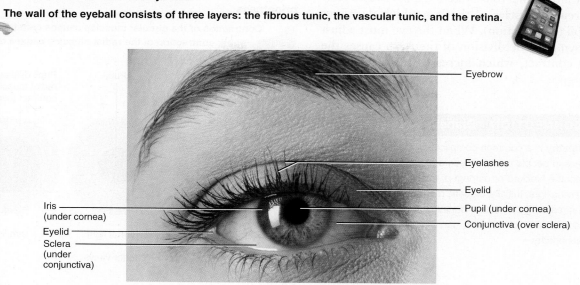

Eyebrow

Eyelashes

Eyelid

Iris (under cornea)

Pupil (under cornea)

Eyelid

Conjunctiva (over sclera)

Sclera (under conjunctiva)

(a) Anterior view of right eyeball

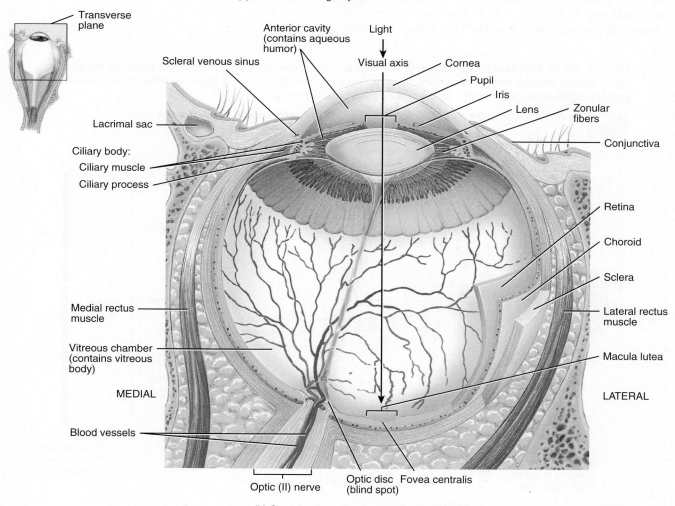

Transverse plane

Anterior cavity (contains aqueous humor)

Light

Visual axis

Scleral venous sinus

Cornea

Pupil

Iris

Lens

Zonular fibers

Lacrimal sac

Conjunctiva

Ciliary body:

Ciliary muscle

Ciliary process

Retina

Choroid

Sclera

Medial rectus muscle

Lateral rectus muscle

Vitreous chamber (contains vitreous body)

Macula lutea

MEDIAL

LATERAL

Blood vessels

Optic (II) nerve

Optic disc (blind spot)

Fovea centralis

(b) Superior view of transverse section of right eyeball

What are the components of the fibrous tunic and vascular tunic?

division of the autonomic nervous system (ANS) causes contraction of the circular muscles of the iris, which decreases the size of the pupil (constriction). When the eye must adjust to dim light, the sympathetic division of the ANS causes the radial muscles to contract, which increases the size of the pupil (dilation) (Figure 12.7).

CLINICAL CONNECTION | Diabetic Retinopathy

Diabetic Retinopathy is a common complication of diabetes and the main cause of blindness in people under 65 years old. In diabetic patients, blood vessels may become blocked or damaged; this damages the retina, preventing it from receiving light. Initially, diabetic retinopathy will not exhibit symptoms but if left untreated it can cause partial or total blindness. Adequate management and control of diabetes can help reduce the risk of developing diabetic retinopathy. •

Retina

The third and inner coat of the eyeball, the *retina*, lines the posterior three-quarters of the eyeball and is the beginning of the visual pathway (Figure 12.8). It has two layers: the neural layer and the pigmented layer. The *neural layer* is a

Figure 12.7 Responses of the pupil to light of varying brightness.

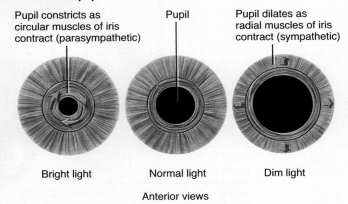

Contraction of the circular muscles causes constriction of the pupil; contraction of the radial muscles causes dilation of the pupil.

Pupil constricts as circular muscles of iris contract (parasympathetic)   Pupil   Pupil dilates as radial muscles of iris contract (sympathetic)

Bright light   Normal light   Dim light

Anterior views

Which division of the autonomic nervous system causes pupillary constriction? Which causes pupillary dilation?

Figure 12.8 Microscopic structure of the retina. The downward blue arrow at right indicates the direction of the signals passing through the neural layer of the retina. Eventually, nerve impulses arise in ganglion cells and propagate along their axons, which make up the optic (II) nerve.

In the retina, visual signals pass from photoreceptors to bipolar cells to ganglion cells.

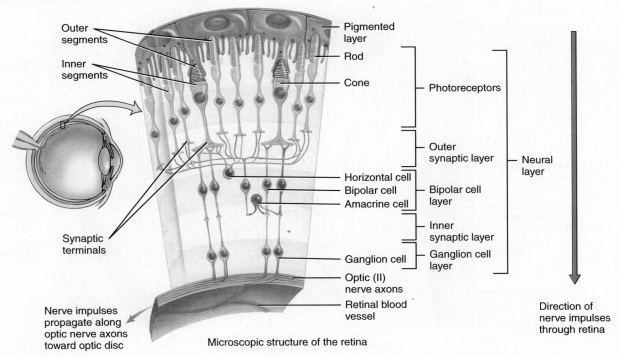

Microscopic structure of the retina

What are the two types of photoreceptors, and how do their functions differ?

multilayered outgrowth of the brain. Three distinct layers of retinal neurons—the **photoreceptor layer**, the **bipolar cell layer**, and the **ganglion cell layer**—are separated by two zones, the outer and inner synaptic layers, where synaptic contacts are made. Light passes through the ganglion and bipolar cell layers and both synaptic layers before it reaches the photoreceptor layer.

The *pigmented layer* of the retina is a sheet of melanin-containing epithelial cells located between the choroid and the neural part of the retina. The melanin in the pigmented layer of the retina, like in the choroid, also helps to absorb stray light rays.

Photoreceptors are specialized cells that begin the process by which light rays are ultimately converted to nerve impulses. There are two types of photoreceptors: rods and cones. **Rods** allow us to see shades of gray in dim light, such as moonlight. Brighter light stimulates the **cones**, giving rise to highly acute, color vision. Three types of cones are present in the retina: (1) *blue cones*, which are sensitive to blue light; (2) *green cones*, which are sensitive to green light; and (3) *red cones*, which are sensitive to red light. Color vision results from the stimulation of various combinations of these three types of cones. Just as an artist can obtain almost any color by mixing them on a palette, the cones can code for different colors by differential stimulation. There are about 6 million cones and 120 million rods. Cones are most densely concentrated in the *fovea centralis*, a small depression in the center of the *macula lutea* (MAK-ū-la LOO-tē-a), or yellow spot, in the exact center of the retina. The fovea centralis is the area of highest *visual acuity* or *resolution* (sharpness of vision) because of its high concentration of cones. The main reason that you move your head and eyes while looking at something, such as the words of this sentence, is to place images of interest on your fovea. Rods are absent from the fovea centralis and macula lutea and increase in numbers toward the periphery of the retina.

From photoreceptors, information flows through the outer synaptic layer to the bipolar cells of the bipolar cell layer, and then from bipolar cells through the inner synaptic layer to the ganglion cells of the ganglion cell layer. Between 6 and 600 rods synapse with a single bipolar cell in the outer synaptic layer; a cone usually synapses with just one bipolar cell. The convergence of many rods onto a single bipolar cell increases the light sensitivity of rod vision but slightly blurs the image that is perceived. Cone vision, although less sensitive, has higher acuity because of the one-to-one synapses between cones and their bipolar cells. The axons of the ganglion cells extend posteriorly to a small area of the retina called the *optic disc (blind spot)*, where they all exit as the optic (II) nerve (see Figure 12.6). Because the optic disc contains no rods or cones, we cannot see an image that strikes the blind spot. Normally, you are not aware of having a blind spot, but you can easily demonstrate its presence. Cover your left eye and gaze directly at the cross near the top of the next column. Then increase

or decrease the distance between the book and your eye. At some point, the square will disappear as its image falls on the blind spot.

Interior of the Eyeball

The lens divides the interior of the eyeball into two cavities, the anterior cavity and the vitreous chamber. The **anterior cavity** lies anterior to the lens and is filled with **aqueous humor** (AK-wē-us HŪ-mer; *aqua* = water), a watery fluid similar to cerebrospinal fluid. Blood capillaries of the ciliary processes of the ciliary body secrete aqueous humor into the anterior cavity. It then drains into the *scleral venous sinus (canal of Schlemm)*, an opening where the sclera and cornea meet (see Figure 12.6), and reenters the blood. The aqueous humor helps maintain the shape of the eye and nourishes the lens and cornea, neither of which has blood vessels. Normally, aqueous humor is completely replaced about every 90 minutes.

Behind the lens is the second, and larger, cavity of the eyeball, the **vitreous chamber**. It contains a clear, jellylike substance called the **vitreous body**, which forms during embryonic life and is not replaced thereafter. This substance helps prevent the eyeball from collapsing and holds the retina flush against the choroid.

The pressure in the eye, called **intraocular pressure**, is produced mainly by the aqueous humor with a smaller contribution from the vitreous body. Intraocular pressure maintains the shape of the eyeball and keeps the retina smoothly pressed against the choroid so the retina is well nourished and forms clear images. Normal intraocular pressure (about 16 mm Hg) is maintained by a balance between production and drainage of the aqueous humor.

Table 12.2 summarizes the structures of the eyeball.

Image Formation and Binocular Vision

In some ways the eye is like a camera: Its optical elements focus an image of some object on a light-sensitive "film"—the retina—while ensuring the correct amount of light makes the proper "exposure." To understand how the eye forms clear images of objects on the retina, we must examine three processes: (1) the refraction or bending of light by the lens and cornea, (2) the change in shape of the lens, and (3) constriction or narrowing of the pupil.

Refraction of Light Rays

When light rays traveling through a transparent substance (such as air) pass into a second transparent substance with a different density (such as water), they bend at the junction between the two substances. This bending is called **refraction**

TABLE 12.2

Summary of the Structures of the Eyeball and Their Functions

STRUCTURE	FUNCTION
Fibrous tunic 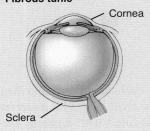	*Cornea:* Admits and refracts (bends) light *Sclera:* Provides shape and protects inner parts
Vascular tunic 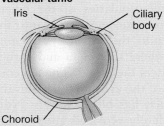	*Iris:* Regulates the amount of light that enters eyeball *Ciliary body:* Secretes aqueous humor and alters the shape of the lens for near or far vision (accommodation) *Choroid:* Provides blood supply and absorbs scattered light
Retina 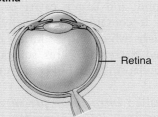	Receives light and converts it into nerve impulses; provides output to brain via axons of ganglion cells, which form the optic (II) nerve
Lens 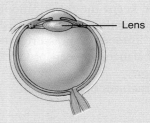	Refracts light
Anterior cavity 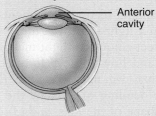	Contains aqueous humor that helps maintain the shape of the eyeball and supplies oxygen and nutrients to the lens and cornea
Vitreous chamber 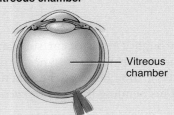	Contains the vitreous body, which helps maintain the shape of eyeball and keeps the retina attached to the choroid

(Figure 12.9a). About 75 percent of the total refraction of light occurs at the cornea. Then, the lens of the eye further refracts the light rays so that they come into exact focus on the retina.

Images focused on the retina are inverted (upside down) (Figure 12.9b, c). They also undergo right-to-left reversal; that is, light from the right side of an object strikes the left side of the retina, and vice versa. The reason the world does not look inverted and reversed is that the brain "learns" early in life to coordinate visual images with the orientations of objects. The brain stores the inverted and reversed images we acquire when we first reach for and touch objects and interprets those visual images as being correctly oriented in space.

Figure 12.9 Refraction of light rays and accommodation.

Refraction is the bending of light rays.

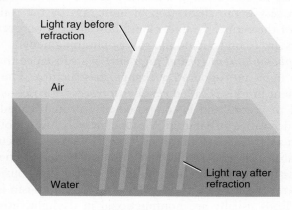

(a) Refraction of light rays

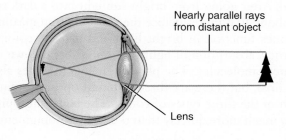

(b) Viewing distant object

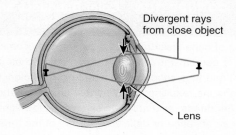

(c) Accommodation

? **What changes occur during accommodation?**

When an object is more than 6 meters (20 ft) away from the viewer, the light rays reflected from the object are nearly parallel to one another, and the curvatures of the cornea and lens exactly focus the image on the retina (Figure 12.9b). However, light rays from objects closer than 6 meters are divergent rather than parallel (Figure 12.9c). The rays must be refracted more if they are to be focused on the retina. This additional refraction is accomplished by changes in the shape of the lens.

Accommodation

A surface that curves outward, like the surface of a ball, is said to be *convex*. The convex surface of a lens refracts incoming light rays toward each other, so that they eventually intersect. The lens of the eye is convex on both its anterior and posterior surfaces, and its ability to refract light increases as its curvature becomes greater. When the eye is focusing on a close object, the lens becomes more convex and refracts the light rays more. This increase in the curvature of the lens for near vision is called ***accommodation*** (Figure 12.9c).

When you are viewing distant objects, the ciliary muscle of the ciliary body is relaxed and the lens is fairly flat because it is stretched in all directions by taut zonular fibers. When you view a close object, the ciliary muscle contracts, which pulls the ciliary process and choroid forward toward the lens. This action releases tension on the lens, allowing it to become rounder (more convex), which increases its focusing power and causes greater convergence of the light rays.

The normal eye, known as an ***emmetropic eye*** (em′-e-TROP-ik), can sufficiently refract light rays from an object 6 m (20 ft) away so that a clear image is focused on the retina (Figure 12.10a). Many people, however, lack this ability because of refraction abnormalities. Among these abnormalities are ***myopia*** (mī-Ō-pē-a), or nearsightedness, which occurs when the eyeball is too long relative to the focusing power of the cornea and lens. Myopic individuals can see nearby objects clearly, but not distant objects. In ***hyperopia*** (hī-per-Ō-pē-a) or farsightedness, also known as ***hypermetropia*** (hī′-per-me-TRŌ-pē-a), the eyeball length is short relative to the focusing power of the cornea and lens. Hyperopic individuals can see distant objects clearly, but not nearby objects. Figure 12.10b–e illustrates these conditions and shows how they are corrected. Another refraction abnormality is ***astigmatism*** (a-STIG-ma-tizm), in which either the cornea or the lens has an irregular curvature.

CLINICAL CONNECTION | Presbyopia

With aging, the lens loses some of its elasticity so its ability to accommodate decreases. At about age 40, people who have not previously worn glasses begin to require them for close vision, such as reading. This condition is called **presbyopia** (prez′-bē-Ō-pē-a; *presby-* = old; *-opia* = pertaining to the eye or vision). •

Figure 12.10 Normal and abnormal refraction in the eyeball. (a) In the normal (emmetropic) eye, light rays from an object are bent sufficiently by the cornea and lens to focus on the fovea centralis. (b) In the nearsighted (myopic) eye, the image is focused in front of the retina. (c) Correction for myopia is by use of a concave lens that diverges entering light rays so that they have to travel further through the eyeball. (d) In the farsighted (hyperopic) eye, the image is focused behind the retina. (e) Correction for hyperopia is by a convex lens that causes entering light rays to converge.

🔑 **In uncorrected myopia, distant objects can't be seen clearly; in uncorrected hyperopia, nearby objects can't be seen clearly.**

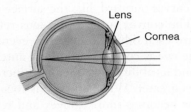

(a) Normal (emmetropic) eye

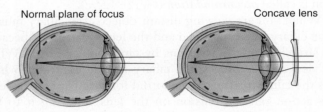

(b) Nearsighted (myopic) eye, uncorrected

(c) Nearsighted (myopic) eye, corrected

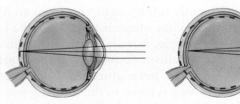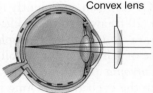

(d) Farsighted (hyperopic) eye, uncorrected

(e) Farsighted (hyperopic) eye, corrected

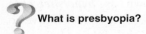 **What is presbyopia?**

Constriction of the Pupil

Constriction of the pupil is a narrowing of the diameter of the hole through which light enters the eye due to contraction of the circular muscles of the iris. This autonomic reflex occurs simultaneously with accommodation and prevents light rays from entering the eye through the periphery of the lens. Light rays entering at the periphery of the lens would not be brought to focus on the retina and would result in blurred vision. The pupil, as noted earlier, also constricts in bright light to limit the amount of light that strikes the retina.

Convergence

In humans, both eyes focus on only one set of objects, a characteristic called **binocular vision**. This feature of our visual system allows the perception of depth and an appreciation of the three-dimensional nature of objects. When you stare straight ahead at a distant object, the incoming light rays are aimed directly at the pupils of both eyes and are refracted to comparable spots on the two retinas. As you move closer to the object, your eyes must rotate toward the nose if the light rays from the object are to strike comparable points on both retinas. **Convergence** is the name for this automatic movement of the two eyeballs toward the midline, which is caused by the coordinated action of the extrinsic eye muscles. The nearer the object, the greater the convergence needed to maintain binocular vision.

Stimulation of Photoreceptors

After an image is formed on the retina by refraction, accommodation, constriction of the pupil, and convergence, light rays must be converted into neural signals. The initial step in this process is the absorption of light rays by the rods and cones of the retina. To understand how absorption occurs, it is necessary to understand the role of photopigments.

A **photopigment (visual pigment)** is a substance that can absorb light and undergo a change in structure. The photopigment in rods is called **rhodopsin** (*rhodo-* = rose; *-opsin* = related to vision) and is composed of a protein called *opsin* and a derivative of vitamin A called *retinal*. Any amount of light in a darkened room causes some rhodopsin molecules to split into opsin and retinal and initiate a series of chemical changes in the rods. When the light level is dim, opsin and retinal recombine into rhodopsin as fast as rhodopsin is split apart. Rods usually are nonfunctional in daylight, however, because rhodopsin is split apart faster than it can be reformed. After going from bright sunlight into a dark room, it takes about 40 minutes before the rods function maximally.

Cones function in bright light and provide color vision. As in rods, absorption of light rays causes breakdown of photopigment molecules. The photopigments in cones also contain retinal, but there are three different opsin proteins—one in each of the three types of cones. The cone photopigments reform much more quickly than the rod photopigment.

⚕️ **CLINICAL CONNECTION | Color Blindness and Night Blindness**

The complete loss of cone vision causes a person to become legally blind. In contrast, a person who loses rod vision mainly has difficulty seeing in dim light and thus should not drive at night. Prolonged vitamin A deficiency and the resulting below-normal amount of rhodopsin may cause **night blindness**, an inability to see well at low light levels. An individual with an absence or deficiency of one of the three types of cones from the retina cannot distinguish some colors from others and is said to be **color-blind**. In the most common type, *red–green color blindness*, either red cones or green cones are missing. Thus, the person cannot distinguish between red and green. The inheritance of color blindness is illustrated in Figure 24.12. •

The Visual Pathway

After stimulation by light, the rods and cones trigger electrical signals in bipolar cells. Bipolar cells transmit both excitatory and inhibitory signals to ganglion cells. The ganglion cells become depolarized and generate nerve impulses. The axons of the ganglion cells exit the eyeball as the *optic (II) nerve* (Figure 12.11, **1**) and extend posteriorly to the *optic chiasm* (KĪ-azm = a crossover, as in the letter X; Figure 12.11, **2**). In the optic chiasm, about half of the axons from each eye cross to the opposite side of the brain. After passing the optic chiasm, the axons, now part of the *optic tract* (Figure 12.11, **3**), terminate in the thalamus. Here they synapse with neurons whose axons project to the primary visual areas in the occipital lobes of the cerebral cortex (Figure 12.11, **4**; see also Figure 10.13). Because of crossing at the optic chiasm, the right side of the brain receives signals from both eyes for interpretation of visual sensations from the left side of an object, and the left side of the brain receives signals from both eyes for interpretation of visual sensations from the right side of an object.

Figure 12.11 Visual pathway.

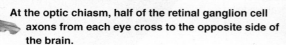

At the optic chiasm, half of the retinal ganglion cell axons from each eye cross to the opposite side of the brain.

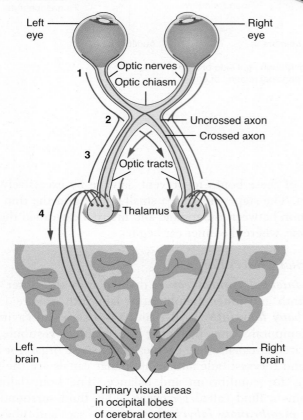

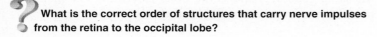
What is the correct order of structures that carry nerve impulses from the retina to the occipital lobe?

✓ CHECKPOINT

10. List and describe the accessory structures of the eye.
11. Describe the layers of the eyeball and their functions.
12. How is an image formed on the retina?
13. How does the shape of the lens change during accommodation?
14. How do photopigments respond to light?
15. By what pathway do nerve impulses triggered by an object in the left half of the visual field of the left eye reach the primary visual area of the cerebral cortex?

12.7 HEARING AND EQUILIBRIUM

OBJECTIVES • **Distinguish the structures of the external, middle, and internal ear.**

• **Describe the receptors for hearing and equilibrium and outline their pathways to the brain.**

The ear is a marvelously sensitive structure. Its sensory receptors can convert sound vibrations into electrical signals 1000 times faster than photoreceptors can respond to light. Beside receptors for sound waves, the ear also contains receptors for equilibrium (balance).

Structure of the Ear

The ear is divided into three main regions: (1) the external ear, which collects sound waves and channels them inward; (2) the middle ear, which conveys sound vibrations to the oval window; and (3) the internal ear, which houses the receptors for hearing and equilibrium.

External (Outer) Ear

The *external (outer) ear* collects sound waves and passes them inward (Figure 12.12). It consists of an auricle, external auditory canal, and eardrum. The *auricle*, the part of the ear that you can see, is a skin-covered flap of elastic cartilage shaped like the flared end of a trumpet. It plays a small role in collecting sound waves and directing them toward the *external auditory canal* (audit- = hearing), a curved tube that extends from the auricle and directs sound waves toward the eardrum. The canal contains a few hairs and *ceruminous glands* (se-RŪ-mi-nus; cer- = wax), which secrete *cerumen* (se-RŪ-men) (earwax). The hairs and cerumen help prevent foreign objects from entering the ear. The *eardrum*, also called the *tympanic membrane* (tim-PAN-ik; tympan- = adrum), is a thin, semitransparent partition between the external auditory canal and the middle ear. Sound waves cause the eardrum to vibrate. Tearing of the tympanic membrane, due to trauma or infection, is called a *perforated eardrum*.

Figure 12.12 Structure of the ear.

The ear has three principal regions: the external ear, the middle ear, and the internal ear.

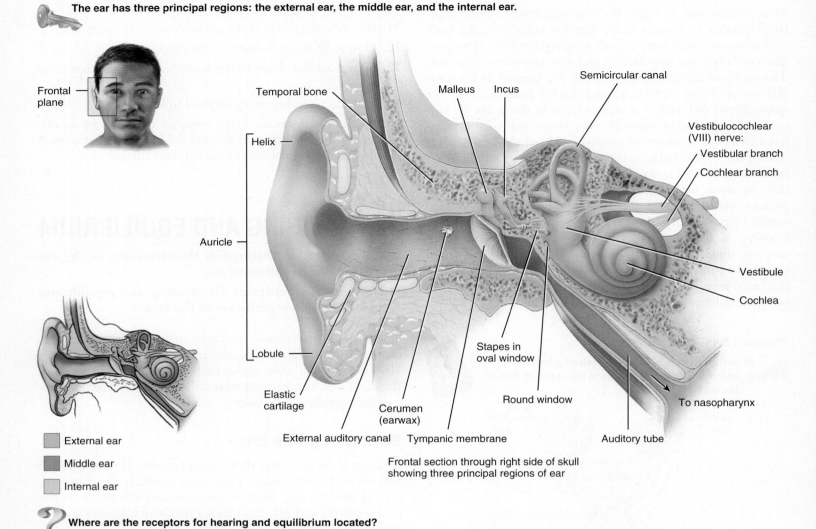

Frontal
plane

Temporal bone

Malleus   Incus

Semicircular canal

Vestibulocochlear
(VIII) nerve:

Vestibular branch

Cochlear branch

Helix

Auricle

Vestibule

Cochlea

Stapes in
oval window

Lobule

Round window

To nasopharynx

Elastic
cartilage

Cerumen
(earwax)

External auditory canal   Tympanic membrane

Auditory tube

Frontal section through right side of skull
showing three principal regions of ear

External ear

Middle ear

Internal ear

? **Where are the receptors for hearing and equilibrium located?**

Middle Ear

The *middle ear* is a small, air-filled cavity between the eardrum and inner ear (Figure 12.12). An opening in the anterior wall of the middle ear leads directly into the *auditory tube*, commonly known as the *eustachian tube*, which connects the middle ear with the upper part of the throat. When the auditory tube is open, air pressure can equalize on both sides of the eardrum. Otherwise, abrupt changes in air pressure on one side of the eardrum might cause it to rupture. During swallowing and yawning, the tube opens, which explains why yawning can help equalize the pressure changes that occur while flying in an airplane.

Extending across the middle ear and attached to it by means of ligaments are three tiny bones called *auditory ossicles* (OS-si-kuls) that are named for their shapes: the *malleus* (MAL-ē-us), *incus* (ING-kus), and *stapes* (STĀ-pēz), commonly called the hammer, anvil, and stirrup (Figure 12.12). Equally tiny skeletal muscles control the amount of move-

ment of these bones to prevent damage by excessively loud noises. The stapes fits into a small opening in the thin bony partition between the middle and internal ear called the *oval window*, where the inner ear begins.

Internal (Inner) Ear

The *internal (inner) ear* is divided into the outer bony labyrinth and inner membranous labyrinth (Figure 12.13). The *bony labyrinth* (LAB-i-rinth) is a series of cavities in the temporal bone, including the cochlea, vestibule, and semicircular canals. The cochlea is the sense organ for hearing, and the vestibule and semicircular canals are the sense organs for equilibrium and balance. The bony labyrinth contains a fluid called *perilymph*. This fluid surrounds the inner *membranous labyrinth*, a series of sacs and tubes with the same general shape as the bony labyrinth. The membranous labyrinth contains a fluid called *endolymph*.

Figure 12.13 Details of the right internal ear. (a) Relationship of the scala tympani, cochlear duct, and scala vestibuli. The arrows indicate the transmission of sound waves. (b) Details of the spiral organ (organ of Corti).

🔑 The three channels in the cochlea are the scala vestibuli, scala tympani, and cochlear duct.

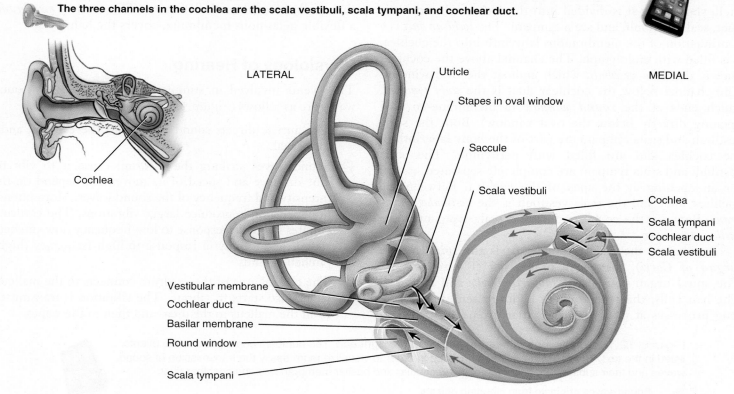

(a) Sections through the cochlea

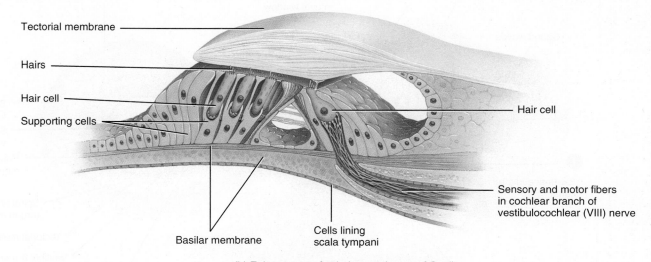

(b) Enlargement of spiral organ (organ of Corti)

❓ **What structures separate the external ear from the middle ear? The middle ear from the internal ear?**

The **vestibule** (VES-ti-būl) is the oval-shaped middle part of the bony labyrinth. The membranous labyrinth in the vestibule consists of two sacs called the **utricle** (Ū-tri-kul = little bag) and **saccule** (SAK-ūl = little sac). Behind the vestibule are the three bony **semicircular canals**. The anterior and posterior semicircular canals are both vertical, and the lateral canal is horizontal. One end of each canal enlarges into a swelling called the **ampulla** (am-PUL-la = little jar). The portions of the membranous labyrinth that lie inside the bony semicircular canals are called the **semicircular ducts**, which connect with the utricle of the vestibule.

A transverse section through the ***cochlea*** (KOK-lē-a = snail's shell), a bony spiral canal that resembles a snail's shell, shows that it is divided into three channels: cochlear duct, scala vestibuli, and scala tympani. The *cochlear duct* is a continuation of the membranous labyrinth into the cochlea; it is filled with endolymph. The channel above the cochlear duct is the *scala vestibuli*, which ends at the oval window. The channel below the cochlear duct is the *scala tympani*, which ends at the ***round window*** (a membrane-covered opening directly below the oval window). Both the scala vestibuli and scala tympani are part of the bony labyrinth of the cochlea and are filled with perilymph. The scala vestibuli and scala tympani are completely separated, except for an opening at the apex of the cochlea. Between the cochlear duct and the scala vestibuli is the *vestibular membrane*. Between the cochlear duct and scala tympani is the *basilar membrane*.

Resting on the basilar membrane is the ***spiral organ (organ of Corti)***, the organ of hearing (Figure 12.13b). The spiral organ consists of *supporting cells* and *hair cells*. The hair cells, the receptors for auditory sensations, have long processes at their free ends that extend into the en-

dolymph of the cochlear duct. The hair cells form synapses with sensory and motor neurons in the cochlear branch of the vestibulocochlear (VIII) nerve. The *tectorial membrane*, a flexible gelatinous membrane, covers the hair cells.

Physiology of Hearing

The events involved in stimulation of hair cells by sound waves are as follows (Figure 12.14):

1 The auricle directs sound waves into the external auditory canal.

2 Sound waves striking the eardrum cause it to vibrate. The distance and speed of its movement depend on the intensity and frequency of the sound waves. More intense (louder) sounds produce larger vibrations. The eardrum vibrates slowly in response to low-frequency (low-pitched) sounds and rapidly in response to high-frequency (high-pitched) sounds.

3 The central area of the eardrum connects to the malleus, which also starts to vibrate. The vibration is transmitted from the malleus to the incus and then to the stapes.

Figure 12.14 Physiology of hearing shown in the right ear. The numbers correspond to the events listed in the text. The cochlea has been uncoiled in order to visualize more easily the transmission of sound waves and their subsequent distortion of the vestibular and basilar membranes of the cochlear duct.

Sound waves originate from vibrating objects.

Malleus Incus Stapes vibrating in oval window Cochlea

Sound waves

Perilymph

External auditory canal

1 **2** **3** **4** **5** **6** **7** **8** **9**

Scala tympani

Scala vestibuli

Basilar membrane

Spiral organ (organ of Corti)

Tectorial membrane

Vestibular membrane

Cochlear duct (contains endolymph)

Tympanic membrane

Round window

Middle ear Auditory tube

? **What is the function of hair cells?**

④ As the stapes moves back and forth, it pushes the oval window in and out.

⑤ The movement of the oval window sets up fluid pressure waves in the perilymph of the cochlea. As the oval window bulges inward, it pushes on the perilymph of the scala vestibuli.

⑥ The fluid pressure waves are transmitted from the scala vestibuli to the scala tympani and eventually to the membrane covering the round window, causing it to bulge outward into the middle ear. (See ⑨ in the figure.)

⑦ As the pressure waves deform the walls of the scala vestibuli and scala tympani, they also push the vestibular membrane back and forth, creating pressure waves in the endolymph inside the cochlear duct.

⑧ The pressure waves in the endolymph cause the basilar membrane to vibrate, which moves the hair cells of the spiral organ against the tectorial membrane. Bending of their hairs stimulates the hair cells to release neurotransmitter molecules at synapses with sensory neurons that are part of the vestibulocochlear (VIII) nerve (see Figure 12.13b). Then, the sensory neurons generate nerve impulses that conduct along the vestibulocochlear (VIII) nerve.

Sound waves of various frequencies cause certain regions of the basilar membrane to vibrate more intensely than other regions. Each segment of the basilar membrane is "tuned" for a particular pitch. Because the membrane is narrower and stiffer at the base of the cochlea (closer to the oval window), high-frequency (high-pitched) sounds induce maximal vibrations in this region. Toward the apex of the cochlea, the basilar membrane is wider and more flexible; low-frequency (low-pitched) sounds cause maximal vibration of the basilar membrane there. Loudness is determined by the intensity of sound waves. High-intensity sound waves cause larger vibrations of the basilar membrane, which leads to a higher frequency of nerve impulses reaching the brain. Louder sounds also may stimulate a larger number of hair cells.

CLINICAL CONNECTION | Perforated Eardrum

Perforated eardrum is a hole or tear in the tympanic membrane. Symptoms include a degree of hearing loss, earache, mucus discharge and ringing in the ear. Any hearing loss is usually temporary. Although there are several possible causes of a perforated eardrum, the most common cause is infection of the middle ear. The eardrum may also be damaged by loud noise, injury to the ear or changes in air pressure. It can heal without treatment within two months although treatment may be recommended for infection and to relieve discomfort. •

Auditory Pathway

Sensory neurons in the cochlear branch of each vestibulocochlear (VIII) nerve terminate in the medulla oblongata on the same side of the brain. From the medulla, axons ascend to the midbrain, then to the thalamus, and finally to the primary auditory area in the temporal lobe (see Figure 10.13). Because many auditory axons cross to the opposite side, the right and left primary auditory areas receive nerve impulses from both ears.

Physiology of Equilibrium

You learned about the anatomy of the internal ear structures for equilibrium in the previous section. In this section we will cover the physiology of balance, or how you are able to stay on your feet after tripping over your roommate's shoes. There are two types of *equilibrium* (balance). One kind, called *static equilibrium*, refers to the maintenance of the position of the body (mainly the head) relative to the force of gravity. Body movements that stimulate the receptors for static equilibrium include tilting the head and *linear* acceleration or deceleration, such as when the body is being moved in an elevator or in a car that is speeding up or slowing down. The second kind, *dynamic equilibrium*, is the maintenance of body position (mainly the head) in response to *rotational* acceleration or deceleration. Collectively, the receptor organs for equilibrium, which include the saccule, utricle, and membranous semicircular ducts, are called the *vestibular apparatus* (ves-TIB-ū-lar).

Static Equilibrium

The walls of both the utricle and the saccule contain a small, thickened region called a *macula* (MAK-ū-la; *macula* = spot), not to be confused with the macula lutea of the eye. The two maculae (plural), which are perpendicular to one another, are the receptors for static equilibrium. The maculae provide sensory information on the position of the head in space and help maintain appropriate posture and balance. The maculae also contribute to some aspects of dynamic equilibrium by detecting linear acceleration and deceleration.

The two maculae consist of two kinds of cells: *hair cells*, which are the sensory receptors, and *supporting cells* (Figure 12.15). The hairs of the hair cells protrude into a thick, jellylike substance called the *otolithic membrane*. A layer of dense calcium carbonate crystals, called *otoliths* (*oto-* = ear; *-liths* = stones), extends over the entire surface of the otolithic membrane. If you tilt your head forward, gravity pulls the membrane (and the otoliths) so it slides over the hair cells in the direction of the tilt. This stimulates the hair cells and triggers nerve impulses that conduct along the *vestibular branch* of the vestibulocochlear (VIII) nerve (see Figure 12.12).

Dynamic Equilibrium

The three membranous semicircular ducts lie at right angles to one another in three planes (see Figure 12.13a). This positioning permits detection of rotational acceleration or deceleration. The dilated portion of each duct, the ampulla, contains a small elevation called the *crista* (KRIS-ta = crest; plural is *cristae*)

Figure 12.15 Location and structure of receptors in the maculae of the right ear. Both sensory neurons (blue) and motor neurons (red) synapse with the hair cells.

Movements of the otolithic membrane stimulate the hair cells.

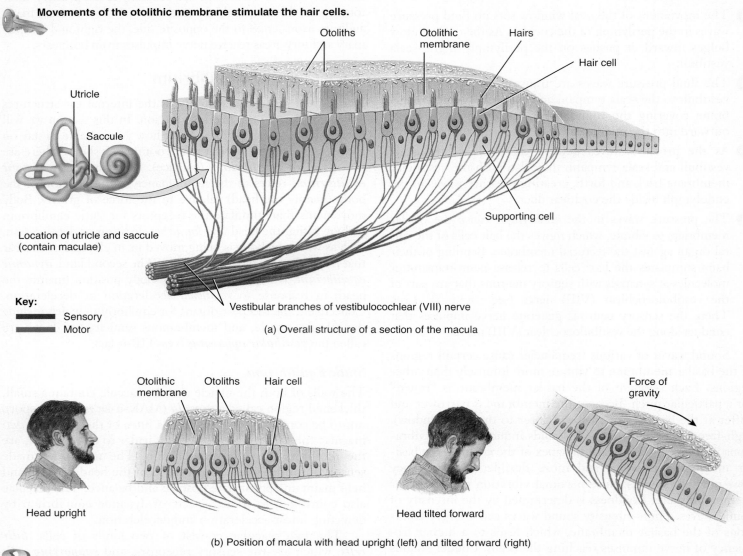

(a) Overall structure of a section of the macula

(b) Position of macula with head upright (left) and tilted forward (right)

? **What is the function of the maculae?**

(Figure 12.16). Each *crista* contains a group of **hair cells** and **supporting cells**. Covering the crista is a mass of gelatinous material called the **cupula** (KŪ-pū-la). When the head moves, the attached membranous semicircular ducts and hair cells move with it. However, the endolymph within the membranous semicircular ducts is not attached and lags behind due to its inertia. As the moving hair cells drag along the stationary endolymph, the hairs bend. Bending of the hairs causes electrical signals in the hair cells. In turn, these signals trigger nerve impulses in sensory neurons that are part of the vestibular branch of the vestibulocochlear (VIII) nerve.

Equilibrium Pathways

Most of the vestibular branch axons of the vestibulocochlear

(VIII) nerve enter the brain stem and then extend to the medulla or the cerebellum, where they synapse with the next neurons in the equilibrium pathways. From the medulla, some axons conduct nerve impulses along the cranial nerves that control eye movements and head and neck movements. Other axons form a spinal cord tract that conveys impulses for regulation of muscle tone in response to head movements. Various pathways among the medulla, cerebellum, and cerebrum enable the cerebellum to play a key role in maintaining equilibrium. In response to continuous reception of sensory information from the utricle and saccule, the cerebellum makes adjustments to the signals going from the motor cortex to specific skeletal muscles to maintain equilibrium.

Figure 12.16 Location and structure of the membranous semicircular ducts of the right ear. Both sensory neurons (blue) and motor neurons (red) synapse with the hair cells. The ampullary nerves are branches of the vestibular division of the vestibulocochlear (VIII) nerve.

 The positions of the membranous semicircular ducts permit detection of rotational movements.

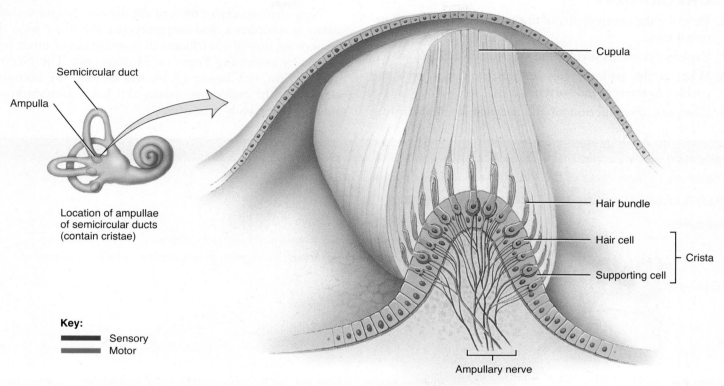

Semicircular duct

Ampulla

Location of ampullae
of semicircular ducts
(contain cristae)

Cupula

Hair bundle

Hair cell

Supporting cell

Crista

Key:
— Sensory
— Motor

Ampullary nerve

(a) Details of a crista

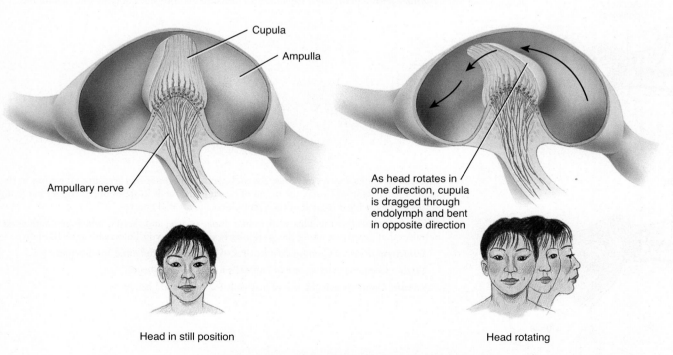

Cupula

Ampulla

Ampullary nerve

As head rotates in
one direction, cupula
is dragged through
endolymph and bent
in opposite direction

Head in still position

Head rotating

(b) Position of a cupula with the head in the still position (left)
and when the head rotates (right)

With which type of equilibrium are the membranous semicircular ducts associated?

Table 12.3 summarizes the structures of the ear related to hearing and equilibrium.

static equilibrium with the role of the cristae in maintaining dynamic equilibrium.

✓ CHECKPOINT

16. Describe the components of the external, middle, and internal ears.

17. Explain the events involved in stimulation of hair cells.

18. What is the pathway for auditory impulses from the cochlea to the cerebral cortex?

19. Compare the function of the maculae in maintaining

• • •

Now that our exploration of the nervous system and sensations is completed, you can appreciate the many ways that the nervous system contributes to homeostasis of other body systems by examining Focus on Homeostasis: The Nervous System. Next, in Chapter 13, we will see how the hormones released by the endocrine system also help maintain homeostasis of many body processes.

TABLE 12.3

Summary of Structures of the Ear Related to Hearing and Equilibrium

REGIONS OF THE EAR AND KEY STRUCTURES	FUNCTIONS
External Ear	*Auricle:* Collects sound waves *External auditory canal:* Directs sound waves to the eardrum *Eardrum (tympanic membrane):* Sound waves cause it to vibrate, which in turn causes the malleus to vibrate
Middle Ear	*Auditory ossicles:* Transmit and amplify vibrations from tympanic membrane to oval window *Auditory (eustachian) tube:* Equalizes air pressure on both sides of tympanic membrane

Internal Ear	*Cochlea:* Contains a series of fluids, channels, and membranes that transmit vibrations to the spiral organ (organ of Corti), the organ of hearing; hair cells in spiral organ trigger nerve impulses in the cochlear branch of the vestibulocochlear (VIII) nerve
	 Vestibular apparatus: Includes semicircular ducts, utricle, and saccule, which generate nerve impulses that propagate along the vestibular branch of the vestibulocochlear (VIII) nerve
	   Semicircular ducts: Contain cristae, sites of hair cells for dynamic equilibrium
	   Utricle: Contains macula, site of hair cells for static equilibrium
	   Saccule: Contains macula, site of hair cells for static equilibrium

BODY SYSTEM	CONTRIBUTION OF THE NERVOUS SYSTEM

For all body systems
Together with hormones from the endocrine system, nerve impulses provide communication and regulation of most body tissues.

Integumentary system
Sympathetic nerves of the autonomic nervous system (ANS) control contraction of smooth muscles attached to hair follicles and secretion of perspiration from sweat glands.

Skeletal system
Nociceptors (pain receptors) in bone tissue warn of bone trauma or damage.

Muscular system
Somatic motor neurons receive instructions from motor areas of the brain and stimulate contraction of skeletal muscles to bring about body movements. The basal ganglia and reticular formation set the level of muscle tone. The cerebellum coordinates skilled movements.

Endocrine system
The hypothalamus regulates secretion of hormones from the anterior and posterior pituitary. The ANS regulates secretion of hormones from the adrenal medulla and pancreas.

Cardiovascular system
The cardiovascular center in the medulla oblongata provides nerve impulses to the ANS that govern heart rate and the forcefulness of the heartbeat. Nerve impulses from the ANS also regulate blood pressure and blood flow through blood vessels.

Lymphatic system and immunity
Certain neurotransmitters help regulate immune responses. Activity in the nervous system may increase or decrease immune responses.

Respiratory system
Respiratory areas in the brain stem control breathing rate and depth. The ANS helps regulate the diameter of airways.

Digestive system
The ANS and enteric nervous system (ENS) help regulate digestion. The parasympathetic division of the ANS stimulates many digestive processes.

Urinary system
The ANS helps regulate blood flow to kidneys, thereby influencing the rate of urine formation; brain and spinal cord centers govern emptying of urinary bladder.

Reproductive systems
The hypothalamus and limbic system govern a variety of sexual behaviors; the ANS brings about erection of the penis in males and the clitoris in females and ejaculation of semen in males. The hypothalamus regulates release of anterior pituitary hormones that control the gonads (ovaries and testes). Nerve impulses elicited by touch stimuli from suckling infants cause release of oxytocin and milk ejection in nursing mothers.

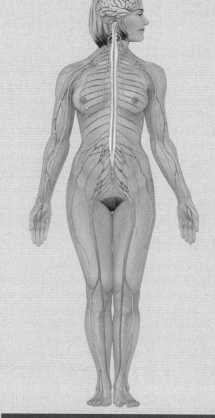

THE NERVOUS SYSTEM

COMMON DISORDERS

Cataracts

A common cause of blindness is a loss of transparency of the lens known as a *cataract* (CAT-a-rakt). The lens becomes cloudy (less transparent) due to changes in the structure of the lens proteins. Cataracts often occur with aging but may also be caused by injury, excessive exposure to ultraviolet rays, certain medications (such as long-term use of steroids), or complications of other diseases (for example, diabetes). People who smoke also have increased risk of developing cataracts. Fortunately, sight can usually be restored by surgical removal of the old lens and implantation of an artificial one.

Glaucoma

In *glaucoma* (glaw-KŌ-ma), the most common cause of blindness in the United States, a buildup of aqueous humor within the anterior cavity causes an abnormally high intraocular pressure. Persistent pressure results in a progression from mild visual impairment to irreversible destruction of the retina, damage to the optic nerve, and blindness. Because glaucoma is painless, and because the other eye initially compensates to a large extent for the loss of vision, a person may experience considerable retinal damage and loss of vision before the condition is diagnosed.

Some individuals have another form of glaucoma called *normal-tension (low-tension) glaucoma*. In this condition, there is damage to the optic nerve with a corresponding loss of vision, even though intraocular pressure is normal. Although the cause is unknown, it appears to be related to a fragile optic nerve, vasospasm of blood vessels around the optic nerve, and ischemia due to narrowed or obstructed blood vessels around the optic nerve. The incidence of normal-tension glaucoma is higher among Japanese and Koreans and females.

Deafness

Deafness is significant or total hearing loss. *Sensorineural deafness* is caused by either impairment of hair cells in the cochlea or damage of the cochlear branch of the vestibulocochlear nerve. This type of deafness may be caused by atherosclerosis, which reduces blood supply to the ears; repeated exposure to loud noise, which destroys hair cells of the spiral organ; or certain drugs such as aspirin and streptomycin. *Conduction deafness* is caused by impairment of the outer and middle ear mechanisms for transmitting sounds to the cochlea. It may be caused by otosclerosis, the deposition of new bone around the oval window; impacted cerumen; injury to the eardrum; or aging, which often results in thickening of the eardrum and stiffening of the joints of the auditory ossicles.

Ménière's Disease

Ménière's disease (men'-ē-ĀRZ) results from an increased amount of endolymph that enlarges the membranous labyrinth. Among the symptoms are fluctuating hearing loss (caused by distortion of the basilar membrane of the cochlea) and roaring tinnitus (ringing). Vertigo (a sensation of spinning or whirling) is characteristic of Ménière's disease. Almost total destruction of hearing may occur over a period of years.

Otitis Media

Otitis media is an acute infection of the middle ear caused primarily by bacteria and associated with infections of the nose and throat. Symptoms include pain; malaise (discomfort or uneasiness); fever; and a reddening and outward bulging of the eardrum, which may rupture unless prompt treatment is received (this may involve draining pus from the middle ear). Bacteria from the nasopharynx passing into the auditory tube is the primary cause of all middle ear infections. Children are more susceptible than adults to middle ear infections because their auditory tubes are almost horizontal, which decreases drainage.

MEDICAL TERMINOLOGY AND CONDITIONS

Age-related macular disease (AMD) Degeneration of the macula lutea of the retina in persons 50 years of age and older.

Anosmia (an-OZ-mē-a; *a-* = without; *-osmi* = smell, odor) Total lack of the sense of smell.

Barotrauma (bar'-ō-TRAW-ma; *baros-* = weight) Damage or pain, mainly affecting the middle ear, as a result of pressure changes. It occurs when pressure on the outer side of the tympanic membrane is higher than on the inner side, for example, when flying in an airplane or diving. Swallowing or holding your nose and exhaling with your mouth closed usually opens the auditory tubes, allowing air into the middle ear to equalize the pressure.

Cochlear implant A device that translates sounds into electrical signals that can be interpreted by the brain. It is especially useful for people with deafness caused by damage to hair cells in the cochlea.

Conjunctivitis (pinkeye) An inflammation of the conjunctiva; when caused by bacteria such as pneumococci, staphylococci, or *Hemophilus influenzae*, it is very contagious and more common in children. May also be caused by irritants, such as dust, smoke, or pollutants in the air, in which case it is not contagious.

Corneal transplant A procedure in which a defective cornea is removed and a donor cornea of similar diameter is sewn in. It is the most common and most successful transplant operation. Since the cornea is avascular, antibodies in the blood that might cause rejection do not enter the transplanted tissue, and rejection rarely occurs. The shortage of donor corneas has been partially overcome by the development of artificial corneas made of plastic.

Detached retina Detachment of the neural portion of the retina from the pigment epithelium due to trauma, disease, or age-related degeneration. The result is distorted vision and blindness.

Diabetic retinopathy (ret-i-NOP-a-thē; *retino-* = retina; *-pathos* = suffering) Degenerative disease of the retina due to diabetes mellitus, in which blood vessels in the retina are damaged or new ones grow and interfere with vision.

LASIK (laser-assisted in-situ keratomileusis) Surgery with a laser to correct the curvature of the cornea for conditions such as nearsightedness, farsightedness, and astigmatism.

Nystagmus (nis-TAG-mus; *nystagm-* = nodding or drowsy) A rapid involuntary movement of the eyeballs, possibly caused by a disease of the central nervous system. It is associated with conditions that cause vertigo.

Otalgia (ō-TAL-jē-a; *ot-* = ear; *-algia* = pain) Earache.

Retinoblastoma (ret-i-nō-blas-TŌ-ma; *-oma* = tumor) A tumor arising from immature retinal cells; it accounts for 2% of childhood cancers.

Scotoma (skō-TŌ-ma = darkness) An area of reduced or lost vision in the visual field.

Strabismus (stra-BIZ-mus) An imbalance in the extrinsic eye muscles that causes a misalignment of one eye so that its line of vision is not parallel with that of the other eye (cross-eyes) and both eyes are not pointed at the same object at the same time; the condition produces a squint.

Tinnitus (ti-NĪ-tus) A ringing, roaring, or clicking in the ears.

Trachoma (tra-KŌ-ma) A serious form of conjunctivitis and the greatest single cause of blindness in the world. It is caused by the bacterium *Chlamydia trachomatis*. The disease produces an excessive growth of subconjunctival tissue and invasion of blood vessels into the cornea, which progresses until the entire cornea is opaque, causing blindness.

Vertigo (VER-ti-gō = dizziness) A sensation of spinning or movement in which the world seems to revolve or the person seems to revolve in space.

CHAPTER REVIEW AND RESOURCE SUMMARY

REVIEW	RESOURCES
12.1 Overview of Sensations	
1. **Sensation** is the conscious or subconscious awareness of external and internal stimuli.	Anatomy Overview—Special Senses
2. Two general classes of senses are (1) **general senses**, which include **somatic senses** and **visceral senses**, and (2) **special senses**, which include smell, taste, vision, hearing, and equilibrium (balance).	
3. The conditions for a sensation to occur are reception of a stimulus by a sensory receptor, conversion of the stimulus into one or more nerve impulses, conduction of the impulses to the brain, and integration of the impulses by a region of the brain.	
4. Sensory impulses from each part of the body arrive in specific regions of the cerebral cortex.	
5. **Adaptation** is a decrease in sensation during a prolonged stimulus. Some receptors are rapidly adapting; others are slowly adapting.	
6. Receptors can be classified structurally by their microscopic features as free nerve endings, encapsulated nerve endings, or separate cells. Functionally, receptors are classified by the type of stimulus they detect as mechanoreceptors, thermoreceptors, nociceptors, photoreceptors, osmoreceptors, and chemoreceptors.	
12.2 Somatic Senses	
1. Somatic sensations include **tactile sensations** (**touch**, **pressure**, **vibration**, **itch**, and **tickle**), **thermal sensations** (heat and cold), pain sensations, and **proprioceptive sensations** (joint and muscle position sense and movements of the limbs). Receptors for these sensations are located in the skin, mucous membranes, muscles, tendons, and joints.	Anatomy Overview—Visceral Receptors Anatomy Overview—Visceral Effectors Anatomy Overview—Sensory Receptors of the Skin
2. Receptors for touch include **Meissner corpuscles**, **hair root plexuses**, **Merkel discs**, and **Ruffini corpuscles**. Receptors for pressure and vibration are **pacinian corpuscles**. Tickle and itch sensations result from stimulation of free nerve endings.	**Figure 12.2** Distribution of referred pain
3. **Thermoreceptors**, free nerve endings in the epidermis and dermis, adapt to continuous stimulation.	
4. **Nociceptors** are free nerve endings that are located in nearly every body tissue; they provide pain sensations.	
5. **Proprioceptors** inform us of the degree to which muscles are contracted, the amount of tension present in tendons, the positions of joints, and the orientation of the head.	

REVIEW	RESOURCES

12.3 Special Senses

1. The special senses include smell, sight, taste, hearing, and equilibrium.
2. Like the general senses, the special senses allow us to detect changes in the environment.

12.4 Olfaction: Sense of Smell

1. The olfactory epithelium in the upper portion of the nasal cavity contains **olfactory receptors**, **supporting cells**, and **basal cells**.
2. Individual olfactory receptors respond to hundreds of different odorant molecules by producing an electrical signal that triggers one or more nerve impulses. Adaptation (decreasing sensitivity) to odors occurs rapidly.
3. Axons of olfactory receptors form the **olfactory (I) nerves**, which convey nerve impulses to the **olfactory bulbs**. From there, impulses conduct via the **olfactory tract** to the limbic system, hypothalamus, and cerebral cortex (temporal lobe).

Anatomy Overview—Special Senses–Smell and Taste

12.5 Gustation: Sense of Taste

1. The receptors for **gustation**, the **gustatory receptor cells**, are located in **taste buds**.
2. To be tasted, substances must be dissolved in saliva.
3. The five primary tastes are salty, sweet, sour, bitter, and umami.
4. Gustatory receptor cells trigger impulses in the following cranial nerves: facial (VII), glossopharyngeal (IX), and vagus (X). Impulses for taste conduct to the medulla oblongata, limbic system, hypothalamus, thalamus, and the primary gustatory area in the parietal lobe of the cerebral cortex.

Anatomy Overview—Special Senses–Smell and Taste

12.6 Vision

1. **Accessory structures** of the eyes include the **eyebrows**, **eyelids**, **eyelashes**, the **lacrimal apparatus** (which produces and drains tears), and extrinsic eye muscles (which move the eyes).
2. The eyeball has three layers: (a) **fibrous tunic** (**sclera** and **cornea**), (b) **vascular tunic** (**choroid**, **ciliary body**, and **iris**), and (c) **retina**.
3. The retina consists of a neural layer (**photoreceptor layer**, **bipolar cell layer**, and **ganglion cell layer**) and a pigmented layer (a sheet of melanin-containing epithelial cells).
4. The **anterior cavity** contains **aqueous humor**; the **vitreous chamber** contains the **vitreous body**.
5. Image formation on the retina involves **refraction** of light rays by the cornea and lens, which focus an inverted image on the central fovea of the retina.
6. For viewing close objects, the lens increases its curvature (**accommodation**), and the pupil constricts to prevent light rays from entering the eye through the periphery of the lens.
7. Improper refraction may result from **myopia** (nearsightedness), **hyperopia** (farsightedness), or **astigmatism** (irregular curvature of the cornea or lens).
8. Movement of the eyeballs toward the nose to view an object is called **convergence**.
9. The first step in vision is the absorption of light rays by **photopigments** in **rods** and **cones** (photoreceptors). Stimulation of the rods and cones then activates bipolar cells, which in turn activate the ganglion cells.
10. Nerve impulses arise in ganglion cells and conduct along the **optic (II) nerve**, through the **optic chiasm** and **optic tract** to the thalamus. From the thalamus, impulses extend to the primary visual area in the occipital lobe of the cerebral cortex.

Anatomy Overview—Special Senses-Vision

Figure 12.6 Structure of the eyeball

Figure 12.11 Visual pathway

12.7 Hearing and Equilibrium

1. The **external ear** consists of the **auricle**, **external auditory canal**, and **eardrum**.
2. The **middle ear** consists of the **auditory (eustachian) tube**, **auditory ossicles**, and **oval window**.
3. The **internal ear** consists of the **bony labyrinth** and **membranous labyrinth**. The internal ear contains the **spiral organ (organ of Corti)**, the organ of hearing.

Anatomy Overview—Special Senses-Hearing
Anatomy Overview—Special Senses-Equilibrium

REVIEW

4. Sound waves enter the external auditory canal, strike the eardrum, pass through the ossicles, strike the oval window, set up pressure waves in the perilymph, strike the vestibular membrane and scala tympani, increase pressure in the endolymph, vibrate the basilar membrane, and stimulate hair cells in the spiral organ.

5. Hair cells release neurotransmitter molecules that can initiate nerve impulses in sensory neurons.

6. Sensory neurons in the cochlear branch of the vestibulocochlear (VIII) nerve terminate in the medulla oblongata. Auditory signals then pass to the midbrain, thalamus, and temporal lobes.

7. **Static equilibrium** is the orientation of the body relative to the pull of gravity. The maculae of the utricle and saccule are the sense organs of static equilibrium.

8. **Dynamic equilibrium** is the maintenance of body position in response to rotational acceleration and deceleration.

9. Most **vestibular branch** axons of the vestibulocochlear (VIII) nerve enter the brain stem and terminate in the medulla and pons; other axons extend to the cerebellum.

RESOURCES

Figure 12.13 Details of the right internal ear

Figure 12.14 Physiology of hearing shown in the right ear

SELF-QUIZ

1. You enter a sauna and it feels awfully hot, but soon the temperature feels comfortably warm. What have you have experienced?
   a. damage to your thermoreceptors
   b. sensory adaptation
   c. a change in the temperature of the sauna
   d. inactivation of your thermoreceptors
   e. temporary damage to sensory neurons

2. The lacrimal glands produce _____, which drain(s) into the _____.
   a. tears; anterior cavity
   b. tears; nasal cavity
   c. the vitreous body; vitreous chamber
   d. aqueous humor; anterior cavity
   e. aqueous humor; scleral venous sinus

3. The spiral organ (organ of Corti)
   a. contains hair cells.
   b. is responsible for equilibrium.
   c. is filled with perilymph.
   d. is another name for the auditory (eustachian) tube.
   e. transmits auditory nerve impulses to the brain.

4. Equilibrium and the activities of muscles and joints are monitored by
   a. olfactory receptors.
   b. nociceptors.
   c. tactile receptors.
   d. proprioceptors.
   e. thermoreceptors

5. In the retina, cone photoreceptors
   a. are more numerous than rods.
   b. contain the photopigment rhodopsin.
   c. are more sensitive to low light levels than are rods.
   d. tend to be highly concentrated in the optic disc.
   e. provide higher acuity of vision than do rods.

6. Which of the following is NOT required for a sensation to occur?
   a. the presence of a stimulus
   b. a receptor specialized to detect a stimulus

c. the presence of slowly adapting receptors
   d. a sensory neuron to conduct impulses
   e. a region of the brain for integration of the nerve impulse

7. Match each receptor with its function.
   ____ a. color vision   A. pacinian corpuscle
   ____ b. taste   B. Meissner corpuscle
   ____ c. smell   C. rod photoreceptor
   ____ d. dynamic equilibrium   D. nociceptor
   ____ e. vision in dim light   E. gustatory receptor cell
   ____ f. stretch in a muscle   F. olfactory receptor
   ____ g. static equilibrium   G. muscle spindle
   ____ h. pressure   H. maculae
   ____ i. touch   I. cristae
   ____ j. detection of pain   J. cone photoreceptors

8. The type of pain that can be precisely localized is
   a. referred pain.   b. fast pain.
   c. slow pain.   d. chronic pain.
   e. none of the above because pain cannot be localized.

9. Which of the following characteristics of taste is NOT true?
   a. Olfaction can affect taste.
   b. Three cranial nerves conduct the impulses for taste to the brain.
   c. Taste adaptation occurs quickly.
   d. Humans can recognize about 10 primary tastes.
   e. Taste receptors are located in taste buds on the tongue, on the roof of the mouth, in the throat, and in the epiglottis.

10. An ice skater is spinning rapidly on the ice. What is occurring in her inner ear?
   a. The hair cells on the macula are responding to changes in static equilibrium.
   b. The hair cells in the cochlea are responding to changes in dynamic equilibrium.
   c. The cristae of each semicircular duct are responding to changes in dynamic equilibrium.

   d. The cochlear branch of the vestibulocochlear (VIII) nerve is transmitting nerve impulses to the brain related to equilibrium.

   e. The auditory (eustachian) tube is adjusting for varying air pressures.

11. Kinesthesia is the
   a. perception of body movements.
   b. ability to identify an object by feeling it.
   c. sensation of weightlessness that occurs in outer space.
   d. decrease in sensitivity of receptors to a prolonged stimulus.
   e. movement of body parts in a rhythmic manner.

12. Which of the following is NOT true about nociceptors?
   a. They respond to stimuli that may cause tissue damage.
   b. They consist of free nerve endings.
   c. They can be activated by excessive stimuli from other sensations.
   d. They are found in virtually every body tissue except the brain.
   e. They adapt very rapidly.

13. The sense of smell
   a. requires the presence of dissolved odorants.
   b. is transmitted through olfactory nerves and olfactory tracts.
   c. evokes emotional responses because of limbic system involvement.
   d. is initiated by stimulating olfactory hairs.
   e. is described by all of the above characteristics.

14. Transmission of vibrations (sound waves) from the tympanic membrane to the oval window is accomplished by
   a. neurons.   **b.** the tectorial membrane.
   c. the auditory ossicles.   **d.** the endolymph.
   e. the auditory (eustachian) tube.

15. Match the following:
   ____ **a.** focuses light rays onto the retina   **A.** sclera
   ____ **b.** regulates the amount of light entering the eye   **B.** choroid
   C. pupil
   ____ **c.** contains aqueous humor   **D.** lens
   ____ **d.** contains blood vessels that help nourish the retina   **E.** retina
   F. iris
   ____ **e.** hole in the middle of the iris   **G.** anterior cavity
   ____ **f.** dense connective tissue that provides shape to the eye
   ____ **g.** contains photoreceptors

16. Which of the following structures refracts light rays entering the eye?
   a. cornea   **b.** sclera   **c.** pupil
   d. retina   **e.** conjunctiva

17. Your 45-year-old neighbor has recently begun to have difficulty reading the morning newspaper. You explain that this condition is known as _____ and is due to _____.
   a. myopia, inability of his eyes to properly focus light on his retinas
   b. night blindness, a vitamin A deficiency
   c. binocular vision, the eyes focusing on two different objects
   d. astigmatism, an irregularity in the curvature of the lens
   e. presbyopia, the loss of elasticity in the lens

18. Damage to cells in the fovea centralis would interfere with
   a. convergence.   **b.** accommodation.   **c.** visual acuity.
   d. ability to see in dim light.   **e.** intraocular pressure.

19. Place the following events concerning the visual pathway in the correct order:
   1. Nerve impulses exit the eye via the optic nerve.
   2. Optic tract axons terminate in the thalamus.
   3. Light reaches the retina.
   4. Rods and cones are stimulated.
   5. Synapses occur in the thalamus and continue to the primary visual area in the occipital lobe.
   6. Ganglion cells generate nerve impulses.
   a. 4, 1, 2, 5, 6, 3   **b.** 5, 4, 1, 3, 2, 6   **c.** 3, 4, 6, 1, 5, 2
   d. 3, 4, 6, 1, 2, 5   **e.** 3, 4, 5, 6, 1, 2

20. Place the following events of the auditory pathway in the correct order:
   1. Hair cells in the spiral organ bend as they rub against the tectorial membrane.
   2. Movement in the oval window begins movement in the perilymph.
   3. Nerve impulses exit the ear via the vestibulocochlear (VIII) nerve.
   4. The eardrum and auditory ossicles transmit vibrations from sound waves.
   5. Pressure waves from the perilymph cause bulging of the round window and formation of pressure waves in the endolymph.
   a. 4, 2, 5, 1, 3   **b.** 4, 5, 2, 3, 1   **c.** 5, 3, 2, 4, 1
   d. 3, 4, 5, 1, 2   **e.** 2, 4, 1, 5, 3

CRITICAL THINKING APPLICATIONS

1. Kate is preparing her six-month-old baby for bed. She gives him a warm bath, dries him off, dresses him and gives him a quick tickle to make him smile. As she lays him in his crib she gives him a light kiss on his lips and gently strokes his arms until he dozes off. What are some of the baby's receptors that have been activated by his mother's actions?

2. Jim works the night shift and sometimes falls asleep in A&P class. What is the effect on the structures in his internal ear when his head falls backward as he slumps in his seat?

3. A medical procedure used to improve visual acuity involves shaving of a thin layer off the cornea. How could this procedure improve vision?

4. The optometrist put drops in Louise's eyes during her eye exam. When Louise looked in the mirror after the exam, her pupils were very large and her eyes were sensitive to the bright light. How did the eyedrops produce this effect on Louise's eyes?

ANSWERS TO FIGURE QUESTIONS

12.1 Meissner corpuscles are abundant in the fingertips, palms, and soles.

12.2 The kidneys have the broadest area for referred pain.

12.3 Basal stem cells undergo cell division to produce new olfactory receptors.

12.4 The gustatory pathway: gustatory receptor cells → cranial nerves VII, IX, and X → medulla oblongata → thalamus → primary gustatory area in the parietal lobe of the cerebral cortex.

12.5 Tears clean, lubricate, and moisten the eyeball.

12.6 The fibrous tunic consists of the cornea and sclera; the vascular tunic consists of the choroid, ciliary body, and iris.

12.7 The parasympathetic division of the autonomic nervous system causes pupillary constriction; the sympathetic division causes pupillary dilation.

12.8 The two types of photoreceptors are rods and cones. Rods provide black-and-white vision in dim light; cones provide high visual acuity and color vision in bright light.

12.9 During accommodation, the ciliary muscle contracts, zonular fibers slacken, and the lens becomes more rounded (convex) and refracts light more.

12.10 Presbyopia is the loss of elasticity in the lens that occurs with aging.

12.11 Structures carrying visual impulses from the retina to the occipital lobe: axons of ganglion cells → optic (II) nerve → optic chiasm → optic tract → thalamus → primary visual area in occipital lobe of the cerebral cortex.

12.12 The receptors for hearing and equilibrium are located in the inner ear: cochlea (hearing) and semicircular ducts (equilibrium).

12.13 The eardrum (tympanic membrane) separates the outer ear from the middle ear. The oval and round windows separate the middle ear from the internal ear.

12.14 Hair cells convert a mechanical force (stimulus) into an electrical signal (depolarization and repolarization of the hair cell membrane).

12.15 The maculae are the receptors for static equilibrium.

12.16 The membranous semicircular ducts function in dynamic equilibrium.

CHAPTER 13

THE ENDOCRINE SYSTEM

As they mature, boys and girls develop striking differences in physical appearance and behavior. In girls, estrogens (female sex hormones) promote accumulation of adipose tissue in the breasts and hips, sculpting a feminine shape. In boys, testosterone (a male sex hormone) enlarges the vocal cords, producing a lower-pitched voice, and begins to help build muscle mass. These changes are examples of the powerful influence of **hormones** (*hormon* = excite or get moving), secretions of the endocrine system. Less dramatically, but just as importantly, hormones help maintain homeostasis on a daily basis. They regulate the activity of smooth muscle, cardiac muscle, and some glands; alter metabolism; spur growth and development; influence reproductive processes; and participate in circadian (daily) rhythms established by the hypothalamus.

Did you know?   *Research suggests that working the graveyard shift is associated with increased risk of cancers of the breast, prostate, colon, and endometrium (uterine lining). In addition to lack of sleep and the stress of irregular schedules, a deficiency of melatonin production may be partly to blame. Melatonin levels rise during darkness, and help to regulate circadian rhythm.*

LOOKING BACK TO MOVE AHEAD...

Steroids (Section 2.2)

The Plasma Membrane (Section 3.2)

Neurons (Section 9.2)

Negative and Positive Feedback Systems (Section 1.4)

FOCUS ON WELLNESS

Shedding Light on Shift Work Cancer Risk

13.1 INTRODUCTION

OBJECTIVE • **List the components of the endocrine system.**

The *endocrine system* consists of several endocrine glands plus many hormone-secreting cells in organs that have functions besides secreting hormones (Figure 13.1). In contrast to the nervous system, which controls body activities through the release of neurotransmitters at synapses, the endocrine system releases hormones into interstitial fluid (fluid that surrounds cells) and then into the bloodstream. The circulating blood then delivers hormones to virtually all cells throughout the body, and cells that recognize a particular hormone will respond. The nervous system and endocrine system often work together. For example, certain parts of the nervous

Figure 13.1 Location of endocrine glands and other organs that contain endocrine cells. Some nearby structures are shown for orientation (trachea, lungs, scrotum, and uterus).

 Endocrine glands secrete hormones, which circulating blood delivers to target tissues.

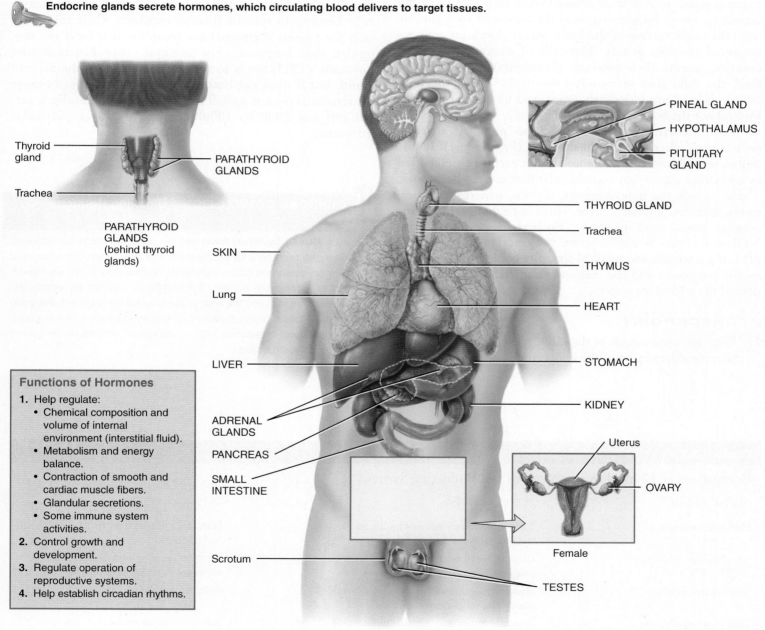

Thyroid gland

Trachea

PARATHYROID GLANDS

PARATHYROID GLANDS (behind thyroid glands)

PINEAL GLAND

HYPOTHALAMUS

PITUITARY GLAND

THYROID GLAND

Trachea

THYMUS

HEART

STOMACH

KIDNEY

SKIN

Lung

LIVER

ADRENAL GLANDS

PANCREAS

SMALL INTESTINE

Scrotum

Uterus

OVARY

Female

TESTES

Male

Functions of Hormones

1. Help regulate:
   - Chemical composition and volume of internal environment (interstitial fluid).
   - Metabolism and energy balance.
   - Contraction of smooth and cardiac muscle fibers.
   - Glandular secretions.
   - Some immune system activities.
2. Control growth and development.
3. Regulate operation of reproductive systems.
4. Help establish circadian rhythms.

? **What is the basic difference between endocrine glands and exocrine glands?**

system stimulate or inhibit the release of hormones by the endocrine system. Typically, the endocrine system acts more slowly than the nervous system, which often produces an effect within a fraction of a second. Moreover, the effects of hormones linger until they are cleared from the blood. The liver inactivates some hormones, and the kidneys excrete others in the urine.

Table 13.1 compares the characteristics of the nervous and endocrine systems.

As you learned in Chapter 4, two types of glands are present in the body: exocrine glands and endocrine glands. *Exocrine glands* secrete their products into *ducts* that carry the secretions into a body cavity, into the lumen of an organ, or onto the outer surface of the body. Sweat glands are one example of exocrine glands. The cells of *endocrine glands*, by contrast, secrete their products (hormones) into *interstitial fluid*, the fluid that surrounds tissue cells. Then, the hormones diffuse into blood capillaries, and blood carries them throughout the body.

The endocrine glands include the pituitary, thyroid, parathyroid, adrenal, and pineal glands (Figure 13.1). In addition, several organs and tissues are not exclusively classified as endocrine glands but contain cells that secrete hormones. These include the hypothalamus, thymus, pancreas, ovaries, testes, kidneys, stomach, liver, small intestine, skin, heart, adipose tissue, and placenta. **Endocrinology** (en′-dō-kri-NOL-ō-jē; *endo-* = within; *-crino* = to secrete; *-logy* = study of) is the scientific and medical specialty concerned with hormonal secretions and the diagnosis and treatment of disorders of the endocrine system.

✓ **CHECKPOINT**

1. Why are organs such as the kidneys, stomach, heart, and skin considered part of the endocrine system?

13.2 HORMONE ACTION

OBJECTIVES • **Define target cells and describe the role of hormone receptors.**

• **Describe the two general mechanisms of action of hormones.**

Target Cells and Hormone Receptors

Although a given hormone travels throughout the body in the blood, it affects only specific *target cells*. Hormones, like neurotransmitters, influence their target cells by chemically binding to specific protein *receptors*. Only the target cells for a given hormone have receptors that bind and recognize that hormone. For example, thyroid-stimulating hormone (TSH) binds to receptors on cells of the thyroid gland, but it does not bind to cells of the ovaries because ovarian cells do not have TSH receptors. Generally, a target cell has 2000 to 100,000 receptors for a particular hormone.

CLINICAL CONNECTION | **Blocking Hormone Receptors**

The drug **RU486 (mifepristone)** can be used to induce an abortion. It binds to the receptors for progesterone (a female sex hormone) and prevents progesterone from exerting its normal effects. When RU486 is given to a pregnant woman, the conditions needed for embryonic development are lost, and the embryo is sloughed off along with the lining of the uterus. This example illustrates an important principle: If a hormone is prevented from interacting with its receptors, the hormone cannot perform its normal functions. •

Table 13.1		
Comparison of Control by the Nervous and Endocrine Systems		
CHARACTERISTIC	NERVOUS SYSTEM	ENDOCRINE SYSTEM
Mediator molecules	Neurotransmitters released locally in response to nerve impulses	Hormones delivered to tissues throughout body by blood
Site of mediator action	Close to site of release, at synapse; binds to receptors in postsynaptic membrane	Far from site of release (usually); binds to receptors on or in target cells
Types of target cells	Muscle (smooth, cardiac, and skeletal) cells, or gland cells, other neurons	Cells throughout body
Time to onset of action	Typically within milliseconds (thousandths of a second)	Seconds to hours or days
Duration of action	Generally briefer (milliseconds)	Generally longer (seconds to days)

Chemistry of Hormones

Chemically, some hormones are soluble in lipids (fats) and others are soluble in water. The lipid-soluble hormones include steroid hormones, thyroid hormones, and nitric oxide. *Steroid hormones* are made from cholesterol. The two *thyroid hormones* (T_3 and T_4) are made by attaching iodine atoms to the amino acid tyrosine. The gas *nitric oxide (NO)* functions as both a hormone and a neurotransmitter.

Most of the water-soluble hormones are made from amino acids. For instance, the amino acid tyrosine is modified to form the hormones epinephrine and norepinephrine (which are also neurotransmitters). Other water-soluble hormones consist of short chains of amino acids (peptide hormones), such as antidiuretic hormone (ADH) and oxytocin, or longer chains of amino acids (protein hormones), such as insulin and human growth hormone.

Mechanisms of Hormone Action

The response to a hormone depends on both the hormone and the target cell. Various target cells respond differently to the same hormone. Insulin, for example, stimulates the synthesis of glycogen in liver cells but the synthesis of triglycerides in adipose cells. To exert an effect, a hormone first must "announce its arrival" to a target cell by binding to its receptors. The receptors for lipid-soluble hormones are located inside target cells, and the receptors for water-soluble hormones are part of the plasma membrane of target cells.

Action of Lipid-soluble Hormones

Lipid-soluble hormones are transported in blood by being attached to *transport proteins*. These proteins make the lipid-soluble hormones in blood temporarily water-soluble, thus increasing their solubility in blood. Lipid-soluble hormones diffuse through the lipid bilayer of the plasma membrane and bind to their receptors *within* target cells. They exert their effects in the following way (Figure 13.2):

1 A lipid-soluble hormone detaches from its transport protein in the bloodstream. Then, the free hormone diffuses from blood into interstitial fluid, and through the plasma membrane into a cell.

2 The hormone binds to and activates receptors within the cell. The activated receptor–hormone complex then alters gene expression: It turns specific genes on or off.

3 As the DNA is transcribed, new messenger RNA (mRNA) forms, leaves the nucleus, and enters the cytosol. There it directs synthesis of a new protein, often an enzyme, on the ribosomes.

Figure 13.2 Mechanism of action of lipid-soluble hormones.

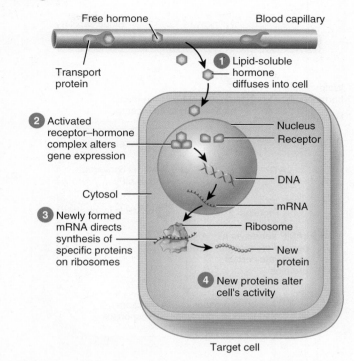

Lipid-soluble hormones bind to their receptors inside target cells.

Target cell

? What types of molecules are synthesized after lipid-soluble hormones bind to their receptors?

4 The new proteins alter the cell's activity and cause the responses typical of that specific hormone.

Action of Water-soluble Hormones

Because most amino acid–based hormones are not lipid-soluble, they cannot diffuse through the lipid bilayer of the plasma membrane. Instead, water-soluble hormones bind to receptors that protrude from the target cell surface. When a water-soluble hormone binds to its receptor at the outer surface of the plasma membrane, it acts as the *first messenger*. The first messenger (the hormone) then causes production of a *second messenger* inside the cell, where specific hormone-stimulated responses take place. One common second messenger is *cyclic AMP (cAMP)*, which is synthesized from ATP.

Water-soluble hormones exert their effects as follows (Figure 13.3):

1 A water-soluble hormone (the first messenger) diffuses from the blood and binds to its receptor in a target cell's plasma membrane.

2 As a result of the binding, a reaction starts inside the cell that converts ATP into cyclic AMP.

Figure 13.3 Mechanism of action of water-soluble hormones.

Water-soluble hormones bind to receptors embedded in the plasma membrane of target cells.

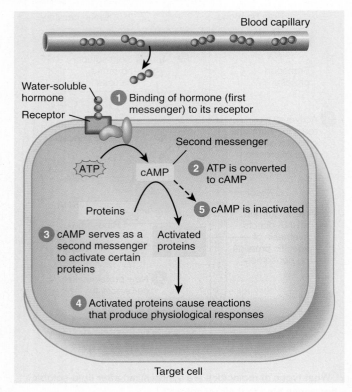

Target cell

? Why is cAMP called a "second messenger"?

③ Cyclic AMP (the second messenger) causes the activation of several proteins (such as enzymes).

④ Activated proteins cause reactions that produce physiological responses.

⑤ After a brief time, cyclic AMP is inactivated. Thus, the cell's response is turned off unless new hormone molecules continue to bind to their receptors in the plasma membrane.

Control of Hormone Secretions

The release of most hormones occurs in short bursts, with little or no secretion between bursts. When stimulated, an endocrine gland releases its hormone in more frequent bursts, increasing the concentration of the hormone in the blood. In the absence of stimulation, the blood level of the hormone decreases as the hormone is inactivated or excreted. Regulation of secretion normally prevents overproduction or underproduction of any given hormone.

Hormone secretion is regulated by (1) signals from the nervous system, (2) chemical changes in the blood, and (3) other hormones. For example, nerve impulses to the adrenal medullae regulate the release of epinephrine and norepinephrine; blood Ca^{2+} level regulates the secretion of parathyroid hormone; and a hormone from the anterior pituitary (adrenocorticotropic hormone or ACTH) stimulates the release of cortisol by the adrenal cortex. ACTH is an example of a tropic hormone. **Tropic hormones** (TRŌ-pik), or **tropins**, are hormones that act on other endocrine glands or tissues to regulate the secretion of another hormone.

Most hormonal regulatory systems work via negative feedback, but a few operate via positive feedback. For example, during childbirth, the hormone oxytocin stimulates contractions of the uterus, and uterine contractions in turn stimulate more oxytocin release, a positive feedback effect.

✓ CHECKPOINT

2. Why are target cell receptors important?

3. Chemically, what types of molecules are hormones?

4. What are the general ways in which blood hormone levels are regulated?

13.3 HYPOTHALAMUS AND PITUITARY GLAND

OBJECTIVES • **Describe the locations of and relationship between the hypothalamus and the pituitary gland.**

• **Describe the functions of each hormone secreted by the pituitary gland.**

For many years, the **pituitary gland** (pi-TOO-i-tār-ē) or *hypophysis* (hī-POF-i-sis) was considered the "master" endocrine gland because it secretes several hormones that control other endocrine glands. We now know that the pituitary gland itself has a master—the **hypothalamus**. This small region of the brain is the major link between the nervous and endocrine systems. Cells in the hypothalamus synthesize at least nine hormones, and the pituitary gland synthesizes seven. Together, these hormones play important roles in the regulation of virtually all aspects of growth, development, metabolism, and homeostasis.

The pituitary gland is about the size of a small grape and has two lobes: a larger **anterior pituitary** or *anterior lobe* and a smaller **posterior pituitary** or *posterior lobe* (Figure 13.4). Both lobes of the pituitary gland rest in the *hypophyseal fossa*, a cup-shaped depression in the sphenoid bone (see Figure 6.9). A stalklike structure, the **infundibulum**, attaches the pituitary gland to the hypothalamus. Within the infundibulum, blood vessels termed **hypophyseal portal veins** (hī′-pō-FIZ-ē-al) connect capillaries in the hypothalamus to capillaries in the anterior pituitary. Axons of hypothalamic neurons called **neurosecretory cells** end near the capillaries of the hypothalamus (Figure 13.4), where they release several hormones into the blood.

Figure 13.4 The pituitary gland and its blood supply. As shown in the inset, to the right, releasing and inhibiting hormones synthesized by hypothalamic neurosecretory cells diffuse into capillaries of the hypothalamus and are carried by the hypophyseal portal veins to the anterior pituitary.

🔑 **Hypothalamic releasing and inhibiting hormones are an important link between the nervous and endocrine systems.**

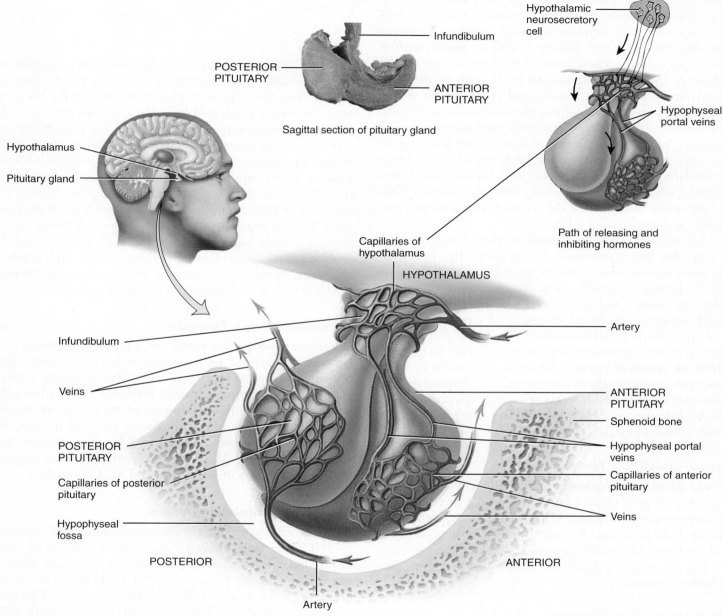

Sagittal section of pituitary gland

Path of releasing and inhibiting hormones

❓ **Which lobe of the pituitary gland does not synthesize the hormones it releases? Where are its hormones produced?**

Anterior Pituitary Hormones

The anterior pituitary synthesizes and secretes hormones that regulate a wide range of bodily activities, from growth to reproduction. Secretion of anterior pituitary hormones is stimulated by **releasing hormones** and suppressed by **inhibiting hormones**, both produced by neurosecretory cells of the hypothalamus. The hypophyseal portal veins deliver the hypothalamic releasing and inhibiting hormones from the hypothalamus to the anterior pituitary (Figure 13.4). This direct route allows the releasing and inhibiting hormones to act quickly on cells of the anterior pituitary before the hormones are diluted or destroyed in the general circulation.

Human Growth Hormone and Insulinlike Growth Factors

Human growth hormone (hGH) is the most abundant anterior pituitary hormone. The main function of hGH is to promote synthesis and secretion of small protein hormones called **insulinlike growth factors (IGFs)** or **somatomedins**. IGFs are so named because some of their actions are similar to those of insulin. In response to hGH, cells in the liver, skeletal muscles, cartilage, bones, and other tissues secrete IGFs, which may either enter the bloodstream or act locally. IGFs stimulate protein synthesis, help maintain muscle and bone mass, and promote healing of injuries and tissue repair. They also enhance breakdown of triglycerides (fats), which releases fatty acids into the blood, and breakdown of liver glycogen, which releases glucose into the blood. Cells throughout the body can use the released fatty acids and glucose for the production of ATP.

The anterior pituitary releases hGH in bursts that occur every few hours, especially during sleep. Two hypothalamic hormones control secretion of hGH: *Growth hormone-releasing hormone (GHRH)* promotes secretion of human growth hormone, and *growth hormone-inhibiting hormone (GHIH)* suppresses it. Blood glucose level is a major regulator of GHRH and GHIH secretion. Low blood glucose level (hypoglycemia) stimulates the hypothalamus to secrete GHRH. By means of negative feedback, an increase in blood glucose concentration above the normal level (hyperglycemia) inhibits release of GHRH. By contrast, hyperglycemia stimulates the hypothalamus to secrete GHIH and hypoglycemia inhibits release of GHIH.

Thyroid-stimulating Hormone

Thyroid-stimulating hormone (TSH) stimulates the synthesis and secretion of thyroid hormones by the thyroid gland. *Thyrotropin-releasing hormone (TRH)* from the hypothalamus controls TSH secretion. Release of TRH, in turn, depends on blood levels of thyroid hormones, which inhibit secretion of TRH via negative feedback. There is no thyrotropin-inhibiting hormone.

Follicle-stimulating Hormone and Luteinizing Hormone

In females, the ovaries are the targets for **follicle-stimulating hormone (FSH)** and **luteinizing hormone (LH)**. Each month FSH initiates the development of several ovarian follicles and LH triggers ovulation (described in Section 23.3). After ovulation, LH stimulates formation of the corpus luteum in the ovary and the secretion of progesterone (another female sex hormone) by the corpus luteum. FSH and LH also stimulate follicular cells to secrete estrogens. In males, FSH stimulates

sperm production in the testes, and LH stimulates the testes to secrete testosterone. *Gonadotropin-releasing hormone (GnRH)* from the hypothalamus stimulates release of FSH and LH. The release of GnRH, FSH, and LH is suppressed by estrogens in females and by testosterone in males through negative feedback systems. There is no gonadotropin-inhibiting hormone.

Prolactin

Prolactin (PRL), together with other hormones, initiates and maintains milk production by the mammary glands. Ejection of milk from the mammary glands depends on the hormone oxytocin, which is released from the posterior pituitary. The function of prolactin is unknown in males, but prolactin hypersecretion causes erectile dysfunction (impotence, the inability to have an erection of the penis). In females, *prolactin-inhibiting hormone (PIH)* suppresses release of prolactin most of the time. Each month, just before menstruation begins, the secretion of PIH diminishes and the blood level of prolactin rises, but not enough to stimulate milk production. As the menstrual cycle begins anew, PIH is again secreted and the prolactin level drops. During pregnancy, very high levels of estrogens promote secretion of *prolactin-releasing hormone (PRH)*, which in turn stimulates release of prolactin.

Adrenocorticotropic Hormone

Adrenocorticotropic hormone (ACTH) or *corticotropin* controls the production and secretion of hormones called glucocorticoids by the cortex (outer portion) of the adrenal glands. Corticotropin-releasing hormone (CRH) from the hypothalamus stimulates secretion of ACTH. Stress-related stimuli, such as low blood glucose or physical trauma, and interleukin-1, a substance produced by macrophages, also stimulate release of ACTH. Glucocorticoids cause negative feedback inhibition of both CRH and ACTH release.

Melanocyte-stimulating Hormone

There is little circulating melanocyte-stimulating hormone (MSH) in humans. Although an excessive amount of MSH causes darkening of the skin, the function of normal levels of MSH is unknown. The presence of MSH receptors in the brain suggests it may influence brain activity. Excessive corticotropin-releasing hormone (CRH) can stimulate MSH release, and dopamine inhibits MSH release.

Posterior Pituitary Hormones

The **posterior pituitary** contains the axons and axon terminals of more than 10,000 neurosecretory cells whose cell bodies are in the hypothalamus (Figure 13.5). Although the posterior pituitary does not *synthesize* hormones, it does *store* and *release* two hormones. In the hypothalamus, the

Figure 13.5 Hypothalamic neurosecretory cells synthesize oxytocin and antidiuretic hormone. Their axons extend from the hypothalamus to the posterior pituitary. Nerve impulses trigger release of the hormones from vesicles in the axon terminals in the posterior pituitary.

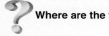 Oxytocin and antidiuretic hormone are synthesized in the hypothalamus and released into capillaries of the posterior pituitary.

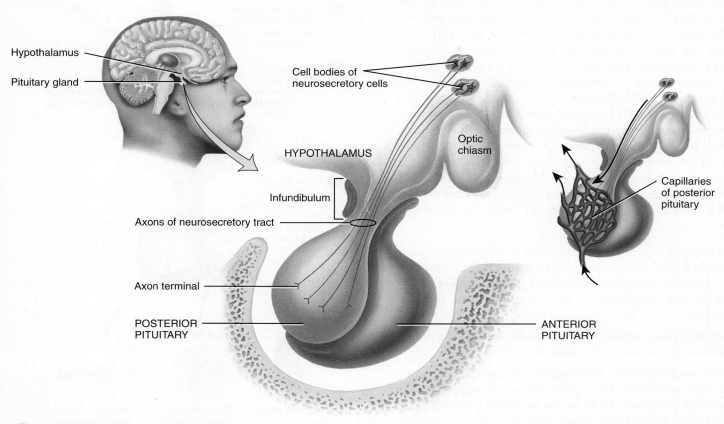

? Where are the target cells of oxytocin located?

hormones ***oxytocin*** (ok′-sē-TŌ-sin; *oxytoc-* = quick birth) and ***antidiuretic hormone (ADH)*** are synthesized and packaged into secretory vesicles within the cell bodies of different neurosecretory cells. The vesicles then move down the axons to the axon terminals in the posterior pituitary. Nerve impulses that arrive at the axon terminals trigger release of these hormones into the capillaries of the posterior pituitary.

Oxytocin

During and after delivery of a baby, oxytocin has two target organs: the mother's uterus and breasts. During delivery, oxytocin enhances contraction of smooth muscle cells in the wall of the uterus; after delivery, it stimulates milk ejection ("letdown") from the mammary glands in response to the mechanical stimulus provided by a suckling infant. Together, milk production and ejection constitute *lactation*.

The function of oxytocin in males and in nonpregnant females is not clear. Experiments with animals have suggested actions within the brain that foster parental caretaking behavior toward young offspring. Oxytocin also may be partly responsible for the feelings of sexual pleasure during and after intercourse.

CLINICAL CONNECTION | Synthetic Oxytocin

Synthetic oxytocin is administered intravenously to induce labour that has not started naturally or to stimulate natural labor when uterine contractions are weak. After birth, synthetic oxytocin may be given to stimulate contractions that expel the placenta and prevent haemorrhaging. It helps the uterus contract to its normal size by mimicking the release of natural oxytocin which occurs when mothers breast-feed immediately after birth. This is one reason why midwives encourage mothers to breast-feed as soon as possible after giving birth. •

Antidiuretic Hormone

An *antidiuretic* (an'-tī-dī-ū-RET-ik; *anti-* = against; *diuretic* = urine-producing agent) is a substance that decreases urine production. *Antidiuretic hormone (ADH)* causes the kidneys to retain more water, thus decreasing urine volume. In the absence of ADH, urine output increases more than tenfold, from the normal 1–2 liters to about 20 liters a day. ADH also decreases the water lost through sweating and causes constriction of arterioles. This hormone's other name, *vasopressin* (*vaso-* = vessel; *pressin-* = pressing or constricting), reflects its effect on increasing blood pressure.

The amount of ADH secreted varies with blood osmotic pressure and blood volume. Blood osmotic pressure is proportional to the concentration of solutes in the blood plasma. When body water is lost faster than it is taken in, a condition termed *dehydration*, the blood volume falls and blood osmotic pressure rises. Figure 13.6 shows regulation of ADH secretion and the actions of ADH on its target tissues:

1️⃣ High blood osmotic pressure—due to dehydration or a drop in blood volume because of hemorrhage, diarrhea, or excessive sweating—stimulates **osmoreceptors**, neurons in the hypothalamus that monitor blood osmotic pressure.

2️⃣ Osmoreceptors activate the hypothalamic neurosecretory cells that synthesize and release ADH.

3️⃣ When neurosecretory cells receive excitatory input from the osmoreceptors, they generate nerve impulses that cause the release of ADH in the posterior pituitary. The ADH then diffuses into blood capillaries of the posterior pituitary.

4️⃣ The blood carries ADH to three targets: the kidneys, sweat glands, and smooth muscle in blood vessel walls. The kidneys respond by retaining more water, which decreases urine output. Secretory activity of sweat glands decreases, which lowers the rate of water loss by perspiration from the skin. Smooth muscle in the walls of arterioles (small arteries) contracts in response to high levels of ADH, which constricts (narrows) the lumen of these blood vessels and increases blood pressure.

5️⃣ Low blood osmotic pressure or increased blood volume inhibits the osmoreceptors.

6️⃣ Inhibition of osmoreceptors reduces or stops ADH secretion. The kidneys then retain less water by forming a larger volume of urine, secretory activity of sweat glands increases, and arterioles dilate. The blood volume and osmotic pressure of body fluids return to normal.

Secretion of ADH can also be altered in other ways. Pain, stress, trauma, anxiety, acetylcholine, nicotine, and

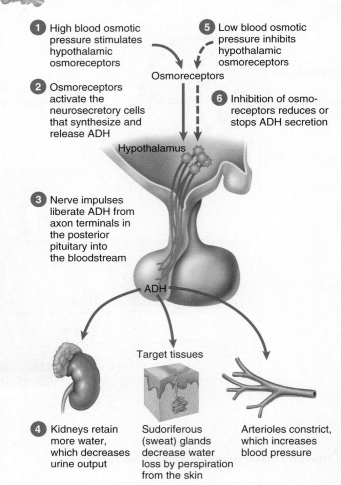

Figure 13.6 Regulation of secretion and actions of antidiuretic hormone (ADH).

🔑 **ADH acts to retain body water and increase blood pressure.**

1️⃣ High blood osmotic pressure stimulates hypothalamic osmoreceptors

2️⃣ Osmoreceptors activate the neurosecretory cells that synthesize and release ADH

3️⃣ Nerve impulses liberate ADH from axon terminals in the posterior pituitary into the bloodstream

5️⃣ Low blood osmotic pressure inhibits hypothalamic osmoreceptors

6️⃣ Inhibition of osmoreceptors reduces or stops ADH secretion

Osmoreceptors

Hypothalamus

ADH

Target tissues

4️⃣ Kidneys retain more water, which decreases urine output

Sudoriferous (sweat) glands decrease water loss by perspiration from the skin

Arterioles constrict, which increases blood pressure

❓ **What effect would drinking a large glass of water have on the osmotic pressure of your blood, and how would the level of ADH change in your blood?**

drugs such as morphine, tranquilizers, and some anesthetics stimulate ADH secretion. Alcohol inhibits ADH secretion, thereby increasing urine output. The resulting dehydration may cause both the thirst and the headache typical of a hangover.

Table 13.2 lists the pituitary gland hormones and summarizes their actions.

✓ CHECKPOINT

5. In what respect is the pituitary gland actually two glands?

6. How do hypothalamic releasing and inhibiting hormones influence secretions of anterior pituitary hormones?

TABLE 13.2

Summary of Pituitary Gland Hormones and Their Actions

HORMONE	ACTIONS
Anterior Pituitary Hormones	
Human growth hormone (hGH)	Stimulates liver, muscle, cartilage, bone, and other tissues to synthesize and secrete insulinlike growth factors (IGFs); IGFs promote growth of body cells, protein synthesis, tissue repair, breakdown of triglycerides, and elevation of blood glucose level
Thyroid-stimulating hormone (TSH)	Stimulates synthesis and secretion of thyroid hormones by the thyroid gland
Follicle-stimulating hormone (FSH)	In females, initiates development of oocytes and induces secretion of estrogens by the ovaries; in males, stimulates testes to produce sperm
Luteinizing hormone (LH)	In females, stimulates secretion of estrogens and progesterone, ovulation, and formation of corpus luteum; in males, stimulates testes to produce testosterone
Prolactin (PRL)	In females, stimulates milk production by the mammary glands
Adrenocorticotropic hormone (ACTH), also known as **corticotropin**	Stimulates secretion of glucocorticoids (mainly cortisol) by the adrenal cortex
Melanocyte-stimulating hormone (MSH)	Exact role in humans is unknown but may influence brain activity; when present in excess, can cause darkening of skin
Posterior Pituitary Hormones	
Oxytocin	Stimulates contraction of smooth muscle cells of uterus during childbirth; stimulates milk ejection from mammary glands
Antidiuretic hormone (ADH), also known as **vasopressin**	Conserves body water by decreasing urine output; decreases water loss through sweating; raises blood pressure by constricting (narrowing) arterioles

13.4 THYROID GLAND

OBJECTIVE • **Describe the location, hormones, and functions of the thyroid gland.**

The butterfly-shaped *thyroid gland* is located just below the larynx (voice box). It is composed of right and left lobes, one on either side of the trachea (Figure 13.7a).

Microscopic spherical sacs called *thyroid follicles* (Figure 13.7b) make up most of the thyroid gland. The wall of each thyroid follicle consists primarily of cells called *follicular cells*, which produce two hormones: *thyroxine* (thī-ROK-sēn), also called T_4 because it contains four atoms of iodine, and *triiodothyronine* (T_3) (trī-ī′-ō-dō-THĪ-rō-nēn), which contains three atoms of iodine. T_3 and T_4 are also known as *thyroid hormones*. The central cavity of each thyroid follicle contains stored thyroid hormones. As T_4 circulates in the blood and enters cells throughout the body, most of it is converted to T_3 by removal of one iodine atom. T_3 is the more potent form of the thyroid hormones.

A smaller number of cells called *parafollicular cells* lie between the follicles (Figure 13.7b). They produce the hormone calcitonin.

Actions of Thyroid Hormones

Because most body cells have receptors for thyroid hormones, T_3 and T_4 exert their effects throughout the body. Thyroid hormones increase *basal metabolic rate (BMR)*, the rate of oxygen consumption under standard or basal conditions (awake, at rest, and fasting). The BMR rises due to increased synthesis and use of ATP. As cells use more oxygen to produce the ATP, more heat is given off, and body temperature rises. In this way, thyroid hormones play an important role in the maintenance of normal body temperature. The thyroid hormones also stimulate protein synthesis, increase the use of glucose and fatty acids for ATP production, increase the breakdown of triglycerides, and enhance cholesterol excretion, thus reducing blood cholesterol level. Together with human growth hormone and insulin, thyroid hormones stimulate body growth, particularly the growth of the nervous and skeletal systems.

CLINICAL CONNECTION | Hyperthyroidism

Excess secretion of thyroid hormones is known as **hyperthyroidism**. Symptoms of hyperthyroidism include increased heart rate and more forceful heartbeats, increased blood pressure, and increased nervousness. •

Figure 13.7 Location and histology of the thyroid gland.

Thyroid hormones regulate (1) oxygen use and basal metabolic rate, (2) cellular metabolism, and (3) growth and development.

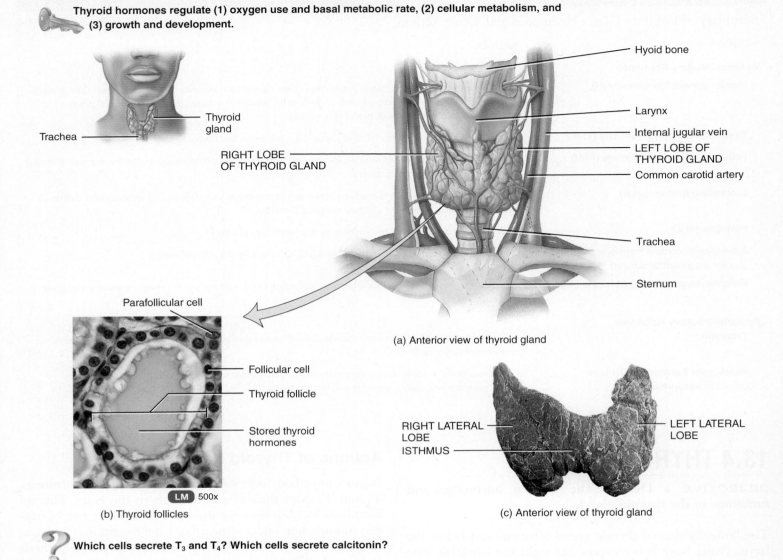

(a) Anterior view of thyroid gland

(b) Thyroid follicles

(c) Anterior view of thyroid gland

Which cells secrete T₃ and T₄? Which cells secrete calcitonin?

Control of Thyroid Hormone Secretion

Thyrotropin-releasing hormone (TRH) from the hypothalamus and thyroid-stimulating hormone (TSH) from the anterior pituitary stimulate synthesis and release of thyroid hormones, as shown in Figure 13.8:

1 Low blood level of thyroid hormones or low metabolic rate stimulates the hypothalamus to secrete TRH.

2 TRH is carried to the anterior pituitary, where it stimulates secretion of thyroid-stimulating hormone (TSH).

3 TSH stimulates thyroid follicular cell activity, including thyroid hormone synthesis and secretion, and growth of the follicular cells.

4 The thyroid follicular cells release thyroid hormones into the blood until the metabolic rate returns to normal.

5 An elevated level of thyroid hormones inhibits release of TRH and TSH (negative feedback).

Conditions that increase ATP demand—a cold environment, low blood glucose, high altitude, and pregnancy—also increase secretion of the thyroid hormones.

Calcitonin

The hormone produced by the parafollicular cells of the thyroid gland is *calcitonin (CT)* (kal-si-TŌ-nin). Calcitonin can decrease the level of calcium in the blood by inhibiting the action of osteoclasts, the cells that break down bone. The secretion of calcitonin is controlled by a negative feedback system (see Figure 13.10). Calcitonin's importance in normal physiology is unclear because it can be present in excess or completely absent without causing clinical symptoms.

Figure 13.8 Regulation of secretion of thyroid hormones.

TSH promotes release of thyroid hormones.

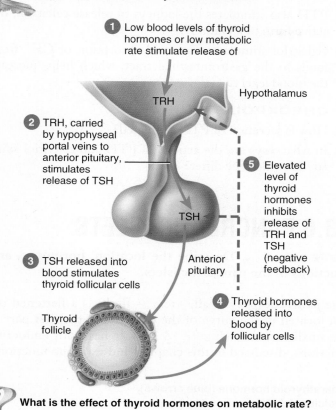

1 Low blood levels of thyroid hormones or low metabolic rate stimulate release of

TRH

Hypothalamus

2 TRH, carried by hypophyseal portal veins to anterior pituitary, stimulates release of TSH

5 Elevated level of thyroid hormones inhibits release of TRH and TSH (negative feedback)

TSH

Anterior pituitary

3 TSH released into blood stimulates thyroid follicular cells

4 Thyroid hormones released into blood by follicular cells

Thyroid follicle

What is the effect of thyroid hormones on metabolic rate?

✓ **CHECKPOINT**

7. How is the secretion of T_3 and T_4 regulated?

8. What are the actions of the thyroid hormones and calcitonin?

13.5 PARATHYROID GLANDS

OBJECTIVE • **Describe the location, hormones, and functions of the parathyroid glands.**

The **parathyroid glands** (*para-* = beside) are small, round masses of glandular tissue that are partially embedded in the posterior surface of the thyroid gland (Figure 13.9). Usually, one superior and one inferior parathyroid gland are attached to each thyroid lobe. Within the parathyroid glands are secretory cells called **chief cells** that release **parathyroid hormone (PTH)**.

PTH is the major regulator of the levels of calcium (Ca^{2+}), magnesium (Mg^{2+}), and phosphate (HPO_4^{2-}) ions in the blood. PTH increases the number and activity of osteoclasts, which break down bone extracellular matrix and release Ca^{2+}

Figure 13.9 Location of the parathyroid glands.

The four parathyroid glands are attached to the posterior surface of the thyroid gland.

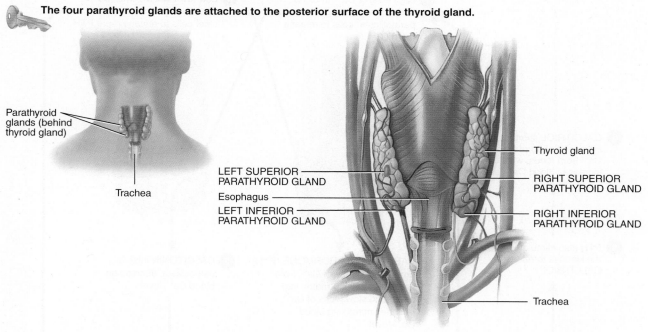

Parathyroid glands (behind thyroid gland)

Trachea

LEFT SUPERIOR PARATHYROID GLAND

Esophagus

LEFT INFERIOR PARATHYROID GLAND

Thyroid gland

RIGHT SUPERIOR PARATHYROID GLAND

RIGHT INFERIOR PARATHYROID GLAND

Trachea

Posterior view

What effect does parathyroid hormone have on osteoclasts?

and HPO_4^{2-} into the blood. PTH also produces three changes in the kidneys. First, it slows the rate at which Ca^{2+} and Mg^{2+} are lost from blood into the urine. Second, it increases loss of HPO_4^{2-} from blood in urine. Because more is lost in the urine than is gained from the bones, PTH decreases blood HPO_4^{2-} level and increases blood Ca^{2+} and Mg^{2+} levels. Third, PTH promotes formation in the kidneys of the hormone *calcitriol*, the active form of vitamin D. Calcitriol acts on the gastrointestinal tract to increase the rate of Ca^{2+}, Mg^{2+}, and HPO_4^{2-} absorption from foods into the blood.

The blood calcium level directly controls the secretion of calcitonin and parathyroid hormone via negative feedback, and the two hormones have opposite effects on blood Ca^{2+} level (Figure 13.10):

1 A higher-than-normal level of calcium ions (Ca^{2+}) in the blood stimulates parafollicular cells of the thyroid gland to release more calcitonin.

2 CT inhibits the activity of osteoclasts, decreasing blood Ca^{2+} level.

3 A lower-than-normal level of Ca^{2+} in the blood stimulates chief cells of the parathyroid gland to release more PTH.

4 PTH increases the number and activity of osteoclasts, which break down bone and release Ca^{2+} into the blood.

PTH also slows loss of Ca^{2+} in the urine. Both actions of PTH raise the blood level of Ca^{2+}.

5 PTH also stimulates the kidneys to release calcitriol, the active form of vitamin D.

6 Calcitriol stimulates increased absorption of Ca^{2+} from foods in the gastrointestinal tract, which helps increase the blood level of Ca^{2+}.

✓ CHECKPOINT

9. How is secretion of PTH regulated?

10. In what ways are the actions of PTH and calcitriol similar? How do they differ?

13.6 PANCREATIC ISLETS

OBJECTIVE • **Describe the location, hormones, and functions of the pancreatic islets.**

The *pancreas* (*pan-* = all; *-creas* = flesh) is a flattened organ located in the curve of the duodenum, the first part of the small intestine (Figure 13.11a). It has both endocrine functions, discussed in this chapter, and exocrine functions,

Figure 13.10 The roles of calcitonin (green arrows), parathyroid hormone (blue arrows), and calcitriol (orange arrows) in homeostasis of blood calcium level.

🔑 **PTH and calcitonin have opposite effects on the level of calcium ions (Ca^{2+}) in the blood.**

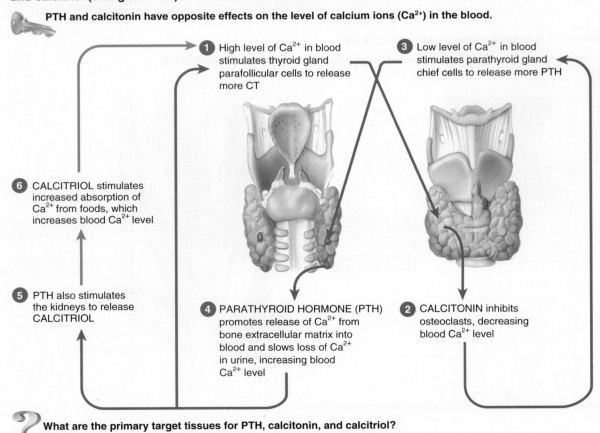

1 High level of Ca^{2+} in blood stimulates thyroid gland parafollicular cells to release more CT

3 Low level of Ca^{2+} in blood stimulates parathyroid gland chief cells to release more PTH

6 CALCITRIOL stimulates increased absorption of Ca^{2+} from foods, which increases blood Ca^{2+} level

5 PTH also stimulates the kidneys to release CALCITRIOL

4 PARATHYROID HORMONE (PTH) promotes release of Ca^{2+} from bone extracellular matrix into blood and slows loss of Ca^{2+} in urine, increasing blood Ca^{2+} level

2 CALCITONIN inhibits osteoclasts, decreasing blood Ca^{2+} level

❓ **What are the primary target tissues for PTH, calcitonin, and calcitriol?**

Figure 13.11 Location and histology of the pancreas.

Hormones released by pancreatic islets regulate blood glucose level.

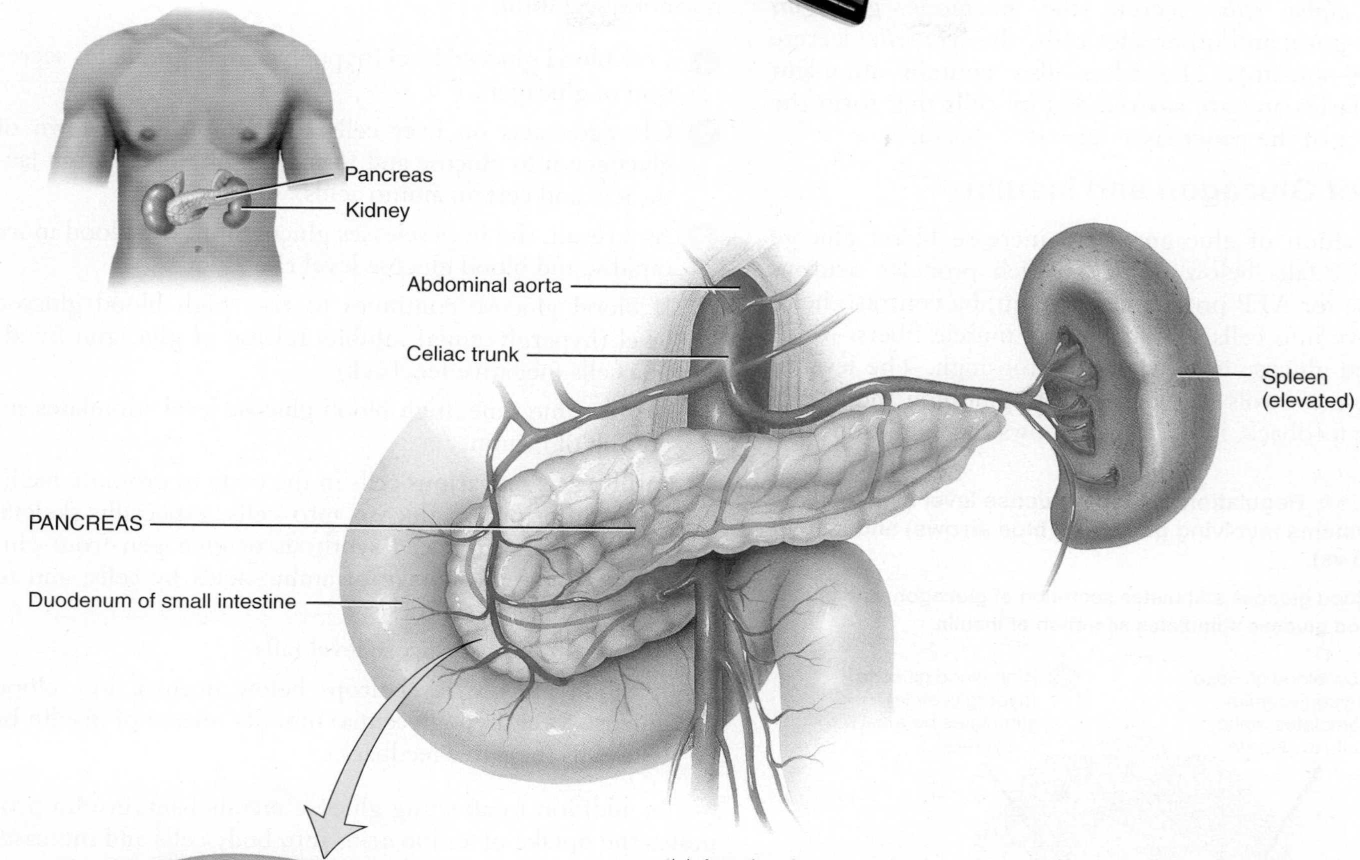

Pancreas

Kidney

Abdominal aorta

Celiac trunk

Spleen (elevated)

PANCREAS

Duodenum of small intestine

(a) Anterior view

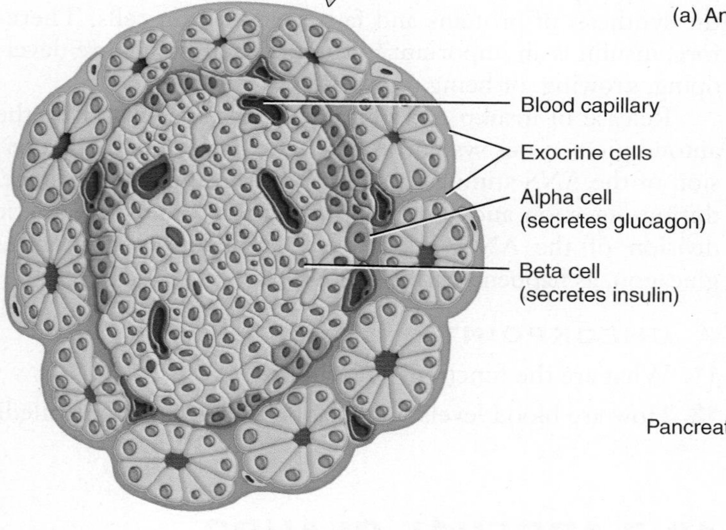

Blood capillary

Exocrine cells

Alpha cell (secretes glucagon)

Beta cell (secretes insulin)

(b) Pancreatic islet and surrounding acini

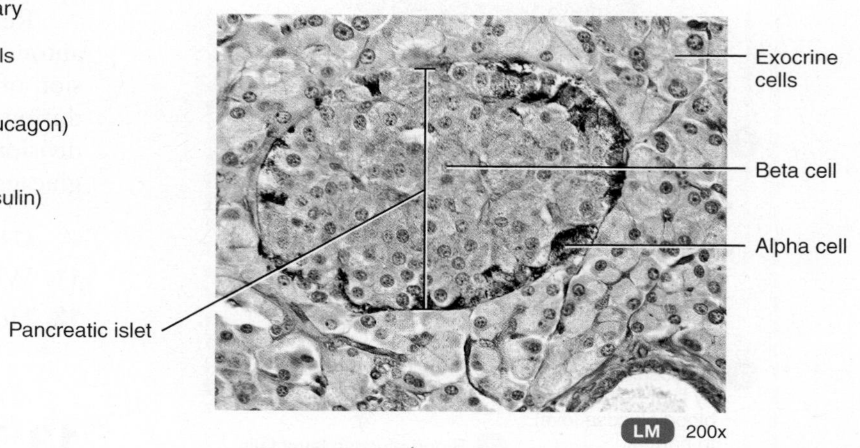

Exocrine cells

Beta cell

Alpha cell

Pancreatic islet

LM 200x

(c) Pancreatic islet and surrounding exocrine cells

 Is the pancreas an exocrine gland or an endocrine gland?

discussed in Section 19.6. The endocrine part of the pancreas consists of clusters of cells called *pancreatic islets* or *islets of Langerhans* (LAHNG-er-hanz). Some of the islet cells, the *alpha cells*, secrete the hormone *glucagon* (GLOO-ka-gon), and other islet cells, the *beta cells*, secrete *insulin* (IN-soo-lin). The islets also contain abundant blood capillaries and are surrounded by cells that form the exocrine part of the pancreas (Figure 13.11b, c).

Actions of Glucagon and Insulin

The main action of glucagon is to increase blood glucose level when it falls below normal, which provides neurons with glucose for ATP production. Insulin, by contrast, helps glucose move into cells, especially into muscle fibers, which lowers blood glucose level when it is too high. The level of blood glucose controls secretion of both glucagon and insulin via negative feedback. Figure 13.12 shows the conditions that

stimulate the pancreatic islets to secrete their hormones, the ways in which glucagon and insulin produce their effects on blood glucose level, and the negative feedback control of hormone secretion:

1. Low blood glucose level (hypoglycemia) stimulates secretion of glucagon.

2. Glucagon acts on liver cells to promote breakdown of glycogen into glucose and formation of glucose from lactic acid and certain amino acids.

3. As a result, the liver releases glucose into the blood more rapidly, and blood glucose level rises.

4. If blood glucose continues to rise, high blood glucose level (hyperglycemia) inhibits release of glucagon by alpha cells (negative feedback).

5. At the same time, high blood glucose level stimulates secretion of insulin.

6. Insulin acts on various cells in the body to promote facilitated diffusion of glucose into cells, especially skeletal muscle fibers; to speed synthesis of glycogen from glucose; to increase uptake of amino acids by cells; and to increase protein synthesis.

7. As a result, blood glucose level falls.

8. If blood glucose level drops below normal, low blood glucose level (hypoglycemia) inhibits release of insulin by beta cells (negative feedback).

In addition to affecting glucose metabolism, insulin promotes the uptake of amino acids into body cells and increases the synthesis of proteins and fatty acids within cells. Therefore, insulin is an important hormone when tissues are developing, growing, or being repaired.

Release of insulin and glucagon is also regulated by the autonomic nervous system (ANS). The parasympathetic division of the ANS stimulates secretion of insulin, for instance, during digestion and absorption of a meal. The sympathetic division of the ANS, by contrast, stimulates secretion of glucagon, as happens during exercise.

Figure 13.12 Regulation of blood glucose level by negative feedback systems involving glucagon (blue arrows) and insulin (orange arrows).

🔑 Low blood glucose stimulates secretion of glucagon, and high blood glucose stimulates secretion of insulin.

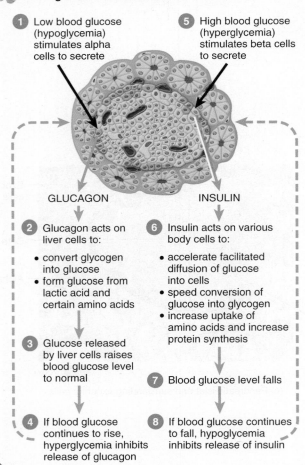

① Low blood glucose (hypoglycemia) stimulates alpha cells to secrete

⑤ High blood glucose (hyperglycemia) stimulates beta cells to secrete

GLUCAGON   INSULIN

② Glucagon acts on liver cells to:
- convert glycogen into glucose
- form glucose from lactic acid and certain amino acids

⑥ Insulin acts on various body cells to:
- accelerate facilitated diffusion of glucose into cells
- speed conversion of glucose into glycogen
- increase uptake of amino acids and increase protein synthesis

③ Glucose released by liver cells raises blood glucose level to normal

⑦ Blood glucose level falls

④ If blood glucose continues to rise, hyperglycemia inhibits release of glucagon

⑧ If blood glucose continues to fall, hypoglycemia inhibits release of insulin

❓ Why is glucagon sometimes called an "anti-insulin" hormone?

✓ CHECKPOINT

11. What are the functions of insulin?

12. How are blood levels of glucagon and insulin controlled?

13.7 ADRENAL GLANDS

OBJECTIVE • **Describe the location, hormones, and functions of the adrenal glands.**

There are two *adrenal glands*, one lying atop each kidney (Figure 13.13). Each adrenal gland has regions that produce different hormones: the outer *adrenal cortex*, which

Figure 13.13 Location and histology of the adrenal glands.

 The adrenal cortex secretes steroid hormones, and the adrenal medulla secretes epinephrine and norepinephrine.

Adrenal glands

Kidney

RIGHT ADRENAL GLAND

LEFT ADRENAL GLAND

Right renal artery
Right renal vein

Left renal artery
Left renal vein

Inferior vena cava

Abdominal aorta

(a) Anterior view

Capsule

Adrenal cortex

Adrenal medulla

(b) Section through left adrenal gland

ADRENAL GLAND

Kidney

(c) Anterior view of adrenal gland and kidney

Capsule

Adrenal cortex:

Zona glomerulosa secretes mineralocorticoids, mainly aldosterone

Zona fasciculata secretes glucocorticoids, mainly cortisol

Zona reticularis secretes androgens

Adrenal medulla chromaffin cells secrete epinephrine and norepinephrine (NE)

LM 50x

(d) Subdivisions of adrenal gland

? What hormones are secreted by the three zones of the adrenal cortex?

makes up 85 percent of the gland, and the inner *adrenal medulla*.

Adrenal Cortex Hormones

The adrenal cortex consists of three zones, each of which synthesizes and secretes different steroid hormones. The outer zone releases hormones called mineralocorticoids because they affect mineral homeostasis. The middle zone releases hormones called glucocorticoids because they affect glucose homeostasis. The inner zone releases androgens (steroid hormones that have masculinizing effects).

Mineralocorticoids

Aldosterone (al-DO-ster-ōn) is the major *mineralocorticoid* (min′-er-al-ō-KOR-ti-koyd). It regulates homeostasis of two mineral ions, namely, sodium ions (Na^+) and potassium ions (K^+). Aldosterone increases reabsorption of Na^+ from the fluid that will become urine into the blood, and it stimulates excretion of K^+ into the fluid that will become urine. It also helps adjust blood pressure and blood volume, and promotes excretion of H^+ in the urine. Such removal of acids from the body can help prevent acidosis (blood pH below 7.35).

Secretion of aldosterone occurs as part of the *renin–angiotensin–aldosterone pathway* (RĒ-nin an′-jē-ō-TEN-sin) (Figure 13.14). Conditions that initiate this pathway include dehydration, Na^+ deficiency, or hemorrhage, which decrease blood volume and blood pressure. Lowered blood pressure stimulates the kidneys to secrete the enzyme *renin*, which promotes a reaction in the blood that forms *angiotensin I*. As blood flows through the lungs, another enzyme called *angiotensin converting enzyme (ACE)* converts inactive angiotensin I into the active hormone *angiotensin II*. Angiotensin II stimulates the adrenal cortex to secrete aldosterone. Aldosterone, in turn, acts on the kidneys to promote the return of Na^+ and water to the blood. As more water returns to the blood (and less is lost in the urine), blood volume increases, increasing blood pressure to normal.

Glucocorticoids

The most abundant *glucocorticoid* (gloo′-kō-KOR-ti-koyd; *gluco-* = sugar; *cortic-* = the bark, shell) is *cortisol*. Cortisol and other glucocorticoids have the following actions:

- **Protein breakdown**. Glucocorticoids increase the rate of protein breakdown, mainly in muscle fibers, and thus increase the liberation of amino acids into the bloodstream. The amino acids may be used by body cells for synthesis of new proteins or for ATP production.

- **Glucose formation**. Upon stimulation by glucocorticoids, liver cells may convert certain amino acids or lactic acid to glucose, which neurons and other cells can use for ATP production.

Figure 13.14 The renin–angiotensin–aldosterone pathway.

Aldosterone helps regulate blood volume, blood pressure, and levels of Na^+ and K^+ in the blood.

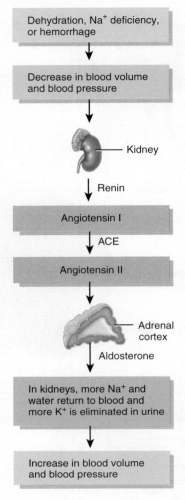

Could a drug that *blocks* the action of the enzyme ACE be used to raise or to lower blood pressure? Why?

- **Breakdown of triglycerides**. Glucocorticoids stimulate the breakdown of triglycerides in adipose tissue. The fatty acids thus released into the blood can be used for ATP production by many body cells.

- **Anti-inflammatory effects**. Although inflammation and immune responses are important defense mechanisms, when these responses become exaggerated during a stressful situation, the body may experience a lot of pain. Glucocorticoids inhibit white blood cells that participate in inflammatory responses. They are often used in the treatment of chronic inflammatory disorders such as rheumatoid arthritis. Unfortunately, glucocorticoids also retard tissue repair, which slows wound healing.

- **Depression of immune responses**. High doses of glucocorticoids depress immune responses. For this reason, glucocorticoids are prescribed for organ transplant recipients to decrease the risk of tissue rejection by the immune system.

The control of secretion of cortisol (and other glucocorticoids) occurs by negative feedback. A low blood level of cortisol stimulates neurosecretory cells in the hypothalamus to secrete *corticotropin-releasing hormone (CRH)*. The hypophyseal portal veins carry CRH to the anterior pituitary, where it stimulates release of *adrenocorticotropic hormone (ACTH)*. ACTH, in turn, stimulates cells of the adrenal cortex to secrete cortisol. As the level of cortisol rises, it exerts negative feedback inhibition both on the anterior pituitary to reduce release of ACTH and on the hypothalamus to reduce release of CRH.

Androgens

In both males and females, the adrenal cortex secretes small amounts of weak androgens. After puberty in males, androgens are also released in much greater quantity by the testes. Thus, the amount of androgens secreted by the adrenal gland in males is usually so low that their effects are insignificant. In females, however, adrenal androgens play important roles: They contribute to libido (sex drive) and are converted into estrogens (feminizing sex steroids) by other body tissues. After menopause, when ovarian secretion of estrogens ceases, all female estrogens come from conversion of adrenal androgens. Adrenal androgens also stimulate growth of axillary (armpit) and pubic hair in boys and girls and contribute to the growth spurt before puberty. Although control of adrenal androgen secretion is not fully understood, the main hormone that stimulates its secretion is ACTH.

CLINICAL CONNECTION | Congenital Adrenal Hyperplasia (CAH)

Congenital adrenal hyperplasia (CAH) is a group of genetic disorders in which one or more enzymes needed for the production of cortisol, or aldosterone, or both are absent. Because the cortisol level is low, secretion of ACTH by the anterior pituitary is high due to lack of negative feedback inhibition. ACTH, in turn, stimulates growth and secretory activity of the adrenal cortex. As a result, both adrenal glands are enlarged. However, certain steps leading to synthesis of cortisol are blocked. Thus, precursor molecules build up, and some of these are weak androgens that can be converted to testosterone. The result is **virilism**, or masculinization. In a female, virile characteristics include growth of a beard, development of a much deeper voice and a masculine distribution of body hair, growth of the clitoris so it may resemble a penis, atrophy of the breasts, and increased muscularity that produces a masculine physique. In males, virilism causes the same characteristics as in females, plus rapid development of the male sexual organs and emergence of male sexual desires. •

Adrenal Medulla Hormones

The innermost region of each adrenal gland, the adrenal medulla, consists of sympathetic postganglionic cells of the autonomic nervous system (ANS) that are specialized to secrete hormones. The two main hormones of the adrenal medullae are *epinephrine* and *norepinephrine (NE)*, also called adrenaline and noradrenaline, respectively.

In stressful situations and during exercise, impulses from the hypothalamus stimulate sympathetic preganglionic neurons, which in turn stimulate the cells of the adrenal medullae to secrete epinephrine and norepinephrine. These two hormones greatly augment the fight-or-flight response (see Section 13.11). By increasing heart rate and force of contraction, epinephrine and norepinephrine increase the pumping output of the heart, which increases blood pressure. They also increase blood flow to the heart, liver, skeletal muscles, and adipose tissue; dilate airways to the lungs; and increase blood levels of glucose and fatty acids. Like the glucocorticoids of the adrenal cortex, epinephrine and norepinephrine also help the body resist acute stress (fight-or-flight).

✓ CHECKPOINT

13. How do the adrenal cortex and adrenal medulla compare with regard to their location and histology?
14. How is secretion of adrenal cortex hormones regulated?

13.8 OVARIES AND TESTES

OBJECTIVE • Describe the location, hormones, and functions of the ovaries and testes.

Gonads are the organs that produce gametes—sperm in males and oocytes in females. The female gonads, the *ovaries*, are paired oval bodies located in the pelvic cavity. They produce the female sex hormones *estrogens* and *progesterone*. Along with FSH and LH from the anterior pituitary, the female sex hormones regulate the menstrual cycle, maintain pregnancy, and prepare the mammary glands for lactation. They also help establish and maintain the feminine body shape.

The ovaries also produce *inhibin*, a protein hormone that inhibits secretion of follicle-stimulating hormone (FSH). During pregnancy, the ovaries and placenta produce a peptide hormone called *relaxin*, which increases the flexibility of the pubic symphysis during pregnancy and helps dilate the uterine cervix during labor and delivery. These actions enlarge the birth canal, which helps ease the baby's passage.

The male gonads, the *testes*, are oval glands that lie in the scrotum. They produce *testosterone*, the primary androgen or male sex hormone. Testosterone regulates production of sperm and stimulates the development and maintenance of masculine characteristics such as beard growth and deepening of the

Shedding Light on Shift Work Cancer Risk

Once upon a time, it got dark at night. With the ability to burn torches and candles, and more recently, turn on electric lights, people could lengthen the day and be active longer. In our industrialized society, some types of work continue around the clock. Hospital staff members aid the sick, transportation workers get people where they need to go, and security guards, public safety workers, and military personnel deal with emergencies and keep watch over the rest of us. Many factories and other work places operate continuous shifts to maximize production or accomplish work that must be done overnight.

Endocrine Disruption

Our bodies evolved in synchrony with the 24-hour light–dark cycle that accompanies the turning of the earth. Until recently, humans have primarily used the nights for sleeping. The pineal gland helps to regulate the body's circadian rhythm. It releases melatonin as darkness sets in, telling the body it is time to sleep. What happens when darkness never sets in? Melatonin production declines, and circadian rhythm is disrupted.

Take decreased exposure to darkness, add the need to be alert, awake, and active all night, and risk of several health problems rises. The most serious of these appears to be cancer. People who work the night shift consistently, as well as those whose shifts vary and include night work, have elevated risks of reproductive cancers, including cancers of the breast, prostate, colon, and endometrium (uterine lining).

Think It Over . . . What can people who work the night shift do to create a more normal circadian rhythm in their bodies?

Normal melatonin levels seem to prevent cancer in several ways. In the lab, melatonin suppresses the activity of tumor cells. It may also act as an antioxidant. Melatonin also seems to increase the production of cytokines (one component of a healthy immune response), which may in turn stimulate immune cells to recognize and destroy cancer cells. High levels of estrogen can promote the growth of cancerous cells; melatonin can bind with some receptors for estrogens, thus reducing the effects of estrogens on cells.

Are Melatonin Supplements the Answer?

Medical professionals discourage the use of melatonin supplements. Some people may find that occasional use of melatonin supplements is helpful for combating jet lag or adjusting to a new schedule. But researchers do not yet understand how melatonin supplements affect the activity of the pineal gland Some evidence suggests melatonin supplements may increase risk for depression or reduce the pineal gland's natural ability to regulate circadian rhythm. Too much melatonin may also interfere with normal sex hormone levels.

---

voice. The testes also produce inhibin, which inhibits secretion of FSH. The detailed structure of the ovaries and testes and the specific roles of sex hormones will be discussed in Chapter 23.

✓ CHECKPOINT

15. Why are the ovaries and testes included among the endocrine glands?

13.9 PINEAL GLAND

OBJECTIVE • Describe the location, hormone, and functions of the pineal gland.

The *pineal gland* (PIN-ē-al = pinecone shape) is a small endocrine gland attached to the roof of the third ventricle of the brain at the midline (see Figures 13.1 and 10.6). One hormone secreted by the pineal gland is *melatonin*, which

contributes to setting the body's biological clock. More melatonin is released in darkness and during sleep; less melatonin is liberated in strong sunlight. In animals that breed during specific seasons, melatonin inhibits reproductive functions. Whether melatonin influences human reproductive function, however, is still unclear. Melatonin levels are higher in children and decline with age into adulthood, but there is no evidence that changes in melatonin secretion correlate with the onset of puberty and sexual maturation.

CLINICAL CONNECTION | Seasonal Affective Disorder (SAD)

Seasonal affective disorder (SAD) is a type of depression that afflicts some people during the winter months, when day length is short. It is thought to be due, in part, to overproduction of melatonin. Bright light therapy—repeated exposure to artificial light—can provide relief. •

✓ CHECKPOINT

16. What is the relationship between melatonin secretion and sleep?

13.10 OTHER HORMONES

OBJECTIVE • **List the hormones secreted by cells in tissues and organs other than endocrine glands, and describe their functions.**

Hormones from Other Endocrine Tissues and Organs

Cells in organs other than those usually classified as endocrine glands have an endocrine function and secrete hormones. Table 13.3 provides an overview of these organs and tissues and their hormones and actions.

Prostaglandins and Leukotrienes

Two families of molecules derived from fatty acids, the *prostaglandins* (*PGs*) (pros′-ta-GLAN-dins), and the *leukotrienes* (*LTs*) (loo-kō-TRĪ-ēns), act locally as hormones in most tissues of the body. Virtually all body cells except red blood cells release these local hormones in response to chemical and mechanical stimuli. Because the PGs and LTs act close to their sites of release, they appear in only tiny quantities in the blood.

Leukotrienes stimulate movement of white blood cells and mediate inflammation. The prostaglandins alter smooth muscle contraction, glandular secretions, blood flow, reproductive processes, platelet function, respiration, nerve impulse transmission, fat metabolism, and immune responses. PGs also have roles in inflammation, promoting fever, and intensifying pain.

CLINICAL CONNECTION | Nonsteroidal Anti-Inflammatory Drugs (NSAIDs)

Aspirin and related **nonsteroidal anti-inflammatory drugs (NSAIDs)**, such as ibuprofen (Advil®, Motrin®), inhibit a key enzyme in prostaglandin synthesis without affecting synthesis of leukotrienes. They are used to treat a wide variety of inflammatory disorders, from rheumatoid arthritis to tennis elbow. •

TABLE 13.3

Summary of Hormones Produced by Other Organs and Tissues that Contain Endocrine Cells

SOURCE AND HORMONE	ACTIONS
Thymus	
Thymosin	Promotes the maturation of T cells (a type of white blood cell that destroys microbes and foreign substances) and may retard the aging process (discussed in Chapter 17)
Gastrointestinal Tract	
Gastrin	Promotes secretion of gastric juice and increases movements of the stomach (discussed in Chapter 19)
Glucose-dependent insulinotropic peptide (GIP)	Stimulates release of insulin by pancreatic beta cells (discussed in Chapter 19)
Secretin	Stimulates secretion of pancreatic juice and bile (discussed in Chapter 19)
Cholecystokinin (CCK)	Stimulates secretion of pancreatic juice, regulates release of bile from the gallbladder, and brings about a feeling of fullness after eating (discussed in Chapter 19)
Kidney	
Erythropoietin (EPO)	Increases rate of red blood cell production (discussed in Chapter 14)
Heart	
Atrial natriuretic peptide (ANP)	Decreases blood pressure (discussed in Chapter 16)
Adipose Tissue	
Leptin	Suppresses appetite and may increase the activity of FSH and LH (discussed in Chapter 20)
Placenta	
Human chorionic gonadotropin (hCG)	Stimulates the ovary to continue production of estrogens and progesterone during pregnancy (discussed in Chapter 24)

17. What hormones are secreted by the gastrointestinal tract, placenta, kidneys, skin, adipose tissue, and heart?

18. What are some functions of prostaglandins and leuko-trienes?

13.11 THE STRESS RESPONSE

OBJECTIVE • **Describe how the body responds to stress.**

It is impossible to remove all stress from our everyday lives. Any stimulus that produces a stress response is called a *stressor*. A stressor may be almost any disturbance—heat or cold, environmental poisons, toxins given off by bacteria, heavy bleeding from a wound or surgery, or a strong emotional reaction. Stressors may be pleasant or unpleasant, and they vary among people and even within the same person at different times. When homeostatic mechanisms are successful in counteracting stress, the internal environment remains within normal physiological limits. If stress is extreme, unusual, or long lasting, it elicits the *stress response*, a sequence of bodily changes that can progress through three stages: (1) an initial fight-or-flight response, (2) a slower resistance reaction, and eventually (3) exhaustion.

The *fight-or-flight response*, initiated by nerve impulses from the hypothalamus to the sympathetic division of the autonomic nervous system (ANS), including the adrenal medullae, quickly mobilizes the body's resources for immediate physical activity. It brings huge amounts of glucose and oxygen to the organs that are most active in warding off danger: the brain, which must become highly alert; the skeletal muscles, which may have to fight off an attacker or flee; and the heart, which must work vigorously to pump enough blood to the brain and muscles. Reduction of blood flow to the kidneys, however, promotes the release of renin, which sets into motion the renin–angiotensin–aldosterone pathway (see Figure 13.14). Aldosterone causes the kidneys to retain Na⁺, which leads to water retention and elevated blood pressure. Water retention also helps preserve body fluid volume in the case of severe bleeding.

The second stage in the stress response is the *resistance reaction*. Unlike the short-lived fight-or-flight response, which is initiated by nerve impulses from the hypothalamus, the resistance reaction is initiated in large part by hypothalamic releasing hormones and is a longer-lasting response. The hormones involved are corticotropin-releasing hormone (CRH), growth hormone-releasing hormone (GHRH), and thyrotropin-releasing hormone (TRH).

CRH stimulates the anterior pituitary to secrete ACTH, which in turn stimulates the adrenal cortex to release more cortisol. Cortisol then stimulates release of glucose by liver cells, breakdown of triglycerides into fatty acids, and catabolism of proteins into amino acids. Tissues throughout the body can use the resulting glucose, fatty acids, and amino acids to produce ATP or to repair damaged cells. Cortisol also reduces inflammation. A second hypothalamic releasing hormone, GHRH, causes the anterior pituitary to secrete human growth hormone (hGH). Acting via insulinlike growth factors, hGH stimulates breakdown of triglycerides and glycogen. A third hypothalamic releasing hormone, TRH, stimulates the anterior pituitary to secrete thyroid-stimulating hormone (TSH). TSH promotes secretion of thyroid hormones, which stimulate the increased use of glucose for ATP production. The combined actions of hGH and TSH thereby supply additional ATP for metabolically active cells.

The resistance stage helps the body continue fighting a stressor long after the fight-or-flight response dissipates. Generally, it is successful in seeing us through a stressful episode, and our bodies then return to normal. Occasionally, however, the resistance stage fails to combat the stressor: The resources of the body may eventually become so depleted that they cannot sustain the resistance stage, and *exhaustion* ensues. Prolonged exposure to high levels of cortisol and other hormones involved in the resistance reaction causes wasting of muscles, suppression of the immune system, ulceration of the gastrointestinal tract, and failure of pancreatic beta cells. In addition, pathological changes may occur because resistance reactions persist after the stressor has been removed.

Although the exact role of stress in human diseases is not known, it is clear that stress can temporarily inhibit certain components of the immune system. Stress-related disorders include gastritis, ulcerative colitis, irritable bowel syndrome, hypertension, asthma, rheumatoid arthritis, migraine headaches, anxiety, and depression. People under stress also are at a greater risk of developing a chronic disease or dying prematurely.

CLINICAL CONNECTION | Posttraumatic Stress Disorder (PTSD)

Posttraumatic stress disorder (PTSD) may develop in someone who has experienced, witnessed, or learned about a physically or psychologically distressing event. The immediate cause of PTSD appears to be the specific stressors associated with the events. Among the stressors are terrorism, hostage taking, imprisonment, serious accidents, torture, sexual or physical abuse, violent crimes, and natural disasters. In the United States, PTSD affects 10 percent of females and 5 percent of males. Symptoms of PTSD include reliving the event through nightmares or flashbacks; loss of interest and lack of motivation; poor concentration; irritability; and insomnia. •

✓ **CHECKPOINT**

19. What is the role of the hypothalamus during stress?
20. How are stress and immunity related?

13.12 AGING AND THE ENDOCRINE SYSTEM

OBJECTIVE • **Describe the effects of aging on the endocrine system.**

Although some endocrine glands shrink as we get older, their performance may or may not be compromised. Production of human growth hormone by the anterior pituitary decreases, which is one cause of muscle atrophy as aging proceeds. The thyroid gland often decreases its output of thyroid hormones with age, causing a decrease in metabolic rate, an increase in body fat, and hypothyroidism, which is seen more often in older people. Because there is less negative feedback (to lower levels of thyroid hormones), the level of thyroid-stimulating hormone increases with age.

With aging, the blood level of PTH rises, perhaps due to inadequate dietary intake of calcium. In a study of older women who took 2,400 mg/day of supplemental calcium, blood levels of PTH were as low as those in younger women. Both calcitriol and calcitonin levels are lower in older persons. Together, the rise in PTH and the fall in calcitonin heighten the age-related decrease in bone mass that leads to osteoporosis and increased risk of fractures.

The adrenal glands contain increasingly more fibrous tissue and produce less cortisol and aldosterone with advancing age. However, production of epinephrine and norepinephrine remains normal. The pancreas releases insulin more slowly with age, and receptor sensitivity to glucose declines. As a result, blood glucose levels in older people increase faster and return to normal more slowly than in younger individuals.

The thymus is largest in infancy. After puberty, its size begins to decrease, and thymic tissue is replaced by adipose and areolar connective tissue. In older adults, the thymus has atrophied significantly. However, it still produces new T cells for immune responses.

The ovaries decrease in size with age, and they no longer respond to gonadotropins. The resultant decreased output of estrogens leads to conditions such as osteoporosis, high blood cholesterol, and atherosclerosis. FSH and LH levels are high due to less negative feedback inhibition of estrogens. Although testosterone production by the testes decreases with age, the effects are not usually apparent until very old age, and many elderly males can still produce active sperm in normal numbers.

✓ **CHECKPOINT**

21. Which hormone is related to the muscle atrophy that occurs with aging?

• • •

To appreciate the many ways the endocrine system contributes to homeostasis of other body systems, examine Focus on Homeostasis: The Endocrine System. Next, in Chapter 14, we will begin to explore the cardiovascular system, starting with a description of the composition and functions of the blood.

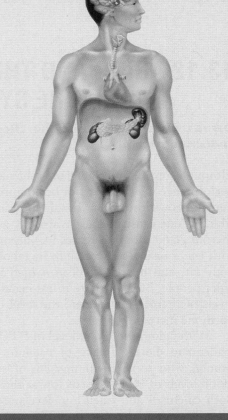

THE ENDOCRINE SYSTEM

For all body systems

Together with the nervous system, hormones of the endocrine system regulate the activity and growth of target cells throughout the body. Several hormones regulate metabolism, uptake of glucose, and molecules used for ATP production by body cells.

Integumentary system

Androgens stimulate the growth of axillary and pubic hair and activation of sebaceous glands. Melanocyte-stimulating hormone (MSH) can cause darkening of the skin.

Skeletal system

Human growth hormone (hGH) and insulinlike growth factors (IGFs) stimulate bone growth. Estrogens cause closure of epiphyseal plates at the end of puberty and help maintain bone mass in adults. Parathyroid hormone (PTH) promotes the release of calcium and other minerals from bone extracellular matrix into the blood. Thyroid hormones are needed for normal development and growth of the skeleton.

Muscular system

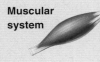

Epinephrine and norepinephrine help increase blood flow to exercising muscles. PTH maintains the proper level of Ca^{2+} in blood and interstitial fluid, which is needed for muscle contraction. Glucagon, insulin, and other hormones regulate metabolism in muscle fibers. IGFs, thyroid hormones, and insulin stimulate protein synthesis and thereby help maintain muscle mass.

Nervous system

Several hormones, especially thyroid hormones, insulin, and IGFs, influence growth and development of the nervous system.

Cardiovascular system

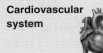

Erythropoietin (EPO) promotes the production of red blood cells. Aldosterone and antidiuretic hormone (ADH) increase blood volume. Epinephrine and norepinephrine increase the heart's rate and force of contraction. Several hormones elevate blood pressure during exercise and other stresses.

Lymphatic system and immunity

Glucocorticoids such as cortisol depress inflammation and immune responses. Hormones from the thymus promote maturation of T cells, a type of white blood cell that participates in immune responses.

Respiratory system

Epinephrine and norepinephrine dilate (widen) the airways during exercise and other stresses. Erythropoietin regulates the amount of oxygen carried in the blood by adjusting the number of red blood cells.

Digestive system

Epinephrine and norepinephrine depress activity of the digestive system. Gastrin, cholecystokinin, secretin, and glucose-dependent insulinotropic peptide (GIP) help regulate digestion. Calcitriol promotes absorption of dietary calcium. Leptin suppresses appetite.

Urinary system

ADH, aldosterone, and atrial natriuretic peptide (ANP) adjust the rate of loss of water and ions in the urine, thereby regulating blood volume and ion levels in the blood.

Reproductive systems

Hypothalamic releasing and inhibiting hormones, follicle-stimulating hormone (FSH), and luteinizing hormone (LH) regulate the development, growth, and secretions of the gonads (ovaries and testes). Estrogens and testosterone contribute to the development of oocytes and sperm and stimulate the development of sexual characteristics. Prolactin promotes milk production in the mammary glands. Oxytocin causes contraction of the uterus and ejection of milk from the mammary glands.

COMMON DISORDERS

Disorders of the endocrine system often involve either **hyposecretion** (*hypo-* = too little or under), inadequate release of a hormone, or **hypersecretion** (*hyper-* = too much or above), excessive release of a hormone. In other cases, the problem is faulty hormone receptors or an inadequate number of receptors.

Pituitary Gland Disorders

Several disorders of the anterior pituitary involve human growth hormone (hGH). Undersecretion of hGH during the growth years slows bone growth, and the epiphyseal plates close before normal height is reached. This condition is called **pituitary dwarfism**. Other organs of the body also fail to grow, and the body proportions are childlike. A dwarf has a normal-sized head and torso but small limbs; a *midget* has a proportional head, torso, and limbs.

Oversecretion of hGH during childhood results in **giantism (gigantism)**, an abnormal increase in the length of long bones. The person grows to be very tall, but body proportions are about normal. Figure 13.15a shows identical twins; one brother developed giantism due to a pituitary tumor. Oversecretion of hGH during adulthood is called **acromegaly** (ak'-rō-MEG-a-lē). Although hGH cannot produce further lengthening of the long bones because the epiphyseal plates are already closed, the bones of the hands, feet, cheeks, and jaws thicken and other tissues enlarge (Figure 13.15b).

The most common abnormality of the posterior pituitary is **diabetes insipidus** (dī-a-BĒ-tēs in-SIP-i-dus; *diabetes* = overflow; insipidus = tasteless). This disorder is due to defects in antidiuretic

Figure 13.15 Photographs of people with various endocrine disorders.

 Disorders of the endocrine system often involve hyposecretion or hypersecretion of various hormones.

(b) Acromegaly (excess hGH during adulthood)

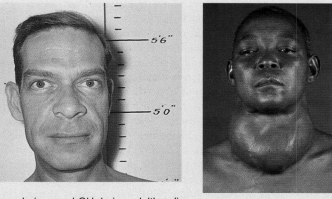

(c) Goiter (enlargement of thyroid gland)

(d) Exophthalmos (excess thyroid hormones, as in Graves disease)

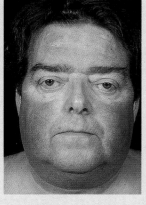

(a) A 22-year old man with pituitary giantism shown beside his identical twin

(e) Cushing's syndrome (excess glucocorticoids)

Which of the disorders shown here is due to antibodies that mimic the action of TSH?

hormone (ADH) receptors or an inability to secrete ADH. Usually the disorder is caused by a brain tumor, head trauma, or brain surgery that damages the posterior pituitary or the hypothalamus. A common symptom is excretion of large volumes of urine, with resulting dehydration and thirst. Because so much water is lost in the urine, a person with diabetes insipidus may die of dehydration if deprived of water for only a day or so.

Thyroid Gland Disorders

Thyroid gland disorders affect all major body systems and are among the most common endocrine disorders. ***Congenital hypothyroidism***, hyposecretion of thyroid hormones that is present at birth, has devastating consequences if not treated promptly. Previously termed *cretinism*, this condition causes severe mental retardation. At birth, the baby typically is normal because lipid-soluble maternal thyroid hormones crossed the placenta during pregnancy and allowed normal development. Most states require testing of all newborns to ensure adequate thyroid function. If congenital hypothyroidism exists, oral thyroid hormone treatment must be started soon after birth and continued for life.

Hypothyroidism during the adult years produces ***myxedema*** (mix-e-DĒ-ma), which occurs about five times more often in females than in males. A hallmark of this disorder is edema (accumulation of interstitial fluid) that causes the facial tissues to swell and look puffy. A person with myxedema has a slow heart rate, low body temperature, sensitivity to cold, dry hair and skin, muscular weakness, general lethargy, and a tendency to gain weight easily.

The most common form of hyperthyroidism is ***Graves disease***, which also occurs much more often in females than in males, usually before age 40. Graves disease is an autoimmune disorder in which the person produces antibodies that mimic the action of thyroid-stimulating hormone (TSH). The antibodies continually stimulate the thyroid gland to grow and produce thyroid hormones. Thus, the thyroid gland may enlarge to two to three times its normal size, a condition called ***goiter*** (GOY-ter; *guttur* = throat) (Figure 13.15c). Goiter also occurs in other thyroid diseases and if dietary intake of iodine is inadequate. Graves patients often have a peculiar edema behind the eyes, called ***exophthalmos*** (ek′-sof-THAL-mos), which causes the eyes to protrude (Figure 13.15d).

Parathyroid Gland Disorders

Hypoparathyroidism—too little parathyroid hormone—leads to a deficiency of Ca^{2+}, which causes neurons and muscle fibers to depolarize and produce action potentials spontaneously. This leads to twitches, spasms, and ***tetany*** (maintained contraction) of skeletal muscle. The leading cause of hypoparathyroidism is accidental damage to the parathyroid glands or to their blood supply during surgery to remove the thyroid gland.

Adrenal Gland Disorders

Hypersecretion of cortisol by the adrenal cortex produces ***Cushing's syndrome***. The condition is characterized by the breakdown of muscle proteins and redistribution of body fat, resulting in spindly arms and legs accompanied by a rounded "moon face" (Figure 13.15e), "buffalo hump" on the back, and pendulous (hanging) abdomen. The elevated level of cortisol causes hyperglycemia, osteoporosis, weakness, hypertension, increased susceptibility to infection, decreased resistance to stress, and mood swings.

Hyposecretion of glucocorticoids and aldosterone causes ***Addison's disease***. Symptoms include mental lethargy, anorexia, nausea and vomiting, weight loss, hypoglycemia, and muscular weakness. Loss of aldosterone leads to elevated potassium and decreased sodium in the blood; low blood pressure; dehydration; and decreased cardiac output, cardiac arrhythmias, and even cardiac arrest. The skin may have a "bronzed" appearance that often is mistaken for a suntan. Such was true in the case of President John F. Kennedy, whose Addison's disease was known to only a few while he was alive.

Usually benign tumors of the adrenal medulla, called ***pheochromocytomas*** (fē′-ō-krō′-mō-sī-TŌ-mas; *pheo-* = dusky; *chromo-* = color; *cyto-* = cell), cause oversecretion of epinephrine and norepinephrine. The result is a prolonged version of the fight-or-flight response: rapid heart rate, headache, high blood pressure, high levels of glucose in blood and urine, an elevated basal metabolic rate (BMR), flushed face, nervousness, sweating, and decreased gastrointestinal motility.

Pancreatic Islet Disorders

The most common endocrine disorder is ***diabetes mellitus*** (MEL-i-tus; *melli-* = honey sweetened), caused by an inability either to produce or to use insulin. Diabetes mellitus is the fourth leading cause of death by disease in the United States, primarily because of its damage to the cardiovascular system. Because insulin is unavailable to aid the movement of glucose into body cells, blood glucose level is high and glucose "spills" into the urine (glucosuria). Hallmarks of diabetes mellitus are the three "polys"; *polyuria*, excessive urine production due to an inability of the kidneys to reabsorb water; *polydipsia*, excessive thirst; and *polyphagia*, excessive eating.

Both genetic and environmental factors contribute to onset of the two types of diabetes mellitus—Type 1 and Type 2—but the exact mechanisms are still unknown. In ***type 1 diabetes*** insulin level is low because the person's immune system destroys the pancreatic beta cells. Most commonly, type 1 diabetes develops in people younger than age 20, though it persists throughout life. By the time symptoms arise, 80–90% of the islet beta cells have been destroyed.

Because insulin is not present to aid the entry of glucose into body cells, most cells use fatty acids to produce ATP. Stores of triglycerides in adipose tissue are broken down to fatty acids and glycerol. The byproducts of fatty acid breakdown—organic acids called ketones or ketone bodies—accumulate. Buildup of ketones causes blood pH to fall, a condition known as ***ketoacidosis***. Unless treated quickly, ketoacidosis can cause death.

Type 2 diabetes is much more common than type 1. It most often occurs in people who are over 35 and overweight. The high glucose levels in the blood often can be controlled by diet, exercise, and weight loss. Sometimes, an antidiabetic drug such as *glyburide* (Diabeta®) is used to stimulate secretion of insulin by pancreatic beta cells. Although some type 2 diabetics need insulin, many have a sufficient amount (or even a surplus) of insulin in the blood. For these people, diabetes arises not from a shortage of insulin but because target cells become less sensitive to it.

Hyperinsulinism most often results when a diabetic injects too much insulin. The main symptom is ***hypoglycemia***, decreased blood glucose level, which occurs because the excess insulin stimulates too much uptake of glucose by body cells. When blood glucose falls, neu-rons are deprived of the steady supply of glucose they need to function effectively. Severe hypoglycemia leads to mental disorientation, convulsions, unconsciousness, and shock and is termed ***insulin shock***. Death can occur quickly unless blood glucose is restored to normal levels.

MEDICAL TERMINOLOGY AND CONDITIONS

Gynecomastia (gī-ne′-kō-MAS-tē-a; *gyneco-* = woman; *mast-* = breast) Excessive development of mammary glands in a male. Sometimes a tumor of the adrenal gland may secrete sufficient amounts of estrogen to cause the condition.

Hirsutism (HER-soo-tizm; *hirsut-* = shaggy) Presence of excessive bodily and facial hair in a male pattern, especially in women; may be due to excess androgen production caused by tumors or drugs.

Thyroid crisis (storm) A severe state of hyperthyroidism that can be life-threatening. It is characterized by high body temperature, rapid heart rate, high blood pressure, gastrointestinal symptoms (abdominal pain, vomiting, diarrhea), agitation, tremors, confusion, seizures, and possibly coma.

Virilizing adenoma (*aden-* = gland; *-oma* = tumor) Tumor of the adrenal gland that liberates excessive androgens, causing virilism (masculinization) in females. Occasionally, adrenal tumor cells liberate estrogens to the extent that a male patient develops gynecomastia. Such a tumor is called a ***feminizing adenoma***.

CHAPTER REVIEW AND RESOURCE SUMMARY

REVIEW	RESOURCES
13.1 Introduction	
1. The nervous system controls homeostasis through the release of neurotransmitters; the endocrine system uses hormones. The nervous system causes muscles to contract and glands to secrete; the endocrine system affects virtually all body tissues. Table 13.1 compares the characteristics of the nervous and endocrine systems.	Anatomy Overview—The Endocrine System
2. Exocrine glands (sweat, oil, mucous, digestive) secrete their products through ducts into body cavities or onto body surfaces.	Animation—Introduction to Hormone Feedback Loops
3. The **endocrine system** consists of endocrine glands and several organs that contain endocrine tissue.	Exercise—Concentrate on Endocrine Activities
13.2 Hormone Action	
1. Endocrine glands secrete hormones into interstitial fluid. Then, the hormones diffuse into the blood.	Anatomy Overview—Hormones
2. Hormones affect only specific **target cells** that have the proper **receptors** to bind a given hormone.	Animation—Introduction to Hormonal Regulation, Secretion, and Concentration
3. Chemically, hormones are either lipid-soluble (**steroids, thyroid hormones**, and **nitric oxide**) or water-soluble (modified amino acids, peptides, and proteins).	Animation—Mechanisms of Hormone Action
4. Lipid-soluble hormones affect cell function by altering gene expression.	Exercise—Produce that Hormone!
5. Water-soluble hormones alter cell function by activating plasma membrane receptors, which elicit production of a **second messenger** that activates various proteins inside the cell.	Exercise—Hormone Actions
6. Hormone secretion is controlled by signals from the nervous system, chemical changes in the blood, and other hormones.	
13.3 Hypothalamus and Pituitary Gland	
1. The **pituitary gland** is attached to the **hypothalamus** and consists of two lobes: the **anterior pituitary** and the **posterior pituitary**. Hormones of the pituitary gland are controlled by inhibiting and releasing hormones produced by the hypothalamus. The **hypophyseal portal veins** carry hypothalamic **releasing** and **inhibiting hormones** from the hypothalamus to the anterior pituitary.	Animation—GHRH /hGH
	Animation—hGH—Growth and Development
	Animation—hGH—Glycogenolysis and Lipolysis

REVIEW

2. The anterior pituitary consists of cells that produce **human growth hormone (hGH)**, **prolactin (PRL)**, **thyroid-stimulating hormone (TSH)**, **follicle-stimulating hormone (FSH)**, **luteinizing hormone (LH)**, **adrenocorticotropic hormone (ACTH)**, and **melanocyte-stimulating hormone (MSH)**.

3. Human growth hormone (hGH) stimulates body growth through **insulinlike growth factors (IGFs)** and is controlled by growth hormone-releasing hormone (GHRH) and growth hormone-inhibiting hormone (GHIH).

4. TSH regulates thyroid gland activities and is controlled by thyrotropin-releasing hormone (TRH).

5. FSH and LH regulate activities of the gonads—ovaries and testes—and are controlled by gonadotropin-releasing hormone (GnRH).

6. PRL helps stimulate milk production. Prolactin-inhibiting hormone (PIH) suppresses release of prolactin. Prolactin-releasing hormone (PRH) stimulates a rise in prolactin level during pregnancy.

7. ACTH regulates activities of the adrenal cortex and is controlled by corticotropin-releasing hormone (CRH).

8. The posterior pituitary contains axon terminals of neurosecretory cells whose cell bodies are in the hypothalamus. Hormones made in the hypothalamus and released in the posterior pituitary include **oxytocin**, which stimulates contraction of the uterus and ejection of milk from the breasts, and **antidiuretic hormone (ADH)**, which stimulates water reabsorption by the kidneys and constriction of arterioles.

9. Oxytocin secretion is stimulated by uterine stretching and by suckling during nursing; ADH secretion is controlled by the osmotic pressure of the blood and blood volume.

10. Table 13.2 summarizes the hormones of the anterior and posterior pituitary.

Animation—Antidiuretic
  Hormone

Figure 13.4 The pituitary
gland and its blood supply

13.4 Thyroid Gland

1. The **thyroid gland**, located below the larynx, consists of **thyroid follicles** composed of **follicular cells**, which secrete the thyroid hormones **thyroxine (T₄)** and **triiodothyronine (T₃)**, and **parafollicular cells**, which secrete calcitonin.

2. Thyroid hormones regulate oxygen use and metabolic rate, cellular metabolism, and growth and development. Secretion is controlled by TRH from the hypothalamus and thyroid-stimulating hormone (TSH) from the anterior pituitary.

3. **Calcitonin (CT)** can lower the blood level of calcium; its secretion is controlled by the level of calcium in the blood.

Animation—TRH/TSH—
  Production
Animation—TRH/TSH Actions
Animation—Calcitonin
Animation—Thyroid Hormones
  and Glucose and Lipid
  Catabolism

13.5 Parathyroid Glands

1. The **parathyroid glands** are embedded on the posterior surfaces of the thyroid.

2. **Parathyroid hormone (PTH)** regulates the homeostasis of calcium, magnesium, and phosphate by increasing blood calcium and magnesium levels and decreasing blood phosphate level. PTH secretion is controlled by the level of calcium in the blood.

Animation—Parathyroid
  Hormone
Exercise—Calcium Homeostasis

13.6 Pancreatic Islets

1. The **pancreas** lies in the curve of the duodenum. It has both endocrine and exocrine functions.

2. The endocrine portion consists of **pancreatic islets** or **islets of Langerhans**, which are made up of alpha and beta cells.

3. Alpha cells secrete **glucagon** and beta cells secrete **insulin**.

4. Glucagon increases blood glucose level, and insulin decreases blood glucose level. Secretion of both hormones is controlled by the level of glucose in the blood.

Animation—Glucagon
Animation—Insulin

Figure 13.11 Location
and histology of the
pancreas

Exercise—Glucose
  Regulation Feedback Loop
Concepts and Connections—
  Blood Glucose Regulation

13.7 Adrenal Glands

1. The **adrenal glands** are located above the kidneys. They consist of an outer **adrenal cortex** and inner **adrenal medulla**.

Animation—ACTH/Cortisol-
  Glycogenolysis
Animation—ACTH/Cortisol

REVIEW	RESOURCES

REVIEW

2. The adrenal cortex is divided into three zones: The outer zone secretes mineralocorticoids; the middle zone secretes glucocorticoids; and the inner zone secretes androgens.

3. **Mineralocorticoids** (mainly **aldosterone**) increase sodium and water reabsorption and decrease potassium reabsorption. Secretion is controlled by the **renin–angiotensin–aldosterone pathway**.

4. **Glucocorticoids** (mainly **cortisol**) promote normal metabolism, help resist stress, and decrease inflammation. Secretion is controlled by ACTH.

5. Androgens secreted by the adrenal cortex stimulate growth of axillary and pubic hair, aid the prepubertal growth spurt, and contribute to libido.

6. The adrenal medullae secrete **epinephrine** and **norepinephrine (NE)**, which are released under stress.

RESOURCES

Animation—Epinephrine/NE
Animation—Cortisol

Figure 13.13 Location and histology of the adrenal glands

13.8 Ovaries and Testes

1. The **ovaries** are located in the pelvic cavity and produce **estrogens**, **progesterone**, and **inhibin**. These sex hormones regulate the menstrual cycle, maintain pregnancy, and prepare the mammary glands for lactation. They also help establish and maintain the feminine body shape.

2. The **testes** lie inside the scrotum and produce **testosterone** and inhibin. Testosterone regulates production of sperm and stimulates the development and maintenance of masculine characteristics such as beard growth and deepening of the voice.

Animation—Hormonal Regulation of Female Reproductive System
Animation—Hormonal Regulation of Male Reproductive System

13.9 Pineal Gland

1. The **pineal gland**, attached to the roof of the third ventricle in the brain, secretes **melatonin**, which contributes to setting the body's biological clock.

Anatomy Overview—The Pineal Gland

13.10 Other Hormones

1. Body tissues other than those normally classified as endocrine glands contain endocrine tissue and secrete hormones. These include the thymus, gastrointestinal tract, placenta, kidneys, adipose tissue, and heart. (See Table 13.3.)

2. **Prostaglandins** and **leukotrienes** act locally in most body tissues.

Anatomy Overview—Hormones

13.11 The Stress Response

1. **Stressors** include surgical operations, poisons, infections, fever, and strong emotional responses.

2. If stress is extreme, it triggers the **stress response**, which occurs in three stages: the fight-or-flight response, resistance reaction, and exhaustion.

3. The **fight-or-flight response** is initiated by nerve impulses from the hypothalamus to the sympathetic division of the autonomic nervous system and the adrenal medullae. This response rapidly increases circulation and promotes ATP production.

4. The **resistance reaction** is initiated by releasing hormones secreted by the hypothalamus. Resistance reactions are longer lasting and accelerate breakdown reactions to provide ATP for counteracting stress.

5. **Exhaustion** results from depletion of body resources during the resistance stage.

6. Stress may trigger certain diseases by inhibiting the immune system.

Animation—General Adaptation Syndrome
Concepts and Connections—General Adaptation Syndrome

13.12 Aging and the Endocrine System

1. Although some endocrine glands shrink as we get older, their performance may or may not be compromised.

2. Production of human growth hormone, thyroid hormones, cortisol, aldosterone, and estrogens decrease with advancing age.

Homeostatic Imbalance—The Case of the Girl with Fruity Breath

REVIEW

3. With aging, the blood levels of TSH, LH, FSH, and PTH rise.

4. The pancreas releases insulin more slowly with age, and receptor sensitivity to glucose declines.

5. After puberty, thymus size begins to decrease, and thymic tissue is replaced by adipose and areolar connective tissue.

SELF-QUIZ

1. Which of the following is NOT true concerning hormones?
   a. Responses to hormones are generally slower and longer lasting than the responses stimulated by the nervous system.
   b. Hormones are generally controlled by negative feedback systems.
   c. The hypothalamus inhibits the release of some hormones.
   d. Most hormones are released steadily throughout the day.
   e. Hormone secretion is determined by the body's need to maintain homeostasis.

2. Which of the following statements concerning hormone action is NOT true?
   a. Hormones bring about changes in metabolic activities of cells.
   b. Target cells must have receptors for a hormone.
   c. Lipid-soluble hormones may directly enter target cells and affect specific genes.
   d. When a hormone attaches to a membrane receptor, it is termed the first messenger.
   e. ATP is a common second messenger in target cells.

3. Which of the following statements is NOT true?
   a. The secretion of hormones by the anterior pituitary is controlled by hypothalamic releasing hormones.
   b. The pituitary is attached to the hypothalamus by the infundibulum.
   c. Hypophyseal portal veins connect the posterior pituitary to the hypothalamus.
   d. The anterior pituitary constitutes the majority of the pituitary gland.
   e. The posterior pituitary releases hormones produced by neurosecretory cells of the hypothalamus.

4. The hormone that promotes milk release from the mammary glands and that stimulates the uterus to contract is
   a. oxytocin.   b. prolactin.   c. relaxin.
   d. calcitonin.   e. follicle-stimulating hormone.

5. The gland that prepares the body to react to stress by releasing epinephrine is the
   a. posterior pituitary.   b. anterior pituitary.
   c. pineal.   d. adrenal.   e. pancreas.

6. Place the following steps of the feedback loop concerning thyroid gland secretion in the correct order.
   1. TRH released from hypothalamus
   2. increased activity of thyroid follicular cells

3. normal blood levels of thyroid hormones
4. TSH released from anterior pituitary
5. low blood levels of thyroid hormones
   a. 2, 3, 1, 4, 5   b. 5, 1, 4, 2, 3
   c. 5, 4, 1, 2, 3   d. 3, 4, 1, 2, 5
   e. 5, 4, 2, 1, 3

7. A female who is sluggish, gaining weight, and has a low body temperature may be having problems with her
   a. pancreas.   b. parathyroid glands.
   c. adrenal medullae.   d. ovaries.
   e. thyroid gland.

8. Destruction of the alpha cells of the pancreas might result in
   a. hypoglycemia.   b. seasonal affective disorder.
   c. low blood calcium levels.   d. hyperglycemia.
   e. decreased urine output.

9. Which of the following is NOT true concerning human growth hormone (hGH) and insulinlike growth factors?
   a. They stimulate protein synthesis.
   b. They have one primary target tissue in the body.
   c. They stimulate skeletal muscle growth.
   d. hGH levels rise during sleep.
   e. Hypoglycemia can stimulate the release of hGH from the pituitary gland.

10. Follicle-stimulating hormone (FSH) acts on the _____ and luteinizing hormone (LH) acts on the _____.
   a. ovaries, testes
   b. testes, ovaries
   c. ovaries and testes, ovaries and testes
   d. ovaries, mammary glands
   e. ovaries and uterus, testes

11. An injection of adrenocorticotropic hormone (ACTH) would
   a. stimulate the ovaries.
   b. influence thyroid gland activity.
   c. stimulate the release of cortisol.
   d. cause uterine contractions.
   e. decrease urine output.

12. Which of the following is NOT true concerning glucocorticoids?
   a. They help to control electrolyte balance.
   b. They help provide resistance to stress.
   c. They help promote normal metabolism.

    d. They are anti-inflammatory hormones.

    e. They provide the body with energy.

13. Mineralocorticoids

    a. help prevent the loss of potassium from the body.

    b. are regulated by the renin–angiotensin–aldosterone pathway.

    c. increase the rate of sodium loss in the urine.

    d. decrease the body's blood pressure.

    e. increase water loss from the body by increasing urine production.

14. Parathyroid hormone

    a. decreases the loss of calcium in the kidneys.

    b. promotes the formation of calcitriol.

    c. encourages calcium absorption in the gastrointestinal tract.

    d. stimulates the release of calcium and phosphate from bone.

    e. is responsible for all of the above.

15. Which of the following pairs of hormones have opposite effects?

    a. parathyroid hormone, thyroid hormones

    b. parathyroid hormone, calcitonin

    c. oxytocin, glucocorticoids

    d. aldosterone, oxytocin

    e. thyroid hormones, thymosin

16. A hormone that stops the release of FSH is

    a. relaxin.   **b.** progesterone.   **c.** oxytocin.

    d. inhibin.   **e.** insulin.

17. Match the following:

    ____ **a.** produce thyroid hormones   **A.** posterior pituitary

    ____ **b.** secrete insulin   **B.** adrenal cortex

    ____ **c.** release hormones into capillaries of the posterior pituitary   **C.** follicular cells

        D. alpha cells

    ____ **d.** store oxytocin   **E.** parafollicular cells

    ____ **e.** secrete glucagon   **F.** beta cells

    ____ **f.** produce calcitonin   **G.** neurosecretory cells

    ____ **g.** secrete steroid hormones

18. In a dehydrated person, you would expect to see an increased release of

    a. parathyroid hormone.   **b.** aldosterone.

    c. insulin.   **d.** melatonin.   **e.** calcitonin.

19. Match the following:

    ____ **a.** diabetes insipidus   **A.** hypersecretion of glucocorticoids

    ____ **b.** diabetes mellitus

    ____ **c.** myxedema   **B.** hyposecretion of antidiuretic hormone

    ____ **d.** Cushing's syndrome   **C.** hyposecretion of insulin

    ____ **e.** acromegaly   **D.** hyposecretion of parathyroid hormone

    ____ **f.** tetany

    ____ **g.** seasonal affective disorder   **E.** hyposecretion of thyroid hormone

    ____ **h.** Graves disease   **F.** hyposecretion of glucocorticoids

    ____ **i.** Addison's disease   **G.** hypersecretion of thyroid hormones

        H. hypersecretion of hGH

        I. hypersecretion of melatonin

20. For each of the following, indicate at which stage they would occur as part of the stress response. Use **F** to indicate fight-or-flight response, **R** to indicate resistance reaction, and **E** to indicate exhaustion.

    ____ **a.** initiated by hypothalamic releasing hormones

    ____ **b.** initiated by the sympathetic division of the autonomic nervous system

    ____ **c.** immediately prepares the body for action

    ____ **d.** increases cortisol release

    ____ **e.** short-lived response

    ____ **f.** body resources become depleted

    ____ **g.** increased release of many hormones that ensure a continued ATP supply

    ____ **h.** failure of pancreatic beta cells; wasting of muscles

    ____ **i.** nonessential body functions inhibited

CRITICAL THINKING APPLICATIONS

1. Sam was diagnosed with diabetes mellitus aged nine. His 63-year-old aunt was just diagnosed with diabetes also. Sam is having a hard time understanding why he needs injections, while his aunt controls her blood sugar with diet and oral medication. Why is his aunt's treatment different from his?

2. Even though he is fairly physically active, 62-year-old Mark has noticed that his muscles aren't as large as they were in his youth. He heard that there is a "special hormone pill" that can help rebuild his muscles. What is one cause of his muscle loss and what hormone could be in the medication?

3. Melatonin has been suggested as a possible aid for sleeping problems due to jet lag and rotating work schedules (shift work). It may also be involved in seasonal affective disorder (SAD). Explain how melatonin may affect sleeping.

4. Belinda is in a 30-mile charity bike race on a hot summer day. She's breathing dust at the back of the pack, sweating profusely, and now she's lost her water bottle. Belinda is not having a good time. How will her hormones respond to decreased intake of water and the stress of the situation?

ANSWERS TO FIGURE QUESTIONS

13.1 Secretions of endocrine glands diffuse into interstitial fluid and then into the blood; exocrine secretions flow into ducts that lead into body cavities or to the body surface.

13.2 RNA molecules are synthesized when genes are expressed (transcribed), and then mRNA codes for the synthesis of protein molecules.

13.3 It brings the message of the first messenger, the water-soluble hormone, into the cell.

13.4 The posterior pituitary releases hormones synthesized in the hypothalamus.

13.5 Oxytocin's target cells are in the uterus and mammary glands.

13.6 Absorption of a large glass of water in the intestines would decrease the osmotic pressure (concentration of solutes) of your blood plasma, turning off secretion of ADH and decreasing the ADH level in your blood.

13.7 Follicular cells secrete T_3 and T_4; parafollicular cells secrete calcitonin.

13.8 Thyroid hormones increase metabolic rate.

13.9 PTH increases the number and activity of osteoclasts.

13.10 Target tissues for PTH are bone and kidneys; the target tissue for calcitonin is bone; the target tissue for calcitriol is the gastrointestinal (GI) tract.

13.11 The pancreas is both an endocrine gland and an exocrine gland.

13.12 Glucagon is considered an "anti-insulin" hormone because it has several effects that are opposite to those of insulin.

13.13 The outer zone of the adrenal cortex secretes mineralocorticoids, the middle zone secretes glucocorticoids, and the inner zone secretes adrenal androgens.

13.14 Because drugs that block ACE lower blood pressure, they are used to treat high blood pressure (hypertension).

13.15 In Graves disease, antibodies are produced that mimic the action of TSH.

CHAPTER 14

THE CARDIOVASCULAR SYSTEM: BLOOD

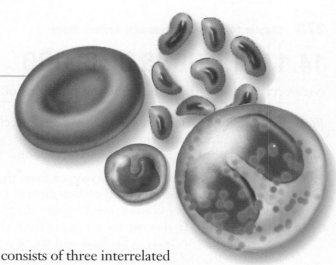

The *cardiovascular system* (*cardio-* = heart; *-vascular* = blood vessels) consists of three interrelated components: blood, the heart, and blood vessels. The focus of this chapter is blood; the next two chapters will cover the heart and blood vessels, respectively.

Functionally, the cardiovascular system transports substances to and from body cells. To perform its functions, blood must circulate throughout the body. The heart serves as the pump for circulation, and blood vessels carry blood from the heart to body cells and from body cells back to the heart.

The branch of science concerned with the study of blood, blood-forming tissues, and the disorders associated with them is *hematology* (hēm-a-TOL-ō-jē; *hemo-* or *hemato-* = blood; *-logy* = study of).

Did you know? Red blood cells carry oxygen, which is vital for the energy production processes that fuel all of the body's energy needs. No oxygen, no energy, no life. Athletes use a great deal of energy to meet the increased demands of muscular contractions on the body. Imagine the emotional devastation that endurance athletes experience from a diagnosis of iron-deficiency anemia (low levels of red blood cells or hemoglobin, the oxygen-carrying molecule). While most cases of iron-deficiency anemia can be treated with iron supplements, recovery to normal iron levels can take weeks, or even months, enough time to cause the athlete to miss an entire sports season.

FOCUS ON WELLNESS
Athletes and Iron-deficiency Anemia

LOOKING BACK TO MOVE AHEAD...

Blood Tissue (Section 4.3)

Positive Feedback System (Section 1.4)

Phagocytosis (Section 3.3)

14.1 FUNCTIONS OF BLOOD

OBJECTIVE • **List and describe the functions of blood.**

Blood is a liquid connective tissue that consists of cells surrounded by extracellular matrix. Blood has three general functions: transportation, regulation, and protection.

1. **Transportation**. Blood transports oxygen from the lungs to cells throughout the body and carbon dioxide (a waste product of cellular respiration; see Chapter 20) from the cells to the lungs. It also carries nutrients from the gastrointestinal tract to body cells, heat and waste products away from cells, and hormones from endocrine glands to other body cells.

2. **Regulation**. Blood helps regulate the pH of body fluids. The heat-absorbing and coolant properties of the water in blood plasma (see Section 2.2) and its variable rate of flow through the skin help adjust body temperature. Blood osmotic pressure also influences the water content of cells.

3. **Protection**. Blood clots (becomes gel-like) in response to an injury, which protects against its excessive loss from the cardiovascular system. In addition, white blood cells protect against disease by carrying on phagocytosis and producing proteins called antibodies. Blood contains additional proteins, called interferons and complement, that also help protect against disease.

✓ **CHECKPOINT**

1. Name several substances transported by blood.
2. How is blood protective?

14.2 COMPONENTS OF WHOLE BLOOD

OBJECTIVE • **Discuss the formation, components, and functions of whole blood.**

Blood is denser and more viscous (thicker) than water. The temperature of blood is about 38°C (100.4°F). Its pH is slightly alkaline, ranging from 7.35 to 7.45. Blood constitutes about 8 percent of the total body weight. The blood volume is 5 to 6 liters (1.5 gal) in an average-sized adult male and 4 to 5 liters (1.2 gal) in an average-sized adult female. The difference in volume is due to differences in body size.

Whole blood is composed of two portions: (1) *blood plasma*, a liquid extracellular matrix that contains dissolved substances, and (2) *formed elements*, which are cells and cell fragments. If a sample of blood is centrifuged (spun at high speed) in a small glass tube, the cells (which are more dense) sink to the bottom of the tube and the lighter-weight blood plasma (which is less dense) forms a layer on top (Figure 14.1a). Blood is about 45 percent formed elements and 55 percent plasma. Normally, more than 99 percent of the formed elements are red blood cells (RBCs) since they are the most dense. The percentage of total blood volume occupied by red blood cells is termed the *hematocrit* (he-MAT-ō-krit). Pale, colorless white blood cells (WBCs) and platelets occupy less than 1 percent of total blood volume. They form a very thin layer, called the *buffy coat*, between the packed RBCs and blood plasma in centrifuged blood. Figure 14.1b shows the composition of blood plasma and the numbers of the various types of formed elements in blood.

Blood Plasma

When the formed elements are removed from blood, a straw-colored liquid called *blood plasma* (or simply *plasma*) remains. Plasma is about 91.5 percent water, 7 percent proteins, and 1.5 percent solutes other than proteins. Proteins in the blood, the *plasma proteins*, are synthesized mainly by the liver. The most plentiful plasma proteins are the *albumins*, which account for about 54 percent of all plasma proteins. Among other functions, albumins help maintain proper blood osmotic pressure, which is an important factor in the exchange of fluids across capillary walls. *Globulins*, which compose 38 percent of plasma proteins, include *antibodies*, defensive proteins produced during certain immune responses. *Fibrinogen* makes up about 7 percent of plasma proteins and is a key protein in formation of blood clots. Other solutes in plasma include electrolytes, nutrients, gases, regulatory substances such as enzymes and hormones, vitamins, and waste products.

Formed Elements

The *formed elements* of the blood are the following (see Figure 14.2):

I. Red blood cells (erythrocytes)

II. White blood cells (leukocytes)

   A. Granular leukocytes (contain conspicuous granules that are visible under a light microscope after staining)

     1. Neutrophils

     2. Eosinophils

     3. Basophils

   B. Agranular leukocytes (no granules are visible under a light microscope after staining)

     1. T and B lymphocytes and natural killer cells

     2. Monocytes

III. Platelets

Figure 14.1 Components of blood in a normal adult.

Blood is a connective tissue that consists of blood plasma (liquid) plus formed elements: red blood cells, white blood cells, and platelets.

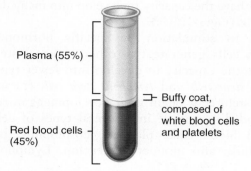

Plasma (55%)

Buffy coat, composed of white blood cells and platelets

Red blood cells (45%)

(a) Appearance of centrifuged blood

Functions of Blood

1. Transports oxygen, carbon dioxide, nutrients, hormones, heat, and wastes.
2. Regulates pH, body temperature, and water content of cells.
3. Protects against blood loss through clotting, and against disease through phagocytic white blood cells and proteins such as antibodies, interferons, and complement.

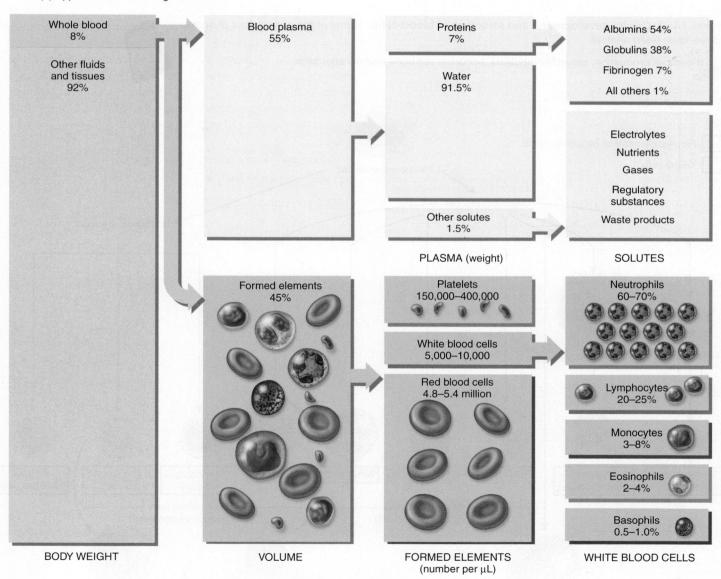

Whole blood 8%

Other fluids and tissues 92%

BODY WEIGHT

Blood plasma 55%

Formed elements 45%

VOLUME

Proteins 7%

Water 91.5%

Other solutes 1.5%

PLASMA (weight)

Platelets 150,000–400,000

White blood cells 5,000–10,000

Red blood cells 4.8–5.4 million

FORMED ELEMENTS (number per μL)

Albumins 54%

Globulins 38%

Fibrinogen 7%

All others 1%

Electrolytes

Nutrients

Gases

Regulatory substances

Waste products

SOLUTES

Neutrophils 60–70%

Lymphocytes 20–25%

Monocytes 3–8%

Eosinophils 2–4%

Basophils 0.5–1.0%

WHITE BLOOD CELLS

(b) Components of blood

 Which formed elements of blood are most numerous?

Formation of Blood Cells

The process by which the formed elements of blood develop is called *hemopoiesis* (hē-mō-poy-Ē-sis; *-poiesis* = making), also called *hematopoiesis*. Before birth, hemopoiesis first occurs in the yolk sac of an embryo and later in the liver, spleen, thymus, and lymph nodes of a fetus. In the last three months before birth, red bone marrow becomes the primary site of hemopoiesis and continues as the source of blood cells after birth and throughout life.

Red bone marrow is a highly vascularized connective tissue located in the microscopic spaces between trabeculae of spongy bone tissue. It is present chiefly in bones of the axial skeleton, pectoral and pelvic girdles, and the proximal epiphyses of the humerus and femur. About 0.05–0.1 percent of red bone marrow cells are cells called *pluripotent stem cells* (ploo-RIP-ō-tent; *pluri-* = several). Pluripotent stem cells are cells that have the capacity to develop into many different types of cells (Figure 14.2a).

In response to stimulation by specific hormones, pluripotent stem cells generate two other types of stem cells, which have the capacity to develop into fewer types of cells: *myeloid stem cells* and *lymphoid stem cells* (Figure 14.2a). Myeloid stem cells begin their development in red bone marrow and differentiate into several types of cells from which red blood cells, platelets, eosinophils, basophils, neutrophils, and monocytes develop. Lymphoid

Figure 14.2 Origin, development, and structure of blood cells. Some of the generations of some cell lines have been omitted.

Blood cell production, called hemopoiesis, occurs in red bone marrow after birth.

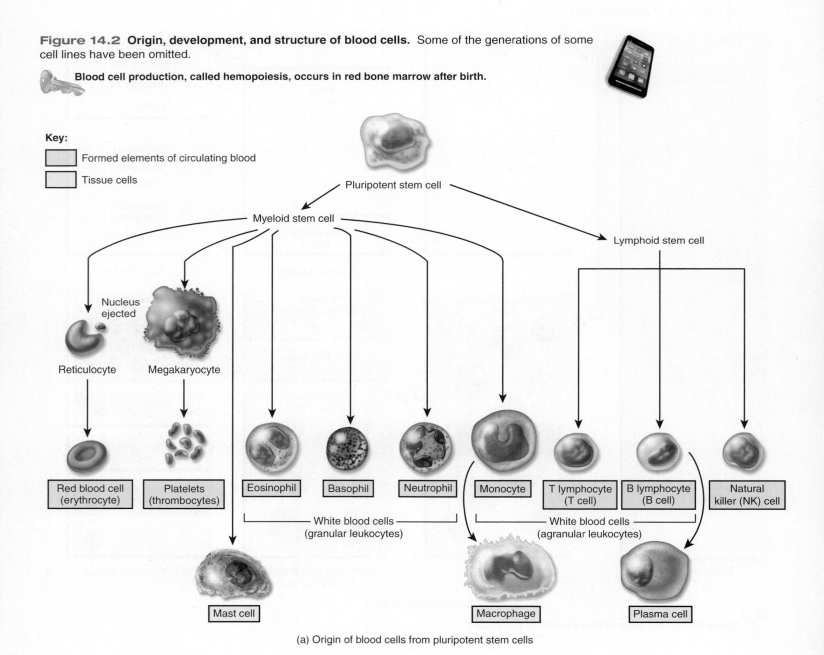

Key:

Formed elements of circulating blood

Tissue cells

Pluripotent stem cell

Myeloid stem cell

Lymphoid stem cell

Nucleus ejected

Reticulocyte

Megakaryocyte

Red blood cell (erythrocyte)

Platelets (thrombocytes)

Eosinophil

Basophil

Neutrophil

Monocyte

T lymphocyte (T cell)

B lymphocyte (B cell)

Natural killer (NK) cell

White blood cells (granular leukocytes)

White blood cells (agranular leukocytes)

Mast cell

Macrophage

Plasma cell

(a) Origin of blood cells from pluripotent stem cells

stem cells begin their development in red bone marrow but complete it in lymphatic tissues. They differentiate into cells from which the T and B lymphocytes and natural killer (NK) cells develop.

Red Blood Cells

Red blood cells (RBCs) or **erythrocytes** (e-RITH-rō-sīts; *erythro-* = red; *-cyte* = cell) contain the oxygen-carrying pro-

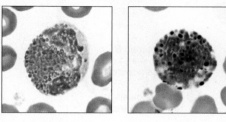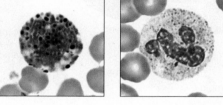

| Eosinophil | Basophil | Neutrophil |

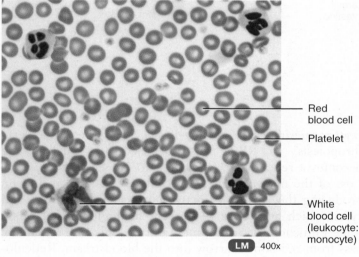

Red blood cell
Platelet

White blood cell (leukocyte: monocyte)

LM 400x

Blood Smear

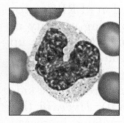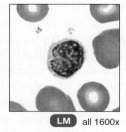

LM all 1600x

Monocyte
Lymphocyte

(b) Photomicrographs

? **What percentage of body weight is made up of blood?**

tein **hemoglobin**, which is a pigment that gives whole blood its red color. Hemoglobin also transports about 23 percent of the carbon dioxide in the blood. A healthy adult male has about 5.4 million red blood cells per microliter (μL) of blood, and a healthy adult female has about 4.8 million. (One drop of blood is about 50μL.) Again this difference reflects differences in body size. To maintain normal numbers of RBCs, new mature cells must enter the circulation at the astonishing rate of at least 2 million per second, a pace that balances the equally high rate of RBC destruction. RBCs are biconcave (concave on both sides) discs averaging about 8μm* in diameter. Mature RBCs lack a nucleus and other organelles and can neither reproduce nor carry on extensive metabolic activities. However, all of their internal space is available for oxygen and carbon dioxide transport. Essentially, RBCs consist of a selectively permeable plasma membrane, cytosol, and hemoglobin.

Since a biconcave disc has a much greater surface area for its volume (compared to a sphere or a cube), this shape provides a large surface area for the diffusion of gas molecules into and out of a RBC.

RBC Life Cycle

Red blood cells live only about 120 days because of wear and tear on their plasma membranes as they squeeze through blood capillaries. Worn-out red blood cells are removed from circulation as follows (Figure 14.3):

1 Macrophages in the spleen, liver, and red bone marrow phagocytize ruptured and worn-out red blood cells, splitting apart the heme and globin portions of hemoglobin.

2 The protein globin is broken down into amino acids, which can be reused by body cells to synthesize other proteins.

3 Iron removed from the heme portion associates with the plasma protein **transferrin** (trans-FER-in; *trans-* = across; *ferr-* = iron), which acts as a transporter.

4 The iron–transferrin complex is then carried to red bone marrow, where RBC precursor cells use it in hemoglobin synthesis. Iron is needed for the heme portion of the hemoglobin molecule, and amino acids are needed for the globin portion. Vitamin B_{12} is also needed for synthesis of hemoglobin. (The lining of the stomach must produce a protein called *intrinsic factor* for absorption of dietary vitamin B_{12} from the GI tract into the blood.)

*1μm = 1/25,000 of an inch or 1/10,000 of a centimeter (cm), which is 1/1000 of a millimeter (mm).

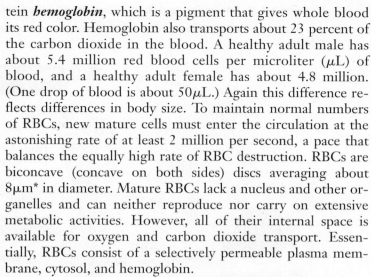
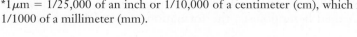

Figure 14.3 Formation and destruction of red blood cells, and the recycling of hemoglobin components.

The rate of RBC formation by red bone marrow equals the rate of RBC destruction by macrophages.

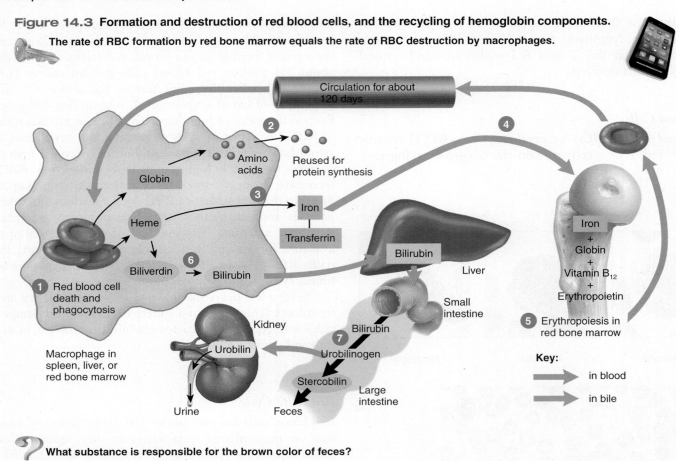

What substance is responsible for the brown color of feces?

⑤ Erythropoiesis in red bone marrow results in the production of red blood cells, which enter the circulation.

⑥ When iron is removed from heme, the non-iron portion of heme is converted to **biliverdin** (bil′-i-VER-din), a green pigment, and then into **bilirubin** (bil′-i-ROO-bin), a yellow-orange pigment. Bilirubin enters the blood and is transported to the liver. Within the liver, bilirubin is secreted by liver cells into bile, which passes into the small intestine and then into the large intestine.

⑦ In the large intestine, bacteria convert bilirubin into **urobilinogen** (ūr-ō-bī-LIN-ō-jen). Some urobilinogen is absorbed back into the blood, converted to a yellow pigment called **urobilin** (ūr-ō-BĪ-lin), and excreted in urine. Most urobilinogen is eliminated in feces in the form of a brown pigment called **stercobilin** (ster′-kō-BĪ-lin), which gives feces its characteristic color.

Because free iron ions bind to and damage molecules in cells or in the blood, transferrin acts as a protective "protein escort" during transport of iron ions. As a result, plasma contains virtually no free iron.

RBC PRODUCTION   The formation of blood cells in general is called hemopoiesis; the formation of just RBCs is termed

erythropoiesis (e-rith′-rō-poy-Ē-sis). Near the end of erythropoiesis, an RBC precursor ejects its nucleus and becomes a reticulocyte (re-TIK-ū-lō-sīt; see Figure 14.2a). Loss of the nucleus causes the center of the cell to indent, producing the RBC's distinctive biconcave shape. Reticulocytes, which are about 34 percent hemoglobin and retain some mitochondria, ribosomes, and endoplasmic reticulum, pass from red bone marrow into the bloodstream. Reticulocytes usually develop into mature RBCs within 1 to 2 days after their release from bone marrow.

Normally, erythropoiesis and destruction of RBCs proceed at the same pace. If the oxygen-carrying capacity of the blood falls because erythropoiesis is not keeping up with RBC destruction, RBC production increases (Figure 14.4). The controlled condition in this particular negative feedback loop is the amount of oxygen delivered to the kidneys (and thus to body tissues in general). **Hypoxia** (hī-POKS-ē-a), a deficiency of oxygen, stimulates increased release of **erythropoietin** (**EPO**) (e-rith′-rō-POY-ē-tin), a hormone made by the kidneys. EPO circulates through the blood to the red bone marrow, where it stimulates erythropoiesis. The larger the number of RBCs in the blood, the higher the oxygen delivery to the tissues (Figure 14.4). A

Figure 14.4 Negative feedback regulation of erythropoiesis (red blood cell formation).

The main stimulus for erythropoiesis is hypoxia, a decrease in the oxygen-carrying capacity of the blood.

Some stimulus disrupts homeostasis by

Decreasing

Oxygen delivery to kidneys (and other tissues)

Receptors

Kidney cells detect low oxygen level

Input — Increased erythropoietin secreted into blood

Control center

Proerythroblasts in red bone marrow mature more quickly into reticulocytes

Return to homeostasis when oxygen delivery to kidneys increases to normal

Output — More reticulocytes enter circulating blood

Effectors

Larger number of RBCs in circulation

Increased oxygen delivery to tissues

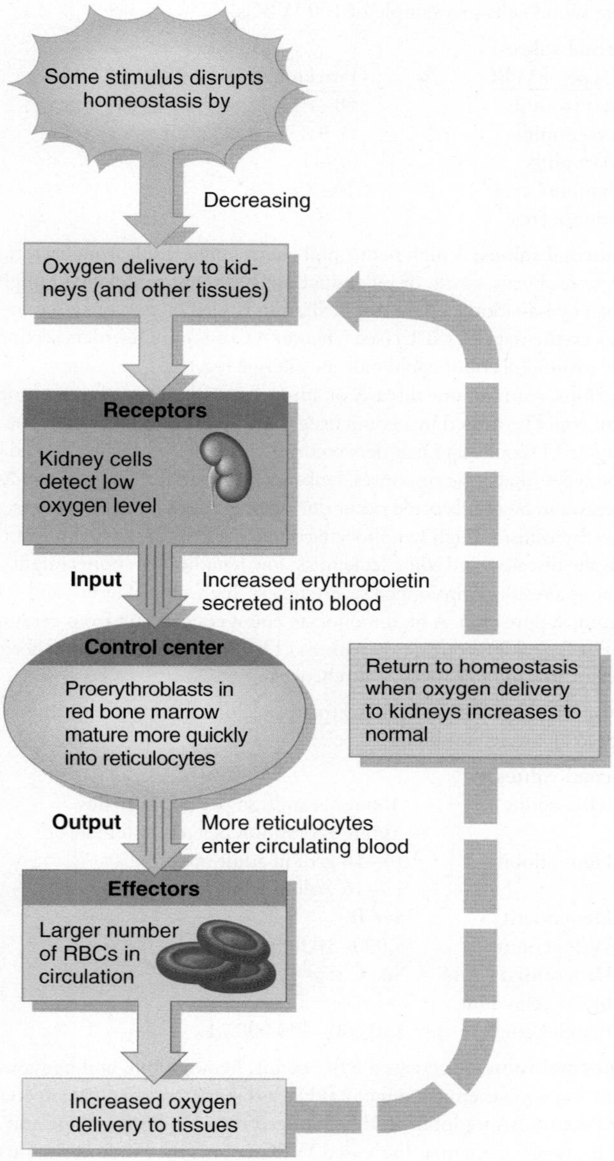

What is the term for cellular oxygen deficiency?

CLINICAL CONNECTION | Induced Polycythemia

Delivery of oxygen to muscles is a limiting factor in muscular feats. As a result, increasing the oxygen-carrying capacity of the blood enhances athletic performance, especially in endurance events. Because RBCs are the main transport vehicle for oxygen, athletes have tried several means of increasing their RBC count, causing **induced polycythemia**, to gain a competitive edge. Athletes have enhanced their RBC production by injecting *Epoetin alfa* (*Procrit*® or *Epogen*®), a synthetic form of erythropoietin that is used to treat anemia by stimulating the production of RBCs by red bone marrow. Practices that increase the number of RBCs are dangerous because they raise the viscosity of the blood, which increases the resistance to blood flow and makes the blood more difficult for the heart to pump. Increased viscosity also contributes to high blood pressure and increased risk of stroke. During the 1980s, at least 15 competitive cyclists died from heart attacks or strokes linked to suspected use of Epoetin alfa. Although the International Olympics Committee bans Epoetin alfa use, enforcement is difficult because the drug is identical to naturally occurring EPO. •

A test that measures the rate of erythropoiesis is called a reticulocyte count. This and several other tests related to red blood cells are explained in Table 14.1.

White Blood Cells

WBC STRUCTURE AND TYPES Unlike red blood cells, **white blood cells (WBCs)** or **leukocytes** (LOO-kō-sīts; *leuko-* = white) have nuclei and a full complement of other organelles but they do not contain hemoglobin. WBCs are classified as either granular or agranular, depending on whether they contain chemical-filled cytoplasmic granules (vesicles) that are made visible by staining when viewed through a light microscope (see Figure 14.2b). The *granular leukocytes* include **neutrophils** (NOO-trō-fils), **eosinophils** (ē′-ō-SIN-ō-fils), and **basophils** (BĀ-sō-fils). The *agranular leukocytes* include **lymphocytes** and **monocytes** (MON-ō-sīts). (See Table 14.2 for the sizes and microscopic characteristics of WBCs.)

WBC FUNCTIONS The skin and mucous membranes of the body are continuously exposed to microbes (microscopic organisms), such as bacteria, some of which are capable of invading deeper tissues and causing disease. Once microbes enter the body, some WBCs combat them by **phagocytosis**, and others produce antibodies. Neutrophils respond first to bacterial invasion, carrying on phagocytosis and releasing enzymes such as lysozyme that destroy certain bacteria. Monocytes take longer to reach the site of infection than neutrophils, but they eventually arrive in larger numbers. Monocytes that migrate into infected tissues develop into cells called **wandering macrophages** (*macro-* = large; *-phages* = eaters), which can phagocytize many more microbes than neutrophils. They also clean up cellular debris following an infection.

Eosinophils leave the capillaries and enter interstitial fluid. They release enzymes that combat inflammation in

person with prolonged hypoxia may develop a life-threatening condition called **cyanosis** (sī′-a-NŌ-sis), characterized by a bluish-purple skin coloration most easily seen in the nails and mucous membranes. Oxygen delivery may fall due to **anemia** (a lower-than-normal number of RBCs or reduced quantity of hemoglobin) or circulatory problems that reduce blood flow to tissues.

TABLE 14.1

Obtaining Blood Samples and Common Medical Tests Involving Blood

I. Obtaining Blood Samples

A. Venipuncture. This most frequently used procedure involves withdrawal of blood from a vein using a sterile needle and syringe. (Veins are used instead of arteries because they are closer to the skin, more readily accessible, and contain blood at a much lower pressure.) A commonly used vein is the median cubital vein in front of the elbow (see Figure 16.14b). A tourniquet is wrapped around the arm, which stops blood flow through the veins and makes the veins below the tourniquet stand out.

B. Fingerstick. Using a sterile needle or lancet, a drop or two of capillary blood is taken from a finger, earlobe, or heel.

C. Arterial stick. Sample is most often taken from radial artery in the wrist or femoral artery in the thigh (see Figure 16.9).

II. Testing Blood Samples

A. Reticulocyte count (indicates the rate of erythropoiesis)

Normal value: 0.5% to 1.5%.

Abnormal values: A high reticulocyte count might indicate the presence of bleeding or hemolysis (rupture of erythrocytes), or it may be the response of someone who is iron deficient. Low reticulocyte count in the presence of anemia might indicate a malfunction of the red bone marrow, owing to a nutritional deficiency, pernicious anemia, or leukemia.

B. Hematocrit (the percentage of red blood cells in blood). A hematocrit of 40 means that 40% of the volume of blood is composed of RBCs.

Normal values:
  Females: 38 to 46 (average 42)
  Males: 40 to 54 (average 47)

Abnormal values: The test is used to diagnose anemia, polycythemia (an increased percentage of red blood cells above 55), and abnormal states of hydration. Anemia may vary from mild (hematocrit of 35) to severe (hematocrit of less than 15). Athletes often have a higher-than-average hematocrit, and the average hematocrit of persons living at high altitude is greater than that of persons living at sea level.

C. Differential white blood cell count (the percentage of each type of white blood cells in a sample of 100 WBCs)

Normal values:

Type of WBC	Percentage
neutrophils	60–70
eosinophils	2–4
basophils	0.5–1
lymphocytes	20–25
monocytes	3–8

Abnormal values: A high neutrophil count might result from bacterial infections, burns, stress, or inflammation; a low neutrophil count might be caused by radiation, certain drugs, vitamin B_{12} deficiency, or systemic lupus erythematosus (SLE) (see Chapter 4 Common Disorders section). A high eosinophil count could indicate allergic reactions, parasitic infections, autoimmune disease, or adrenal insufficiency; a low eosinophil count could be caused by certain drugs, stress, or Cushing's syndrome (see Chapter 13 Common Disorders section). Basophils could be elevated in some types of allergic responses, leukemias, cancers, and hypothyroidism; decreases in basophils could occur during pregnancy, ovulation, stress, and hyperthyroidism. High lymphocyte counts could indicate viral infections, immune diseases, and some leukemias; low lymphocyte counts might occur as a result of prolonged severe illness, high steroid levels, and immunosuppression. A high monocyte count could result from certain viral or fungal infections, tuberculosis (TB), some leukemias, and chronic diseases; low monocyte levels rarely occur.

D. Complete blood count (CBC) (provides information about the formed elements in blood)*

Normal values:

RBC count	About 5.4 million per μL in males
	About 4.8 million per μL in females
Hemoglobin	14–18 g/dl in adult males
	12–16 g/dl in adult females
Hematocrit	See **B**
WBC count	5,000–10,000 per μL
Differential white blood cell count	See **C**
Platelet count	150,000–400,000 μL

Abnormal values: Increased RBC count, hemoglobin, and hematocrit occur in polycythemia, congenital heart disease, and hypoxia; decreased RBC count, hemoglobin, and hematocrit occur in hemorrhage and certain types of anemia. Increased WBC counts may indicate acute or chronic infections, trauma, leukemia, or stress (see also **C**). Decreased WBC counts could indicate anemia and viral infections (see also **C**). High platelet counts may indicate cancer, trauma, or cirrhosis. Low platelet counts could indicate anemia, allergic conditions, or hemorrhage.

*Not all components of a CBC have been included.

allergic reactions. Eosinophils also phagocytize antigen–antibody complexes and are effective against certain parasitic worms. A high eosinophil count often indicates an allergic condition or a parasitic infection.

Basophils are also involved in inflammatory and allergic reactions. They leave capillaries, enter tissues, and can liber-ate heparin, histamine, and serotonin. These substances intensify the inflammatory reaction and are involved in allergic reactions.

Three types of lymphocytes—B cells, T cells, and natural killer (NK) cells—are the major combatants in immune responses, which are described in detail in Chapter 17.

B cells develop into plasma cells, which produce antibodies that help destroy bacteria and inactivate their toxins. T cells attack viruses, fungi, transplanted cells, cancer cells, and some bacteria. Natural killer cells attack a wide variety of infectious microbes and certain spontaneously arising tumor cells.

White blood cells and other nucleated body cells have proteins, called *major histocompatibility (MHC) antigens*, protruding from their plasma membrane into the extracellular fluid. These "cell identity markers" are unique for each person (except identical twins). Although RBCs (which do not possess nuclei) possess blood group antigens, they lack the MHC antigens. An incompatible tissue transplant is rejected by the recipient due, in part, to differences in donor and recipient MHC antigens. The MHC antigens are used to type tissues to identify compatible donors and recipients and thus reduce the chance of tissue rejection.

WBC Life Span   Red blood cells outnumber white blood cells about 700 to 1. There are normally about 5000 to 10,000 WBCs per μL of blood. Bacteria have continuous access to the body through the mouth, nose, and pores of the skin. Furthermore, many cells, especially those of epithelial tissue, age and die daily, and their remains must be removed. However, a WBC can phagocytize only a certain amount of material before it interferes with the WBC's own metabolic activities. Thus, the life span of most WBCs is only a few days. During a period of infection, many WBCs live only a few hours. However, some B and T cells remain in the body for years.

Leukocytosis (loo′-kō-sī-TŌ-sis), an increase in the number of WBCs, is a normal, protective response to stresses such as invading microbes, strenuous exercise, anesthesia, and surgery. Leukocytosis usually indicates an inflammation or infection. Because each type of white blood cell plays a different role, determining the percentage of each type in the blood assists in diagnosing the condition. This test, called a *differential white blood cell count*, measures the number of each kind of white cell in a sample of 100 white blood cells (see Table 14.1). An abnormally low level of white blood cells (below 5000 cells/μL), called *leukopenia* (loo′-kō-PĒ-nē-a), is never beneficial; it may be caused by exposure to radiation, shock, and certain chemotherapeutic agents.

WBC Production   Leukocytes develop in red bone marrow. As shown in Figure 14.2a, monocytes and granular leukocytes develop from a myeloid stem cell. T cells, B cells, and NK cells develop from a lymphoid stem cell.

Platelets

Pluripotent stem cells also differentiate into cells that produce platelets (see Figure 14.2a). Some myeloid stem cells develop into cells called *megakaryoblasts*, which in turn transform into megakaryocytes, huge cells that splinter into 2000–3000 fragments in the red bone marrow and then enter the bloodstream. Each fragment, enclosed by a piece of the megakaryocyte cell membrane, is a *platelet*. Between 150,000 and 400,000 platelets are present in each μL of blood. Platelets are disc-shaped, have a diameter of 2–4 μm, and exhibit many vesicles but no nucleus. When blood vessels are damaged, platelets help stop blood loss by forming a platelet plug. Their vesicles also contain chemicals that promote blood clotting. (Both processes are described shortly.) After their short life span of five to nine days, platelets are removed by macrophages in the spleen and liver.

CLINICAL CONNECTION | Bone Marrow Transplant

A **bone marrow transplant** is the replacement of cancerous or abnormal red bone marrow with healthy red bone marrow in order to establish normal blood cell counts. The defective red bone marrow is destroyed by high doses of chemotherapy and whole body radiation just before the transplant takes place. These treatments kill the cancer cells and destroy the patient's immune system in order to decrease the chance of transplant rejection. The red bone marrow from a donor is usually removed from the hip bone under general anesthesia with a syringe and is then injected into the recipient's vein, much like a blood transfusion. The injected marrow migrates to the recipient's red bone marrow cavities, and the stem cells in the marrow multiply. If all goes well, the recipient's red bone marrow is replaced entirely by healthy, noncancerous cells.

Bone marrow transplants have been used to treat aplastic anemia, certain types of leukemia, severe combined immunodeficiency disease (SCID), Hodgkin's disease, non-Hodgkin's lymphoma, multiple myeloma, thalassemia, sickle-cell disease, breast cancer, ovarian cancer, testicular cancer, and hemolytic anemia. However, there are some drawbacks. Since the recipient's white blood cells have been completely destroyed by chemotherapy and radiation, the patient is extremely vulnerable to infection. (It takes about 2–3 weeks for transplanted red bone marrow to produce enough white blood cells to protect against infection.) In addition, transplanted red bone marrow may produce T lymphocytes that attack the recipient's tissues. Another drawback is that patients must take immunosuppressive drugs for life. Because these drugs reduce the level of immune system activity, they increase the risk of infection. •

Table 14.2 presents a summary of the formed elements in blood.

✓ CHECKPOINT

3. Briefly outline the process of hemopoiesis.

4. What is erythropoiesis? How does erythropoiesis affect hematocrit? What factors speed up and slow down erythropoiesis?

5. What functions do neutrophils, eosinophils, basophils, monocytes, B cells, T cells, and natural killer cells perform?

6. How are leukocytosis and leukopenia different? What is a differential white blood cell count?

TABLE 14.2

Summary of Formed Elements in Blood

NAME AND APPEARANCE	NUMBER	CHARACTERISTICS*	FUNCTIONS
Red Blood Cells (RBCs) or **Erythrocytes**	4.8 million/μL in females; 5.4 million/μL in males	7–8 μm diameter, biconcave discs, without nuclei; live for about 120 days	Contain hemoglobin, which transports most of the oxygen and some of the carbon dioxide in the blood
White Blood Cells (WBCs) or **Leukocytes**	5000–10,000/μL	Most live for a few hours to a few days[†]	Combat pathogens and other foreign substances that enter the body
Granular Leukocytes			
Neutrophils	60–70% of all WBCs	10–12 μm diameter; nucleus has 2–5 lobes connected by thin strands of chromatin; cytoplasm has very fine, pale lilac granules	Phagocytosis; destruction of bacteria with lysozyme, defensins, and strong oxidants, such as superoxide anion, hydrogen peroxide, and hypochlorite anion
Eosinophils	2–4% of all WBCs	10–12 μm diameter; nucleus usually has 2 lobes connected by a thick strand of chromatin; large, red-orange granules fill the cytoplasm	Combat the effects of histamine in allergic reactions, phagocytize antigen–antibody complexes, and destroy certain parasitic worms
Basophils	0.5–1% of all WBCs	8–10 μm diameter; nucleus has 2 lobes; large cytoplasmic granules appear deep blue-purple	Liberate heparin, histamine, and serotonin in allergic reactions that intensify the overall inflammatory response
Agranular Leukocytes			
Lymphocytes (T cells, B cells, and natural killer cells)	20–25% of all WBCs	Small lymphocytes are 6–9 μm in diameter; large lymphocytes are 10–14 μm in diameter; nucleus is round or slightly indented; cytoplasm forms a rim around the nucleus that looks sky blue; the larger the cell, the more cytoplasm is visible	Mediate immune responses, including antigen–antibody reactions. B cells develop into plasma cells, which secrete antibodies. T cells attack invading viruses, cancer cells, and transplanted tissue cells. Natural killer cells attack a wide variety of infectious microbes and certain spontaneously arising tumor cells
Monocytes	3–8% of all WBCs	12–20 μm diameter; nucleus is kidney shaped or horseshoe shaped; cytoplasm is blue-gray and has foamy appearance	Phagocytosis (after transforming into fixed or wandering macrophages)
Platelets (Thrombocytes)	150,000–400,000/μL	2–4 μm diameter cell fragments that live for 5–9 days; contain many vesicles but no nucleus	Form platelet plug in hemostasis; release chemicals that promote vascular spasm and blood clotting

*Colors are those seen when using Wright's stain.
[†]Some lymphocytes, called T and B memory cells, can live for many years once they are established.

14.3 HEMOSTASIS

OBJECTIVE • **Describe the various mechanisms that prevent blood loss.**

Hemostasis (hē′-mō-STĀ-sis; -*stasis* = standing still) is a sequence of responses that stops bleeding when blood vessels are injured. (Be sure not to confuse the two words *hemostasis* and *homeostasis*.) The hemostatic response must be quick, localized to the region of damage, and carefully controlled. Three mechanisms can reduce loss of blood from blood vessels: (1) vascular spasm, (2) platelet plug formation, and (3) blood clotting (coagulation). When successful, hemostasis prevents *hemorrhage* (HEM-o-rij; -*rhage* = burst forth), the loss of a large amount of blood from the vessels. Hemostasis can prevent hemorrhage from smaller blood vessels, but extensive hemorrhage from larger vessels usually requires medical intervention.

Vascular Spasm

When a blood vessel is damaged, the smooth muscle in its wall contracts immediately, a response called a *vascular spasm*. Vascular spasm reduces blood loss for several minutes to several hours, during which time the other hemostatic mechanisms begin to operate. The spasm is probably caused by damage to the smooth muscle and by reflexes initiated by pain receptors. As platelets accumulate at the damaged site, they release chemicals that enhance vasoconstriction (narrowing of a blood vessel), thus maintaining the vascular spasm.

Platelet Plug Formation

When platelets come into contact with parts of a damaged blood vessel, their characteristics change drastically and they quickly come together to form a mass called a *platelet plug* that helps fill the gap in the injured blood vessel wall. Platelet plug formation occurs as follows.

Initially, platelets contact and stick to parts of a damaged blood vessel, such as collagen fibers. Then, they interact with one another and begin to liberate the chemicals. The chemicals activate nearby platelets and sustain the vascular spasm, which decreases blood flow through the injured vessel. The release of platelet chemicals makes other platelets in the area sticky, and the stickiness of the newly recruited and activated platelets causes them to stick to the originally activated platelets. Eventually, a large number of platelets forms a platelet plug, which can stop blood loss completely if the hole in a blood vessel is small enough.

Blood Clotting

Normally, blood remains in its liquid form as long as it stays within its vessels. If it is withdrawn from the body, however, it thickens and forms a gel. Eventually, the gel separates from the liquid. The straw-colored liquid, called *serum*, is simply plasma minus the clotting proteins. The gel is called a *clot* and consists of a network of insoluble protein fibers called *fibrin* in which the formed elements of blood are trapped (see Figure 14.5).

The process of clot formation, called *clotting (coagulation)*, is a series of chemical reactions that culminates in the formation of fibrin threads. If blood clots too easily, the result can be *thrombosis*, clotting in an unbroken blood vessel. If it takes too long to clot, hemorrhage can result.

Clotting is a complex process in which various chemicals known as *clotting factors* activate each other. Clotting (coagulation) factors include calcium ions (Ca^{2+}), several enzymes that are made by liver cells and released into the blood, and various molecules associated with platelets or released by damaged tissues. Many clotting factors are identified by Roman numerals. Clotting occurs in three stages (Figure 14.5):

1 *Prothrombinase* is formed.

2 Prothrombinase converts *prothrombin* (a plasma protein formed by the liver with the help of vitamin K) into the enzyme *thrombin*.

Figure 14.5 Blood clotting.

During blood clotting, the clotting factors activate each other, resulting in a cascade of reactions that includes positive feedback cycles (green arrows).

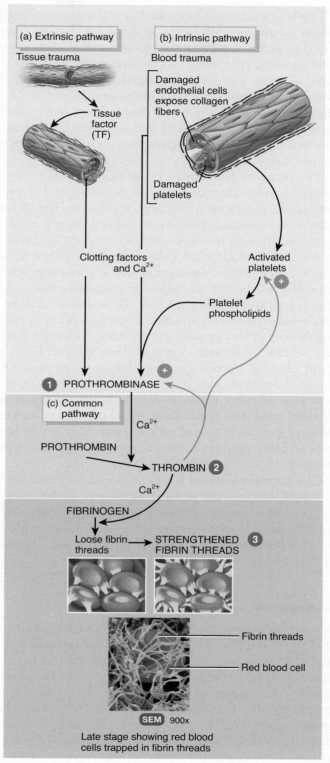

(a) Extrinsic pathway
Tissue trauma

Tissue factor (TF)

(b) Intrinsic pathway
Blood trauma

Damaged endothelial cells expose collagen fibers

Damaged platelets

Clotting factors and Ca^{2+}

Activated platelets

Platelet phospholipids

1 PROTHROMBINASE

(c) Common pathway

Ca^{2+}

PROTHROMBIN

THROMBIN **2**

Ca^{2+}

FIBRINOGEN

Loose fibrin threads → STRENGTHENED FIBRIN THREADS **3**

Fibrin threads

Red blood cell

SEM 900x

Late stage showing red blood cells trapped in fibrin threads

What is the outcome of stage 1 of clotting?

3 Thrombin converts soluble *fibrinogen* (another plasma protein formed by the liver) into insoluble fibrin. Fibrin forms the threads of the clot. (Cigarette smoke contains substances that interfere with fibrin formation.)

Prothrombinase can be formed in two ways, by either the extrinsic or the intrinsic pathway of blood clotting (Figure 14.5). The *extrinsic pathway* of blood clotting occurs rapidly, within seconds. It is so-named because damaged tissue cells release a tissue protein called *tissue factor (TF)* into the blood from *outside* (extrinsic to) blood vessels (Figure 14.5a). Following several additional reactions that require calcium ions (Ca^{2+}) and several clotting factors, tissue factor is eventually converted into prothrombinase. This completes the extrinsic pathway.

The *intrinsic pathway* of blood clotting (Figure 14.5b) is more complex than the extrinsic pathway, and it occurs more slowly, usually requiring several minutes. The intrinsic pathway is so-named because its activators are either in direct contact with blood or contained *within* (intrinsic to) the blood. If endothelial cells lining the blood vessels become roughened or damaged, blood can come in contact with collagen fibers in the adjacent connective tissue. Such contact activates clotting factors. In addition, trauma to endothelial cells activates platelets, causing them to release phospholipids that can also activate certain clotting factors. After several additional reactions that require Ca^{2+} and several clotting factors, prothrombinase is formed. Once formed, thrombin activates more platelets, resulting in the release of more platelet phospholipids, an example of a positive feedback cycle. Both the extrinsic and intrinsic pathways can be activated at the same time since blood vessel and surrounding tissue damage usually occur simultaneously.

Clot formation occurs locally; it does not extend beyond the wound site into the general circulation. One reason for this is that fibrin has the ability to absorb and inactivate up to nearly 90 percent of the thrombin formed from prothrombin. This helps stop the spread of thrombin into the blood and thus inhibits clotting except at the wound.

Clot Retraction and Blood Vessel Repair

Once a clot is formed, it plugs the ruptured area of the blood vessel and thus stops blood loss. *Clot retraction* is the consolidation or tightening of the fibrin clot. The fibrin threads attached to the damaged surfaces of the blood vessel gradually contract as platelets pull on them. As the clot retracts, it pulls the edges of the damaged vessel closer together, decreasing the risk of further damage. Permanent repair of the blood vessel can then take place. In time, fibroblasts form connective tissue in the ruptured area, and new endothelial cells repair the vessel lining.

Hemostatic Control Mechanisms

Many times a day little clots start to form, often at a site of minor roughness inside a blood vessel. Usually, small, inappropriate clots dissolve in a process called *fibrinolysis* (fī′-bri-NOL-i-sis). When a clot is formed, an inactive plasma enzyme called *plasminogen* is incorporated into the clot. Both body tissues and blood contain substances that can activate plasminogen to *plasmin*, an active plasma enzyme. Once plasmin is formed, it can dissolve the clot by digesting fibrin threads. Plasmin also dissolves clots at sites of damage once the damage is repaired. Among the substances that can activate plasminogen are thrombin and tissue plasminogen activator (tPA), which is normally found in many body tissues and is liberated into the blood after a vascular injury.

CLINICAL CONNECTION | Anticoagulant Drug

Patients who are at increased risk of forming blood clots may receive an **anticoagulant drug**, a substance that delays, suppresses, or prevents blood clotting. Examples are heparin or warfarin. *Heparin*, an anticoagulant that is produced by mast cells and basophils, inhibits the conversion of prothrombin to thrombin, thereby preventing blood clot formation. Heparin extracted from animal tissues is often used to prevent clotting during hemodialysis and after open heart surgery. *Coumadin®* *(warfarin sodium)* acts as an antagonist to vitamin K and thus blocks synthesis of several clotting factors. To prevent clotting in donated blood, blood banks and laboratories often add a substance that removes Ca^{2+}, for example, CPD (citrate phosphate dextrose). •

Clotting in Blood Vessels

Despite fibrinolysis and the action of anticoagulants, blood clots sometimes form within blood vessels. The endothelial surfaces of a blood vessel may be roughened as a result of *atherosclerosis* (accumulation of fatty substances on arterial walls), trauma, or infection. These conditions also make the platelets that are attracted to the rough spots more sticky. Clots may also form in blood vessels when blood flows too slowly, allowing clotting factors to accumulate in high enough concentrations to initiate a clot.

Clotting in an unbroken blood vessel is called *thrombosis* (*thromb-* = clot; *-osis* = a condition of). The clot itself, called a *thrombus*, may dissolve spontaneously. If it remains intact, however, the thrombus may become dislodged and be swept away in the blood. A blood clot, bubble of air, fat from broken bones, or a piece of debris transported by the bloodstream is called an *embolus* (*em-* = in; *-bolus* = a mass; plural is *emboli*). Because emboli often form in veins, where blood flow is slower, the most common site for the embolus to become lodged is in the lungs, a condition called *pulmonary embolism*. Massive emboli in the lungs may result in right ventricular failure and death in a few minutes or hours. An embolus that breaks away from an arterial wall may lodge in a smaller-diameter artery downstream. If it blocks blood flow to the brain, kidney, or heart, the embolus can cause a stroke, kidney failure, or heart attack, respectively.

 CLINICAL CONNECTION | Aspirin and Thrombolytic Agents

In patients with heart and blood vessel disease, the events of hemostasis may occur even without external injury to a blood vessel. At low doses, **aspirin** inhibits vasoconstriction and platelet aggregation. It also reduces the chance of thrombus formation. Due to these effects, aspirin reduces the risk of transient ischemic attacks (TIA), strokes, myocardial infarction, and blockage of peripheral arteries.

Thrombolytic agents are chemical substances that are injected into the body to dissolve blood clots that have already formed to restore circulation. They either directly or indirectly activate plasminogen. The first thrombolytic agent, approved in 1982 for dissolving clots in the coronary arteries of the heart, was **streptokinase**, which is produced by streptococcal bacteria. A genetically engineered version of human **tissue plasminogen activator (tPA)** is now used to treat both heart attacks and brain attacks (strokes) that are caused by blood clots. •

✓ **CHECKPOINT**

7. What is hemostasis?

8. How do vascular spasm and platelet plug formation occur?

9. What is fibrinolysis? Why does blood rarely remain clotted inside blood vessels?

14.4 BLOOD GROUPS AND BLOOD TYPES

OBJECTIVE • **Describe the ABO and Rh blood groups.**

The surfaces of red blood cells contain a genetically determined assortment of *antigens* composed of glycolipids and glycoproteins. These antigens, called *agglutinogens* (ag′-loo-TIN-ō-jenz), occur in characteristic combinations. Based on the presence or absence of various antigens, blood is categorized into different *blood groups*. Within a given blood group there may be two or more different *blood types*. There are at least 24 blood groups and more than 100 antigens that can be detected on the surface of red blood cells. Here we discuss two major blood groups: ABO and Rh.

ABO Blood Group

The *ABO blood group* is based on two antigens called *A* and *B* (Figure 14.6). People whose RBCs display only antigen A

Figure 14.6 Antigens and antibodies involved in the ABO blood grouping system.

Your plasma does not contain antibodies that could react with the antigens on your red blood cells.

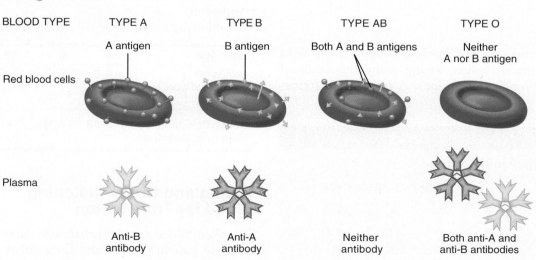

 Which antibodies are found in type O blood?

have type A blood. Those who have only antigen B are type B. Individuals who have both A and B antigens are type AB, and those who have neither antigen A nor B are type O. In about 80 percent of the population, soluble antigens of the ABO type appear in saliva and other body fluids, in which case blood type can be identified from a sample of saliva. The incidence of ABO blood types varies among different population groups, as indicated in Table 14.3.

In addition to antigens on RBCs, blood plasma usually contains **antibodies** or **agglutinins** (a-GLOO-ti-nins) that react with the A or B antigens if the two are mixed. These are the **anti-A antibody**, which reacts with antigen A, and the **anti-B antibody**, which reacts with antigen B. The antibodies present in each of the four ABO blood types are also shown in Figure 14.6. You do not have antibodies that react with your own antigens, but you do have antibodies for any antigens that your RBCs lack. For example, if you have type A blood, it means that you have A antigens on the surfaces of your RBCs, but anti-B antibodies in your blood plasma. If you had anti-A antibodies in your blood plasma, they would attack your RBCs.

Rh Blood Group

The **Rh blood group** is so named because the Rh antigen, called **Rh factor**, was first found in the blood of the rhesus monkey. People whose RBCs have the Rh antigen are designated Rh^+ (Rh positive); those who lack the Rh antigen are designated Rh^- (Rh negative). The percentages of Rh^+ and Rh^- individuals in various populations are shown in Table 14.3. Under normal circumstances, plasma does not contain anti-Rh antibodies. If an Rh^- person receives an Rh^+ blood transfusion, however, the immune system starts to make anti-Rh antibodies that do remain in the blood.

Transfusions

Despite the differences in RBC antigens, blood is the most easily shared of human tissues, saving many thousands of lives every year through transfusions. A **transfusion** (trans-

FŪ-zhun) is the transfer of whole blood or blood components (red blood cells only or plasma only) into the bloodstream. Most often a transfusion is given to alleviate anemia or when blood volume is low, for example, after a severe hemorrhage.

In an incompatible blood transfusion, antibodies in the recipient's plasma bind to the antigens on the donated RBCs. When these antigen–antibody complexes form, they cause hemolysis and release hemoglobin into the plasma. Consider what happens if a person with type A blood receives a transfusion of type B blood. In this situation, two things can happen. First, the anti-B antibodies in the recipient's plasma can bind to the B antigens on the donor's RBCs, causing hemolysis. Second, the anti-A antibodies in the donor's plasma can bind to the A antigens on the recipient's RBCs. The second reaction is usually not serious because the donor's anti-A antibodies become so diluted in the recipient's plasma that they do not cause any significant hemolysis of the recipient's RBCs.

People with type AB blood do not have any anti-A or anti-B antibodies in their plasma. They are sometimes called "universal recipients" because theoretically they can receive blood from donors of all four ABO blood types. People with type O blood have neither A nor B antigens on their RBCs and are sometimes called "universal donors." Theoretically, because there are no antigens on their RBCs for antibodies to attack, they can donate blood to all four ABO blood types. Type O persons requiring blood may receive only type O blood, as they have antibodies to both A and B antigens in their plasma. In practice, use of the terms *universal recipient* and *universal donor* is misleading and dangerous. Blood contains antigens and antibodies other than those associated with the ABO system, and they can cause transfusion problems. Thus, blood should always be carefully matched before transfusion.

Following is a summary of ABO blood group interactions:

Blood type	A	B	AB	O
Compatible donor blood types (no hemolysis)	A, O	B, O	A, B, AB, O	O
Incompatible donor blood types (hemolysis)	B, AB	A, AB	—	A, B, AB

Typing and Cross-matching Blood for Transfusion

To avoid blood-type mismatches, laboratory technicians type the patient's blood and then either cross-match it to potential donor blood or screen it for the presence of antibodies. In the procedure for ABO blood typing, single drops of blood are mixed with different *antisera*, solutions that contain antibodies (Figure 14.7). One drop of blood is mixed with anti-A serum, which contains anti-A antibodies

TABLE 14.3

Blood Types in the United States

| POPULATION GROUP | BLOOD TYPE (PERCENTAGE) | | | | |
	O	A	B	AB	Rh⁺
European American	45	40	11	4	85
African American	49	27	20	4	95
Korean	32	28	30	10	100
Japanese	31	38	21	10	100
Chinese	42	27	25	6	100
Native American	79	16	4	1	100

Athletes and Iron-deficiency Anemia

A male athlete in the prime of good health goes to donate blood, but is rejected as a donor because he is diagnosed with iron-deficiency anemia. This has never happened to him before, and he has given blood on several occasions. What could be the cause? He just began an intense conditioning program to prepare for his upcoming sports season. Did the heavy workouts interfere with his red blood cell production?

Sports Anemia

Probably not. This athlete may have a condition called *sports anemia*. When an athlete suddenly increases training volume, one of the first adaptations that occurs is an increase in blood volume. This increase happens within just two or three days of the start of training, and results from an increase in plasma volume. When blood is tested in these early days of training, the blood has fewer red blood cells per unit volume. As a result, both the hematocrit and the hemoglobin measures appear to be low. But the number of red blood cells has not necessarily declined; the blood is just more dilute, a condition called *hemodilution*.

Sports anemia does not appear to impair performance. It has been proposed that the increased plasma volume and "thinner" blood may even facilitate oxygen delivery to working muscles because the blood can flow more easily. Increased blood flow also improves the ability of the blood to carry heat from the working muscles to the skin, where heat is given off to the environment (except on days that are warmer than body temperature). In any case, with no special treatment red blood cell production accelerates and soon the athlete's blood measures return to normal.

Iron-deficiency Anemia in Athletes

Unlike the athlete described previously, some athletes do develop true iron-deficiency anemia. Multiple causes usually contribute to this condition, some of them dietary (not consuming enough foods high in iron) and some from participation in physical activity. Women of childbearing age are more prone to iron-deficiency anemia than men, since they lose blood during their menstrual cycles.

Sports may accelerate the loss of red blood cells in a few different ways. Athletes who run long distances may rupture red blood cells when the cells are pounded between the heel bone and the ground, a condition called *foot-strike hemolysis*. In addition, tiny amounts of iron may be lost in sweat. Although small, these losses may add up over time if training is demanding and sweating is heavy.

Think It Over . . . **Why do athletes with true iron-deficiency anemia usually see a decline in their aerobic endurance?**

Figure 14.7 ABO blood typing.

 In the procedure for ABO blood typing, blood is mixed with anti-A serum and anti-B serum.

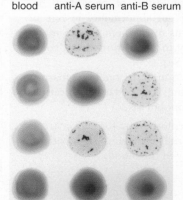

	Untreated blood	Treated with anti-A serum	Treated with anti-B serum	Blood type
				A
				B
				AB
				O

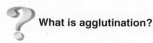

Anti-A serum   Anti-B serum

? What is agglutination?

that will agglutinate (clump together) red blood cells that possess A antigens. Another drop is mixed with anti-B serum, which contains anti-B antibodies that will agglutinate red blood cells that possess B antigens. If the red blood cells agglutinate only when mixed with anti-A serum, the blood is type A. If the red blood cells agglutinate only when mixed with anti-B serum, the blood is type B. The blood is type AB if both drops agglutinate; if neither drop agglutinates, the blood is type O.

✓ CHECKPOINT

10. What is the basis for distinguishing the various blood groups?

11. What precautions must be taken before giving a blood transfusion?

• • •

We will next direct our attention to the heart, the second major component of the cardiovascular system.

COMMON DISORDERS

Anemia

Anemia (a-NĒ-mē-a) is a condition in which the oxygen-carrying capacity of blood is reduced. Many types of anemia exist; all are characterized by reduced numbers of RBCs or a decreased amount of hemoglobin in the blood. The person feels fatigued and is intolerant of cold, both of which are related to lack of oxygen needed for ATP and heat production. Also, the skin appears pale, due to the low content of red-colored hemoglobin circulating in skin blood vessels. Among the most important types of anemia are the following:

• *Iron-deficiency anemia*, the most prevalent kind of anemia, is caused by inadequate absorption of iron, excessive loss of iron, or insufficient intake of iron. Women are at greater risk for iron-deficiency anemia due to monthly menstrual blood loss.

• *Pernicious anemia* is caused by insufficient hemopoiesis resulting from an inability of the stomach to produce intrinsic factor (needed for absorption of dietary vitamin B_{12}).

• *Hemorrhagic anemia* is due to an excessive loss of RBCs through bleeding resulting from large wounds, stomach ulcers, or especially heavy menstruation.

• In *hemolytic anemia*, RBC plasma membranes rupture prematurely. The condition may result from inherited defects or from outside agents such as parasites, toxins, or antibodies from incompatible transfused blood.

• *Thalassemia* (thal′-a-SĒ-mē-a) is a group of hereditary hemolytic anemias in which there is an abnormality in one or more of the four polypeptide chains of the hemoglobin molecule. Thalassemia occurs primarily in populations from countries bordering the Mediterranean Sea.

• *Aplastic anemia* results from destruction of the red bone marrow caused by toxins, gamma radiation, and certain medications that inhibit enzymes needed for hemopoiesis.

Anemia may also be caused by chemotherapy for cancer treatment. A synthetic erythropoietin (Procrit®) is given to such patients to increase the oxygen-carrying capacity of the blood.

Sickle Cell Disease

The RBCs of a person with **sickle cell disease (SCD)** contain Hb-S, an abnormal kind of hemoglobin. When Hb-S gives up oxygen to the interstitial fluid, it forms long, stiff, rodlike structures that bend the erythrocyte into a sickle shape. The sickled cells rupture easily. Even though the loss of RBCs stimulates erythropoiesis, it cannot keep pace with hemolysis. People with sickle-cell disease always have some degree of anemia and mild jaundice and may experience joint or bone pain, breathlessness, rapid heart rate, abdominal pain, fever, and fatigue as a result of tissue damage caused by prolonged recovery oxygen uptake (oxygen debt). Any activity that reduces the amount of oxygen in the blood, such as vigorous exercise, may produce a sickle-cell crisis (worsening of the anemia, pain in the abdomen and long bones of the limbs, fever, and shortness of breath).

Hemolytic Disease of the Newborn

Hemolytic disease of the newborn (HDN) is a problem that results from Rh incompatibility between a mother and her fetus. Normally, no direct contact occurs between maternal and fetal blood while a woman is pregnant. However, if a small amount of Rh^+ blood leaks from the fetus through the placenta into the bloodstream of an Rh^- mother, her body starts to make anti-Rh antibodies. Because the greatest possibility of fetal blood transfer occurs at delivery, the firstborn baby typically is not affected. If the mother becomes pregnant again, however, her anti-Rh antibodies, made after delivery of the first baby, can cross the placenta and enter the bloodstream of the fetus. If the fetus is Rh^-, there is no problem, because Rh^- blood does not have the Rh antigen. If, however, the fetus is Rh^+, life-threatening *hemolysis* (rupture of RBCs) is likely to occur in the fetal blood. By contrast, ABO incompatibility between a mother and her fetus rarely causes problems because the anti-A and anti-B antibodies do not cross the placenta.

HDN is prevented by giving all Rh^- women an injection of anti-Rh antibodies called anti-Rh gamma globulin (RhoGAM) soon after every delivery, miscarriage, or abortion. These antibodies destroy any Rh antigens that are present so the mother doesn't produce her own antibodies to them. In the case of an Rh^+ mother, there are no complications, because she cannot make anti-Rh antibodies.

Leukemia

The term **leukemia** (loo-KĒ-mē-a; *leuko-* = white) refers to a group of red bone marrow cancers in which abnormal white blood cells multiply uncontrollably. The accumulation of the cancerous white blood cells in red bone marrow interferes with the production of red blood cells, white blood cells, and platelets. As a result, the oxygen-carrying capacity of the blood is reduced, an individual is more susceptible to infection, and blood clotting is abnormal. In most leukemias, the cancerous white blood cells spread to the lymph nodes, liver, and spleen, causing them to enlarge. All leukemias produce the usual symptoms of anemia (fatigue, intolerance to cold, and pale skin). In addition, weight loss, fever, night sweats, excessive bleeding, and recurrent infections may also occur.

MEDICAL TERMINOLOGY AND CONDITIONS

Acute normovolemic hemodilution (nor-mō-vō-LĒ-mik hē-mō-di-LOO-shun)   Removal of blood immediately before surgery and its replacement with a cell-free solution to maintain sufficient blood volume for adequate circulation. At the end of surgery, once bleeding has been controlled, the collected blood is returned to the body.

Autologous preoperative transfusion (aw-TOL-o-gus trans-FŪ-zhun; *auto-* = self)   Donating one's own blood in preparation for surgery; can be done up to six weeks before elective surgery. Also called *predonation*.

Blood bank   A facility that collects and stores a supply of blood for future use by the donor or others. Because blood banks have now assumed additional and diverse functions (immunohematology reference work, continuing medical education, bone and tissue storage, and clinical consultation), they are more appropriately referred to as *centers of transfusion medicine*.

Cyanosis (sī-a-NŌ-sis; *cyano-* = blue)   Slightly bluish/dark-purple skin discoloration, most easily seen in the nail beds and mucous membranes, due to an increased quantity of reduced hemoglobin (hemoglobin not combined with oxygen) in systemic blood.

Hemochromatosis (hē′-mō-krō′-ma-TŌ-sis; *chroma* = color)   Disorder of iron metabolism characterized by excess deposits of iron in tissues (especially the liver, heart, pituitary gland, gonads, and pancreas) that result in discoloration (bronzing) of the skin, cirrhosis, diabetes mellitus, and bone and joint abnormalities.

Hemophilia (hē′-mō-FIL-ē-a; *-philia* = loving)   An inherited deficiency of clotting in which bleeding may occur spontaneously or after only minor trauma.

Hemorrhage (HEM-or-ij; *rhegnynai* = bursting forth)   Loss of a large amount of blood; can be either internal (from blood vessels into tissues) or external (from blood vessels directly to the surface of the body).

Jaundice (*jaund-* = yellow)   An abnormal yellowish discoloration of the sclerae of the eyes, skin, and mucous membranes due to excess bilirubin (yellow-orange pigment) in the blood that is produced when the heme pigment in aged red blood cells is broken down.

Phlebotomist (fle-BOT-ō-mist; *phlebo-* = vein; *-tom* = cut)   A technician who specializes in withdrawing blood.

Polycythemia (pol′-ē-sī-THĒ-mē-a)   An abnormal increase in the number of red blood cells in which hematocrit is above 55%, the upper limit of normal.

Septicemia (sep′-ti-SĒ-mē-a; *septic-* = decay; *-emia* = condition of blood)   An accumulation of toxins or disease-causing bacteria in the blood. Also called *blood poisoning*.

Thrombocytopenia (throm′-bō-sī′-tō-PĒ-nē-a; *-penia* = poverty)   Very low platelet count that results in a tendency to bleed from capillaries.

Whole blood   Blood containing all formed elements, plasma, and plasma solutes in natural concentrations.

CHAPTER REVIEW AND RESOURCE SUMMARY

REVIEW	RESOURCES
### 14.1 Functions of Blood	
1. Blood transports oxygen, carbon dioxide, nutrients, wastes, and hormones.	Anatomy Overview—Blood
2. It helps to regulate pH, body temperature, and water content of cells.	
3. It prevents blood loss through clotting and combats microbes and toxins through the action of certain phagocytic white blood cells or specialized plasma proteins.	
### 14.2 Components of Whole Blood	Animation—Erythropoietin
1. Physical characteristics of whole blood include a viscosity greater than that of water, a temperature of 38° (100.4°F), and a pH range between 7.35 and 7.45. Blood constitutes about 8% of body weight in an adult, and consists of 55% plasma and 45% formed elements.	**Figure 14.2** Origin, development, and structure of blood cells

REVIEW	RESOURCES

2. The **formed elements** in blood include red blood cells (erythrocytes), white blood cells (leukocytes), and platelets. **Hematocrit** is the percentage of red blood cells in whole blood.

3. **Plasma** contains 91.5% water, 7% proteins, and 1.5% solutes other than proteins. Principal solutes include proteins (**albumins, globulins, fibrinogen**), nutrients, hormones, respiratory gases, electrolytes, and waste products.

4. **Hemopoiesis**, the formation of blood cells from pluripotent stem cells, occurs in **red bone marrow**.

5. **Red blood cells (RBCs)** are biconcave discs without nuclei that contain hemoglobin. The function of the hemoglobin in red blood cells is to transport oxygen. Red blood cells live about 120 days. A healthy male has about 5.4 million RBCs/mL of blood and a healthy female has about 4.8 million RBCs/mL. After phagocytosis of aged red blood cells by macrophages, hemoglobin is recycled.

6. RBC formation, called **erythropoiesis**, occurs in adult red bone marrow. It is stimulated by **hypoxia**, which stimulates release of **erythropoietin** by the kidneys. A reticulocyte count is a diagnostic test that indicates the rate of erythropoiesis.

7. **White blood cells (WBCs)** are nucleated cells. The two principal types are granular leukocytes (**neutrophils, eosinophils, basophils**) and agranular leukocytes (**lymphocytes** and **monocytes**). The general function of WBCs is to combat inflammation and infection. Neutrophils and macrophages (which develop from monocytes) do so through **phagocytosis**.

8. Eosinophils combat inflammation in allergic reactions, phagocytize antigen–antibody complexes, and combat parasitic worms; basophils liberate heparin, histamine, and serotonin in allergic reactions that intensify the inflammatory response.

9. B cells (B lymphocytes) are effective against bacteria and other toxins. T cells (T lymphocytes) are effective against viruses, fungi, and cancer cells. Natural killer cells attack microbes and tumor cells.

10. White blood cells usually live for only a few hours or a few days. Normal blood contains 5000 to 10,000 WBCs/mL.

11. **Platelets** are disc-shaped cell fragments without nuclei that are formed from megakaryocytes and take part in hemostasis by forming a platelet plug. Normal blood contains 150,000 to 400,000 platelets/mL.

Figure 14.3 Formation and destruction of red blood cells, and the recycling of hemoglobin components

Concepts and Connections—Blood Homeostatic Imbalances— The Case of the Man with Yellow Eyes

14.3 Hemostasis

1. **Hemostasis**, the stoppage of bleeding, involves vascular spasm, platelet plug formation, and blood clotting. In **vascular spasm**, the smooth muscle of a blood vessel wall contracts. **Platelet plug** formation is the aggregation of platelets to stop bleeding. A **clot** is a network of insoluble protein fibers (**fibrin**) in which formed elements of blood are trapped. The chemicals involved in clotting are known as clotting factors.

2. Blood **clotting** involves a series of reactions that may be divided into three stages: formation of **prothrombinase** by either the **extrinsic pathway** or the **intrinsic pathway**, conversion of **prothrombin** into **thrombin**, and conversion of soluble **fibrinogen** into insoluble fibrin.

3. Normal coagulation involves **clot retraction** (tightening of the clot) and **fibrinolysis** (dissolution of the clot).

4. Anticoagulants (for example, heparin) prevent clotting.

5. Clotting in an unbroken blood vessel is called **thrombosis**. A **thrombus** that moves from its site of origin is called an **embolus**.

Figure 14.5 Blood clotting

14.4 Blood Groups and Blood Types

1. In the ABO system, the **antigens** on RBCs, called **A** and **B**, determine blood type. Plasma contains antibodies termed **anti-A antibodies** and **anti-B antibodies**.

2. In the Rh system, individuals whose erythrocytes have Rh antigens (the **Rh factor**) are classified as Rh^+. Those who lack the antigen are Rh^-.

SELF-QUIZ

1. A hematocrit is
   a. used to measure the quantity of the five types of white blood cells.
   b. essential for determining a person's blood type.
   c. the percentage of red blood cells in whole blood.
   d. also known as a platelet count.
   e. involved in blood clotting.

2. Match the following:
   _____ a. involved in certain immune responses
   _____ b. develop into mature red blood cells
   _____ c. most abundant plasma protein
   _____ d. blood after formed elements are removed
   _____ e. plasma without clotting proteins
   _____ f. needed for blood clotting

   A. albumin
   B. fibrinogen
   C. globulins
   D. plasma
   E. serum
   F. reticulocytes

3. In adults, erythropoiesis takes place in
   a. the liver.   b. yellow bone marrow.
   c. red bone marrow.   d. lymphatic tissue.   e. the kidneys.

4. Match the following substances in RBC breakdown.
   _____ a. transports iron to red bone marrow
   _____ b. converted from non-iron portion of heme; green pigment
   _____ c. required for vitamin B_{12} absorption
   _____ d. broken down into amino acids
   _____ e. contributes to yellow color in urine
   _____ f. intestinal conversion of bilirubin
   _____ g. contributes to brown color of feces
   _____ h. yellow-orange pigment; becomes part of bile

   A. globin
   B. stercobilin
   C. biliverdin
   D. urobilin
   E. bilirubin
   F. transferrin
   G. intrinsic factor
   H. urobilinogen

5. Which of the following statements is NOT true about red blood cells?
   a. The production of red blood cells is known as erythropoiesis.
   b. Red blood cells originate from pluripotent stem cells.
   c. Hypoxia increases the production of red blood cells.
   d. The liver takes part in the destruction and recycling of red blood cell components.
   e. Red blood cells have a lobed nucleus and granular cytoplasm.

6. The presence of erythropoietin in blood
   a. maintains blood volume.
   b. stimulates blood clotting.
   c. results in anemia.
   d. stimulates RBC production.
   e. increases the amount of free iron in the blood.

7. If a differential white blood cell count indicated higher than normal numbers of basophils, what may be occurring in the body?
   a. chronic infection   b. allergic reaction   c. leukopenia
   d. initial response to invading bacteria   e. hemostasis

8. In a person with blood type A, the antibodies that would normally be present in the plasma is (are)
   a. anti-A antibody.   b. anti-B antibody.
   c. both anti-A and anti-B antibodies.
   d. neither anti-A nor anti-B.   e. anti-O antibodies.

9. Hemolytic disease of the newborn (HDN) may occur in the fetus of a second pregnancy if
   a. the mother is Rh^+ and the baby is Rh^-.
   b. the mother is Rh^+ and the baby is Rh^+.
   c. the mother is Rh^- and the baby is Rh^-.
   d. the mother is Rh^- and the baby is Rh^+.
   e. the father is Rh^- and the mother is Rh^+.

10. Place the following steps of hemostasis in the correct order.
    1. clot retraction
    2. prothrombinase formed
    3. fibrinolysis by plasmin
    4. vascular spasm
    5. conversion of prothrombin into thrombin
    6. platelet plug formation
    7. conversion of fibrinogen into fibrin
       a. 4, 6, 2, 5, 7, 1, 3   b. 5, 4, 7, 6, 2, 3, 1
       c. 2, 5, 6, 7, 1, 4, 3   d. 4, 6, 5, 2, 7, 1, 3
       e. 4, 2, 6, 5, 3, 7, 1

11. The intrinsic pathway of clotting
    a. involves the release of tissue factor.
    b. does not require the production of prothrombinase.
    c. is initiated by damage to blood vessel linings.
    d. takes less time than the extrinsic pathway.
    e. is described by all of the above.

12. How does aspirin prevent thrombosis?
    a. It inhibits platelet aggregation.
    b. It activates plasmin.
    c. It inhibits the conversion of prothrombin to thrombin.
    d. It acts as an enzyme to dissolve the thrombus.
    e. It prevents the accumulation of fatty substances on blood vessel walls.

13. Match the following:
    _____ a. become wandering macrophages
    _____ b. produce antibodies
    _____ c. are involved in allergic reactions
    _____ d. first to respond to bacterial invasion
    _____ e. destroy antigen–antibody complexes; combat inflammation

    A. neutrophils
    B. eosinophils
    C. basophils
    D. lymphocytes
    E. monocytes

14. Hemostasis is
    a. maintenance of a steady state in the body.
    b. an abnormal increase in leukocytes.
    c. a hereditary condition in which spontaneous hemorrhaging occurs.
    d. an anticoagulant produced by some leukocytes.
    e. a series of events that stops bleeding.

15. Which of the following are mismatched?
    a. white blood cell count below 5000 cells/μL, leukopenia
    b. red blood cell count of 250,000 cells/μL, normal adult male
    c. white blood cell count above 10,000 cells/μL, leukocytosis
    d. platelet count of 300,000 cells/μL, normal adult
    e. pH 7.4, normal blood

16. An individual with type A blood has _____ in the plasma membranes of red blood cells.
    a. antigen A   b. antigen B   c. antigen O
    d. antigen A and antigen Rh   e. antigen B and antigen Rh

17. Mrs. Smith arrives at a health clinic with her ill daughter Beth. It is suspected that Beth has recently developed a bacterial infection. It is likely that Beth's leukocyte count will be _____ cells/μL of blood, a condition known as _____. A differential white blood cell count shows an abnormally high percentage of _____.
    a. 20,000, leukopenia, neutrophils
    b. 5000, leukocytosis, monocytes
    c. 7000, leukocytosis, basophils
    d. 2000, leukopenia, platelets
    e. 20,000, leukocytosis, neutrophils

18. Clot retraction
    a. draws torn edges of the damaged vessel closer together.
    b. dissolves clots.
    c. is also known as the intrinsic pathway.
    d. involves the formation of fibrin from fibrinogen.
    e. helps prevent the formation of an embolus.

19. Persons with blood type AB are sometimes referred to as universal recipients because their blood
    a. lacks A and B antigens.
    b. lacks anti-A and anti-B antibodies.
    c. possesses type O antigens and anti-O antibodies.
    d. has natural immunity to disease.
    e. contains A and B antigens.

20. A clot that is being transported by the bloodstream is called
    a. a plasma protein.   b. a thrombus.
    c. an embolus.   d. a wandering macrophage.
    e. a reticulocyte.

CRITICAL THINKING APPLICATIONS

1. Biliary atresia is a condition in which the ducts that transport bile out of the liver do not function properly. The whites of the eyes in a baby with this condition have a yellow color. What is the name of the yellow color and what is its cause?

2. While working as an intern in a medical lab, you have been assigned to determine the ABO blood type of three individuals. You've mixed antisera with the blood, with the following results.
   Person 1: blood agglutinates with anti-A sera but not anti-B sera.
   Person 2: blood agglutinates with anti-A and anti-B sera.
   Person 3: blood does not agglutinate with anti-A or anti-B sera.
   What blood type does each individual have?

3. The school nurse sighed, "I just can't get used to the blue nail polish the kids are wearing. I keep thinking there's a medical problem." What type of problem might result in blue fingernails? How might it occur?

4. Very small numbers of pluripotent stem cells occur normally in blood. If these cells could be isolated and grown in sufficient numbers, what medically useful products could they produce?

ANSWERS TO FIGURE QUESTIONS

14.1 Red blood cells are the most numerous formed elements in blood.

14.2 Blood makes up 8 percent of body weight.

14.3 Stercobilin is responsible for the brown color of feces.

14.4 Hypoxia means cellular oxygen deficiency.

14.5 Prothrombinase is formed during stage 1 of clotting.

14.6 Type O blood has anti-A and anti-B antibodies.

14.7 Agglutination refers to the clumping of red blood cells.

CHAPTER 15

THE CARDIOVASCULAR SYSTEM: HEART

In the last chapter we examined the composition and functions of blood. For blood to reach body cells and exchange materials with them, it must be constantly pumped by the heart through the body's blood vessels. The heart beats about 100,000 times every day, which adds up to about 35 million beats in a year. The left side of the heart pumps blood through an estimated 100,000 km (60,000 mi) of blood vessels. The right side of the heart pumps blood through the lungs, enabling blood to pick up oxygen and unload carbon dioxide. Even while you are sleeping, your heart pumps 30 times its own weight each minute, which amounts to about 5 liters (5.3 qt) to the lungs and the same volume to the rest of the body. At this rate, the heart pumps more than 14,000 liters (3600 gal) of blood in a day, or 10 million liters (2.6 million gal) in a year. You don't spend all your time sleeping, however, and your heart pumps more vigorously when you are active. Thus, the actual blood volume the heart pumps in a single day is much larger.

The scientific study of the normal heart and the diseases associated with it is **cardiology** (kar'-dē-OL-ō-jē; *cardio-* = heart; *-logy* = study of). This chapter explores the design of the heart and the unique properties that permit it to pump for a lifetime without a moment of rest.

Did you know?   Hypertension is the leading cause of cardiovascular disease, increasing the risks for coronary artery disease, heart failure, and stroke. Many factors raise a person's risk of developing hypertension. While factors such as age and genetic predisposition are not under an individual's control, many others are. Leading lifestyle risk factors for hypertension include the usual deadly sins: smoking, drinking too much alcohol, gluttony, and sloth. Abdominal obesity that causes type 2 diabetes increases a person's risk of developing hypertension. But here's the good news: as little as 30 minutes of exercise per day along with stress reduction and a diet that includes plenty of fruits and vegetables and fewer empty and fatty calories reduces hypertension risk.

FOCUS ON WELLNESS
Hypertension Prevention

LOOKING BACK TO MOVE AHEAD...

Functions of Blood (Section 14.1)

Membranes (Section 4.4)

Muscular Tissue (Section 4.5)

Cardiac Muscle Tissue (Section 8.7)

Action Potential (Section 9.3)

Free Radicals (Section 2.1)

ANS Neurotransmitters (Section 11.3)

15.1 STRUCTURE AND ORGANIZATION OF THE HEART

OBJECTIVES ● **Identify the location of the heart and the structure and functions of the pericardium.**

● **Describe the layers of the heart wall and the chambers of the heart.**

● **Identify the major blood vessels that enter and exit the heart.**

● **Explain the structure and functions of the valves of the heart.**

Location and Coverings of the Heart

The *heart* is situated between the two lungs in the thoracic cavity, with about two-thirds of its mass lying to the left of the body's midline (Figure 15.1). Your heart is about the size of your closed fist. The pointed end, the *apex*, is formed by the tip of the left ventricle, a lower chamber of the heart, and rests on the diaphragm. The *base* of the heart is its posterior surface. It is formed by the atria (upper chambers of the heart), mostly the left atrium, into which the four pulmonary veins open, and a portion of the right atrium that receives the superior and inferior vena cavae (see Figure 15.3b). The base lies opposite the apex.

The membrane that surrounds and protects the heart and holds it in place is the *pericardium* (*peri-* = around). It consists of two parts: the fibrous pericardium and the serous pericardium (see Figure 15.2). The outer *fibrous pericardium* is a tough, inelastic, dense irregular connective tissue layer. It prevents overstretching of the heart, provides protection, and anchors the heart in place.

The inner *serous pericardium* is a thinner, more delicate membrane that forms a double layer around the heart. The outer *parietal layer* of the serous pericardium is fused to the

Figure 15.1 Position of the heart and associated blood vessels in the thoracic cavity. In this and subsequent illustrations, vessels that carry oxygenated blood are colored red; vessels that carry deoxygenated blood are colored blue. The borders of the mediastinum in (b) are indicated by a dashed line.

🔑 The heart is located between the lungs, with about two-thirds of its mass to the left of the midline.

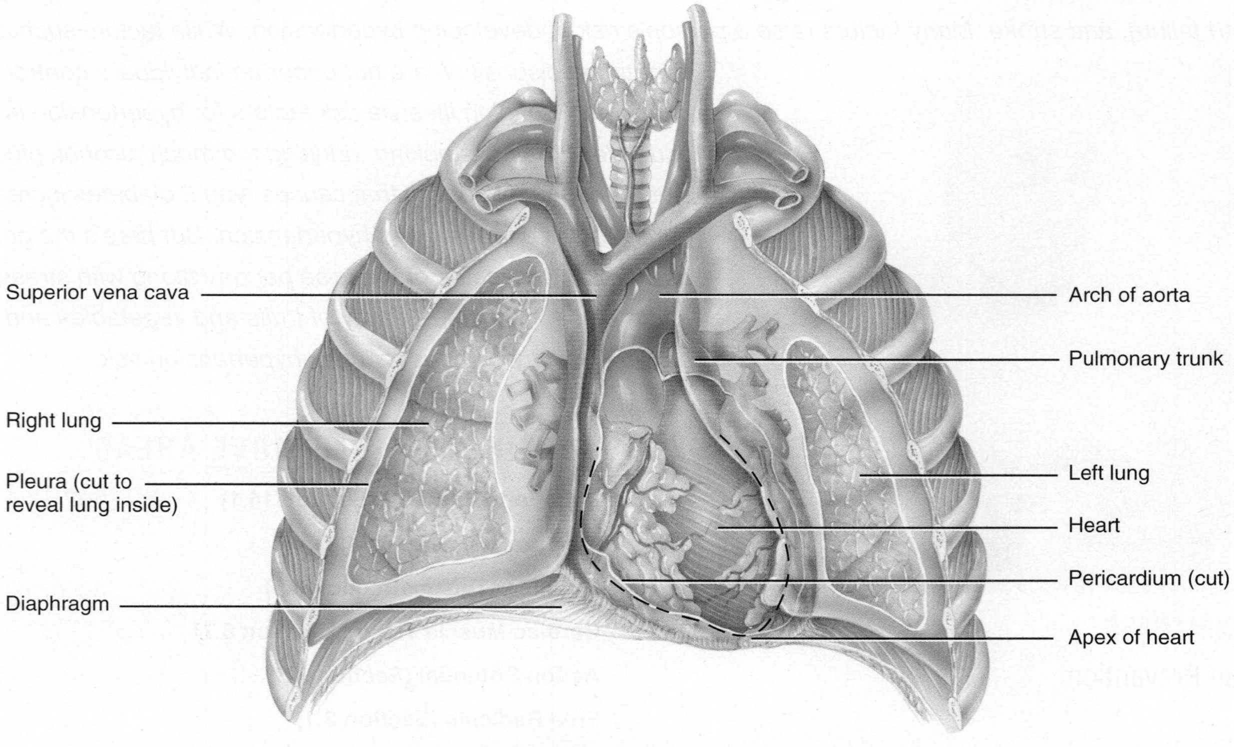

Superior vena cava

Right lung

Pleura (cut to reveal lung inside)

Diaphragm

Arch of aorta

Pulmonary trunk

Left lung

Heart

Pericardium (cut)

Apex of heart

(a) Anterior view of the heart in the thoracic cavity

fibrous pericardium, and the inner *visceral layer* of the serous pericardium, also called the *epicardium* (*epi-* = on top of), adheres tightly to the surface of the heart. Between the parietal and visceral layers of the serous pericardium is a thin film of fluid. This fluid, known as *pericardial fluid*, reduces friction between the membranes as the heart moves. The *pericardial cavity* is the space that contains the pericardial fluid.

CLINICAL CONNECTION | Pericarditis

Inflammation of the pericardium is called **pericarditis** (per'-i-kar-DĪ-tis). The most common type, *acute pericarditis*, begins suddenly and has no known cause in most cases but is sometimes linked to a viral infection. As a result of irritation to the pericardium, there is chest pain that may extend to the left shoulder and down the left arm (often mistaken for a heart attack) and pericardial friction rub (a scratchy or creaking sound heard through a stethoscope as the visceral layer of the serous pericardium rubs against the parietal layer of the serous pericardium). Acute pericarditis usually lasts for about one week and is treated with drugs that reduce inflammation and pain, such as ibuprofen or aspirin.

Chronic pericarditis begins gradually and is long-lasting. In one form of this condition, there is a buildup of pericardial fluid. If a great deal of fluid accumulates, this is a life-threatening condition because the fluid compresses the heart, a condition called *cardiac tamponade* (tam'-pon-ĀD). As a result of the compression, ventricular filling is decreased, cardiac output is reduced, venous return to the heart is diminished, blood pressure falls, and breathing is difficult. Most causes of chronic pericarditis involving cardiac tamponade are unknown, but it is sometimes caused by conditions such as cancer and tuberculosis. Treatment consists of draining the excess fluid through a needle passed into the pericardial cavity. •

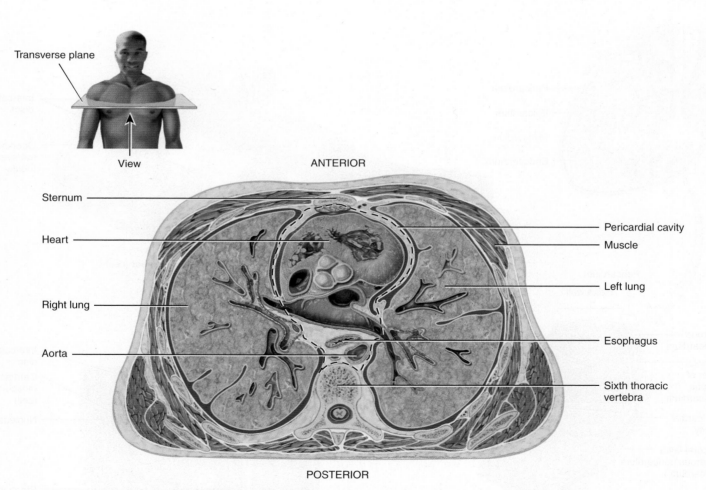

(b) Inferior view of transverse section of thoracic cavity
showing the heart in the mediastinum

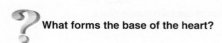

What forms the base of the heart?

Heart Wall

The wall of the heart (Figure 15.2a) is composed of three layers: epicardium (external layer), myocardium (middle layer), and endocardium (inner layer). The **epicardium**, which is also known as the visceral layer of serous pericardium, is the thin, transparent outer layer of the wall. It is composed of mesothelium and connective tissue.

The **myocardium** (*myo-* = muscle) consists of cardiac muscle tissue, which constitutes the bulk of the heart. This tissue is found only in the heart and is specialized in structure and function. The myocardium is responsible for the pumping action of the heart. Cardiac muscle fibers (cells) are involuntary, striated, and branched, and the tissue is arranged in interlacing bundles of fibers (Figure 15.2b).

Cardiac muscle fibers form two separate networks—one atrial and one ventricular. Each cardiac muscle fiber connects with other fibers in the networks by thickenings of the sarcolemma (plasma membrane) called **intercalated discs**. Within the discs are **gap junctions** that allow action potentials to conduct from one cardiac muscle fiber to the next. The intercalated discs also link cardiac muscle fibers to one another so they do not pull apart. Each network contracts as a functional unit, so the atria contract separately from the ventricles. In response to a single action potential, cardiac muscle fibers develop a prolonged contraction, 10–15 times longer than the contraction observed in skeletal muscle fibers. Also, the refractory period of a cardiac fiber lasts longer than the contraction itself. Thus, another contraction

Figure 15.2 Pericardium and heart wall.

The pericardium is a sac that surrounds and protects the heart.

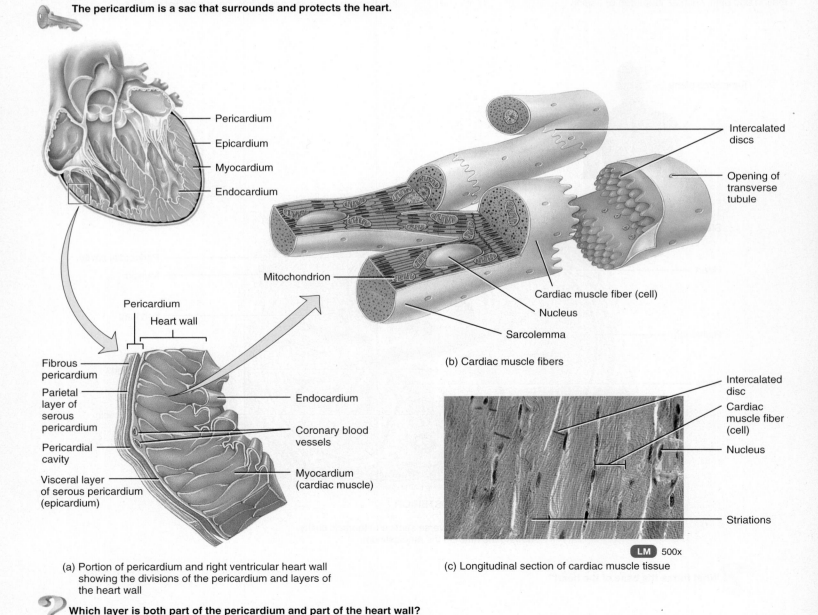

(b) Cardiac muscle fibers

(a) Portion of pericardium and right ventricular heart wall showing the divisions of the pericardium and layers of the heart wall

(c) Longitudinal section of cardiac muscle tissue

? Which layer is both part of the pericardium and part of the heart wall?

of cardiac muscle cannot begin until relaxation is well underway. For this reason, tetanus (maintained contraction) cannot occur in cardiac muscle tissue.

The ***endocardium*** (*endo-* = within) is a thin layer of simple squamous epithelium that lines the inside of the myocardium and covers the valves of the heart and the tendons

attached to the valves. It is continuous with the epithelial lining of the large blood vessels.

Chambers of the Heart

The heart contains four chambers (Figure 15.3). The two upper chambers are the ***atria*** (= entry halls or chambers), and

Figure 15.3 **Structure of the heart.**

The four chambers of the heart are the two upper atria and two lower ventricles.

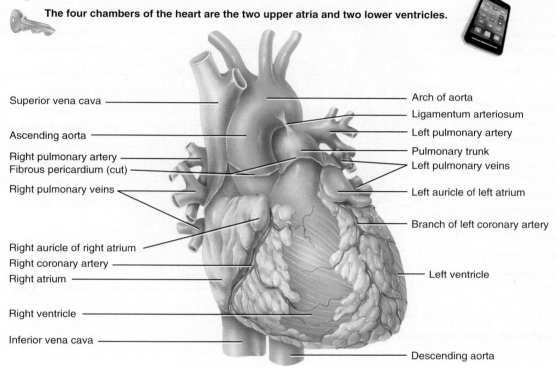

Superior vena cava
Ascending aorta
Right pulmonary artery
Fibrous pericardium (cut)
Right pulmonary veins
Right auricle of right atrium
Right coronary artery
Right atrium
Right ventricle
Inferior vena cava

Arch of aorta
Ligamentum arteriosum
Left pulmonary artery
Pulmonary trunk
Left pulmonary veins
Left auricle of left atrium
Branch of left coronary artery
Left ventricle
Descending aorta

(a) Anterior external view showing surface features

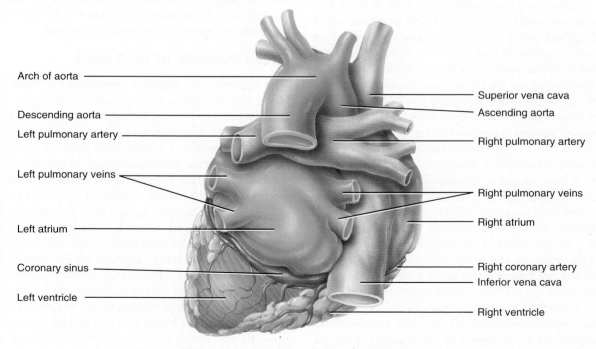

Arch of aorta
Descending aorta
Left pulmonary artery
Left pulmonary veins
Left atrium
Coronary sinus
Left ventricle

Superior vena cava
Ascending aorta
Right pulmonary artery
Right pulmonary veins
Right atrium
Right coronary artery
Inferior vena cava
Right ventricle

(b) Posterior external view showing surface features

Figure 15.3 CONTINUES

Figure 15.3 CONTINUED

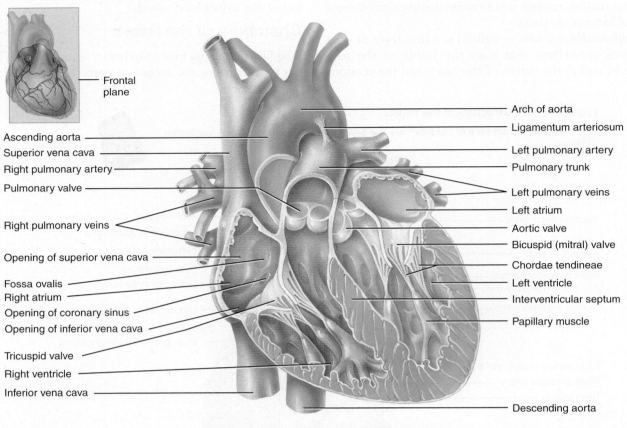

Frontal plane

Ascending aorta
Superior vena cava
Right pulmonary artery
Pulmonary valve

Right pulmonary veins

Opening of superior vena cava

Fossa ovalis
Right atrium
Opening of coronary sinus
Opening of inferior vena cava

Tricuspid valve

Right ventricle

Inferior vena cava

Arch of aorta
Ligamentum arteriosum
Left pulmonary artery
Pulmonary trunk
Left pulmonary veins
Left atrium
Aortic valve
Bicuspid (mitral) valve
Chordae tendineae
Left ventricle
Interventricular septum
Papillary muscle

Descending aorta

(c) Anterior view of frontal section showing internal anatomy

 Through which type of vessel does blood flow away from the heart?

the two lower chambers are the ***ventricles*** (= little bellies). Between the right atrium and left atrium is a thin partition called the ***interatrial septum*** (*inter-* = between; *septum* = a dividing wall or partition); a prominent feature of this septum is an oval depression called the ***fossa ovalis***. It is the remnant of the ***foramen ovale***, an opening in the fetal heart that directs blood from the right to left atrium in order to bypass the nonfunctioning fetal lungs. The foramen ovale normally closes soon after birth. An ***interventricular septum*** separates the right ventricle from the left ventricle (Figure 15.3c). On the anterior surface of each atrium is a wrinkled pouchlike structure called an ***auricle*** (OR-i-kul; *auri-* = ear), so named because of its resemblance to a dog's ear. Each auricle slightly increases the capacity of an atrium so that it can hold a greater volume of blood.

The thickness of the myocardium of the chambers varies according to the amount of work each chamber has to perform. The walls of the atria are thin compared to those of the ventricles because the atria need only enough cardiac muscle tissue to deliver blood into the ventricles (Figure 15.3c). The right ventricle pumps blood only to the lungs (pulmonary circulation); the left ventricle pumps blood to all other parts of the body (systemic circulation). The left ventricle must work harder than the right ventricle to main-

tain the same rate of blood flow, so the muscular wall of the left ventricle is considerably thicker than the wall of the right ventricle to overcome the greater pressure.

Great Vessels of the Heart

The right atrium receives ***deoxygenated blood*** (oxygen-poor blood that has given up some of its oxygen to cells) through three ***veins***, blood vessels that return blood to the heart. The ***superior vena cava*** (VĒ-na CĀ-va; *vena* = vein; *cava* = hollow, a cave) brings blood mainly from parts of the body above the heart; the ***inferior vena cava*** brings blood mostly from parts of the body below the heart; and the ***coronary sinus*** drains blood from most of the vessels supplying the wall of the heart (Figure 15.3b, c). The right atrium then delivers the deoxygenated blood into the right ventricle, which pumps it into the ***pulmonary trunk***. The pulmonary trunk divides into a ***right*** and ***left pulmonary artery***, each of which carries blood to the corresponding lung. ***Arteries*** are blood vessels that carry blood away from the heart. In the lungs, the deoxygenated blood unloads carbon dioxide and picks up oxygen. This ***oxygenated blood*** (oxygen-rich blood that has picked up oxygen as it flows through the lungs) then enters the left atrium via four ***pulmonary veins***. The blood then passes into the left ventricle, which pumps the blood into the ***ascending aorta***. From here the oxygenated blood is carried to all parts of the body.

Between the pulmonary trunk and arch of the aorta is a structure called the ***ligamentum arteriosum***. It is the remnant of the ***ductus arteriosus***, a blood vessel in fetal circulation that allows most blood to bypass the nonfunctional fetal lungs (see Section 16.3).

Valves of the Heart

As each chamber of the heart contracts, it pushes a volume of blood into a ventricle or out of the heart into an artery. To prevent the blood from flowing backward, the heart has four ***valves*** composed of dense connective tissue covered by endothelium. These valves open and close in response to pressure changes as the heart contracts and relaxes.

As their names imply, ***atrioventricular (AV) valves*** lie between the atria and ventricles (Figure 15.3c). The atrioventricular valve between the right atrium and right ventricle is

called the ***tricuspid valve*** because it consists of three cusps (leaflets). The pointed ends of the cusps project into the ventricle. Tendonlike cords, called ***chordae tendineae*** (KOR-dē ten-DI-nē-ē; *chord-* = cord; *tend-* = tendon), connect the pointed ends to ***papillary muscles*** (*papill-* = nipple), cardiac muscle projections located on the inner surface of the ventricles. The chordae tendineae prevent the valve cusps from pushing up into the atria when the ventricles contract and are aligned to allow the valve cusps to tightly close the valve.

The atrioventricular valve between the left atrium and left ventricle is called the ***bicuspid (mitral) valve.*** It has two cusps that work in the same way as the cusps of the tricuspid valve. For blood to pass from an atrium to a ventricle, an atrioventricular valve must open.

The opening and closing of the valves are due to pressure differences across the valves. When blood moves from an atrium to a ventricle, the valve is pushed open, the papillary muscles relax, and the chordae tendineae slacken (Figure 15.4a). When a ventricle contracts, the pressure of the

Figure 15.4 Atrioventricular (AV) valves. The bicuspid and tricuspid valves operate in a similar manner.

Heart valves open and close in response to pressure changes as the heart contracts and relaxes.

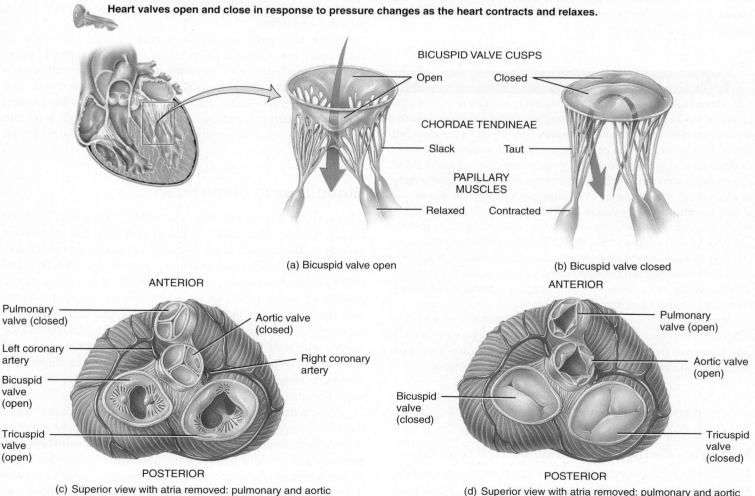

(a) Bicuspid valve open

(b) Bicuspid valve closed

(c) Superior view with atria removed: pulmonary and aortic valves closed, bicuspid and tricuspid valves open

(d) Superior view with atria removed: pulmonary and aortic valves open, bicuspid and tricuspid valves closed

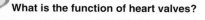 **What is the function of heart valves?**

ventricular blood drives the cusps upward until their edges meet and close the opening (Figure 15.4b). At the same time, contraction of the papillary muscles and tightening of the chordae tendineae help prevent the cusps from swinging upward into the atrium.

Near the origin of the pulmonary trunk and aorta are *semilunar valves* called the *pulmonary valve* and the *aortic valve* that prevent blood from flowing back into the heart (see Figure 15.3c). The pulmonary valve lies in the opening where the pulmonary trunk leaves the right ventricle. The aortic valve is situated at the opening between the left ventricle and the aorta. Each valve consists of three semilunar (half-moon-shaped) cusps that attach to the artery wall. Like the atrioventricular valves, the semilunar valves permit blood to flow in one direction only—in this case, from the ventricles into the arteries.

When the ventricles contract, pressure builds up within them. The semilunar valves open when pressure in the ventricles exceeds the pressure in the arteries, permitting ejection of blood from the ventricles into the pulmonary trunk and aorta (see Figure 15.4d). As the ventricles relax, blood starts to flow back toward the heart. This back-flowing blood fills the valve cusps, which tightly closes the semilunar valves (see Figure 15.4c).

CLINICAL CONNECTION | Heart Valve Disorders

When heart valves operate normally, they open fully and close completely at the proper times. A narrowing of a heart valve opening that restricts blood flow is known as **stenosis** (ste-NŌ-sis = a narrowing); failure of a valve to close completely is termed **insufficiency** or **incompetence**. In **mitral stenosis**, scar formation or a congenital defect causes narrowing of the mitral valve. One cause of **mitral insufficiency**, in which there is backflow of blood from the left ventricle into the left atrium, is **mitral valve prolapse (MVP)**. In MVP, one or both cusps of the mitral valve protrude into the left atrium during ventricular contraction. Mitral valve prolapse is one of the most common valvular disorders, affecting as much as 30 percent of the population. It is more prevalent in women than in men, and does not always pose a serious threat. In **aortic stenosis**, the aortic valve is narrowed, and in **aortic insufficiency**, there is backflow of blood from the aorta into the left ventricle.

If a heart valve cannot be repaired surgically, then the valve must be replaced. Tissue (biologic) valves may be provided by human donors or pigs; sometimes mechanical (artificial) valves made of plastic or metal are used. The aortic valve is the most commonly replaced heart valve. •

✓ CHECKPOINT

1. Identify the location of the heart.
2. Describe the layers of the pericardium and heart wall.
3. How do atria and ventricles differ in structure and function?
4. Which blood vessels that enter and exit the heart carry oxygenated blood? Which carry deoxygenated blood?
5. In correct sequence, which heart chambers, heart valves, and blood vessels would a drop of blood encounter from the time it flows out of the right atrium until it reaches the aorta?

15.2 BLOOD FLOW AND BLOOD SUPPLY OF THE HEART

OBJECTIVES • **Explain how blood flows through the heart.**

• **Describe the clinical importance of the blood supply of the heart.**

Blood Flow Through the Heart

Blood flows through the heart from areas of higher blood pressure to areas of lower blood pressure. As the walls of the atria contract, the pressure of the blood within them increases. This increased blood pressure forces the AV valves open, allowing atrial blood to flow through the AV valves into the ventricles.

After the atria are finished contracting, the walls of the ventricles contract, increasing ventricular blood pressure and pushing blood through the semilunar valves into the pulmonary trunk and aorta. At the same time, the shape of the AV valve cusps causes them to be pushed shut, preventing backflow of ventricular blood into the atria. Figure 15.5 summarizes the flow of blood through the heart.

Blood Supply of the Heart

The wall of the heart, like any other tissue, has its own blood vessels. The flow of blood through the numerous vessels in the myocardium is called *coronary (cardiac) circulation*. The principal coronary vessels are the *left* and *right coronary arteries*, which originate as branches of the ascending aorta (see Figure 15.3a). Each artery branches and then branches again to deliver oxygen and nutrients throughout the heart muscle. Most of the deoxygenated blood, which carries carbon dioxide and wastes, is collected by a large vein on the posterior surface of the heart, the *coronary sinus* (see Figure 15.3b), which empties into the right atrium.

Most parts of the body receive blood from branches of more than one artery, and where two or more arteries supply the same region, they usually connect. These connections, called *anastomoses* (a-nas′-tō-MŌ-sēs), provide alternative routes for blood to reach a particular organ or tissue. The myocardium contains many anastomoses that connect branches of a given coronary artery or extend between branches of different coronary arteries.

Figure 15.5 Blood flow through the heart.

The right and left coronary arteries deliver blood to the heart; the coronary veins drain blood from the heart into the coronary sinus.

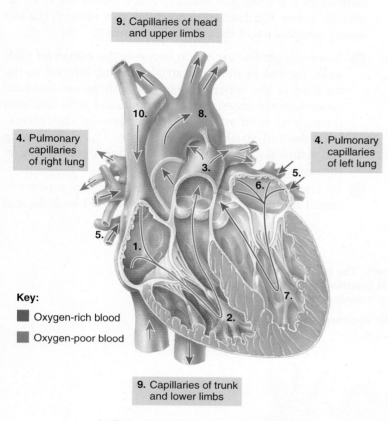

9. Capillaries of head and upper limbs

10.   8.

4. Pulmonary capillaries of right lung

4. Pulmonary capillaries of left lung

3.

5.

6.

5.

1.

7.

Key:

Oxygen-rich blood

Oxygen-poor blood

2.

9. Capillaries of trunk and lower limbs

(a) Path of blood flow through heart

4. In pulmonary capillaries, blood loses CO_2 and gains O_2

3. Pulmonary trunk and pulmonary arteries

5. Pulmonary veins (oxygenated blood)

Pulmonary valve

2. Right ventricle

6. Left atrium

Tricuspid valve

Bicuspid valve

1. Right atrium (deoxygenated blood)

7. Left ventricle

Aortic valve

10. Superior vena cava | Inferior vena cava | Coronary sinus

8. Aorta and systemic arteries

9. In systemic capillaries, blood loses O_2 and gains CO_2

(b) Diagram of blood flow

Which veins deliver deoxygenated blood into the right atrium?

They provide detours for arterial blood if a main route becomes obstructed. Thus, heart muscle may receive sufficient oxygen even if one of its coronary arteries is partially blocked.

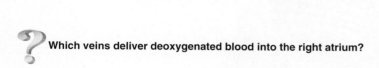

CLINICAL CONNECTION | Reperfusion and Free Radicals

When blockage of a coronary artery deprives the heart muscle of oxygen, **reperfusion**, the reestablishment of blood flow, may damage the tissue further. This surprising effect is due to the formation of oxy-gen **free radicals** from the reintroduced oxygen. Free radicals are electrically charged molecules that have an unpaired electron. Such molecules are unstable and highly reactive. They cause chain reactions that lead to cellular damage and death. To counter the effects of oxygen free radicals, body cells produce enzymes that convert free radicals to less reactive substances. In addition, some nutrients, such as vitamin E, vitamin C, beta-carotene, zinc, and selenium, are antioxidants, which remove oxygen free radicals. Drugs that lessen reperfusion damage after a heart attack or stroke are currently under development. •

✓ **CHECKPOINT**

6. Describe the main force that causes blood to flow through the heart.

7. Why is it that blood flowing through the chambers within the heart cannot supply sufficient oxygen or remove enough carbon dioxide from the myocardium?

15.3 CONDUCTION SYSTEM OF THE HEART

OBJECTIVE • **Explain how each heartbeat is initiated and maintained.**

About 1 percent of the cardiac muscle fibers are different from all others because they can generate action potentials over and over and do so in a rhythmical pattern. They continue to stimulate a heart to beat even after it is removed from the body—

for example, to be transplanted into another person—and all of its nerves have been cut. The nerves regulate the heart rate, but do not determine it. These cells have two important functions: They act as a *pacemaker*, setting the rhythm for the entire heart, and they form the *conduction system*, the route for action potentials throughout the heart muscle. The conduction system ensures that cardiac chambers are stimulated to contract in a coordinated manner, which makes the heart an effective pump. Cardiac action potentials pass through the following components of the conduction system (Figure 15.6):

① Normally, cardiac excitation begins in the *sinoatrial (SA) node*, located in the right atrial wall just inferior to the opening of the superior vena cava. An action potential spontaneously arises in the SA node and then conducts throughout both atria via gap junctions in the intercalated discs of atrial fibers (see Figure 15.2b). Following the action potential, the two atria finish contracting at the same time.

② By conducting along atrial muscle fibers, the action potential also reaches the *atrioventricular (AV) node*, located

Figure 15.6 Conduction system of the heart. The SA node, located in the right atrial wall, is the heart's pacemaker, initiating cardiac action potentials that cause contraction of the heart's chambers. The arrows indicate the flow of action potentials through the atria.

The conduction system ensures that cardiac chambers contract in a coordinated manner.

Frontal plane

Left atrium

Right atrium

① SINOATRIAL (SA) NODE

② ATRIOVENTRICULAR (AV) NODE

③ ATRIOVENTRICULAR (AV) BUNDLE (BUNDLE OF HIS)

④ RIGHT AND LEFT BUNDLE BRANCHES

Right ventricle

Left ventricle

⑤ PURKINJE FIBERS

Anterior view of frontal section

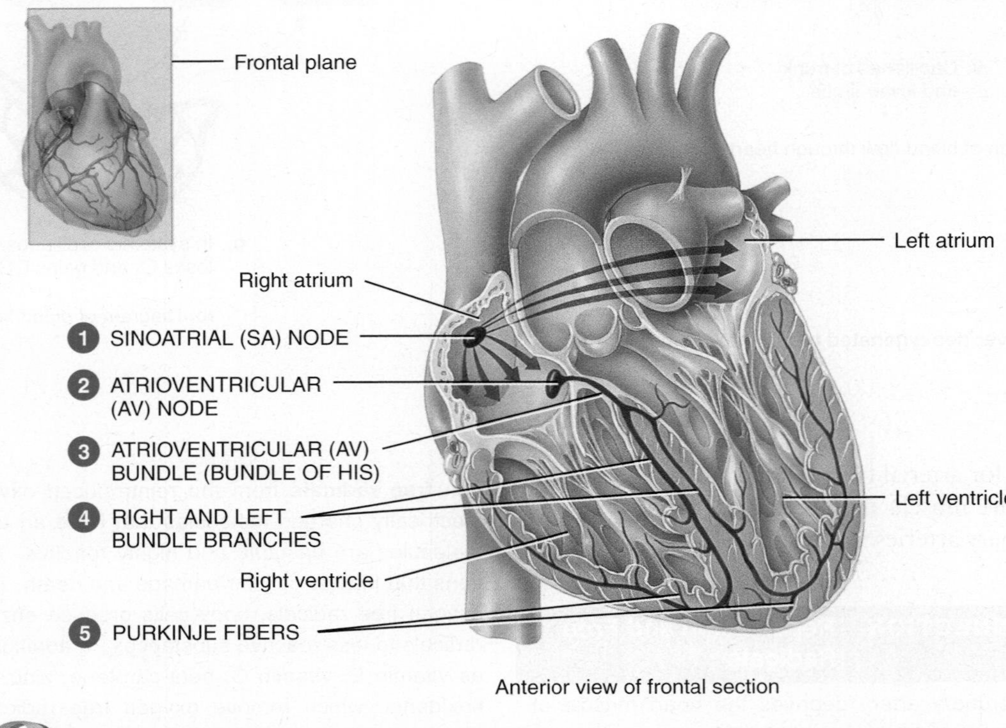

? Which component of the conduction system provides the only route for action potentials to conduct between the atria and the ventricles?

in the interatrial septum, just anterior to the opening of the coronary sinus. At the AV node, the action potential slows considerably, providing time for the atria to empty their blood into the ventricles.

3 From the AV node, the action potential enters the ***atrioventricular (AV) bundle*** (also known as the ***bundle of His***), in the interventricular septum. The AV bundle is the only site where action potentials can conduct from the atria to the ventricles.

4 After conducting along the AV bundle, the action potential then enters both the ***right*** and ***left bundle branches*** that course through the interventricular septum toward the apex of the heart.

5 Finally, large-diameter ***Purkinje fibers*** (pur-KIN-jē) rapidly conduct the action potential, first to the apex of the ventricles and then upward to the remainder of the ventricular myocardium. Then, a fraction of a second after the atria contract, the ventricles contract.

The SA node initiates action potentials about 100 times per minute, faster than any other region of the conducting system. Thus, the SA node sets the rhythm for contraction of the heart—it is the *pacemaker* of the heart. Various hormones and neurotransmitters can speed or slow pacing of the heart by SA node fibers. In a person at rest, for example, acetylcholine released by the parasympathetic division of the autonomic nervous system (ANS) typically slows SA node pacing to about 75 action potentials per minute, causing 75 heartbeats per minute. If the SA node becomes diseased or damaged, the slower AV node fibers can become the pacemaker. With pacing by the AV node, however, heart rate is slower, only 40 to 60 beats/min. If the activity of both nodes is suppressed, the heartbeat may still be maintained by the AV bundle, a bundle branch, or Purkinje fibers. These fibers generate action potentials very slowly, about 20 to 35 times per minute. At such a low heart rate, blood flow to the brain is inadequate.

CLINICAL CONNECTION | Artificial Pacemaker

When the heart rate is too low, normal heart rhythm can be restored and maintained by surgically implanting an **artificial pacemaker**, a device that sends out small electrical currents to stimulate the heart to contract. A pacemaker consists of a battery and impulse generator and is usually implanted beneath the skin just inferior to the clavicle. The pacemaker is connected to one or two flexible wires (leads) that are threaded through the superior vena cava and then passed into the right atrium and right ventricle. Many of the newer pacemakers, called *activity-adjusted pacemakers*, automatically speed up the heartbeat during exercise. •

✓ **CHECKPOINT**

8. Describe the path of an action potential through the conduction system.

15.4 ELECTROCARDIOGRAM

OBJECTIVE • **Describe the meaning and diagnostic value of an electrocardiogram.**

Conduction of action potentials through the heart generates electrical currents that can be picked up by electrodes placed on the skin. A recording of the electrical changes that accompany the heartbeat is called an ***electrocardiogram*** (e-lek′-trō-KAR-dē-ō-gram), which is abbreviated as either ***ECG*** or ***EKG***.

Three clearly recognizable waves accompany each heartbeat. The first, called the ***P wave***, is a small upward deflection on the ECG (Figure 15.7); it represents atrial depolarization, the depolarizing phase of the cardiac action potential as it spreads from the SA node throughout both atria. Depolarization causes contraction. Thus, a fraction of a second after the P wave begins, the atria contract. The second wave, called the ***QRS complex***, begins as a downward deflection (Q); continues as a large, upright, triangular wave (R); and ends as a downward wave (S). The QRS complex represents the onset of

Figure 15.7 Normal electrocardiogram (ECG) of a single heartbeat. P wave = atrial depolarization; QRS complex = onset of ventricular depolarization; T wave = ventricular repolarization.

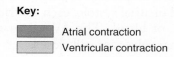 An electrocardiogram is a recording of the electrical activity that initiates each heartbeat.

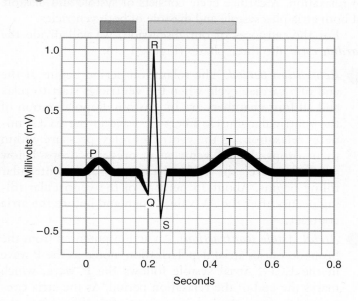

Key:

▮ Atrial contraction
▮ Ventricular contraction

 What event occurs in response to atrial depolarization?

ventricular depolarization, as the cardiac action potential spreads through the ventricles. Shortly after the QRS complex begins, the ventricles start to contract. The third wave is the *T wave*, a dome-shaped upward deflection that indicates ventricular repolarization and occurs just before the ventricles start to relax. Repolarization of the atria is not usually evident in an ECG because it is masked by the larger QRS complex.

Variations in the size and duration of the waves of an ECG are useful in diagnosing abnormal cardiac rhythms and conduction patterns and in following the course of recovery from a heart attack. An ECG can also reveal the presence of a living fetus.

✓ CHECKPOINT

9. What is the significance of the P wave, QRS complex, and T wave?

15.5 THE CARDIAC CYCLE

OBJECTIVE • **Describe the phases of the cardiac cycle.**

A single *cardiac cycle* includes all of the events associated with one heartbeat. In a normal cardiac cycle, the two atria contract while the two ventricles relax; then, while the two ventricles contract, the two atria relax. The term *systole* (SIS-tō-lē = contraction) refers to the phase of contraction; *diastole* (dī-AS-tō-lē = dilation or expansion) refers to the phase of relaxation. A cardiac cycle consists of systole and diastole of both atria plus systole and diastole of both ventricles.

For the purposes of our discussion, we will divide the *cardiac cycle* into three phases (Figure 15.8):

❶ *Relaxation period*. The relaxation period begins at the end of a cardiac cycle when the ventricles start to relax and all four chambers are in diastole. Repolarization of the ventricular muscle fibers (T wave in the ECG) initiates relaxation. As the ventricles relax, pressure within them drops. When ventricular pressure drops below atrial pressure, the AV valves open and ventricular filling begins. About 75 percent of the ventricular filling occurs after the AV valves open and before the atria contract.

❷ *Atrial systole (contraction)*. An action potential from the SA node causes atrial depolarization, noted as the P wave in the ECG. Atrial systole follows the P wave, which marks the end of the relaxation period. As the atria contract, they force the last 25 percent of the blood into the ventricles. At the end of atrial systole, each ventricle contains about 130 mL of blood. The AV valves are still open and the semilunar valves are still closed.

❸ *Ventricular systole (contraction)*. The QRS complex in the ECG indicates ventricular depolarization, which

Figure 15.8 Cardiac cycle.

🔑 **A cardiac cycle is composed of all of the events associated with one heartbeat.**

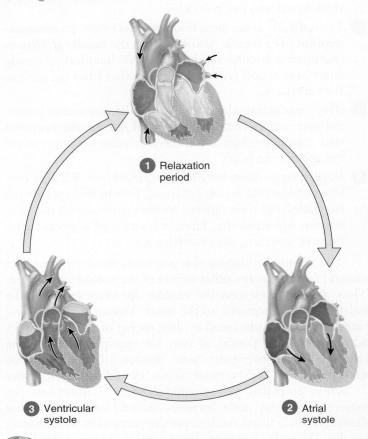

❶ Relaxation period

❸ Ventricular systole

❷ Atrial systole

❓ **What is the term used for the contraction phase of the cardiac cycle? The relaxation phase?**

leads to contraction of the ventricles. Ventricular contraction pushes blood against the AV valves, forcing them shut. As ventricular contraction continues, pressure inside the chambers quickly rises. When left ventricular pressure surpasses aortic pressure and right ventricular pressure rises above the pressure in the pulmonary trunk, both semilunar valves open, and ejection of blood from the heart begins. Ejection continues until the ventricles start to relax. At rest, the volume of blood ejected from each ventricle during ventricular systole is about 70 mL (a little more than 2 oz.). When the ventricles begin to relax, ventricular pressure drops, the semilunar valves close, and another relaxation period begins.

At rest, each cardiac cycle lasts about 0.8 sec. In one complete cycle, the first 0.4 sec of the cycle is the relaxation period, when all four chambers are in diastole. Then, the atria are in systole for 0.1 sec and in diastole for the next 0.7 sec. After atrial systole, the ventricles are in systole for 0.3 sec and in diastole for 0.5 sec. When the heart beats faster, for instance during exercise, the relaxation period is shorter.

Heart Sounds

The sound of the heartbeat comes primarily from turbulence in blood flow created by the closure of the valves, not from the contraction of the heart muscle. The first sound, *lubb*, is a long, booming sound from the AV valves closing after ventricular systole begins. The second sound, a short, sharp sound, *dupp*, is from the semilunar valves closing at the end of ventricular systole. There is a pause during the relaxation period. Thus, the cardiac cycle is heard as: lubb, dupp, pause; lubb, dupp, pause; lubb, dupp, pause.

> **CLINICAL CONNECTION | Heart Murmurs**
>
> Heart sounds provide valuable information about the mechanical operation of the heart. A **heart murmur** is an abnormal sound consisting of a clicking, rushing, or gurgling noise that is heard before, between, or after the normal heart sounds, or that may mask the normal heart sounds. Heart murmurs in children are extremely common and usually do not represent a health condition. These types of heart murmurs often subside or disappear with growth. Although some heart murmurs in adults are innocent, most often a murmur indicates a valve disorder. •

✓ CHECKPOINT

10. Explain the events that occur during each of the three phases of the cardiac cycle.
11. What causes the heart sounds?

15.6 CARDIAC OUTPUT

OBJECTIVE • **Define cardiac output, explain how it is calculated, and describe how it is regulated.**

The volume of blood ejected per minute from the left ventricle into the aorta is called the *cardiac output (CO)*. (Note that the same amount of blood is also ejected from the right ventricle into the pulmonary trunk.) Cardiac output is determined by (1) the *stroke volume (SV)*, the amount of blood ejected by the left ventricle during each beat (contraction), and (2) *heart rate (HR)*, the number of heartbeats per minute. In a resting adult, stroke volume averages 70 mL, and heart

rate is about 75 beats per minute. Thus the average cardiac output in a resting adult is

$$\text{Cardiac output} = \text{stroke volume} \times \text{heart rate}$$
$$= 70 \text{ mL/beat} \times 75 \text{ beats/min}$$
$$= 5250 \text{ mL/min or } 5.25 \text{ liters/min}$$

Factors that increase stroke volume or heart rate, such as exercise, increase cardiac output.

Regulation of Stroke Volume

Although some blood is always left in the ventricles at the end of their contraction, a healthy heart pumps out the blood that has entered its chambers during the previous diastole. The more blood that returns to the heart during diastole, the more blood that is ejected during the next systole. Three factors regulate stroke volume and ensure that the left and right ventricles pump equal volumes of blood:

1. **The degree of stretch in the heart before it contracts.** Within limits, the more the heart is stretched as it fills during diastole, the greater the force of contraction during systole, a relationship known as the *Frank–Starling law of the heart*. The situation is somewhat like stretching a rubber band: The more you stretch the heart, the more forcefully it contracts. In other words, within physiological limits, the heart pumps all the blood it receives. If the left side of the heart pumps a little more blood than the right side, a larger volume of blood returns to the right ventricle. On the next beat the right ventricle contracts more forcefully, and the two sides are again in balance.

2. **The forcefulness of contraction of individual ventricular muscle fibers.** Even at a constant degree of stretch, the heart can contract more or less forcefully when certain substances are present. Stimulation of the sympathetic division of the autonomic nervous system (ANS), hormones such as epinephrine and norepinephrine, increased Ca^{2+} level in the interstitial fluid, and the drug digitalis all increase the force of contraction of cardiac muscle fibers. In contrast, inhibition of the sympathetic division of the ANS, anoxia, acidosis, some anesthetics, and increased K^+ level in the extracellular fluid decrease contraction force.

3. **The pressure required to eject blood from the ventricles.** The semilunar valves open and ejection of blood from the heart begins when pressure in the right ventricle exceeds the pressure in the pulmonary trunk and when the pressure in the left ventricle exceeds the pressure in the aorta. When the required pressure is higher than normal, the valves open later than normal, stroke volume decreases, and more blood remains in the ventricles at the end of systole.

Hypertension Prevention

Hypertension is a leading cause of death around the world, increasing the risks of heart disease, kidney disease, and stroke. A majority of North Americans who reach their sixties develop hypertension. Many researchers believe that hypertension is caused by our modern lifestyles, which include too much of the wrong kinds of food, too little physical activity, and high levels of chronic stress. While hypertension can be treated with drugs, these are often expensive and may have unwanted side effects. Fortunately, a healthful lifestyle can often prevent the development of hypertension or at least delay its onset.

Calming the Fight-or-flight Response

The sympathetic nervous system sets the fight-or-flight stress response into motion, increasing heart rate, peripheral resistance, and blood pressure. While this response can save your life, most people spend too much time in an overly wound-up state. Many people rely on stress arousal and anxiety to motivate their productivity in school and at work.

This is a hard pattern to change, requiring the cultivation of new habits. At the very least, it is important to balance the hours spent in a stress buzz with more relaxing times. Unwinding with alcohol, caffeine, or cigarettes does not count. All three stimulate the sympathetic nervous system, elevating heart rate and blood pressure, thus increasing risk for hypertension. Many people learn and practice relaxation techniques to increase parasympathetic nervous system output, also called parasympathetic tone. Even just ten or fifteen minutes a day of breathing exercises, meditation, or other practices can help lower resting blood pressure.

One of the best ways to increase parasympathetic tone is with at least 30 minutes of regular exercise each day. Exercise can also improve blood sugar regulation and body composition, which themselves affect blood pressure. Excess body fat and higher levels of insulin activate the sympathetic nervous system response.

Reducing Blood Volume

Excess sodium is normally secreted by the kidneys, but when the sympathetic nervous system is activated or excessive amounts of salt are consumed, the kidneys retain too much sodium, which causes water retention and higher blood volume. More blood in the circulatory system means higher blood pressure. Reducing stress and consuming a healthful diet can lower blood volume and reduce blood pressure. Diets high in fruits and vegetables, low in salt, and with adequate low-fat sources of calcium have been shown to significantly reduce high blood pressure.

Think It Over . . . **The DASH (Dietary Approaches to Stop Hypertension) Diet recommends four to five servings of fruit and four to five servings of vegetables each day. Outline a day's meals (including snacks) that would include four servings of fruits and five servings of vegetables. (Serving sizes are approximately $1/2$ cup of cooked fruits or vegetables, or 1 cup raw.)**

CLINICAL CONNECTION | Congestive Heart Failure

In **congestive heart failure (CHF),** the heart is a failing pump. It pumps blood less and less effectively, leaving more blood in the ventricles at the end of each cycle. The result is a positive feedback cycle: Less-effective pumping leads to even lower pumping capability. Often, one side of the heart starts to fail before the other. If the left ventricle fails first, it can't pump out all the blood it receives, and blood backs up in the lungs. The result is *pulmonary edema*, fluid accumulation in the lungs that can lead to suffocation. If the right ventricle fails first, blood backs up in the systemic blood vessels. In this case, the resulting *peripheral edema* is usually most noticeable as swelling in the feet and ankles. Common causes of CHF are coronary artery disease (see Common Disorders), long-term high blood pressure, myocardial infarctions, and valve disorders. •

Regulation of Heart Rate

Adjustments to the heart rate are important in the short-term control of cardiac output and blood pressure. If left to itself, the sinoatrial node would set a constant heart rate of about 100 beats/min. However, tissues require different volumes of blood flow under different conditions. During exercise, for example, cardiac output rises to supply working tissues with increased amounts of oxygen and nutrients. The most important factors in the regulation of heart rate are the autonomic nervous system and the hormones epinephrine and norepinephrine, released by the adrenal glands.

Autonomic Regulation of Heart Rate

The nervous system regulation of the heart originates in the **cardiovascular (CV) center** in the medulla oblongata. This region of the brain stem receives input from a variety of sensory receptors and from higher brain centers, such as the limbic system and cerebral cortex. The cardiovascular center then directs appropriate output by increasing or decreasing the frequency of nerve impulses sent out to both the sympathetic and parasympathetic branches of the ANS (Figure 15.9).

Arising from the CV center are sympathetic neurons that reach the heart via *cardiac accelerator nerves*. They innervate the conduction system, atria, and ventricles. The norepinephrine released by cardiac accelerator nerves increases the heart rate. Also arising from the CV center are parasympathetic neurons that reach the heart via the *vagus (X) nerves*. These parasympathetic neurons extend to the conduction system and atria. The neurotransmitter they release—acetylcholine (ACh)—decreases the heart rate by slowing the pacemaking activity of the SA node.

Several types of sensory receptors provide input to the cardiovascular center. For example, *baroreceptors* (*baro-* = pressure), neurons sensitive to blood pressure changes, are strategically located in the arch of the aorta and carotid arteries (arteries in the neck that supply blood to the brain). If there is an increase in blood pressure, the baroreceptors send nerve impulses along sensory neurons that are part of the glossopharyn-

geal (IX) and vagus (X) nerves to the CV center (Figure 15.9). The cardiovascular center responds by putting out more nerve impulses along the parasympathetic (motor) neurons that are also part of the vagus (X) nerves and by decreasing cardiac accelerator output. The resulting decrease in heart rate lowers cardiac output and thus lowers blood pressure. If blood pressure falls, baroreceptors do not stimulate the cardiovascular center. As a result of this lack of stimulation, heart rate increases, cardiac output increases, and blood pressure increases to the normal level. *Chemoreceptors*, neurons sensitive to chemical changes in the blood, detect changes in blood levels of chemicals such as O_2, CO_2, and H^+. Their relationship to the cardiovascular center is considered in Chapter 16 with regard to blood pressure (see Section 16.2).

Chemical Regulation of Heart Rate

Certain chemicals influence both the basic physiology of cardiac muscle and its rate of contraction. Chemicals with major effects on the heart fall into one of two categories:

1. **Hormones.** Epinephrine and norepinephrine (from the adrenal medullae) enhance the heart's pumping effectiveness by increasing both heart rate and contraction force.

Figure 15.9 Autonomic nervous system regulation of heart rate.

The cardiovascular center in the medulla oblongata controls both sympathetic and parasympathetic nerves that innervate the heart.

Cardiovascular (CV) center

Glossopharyngeal (IX) nerves

Baroreceptors in carotid sinus

Baroreceptors in arch of aorta

Vagus (X) nerves (parasympathetic)

Medulla oblongata

SA node

AV node

Spinal cord

Ventricular myocardium

Cardiac accelerator nerves (sympathetic)

Sympathetic trunk ganglion

Key:
Sensory neurons ←
Motor neurons →

 What effect does acetylcholine, released by parasympathetic nerves, have on heart rate?

Exercise, stress, and excitement cause the adrenal medullae to release more hormones. Thyroid hormones also increase heart rate. One sign of hyperthyroidism (excessive levels of thyroid hormone) is tachycardia (elevated resting heart rate).

2. **Ions.** Elevated blood levels of K^+ or Na^+ decrease heart rate and contraction force. A moderate increase in extracellular and intracellular Ca^{2+} level increases heart rate and contraction force.

Other Factors in Heart Rate Regulation

Age, gender, physical fitness, and body temperature also influence resting heart rate. A newborn baby is likely to have a resting heart rate over 120 beats per minute; the rate then declines throughout childhood to the adult level of 75 beats per minute. Adult females generally have slightly higher resting heart rates than adult males, although regular exercise tends to bring resting heart rate down in both sexes. As adults age, their heart rates may increase.

Increased body temperature, such as occurs during fever or strenuous exercise, increases heart rate by causing the SA node to discharge more rapidly. Decreased body temperature decreases heart rate and force of contraction. During surgical repair of certain heart abnormalities, it is helpful to slow a patient's heart rate by deliberately cooling the body.

✓ CHECKPOINT

12. Describe how stroke volume is regulated.

13. How does the autonomic nervous system help regulate heart rate?

15.7 EXERCISE AND THE HEART

OBJECTIVE • **Explain the relationship between exercise and the heart.**

A person's cardiovascular fitness can be improved at any age with regular exercise. Some types of exercise are more effective than others for improving the health of the cardiovascular system. **Aerobic exercise,** any activity that works large body muscles for at least 20 minutes, elevates cardiac output and accelerates metabolic rate. Three to five such sessions a week are usually recommended for improving the health of the cardiovascular system. Brisk walking, running, bicycling, cross-country skiing, and swimming are examples of aerobic activities.

Sustained exercise increases the oxygen demand of the muscles. Whether the demand is met depends mainly on the adequacy of cardiac output and proper functioning of the respiratory system. After several weeks of training, a healthy person increases maximal cardiac output (the amount of blood ejected from the ventricles into their respective arteries per minute), thereby increasing the maximal rate of oxygen delivery to the tissues. Oxygen delivery also rises because skeletal muscles develop more capillary networks in response to long-term training.

During strenuous activity, a well-trained athlete can achieve a cardiac output double that of a sedentary person, in part because training causes hypertrophy (enlargement) of the heart. This condition is referred to as **physiological cardiomegaly** (kar′-dē-ō-MEG-a-lē; *mega* = large). A **pathological cardiomegaly** is related to significant heart disease. Even though the heart of a well-trained athlete is larger, *resting* cardiac output is about the same as in a healthy untrained person, because *stroke volume* (volume of blood pumped by each beat of a ventricle) is increased while heart rate is decreased. The resting heart rate of a trained athlete often is only 40–60 beats per minute *(resting bradycardia)*. Regular exercise also helps to reduce blood pressure, anxiety, and depression; control weight; and increase the body's ability to dissolve blood clots.

✓ CHECKPOINT

14. What is aerobic exercise? Why are aerobic exercises beneficial?

• • •

The heart is the blood pump for the cardiovascular system, but it is the blood vessels that distribute blood to all parts of the body and collect blood from them. In the next chapter we will see how blood vessels accomplish this essential task.

COMMON DISORDERS

Coronary Artery Disease

Coronary artery disease (CAD) is a serious medical problem that affects about 7 million people and causes nearly 750,000 deaths in the United States each year. CAD is defined as the effects of the accumulation of atherosclerotic plaques (described shortly) in coronary arteries that lead to a reduction in blood flow to the myocardium. Some individuals have no signs or symptoms, others experience angina pectoris (chest pain), and still others suffer a heart attack.

People who possess combinations of certain risk factors are more likely to develop CAD. *Risk factors* (characteristics, symptoms, or signs that are statistically associated with a greater chance of developing a disease) include smoking, high blood pressure, diabetes, high cholesterol levels, obesity, "type A" personality, sedentary lifestyle, and a family history of CAD. Most of these can be modified by changing diet and other habits or can be controlled by taking medications. However, other risk factors are unmodifiable—that is, beyond our control—including genetic predisposition (family history of CAD at an early age), age, and gender. For example, adult males are more likely than adult females to develop CAD; after age 70 the risks are roughly equal. Smoking is undoubtedly the number-one risk factor in all CAD-associated diseases, roughly doubling the risk of morbidity and mortality.

A number of other risk factors (all modifiable) have also been identified as significant predictors of CAD. *C-reactive proteins (CRPs)* are proteins produced by the liver or present in blood in an inactive form that are converted to an active form during inflammation. CRPs may play a direct role in the development of atherosclerosis by promoting the uptake of LDLs by macrophages. *Lipoprotein (a)* is an LDL-like particle that binds to endothelial cells, macrophages, and blood platelets, may promote the proliferation of smooth muscle fibers, and inhibits the breakdown of blood clots. *Fibrinogen* is a glycoprotein involved in blood clotting that may help regulate cellular proliferation, vasoconstriction, and platelet aggregation. *Homocysteine* is an amino acid that may induce blood vessel damage by promoting platelet aggregation and smooth muscle fiber proliferation.

Atherosclerosis (ath′-er-ō-skler-Ō-sis) is a progressive disease characterized by the formation in the walls of large- and medium-sized arteries of lesions called ***atherosclerotic plaques*** (Figure 15.10).

To understand how atherosclerotic plaques develop, you will need to know about molecules produced by the liver and small intestine called ***lipoproteins***. These spherical particles consist of an inner core of triglycerides and other lipids and an outer shell of proteins, phospholipids, and cholesterol. Two major lipoproteins are ***low-density lipoproteins*** or ***LDLs*** and ***high-density lipoproteins*** or ***HDLs***. LDLs transport cholesterol from the liver to body cells for use in cell membrane repair and the production of steroid hormones and bile salts. However, excessive amounts of LDLs promote atherosclerosis, so the cholesterol in these particles is known as "bad cholesterol." HDLs, on the other hand, remove excess cholesterol from body cells and transport it to the liver for elimination. Because HDLs decrease blood cholesterol level, the cholesterol in HDLs is known as "good cholesterol." Basically, you want your LDL to be low and your HDL to be high.

It has recently been learned that inflammation, a defensive response of the body to tissue damage, plays a key role in the development of atherosclerotic plaques. As a result of the damage, blood vessels dilate and increase their permeability. The formation of atherosclerotic plaques begins when excess LDLs from the blood accumulate in the artery wall and undergo oxidation. In response, endothelial and smooth muscle cells of the artery secrete substances that attract monocytes from the blood and convert them into macrophages. The macrophages then ingest and become so filled with the oxidized LDL particles that they have a foamy appearance when viewed microscopically (***foam cells***). Together with T cells (lymphocytes), foam cells form a ***fatty streak***, the beginning of an atherosclerotic plaque. Following fatty streak formation, smooth muscle cells of the artery migrate to the top of the atherosclerotic plaque, forming a cap over it and thus walling it off from the blood.

Because most atherosclerotic plaques expand away from the bloodstream rather than into it, blood can flow through an artery with relative ease, often for decades. Most heart attacks occur when the cap over the plaque breaks open in response to chemicals produced by foam cells, causing a clot to form. If the clot in a coronary artery is large enough, it can significantly decrease or stop the flow of blood and result in a heart attack.

Treatment options for CAD include drugs (antihypertensive drugs, nitroglycerin, beta-blockers, and cholesterol-lowering and clot-dissolving agents) and various surgical and nonsurgical procedures designed to increase the blood supply to the heart.

Figure 15.10 Photomicrographs of transverse sections of (a) a normal artery and (b) an artery partially obstructed by an atherosclerotic plaque.

 Atherosclerosis is a progressive disease caused by the formation of atherosclerotic plaques.

Atherosclerotic plaque

Partially obstructed space through which blood flows

LM 20x

(a) Normal artery

LM 20x

(b) Obstructed artery

 What substances are part of an atherosclerotic plaque?

Myocardial Ischemia and Infarction

Partial obstruction of blood flow in the coronary arteries may cause *myocardial ischemia* (is-KĒ-mē-a; *ische-* = to obstruct; *-emia* = in the blood), a condition of reduced blood flow to the myocardium. Usually, ischemia causes *hypoxia* (reduced oxygen supply), which may weaken cells without killing them. *Angina pectoris* (an-JĪ-na or AN-ji-na PEK-to-ris), which literally means "strangled chest," is a severe pain that usually accompanies myocardial ischemia. Typically, sufferers describe it as a tightness or squeezing sensation, as though the chest were in a vise. The pain associated with angina pectoris is often referred to the neck, chin, or down the left arm to the elbow. *Silent myocardial ischemia*, ischemic episodes without pain, is particulary dangerous because the person has no forewarning of an impending heart attack.

A complete obstruction to blood flow in a coronary artery may result in a *myocardial infarction* (*MI*) (in-FARK-shun), commonly called a *heart attack*. *Infarction* means the death of an area of tissue because of interrupted blood supply. Because the heart tissue distal to the obstruction dies and is replaced by non-contractile scar tissue, the heart muscle loses some of its strength. Depending on the size and location of the infarcted (dead) area, an infarction may disrupt the conduction system of the heart and cause sudden death by triggering ventricular fibrillation. Treatment for a myocardial infarction may involve injection of a thrombolytic (clot-dissolving) agent such as streptokinase or tPA, plus heparin (an anticoagulant), or performing coronary angioplasty or coronary artery bypass grafting. Fortunately, heart muscle can remain alive in a resting person if it receives as little as 10–15% of its normal blood supply.

Congenital Defects

A defect that exists at birth (and usually before) is a *congenital defect*. Among the several congenital defects that affect the heart are the following:

- In *patent ductus arteriosus*, the ductus arteriosus (temporary blood vessel) between the aorta and the pulmonary trunk, which normally closes shortly after birth, remains open (see Figure 16.17). Closure of the ductus arteriosus leaves a remnant called the ligamentum arteriosum (see Figure 15.3a).
- *Atrial septal defect (ASD)* is caused by incomplete closure of the interatrial septum. The most common type involves the foramen ovale, which normally closes shortly after birth (see Figure 16.17).
- *Ventricular septal defect (VSD)* is caused by an incomplete closure of the interventricular septum.
- *Valvular stenosis* is a narrowing of one of the valves associated with blood flow through the heart.
- *Tetralogy of Fallot* (te-TRAL-Ō-jē of fa-LŌ) is a combination of four defects: an interventricular septal defect, an aorta that emerges from both ventricles instead of from the left ventricle only, a narrowed pulmonary semilunar valve, and an enlarged right ventricle.

Some congenital heart defects are being surgically corrected prior to birth in order to prevent complications at the time of birth or following the birth of an infant.

Arrhythmias

The usual rhythm of heartbeats, established by the SA node, is called *normal sinus rhythm*. The term *arrhythmia* (a-RITH-mē-a; *a-* = without) or *dysrhythmia* refers to an abnormal rhythm as a result of a defect in the conduction system of the heart. The heart may beat irregularly, too fast, or too slowly. Symptoms include chest pain, shortness of breath, lightheadedness, dizziness, and fainting. Arrhythmias may be caused by factors that stimulate the heart, such as stress, caffeine, alcohol, nicotine, cocaine, and certain drugs that contain caffeine or other stimulants. Arrhythmias may also be caused by a congenital defect, coronary artery disease, myocardial infarction, hypertension, defective heart valves, rheumatic heart disease, hyperthyroidism, and potassium deficiency.

Following are some types of arrhythmias:

- *Supraventricular tachycardia (SVT)* is a rapid but regular heart rate (160–200 beats per minute) that originates in the atria. The episodes begin and end suddenly and may last from a few minutes to many hours.
- *Heart block* is an arrhythmia that occurs when the electrical pathways between the atria and ventricles are blocked, slowing the transmission of nerve impulses. The most common site of blockage is the atrioventricular node, a condition called *atrioventricular (AV) block*.
- *Atrial premature contraction (APC)* is a heartbeat that occurs earlier than expected and briefly interrupts the normal heart rhythm. It often causes a sensation of a skipped heartbeat followed by a more forceful heartbeat. APCs originate in the atrial myocardium and are common in healthy individuals.
- *Atrial flutter* consists of rapid, regular atrial contractions (240–360 beats/min) accompanied by an atrioventricular (AV) block in which some of the nerve impulses from the SA node are not conducted through the AV node.
- *Atrial fibrillation (AF)* is a common arrhythmia, affecting mostly older adults, in which contraction of the atrial fibers is asynchronous (not in unison) so that atrial pumping ceases altogether. The atria may beat 300 to 600 beats per minute. The ventricles may also speed up, resulting in a rapid heartbeat (up to 160 beats/min).
- *Ventricular premature contraction (VPC)*. Another form of arrhythmia arises when an *ectopic focus* (ek-TOP-ik), a region of the heart other than the conduction system, becomes more excitable than normal and causes an occasional abnormal action potential to occur. As a wave of depolarization spreads outward from the ectopic focus, it causes a *ventricular premature contraction (beat)*. The contraction occurs early in diastole before the SA node is normally scheduled to discharge its action potential. Ventricular premature contractions may be relatively benign and may be caused by emotional stress, excessive intake of stimulants such as caffeine, alcohol, or nicotine, and lack of sleep. In other cases, the premature beats may reflect an underlying pathology.
- *Ventricular tachycardia (VT or V-tach)* is an arrhythmia that originates in the ventricles and is characterized by four or more ventricular premature contractions. It causes the ventricles to beat too fast (at least 120 beats/min). VT is almost always associated with heart disease or a recent myocardial infarction and may develop into a very serious arrhythmia called ventricular fibrillation

(described shortly). Sustained VT is dangerous because the ventricles do not fill properly and thus do not pump sufficient blood. The result may be low blood pressure and heart failure.

- *Ventricular fibrillation (VF or V-fib)* is the most deadly arrhythmia, in which contractions of the ventricular fibers are completely asynchronous so that the ventricles quiver rather than contract in a coordinated way. As a result, ventricular pumping stops, blood ejection ceases, and circulatory failure and death occur unless there is immediate medical intervention. Ventricular fibrillation causes unconsciousness in seconds

and, if untreated, seizures occur and irreversible brain damage may occur after five minutes. Death soon follows. Treatment involves cardiopulmonary resuscitation (CPR) and defibrillation. In *defibrillation* (dē-fib-ri-LĀ-shun), also called *cardioversion* (kar′-dē-ō-VER-shun), a strong, brief electrical current is passed to the heart and often can stop the ventricular fibrillation. The electrical shock is generated by a device called a *defibrillator* (de-FIB-ri-lā-tor) and applied via two large paddle-shaped electrodes pressed against the skin of the chest.

MEDICAL TERMINOLOGY AND CONDITIONS

Angiocardiography (an′-jē-ō-kar′-dē-OG-ra-fē; *angio-* = vessel; *cardio-* = heart)   X-ray examination of the heart and great blood vessels after injection of a radiopaque dye into the bloodstream.

Asystole (ā-SIS-tō-lē; *a-* = without)   Failure of the myocardium to contract.

Cardiac arrest (KAR-dē-ak a-REST)   A clinical term meaning cessation of an effective heartbeat. The heart may be completely stopped or in ventricular fibrillation.

Cardiac catheterization (kath′-e-ter-i-ZĀ-shun)   Procedure that is used to visualize the heart's coronary arteries, chambers, valves, and great vessels. It may also be used to measure pressure in the heart and blood vessels; to assess cardiac output; and to measure the flow of blood through the heart and blood vessels, the oxygen content of blood, and the status of the heart valves and conduction system. The basic procedure involves inserting a catheter into a peripheral vein (for right heart catheterization) or artery (for left heart catheterization) and guiding it under fluoroscopy (x-ray observation).

Cardiac rehabilitation (rē-ha-bil-i-TĀ-shun)   A supervised program of progressive exercise, psychological support, education, and training to enable a patient to resume normal activities following a myocardial infarction.

Cardiomegaly (kar′-dē-ō-MEG-a-lē; *mega-* = large)   Heart enlargement.

Cor pulmonale (CP) (kor pul-mōn-ĀL; *cor-* = heart; *pulmon-* = lung)   Right ventricular hypertrophy caused by hypertension (high blood pressure) in the pulmonary circulation.

Cardiopulmonary resuscitation (kar′-dē-ō-PUL-mō-ner-ē re-sus′-i-TĀ-shun) **(CPR)**   The artificial establishment of normal or

near-normal respiration and circulation. The *ABCs* of cardiopulmonary resuscitation are *Airway*, *Breathing*, and *Circulation*, meaning the rescuer must establish an airway, provide artificial ventilation if breathing has stopped, and reestablish circulation if there is inadequate cardiac action.

Endocarditis (en′-dō-kar-DĪ-tis)   Inflammation of the endocardium that typically involves the heart valves. Most cases are caused by bacteria (bacterial endocarditis).

Myocarditis (mī-ō-kar-DĪ-tis)   Inflammation of the myocardium that usually occurs as a complication of a viral infection, rheumatic fever, or exposure to radiation or certain chemicals or medications.

Palpitation (pal′-pi-TĀ-shun)   A fluttering of the heart or abnormal rate or rhythm of the heart.

Paroxysmal tachycardia (par′-ok-SIZ-mal tak′-e-KAR-dē-a)   A period of rapid heartbeats that begins and ends suddenly.

Rheumatic fever (roo-MAT-ik)   An acute systemic inflammatory disease that usually occurs after a streptococcal infection of the throat. The bacteria trigger an immune response in which antibodies that are produced to destroy the bacteria attack and inflame the connective tissues in joints, heart valves, and other organs. Even though rheumatic fever may weaken the entire heart wall, most often it damages the bicuspid (mitral) and aortic valves.

Sudden cardiac death   The unexpected cessation of circulation and breathing due to an underlying heart disease such as ischemia, myocardial infarction, or a disturbance in cardiac rhythm.

CHAPTER REVIEW AND RESOURCE SUMMARY

REVIEW

RESOURCES

15.1 Structure and Organization of the Heart

1. The **heart** is situated between the lungs, with about two-thirds of its mass to the left of the midline.
2. The **pericardium** consists of an outer fibrous layer (**fibrous pericardium**) and an inner **serous pericardium**. The serous pericardium is composed of a **parietal layer** and a **visceral layer**. Between the parietal and visceral layers of the serous pericardium is the **pericardial cavity**, a space filled with **pericardial fluid** that reduces friction between the two membranes.

Anatomy Overview—Cardiac Muscle

Figure 15.3 Structure of the heart

REVIEW	RESOURCES
3. The wall of the heart has three layers: **epicardium**, **myocardium**, and **endocardium**.	Exercise—Paint the heart
4. The chambers include two upper **atria** and two lower **ventricles**.	
5. The blood flows through the heart from the **superior** and **inferior venae cavae** and the **coronary sinus** to the right atrium, through the right ventricle, and through the **pulmonary trunk** to the lungs.	
6. From the lungs, blood flows through the **pulmonary veins** into the left atrium, through the left ventricle, and out through the aorta.	
7. Four valves prevent the backflow of blood in the heart. **Atrioventricular (AV) valves**, between the atria and their ventricles, are the **tricuspid valve** on the right side of the heart and the **bicuspid (mitral) valve** on the left. The atrioventricular valves, **chordae tendineae**, and their **papillary muscles** stop blood from flowing back into the atria. Each of the two arteries that leave the heart has a **semilunar valve**.	

15.2 Blood Flow and Blood Supply of the Heart

1. Blood flows through the heart from areas of higher pressure to areas of lower pressure. The pressure is related to the size and volume of a chamber.	Exercise—Drag and Drop Blood Flow
2. The movement of blood through the heart is controlled by the opening and closing of the valves and the contraction and relaxation of the myocardium.	**Figure 15.5** Blood flow through the heart
3. **Coronary (cardiac) circulation** delivers oxygenated blood to the myocardium and removes carbon dioxide from it.	
4. Deoxygenated blood returns to the right atrium via the **coronary sinus**.	

15.3 Conduction System of the Heart

1. The **conduction system** consists of specialized cardiac muscle tissue that generates and distributes action potentials.	Animation—Cardiac conduction Exercise—Sequence Cardiac Conduction
2. Components of this system are the **sinoatrial (SA) node (pacemaker)**, **atrioventricular (AV) node**, **atrioventricular (AV) bundle (bundle of His)**, **bundle branches**, and **Purkinje fibers**.	

15.4 Electrocardiogram

1. The record of electrical changes during each cardiac cycle is referred to as an **electrocardiogram (ECG)**.	Animation—Cardiac Cycle and ECG Exercise—ECG Jigsaw Puzzle
2. A normal ECG consists of a **P wave** (depolarization of atria), **QRS complex** (onset of ventricular depolarization), and **T wave** (ventricular repolarization).	
3. The ECG is used to diagnose abnormal cardiac rhythms and conduction patterns.	

15.5 The Cardiac Cycle

1. A **cardiac cycle** consists of **systole** (contraction) and **diastole** (relaxation) of the chambers of the heart.	Animation—Cardiac Cycle Exercise—Cardiac Cycle
2. The phases of the cardiac cycle are (a) the **relaxation period**, (b) **atrial systole**, and (c) **ventricular systole**.	
3. A complete cardiac cycle takes 0.8 sec at an average heartbeat of 75 beats per minute.	
4. The first heart sound (**lubb**) represents the closing of the atrioventricular valves. The second sound (**dupp**) represents the closing of semilunar valves.	

15.6 Cardiac Output

1. **Cardiac output (CO)** is the amount of blood ejected by the left ventricle into the aorta each minute: CO = stroke volume × beats per minute.	Animation—Cardiac output Animation—ANS Effects on Cardiac Cycle Exercise—Cardiac Output Factors
2. **Stroke volume (SV)** is the amount of blood ejected by a ventricle during ventricular systole. It is related to stretch on the heart before it contracts, forcefulness of contraction, and the amount of pressure required to eject blood from the ventricles.	

REVIEW

3. Nervous control of the cardiovascular system originates in the **cardiovascular (CV) center** in the medulla oblongata. Sympathetic impulses increase heart rate and force of contraction; parasympathetic impulses decrease heart rate.

4. Heart rate is affected by hormones (epinephrine, norepinephrine, thyroid hormones), ions (Na^+, K^+, Ca^{2+}), age, gender, physical fitness, and body temperature.

15.7 Exercise and the Heart

1. Sustained exercise increases oxygen demand on muscles.

2. Among the benefits of **aerobic exercise** are increased maximal cardiac output, decreased blood pressure, weight control, and increased ability to dissolve clots.

RESOURCES

Animation—Exercise and Cardiac Output

SELF-QUIZ

1. Match the following:

_____ **a.** valve between the left atrium and left ventricle

_____ **b.** valve between the right atrium and right ventricle

_____ **c.** chamber that pumps blood to the lungs

_____ **d.** chamber that pumps blood into the aorta

_____ **e.** chamber that receives oxygenated blood from lungs

_____ **f.** chamber that receives deoxygenated blood from body

_____ **g.** valve between the left ventricle and aorta

_____ **h.** valve between right ventricle and pulmonary trunk

_____ **i.** wall between lower heart chambers

_____ **j.** sac-like structure located on anterior surface of atrium

_____ **k.** wall between upper heart chambers

A. aortic valve
B. right atrium
C. left atrium
D. bicuspid (mitral) valve
E. pulmonary valve
F. auricle
G. interatrial septum
H. right ventricle
I. left ventricle
J. tricuspid valve
K. interventricular septum

2. Which of the following statements describes the pericardium?

a. It is a layer of nervous tissue.

b. It lines the inside of the myocardium.

c. It is continuous with the epithelial lining of the large blood vessels.

d. It is responsible for the contraction of the heart.

e. It is a two-part membrane that surrounds and protects the heart.

3. Which blood vessel delivers deoxygenated blood from the head and neck to the heart?

a. pulmonary vein
b. thoracic aorta
c. pulmonary artery
d. inferior vena cava
e. superior vena cava

4. Cardiac muscle fibers

a. are connected by intercalated discs.

b. form networks that contract as functional units.

c. have longer contraction and refractory periods than skeletal muscle fibers.

d. conduct action potentials through gap junctions.

e. are described by all of the above.

5. The chordae tendineae and papillary muscles of the heart

a. are responsible for connecting cardiac muscle fibers.

b. can develop self-excitability and stimulate contraction.

c. help prevent the atrioventricular valves from protruding into the atria when the ventricles contract.

d. help anchor and protect the heart.

e. form the cusps (flaps) of the heart valves.

6. Which chamber of the heart has the thickest layer of myocardium?

a. right ventricle
b. right atrium
c. left ventricle
d. left atrium
e. coronary sinus

7. The normal "pacemaker" of the heart is the

a. sinoatrial (SA) node.
b. atrioventricular (AV) node.
c. Purkinje fibers.
d. atrioventricular (AV) bundle.
e. right bundle branch.

8. In normal heart action,

a. the right atrium and ventricle contract, followed by the contraction of the left atrium and ventricle.

b. the order of contraction is right atrium, then right ventricle, then left atrium, then left ventricle.

c. the two atria contract together, and then the two ventricles contract together.

d. the right atrium and left ventricle contract, followed by the contraction of the left atrium and right ventricle.

e. all four chambers of the heart contract and then relax simultaneously.

9. Heart sounds are produced by
   a. contraction of the myocardium.
   b. closure of the heart valves.
   c. the flow of blood in the coronary arteries.
   d. the flow of blood in the ventricles.
   e. the transmission of action potentials through the conduction system.

10. Heart rate and strength of contraction are controlled by the cardiovascular center, which is located in the
    a. cerebrum.   b. pons.   c. right atrium.
    d. medulla.   e. atrioventricular node.

11. The portion of the ECG that corresponds to atrial depolarization is the
    a. R peak.   b. space between the T wave and P wave.
    c. T wave.   d. P wave.   e. QRS complex

12. The opening of the semilunar valves is due to the pressure in the
    a. ventricles exceeding the pressure in the aorta and pulmonary trunk.
    b. ventricles exceeding the pressure in the atria.
    c. atria exceeding the pressure in the ventricles.
    d. atria exceeding the pressure in the aorta and pulmonary trunk.
    e. aorta and pulmonary trunk exceeding the pressure in the ventricles.

13. The blood supply to the myocardium is the
    a. coronary circulation.   b. pulmonary circulation.
    c. coronary conduction system. d. cardiac cycle.
    e. cardiac output.

14. The Frank–Starling law of the heart
    a. is important in maintaining equal blood output from both ventricles.
    b. is used in reference to the force of contraction of the atria.
    c. results in a decreased heart rate.
    d. causes blood to accumulate in the lungs.
    e. is related to the stretching of the cardiac muscle cells in the atria.

15. Which of the following sequences best represents the pathway of an action potential through the heart's conduction system?
    1. sinoatrial (SA) node
    2. Purkinje fibers
    3. atrioventricular (AV) bundle
    4. atrioventricular (AV) node
    5. right and left bundle branches
       a. 1, 4, 3, 2, 5   b. 4, 1, 3, 5, 2   c. 3, 4, 1, 2, 5
       d. 1, 4, 3, 5, 2   e. 2, 5, 3, 4, 1

16. Which of the following is NOT true concerning ventricular filling during the cardiac cycle?
    a. The atrioventricular (AV) valves are open.
    b. The ventricles fill to 75% of their capacity before the atria contract.
    c. The remaining 25% of the ventricular blood is forced into the ventricles when the atria contract.

d. The semilunar valves are open.
e. Ventricular filling begins when the ventricular pressure drops below the atrial pressure, causing the AV valves to open.

17. Cardiac output
    a. equals stroke volume (SV) × blood pressure (BP).
    b. equals stroke volume (SV) × heart rate (HR).
    c. is calculated using the formula for the Frank–Starling law of the heart.
    d. is about 70 mL in the average adult male.
    e. equals blood pressure (BP) × heart rate (HR).

18. Which of the following statements is NOT true?
    a. An increase in blood pressure would activate baroreceptors.
    b. Chemoreceptors monitor changes in carbon dioxide and oxygen levels.
    c. Baroreceptors monitor blood pressure in the aorta and carotid arteries.
    d. A drop in blood pressure would cause baroreceptors to stimulate the cardiovascular center to increase parasympathetic impulses.
    e. Both baroreceptors and chemoreceptors can affect heart rate by influencing the cardiovascular center.

19. Using the situations that follow, indicate if the heart rate would speed up (**A**) or slow down (**B**).
    ____ a. sympathetic stimulation of the sinoatrial (SA) node
    ____ b. decrease in blood pressure
    ____ c. fever
    ____ d. parasympathetic stimulation of the heart's conduction system
    ____ e. release of epinephrine
    ____ f. elevated K^+ level
    ____ g. release of acetylcholine
    ____ h. strenuous exercise
    ____ i. stimulation by the vagus (X) nerve
    ____ j. stress, excitement
    ____ k. cooling the body
    ____ l. excessive thyroid hormones

20. Match the following:
    ____ a. may cause a heart murmur
    ____ b. heart compression
    ____ c. inflammation of heart covering
    ____ d. heart chamber contraction
    ____ e. chest pain from ischemia
    ____ f. heart attack
    ____ g. heart chamber relaxation

    A. pericarditis
    B. valve disorder
    C. myocardial infarction
    D. angina pectoris
    E. diastole
    F. systole
    G. cardiac tamponade

CRITICAL THINKING APPLICATIONS

1. Gareth had an artificial pacemaker inserted after his last bout with heart trouble. What is the function of a pacemaker? For which heart structure does the pacemaker substitute?

2. David was strolling across the bend in a road when a car suddenly appeared out of nowhere. As he finished sprinting across the road, he felt his heart racing. Trace the route of the signal from his brain to his heart.

3. Samantha, a member of the college's tennis team, volunteered to have her heart function evaluated by the exercise physiology class. Her resting pulse rate was 40 beats per minute. Assuming that she has an average cardiac output (CO), determine Samantha's stroke volume (SV). Next, Samantha rode an exercise bike until her heart rate had risen to 60 beats per minute. Assuming that her SV stayed constant, calculate Samantha's CO during this moderate exercise.

4. Tricia called you with great anxiety because her husband told her that his HDL levels were high and his LDL levels were low. She knows that these blood measurements have something to do with "heart health" and cholesterol levels. Should Tricia be concerned about her husband's HDL and LDL levels?

ANSWERS TO FIGURE QUESTIONS

15.1 The base of the heart consists mainly of the left atrium.

15.2 The visceral layer of the serous pericardium is also part of the heart wall (epicardium).

15.3 Blood flows away from the heart in arteries.

15.4 Heart valves prevent the backflow of blood.

15.5 The superior vena cava, inferior vena cava, and coronary sinus deliver deoxygenated blood into the right atrium.

15.6 The only electrical connection between the atria and the ventricles is the atrioventricular (AV) bundle.

15.7 Atrial depolarization causes contraction of the atria.

15.8 The contraction phase is called systole; the relaxation phase is called diastole.

15.9 Acetylcholine decreases heart rate.

15.10 Fatty substances, cholesterol, and smooth muscle fibers make up atherosclerotic plaques.

CHAPTER 16

THE CARDIOVASCULAR SYSTEM: BLOOD VESSELS AND CIRCULATION

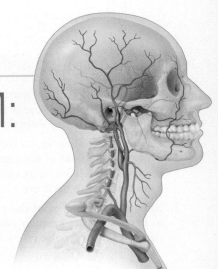

The cardiovascular system contributes to the homeostasis of other body systems by transporting and distributing blood throughout the body to deliver materials such as oxygen, nutrients, and hormones and to carry away wastes. This transport is accomplished by blood vessels, which form closed circulatory routes for blood to travel from the heart to body organs and back again. In Chapters 14 and 15 we discussed the composition and functions of blood and the structure and function of the heart. In this chapter, we examine the structure and functions of the different types of blood vessels that carry the blood to and from the heart, as well as factors that contribute to blood flow and regulation of blood pressure.

Did you know? *Physical activity helps to protect the cardiovascular system in many ways. It improves blood cholesterol levels and blood sugar regulation. Regular exercise leads to increased output from the parasympathetic nervous system (not during exercise, but during the rest of the day), which leads to a lower resting heart rate and lower resting blood pressure. Physical activity burns calories, thus helping to prevent obesity. People who exercise regularly have lower rates of inflammation markers, such as C-reactive protein. Less inflammation suggests lower risk of artery disease. To maximize cardiovascular health, researchers recommend about an hour of brisk activity each day.*

FOCUS ON WELLNESS
Reversing Artery Disease

LOOKING BACK TO MOVE AHEAD...

Diffusion (Section 3.3)

Medulla Oblongata (Section 10.4)

Antidiuretic Hormone (Section 13.3)

Mineralocorticoids (Section 13.7)

Great Vessels of the Heart (Section 15.1)

16.1 BLOOD VESSEL STRUCTURE AND FUNCTION

OBJECTIVES • **Compare the structure and function of the different types of blood vessels.**

- **Describe how substances enter and leave the blood in capillaries.**

- **Explain how venous blood returns to the heart.**

There are five types of blood vessels: arteries, arterioles, capillaries, venules, and veins (Figure 16.1). *Arteries* (AR-ter-ēz) carry blood *away from the heart* to body tissues. Two large arteries—the aorta and the pulmonary trunk—emerge from the heart and branch out into medium-sized arteries that serve various regions of the body. These medium-sized arter-

ies then divide into small arteries, which, in turn, divide into still smaller arteries called *arterioles* (ar-TER-ē-ōls). Arterioles within a tissue or organ branch into numerous microscopic vessels called *capillaries* (KAP-i-lar'-ēz). Groups of capillaries within a tissue reunite to form small veins called *venules* (VEN-ūls). These, in turn, merge to form progressively larger vessels called veins. *Veins* (VĀNZ) are the blood vessels that convey blood from the tissues *back to the heart*.

At any one time, systemic veins and venules contain about 64 percent of the total volume of blood in the system, systemic arteries and arterioles about 13 percent, systemic capillaries about 7 percent, pulmonary blood vessels about 9 percent, and the heart chambers about 7 percent. Because veins contain so much of the blood, certain veins function as *blood reservoirs*. The main blood reservoirs are the veins of the abdominal organs (especially the liver and spleen) and the

Figure 16.1 Comparative structure of blood vessels. The relative size of the capillary in (c) is enlarged for emphasis. Note the valve in the vein.

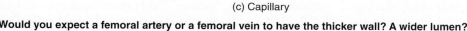
Arteries carry blood away from the heart to tissues. Veins carry blood from tissues back to the heart.

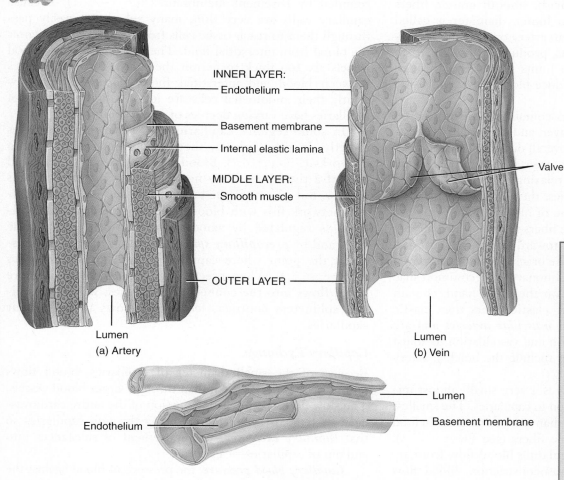

INNER LAYER:
Endothelium

Basement membrane

Internal elastic lamina

MIDDLE LAYER:
Smooth muscle

OUTER LAYER

Lumen
(a) Artery

Valve

Lumen
(b) Vein

Lumen

Basement membrane

Endothelium

(c) Capillary

Functions of Blood Vessels

1. Blood vessels form a closed system of tubes that carries blood away from the heart (in arteries), transports it through the tissues of the body (in arterioles, capillaries, and venules), and then returns it to the heart (in veins).
2. Exchange of substances between the blood and body tissue cells occurs as blood flows through the capillaries.
3. Nutrients and oxygen diffuse from the blood through interstitial fluid into tissue cells. Waste products, including carbon dioxide, diffuse from tissue cells through interstitial fluid into the blood.

? Would you expect a femoral artery or a femoral vein to have the thicker wall? A wider lumen?

skin. Blood can be diverted quickly from its reservoirs to other parts of the body, for example, to skeletal muscles to support increased muscular activity.

Arteries and Arterioles

The walls of arteries have three layers of tissue surrounding a hollow space, the **lumen**, through which the blood flows (Figure 16.1a). The inner layer is composed of **endothelium**, a type of simple squamous epithelium; a basement membrane; and an elastic tissue called the internal elastic lamina. The middle layer consists of smooth muscle and elastic tissue. The outer layer is composed mainly of elastic and collagen fibers.

Sympathetic fibers of the autonomic nervous system innervate vascular smooth muscle. An increase in sympathetic stimulation typically causes the smooth muscle to contract, squeezing the vessel wall and narrowing the lumen. Such a decrease in the diameter of the lumen of a blood vessel is called **vasoconstriction**. In contrast, when sympathetic stimulation decreases, or in the presence of certain chemicals (such as nitric oxide or lactic acid), smooth muscle fibers relax. The resulting increase in lumen diameter is called **vasodilation**. Additionally, when an artery or arteriole is damaged, its smooth muscle contracts, producing vascular spasm of the vessel. Such a vasospasm limits blood flow through the damaged vessel and helps reduce blood loss if the vessel is small.

The largest-diameter arteries contain a high proportion of elastic fibers in their middle layer, and their walls are relatively thin in proportion to their overall diameter. Such arteries are called **elastic arteries**. These arteries help propel blood onward while the ventricles are relaxing. As blood is ejected from the heart into elastic arteries, their highly elastic walls stretch, accommodating the surge of blood. Then, while the ventricles are relaxing, the elastic fibers in the artery walls recoil, which forces blood onward toward the smaller arteries. Examples include the aorta and the brachiocephalic, common carotid, subclavian, vertebral, pulmonary, and common iliac arteries. Medium-sized arteries, on the other hand, contain more smooth muscle and fewer elastic fibers than elastic arteries. Such arteries are called **muscular arteries** and are capable of greater vasoconstriction and vasodilation to adjust the rate of blood flow. Examples include the brachial artery (arm) and radial artery (forearm).

An **arteriole** (= small artery) is a very small, almost microscopic, artery that delivers blood to capillaries. The smallest arterioles consist of little more than a layer of endothelium covered by a few smooth muscle fibers (see Figure 16.2a). Arterioles play a key role in regulating blood flow from arteries into capillaries. During vasoconstriction, blood flow from the arterioles to the capillaries is restricted; during vasodilation, the flow is significantly increased. A change in diameter of arterioles can also significantly alter blood pres-

sure; vasodilation decreases blood pressure and vasoconstriction increases blood pressure.

Capillaries

Capillaries (*capillar-* = hairlike) are microscopic vessels that connect arterioles to venules (Figure 16.1c). Capillaries are present near almost every body cell, and they are known as *exchange vessels* because they permit the exchange of nutrients and wastes between the body's cells and the blood. The number of capillaries varies with the metabolic activity of the tissue they serve. Body tissues with high metabolic requirements, such as muscles, the liver, the kidneys, and the nervous system, have extensive capillary networks. Tissues with lower metabolic requirements, such as tendons and ligaments, contain fewer capillaries. A few tissues—all covering and lining epithelia, the cornea and lens of the eye, and cartilage—lack capillaries completely.

Structure of Capillaries

A capillary consists of a layer of endothelium that is surrounded by basement membrane (Figure 16.1c). Because capillary walls are very thin, many substances easily pass through them to reach tissue cells from the blood or to enter the blood from interstitial fluid. The walls of all other blood vessels are too thick to permit the exchange of substances between blood and interstitial fluid. Depending on how tightly their endothelial cells are joined, different types of capillaries have varying degrees of permeability.

In some regions, capillaries link arterioles to venules directly. In other places, they form extensive branching networks (Figure 16.2). Blood flows through only a small part of a tissue's capillary network when metabolic needs are low. But when a tissue becomes active, the entire capillary network fills with blood. The flow of blood in capillaries is regulated by smooth muscle fibers in arteriole walls and by **precapillary sphincters**, rings of smooth muscle at the point where capillaries branch from arterioles (Figure 16.2a). When precapillary sphincters relax, more blood flows into the connected capillaries; when precapillary sphincters contract, less blood flows through their capillaries.

Capillary Exchange

Because of the small diameter of capillaries, blood flows more slowly through them than through larger blood vessels. The slow flow aids the prime mission of the entire cardiovascular system: to keep blood flowing through capillaries so that **capillary exchange**—the movement of substances into and out of capillaries—can occur.

Capillary blood pressure, the pressure of blood against the walls of capillaries, "pushes" fluid out of capillaries into interstitial fluid. An opposing pressure, termed **blood colloid osmotic pressure**, "pulls" fluid into capillaries. (Recall from Chapter 3

Figure 16.2 Capillaries. Because red blood cells and capillaries are nearly the same size, red blood cells squeeze through capillaries in single file.

 Arterioles regulate blood flow into capillaries, where nutrients, gases, and wastes are exchanged between blood and interstitial fluid.

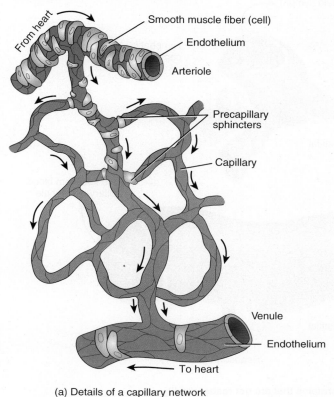

Smooth muscle fiber (cell)
From heart
Endothelium
Arteriole
Precapillary sphincters
Capillary
Venule
Endothelium
To heart

(a) Details of a capillary network

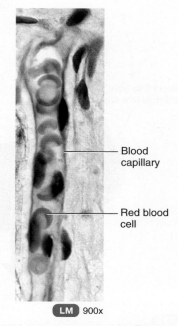

Blood capillary

Red blood cell

LM 900x

(b) Photomicrograph showing red blood cells squeezing through a blood capillary

Why do metabolically active tissues have extensive capillary networks?

that osmotic pressure is the pressure of a fluid due to its solute concentration. The higher the solute concentration, the greater the osmotic pressure.) Most solutes are present in nearly equal concentrations in blood and interstitial fluid. But the presence of proteins in plasma and their virtual absence in interstitial fluid gives blood the higher osmotic pressure. Blood colloid osmotic pressure is osmotic pressure due mainly to plasma proteins.

Capillary blood pressure is higher than blood colloid osmotic pressure for about the first half of the length of a typical capillary. Thus, water and solutes flow out of the blood capillary into the surrounding interstitial fluid, a movement called *filtration* (Figure 16.3). Because capillary blood pressure decreases progressively as blood flows along a capillary, at about the capillary's midpoint, blood pressure drops below blood colloid osmotic pressure. Then, water and solutes move from interstitial fluid into the blood capillary, a process termed *reabsorption*. Normally, about 85 percent of the filtered fluid is reabsorbed. The excess filtered fluid and the few plasma proteins that do escape enter lymphatic capillaries and eventually are returned by the lymphatic system to the cardiovascular system.

Localized changes in each capillary network can regulate vasodilation and vasoconstriction. When vasodilators are released by tissue cells, they cause dilation of nearby arterioles and relaxation of precapillary sphincters. Then, blood flow into the capillary networks increases, and O_2 delivery to the tissue rises. Vasoconstrictors have the opposite effect. The ability of a tissue to automatically adjust its blood flow to match its metabolic demands is called *autoregulation*.

Venules and Veins

When several capillaries unite, they form venules. Venules receive blood from capillaries and empty blood into veins, which return blood to the heart.

Structure of Venules and Veins

Venules (= little veins) are similar in structure to arterioles; their walls are thinner near the capillary end and thicker as they progress toward the heart. *Veins* are structurally similar to arteries, but their middle and inner layers are thinner

Figure 16.3 Capillary exchange.

Capillary blood pressure pushes fluid out of capillaries (filtration); blood colloid osmotic pressure pulls fluid into capillaries (reabsorption).

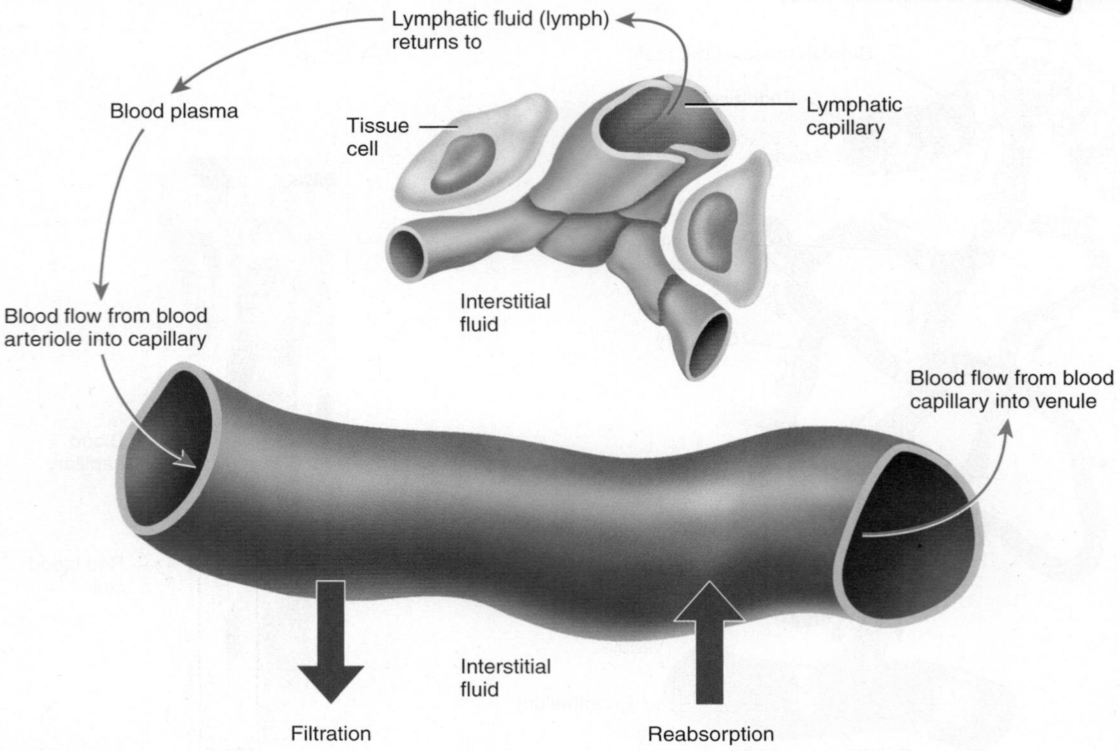

Lymphatic fluid (lymph) returns to

Blood plasma

Tissue cell

Lymphatic capillary

Interstitial fluid

Blood flow from blood arteriole into capillary

Blood flow from blood capillary into venule

Interstitial fluid

Filtration

Reabsorption

What happens to excess filtered fluid and proteins that are not reabsorbed?

(see Figure 16.1b). The outer layer of veins is the thickest layer. The lumen of a vein is wider than that of a corresponding artery.

In some veins, the inner layer folds inward to form *valves* that prevent the backflow of blood. In people with weak venous valves, gravity forces blood backward through the valve. This increases venous blood pressure, which pushes the vein's wall outward. After repeated overloading, the walls lose their elasticity and become stretched and flabby, a condition called *varicose veins*.

By the time blood leaves the capillaries and moves into veins, it has lost a great deal of pressure. This can be observed in the blood leaving a cut vessel: Blood flows from a cut vein slowly and evenly, but it gushes out of a cut artery in rapid spurts. When a blood sample is needed, it is usually collected from a vein because pressure is low in veins and many of them are close to the skin surface.

Venous Return

Venous return, the volume of blood flowing back to the heart through systemic veins, occurs due to pressure generated in three ways: (1) contractions of the heart, (2) the skeletal muscle pump, and (3) the respiratory pump. Blood pressure is generated by contraction of the heart's ventricles and is measured in millimeters of mercury, abbreviated mm Hg. The pressure difference from venules (averaging about 16 mm Hg) to the right atrium (0 mm Hg), although small, normally is sufficient to cause venous return to the heart. When you stand, the pressure pushing blood up the veins in your lower limbs is barely enough to overcome the force of gravity pushing it back down.

The *skeletal muscle pump* operates as follows (Figure 16.4):

1 While you are standing at rest, both the venous valve closer to the heart and the one farther from the heart in this part of the leg are open, and blood flows upward toward the heart.

2 Contraction of leg muscles, such as when you stand on tiptoes or take a step, compresses the vein. The compression pushes blood through the valve closer to the heart, an action called *milking*. At the same time, the valve farther from the heart in the uncompressed segment

Figure 16.4 Action of the skeletal muscle pump in returning blood to the heart.

 Milking refers to skeletal muscle contractions that drive venous blood toward the heart.

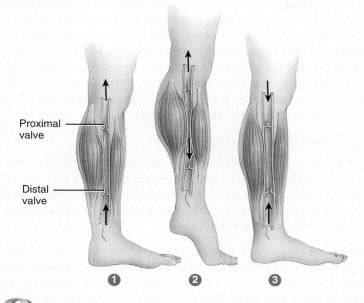

Proximal valve

Distal valve

① ② ③

 What mechanisms, besides cardiac contractions, act as pumps to boost venous return?

of the vein closes as some blood is pushed against it. People who are immobilized through injury or disease lack these contractions of leg muscles. As a result, their venous return is slower and they may develop circulation problems.

③ Just after muscle relaxation, pressure falls in the previously compressed section of vein, which causes the valve closer to the heart to close. The valve farther from the heart now opens because blood pressure in the foot is higher than in the leg, and the vein fills with blood from the foot.

The *respiratory pump* is also based on alternating compression and decompression of veins. During inhalation (breathing in) the diaphragm moves downward, which causes a decrease in pressure in the thoracic cavity and an increase in pressure in the abdominal cavity. As a result, abdominal veins are compressed, and a greater volume of blood moves from the compressed abdominal veins into the decompressed thoracic veins and then into the right atrium. When the pressures reverse during exhalation (breathing out), the valves in the veins prevent backflow of blood from the thoracic veins to the abdominal veins.

✓ CHECKPOINT

1. How do arteries, capillaries, and veins differ in function?

2. Distinguish between filtration and reabsorption.

3. What factors contribute to blood flow back to the heart?

16.2 BLOOD FLOW THROUGH BLOOD VESSELS

OBJECTIVES • **Define blood pressure and describe how it varies throughout the systemic circulation.**

• **Identify the factors that affect blood pressure and vascular resistance.**

• **Describe how blood pressure and blood flow are regulated.**

We saw in Chapter 15 that cardiac output (CO) depends on stroke volume and heart rate. Two other factors influencing cardiac output and the proportion of blood that flows through specific circulatory routes are blood pressure and vascular resistance.

Blood Pressure

As you have just learned, blood flows from regions of higher pressure to regions of lower pressure; the greater the pressure difference, the greater the blood flow. Contraction of the ventricles generates ***blood pressure (BP)***, the pressure exerted by blood on the walls of a blood vessel. BP is highest in the aorta and large systemic arteries; in a resting, young adult, BP rises to about 110 mm Hg during systole (contraction) and drops to about 70 mm Hg during diastole (relaxation). Blood pressure falls progressively as the distance from the left ventricle increases (Figure 16.5), to about

Figure 16.5 Blood pressure changes as blood flows through the systemic circulation. The dashed line is the mean (average) pressure in the aorta, arteries, and arterioles.

Blood pressure falls progressively as blood flows from systemic arteries through capillaries and back to the right atrium. The greatest drop in blood pressure occurs in the arterioles.

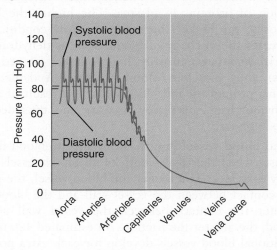

 What is the relationship between blood pressure and blood flow?

35 mm Hg as blood passes into systemic capillaries. At the venous end of capillaries, blood pressure drops to about 16 mm Hg. Blood pressure continues to drop as blood enters systemic venules and then veins, and it reaches 0 mm Hg as blood returns to the right atrium.

Blood pressure depends in part on the total volume of blood in the cardiovascular system. The normal volume of blood in an adult is about 5 liters (5.3 qt). Any decrease in this volume, as from hemorrhage, decreases the amount of blood that is circulated through the arteries. A modest decrease can be compensated by homeostatic mechanisms that help maintain blood pressure, but if the decrease in blood volume is greater than 10 percent of total blood volume, blood pressure drops, with potentially life-threatening results. Conversely, anything that increases blood volume, such as water retention in the body, tends to increase blood pressure.

Resistance

Vascular resistance is the opposition to blood flow due to friction between blood and the walls of blood vessels. An increase in vascular resistance increases blood pressure; a decrease in vascular resistance has the opposite effect. Vascular resistance depends on the following three factors:

1. **Size of the lumen.** The smaller the lumen of a blood vessel, the greater its resistance to blood flow. Vasoconstriction narrows the lumen, and vasodilation widens it. Normally, moment-to-moment fluctuations in blood flow through a given tissue are due to vasoconstriction and vasodilation of the tissue's arterioles. As arterioles dilate, resistance decreases, and blood pressure falls. As arterioles constrict, resistance increases, and blood pressure rises.

2. **Blood viscosity.** The viscosity (thickness) of blood depends mostly on the ratio of red blood cells to plasma (fluid) volume, and to a smaller extent on the concentration of proteins in plasma. The higher the blood's viscosity, the higher the resistance. Any condition that increases the viscosity of blood, such as dehydration or *polycythemia* (an unusually high number of red blood cells), thus increases blood pressure. A depletion of plasma proteins or red blood cells, as a result of anemia or hemorrhage, decreases viscosity and thus decreases blood pressure.

3. **Total blood vessel length.** Resistance to blood flow increases when the total length of all blood vessels in the body increases. The longer the blood vessel, the greater the contact between the vessel wall and the blood. The greater the contact between the vessel wall and the blood, the greater the friction. An estimated 400 miles of additional blood vessels develop for each extra pound of fat, one reason why overweight individuals may have higher blood pressure.

Regulation of Blood Pressure and Blood Flow

Several interconnected negative feedback systems control blood pressure and blood flow by adjusting heart rate, stroke volume, vascular resistance, and blood volume. Some systems allow rapid adjustments to cope with sudden changes, such as the drop in blood pressure in the brain that occurs when you stand up; others provide long-term regulation. The body may also require adjustments to the distribution of blood flow. During exercise, for example, a greater percentage of blood flow is diverted to skeletal muscles.

Role of the Cardiovascular Center

In Chapter 15 we noted how the **cardiovascular (CV) center** in the medulla oblongata helps regulate heart rate and stroke volume. The CV center also controls the neural and hormonal negative feedback systems that regulate blood pressure and blood flow to specific tissues.

INPUT The cardiovascular center receives input from higher brain regions: the cerebral cortex, limbic system, and hypothalamus (Figure 16.6). For example, even before you start to run a race, your heart rate may increase due to nerve impulses conveyed from the limbic system to the CV center. If your body temperature rises during a race, the hypothalamus sends nerve impulses to the CV center. The resulting vasodilation of skin blood vessels allows heat to dissipate more rapidly from the surface of the skin.

The CV center also receives input from three main types of sensory receptors: proprioceptors, baroreceptors, and chemoreceptors. *Proprioceptors*, which monitor movements of joints and muscles, provide input to the cardiovascular center during physical activity, such as playing tennis, and cause the rapid increase in heart rate at the beginning of exercise.

Baroreceptors (pressure receptors) are located in the aorta, internal carotid arteries (arteries in the neck that supply blood to the brain), and other large arteries in the neck and chest. They send impulses continuously to the cardiovascular center to help regulate blood pressure. If blood pressure falls, the baroreceptors are stretched less, and they send nerve impulses at a slower rate to the cardiovascular center (Figure 16.7). In response, the cardiovascular center decreases parasympathetic stimulation of the heart and increases sympathetic stimulation of the heart. As the heart beats faster and more forcefully, and as vascular resistance increases, blood pressure increases to the normal level.

By contrast, when an increase in pressure is detected, the baroreceptors send impulses at a faster rate. The cardiovascular center responds by increasing parasympathetic stimulation and decreasing sympathetic stimulation. The resulting decreases in heart rate and force of contraction lower cardiac output, and vasodilation lowers vascular resistance. Decreased cardiac output and decreased vascular resistance both lower blood pressure.

Figure 16.6 **The cardiovascular (CV) center.** Located in the medulla oblongata, the CV center receives input from higher brain centers, proprioceptors, baroreceptors, and chemoreceptors. It provides output to both the sympathetic and parasympathetic divisions of the autonomic nervous system.

> The cardiovascular center is the main region for the nervous system regulation of heart rate, force of heart contractions, and vasodilation or vasoconstriction of blood vessels.

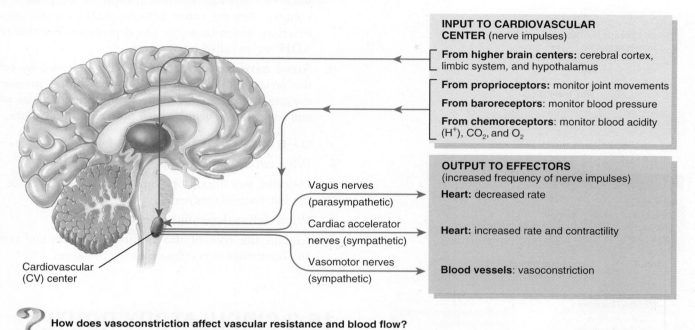

INPUT TO CARDIOVASCULAR CENTER (nerve impulses)

From higher brain centers: cerebral cortex, limbic system, and hypothalamus

From proprioceptors: monitor joint movements

From baroreceptors: monitor blood pressure

From chemoreceptors: monitor blood acidity (H^+), CO_2, and O_2

OUTPUT TO EFFECTORS (increased frequency of nerve impulses)

Heart: decreased rate

Heart: increased rate and contractility

Blood vessels: vasoconstriction

Vagus nerves (parasympathetic)

Cardiac accelerator nerves (sympathetic)

Vasomotor nerves (sympathetic)

Cardiovascular (CV) center

? **How does vasoconstriction affect vascular resistance and blood flow?**

Moving from a prone (lying down) position to an erect position decreases blood pressure and blood flow in the head and upper part of the body. The drop in pressure, however, is quickly counteracted by the **baroreceptor reflexes**. Sometimes these reflexes operate more slowly than normal, especially in older people. As a result, a person can faint due to reduced brain blood flow upon standing up too quickly. •

CHEMORECEPTOR REFLEXES *Chemoreceptors* (chemical receptors) that monitor blood levels of O_2, CO_2, and H^+ are located in the two *carotid bodies* in the common carotid arteries and in the *aortic body* in the arch of the aorta. *Hypoxia* (lowered O_2 availability), *acidosis* (an increase in H^+ concentration), or *hypercapnia* (excess CO_2) stimulates the chemoreceptors to send impulses to the cardiovascular center. In response, the CV center increases sympathetic stimulation of arterioles and veins, producing vasoconstriction and an increase in blood pressure.

OUTPUT Output from the cardiovascular center flows along sympathetic and parasympathetic fibers of the ANS (see Figure 16.6). An increase in sympathetic stimulation increases heart rate and the forcefulness of contraction, and a decrease in sympathetic stimulation decreases heart rate and contraction force. The vasomotor region of the cardiovascular center also

sends impulses to arterioles throughout the body. The result is a moderate state of vasoconstriction, called *vasomotor tone*, that sets the resting level of vascular resistance. Sympathetic stimulation of most veins results in movement of blood out of venous blood reservoirs, which increases blood pressure.

Hormonal Regulation of Blood Pressure and Blood Flow

Several hormones help regulate blood pressure and blood flow by altering cardiac output, changing vascular resistance, or adjusting the total blood volume.

1. **Renin–angiotensin–aldosterone (RAA) system.** When blood volume falls or blood flow to the kidneys decreases, certain cells in the kidneys secrete the enzyme *renin* into the bloodstream (see Figure 13.14). Together, renin and angiotensin converting enzyme (ACE) produce the active hormone *angiotensin II*, which raises blood pressure by causing vasoconstriction. Angiotensin II also stimulates secretion of *aldosterone*, which increases reabsorption of sodium ions (Na^+) and water by the kidneys. The water reabsorption increases total blood volume, which increases blood pressure.

2. **Epinephrine and norepinephrine.** In response to sympathetic stimulation, the adrenal medulla releases epinephrine and norepinephrine. These hormones increase cardiac output by increasing the rate and force of heart

Figure 16.7 Negative feedback regulation of blood pressure via baroreceptor reflexes.

The baroreceptor reflex is a neural mechanism for rapid regulation of blood pressure.

Some stimulus disrupts homeostasis by

Decreasing

Blood pressure

Receptors

Baroreceptors in arch of aorta and carotid sinus are stretched less

Input — Decreased rate of nerve impulses

Control centers

CV center in medulla oblongata

and adrenal medulla

Return to homeostasis when increased cardiac output and increased vascular resistance bring blood pressure back to normal

Output — Increased sympathetic, decreased parasympathetic stimulation

Increased secretion of epinephrine and norepinephrine from adrenal medulla

Effectors

Increased stroke volume and heart rate lead to increased cardiac output (CO)

Constriction of blood vessels increases systemic vascular resistance (SVR)

Increased blood pressure

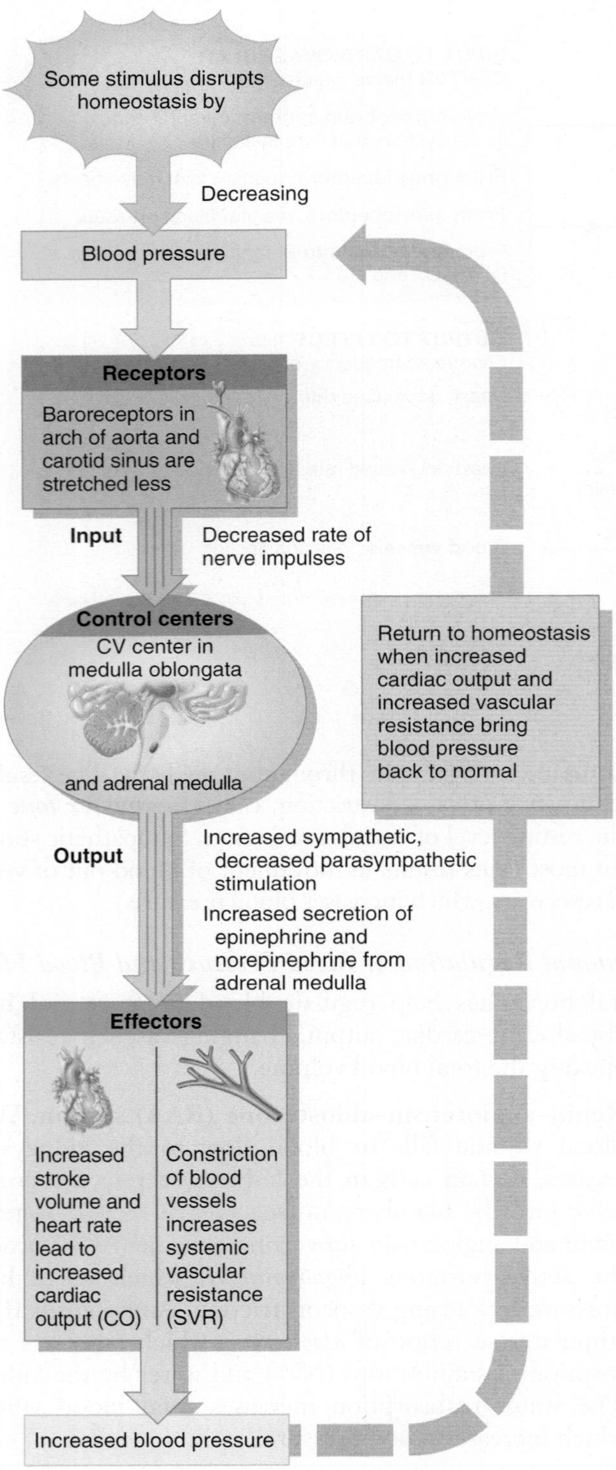

Does this negative feedback cycle happen when you lie down or when you stand up?

contractions; they also cause vasoconstriction of arterioles and veins in the skin and abdominal organs.

3. **Antidiuretic hormone (ADH).** ADH is produced by the hypothalamus and released from the posterior pituitary in response to dehydration or decreased blood volume. Among other actions, ADH causes vasoconstriction, which increases blood pressure. For this reason ADH is also called **vasopressin**.

4. **Atrial natriuretic peptide (ANP).** Released by cells in the atria of the heart, ANP lowers blood pressure by causing vasodilation and by promoting the loss of salt and water in the urine, which reduces blood volume.

✓ CHECKPOINT

4. What two factors influence cardiac output?

5. Describe how blood pressure decreases as distance from the left ventricle increases.

6. What factors determine vascular resistance?

7. Explain the role of the cardiovascular center, reflexes, and hormones in regulating blood pressure.

16.3 CIRCULATORY ROUTES

OBJECTIVE • **Compare the major routes that blood takes through various regions of the body.**

Blood vessels are organized into **circulatory routes** that carry blood throughout the body (Figure 16.8). As noted earlier, the two main circulatory routes are the systemic circulation and the pulmonary circulation.

Systemic Circulation

The **systemic circulation** includes the arteries and arterioles that carry blood containing oxygen and nutrients from the left ventricle to systemic capillaries throughout the body, plus the veins and venules that carry blood containing carbon dioxide and wastes to the right atrium. Blood leaving the aorta and traveling through the systemic arteries is a bright red color. As it moves through the capillaries, it loses some of its oxygen and takes on carbon dioxide, so that the blood in the systemic veins is a dark red color.

All systemic arteries branch from the **aorta**, which arises from the left ventricle of the heart (see Figure 16.9). Deoxygenated blood returns to the heart through the systemic veins. All veins of the systemic circulation empty into the **superior vena cava, inferior vena cava,** or the **coronary sinus,** which, in turn, empty into the right atrium. The principal blood vessels of the systemic circulation are described and illustrated in Exhibits 16.A through 16.G and Figures 16.9 through 16.15.

Figure 16.8 Circulatory routes. Red arrows indicate hepatic portal circulation. The details of the pulmonary circulation are shown here, and the details of the hepatic portal circulation are shown in Figure 16.16.

Blood vessels are organized into routes that deliver blood to various tissues of the body.

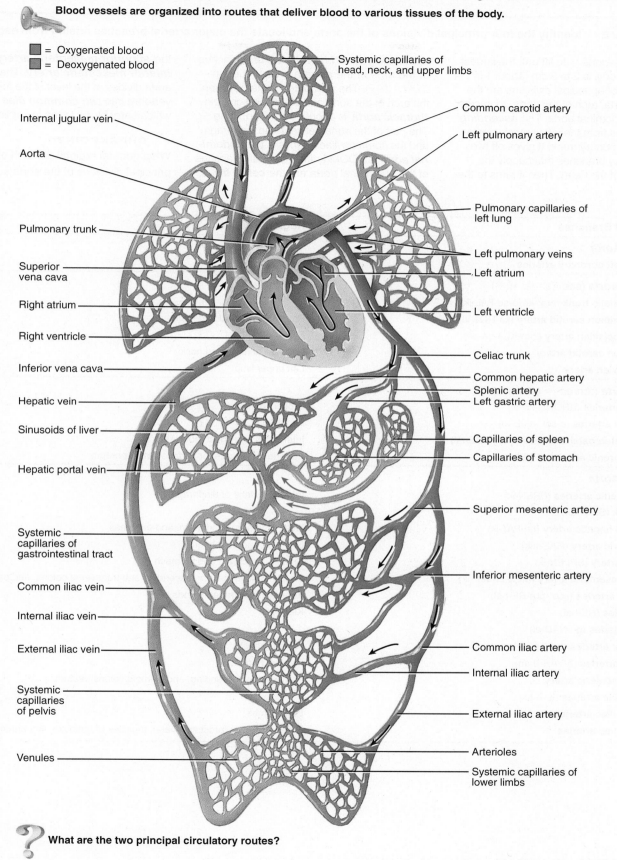

■ = Oxygenated blood
■ = Deoxygenated blood

Systemic capillaries of head, neck, and upper limbs

Internal jugular vein

Common carotid artery

Aorta

Left pulmonary artery

Pulmonary trunk

Pulmonary capillaries of left lung

Superior vena cava

Left pulmonary veins

Right atrium

Left atrium

Right ventricle

Left ventricle

Inferior vena cava

Celiac trunk

Hepatic vein

Common hepatic artery
Splenic artery
Left gastric artery

Sinusoids of liver

Capillaries of spleen

Hepatic portal vein

Capillaries of stomach

Superior mesenteric artery

Systemic capillaries of gastrointestinal tract

Common iliac vein

Inferior mesenteric artery

Internal iliac vein

External iliac vein

Common iliac artery

Internal iliac artery

Systemic capillaries of pelvis

External iliac artery

Venules

Arterioles

Systemic capillaries of lower limbs

? What are the two principal circulatory routes?

429

EXHIBIT 16.A The Aorta and Its Branches *(Figure 16.9)*

OBJECTIVE • Identify the four principal divisions of the aorta and locate the major arterial branches arising from each.

• The **aorta** (*aortae* = to lift up), the largest artery of the body, is 2 to 3 cm (about 1 in.) in diameter. Its four principal divisions are the ascending aorta, arch of the aorta, thoracic aorta, and abdominal aorta. The **ascending aorta** emerges from the left ventricle posterior to the pulmonary trunk. It gives off two coronary artery branches that supply the myocardium of the heart. Then it turns to the left, forming the **arch of the aorta**. Branches of the arch of the aorta are described in Exhibit 16.B. The part of the aorta between the arch of the aorta and the diaphragm, the **thoracic aorta**, is about 20 cm (8 in.) long. The part of the aorta between the diaphragm and the common iliac arteries is the **abdominal aorta** (ab-DOM-i-nal). The main branches of the abdominal aorta are the **celiac trunk**, the **superior mesenteric artery**, and the **inferior mesenteric artery**. The abdominal aorta divides at the level of the fourth lumbar vertebra into two **common iliac arteries**, which carry blood to the lower limbs.

✓ **CHECKPOINT**

What general regions do each of the four principal divisions of the aorta supply?

Division and Branches	Region Supplied
Ascending Aorta	
Right and left coronary arteries	Heart
Arch of the Aorta (see Exhibit 16.B)	
Brachiocephalic trunk (brā′-kē-ō-se-FAL-ik)	
Right common carotid artery (ka-ROT-id)	Right side of head and neck
Right subclavian artery (sub-KLĀ-vē-an)	Right upper limb
Left common carotid artery	Left side of head and neck
Left subclavian artery	Left upper limb
Thoracic Aorta (*thorac-* = chest)	
Bronchial arteries (BRONG-kē-al)	Bronchi of lungs
Esophageal arteries (e-sof′-a-JĒ-al)	Esophagus
Posterior intercostal arteries (in′-ter-KOS-tal)	Intercostal and chest muscles
Superior phrenic arteries (FREN-ik)	Superior and posterior surfaces of diaphragm
Abdominal Aorta	
Inferior phrenic arteries (FREN-ik)	Inferior surface of diaphragm
Celiac trunk (SĒ-lē-ak)	
Common hepatic artery (he-PAT-ik)	Liver, stomach, duodenum, and pancreas
Left gastric artery (GAS-trik)	Stomach and esophagus
Splenic artery (SPLĒN-ik)	Spleen, pancreas, and stomach
Superior mesenteric artery (MES-en-ter′-ik)	Small intestine, cecum, ascending and transverse colons, and pancreas
Suprarenal arteries (soo′-pra-RĒ-nal)	Adrenal (suprarenal) glands
Renal arteries (RĒ-nal)	Kidneys
Gonadal arteries (gō-NAD-al)	
Testicular arteries (tes-TIK-ū-lar)	Testes (male)
Ovarian arteries (ō-VAR-ē-an)	Ovaries (female)
Inferior mesenteric artery	Transverse, descending, and sigmoid colons; rectum
Common iliac arteries (IL-ē-ak)	
External iliac arteries	Lower limbs
Internal iliac arteries	Uterus (female), prostate (male), muscles of buttocks, and urinary bladder

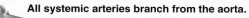

Figure 16.9 Aorta and its principal branches.

All systemic arteries branch from the aorta.

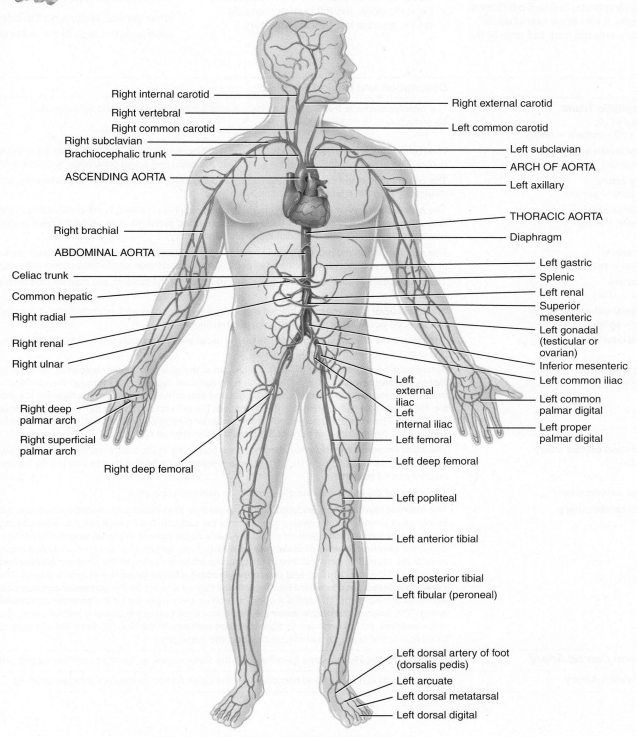

Overall anterior view of the principal branches of the aorta

After blood is ejected from the heart, what are the names of the four divisions of the aorta that it passes through?

EXHIBIT 16.B   The Arch of the Aorta   *(Figure 16.10)*

OBJECTIVE • **Identify the three arteries that branch from the arch of the aorta.**

• The *arch of the aorta*, the continuation of the ascending aorta, is 4 to 5 cm (almost 2 in.) in length. It has three branches. In order, as they emerge from the arch of the aorta, the three branches are the brachiocephalic trunk, the left common carotid artery, and the left subclavian artery.

✓ **CHECKPOINT**

What general regions do the arteries that arise from the arch of the aorta supply?

Artery	Description and Region Supplied
Brachiocephalic Trunk (brā′-kē-ō-se-FAL-ik; *brachio-* = arm; *-cephalic* = head)	The *brachiocephalic trunk* divides to form the right subclavian artery and right common carotid artery (Figure 16.10a).
Right subclavian artery (sub-KLĀ-vē-an)	The *right subclavian artery* extends from the brachiocephalic trunk and then passes into the armpit (axilla). The general distribution of the artery is to the brain and spinal cord, neck, shoulder, and chest.
Axillary artery (AK-si-ler-ē = armpit)	The continuation of the right subclavian artery into the axilla is called the *axillary artery*. Its general distribution is to the shoulder.
Brachial artery (BRĀ-kē-al = arm)	The *brachial artery*, which provides the main blood supply to the arm, is the continuation of the axillary artery into the arm. It is commonly used to measure blood pressure. Just below the bend in the elbow, the brachial artery divides into the radial artery and ulnar artery.
Radial artery (RĀ-dē-al = radius)	The *radial artery* is a direct continuation of the brachial artery. It passes along the lateral (radial) aspect of the forearm and then through the wrist and hand; it is a common site for measuring radial pulse.
Ulnar artery (UL-nar = ulna)	The *ulnar artery* passes along the medial (ulnar) aspect of the forearm and then into the wrist and hand.
Superficial palmar arch (*palma* = palm)	The *superficial palmar arch* is formed mainly by the ulnar artery and extends across the palm. It gives rise to blood vessels that supply the palm and the fingers.
Deep palmar arch	The *deep palmar arch* is formed mainly by the radial artery. The arch extends across the palm and gives rise to blood vessels that supply the palm.
Vertebral artery (VER-te-bral)	Before passing into the axilla, the right subclavian artery gives off a major branch to the brain called the *right vertebral artery* (Figure 16.10b). The right vertebral artery passes through the foramina of the transverse processes of the cervical vertebrae and enters the skull through the foramen magnum to reach the inferior surface of the brain. Here it unites with the left vertebral artery to form the *basilar artery* (BAS-i-lar). The vertebral artery supplies the posterior portion of the brain with blood. The basilar artery supplies the cerebellum and pons of the brain and the internal ear.
Right common carotid artery (ka-ROT-id)	The *right common carotid artery* begins at the branching of the brachiocephalic trunk and supplies structures in the head (Figure 16.10b). Near the larynx (voice box), it divides into the right external and right internal carotid arteries.
External carotid artery	The *external carotid artery* supplies structures *external* to the skull.
Internal carotid artery	The *internal carotid artery* supplies structures *internal* to the skull such as the eyeball, ear, most of the cerebrum of the brain, and pituitary gland. Inside the cranium, the internal carotid arteries along with the basilar artery form an arrangement of blood vessels at the base of the brain near the hypophyseal fossa called the *cerebral arterial circle (circle of Willis)*. From this circle (Figure 16.10c) arise arteries supplying most of the brain. The cerebral arterial circle is formed by the union of the *anterior cerebral arteries* (branches of internal carotids) and *posterior cerebral arteries* (branches of basilar artery). The *posterior cerebral arteries* are connected with the internal carotid arteries by the *posterior communicating arteries* (ko-MŪ-ni-kā′-ting). The anterior cerebral arteries are connected by the *anterior communicating artery*. The *internal carotid arteries* are also considered part of the cerebral arterial circle. The functions of the cerebral arterial circle are to equalize blood pressure to the brain and provide alternate routes for blood flow to the brain, should the arteries become damaged.
Left Common Carotid Artery	Divides into basically the same branches with the same names as the right common carotid artery.
Left Subclavian Artery	Divides into basically the same branches with the same names as the right subclavian artery.

Figure 16.10 Arch of the aorta and its branches.

The arch of the aorta is the continuation of the ascending aorta.

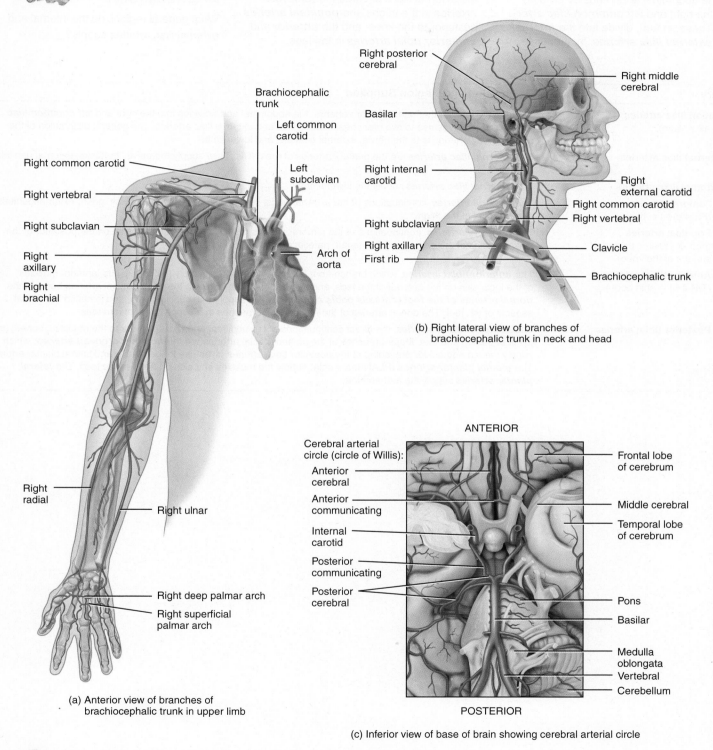

(a) Anterior view of branches of brachiocephalic trunk in upper limb

(b) Right lateral view of branches of brachiocephalic trunk in neck and head

(c) Inferior view of base of brain showing cerebral arterial circle

 What are the three major branches of the arch of the aorta, in order of their origination?

EXHIBIT 16.C Arteries of the Pelvis and Lower Limbs *(Figure 16.11)*

OBJECTIVE • Identify the two major branches of the common iliac arteries.

• The abdominal aorta ends by dividing into the right and left *common iliac arteries*. These, in turn, divide into the *internal* and *external iliac arteries*. In sequence, the external iliacs become the *femoral arteries* in the thighs, the *popliteal arteries* posterior to the knee, and the *anterior* and *posterior tibial arteries* in the legs.

✓ **CHECKPOINT**

What general regions do the internal and external iliac arteries supply?

Artery	Description and Region Supplied
Common iliac arteries (IL-ē-ak = ilium)	At about the level of the fourth lumbar vertebra, the abdominal aorta divides into the right and left *common iliac arteries*. Each gives rise to two branches: internal iliac and external iliac arteries. The general distribution of the common iliac arteries is to the pelvis, external genitals, and lower limbs.
Internal iliac arteries	The *internal iliac arteries* are the primary arteries of the pelvis. They supply the pelvis, buttocks, external genitals, and thigh.
External iliac arteries	The *external iliac arteries* supply the lower limbs.
Femoral arteries (FEM-o-ral = thigh)	The *femoral arteries*, continuations of the external iliacs, supply the lower abdominal wall, groin, external genitals, and muscles of the thigh.
Popliteal arteries (pop′-li-TĒ-al = posterior surface of the knee)	The *popliteal arteries*, continuations of the femoral arteries, supply muscles and the skin on the posterior of the legs; muscles of the calf; knee joint; femur; patella; and fibula.
Anterior tibial arteries (TIB-ē-al = shin bone)	The *anterior tibial arteries*, which branch from the popliteal arteries, supply the knee joints, anterior muscles of the legs, skin on the anterior of the legs, and ankle joints. At the ankles, the anterior tibial arteries become the *dorsal arteries of the foot (dorsalis pedis arteries)*, which supply the muscles, skin, and joints on the dorsal aspects of the feet. The dorsal arteries of the foot give off branches that supply the feet and toes.
Posterior tibial arteries	The *posterior tibial arteries*, the direct continuations of the popliteal arteries, distribute to the muscles, bones, and joints of the leg and foot. Major branches of the posterior tibial arteries are the *fibular (peroneal) arteries*, which supply the leg and ankle. Branching of the posterior tibial arteries gives rise to the medial and lateral plantar arteries. The *medial plantar arteries* (PLAN-tar = sole) supply the muscles and skin of the feet and toes. The *lateral plantar arteries* supply the feet and toes.

Figure 16.11 Arteries of the pelvis and right lower limb.

 The internal iliac arteries carry most of the blood supply to the pelvis, buttocks, external genitals, and thigh.

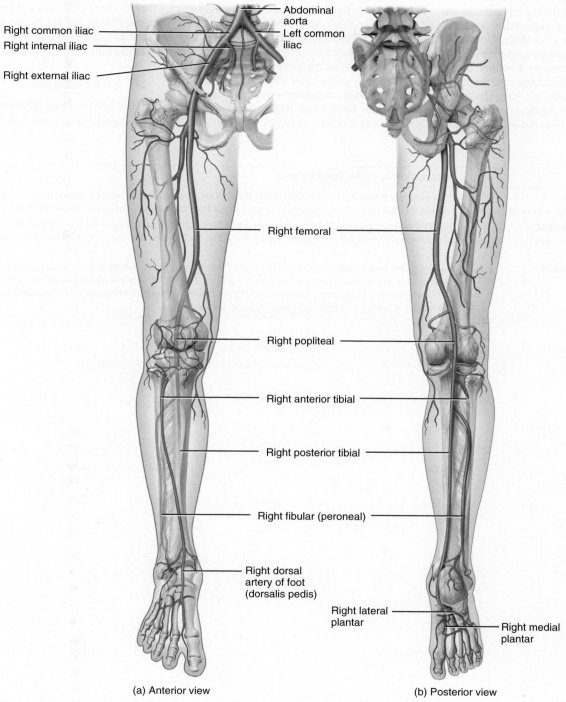

Abdominal aorta

Right common iliac

Left common iliac

Right internal iliac

Right external iliac

Right femoral

Right popliteal

Right anterior tibial

Right posterior tibial

Right fibular (peroneal)

Right dorsal artery of foot (dorsalis pedis)

Right lateral plantar

Right medial plantar

(a) Anterior view

(b) Posterior view

At what point does the abdominal aorta divide into the common iliac arteries?

EXHIBIT 16.D Veins of the Systemic Circulation *(Figure 16.12)*

OBJECTIVE • **Identify the three systemic veins that return deoxygenated blood to the heart.**

• Arteries distribute blood to various parts of the body, and veins drain blood away from them. For the most part, arteries are deep. Veins may be *superficial* (located just beneath the skin) or *deep*. Deep veins generally travel alongside arteries and usually bear the same name. Because there are no large superficial arteries, the names of superficial veins do not correspond to those of arteries. Superficial veins are clinically important as sites for withdrawing blood or giving injections. Arteries usually follow definite pathways. Veins are more difficult to follow because they connect in irregular networks in which many smaller veins merge to form a larger vein. Although only one systemic artery, the aorta, takes oxygenated blood away from the heart (left ventricle), three systemic veins, the *coronary sinus, superior vena cava*, and *inferior vena cava*, deliver deoxygenated blood to the right atrium of the heart. The coronary sinus receives blood from the cardiac veins; the superior vena cava receives blood from other veins superior to the diaphragm, except the air sacs (alveoli) of the lungs; the inferior vena cava receives blood from veins inferior to the diaphragm.

✓ CHECKPOINT

What are the basic differences between systemic arteries and veins?

Vein	Description and Region Drained
Coronary sinus (KOR-ō-nar-ē; *corona* = crown)	The *coronary sinus* is the main vein of the heart; it receives almost all venous blood from the myocardium. It opens into the right atrium between the opening of the inferior vena cava and the tricuspid valve.
Superior vena cava (SVC) (VĒ-na CĀ-va; *vena* = vein; *cava* = cavelike)	The *SVC* empties its blood into the superior part of the right atrium. It begins by the union of the right and left brachiocephalic veins and enters the right atrium. The SVC drains the head, neck, chest, and upper limbs.
Inferior vena cava (IVC)	The *IVC* is the largest vein in the body. It begins by the union of the common iliac veins, passes through the diaphragm, and enters the inferior part of the right atrium. The IVC drains the abdomen, pelvis, and lower limbs. The inferior vena cava is commonly compressed during the later stages of pregnancy by the enlarging uterus, producing edema of the ankles and feet and temporary varicose veins.

Figure 16.12 Principal veins.

 Deoxygenated blood returns to the heart via the superior and inferior venae cavae and the coronary sinus.

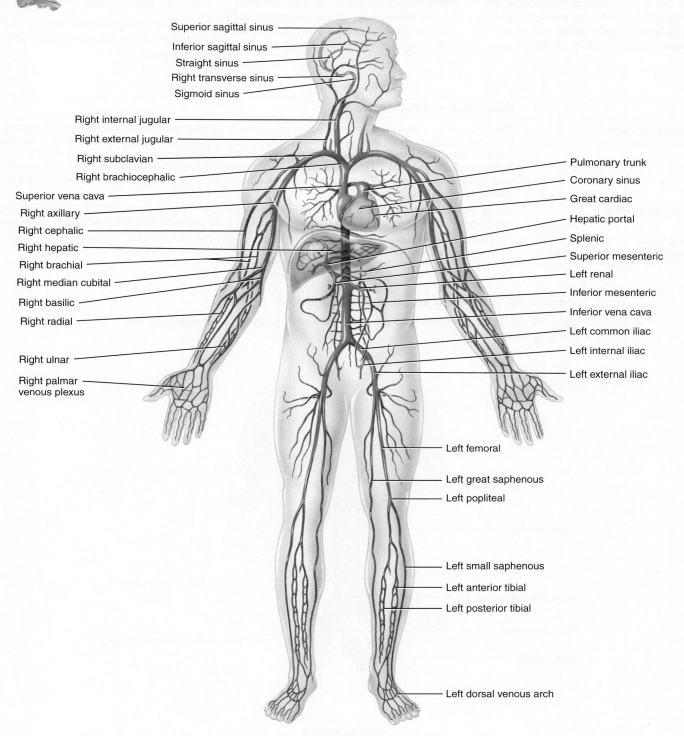

Superior sagittal sinus
Inferior sagittal sinus
Straight sinus
Right transverse sinus
Sigmoid sinus

Right internal jugular
Right external jugular
Right subclavian
Right brachiocephalic
Superior vena cava
Right axillary
Right cephalic
Right hepatic
Right brachial
Right median cubital
Right basilic
Right radial
Right ulnar
Right palmar venous plexus

Pulmonary trunk
Coronary sinus
Great cardiac
Hepatic portal
Splenic
Superior mesenteric
Left renal
Inferior mesenteric
Inferior vena cava
Left common iliac
Left internal iliac
Left external iliac

Left femoral
Left great saphenous
Left popliteal
Left small saphenous
Left anterior tibial
Left posterior tibial
Left dorsal venous arch

Overall anterior view of the principal veins

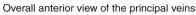 **Which general regions of the body are drained by the superior vena cava and the inferior vena cava?**

EXHIBIT 16.E Veins of the Head and Neck *(Figure 16.13)*

OBJECTIVE • **Identify the three major veins that drain blood from the head.**

• Most blood draining from the head passes into three pairs of veins: the *internal jugular veins*, *external jugular veins*, and *vertebral veins*. Within the brain, all veins drain into dural venous sinuses and then into the internal jugular veins. *Dural venous sinuses* are endothelium-lined venous channels between layers of the cranial dura mater.

✓ **CHECKPOINT**

Which general areas are drained by the internal jugular, external jugular, and vertebral veins?

Vein	Description and Region Drained
Internal jugular veins (JUG-ū-lar; *jugular* = throat)	The dural venous sinuses (the light blue vessels in Figure 16.13) drain blood from the cranial bones, meninges, and brain. The right and left *internal jugular veins* pass inferiorly on either side of the neck lateral to the internal carotid and common carotid arteries. They then unite with the subclavian veins to form the right and left *brachiocephalic veins* (brā′-kē-ō-se-FAL-ik; *brachio-* = arm; *-cephalic* = head). From here blood flows into the superior vena cava. The general structures drained by the internal jugular veins are the brain (through the dural venous sinuses), face, and neck.
External jugular veins	The right and left *external jugular veins* empty into the subclavian veins. The general structures drained by the external jugular veins are external to the cranium, such as the scalp and superficial and deep regions of the face.
Vertebral veins (VER-te-bral; *vertebra* = vertebrae)	The right and left *vertebral veins* empty into the brachiocephalic veins in the neck. They drain deep structures in the neck such as the cervical vertebrae, cervical spinal cord, and some neck muscles.

Figure 16.13 Principal veins of the head and neck.

Blood draining from the head passes into the internal jugular, external jugular, and vertebral veins.

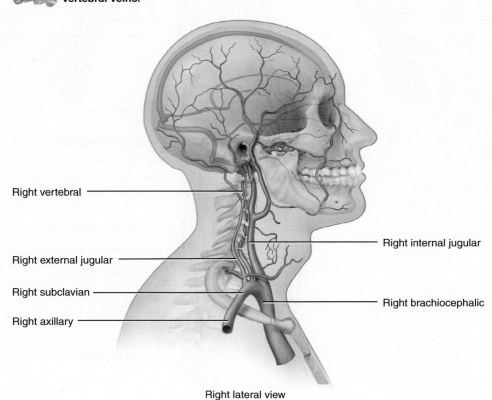

Right vertebral

Right external jugular

Right subclavian

Right axillary

Right internal jugular

Right brachiocephalic

Right lateral view

 Into which veins in the neck does all venous blood from the brain drain?

EXHIBIT 16.F Veins of the Upper Limbs *(Figure 16.14)*

OBJECTIVE • **Identify the principal veins that drain the upper limbs.**

• Blood from the upper limbs is returned to the heart by both *superficial* and *deep veins*. Both sets of veins have valves, which are more numerous in the deep veins.

• Superficial veins are larger than deep veins and return most of the blood from the upper limbs.

✓ **CHECKPOINT**

Where do the cephalic, basilic, median antebrachial, radial, and ulnar veins originate?

Vein	Description and Region Drained
Superficial Veins	
Cephalic veins (se-FAL-ik = head)	The principal superficial veins that drain the upper limbs originate in the hand and convey blood from the smaller superficial veins into the axillary veins. The **cephalic veins** begin on the lateral aspect of the **dorsal venous networks of the hands (dorsal venous arches)**, networks of veins on the dorsum of the hands (Figure 16.14a) that drain the fingers. The cephalic veins drain blood from the lateral aspect of the upper limbs.
Basilic veins (ba-SIL-ik = royal)	The **basilic veins** begin on the medial aspects of the dorsal venous networks of the hands (Figure 16.14b) and drain blood from the medial aspects of the upper limbs. Anterior to the elbow, the basilic veins are connected to the cephalic veins by the **median cubital veins** (*cubitus* = elbow), which drain the forearm. If a vein must be punctured for an injection, transfusion, or removal of a blood sample, the median cubital vein is preferred. The basilic veins continue ascending until they join the brachial veins. As the basilic and brachial veins merge in the axillary area, they form the axillary veins.
Median antebrachial veins (an′-tē-BRĀ-kē-al; *ante-* = before, in front of; *brachi-* = arm)	The **median antebrachial veins (median veins of the forearm)** begin in the **palmar venous plexuses**, networks of veins on the palms. The plexuses drain the fingers. The median antebrachial veins ascend in the forearms to join the basilic or median cubital veins, sometimes both. They drain the palms and forearms.
Deep Veins	
Radial veins (RĀ-dē-al = pertaining to the radius)	The paired **radial veins** begin at the **deep palmar venous arches** (Figure 16.14c). These arches drain the palms. The radial veins drain the lateral aspects of the forearms and pass alongside each radial artery. Just below the elbow joint, the radial veins unite with the ulnar veins to form the brachial veins.
Ulnar veins (UL-nar = pertaining to the ulna)	The paired **ulnar veins** begin at the **superficial palmar venous arches**, which drain the palms and the fingers. The ulnar veins drain the medial aspect of the forearms, pass alongside each ulnar artery, and join with the radial veins to form the brachial veins.
Brachial veins (BRĀ-kē-al)	The paired **brachial veins** accompany the brachial arteries. They drain the forearms, elbow joints, and arms. They join with the basilic veins to form the axillary veins.
Axillary veins (AK-si-ler′-ē; *axilla* = armpit)	The **axillary veins** ascend to become the subclavian veins. They drain the arms, axillae, and upper part of the chest wall.
Subclavian veins (sub-KLĀ-vē-an; *sub-* = under; *-clavian* = pertaining to the clavicle)	The **subclavian veins** are continuations of the axillary veins that unite with the internal jugular veins to form the brachiocephalic veins. The brachiocephalic veins unite to form the superior vena cava. The subclavian veins drain the arms, neck, and thoracic wall.

EXHIBIT 16.F CONTINUES

EXHIBIT 16.F   Veins of the Upper Limbs   *(Figure 16.14)*   **CONTINUED**

Figure 16.14 Principal veins of the right upper limb.

  Deep veins usually accompany arteries that have similar names.

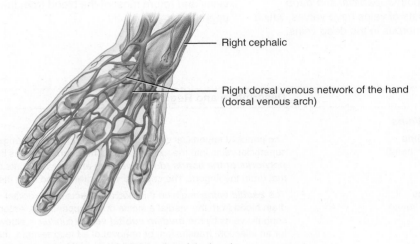

— Right cephalic

— Right dorsal venous network of the hand
(dorsal venous arch)

(a) Posterior view of superficial veins of the hand

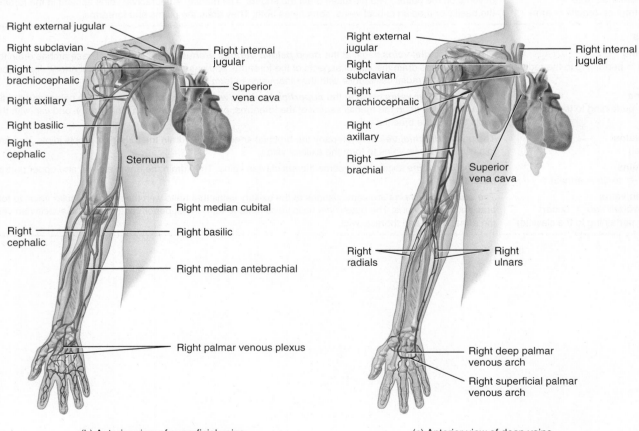

Right external jugular —

Right subclavian —

Right brachiocephalic —

Right axillary —

Right basilic —

Right cephalic —

— Right internal jugular

— Superior vena cava

Sternum —

— Right median cubital

Right cephalic —

— Right basilic

— Right median antebrachial

— Right palmar venous plexus

(b) Anterior view of superficial veins

Right external jugular —

Right subclavian —

Right brachiocephalic —

Right axillary —

Right brachial —

— Right internal jugular

— Superior vena cava

Right radials —

— Right ulnars

— Right deep palmar venous arch

— Right superficial palmar venous arch

(c) Anterior view of deep veins

? From which vein in the upper limb is a blood sample often taken?

EXHIBIT 16.G Veins of the Lower Limbs *(Figure 16.15)*

OBJECTIVE • **Identify the principal veins that drain the lower limbs.**

• As with the upper limbs, blood from the lower limbs is drained by both *superficial* and *deep veins*. The superficial veins often branch with each other and with deep veins along their length. All veins of the lower limbs have valves, which are more numerous than in veins of the upper limbs.

✓ **CHECKPOINT**

Why are the great saphenous veins clinically important?

Vein	Description and Region Drained
Superficial Veins	
Great saphenous veins (sa-FĒ-nus = clearly visible)	The **great saphenous veins**, the longest veins in the body, begin at the medial side of the **dorsal venous arches** (VĒ-nus) of the foot, networks of veins on the top of the foot that collect blood from the toes. The great saphenous veins empty into the femoral veins, and they mainly drain the leg and thigh, the groin, external genitals, and abdominal wall. Along their length, the great saphenous veins have from 10 to 20 valves, with more located in the leg than the thigh. The great saphenous veins are often used for prolonged administration of intravenous fluids. This is particularly important in very young children and in patients of any age who are in shock and whose veins are collapsed. The great saphenous veins are also often used as a source of vascular grafts, especially for coronary bypass surgery. In the procedure, the vein is removed and then reversed so that the valves do not obstruct the flow of blood.
Small saphenous veins	The **small saphenous veins** begin at the lateral side of the dorsal venous arches of the foot. They empty into the popliteal veins behind the knee. Along their length, the small saphenous veins have from 9 to 12 valves. The small saphenous veins drain the foot and leg.
Deep Veins	
Posterior tibial veins (TIB-ē-al)	The **deep plantar venous arches** on the soles drain the toes and ultimately give rise to the paired **posterior tibial veins**. They accompany the posterior tibial arteries through the leg and drain the foot and posterior leg muscles. About two-thirds the way up the leg, the posterior tibial veins drain blood from the **fibular (peroneal) veins**, which serve the lateral and posterior leg muscles.
Anterior tibial veins	The paired **anterior tibial veins** arise in the dorsal venous arch and accompany each anterior tibial artery. They unite with the posterior tibial veins to form the popliteal vein. The anterior tibial veins drain the ankle joint, knee joint, tibiofibular joint, and anterior portion of the leg.
Popliteal veins (pop′-li-TĒ-al; *popliteus* = hollow behind knee)	The **popliteal veins** are formed by the union of the anterior and posterior tibial veins. They drain the skin, muscles, and bones of the knee joint.
Femoral veins (FEM-o-ral)	The **femoral veins** accompany each femoral artery and are the continuations of the popliteal veins. They drain the muscles of the thigh, femur, external genitals, and superficial lymph nodes. The femoral veins enter the pelvic cavity, where they are known as the **external iliac veins**. The external and internal iliac veins unite to form the common iliac veins, which unite to form the inferior vena cava.

EXHIBIT 16.G CONTINUES

EXHIBIT 16.G Veins of the Lower Limbs *(Figure 16.15)* **CONTINUED**

Figure 16.15 Principal veins of the pelvis and lower limbs.

 All veins of the lower limbs have valves.

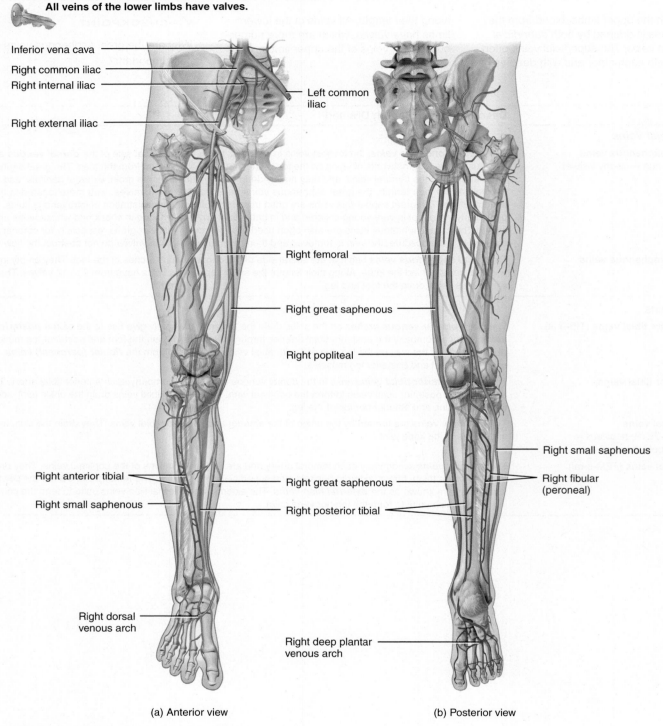

(a) Anterior view

(b) Posterior view

Which veins of the lower limb are superficial?

Reversing Artery Disease

Scientists once believed that when plaque formed in an artery, it never went away. Medical researchers thought that lifestyle changes and drugs could slow the process of atherosclerosis, but they could not undo damage already done. In recent years, however, researchers have discovered that, if given half a chance, the body's own healing processes can reverse arterial plaque buildup. Lifestyle changes and drug treatments appear to help stabilize dangerous atherosclerotic plaques and in some people may even eliminate the need for surgical interventions such as bypass surgery.

What Makes Plaque Dangerous?

The health risk imposed by plaque that accumulates within the artery lining depends on several factors. Some plaque is fairly stable: It has a low lipid content, does not increase much in size, and has a strong fibrous cap that keeps it from rupturing when blood pressure rises. Unstable plaque is characterized by a large accumulation of lipid in its core covered only by a thin fibrous cap. In addition, unstable plaques contain a large number of macrophages, indicating active inflammation in the plaque area as the body tries to cope with the damage to the artery lining. In a misguided attempt to heal endothelial damage, macrophages ingest plaque lipids; the net result is increased arterial injury and lipid accumulation. An unstable plaque is apt to rupture, triggering formation of a life-threatening blood clot at the plaque site.

Stabilizing Influences

The first step in preventing, slowing, and reversing artery disease is to control the risk factors associated with its progression. Artery disease is aggravated by oxidized low-density lipoprotein (LDL) particles that stick to the artery lining. Reducing blood levels of LDL through a heart-healthy diet is a good first step. Such a diet involves reduced intake of animal fats and increased intake of plant foods. Quitting smoking is a must, as smoking can cause the oxidation of the LDL particles.

Regular physical activity improves artery health in several ways. Exercise increases high-density lipoprotein (HDL) levels; HDLs help shuttle harmful fats away from the artery lining. Exercise decreases platelet stickiness and the tendency of the blood to form clots around the plaque. It also helps reduce high blood pressure, which can damage the artery lining. By improving blood sugar control, exercise reduces the risk of type 2 diabetes and the damage that high levels of blood glucose and insulin can do to the artery lining.

Think It Over . . . Lifestyle programs designed to reverse atherosclerosis damage usually include a healthful diet, physical activity, and stress management. Why is stress management seen as an important part of these programs?

Pulmonary Circulation

When deoxygenated blood returns to the heart from the systemic route, it is pumped out of the right ventricle into the lungs. In the lungs, it loses carbon dioxide and picks up oxygen. Now bright red again, the blood returns to the left atrium of the heart and is pumped again into the systemic circulation. The flow of deoxygenated blood from the right ventricle to the air sacs of the lungs and the return of oxygenated blood from the air sacs to the left atrium is called the *pulmonary circulation* (see Figure 16.8). The *pulmonary trunk* emerges from the right ventricle and then divides into two branches. The *right pulmonary artery* runs to the right lung; the *left pulmonary artery* goes to the left lung. After birth, the pulmonary arteries are the only arteries that carry deoxygenated blood. On entering the lungs, the branches divide and subdivide until ultimately they form capillaries around the air sacs in the lungs. Carbon dioxide passes from the blood into the air sacs and is exhaled, while inhaled oxygen passes from the air sacs into the blood. The capillaries unite, venules and veins are formed, and, eventually, two *pulmonary veins* from each lung transport the oxygenated blood to the left atrium. (After birth, the pulmonary veins are the only veins that carry oxygenated blood.) Contractions of the left ventricle then send the blood into the systemic circulation.

Hepatic Portal Circulation

A vein that carries blood from one capillary network to another is called a *portal vein*. The hepatic portal vein, formed by the union of the splenic and superior mesenteric veins (Figure 16.16), receives blood from capillaries of digestive organs and delivers it to capillary-like structures in the liver called sinusoids. In the *hepatic portal circulation* (*hepat-* = liver), venous blood from the gastrointestinal organs and spleen, rich with substances absorbed from the gastrointestinal tract, is delivered

Figure 16.16 Hepatic portal circulation.

The hepatic portal circulation delivers venous blood from the gastrointestinal organs and spleen to the liver.

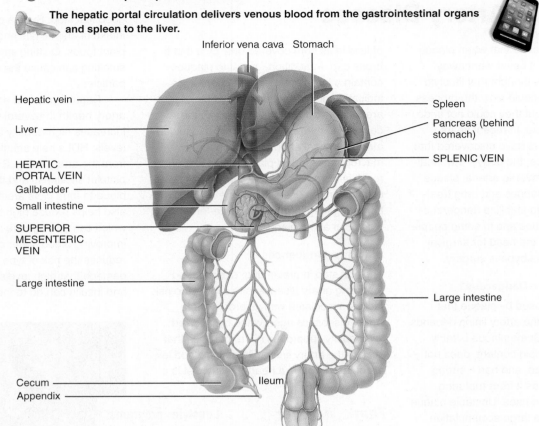

(a) Anterior view of veins draining into the hepatic portal vein

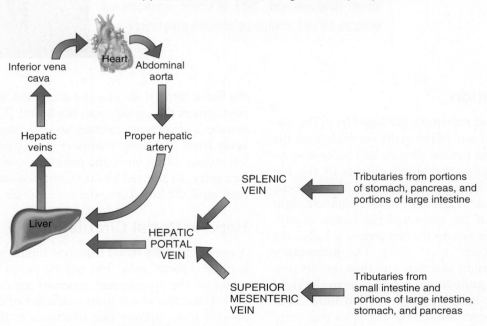

(b) Scheme of principal blood vessels of hepatic portal circulation and arterial supply and venous drainage of liver

 **Which veins carry blood away from the liver?**

to the hepatic portal vein and enters the liver. The liver processes these substances before they pass into the general circulation. At the same time, the liver receives oxygenated blood from the systemic circulation via the hepatic artery. The oxygenated blood mixes with the deoxygenated blood in sinusoids. Ultimately, all blood leaves the sinusoids of the liver through the hepatic veins, which drain into the inferior vena cava.

Fetal Circulation

The circulatory system of a fetus, called *fetal circulation*, exists only in the fetus and contains special structures that allow the developing fetus to exchange materials with its mother (Figure 16.17). It differs from the postnatal (after birth) circulation because the lungs, kidneys, and gastrointestinal organs

Figure 16.17 Fetal circulation and changes at birth. The boxes between parts (a) and (b) describe the fate of certain fetal structures once postnatal circulation is established.

The lungs and gastrointestinal organs do not begin to function until birth.

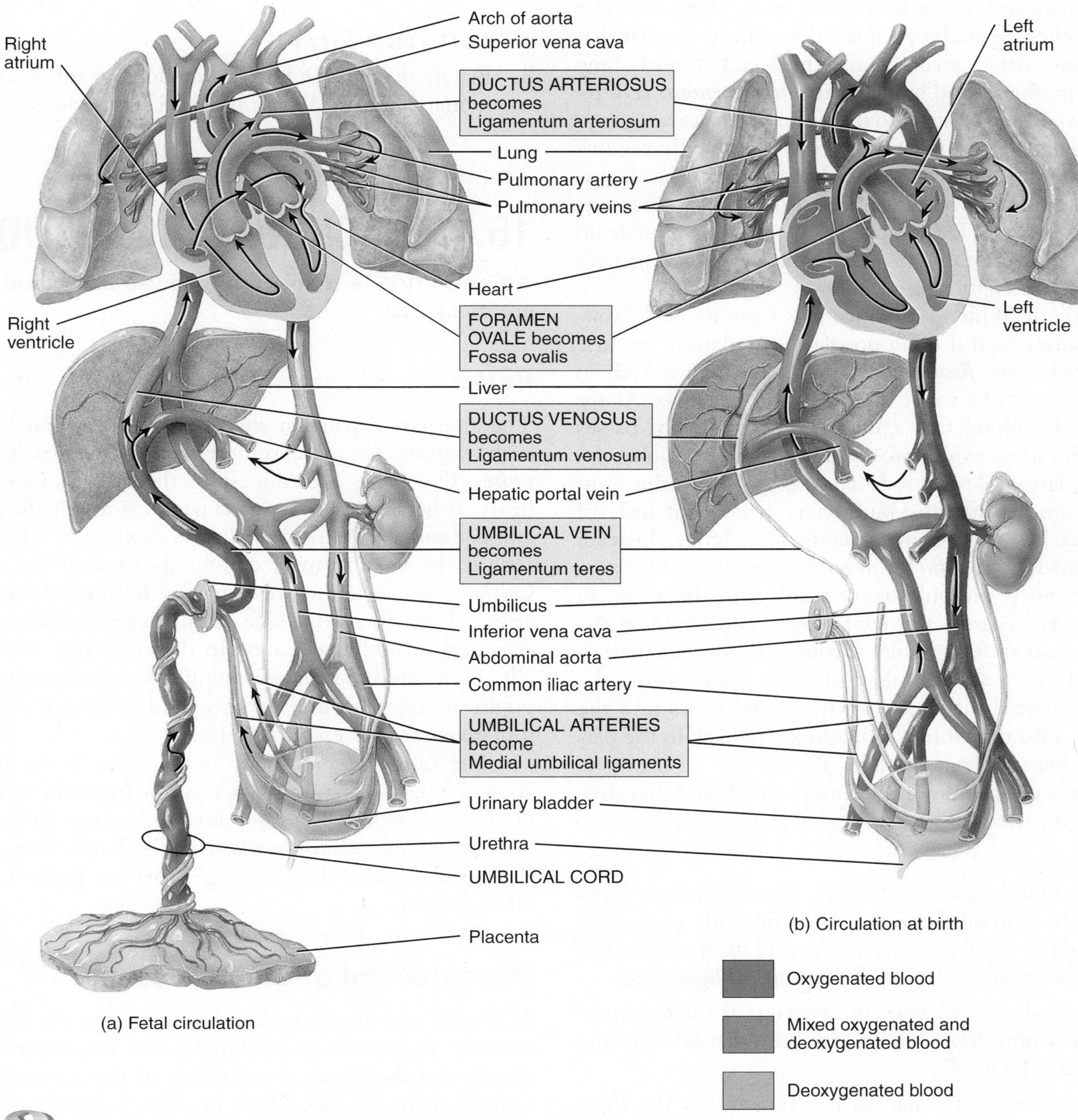

Arch of aorta
Superior vena cava
Right atrium
Left atrium

DUCTUS ARTERIOSUS becomes Ligamentum arteriosum

Lung
Pulmonary artery
Pulmonary veins

Heart
Right ventricle
Left ventricle

FORAMEN OVALE becomes Fossa ovalis

Liver

DUCTUS VENOSUS becomes Ligamentum venosum

Hepatic portal vein

UMBILICAL VEIN becomes Ligamentum teres

Umbilicus
Inferior vena cava
Abdominal aorta
Common iliac artery

UMBILICAL ARTERIES become Medial umbilical ligaments

Urinary bladder
Urethra
UMBILICAL CORD

Placenta

(a) Fetal circulation

(b) Circulation at birth

Oxygenated blood

Mixed oxygenated and deoxygenated blood

Deoxygenated blood

Which structure provides for exchange of materials between mother and fetus?

do not begin to function until birth. The fetus obtains O_2 and nutrients from and eliminates CO_2 and other wastes into the maternal blood.

The exchange of materials between fetal and maternal circulations occurs through the *placenta* (pla-SEN-ta), which forms inside the mother's uterus and attaches to the umbilicus (navel) of the fetus by the *umbilical cord* (um-BIL-i-kal). Blood passes from the fetus to the placenta via two *umbilical arteries* (Figure 16.17a). These branches of the internal iliac arteries are within the umbilical cord. At the placenta, fetal blood picks up O_2 and nutrients and eliminates CO_2 and wastes. The oxygenated blood returns from the placenta via a single *umbilical vein*. This vein ascends to the liver of the fetus, where it divides into two branches. Some blood flows through the branch that joins the hepatic portal vein and enters the liver, but most of the blood flows into the second branch, the *ductus venosus* (DUK-tus ve-NŌ-sus), which drains into the inferior vena cava.

Deoxygenated blood returning from lower body regions of the fetus mingles with oxygenated blood from the ductus venosus in the inferior vena cava. This mixed blood then enters the right atrium. Deoxygenated blood returning from upper body regions of the fetus enters the superior vena cava and passes into the right atrium.

Most of the fetal blood does not pass from the right ventricle to the lungs, as it does in postnatal circulation, because an opening called the *foramen ovale* (fō-RĀ-men ō-VAL-ē) exists in the septum between the right and left atria. About one-third of the blood that enters the right atrium passes through the foramen ovale into the left atrium and joins the systemic circulation. The blood that does pass into the right ventricle is pumped into the pulmonary trunk, but little of this blood reaches the nonfunctioning fetal lungs. Instead, most is sent through the *ductus arteriosus* (ar-tē-rē-Ō-sus), a vessel that connects the pulmonary trunk with the aorta, so that most blood bypasses the fetal lungs. The blood in the aorta is carried to all fetal tissues through the systemic circulation. When the common iliac arteries branch into the external and internal iliacs, part of the blood flows into the internal iliacs, into the umbilical arteries, and back to the placenta for another exchange of materials.

After birth, when pulmonary (lung), renal, and digestive functions begin, the following vascular changes occur (Figure 16.17b):

1. When the umbilical cord is tied off, blood no longer flows through the umbilical arteries, they fill with connective tissue, and the distal portions of the umbilical arteries become fibrous cords called *medial umbilical ligaments*.

2. The umbilical vein collapses but remains as the *ligamentum teres (round ligament)*, a structure that attaches the umbilicus to the liver.

3. The ductus venosus collapses but remains as the *ligamentum venosum*, a fibrous cord on the inferior surface of the liver.

4. The placenta is expelled as the "*afterbirth*."

5. The foramen ovale normally closes shortly after birth to become the *fossa ovalis*, a depression in the interatrial septum. When an infant takes its first breath, the lungs expand and blood flow to the lungs increases. Blood returning from the lungs to the heart increases pressure in the left atrium. This closes the foramen ovale by pushing the valve that guards it against the interatrial septum. Permanent closure occurs in about a year.

6. The ductus arteriosus closes by vasoconstriction almost immediately after birth and becomes the *ligamentum arteriosum*.

✓ **CHECKPOINT**

8. Briefly describe the main functions of the systemic, pulmonary, hepatic portal, and fetal circulations.

16.4 CHECKING CIRCULATION

OBJECTIVE • **Explain how pulse and blood pressure are measured.**

Pulse

The alternate expansion and elastic recoil of an artery after each contraction and relaxation of the left ventricle is called a *pulse*. The pulse is strongest in the arteries closest to the heart. It becomes weaker as it passes through the arterioles, and it disappears altogether in the capillaries. The radial artery at the wrist is most commonly used to feel the pulse. Other sites where the pulse may be felt include the brachial artery along the medial side of the biceps brachii muscle; the common carotid artery, next to the voice box, which is usually monitored during cardiopulmonary resuscitation; the popliteal artery behind the knee; and the dorsal artery of the foot above the instep of the foot.

The pulse rate normally is the same as the heart rate, about 75 beats per minute at rest. *Tachycardia* (tak'-i-KAR-dē-a; *tachy-* = fast) is a rapid resting heart or pulse rate over 100 beats/min. *Bradycardia* (brād'-i-KAR-dē-a; *brady-* = slow) indicates a slow resting heart or pulse rate under 50 beats/min.

Measurement of Blood Pressure

In clinical use, the term *blood pressure* usually refers to the pressure in arteries generated by the left ventricle during systole and the pressure remaining in the arteries when the ventricle is in diastole. Blood pressure is usually measured in the brachial artery in the left arm (see Figure 16.10a). The device used to measure blood pressure is a *sphygmomanometer* (sfig′-mō-ma-NOM-e-ter; *sphygmo-* = pulse; *manometer* = instrument used to measure pressure). When the pressure cuff is inflated above the blood pressure attained during systole, the artery is compressed so that blood flow stops. The technician places a stethoscope below the cuff over the brachial artery and then slowly deflates the cuff. When the cuff is deflated enough to allow the artery to open, a spurt of blood passes through, resulting in the first sound heard through the stethoscope. This sound corresponds to **systolic blood pressure (SBP)**—the force with which blood is pushing against arterial walls during ventricular contraction. As the cuff is deflated further, the sounds suddenly become faint. This level, called the **diastolic blood pressure (DBP)**, represents the force exerted by the blood remaining in arteries during ventricular relaxation.

The normal blood pressure of a young adult male is less than 120 mm Hg systolic and less than 80 mm Hg diastolic, reported, for example, as "110 over 70" and written as 110/70. In young adult females, the pressures are 8 to 10 mm Hg less. People who exercise regularly and are in good physical condition may have even lower blood pressures.

✓ CHECKPOINT

9. What causes pulse?

10. Distinguish between systolic and diastolic blood pressure.

16.5 AGING AND THE CARDIOVASCULAR SYSTEM

OBJECTIVE • **Describe the effects of aging on the cardiovascular system.**

General changes in the cardiovascular system associated with aging include increased stiffness of the aorta, reduction in cardiac muscle fiber size, progressive loss of cardiac muscular strength, reduced cardiac output, a decline in maximum heart rate, and an increase in systolic blood pressure. Coronary artery disease (CAD) is the major cause of heart disease and death in older Americans. Congestive heart failure (CHF), a set of symptoms associated with impaired pumping of the heart, is also prevalent in older individuals. Changes in blood vessels that serve brain tissue—for example, atherosclerosis—reduce nourishment to the brain and result in the malfunction or death of brain cells. By age 80, blood flow to the brain is 20 percent less, and blood flow to the kidneys is 50 percent less than it was in the same person at age 30 because of the effects of aging on blood vessels.

✓ CHECKPOINT

11. What are some of the signs that the cardiovascular system is aging?

• • •

To appreciate the many ways the cardiovascular system contributes to homeostasis of other body systems, examine Focus on Homeostasis: The Cardiovascular System. Next, in Chapter 17, we will examine the structure and function of the lymphatic system, seeing how it returns excess fluid filtered from capillaries to the cardiovascular system. We will also take a more detailed look at how some white blood cells function as defenders of the body by carrying out immune responses.

Focus on Homeostasis

BODY SYSTEM	CONTRIBUTION OF THE CARDIOVASCULAR SYSTEM

For all body systems

The heart pumps blood through blood vessels to body tissues, delivering oxygen and nutrients and removing wastes by means of capillary exchange. Circulating blood keeps body tissues at a proper temperature.

Integumentary system

Blood delivers clotting factors and white blood cells that aid in hemostasis when skin is damaged and contribute to repair of injured skin. Changes in skin blood flow contribute to body temperature regulation by adjusting the amount of heat loss via the skin. Blood flowing in skin may give skin a pink hue.

Skeletal system

Blood delivers calcium and phosphate ions that are needed for building bone extracellular matrix, hormones that govern building and breakdown of bone extracellular matrix, and erythropoietin that stimulates production of red blood cells by red bone marrow.

Muscular system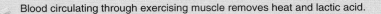

Blood circulating through exercising muscle removes heat and lactic acid.

Nervous system

Endothelial cells lining choroid plexuses in brain ventricles help produce cerebrospinal fluid (CSF) and contribute to the blood–brain barrier.

Endocrine system

Circulating blood delivers most hormones to their target tissues. Atrial cells of the heart secrete atrial natriuretic peptide.

Lymphatic system and immunity

Circulating blood distributes lymphocytes, antibodies, and macrophages that carry out immune functions. Lymph forms from excess interstitial fluid, which filters from blood plasma due to blood pressure generated by the heart.

Respiratory system

Circulating blood transports oxygen from the lungs to body tissues and carbon dioxide to the lungs for exhalation.

Digestive system

Blood carries newly absorbed nutrients and water to the liver. Blood distributes hormones that aid digestion.

Urinary system

The heart and blood vessels deliver 20 percent of the resting cardiac output to the kidneys, where blood is filtered, needed substances are reabsorbed, and unneeded substances are eliminated as part of urine, which is excreted.

Reproductive systems

Vasodilation of arterioles in penis and clitoris cause erection during sexual intercourse. Blood distributes hormones that regulate reproductive functions.

THE CARDIOVASCULAR SYSTEM

COMMON DISORDERS

Hypertension

About 50 million Americans have **hypertension**, or persistently high blood pressure. It is the most common disorder affecting the heart and blood vessels and is the major cause of heart failure, kidney disease, and stroke. In May 2003, the Joint National Committee on Prevention, Detection, Evaluation, and Treatment of High Blood Pressure published new guidelines for hypertension because clinical studies have linked what were once considered fairly low pressure readings to an increased risk of cardiovascular disease. The new guidelines are as follows:

Category	Systolic (mm Hg)	Diastolic (mm Hg)
Normal	Less than 120 *and*	Less than 80
Prehypertension	120–139 *or*	80–89
Stage 1 hypertension	140–159 *or*	90–99
Stage 2 hypertension	Greater than 160 *or*	Greater than 100

Using the new guidelines, the normal classification was previously considered optimal; prehypertension now includes many more individuals previously classified as normal or high-normal; stage 1 hypertension is the same as in previous guidelines; and stage 2 hypertension now combines the previous stage 2 and stage 3 categories since treatment options are the same for the former stages 2 and 3.

Although several categories of drugs can reduce elevated blood pressure, the following lifestyle changes are also effective in managing hypertension:

- *Lose weight.* This is the best treatment for high blood pressure short of using drugs. Loss of even a few pounds helps reduce blood pressure in overweight hypertensive individuals.

- *Limit alcohol intake.* Drinking in moderation may lower the risk of coronary heart disease, mainly among males over 45 and females over 55. Moderation is defined as no more than one 12-oz beer per day for females and no more than two 12-oz beers per day for males.

- *Exercise.* Becoming more physically fit by engaging in moderate activity (such as brisk walking) several times a week for 30 to 45 minutes can lower systolic blood pressure by about 10 mm Hg.

- *Reduce intake of sodium (salt).* Roughly half the people with hypertension are "salt sensitive." For them, a high-salt diet appears to promote hypertension, and a low-salt diet can lower their blood pressure.

- *Maintain recommended dietary intake of potassium, calcium, and magnesium.* Higher levels of potassium, calcium, and magnesium in the diet are associated with a lower risk of hypertension.

- *Don't smoke.* Smoking has devastating effects on the heart and can augment the damaging effects of high blood pressure by promoting vasoconstriction.

- *Manage stress.* Various meditation and biofeedback techniques help some people reduce high blood pressure. These methods may work by decreasing the daily release of epinephrine and norepinephrine by the adrenal medulla.

Shock

Shock is a failure of the cardiovascular system to deliver enough O_2 and nutrients to meet cellular metabolic needs. The causes of shock are many and varied, but all are characterized by inadequate blood flow to body tissues. Common causes of shock include loss of body fluids, as occurs in hemorrhage, dehydration, burns, excessive vomiting, diarrhea, or sweating. If shock persists, cells and organs become damaged, and cells may die unless proper treatment begins quickly.

Although the symptoms of shock vary with the severity of the condition, the following are commonly observed: systolic blood pressure lower than 90 mm Hg; rapid resting heart rate due to sympathetic stimulation and increased blood levels of epinephrine and norepinephrine; weak, rapid pulse due to reduced cardiac output and fast heart rate; cool, pale skin due to vasoconstriction of skin blood vessels; sweating due to sympathetic stimulation; reduced urine formation and output due to increased levels of aldosterone and antidiuretic hormone (ADH); altered mental state due to reduced oxygen supply to the brain; thirst due to loss of extracellular fluid; and nausea due to impaired circulation to digestive organs.

Aneurysm

An **aneurysm** (AN-ū-rizm) is a thin, weakened section of the wall of an artery or a vein that bulges outward, forming a balloonlike sac. Common causes are atherosclerosis, syphilis, congenital blood vessel defects, and trauma. If untreated, the aneurysm enlarges and the blood vessel wall becomes so thin that it bursts. The result is massive hemorrhage along with shock, severe pain, stroke, or death.

MEDICAL TERMINOLOGY AND CONDITIONS

Angiogenesis (an′-jē-ō-JEN-e-sis)   Formation of new blood vessels.

Aortography (a′-or-TOG-ra-fē)   X-ray examination of the aorta and its main branches after injection of a dye.

Circulation time   The time required for a drop of blood to pass from the right atrium, through the pulmonary circulation, back to the left atrium, through the systemic circulation, down to the foot, and back again to the right atrium; normally about 1 minute in a resting person.

Claudication (klaw′-di-KĀ-shun)   Pain and lameness or limping caused by defective circulation of the blood in the vessels of the limbs.

Deep vein thrombosis (DVT)   The presence of a thrombus (blood clot) in a deep vein of the lower limbs.

Doppler ultrasound scanning   Imaging technique commonly used to measure blood flow. A transducer is placed on the skin and an image is displayed on a monitor that provides the exact position and severity of a blockage.

Hypotension (hī-pō-TEN-shun)   Low blood pressure; most commonly used to describe an acute drop in blood pressure, as occurs during excessive blood loss.

Occlusion (ō-KLOO-zhun) The closure or obstruction of the lumen of a structure such as a blood vessel. An example is an atherosclerotic plaque in an artery.

Orthostatic hypotension (or′-thō-STAT-ik; *ortho-* = straight; *-static* = causing to stand) An excessive lowering of systemic blood pressure when a person stands up; usually a sign of disease. May be caused by excessive fluid loss, certain drugs, and cardiovascular or neurogenic factors. Also called *postural hypotension.*

Phlebitis (fle-BĪ-tis; *phleb-* = vein) Inflammation of a vein, often in a leg. The condition is often accompanied by pain and red-

ness of the skin over the inflamed vein. It is frequently caused by trauma or bacterial infection.

Syncope (SIN-kō-pē) A temporary cessation of consciousness; a faint. One cause is insufficient blood supply to the brain.

Thrombophlebitis (throm′-bō-fle-BĪ-tis) Inflammation of a vein involving clot formation. Superficial thrombophlebitis occurs in veins under the skin, especially in the calf.

White coat (office) hypertension A stress-induced syndrome found in patients who have elevated blood pressure when being examined by health-care personnel, but otherwise have normal blood pressure.

CHAPTER REVIEW AND RESOURCE SUMMARY

REVIEW

RESOURCES

16.1 Blood Vessel Structure and Function

1. **Arteries** carry blood away from the heart. Their walls consist of three layers. The structure of the middle layer gives arteries their two major properties, elasticity and contractility.

2. **Arterioles** are small arteries that deliver blood to capillaries. Through constriction and dilation, arterioles play a key role in regulating blood flow from arteries into capillaries.

3. **Capillaries** are microscopic blood vessels through which materials are exchanged between blood and interstitial fluid. **Precapillary sphincters** regulate blood flow through capillaries.

4. **Capillary blood pressure** "pushes" fluid out of capillaries into interstitial fluid (**filtration**). **Blood colloid osmotic pressure** "pulls" fluid into capillaries from interstitial fluid (**reabsorption**).

5. **Autoregulation** refers to local adjustments of blood flow in response to physical and chemical changes in a tissue.

6. **Venules** are small vessels that emerge from capillaries and merge to form veins. They drain blood from capillaries into veins.

7. **Veins** consist of the same three layers as arteries but have less elastic tissue and smooth muscle. They contain **valves** that prevent backflow of blood. Weak venous valves can lead to **varicose veins**.

8. **Venous return**, the volume of blood flowing back to the heart through systemic veins, occurs due to the pumping action of the heart, aided by skeletal muscle contractions (the **skeletal muscle pump**), and breathing (the **respiratory pump**).

Anatomy Overview—The
  Cardiovascular System
Animation—Capillary Exchange
Animation—Vascular
  Regulation

Figure 16.1 Comparative structure of blood vessels

Figure 16.3 Capillary exchange

Exercise—Concentrate on
  Blood Vessels
Exercise—Capillary Exchange
  Pick 'em
Concepts and Connections—
  Capillary Exchange

16.2 Blood Flow Through Blood Vessels

1. Blood flow is determined by blood pressure and vascular resistance.

2. Blood flows from regions of higher pressure to regions of lower pressure. **Blood pressure** is highest in the aorta and large systemic arteries; it drops progressively as distance from the left ventricle increases. Blood pressure in the right atrium is close to 0 mm Hg.

3. An increase in blood volume increases blood pressure, and a decrease in blood volume decreases it.

4. **Vascular resistance** is the opposition to blood flow mainly as a result of friction between blood and the walls of blood vessels. It depends on the size of the blood vessel lumen, blood viscosity, and total blood vessel length.

5. Blood pressure and blood flow are regulated by neural and hormonal negative feedback systems and by autoregulation.

6. The **cardiovascular (CV) center** in the medulla oblongata helps regulate heart rate, stroke volume, and size of blood vessel lumen.

7. Vasomotor nerves (sympathetic) control vasoconstriction and vasodilation.

8. Baroreceptors (pressure-sensitive receptors) send impulses to the cardiovascular center to regulate blood pressure.

Animation—Regulating Blood
  Pressure
Exercise—Regulating BP with
  Hormones
Exercise—Regulating BP with
  Nervous Impluses
Concepts and Connections—
  Blood Flow
Concepts and Connections—
  Blood Pressure Regulation

REVIEW	**RESOURCES**

9. Chemoreceptors (receptors sensitive to concentrations of oxygen, carbon dioxide, and hydrogen ions) also send impulses to the cardiovascular center to regulate blood pressure.

10. Hormones such as angiotensin II, aldosterone, epinephrine, norepinephrine, and antidiuretic hormone raise blood pressure; atrial natriuretic peptide lowers it.

16.3 Circulatory Routes

1. The two major **circulatory routes** are the systemic circulation and the pulmonary circulation.

2. The **systemic circulation** takes oxygenated blood from the left ventricle through the **aorta** to all parts of the body and returns deoxygenated blood to the right atrium.

3. The parts of the aorta include the **ascending aorta**, the **arch of the aorta**, the **thoracic aorta**, and the **abdominal aorta** (see Exhibit 16.A). Each part gives off arteries that branch to supply the whole body (see Exhibits 16.B–16.C).

4. Deoxygenated blood is returned to the heart through the systemic veins (see Exhibit 16.D). All veins of the systemic circulation flow into either the **superior** or **inferior vena cava** or the **coronary sinus**, which empty into the right atrium (see Exhibits 16.E–16.G).

5. The **pulmonary circulation** takes deoxygenated blood from the right ventricle to the air sacs of the lungs and returns oxygenated blood from the air sacs to the left atrium. It allows blood to be oxygenated for systemic circulation.

6. The **hepatic portal circulation** collects deoxygenated blood from the veins of the gastrointestinal tract and spleen and directs it into the hepatic portal vein of the liver. This routing allows the liver to extract and modify nutrients and detoxify harmful substances in the blood. The liver also receives oxygenated blood from the hepatic artery.

7. **Fetal circulation** exists only in the fetus. It involves the exchange of materials between fetus and mother via the **placenta**. The fetus derives O_2 and nutrients from and eliminates CO_2 and wastes into maternal blood. At birth, when pulmonary (lung), digestive, and liver functions begin, the special structures of fetal circulation are no longer needed.

Resources:
Anatomy Overview— Comparison of Circulatory Routes

Figure 16.16 Hepatic portal circulation

Figure 16.17 Fetal circulation and changes at birth

Exercise—Artery Archery
Exercise—Drag and Drop Blood Flow
Exercise—Vein Archery

16.4 Checking Circulation

1. **Pulse** is the alternate expansion and elastic recoil of an artery with each heartbeat. It may be felt in any artery that lies near the surface or over a hard tissue.

2. A normal pulse rate is about 75 beats per minute.

3. **Blood pressure** is the pressure exerted by blood on the wall of an artery when the left ventricle undergoes systole and then diastole. It is measured by a **sphygmomanometer**.

4. **Systolic blood pressure (SBP)** is the force of blood recorded during ventricular contraction. **Diastolic blood pressure (DBP)** is the force of blood recorded during ventricular relaxation. The normal blood pressure of a young adult male is less than 120/80.

Resources:
Animation—Regulating Blood Pressure

16.5 Aging and the Cardiovascular System

1. General changes associated with aging include reduced elasticity of blood vessels, reduction in cardiac muscle size, reduced cardiac output, and increased systolic blood pressure.

2. The incidence of coronary artery disease (CAD), congestive heart failure (CHF), and atherosclerosis increases with age.

Resources:
Homeostatic Imbalances— The Case of the Anxious Man with Chest Pains

SELF-QUIZ

1. Sensory receptors that monitor changes in the blood pressure to the brain are
   a. chemoreceptors in the aorta.
   b. baroreceptors in the carotid arteries.
   c. the aortic bodies.
   d. precapillary sphincters in the arterioles.
   e. proprioceptors in the muscles.

2. The blood vessels that allow the exchange of nutrients, wastes, oxygen, and carbon dioxide between the blood and tissues are the
   a. capillaries.
   b. arteries.
   c. venules.
   d. arterioles.
   e. veins.

3. Autoregulation is
   a. the reabsorption of fluids in capillaries.
   b. an ability of a tissue to adjust its blood flow.
   c. a type of capillary exchange mechanism.
   d. a measure of capillary blood pressure.
   e. the slowing of blood flow in capillaries.

4. Blood flows through the blood vessels because of the
   a. establishment of a concentration gradient.
   b. elastic recoil of the veins.
   c. establishment of a pressure gradient.
   d. viscosity of the blood.
   e. thinness of the walls of capillaries.

5. Which of the following represents pulmonary circulation as the blood flows from the right ventricle?
   a. pulmonary trunk → pulmonary veins → pulmonary capillaries → pulmonary arteries
   b. pulmonary arteries → pulmonary capillaries → pulmonary trunk → pulmonary veins
   c. pulmonary capillaries → pulmonary trunk → pulmonary arteries → pulmonary veins
   d. pulmonary trunk → pulmonary arteries → pulmonary capillaries → pulmonary veins
   e. pulmonary veins → pulmonary capillaries → pulmonary arteries → pulmonary trunk

6. Supplying additional fat tissue with blood may raise blood pressure because
   a. the blood's viscosity will increase.
   b. arterioles will constrict in the remainder of the body's vessels.
   c. the respiratory pump will be inhibited.
   d. baroreceptors become less responsive to pressure changes.
   e. of an increase in total vessel length.

7. Match the following descriptions to the appropriate blood vessel.
   ____ a. composed of a single layer of endothelium and a basement membrane
   ____ b. formed by reuniting capillaries
   ____ c. carry blood away from heart
   ____ d. medium-sized vessels capable of high degree of vasoconstriction and vasodilation
   ____ e. may contain valves
   ____ f. very stretchable and able to recoil
   ____ g. regulate blood flow to capillaries

   A. arteries
   B. arterioles
   C. veins
   D. venules
   E. capillaries
   F. elastic arteries
   G. muscular arteries

8. Filtration of substances out of capillaries occurs when the capillary blood pressure is ____ and blood colloid osmotic pressure is ____.
   a. low, low   b. low, high
   c. high, high   d. high, low
   e. low, not changing

9. Which of the following pairs of hormones have opposite effects on blood pressure?
   a. aldosterone, ADH
   b. epinephrine, angiotensin II
   c. ADH, ANP
   d. epinephrine, norepinephrine
   e. angiotensin II, aldosterone

10. Which of the following statements about blood vessels is true?
    a. Capillaries contain valves.
    b. Walls of arteries are generally thicker and contain more elastic tissue than walls of veins.
    c. Arteries generally contain most of the body's total blood volume.
    d. Blood flows most rapidly through capillaries.
    e. Blood pressure in arteries is always lower than in veins.

11. Why is it important that blood flows slowly through the capillaries?
    a. It allows time for the materials in the blood to pass through the thick capillary walls.
    b. It prevents damage to the capillaries.
    c. It permits the efficient exchange of nutrients and wastes between the blood and body cells.
    d. It allows the heart time to rest.
    e. It allows the blood pressure in capillaries to rise above the blood pressure in the veins.

12. Match the following:
    ____ a. source of all systemic arteries
    ____ b. supplies a lower limb
    ____ c. heart's blood system
    ____ d. returns blood to heart from lower limbs
    ____ e. carries blood to liver
    ____ f. leads to lungs
    ____ g. returns blood from lungs to heart
    ____ h. supplies blood to brain
    ____ i. returns blood to heart from head and upper body

    A. hepatic portal vein
    B. pulmonary trunk
    C. pulmonary vein
    D. common iliac artery
    E. coronary circulation
    F. inferior vena cava
    G. superior vena cava
    H. aorta
    I. cerebral arterial circle

13. For each of the following factors, indicate if it increases (I) or decreases (D) blood pressure:
    ____ a. an increase in cardiac output
    ____ b. hemorrhage
    ____ c. vasodilation
    ____ d. vasoconstriction
    ____ e. stimulation of the heart by the sympathetic nervous system
    ____ f. hypoxia
    ____ g. epinephrine
    ____ h. decrease in blood viscosity
    ____ i. bradycardia
    ____ j. increase in blood volume

14. Aldosterone affects blood pressure by
    a. increasing heart rate.
    b. increasing vasoconstriction of arterioles.
    c. reducing blood volume.
    d. stimulating the release of atrial natriuretic peptide by the heart.
    e. increasing reabsorption of sodium ions and water by the kidneys.

15. In a blood pressure reading of 110/70,
    a. 110 represents the diastolic pressure.
    b. 70 represents the pressure of the blood against the arteries during ventricular relaxation.
    c. 110 represents the blood pressure and 70 represents the heart rate.
    d. 70 is the reading taken when the first sound is heard.
    e. the patient has a severe problem with hypertension.

16. Which of the following statements is NOT true?
    a. Regulation of blood vessel diameter originates from the hypothalamus.
    b. The cerebral cortex may provide input to the cardiovascular center.
    c. Baroreceptors may stimulate the cardiovascular center.
    d. Activation of proprioceptors increases heart rate at the beginning of exercise.
    e. Vasomotor tone is due to a moderate level of vasoconstriction.

17. Venous return to the heart is enhanced by all of the following EXCEPT
    a. skeletal muscle pump.
    b. valves in veins.
    c. the pressure difference from venules to the right ventricle.
    d. vasodilation.
    e. inhalation during breathing.

CRITICAL THINKING APPLICATIONS

1. The local anesthetic injected by a dentist often contains a small amount of adrenaline. What effect would adrenaline have on the blood vessels in the vicinity of the dental work? Why might this effect be desired?

2. In this chapter, you've read about varicose veins. Why didn't you read about varicose arteries?

3. Kate is expecting her first child and has had cravings for chocolate throughout her pregnancy. She visits her favorite shop and buys a large bag of chocolate bars. As the shop assistant hands her the bag, Kate rubs her swollen abdomen and proclaims she's "eating and breathing for two." Exactly how is her unborn child "eating and breathing"?

4. Philip spent 10 minutes sharpening his favorite knife before carving the roast. Unfortunately, he sliced his finger along with the roast. His wife pressed a towel over the spurting cut and drove him to the Accident and Emergency department. What type of vessel did Philip cut, and how do you know?

ANSWERS TO FIGURE QUESTIONS

16.1 The femoral artery has the thicker wall; the femoral vein has the wider lumen.

16.2 Metabolically active tissues have more capillaries because they use oxygen and produce wastes more rapidly than inactive tissues.

16.3 Excess filtered fluid and proteins that escape from plasma drain into lymphatic capillaries and are returned by the lymphatic system to the cardiovascular system.

16.4 The skeletal muscle pump and the respiratory pump help boost venous return.

16.5 As blood pressure increases, blood flow increases.

16.6 Vasoconstriction increases vascular resistance, which decreases blood flow through the vasoconstricted blood vessels.

16.7 It happens when you stand up because gravity causes pooling of blood in leg veins as you stand upright, decreasing the blood pressure in your upper body.

16.8 The principal circulatory routes are the systemic and pulmonary circulations.

16.9 The four parts of the aorta are the ascending aorta, arch of the aorta, thoracic aorta, and abdominal aorta.

16.10 Branches of the arch of the aorta are the brachiocephalic trunk, left common carotid artery, and left subclavian artery.

16.11 The abdominal aorta divides into the common iliac arteries at about the level of the fourth lumbar vertebra.

16.12 The superior vena cava drains regions above the diaphragm (except the cardiac veins and the alveoli of the lungs), and the inferior vena cava drains regions below the diaphragm.

16.13 All venous blood in the brain drains into the internal jugular veins.

16.14 The median cubital vein is often used for withdrawing blood.

16.15 Superficial veins of the lower limbs include the dorsal venous arch and the great saphenous and small saphenous veins.

16.16 The hepatic veins carry blood away from the liver.

16.17 The exchange of materials between mother and fetus occurs across the placenta.

CHAPTER 17

THE LYMPHATIC SYSTEM AND IMMUNITY

Maintaining homeostasis in the body requires continual combat against harmful agents in our environment. Despite constant exposure to a variety of *pathogens* (PATH-ō-jens), disease-producing microbes such as bacteria and viruses, most people remain healthy. The body surface also endures cuts and bumps, exposure to ultraviolet rays in sunlight, chemical toxins, and minor burns with an array of defenses. In this chapter, we will explore the mechanisms that provide defenses against intruders and promote the repair of damaged body tissues.

Immunity (i-MŪ-ni-tē) or *resistance* is the ability to use our body's defenses to ward off damage or disease. The two types of immunity are (1) innate and (2) adaptive. *Innate (nonspecific) immunity* refers to defenses that are present at birth. They are always present and available to provide rapid responses to protect us against disease. Innate immunity does not involve specific recognition of a microbe and acts against all microbes in the same way. However, innate immunity does not have a memory component, that is, it cannot recall a previous contact with a foreign molecule. Among the components of innate immunity are the first line of defense (skin and mucous membranes) and the second line of defense (antimicrobial substances, natural killer cells, phagocytes, inflammation, and fever). Innate immune responses represent immunity's early-warning system and are designed to prevent microbes from gaining access into the body and to help eliminate those that do gain access.

Adaptive (specific) immunity refers to defenses that involve specific recognition of a microbe once it has breached the innate immunity defenses. Adaptive immunity is based on a specific response to a specific microbe; that is, it adapts or adjusts to handle a specific microbe. Unlike innate immunity, adaptive immunity is slower to respond but it does have a memory component. Adaptive immunity involves lymphocytes (a type of white blood cell) called T lymphocytes (T cells) and B lymphocytes (B cells). The body system responsible for adaptive immunity (and some aspects of innate immunity) is the *lymphatic system* (lim-FAT-ik) (Figure 17.1).

Did you know?　*The immune system defends us from viruses, bacteria, and other pathogens. It provides barriers that keep these potential invaders out of our bodies, and attacks the pathogens that manage to break through these barriers. But sometimes the immune system makes mistakes, identifying the body's own cells as foreign and initiating an attack. In addition, the process of inflammation, which normally stimulates the healing process, can sometimes become a chronic and damaging condition.*

LOOKING BACK TO MOVE AHEAD...

Veins (Section 16.3)

Cancer (Chapter 3, Common Disorders)

Epidermis of Skin (Section 5.1)

Mucous Membranes (Section 4.4)

Phagocytosis (Section 3.3)

FOCUS ON WELLNESS

Inflammatory Messages—Fatness versus Fitness

454

Figure 17.1 Components of the lymphatic system.

 The lymphatic system consists of lymph, lymphatic vessels, lymphatic tissues, and red bone marrow.

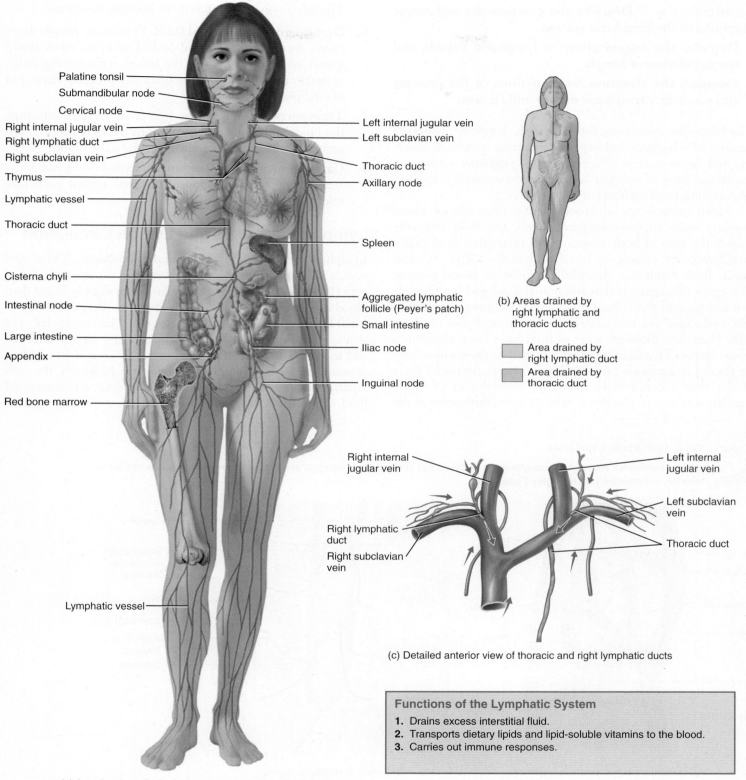

Palatine tonsil
Submandibular node
Cervical node
Right internal jugular vein
Right lymphatic duct
Right subclavian vein
Thymus
Lymphatic vessel
Thoracic duct
Cisterna chyli
Intestinal node
Large intestine
Appendix
Red bone marrow

Left internal jugular vein
Left subclavian vein
Thoracic duct
Axillary node
Spleen
Aggregated lymphatic follicle (Peyer's patch)
Small intestine
Iliac node
Inguinal node

Lymphatic vessel

(a) Anterior view of principal components of lymphatic system

(b) Areas drained by right lymphatic and thoracic ducts

Area drained by right lymphatic duct
Area drained by thoracic duct

Right internal jugular vein
Right lymphatic duct
Right subclavian vein

Left internal jugular vein
Left subclavian vein
Thoracic duct

(c) Detailed anterior view of thoracic and right lymphatic ducts

Functions of the Lymphatic System

1. Drains excess interstitial fluid.
2. Transports dietary lipids and lipid-soluble vitamins to the blood.
3. Carries out immune responses.

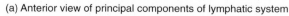 **What is lymphatic tissue?**

17.1 LYMPHATIC SYSTEM

OBJECTIVES • **Describe the components and major functions of the lymphatic system.**

• **Describe the organization of lymphatic vessels and the circulation of lymph.**

• **Compare the structure and functions of the primary and secondary lymphatic organs and tissues.**

The lymphatic system consists of lymph, lymphatic vessels, a number of structures and organs containing lymphatic tissue, and red bone marrow (Figure 17.1). *Lymphatic tissue* is a specialized form of reticular connective tissue (see Table 4.3C) that contains large numbers of lymphocytes.

Most components of blood plasma filter out of blood capillary walls to form *interstitial fluid*, the fluid that surrounds the cells of body tissues. After interstitial fluid passes into lymphatic vessels, it is called *lymph* (LIMF = clear fluid). Both fluids are chemically similar to blood plasma. The main difference is that interstitial fluid and lymph contain less protein than blood plasma because most plasma protein molecules are too large to filter through the capillary wall. Each day, about 20 liters of fluid filter from blood into tissue spaces. This fluid must be returned to the cardiovascular system to maintain normal blood volume. About 17 liters of the fluid filtered daily from the arterial end of blood capillaries return to the blood directly by reabsorption at the venous end of the capillaries. The remaining 3 liters per day pass first into lymphatic vessels and are then returned to the blood.

The lymphatic system has three primary functions:

1. **Drains excess interstitial fluid.** Lymphatic vessels drain excess interstitial fluid and leaked proteins from tissue spaces and return them to the blood. This activity helps maintain fluid balance in the body and prevents depletion of vital plasma proteins.

2. **Transports dietary lipids.** Lymphatic vessels transport the lipids and lipid-soluble vitamins (A, D, E, and K) absorbed by the gastrointestinal tract into the blood.

3. **Carries out immune responses.** Lymphatic tissue initiates highly specific responses directed against particular microbes or abnormal cells.

Lymphatic Vessels and Lymph Circulation

Lymphatic vessels begin as *lymphatic capillaries*. These tiny vessels are closed at one end and located in the spaces between cells (Figure 17.2). Lymphatic capillaries are slightly larger than blood capillaries and have a unique structure that permits interstitial fluid to flow into them, but not out. The endothelial cells that make up the wall of a lymphatic capillary are not attached end to end, but rather, the ends overlap (Figure 17.2b). When pressure is greater in interstitial fluid than in lymph, the cells separate slightly, like a one-way swinging door, and interstitial fluid enters the lymphatic capillary. When pressure is greater

Figure 17.2 Lymphatic capillaries.

Lymphatic capillaries are found throughout the body except in the central nervous system, portions of the spleen, red bone marrow, and tissues that lack blood capillaries.

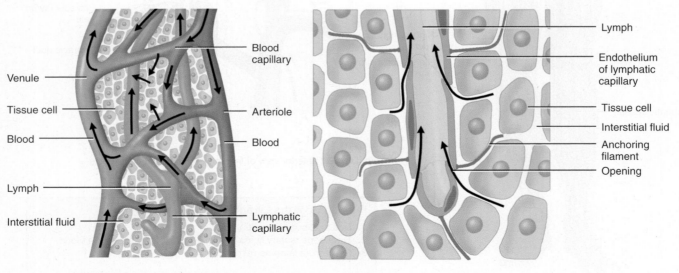

(a) Relationship of lymphatic capillaries to tissue cells and blood capillaries

(b) Details of lymphatic capillary

Labels (a): Venule; Tissue cell; Blood; Lymph; Interstitial fluid; Blood capillary; Arteriole; Blood; Lymphatic capillary

Labels (b): Lymph; Endothelium of lymphatic capillary; Tissue cell; Interstitial fluid; Anchoring filament; Opening

Why is lymph more similar to interstitial fluid than it is to blood plasma?

inside the lymphatic capillary, the cells adhere more closely and lymph cannot escape back into interstitial fluid.

Unlike blood capillaries, which link two larger blood vessels that form part of a circuit, lymphatic capillaries begin in the tissues and carry the lymph that forms there toward a larger lymphatic vessel. Just as blood capillaries unite to form venules and veins, lymphatic capillaries unite to form larger and larger *lymphatic vessels* (see Figure 17.1a). Lymphatic vessels resemble veins in structure but have thinner walls and more valves. Located along lymphatic vessels are *lymph nodes*, masses of B cells and T cells that are surrounded by a capsule. Lymph flows through lymph nodes.

From the lymphatic vessels, lymph eventually passes into one of two main channels: the thoracic duct or the right lymphatic duct. The ***thoracic duct***, the main lymph-collecting duct, receives lymph from the left side of the head, neck, and chest; the left upper limb; and the entire body below the ribs. The ***right lymphatic duct*** drains lymph from the upper right side of the body (see Figure 17.1b, c).

Ultimately, the thoracic duct empties its lymph into the junction of the left internal jugular and left subclavian veins, and the right lymphatic duct empties its lymph into the junction of the right internal jugular and right subclavian veins. Thus, lymph drains back into the blood (Figure 17.3).

The same two pumps that aid return of venous blood to the heart maintain the flow of lymph:

1. **Skeletal muscle pump.** The "milking action" of skeletal muscle contractions (see Figure 16.4) compresses lymphatic vessels (as well as veins) and forces lymph toward the subclavian veins.

2. **Respiratory pump.** Lymph flow is also maintained by pressure changes that occur during inhalation (breathing in). Lymph flows from the abdominal region, where the pressure is higher, toward the thoracic region, where it is lower. When the pressures reverse during exhalation (breathing out), the valves prevent backflow of lymph.

CLINICAL CONNECTION | Edema

Edema (e-DĒ-ma) is an excessive accumulation of interstitial fluid in tissue spaces. It may be caused by a lymphatic system obstruction, such as an infected lymph node or a blocked lymphatic vessel. Edema may also result from increased capillary blood pressure, which causes excess interstitial fluid to form faster than it can pass into lymphatic vessels or be reabsorbed back into the capillaries. Another cause is lack of skeletal muscle contractions, as in individuals who are paralyzed. •

Figure 17.3 Relationship of lymphatic vessels and lymph nodes to the cardiovascular system. Arrows show the direction of flow of lymph and blood.

The sequence of fluid flow is: blood capillaries (blood plasma) → interstitial spaces (interstitial fluid) → lymphatic capillaries (lymph) → lymphatic vessels and lymph nodes (lymph) → lymphatic ducts (lymph) → junction of jugular and subclavian veins (blood plasma).

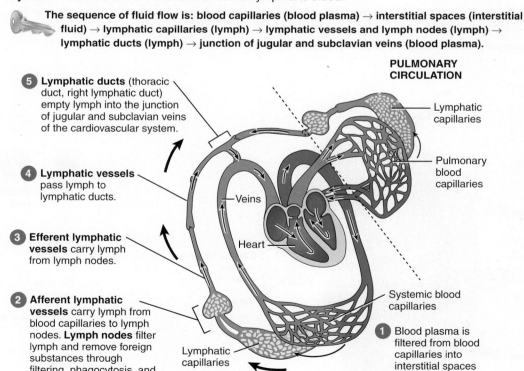

5 **Lymphatic ducts** (thoracic duct, right lymphatic duct) empty lymph into the junction of jugular and subclavian veins of the cardiovascular system.

4 **Lymphatic vessels** pass lymph to lymphatic ducts.

3 **Efferent lymphatic vessels** carry lymph from lymph nodes.

2 **Afferent lymphatic vessels** carry lymph from blood capillaries to lymph nodes. **Lymph nodes** filter lymph and remove foreign substances through filtering, phagocytosis, and immune reactions.

PULMONARY CIRCULATION

Lymphatic capillaries

Pulmonary blood capillaries

Veins

Heart

Systemic blood capillaries

1 Blood plasma is filtered from blood capillaries into interstitial spaces to become interstitial fluid.

Lymphatic capillaries

SYSTEMIC CIRCULATION

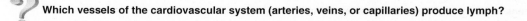

Which vessels of the cardiovascular system (arteries, veins, or capillaries) produce lymph?

Lymphatic Organs and Tissues

Lymphatic organs and tissues, which are widely distributed throughout the body, are classified into two groups based on their functions. *Primary lymphatic organs*, the sites where stem cells divide and develop into mature B cells and T cells, include the *red bone marrow* (in flat bones and the ends of the long bones of adults) and the *thymus*. The *secondary lymphatic organs* and *tissues*, the sites where most immune responses occur, include lymph nodes, the spleen, and lymphatic nodules.

Thymus

The thymus is a two-lobed organ located posterior to the sternum, medial to the lungs, and superior to the heart (see Figure 17.1). It contains large numbers of T cells and scattered dendritic cells (so named for their long, branchlike projections), epithelial cells, and macrophages. Immature T cells migrate from red bone marrow to the thymus, where they multiply and begin to mature. Only about 2 percent of the immature T cells that arrive in the thymus achieve the proper "education" to "graduate" into mature T cells. The remaining cells die via apoptosis (programmed cell death). Thymic macrophages help clear out the debris of dead and dying cells. Mature T cells leave the thymus via the blood and are carried to lymph nodes, the spleen, and other lymphatic tissues where they populate parts of these organs and tissues.

Lymph Nodes

Located along lymphatic vessels are about 600 bean-shaped *lymph nodes*. They are scattered throughout the body, both superficially and deep, and usually occur in groups (see Figure 17.1). Lymph nodes are heavily concentrated near the mammary glands and in the axillae and groin. Each node is covered by a capsule of dense connective tissue (Figure 17.4). Internally, different regions of a lymph node may contain B cells that develop into plasma cells, as well as T cells, dendritic cells, and macrophages.

Lymph enters a node through one of several *afferent lymphatic vessels* (*af-* = toward; *-ferrent* = to carry). As lymph flows through the node, foreign substances are trapped by *reticular fibers* within the spaces between cells. Macrophages destroy some foreign substances by phagocytosis, and lymphocytes destroy others by a variety of immune responses. Filtered lymph leaves the other end of the node through one or two *efferent lymphatic vessels* (*ef-* = away). Plasma cells and T cells that have divided many times within a lymph node can also leave the node and circulate to other parts of the body. Valves direct the flow of lymph inward through the afferent lymphatic vessels and outward through the efferent lymphatic vessels (Figure 17.4).

Lymph nodes function as filters. As lymph enters one end of a lymph node, foreign substances are trapped by the

Figure 17.4 Structure of a lymph node (partially sectioned). Green arrows indicate direction of lymph flow into and out of the lymph node.

Lymph nodes are present throughout the body, usually clustered in groups.

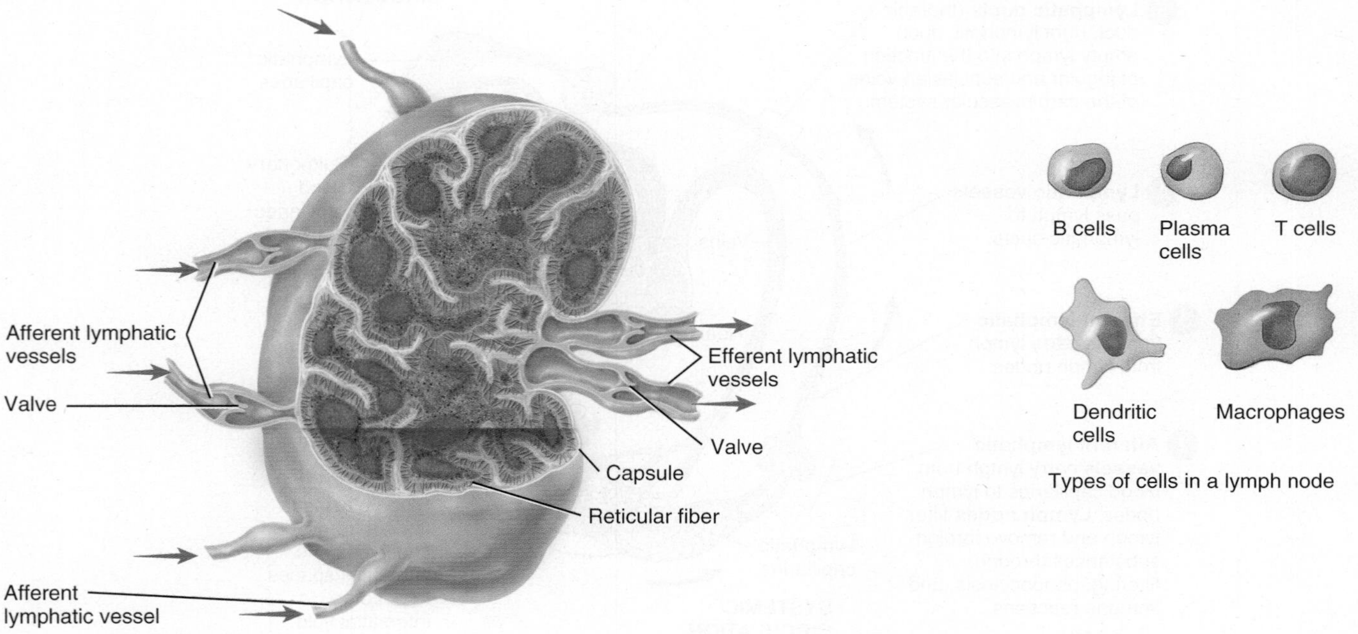

Afferent lymphatic vessels

Valve

Efferent lymphatic vessels

Valve

Capsule

Reticular fiber

Afferent lymphatic vessel

B cells   Plasma cells   T cells

Dendritic cells   Macrophages

Types of cells in a lymph node

? What happens to foreign substances that enter a lymph node in lymph?

reticular fibers within the sinuses of the node. Then macrophages destroy some foreign substances by phagocytosis, while lymphocytes destroy others by immune responses. The filtered lymph then leaves the other end of the lymph node. Since there are many afferent lymphatic vessels that bring lymph into a lymph node and only one or two efferent lymphatic vessels that transport lymph out of a lymph node, the slow flow of lymph within the lymph nodes allows additional time for lymph to be filtered. Additionally, all lymph flows through multiple lymph nodes on its path through the lymph vessels. This exposes the lymph to multiple filtering events before returning to the blood.

CLINICAL CONNECTION | Metastasis

Metastasis (me-TAS-ta-sis; *meta-* = beyond; *stasis* = to stand), the spread of a disease from one part of the body to another, can occur via lymphatic vessels. All malignant tumors eventually metastasize. Cancer cells may travel in the blood or lymph and establish new tumors where they lodge. When metastasis occurs via lymphatic vessels, secondary tumor sites can be predicted according to the direction of lymph flow from the primary tumor site. Cancerous lymph nodes feel enlarged, firm, nontender, and fixed to underlying structures. By contrast, most lymph nodes that are enlarged due to an infection are softer, tender, and movable. •

Spleen

The *spleen* is the largest single mass of lymphatic tissue in the body (see Figure 17.1). It lies between the stomach and diaphragm and is covered by a capsule of dense connective tissue. The spleen contains two types of tissue, called white pulp and red pulp. *White pulp* is lymphatic tissue, consisting mostly of lymphocytes and macrophages. *Red pulp* consists of blood-filled *venous sinuses* and cords of *splenic tissue* consisting of red blood cells, macrophages, lymphocytes, plasma cells, and granular leukocytes.

Blood flowing into the spleen through the splenic artery enters the white pulp. Within the white pulp, B cells and T cells carry out immune responses, while macrophages destroy pathogens by phagocytosis. Within the red pulp, the spleen performs three functions related to blood cells: (1) removal by macrophages of worn-out or defective blood cells and platelets; (2) storage of platelets, perhaps up to one-third of the body's supply; and (3) production of blood cells (hemopoiesis) during fetal life.

CLINICAL CONNECTION | Splenectomy

The spleen is the organ most often damaged in cases of abdominal trauma. A ruptured spleen causes severe internal hemorrhage and shock. Prompt **splenectomy**, removal of the spleen, is needed to prevent bleeding to death. After a splenectomy, other structures, particularly red bone marrow and the liver, can take over functions normally carried out by the spleen. •

Lymphatic Nodules

Lymphatic nodules are egg-shaped masses of lymphatic tissue that are not surrounded by a capsule. They are plentiful in the connective tissue of mucous membranes lining the gastrointestinal, urinary, and reproductive tracts and the respiratory airways. Although many lymphatic nodules are small and solitary, some occur as large aggregations in specific parts of the body. Among these are the tonsils in the pharyngeal region and the aggregated lymphatic follicles (Peyer's patches) in the ileum of the small intestine (see Figure 17.1). Aggregations of lymphatic nodules also occur in the appendix. The five *tonsils*, which form a ring at the junction of the oral cavity, nasal cavity, and throat, are strategically positioned to participate in immune responses against inhaled or ingested foreign substances. The single *pharyngeal tonsil* (fa-RIN-jē-al) or *adenoid* is embedded in the posterior wall of the upper part of the throat (see Figure 18.2). The two *palatine tonsils* (PAL-a-tīn) lie at the back of the mouth, one on either side; these are the tonsils commonly removed in a tonsillectomy. The paired *lingual tonsils* (LIN-gwal), located at the base of the tongue, may also require removal during a tonsillectomy.

✓ CHECKPOINT

1. How are interstitial fluid and lymph similar, and how do they differ?
2. What are the roles of the thymus and the lymph nodes in immunity?
3. Describe the functions of the spleen and tonsils.

17.2 INNATE IMMUNITY

OBJECTIVE • Describe the various components of innate immunity.

Innate (nonspecific) immunity includes the external physical and chemical barriers provided by the skin and mucous membranes. It also includes various internal defenses, such as antimicrobial substances, natural killer cells, phagocytes, inflammation, and fever.

First Line of Defense: Skin and Mucous Membranes

The skin and mucous membranes of the body are the **first line of defense** against pathogens. These structures provide both physical and chemical barriers that discourage pathogens and foreign substances from penetrating the body and causing disease.

With its many layers of closely packed, keratinized cells, the outer epithelial layer of the skin—the *epidermis*—provides a formidable physical barrier to the entrance of microbes (see Figure 5.1). In addition, periodic shedding of epidermal cells helps remove microbes at the skin surface. Bacteria rarely penetrate the intact surface of healthy epidermis. If this surface is broken by cuts, burns, or punctures, however, pathogens can penetrate the epidermis and invade adjacent tissues or circulate in the blood to other parts of the body.

The epithelial layer of *mucous membranes*, which line body cavities, secretes a fluid called *mucus* that lubricates and moistens the cavity surface. Because mucus is slightly viscous, it traps many microbes and foreign substances. The mucous membrane of the nose has mucus-coated *hairs* that trap and filter microbes, dust, and pollutants from inhaled air. The mucous membrane of the upper respiratory tract contains *cilia*, microscopic hairlike projections on the surface of the epithelial cells. The waving action of cilia propels inhaled dust and microbes that have become trapped in mucus toward the throat. Coughing and sneezing accelerate movement of mucus and its entrapped pathogens out of the body. Swallowing mucus sends pathogens to the stomach, where gastric juice destroys them.

Other fluids produced by various organs also help protect epithelial surfaces of the skin and mucous membranes. The *lacrimal apparatus* (LAK-ri-mal) of the eyes (see Figure 12.5) manufactures and drains away tears in response to irritants. Blinking spreads tears over the surface of the eyeball, and the continual washing action of tears helps to dilute microbes and keep them from settling on the surface of the eye. Tears also contain *lysozyme* (LĪ-so-zīm), an enzyme capable of breaking down the cell walls of certain bacteria. Besides tears, lysozyme is present in saliva, perspiration, nasal secretions, and tissue fluids. *Saliva*, produced by the salivary glands, washes microbes from the surfaces of the teeth and from the mucous membrane of the mouth, much as tears wash the eyes. The flow of saliva reduces colonization of the mouth by microbes.

The cleansing of the urethra by the *flow of urine* retards microbial colonization of the urinary system. *Vaginal secretions* likewise move microbes out of the body in females. *Defecation* and *vomiting* also expel microbes. For example, in response to some microbial toxins, the smooth muscle of the lower gastrointestinal tract contracts vigorously; the resulting diarrhea rapidly expels many of the microbes.

Certain chemicals also contribute to the high degree of resistance of the skin and mucous membranes to microbial invasion. Sebaceous (oil) glands of the skin secrete an oily substance called *sebum* that forms a protective film over the surface of the skin. The unsaturated fatty acids in sebum inhibit the growth of certain pathogenic bacteria and fungi. The acidity of the skin (pH 3–5) is caused in part by the secretion of fatty acids and lactic acid. *Perspiration* helps flush microbes from the surface of the skin. *Gastric juice*, produced by the glands of the stomach, is a mixture of hydrochloric acid, enzymes, and mucus. The strong acidity of gastric juice (pH 1.2–3.0) destroys many

bacteria and most bacterial toxins. Vaginal secretions also are slightly acidic, which discourages bacterial growth.

Second Line of Defense: Internal Defenses

Although the skin and mucous membranes are very effective barriers in preventing invasion by pathogens, they may be broken by injuries or everyday activities such as brushing the teeth or shaving. Any pathogens that get past the surface barriers encounter a **second line of defense** consisting of internal antimicrobial substances, phagocytes, natural killer cells, inflammation, and fever.

Antimicrobial Substances

Various body fluids contain four main types of *antimicrobial substances* that discourage microbial growth:

1. Lymphocytes, macrophages, and fibroblasts infected with viruses produce proteins called *interferons* (*IFNs*) (in′-ter-FĒR-ons). After their release by virus-infected cells, IFNs diffuse to uninfected neighboring cells, where they stimulate synthesis of proteins that interfere with viral replication. Viruses can cause disease only if they can replicate within body cells.

2. A group of normally inactive proteins in blood plasma and on plasma membranes makes up the *complement system*. When activated, these proteins "complement" or enhance certain immune, allergic, and inflammatory reactions. One effect of complement proteins is to create holes in the plasma membrane of the microbe. As a result, extracellular fluid moves into the holes, causing the microbe to burst, a process called *cytolysis*. Another effect of complement is to cause *chemotaxis* (kē′-mō-TAK-sis), the chemical attraction of phagocytes to a site. Some complement proteins cause *opsonization* (op′-son-i-ZĀ-shun), a process in which complement proteins bind to the surface of a microbe and promote phagocytosis.

3. **Iron-binding proteins** inhibit the growth of certain bacteria by reducing the amount of available iron. Examples include *transferrin* (found in blood and tissue fluids), *lactoferrin* (found in milk, saliva, and mucus), *ferretin* (found in the liver, spleen, and red bone marrow), and *hemoglobin* (found in red blood cells).

4. **Antimicrobial proteins (AMPs)** are short peptides that have a broad spectrum of antimicrobial activity. Examples of AMPs are *dermicidin* (der-ma-SĪ-din) (produced by sweat glands), *defensins* and *cathelicidins* (cath-el-i-SĪ-dins) (produced by neutrophils, macrophages, and epithelia), and *thrombocidin* (throm′-bō-SĪ-din) (produced by platelets). Besides killing a wide range of microbes, AMPs can attract dendritic cells and mast cells, which participate in immune responses. Interestingly enough, microbes exposed to AMPs do not appear to develop resistance, as often happens with antibiotics.

Phagocytes and Natural Killer Cells

When microbes penetrate the skin and mucous membranes or bypass the antimicrobial substances in blood, the next nonspecific defense consists of phagocytes and natural killer cells.

Phagocytes (*phago-* = eat; *-cytes* = cells) are specialized cells that perform **phagocytosis** (*-osis* = process), the ingestion of microbes or other particles such as cellular debris. The two main types of phagocytes are neutrophils and macrophages. When an infection occurs, neutrophils and monocytes migrate to the infected area. During this migration, the monocytes enlarge and develop into actively phagocytic cells called **macrophages** (MAK-rō-fā-jez) (see Figure 14.2a). Some are *wandering macrophages*, which migrate to infected areas. Others are *fixed macrophages*, which remain in certain locations, including the skin and subcutaneous layer, liver, lungs, brain, spleen, lymph nodes, and red bone marrow.

About 5–10 percent of lymphocytes in the blood are **natural killer (NK) cells**, which have the ability to kill a wide variety of microbes and certain tumor cells. NK cells also are present in the spleen, lymph nodes, and red bone marrow. They cause cellular destruction by releasing proteins that destroy the target cell's membrane.

Inflammation

Inflammation is a defensive response of the body to tissue damage. Because inflammation is one of the body's innate defenses, the response of a tissue to a cut is similar to the response to damage caused by burns, radiation, or invasion of bacteria or viruses. The events of inflammation dispose of microbes, toxins, or foreign material at the site of injury, prevent their spread to other tissues, and prepare the site for tissue repair. Thus, inflammation helps restore tissue homeostasis. The four signs and symptoms of inflammation are redness, pain, heat, and swelling. Inflammation can also cause the loss of function in the injured area, depending on the site and extent of the injury.

The stages of inflammation are as follows:

1. In a region of tissue injury, mast cells in connective tissue and basophils and platelets in blood release *histamine*. In response to histamine, two immediate changes occur in the blood vessels: *increased permeability* and *vasodilation*, an increase in the diameter of the blood vessels (Figure 17.5). Increased permeability means that substances normally retained in blood are permitted to pass out of the blood vessels. Vasodilation allows more blood to flow to the damaged area and helps remove microbial toxins and dead cells. Increased permeability also permits defensive substances such as antibodies and clot-forming chemicals to enter the injured area from the blood.

   From the events that occur during inflammation, it's easy to understand the signs and symptoms. Heat and redness result from the large amount of blood that accumulates in the damaged area. The area swells due to an increased amount of interstitial fluid that has leaked out

Figure 17.5 Inflammation. Several substances stimulate vasodilation, increased permeability of blood vessels, chemotaxis, emigration, and phagocytosis. Phagocytes migrate from blood to the site of tissue injury.

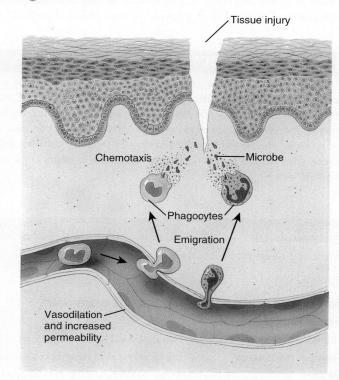

Inflammation is an innate immune response of the body to tissue damage.

Phagocytes migrate from blood to site of tissue injury

 What causes redness at a site of inflammation?

of the capillaries (edema). Pain results from injury to neurons, from toxic chemicals released by microbes, and from the increased pressure of edema.

2. The increased permeability of capillaries causes leakage of clotting proteins into tissues. Fibrinogen is converted to an insoluble, thick network of fibrin threads, which traps the invading organisms and prevents their spread. The resulting clot isolates the invading microbes and their toxins.

3. Shortly after the inflammatory process starts, phagocytes are attracted to the site of injury by chemotaxis (Figure 17.5). Near the damaged area, neutrophils begin to squeeze through the wall of the blood vessel, a process called *emigration*. Neutrophils predominate in the early stages of infection, but they die off rapidly along with the microbes they have eaten. Within a few hours, monocytes arrive in the infected area. Once in the tissue, they turn into wandering macrophages that engulf damaged tissue, worn-out neutrophils, and invading microbes.

4. Eventually, macrophages also die. Within a few days, a pocket of dead phagocytes and damaged tissue forms; this collection of dead cells and fluid is called *pus*. At times, pus reaches the surface of the body or drains into an internal cavity and is dispersed; on other occasions the pus remains even after the infection is terminated. In this case, the pus is gradually destroyed over a period of days and is absorbed.

CLINICAL CONNECTION | Abscesses and Ulcers

If pus cannot drain out of an inflamed region, the result is an **abscess**—an excessive accumulation of pus in a confined space. Common examples are pimples and boils. When superficial inflamed tissue sloughs off the surface of an organ or tissue, the resulting open sore is called an **ulcer**. People with poor circulation—for instance, diabetics with advanced atherosclerosis—are particularly susceptible to ulcers in the tissues of their legs. •

Fever

Fever is an abnormally high body temperature that occurs because the hypothalamic thermostat is reset. It commonly occurs during infection and inflammation. Many bacterial toxins elevate body temperature, sometimes by triggering release of fever-causing substances such as interleukin-1 from macrophages. Elevated body temperature intensifies the effects of interferons, inhibits the growth of some microbes, and speeds up body reactions that aid repair.

Table 17.1 summarizes the components of innate defenses.

✓ CHECKPOINT

4. What physical and chemical factors provide protection from disease in the skin and mucous membranes?
5. What internal defenses provide protection against microbes that penetrate the skin and mucous membranes?
6. What are the main signs and symptoms of inflammation?

TABLE 17.1

Summary of Innate Defenses

COMPONENT	FUNCTIONS
FIRST LINE OF DEFENSE: SKIN AND MUCOUS MEMBRANES	
Physical Factors	
Epidermis of skin	Forms physical barrier to entrance of microbes
Mucous membranes	Inhibit entrance of many microbes, but not as effective as intact skin
Mucus	Traps microbes in respiratory and gastrointestinal tracts
Hairs	Filter out microbes and dust in nose
Cilia	Together with mucus, trap and remove microbes and dust from upper respiratory tract
Lacrimal apparatus	Tears dilute and wash away irritating substances and microbes
Saliva	Washes microbes from surfaces of teeth and mucous membranes of mouth
Urine	Washes microbes from urethra
Defecation and vomiting	Expel microbes from body
Chemical Factors	
Sebum	Forms protective acidic film over skin surface that inhibits growth of many microbes
Lysozyme	Antimicrobial substance in perspiration, tears, saliva, nasal secretions, and tissue fluids
Gastric juice	Destroys bacteria and most toxins in stomach
Vaginal secretions	Slight acidity discourages bacterial growth; flush microbes out of vagina
SECOND LINE OF DEFENSE: INTERNAL DEFENSES	
Antimicrobial Substances	
Interferons (IFNs)	Protect uninfected host cells from viral infection
Complement system	Causes cytolysis of microbes; promotes phagocytosis; contributes to inflammation
Iron-binding proteins	Inhibit growth of certain bacteria by reducing amount of available iron
Antimicrobial proteins (AMPs)	Have broad-spectrum antimicrobial activities and attract dendritic cells and mast cells
Natural killer (NK) cells	Kill infected target cells by releasing granules that contain perforin and granzymes; phagocytes then kill released microbes
Phagocytes	Ingest foreign particulate matter
Inflammation	Confines and destroys microbes; initiates tissue repair
Fever	Intensifies effects of interferons; inhibits growth of some microbes; speeds up body reactions that aid repair

17.3 ADAPTIVE IMMUNITY

OBJECTIVES • **Define adaptive immunity and compare it with innate immunity.**

• **Explain the relationship between an antigen and an antibody.**

• **Compare the functions of cell-mediated immunity and antibody-mediated immunity.**

The various aspects of innate immunity have one thing in common: They are not specifically directed against a particular type of invader. Adaptive (specific) immunity involves the production of specific types of cells or specific antibodies to destroy a particular antigen. An *antigen* is any substance—such as microbes, foods, drugs, pollen, or tissue—that the immune system recognizes as foreign (nonself). The branch of science that deals with the responses of the body to antigens is called *immunology* (im′-ū-NOL-ō-jē). The *immune system* includes the cells and tissues that carry out immune responses. Normally, a person's adaptive immune system cells recognize and do not attack their own tissues and chemicals. Such lack of reaction against self-tissues is called *self-tolerance*.

CLINICAL CONNECTION | Autoimmune Disease

At times, self-tolerance breaks down, which leads to an **autoimmune disease**. Sometimes tissues undergo changes that cause the adaptive immune system to recognize them as foreign antigens and attack them. Among human autoimmune diseases are systemic lupus erythematosus (SLE), Addison's disease, Graves disease, type 1 diabetes mellitus, myasthenia gravis, multiple sclerosis (MS), and ulcerative colitis. •

Maturation of T Cells and B Cells

Adaptive immunity involves lymphocytes called B cells and T cells. Both develop in primary lymphatic organs (red bone marrow and the thymus) from stem cells that originate in red bone marrow (see Figure 14.2). B cells complete their development in red bone marrow. T cells develop from pre-T cells that migrate from red bone marrow into the thymus, where they mature (Figure 17.6). Before T cells leave the thymus or B cells leave red bone marrow, they begin to make several distinctive proteins that are inserted into their plasma membranes. Some of these proteins function as *antigen receptors*—molecules capable of recognizing specific antigens (Figure 17.6). There are two major types of mature T cells that exit the thymus: *helper T cells* and *cytotoxic T cells* (Figure 17.6). As we will see later in this chapter, these two types of T cells have very different functions.

Types of Adaptive Immunity

There are two types of adaptive immunity: cell-mediated immunity and antibody-mediated immunity. Both types of adaptive immunity are triggered by antigens. In *cell-mediated immunity*, cytotoxic T cells directly attack invading antigens. In *antibody-mediated immunity*, B cells transform into plasma cells, which synthesize and secrete specific proteins called *antibodies*. A given antibody can bind to and inactivate a specific antigen. Helper T cells aid the immune responses of both cell-mediated and antibody-mediated immunity.

Cell-mediated immunity is particularly effective against (1) intracellular pathogens, which include any viruses, bacteria, or fungi that are inside cells; (2) some cancer cells; and (3) foreign tissue transplants. Thus, cell-mediated immunity always involves cells attacking cells. Antibody-mediated immunity works mainly against extracellular pathogens, which include any viruses, bacteria, or fungi that are in body fluids outside cells. Since antibody-mediated immunity involves antibodies that bind to antigens in body *humors* or fluids (such as blood and lymph), it is also referred to as *humoral immunity*.

In most cases, when a particular antigen initially enters the body, there is only a small group of lymphocytes with the correct antigen receptors to respond to that antigen; this small group of cells include a few helper T cells, cytotoxic T cells, and B cells. Depending on its location, a given antigen can provoke both types of adaptive immune responses. This is due to the fact that when a specific antigen invades the body, there are usually many copies of that antigen spread throughout the body's tissues and fluids. Some copies of the antigen may be present inside body cells (which provokes a cell-mediated immune response by cytotoxic T cells), while other copies of the antigen may be present in extracellular fluid (which provokes an antibody-mediated immune response by B cells). Thus, cell-mediated and antibody-mediated immune responses often work together to get rid of the large number of copies of a particular antigen from the body.

Clonal Selection: The Principle

As you just learned, when a specific antigen is present in the body, there are usually many copies of that antigen located throughout the body's tissues and fluids. The numerous copies of the antigen initially outnumber the small group of helper T cells, cytotoxic T cells, and B cells with the correct antigen receptors to respond to that antigen. Therefore, once each of these lymphocytes encounters a copy of the antigen and receives stimulatory cues, it subsequently undergoes clonal selection. *Clonal selection* is the process by which a lymphocyte *proliferates* (divides) and *differentiates* (forms more highly specialized cells) in response to a specific antigen. The result of clonal selection is the formation of a population of identical cells, called a *clone*, that can recognize the same specific antigen as the original lymphocyte (Figure 17.6). Before the first exposure to a given antigen, only a few lymphocytes are able to recognize it, but once clonal selection occurs, there are

Figure 17.6 Origin of B cells and pre-T cells from stem cells in red bone marrow.
B cells and T cells develop in primary lymphatic tissues (red bone marrow and the thymus) and are activated in secondary lymphatic organs and tissues (lymph nodes, spleen, and lymphatic nodules). Once activated, each type of lymphocyte forms a clone of cells that can recognize a specific antigen. For simplicity, antigen receptors are not shown in the plasma membranes of the cells of the lymphocyte clones.

The two types of adaptive immunity are cell-mediated immunity and antibody-mediated immunity.

CELL-MEDIATED IMMUNITY
Directed against intracellular pathogens, some cancer cells, and tissue transplants

ANTIBODY-MEDIATED IMMUNITY
Directed against extracellular pathogens

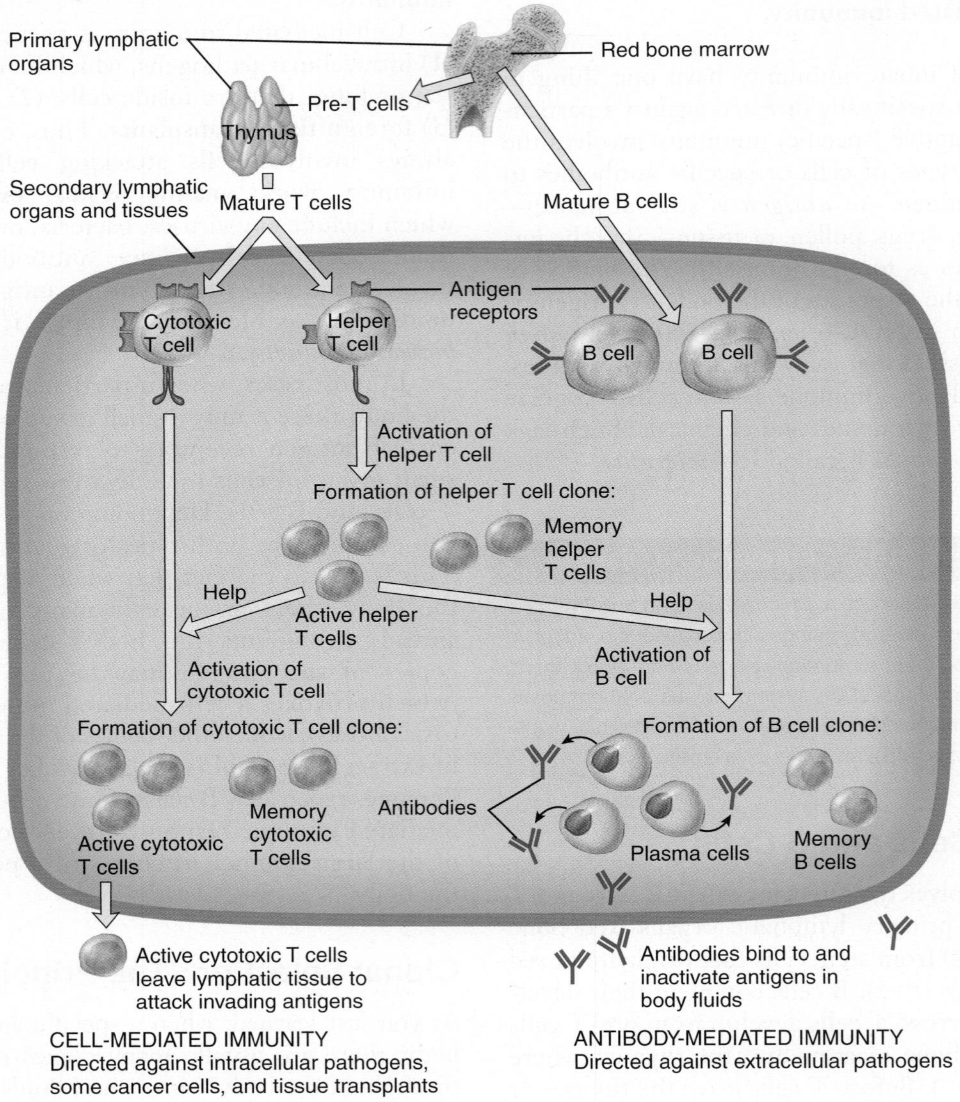

Which type of T cell participates in both cell-mediated and antibody-mediated immune responses?

thousands of lymphocytes that can respond to that antigen. Clonal selection of lymphocytes occurs in the secondary lymphatic organs and tissues. The swollen tonsils or lymph nodes in your neck you experienced the last time you were sick were probably caused by clonal selection of lymphocytes participating in an immune response.

A lymphocyte that undergoes clonal selection gives rise to two major types of cells in the clone: effector cells and memory cells. The thousands of *effector cells* of a lymphocyte

clone carry out immune responses that ultimately result in the destruction or inactivation of the antigen. Effector cells include *active helper T cells*, which are part of a helper T cell clone; *active cytotoxic T cells*, which are part of a cytotoxic T cell clone; and *plasma cells*, which are part of a B cell clone. Most effector cells eventually die after the immune response has been completed.

Memory cells do not actively participate in the initial immune response to the antigen. However, if the same antigen enters the body again in the future, the thousands of memory cells of a lymphocyte clone are available to initiate a far swifter reaction than occurred during the first invasion. The memory cells respond to the antigen by proliferating and differentiating into more effector cells and more memory cells. Consequently, the second response to the antigen is usually so fast and so vigorous that the antigen is destroyed before any signs or symptoms of disease can occur. Memory cells include *memory helper T cells*, which are part of a helper T cell clone; *memory cytotoxic T cells*, which are part of a cytotoxic T cell clone; and *memory B cells*, which are part of a B cell clone. Most memory cells do not die at the end of an immune response. Instead, they have long life spans (often lasting for decades). The functions of effector cells and memory cells are described in more detail later in this chapter.

Antigens and Antibodies

An antigen (meaning *anti*body *gen*erator) causes the body to produce specific antibodies and/or specific T cells that react with it. Entire microbes or parts of microbes may act as antigens. Chemical components of bacterial structures such as flagella, capsules, and cell walls are antigenic, as are bacterial toxins and viral proteins. Other examples of antigens include chemical components of pollen, egg white, incompatible blood cells, and transplanted tissues and organs. The huge variety of antigens in the environment provides myriad opportunities for provoking immune responses.

Located at the plasma membrane surface of most body cells are protein "self-antigens" known as *major histocompatibility complex (MHC)* proteins. Unless you have an identical twin, your MHC proteins are unique. Thousands to several hundred thousand MHC molecules mark the surface of each of your body cells except red blood cells. MHC proteins are the reason that tissues may be rejected when they are transplanted from one person to another, but their normal function is to help T cells recognize that an antigen is foreign, not self. This recognition is an important first step in any adaptive immune response.

CLINICAL CONNECTION | Tissue Typing

Tissue typing (histocompatibility) examines human leukocyte antigens (HLA) on the surface of blood cells in patients and donors prior to organ transplantation, to match compatible donors and recipients. The more HLAs that donors and recipients have in common, the less likely the transplant will be rejected. Tissue type is inherited, with three HLAs passed from each parent; therefore compatibility with a relative is more likely than a non-related donor. •

Figure 17.7 Structure of an antibody and relationship of an antigen to an antibody.

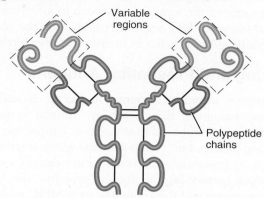

An antigen stimulates plasma cells to secrete specific antibodies that combine with the antigen.

Variable regions

Polypeptide chains

(a) Diagram of an antibody molecule

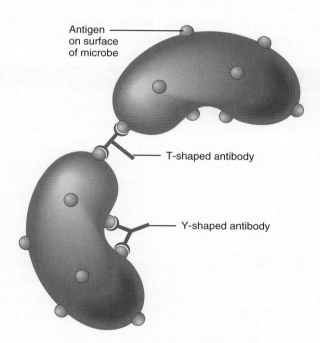

Antigen on surface of microbe

T-shaped antibody

Y-shaped antibody

(b) Antibody molecules binding to antigens

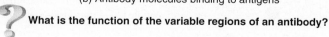

What is the function of the variable regions of an antibody?

body that "fit" and bind to a particular antigen, much like your house key fits into its lock. Because the antibody "arms" can move somewhat, an antibody can assume either a T shape or a Y shape. This flexibility enhances the ability of the antibody to bind to two identical antigens at the same time—for example, on the surface of two adjacent microbes (Figure 17.7b).

Antibodies belong to a group of plasma proteins called globulins, and for this reason they are also known as *immunoglobulins* (im′-ū-nō-GLOB-ū-lins). Immunoglobulins are grouped in five different classes, designated IgG, IgA, IgM, IgD, and IgE. Each class has a distinct chemical

Antigens induce plasma cells to secrete proteins known as *antibodies*. Most antibodies contain four polypeptide chains (Figure 17.7a). At two tips of the chains are *variable regions*, so named because the sequence of amino acids there varies for each different antibody. The variable regions are the *antigen-binding sites*, the parts of an anti-structure and different functions (Table 17.2). Because they appear first and are relatively short-lived, IgM antibodies indicate a recent invasion. In a sick patient, a high level of IgM against a particular pathogen helps identify the cause of the illness. Resistance of the fetus and newborn to infection stems mainly from maternal IgG antibodies that cross the placenta before birth and IgA antibodies in breast milk after birth.

Processing and Presenting Antigens

For an adaptive immune response to occur, B cells and T cells must recognize that a foreign antigen is present. B cells can recognize and bind to antigens in lymph, interstitial fluid, or blood plasma, but T cells only recognize fragments of antigens that are processed and presented in a certain way.

In *antigen processing*, antigenic proteins are broken down into fragments and then combine with MHC molecules. Next, the antigen–MHC complex is inserted into the plasma membrane of a body cell. The insertion of the complex into the plasma membrane is called *antigen presentation*. When an antigenic fragment comes from a *self protein*, T cells ignore the antigen–MHC complex. However, if the fragment comes from a *foreign protein*, T cells recognize the antigen–MHC as an intruder, and an adaptive immune response takes place.

A special class of cells called *antigen-presenting cells (APCs)* process and present antigens. APCs include dendritic cells, macrophages, and B cells. They are strategically located in places where antigens are likely to penetrate innate defenses and enter the body, such as the epidermis and dermis of the skin (Langerhans cells are a type of dendritic cell); mucous membranes that line the respiratory, gastrointestinal, urinary, and reproductive tracts; and lymph nodes. After processing and presenting an antigen, APCs migrate from tissues via lymphatic vessels to lymph nodes.

The steps in the processing and presenting of an antigen by an APC occur as follows (Figure 17.8):

1. **Ingestion of the antigen.** Antigen-presenting cells ingest antigens by phagocytosis. Ingestion could occur almost anywhere in the body that invaders, such as microbes, have penetrated the nonspecific defenses.

TABLE 17.2

Classes of Immunoglobulins

NAME AND STRUCTURE	CHARACTERISTICS AND FUNCTIONS
IgG	About 80 percent of all antibodies in the blood; also found in lymph and the intestines
	Protects against bacteria and viruses by enhancing phagocytosis, neutralizing toxins, and triggering the complement system
	Only class of antibody to cross the placenta from mother to fetus, conferring considerable immune protection in newborns
IgA	About 10–15 percent of all antibodies in the blood; found mainly in sweat, tears, saliva, mucus, breast milk, and gastrointestinal secretions
	Levels decrease during stress, lowering resistance to infection
	Provides localized protection against bacteria and viruses on mucous membranes
IgM	About 5–10 percent of all antibodies in the blood; also found in lymph
	First antibody class to be secreted by plasma cells after an initial exposure to any antigen
	Activates complement and causes agglutination and lysis of microbes
	In blood plasma, anti-A and anti-B antibodies of ABO blood group, which bind to A and B antigens during incompatible blood transfusions, are also IgM antibodies (see Figure 14.6)
IgD	About 0.2 percent of all antibodies in the blood; also found in lymph and on the surfaces of B cells as antigen receptors
	Involved in activation of B cells
IgE	Less than 0.1 percent of all antibodies in the blood; also located on mast cells and basophils
	Involved in allergic and hypersensitivity reactions and provides protection against parasitic worms

Figure 17.8 Processing and presenting of antigen by an antigen-presenting cell (APC).

 An APC migrates to a lymphatic tissue where it "presents" a processed antigen to T cells having receptors that fit that particular antigen fragment.

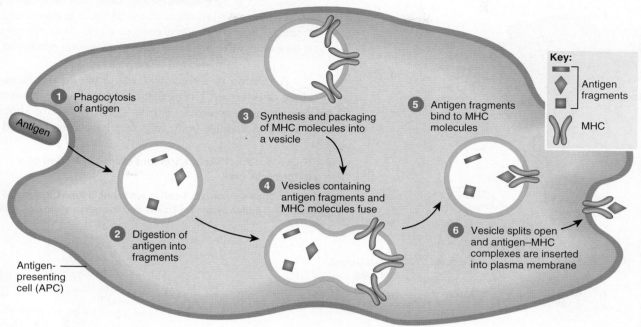

APCs present antigens in association with MHC molecules

? Which types of cells can function as APCs?

2 **Digestion of antigen into fragments.** Within the APC, protein-digesting enzymes split large antigens into short peptide fragments.

3 **Synthesis of MHC molecules.** At the same time, the APC synthesizes MHC molecules and packages them into vesicles.

4 **Fusion of vesicles.** The vesicles containing antigen fragments and MHC molecules merge and fuse.

5 **Binding of fragments to MHC molecules.** After fusion of the two vesicles, antigen fragments bind to MHC molecules.

6 **Insertion of antigen–MHC complexes into the plasma membrane.** The combined vesicle that contains antigen–MHC complexes splits open and the antigen–MHC complexes are inserted into the plasma membrane.

After processing an antigen, the APC migrates to lymphatic tissue to present the antigen to T cells. Within lymphatic tissue, a small number of T cells that have the correct antigen receptors recognize and bind to the antigen fragment–MHC complex, triggering either a cell-mediated immune response or an antibody-mediated immune response.

T Cells and Cell-mediated Immunity

The presentation of an antigen together with MHC molecules by APCs informs T cells that intruders are present in the body and that combative action should begin. But a T cell becomes activated only if its antigen receptor binds to the foreign antigen (antigen recognition) and at the same time it receives a second stimulating signal, a process known as *costimulation* (Figure 17.9). A common costimulator is *interleukin-2 (IL-2)*. The need for two signals is a little like starting and driving a car. When you insert the correct key (antigen) in the ignition (T cell receptor) and turn it, the car starts (recognition of specific antigen), but it cannot move forward until you move the gear shift into drive (costimulation). The need for costimulation probably helps prevent immune responses from occurring accidentally.

Once a T cell has been activated, it undergoes clonal selection. Recall that clonal selection is the process by which a lymphocyte proliferates (divides several times) and differentiates (forms more highly specialized cells) in response to a specific antigen. The result of clonal selection is the formation of a clone of cells that can recognize the same antigen as the original lymphocyte (see Figure 17.6). Some of the cells of a T cell clone become effector cells, while other cells of the clone become memory cells. The effector cells of a T cell clone carry out immune responses that ultimately result in *elimination* of the intruder.

As you have already learned, there are two major types of mature T cells: helper T cells and cytotoxic T cells. Activation of a *helper T cell* results in the formation of a clone of active helper T cells and memory helper T cells

Figure 17.9 Activation and clonal selection of a helper T cell.

Once a helper T cell is activated, it forms a clone of active helper T cells and memory helper T cells.

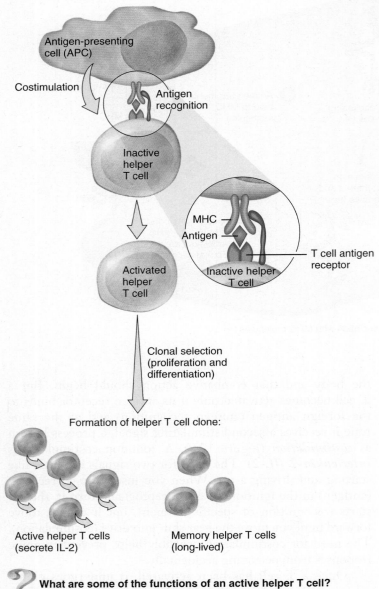

What are some of the functions of an active helper T cell?

To reduce the risk of rejection, recipients of **organ transplants** receive immunosuppressive drugs. One such drug is *cyclosporine*, derived from a fungus, which inhibits secretion of interleukin-2 by helper T cells but has only a minimal effect on B cells. Thus, the risk of rejection is diminished while resistance to some diseases is maintained. •

Activation of a *cytotoxic T cell* results in the formation of a clone of cytotoxic T cells that consists of active cytotoxic T cells and memory cytotoxic T cells (Figure 17.10). *Active cytotoxic T cells* attack other body cells that have been infected with the

Figure 17.10 Activation and clonal selection of a cytotoxic T cell.

Once a cytotoxic T cell is activated, it forms a clone of active cytotoxic T cells and memory cytotoxic T cells.

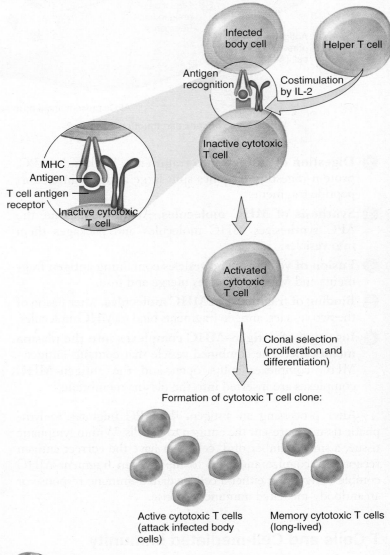

What is the function of a memory cytotoxic T cell?

(Figure 17.9). The *active helper T cells* help other cells of the adaptive immune system combat intruders. For instance, helper T cells release the protein interleukin-2, which acts as a costimulator for resting helper T cells or cytotoxic T cells, and it enhances activation and proliferation of T cells, B cells, and natural killer cells. The *memory helper T cells* of a helper T cell clone are not active cells. However, if the same antigen enters the body again in the future, memory helper T cells can quickly proliferate and differentiate into more active helper T cells and more memory helper T cells.

antigen. *Memory cytotoxic T cells* do not attack infected body cells. Instead, they can quickly proliferate and differentiate into more active cytotoxic T cells and more memory cytotoxic T cells if the same antigen enters the body at a future time.

Elimination of Invaders

Cytotoxic T cells are the soldiers that march forth to do battle with foreign invaders in cell-mediated immune responses. The name *cytotoxic* reflects their function—killing cells. They leave secondary lymphatic organs and tissues and migrate to seek out and destroy infected target cells, cancer cells, and transplanted cells (Figure 17.11). Cytotoxic T cells recognize and attach to target cells. Then, the cytotoxic T cells deliver a "lethal hit" that kills the target cells.

Cytotoxic T cells kill infected target body cells much like natural killer cells do. The major difference is that cytotoxic T cells have receptors specific for a particular microbe and thus kill only target body cells infected with *one* particular type of microbe; natural killer cells can destroy a wide variety of microbe-infected body cells. Cytotoxic T cells have two principal mechanisms for killing infected target cells:

1. Cytotoxic T cells, using receptors on their surfaces, recognize and bind to infected target cells that have microbial antigens displayed on their surface. The cytotoxic T cell then releases *granzymes*, protein-digesting enzymes that trigger apoptosis, the fragmentation of cellular contents (Figure 17.11a). Once the infected cell is destroyed, the released microbes are killed by phagocytes.

Figure 17.11 Action of a cytotoxic T cell. After delivering a "lethal hit," a cytotoxic T cell can detach and attack another target cell displaying the same antigen.

Cytotoxic T cells kill their targets directly by secreting granzymes, which trigger apoptosis, and perforin, which triggers cytolysis of infected target cells.

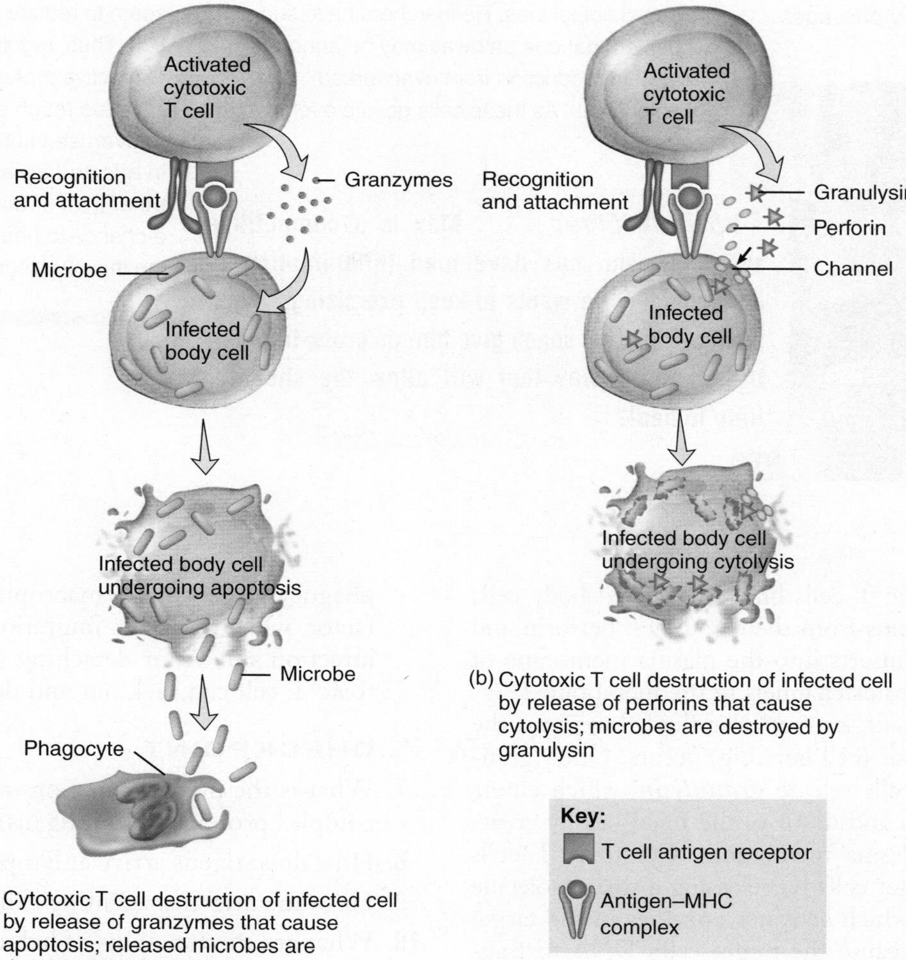

(a) Cytotoxic T cell destruction of infected cell by release of granzymes that cause apoptosis; released microbes are destroyed by phagocyte

(b) Cytotoxic T cell destruction of infected cell by release of perforins that cause cytolysis; microbes are destroyed by granulysin

Key:

T cell antigen receptor

Antigen–MHC complex

? Besides cells infected by microbes, what other types of cells do cytotoxic T cells attack?

Inflammatory Messages—Fatness versus Fitness

The process of inflammation helps the body fight infection and heal wounds. The immune cells attack foreign invaders, clean up the debris, and stimulate the growth of new cells. Once the healing has occurred, the inflammation resolves, and the body returns to homeostasis.

Sometimes, however, the immune system sets up an inflammatory response that makes matters worse rather than better. An inappropriate immune response can lead to low-level, chronic inflammation that can affect many parts of the body, including joints (as in rheumatoid arthritis), the skin (eczema), the digestive system (inflammatory bowel disorders), the artery lining (artery disease), and respiratory passages (asthma). Chronic inflammation can be estimated from a blood test for a substance called C-reactive protein.

Researchers have puzzled over the observation that inflammatory disorders are much more common in affluent nations than in poorer countries. Several lifestyle factors may be related to this difference, and adopting a healthful lifestyle may help reduce your risk of chronic low-level inflammation.

Fatness

Rich nations have higher rates of obesity. Excess body fat appears to stimulate the production of immune system messengers called adipokines. Researchers have suggested that one pathway may be abnormal protein production from overworked adipocytes. As these cells go into overdrive to deal with the excess calories that the liver has converted to triglycerides, errors occur in the cellular processes that produce proteins.

These errors in turn may trigger a distress call from the adipocytes, which release adipokines that carry messages from the adipose tissue to the rest of the body. This distress signal calls in the immune system to cope with the emergency. The adipokines in this case may trigger chronic inflammation throughout the body, inflammation that does not resolve but instead becomes the new norm.

Physical Activity Level

Appropriate levels of physical activity stimulate muscle tissue to produce protein messengers that talk to the immune system. Unlike the messenger molecules produced by adipose tissue, those from muscular contraction seem to reduce levels of chronic inflammation. Thus, regular physical activity reduces C-reactive protein levels in most people.

Too much exercise, however, can lead to overuse injuries and chronic inflammation in affected areas, especially tendons and other joint structures. Injuries may never get a chance to heal if the type of exercise causing the injury continues to be performed.

Think It Over . . . **Max is a competitive swimmer who has developed inflammation in his shoulder. He wants to keep exercising. What advice might his coach give him on cross-training, training in a way that will allow the shoulder time to heal?**

2. Alternatively, cytotoxic T cells bind to infected body cells and release two proteins from their granules: perforin and granulysin. *Perforin* inserts into the plasma membrane of the target cell and creates channels in the membrane (Figure 17.11b). As a result, extracellular fluid flows into the target cell and cytolysis (cell bursting) occurs. Other granules in cytotoxic T cells release *granulysin*, which enters through the channels and destroys the microbes by creating holes in their plasma membranes. Cytotoxic T cells may also destroy target cells by releasing a toxic molecule called *lymphotoxin*, which activates enzymes in the target cell. These enzymes cause the target cell's DNA to fragment, and the cell dies. In addition, cytotoxic T cells secrete gamma-interferon, which attracts and activates phagocytic cells, and macrophage migration inhibition factor, which prevents migration of phagocytes from the infection site. After detaching from a target cell, a cytotoxic T cell can seek out and destroy another target cell.

✓ CHECKPOINT

7. What is the normal function of major histocompatibility complex proteins (self-antigens)?

8. How do antigens arrive at lymphatic tissues?

9. How do antigen-presenting cells process antigens?

10. What are the functions of helper, cytotoxic, and memory T cells?

11. How do cytotoxic T cells kill their targets?

B Cells and Antibody-mediated Immunity

The body contains not only millions of different T cells, but also millions of different B cells, each capable of responding to a specific antigen. Cytotoxic T cells leave lymphatic tissues to seek out and destroy a foreign antigen, but B cells stay put. In the presence of a foreign antigen, a specific B cell in a lymph node, the spleen, or mucosa-associated lymphatic tissue becomes activated. Then it undergoes clonal selection, forming a clone of plasma cells and memory cells. Plasma cells are the effector cells of a B cell clone; they secrete spe-

cific antibodies, which in turn circulate in the lymph and blood to reach the site of invasion.

During activation of a B cell, antigen receptors on the cell surface of a B cell bind to an antigen (Figure 17.12). B cell antigen receptors are chemically similar to the antibodies that eventually are secreted by the plasma cells. Although B cells can respond to an unprocessed antigen present in lymph or interstitial fluid, their response is much more intense when they process the antigen. Antigen processing in a B cell occurs in the following way: the antigen is taken into the B cell, broken down into fragments and combined with MHC protein, and

Figure 17.12 Activation and clonal selection of B cells. Plasma cells are actually much larger than B cells.

Plasma cells secrete antibodies.

B-cell antigen receptor

Inactive B cell

Antigen

Microbe   Microbe   Microbe

Activated B cell

Activated B cell

Helper T cell

B cell recognizing unprocessed antigen

Costimulation by interleukin-2 and other proteins

B cell displaying processed antigen is recognized by helper T cell, which releases costimulators

Clonal selection (proliferation and differentiation)

Formation of B cell clone:

Antibodies

Plasma cells (secrete antibodies)

Memory B cells (long-lived)

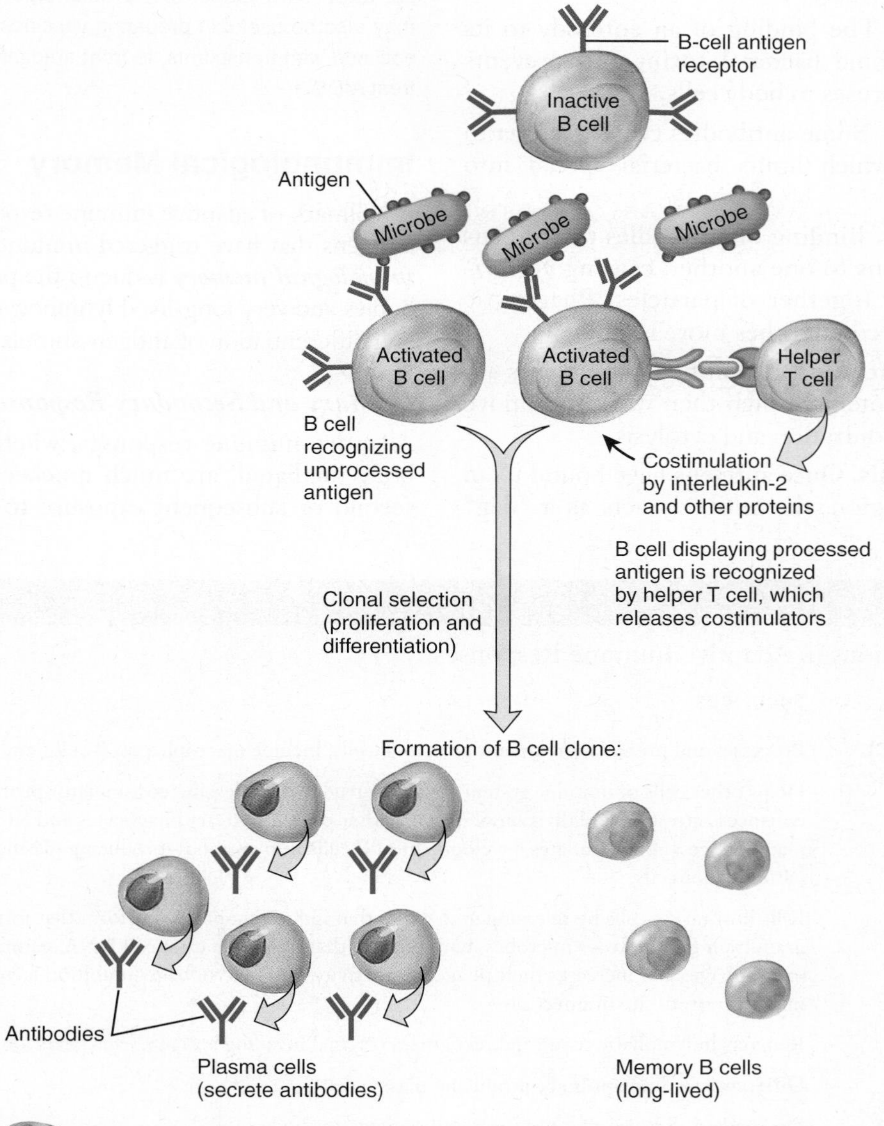

How many different kinds of antibodies will be secreted by the plasma cells in the clone shown here?

moved to the B cell surface. Helper T cells recognize the processed antigen–MHC protein complex and deliver the co-stimulation needed for B cell division and differentiation. The helper T cell releases interleukin-2 and other proteins that function as costimulators to activate B cells.

Once activated, a B cell undergoes clonal selection (Figure 17.12). The result is the formation of a clone of B cells that consists of plasma cells and memory B cells. *Plasma cells* secrete antibodies. A few days after exposure to an antigen, a plasma cell secretes hundreds of millions of antibodies each day for about 4 or 5 days, until the plasma cell dies. Most antibodies travel in lymph and blood to the invasion site. *Memory B cells* do not secrete antibodies. Instead, they can quickly proliferate and differentiate into more plasma cells and more memory B cells should the same antigen reappear at a future time.

Although the functions of the five classes of antibodies differ somewhat, all attack antigens in several ways:

1. **Neutralize antigen.** The binding of an antibody to its antigen neutralizes some bacterial toxins and prevents attachment of some viruses to body cells.

2. **Immobilize bacteria.** Some antibodies cause bacteria to lose their motility, which limits bacterial spread into nearby tissues.

3. **Agglutinate antigen.** Binding of antibodies to antigens may connect pathogens to one another, causing *agglutination*, the clumping together of particles. Phagocytic cells ingest agglutinated microbes more readily.

4. **Activate complement.** Antigen–antibody complexes activate complement proteins, which then work to remove microbes through opsonization and cytolysis.

5. **Enhance phagocytosis.** Once antigens have bound to an antibody's variable region, the antibody acts as a "flag" that attracts phagocytes. Antibodies enhance the activity of phagocytes by causing agglutination, by activating complement, and by coating microbes so that they are more susceptible to phagocytosis (opsonization).

Table 17.3 summarizes the functions of cells that participate in adaptive immune responses.

CLINICAL CONNECTION | Monoclonal Antibodies

An antibody-mediated response typically produces many different antibodies that recognize different parts of an antigen or different antigens of a foreign cell. By contrast, a **monoclonal antibody (MAb)** is a pure antibody produced from a single clone of identical cells grown in the laboratory. Clinical uses of MAbs include the diagnosis of pregnancy, allergies, and diseases such as strep throat, hepatitis, rabies, and some sexually transmitted diseases. MAbs have also been used to detect cancer at an early stage and to determine the extent of metastasis. They may also be useful in preparing vaccines to counteract the rejection associated with transplants, to treat autoimmune diseases, and perhaps to treat AIDS. •

Immunological Memory

A hallmark of adaptive immune responses is memory for specific antigens that have triggered immune responses in the past. *Immunological memory* is due to the presence of long-lasting antibodies and very long-lived lymphocytes that arise during division and differentiation of antigen-stimulated B cells and T cells.

Primary and Secondary Responses

Adaptive immune responses, whether cell-mediated or antibody-mediated, are much quicker and more intense after a second or subsequent exposure to an antigen than after the

---

TABLE 17.3

Summary of Cell Functions in Adaptive Immune Responses

CELL	FUNCTIONS
Antigen-presenting Cell (APC)	Processes and presents foreign antigens to T cells; include macrophages, B cells, and dendritic cells
Helper T Cell	Helps other cells of immune system combat intruders by releasing costimulator protein interleukin-2 (IL-2), which enhances activation and division of T cells; other proteins attract phagocytes and enhance phagocytic ability of macrophages; also stimulates development of B cells into antibody-producing plasma cells and development of natural killer cells
Cytotoxic T Cell	Kills host target cells by releasing granzymes that induce apoptosis, perforin that forms channels to cause cytolysis, granulysin that destroys microbes, lymphotoxin that destroys target cell DNA, gamma-interferon that attracts macrophages and increases their phagocytic activity, and macrophage inhibition factor that prevents macrophage migration from site of infection
Memory T Cell	Remains in lymphatic tissue and recognizes original invading antigen, even years after the first encounter
B Cell	Differentiates into antibody-producing plasma cell
Plasma Cell	Descendant of B cell that produces and secretes antibodies
Memory B Cell	Remains ready to produce a more rapid and forceful secondary response should the same antigen enter the body in the future

first exposure. Initially, only a few cells have the correct antigen receptors to respond, and the immune response may take several days to build to maximum intensity. Because thousands of memory cells exist after an initial encounter with an antigen, they can divide and differentiate into helper T cells, cytotoxic T cells, or plasma cells within hours the next time the same antigen appears.

One measure of immunological memory is the amount of antibody in blood plasma. After an initial contact with an antigen, no antibodies are present for a few days. Then, the levels of antibodies slowly rise, first IgM and then IgG, followed by a gradual decline (Figure 17.13). This is the ***primary response***. Memory cells may live for decades. Every new encounter with the same antigen causes a rapid division of memory cells. The antibody level after subsequent encounters is far greater than during a primary response and consists mainly of IgG antibodies. This accelerated, more intense response is called the ***secondary response***. Antibodies produced during a secondary response are even more effective than those produced during a primary response. Thus, they are more successful in disposing of the invaders.

Primary and secondary responses occur during microbial infection. When you recover from an infection without taking antimicrobial drugs, it is usually because of the primary response. If the same microbe infects you later, the secondary response could be so swift that the microbes are destroyed before you exhibit any signs or symptoms of infection.

Naturally Acquired Immunity and Artificially Acquired Immunity

Immunological memory provides the basis for immunization by vaccination against certain diseases, for instance, polio.

Figure 17.13 Secretion of antibodies. The primary response (after first exposure) is milder than the secondary response (after second or subsequent exposure) to a given antigen.

🔑 **Immunological memory is the basis for successful immunization by vaccination.**

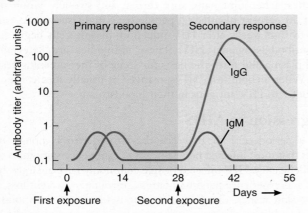

❓ **Which type of antibody responds most strongly during the secondary response?**

When you receive the *vaccine*, which may contain weakened or killed whole microbes or parts of microbes, your B cells and T cells are activated. Should you subsequently encounter the living pathogen as an infecting microbe, your body initiates a secondary response. However, booster doses of some immunizing agents must be given periodically to maintain adequate protection against the pathogen. Table 17.4 summarizes the various types of antigen encounters that provide naturally and artificially acquired immunity.

✓ CHECKPOINT

12. How are cell-mediated and antibody-mediated immune responses similar? How do they differ?

13. What is the difference between the secondary response to an antigen and the primary response?

17.4 AGING AND THE IMMUNE SYSTEM

OBJECTIVE • **Describe the effects of aging on the immune system.**

With advancing age, most people become more susceptible to all types of infections and malignancies. Their response to vaccines is decreased, and they tend to produce more autoantibodies (antibodies against their body's own molecules). In addition, the immune system exhibits lowered levels of

TABLE 17.4	
Types of Adaptive Immunity	
TYPE	**HOW ACQUIRED**
Naturally acquired active immunity	Following exposure to a microbe, antigen recognition by B cells and T cells and costimulation lead to antibody-secreting plasma cells, cytotoxic T cells, and B and T memory cells
Naturally acquired passive immunity	Transfer of IgG antibodies from mother to fetus across the placenta, or of IgA antibodies from mother to baby in milk during breast-feeding
Artificially acquired active immunity	Antigens introduced during a vaccination stimulate cell-mediated and antibody-mediated immune responses, leading to production of memory cells. The antigens are pretreated to be immunogenic but not pathogenic; that is, they will trigger an immune response but not cause significant illness
Artificially acquired passive immunity	Intravenous injection of immunoglobulins (antibodies)

function. For example, T cells become less responsive to antigens, and fewer T cells respond to infections. This may result from age-related atrophy of the thymus or decreased production of thymic hormones. Because the T cell population decreases with age, B cells are also less responsive. Consequently, antibody levels do not increase as rapidly in response to a challenge by an antigen, resulting in increased susceptibility to various infections. It is for this key reason that elderly individuals are encouraged to get influenza (flu) vaccinations each year.

✓ CHECKPOINT

14. What are the consequences of decreases in the number of T cells and B cells with advancing age?

• • •

To appreciate the many ways that the lymphatic system and immunity contribute to homeostasis of other body systems, examine Focus on Homeostasis: The Lymphatic System and Immunity. Next, in Chapter 18, we will explore the structure and function of the respiratory system and see how its operation is regulated by the nervous system. Most importantly, the respiratory system provides for gas exchange—taking in oxygen and blowing off carbon dioxide. The cardiovascular system aids gas exchange by transporting blood containing the gases between the lungs and tissue cells.

COMMON DISORDERS

AIDS: Acquired Immunodeficiency Syndrome

Acquired immunodeficiency syndrome (AIDS) is a condition in which a person experiences a telltale assortment of infections due to the progressive destruction of immune system cells by the *human immunodeficiency virus (HIV)*. AIDS represents the end stage of infection by HIV. A person who is infected with HIV may be symptom-free for many years, even while the virus is actively attacking the immune system. In the two decades after the first five cases were reported in 1981, 22 million people died of AIDS. Worldwide, about 40 million people are currently infected with HIV.

HIV Transmission

Because HIV is present in the blood and some body fluids, it is most effectively transmitted (spread from one person to another) by practices that involve the exchange of blood or body fluids. HIV is transmitted in semen or vaginal fluid during unprotected (without a condom) anal, vaginal, or oral sex. HIV also is transmitted by direct blood-to-blood contact, such as occurs in intravenous drug users who share hypodermic needles or health-care professionals who may be accidentally stuck by HIV-contaminated hypodermic needles. In addition, HIV can be transmitted from an HIV-infected mother to her baby at birth or during breast feeding.

The chances of transmitting or of being infected by HIV during vaginal or anal intercourse can be greatly reduced—although not eliminated—by the use of latex condoms. Public health programs aimed at encouraging drug users not to share needles have proved effective at checking the increase in new HIV infections in this population. Also, giving certain drugs to pregnant HIV-infected women greatly reduces the risk of transmission of the virus to their babies.

HIV is a very fragile virus; it cannot survive for long outside the human body. The virus is not transmitted by insect bites. A person cannot become infected by casual physical contact with an HIV-infected person, such as by hugging or sharing household items. The virus can be eliminated from personal care items and medical equipment by exposing them to heat (135°F for 10 min-utes) or by cleaning them with common disinfectants such as hydrogen peroxide, rubbing alcohol, household bleach, or germicidal cleansers such as Betadine® or Hibiclens®. Standard dishwashing and clothes washing also kill HIV.

HIV: Structure and Infection

HIV consists of an inner core of ribonucleic acid (RNA) covered by a protein coat (capsid) surrounded by an outer layer (envelope), composed of a lipid bilayer penetrated by proteins. Outside a living host cell, a virus is unable to replicate. However, when the virus infects and enters a host cell, its RNA uses the host cell's resources to make thousands of copies of the virus. New viruses eventually leave and then infect other cells.

HIV mainly damages helper T cells. Over 10 billion viral copies may be made each day. The viruses bud so rapidly from an infected cell's plasma membrane that the cell ruptures and dies. In most HIV-infected people, helper T cells are initially replaced as fast as they are destroyed. After several years, however, the body's ability to replace helper T cells is slowly exhausted, and the number of helper T cells in circulation gradually declines.

Signs, Symptoms, and Diagnosis of HIV Infection

Soon after being infected with HIV, most people experience a brief flu-like illness. Common signs and symptoms are fever, fatigue, rash, headache, joint pain, sore throat, and swollen lymph nodes. About 50 percent of infected people have night sweats. As early as three to four weeks after HIV infection, plasma cells begin secreting antibodies against HIV. These antibodies are detectable in blood plasma and form the basis for some of the screening tests for HIV. When people test "HIV-positive," it usually means they have antibodies to HIV antigens in their bloodstream.

Progression to AIDS

After a period of 2 to 10 years, the virus destroys enough helper T cells that most infected people begin to experience symptoms of immunodeficiency. HIV-infected people commonly have enlarged lymph nodes and experience persistent fatigue, involuntary weight loss, night sweats, skin rashes, diarrhea, and various lesions of the mouth and gums. In addition, the virus may begin to infect neurons in the brain, affecting the person's memory and producing visual disturbances.

BODY SYSTEM	CONTRIBUTION OF THE LYMPHATIC SYSTEM AND IMMUNITY

For all body systems
B cells, T cells, and antibodies protect all body systems from attack by harmful foreign microbes (pathogens), foreign cells, and cancer cells.

Integumentary system
Lymphatic vessels drain excess interstitial fluid and leaked plasma proteins from the dermis of the skin. Immune system cells (Langerhans cells) in the skin help protect skin. Lymphatic tissue provides IgA antibodies in sweat.

Skeletal system
Lymphatic vessels drain excess interstitial fluid and leaked plasma proteins from connective tissue around bones.

Muscular system
Lymphatic vessels drain excess interstitial fluid and leaked plasma proteins from muscles.

Nervous system
Lymphatic vessels drain excess interstitial fluid and leaked plasma proteins from the peripheral nervous system.

Endocrine system
Flow of lymph helps distribute some hormones. Lymphatic vessels drain excess interstitial fluid and leaked plasma proteins from endocrine glands.

Cardiovascular system
Lymph returns excess fluid filtered from blood capillaries and leaked plasma proteins to venous blood. Macrophages in spleen destroy aged red blood cells and remove debris in blood.

Respiratory system
Tonsils, lymphatic nodules in the mucosa, and alveolar macrophages in the lungs help protect airways and lungs from pathogens. Lymphatic vessels drain excess interstitial fluid from the lungs.

Digestive system
Tonsils and lymphatic nodules in the mucosa help defend against toxins and pathogens that penetrate the body from the gastrointestinal tract. The immune system provides IgA antibodies in saliva and gastrointestinal secretions. Lymphatic vessels pick up absorbed dietary lipids and fat-soluble vitamins from the small intestine and transport them to the blood. Lymphatic vessels drain excess interstitial fluid and leaked plasma proteins from organs of the digestive system.

Urinary system
Lymphatic vessels drain excess interstitial fluid and leaked plasma proteins from organs of the urinary system. Lymphatic nodules in the mucosa help defend against toxins and pathogens that penetrate the body via the urethra.

Reproductive systems
Lymphatic vessels drain excess interstitial fluid and leaked plasma proteins from organs of the reproductive systems. Lymphatic nodules in the mucosa help defend against toxins and pathogens that penetrate the body via the vagina and penis. In females, sperm deposited in the vagina are not attacked as foreign invaders due to components in seminal fluid that inhibit immune responses. IgG antibodies can cross the placenta to provide protection to a developing fetus. Lymphatic tissue provides IgA antibodies in the milk of the nursing mother.

THE LYMPHATIC SYSTEM AND IMMUNITY

As the immune system slowly collapses, an HIV-infected person becomes susceptible to a host of *opportunistic infections*. These are diseases caused by microorganisms that are normally held in check but now proliferate because of the defective immune system. AIDS is diagnosed when the helper T cell count drops below 200 cells per microliter (= cubic millimeter) of blood or when opportunistic infections arise, whichever occurs first. In time, opportunistic infections usually are the cause of death.

Treatment of HIV Infection

At present, infection with HIV cannot be cured. Vaccines designed to block new HIV infections and to reduce the viral load (the number of copies of HIV RNA in a microliter of blood plasma) in those who are already infected are in clinical trials. Meanwhile, two categories of drugs have proved successful in extending the life of many HIV-infected people:

1. *Reverse transcriptase inhibitors* interfere with the action of reverse transcriptase, the enzyme that the virus uses to convert its RNA into a DNA copy. Among the drugs in this category are zidovudine (ZDV, previously called AZT), didanosine (ddl), and stavudine (d4T). Trizivir®, approved in 2000 for treatment of HIV infection, combines three reverse transcriptase inhibitors in one pill.

2. *Protease inhibitors* interfere with the action of protease, a viral enzyme that cuts proteins into pieces to assemble the coat of newly produced HIV particles. Drugs in this category include nelfinavir, saquinavir, ritonavir, and indinavir.

In 1996, physicians treating HIV-infected patients widely adopted *highly active antiretroviral therapy (HAART)*—a combination of two differently acting reverse transcriptase inhibitors and one protease inhibitor. Most HIV-infected individuals receiving HAART experience a drastic reduction in viral load and an increase in the number of helper T cells in their blood. Not only does HAART delay the progression of HIV infection to AIDS, but many people with AIDS have seen the remission or disappearance of opportunistic infections and an apparent return to health. Unfortunately, HAART is very costly (exceeding $10,000 per year), the dosing schedule is grueling, and not all people can tolerate the toxic side effects of these drugs. Although HIV may virtually disappear from the blood with drug treatment (and thus a blood test may be "negative" for HIV), the virus typically still lurks in various lymphatic tissues. In such cases, the infected person can still transmit the virus to another person.

Allergic Reactions

A person who is overly reactive to a substance that is tolerated by most other people is said to be *allergic*. Whenever an allergic reaction takes place, some tissue injury occurs. The antigens that induce an allergic reaction are termed *allergens*. Common allergens include certain foods (milk, peanuts, shellfish, eggs), antibiotics (penicillin, tetracycline), vaccines (pertussis, typhoid), venoms (honeybee, wasp, snake), cosmetics, chemicals in plants such as poison ivy, pollens, dust, molds, iodine-containing dyes used in certain x-ray procedures, and even microbes.

Type I (anaphylactic) reactions are the most common and typically occur within a few minutes after a person who was previously sensitized to an allergen is reexposed to it. In response to certain allergens, some people produce IgE antibodies that bind to the surface of mast cells and basophils. The next time the same allergen enters the body, it attaches to the IgE antibodies already present. In response, both the mast cells and basophils release histamine, prostaglandins, and other chemicals. Collectively, these chemicals cause vasodilation, increased blood capillary permeability, increased smooth muscle contraction in the airways of the lungs, and increased mucus secretion. As a result, a person may experience inflammatory responses, difficulty in breathing through the narrowed airways, and a runny nose from excess mucus secretion. In *anaphylactic shock*, which may occur in a susceptible individual who has just received a triggering drug or been stung by a wasp, wheezing and shortness of breath as airways constrict are usually accompanied by shock due to vasodilation and fluid loss from blood. Injecting epinephrine to dilate the airways and strengthen the heartbeat usually is effective in this life-threatening emergency.

Type II (cytotoxic) reactions are caused by antibodies directed against antigens on a person's blood cells or tissue cells. Type II reactions, which may occur in incompatible blood transfusion reactions, damage cells by causing lysis.

Type III (immune-complex) reactions involve antigens, antibodies, and complement. Glomerulonephritis and rheumatoid arthritis (RA) arise in this way.

Type IV (cell-mediated) reactions or *delayed hypersensitivity reactions* usually appear 12–72 hours after exposure to an allergen. Type IV reactions occur when allergens are taken up by antigen-presenting cells (such as Langerhans cells in the skin) that migrate to lymph nodes and present the allergen to T cells, which then divide. Some of the new T cells return to the site of allergen entry into the body, where they produce gamma-interferon, which activates macrophages, and tumor necrosis factor, which stimulates an inflammatory response. Intracellular bacteria such as *Mycobacterium tuberculosis* trigger this type of cell-mediated immune response, as do certain haptens, such as poison ivy toxin. The skin test for tuberculosis also is a delayed hypersensitivity reaction.

Infectious Mononucleosis

Infectious mononucleosis or glandular fever is a contagious disease caused by the *Epstein–Barr virus (EBV)*. It occurs mainly in children and young adults, and more often in females than in males. The virus commonly enters the body through intimate oral contact such as kissing, which accounts for its being called the "kissing disease." EBV then multiplies in lymphatic tissues and spreads into the blood, where it infects and multiplies in B cells, the primary host cells. Because of this infection, the B cells become enlarged and abnormal in appearance so that they resemble monocytes, the primary reason for the term *mononucleosis*. Besides an elevated white blood cell count with an abnormally high percentage of lymphocytes, signs and symptoms include fatigue, headache, dizziness, sore throat, enlarged and tender lymph nodes, and fever. There is no cure for infectious mononucleosis, but the disease usually runs its course in a few weeks.

Lymphomas

Lymphomas (lim-FŌ-mas; *lymph-* = clear water; *-oma* = tumor) are cancers of the lymphatic organs, especially the lymph nodes. Most have no known cause. The two main types of lymphomas are Hodgkin disease and non-Hodgkin lymphoma.

Hodgkin disease (HD) is characterized by painless, nontender enlargement of one or more lymph nodes, most commonly in the neck, chest, and axillae (armpits). If the disease has metastasized from these sites, fevers, night sweats, weight loss, and bone pain also occur. HD primarily affects individuals between ages 15 and 35 and those over 60; it is more common in males. If diagnosed early, HD has a 90–95% cure rate.

Non-Hodgkin lymphoma (NHL), which is more common than HD, occurs in all age groups. NHL may start the same way as HD but may also include an enlarged spleen, anemia, and general malaise. Up to half of all individuals with NHL are cured or survive for a lengthy period. Treatment options for both HD and NHL include radiation therapy, chemotherapy, and red bone marrow transplantation.

Systemic Lupus Erythematosus

Systemic lupus erythematosus (*SLE*) (er-e-thēm-a-TŌ-sus), or *lupus* (*lupus* = wolf), is a chronic autoimmune disease that affects multiple body systems. Most cases of SLE occur in women between the ages of 15 and 25, more often in blacks than in whites. Although the cause of SLE is not known, both a genetic predisposition to the disease and environmental factors contribute. Females are nine times more likely than males to suffer from SLE. The disorder often occurs in females who exhibit extremely low levels of androgens (male sex hormones).

Signs and symptoms of SLE include joint pain, slight fever, fatigue, oral ulcers, weight loss, enlarged lymph nodes and spleen, photosensitivity, rapid loss of large amounts of scalp hair, and sometimes an eruption across the bridge of the nose and cheeks called a "butterfly rash." The erosive nature of some of the SLE skin lesions was thought to resemble the damage inflicted by the bite of a wolf—thus, the term lupus. Kidney damage occurs as antigen–antibody complexes become trapped in kidney capillaries, thereby obstructing blood filtering. Renal failure is the most common cause of death.

MEDICAL TERMINOLOGY AND CONDITIONS

Adenitis (ad'-e-NĪ-tis; *aden-* = gland; *-itis* = inflammation of) Enlarged, tender, and inflamed lymph nodes resulting from an infection.

Allograft (AL-ō-graft; *allo-* = other) A transplant between genetically different individuals of the same species. Skin transplants from other people and blood transfusions are allografts.

Autograft (AW-tō-graft; *auto-* = self) A transplant in which one's own tissue is grafted to another part of the body (such as skin grafts for burn treatment or plastic surgery).

Chronic fatigue syndrome (CFS) A disorder, usually occurring in young female adults, characterized by (1) extreme fatigue that impairs normal activities for at least six months and (2) the absence of other known diseases (cancer, infections, drug abuse, toxicity, or psychiatric disorders) that might produce similar symptoms.

Gamma globulin (GLOB-ū-lin) Suspension of immunoglobulins from blood consisting of antibodies that react with a specific pathogen. It is prepared by injecting the pathogen into animals, removing blood from the animals after antibodies have been produced, isolating the antibodies, and injecting them into a human to provide short-term immunity.

Graft Any tissue or organ used for transplantation or a transplant of such structures.

Hypersplenism (hī-per-SPLĒN-izm; *hyper-* = over) Abnormal splenic activity due to splenic enlargement and associated with an increased rate of destruction of normal blood cells.

Lymphadenopathy (lim-fad'-e-NOP-a-thē; *lymph-* = clear fluid; *-pathy* = disease) Enlarged, sometimes tender lymph glands as a response to infection; also called *swollen glands*.

Splenomegaly (splē'-nō-MEG-a-lē; *mega-* = large) Enlarged spleen.

Tonsillectomy (ton'-si-LEK-tō-mē; *-ectomy* = excision) Removal of a tonsil.

Xenograft (ZEN-ō-graft; *xeno-* = strange or foreign) A transplant between animals of different species. Xenografts from porcine (pig) or bovine (cow) tissue may be used in people as a physiological dressing for severe burns.

CHAPTER REVIEW AND RESOURCE SUMMARY

REVIEW	RESOURCES

Introduction

1. Despite constant exposure to a variety of **pathogens** (disease-producing microbes such as bacteria and viruses), most people remain healthy.

2. **Immunity** or **resistance** is the ability to ward off damage or disease. **Innate (nonspecific) immunity** refers to defenses that are present at birth; they are always present and provide immediate but general protection against invasion by a wide range of pathogens. **Adaptive (specific) immunity** refers to defenses that respond to a particular invader; it involves activation of specific lymphocytes that can combat a specific invader.

REVIEW	RESOURCES

17.1 Lymphatic System

Anatomy Overview—The Lymphatic System and Disease Resistance
Animation—Lymph Formation and Flow

1. The body system responsible for adaptive immunity (and some aspects of innate immunity) is the **lymphatic system**, which consists of lymph, lymphatic vessels, structures and organs that contain **lymphatic tissue**, and red bone marrow.

2. Components of blood plasma filter through blood capillary walls to form interstitial fluid, the fluid that bathes the cells of body tissues. After interstitial fluid passes into lymphatic vessels, it is called **lymph**. Interstitial fluid and lymph are chemically similar to blood plasma.

Figure 17.4 Structure of a lymph node (partially sectioned)

3. The lymphatic system drains tissue spaces of excess fluid and returns proteins that have escaped from blood to the cardiovascular system. It also transports lipids and lipid-soluble vitamins from the gastrointestinal tract to the blood, and protects the body against invasion.

4. Lymphatic vessels begin as **lymphatic capillaries** in tissue spaces between cells. The lymphatic capillaries merge to form larger **lymphatic vessels**, which ultimately drain into the **thoracic duct** or **right lymphatic duct**. Located at intervals along lymphatic vessels are **lymph nodes**, masses of B cells and T cells surrounded by a capsule.

Exercise—Disease Resistance Fighters
Exercise—Lymphatic Highway
Concepts and Connections— Lymphatic System

5. The passage of lymph is from interstitial fluid, to lymphatic capillaries, to lymphatic vessels and lymph nodes, to the thoracic duct or right lymphatic duct, to the junction of the internal jugular and subclavian veins. Lymph flows due to the "milking action" of skeletal muscle contractions and pressure changes that occur during inhalation. Valves in the lymphatic vessels prevent backflow of lymph.

6. **Primary lymphatic organs** are the sites where stem cells divide and develop into mature B cells and T cells. They include the **red bone marrow** (in flat bones and the ends of the long bones of adults) and the **thymus**. Stem cells in red bone marrow give rise to mature B cells and to immature T cells that migrate to the thymus, where they mature into functional T cells.

7. The **secondary lymphatic organs** and **tissues** are the sites where most immune responses occur. They include lymph nodes, the spleen, and lymphatic nodules.

8. Lymph nodes contain B cells that develop into plasma cells, T cells, dendritic cells, and macrophages. Lymph enters nodes through afferent lymphatic vessels and exits through efferent lymphatic vessels.

9. The **spleen** is the single largest mass of lymphatic tissue in the body. It is where B cells divide into plasma cells and macrophages phagocytize worn-out red blood cells and platelets.

10. **Lymphatic nodules** are oval-shaped concentrations of lymphatic tissue that are not surrounded by a capsule. They are scattered throughout the mucosa of the gastrointestinal, respiratory, urinary, and reproductive tracts.

17.2 Innate Immunity

Animation—Introduction to Disease Resistance
Animation—Nonspecific Disease Resistance
Exercise—Integument vs. Infection
Exercise—Microbe Massacre
Concepts and Connections— Nonspecific Disease Resistance
Concepts and Connections— Complement Proteins

1. The defenses of innate (nonspecific) immunity include barriers provided by the skin and mucous membranes (**first line of defense**). They also include various internal defenses (**second line of defense**): internal antimicrobial substances (**interferons**, **complement system**, **iron-binding proteins**, and **antimicrobial proteins**), phagocytes (neutrophils and **macrophages**), **natural killer (NK) cells** (which have the ability to kill a wide variety of infectious microbes and certain tumor cells), **inflammation**, and **fever**.

2. Table 17.1 summarizes the components of innate immunity.

17.3 Adaptive Immunity

Anatomy Overview—Antigens and Antibodies
Animation—Cell-mediated Immunity
Animation—Antibody-mediated Immunity

1. Adaptive (specific) immunity involves the production of specific types of cells or specific antibodies to destroy a particular antigen. An **antigen** is any substance that the adaptive immune system recognizes as foreign (nonself). Normally, a person's immune system cells exhibit **self-tolerance**: They recognize and do not attack their own tissues and cells.

2. B cells complete their development in red bone marrow, but mature T cells develop in the thymus from immature T cells that migrate from bone marrow.

REVIEW

3. There are two types of adaptive immunity: **cell-mediated immunity** and **antibody-mediated immunity**. In cell-mediated immune responses, cytotoxic T cells directly attack invading antigens; in antibody-mediated immune responses, B cells transform into plasma cells that secrete antibodies.

4. **Clonal selection** is the process by which a lymphocyte **proliferates** and **differentiates** in response to a specific antigen. The result of clonal selection is the formation of a **clone** of cells that can recognize the same specific antigen as the original lymphocyte. A lymphocyte that undergoes clonal selection gives rise to two major types of cells in the clone: effector cells and memory cells.

5. The **effector cells** of a lymphocyte clone carry out immune responses that ultimately result in the destruction or inactivation of the antigen. Effector cells include **active helper T cells**, which are part of a helper T cell clone; **active cytotoxic T cells**, which are part of a cytotoxic T cell clone; and **plasma cells**, which are part of a B cell clone.

6. The **memory cells** of a lymphocyte clone do not actively participate in the initial immune response. However, if the antigen reappears in the body in the future, the memory cells can quickly respond to the antigen by proliferating and differentiating into more effector cells and more memory cells. Memory cells include **memory helper T cells**, which are part of a helper T cell clone; **memory cytotoxic T cells**, which are part of a cytotoxic T cell clone; and **memory B cells**, which are part of a B cell clone.

7. The major histocompatibility complex (MHC) proteins are protein "self-antigens" unique to each person's body cells. All cells except red blood cells display MHC molecules. Antigens induce plasma cells to secrete **antibodies**, proteins that typically contain four polypeptide chains. The **variable regions** of an antibody are the **antigen-binding sites**, where the antibody can bind to a particular antigen. Based on chemistry and structure, antibodies, also known as **immunoglobulins**, are grouped in five classes, each with specific functions: IgG, IgA, IgM, IgD, and IgE (see Table 17.2). Functionally, antibodies neutralize antigens, immobilize bacteria, agglutinate antigens, activate complement, and enhance phagocytosis.

8. **Antigen-presenting cells (APCs)** process and present antigens to activate T cells, and they secrete substances that stimulate division of T cells and B cells. A cell-mediated immune response begins with activation of a small number of T cells by a specific antigen. There are two major types of mature T cells that exit the thymus: helper T cells and cytotoxic T cells. Activation of a helper T cell results in the formation of a clone of active helper T cells and memory helper T cells. Active helper T cells secrete **interleukin-2**, which provides **costimulation** for other helper T cells, cytotoxic T cells, and B cells. Activation of a cytotoxic T cell results in the formation of a clone of active cytotoxic T cells and memory cytotoxic T cells. Active cytotoxic T cells eliminate invaders by (1) releasing **granzymes** that cause target cell apoptosis (phagocytes then kill the microbes) and (2) releasing **perforin**, which causes cytolysis, and **granulysin**, which destroys the microbes.

9. An antibody-mediated immune response begins with activation of a B cell by a specific antigen. B cells can respond to unprocessed antigens, but their response is more intense when they process the antigen. Interleukin-2 and other cytokines secreted by helper T cells provide costimulation for activation of B cells. Once activated, a B cell undergoes clonal selection, forming a clone of plasma cells and memory cells. Plasma cells are the effector cells of a B cell clone; they secrete antibodies. Table 17.3 summarizes the functions of cells that participate in adaptive immune responses.

10. Immunization against certain microbes is possible because memory B cells and memory T cells remain after a **primary response** to an antigen, providing **immunological memory**. The **secondary response** provides protection should the same microbe enter the body again. Table 17.4 summarizes the various types of antigen encounters that provide naturally and artificially acquired immunity.

17.4 Aging and the Immune System

1. With advancing age, individuals become more susceptible to infections and malignancies, respond less well to vaccines, and produce more autoantibodies.

2. T cell responses also diminish with age.

RESOURCES

Figure 17.10 Activation and clonal selection of a cytotoxic T cell

Exercise—Activate the Attack
Exercise—Antibody Ambush
Concepts and Connections—
    Specific Disease Resistance

Homeostatic Imbalance—
    The Case of the Man with
    Opportunistic Infections

SELF-QUIZ

1. Which of the following is NOT true concerning the lymphatic system?

   a. Lymphatic vessels transport lipids from the gastrointestinal tract to the blood.

   b. Lymph is more similar to interstitial fluid than to blood.

   c. Lymphatic tissue is present in only a few isolated organs in the body.

   d. The unique structure of lymphatic capillaries allows fluid to flow into them but not out of them.

   e. Lymphatic vessels resemble veins in structure.

2. Which of the following are produced by virus-infected cells to protect uninfected cells from viral invasion?

   a. complement proteins   b. prostaglandins

   c. transferrins   d. interferons

   e. histamines

3. Lymph flow is aided by

   a. one-way valves.   b. the skeletal muscle pump.

   c. the respiratory pump.   d. pressure changes.

   e. all of the above.

4. Which of the following best represents lymph flow from the interstitial spaces back to the blood?

   a. lymphatic capillaries → lymphatic ducts → lymphatic vessels → junction of internal jugular and subclavian veins

   b. junction of internal jugular and subclavian veins → lymphatic capillaries → lymphatic vessels → lymphatic ducts

   c. lymphatic capillaries → lymphatic vessels → lymphatic ducts → junction of internal jugular and subclavian veins

   d. lymphatic ducts → lymphatic vessels → lymphatic capillaries → junction of internal jugular and subclavian veins

   e. lymphatic capillaries → lymphatic vessels → junction of internal jugular and subclavian veins → lymphatic ducts

5. Lymph nodes

   a. filter lymph.   b. are also called tonsils.

   c. produce lymph.   d. are a primary storage site for blood.

   e. are composed of ciliated epithelial cells.

6. Which of the following is NOT true about the role of skin in innate immunity?

   a. Sebum inhibits the growth of certain bacteria by forming a protective film.

   b. Epidermal cells produce interferons to destroy viruses.

   c. Shedding of epidermal cells helps remove microbes.

   d. Lysozyme in sweat destroys some bacteria.

   e. The skin forms a physical barrier to prevent entry of microbes.

7. Which of the following statements about B cells is true?

   a. They become functional while in the thymus.

   b. Some develop into plasma cells that secrete antibodies.

   c. Some B cells become natural killer cells.

   d. Cytotoxic B cells travel in lymph and blood to react with foreign antigens.

   e. They kill virus-infected cells by secreting perforin.

8. The cells that release granzymes, perforin, granulysin, and lymphotoxin are

   a. cytotoxic T cells.   b. plasma cells.   c. B cells.

   d. natural killer cells.   e. helper T cells.

9. The secondary response in antibody-mediated immunity

   a. is characterized by a slow rise in antibody levels and then a gradual decline.

   b. occurs when you first receive a vaccination against some disease.

   c. produces fewer but more responsive antibodies than are produced during the primary response.

   d. is an intense response by memory cells to produce antibodies when an antigen is contacted again.

   e. is rarely seen except in autoimmune disorders.

10. The ability of the body's immune system to recognize its own tissues is known as

    a. immunodeficiency.   b. autoimmunity.

    c. innate immunity.   d. hypersensitivity.

    e. self-tolerance.

11. A disease that causes destruction of helper T cells would result in all of the following effects EXCEPT

    a. inability to produce cytotoxic T cells.

    b. alteration of lymph flow.

    c. lack of development of plasma cells.

    d. decreased production of antibodies.

    e. increased risk of developing infections.

12. In which lymphatic organ do T cells mature?

    a. thyroid gland   b. spleen   c. thymus

    d. red bone marrow   e. lymph node

13. Place the following steps involved in the process of inflammation in the correct order.

    1. arrival of large numbers of neutrophils

    2. vasodilation and increased permeability of blood vessels

    3. formation of pus

    4. increased migration of monocytes

    5. formation of fibrin network to form a clot and isolate microbes

    6. release of histamine

       a. 6, 2, 4, 1, 5, 3   b. 3, 6, 1, 4, 2, 5   c. 5, 1, 4, 2, 6, 3

       d. 6, 2, 5, 1, 4, 3   e. 4, 6, 1, 3, 2, 5

14. Match the following:

____	**a.**	destroy antigens by cytolysis	**A.** natural killer cells
____	**b.**	stimulate other cells of the adaptive immune response	**B.** helper T cells
			C. B cells
____	**c.**	are programmed to recognize the original invading antigen; allow immunity to last for years	**D.** memory T cells
			E. cytotoxic T cells
____	**d.**	function in innate immunity	
____	**e.**	can develop into plasma cells	

15. What happens during opsonization?
    a. engulfment of a microbe by a phagocyte
    b. chemical attraction of a phagocyte
    c. binding of complement to a microbe
    d. attachment of a phagocyte to a microbe
    e. breakdown of a microbe by enzymes

16. All of the following contribute to innate immunity EXCEPT
    a. gastric and vaginal secretions.   b. immunoglobulins.
    c. natural killer cells.   d. fever.   e. interferons.

17. Vasodilation and increased vessel permeability occur with the release of
    a. lysozyme.   b. antimicrobial peptides.
    c. major histocompatibility complex proteins.
    d. antibodies.   e. histamine.

18. Place the phases of antigen processing in the correct order.
    1. digestion of antigen into fragments
    2. fusion of vesicles containing MHC molecules and antigen fragments
    3. insertion of complex into plasma membrane for presentation
    4. ingestion of foreign material
    5. formation of antigen–MHC complex
    6. synthesis and packaging of MHC molecules
       a. 1, 2, 3, 4, 6, 5   **b.** 4, 1, 6, 2, 5, 3   **c.** 1, 4, 5, 6, 2, 3
       d. 4, 1, 2, 6, 3, 5   **e.** 3, 1, 5, 4, 6, 2

19. Antibodies attack antigens by all of the following methods EXCEPT
    a. agglutination of antigens.
    b. activation of complement.
    c. opsonization to enhance phagocytosis.
    d. prevention of attachment to body cells.
    e. production of acid secretions.

20. What is the importance of tonsils in the body's defenses?
    a. They help destroy microbes that are inhaled or ingested.
    b. They contain ciliated cells that move trapped pathogens from the breathing passages.
    c. They are needed for T cell maturation.
    d. They are needed for B cell maturation.
    e. They filter lymph.

CRITICAL THINKING APPLICATIONS

1. Ruth found a lump in her right breast during a self-examination. The lump was found to be cancerous. The surgeon removed the breast lump, the surrounding tissue, and some lymph nodes. Which nodes were probably removed and why?

2. While driving home from work, Simon was in a car accident. The doctor in A&E had Simon rushed to surgery to remove his ruptured spleen. What is the role of Simon's spleen, and how will his body be affected by its absence?

3. Adrian stepped on a rusty nail while decorating at home. The Accident and Emergency nurse removed the nail and gave Adrian a tetanus booster. Why?

4. You learned in Chapter 16 that the cornea and lens of the eye are completely lacking in capillaries. How is this fact related to the high success of corneal transplants?

ANSWERS TO FIGURE QUESTIONS

17.1 Lymphatic tissue is reticular connective tissue that contains large numbers of lymphocytes.

17.2 Lymph is more similar to interstitial fluid because its protein content is low.

17.3 Capillaries produce lymph.

17.4 Foreign substances in lymph may be phagocytized by macrophages or destroyed by T cells or antibodies produced by plasma cells.

17.5 Redness is caused by increased blood flow due to vasodilation.

17.6 Helper T cells participate in both cell-mediated and antibody-mediated immune responses.

17.7 The variable regions of an antibody can bind specifically to the antigen that triggered its production.

17.8 APCs include macrophages, B cells, and dendritic cells.

17.9 Active helper T cells release the protein interleukin-2, which acts as a costimulator for resting helper T cells or cytotoxic T cells; and enhance activation and proliferation of T cells, B cells, and natural killer cells.

17.10 Memory cytotoxic T cells quickly proliferate and differentiate into more active cytotoxic T cells and more memory T cells if the same antigen enters the body at a future time.

17.11 Cytotoxic T cells can also attack some tumor cells and transplanted tissue cells.

17.12 Since all of the plasma cells in this figure are part of the same clone, they secrete just one kind of antibody.

17.13 IgG is the antibody secreted in the greatest amount during a secondary response.

## CHAPTER 18

THE RESPIRATORY SYSTEM

Body cells continually use oxygen (O_2) for the metabolic reactions that release energy from nutrient molecules and produce ATP. These same reactions produce carbon dioxide (CO_2). Because an excessive amount of CO_2 produces acidity that can be toxic to cells, excess CO_2 must be eliminated quickly and efficiently. The *respiratory system*, which includes the nose, pharynx (throat), larynx (voice box), trachea (windpipe), bronchi, and lungs (Figure 18.1), provides for gas exchange, the intake of O_2, and the removal of CO_2. The respiratory system also helps regulate blood pH; contains receptors for the sense of smell; filters, warms, and moistens inspired air; produces sounds; and rids the body of some water and heat in exhaled air.

Did you know?   *Cigarette smoking is the single most preventable cause of death and disability worldwide. Smoking disrupts the body's ability to maintain homeostasis and health because it introduces many harmful substances into the body, and wreaks havoc on the fragile tissues of the respiratory system. For example, smoking causes emphysema by progressively destroying the alveoli and bronchioles. The irritation produced by cigarette smoke often leads to chronic bronchitis. Smoking causes many types of cancers, including cancers of the mouth, throat, lungs, esophagus, stomach, kidneys, pancreas, colon, and urinary bladder. The chemicals in cigarette smoke also raise blood pressure and accelerate the process of atherosclerosis.*

FOCUS ON WELLNESS
Smoking—A Breathtaking Experience

LOOKING BACK TO MOVE AHEAD...

Cartilage (Section 4.3)

Pseudostratified Ciliated Columnar Epithelium (Section 4.2)

Simple Squamous Epithelium (Section 4.2)

Muscles Used in Breathing (Section 8.11)

Diffusion (Section 3.3)

Ions (Section 2.1)

Medulla and Pons (Section 10.4)

Figure 18.1 Organs of the respiratory system.

 The upper respiratory system includes the nose, pharynx, and associated structures. The lower respiratory system includes the larynx, trachea, bronchi, and lungs.

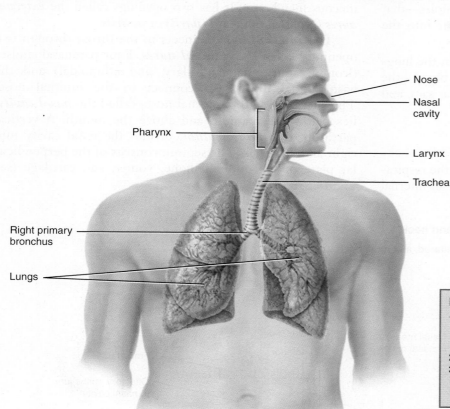

Nose

Nasal cavity

Pharynx

Larynx

Trachea

Right primary bronchus

Lungs

Functions of the Respiratory System

1. Provides for gas exchange—intake of O_2 for delivery to body cells and removal of CO_2 produced by body cells.
2. Helps regulate blood pH.
3. Contains receptors for the sense of smell, filters inspired air, produces sounds, and excretes small amounts of water and heat.

Anterior view showing organs of respiration

 **Which structures comprise the conducting zone of the respiratory system?**

The branch of medicine that deals with the diagnosis and treatment of diseases of the ears, nose, and throat (ENT) is called **otorhinolaryngology** (ō′-tō-rī′-nō-lar′-in-GOL-ō-jē; *oto-* = ear; *rhino-* = nose; *laryngo-* = voice box; *-logy* = study of). A **pulmonologist** (*pulmon-* = lung) is a specialist in the diagnosis and treatment of diseases of the lungs.

The entire process of gas exchange in the body, called *respiration*, occurs in three basic steps:

1. **Pulmonary ventilation**, or **breathing**, is the flow of air into and out of the lungs.

2. **External respiration** is the exchange of gases between the air spaces (alveoli) of the lungs and the blood in pulmonary capillaries. In this process, pulmonary capillary blood gains O_2 and loses CO_2.

3. **Internal respiration** is the exchange of gases between blood in systemic capillaries and tissue cells. The blood loses O_2 and gains CO_2. Within cells, the metabolic reactions that consume O_2 and give off CO_2 during the production of ATP are termed *cellular respiration* (discussed in Chapter 20).

As you can see, two systems are cooperating to supply O_2 and eliminate CO_2—the cardiovascular and respiratory systems. The first two steps are the responsibility of the respiratory system, while the third step is a function of the cardiovascular system.

18.1 ORGANS OF THE RESPIRATORY SYSTEM

OBJECTIVE • **Describe the structure and functions of the nose, pharynx, larynx, trachea, bronchi, bronchioles, and lungs.**

Structurally, the respiratory system consists of two parts: The **upper respiratory system** includes the nose, nasal cavity, pharynx (throat), and associated structures; the **lower respiratory system** consists of the larynx (voice box), trachea (windpipe), bronchi, and lungs. *Functionally*, the respiratory system can also be divided into two parts:

- The ***conducting zone*** consists of a series of interconnecting cavities and tubes both outside and within the lungs—nose, nasal cavity, pharynx, larynx, trachea, bronchi, bronchioles, and terminal bronchioles—that filter, warm, and moisten air and conduct air into the lungs.

- The ***respiratory zone*** consists of tissues within the lungs where gas exchange occurs between air and blood—the respiratory bronchioles, alveolar ducts, alveolar sacs, and alveoli.

Nose

The ***nose*** is a specialized organ at the entrance to the respiratory system and has a visible external portion and an internal portion inside the skull called the nasal cavity (Figure 18.2). The ***external nose*** is the portion visible on the face and consists of bone and cartilage covered with skin and lined with mucous membrane. It has two openings called the ***external nares*** (NA-rēz; singular is ***naris***) or ***nostrils***.

The ***internal nose*** connects to the throat through two openings called the ***internal nares***. Four paranasal sinuses (frontal, sphenoidal, maxillary, and ethmoidal) and the nasolacrimal ducts also connect to the internal nose. The space inside the internal nose, called the ***nasal cavity***, lies below the cranium and above the mouth. A vertical partition, the ***nasal septum***, divides the nasal cavity into right and left sides. The septum consists of the perpendicular plate of the ethmoid bone, vomer, and cartilage (see Figure 6.7a).

Figure 18.2 Respiratory organs in the head and neck.

 As air passes through the nose, it is warmed, filtered, and moistened.

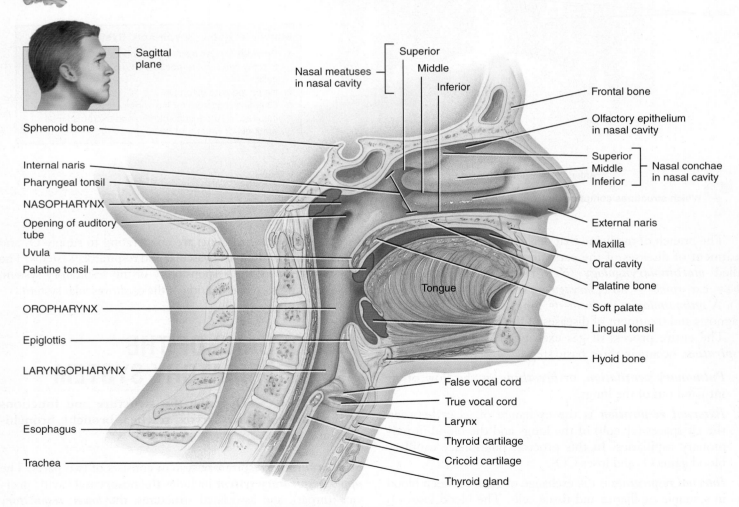

Sagittal section of the left side of the head and neck

? **What is the path taken by air molecules into and through the nose?**

Rhinoplasty (RĪ-nō-plas′-tē; -*plasty* = to mold or to shape), commonly called a "nose job," is a surgical procedure to alter the shape of the external nose. Although rhinoplasty is often done for cosmetic reasons, it is sometimes performed to repair a fractured nose or a deviated nasal septum. With anesthesia, instruments inserted through the nostrils are used to reshape the nasal cartilage and fracture and reposition the nasal bones, to achieve the desired shape. An internal packing and splint keep the nose in the desired position while it heals. •

The interior structures of the nose are specialized for three basic functions: (1) filtering, warming, and moistening incoming air; (2) detecting olfactory (smell) stimuli; and (3) modifying the vibrations of speech sounds. When air enters the nostrils, it passes coarse hairs that trap large dust particles. The air then flows over three bones called the superior, middle, and inferior **nasal conchae** (KONG-kē) or **turbinates** that extend out of the wall of the cavity. Mucous membrane lines the nasal cavity and the three conchae. As inspired air whirls around the conchae, it is warmed by blood circulating in abundant capillaries. Some individuals experience a loss of airflow sensation and a feeling of suffocation (empty nose syndrome) as a result of surgical procedures that reduce the size of the nasal conchae. The olfactory receptors lie in the membrane lining the superior nasal conchae and adjacent septum. This region is called the **olfactory epithelium**.

Pseudostratified ciliated columnar epithelial cells and goblet cells line the nasal cavity. Mucus secreted by goblet cells moistens the air and traps dust particles. Cilia move the dust-laden mucus toward the pharynx (mucociliary escalator), at which point it can be swallowed or spit out, thus removing particles from the respiratory tract.

Substances in cigarette smoke **inhibit movement of cilia**. If the cilia are paralyzed, only coughing can remove mucus–dust packages from the airways. This is why smokers cough so much and are more prone to respiratory infections. •

Pharynx

The **pharynx** (FAR-inks), or throat, is a funnel-shaped tube that starts at the internal nares and extends partway down the neck (Figure 18.2). It lies just posterior to the nasal and oral cavities and just anterior to the cervical (neck) vertebrae. Its wall is composed of skeletal muscle and lined with mucous membrane. The pharynx functions as a passageway for air and food, provides a resonating chamber for speech sounds, and houses the tonsils, which participate in immunological responses to foreign invaders.

The upper part of the pharynx, called the **nasopharynx**, connects with the two internal nares and has two openings that lead into the auditory (eustachian) tubes. The posterior wall contains the **pharyngeal tonsil**. The nasopharynx exchanges air with the nasal cavities and receives mucus–dust packages. The cilia of its pseudostratified ciliated columnar epithelium move the mucus–dust packages toward the mouth. The nasopharynx also exchanges small amounts of air with the auditory tubes to equalize air pressure between the pharynx and middle ear. The middle portion of the pharynx, the **oropharynx**, opens into the mouth and nasopharynx. Two pairs of tonsils, the **palatine tonsils** and **lingual tonsils**, are found in the oropharynx. The lowest portion of the pharynx, the **laryngopharynx** (la-rin′-gō-FAIR-inks), connects with both the esophagus (food tube) and the larynx (voice box). Thus, the oropharynx and laryngopharynx both serve as passageways for air as well as for food and drink.

Larynx

The **larynx** (LAR-inks), or voice box, is a short tube of cartilage lined by mucous membrane that connects the pharynx with the trachea (Figure 18.3). It lies in the midline of the neck anterior to the fourth, fifth, and sixth cervical vertebrae (C4 to C6).

The **thyroid cartilage**, which consists of hyaline cartilage, forms the anterior wall of the larynx. Its common name (Adam's apple) reflects the fact that it is often larger in males than in females due to the influence of male sex hormones during puberty.

The **epiglottis** (*epi-* = over; *glottis* = tongue) is a large, leaf-shaped piece of elastic cartilage that is covered with epithelium (see also Figure 18.2). The "stem" of the epiglottis is attached to the anterior rim of the thyroid cartilage and hyoid bone. The broad superior "leaf" portion of the epiglottis is unattached and is free to move up and down like a trap door. During swallowing, the pharynx and larynx rise. Elevation of the pharynx widens it to receive food or drink; elevation of the larynx causes the epiglottis to move down and form a lid over the larynx, closing it off. The closing of the larynx in this way during swallowing routes liquids and foods into the esophagus and keeps them out of the airways below. When anything but air passes into the larynx, a cough reflex attempts to expel the material.

The **cricoid cartilage** (KRĪ-koyd) is a ring of hyaline cartilage that forms the inferior wall of the larynx and is attached to the first tracheal cartilage. The paired **arytenoid cartilages** (ar′-i-TĒ-noyd), consisting mostly of hyaline cartilage, are located above the cricoid cartilage. They attach to the true vocal cords and pharyngeal muscles and function in voice production. The cricoid cartilage is the landmark for making an emergency airway (a tracheotomy; see Clinical Connection later in this section).

Figure 18.3 Larynx.

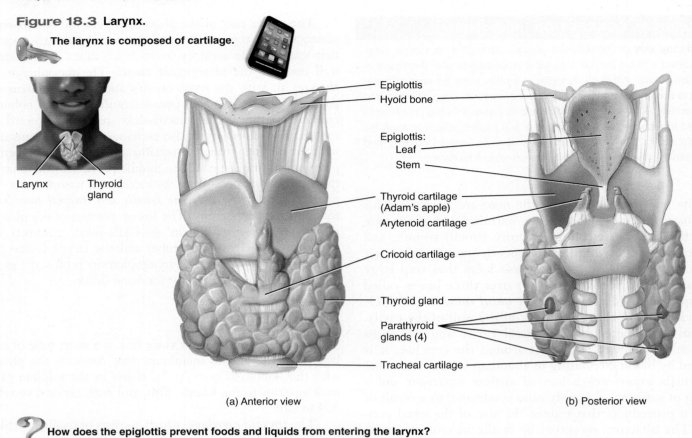

The larynx is composed of cartilage.

Epiglottis
Hyoid bone

Epiglottis:
  Leaf
  Stem

Thyroid cartilage
(Adam's apple)

Arytenoid cartilage

Cricoid cartilage

Thyroid gland

Parathyroid
glands (4)

Tracheal cartilage

Larynx   Thyroid
                gland

(a) Anterior view   (b) Posterior view

How does the epiglottis prevent foods and liquids from entering the larynx?

The Structures of Voice Production

The mucous membrane of the larynx forms two pairs of folds: an upper pair called the *false vocal cords* and a lower pair called the *true vocal cords* (see Figure 18.2). The false vocal cords hold the breath against pressure in the thoracic cavity when you strain to lift a heavy object, such as a backpack filled with textbooks. They do not produce sound.

The true vocal cords produce sounds during speaking and singing. They contain elastic ligaments stretched between pieces of rigid cartilage like the strings on a guitar. Muscles attach both to the cartilage and to the true vocal cords. When the muscles contract, they pull the elastic ligaments tight, which moves the true vocal cords out into the air passageway. The air pushed against the true vocal cords causes them to vibrate and sets up sound waves in the air in the pharynx, nose, and mouth. The greater the air pressure, the louder the sound.

Pitch is controlled by the tension of the true vocal cords. If they are pulled taut, they vibrate more rapidly and a higher pitch results. Lower sounds are produced by decreasing the muscular tension. Due to the influence of male sex hormones, vocal cords are usually thicker and longer in males than in females. They therefore vibrate more slowly, giving men a lower range of pitch than women.

✓ CHECKPOINT

1. What functions do the respiratory and cardiovascular systems have in common?

2. Compare the structure and functions of the external and internal nose.

3. How does the larynx function in respiration and voice production?

Trachea

The *trachea* (TRĀ-kē-a), or windpipe, is a tubular passageway for air that is located anterior to the esophagus. It extends from the larynx to the upper part of the fifth thoracic vertebra (T5), where it divides into right and left primary bronchi (Figure 18.4).

The wall of the trachea is lined with mucous membrane and is supported by cartilage. The mucous membrane is composed of pseudostratified ciliated columnar epithelium, consisting of ciliated columnar cells, goblet cells, and basal cells (see Table 4.1E), and provides the same protection against dust as the membrane lining the nasal cavity and larynx. The cilia in the upper respiratory tract move mucus and trapped particles *down* toward the pharynx, but the cilia in the lower respiratory tract move mucus and trapped particles *up* toward the pharynx. The cartilage layer consists of 16 to 20 C-shaped rings of hyaline cartilage stacked one on top of another. The open part of each C-shaped cartilage ring faces the esophagus and permits it to expand slightly into the trachea during swallowing. The solid parts of the C-shaped cartilage rings provide a rigid support so

the tracheal wall does not collapse inward and obstruct the air passageway. The rings of cartilage may be felt under the skin below the larynx.

CLINICAL CONNECTION | Tracheotomy

Several conditions may block airflow by obstructing the trachea. The rings of cartilage that support the trachea may be accidentally crushed, the mucous membrane may become inflamed and swell so much that it closes off the passageway, excess mucus secreted by inflamed membranes may clog the lower respiratory passages, or a large object may be aspirated (breathed in). If the obstruction is above the level of the larynx, a **tracheotomy** (trā-kē-O-tō-mē) may be performed. In this procedure, also called a *tracheostomy*, an incision is made in the trachea below the cricoid cartilage and a tracheal tube is inserted to create an emergency air passageway. •

Figure 18.4 Branching of airways from the trachea and lobes of the lungs.

The bronchial tree consists of airways that begin at the trachea and end at the terminal bronchioles.

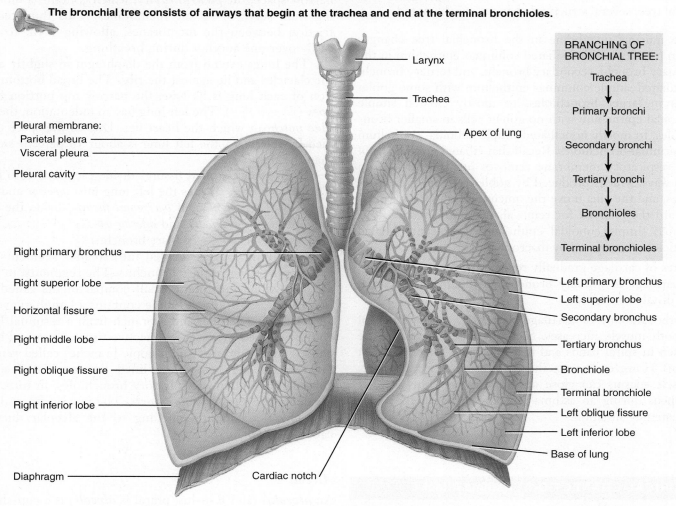

BRANCHING OF BRONCHIAL TREE:

Trachea
↓
Primary bronchi
↓
Secondary bronchi
↓
Tertiary bronchi
↓
Bronchioles
↓
Terminal bronchioles

Larynx

Trachea

Apex of lung

Pleural membrane:
Parietal pleura
Visceral pleura

Pleural cavity

Right primary bronchus

Right superior lobe

Horizontal fissure

Right middle lobe

Right oblique fissure

Right inferior lobe

Diaphragm

Cardiac notch

Left primary bronchus
Left superior lobe
Secondary bronchus
Tertiary bronchus
Bronchiole
Terminal bronchiole
Left oblique fissure
Left inferior lobe
Base of lung

Anterior view

How many lobes and secondary bronchi are present in each lung?

Bronchi and Bronchioles

The trachea divides into a *right primary bronchus* (BRON-kus = windpipe), which goes to the right lung, and a *left primary bronchus*, which goes to the left lung (Figure 18.4). Like the trachea, the primary bronchi (BRONG-kē) contain incomplete rings of cartilage and are lined by pseudostratified ciliated columnar epithelium. Pulmonary blood vessels, lymphatic vessels, and nerves enter and exit the lungs with the two bronchi.

On entering the lungs, the primary bronchi divide to form the *secondary bronchi*, one for each lobe of the lung. (The right lung has three lobes; the left lung has two.) The secondary bronchi continue to branch, forming still smaller bronchi, called *tertiary bronchi*, that divide several times, ultimately giving rise to smaller *bronchioles*. Bronchioles, in turn, branch into even smaller tubes called *terminal bronchioles*. Because all of the airways resemble an upside-down tree with many branches, their arrangement is known as the *bronchial tree*.

As the branching becomes more extensive in the bronchial tree, several structural changes may be noted.

1. The mucous membrane in the bronchial tree changes from pseudostratified ciliated columnar epithelium in the primary bronchi, secondary bronchi, and tertiary bronchi to ciliated simple columnar epithelium with some goblet cells in larger bronchioles, to mostly ciliated simple cuboidal epithelium with no goblet cells in smaller bronchioles, to mostly nonciliated simple cuboidal epithelium in terminal bronchioles. Recall that ciliated epithelium of the respiratory membrane removes inhaled particles in two ways. Mucus produced by goblet cells traps the particles, and the cilia move the mucus and trapped particles toward the pharynx for removal. In regions where nonciliated simple cuboidal epithelium is present, inhaled particles are removed by macrophages.

2. Plates of cartilage gradually replace the incomplete rings of cartilage in primary bronchi and finally disappear in the distal bronchioles.

3. As the amount of cartilage decreases, the amount of smooth muscle increases. Smooth muscle encircles the lumen in spiral bands and helps *maintain patency* (keep it open). However, because there is no supporting cartilage, muscle spasms can close off the airways. This is what happens during an asthma attack, which can be a life-threatening situation.

CLINICAL CONNECTION | Asthma Attack

During an **asthma attack**, bronchiolar smooth muscle goes into spasm. Because there is no supporting cartilage, the spasms can close off the air passageways. Movement of air through constricted bronchi-oles causes breathing to be more labored. The parasympathetic division of the ANS and mediators of allergic reactions such as histamine also cause narrowing of bronchioles (bronchoconstriction) due to contraction of bronchiolar smooth muscle. •

Lungs

The *lungs* (= lightweights, because they float) are two spongy, cone-shaped organs in the thoracic cavity. They are separated from each other by the heart and other structures in the mediastinum (see Figure 15.1). The *pleural membrane* (PLOOR-al; *pleur* = side) is a double-layered serous membrane that encloses and protects each lung (Figure 18.4). The outer layer is attached to the wall of the thoracic cavity and diaphragm and is called the *parietal pleura*. The inner layer, the *visceral pleura*, is attached to the lungs. Between the visceral and parietal pleurae is a narrow space, the *pleural cavity*, which contains a lubricating fluid secreted by the membranes. This fluid reduces friction between the membranes, allowing them to slide easily over one another during breathing.

The lungs extend from the diaphragm to slightly above the clavicles and lie against the ribs. The broad bottom portion of each lung is its *base*; the narrow top portion is the *apex* (Figure 18.4). The left lung has an indentation, the *cardiac notch*, in which the heart lies. Due to the space occupied by the heart, the left lung is about 10 percent smaller than the right lung.

Deep grooves called fissures divide each lung into *lobes*. The *oblique fissure* divides the left lung into *superior* and *inferior lobes*. The *oblique* and *horizontal fissures* divide the right lung into *superior, middle,* and *inferior lobes* (Figure 18.4). Each lobe receives its own secondary bronchus.

Each lung lobe is divided into smaller segments that are supplied by a tertiary bronchus. The segments, in turn, are subdivided into many small compartments called *lobules* (Figure 18.5). Each lobule contains a lymphatic vessel, an arteriole, a venule, and a branch from a terminal bronchiole wrapped in elastic connective tissue. Terminal bronchioles subdivide into microscopic branches called *respiratory bronchioles*, which are lined by nonciliated simple cuboidal epithelium. Respiratory bronchioles, in turn, subdivide into several *alveolar ducts*. The two or more alveoli that share a common opening to the alveolar duct are called *alveolar sacs*.

Alveoli

An *alveolus* (al-VĒ-ō-lus; plural is *alveoli*) is a cup-shaped outpouching of an alveolar sac. Many alveoli and alveolar sacs surround each alveolar duct. The walls of alveoli consist mainly of thin *alveolar cells*, which are simple squamous

Figure 18.5 Lobule of the lung.

Alveolar sacs are two or more alveoli that share a common opening into an alveolar duct.

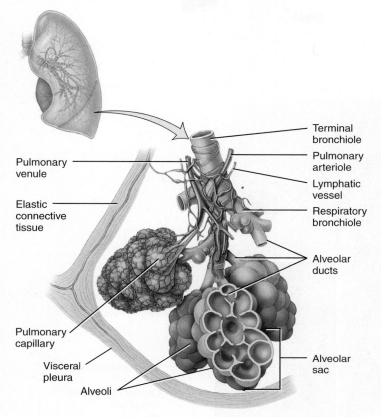

(a) Diagram of a portion of a lobule of the lung

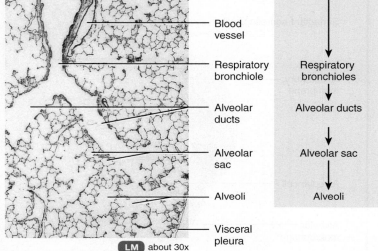

(b) Lung lobule

? What are the major parts of a lobule of a lung?

---

epithelial cells (Figure 18.6). They are the main sites of gas exchange. Scattered among them are *surfactant-secreting cells* that secrete *alveolar fluid*, which keeps the surface between the cells and the air moist. Included in the alveolar fluid is **surfactant** (sur-FAK-tant), a mixture of phospholipids and lipoproteins that reduces the tendency of alveoli to collapse. Also present are *alveolar macrophages (dust cells)*, wandering phagocytes that remove fine dust particles and other debris in the alveolar spaces. Underlying the layer of alveolar cells is an elastic basement membrane and a thin layer of connective tissue containing plentiful elastic and reticular fibers (described shortly). Around the alveoli, the pulmonary arteriole and venule form lush networks of blood capillaries (see Figure 18.5a).

Exchange of O_2 and CO_2 between air spaces in the lungs and the blood takes place by diffusion across alveolar and capillary walls, which together form the **respiratory membrane**. It consists of the following layers (Figure 18.6b):

1. *Alveolar cells* that form the wall of an alveolus.
2. *Epithelial basement membrane* underlying the alveolar cells.

3. *Capillary basement membrane* that is often fused to the epithelial basement membrane.
4. *Endothelial cells* of a capillary wall.

Despite having several layers, the respiratory membrane is only 0.5 μm* wide. This thin width, far less than the thickness of a sheet of tissue paper, permits O_2 and CO_2 to diffuse efficiently between the blood and alveolar air spaces. Moreover, the lungs contain roughly 300 million alveoli. They provide a huge surface area for the exchange of O_2 and CO_2—about 30 to 40 times greater than the surface area of your skin or half the size of a tennis court!

✓ **CHECKPOINT**

4. What is the bronchial tree? Describe its structure.
5. Where are the lungs located? Distinguish the parietal pleura from the visceral pleura.
6. Where in the lungs does the exchange of O_2 and CO_2 take place?

---

*1 μm (micrometer) = 1/1,000,000 of a meter or 1/25,000 of an inch.

Figure 18.6 Structure of an alveolus.

The exchange of respiratory gases occurs by diffusion across the respiratory membrane.

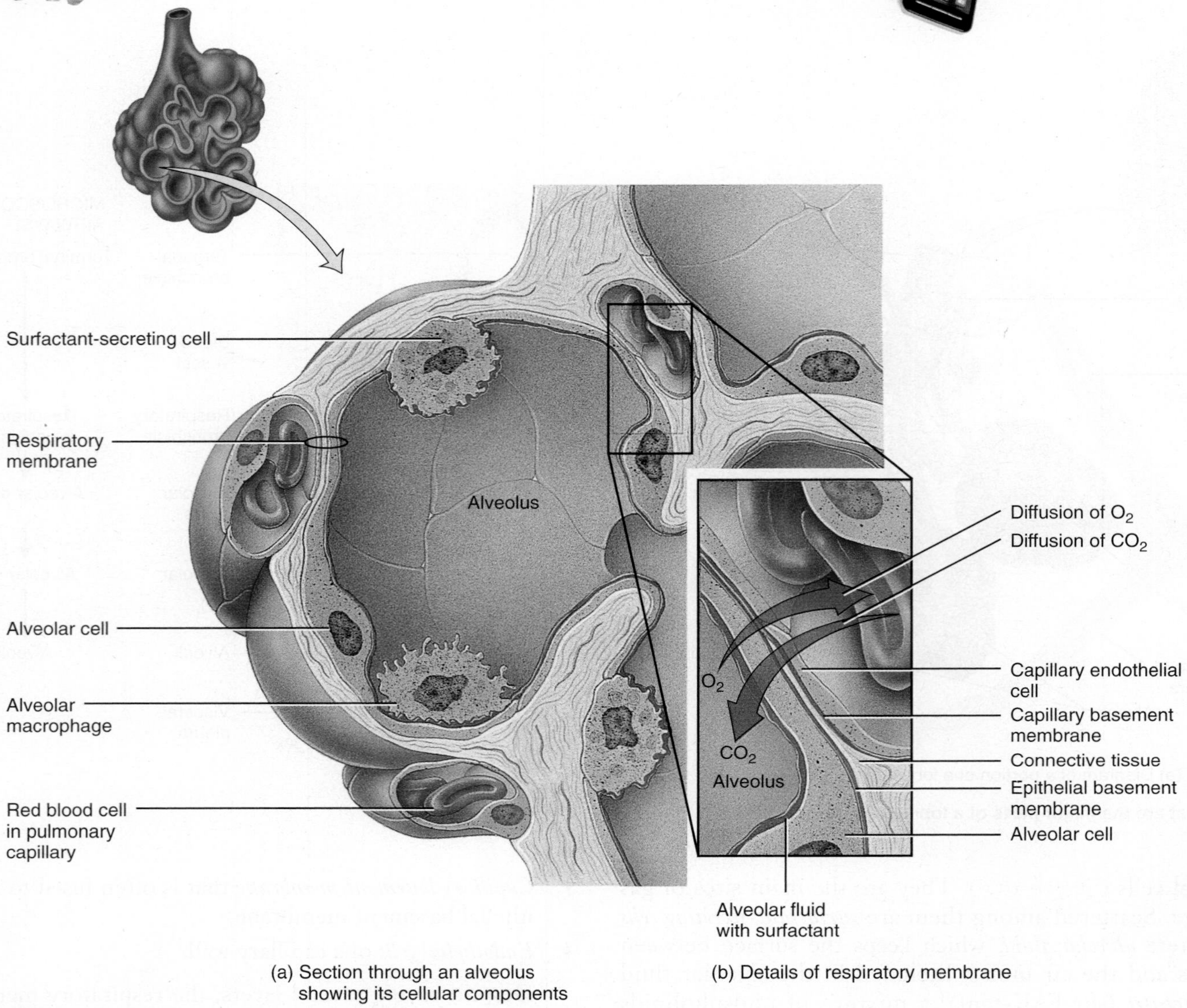

Surfactant-secreting cell

Respiratory membrane

Alveolus

Alveolar cell

Alveolar macrophage

Red blood cell in pulmonary capillary

Diffusion of O_2

Diffusion of CO_2

O_2

CO_2
Alveolus

Capillary endothelial cell

Capillary basement membrane

Connective tissue

Epithelial basement membrane

Alveolar cell

Alveolar fluid with surfactant

(a) Section through an alveolus showing its cellular components

(b) Details of respiratory membrane

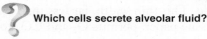 **Which cells secrete alveolar fluid?**

18.2 PULMONARY VENTILATION

OBJECTIVES • **Explain how inhalation and exhalation take place.**

• **Define the various lung volumes and capacities.**

Pulmonary ventilation, the flow of air between the atmosphere and the lungs, occurs due to differences in air pressure. We inhale or breathe in when the pressure inside the lungs is less than the atmospheric air pressure. We exhale or breathe out when the pressure inside the lungs is greater than the atmospheric air pressure. Contraction and relaxation of skeletal muscles create the air pressure changes that power breathing.

Muscles of Inhalation and Exhalation

Breathing in is called *inhalation* or *inspiration*. The muscles of quiet (unforced) inhalation are the diaphragm, the dome-shaped skeletal muscle that forms the floor of the thoracic cavity, and the external intercostals, which extend between the ribs (Figure 18.7). The diaphragm contracts when it receives

Figure 18.7 Muscles of inhalation and exhalation and their actions. The pectoralis minor muscle (not shown here) is illustrated in Figures 8.17 and 8.18.

🔑 **During quiet inhalation, the diaphragm and external intercostals contract, the lungs expand, and air moves into the lungs. During exhalation, the diaphragm relaxes and the lungs recoil inward, forcing air out of the lungs.**

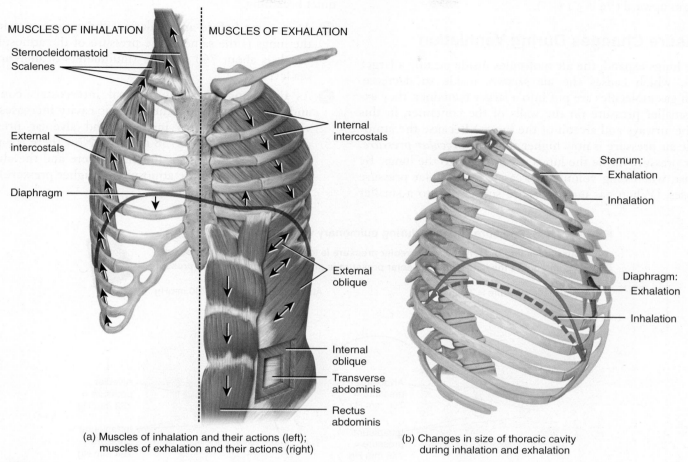

(a) Muscles of inhalation and their actions (left); muscles of exhalation and their actions (right)

(b) Changes in size of thoracic cavity during inhalation and exhalation

❓ **What are the main muscles that power your quiet breathing?**

nerve impulses from the phrenic nerves. As the diaphragm contracts, it descends and becomes flatter, which causes the volume of the attached lungs to expand. As the external intercostals contract, they pull the ribs upward and outward; the attached lungs follow, further increasing lung volume. Contraction of the diaphragm is responsible for about 75 percent of the air that enters the lungs during quiet breathing. Advanced pregnancy, obesity, confining clothing, or increased size of the stomach after eating a large meal can impede descent of the diaphragm and may cause shortness of breath.

During deep, labored inhalations, the sternocleidomastoid muscles elevate the sternum, the scalene muscles elevate the two uppermost ribs, and the pectoralis minor muscles elevate the third through fifth ribs. As the ribs and sternum are elevated, the size of the lungs increases (Figure 18.7b). Movements of the pleural membrane aid expansion of the lungs. The parietal and visceral pleurae normally adhere

tightly because of the surface tension created by their moist adjoining surfaces. Whenever the thoracic cavity expands, the parietal pleura lining the cavity follows, and the visceral pleura and lungs are pulled along with it.

Breathing out, called **exhalation** or *expiration*, begins when the diaphragm and external intercostals relax. Exhalation occurs due to *elastic recoil* of the chest wall and lungs, both of which have a natural tendency to spring back after they have been stretched. Although the alveoli and airways recoil, they don't completely collapse. Because surfactant in alveolar fluid *reduces* elastic recoil, a lack of surfactant causes breathing difficulty by increasing the chance of alveolar collapse.

Because no muscular contractions are involved, quiet exhalation, unlike quiet inhalation, is a *passive process*. Exhalation becomes *active* only during forceful breathing, such as in playing a wind instrument or during exercise. During these times, muscles of exhalation—the internal intercostals, external oblique, internal oblique, transverse abdominis, and rectus abdominis—contract to move the lower ribs downward and compress the abdominal viscera, thus forcing the diaphragm upward (Figure 18.7a).

Pressure Changes During Ventilation

As the lungs expand, the air molecules inside occupy a larger *volume*, which causes the air *pressure* inside to decrease. (When gas molecules are put into a larger container, they exert a smaller pressure on the walls of the container, in this case the airways and alveoli of the lungs.) Because the atmospheric air pressure is now higher than the **alveolar pressure**, the air pressure inside the lungs, air moves into the lungs. By contrast, when lung volume decreases, the alveolar pressure increases. (When gas molecules are squeezed into a smaller

container, they exert a larger pressure on the walls of the container.) Air then flows from the area of higher pressure in the alveoli to the area of lower pressure in the atmosphere. Figure 18.8 shows the sequence of pressure changes during quiet breathing.

1 At rest just before an inhalation, the air pressure inside the lungs is the same as the pressure of the atmosphere, which is about 760 mm Hg (millimeters of mercury) at sea level.

2 As the diaphragm and external intercostals contract and the overall size of the thoracic cavity increases, the volume of the lungs increases and alveolar pressure decreases from 760 to 758 mm Hg. Now there is a pressure difference between the atmosphere and the alveoli, and air flows from the atmosphere (higher pressure) into the lungs (lower pressure).

Figure 18.8 Pressure changes during pulmonary ventilation.

🔑 **Air moves into the lungs when alveolar pressure is less than atmospheric pressure, and out of the lungs when alveolar pressure is greater than atmospheric pressure.**

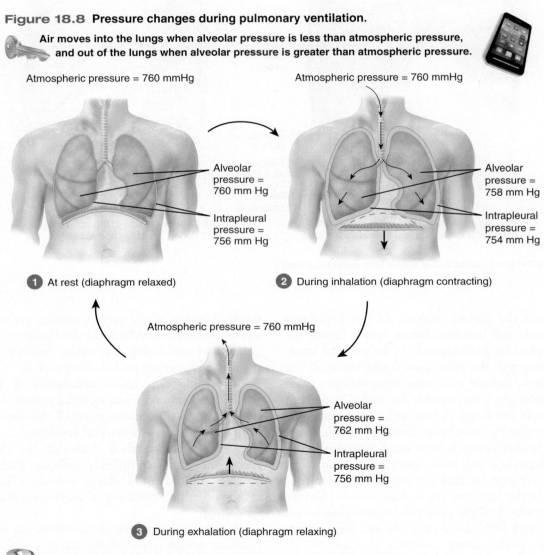

Atmospheric pressure = 760 mmHg

Alveolar pressure = 760 mm Hg

Intrapleural pressure = 756 mm Hg

1 At rest (diaphragm relaxed)

Atmospheric pressure = 760 mmHg

Alveolar pressure = 758 mm Hg

Intrapleural pressure = 754 mm Hg

2 During inhalation (diaphragm contracting)

Atmospheric pressure = 760 mmHg

Alveolar pressure = 762 mm Hg

Intrapleural pressure = 756 mm Hg

3 During exhalation (diaphragm relaxing)

❓ **How does the alveolar pressure change during a normal, quiet breath?**

③ When the diaphragm and external intercostals relax, lung elastic recoil causes the lung volume to decrease, and alveolar pressure rises from 760 to 762 mm Hg. Air then flows from the area of higher pressure in the alveoli to the area of lower pressure in the atmosphere.

Lung Volumes and Capacities

While at rest, a healthy adult breathes about 12 times a minute, with each inhalation and exhalation moving about 500 mL of air into and out of the lungs. The volume of one breath is called the ***tidal volume***. The ***minute ventilation (MV)***—the total volume of air inhaled and exhaled each minute—is equal to breathing rate multiplied by tidal volume:

$$MV = 12 \text{ breaths/min} \times 500 \text{ mL/breath}$$
$$= 6000 \text{ mL/min or 6 liters/min}$$

Tidal volume varies considerably from one person to another and in the same person at different times. About 70 percent of the tidal volume (350 mL) actually reaches the respiratory bronchioles and alveolar sacs and thus participates in gas exchange. The other 30 percent (150 mL) does not participate in gas exchange because it remains in the conducting airways of the nose, pharynx, larynx, trachea, bronchi, bronchioles, and terminal bronchioles. Collectively, these conducting airways are known as the ***anatomic dead space***.

The apparatus commonly used to measure respiratory rate and the amount of air inhaled and exhaled during breathing is a ***spirometer*** (*spiro-* = breathe; *meter* = measuring device). The record produced by a spirometer is called a ***spirogram***. Inhalation is recorded as an upward deflection, and exhalation is recorded as a downward deflection (Figure 18.9).

By taking a very deep breath, you can inhale a good deal more than 500 mL. This additional inhaled air, called the ***inspiratory reserve volume***, is about 3100 mL in an average adult male and 1900 mL in an average adult female (Figure 18.9). Even more air can be inhaled if inhalation follows forced exhalation. If you inhale normally and then exhale as forcibly as possible, you should be able to push out considerably more air in addition to the 500 mL of tidal volume. The extra 1200 mL in males and 700 mL in females is called the ***expiratory reserve volume***. Even after the expiratory reserve volume is expelled, considerable air remains

Figure 18.9 Spirogram showing lung volumes and capacities in milliliters (mL). The average values for a healthy adult male and female are indicated, with the values for a female in parentheses. Note that the spirogram is read from right (start of record) to left (end of record).

Lung capacities are combinations of various lung volumes.

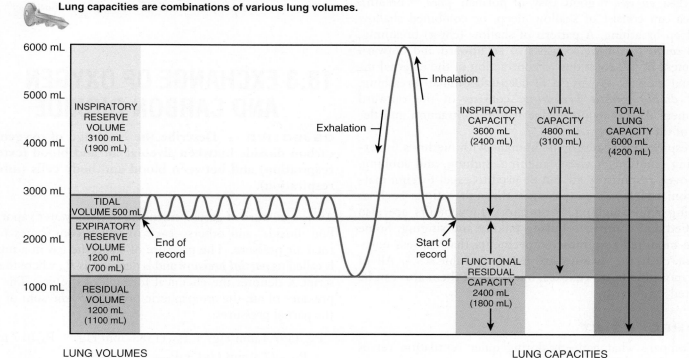

LUNG VOLUMES　　　　　　　　　　　　　　　　　　　LUNG CAPACITIES

? Breathe in as deeply as possible and then exhale as much air as you can. Which lung capacity have you demonstrated?

in the lungs and airways. This volume, called the *residual volume*, amounts to about 1200 mL in males and 1100 mL in females.

Lung *capacities* are combinations of specific lung *volumes* (Figure 18.9). *Inspiratory capacity* is the sum of tidal volume and inspiratory reserve volume (500 mL + 3100 mL = 3600 mL in males and 500 mL + 1900 mL = 2400 mL in females). *Functional residual capacity* is the sum of residual volume and expiratory reserve volume (1200 mL + 1200 mL = 2400 mL in males and 1100 mL + 700 mL = 1800 mL in females). *Vital capacity* is the sum of inspiratory reserve volume, tidal volume, and expiratory reserve volume (4800 mL in males and 3100 mL in females). Finally, *total lung capacity* is the sum of vital capacity and residual volume (4800 mL + 1200 mL = 6000 mL in males and 3100 mL + 1100 mL = 4200 mL in females). The values given here are typical for young adults. Lung volumes and capacities vary with age (smaller in older people), gender (generally smaller in females), and body size (smaller in shorter people). Lung volumes and capacities provide information about an individual's respiratory status since they are usually abnormal in people with pulmonary disorders.

Breathing Patterns and Modified Respiratory Movements

The term for the normal pattern of quiet breathing is *eupnea* (ūp-NĒ-a; *eu- eu-* = good, easy, or normal; *-pnea* = breath). Eupnea can consist of shallow, deep, or combined shallow and deep breathing. A pattern of shallow (chest) breathing, called *costal breathing*, consists of an upward and outward movement of the chest due to contraction of the external intercostal muscles. A pattern of deep (abdominal) breathing, called *diaphragmatic breathing*, consists of the outward movement of the abdomen due to the contraction and descent of the diaphragm.

Respirations also provide humans with methods for expressing emotions such as laughing, sighing, and sobbing. Moreover, respiratory air can be used to expel foreign matter from the lower air passages through actions such as sneezing and coughing. Respiratory movements are also modified and controlled during talking and singing. Some of the modified respiratory movements that express emotion or clear the airways are listed in Table 18.1. All of these movements are reflexes, but some of them also can be initiated voluntarily.

✓ CHECKPOINT

7. Compare what happens during quiet ventilation versus labored ventilation.

8. What is the basic difference between a lung volume and a lung capacity?

TABLE 18.1

Modified Respiratory Movements

MOVEMENT	DESCRIPTION
Coughing	A long-drawn and deep inhalation followed by a strong exhalation that suddenly sends a blast of air through the upper respiratory passages; stimulus for this reflex act may be a foreign body lodged in the larynx, trachea, or epiglottis
Sneezing	Spasmodic contraction of muscles of exhalation that forcefully expels air through the nose and mouth; stimulus may be an irritation of the nasal mucosa
Sighing	A long-drawn and deep inhalation immediately followed by a shorter but forceful exhalation
Yawning	A deep inhalation through the widely opened mouth producing an exaggerated depression of the mandible; may be stimulated by drowsiness, fatigue, or someone else's yawning, but precise cause is unknown
Sobbing	A series of convulsive inhalations followed by a single prolonged exhalation
Crying	An inhalation followed by many short convulsive exhalations, during which the vocal folds vibrate; accompanied by characteristic facial expressions and tears
Laughing	The same basic movements as crying, but the rhythm of the movements and the facial expressions usually differ from those of crying
Hiccupping	Spasmodic contraction of the diaphragm followed by a spasmodic closure of the larynx, which produces a sharp sound on inhalation; stimulus is usually irritation of the sensory nerve endings of the gastrointestinal tract

18.3 EXCHANGE OF OXYGEN AND CARBON DIOXIDE

OBJECTIVE • Describe the exchange of oxygen and carbon dioxide between alveolar air and blood (external respiration) and between blood and body cells (internal respiration).

Air is a mixture of gases—nitrogen, oxygen, water vapor, carbon dioxide, and others—each of which contributes to the total air pressure. The pressure of a specific gas in a mixture is called its *partial pressure* and is denoted as P_x, where the subscript X denotes the chemical formula of the gas. The total pressure of air, the atmospheric pressure, is the sum of all of the partial pressures:

P_{N_2} (597.4 mm Hg) + P_{O_2} (158.8 mm Hg) + P_{Ar} (0.7 mm)
 + P_{H_2O} (2.3 mm Hg) + P_{CO_2} (0.3 mm Hg)

 + $P_{other\ gases}$ (0.5 mm Hg)

 = atmospheric pressure (760 mm Hg)

Partial pressures are important because each gas diffuses from areas where its partial pressure is higher to areas where its partial pressure is lower in the body.

In body fluids, the ability of a gas to stay in solution is greater when its partial pressure is higher and when it has a high solubility in water. The higher the partial pressure of a gas over a liquid and the higher its solubility, the more gas will stay in solution. In comparison to oxygen, much more CO_2 is dissolved in blood plasma because the solubility of CO_2 is 24 times greater than that of O_2. Even though the air we breathe contains mostly N_2, this gas has no known effect on bodily functions, and at sea level pressure very little of it dissolves in blood plasma because its solubility is very low.

CLINICAL CONNECTION | Hyperbaric Oxygenation

Hyperbaric oxygenation (*hyper* = over; *baros* = pressure) is the use of pressure to cause more O_2 to dissolve in the blood. It is an effective technique in treating patients infected by anaerobic bacteria, such as those that cause tetanus and gangrene. (Anaerobic bacteria cannot live in the presence of free O_2.) A person undergoing hyperbaric oxygenation is placed in a hyperbaric chamber, which contains O_2 at a pressure greater than one atmosphere (760 mmHg). As body tissues pick up the O_2, the bacteria are killed. Hyperbaric chambers may also be used for treating certain heart disorders, carbon monoxide poisoning, gas embolisms, crush injuries, cerebral edema, certain hard-to-treat bone infections caused by anaerobic bacteria, smoke inhalation, near-drowning, asphyxia, vascular insufficiencies, and burns. •

External Respiration: Pulmonary Gas Exchange

External respiration, also termed *pulmonary gas exchange*, is the diffusion of O_2 from air in the alveoli of the lungs to blood in pulmonary capillaries and the diffusion of CO_2 in the opposite direction (Figure 18.10a). External respiration in the lungs converts *deoxygenated* (low-oxygen) *blood* that comes from the right side of the heart to *oxygenated* (high-oxygen) *blood* that returns to the left side of the heart. As blood flows through the pulmonary capillaries, it picks up O_2 from alveolar air and unloads CO_2 into alveolar air. Although this process is commonly called an "exchange" of gases, each gas diffuses *independently* from an area where its partial pressure is higher to an area where its partial pressure is lower. An important factor that affects the rate of external respiration is the total surface area available for gas exchange. Any pulmonary disorder that decreases the functional surface area of the respiratory membrane, such as emphysema (see Common Disorders), decreases the rate of gas exchange.

O_2 diffuses from alveolar air, where its partial pressure (P_{O_2}) is 105 mm Hg, into the blood in pulmonary capillaries, where P_{O_2} is about 40 mm Hg in a resting person. During exercise, the P_{O_2} of blood entering the pulmonary capillaries is even lower because contracting muscle fibers are using more O_2. Diffusion continues until the P_{O_2} of pulmonary capillary blood increases to 105 mm Hg, matching the P_{O_2} of alveolar air. Blood leaving pulmonary capillaries near alveolar air spaces mixes with a small volume of blood that has flowed through conducting portions of the respiratory system, where gas exchange does not occur. Thus, the P_{O_2} of blood in the pulmonary veins is about 100 mm Hg, slightly less than the P_{O_2} in pulmonary capillaries.

While O_2 is diffusing from alveolar air into deoxygenated blood, CO_2 is diffusing in the opposite direction. The P_{CO_2} of deoxygenated blood is 45 mm Hg in a resting person, compared to the P_{CO_2} of alveolar air, which is 40 mm Hg. Because of this difference in P_{CO_2}, carbon dioxide diffuses from deoxygenated blood into the alveoli until the P_{CO_2} of the blood decreases to 40 mm Hg. Exhalation keeps alveolar P_{CO_2} at 40 mm Hg. Oxygenated blood returning to the left side of the heart in the pulmonary veins thus has a P_{CO_2} of 40 mm Hg.

CLINICAL CONNECTION | High Altitude Sickness

As a person ascends in altitude, the total atmospheric pressure decreases, with a parallel decrease in the partial pressure of oxygen. P_{O_2} decreases from 159 mm Hg at sea level to 73 mm Hg at 6000 meters (about 20,000 ft). Alveolar P_{O_2} decreases correspondingly, and less oxygen diffuses into the blood. The common symptoms of **high altitude sickness**—shortness of breath, nausea, and dizziness—are due to a lower level of oxygen in the blood. •

Internal Respiration: Systemic Gas Exchange

The left ventricle pumps oxygenated blood into the aorta and through the systemic arteries to systemic capillaries. The exchange of O_2 and CO_2 between systemic capillaries and tissue cells is called *internal respiration* or *systemic gas exchange* (Figure 18.10b). As O_2 leaves the bloodstream, oxygenated blood is converted into deoxygenated blood. Unlike external respiration, which occurs only in the lungs, internal respiration occurs in tissues throughout the body.

The P_{O_2} of blood pumped into systemic capillaries is higher (100 mm Hg) than the P_{O_2} in tissue cells (about 40 mm Hg at rest) because cells constantly use up O_2 to produce ATP. Due to this pressure difference, oxygen diffuses out of the capillaries into tissue cells, and blood P_{O_2} decreases. While O_2 diffuses from the systemic capillaries into tissue cells, CO_2 diffuses in the opposite direction. Because tissue cells are constantly producing CO_2, the P_{CO_2} of cells (45 mm Hg at rest) is higher than that of systemic capillary blood (40 mm Hg). As a result, CO_2 diffuses from tissue cells through interstitial fluid into systemic capillaries until the P_{CO_2} in the blood increases. The deoxygenated blood then returns to the heart and is pumped to the lungs for another cycle of external respiration.

Figure 18.10 Changes in partial pressures of oxygen (O₂) and carbon dioxide (CO₂) in mm Hg during external and internal respiration.

Each gas in a mixture of gases diffuses from an area of higher partial pressure of that gas to an area of lower partial pressure of that gas.

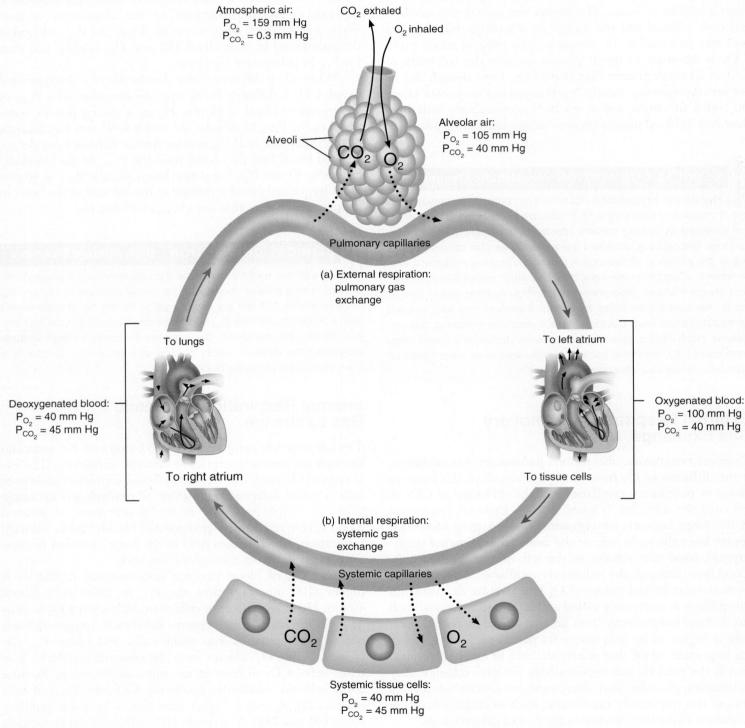

Atmospheric air:
P_{O_2} = 159 mm Hg
P_{CO_2} = 0.3 mm Hg

CO₂ exhaled

O₂ inhaled

Alveolar air:
P_{O_2} = 105 mm Hg
P_{CO_2} = 40 mm Hg

Alveoli

CO₂   O₂

Pulmonary capillaries

(a) External respiration:
pulmonary gas
exchange

To lungs

To left atrium

Deoxygenated blood:
P_{O_2} = 40 mm Hg
P_{CO_2} = 45 mm Hg

Oxygenated blood:
P_{O_2} = 100 mm Hg
P_{CO_2} = 40 mm Hg

To right atrium

To tissue cells

(b) Internal respiration:
systemic gas
exchange

Systemic capillaries

CO₂   O₂

Systemic tissue cells:
P_{O_2} = 40 mm Hg
P_{CO_2} = 45 mm Hg

What causes O₂ to enter pulmonary capillaries from alveolar air and to enter tissue cells from systemic capillaries?

✓ **CHECKPOINT**

9. What are the basic differences among pulmonary ventilation, external respiration, and internal respiration?

10. In a person at rest, what is the partial pressure difference that drives diffusion of oxygen into the blood in pulmonary capillaries?

18.4 TRANSPORT OF RESPIRATORY GASES

OBJECTIVE • Describe how the blood transports oxygen and carbon dioxide.

The blood transports gases between the lungs and body tissues. When O_2 and CO_2 enter the blood, certain physical and chemical changes occur that aid in gas transport and exchange.

Oxygen Transport

Oxygen does not dissolve easily in water, and therefore only about 1.5 percent of the O_2 in blood is dissolved in blood plasma, which is mostly water. About 98.5 percent of blood O_2 is bound to hemoglobin in red blood cells (Figure 18.11).

The heme part of hemoglobin contains four ions of iron, each capable of binding to a molecule of O_2. Oxygen and *deoxyhemoglobin* (Hb) bind in an easily reversible reaction to form *oxyhemoglobin* (Hb–O_2):

$$\text{Hb} + \text{O}_2 \underset{\text{Release of O}_2}{\overset{\text{Binding of O}_2}{\rightleftharpoons}} \text{Hb–O}_2$$

Deoxyhemoglobin   Oxygen   Oxyhemoglobin

When blood P_{O_2} is high, hemoglobin binds with large amounts of O_2 and is *fully saturated*; that is, every available iron atom has combined with a molecule of O_2. When blood P_{O_2} is low, hemoglobin releases O_2. Therefore, in systemic capillaries, where the P_{O_2} is lower, hemoglobin releases O_2, which then can diffuse from blood plasma into interstitial fluid and into tissue cells (Figure 18.11b).

Besides P_{O_2}, several other factors influence the amount of O_2 released by hemoglobin:

■ **Carbon dioxide.** As the P_{CO_2} rises in any tissue, hemoglobin releases O_2 more readily. Thus, hemoglobin releases more O_2 as blood flows through active tissues that are producing more CO_2, such as muscular tissue during exercise.

■ **Acidity.** In an acidic environment, hemoglobin releases O_2 more readily. During exercise, muscles produce lactic acid, which promotes release of O_2 from hemoglobin.

■ **Temperature.** Within limits, as temperature increases, so does the amount of O_2 released from hemoglobin. Active tissues produce more heat, which elevates the local temperature and promotes release of O_2.

Carbon Dioxide Transport

Carbon dioxide is transported in the blood in three main forms (Figure 18.11):

1. **Dissolved CO_2.** The smallest percentage—about 7 percent—is dissolved in blood plasma. Upon reaching the lungs, it diffuses into alveolar air and is exhaled.

2. **Bound to amino acids.** A somewhat higher percentage, about 23 percent, combines with the amino groups of amino acids and proteins in blood. Because the most prevalent protein in blood is hemoglobin (inside red blood cells), most of the CO_2 transported in this manner is bound to hemoglobin. Hemoglobin that has bound CO_2 is termed *carbaminohemoglobin (Hb–CO_2)*:

$$\text{Hb} + \text{CO}_2 \rightleftharpoons \text{Hb–CO}_2$$

Hemoglobin   Carbon dioxide   Carbaminohemoglobin

In tissue capillaries P_{CO_2} is relatively high, which promotes formation of carbaminohemoglobin. But in pulmonary capillaries, P_{CO_2} is relatively low, and the CO_2 readily splits apart from hemoglobin and enters the alveoli by diffusion.

3. **Bicarbonate ions.** The greatest percentage of CO_2—about 70 percent—is transported in blood plasma as *bicarbonate ions (HCO$_3^-$)*. As CO_2 diffuses into tissue capillaries and enters the red blood cells, it combines with water to form carbonic acid (H_2CO_3). The enzyme inside red blood cells that drives this reaction is *carbonic anhydrase (CA)*. The carbonic acid then breaks down into hydrogen ions (H^+) and HCO_3^-:

$$\text{CO}_2 + \text{H}_2\text{O} \overset{\text{CA}}{\rightleftharpoons} \text{H}_2\text{CO}_3 \rightleftharpoons \text{H}^+ + \text{HCO}_3^-$$

Carbon dioxide   Water   Carbonic acid   Hydrogen ion   Bicarbonate ion

Thus, as blood picks up CO_2, HCO_3^- accumulates inside RBCs. Some HCO_3^- moves out into the blood

Figure 18.11 Transport of oxygen and carbon dioxide in the blood.

Most O_2 is transported by hemoglobin as oxyhemoglobin within red blood cells; most CO_2 is transported in blood plasma as bicarbonate ions.

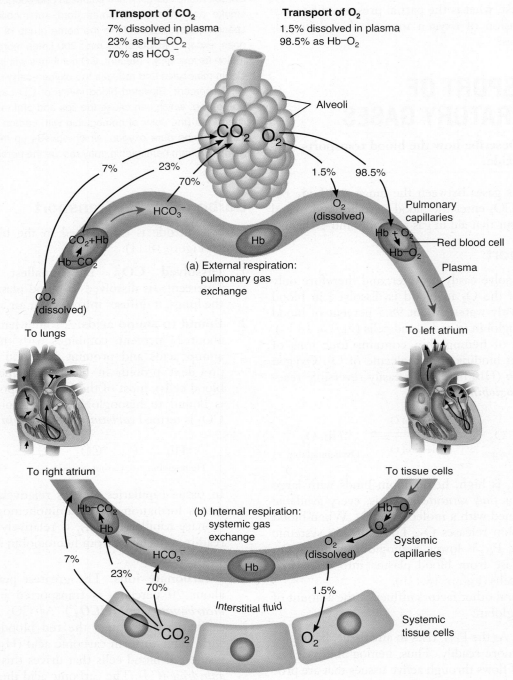

Transport of CO_2
7% dissolved in plasma
23% as $Hb–CO_2$
70% as HCO_3^-

Transport of O_2
1.5% dissolved in plasma
98.5% as $Hb–O_2$

Alveoli

7%   23%   70%   1.5%   98.5%

CO_2 O_2

HCO_3^- O_2
(dissolved)

Pulmonary capillaries

$CO_2 + Hb$ Hb $Hb + O_2$
$Hb–CO_2$ $Hb–O_2$ Red blood cell

Plasma

CO_2
(dissolved)

(a) External respiration: pulmonary gas exchange

To lungs   To left atrium

To right atrium   To tissue cells

$Hb–CO_2$ $Hb–O_2$
Hb O_2

(b) Internal respiration: systemic gas exchange   Systemic capillaries

7%   23% HCO_3^- Hb O_2
(dissolved)

70%   1.5%

Interstitial fluid   Systemic tissue cells

CO_2 O_2

? What percentage of oxygen is transported in blood by hemoglobin?

plasma, down its concentration gradient. In exchange, chloride ions (Cl^-) move from plasma into the RBCs. This exchange of negative ions, which maintains the electrical balance between blood plasma and RBC cytosol, is known as the *chloride shift*. As a result of these

chemical reactions, CO_2 is removed from tissue cells and transported in blood plasma as HCO_3^-.

As blood passes through pulmonary capillaries in the lungs, all these reactions reverse. The CO_2 that was dissolved in plasma diffuses into alveolar air. The CO_2 that was combined with hemoglobin splits and diffuses into the alveoli. The bicarbonate ions (HCO_3^-) reenter the red blood cells from the blood plasma and recombine with H^+ to form H_2CO_3, which splits into CO_2 and H_2O. This CO_2 leaves the red blood cells, diffuses into alveolar air, and is exhaled (Figure 18.11a).

✓ **CHECKPOINT**

11. What is the relationship between hemoglobin and P_{O_2}?

12. What factors cause hemoglobin to unload more oxygen as blood flows through capillaries of metabolically active tissues, such as skeletal muscle during exercise?

18.5 CONTROL OF RESPIRATION

OBJECTIVES ● **Explain how the nervous system controls breathing.**

● **List the factors that can alter the rate and depth of breathing.**

At rest, about 200 mL of O_2 are used each minute by body cells. During strenuous exercise, however, O_2 use typically increases 15- to 20-fold in normal healthy adults, and as much as 30-fold in elite endurance-trained athletes. Several mechanisms help match respiratory effort to metabolic demand.

Respiratory Center

The basic rhythm of respiration is controlled by groups of neurons in the brain stem. The area from which nerve impulses are sent to respiratory muscles is called the ***respiratory center*** and consists of groups of neurons in both the medulla oblongata and the pons.

The ***medullary rhythmicity area*** (rith-MIS-i-tē) in the medulla oblongata controls the basic rhythm of respiration. Within the medullary rhythmicity area are both inspiratory and expiratory areas. Figure 18.12 shows the relationships of the inspiratory and expiratory areas during normal quiet breathing and forceful breathing.

During quiet breathing, inhalation lasts for about 2 seconds and exhalation lasts for about 3 seconds. Nerve impulses generated in the ***inspiratory area*** establish the basic rhythm of breathing. While the inspiratory area is active, it generates nerve impulses for about 2 seconds (Figure 18.12a). The impulses propagate to the external intercostal muscles via intercostal nerves and to the diaphragm via the phrenic nerves. When the nerve impulses reach the diaphragm and external intercostal muscles, the muscles contract and inhalation occurs. Even when all incoming nerve connections to the inspi-

Figure 18.12 Roles of the medullary rhythmicity area in controlling (a) the basic rhythm of quiet respiration and (b) forceful breathing.

 During normal, quiet breathing, the expiratory area is inactive. During forceful breathing, the inspiratory area activates the expiratory area.

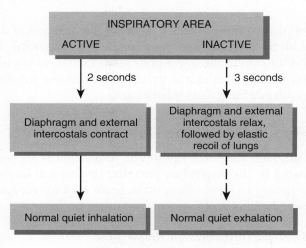

(a) During normal quiet breathing

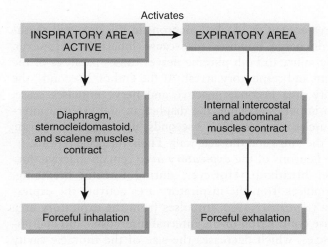

(b) During forceful breathing

? **Which nerves convey impulses from the respiratory center to the diaphragm?**

Smoking—A Breathtaking Experience

Cigarette smoking is the single most preventable cause of death and disability worldwide. All forms of tobacco use disrupt the body's ability to maintain homeostasis and health. Here are a few of smoking's most deadly effects on the respiratory system.

Where There's Smoke . . .

. . . there is injury. The delicate structure of the alveoli allows you to extract life-giving oxygen and cleanse your body of metabolic

wastes. Chronic exposure to smoke gradually destroys lung elasticity. The result is emphysema, a progressive destruction of the alveoli and collapse of respiratory bronchioles. As a result, oxygen uptake is increasingly more difficult.

Bronchitis is an inflammation of the upper respiratory tract. In smokers, chronic bronchitis may result from the irritation and inflammation caused by cigarette smoke and its chemicals. "Smoker's cough" is a symptom of chronic bronchitis. Chronic bronchitis can cause permanent airway damage.

Even in the short term, several factors decrease respiratory function in smokers. Nicotine constricts terminal bronchioles, decreasing airflow into and out of the lungs. Carbon monoxide in smoke binds to hemoglobin and reduces its oxygen-carrying capacity. Irritants in smoke cause increased mucus secretion by the mucosa of the bronchial tree

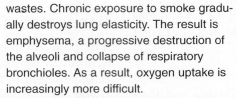

Think It Over . . . Smoking reduces the amount of oxygen reaching body tissues, which interferes with the body's ability to make collagen. How does this help explain why smokers heal more slowly from surgery and fractures?

and swelling of the mucosal lining, both of which impede airflow into and out of the lungs. Irritants in smoke also inhibit the cilia and destroy cilia in the lining of the airways. Thus, excess mucus and foreign debris are not easily removed, making breathing more difficult and setting the stage for infection.

Cancerous Genetic Changes with a Single Puff

Researchers recently discovered that carcinogenic genetic changes begin to occur after a single puff of tobacco smoke into the lungs. Cigarette smoke contains dozens of known carcinogens. Smoking increases cancers of the oral cavity (mouth, tongue, lip, cheek, and throat), larynx, and lungs. When smoke is inhaled, carcinogens are absorbed from the airways into the bloodstream and can thus contribute to cancers at sites outside the lungs. Cancers of the esophagus, stomach, kidney, pancreas, colon, and urinary bladder are more common in smokers than in nonsmokers. In addition, women who smoke have higher rates of breast and cervical cancers.

---

ratory area are cut or blocked, neurons in this area still rhythmically discharge impulses that cause inhalation. However, traumatic injury to both phrenic nerves causes paralysis of the diaphragm and respiratory arrest. At the end of 2 seconds, the inspiratory area becomes inactive and nerve impulses cease. With no impulses arriving, the diaphragm and external intercostal muscles relax for about 3 seconds, allowing passive elastic recoil of the lungs and thoracic wall. Then, the cycle repeats.

The neurons of the *expiratory area* remain inactive during quiet breathing. However, during forceful breathing, nerve impulses from the inspiratory area activate the expiratory area (Figure 18.12b). Impulses from the expiratory area then cause contraction of the internal intercostal and abdominal muscles, which decreases the size of the thoracic cavity and causes forceful exhalation.

The *pneumotaxic area* (noo-mō-TAK-sik; *pneumo-* = air or breath; *-taxic* = arrangement) in the upper pons helps turn off the inspiratory area to shorten the duration of inhalations and to increase breathing rate. The *apneustic area* (ap-NOO-stik) in the lower pons sends excitatory impulses to the inspiratory area that activate it and prolong inhalation. The result is a long, deep inhalation.

Regulation of the Respiratory Center

Although the basic rhythm of respiration is set and coordinated by the inspiratory area, the rhythm can be modified in response to inputs from other brain regions, receptors in the peripheral nervous system, and other factors.

Cortical Influences on Respiration

Because the cerebral cortex has connections with the respiratory center, we can voluntarily alter our pattern of breathing. We can even refuse to breathe at all for a short time. Voluntary control is protective because it enables us to prevent water or irritating gases from entering the lungs. The ability to not breathe, however, is limited by the buildup of CO_2 and H^+ in body fluids. When the P_{CO_2} and H^+ concentration reach a certain level, the inspiratory area is strongly stimulated and breathing resumes, whether the person wants it or not. Despite threats to the contrary by some small children, it is impossible for people to kill themselves by voluntarily holding their breath. Even if breath is held long enough to cause fainting, breathing resumes when consciousness is lost. Nerve impulses from the hypothalamus and limbic system also stimulate the respiratory center, allowing emotional stimuli to alter respirations as, for example, in laughing and crying.

Chemoreceptor Regulation of Respiration

Certain chemical stimuli determine how quickly and how deeply we breathe. The respiratory system functions to maintain proper levels of CO_2 and O_2 and is very responsive to changes in the levels of either in body fluids. Sensory neurons that are responsive to chemicals are termed *chemoreceptors*. *Central chemoreceptors*, located within the medulla oblongata, respond to changes in H^+ level or P_{CO_2} or both, in cerebrospinal fluid. *Peripheral chemoreceptors*, located within the arch of the aorta and common carotid arteries, are especially sensitive to changes in P_{O_2}, H^+, and P_{CO_2}, in the blood.

Because CO_2 is lipid-soluble, it easily diffuses through the plasma membrane into cells, where it combines with water (H_2O) to form carbonic acid (H_2CO_3). Carbonic acid quickly breaks down into H^+ and HCO_3^-. Any increase in CO_2 in the blood thus causes an increase in H^+ inside cells, and any decrease in CO_2 causes a decrease in H^+.

CLINICAL CONNECTION | Hypercapnia and Hypoxia

Normally, the P_{CO_2} in arterial blood is 40 mm Hg. If even a slight increase in P_{CO_2} occurs—a condition called **hypercapnia**—the central chemoreceptors are stimulated and respond vigorously to the resulting increase in H^+ level. The peripheral chemoreceptors also are stimulated by both the high P_{CO_2} and the rise in H^+. In addition, the peripheral chemoreceptors respond to severe **hypoxia**, a deficiency of O_2. If P_{O_2} in arterial blood falls from a normal level of 100 mm Hg to about 50 mm Hg, the peripheral chemoreceptors are strongly stimulated. •

The chemoreceptors participate in a negative feedback system that regulates the levels of CO_2, O_2, and H^+ in the blood (Figure 18.13). As a result of increased P_{CO_2}, decreased pH (increased H^+), or decreased P_{O_2}, input from the central and peripheral chemoreceptors causes the inspiratory area to become highly active. Then, the rate and depth of breathing increase. Rapid and deep breathing, called *hyperventilation*, allows the exhalation of more CO_2 until P_{CO_2} and H^+ are lowered to normal.

Figure 18.13 Negative feedback control of breathing in response to changes in blood P_{CO_2}, pH (H^+ level), and P_{O_2}.

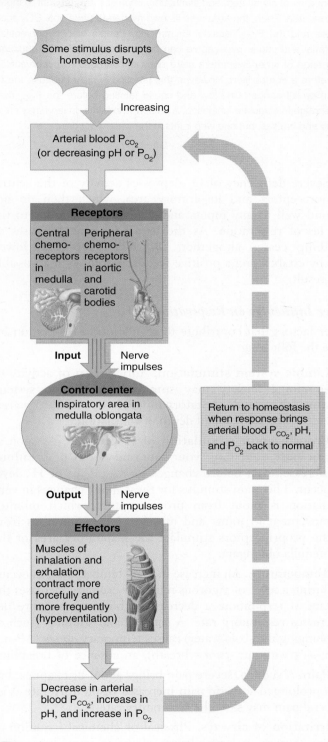

An increase in blood P_{CO_2} stimulates the inspiratory center.

Some stimulus disrupts homeostasis by

Increasing

Arterial blood P_{CO_2} (or decreasing pH or P_{O_2})

Receptors

Central chemoreceptors in medulla   Peripheral chemoreceptors in aortic and carotid bodies

Input   Nerve impulses

Control center
Inspiratory area in medulla oblongata

Return to homeostasis when response brings arterial blood P_{CO_2}, pH, and P_{O_2} back to normal

Output   Nerve impulses

Effectors
Muscles of inhalation and exhalation contract more forcefully and more frequently (hyperventilation)

Decrease in arterial blood P_{CO_2}, increase in pH, and increase in P_{O_2}

What is the normal arterial blood P_{CO_2}?

When these receptors become stretched during overinflation of the lungs, the inspiratory area is inhibited. As a result, exhalation begins. This reflex is mainly a protective mechanism for preventing excessive inflation of the lungs.

✓ **CHECKPOINT**

13. How does the medullary rhythmicity area function in regulating respiration?

14. How do the cerebral cortex, levels of CO_2 and O_2, proprioceptors, inflation reflex, temperature changes, pain, and irritation of the airways modify respiration?

Severe deficiency of O_2 depresses activity of the central chemoreceptors and inspiratory area, which then do not respond well to any inputs and send fewer impulses to the muscles of respiration. As the breathing rate decreases or breathing ceases altogether, P_{O_2} falls lower and lower, thereby establishing a positive feedback cycle with a possibly fatal result.

Other Influences on Respiration

Other factors that contribute to regulation of respiration include the following:

■ **Limbic system stimulation.** Anticipation of activity or emotional anxiety may stimulate the limbic system, which then sends excitatory input to the inspiratory area, increasing the rate and depth of ventilation.

■ **Proprioceptor stimulation of respiration.** As soon as you start exercising, your rate and depth of breathing increase, even before changes in P_{O_2}, P_{CO_2}, or H^+ level occur. The main stimulus for these quick changes in ventilation is input from proprioceptors, which monitor movement of joints and muscles. Nerve impulses from the proprioceptors stimulate the inspiratory area of the medulla oblongata.

■ **Temperature.** An increase in body temperature, as occurs during a fever or vigorous muscular exercise, increases the rate of respiration; a decrease in body temperature decreases respiratory rate. A sudden cold stimulus (such as plunging into cold water) causes temporary *apnea* (AP-nē-a; *a-* = without; *-pnea* = breath), an absence of breathing.

■ **Pain.** A sudden, severe pain brings about brief apnea, but a prolonged somatic pain increases respiratory rate. Visceral pain may slow the respiratory rate.

■ **Irritation of airways.** Physical or chemical irritation of the pharynx or larynx brings about an immediate cessation of breathing followed by coughing or sneezing.

■ **The inflation reflex.** Located in the walls of bronchi and bronchioles are pressure-sensitive *stretch receptors*.

18.6 EXERCISE AND THE RESPIRATORY SYSTEM

OBJECTIVE • **Describe the effects of exercise on the respiratory system.**

During exercise, the respiratory and cardiovascular systems make adjustments in response to both the intensity and duration of the exercise. The effects of exercise on the heart were discussed in Chapter 15; here we focus on how exercise affects the respiratory system.

Recall that the heart pumps the same amount of blood to the lungs as to all the rest of the body. Thus, as cardiac output rises, the rate of blood flow through the lungs also increases. If blood flows through the lungs twice as fast as at rest, it picks up twice as much oxygen per minute. In addition, the rate at which O_2 diffuses from alveolar air into the blood increases during maximal exercise because blood flows through a larger percentage of the pulmonary capillaries, providing a greater surface area for diffusion of O_2 into the blood.

When muscles contract during exercise, they consume large amounts of O_2 and produce large amounts of CO_2, forcing the respiratory system to work harder to maintain normal blood gas levels. During vigorous exercise, O_2 consumption and ventilation increase dramatically. At the onset of exercise, an abrupt increase in ventilation, due to activation of proprioceptors, is followed by a more gradual increase. With moderate exercise, the depth of ventilation rather than breathing rate is increased. When exercise is more strenuous, breathing rate also increases.

At the end of an exercise session, an abrupt decrease in ventilation rate is followed by a more gradual decline to the resting level. The initial decrease is due mainly to decreased stimulation of proprioceptors when movement stops or slows. The more gradual decrease reflects the slower return of blood chemistry and blood temperature to resting levels.

✓ **CHECKPOINT**

15. How does exercise affect the inspiratory area?

18.7 AGING AND THE RESPIRATORY SYSTEM

OBJECTIVE • **Describe the effects of aging on the respiratory system.**

With advancing age, the airways and tissues of the respiratory tract, including the alveoli, become less elastic and more rigid; the chest wall becomes more rigid as well. The result is a decrease in lung capacity. In fact, vital capacity (the maximum amount of air that can be expired after maximal inhalation) can decrease as much as 35 percent by age 70. A decrease in blood level of O_2, decreased activity of alveo-lar macrophages, and diminished ciliary action of the epithelium lining the respiratory tract occur. Owing to all these age-related factors, elderly people are more susceptible to pneumonia, bronchitis, emphysema, and other pulmonary disorders. Age-related changes in the structure and functions of the lung can also contribute to an older person's reduced ability to perform vigorous exercises, such as running.

✓ **CHECKPOINT**

16. What accounts for the decrease in vital capacity with aging?

• • •

To appreciate the many ways that the respiratory system contributes to homeostasis of other body systems, examine Focus on Homeostasis: The Respiratory System. Next, in Chapter 19, we will see how the digestive system makes nutrients available to body cells so that oxygen provided by the respiratory system can be used for ATP production.

Focus on Homeostasis

BODY SYSTEM	CONTRIBUTION OF THE RESPIRATORY SYSTEM

For all body systems

Provides oxygen and removes carbon dioxide. Helps adjust the pH of body fluids through exhalation of carbon dioxide.

Muscular system

Increased rate and depth of breathing support increased activity of skeletal muscles during exercise.

Nervous system

Nose contains receptors for the sense of smell (olfaction). Vibrations of air flowing across the vocal cords produce sounds for speech.

Endocrine system
Angiotensin converting enzyme (ACE) in the lungs promotes formation of the hormone angiotensin II, which in turn stimulates the adrenal gland to release the hormone aldosterone.

Cardiovascular system

During inhalations, the respiratory pump aids the return of venous blood to the heart.

Lymphatic system and immunity

Hairs in the nose, cilia and mucus in the trachea, bronchi, and smaller airways, and alveolar macrophages contribute to nonspecific immunity to disease. The pharynx (throat) contains lymphatic tissue (tonsils). During inhalation, the respiratory pump promotes the flow of lymph.

Digestive system

Forceful contraction of the respiratory muscles can assist in defecation.

Urinary system

Together, the respiratory and urinary systems regulate the pH of body fluids.

Reproductive systems

Increased rate and depth of breathing support activity during sexual intercourse. Internal respiration provides oxygen to the developing fetus.

THE RESPIRATORY SYSTEM

COMMON DISORDERS

Asthma

Asthma (AZ-ma = panting) is a disorder characterized by chronic airway inflammation, airway hypersensitivity to a variety of stimuli, and airway obstruction. The airway obstruction may be due to smooth muscle spasms in the walls of smaller bronchi and bronchioles, swelling of the mucosa of the airways, increased mucus secretion, or damage to the epithelium of the airway. Asthma is at least partially reversible, either spontaneously or with treatment. It affects 3–5% of the U.S. population and is becoming increasingly common in children.

Asthmatics typically react to low concentrations of stimuli that do not normally cause symptoms in people without asthma. Sometimes the trigger is an allergen such as pollen, dust mites, molds, or a particular food. Other common triggers include emotional upset, aspirin, sulfiting agents (used in wine and beer and to keep greens fresh in salad bars), exercise, and breathing cold air or cigarette smoke. Symptoms include difficult breathing, coughing, wheezing, chest tightness, tachycardia, fatigue, moist skin, and anxiety.

Chronic Obstructive Pulmonary Disease

Chronic obstructive pulmonary disease (COPD) is a respiratory disorder characterized by chronic obstruction of airflow. The principal types of COPD are emphysema and chronic bronchitis. In most cases, COPD is preventable because its most common cause is cigarette smoking or breathing secondhand smoke. Other causes include air pollution, pulmonary infection, occupational exposure to dusts and gases, and genetic factors.

Emphysema

Emphysema (em′-fi-SĒ-ma = blown up or full of air) is a disorder characterized by destruction of the walls of the alveoli, which produces abnormally large air spaces that remain filled with air during exhalation. With less surface area for gas exchange, O_2 diffusion across the respiratory membrane is reduced. Blood O_2 level is somewhat lowered, and any mild exercise that raises the O_2 requirements of the cells leaves the patient breathless. As increasing numbers of alveolar walls are damaged, lung elastic recoil decreases due to loss of elastic fibers, and an increasing amount of air becomes trapped in the lungs at the end of exhalation. Over several years, added respiratory exertion increases the size of the chest cage, resulting in a "barrel chest." Emphysema is a common precursor to the development of lung cancer.

Chronic Bronchitis

Chronic bronchitis (brong-KĪ-tis) is a disorder characterized by excessive secretion of bronchial mucus accompanied by a cough. Inhaled irritants lead to chronic inflammation with an increase in the size and number of mucous glands and goblet cells in the airway epithelium. The thickened and excessive mucus produced narrows the airway and impairs the action of cilia. Thus, inhaled pathogens become embedded in airway secretions and multiply rapidly. Besides a cough, symptoms of chronic bronchitis are shortness of breath, wheezing, cyanosis, and pulmonary hypertension.

Lung Cancer

In the United States, *lung cancer* is the leading cause of cancer death in both males and females. At the time of diagnosis, lung cancer is usually well advanced. Most people with lung cancer die within a year of the diagnosis, and the overall survival rate is only 10–15%. About 85% of lung cancer cases are due to smoking, and the disease is 10 to 30 times more common in smokers than nonsmokers. Exposure to secondhand smoke also causes lung cancer and heart disease. Other causes of lung cancer are ionizing radiation, such as x-rays, and inhaled irritants, such as asbestos and radon gas.

Symptoms of lung cancer may include a chronic cough, spitting blood from the respiratory tract, wheezing, shortness of breath, chest pain, hoarseness, difficulty swallowing, weight loss, anorexia, fatigue, bone pain, confusion, problems with balance, headache, anemia, low blood platelet count, and jaundice.

Pneumonia

Pneumonia or *pneumonitis* (nū′-mō-NĪ-tis) is an acute infection or inflammation of the alveoli. It is the most common infectious cause of death in the United States, where an estimated 4 million cases occur annually. When certain microbes enter the lungs of susceptible individuals, they release harmful toxins, stimulating inflammation and immune responses that have damaging side effects. The toxins and immune response damage alveoli and bronchial mucous membranes; inflammation and edema cause the alveoli to fill with debris and fluid, interfering with ventilation and gas exchange. The most common cause is the bacterium *Streptococcus pneumoniae*, but other bacteria, viruses, or fungi may also cause pneumonia.

Tuberculosis

The bacterium *Mycobacterium tuberculosis* produces an infectious, communicable disease called *tuberculosis (TB)* that most often affects the lungs and the pleurae but may involve other parts of the body. Once the bacteria are inside the lungs, they multiply and cause inflammation, which stimulates neutrophils and macrophages to migrate to the area and engulf the bacteria to prevent their spread. If the immune system is not impaired, the bacteria may remain dormant for life. Impaired immunity may enable the bacteria to escape into blood and lymph to infect other organs. In many people, symptoms—fatigue, weight loss, lethargy, anorexia, a low-grade fever, night sweats, cough, dyspnea, chest pain, and spitting blood (hemoptysis)—do not develop until the disease is advanced.

Upper Respiratory Infections, Seasonal Influenza, and H1N1 Influenza

Hundreds of viruses can cause **upper respiratory infections** or the **common cold,** but a group of viruses called *rhinoviruses* (RĪ-nō-vī-rus-es) is responsible for about 40 percent of all colds in adults. Typical symptoms include sneezing, excessive nasal secretion, dry cough, and congestion. The uncomplicated common cold is not usually accompanied by a fever. Complications include sinusitis, asthma, bronchitis, ear infections, and laryngitis. Recent investigations suggest an association between emotional stress and the common cold. The higher the stress level, the greater the frequency and duration of colds.

Seasonal influenza (flu) is also caused by a virus. Its symptoms include chills, fever (usually higher than 101°F = 39°C), headache, and muscular aches. Seasonal influenza can become life-threatening and may develop into pneumonia. It is important to recognize that influenza is a respiratory disease, not a gastrointestinal (GI) disease. Many people mistakenly report having seasonal flu when they are suffering from a GI illness.

H1N1 influenza (flu), also known as *swine flu*, is a type of influenza caused by a new virus called *influenza H1N1*. The virus is spread in the same way that seasonal flu spreads: from person to person through coughing or sneezing or by touching infected objects and then touching one's mouth or nose. Most individuals infected with the virus have mild disease and recover without medical treatment, but some people have severe disease and have even died. The symptoms of H1N1 flu include fever, cough, runny or stuffy nose, headache, body aches, chills, and fatigue. Some people also have vomiting and diarrhea. Most people who have been hospitalized for H1N1 flu have had one or more preexisting medical conditions such as diabetes, heart disease, asthma, kidney disease, or pregnancy. People infected with the virus can infect others from 1 day before symptoms occur to 5–7 days or more after they occur. Treatment of H1N1 flu involves taking an-

tiviral drugs, such as Tamiflu® and Relenza®. A vaccine is also available, but the H1N1 flu vaccine is not a substitute for seasonal flu vaccines. In order to prevent infection, the Centers for Disease Control and Prevention (CDC) recommends washing your hands often with soap and water or with an alcohol-based hand cleaner; covering your mouth and nose with a tissue when coughing or sneezing and disposing of the tissue; avoiding touching your mouth, nose, or eyes; avoiding close contact (within 6 feet) with people who have flulike symptoms; and staying home for 7 days after symptoms begin or for 24 hours after being symptom-free, whichever is longer.

Pulmonary Edema

Pulmonary edema is an abnormal accumulation of interstitial fluid in the interstitial spaces and alveoli of the lungs. The edema may arise from increased pulmonary capillary permeability (pulmonary origin) or increased pulmonary capillary pressure due to congestive heart failure (cardiac origin). The most common symptom is painful or labored breathing. Other symptoms include wheezing, rapid breathing rate, restlessness, a feeling of suffocation, cyanosis, paleness, and excessive perspiration.

MEDICAL TERMINOLOGY AND CONDITIONS

Abdominal thrust maneuver   First-aid procedure to clear the airways of obstructing objects. It is performed by applying a quick upward thrust between the navel and lower ribs that causes sudden elevation of the diaphragm and forceful, rapid expulsion of air from the lungs, forcing air out of the trachea to eject the obstructing object. Also used to expel water from the lungs of near-drowning victims before resuscitation is begun. Also known as the *Heimlich maneuver* (HĪM-lik ma-NOO-ver).

Asphyxia (as-FIK-sē-a; *sphyxia* = pulse)   Oxygen starvation due to low atmospheric oxygen or interference with ventilation, external respiration, or internal respiration.

Aspiration (as'-pi-RĀ-shun)   Inhalation into the bronchial tree of a substance other than air, for instance, water, food, or a foreign body.

Bronchoscopy (brong-KOS-kō-pē)   The visual examination of the bronchi through a *bronchoscope*, an illuminated, tubular instrument that is passed through the mouth (or nose), larynx, and trachea into the bronchi.

Cystic fibrosis (CF)   An inherited disease of secretory epithelia that affects the airways, liver, pancreas, small intestine, and sweat glands. Clogging and infection of the airways leads to difficulty in breathing and eventual destruction of lung tissue.

Dyspnea (DISP-nē-a; *dys-* = painful, difficult)   Painful or labored breathing.

Epistaxis (ep'-i-STAK-sis)   Loss of blood from the nose due to trauma, infection, allergy, malignant growths, or bleeding disorders. It can be arrested by cautery with silver nitrate, electrocautery, or firm packing. Also called *nosebleed*.

Hypoxia (hī-POK-sē-a; *hypo-* = below or under)   A deficiency of O_2 at the tissue level that may be caused by a low P_{O_2} in arterial blood, as from high altitudes; too little functioning hemoglobin in the blood, as in anemia; inability of the blood to carry O_2 to tissues fast enough to sustain their needs, as in heart failure; or inability of tissues to use O_2 properly, as in cyanide poisoning.

Mechanical ventilation   The use of an automatically cycling device (ventilator or respirator) to assist breathing. A plastic tube is inserted into the nose or mouth and the tube is attached to a device that forces air into the lungs. Exhalation occurs passively due to the elastic recoil of the lungs.

Pleurisy (PLOOR-i-sē)   Inflammation of the pleural membranes, which causes friction during breathing that can be quite painful when the swollen membranes rub against each other. Also known as *pleuritis*.

Rales (RĀLS)   Sounds sometimes heard in the lungs that resemble bubbling or rattling. Different types are due to the presence of an abnormal type or amount of fluid or mucus within the bronchi or alveoli, or to bronchoconstriction that causes turbulent airflow.

Respiratory distress syndrome (RDS)   A breathing disorder of premature newborns in which the alveoli do not remain open due to a lack of surfactant. Surfactant reduces surface tension and is necessary to prevent the collapse of alveoli during exhalation.

Respiratory failure   A condition in which the respiratory system either cannot supply enough O_2 to maintain metabolism or cannot eliminate enough CO_2 to prevent respiratory acidosis (a higher-than-normal H^+ level in interstitial fluid).

Rhinitis (rī-NĪ-tis; *rhin-* = nose)   Chronic or acute inflammation of the mucous membrane of the nose.

Sudden infant death syndrome (SIDS)   Death of infants between the ages of 1 week and 12 months thought to be due to hypoxia that occurs while sleeping in a prone position (on the stomach) and rebreathing exhaled air trapped in a depression of the mattress. It is now recommended that normal newborns be placed on their backs for sleeping (remember: "back to sleep").

Tachypnea (tak'-ip-NĒ-a; *tachy-* = rapid)   Rapid breathing rate.

Wheeze (HWĒZ)   A whistling, squeaking, or musical high-pitched sound during breathing resulting from a partially obstructed airway.

CHAPTER REVIEW AND RESOURCE SUMMARY

REVIEW

RESOURCES

18.1 Organs of the Respiratory System

1. Respiratory organs include the nose, pharynx, larynx, trachea, bronchi, and lungs, and they act with the cardiovascular system to supply oxygen and remove carbon dioxide from the blood.

2. The external portion of the **nose** is made of cartilage and skin and is lined with mucous membrane. Openings to the exterior are the **external nares**. The internal portion of the nose, divided from the external portion by the **nasal septum**, communicates with the paranasal sinuses and nasopharynx through the internal nares. The nose is adapted for warming, moistening, and filtering air; olfaction; and serving as a resonating chamber.

3. The **pharynx** (throat), a muscular tube lined by a mucous membrane, is divided into the nasopharynx, oropharynx, and laryngopharynx. The **nasopharynx** functions in respiration. The **oropharynx** and **laryngopharynx** function both in digestion and in respiration.

4. The **larynx** connects the pharynx and the trachea. It contains the **thyroid cartilage** (Adam's apple), the **epiglottis**, the **cricoid cartilage**, **arytenoid cartilages**, **false vocal cords**, and **true vocal cords**. Taut true vocal cords produce high pitches; relaxed ones produce low pitches.

5. The **trachea** (windpipe) extends from the larynx to the primary bronchi. It is composed of smooth muscle and C-shaped rings of cartilage and is lined with pseudostratified ciliated columnar epithelium.

6. The **bronchial tree** consists of the trachea, primary bronchi (**right primary bronchus** and **left primary bronchus**), **secondary bronchi**, **tertiary bronchi**, **bronchioles**, and **terminal bronchioles**.

7. **Lungs** are paired organs in the thoracic cavity enclosed by the **pleural membrane**. The **parietal pleura** is the outer layer; the **visceral pleura** is the inner layer. The right lung has three lobes separated by two fissures; the left lung has two lobes separated by one fissure plus a depression, the cardiac notch.

8. Each lobe consists of **lobules**, which contain lymphatic vessels, arterioles, venules, terminal bronchioles, **respiratory bronchioles**, **alveolar ducts**, **alveolar sacs**, and **alveoli**.

9. Exchange of gases (oxygen and carbon dioxide) in the lungs occurs across the **respiratory membrane**, a thin "sandwich" consisting of alveolar cells, basement membrane, and endothelial cells of a capillary.

Anatomy Overview—Overview of Respiratory Organs
Anatomy Overview—Respiratory Tissues
Figure 18.3 Larynx
Figure 18.6 Structure of an alveolus

Exercise—Build an Airway
Exercise—Paint the Lung
Exercise—The Airway
Concepts and Connections—Functional Anatomy of the Respiratory System

18.2 Pulmonary Ventilation

1. **Pulmonary ventilation** (breathing) consists of inhalation and exhalation, the movement of air into and out of the lungs. Air flows from higher to lower pressure.

2. **Inhalation** occurs when **alveolar pressure** falls below atmospheric pressure. Contraction of the diaphragm and external intercostals expands lung volume. Increased volume of the lungs decreases alveolar pressure, and air moves from higher to lower pressure, from the atmosphere into the lungs.

3. **Exhalation** occurs when alveolar pressure is higher than atmospheric pressure. Relaxation of the diaphragm and external intercostals decreases lung volume, and alveolar pressure increases so that air moves from the lungs to the atmosphere.

4. The sternocleidomastoids, scalenes, and pectoralis minors contribute to forced inhalation. Forced exhalation involves contraction of the internal intercostals, external oblique, internal oblique, transverse abdominis, and rectus abdominis.

5. The **minute ventilation** is the total air taken in during 1 minute (breathing rate per minute multiplied by **tidal volume**).

6. The lung volumes are tidal volume, **inspiratory reserve volume**, **expiratory reserve volume**, and **residual volume**.

7. Lung capacities, the sum of two or more lung volumes, include **inspiratory capacity**, **functional residual capacity**, **vital capacity**, and **total capacity**.

Animation—Pulmonary Ventilation
Figure 18.8 Pressure changes during pulmonary ventilation

Concepts and Connections—Ventilation

REVIEW	RESOURCES

18.3 Exchange of Oxygen and Carbon Dioxide

1. The **partial pressure** of a gas (P_x) is the pressure exerted by that gas in a mixture of gases.
2. Each gas in a mixture of gases exerts its own pressure and behaves as if no other gases are present.
3. In external and internal respiration, O_2 and CO_2 move from areas of higher partial pressure to areas of lower partial pressure.
4. **External respiration**, the exchange of gases between alveolar air and pulmonary blood capillaries, is aided by a thin respiratory membrane, a large alveolar surface area, and a rich blood supply.
5. **Internal respiration** is the exchange of gases between systemic tissue capillaries and systemic tissue cells.

Resources:
Animation—Gas Exchange
Figure 18.10 Changes in partial pressures of oxygen (O_2) and carbon dioxide (CO_2) in mm Hg during external and internal respiration
Exercise—Gas Exchange Match-up
Concepts and Connections—Gas Exchange

18.4 Transport of Respiratory Gases

1. Most oxygen, 98.5 percent, is carried by the iron ions of the heme in hemoglobin; 1.5 percent is dissolved in plasma.
2. The association of O_2 and hemoglobin is affected by P_{O_2}, pH, temperature, and P_{CO_2}.
3. Carbon dioxide is transported in three ways. About 7 percent is dissolved in plasma, 23 percent combines with the globin of hemoglobin, and 70 percent is converted to bicarbonate ions (HCO_3^-).

Resources:
Animation—Gas Transport
Concepts and Connections—Oxygen Transport

18.5 Control of Respiration

1. The **respiratory center** consists of a **medullary rhythmicity area** (**inspiratory** and **expiratory areas**) in the medulla oblongata and groups of neurons in the pons (**pneumotaxic area** and **apneustic area**).
2. The inspiratory area sets the basic rhythm of respiration.
3. Respirations may be modified by several factors, including cortical influences; chemical stimuli, such as levels of O_2, CO_2, and H^+; limbic system stimulation; proprioceptor input; temperature; pain; the inflation reflex; and irritation to the airways.

Resources:
Anatomy Overview—Structures that Control Respiration
Animation—Regulation of Ventilation
Exercise—Respiration and pH Reflex
Concepts and Connections—Respiratory Rate

18.6 Exercise and the Respiratory System

1. The rate and depth of ventilation change in response to both the intensity and duration of exercise.
2. The abrupt increase in ventilation at the start of exercise is due to neural changes that send excitatory impulses to the inspiratory area in the medulla oblongata. The more gradual increase in ventilation during moderate exercise is due to chemical and physical changes in the bloodstream.

18.7 Aging and the Respiratory System

1. Aging results in decreased vital capacity, decreased blood level of O_2, and diminished alveolar macrophage activity.
2. Elderly people are more susceptible to pneumonia, emphysema, bronchitis, and other pulmonary disorders.

Resources:
Homeostatic Imbalance—The Case of the Worried Smoker

SELF-QUIZ

1. Which of the following is NOT true concerning the pharynx?
   a. Food, drink, and air pass through the oropharynx and laryngopharynx.
   b. The auditory (eustachian) tubes have openings in the nasopharynx.
   c. The pseudostratified ciliated epithelium of the nasopharynx helps move dust-laden mucus toward the mouth.
   d. The palatine and lingual tonsils are located in the laryngopharynx.
   e. The wall of the pharynx is composed of skeletal muscle lined with mucous membranes.

2. During speaking, you raise your voice's pitch. This is possible because
   a. the epiglottis vibrates rapidly.
   b. you have increased the air pressure pushing against the vocal cords.
   c. you have increased the tension on the true vocal cords.
   d. your true vocal cords have become thicker and longer.
   e. the true vocal cords begin to vibrate more slowly.

3. Johnny is having an asthma attack and feels as if he cannot breathe. Why?

a. His diaphragm is not contracting.

b. Spasms in the bronchiole smooth muscle have blocked airflow to the alveoli.

c. Excess mucus production is interfering with airflow into the lungs.

d. The epiglottis has closed and air is not entering the lungs.

e. Insufficient surfactant is being produced.

4. Which sequence of events best describes inhalation?

a. contraction of diaphragm and external intercostals → increase in size of thoracic cavity → decrease in alveolar pressure

b. relaxation of diaphragm and external intercostals → decrease in size of thoracic cavity → increase in alveolar pressure

c. contraction of diaphragm and external intercostals → decrease in size of thoracic cavity → decrease in alveolar pressure

d. relaxation of diaphragm and external intercostals → increase in size of thoracic cavity → increase in alveolar pressure

e. contraction of diaphragm and external intercostals → decrease in size of thoracic cavity → increase in alveolar pressure

5. Which of the following does NOT help keep air passages clear?

a. nostril hairs

b. alveolar macrophages

c. capillaries in the nasal cavities

d. cilia in the upper and lower respiratory tracts

e. mucus produced by goblet cells

6. The inflation reflex

a. is initiated by overinflation of the lungs.

b. causes temporary apnea.

c. prolongs inspiration.

d. results from irritants in the breathing passages.

e. involves input from muscle and joint proprioceptors.

7. How does hypercapnia affect respiration?

a. It increases the rate of respiration.

b. It decreases the rate of respiration.

c. It causes hypoventilation.

d. It does not change the rate of respiration.

e. It activates stretch receptors in the lungs.

8. Air would flow into the lungs along the following route:

1. bronchioles   **2.** primary bronchi

3. secondary bronchi   **4.** terminal bronchioles

5. tertiary bronchi   **6.** trachea

  a. 6, 1, 2, 3, 5, 4   **b.** 6, 5, 3, 4, 2, 1   **c.** 6, 2, 3, 5, 4, 1

  d. 6, 2, 3, 5, 1, 4   **e.** 6, 1, 4, 5, 3, 2

9. Match the following:

____ **a.** normally inactive; when activated, causes contraction of internal intercostals and abdominal muscles and forced exhalation

A. inspiratory area
B. expiratory area
C. pneumotaxic area
D. apneustic area
E. cerebral cortex

____ **b.** located in pons; stimulates inspiratory area to prolong inhalation

____ **c.** sets basic rhythm of respiration; located in medulla

____ **d.** transmits inhibitory impulses to inspiratory area; located in pons

____ **e.** allows voluntary alteration of breathing patterns

10. Under normal body conditions, hemoglobin releases oxygen more readily when

a. body temperature increases.   **b.** body tissues are less active.

c. blood pH increases.   **d.** blood P_{O_2} is high.

e. blood P_{CO_2} is low.

11. Match the following:

____ **a.** decreased carbon dioxide levels

____ **b.** normal, quiet breathing

____ **c.** rapid breathing

____ **d.** exchange of gases between the pulmonary capillaries and alveoli

____ **e.** inhalation and exhalation

____ **f.** increased carbon dioxide levels

____ **g.** exchange of gases between systemic capillaries and tissue cells

____ **h.** absence of breathing

A. external respiration
B. apnea
C. hypercapnia
D. eupnea
E. internal respiration
F. hypocapnia
G. pulmonary ventilation
H. hyperventilation

12. Which of the following statements is NOT true concerning the lungs?

a. The alveoli are composed of simple squamous epithelial cells.

b. The right lung contains the cardiac notch.

c. The left lung is composed of two lobes, separated by the oblique fissure.

d. The top portion of the lung is the apex.

e. The lungs are surrounded by a serous membrane.

13. Exhalation

a. occurs when alveolar pressure reaches 758 mm Hg.

b. is normally considered an active process requiring muscle contraction.

c. occurs when alveolar pressure is greater than atmospheric pressure.

d. involves the expansion of the pleural membranes.

e. occurs when the atmospheric pressure is equal to the pressure in the lungs.

14. Where is the cricoid cartilage located?

a. larynx   **b.** pharynx   **c.** trachea

d. primary bronchi   **e.** alveoli

15. What is a primary factor that allows bronchioles to collapse more readily than the rest of the bronchial tree?

a. The bronchioles are so small.

b. Bronchioles don't contain smooth muscle.

c. There is more air present in the larger breathing tubes.

d. The rest of the bronchial tree contains supporting cartilage.

e. Bronchioles lack lubricating mucus.

16. The primary form in which oxygen is transported in the blood is
    a. carbaminohemoglobin.   b. oxyhemoglobin.
    c. attached to carbon dioxide.   d. deoxyhemoglobin.
    e. dissolved in the blood plasma.

17. Decreasing the surface area of the respiratory membrane would affect
    a. internal respiration.   b. inhalation.   c. cellular respiration.
    d. external respiration.   e. mucus production.

18. In which form is carbon dioxide NOT carried in the blood?
    a. bicarbonate ion   b. bound to amino acids
    c. oxyhemoglobin   d. carbaminohemoglobin
    e. dissolved in plasma

19. Fill in the blanks in the following chemical reactions.
    $CO_2 +$ _____ $\rightleftharpoons H_2CO_3 \rightleftharpoons H^+ +$ _____
    a. HCO_3^-, O_2
    b. HCO_3^-, H_2O
    c. H^+, H_2O
    d. O_2, HCO_3^-
    e. H_2O, HCO_3^-

20. Of the following, which would have the highest partial pressure of oxygen?
    a. alveolar air at the end of exhalation
    b. rapidly contracting skeletal muscle fibers
    c. alveolar air immediately after inhalation
    d. blood flowing into the lungs from the right side of the heart
    e. blood returning to the heart from the tissue cells

21. Match the following:
    ____ a. forceful exhalation of air
    ____ b. inspiratory reserve volume + tidal volume + expiratory reserve volume
    ____ c. volume of air moved during normal quiet breathing
    ____ d. air remaining after forced exhalation
    ____ e. forceful inhalation of air
    ____ f. breathing rate × tidal volume
    ____ g. residual volume + expiratory reserve volume

    A. vital capacity
    B. inspiratory reserve volume
    C. residual volume
    D. expiratory reserve volume
    E. functional residual capacity
    F. tidal volume
    G. minute volume

CRITICAL THINKING APPLICATIONS

1. Your three-year-old nephew Toby likes to get his own way all the time! Toby wants to eat 10 chocolate biscuits (1 for each finger), but you'll only give him one for each year of his age. He is at this moment "holding my breath until I turn blue and won't you be sorry!" Is he in danger of death?

2. Emily was diagnosed with exercise-induced asthma after she reported trouble catching her breath during a swim meet. Exercise-induced asthma is a particularly annoying condition for an athlete because the body's response to exercise is the exact opposite of the body's need. Explain this statement.

3. Zoe has a flair for being dramatic. "I can't come to work today," she whispered, "I've got laryngitis and a horrible rhinovirus." What is wrong with Zoe?

4. Your friend told a hilarious joke just as you were swallowing a mouthful of beer. Instead of laughing, you began to choke and then cough. Your friend expected laughter, but got sprayed with beer. What happened?

ANSWERS TO FIGURE QUESTIONS

18.1   The conducting zone of the respiratory system includes the nose, pharynx, larynx, trachea, bronchi, and bronchioles (except the respiratory bronchioles).

18.2   Air molecules flow through the external nares, the nasal cavity, and then the internal nares.

18.3   During swallowing, the epiglottis closes over the larynx to block food and liquids from entering.

18.4   There are two lobes and two secondary bronchi in the left lung and three lobes and three secondary bronchi in the right lung.

18.5   A lung lobule includes a lymphatic vessel, arteriole, venule, and branch of a terminal bronchiole wrapped in elastic connective tissue.

18.6   Surfactant-secreting cells secrete alveolar fluid, which includes surfactant.

18.7   The main muscles that power quiet breathing are the diaphragm and external intercostals.

18.8   Alveolar pressure is 758 mm Hg during inhalation; alveolar pressure during exhalation is 762 mm Hg.

18.9   You demonstrate vital capacity when you breathe in as deeply as possible and then exhale as much air as you can.

18.10   Oxygen enters pulmonary capillaries from alveolar air and enters tissue cells from systemic capillaries due to differences in P_{O_2}.

18.11   Hemoglobin transports about 98.5 percent of the oxygen carried in blood.

18.12   The phrenic nerves stimulate the diaphragm to contract.

18.13   Normal arterial blood P_{CO_2} is 40 mm Hg.

## CHAPTER 19

THE DIGESTIVE SYSTEM

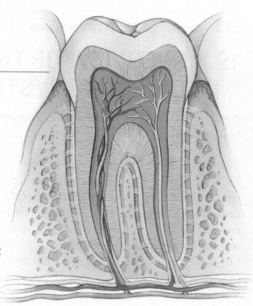

The food we eat contains a variety of nutrients, which are used for building new body tissues and repairing damaged tissues. However, most of the food we eat consists of molecules that are too large to be used by body cells. Therefore, food must be broken down into molecules that are small enough to enter body cells, a process known as *digestion*. Collectively, the organs that perform these functions are known as the *digestive system*.

The medical specialty that deals with the structure, function, diagnosis, and treatment of diseases of the stomach and intestines is *gastroenterology* (gas′-trō-en′-ter-OL-ō-jē; *gastro-* = stomach; *entero-* = intestines; *-logy* = study of). The medical specialty that deals with the diagnosis and treatment of disorders of the rectum and anus is *proctology* (prok-TOL-ō-jē; *proct-* = rectum).

Did you know?   The drive to seek and consume food is called hunger. Many factors influence a person's feelings of hunger. Needing calories is an obvious cause; if people don't eat they die, so nature has endowed people with strong hunger drives. But hunger can also be influenced by other factors. For example, a lack of sleep raises levels of an important hunger-regulating hormone called ghrelin. Perhaps when the body lacks energy because of a lack of sleep, it looks to food instead. Hunger can also be triggered by emotions such as boredom, sadness, anxiety, anger, and happiness. When emotional eating gets out of hand, the result can be obesity, or even eating disorders.

FOCUS ON WELLNESS
Emotional Eating—Consumed by Food

LOOKING BACK TO MOVE AHEAD...

Mucous Membranes (Section 4.4)

Serous Membranes (Section 4.4)

Smooth Muscle Tissue (Section 8.8)

Muscles That Move the Mandible (Section 8.11)

Negative Feedback System (Section 1.4)

Simple Columnar Epithelium (Section 4.2)

Carbohydrates, Lipids, Proteins (Section 2.2)

Enzymes (Section 2.2)

19.1 OVERVIEW OF THE DIGESTIVE SYSTEM

OBJECTIVE • **Identify the organs of the digestive system and their basic functions.**

Two groups of organs compose the digestive system (Figure 19.1): the gastrointestinal tract and the accessory digestive organs. The *gastrointestinal (GI) tract* or *alimentary canal* is a continuous tube that extends from the mouth to the anus. The GI tract contains food from the time it is eaten until it is digested and absorbed or eliminated from the body. Organs of the gastrointestinal tract include the mouth, pharynx, esophagus, stomach, small intestine, and large intestine. The length of the GI tract is about 5–7 meters (16.5–23 ft) in a living person. It is longer in a cadaver (about 7–9 meters or 23–29.5 ft) because the muscles along the wall of the GI tract organs no longer are in a state of tonus (sustained contraction). The teeth, tongue, salivary glands, liver, gallbladder, and pancreas serve as *accessory digestive organs*. Teeth aid in the physical breakdown of food, and the tongue assists in chewing and swallowing. The other accessory digestive organs never come into direct contact with food. The secretions that they produce or store flow into the GI tract through ducts and aid in the chemical breakdown of food.

Overall, the digestive system performs six basic processes:

1. **Ingestion.** This process involves taking foods and liquids into the mouth (eating).

2. **Secretion.** Each day, cells within the walls of the GI tract and accessory organs secrete a total of about 7 liters of water, acid, buffers, and enzymes into the lumen of the tract.

3. **Mixing and propulsion.** Alternating contraction and relaxation of smooth muscle in the walls of the GI tract mix food and secretions and move them toward the anus. The ability of the GI tract to mix and move material along its length is termed *motility*.

4. **Digestion.** Mechanical and chemical processes break down ingested food into small molecules. In *mechanical digestion* the teeth cut and grind food before it is swallowed, and then smooth muscles of the stomach and small intestine churn the food to further assist the process. As a result, food molecules become dissolved and thoroughly mixed with digestive enzymes. In *chemical digestion* the large carbohydrate, lipid, protein, and nucleic acid molecules in food are broken down into smaller molecules by digestive enzymes.

5. **Absorption.** The entrance of ingested and secreted fluids, ions, and small molecules that are products of digestion into the epithelial cells lining the lumen of the GI tract is called *absorption*. The absorbed substances pass into interstitial fluid and then into blood or lymph and circulate to cells throughout the body.

6. **Defecation.** Wastes, indigestible substances, bacteria, cells shed from the lining of the GI tract, and digested materials that were not absorbed leave the body through the anus in a process called *defecation*. The eliminated material is termed *feces (stool)*.

✓ CHECKPOINT

1. Which components of the digestive system are GI tract organs and which are accessory digestive organs?

2. Which organs of the digestive system come in contact with food?

19.2 LAYERS OF THE GI TRACT AND THE OMENTUM

OBJECTIVE • **Describe the four layers that form the wall of the gastrointestinal tract.**

The wall of the GI tract, from the lower esophagus to the anal canal, has the same basic, four-layered arrangement of tissues. The four layers of the tract, from the inside out, are the mucosa, submucosa, muscularis, and serosa (Figure 19.2).

1. **Mucosa.** The *mucosa*, or inner lining of the tract, is a mucous membrane. It is composed of a layer of epithelium in direct contact with the contents of the GI tract, a layer of areolar connective tissue called the *lamina propria*, and a thin layer of smooth muscle called the *muscularis mucosae*. Contractions of the muscularis mucosae create folds in the mucosa that increase the surface area for digestion and absorption. The mucosa also contains prominent lymphatic nodules that protect against the entry of pathogens through the GI tract.

2. **Submucosa.** The *submucosa* consists of areolar connective tissue that binds the mucosa to the muscularis. It contains many blood and lymphatic vessels that receive absorbed food molecules. Also located in the submucosa are networks of neurons subject to regulation by the autonomic nervous system (ANS) called the *enteric nervous system (ENS)*, the "brain of the gut." ENS neurons within the submucosa control the secretions of the organs of the GI tract.

3. **Muscularis.** As its name implies, the *muscularis* of the GI tract is a thick layer of muscle. In the mouth, pharynx, and upper esophagus, it consists in part of *skeletal muscle* that produces voluntary swallowing. Skeletal muscle also forms the external anal sphincter, which permits voluntary control of defecation. Recall that a sphincter is a thick circle of muscle around an opening. In the rest of the tract, the muscularis consists of *smooth muscle*, usually

Figure 19.1 Organs of the digestive system and related structures. Accessory digestive organs are the teeth, tongue, salivary glands, liver, gallbladder, and pancreas and are indicated with an asterisk (*).

🔑 **Organs of the gastrointestinal (GI) tract are the mouth, pharynx, esophagus, stomach, small intestine, and large intestine.**

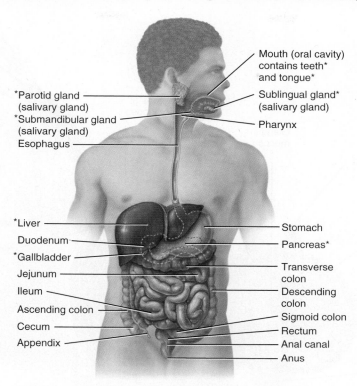

Mouth (oral cavity) contains teeth* and tongue*

*Parotid gland (salivary gland)

Sublingual gland* (salivary gland)

*Submandibular gland (salivary gland)

Pharynx

Esophagus

*Liver

Stomach

Duodenum

Pancreas*

*Gallbladder

Transverse colon

Jejunum

Ileum

Descending colon

Ascending colon

Sigmoid colon

Cecum

Rectum

Appendix

Anal canal

Anus

(a) Diagram of right lateral view of head and neck and anterior view of trunk

SUPERIOR

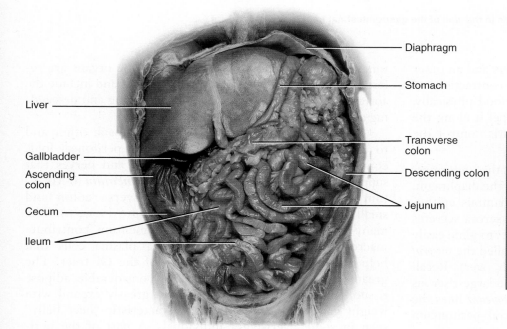

Diaphragm

Stomach

Liver

Transverse colon

Gallbladder

Descending colon

Ascending colon

Jejunum

Cecum

Ileum

Functions of the Digestive System

1. Ingestion: taking food into the mouth.
2. Secretion: release of water, acid, buffers, and enzymes into the lumen of the GI tract.
3. Mixing and propulsion: churning and pushing food through the GI tract.
4. Digestion: mechanical and chemical breakdown of food.
5. Absorption: passage of digested products from the GI tract into the blood and lymph.
6. Defecation: elimination of feces from the GI tract.

(b) Anterior view

Which accessory digestive organs assist in the mechanical breakdown of food?

Figure 19.2 Layers of the gastrointestinal tract. Variations in this basic plan may be seen in the stomach (Figure 19.8), small intestine (Figure 19.12), and large intestine (Figure 19.15).

🔑 **The four layers of the GI tract from inside to outside are the mucosa, submucosa, muscularis, and serosa.**

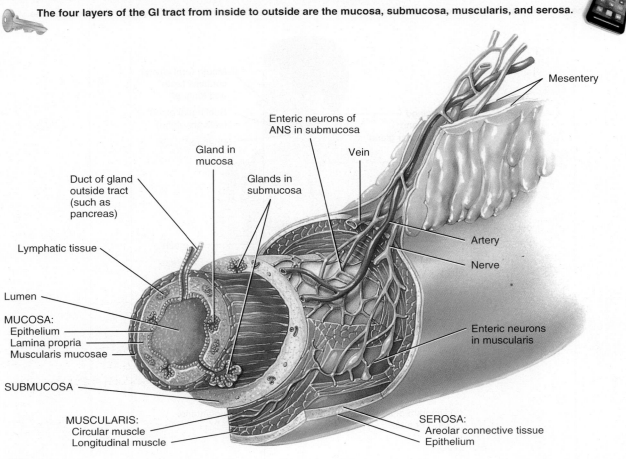

❓ **What is the function of the nerves in the wall of the gastrointestinal tract?**

arranged as an inner sheet of circular fibers and an outer sheet of longitudinal fibers. Involuntary contractions of these smooth muscles help break down food physically, mix it with digestive secretions, and propel it along the tract. ENS neurons within the muscularis control the frequency and strength of its contractions.

4. **Serosa and peritoneum.** The *serosa*, the outermost layer around organs of the GI tract below the diaphragm, is a membrane composed of simple squamous epithelium and areolar connective tissue. The serosa secretes a slippery, watery fluid that allows the tract to glide easily against other organs. The serosa is also called the *visceral peritoneum* (per′-i-tō-NE-um = to stretch over). Recall from Section 4.4 that the *peritoneum* is the largest serous membrane of the body. The *parietal peritoneum* lines the wall of the abdominal cavity; the visceral peritoneum covers organs in the cavity.

Some organs of the body lie on the posterior abdominal wall *behind* the parietal peritoneum, and are covered by peritoneum only on their anterior surfaces. The organs are referred to as *retroperitoneal* (*retro-* = behind) and include the aorta, inferior vena cava, duodenum, ascending and descending colons, kidneys, and ureters.

In addition to binding the organs to each other and to the walls of the abdominal cavity, the peritoneal folds contain blood vessels, lymphatic vessels, and nerves that supply the abdominal organs. The *greater omentum* (ō-MENtum = fat skin) drapes over the transverse colon and small intestine like a "fatty apron" (Figure 19.3a, b). The many lymph nodes of the greater omentum contribute macrophages and antibody-producing plasma cells that help combat and contain infections of the GI tract. The greater omentum normally contains considerable adipose tissue. Its adipose tissue content can greatly expand with weight gain, giving rise to the characteristic "beer belly" seen in some overweight individuals. A part of the peritoneum, the *mesentery* (MEZ-en-ter′-ē; *mes-* = middle), binds the small intestine to the posterior abdominal wall (Figure 19.3b).

Figure 19.3 Views of the abdomen and pelvis. The relationship of the parts of the peritoneum (greater omentum and mesentery) to each other and to organs of the digestive system is shown.

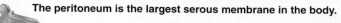

The peritoneum is the largest serous membrane in the body.

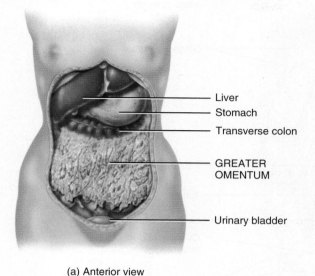

Liver
Stomach
Transverse colon
GREATER OMENTUM
Urinary bladder

(a) Anterior view

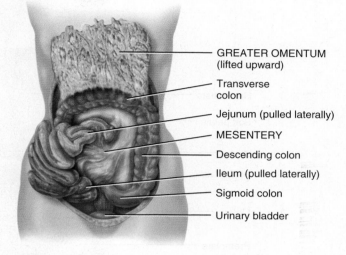

GREATER OMENTUM (lifted upward)
Transverse colon
Jejunum (pulled laterally)
MESENTERY
Descending colon
Ileum (pulled laterally)
Sigmoid colon
Urinary bladder

(b) Anterior view (greater omentum lifted and small intestine reflected to right side)

 Which part of the peritoneum binds the small intestine to the posterior abdominal wall?

CLINICAL CONNECTION | Peritonitis

A common cause of **peritonitis**, an acute inflammation of the peritoneum, is contamination of the peritoneum by infectious microbes, which can result from accidental or surgical wounds in the abdominal wall, or from perforation or rupture of abdominal organs. •

✓ CHECKPOINT

3. Where along the GI tract is the muscularis composed of skeletal muscle? Is control of this skeletal muscle voluntary or involuntary?

4. Where are the visceral peritoneum and parietal peritoneum located?

19.3 MOUTH

OBJECTIVES • **Identify the locations of the salivary glands, and describe the functions of their secretions.**

• **Outline the structure and functions of the tongue.**

• **List the parts of a typical tooth, and compare the deciduous and permanent dentitions.**

The **mouth** or **oral cavity** is formed by the cheeks, hard and soft palates, and tongue (Figure 19.4). The cheeks form the lateral walls of the oral cavity. The *lips* are fleshy folds around the opening of the mouth. Both the cheeks and lips are cov-

ered on the outside by skin and on the inside by a mucous membrane. During chewing, the lips and cheeks help keep food between the upper and lower teeth. They also assist in speech.

The *hard palate*, consisting of the maxillae and palatine bones, forms most of the roof of the mouth. The rest is formed by the muscular *soft palate*. Hanging from the soft palate is a finger-like structure called the *uvula* (Ū-vū-la). During swallowing, the uvula moves upward with the soft palate, which prevents entry of swallowed foods and liquids into the nasal cavity. At the back of the soft palate, the mouth opens into the oropharynx. The *palatine tonsils* are just posterior to the opening.

Tongue

The **tongue** forms the floor of the oral cavity. It is an accessory digestive organ composed of skeletal muscle covered with mucous membrane (see Figure 12.4).

The muscles of the tongue move food for chewing, shape the food into a rounded mass, force the food to the back of the mouth for swallowing, and alter the shape and size of the tongue for swallowing and speech. The **lingual frenulum** (LING-gwal FREN-ū-lum; *lingua* = tongue; *frenum* = bridle), a fold of mucous membrane in the midline of the undersurface of the tongue, limits the movement of the tongue posteriorly (Figure 19.4a). If a person's lingual frenulum is abnormally short or rigid, the person is said to be

Figure 19.4 Structures of the mouth (oral cavity).

The cheeks, hard and soft palates, and tongue form the mouth.

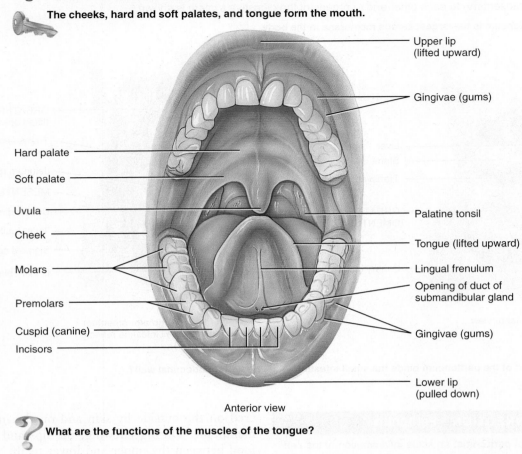

Upper lip (lifted upward)

Gingivae (gums)

Hard palate

Soft palate

Uvula

Cheek

Molars

Premolars

Cuspid (canine)

Incisors

Palatine tonsil

Tongue (lifted upward)

Lingual frenulum

Opening of duct of submandibular gland

Gingivae (gums)

Lower lip (pulled down)

Anterior view

What are the functions of the muscles of the tongue?

"tongue-tied" because of the resulting impairment to speech. It can be corrected surgically. The lingual tonsils lie at the base of the tongue (see Figure 12.4a). The upper surface and sides of the tongue are covered with projections called *papillae* (pa-PIL-ē), some of which contain taste buds. Glands in the tongue secrete an enzyme called *lingual lipase* which begins the digestion of triglycerides into fatty acids and diglycerides (glycerol plus two fatty acids) once in the acid environment of the stomach.

Salivary Glands

The three pairs of *salivary glands* (SAL-i-var-ē) are accessory organs of digestion that lie outside the mouth and release their secretion (saliva) into ducts emptying into the oral cavity (see Figure 19.1). The *parotid glands* (pa-ROT-id) are located inferior and anterior to the ears between the skin and the masseter muscle. The *submandibular glands* (sub'-man-DIB-ū-lar) are found in the floor of the mouth; they are medial and partly inferior to the mandible. The *sublingual glands* (sub-LING-gwal) are beneath the tongue and superior to the submandibular glands.

The fluid secreted by the salivary glands, called *saliva*, is composed of 99.5 percent water and 0.5 percent solutes. The water in saliva helps dissolve foods so they can be tasted and digestive reactions can begin. One of the solutes, the digestive enzyme *salivary amylase*, begins the digestion of starches in the mouth. Mucus in saliva lubricates food so it can be swallowed easily. The enzyme lysozyme kills bacteria, thereby protecting the mouth's mucous membrane from infection and the teeth from decay.

Secretion of saliva, called *salivation* (sal-i-VĀ-shun), is controlled by the autonomic nervous system. Normally, parasympathetic stimulation promotes continuous secretion of a moderate amount of saliva, which keeps the mucous membranes moist and lubricates the movements of the tongue and lips during speech. Sympathetic stimulation dominates during stress, resulting in dryness of the mouth.

Teeth

The *teeth (dentes)* are accessory digestive organs located in bony sockets of the mandible and maxillae. The sockets are covered by the *gingivae* (JIN-ji-vē; singular is *gingiva*) or *gums* and are lined with the *periodontal ligament* (per'-ē-ō-DON-tal; *peri-* = around; *odont-* = tooth). This dense fibrous connective tissue anchors the teeth to bone (Figure 19.5a).

Figure 19.5 Parts of a typical tooth.

There are 20 teeth in a complete deciduous set and 32 teeth in a complete permanent set.

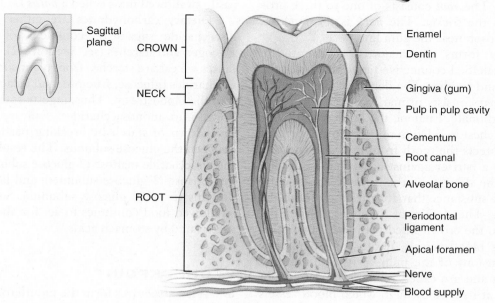

(a) Sagittal section of a mandibular (lower) molar

CROWN
NECK
ROOT

Sagittal plane

Enamel
Dentin
Gingiva (gum)
Pulp in pulp cavity
Cementum
Root canal
Alveolar bone
Periodontal ligament
Apical foramen
Nerve
Blood supply

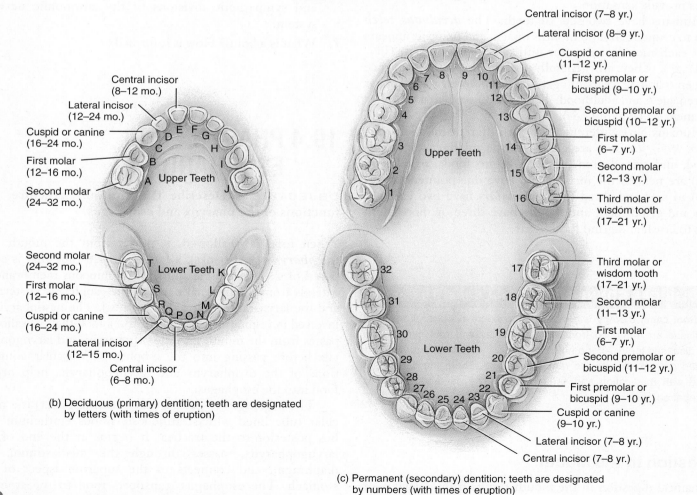

(b) Deciduous (primary) dentition; teeth are designated by letters (with times of eruption)

Central incisor (8–12 mo.)
Lateral incisor (12–24 mo.)
Cuspid or canine (16–24 mo.)
First molar (12–16 mo.)
Second molar (24–32 mo.)

Upper Teeth

Second molar (24–32 mo.)
First molar (12–16 mo.)
Cuspid or canine (16–24 mo.)
Lateral incisor (12–15 mo.)
Central incisor (6–8 mo.)

Lower Teeth

(c) Permanent (secondary) dentition; teeth are designated by numbers (with times of eruption)

Central incisor (7–8 yr.)
Lateral incisor (8–9 yr.)
Cuspid or canine (11–12 yr.)
First premolar or bicuspid (9–10 yr.)
Second premolar or bicuspid (10–12 yr.)
First molar (6–7 yr.)
Second molar (12–13 yr.)
Third molar or wisdom tooth (17–21 yr.)

Upper Teeth

Third molar or wisdom tooth (17–21 yr.)
Second molar (11–13 yr.)
First molar (6–7 yr.)
Second premolar or bicuspid (11–12 yr.)
First premolar or bicuspid (9–10 yr.)
Cuspid or canine (9–10 yr.)
Lateral incisor (7–8 yr.)
Central incisor (7–8 yr.)

Lower Teeth

What type of tissue is the main component of teeth?

A typical tooth has three major external regions: the crown, root, and neck. The *crown* is the visible portion above the level of the gums. The *root* consists of one to three projections embedded in the socket. The *neck* is the junction line of the crown and root, near the gum line.

Internally, *dentin* forms the majority of the tooth. Dentin consists of a calcified connective tissue that gives the tooth its basic shape and rigidity. The dentin of the crown is covered by *enamel* that consists primarily of calcium phosphate and calcium carbonate. Enamel, the hardest substance in the body and the richest in calcium salts (about 95 percent of its dry weight), protects the tooth from the wear and tear of chewing. It is also a barrier against acids that easily dissolve the dentin. The dentin of the root is covered by *cementum*, a bonelike substance that attaches the root to the periodontal ligament. The dentin of a tooth encloses the *pulp cavity*, a space in the crown filled with *pulp*, a connective tissue containing blood vessels, nerves, and lymphatic vessels. Narrow extensions of the pulp cavity run through the root of the tooth and are called *root canals*. Each root canal has an opening at its base through which blood vessels bring nourishment, lymphatic vessels offer protection, and nerves provide sensation.

Humans have two sets of teeth. The *deciduous teeth* begin to erupt at about 6 months of age, and one pair appears about each month thereafter until all 20 are present (Figure 19.5b). They are generally lost in the same sequence between 6 and 12 years of age. The *permanent teeth* appear between age 6 and adulthood. There are 32 teeth in a complete permanent set (Figure 19.5c).

Humans have different teeth for different functions (see Figure 19.4). *Incisors* are closest to the midline, are chisel-shaped, and are adapted for cutting into food; *cuspids* (canines) are next to the incisors and have one pointed surface (cusp) to tear and shred food; *premolars* have two cusps to crush and grind food; and *molars* have three or more blunt cusps to crush and grind food.

CLINICAL CONNECTION | Root Canal Therapy

Root canal therapy is a multistep procedure in which all traces of pulp tissue are removed from the pulp cavity and root canals of a badly diseased tooth. After a hole is made in the tooth, the root canals are filed out and irrigated to remove bacteria. Then, the canals are treated with medication and sealed tightly. The damaged crown is then repaired. •

Digestion in the Mouth

Mechanical digestion in the mouth results from chewing, or *mastication* (mas'-ti-KĀ-shun = to chew), in which food is manipulated by the tongue, ground by the teeth, and mixed with saliva. As a result, the food is reduced to a soft, flexible, easily swallowed mass called a *bolus* (= lump).

Dietary carbohydrates are either monosaccharide and disaccharide sugars or complex polysaccharides such as glycogen and starches (see Section 2.2). Most of the carbohydrates we eat are starches from plant sources, but only monosaccharides (glucose, fructose, and galactose) can be absorbed into the bloodstream. Thus, ingested starches must be broken down into monosaccharides. Salivary amylase begins the breakdown of starch by breaking particular chemical bonds between the glucose subunits. The resulting products include the disaccharide maltose (2 glucose subunits), the trisaccharide maltotriose (3 glucose subunits), and larger fragments called dextrins (5 to 10 glucose subunits). Salivary amylase in the swallowed food continues to act for about an hour until it is inactivated by stomach acids.

✓ CHECKPOINT

5. What structures form the mouth (oral cavity)?

6. How is saliva secretion regulated by the parasympathetic and sympathetic divisions of the autonomic nervous system?

7. What is a bolus? How is it formed?

19.4 PHARYNX AND ESOPHAGUS

OBJECTIVE • **Describe the location, structure, and functions of the pharynx and esophagus.**

When food is swallowed, it passes from the mouth into the *pharynx* (FAIR-inks), a funnel-shaped tube that is composed of skeletal muscle and lined by mucous membrane. It extends from the internal nares to the esophagus posteriorly and the larynx anteriorly (Figure 19.6a). The nasopharynx is involved in respiration (see Figure 18.2); food that is swallowed passes from the mouth into the oropharynx and laryngopharynx before passing into the esophagus. Muscular contractions of the oropharynx and laryngopharynx help propel food into the esophagus.

The *esophagus* (e-SOF-a-gus = eating gullet) is a muscular tube lined with stratified squamous epithelium that lies posterior to the trachea. It begins at the end of the laryngopharynx, passes through the mediastinum and diaphragm, and connects to the superior aspect of the stomach. The esophagus transports food to the stomach and secretes mucus. At each end of the esophagus, the muscularis forms two sphincters—the ***upper esophageal sphincter (UES)*** (e-sof′-a-JĒ-al), which consists of skeletal muscle, and the ***lower esophageal sphincter (LES)*** or ***cardiac sphincter***, which consists of smooth muscle and is near the heart. The upper esophageal sphincter regulates the movement of food from the pharynx into the esophagus; the lower esophageal sphincter regulates the movement of food from the esophagus into the stomach.

> ### CLINICAL CONNECTION | Gastroesophageal Reflux Disease (GERD)
>
> If the lower esophageal sphincter fails to close adequately after food has entered the stomach, the stomach contents can reflux (back up) into the inferior portion of the esophagus. This condition is known as **gastroesophageal reflux disease (GERD)**. Hydrochloric acid (HCl) from the stomach contents can irritate the esophageal wall, resulting in a burning sensation that is called **heartburn** because it is experienced in a region very near the heart; however, it is unrelated to any cardiac problem. Drinking alcohol and smoking can cause the sphincter to relax, worsening the problem. The symptoms of GERD often can be controlled by avoiding foods that strongly stimulate stomach acid secretion (coffee, chocolate, tomatoes, fatty foods, orange juice, peppermint, spearmint, and onions). Other acid-reducing strategies include taking over-the-counter histamine-2 (H_2) blockers such as Tagamet HB® or Pepcid AC® 30 to 60 minutes before eating to block acid secretion, and neutralizing acid that has already been secreted with antacids such as Tums® or Maalox®. GERD may be associated with cancer of the esophagus. •

Swallowing or ***deglutition*** (dē′-glū-TISH-un), the movement of food from the mouth to the stomach, involves the mouth, pharynx, and esophagus and is helped by saliva and mucus. Swallowing is divided into three stages: voluntary, pharyngeal, and esophageal.

In the ***voluntary stage*** of swallowing, the bolus is forced to the back of the mouth cavity and into the oropharynx by the movement of the tongue upward and backward against the palate. With the passage of the bolus into the oropharynx, the involuntary ***pharyngeal stage*** of swallowing begins (Figure 19.6b). Breathing is temporarily interrupted when the soft palate and uvula move upward to close off the nasopharynx, the epiglottis seals off the larynx, and the vocal cords come together. After the bolus passes through the oropharynx, the respiratory passageways reopen and breathing resumes. Once the upper esophageal sphincter relaxes, the bolus moves into the esophagus.

Figure 19.6 Swallowing. During the pharyngeal stage of swallowing (b), the tongue rises against the palate, the nasopharynx is closed off, the larynx rises, the epiglottis seals off the larynx, and the bolus passes into the esophagus. During the esophageal stage of swallowing (c), food moves through the esophagus into the stomach via peristalsis.

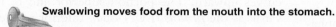 **Swallowing moves food from the mouth into the stomach.**

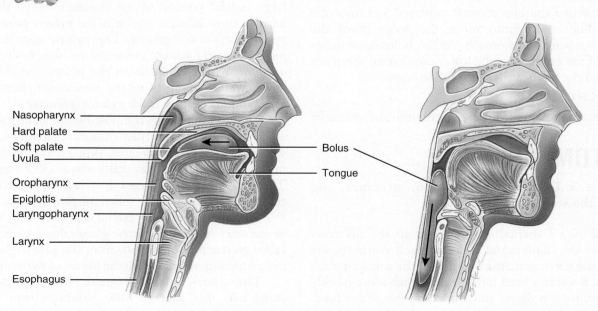

(a) Position of structures before swallowing

(b) During the pharyngeal stage of swallowing

FIGURE 19.6 CONTINUES

Figure 19.6 CONTINUED

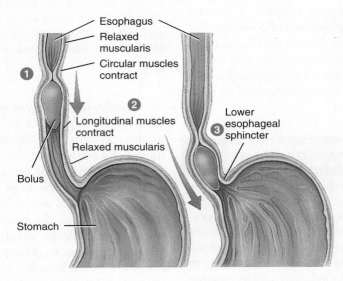

(c) Anterior view of frontal sections of peristalsis in esophagus

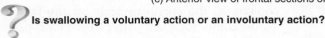

Is swallowing a voluntary action or an involuntary action?

In the *esophageal stage*, food is pushed through the esophagus by a process called *peristalsis* (per′-i-STAL-sis; *stalis* = constriction) (Figure 19.6c):

1. The circular muscle fibers in the section of esophagus above the bolus contract, constricting the wall of the esophagus and squeezing the bolus downward.
2. Longitudinal muscle fibers around the bottom of the bolus contract, shortening the section of the esophagus below the bolus and pushing its walls outward.
3. After the bolus moves into the new section of the esophagus, the circular muscles above it contract, and the cycle repeats. The contractions move the bolus down the esophagus toward the stomach. As the bolus approaches the end of the esophagus, the lower esophageal sphincter relaxes and the bolus moves into the stomach.

✓ **CHECKPOINT**

8. How does a bolus pass from the mouth into the stomach?

19.5 STOMACH

OBJECTIVE • **Describe the location, structure, and functions of the stomach.**

The *stomach* is a J-shaped enlargement of the GI tract directly below the diaphragm. The stomach connects the esophagus to the duodenum, the first part of the small intestine (Figure 19.7). Because a meal can be eaten much more quickly than the intestines can digest and absorb it, one of the functions of the stomach is to serve as a mixing chamber and holding reservoir. At appropriate intervals after food is ingested, the stomach forces a small quantity of material into the duodenum. The position and size of the stomach vary continually; the diaphragm pushes it inferiorly with each inhalation and

pulls it superiorly with each exhalation. The stomach is the most elastic part of the GI tract and can accommodate a large quantity of food, up to about 6.4 liters (6 qt).

Structure of the Stomach

The stomach has four main regions: cardia, fundus, body, and pyloric part (Figure 19.7). The *cardia* (CAR-dē-a) surrounds the superior opening of the stomach. The stomach then curves upward. The portion superior and to the left of the cardia is the *fundus* (FUN-dus). Inferior to the fundus is the large central portion of the stomach, called the *body*. The narrow, most inferior region is the *pyloric part* (pĪ-LOR-ik; *pyl-* = gate; *-orus* = guard). The pyloric part consists of the *pyloric canal*, which connects to the body; the *pyloric antrum*, which connects to the pyloric canal; and the *pylorus*, which connects to the duodenum. Between the pylorus and duodenum is the *pyloric sphincter*.

The stomach wall is composed of the same four basic layers as the rest of the GI tract (mucosa, submucosa, muscularis, serosa), with certain differences. When the stomach is empty, the mucosa lies in large folds, called *rugae* (ROO-gē = wrinkles). The surface of the mucosa is a layer of nonciliated simple columnar epithelial cells called *surface mucous cells* (Figure 19.8). Epithelial cells also extend downward and form columns of secretory cells called *gastric glands* that line narrow channels called *gastric pits*. Secretions from the gastric glands flow into the gastric pits and then into the lumen of the stomach.

The gastric glands contain three types of *exocrine gland cells* that secrete their products into the stomach lumen: mucous neck cells, chief cells, and parietal cells (Figure 19.8). Both surface mucous cells and *mucous neck cells* secrete mucus. The *chief cells* secrete an inactive gastric enzyme called *pepsinogen*. *Parietal cells* produce hydrochloric acid, which kills many microbes in food and

Figure 19.7 External and internal anatomy of the stomach. The dashed lines indicate the approximate borders of the regions of the stomach.

The four regions of the stomach are the cardia, fundus, body, and pylorus.

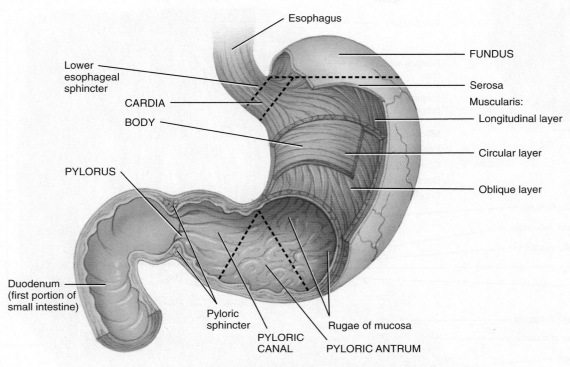

(a) Anterior view of external and internal anatomy

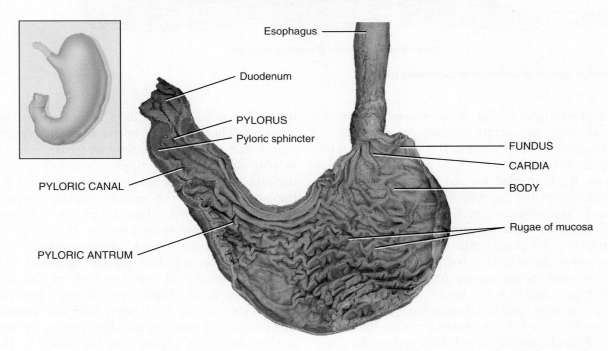

(b) Anterior view of internal anatomy

 Does your stomach still have rugae after a very big meal?

Figure 19.8 Layers of the stomach.

Secretions from the gastric glands flow into the gastric pits and then into the lumen of the stomach.

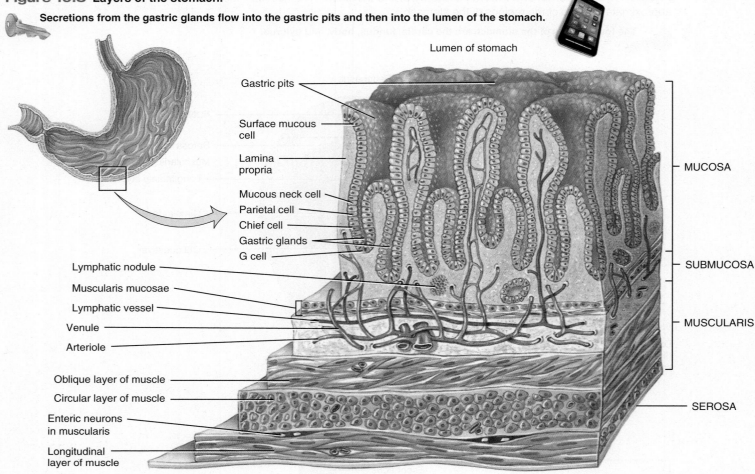

Three-dimensional view of layers of stomach

 Which stomach layer is in contact with swallowed food?

helps convert pepsinogen to the active digestive enzyme *pepsin*. Parietal cells also secrete *intrinsic factor*, which is involved in the absorption of vitamin B$_{12}$. Inadequate production of intrinsic factor can result in pernicious anemia because vitamin B$_{12}$ is needed for red blood cell production. The secretions of the mucous neck cells, chief cells, and parietal cells are collectively called *gastric juice*. The *G cells*, a fourth type of cell in the gastric glands, secrete the hormone *gastrin* into the bloodstream.

The submucosa of the stomach is composed of areolar connective tissue that connects the mucosa to the muscularis. The muscularis has three rather than two layers of smooth muscle: an outer longitudinal layer, a middle circular layer, and an inner oblique layer (see Figure 19.7). The serosa covering the stomach, composed of simple squamous epithelium and areolar connective tissue, is part of the visceral peritoneum.

Digestion and Absorption in the Stomach

Once food reaches the stomach, the stomach wall is stretched and the pH of the stomach contents increases because proteins in food have buffered some of the stomach acid. These changes in the stomach trigger nerve impulses that stimulate the flow of gastric juice and initiate *mixing waves*, gentle, rippling peristaltic movements of the muscularis. These waves macerate food and mix it with the secretions of the gastric glands, producing *chyme* (KĪM = juice), a thick liquid with the consistency of pea soup. Each mixing wave forces a small amount of chyme through the partially closed pyloric sphincter into the duodenum, a process called *gastric emptying*. Most of the chyme is forced back into the body of the stomach. The next mixing wave pushes chyme forward again and forces a little more into the duodenum. After the stomach has emptied some of its contents into the duodenum, reflexes begin to slow the exit of chyme from the stomach. This prevents overloading of the duodenum with more chyme than it can handle. Foods rich in carbohydrate spend the least time in the stomach; high-protein foods remain somewhat longer, and gastric emptying is slowest after a meal containing large amounts of fat.

Vomiting is the forcible expulsion of the contents of the upper GI tract (stomach and sometimes duodenum) through the mouth. The strongest stimuli for vomiting are irritation and excessive distension of the stomach. Other stimuli include unpleasant sights, general anesthesia, dizziness, and certain drugs such as morphine. Prolonged vomiting, especially in infants and elderly people, can be serious because the loss of acidic gastric juice can lead to alkalosis (higher than normal blood pH), dehydration, and damage to the esophagus and teeth. •

The main event of chemical digestion in the stomach is the beginning of protein digestion by the enzyme pepsin, which breaks peptide bonds between the amino acids of proteins. As a result, the proteins become fragmented into *peptides*, smaller strings of amino acids. Pepsin is most effective in the very acidic environment of the stomach, which has a pH of 2. What keeps pepsin from digesting the protein in stomach cells along with the food? First, recall that chief cells secrete pepsin in an inactive form (pepsinogen). It is not converted into active pepsin until it contacts hydrochloric acid in gastric juice. Second, mucus secreted by mucous cells coats the mucosa, forming a thick barrier between the cells of the stomach lining and the gastric juice. Lingual lipase produced by the tongue digests triglycerides into fatty acids and diglycerides in the acid environment of the stomach.

The epithelial cells of the stomach are impermeable to most materials, so little absorption occurs. However, mucous cells of the stomach absorb some water, ions, and short-chain fatty acids, as well as certain drugs (especially aspirin) and alcohol.

✓ CHECKPOINT

9. What are the components of gastric juice?

10. What is the role of pepsin? Why is it secreted in an inactive form?

11. What substances are absorbed in the stomach?

19.6 PANCREAS

OBJECTIVE • **Describe the location, structure, and functions of the pancreas.**

From the stomach, chyme passes into the small intestine. Because chemical digestion in the small intestine depends on activities of the pancreas, liver, and gallbladder, we first consider these accessory digestive organs and their contributions to digestion in the small intestine.

Structure of the Pancreas

The **pancreas** (*pan-* = all; *-creas* = flesh) lies behind the stomach (see Figure 19.1). Secretions pass from the pancreas to the duodenum via the **pancreatic duct**, which unites with the common bile duct from the liver and gallbladder, forming the hepatopancreatic duct, which enters the duodenum (Figure 19.9).

The pancreas is made up of small clusters of glandular epithelial cells, most of which are arranged in clusters called **acini** (AS-i-nī). The acini constitute the *exocrine* portion of the organ (see Figure 13.11). The cells within acini secrete a mixture of fluid and digestive enzymes called **pancreatic juice**. The remaining 1 percent of the cells are organized into clusters called **pancreatic islets (islets of Langerhans)**, the *endocrine* portion of the pancreas. These cells secrete the hormones glucagon, insulin, somatostatin, and pancreatic polypeptide, which are discussed in Section 13.6.

Pancreatic Juice

Pancreatic juice is a clear, colorless liquid that consists mostly of water, some salts, sodium bicarbonate, and enzymes. The bicarbonate ions give pancreatic juice a slightly alkaline pH (7.1 to 8.2), which inactivates pepsin from the stomach and creates the optimal environment for activity of enzymes in the small intestine. The enzymes in pancreatic juice include a starch-digesting enzyme called **pancreatic amylase**; several protein-digesting enzymes including **trypsin** (TRIP-sin), **chymotrypsin** (kī′-mō-TRIP-sin), and **carboxypeptidase** (kar-bok′-sē-PEP-ti-dās); the main triglyceride-digesting enzyme in adults, called **pancreatic lipase**; and nucleic acid–digesting enzymes called **ribonuclease** (rī′-bō-NOO-klē-ās) and **deoxyribonuclease** (dē-oks-ē-rī′-bō-NOO-klē-ās). The protein-digesting enzymes are produced in an inactive form, which prevents them from digesting the pancreas itself. Upon reaching the small intestine, the inactive form of trypsin is activated by an enzyme called **enterokinase**. In turn, trypsin activates the other protein-digesting pancreatic enzymes.

Pancreatic cancer usually affects people over 50 years of age and occurs more frequently in males. Typically, there are few symptoms until the disorder reaches an advanced stage and often not until it has metastasized to other parts of the body such as the lymph nodes, liver, or lungs. The disease is nearly always fatal and is the fourth most common cause of death from cancer in the United States. Pancreatic cancer has been linked to fatty foods, high alcohol consumption, genetic factors, smoking, and chronic *pancreatitis* (inflammation of the pancreas). •

✓ CHECKPOINT

12. What are the pancreatic acini? How do their functions differ from those of the pancreatic islets?

13. What is the role of pancreatic juice in digestion?

Figure 19.9   Relation of the pancreas to the liver, gallbladder, and duodenum. The inset shows details of the common bile duct and pancreatic duct forming the common duct.

🔑 **Pancreatic juice in the pancreatic duct and bile in the common bile duct both flow into the common duct to the duodenum.**

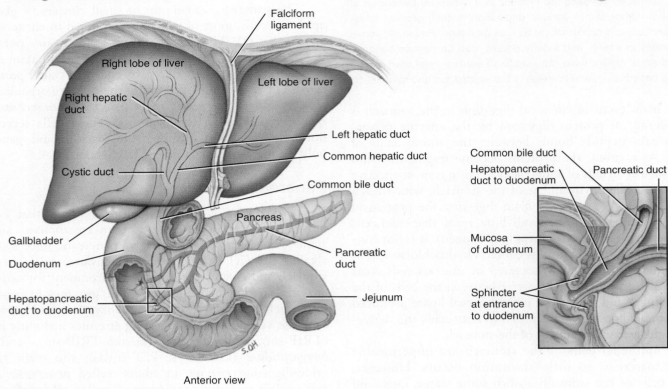

Anterior view

❓ **What substances are present in pancreatic juice?**

19.7 LIVER AND GALLBLADDER

OBJECTIVE • Describe the location, structure, and functions of the liver and gallbladder.

In an average adult, the *liver* weighs 1.4 kg (about 3 lb) and, after the skin, is the second largest organ of the body. It is located below the diaphragm, mostly on the right side of the body. A connective tissue capsule covers the liver, which in turn is covered by peritoneum, the serous membrane that covers all of the viscera. The *gallbladder* (*gall-* = bile) is a pear-shaped sac that hangs from the lower front margin of the liver (Figure 19.9).

Structure of the Liver and Gallbladder

Microscopically, the liver consists of several components (Figure 19.10):

1. *Hepatocytes* (*hepat-* = liver; *-cytes* = cells). These are the major functional cells of the liver; they perform metabolic, secretory, and endocrine functions.

2. *Bile canaliculi* (kan-a-LIK-ū-li = small canals). These are small ducts between hepatocytes that collect bile

produced by the hepatocytes. From bile canaliculi, bile passes into *bile ducts*. The bile ducts merge and eventually form the larger *right* and *left hepatic ducts*, which unite and exit the liver as the *common hepatic duct*. The common hepatic duct joins the *cystic duct* (*cystic* = bladder) from the gallbladder to form the *common bile duct*. From here, bile enters the *hepatopancreatic duct* to enter the duodenum of the small intestine to participate in digestion (see Figure 19.9). When the small intestine is empty, the sphincter around the hepatopancreatic duct at the entrance to the duodenum closes, and bile backs up into the cystic duct to the gallbladder for storage.

3. *Hepatic sinusoids.* These are highly permeable blood capillaries between rows of hepatocytes that receive oxygenated blood from branches of the hepatic artery and nutrient-rich deoxygenated blood from branches of the hepatic portal vein. Recall that the hepatic portal vein brings venous blood from the gastrointestinal organs into the liver. Hepatic sinusoids converge and deliver blood into a *central vein*. From central veins the blood flows into the *hepatic veins*, which drain into the inferior vena cava (see Figure 16.16). Also present in the hepatic sinusoids are fixed phagocytes called *stellate reticuloendothelial*

Figure 19.10 Structure of the liver.

 A liver lobule consists of hepatocytes arranged around a central vein.

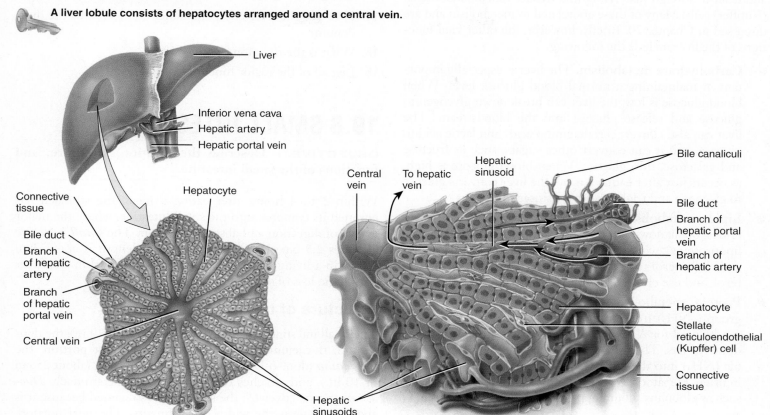

(a) Overview of microscopic components of liver

(b) Details of microscopic components of liver

Which cells in the liver are phagocytes?

(Kupffer) cells (STEL-āt re-tik′-ū-lō-en′-dō-THĒ-lē-al), which destroy worn-out white and red blood cells, bacteria, and other foreign matter in the venous blood draining from the gastrointestinal tract.

Bile

Bile salts in *bile* aid in *emulsification* (ē-mul′-si-fi-KĀ-shun), the breakdown of large lipid globules into a suspension of small lipid globules, and in absorption of lipids following their digestion. The small lipid globules formed as a result of emulsification present a very large surface area so that pancreatic lipase can digest them rapidly. The principal bile pigment is *bilirubin*, which is derived from heme. When worn-out red blood cells are broken down, iron, globin, and bilirubin are released. The iron and globin are recycled, but some of the bilirubin is excreted in bile. Bilirubin eventually is broken down in the intestine, and one of its breakdown products (stercobilin) gives feces their normal brown color (see Figure 14.3). After they have served as emulsifying agents, most bile salts are reabsorbed by active transport in the final portion of the small intestine (ileum) and enter the blood flowing toward the liver.

CLINICAL CONNECTION | Gallstones

If bile contains either insufficient bile salts or lecithin or excessive cholesterol, the cholesterol may crystallize to form **gallstones**. As they grow in size and number, gallstones may cause minimal or complete obstruction to the flow of bile from the gallbladder into the duodenum. Treatment consists of using gallstone-dissolving drugs, lithotripsy (shockwave therapy), or surgery. For people with a history of gallstones or for whom drugs or lithotripsy are not options, **cholecystectomy** (kō′-lē-sis-TEK-tō-mē)—the removal of the gallbladder and its contents— is necessary. More than half a million cholecystectomies are performed each year in the United States. To prevent side effects resulting from a loss of the gallbladder, patients should make lifestyle and dietary changes, including the following: (1) limiting the intake of saturated fat; (2) avoiding the consumption of alcoholic beverages; (3) eating smaller amounts of food during a meal and eating five to six smaller meals per day instead of two to three larger meals; and (4) taking vitamin and mineral supplements. •

Functions of the Liver

The liver performs many other vital functions in addition to the secretion of bile and bile salts and the phagocytosis of bacteria and dead or foreign material by the stellate reticuloendothelial (Kupffer) cells. Many of these are related to metabolism and are discussed in Chapter 20. Briefly, however, the other vital functions of the liver include the following:

- **Carbohydrate metabolism.** The liver is especially important in maintaining a normal blood glucose level. When blood glucose is low, the liver can break down glycogen to glucose and release glucose into the bloodstream. The liver can also convert certain amino acids and lactic acid to glucose, and it can convert other sugars, such as fructose and galactose, into glucose. When blood glucose is high, as occurs just after eating a meal, the liver converts glucose to glycogen and triglycerides for storage.

- **Lipid metabolism.** Hepatocytes store some triglycerides; break down fatty acids to generate ATP; synthesize lipoproteins, which transport fatty acids, triglycerides, and cholesterol to and from body cells; synthesize cholesterol; and use cholesterol to make bile salts.

- **Protein metabolism.** Hepatocytes remove the amino group ($-NH_2$) from amino acids so that the amino acids can be used for ATP production or converted to carbohydrates or fats. They also convert the resulting toxic ammonia (NH_3) into the much less toxic urea, which is excreted in urine. Hepatocytes also synthesize most plasma proteins, such as globulins, albumin, prothrombin, and fibrinogen.

- **Processing of drugs and hormones.** The liver can detoxify substances such as alcohol or secrete drugs such as penicillin, erythromycin, and sulfonamides into bile. It can also inactivate thyroid hormones and steroid hormones such as estrogens and aldosterone.

- **Excretion of bilirubin.** Bilirubin, derived from the heme of aged red blood cells, is absorbed by the liver from the blood and secreted into bile. Most of the bilirubin in bile is metabolized in the small intestine by bacteria and eliminated in feces.

- **Storage of vitamins and minerals.** In addition to storing glycogen, the liver stores certain vitamins (A, D, E, and K) and minerals (iron and copper), which are released from the liver when needed elsewhere in the body.

- **Activation of vitamin D.** The skin, liver, and kidneys participate in synthesizing the active form of vitamin D.

CLINICAL CONNECTION | Liver Function Tests

Liver function tests are blood tests designed to determine the presence of certain chemicals (enzymes and proteins) released by liver cells. These tests are used to evaluate and monitor liver disease or damage. Common causes of elevated liver enzymes include nonsteroidal anti-inflammatory drugs, cholesterol-lowering medications, some antibiotics, alcohol, diabetes, infections (viral hepatitis and mononucleosis), gallstones, tumors of the liver, and excessive use of herbal supplements such as kava, comfrey, pennyroyal, dandelion root, skullcap, and ephedra. •

✓ **CHECKPOINT**

14. How are the liver and gallbladder connected to the duodenum?

15. What is the function of bile?

16. List all of the major functions of the liver.

19.8 SMALL INTESTINE

OBJECTIVE • **Describe the location, structure, and functions of the small intestine.**

Within 2 to 4 hours after eating a meal, the stomach has emptied its contents into the small intestine, where the major events of digestion and absorption occur. The **small intestine** averages 2.5 cm (1 in.) in diameter; its length is about 3 m (10 ft) in a living person and about 6.5 m (21 ft) in a cadaver due to the loss of smooth muscle tone after death.

Structure of the Small Intestine

The small intestine has three portions (Figure 19.11a): the duodenum, the jejunum, and the ileum. The first portion, the **duodenum** (doo′-ō-DĒ-num), is the shortest part (about 25 cm or 10 in.), and attaches to the pylorus of the stomach. *Duodenum* means "twelve"; the structure is so named because it is about as long as the width of 12 fingers. The next portion, the **jejunum** (je-JOO-num = empty), is about 1 m (3 ft) long and is so named because it is empty at death. It is mostly in the left upper quadrant (see Figure 1.11). The final portion of the small intestine, the **ileum** (IL-e-um = twisted), measures about 2 m (6 ft) and joins the large intestine at the **ileocecal sphincter** or **ileocecal valve** (il′-ē-ō-SĒ-kal). The ileum is mostly in the right lower quadrant.

The wall of the small intestine is composed of the same four layers that make up most of the GI tract: mucosa, submucosa, muscularis, and serosa (Figure 19.12). The epithelial layer of the small intestinal mucosa consists of simple columnar epithelium that contains many types of cells. **Absorptive cells** of the epithelium release enzymes to digest food and contain microvilli to absorb nutrients in small intestinal chyme. Also present in the epithelium are **goblet cells**, which secrete mucus. The small intestinal mucosa contains **intestinal glands**, which are deep crevices lined by epithelial cells that secrete intestinal juice. Besides absorptive cells and goblet cells, intestinal glands also contain three types of endocrine cells that secrete hormones into the bloodstream: **S cells**, **CCK cells**, and **K cells**, which secrete **secretin** (se-KRĒ-tin), **cholecystokinin** (**CCK**) (kō-le-sis′-tō-KĪN-in), and **glucose-dependent insulinotropic peptide** (**GIP**), respectively (see Table 19.2 for secretin and CCK and Table 13.3 for GIP). The lamina propria of the small intestinal mucosa contains areolar connective tissue that has an abundance of lymphatic tissue, which helps defend against pathogens in

Figure 19.11 External and internal anatomy of the small intestine. (a) Portions of the small intestine are the duodenum, jejunum, and ileum. (b) Circular folds increase the surface area for digestion and absorption in the small intestine.

Most digestion and absorption occur in the small intestine.

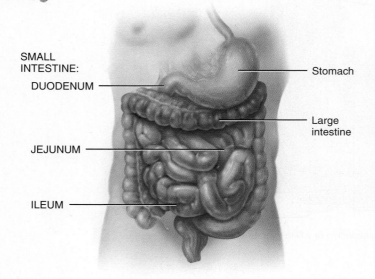

SMALL INTESTINE:
DUODENUM
JEJUNUM
ILEUM

Stomach
Large intestine

(a) Anterior view of external anatomy

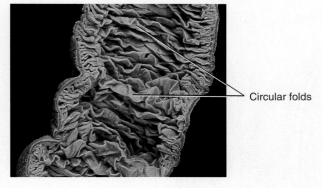

Circular folds

(b) Internal anatomy of the jejunum

Which segment of the small intestine is the longest?

food. The submucosa of the duodenum contains **duodenal glands** that secrete an alkaline mucus. It helps neutralize gastric acid in the chyme. The muscularis of the small intestine consists of two layers of smooth muscle—an outer longitudinal layer and an inner circular layer. The serosa is composed of simple squamous epithelium and areolar connective tissue.

Even though the wall of the small intestine is composed of the same four basic layers as the rest of the GI tract, special structural features of the small intestine facilitate the process of digestion and absorption. These structural features include circular folds, villi, and microvilli. **Circular folds** are permanent ridges of the mucosa and submucosa that enhance absorption by increasing surface area and causing the chyme to spiral, rather than move in a straight line, as it passes through the

small intestine (Figure 19.11b). Also present in the small intestine are numerous **villi** (= tufts of hair; singular is *villus*), fingerlike projections of the mucosa that increase the surface area of the intestinal epithelium (Figure 19.12). Each villus consists of a layer of simple columnar epithelium surrounding a core of lamina propria. Within the core are an arteriole, a venule, a blood capillary network, and a **lacteal** (LAK-tē-al = milky), which is a lymphatic capillary. Nutrients absorbed by the epithelial cells covering the villus pass through the wall of a capillary or a lacteal to enter blood or lymph, respectively. Besides circular folds and villi, the small intestine also has **microvilli** (mī-krō-VIL-ī; *micro-* = small), tiny projections of the plasma membrane of absorptive cells that increase the surface area of these cells (see Figure 19.13a). When viewed through a light microscope, the microvilli are too small to be seen individually; instead they form a fuzzy line, called the **brush border**, extending into the lumen of the small intestine. Because the microvilli greatly increase the surface area of the plasma membrane, larger amounts of digested nutrients can diffuse into absorptive cells in a given period.

Intestinal Juice

Intestinal juice, secreted by the intestinal glands, is a watery clear yellow fluid with a slightly alkaline pH of 7.6 that contains some mucus. The alkaline pH of intestinal juice is due to the high bicarbonate ion (HCO_3^-) content of pancreatic juice. Together, pancreatic and intestinal juices provide a liquid medium that aids absorption of substances from chyme as they come in contact with the microvilli. Intestinal enzymes are synthesized in the absorptive cells that line the villi (brush border enzymes). Most digestion by enzymes of the small intestine occurs in or on the surface of these absorptive cells.

Mechanical Digestion in the Small Intestine

Two types of movements contribute to intestinal motility in the small intestine: segmentations and peristalsis. **Segmentations** are localized contractions that slosh chyme back and forth, mixing it with digestive juices and bringing food particles into contact with the mucosa for absorption. The movements are similar to alternately squeezing the middle and the ends of a capped tube of toothpaste. They do not push the intestinal contents along the tract.

After most of a meal has been absorbed, segmentation stops; peristalsis begins in the lower portion of the stomach and pushes chyme forward along a short stretch of small intestine. The peristaltic wave slowly migrates down the small intestine, reaching the end of the ileum in 90 to 120 minutes. Then another wave of peristalsis begins in the stomach. Altogether, chyme remains in the small intestine for 3 to 5 hours.

Chemical Digestion in the Small Intestine

The chyme entering the small intestine contains partially digested carbohydrates and proteins. The completion of

Figure 19.12 Structure of the small intestine.

🔑 Circular folds, villi, and microvilli increase the surface area for digestion and absorption in the small intestine.

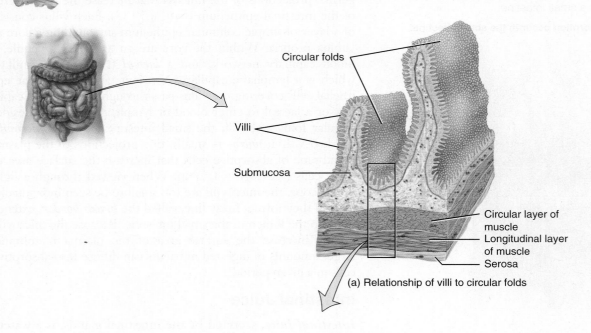

- Circular folds
- Villi
- Submucosa
- Circular layer of muscle
- Longitudinal layer of muscle
- Serosa

(a) Relationship of villi to circular folds

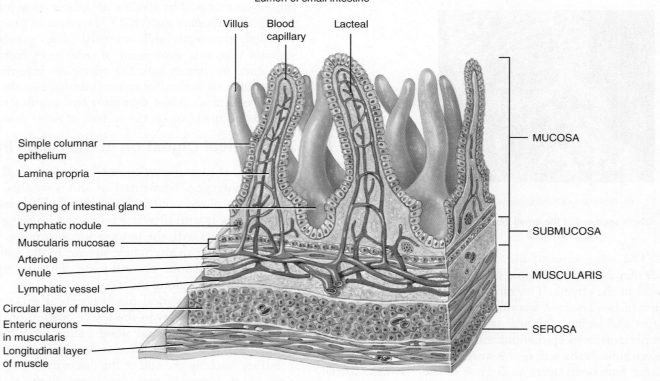

Lumen of small intestine

- Villus
- Blood capillary
- Lacteal
- Simple columnar epithelium
- Lamina propria
- Opening of intestinal gland
- Lymphatic nodule
- Muscularis mucosae
- Arteriole
- Venule
- Lymphatic vessel
- Circular layer of muscle
- Enteric neurons in muscularis
- Longitudinal layer of muscle

- MUCOSA
- SUBMUCOSA
- MUSCULARIS
- SEROSA

(b) Three-dimensional view of layers of the small intestine showing villi

 Where are the cells that absorb dietary nutrients located?

digestion in the small intestine is a collective effort of pancreatic juice, bile, and intestinal juice. Once digestion is completed, the final products of digestion are ready for absorption.

Starches and dextrins not reduced to maltose by the time chyme leaves the stomach are broken down by *pancreatic amylase*, an enzyme in pancreatic juice that acts in the small intestine. Three enzymes located at the surface of small intestinal absorptive cells complete the digestion of disaccharides, breaking them down into monosaccharides, which are small enough to be absorbed. *Maltase* splits maltose into two molecules of glucose. *Sucrase* breaks sucrose into a molecule of glucose and a molecule of fructose. *Lactase* digests lactose into a molecule of glucose and a molecule of galactose.

CLINICAL CONNECTION | Coeliac Disease

Coeliac disease—A common digestive condition associated with intolerance to the protein gluten (found in wheat, barley and rye). Exposure to gluten causes inflammation and damages the villi lining the small intestine, hindering absorption of nutrients. Patients experience a range of symptoms including diarrhoea, bloating and abdominal pain. The cause of coeliac disease is unknown, although a combination of genetic and environmental factors is suggested. There is no cure, but a gluten-free diet is recommended to help control symptoms. •

Enzymes in pancreatic juice (trypsin, chymotrypsin, elastase, and carboxypeptidase) continue the digestion of proteins begun in the stomach, though their actions differ somewhat because each splits the peptide bond between different amino acids. Protein digestion is completed by *peptidases*, enzymes produced by absorptive cells that line the villi. The final products of protein digestion are amino acids, dipeptides, and tripeptides.

In an adult, most lipid digestion occurs in the small intestine. In the first step of lipid digestion, bile salts emulsify large globules of triglycerides and lipids into small lipid globules, giving pancreatic lipase easy access. Recall that triglycerides consist of a molecule of glycerol with three attached fatty acids (see Figure 2.10). In the second step, *pancreatic lipase* (found in pancreatic juice) breaks down each triglyceride molecule by removing two of the three fatty acids from glycerol; the third remains attached to the glycerol. Thus, fatty acids and monoglycerides are the end products of triglyceride digestion.

Pancreatic juice contains two nucleases: ribonuclease, which digests RNA, and deoxyribonuclease, which digests DNA. The nucleotides that result from the action of the two nucleases are further digested by small intestinal enzymes into pentoses, phosphates, and nitrogenous bases.

Table 19.1 summarizes the enzymes that contribute to digestion.

TABLE 19.1

Summary of Digestive Enzymes

ENZYME	SOURCE	SUBSTRATE	PRODUCT
Carbohydrate-digesting			
Salivary amylase	Salivary glands	Starches	Maltose (disaccharide), maltotriose (trisaccharide), and dextrins
Pancreatic amylase	Pancreas	Starches	Maltose, maltotriose, and dextrins
Maltase	Small intestine	Maltose	Glucose
Sucrase	Small intestine	Sucrose	Glucose and fructose
Lactase	Small intestine	Lactose	Glucose and galactose
Protein-digesting			
Pepsin	Stomach (chief cells)	Proteins	Peptides
Trypsin	Pancreas	Proteins	Peptides
Chymotrypsin	Pancreas	Proteins	Peptides
Carboxypeptidase	Pancreas	Amino acid at carboxyl (acid) end of peptides	Peptides and amino acids
Peptidases	Small intestine	Amino acid at amino end of peptides and dipeptides	Peptides and amino acids
Lipid-digesting			
Lingual lipase	Tongue	Triglycerides (fats)	Fatty acids and diglycerides
Pancreatic lipase	Pancreas	Triglycerides (fats) that have been emulsified by bile salts	Fatty acids and monoglycerides
Nucleases			
Ribonuclease	Pancreas	Ribonucleic acid	Nucleotides
Deoxyribonuclease	Pancreas	Deoxyribonucleic acid	Nucleotides

Absorption in the Small Intestine

All of the mechanical and chemical phases of digestion from the mouth down through the small intestine are directed toward changing food into molecules that can undergo *absorption*. Recall that absorption refers to the movement of small molecules through the absorptive epithelial cells of the mucosa into the underlying blood and lymphatic vessels. About 90 percent of all absorption takes place in the small intestine. The other 10 percent occurs in the stomach and large intestine. Absorption in the small intestine occurs by simple diffusion, facilitated diffusion, osmosis, and active transport. Any undigested or unabsorbed material left in the small intestine is passed on to the large intestine.

Absorption of Monosaccharides

All carbohydrates are absorbed as monosaccharides. Glucose and galactose are transported into absorptive cells of the villi by active transport. Fructose is transported by facilitated diffusion (Figure 19.13a). After absorption, monosaccharides are transported out of the epithelial cells by facilitated diffusion into the blood capillaries, which drain into venules of the villi. From here, monosaccharides are carried to the liver via the hepatic portal vein, then through the heart and to the general circulation (Figure 19.13b). Recall from Chapter 16 that the liver processes substances it receives from the hepatic portal vein before they pass into general circulation.

Absorption of Amino Acids

Enzymes break down dietary proteins into amino acids, dipeptides, and tripeptides, which are absorbed mainly in the duodenum and jejunum. About half of the absorbed amino acids are present in food, but half come from proteins in digestive juices and dead cells that slough off your body's own mucosa! Amino acids, dipeptides, and tripeptides enter absorptive cells of the villi via active transport (Figure 19.13a). Inside the epithelial cells, peptides are digested into amino acids, which leave via diffusion and enter blood capillaries. Like monosaccharides, amino acids are carried in hepatic portal blood to the liver (Figure 19.13b). If not removed by liver cells, amino acids enter the general circulation. From there, body cells take up amino acids for use in protein synthesis and ATP production.

Absorption of Ions and Water

Absorptive cells lining the small intestine also absorb most of the ions and water that enter the GI tract in food, drink, and digestive secretions. Major ions absorbed in the small intestine include sodium, potassium, calcium, iron, magnesium, chloride, phosphate, nitrate, and iodide. All water absorption in the GI tract, about 9 liters (a little more than 2 gallons) daily, occurs via osmosis. When monosaccharides, amino acids, peptides, and ions are absorbed, they "pull" water along by osmosis.

Absorption of Lipids and Bile Salts

Lipases break down triglycerides into monoglycerides and fatty acids. The fatty acids can be either short-chain fatty acids (with fewer than 10–12 carbons) or long-chain fatty acids. Small short-chain fatty acids are absorbed via simple diffusion into absorptive cells of the villi and then pass into blood capillaries along with monosaccharides and amino acids (Figure 19.13a). Bile salts emulsify the larger lipids, forming many *micelles* (mī-SELZ = small morsels), tiny droplets that include some bile salt molecules along with the large short-chain and long-chain fatty acids, monoglycerides, cholesterol, and other dietary lipids (Figure 19.13a). From micelles, these lipids diffuse into absorptive cells of the villi where they are packaged into *chylomicrons* (kī-lō-MĪ-krons), large spherical particles that are coated with proteins. Chylomicrons leave the epithelial cells via exocytosis and enter lymphatic fluid within a lacteal. Thus, most absorbed dietary lipids bypass the hepatic portal circulation because they enter lymphatic vessels instead of blood capillaries. Lymphatic fluid carrying chylomicrons from the small intestine passes into the thoracic duct and in due course empties into the left subclavian vein (Figure 19.13b). As blood passes through capillaries in adipose tissue and the liver, chylomicrons are removed and their lipids are stored for future use.

There are several benefits to including some healthy fats in the diet. For example, fats delay gastric emptying and this helps a person to feel full. Fats also enhance the feeling of fullness by triggering the release of a hormone called cholecystokinin (CCK) (see Table 19.2). Finally, fats are necessary for the absorption of fat-soluble vitamins.

When chyme reaches the ileum, most of the bile salts are reabsorbed and returned by the blood to the liver for recycling. Insufficient bile salts, due to either obstruction of the bile ducts or liver disease, can result in the loss of up to 40 percent of dietary lipids in feces due to diminished lipid absorption.

Absorption of Vitamins

Fat-soluble vitamins (A, D, E, and K) are included along with ingested dietary lipids in micelles and are absorbed via simple diffusion. Most water-soluble vitamins, such as the B vitamins and vitamin C, are absorbed by simple diffusion. Vitamin B_{12} must be combined with intrinsic factor (produced by the stomach) for its absorption via active transport in the ileum.

Figure 19.13 Absorption of digested nutrients in the small intestine. For simplicity, all digested foods are shown in the lumen of the small intestine, even though some nutrients are digested at the surface of or within absorptive epithelial cells of the villi.

Long-chain fatty acids and monoglycerides are absorbed into lacteals; other products of digestion enter blood capillaries.

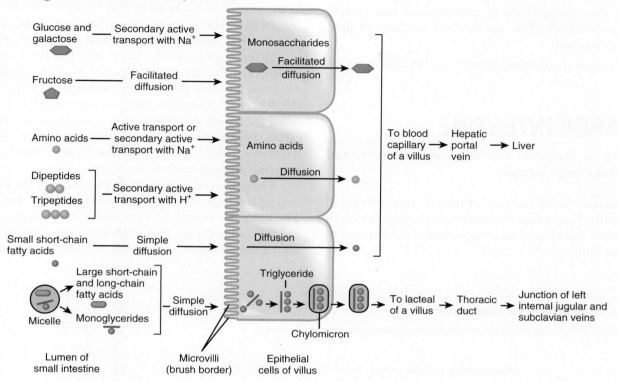

(a) Mechanisms for movement of nutrients through absorptive epithelial cells of villi

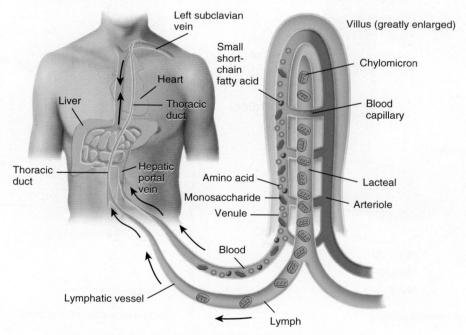

(b) Movement of absorbed nutrients into blood and lymph

 How are fat-soluble vitamins (A, D, E, and K) absorbed?

✓ **CHECKPOINT**

17. In what ways are the mucosa and submucosa of the small intestine adapted for digestion and absorption?

18. Define absorption. Where does most absorption occur?

19. How are the end products of carbohydrate and protein digestion absorbed? How are the end products of lipid digestion absorbed?

20. By what routes do absorbed nutrients reach the liver?

19.9 LARGE INTESTINE

OBJECTIVE • **Describe the location, structure, and functions of the large intestine.**

The large intestine is the last part of the GI tract. Its overall functions are the completion of absorption, the production of certain vitamins, the formation of feces, and the expulsion of feces from the body.

Structure of the Large Intestine

The **large intestine** averages about 6.5 cm (2.5 in.) in diameter and about 1.5 m (5 ft) in length in both living humans and cadavers. It extends from the ileum to the anus and is attached to the posterior abdominal wall by its mesentery (see Figure 19.3b). The large intestine has four principal regions: cecum, colon, rectum, and anal canal (Figure 19.14).

At the opening of the ileum into the large intestine is a valve called the ileocecal sphincter. It allows materials from the small intestine to pass into the large intestine. Inferior to the ileocecal sphincter is the first segment of large intestine, called the **cecum**. Attached to the cecum is a twisted coiled tube called the **appendix**. This structure has highly concentrated lymphatic nodules that control the bacteria entering the large intestine through immune responses.

The open end of the cecum merges with the longest portion of the large intestine, called the **colon** (= food passage). The colon is divided into ascending, transverse, descending, and sigmoid portions. The **ascending colon** ascends on the right side of the abdomen, reaches the undersurface of the liver, and

Figure 19.14 Anatomy of the large intestine.

The regions of the large intestine are the cecum, colon, rectum, and anal canal.

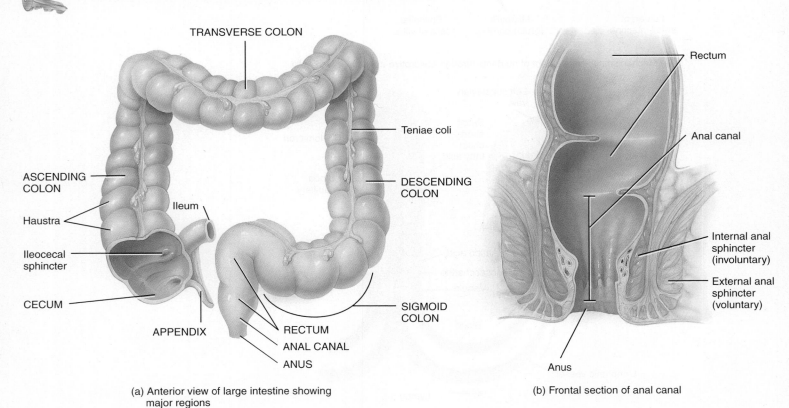

(a) Anterior view of large intestine showing major regions

(b) Frontal section of anal canal

 What are the functions of the large intestine?

turns to the left. The colon continues across the abdomen to the left side as the **transverse colon**. It curves beneath the lower border of the spleen on the left side and passes downward as the **descending colon**. The S-shaped **sigmoid colon** begins near the iliac crest of the left hip bone and ends as the **rectum**.

The last 2 to 3 cm (1 in.) of the rectum is called the **anal canal**. The opening of the anal canal to the exterior is called the **anus**. It has an internal sphincter of smooth (involuntary) muscle and an external sphincter of skeletal (voluntary) muscle. Normally, the anal sphincters are closed except during the elimination of feces.

The wall of the large intestine contains the typical four layers found in the rest of the GI tract: mucosa, submucosa, muscularis, and serosa. The epithelium of the mucosa is simple columnar epithelium that contains mostly absorptive cells and goblet cells (Figure 19.15). The cells form long tubes called *intestinal glands*. The absorptive cells function primarily in ion and water absorption. The goblet cells secrete mucus that lubricates the contents of the colon. Lymphatic nodules also are found in the mucosa. Compared to the small intestine, the mucosa of the large intestine does not have as many structural adaptations that

increase surface area. There are no circular folds or villi; however, microvilli of the absorptive cells are present. Consequently, much more absorption occurs in the small intestine than in the large intestine. The muscularis consists of an external layer of longitudinal muscles and an internal layer of circular muscles. Unlike other parts of the gastrointestinal tract, the outer longitudinal layer of the muscularis is bundled into three longitudinal bands called **teniae coli** (TĒ-nē-ē kō-lī; *teniae* = flat bands) that run most of the length of the large intestine (see Figure 19.14a). Contractions of the bands gather the colon into a series of pouches called **haustra** (= shaped like pouches), which give the colon a puckered appearance.

CLINICAL CONNECTION | Polyps in the Colon

Polyps in the colon are generally slow-developing benign growths that arise from the mucosa of the large intestine. Often, they do not cause symptoms. If symptoms do occur, they include diarrhea, blood in the feces, and mucus discharged from the anus. The polyps are removed by colonoscopy or surgery because some of them may become cancerous. •

Figure 19.15 Structure of the large intestine.

Intestinal glands formed by absorptive cells and goblet cells extend the full thickness of the mucosa.

Lumen of large intestine

Openings of intestinal glands

Simple columnar epithelium

Intestinal gland

Lamina propria

MUCOSA

Lymphatic nodule

Muscularis mucosae

Lymphatic vessel

Arteriole

Venule

Circular layer of muscle

Enteric neurons in muscularis

Longitudinal layer of muscle

SUBMUCOSA

MUSCULARIS

SEROSA

Three-dimensional view of layers of the large intestine

How does the muscularis of the large intestine differ from that of other parts of the GI tract?

Digestion and Absorption in the Large Intestine

The passage of chyme from the ileum into the cecum is regulated by the ileocecal sphincter. The sphincter normally remains slightly contracted so that the passage of chyme is usually a slow process. Immediately after a meal, a reflex intensifies peristalsis, forcing any chyme in the ileum into the cecum. *Peristalsis* occurs in the large intestine at a slower rate than in other portions of the GI tract. Characteristic of the large intestine is *mass peristalsis*, a strong peristaltic wave that begins in the middle of the colon and drives the colonic contents into the rectum. Food in the stomach initiates mass peristalsis, which usually takes place three or four times a day, during or immediately after a meal.

The final stage of digestion occurs in the colon through the activity of bacteria that normally inhabit the lumen. The glands of the large intestine secrete mucus but no enzymes. Bacteria ferment any remaining carbohydrates and release hydrogen, carbon dioxide, and methane gases. These gases contribute to flatus (gas) in the colon, termed *flatulence* when it is excessive. Bacteria also break down the remaining proteins to amino acids and decompose bilirubin to simpler pigments, including stercobilin, which give feces their brown color. Several vitamins needed for normal metabolism, including some B vitamins and vitamin K, are bacterial products that are absorbed in the colon.

Although most water absorption occurs in the small intestine, the large intestine also absorbs a significant amount. The large intestine also absorbs ions, including sodium and chloride, and some dietary vitamins.

By the time chyme has remained in the large intestine 3 to 10 hours, it has become solid or semisolid as a result of water absorption and is now called *feces*. Chemically, feces consist of water, inorganic salts, sloughed-off epithelial cells from the mucosa of the gastrointestinal tract, bacteria, products of bacterial decomposition, unabsorbed digested materials, and indigestible parts of food.

The Defecation Reflex

Mass peristaltic movements push fecal material from the sigmoid colon into the rectum. The resulting distension of the rectal wall stimulates stretch receptors, which initiates a *defecation reflex* that empties the rectum. Impulses from the spinal cord travel along parasympathetic nerves to the descending colon, sigmoid colon, rectum, and anus. The resulting contraction of the longitudinal rectal muscles shortens the rectum, thereby increasing the pressure within it. This pressure plus parasympathetic stimulation opens the internal sphincter. The external sphincter is voluntarily controlled. If it is voluntarily relaxed, **defecation**, the elimination of feces from the rectum through the anus, occurs; if it is voluntarily constricted, defecation can be postponed. Voluntary contractions of the diaphragm and abdominal muscles aid defecation by increasing the pressure within the abdomen, which pushes the walls of the sigmoid colon and rectum inward. If defecation does not occur, the feces back up into the sigmoid colon until the next wave of mass peristalsis stimulates the stretch receptors. In infants, the defecation reflex causes automatic emptying of the rectum because voluntary control of the external anal sphincter has not yet developed.

CLINICAL CONNECTION | Diarrhea and Constipation

Diarrhea (dī-a-RĒ-a; *dia-* = through; *rrhea* = flow) is an increase in the frequency, volume, and fluid content of the feces caused by increased motility of and decreased absorption by the intestines. When chyme passes too quickly through the small intestine and feces pass too quickly through the large intestine, there is not enough time for absorption. Frequent diarrhea can result in dehydration and electrolyte imbalances. Excessive motility may be caused by lactose intolerance, stress, and microbes that irritate the gastrointestinal mucosa.

Constipation (kon′-sti-PĀ-shun; *con-* = together; *stip-* = to press) refers to infrequent or difficult defecation caused by decreased motility of the intestines. Because the feces remain in the colon for prolonged periods, excessive water absorption occurs, and the feces become dry and hard. Constipation may be caused by poor habits (delaying defecation), spasms of the colon, insufficient fiber in the diet, inadequate fluid intake, lack of exercise, emotional stress, or certain drugs. •

✓ CHECKPOINT

21. What activities occur in the large intestine to change its contents into feces?
22. What is defecation and how does it occur?

19.10 PHASES OF DIGESTION

OBJECTIVES • Outline the three phases of digestion.
• **Describe the major hormones that regulate digestive activities.**

Digestive activities occur in three overlapping phases: the cephalic phase, the gastric phase, and the intestinal phase.

Cephalic Phase

During the *cephalic phase* of digestion, the smell, sight, sound, or thought of food activates neural centers in the brain. The brain then activates the facial (VII), glossopharyngeal (IX), and vagus (X) nerves. The facial and glossopharyngeal nerves stimulate the salivary glands to secrete saliva, while the vagus nerves stimulate the gastric glands to

Emotional Eating—Consumed by Food

In addition to keeping us alive, eating serves countless psychological, social, and cultural purposes. We eat to celebrate, punish, comfort, defy, and deny. Eating in response to emotional drives, such as feeling stressed, bored, or tired, rather than in response to true physical hunger, is called *emotional eating*.

Food as Emotional Rescue

Emotional eating is so common that, within limits, it is considered well within the range of normal behavior. Who hasn't at one time or another headed for the refrigerator after a bad day? Problems arise when emotional eating becomes so excessive that it interferes with health. Physical health problems include obesity and associated disorders such as hypertension and heart disease. Psychological health problems include poor self-esteem; an inability to cope effectively with feelings of stress; and in extreme cases, eating disorders.

For emotional eaters, the drive to eat often masks unpleasant feelings such as boredom, loneliness, depression, anxiety, anger, or fatigue. Eating provides comfort and solace, numbing pain and "feeding the hungry heart." Some emotional overeaters say that stuffing themselves with food becomes a metaphor for suppressing undesirable feelings.

Eating may provide a biochemical "fix" as well. Emotional eaters typically overeat carbohydrate foods (sweets and starches), which may raise brain serotonin levels and lead to feelings of relaxation. Food becomes a way to self-medicate when negative emotions arise.

Consumed by Food

In extreme cases, eating becomes an addiction, and the drive to consume excessive amounts of food begins to take over the emotional eater's life. People with bulimia or binge-eating disorder have an overwhelmingly urgent and totally uncontrollable drive to eat, causing them to consume huge volumes of food several times a week, sometimes several times a day. People with bulimia try to purge the calories they have consumed by vomiting, exercising excessively, or using laxatives and diuretics, but people with binge-eating disorder usually do not.

Eating disorders can be very dangerous and even lethal, requiring prompt, comprehensive, and in-depth professional treatment that helps people cope with underlying psychological issues. Therapy for emotional eaters requires the patient to address the emotions that trigger overeating and devise effective coping strategies that eliminate the need to deal with stress by overeating.

Think It Over . . . Why might repeated attempts to lose weight with very restrictive diets lead to emotional overeating?

---

secrete gastric juice. The purpose of the cephalic phase of digestion is to prepare the mouth and stomach for food that is about to be eaten.

Gastric Phase

Once food reaches the stomach, the **gastric phase** of digestion begins. The purpose of this phase of digestion is to continue gastric secretion and to promote gastric motility. Gastric secretion during the gastric phase is regulated by the hormone **gastrin**. Gastrin is released from the G cells of the gastric glands in response to several stimuli: stretching of the stomach by chyme, partially digested proteins in chyme, caffeine in chyme, and the high pH of chyme due to the presence of food in the stomach. Gastrin stimulates gastric glands to secrete large amounts of gastric juice. It also strengthens the contraction of the lower esophageal sphincter to prevent reflux of acid chyme into the esophagus, increases motility of the stomach, and relaxes the pyloric sphincter, which promotes gastric emptying.

Intestinal Phase

The **intestinal phase** of digestion begins once food enters the small intestine. In contrast to the activities initiated during the cephalic and gastric phases, which stimulate stomach secretory activity and motility, those occurring during the intestinal phase have inhibitory effects that slow the exit of chyme from the stomach and prevent overloading of the duodenum with more chyme than it can handle. In addition, responses occurring during the intestinal phase promote the continued digestion of foods that have reached the small intestine.

The activities of the intestinal phase are mediated by two major hormones secreted by the small intestine: cholecystokinin and secretin. *Cholecystokinin (CCK)* is secreted by CCK cells in intestinal glands of the small intestine in response to chyme containing amino acids from partially digested proteins and fatty acids from partially digested triglycerides. CCK stimulates secretion of pancreatic juice that is rich in digestive enzymes. It also causes contraction of the wall of the gallbladder, which squeezes stored bile out of the gallbladder into the cystic duct and through the common bile duct. In addition, CCK slows gastric emptying by promoting contraction of the pyloric sphincter, and it produces *satiety* (feeling full to satisfaction) by acting on the hypothalamus in the brain.

Acidic chyme entering the duodenum stimulates the release of *secretin* from S cells in intestinal glands of the small intestine. In turn, secretin stimulates the flow of pancreatic juice that is rich in bicarbonate (HCO_3^-) ions to buffer the acidic chyme that enters the duodenum from the stomach.

Table 19.2 summarizes the major hormones that control digestion.

✓ **CHECKPOINT**

23. What are the stimuli that cause the cephalic phase of digestion?

24. Compare and contrast the activities that occur during the gastric phase of digestion with those that occur during the intestinal phase of digestion.

19.11 AGING AND THE DIGESTIVE SYSTEM

OBJECTIVE • **Describe the effects of aging on the digestive system.**

Changes in the digestive system associated with aging include decreased secretory mechanisms, decreased motility of the digestive organs, loss of strength and tone of the muscular tissue and its supporting structures, changes in sensory feedback regarding enzyme and hormone release, and diminished response to pain and internal sensations. In the upper portion of the GI tract, common changes include reduced sensitivity to mouth irritations and sores, loss of taste, periodontal disease, difficulty in swallowing, hiatal hernia, gastritis, and peptic ulcer disease. Changes that may appear in the small intestine include duodenal ulcers, maldigestion, and malabsorption. Other pathologies that increase in incidence with age are appendicitis, gallbladder problems, jaundice, cirrhosis of the liver, and acute pancreatitis. Changes in the large intestine such as constipation, hemorrhoids, and diverticular disease may also occur. The incidence of cancer of the colon or rectum increases with age.

✓ **CHECKPOINT**

25. List several changes in the upper and lower portions of the GI tract associated with aging.

• • •

Now that our exploration of the digestive system is complete, you can appreciate the many ways that this system contributes to homeostasis of other body systems by examining Focus on Homeostasis: The Digestive System. Next, in Chapter 20, you will discover how the nutrients absorbed by the GI tract are utilized in metabolic reactions by the body tissues.

TABLE 19.2

Major Hormones That Control Digestion

HORMONE	WHERE PRODUCED	STIMULANT	ACTION
Gastrin	Stomach mucosa (pyloric region)	Stretching of stomach, partially digested proteins and caffeine in stomach, and high pH of stomach chyme	Stimulates secretion of gastric juice, increases motility of GI tract, and relaxes pyloric sphincter
Secretin	Intestinal mucosa	Acidic chyme that enters the small intestine	Stimulates secretion of pancreatic juice rich in bicarbonate ions
Cholecystokinin (CCK)	Intestinal mucosa	Amino acids and fatty acids in chyme in small intestine	Inhibits gastric emptying, stimulates secretion of pancreatic juice rich in digestive enzymes, causes ejection of bile from the gallbladder, and induces a feeling of satiety (feeling full to satisfaction)

BODY SYSTEM	CONTRIBUTION OF THE DIGESTIVE SYSTEM

For all body systems
The digestive system breaks down dietary nutrients into forms that can be absorbed and used by body cells for producing ATP and building body tissues; absorbs water, minerals, and vitamins needed for the growth and functions of body tissues; and eliminates wastes from body tissues in feces.

Integumentary system
The small intestine absorbs vitamin D, which the skin and kidneys modify to produce the hormone calcitriol. Excess dietary calories are stored as triglycerides in adipose cells in the dermis and subcutaneous layer.

Skeletal system
The small intestine absorbs dietary calcium and phosphorus salts needed to build bone extracellular matrix.

Muscular system
The liver can convert lactic acid produced by muscles during exercise to glucose.

Nervous system
Gluconeogenesis (synthesis of new glucose molecules) in the liver plus digestion and absorption of dietary carbohydrates provide glucose, needed for ATP production by neurons.

Endocrine system
The liver inactivates some hormones, ending their activity. Pancreatic islets release insulin and glucagon. Cells in the mucosa of the stomach and small intestine release hormones that regulate digestive activities.

Cardiovascular system
The GI tract absorbs water that helps maintain blood volume and iron that is needed for the synthesis of hemoglobin in red blood cells. Bilirubin from hemoglobin breakdown is partially excreted in feces. The liver synthesizes most plasma proteins.

Lymphatic system and immunity
The acidity of gastric juice destroys bacteria and most toxins in the stomach.

Respiratory system
The pressure of abdominal organs against the diaphragm helps expel air quickly during a forced exhalation.

Urinary system
Absorption of water by the GI tract provides water needed to excrete waste products in urine.

Reproductive systems
Digestion and absorption provides adequate nutrients, including fats, for normal development of reproductive structures, for the production of gametes (oocytes and sperm), and for fetal growth and development during pregnancy.

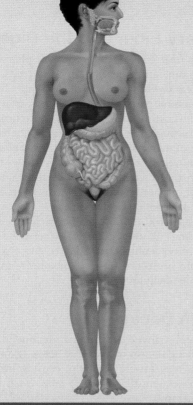

THE DIGESTIVE SYSTEM

COMMON DISORDERS

Dietary Fiber and the Digestive System

Dietary fiber consists of indigestible plant substances, such as cellulose, lignin, and pectin, found in fruits, vegetables, grains, and beans. *Insoluble fiber*, which does not dissolve in water, includes the structural parts of plants such as fruit and vegetable skins and the bran coating around wheat and corn kernels. Insoluble fiber passes through the GI tract largely unchanged and speeds up the passage of material through the tract. *Soluble fiber*, which does dissolve in water, forms a gel that slows the passage of materials through the tract. It is found in abundance in beans, oats, barley, broccoli, prunes, apples, and citrus fruits. It tends to slow the passage of material through the tract.

People who choose a fiber-rich diet may reduce their risk of developing obesity, diabetes, atherosclerosis, gallstones, hemorrhoids, diverticulitis, appendicitis, and colon cancer. Insoluble fiber may help protect against colon cancer, and soluble fiber may help lower blood cholesterol level.

Tooth Decay

Tooth decay involves a gradual demineralization (softening) of the enamel and dentin by bacterial acids. If untreated, various microorganisms may invade the pulp, causing inflammation and infection with subsequent death of the pulp. Such teeth are treated by root canal therapy.

Periodontal Disease

Periodontal disease refers to a variety of conditions characterized by inflammation and degeneration of the gums, bone, periodontal ligament, and cementum. Periodontal diseases are often caused by poor oral hygiene; by local irritants, such as bacteria, impacted food, and cigarette smoke; or by a poor "bite."

Peptic Ulcer Disease

Five to ten percent of the U.S. population develops *peptic ulcer disease (PUD)* each year. An *ulcer* is a craterlike lesion in a membrane; ulcers that develop in areas of the GI tract exposed to acidic gastric juice are called *peptic ulcers*. The most common complication of peptic ulcers is bleeding, which can lead to anemia. In acute cases, peptic ulcers can lead to shock and death. Three distinct causes of PUD are recognized: (1) the bacterium *Helicobacter pylori*, (2) nonsteroidal anti-inflammatory drugs (NSAIDs) such as aspirin, and (3) hypersecretion of HCl.

Helicobacter pylori is the most frequent cause of PUD. The bacterium produces an enzyme that splits urea into ammonia and carbon dioxide. While shielding the bacterium from the acidity of the stomach, the ammonia also damages the protective mucous layer of the stomach and the underlying gastric cells. *H. pylori* also produces several adhesion proteins that allow the bacterium to attach itself to gastric cells.

Several therapeutic approaches are helpful in the treatment of PUD. Cigarette smoke, alcohol, caffeine, and NSAIDs should be avoided because they can impair mucosal defensive mechanisms, which increases mucosal susceptibility to the damaging effects of HCl. In cases associated with *H. pylori*, treatment with an antibiotic drug often resolves the problem. Oral antacids such as Tums® or Maalox® can help temporarily by buffering gastric acid. When hypersecretion of HCl is the cause of PUD, histamine-2 (H_2) blockers (such as Tagamet®) or Prilosec®, which block secretion of H^+ from parietal cells, may be used.

Appendicitis

Appendicitis (a-pen′-di-SĪ-tis) is an inflammation of the appendix. Appendectomy (surgical removal of the appendix) is recommended in all suspected cases because it is safer to operate than to risk gangrene, rupture, and peritonitis.

Colorectal Cancer

Colorectal cancer is among the deadliest of malignancies. An inherited predisposition contributes to more than half of all cases of colorectal cancer. Intake of alcohol and diets high in animal fat and protein are associated with increased risk of colorectal cancer; dietary fiber, retinoids, calcium, and selenium may be protective. Signs and symptoms of colorectal cancer include diarrhea, constipation, cramping, abdominal pain, and rectal bleeding. Screening for colorectal cancer includes testing for blood in the feces, digital rectal examination, sigmoidoscopy, colonoscopy, and barium enema.

Diverticular Disease

Diverticulosis (dī-ver-tik′-ū-LŌ-sis) is the development of diverticula, saclike outpouchings of the wall of the colon in places where the muscularis has become weak. Many people who develop diverticulosis have no symptoms and experience no complications. About 15% of people with diverticulosis eventually develop an inflammation known as *diverticulitis*, characterized by pain, either constipation or increased frequency of defecation, nausea, vomiting, and low-grade fever. Patients who change to high-fiber diets often show marked relief of symptoms.

Hepatitis

Hepatitis is an inflammation of the liver caused by viruses, drugs, and chemicals, including alcohol.

Hepatitis A (infectious hepatitis), caused by the hepatitis A virus, is spread by fecal contamination of food, clothing, toys, eating utensils, and so forth (fecal–oral route). It does not cause lasting liver damage.

Hepatitis B, caused by the hepatitis B virus, is spread primarily by sexual contact and contaminated syringes and transfusion equipment. It can also be spread by any secretion via saliva and tears. Hepatitis B can produce chronic liver inflammation. Vaccines are available for hepatitis B and are required for certain individuals, such as health-care providers.

Hepatitis C, caused by the hepatitis C virus, is clinically similar to hepatitis B. It is often spread by blood transfusions and can cause cirrhosis and liver cancer.

Hepatitis D is caused by the hepatitis D virus. It is transmitted like hepatitis B. A person must be infected with hepatitis B to con-

tract hepatitis D. Hepatitis D results in severe liver damage and has a fatality rate higher than that due to infection with hepatitis B virus alone.

Hepatitis E is caused by the hepatitis E virus and is spread like hepatitis A. Although it does not cause chronic liver disease, the hepatitis E virus is responsible for a very high death rate in pregnant women.

MEDICAL TERMINOLOGY AND CONDITIONS

Anorexia nervosa (an´-ō-REK-sē-a ner-VŌ-sa; *rexis* = appetite) A chronic disorder characterized by self-induced weight loss, negative perception of body image, and physiological changes that result from nutritional depletion. Patients have a fixation on weight control and often abuse laxatives, which worsens their fluid and electrolyte imbalances and nutrient deficiencies. The disorder is found predominantly in young, single females, and it may be inherited. Individuals may become emaciated and may ultimately die of starvation or one of its complications.

Bariatric surgery (bar´-ē-AT-rik; *baros-* = weight; *-iatreia* = medical treatment) A surgical procedure that limits the amount of food that can be ingested and absorbed in order to bring about a significant weight loss in obese individuals. The most commonly performed type of bariatric surgery is called *gastric bypass surgery*. In one variation of this procedure, the stomach is reduced in size by making a small pouch at the top of the stomach about the size of a walnut. The pouch, which is only 5–10 percent of the stomach, is sealed off from the rest of the stomach using surgical staples or a plastic band. The pouch is connected to the jejunum of the small intestine, thus bypassing the rest of the stomach and the duodenum. The result is that smaller amounts of food are ingested and fewer nutrients are absorbed in the small intestine. This leads to weight loss.

Canker sore (KANG-ker) Painful ulcer on the mucous membrane of the mouth that affects females more often than males, usually between ages 10 and 40; it may be an autoimmune reaction or result from a food allergy.

Cholecystitis (kō´-lē-sis-TĪ-tis; *chole-* = bile; *cyst-* = bladder; *-itis* = inflammation of) In some cases, an autoimmune inflammation of the gallbladder; other cases are caused by obstruction of the cystic duct by bile stones.

Cirrhosis (si-RŌ-sis) Distorted or scarred liver as a result of chronic inflammation due to hepatitis, chemicals that destroy hepatocytes, parasites that infect the liver, or alcoholism; the hepatocytes are replaced by fibrous or adipose connective tissue. Symptoms include jaundice, edema in the legs, uncontrolled bleeding, and increased sensitivity to drugs.

Colostomy (kō-LOS-tō-mē; *-stomy* = provide an opening) The diversion of the fecal stream through an opening in the colon, creating a surgical "stoma" (artificial opening) that is affixed to the exterior of the abdominal wall. This opening serves as a substitute anus through which feces are eliminated into a bag worn on the abdomen.

Flatus (FLĀ-tus) Air (gas) in the stomach or intestine, usually expelled through the anus. If the gas is expelled through the mouth, it is called **eructation** or **belching** (burping). Flatus may result from gas released during the breakdown of foods in the stomach or from swallowing air or gas-containing substances such as carbonated drinks.

Food poisoning A sudden illness caused by ingesting food or drink contaminated by an infectious microbe (bacterium, virus, or protozoan) or a toxin (poison). The most common cause of food poisoning is the toxin produced by the bacterium *Staphylococcus aureus*. Most types of food poisoning cause diarrhea and/or vomiting, often associated with abdominal pain.

Inflammatory bowel disease (in-FLAM-a-tō´-rē BOW-el) Disorder that exists in two forms: (1) *Crohn's disease*, an inflammation of the gastrointestinal tract, especially the distal ileum and proximal colon, in which the inflammation may extend from the mucosa through the serosa, and (2) *ulcerative colitis*, an inflammation of the mucosa of the gastrointestinal tract, usually limited to the large intestine and usually accompanied by rectal bleeding.

Irritable bowel syndrome (IBS) Disease of the entire gastrointestinal tract in which a person reacts to stress by developing symptoms (such as cramping and abdominal pain) associated with alternating patterns of diarrhea and constipation. Excessive amounts of mucus may appear in feces; other symptoms include flatulence, nausea, and loss of appetite.

Malocclusion (mal´-ō-KLOO-zhun; *mal-* = bad; *occlusion* = to fit together) Condition in which the surfaces of the maxillary (upper) and mandibular (lower) teeth fit together poorly.

Nausea (NAW-sē-a = seasickness) Discomfort characterized by a loss of appetite and the sensation of impending vomiting. Its causes include local irritation of the gastrointestinal tract, a systemic disease, brain disease or injury, overexertion, or the effects of medication or drug overdose.

Traveler's diarrhea Infectious disease of the gastrointestinal tract that results in loose, urgent bowel movements; cramping; abdominal pain; malaise; nausea; and occasionally fever and dehydration. It is acquired through ingestion of food or water contaminated with fecal material typically containing bacteria (especially *Escherichia coli*); viruses or protozoan parasites are a less common cause.

CHAPTER REVIEW AND RESOURCE SUMMARY

REVIEW	RESOURCES

Introduction

1. The breakdown of larger food molecules into smaller molecules is called **digestion**.
2. The organs that collectively perform digestion and absorption constitute the **digestive system**.

19.1 Overview of the Digestive System

1. The **gastrointestinal (GI) tract** is a continuous tube extending from the mouth to the anus.
2. The **accessory digestive organs** include the teeth, tongue, salivary glands, liver, gallbladder, and pancreas.
3. Digestion includes six basic processes: ingestion, secretion, mixing and propulsion, **mechanical** and **chemical digestion**, **absorption**, and **defecation**.

Resources:
Anatomy Overview—Digestive System
Animation—Chemical Digestion—Enzymes
Exercise—Digestive Enzyme Activity
Exercise—Match the Movement
Exercise—Role of Digestive Chemicals

19.2 Layers of the GI Tract and the Omentum

1. The basic arrangement of layers in most of the gastrointestinal tract, from the inside to the outside, is the **mucosa**, **submucosa**, **muscularis**, and **serosa**.
2. Parts of the peritoneum include the **mesentery** and **greater omentum**.

Resources:
Anatomy Overview—Digestive System Histology

Figure 19.2 Layers of the gastrointestinal tract

19.3 Mouth

1. The **mouth** or **oral cavity** is formed by the cheeks, hard and soft palates, lips, and tongue, which aid mechanical digestion.
2. The **tongue** forms the floor of the oral cavity. It is composed of skeletal muscle covered with mucous membrane. The superior surface and lateral areas of the tongue are covered with **papillae**. Some papillae contain taste buds. Glands in the tongue secrete lingual lipase which digests triglycerides once in the acid environment of the stomach.
3. Most saliva is secreted by the **salivary glands**, which lie outside the mouth and release their secretions into ducts that empty into the oral cavity. There are three pairs of salivary glands: **parotid**, **submandibular**, and **sublingual**. Saliva lubricates food and starts the chemical digestion of carbohydrates. **Salivation** is controlled by the autonomic nervous system.
4. The **teeth**, or **dentes**, project into the mouth and are adapted for mechanical digestion. A typical tooth consists of three principal portions: **crown**, **root**, and **neck**. Teeth are composed primarily of **dentin** and are covered by **enamel**, the hardest substance in the body. Humans have two sets of teeth: **deciduous** and **permanent**.
5. Through **mastication**, food is mixed with saliva and shaped into a **bolus**.
6. **Salivary amylase** begins the digestion of starches in the mouth.

Resources:
Animation—Mastication
Animation—Carbohydrate Digestion in the Mouth
Animation—Lipid Digestion in the Mouth

Figure 19.5b Deciduous (primary) dentition
Figure 19.5c Permanent (secondary) dentition

19.4 Pharynx and Esophagus

1. Food that is swallowed passes from the mouth into the portion of the **pharynx** called the oropharynx. From the oropharynx, food passes into the laryngopharynx.
2. The **esophagus** is a muscular tube that connects the pharynx to the stomach.
3. **Swallowing** moves a bolus from the mouth to the stomach by **peristalsis**. It consists of a **voluntary stage**, **pharyngeal stage** (involuntary), and **esophageal stage** (involuntary).

Resources:
Animation—Deglutition

19.5 Stomach

1. The **stomach** connects the esophagus to the duodenum. The main regions of the stomach are the **cardia**, **fundus**, **body**, and **pylorus**. Between the pylorus and duodenum is the **pyloric sphincter**.

Resources:
Animation—Stomach Peristalsis
Animation—Protein Digestion in the Stomach

REVIEW

2. Adaptations of the stomach for digestion include **rugae**; glands that produce mucus, hydrochloric acid, a protein-digesting enzyme (pepsin), intrinsic factor, and gastrin; and a three-layered muscularis for efficient mechanical movement.

3. Mechanical digestion consists of **mixing waves** that macerate food and mix it with **gastric juice**, forming **chyme**.

4. Chemical digestion consists of the conversion of proteins into peptides by **pepsin**.

5. The stomach wall is impermeable to most substances. Among the substances the stomach can absorb are water, ions, short-chain fatty acids, some drugs, and alcohol.

Animation—Lipid Digestion in the Stomach
Animation—Chemical Digestion-Gastric Acid
Figure 19.8 Layers of the stomach
Figure 19.9 Relation of the pancreas to the liver, gallbladder, and duodenum

19.6 Pancreas

1. Secretions pass from the **pancreas** to the duodenum via the **pancreatic duct**.

2. **Pancreatic islets (islets of Langerhans)** secrete hormones and constitute the endocrine portion of the pancreas.

3. **Acini**, which secrete pancreatic juice, constitute the exocrine portion of the pancreas.

4. **Pancreatic juice** contains enzymes that digest starch (**pancreatic amylase**); proteins (**trypsin**, **chymotrypsin**, and **carboxypeptidase**); triglycerides (**pancreatic lipase**); and nucleic acids (**ribonuclease** and **deoxyribonuclease**).

Animation—Protein Digestion–Pancreatic Juice

19.7 Liver and Gallbladder

1. The **liver** has left and right lobes. The **gallbladder** is a sac located in a depression under the liver that stores and concentrates **bile** produced by the liver.

2. The lobes of the liver are made up of lobules that contain **hepatocytes** (liver cells), **hepatic sinusoids**, **reticuloendothelial (Kupffer) cells**, and a **central vein**.

3. Hepatocytes produce bile, which is carried by a duct system to the gallbladder for concentration and temporary storage.

4. Bile's contribution to digestion is the **emulsification** of dietary lipids.

5. The liver also functions in carbohydrate, lipid, and protein metabolism; processing of drugs and hormones; excretion of bilirubin; synthesis of bile salts; storage of vitamins and minerals; phagocytosis; and activation of vitamin D.

Animation—Lipid Digestion-Bile Salts and Pancreatic Lipase
Animation—Chemical Digestion-Bile

19.8 Small Intestine

1. The **small intestine** extends from the pyloric sphincter to the **ileocecal sphincter**. It is divided into the **duodenum**, the **jejunum**, and the **ileum**.

2. The small intestine is highly adapted for digestion and absorption. Its glands produce enzymes and mucus, and the **microvilli**, **villi**, and **circular folds** of its wall provide a large surface area for digestion and absorption.

3. Mechanical digestion in the small intestine involves **segmentations** and migrating waves of peristalsis.

4. Enzymes in pancreatic juice, bile, and the microvilli of the absorptive cells of the small intestine break down disaccharides to monosaccharides; protein digestion is completed by **peptidases**; triglycerides are broken down into fatty acids and monoglycerides by **pancreatic lipase**; and nucleases break down nucleic acids to pentoses and nitrogenous bases (Table 19.1).

5. **Absorption** is the passage of nutrients from digested food in the gastrointestinal tract into the blood or lymph. It occurs mostly in the small intestine by means of simple diffusion, facilitated diffusion, osmosis, and active transport.

6. Monosaccharides, amino acids, and short-chain fatty acids pass into the blood capillaries.

7. Long-chain fatty acids and monoglycerides are absorbed as part of **micelles**, resynthesized to triglycerides, and transported in **chylomicrons** to the **lacteal** of a villus.

8. The small intestine also absorbs water, electrolytes, and vitamins.

Animation—Segmentation
Animation—Carbohydrate Digestion in the Small Intestine
Animation—Protein Digestion in the Small Intestine
Animation—Nucleic Acid Digestion in the Small Intestine
Animation—Carbohydrate Absorption in the Small Intestine
Animation—Protein Absorption in the Small Intestine
Animation—Nucleic Acid Absorption in the Small Intestine
Animation—Lipid Absorption in the Small Intestine

Figure 19.13 Absorption of digested nutrients in the small intestine

REVIEW	**RESOURCES**

19.9 Large Intestine

1. The **large intestine** extends from the ileocecal sphincter to the **anus**. Its regions include the **cecum**, **colon**, **rectum**, and **anal canal**.

2. The mucosa contains numerous absorptive cells that absorb water and goblet cells that secrete mucus.

3. **Mass peristalsis** is a strong peristaltic wave that drives the contents of the colon into the rectum.

4. In the large intestine, substances are further broken down, and some vitamins are synthesized through bacterial action.

5. The large intestine absorbs water, electrolytes, and vitamins.

6. **Feces** consist of water, inorganic salts, epithelial cells, bacteria, and undigested foods.

7. The elimination of feces from the rectum is called **defecation**. Defecation is a reflex action aided by voluntary contractions of the diaphragm and abdominal muscles and relaxation of the external anal sphincter.

19.10 Phases of Digestion

1. Digestive activities occur in three overlapping phases: cephalic phase, gastric phase, and intestinal phase.

2. During the **cephalic phase** of digestion, salivary glands secrete saliva and gastric glands secrete gastric juice in order to prepare the mouth and stomach for food that is about to be eaten.

3. The presence of food in the stomach causes the **gastric phase** of digestion, which promotes gastric juice secretion and gastric motility.

4. During the **intestinal phase** of digestion, food is digested in the small intestine. In addition, gastric motility and gastric secretion decrease in order to slow the exit of chyme from the stomach, which prevents the small intestine from being overloaded with more chyme than it can handle.

5. The activities that occur during the various phases of digestion are coordinated by hormones. Table 19.2 summarizes the major hormones that control digestion.

19.11 Aging and the Digestive System

1. General changes with age include decreased secretory mechanisms, decreased motility, and loss of tone.

2. Specific changes may include loss of taste, hernias, peptic ulcer disease, constipation, hemorrhoids, and diverticular diseases.

SELF-QUIZ

1. Which of the following is NOT an accessory digestive organ?
   a. teeth **b.** salivary glands **c.** liver
   d. pancreas **e.** esophagus

2. Chewing food is an example of
   a. absorption. **b.** mechanical digestion.
   c. secretion. **d.** chemical digestion. **e.** ingestion.

3. Which of the following is mismatched?
   a. submucosa, enteric nervous system (ENS)
   b. muscularis, lacteal **c.** serosa, greater omentum
   d. mucosa, villi **e.** serosa, visceral peritoneum

4. Most chemical digestion occurs in the
   a. liver. **b.** stomach. **c.** duodenum.
   d. colon. **e.** pancreas.

5. Absorption is defined as
   a. the elimination of solid wastes from the digestive system.
   b. a reflex action controlled by the autonomic nervous system.
   c. the breakdown of foods by enzymes.
   d. the passage of nutrients from the gastrointestinal tract into the bloodstream.
   e. the mechanical breakdown of triglycerides.

6. The exposed portions of the teeth that you clean with a toothbrush are the
   a. crowns. **b.** periodontal ligaments. **c.** roots.
   d. pulp cavities. **e.** gingivae.

7. What causes your mouth to "water" at the sight of your favorite food?

   a. sympathetic stimulation of the salivary glands

   b. the cephalic phase of digestion

   c. activation of the ENS

   d. the release of gastrin

   e. the intestinal phase of digestion

8. Match the following:

   ____ a. carries bile

   ____ b. proteins combined with triglycerides and cholesterol

   ____ c. surrounds the opening between the stomach and duodenum

   ____ d. secrete pancreatic juice

   ____ e. increase surface area in small intestine

   ____ f. bile salts combined with partially digested lipids

   ____ g. location between the opening of the small and large intestine

   ____ h. large mucosal folds in stomach

   A. pyloric sphincter

   B. circular folds

   C. micelles

   D. cystic duct

   E. ileocecal sphincter

   F. rugae

   G. chylomicrons

   H. acini

9. Which of the following correctly describes the esophagus?

   a. Food enters the esophagus from the pyloric region of the stomach.

   b. The movement of food through the entire esophagus is under voluntary control.

   c. It allows the passage of chyme.

   d. It produces several enzymes that aid in the digestion of food.

   e. It is a muscular tube extending from the pharynx to the stomach.

10. If an incision were made into the stomach, the tissue layers would be cut in what order?

    a. mucosa, muscularis, serosa, submucosa

    b. mucosa, muscularis, submucosa, serosa

    c. serosa, muscularis, mucosa, submucosa

    d. muscularis, submucosa, mucosa, serosa

    e. serosa, muscularis, submucosa, mucosa

11. Most water absorption in the digestive tract occurs in the

    a. small intestine.   b. stomach.   c. mouth.

    d. liver.   e. large intestine.

12. Which of the following is NOT true concerning the large intestine?

    a. Its intestinal glands produce several enzymes to complete chemical digestion.

    b. Some necessary vitamins are absorbed in the large intestine.

    c. It contains vital bacterial colonies.

    d. Mass peristalsis in the large intestine is initiated by food in the stomach.

    e. Activation of stretch receptors in the rectal walls stimulates defecation.

13. Match the following.

    ____ a. starch-digesting enzyme

    ____ b. emulsifies lipids

    ____ c. helps neutralize stomach acids

    ____ d. intestinal enzyme that completes protein digestion

    ____ e. digests DNA

    ____ f. triglyceride-digesting enzyme

    ____ g. stimulates release of pancreatic juices

    ____ h. protein-digesting enzyme of the stomach

    ____ i. inactive gastric enzyme

    ____ j. breaks down disaccharides

    ____ k. protein-digesting enzyme from the pancreas

    A. CCK

    B. sodium bicarbonate

    C. trypsin

    D. salivary amylase

    E. deoxyribonuclease

    F. peptidase

    G. bile

    H. pancreatic lipase

    I. pepsin

    J. sucrase; lactase

    K. pepsinogen

14. Which of the following is NOT a function of the liver?

    a. processing newly absorbed nutrients

    b. producing enzymes that digest proteins

    c. breaking down old red blood cells

    d. processing of some drugs and hormones

    e. producing bile

15. The purpose of villi in the small intestine is to

    a aid in the movement of food through the small intestines.

    b. phagocytize microbes.

    c. produce digestive enzymes.

    d. increase the surface area for absorption of digested nutrients.

    e. produce acidic secretions.

16. Which of the following is NOT produced in the stomach?

    a. sodium bicarbonate   b. gastrin

    c. pepsinogen   d. mucus

    e. hydrochloric acid (HCl)

17. Which of the following is NOT correctly paired?

    a. esophagus, peristalsis

    b. mouth, mastication

    c. large intestine, mass peristalsis

    d. small intestine, segmentations

    e. stomach, emulsification

18. Lipases digest triglycerides into

    a. glucose.   b. amino acids.

    c. fatty acids and monoglycerides.

    d. nucleic acids.   e. peptide chains.

19. Place the following in the correct order as food passes from the small intestine:

    1. sigmoid colon   2. transverse colon

    3. ascending colon   4. rectum

    5. cecum   6. descending colon

    a. 1, 3, 2, 6, 5, 4   b. 5, 1, 6, 2, 3, 4

    c. 4, 1, 6, 2, 3, 5   d. 2, 3, 5, 6, 4, 1

    e. 5, 3, 2, 6, 1, 4

20. Lacteals function
   a. in the absorption of lipids in chylomicrons.
   b. to produce bile in the liver.
   c. in the absorption of electrolytes.
   d. in the fermentation of carbohydrates in the large intestine.
   e. to produce salivary amylase.

CRITICAL THINKING APPLICATIONS

1. Four out of five dentists think that you should chew sugarless gum, but all five think that you should brush your teeth. Why?

2. Two friends are having coffee and start to argue. Pam is convinced that lactose intolerance is the cause of her constipation. Jane insists that lactose intolerance has nothing to do with bowel problems but is the cause of her heartburn. Of course, neither of these ladies has eaten dairy products for years (which may help explain their osteoporosis). Please settle the argument.

3. Benjamin put a plastic spider in his sister's drink as a joke. Unfortunately, his mother doesn't think the joke was very funny because his sister swallowed it and now they're all at the Accident and Emergency department. The doctor suspects that the spider may have lodged at the junction of the stomach and the duodenum.

Name the sphincter at this junction. Trace the path taken by the plastic spider on its journey to its new home. What procedure could the doctor use to view the interior of the stomach? What structures may be viewed in the stomach (besides the spider)?

4. Upset about the argument she had with Pam, Jane (in question #2 above) stormed out of the coffee shop and returned home. While still fuming, she warms up leftover spaghetti which she consumes with a glass of wine. She follows that with a piece of chocolate cake and two cups of black coffee. That night, her heartburn is so bad she cannot sleep. Of course, she blames her argument with Pam. What do you think aggravated her heartburn and how could she temporarily relieve it? What might be her long-term solution?

ANSWERS TO FIGURE QUESTIONS

19.1 The teeth cut and grind food.

19.2 Nerves in its wall help regulate secretions and contractions of the gastrointestinal tract.

19.3 Mesentery binds the small intestine to the posterior abdominal wall.

19.4 Muscles of the tongue maneuver food for chewing, shape food into a bolus, force food to the back of the mouth for swallowing, and alter the shape of the tongue for swallowing and speech production.

19.5 The main component of teeth is a connective tissue called dentin.

19.6 Swallowing is both voluntary and involuntary. Initiation of swallowing, carried out by skeletal muscles, is voluntary. Completion of swallowing—moving a bolus along the esophagus and into the stomach—involves peristalsis of smooth muscle and is involuntary.

19.7 After a very large meal, the stomach probably does not have rugae because as the stomach fills, the rugae stretch out.

19.8 The simple columnar epithelial cells of the mucosa are in contact with food in the stomach.

19.9 Pancreatic juice is a mixture of water, salts, bicarbonate ions, and digestive enzymes.

19.10 Stellate reticuloendothelial (Kupffer) cells in the liver are phagocytes.

19.11 The ileum is the longest portion of the small intestine.

19.12 The absorptive cells are located in the mucosal epithelium.

19.13 Fat-soluble vitamins are absorbed by diffusion from micelles.

19.14 Functions of the large intestine include completion of absorption, synthesis of certain vitamins, formation of feces, and elimination of feces.

19.15 The muscularis of the large intestine forms three longitudinal bands that gather the colon into a series of pouches.

CHAPTER 20

NUTRITION AND METABOLISM

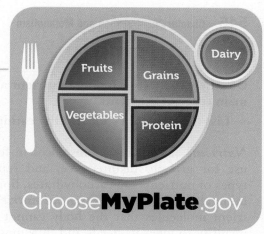

ChooseMyPlate.gov

The food we eat is our only source of energy for performing biological work. Many molecules needed to maintain cells and tissues can be made from building blocks within the body; others must be obtained in food because we cannot make them. Food molecules absorbed by the gastrointestinal (GI) tract have three main fates:

1. *To supply energy* for sustaining life processes, such as active transport, DNA replication, protein synthesis, muscle contraction, maintenance of body temperature, and cell division.
2. *To serve as building blocks* for the synthesis of more complex molecules, such as muscle proteins, hormones, and enzymes.
3. *Storage for future use.* For example, glycogen is stored in liver cells, and triglycerides are stored in adipose cells.

In this chapter we will discuss the major groups of nutrients; guidelines for healthy eating; how each group of nutrients is used for ATP production, growth, and repair of the body; and how various factors affect the body's metabolic rate.

Did you know?   *Metabolism refers to all of the chemical reactions of the body. Metabolic rate refers to the energy expenditure required by these reactions. Resting metabolic rate varies enormously from person to person.*

Many of the factors that contribute to metabolic rate, such as genetics, age, and size, are outside of a person's control. But lifestyle factors such as physical activity level and eating habits also influence resting metabolic rate. Changes in these factors can affect daily energy balance and body composition.

LOOKING BACK TO MOVE AHEAD...

Main Chemical Elements in the Body (Section 2.1)

Enzymes (Section 2.2)

Carbohydrates, Lipids, Proteins (Section 2.2)

Negative Feedback System (Section 1.4)

Functions of the Liver (Section 19.7)

Hypothalamus and Body Temperature Regulation (Section 10.4)

FOCUS ON WELLNESS

Can you "rev" your metabolic rate?

20.1 NUTRIENTS

OBJECTIVES ● **Define a nutrient and identify the six main types of nutrients.**

● **List the guidelines for healthy eating.**

Nutrients are chemical substances in food that body cells use for growth, maintenance, and repair. The six main types of nutrients are carbohydrates, lipids, proteins, water, minerals, and vitamins. *Essential nutrients* are specific nutrient molecules that the body cannot make in sufficient quantity to meet its needs and thus must be obtained from the diet. Some amino acids, some fatty acids, vitamins, and minerals are essential nutrients. The structures and functions of carbohydrates, proteins, lipids, and water were discussed in Chapter 2. In this chapter, we discuss some guidelines for healthy eating and the roles of minerals and vitamins in metabolism.

Guidelines for Healthy Living

Each gram of protein or carbohydrate in food provides about 4 Calories; 1 gram of fat (lipids) provides about 9 Calories.* We do not know with certainty what levels and types of carbohydrate, fat, and protein are optimal in the diet. Different populations around the world eat radically different diets that are adapted to their particular lifestyles.

On June 2, 2011, the United States Department of Agriculture (USDA) introduced a revised icon called *MyPlate* based upon revised guidelines for healthy eating. It replaces the USDA MyPyramid, which first appeared in 2005. As shown in Figure 20.1, the plate is divided into four different-sized colored sections:

- Green (vegetables)
- Red (fruits)
- Orange (grains)
- Purple (protein)

The blue cup (dairy) adjacent to the plate icon is a reminder to include three daily servings of dairy.

The Dietary Guidelines for Americans released in January 2011 are the basis for MyPlate. Among the guidelines are the following:

- Enjoy food but balance calories by eating less.
- Avoid oversized portions, make half of your plate vegetables and fruits.

- Switch to fat-free or low-fat milk.
- Make at least half of your grains whole grains.
- Choose foods that have a lower sodium content.
- Drink water instead of sugary drinks.

MyPlate places a lot of emphasis on proportionality, variety, moderation, and nutrient density in a healthy diet. Proportionality simply means eating more of some types of foods than others. The MyPlate icon shows how much of your plate should be filled with foods from various food groups. Note that the vegetables and fruits take up one half of the plate, while protein and grains take up the other half. Note also that vegetables and grains represent the largest portions.

Variety is important for a healthy diet because no one food or food group provides all of the nutrients and food types that the body needs. Accordingly, a variety of foods should be selected from within each food group. Vegetable choices should be varied to include dark green vegetables such as broccoli, collard greens, and kale; red and orange vegetables such as carrots, sweet potatoes, and red peppers; starchy vegetables such as corn, green peas, and potatoes; other vegetables such as cabbage, asparagus, and artichokes; and beans and peas such as lentils, chickpeas, and black beans. Beans and peas are good sources of the nutrients found in both vegetables and protein foods so they can be counted in either food group. Protein food choices are extremely varied, and include meat, poultry, seafood, beans and peas, eggs, processed soy products, nuts, and seeds. Grains include whole grains such as whole-wheat bread, oatmeal, and brown rice as well as refined grains such as white bread, white rice, and white pasta. Fruits include fresh, canned, or dried fruit and 100 percent fruit juice. Dairy includes all fluid milk products and many foods made from milk such as cheese, yogurt, and pudding, as well as calcium-fortified soy products.

Choosing nutrient-dense foods helps individuals practice moderation to balance calories consumed with calories expended. Tips include making half of your grains whole grains, choosing whole or cut-up fruits more often than juice, selecting fat-free or low-fat dairy products, and keeping meat and poultry portions small and lean.

Minerals

Minerals are inorganic elements that constitute about 4 percent of the total body weight and are concentrated most heavily in the skeleton. Minerals with known functions in the body include calcium, phosphorus, potassium, sulfur, sodium, chloride, magnesium, iron, iodide, manganese, copper, cobalt, zinc, fluoride, selenium, and chromium. Others—aluminum, boron, silicon, and molybdenum—are present but may have no functions. Typical diets supply adequate amounts of potassium, sodium, chloride, and magnesium. Some attention must be paid to eating foods that provide enough calcium, phos-

---

*A *calorie* (with a lowercase c) is the amount of heat required to raise the temperature of 1 gram of water 1°C. Because it is a relatively small unit, the *kilocalorie* or *Calorie* (with an uppercase C) is often used to express the energy content of foods. One kilocalorie equals 1000 calories.

Figure 20.1 MyPlate.

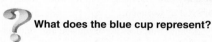 The different colored sections are meant to be visual cues to help make healthier eating choices.

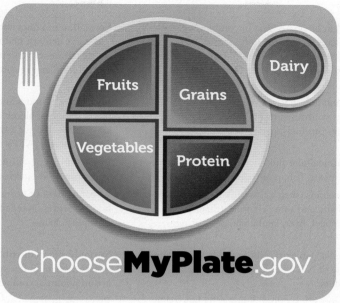

What does the blue cup represent?

phorus, iron, and iodine. Excess amounts of most minerals are excreted in the urine and feces.

A major role of minerals is to help regulate enzymatic reactions. Calcium, iron, magnesium, and manganese are part of some coenzymes. Magnesium also serves as a catalyst for the conversion of ADP to ATP. Minerals such as sodium and phosphorus work in buffer systems, which help control the pH of body fluids. Sodium also helps regulate the osmosis of water and, with other ions, is involved in the generation of nerve impulses. Table 20.1 describes the roles of several minerals in various body functions.

Vitamins

Organic nutrients required in small amounts to maintain growth and normal metabolism are called *vitamins*. Unlike carbohydrates, lipids, or proteins, vitamins do not provide energy or serve as the body's building materials. Most vitamins with known functions serve as coenzymes.

Most vitamins cannot be synthesized by the body and must be ingested. Other vitamins, such as vitamin K, are produced by bacteria in the gastrointestinal (GI) tract and then absorbed. The body can assemble some vitamins if the raw materials, called *provitamins*, are provided. For example, vitamin A is produced by the body from the provitamin beta-carotene, a chemical present in orange and yellow vegetables such as carrots and in dark green vegetables such as spinach. No single food contains all of the vitamins required by the body—one of the best reasons to eat a varied diet.

Vitamins are divided into two main groups: fat-soluble and water-soluble. The *fat-soluble vitamins* are vitamins A, D, E, and K. As you learned in Chapter 19, they are absorbed along with dietary lipids in the small intestine and packaged into chylomicrons (see Section 19.8). They cannot be absorbed in adequate quantity unless they are ingested with other lipids. Fat-soluble vitamins may be stored in cells, particularly in the liver. The *water-soluble vitamins* include the B vitamins and vitamin C. They are dissolved in body fluids. Excess quantities of these vitamins are not stored but instead are excreted in the urine.

Besides their other functions, three vitamins—C, E, and beta-carotene (a provitamin)—are termed *antioxidant vitamins* because they inactivate oxygen free radicals. Recall that free radicals are highly reactive ions or molecules that carry an unpaired electron in their outermost electron shell. Free radicals damage cell membranes, DNA, and other cellular structures and contribute to the formation of atherosclerotic plaques. Some free radicals arise naturally in the body, and others come from environmental hazards such as tobacco smoke and radiation. Antioxidant vitamins are thought to play a role in protecting against some kinds of cancer, reducing the buildup of atherosclerotic plaque, delaying some effects of aging, and decreasing the chance of cataract formation in the lenses of the eyes. Table 20.2 lists the principal vitamins, their sources, their functions, and related deficiency disorders.

TABLE 20.1

Minerals Vital to the Body

MINERAL	COMMENTS	IMPORTANCE
Calcium	Most abundant mineral in body. Appears in combination with phosphates. About 99 percent is stored in bone and teeth. Blood Ca^{2+} level is controlled by parathyroid hormone (PTH). Calcitriol promotes absorption of dietary calcium. Sources are milk, egg yolk, shellfish, and leafy green vegetables.	Formation of bones and teeth, blood clotting, normal muscle and nerve activity, endocytosis and exocytosis, cellular motility, chromosome movement during cell division, glycogen metabolism, and release of neurotransmitters and hormones.
Phosphorus	About 80 percent is found in bones and teeth as phosphate salts. Blood phosphate level is controlled by parathyroid hormone (PTH). Sources are dairy products, meat, fish, poultry, and nuts.	Formation of bones and teeth. Phosphates constitute a major buffer system of blood. Plays important role in muscle contraction and nerve activity. Component of many enzymes. Involved in energy transfer (ATP). Component of DNA and RNA.
Potassium	Major cation (K^+) in intracellular fluid. Excess excreted in urine. Present in most foods (meats, fish, poultry, fruits, and nuts).	Needed for generation and conduction of action potentials in neurons and muscle fibers.
Sulfur	Component of many proteins (such as insulin), electron carriers in electron transport chain, and some vitamins (thiamine and biotin). Sources include beef, liver, lamb, fish, poultry, eggs, cheese, and beans.	As component of hormones and vitamins, regulates various body activities. Needed for ATP production by electron transport chain.
Sodium	Most abundant cation (Na^+) in extracellular fluids; some found in bones. Normal intake of NaCl (table salt) supplies more than the required amounts.	Strongly affects distribution of water through osmosis. Part of bicarbonate buffer system. Functions in nerve and muscle action potential conduction.
Chloride	Major anion (Cl^-) in extracellular fluid. Sources include table salt (NaCl), soy sauce, and processed foods.	Plays role in acid–base balance of blood, water balance, and formation of HCl in stomach.
Magnesium	Important cation (Mg^{2+}) in intracellular fluid. Excreted in urine and feces. Widespread in various foods, such as green leafy vegetables, seafood, and whole-grain cereals.	Required for normal functioning of muscle and nervous tissue. Participates in bone formation. Constituent of many coenzymes.
Iron	About 66 percent found in hemoglobin of blood. Normal losses of iron occur by shedding of hair, epithelial cells, and mucosal cells, and in sweat, urine, feces, bile, and blood lost during menstruation. Sources are meat, liver, shellfish, egg yolk, beans, legumes, dried fruits, nuts, and cereals.	As component of hemoglobin, reversibly binds O_2. Component of cytochromes in electron transport chain.
Iodide	Essential component of thyroid hormones. Sources are seafood, iodized salt, and vegetables grown in iodine-rich soils.	Required by thyroid gland to synthesize thyroid hormones, which regulate metabolic rate.
Manganese	Some stored in liver and spleen. Sources include spinach, romaine lettuce, and pineapple.	Activates several enzymes. Needed for hemoglobin synthesis, urea formation, growth, reproduction, lactation, and bone formation.
Copper	Some stored in liver and spleen. Sources include eggs, whole-wheat flour, beans, beets, liver, fish, spinach, and asparagus.	Required with iron for synthesis of hemoglobin. Component of coenzymes in electron transport chain and enzyme necessary for melanin formation.
Cobalt	Constituent of vitamin B_{12}. Sources include liver, kidney, milk, eggs, cheese, and meat.	As part of vitamin B_{12}, required for erythropoiesis.
Zinc	Important component of certain enzymes. Widespread in many foods, especially meats.	As a component of carbonic anhydrase, important in carbon dioxide metabolism. Necessary for normal growth and wound healing, normal taste sensations and appetite, and normal sperm counts in males. As a component of peptidases, it is involved in protein digestion.
Fluoride	Components of bones, teeth, other tissues. Sources include seafood, tea, and gelatin.	Appears to improve tooth structure and inhibit tooth decay.
Selenium	Important component of certain enzymes. Found in seafood, meat, chicken, tomatoes, egg yolk, milk, mushrooms, and garlic, and cereal grains grown in selenium-rich soil.	Needed for synthesis of thyroid hormones, sperm motility, and proper functioning of the immune system. Also functions as an antioxidant. Prevents chromosome breakage and may play a role in preventing certain birth defects, miscarriage, prostate cancer, and coronary artery disease.
Chromium	Found in high concentrations in brewer's yeast. Also found in wine and some brands of beer.	Needed for normal activity of insulin in carbohydrate and lipid metabolism.

TABLE 20.2

The Principal Vitamins

VITAMIN	COMMENT AND SOURCE	FUNCTIONS	DEFICIENCY SYMPTOMS AND DISORDERS
Fat-soluble Vitamins	**All require bile salts and some dietary lipids for adequate absorption.**		
A	Formed from provitamin beta-carotene (and other provitamins) in GI tract. Stored in liver. Sources of carotene and other provitamins include orange, yellow, and green vegetables; sources of vitamin A include liver and milk.	Maintains general health and vigor of epithelial cells. Beta-carotene acts as an antioxidant to inactivate free radicals.	Atrophy and keratinization of epithelium, leading to dry skin and hair; increased incidence of ear, sinus, respiratory, urinary, and digestive system infections; inability to gain weight; drying of cornea; and skin sores.
		Essential for formation of light-sensitive pigments in photoreceptors of retina.	*Night blindness* or decreased ability for dark adaptation.
		Aids in growth of bones and teeth by helping to regulate activity of osteoblasts and osteoclasts.	Slow and faulty development of bones and teeth.
D	In the presence of sunlight, the skin produces a precursor molecule and then enzymes in the liver and kidneys modify the activated molecule, finally producing the active form of vitamin D (calcitriol). Stored in tissues to slight extent. Most excreted in bile. Dietary sources include fish-liver oils, egg yolk, and fortified milk.	Essential for absorption of calcium and phosphorus from GI tract. Works with parathyroid hormone (PTH) to maintain Ca^{2+} homeostasis.	Defective utilization of calcium by bones leads to *rickets* in children and *osteomalacia* in adults. Possible loss of muscle tone.
E (tocopherols)	Stored in liver, adipose tissue, and muscles. Sources include fresh nuts and wheat germ, seed oils, and green leafy vegetables.	Inhibits catabolism of certain fatty acids that help form cell structures, especially membranes. Involved in formation of DNA, RNA, and red blood cells. May promote wound healing, contribute to the normal structure and functioning of the nervous system, and prevent scarring. Acts as an antioxidant to inactivate free radicals.	May cause oxidation of monoun-saturated fats resulting in abnormal structure and function of mitochondria, lysosomes, and plasma membranes. A possible consequence is *hemolytic anemia*.
K	Produced by intestinal bacteria. Stored in liver and spleen. Dietary sources include spinach, cauliflower, cabbage, and liver.	Coenzyme essential for synthesis of several clotting factors by liver, including prothrombin.	Delayed clotting time results in excessive bleeding.
Water-soluble Vitamins	**Dissolved in body fluids. Most are not stored in body. Excess intake is eliminated in urine.**		
B_1 (thiamine)	Rapidly destroyed by heat. Sources include whole-grain products, eggs, pork, nuts, liver, and yeast.	Acts as a coenzyme for many different enzymes that break carbon-to-carbon bonds and are involved in carbohydrate metabolism of pyruvic acid to CO_2 and H_2O. Essential for synthesis of the neurotransmitter acetylcholine.	Buildup of pyruvic and lactic acids and insufficient production of ATP for muscle and nerve cells leads to: (1) *beriberi*, partial paralysis of smooth muscle of GI tract, causing digestive disturbances, skeletal muscle paralysis, and atrophy of limbs; (2) *polyneuritis*, due to degeneration of myelin sheaths: impaired reflexes, impaired sense of touch, stunted growth in children, and poor appetite.
B_2 (riboflavin)	Small amounts supplied by bacteria of GI tract. Dietary sources include yeast, liver, beef, veal, lamb, eggs, whole-grain products, asparagus, peas, beets, and peanuts.	Component of certain coenzymes (for example, FAD and FMN) in carbohydrate and protein metabolism, especially in cells of the eyes, skin, mucosa of the intestine, and blood.	Faulty use of oxygen resulting in blurred vision, cataracts, and corneal ulcerations. Also dermatitis and cracking of skin, lesions of intestinal mucosa, and one type of anemia.

TABLE 20.2 CONTINUES

TABLE 20.2 CONTINUED

The Principal Vitamins

VITAMIN	COMMENT AND SOURCE	FUNCTIONS	DEFICIENCY SYMPTOMS AND DISORDERS
Water-soluble Vitamins (continued)			
Niacin (nicotinamide)	Derived from amino acid tryptophan. Sources include yeast, meats, liver, fish, whole-grain products, peas, beans, and nuts.	Essential component of NAD and NADP, coenzymes in oxidation–reduction reactions. In lipid metabolism, inhibits production of cholesterol and assists in triglyceride breakdown.	*Pellagra*, characterized by dermatitis, diarrhea, and psychological disturbances.
B$_6$ (pyridoxine)	Synthesized by bacteria of GI tract. Stored in liver, muscles, and brain. Other sources include salmon, yeast, tomatoes, yellow corn, spinach, whole grain products, liver, and yogurt.	Essential coenzyme for normal amino acid metabolism. Assists production of circulating antibodies. May function as coenzyme in triglyceride metabolism.	Dermatitis of eyes, nose, and mouth, retarded growth, and nausea.
B$_{12}$ (cyanocobalamin)	Only B vitamin not found in vegetables; only vitamin containing cobalt. Absorption from GI tract depends on intrinsic factor secreted by the stomach mucosa. Sources include liver, kidney, milk, eggs, cheese, and meat.	Coenzyme necessary for red blood cell formation, formation of the amino acid methionine, entrance of some amino acids into Krebs cycle, and synthesis of choline (used to make acetylcholine).	Pernicious anemia, neuropsychiatric abnormalities (ataxia, memory loss, weakness, personality and mood changes, and abnormal sensations), and impaired activity of osteoblasts.
Pantothenic acid	Some produced by bacteria of GI tract. Stored primarily in liver and kidneys. Other sources include liver, kidneys, yeast, green vegetables, and cereal.	Constituent of coenzyme A, which is used to transfer acetyl groups into Krebs cycle; conversion of lipids and amino acids into glucose; and synthesis of cholesterol and steroid hormones.	Fatigue, muscle spasms, insufficient production of adrenal steroid hormones, vomiting, and insomnia.
Folic acid (folate, folacin)	Synthesized by bacteria of GI tract. Dietary sources include green leafy vegetables, broccoli, asparagus, breads, dried beans, and citrus fruits.	Component of enzyme systems synthesizing nitrogenous bases of DNA and RNA. Essential for normal production of red and white blood cells.	Production of abnormally large red blood cells. Higher risk of neural tube defects in babies born to folic acid–deficient mothers.
Biotin	Synthesized by bacteria of GI tract. Dietary sources include yeast, liver, egg yolk, and kidneys.	Essential coenzyme for conversion of pyruvic acid to oxaloacetic acid and synthesis of fatty acids and purines.	Mental depression, muscular pain, dermatitis, fatigue, and nausea.
C (ascorbic acid)	Rapidly destroyed by heat. Some stored in glandular tissue and plasma. Sources include citrus fruits, strawberries, melons, tomatoes, and green vegetables.	Promotes protein synthesis including synthesis of collagen in connective tissue. As coenzyme, may combine with poisons, rendering them harmless until excreted. Works with antibodies, promotes wound healing, and functions as an antioxidant.	Scurvy; anemia; many symptoms related to poor collagen formation, including tender swollen gums, loosening of teeth, poor wound healing, bleeding, impaired immune responses, and retardation of growth.

CLINICAL CONNECTION | Vitamin and Mineral Supplements

Most nutritionists recommend eating a balanced diet that includes a variety of foods rather than taking **vitamin supplements** and **mineral supplements**, except in special circumstances. Common examples of necessary supplementations include iron for women who have excessive menstrual bleeding; iron and calcium for women who are pregnant or breast-feeding; folic acid (folate) for all women who may become pregnant, to reduce the risk of fetal neural tube defects; calcium for most adults, because they do not receive the recommended amount in their diets; and vitamin B$_{12}$ for strict vegetarians, who eat no meat. Because most North Americans do not ingest in their food the high levels of antioxidant vitamins thought to have beneficial effects, some experts recommend supplementing vitamins C and E. More is not always better; larger doses of vitamins or minerals can be very harmful. •

✓ CHECKPOINT

1. Describe the USDA's MyPlate and give examples of foods from each food group.

2. Briefly describe the functions of the minerals calcium and sodium in the body.

3. Explain how vitamins are different from minerals, and distinguish between a fat-soluble vitamin and a water-soluble vitamin.

20.2 METABOLISM

OBJECTIVES • **Define metabolism and describe its importance in homeostasis.**

• **Explain how the body uses carbohydrates, lipids, and proteins.**

Metabolism (me-TAB-ō-lizm; *metabol-* = change) refers to all of the chemical reactions of the body. Recall from Chapter 2 that chemical reactions occur when chemical bonds between substances are formed or broken, and that *enzymes* serve as catalysts to speed up chemical reactions. Some enzymes require the presence of an ion such as calcium, iron, or zinc. Other enzymes work together with *coenzymes*, which function as temporary carriers of atoms being removed from or added to a substrate during a reaction. Many coenzymes are derived from vitamins. Examples include the coenzyme *NAD⁺*, derived from the B vitamin niacin, and the coenzyme *FAD*, derived from vitamin B_2 (riboflavin).

The body's metabolism may be thought of as an energy-balancing act between anabolic (synthesis) and catabolic (decomposition) reactions. Chemical reactions that combine simple substances into more complex molecules are collectively known as *anabolism* (a-NAB-ō-lizm; *ana-* = upward). Overall, anabolic reactions use more energy than they produce. The energy they use is supplied by catabolic reactions (Figure 20.2). One example of an anabolic process is the formation of peptide bonds between amino acids, combining them into proteins.

The chemical reactions that break down complex organic compounds into simple ones are collectively known as *catabolism* (ka-TAB-ō-lizm; *cata-* = downward). Catabolic reactions release the energy stored in organic molecules. This energy is transferred to molecules of ATP and then used to power anabolic reactions. Important sets of catabolic reactions occur during glycolysis, the Krebs cycle, and the electron transport chain, which are discussed shortly.

About 40 percent of the energy released in catabolism is used for cellular functions; the rest is converted to heat, some of which helps maintain normal body temperature. Excess heat is lost to the environment. Compared with machines, which typically convert only 10–20 percent of energy into work, the 40 percent efficiency of the body's metabolism is impressive. Still, the body has a continuous need to take in and process external sources of energy so that cells can synthesize enough ATP to sustain life.

Carbohydrate Metabolism

During digestion, polysaccharide and disaccharide carbohydrates are catabolized to monosaccharides—glucose, fructose, and galactose—which are absorbed in the small intestine. Shortly after their absorption, however, fructose and galactose are converted to glucose. Thus, the story of carbohydrate metabolism is really the story of glucose metabolism.

Because glucose is the body's preferred source for synthesizing ATP, the fate of glucose absorbed from the diet depends on the needs of body cells. If the cells require ATP immediately, they oxidize the glucose. Glucose not needed for immediate ATP production may be converted to glycogen for storage in liver cells and skeletal muscle fibers. If these glycogen stores are full, the liver cells can transform the glucose to triglycerides for storage in adipose tissue. At a later time, when the cells need more ATP, glycogen and the glycerol part of a triglyceride can be converted back to glucose. Cells throughout the body also can use glucose to make certain amino acids, the building blocks of proteins.

Before glucose can be used by body cells, it must pass through the plasma membrane by facilitated diffusion and enter the cytosol. Insulin increases the rate of facilitated diffusion of glucose.

Glucose Catabolism

The catabolism of glucose to produce ATP is known as *cellular respiration*. Overall, its many reactions can be summarized as follows.

$$1 \text{ glucose} + 6 \text{ oxygen} \longrightarrow 38 \text{ ATP} + 6 \text{ carbon dioxide} + 6 \text{ water}$$

Figure 20.2 Role of ATP in linking anabolic and catabolic reactions. When complex molecules are split apart (catabolism, at left), some of the energy is transferred to form ATP and the rest is given off as heat. When simple molecules are combined to form complex molecules (anabolism, at right), ATP provides the energy for synthesis, and again some energy is given off as heat.

🔑 **The coupling of energy-releasing and energy-requiring reactions is achieved through ATP.**

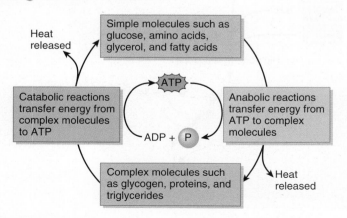

❓In a pancreatic cell that produces digestive enzymes, does anabolism or catabolism predominate?

Four interconnecting sets of chemical reactions contribute to cellular respiration (Figure 20.3):

1 During *glycolysis* (glī-KOL-i-sis; *glyco-* = sugar; *-lysis* = breakdown), reactions that take place in the cytosol convert one six-carbon glucose molecule into two three-carbon pyruvic acid molecules. The reactions of glycolysis directly produce two ATPs. They also transfer some chemical energy, in the form of high-energy electrons, from glucose to the coenzyme NAD^+, forming two $NADH + H^+$. Because glycolysis does not require oxygen, it is a way to produce ATP anaerobically (without oxygen) and is known as *anaerobic cellular respiration*. If oxygen is available, however, most cells next convert pyruvic acid to acetyl coenzyme A.

2 The formation of *acetyl coenzyme A* is a transition step that prepares pyruvic acid for entrance into the Krebs cycle. First, pyruvic acid enters a mitochondrion and is converted to a two-carbon fragment by removing a molecule of carbon dioxide (CO_2). Molecules of CO_2 produced during glucose catabolism diffuse into the blood and are eventually exhaled. Then, the coenzyme NAD^+ is converted to $NADH + H^+$. Finally, the remaining atoms, called an *acetyl group*, are attached to coenzyme A, to form acetyl coenzyme A.

3 The *Krebs cycle* is a series of reactions that transfer the chemical energy from acetyl coenzyme A to two other coenzymes—NAD^+ and FAD—thereby forming $NADH + H^+$ and $FADH_2$. Krebs cycle reactions also produce CO_2

Figure 20.3 Cellular respiration.

The catabolism of glucose to produce ATP involves glycolysis, the formation of acetyl coenzyme A, the Krebs cycle, and the electron transport chain.

1 Glucose

1 GLYCOLYSIS
in cytosol

2 ATP
2 NADH + 2 H⁺

2 Pyruvic acid

Mitochondrion

2 CO₂
2 NADH + 2 H⁺

2 FORMATION OF ACETYL COENZYME A

2 Acetyl coenzyme A

2 ATP
4 CO₂
6 NADH + 6 H⁺
2 FADH₂

3 KREBS CYCLE

High-energy electrons

4 ELECTRON TRANSPORT CHAIN

e⁻
e⁻
e⁻

32 – 34 ATP

6 O₂

6 H₂O

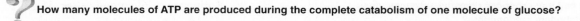

How many molecules of ATP are produced during the complete catabolism of one molecule of glucose?

and one ATP for each acetyl coenzyme A that enters the Krebs cycle. To harvest the energy in NADH and FADH$_2$, their high-energy electrons must first go through the electron transport chain.

4 Through the reactions of the *electron transport chain*, the energy in NADH + H$^+$ and FADH$_2$ is used to synthesize ATP. As the coenzymes pass their high-energy electrons through a series of "electron carriers," ATP is synthesized. Finally, lower-energy electrons are passed to oxygen in a reaction that produces water. Because the Krebs cycle and the electron transport chain together require oxygen to produce ATP, they are known collectively as *aerobic cellular respiration*.

Glucose Anabolism

Even though most of the glucose in the body is catabolized to generate ATP, glucose may take part in or be formed via several anabolic reactions. One is the synthesis of glycogen; another is the synthesis of new glucose molecules from some of the products of protein and lipid breakdown.

If glucose is not needed immediately for ATP production, it combines with many other molecules of glucose to form a long-chain molecule called *glycogen* (Figure 20.4). Synthesis of glycogen is stimulated by insulin. The body can store about 500 grams (about 1.1 lb) of glycogen, roughly 75 percent in skeletal muscle fibers and the rest in liver cells.

If blood glucose level falls below normal, glucagon is released from the pancreas and epinephrine is released from the adrenal medullae. These hormones stimulate breakdown of glycogen into its glucose subunits (Figure 20.4). Liver cells release this glucose into the blood, and body cells pick it up to use for ATP production. Glycogen breakdown usually occurs between meals.

CLINICAL CONNECTION | Carbohydrate Loading

The amount of glycogen stored in the liver and skeletal muscles varies and can be completely used up during long-term athletic endeavors. Thus, many marathon runners and other endurance athletes follow a precise exercise and dietary regimen that includes eating large amounts of complex carbohydrates, such as pasta and potatoes, in the three days before an event. This practice, called **carbohydrate loading**, helps maximize the amount of glycogen available for ATP production in muscles. For athletic events lasting more than an hour, carbohydrate loading has been shown to increase an athlete's endurance. •

When your liver runs low on glycogen, it is time to eat. If you don't, your body starts catabolizing triglycerides (fats) and proteins. Actually, the body normally catabolizes some of its triglycerides and proteins, but large-scale triglyceride and protein catabolism does not happen unless you are starving, eating very few carbohydrates, or suffering from an endocrine disorder.

Figure 20.4 Reactions of glucose anabolism: synthesis of glycogen, breakdown of glycogen, and synthesis of glucose from amino acids, lactic acid, or glycerol.

About 500 grams (1.1 lb) of glycogen are stored in skeletal muscles and the liver.

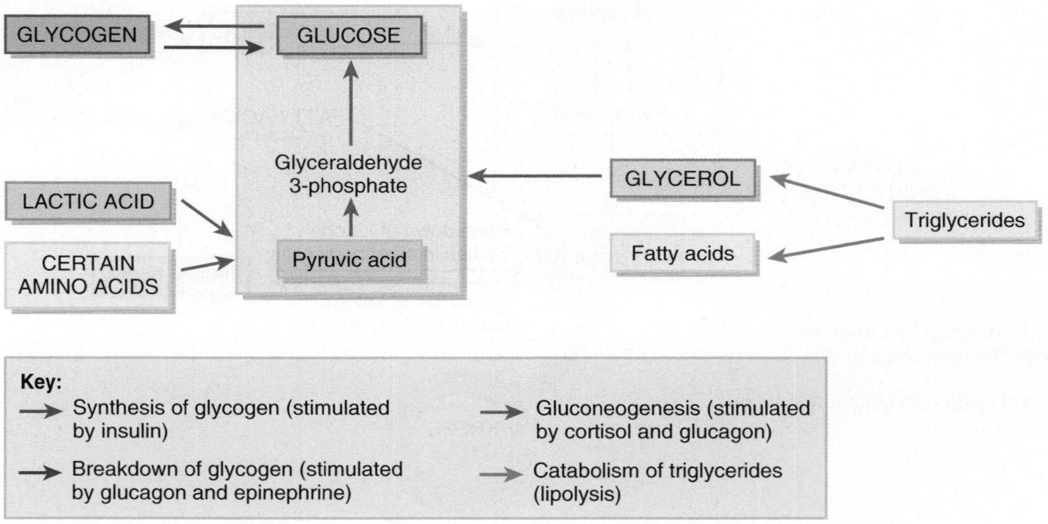

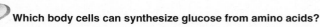

Key:

→ Synthesis of glycogen (stimulated by insulin)

→ Breakdown of glycogen (stimulated by glucagon and epinephrine)

→ Gluconeogenesis (stimulated by cortisol and glucagon)

→ Catabolism of triglycerides (lipolysis)

? Which body cells can synthesize glucose from amino acids?

Liver cells can convert the glycerol part of triglycerides, lactic acid, and certain amino acids to glucose (Figure 20.4). The series of reactions that form glucose from these noncarbohydrate sources is called **gluconeogenesis** (gloo′-kō-nē′-ō-JEN-e-sis; *neo-* = new). This process releases glucose into the blood, thereby keeping blood glucose level normal during the hours between meals when glucose is not being absorbed. Gluconeogenesis occurs when the liver is stimulated by cortisol from the adrenal cortex and glucagon from the pancreas.

Lipid Metabolism

Lipids, like carbohydrates, may be catabolized to produce ATP. If the body has no immediate need to use lipids in this way, they are stored as triglycerides in adipose tissue throughout the body, and in the liver. A few lipids are used as structural molecules or to synthesize other substances. Two essential fatty acids that the body cannot synthesize are linoleic acid and linolenic acid. Dietary sources of these lipids include vegetable oils and leafy vegetables.

Lipid Catabolism

Muscle, liver, and adipose cells routinely catabolize fatty acids from triglycerides to produce ATP. First, the triglycerides are split into glycerol and fatty acids—a process called **lipolysis** (li-POL-i-sis) (Figure 20.5). The hormones epinephrine, norepinephrine, and cortisol enhance lipolysis.

The glycerol and fatty acids that result from lipolysis are catabolized via different pathways. Glycerol is converted by many cells of the body to glyceraldehyde 3-phosphate. If the ATP supply in a cell is high, glyceraldehyde 3-phosphate is converted into glucose, an example of gluconeogenesis. If the ATP supply in a cell is low, glyceraldehyde 3-phosphate enters the catabolic pathway to pyruvic acid.

Fatty acid catabolism begins as enzymes remove two carbon atoms at a time from the fatty acid and attach them to molecules of coenzyme A, forming acetyl coenzyme A (acetyl CoA). Then the acetyl CoA enters the Krebs cycle (Figure 20.5). A 16-carbon fatty acid such as palmitic acid can yield as many as 129 ATPs via the Krebs cycle and the electron transport chain.

As part of normal fatty acid catabolism, the liver converts some acetyl CoA molecules into substances known as **ketone**

Figure 20.5 Metabolism of lipids. Lipolysis is the breakdown of triglycerides into glycerol and fatty acids. Glycerol may be converted to glyceraldehyde 3-phosphate, which can then be converted to glucose or enter the Krebs cycle. Fatty acid fragments enter the Krebs cycle as acetyl coenzyme A. Fatty acids also can be converted into ketone bodies.

 Glycerol and fatty acids are catabolized in separate pathways.

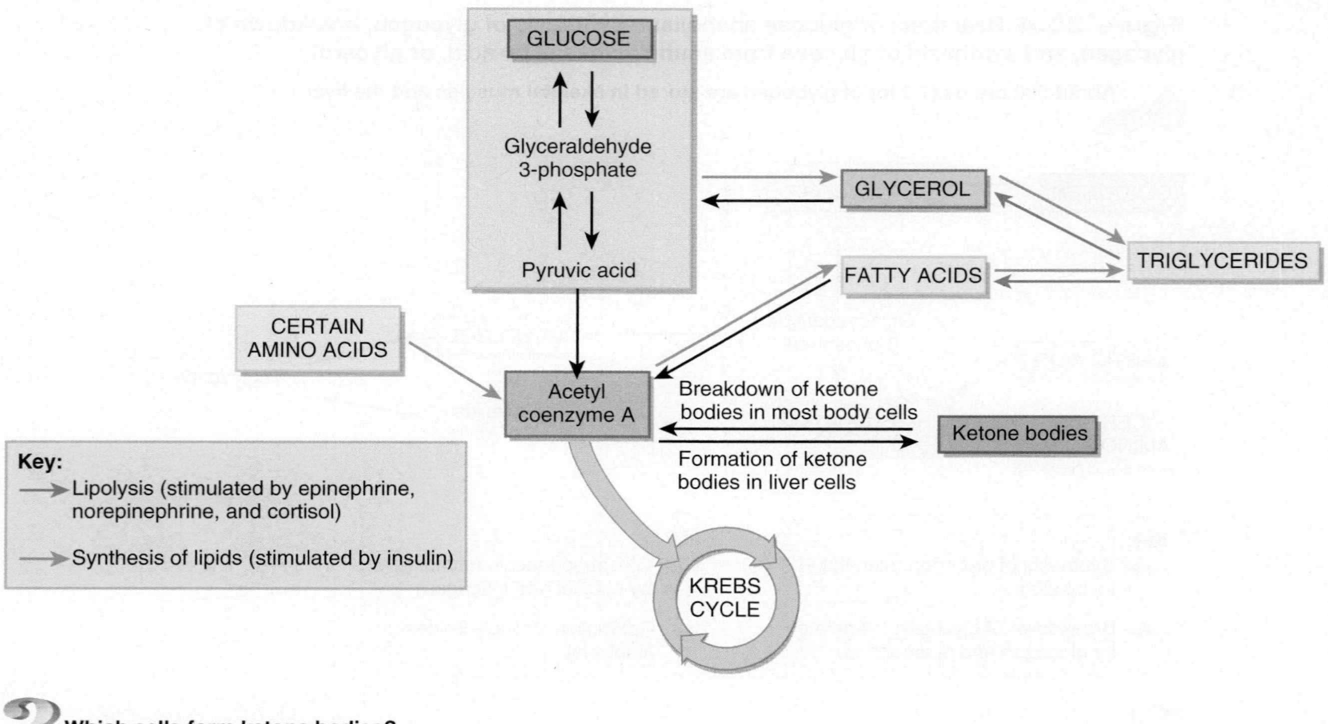

Key:
→ Lipolysis (stimulated by epinephrine, norepinephrine, and cortisol)

→ Synthesis of lipids (stimulated by insulin)

? Which cells form ketone bodies?

bodies (Figure 20.5). Ketone bodies then leave the liver to enter body cells, where they are broken down into acetyl CoA, which enters the Krebs cycle.

Lipid Anabolism

Insulin stimulates liver cells and adipose cells to synthesize triglycerides when more calories are consumed than are needed to satisfy ATP needs (Figure 20.5). Excess dietary carbohydrates, proteins, and fats all have the same fate—they are converted into triglycerides. Certain amino acids can undergo the following reactions: amino acids ➜ acetyl CoA ➜ fatty acids ➜ triglycerides. The use of glucose to form lipids takes place via two pathways:

1. glucose ➜ glyceraldehyde 3-phosphate ➜ glycerol
   or

2. glucose ➜ glyceraldehyde 3-phosphate ➜ acetyl CoA ➜ fatty acids

The resulting glycerol and fatty acids can undergo anabolic reactions to become stored triglycerides, or they can go through a series of anabolic reactions to produce other lipids such as lipoproteins, phospholipids, and cholesterol.

Lipid Transport in Blood

Most lipids, such as triglycerides and cholesterol, are not water-soluble. For transport in watery blood, such molecules first are made more water-soluble by combining them with proteins. Such *lipoproteins* are spherical particles with an outer shell of proteins, phospholipids, and cholesterol molecules surrounding an inner core of triglycerides and other lipids. The proteins in the outer shell help the lipoprotein particles dissolve in body fluids and also have specific functions.

Lipoproteins are transport vehicles: They provide delivery and pickup services so that lipids can be available when cells need them or removed when they are not needed. Lipoproteins are categorized and named mainly according to their size and density. From largest and lightest to smallest and heaviest, the four major types of lipoproteins are chylomicrons, very low-density lipoproteins, low-density lipoproteins, and high-density lipoproteins.

1. *Chylomicrons* form in absorptive epithelial cells of the small intestine and transport dietary lipids to adipose tissue for storage.

2. *Very low-density lipoproteins (VLDLs)* transport triglycerides made in liver cells to adipose cells for storage. After depositing some of their triglycerides in adipose cells, VLDLs are converted to LDLs.

3. *Low-density lipoproteins (LDLs)* carry about 75 percent of the total cholesterol in blood and deliver it to cells throughout the body for use in repair of cell membranes and synthesis of steroid hormones and bile salts.

4. *High-density lipoproteins (HDLs)* remove excess cholesterol from body cells and transport it to the liver for elimination.

Protein Metabolism

During digestion, proteins are broken down into amino acids. Unlike carbohydrates and triglycerides, proteins are not warehoused for future use. Instead, their amino acids are either oxidized to produce ATP or used to synthesize new proteins for growth and repair of body tissues. Excess dietary amino acids are converted into glucose (gluconeogenesis) or triglycerides.

The active transport of amino acids into body cells is stimulated by insulinlike growth factors (IGFs) and insulin. Almost immediately after digestion, amino acids are reassembled into proteins. Many proteins function as enzymes; other proteins are involved in transportation (hemoglobin) or serve as antibodies, clotting factors (fibrinogen), hormones (insulin), or contractile elements in muscle fibers (actin and myosin). Several proteins serve as structural components of the body (collagen, elastin, and keratin).

Protein Catabolism

A certain amount of protein catabolism occurs in the body each day, stimulated mainly by cortisol from the adrenal cortex. Proteins from worn-out cells (such as red blood cells) are broken down into amino acids. Some amino acids are converted into other amino acids, peptide bonds are reformed, and new proteins are made as part of the recycling process. Liver cells convert some amino acids to fatty acids, ketone bodies, or glucose. Figure 20.4 shows the conversion of amino acids into glucose (gluconeogenesis). Figure 20.5 shows the conversion of amino acids into fatty acids or ketone bodies.

Amino acids also are oxidized to generate ATP. Before amino acids can enter the Krebs cycle, however, their amino group ($-NH_2$) must first be removed, a process called **deamination** (dē-am′-i-NĀ-shun). Deamination occurs in liver cells and produces ammonia (NH_3). Liver cells then convert the highly toxic ammonia to urea, a relatively harmless substance that is excreted in the urine.

Protein Anabolism

Protein anabolism, the formation of peptide bonds between amino acids to produce new proteins, is carried out on the ribosomes of almost every cell in the body, directed by the cells' DNA and RNA. Insulinlike growth factors, thyroid hormones, insulin, estrogens, and testosterone stimulate protein synthesis. Because proteins are a main component of most cell structures, adequate dietary protein is especially essential during the growth years, during pregnancy, and when tissue has been damaged by disease or injury. Once dietary intake of protein is adequate, eating more protein does not increase bone or muscle mass; only a regular program of forceful, weight-bearing muscular activity accomplishes that goal.

Of the 20 amino acids in the human body, 10 are **essential amino acids**: They must be present in the diet because they cannot be synthesized in the body in adequate amounts. **Nonessential amino acids** are those synthesized by the body. They are formed by the transfer of an amino group from an amino acid to pyruvic acid or to an acid in the Krebs cycle. Once the appropriate essential and nonessential amino acids are present in cells, protein synthesis occurs rapidly. Table 20.3 summarizes the processes occurring in both catabolism and anabolism of carbohydrates, lipids, and proteins.

TABLE 20.3

Summary of Metabolism

PROCESS	COMMENT
Carbohydrate Metabolism	
Glucose catabolism	Complete catabolism of glucose (cellular respiration) is the chief source of ATP in most cells. It consists of glycolysis, the Krebs cycle, and the electron transport chain. One molecule of glucose yields 36–38 molecules of ATP.
Glycolysis	Conversion of glucose into pyruvic acid, with net production of two ATPs per glucose molecule; reactions do not require oxygen (anaerobic cellular respiration).
Krebs cycle	Series of reactions in which coenzymes (NAD^+ and FAD) pick up high-energy electrons. Some ATP is produced. CO_2, H_2O, and heat are byproducts. Reactions are aerobic (aerobic cellular respiration).
Electron transport chain	Third set of reactions in glucose catabolism in which electrons are passed from one carrier to the next and most of the ATP is produced. Reactions are aerobic (aerobic cellular respiration).
Glucose anabolism	Some glucose is converted into glycogen for storage if not needed immediately for ATP production. Glycogen can be converted back to glucose for use in ATP production. Gluconeogenesis is the synthesis of glucose from amino acids, glycerol, or lactic acid.
Lipid Metabolism	
Triglyceride catabolism	Triglycerides are broken down into glycerol and fatty acids. Glycerol may be converted into glucose (gluconeogenesis) or catabolized via glycolysis. Fatty acids are converted into acetyl CoA that can enter the Krebs cycle for ATP production or be used to form ketone bodies.
Triglyceride anabolism	Synthesis of triglycerides from glucose and amino acids. Triglycerides are stored in adipose tissue.
Protein Metabolism	
Catabolism	Amino acids are deaminated to enter the Krebs cycle. Ammonia formed during deamination is converted to urea in the liver and excreted in the urine. Amino acids may be converted into glucose (gluconeogenesis), fatty acids, or ketone bodies.
Anabolism	Protein synthesis is directed by DNA and uses the cell's RNA and ribosomes.

CLINICAL CONNECTION | Phenylketonuria

Phenylketonuria (PKU) (fen′-il-kē′-tō-NOO-rē-a) is a genetic error of protein metabolism characterized by elevated blood levels of the amino acid phenylalanine. Most children with phenylketonuria have a mutation in the gene that codes for the enzyme needed to convert phenylalanine into the amino acid tyrosine, which can enter the Krebs cycle. Because the enzyme is deficient, phenylalanine cannot be metabolized, and what is not used in protein synthesis builds up in the blood. If untreated, the disorder causes vomiting, rashes, seizures, growth deficiency, and severe mental retardation. Newborns are screened for PKU, and mental retardation can be prevented by restricting the child to a diet that supplies only the amount of phenylalanine needed for growth, although learning disabilities may still ensue. Because the artificial sweetener aspartame (NutraSweet®) contains phenylalanine, its consumption must be restricted in children with PKU. •

✓ CHECKPOINT

4. What happens during glycolysis?

5. What happens in the electron transport chain?

6. Which reactions produce ATP during the complete oxidation of a molecule of glucose?

7. What is gluconeogenesis, and why is it important?

8. What is the difference between anabolism and catabolism?

9. How does ATP provide a link between anabolism and catabolism?

10. What are the functions of the proteins in lipoproteins?

11. Which lipoprotein particles contain "bad" and "good" cholesterol, and why are these terms used?

12. Where are triglycerides stored in the body?

13. What are ketone bodies? What is ketosis?

14. What are the possible fates of the amino acids from protein catabolism?

20.3 METABOLISM AND BODY HEAT

OBJECTIVES • **Explain how body heat is produced and lost.**

• **Describe how body temperature is regulated.**

We now consider the relationship of foods to body heat, heat production and loss, and the regulation of body temperature.

Measuring Heat

Heat is a form of energy that can be measured as *temperature* and expressed in units called calories. As noted earlier in the chapter, a *calorie (cal)*, the amount of heat required to raise the temperature of 1 gram of water 1°C, is a relatively small unit, so the *kilocalorie (kcal)* or *Calorie (Cal)* (always spelled with an uppercase C) is often used to measure the body's metabolic rate and to express the energy content of foods. One kilocalorie equals 1000 Calories. Thus, when we say that a particular food item contains 500 Calories, we are actually referring to kilocalories. Knowing the caloric value of foods is important. If you know the amount of energy the body uses for various activities, you can adjust your food intake by taking in only enough kilocalories to sustain your activities.

Body Temperature Homeostasis

The body produces more or less heat depending on the rates of metabolic reactions. Homeostasis of body temperature can be maintained only if the rate of heat production by metabolism equals the rate of heat loss from the body. Thus, it is important to understand the ways in which heat can be produced and lost.

Body Heat Production

Most of the heat produced by the body comes from the catabolism of the food we eat. The rate at which this heat is produced, the *metabolic rate*, is measured in kilocalories. Because many factors affect metabolic rate, it is measured under standard conditions, with the body in a quiet, resting, and fasting condition called the *basal state*. The measurement obtained is the *basal metabolic rate (BMR)*. BMR is 1200 to 1800 Calories per day in adults, which amounts to about 24 Calories per kilogram of body mass in adult males and 22 Calories per kilogram in adult females.

The added Calories needed to support daily activities, such as digestion and walking, range from 500 Calories for a small, relatively sedentary person to over 3000 Calories for a person in training for Olympic-level competitions. The following seven factors affect metabolic rate:

1. **Exercise.** During strenuous exercise the metabolic rate increases by as much as 15 to 20 times the BMR.

2. **Hormones.** Thyroid hormones are the main regulators of BMR, which increases as the blood levels of thyroid hormones rise. Testosterone, insulin, and human growth hormone can increase the metabolic rate by 5–15 percent.

3. **Nervous system.** During exercise or in a stressful situation, the sympathetic division of the autonomic nervous system releases norepinephrine, and it stimulates release of the hormones epinephrine and norepinephrine by the adrenal medullae. Both epinephrine and norepinephrine increase the metabolic rate of body cells.

4. **Body temperature.** The higher the body temperature, the higher the metabolic rate. As a result, metabolic rate is substantially increased during a fever.

5. **Ingestion of food.** The ingestion of food, especially proteins, can raise metabolic rate by 10–20 percent.

6. **Age.** The metabolic rate of a child, in relation to its size, is about double that of an elderly person due to the high rates of growth-related reactions in children.

7. **Other factors.** Other factors that affect metabolic rate are gender (lower in females, except during pregnancy and lactation), climate (lower in tropical regions), sleep (lower), and malnutrition (lower).

Body Heat Loss

Because body heat is continuously produced by metabolic reactions, heat must also be removed continuously or body temperature would rise steadily. The four principal routes of heat loss from the body to the environment are radiation, conduction, convection, and evaporation.

1. *Radiation* is the transfer of heat in the form of infrared rays between a warmer object and a cooler one without physical contact. Your body loses heat by radiating more infrared waves than it absorbs from cooler objects. If surrounding objects are warmer than you are, you absorb more heat by radiation than you lose.

2. *Conduction* is the heat exchange that occurs between two materials that are in direct contact. Body heat is lost by conduction to solid materials in contact with the body, such as your chair, clothing, and jewelry. Heat can also be gained by conduction, for example, while soaking in a hot tub.

3. *Convection* is the transfer of heat by the movement of a gas or a liquid between areas of different temperatures. The contact of air or water with your body results in heat transfer by both conduction and convection. When cool air makes contact with the body, it becomes warmed and is carried away by convection currents. The faster the air moves—for example, by a breeze or a fan—the faster the rate of convection.

4. *Evaporation* is the conversion of a liquid to a vapor. Under typical resting conditions, about 22 percent of heat loss occurs through evaporation of water—a daily loss of about 300 mL in exhaled air and 400 mL from the skin surface. Evaporation provides the main defense against overheating during exercise. Under extreme conditions, a maximum of about 3 liters of sweat can be produced each hour, removing more than 1700 kcal of heat if all of it evaporates. Sweat that drips off the body rather than evaporating removes very little heat.

Regulation of Body Temperature

If the amount of heat production equals the amount of heat loss, you maintain a nearly constant body temperature near 37°C (98.6°F). If your heat-producing mechanisms generate more heat than is lost by your heat-losing mechanisms, your body temperature rises. For example, strenuous exercise and some infections elevate body temperature. If you lose heat faster than you produce it, your body temperature falls. Immersion in cold water, certain diseases such as hypothyroidism, and some drugs such as alcohol and antidepressants can cause body temperature to fall. An elevated temperature may destroy body proteins, and a depressed temperature may cause cardiac arrhythmias; both can lead to death.

The balance between heat production and heat loss is controlled by neurons in the hypothalamus. These neurons generate more nerve impulses when blood temperature increases and fewer impulses when blood temperature decreases. If body temperature falls, mechanisms that help conserve heat and increase heat production act by means of several negative feedback loops to raise the body temperature to normal (Figure 20.6). Thermoreceptors send nerve impulses to the hypothalamus, which produces a releasing hormone called thyrotropin-releasing hormone (TRH). TRH in turn stimulates the anterior pituitary to release thyroid-stimulating hormone (TSH). Nerve impulses from the hypothalamus and TSH then activate several effectors:

- Sympathetic nerves cause blood vessels of the skin to constrict (vasoconstriction). The decrease of blood flow slows the rate of heat loss from the skin. Because less heat is lost, body temperature increases even if the metabolic rate remains the same.

- Sympathetic nerves stimulate the adrenal medullae to release epinephrine and norepinephrine into the blood. These hormones increase cellular metabolism, which increases heat production.

- The hypothalamus stimulates parts of the brain that increase muscle tone. As muscle tone increases in one muscle (the agonist), the small contractions stretch muscle spindles in its antagonist muscle, initiating a stretch reflex. The resulting contraction in the antagonist stretches muscle spindles in the agonist, and it too develops a stretch reflex. This repetitive cycle—called *shivering*—greatly increases the rate of heat production. During maximal shivering, body heat production can rise to about four times the basal rate in just a few minutes.

- The thyroid gland responds to TSH by releasing more thyroid hormones into the blood, increasing the metabolic rate.

If body temperature rises above normal, a negative feedback system opposite to the one depicted in Figure 20.6 goes into action. The higher temperature of the blood stimulates the hypothalamus. Nerve impulses cause dilation of blood vessels in the skin. The skin becomes warm, and the excess heat is lost to

Figure 20.6 Negative feedback mechanisms that increase heat production.

When stimulated, the heat-promoting center in the hypothalamus raises body temperature.

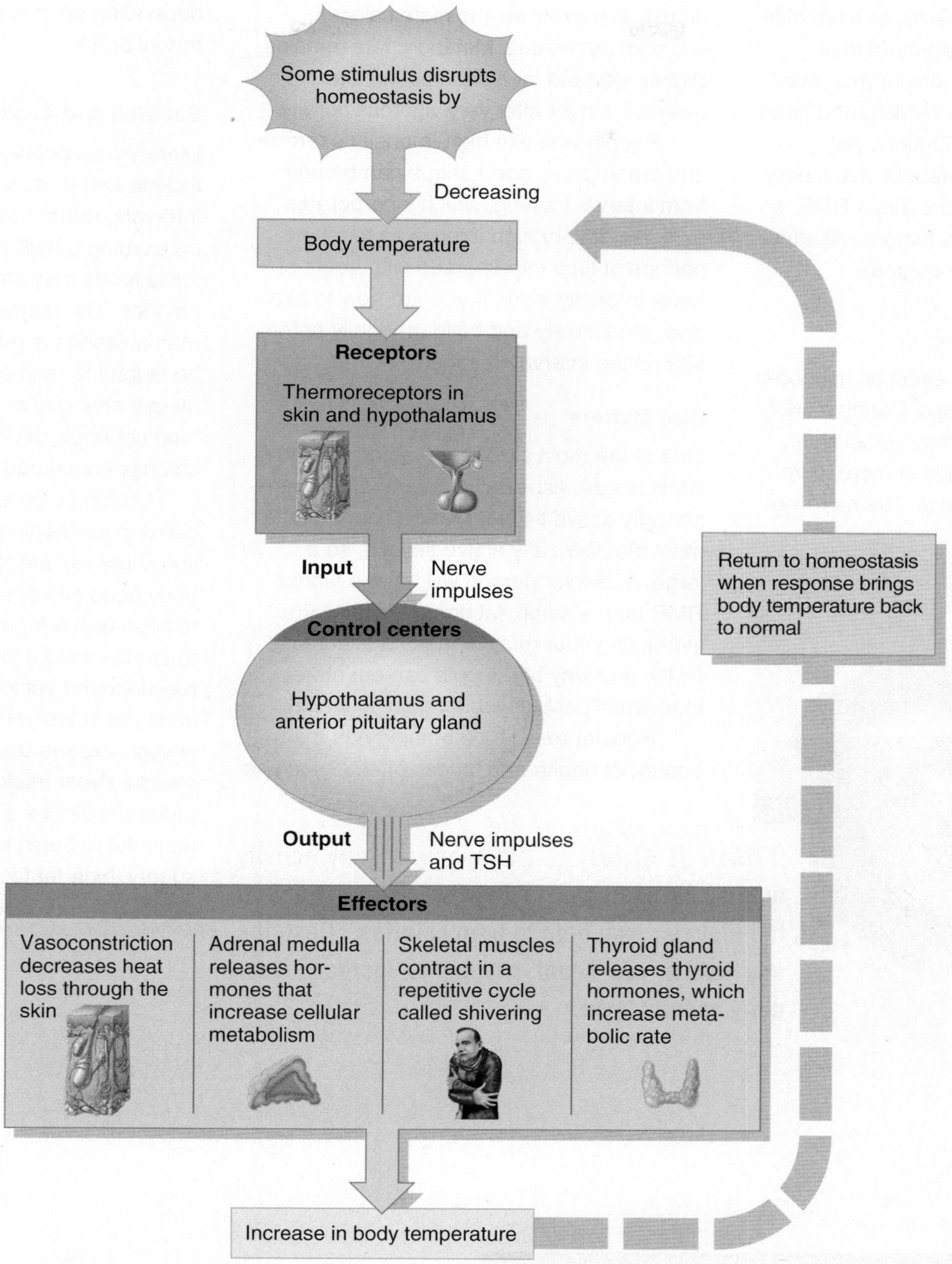

Some stimulus disrupts homeostasis by

Decreasing

Body temperature

Receptors

Thermoreceptors in skin and hypothalamus

Input   Nerve impulses

Control centers

Hypothalamus and anterior pituitary gland

Return to homeostasis when response brings body temperature back to normal

Output   Nerve impulses and TSH

Effectors

| Vasoconstriction decreases heat loss through the skin | Adrenal medulla releases hormones that increase cellular metabolism | Skeletal muscles contract in a repetitive cycle called shivering | Thyroid gland releases thyroid hormones, which increase metabolic rate |

Increase in body temperature

What factors can increase metabolic rate and thus increase heat production?

the environment by radiation and conduction as an increased volume of blood flows from the warmer interior of the body into the cooler skin. At the same time, metabolic rate decreases, and the high temperature of the blood stimulates sweat glands of the skin by means of hypothalamic activation of sympathetic nerves. As the water in sweat evaporates from the surface of the skin, the skin is cooled. All these responses counteract heat-promoting effects and help return body temperature to normal.

Can you "rev" your metabolic rate?

People who gain weight easily, or have difficulty losing weight, often attribute their problems to a sluggish or slower than average resting metabolic rate (RMR), and hope for tricks to rev up the revolutions per minute (RPMs) on the metabolic machinery of their bodies. Many factors affect RMR, so anything that affects these factors will affect the idling of the metabolic engine.

Muscular Contraction

By far, the most significant effect on metabolic rate is achieved with exercise. During moderately vigorous physical activity you elevate your metabolic rate ten times or more, burning hundreds of extra calories. The more vigorously you exercise, the more calories you burn per minute. Metabolic rate remains slightly elevated for a short period after exercise, longer after very vigorous activity.

People who like high-intensity exercise and are in pretty good shape can benefit from interval training, which incorporates exercise at very high intensities for short periods of time interspersed with bouts of lower intensity work. If you are new to exercise, start slowly and build gradually before attempting interval training.

Size Matters

One of the most significant factors affecting RMR is size, especially the amount of metabolically active tissue. Muscle is one of the most metabolically active tissues, so a large, muscular person will have a higher RMR than a small, fat one. This is why increasing your muscle mass will increase RMR, and why big people can eat more than small people without gaining weight.

Regular resistance training has many beneficial health and fitness effects, such as strengthening muscles, joints, and bones. It can also increase muscle mass somewhat, depending upon your age, sex, and hormonal profile.

Eat Well and Avoid Restrictive Diets

Dietary manipulations to increase RMR include eating several small meals at regular intervals, rather than one large meal, and consuming a high-protein diet. Consuming spicy foods may also burn a few extra calories. The cumulative effect of dietary manipulations is relatively small, but may be helpful for some people, especially if the dietary changes also reduce hunger and food cravings, decreasing the number of calories consumed per day.

Limiting calorie intake too severely can depress resting metabolic rate, a reaction known as the "starvation response." Your body goes into energy conservation mode to cope with a food shortage. This response frustrates weight loss attempts as dieters feel deprived yet lose no weight. While you must decrease your food intake to lose weight, experts usually recommend decreasing your intake by only about a few hundred calories a day (if your diet is fairly healthful to begin with) by eliminating sugary, high-fat foods.

Think It Over . . . **Changes in daily activity habits, for example, taking the stairs instead of the elevator, can help to burn calories. Describe any habits you could change to increase your daily activity level.**

CLINICAL CONNECTION | Hypothermia

Hypothermia is a lowering of core body temperature to 35°C (95°F) or below. Causes of hypothermia include an overwhelming cold stress (immersion in icy water), metabolic diseases (hypoglycemia, adrenal insufficiency, or hypothyroidism), drugs (alcohol, antidepressants, sedatives, or tranquilizers), burns, and malnutrition. Symptoms of hypothermia include sensation of cold, shivering, confusion, vasoconstriction, muscle rigidity, slow heart rate, loss of spontaneous movement, and coma. Death is usually caused by cardiac arrhythmias. Because the elderly have reduced metabolic protection against a cold environment coupled with a reduced perception of cold, they are at greater risk for developing hypothermia. •

✓ CHECKPOINT

15. In what ways can a person lose heat to or gain heat from the surroundings? How is it possible for a person to lose heat on a sunny beach when the temperature is 40°C (104°F) and the humidity is 85 percent?

COMMON DISORDERS

Fever

A *fever* is an elevation of body temperature that results from a resetting of the hypothalamic thermostat. The most common causes of fever are viral or bacterial infections and bacterial toxins; other causes are ovulation, excessive secretion of thyroid hormones, tumors, and reactions to vaccines. When phagocytes ingest certain bacteria, they are stimulated to secrete a *pyrogen* (PĪ-rō-gen; *pyro-* = fire; *-gen* = produce), a fever-producing substance. The pyrogen circulates to the hypothalamus and induces secretion of prostaglandins. Some prostaglandins can reset the hypothalamic thermostat at a higher temperature, and temperature-regulating reflex mechanisms then act to bring body temperature up to this new setting. *Antipyretics* are agents that relieve or reduce fever. Examples include aspirin, acetaminophen (Tylenol®), and ibuprofen (Advil®), all of which reduce fever by inhibiting synthesis of certain prostaglandins.

Although death results if core temperature rises above 44–46°C (112–114°F), up to a point, fever is beneficial. A higher temperature intensifies the effect of interferon and the phagocytic activities of macrophages while hindering replication of some pathogens. Because fever increases heart rate, infection-fighting white blood cells are delivered to sites of infection more rapidly. In addition, antibody production and T cell proliferation increase.

Obesity

Obesity is body weight more than 20 percent above a desirable standard due to an excessive accumulation of adipose tissue; it affects one-third of the adult population in the United States. (An athlete may be *overweight* due to a higher-than-normal amount of muscle tissue without being obese.) Even moderate obesity is hazardous to health; it is implicated as a risk factor in cardiovascular disease, hypertension, pulmonary disease, non-insulin-dependent diabetes mellitus, arthritis, certain cancers (breast, uterus, and colon), varicose veins, and gallbladder disease.

In a few cases, obesity may result from trauma to or tumors in the food-regulating centers in the hypothalamus. In most cases of obesity, no specific cause can be identified. Contributing factors include genetic factors, eating habits taught early in life, overeating to relieve tension, and social customs.

MEDICAL TERMINOLOGY AND CONDITIONS

Bulimia (*bu-* = ox; *-limia* = hunger) or *binge–purge syndrome* A disorder that typically affects young, single, middle-class, white females, characterized by overeating at least twice a week followed by purging which can be accomplished by self-induced vomiting, strict dieting or fasting, vigorous exercise, or use of laxatives or diuretics; it occurs in response to fears of being overweight or to stress, depression, and physiological disorders such as hypothalamic tumors.

Heat cramps Cramps that result from profuse sweating. The salt lost in sweat causes painful contractions of muscles; such cramps tend to occur in muscles used while working but do not appear until the person relaxes once the work is done. Drinking salted liquids usually leads to rapid improvement.

Heat exhaustion (heat prostration) A condition in which the core temperature is generally normal, or a little below, and the skin is cool and moist due to profuse perspiration. Heat exhaustion is usually characterized by loss of fluid and electrolytes, especially salt (NaCl). The salt loss results in muscle cramps, dizziness, vomiting, and fainting; fluid loss may cause low blood pressure. Complete rest, rehydration, and electrolyte replacement are recommended.

Heatstroke (sunstroke) A severe and often fatal disorder caused by exposure to high temperatures. Blood flow to the skin is decreased, perspiration is greatly reduced, and body temperature rises sharply because of failure of the hypothalamic thermostat. Body temperature may reach 43°C (110°F). Treatment, which must be undertaken immediately, consists of cooling the body by immersing the victim in cool water and by administering fluids and electrolytes.

Kwashiorkor (kwash′-ē-OR-kor) A disorder in which protein intake is deficient despite normal or nearly normal caloric intake, characterized by edema of the abdomen, enlarged liver, decreased blood pressure, low pulse rate, lower than normal body temperature, and sometimes mental retardation. Because the main protein in corn lacks two essential amino acids, which are needed for growth and tissue repair, many African children whose diet consists largely of cornmeal develop kwashiorkor.

Malnutrition (*mal-* = bad) An imbalance of total caloric intake or intake of specific nutrients, which can be either inadequate or excessive.

Marasmus (mar-AZ-mus) A type of undernutrition that results from inadequate intake of both protein and calories. Its characteristics include retarded growth, low weight, muscle wasting, emaciation, dry skin, and thin, dry, dull hair.

CHAPTER REVIEW AND RESOURCE SUMMARY

REVIEW	RESOURCES

Introduction

1. The food we eat is our only source of energy for performing biological work; it also provides essential substances that we cannot synthesize.

2. Food molecules absorbed by the gastrointestinal tract are used to supply energy for life processes, serve as building blocks during synthesis of complex molecules, or are stored for future use.

20.1 Nutrients

1. **Nutrients** include carbohydrates, lipids, proteins, water, minerals, and vitamins. Each gram of protein or carbohydrate in food provides about 4 Calories; 1 gram of fat provides about 9 Calories. Nutrition experts suggest dietary Calories be 50–60 percent from carbohydrates, 30 percent or less from fats, and 12–15 percent from proteins.

2. **MyPlate** emphasizes proportionality, variety, moderation, and nutrient density. In a healthy diet vegetables and fruits take up half the plate, while protein and grains take up the other half. Vegetables and grains represent the largest portion. Three servings of dairy a day are also recommended.

3. Some **minerals** known to perform essential functions include calcium, phosphorus, potassium, sodium, chloride, magnesium, iron, manganese, copper, and zinc. Their functions are summarized in Table 20.1.

4. **Vitamins** are organic nutrients that maintain growth and normal metabolism. Many function as coenzymes. **Fat-soluble vitamins** are absorbed with fats and include vitamins A, D, E, and K; **water-soluble vitamins** are absorbed with water and include the B vitamins and vitamin C. The functions of the principal vitamins and their deficiency disorders are summarized in Table 20.2.

Anatomy Overview—
Role of Nutrients

20.2 Metabolism

1. **Metabolism** refers to all chemical reactions of the body and has two phases: anabolism and catabolism. **Anabolism** consists of reactions that combine simple substances into more complex molecules. **Catabolism** consists of reactions that break down complex organic compounds into simple ones. Metabolic reactions are catalyzed by **enzymes**, proteins that speed up chemical reactions without being changed. Anabolic reactions require energy, which is supplied by catabolic reactions.

2. During digestion, polysaccharides and disaccharides are converted to glucose. Glucose moves into cells by facilitated diffusion, which is stimulated by insulin. Some glucose is catabolized by cells to produce ATP. Excess glucose can be stored by the liver and skeletal muscles as glycogen or converted to fat. Glucose catabolism is also called **cellular respiration**. The complete catabolism of glucose to produce ATP involves glycolysis, the Krebs cycle, and the electron transport chain. It can be represented as follows: 1 glucose + 6 oxygen ➤ 38 ATP + 6 carbon dioxide + 6 water.

3. **Glycolysis** is also called **anaerobic cellular respiration** because it occurs without oxygen. During glycolysis, which occurs in the cytosol, one glucose molecule is broken down into two molecules of pyruvic acid. Glycolysis yields a net of two ATP and two NADH + H^+.

4. When oxygen is plentiful, most cells convert pyruvic acid to **acetyl coenzyme A**, which enters the Krebs cycle. The **Krebs cycle** occurs in mitochondria. The chemical energy originally contained in glucose, pyruvic acid, and acetyl coenzyme A is transferred to the coenzymes NADH and $FADH_2$.

5. The **electron transport chain** is a series of reactions that occur in mitochondria in which the energy in the reduced coenzymes is transferred to ATP.

6. The conversion of glucose to **glycogen** for storage occurs extensively in liver and skeletal muscle fibers and is stimulated by insulin. The body can store about 500 g of glycogen. The breakdown of glycogen to glucose occurs mainly between meals. **Gluconeogenesis** is the conversion of glycerol, lactic acid, or amino acids to glucose.

Animation—Introduction to
    Metabolism
Animation—Carbohydrate
    Metabolism
Animation—Protein Metabolism
Animation—Lipid Metabolism
Animation—Regulation of
    Metabolism

Figure 20.3 Cellular respiration

REVIEW	RESOURCES

7. Some triglycerides may be catabolized to produce ATP; others are stored in adipose tissue. Other lipids are used as structural molecules or to synthesize other substances. Triglycerides must be split into fatty acids and glycerol before they can be catabolized. Glycerol can be transformed into glucose by conversion into glyceraldehyde 3-phosphate. Fatty acids are catabolized through formation of acetyl coenzyme A, which can enter the Krebs cycle. The formation of **ketone bodies** by the liver is a normal phase of fatty acid catabolism, but an excess of ketone bodies, called ketosis, may cause acidosis.

8. The conversion of glucose or amino acids into lipids is stimulated by insulin. **Lipoproteins** transport lipids in the bloodstream. Types of lipoproteins include **chylomicrons**, which carry dietary lipids to adipose tissue; **very low-density lipoproteins (VLDLs)**, which carry triglycerides from the liver to adipose tissue; **low-density lipoproteins (LDLs)**, which deliver cholesterol to body cells; and **high-density lipoproteins (HDLs)**, which remove excess cholesterol from body cells and transport it to the liver for elimination.

9. Amino acids, under the influence of insulinlike growth factors and insulin, enter body cells by means of active transport. Inside cells, amino acids are reassembled into proteins that function as enzymes, hormones, structural elements, and so forth; stored as fat or glycogen; or used for ATP production. Before amino acids can be catabolized, they must deaminated (**deamination**). Liver cells convert the resulting ammonia to urea, which is excreted in urine. Amino acids may also be converted into glucose, fatty acids, and ketone bodies. Protein synthesis is stimulated by insulinlike growth factors, thyroid hormones, insulin, estrogen, and testosterone. It is directed by DNA and RNA and carried out on ribosomes.

10. Table 20.3 summarizes carbohydrate, lipid, and protein metabolism.

20.3 Metabolism and Body Heat

1. A **calorie** is the amount of energy required to raise the temperature of 1 gram of water 1°C. The **Calorie** is the unit of heat used to express the caloric value of foods and to measure the body's metabolic rate. One Calorie equals 1000 Calories, or 1 **kilocalorie**.

2. Most body **heat** is a result of catabolism of the food we eat. The rate at which this heat is produced is known as the **metabolic rate** and is affected by exercise, hormones, the nervous system, body temperature, ingestion of food, age, gender, climate, sleep, and nutrition. Measurement of the metabolic rate under basal conditions is called the basal **metabolic rate (BMR)**.

3. Mechanisms of heat loss are radiation, conduction, convection, and evaporation. **Radiation** is the transfer of heat from a warmer object to a cooler object without physical contact. **Conduction** is the transfer of heat between two objects in contact with each other. **Convection** is the transfer of heat by the movement of a liquid or gas between areas of different temperatures. **Evaporation** is the conversion of a liquid to a vapor; in the process, heat is lost.

4. A normal body temperature is maintained by negative feedback loops that regulate heat-producing and heat-losing mechanisms. Responses that produce or retain heat when body temperature falls include vasoconstriction; release of epinephrine, norepinephrine, and thyroid hormones; and **shivering**. Responses that increase heat loss when body temperature rises include vasodilation, decreased metabolic rate, and evaporation of sweat.

RESOURCES

Animation—Metabolic Rate, Heat, and Thermoregulation
Animation—Regulation of Metabolism

SELF-QUIZ

1. Glycolysis is an example of _____ while the production of glycogen is an example of _____.
   a. deamination, cellular respiration
   b. anabolism, catabolism   c. catabolism, gluconeogenesis
   d. catabolism, anabolism
   e. cellular respiration, catabolism

2. Free radicals
   a. are a type of provitamin.
   b. are essential amino acids.
   c. can cause damage to cellular structures.
   d. help regulate enzymatic reactions.
   e. are a form of energy.

3. Which of the following statements about vitamins is NOT true?
   a. Most vitamins are synthesized by the body cells.
   b. Vitamins can act as coenzymes.
   c. Vitamin K is produced by bacteria in the GI tract.
   d. Lipid-soluble vitamins may be stored in the liver.
   e. Excess water-soluble vitamins are excreted in urine.

4. Match the following:
   ____ a. precursor for vitamin A
   ____ b. form in which lipids are transported in the blood plasma
   ____ c. needed to convert ADP to ATP
   ____ d. derived from vitamin B_2 (riboflavin)

   A. lipoproteins
   B. FAD
   C. beta-carotene
   D. magnesium

5. Body temperature is controlled by the
   a. pons.   b. thyroid gland.   c. hypothalamus.
   d. adrenal medulla.   e. autonomic nervous system.

6. The removal of an amino group ($-NH_2$) from amino acids entering the Krebs cycle is known as
   a. deamination.   b. convection.   c. ketogenesis.
   d. lipolysis.   e. aerobic respiration.

7. If your diet is low in carbohydrates, which compound(s) does your body begin to catabolize next for ATP production?
   a. vitamins   b. triglycerides   c. minerals
   d. cholesterol   e. amino acids

8. Cellular respiration includes the following steps in order:
   a. Krebs cycle, glycolysis, electron transport chain
   b. Krebs cycle, electron transport chain, glycolysis
   c. glycolysis, electron transport chain, Krebs cycle
   d. electron transport chain, Krebs cycle, glycolysis
   e. glycolysis, Krebs cycle, electron transport chain

9. Which of the following is most often used to synthesize ATP?
   a. galactose   b. triglycerides   c. amino acids
   d. glucose   e. glycerol

10. How does glucose enter the cytosol of cells?
    a. facilitated diffusion   b. simple diffusion
    c. active transport   d. osmosis   e. electron transport

11. Sweat drying from a person's skin surface causes loss of body heat by
    a. radiation.   b. conduction.   c. convection.
    d. evaporation.   e. shivering.

12. Lipid metabolism
    a. can result in the production of ketone bodies.
    b. results in small quantities of ATP.
    c. is a method for storing fatty acids.
    d. requires essential amino acids.
    e. is also known as metabolism.

13. Which of the following statements is NOT true?
    a. Triglycerides are stored in adipose tissue.
    b. Chylomicrons form in the small intestines and transport lipids for storage.

c. Most of the body's cholesterol is carried in low-density lipoproteins.
   d. Lipids can be stored in the liver.
   e. High-density lipoproteins contribute to the formation of fatty plaques.

14. Which of the following equations summarizes the complete catabolism of a molecule of glucose?
    a. glucose + 6 water → 36 or 38 ATP + 6 CO_2 + 6 O_2
    b. glucose + 6 O_2 → 36 or 38 ATP + 6 CO_2 + 6 water
    c. glucose + ATP → 31 or 38 CO_2 + 6 water
    d. glucose + pyruvic acid → 36 or 38 ATP + 6 O_2
    e. glucose + citric acid → 31 or 38 ATP + 6 CO_2

15. Those amino acids that cannot be synthesized by the body and must be obtained from the diet are known as
    a. coenzymes.   b. ketones.
    c. essential amino acids.   d. nonessential amino acids.
    e. polypeptides.

16. Which of the following would NOT increase the metabolic rate?
    a. increased levels of thyroid hormones
    b. epinephrine   c. old age
    d. fever   e. exercise

17. FAD and NAD^+ are examples of
    a. nutrients.   b. antioxidants.   c. pyrogens.
    d. coenzymes.   e. minerals.

18. The process by which glucose is formed from amino acids is
    a. gluconeogenesis.   b. deamination.
    c. anaerobic respiration.   d. ketogenesis.   e. glycolysis.

19. All of the following can contribute to an increase in body temperature EXCEPT
    a. shivering.
    b. release of thyroid hormones.
    c. sympathetic stimulation of the adrenal medulla.
    d. vasodilation of blood vessels in the skin.
    e. activation of the hypothalamus.

20. Match the following (answers may be used multiple times):
    ____ a. conversion of glucose to pyruvic acid
    ____ b. the complete breakdown of glucose
    ____ c. also known as anaerobic cellular respiration
    ____ d. NAD^+ and FAD pick up high-energy electrons
    ____ e. passing low-energy electrons to O_2 resulting in the production of water and ATP
    ____ f. requires oxygen
    ____ g. reactions occur in the cytosol
    ____ h. final products include 36–38 ATP

    A. glycolysis
    B. cellular respiration
    C. Krebs cycle
    D. electron tansport chain

CRITICAL THINKING APPLICATIONS

1. Suzy and Anna want to follow a healthy diet. They recently learned in their Nutrition class that the USDA's MyPyramid had been replaced by MyPlate to reflect new guidelines for healthy eating. Based on your knowledge of MyPlate, what suggestions could you give Suzy and Anna to maintain their healthy lifestyles? Do this by drawing a plate, coloring the four areas, indicating what each means, and giving several examples. Don't forget the blue cup.

2. It's noon on a hot summer day, the sun is directly overhead, and a group of sunbathers lies on the beach. What mechanism causes their body temperature to increase? Several of the sunbathers jump into the cool water. What mechanisms decrease their body temperature?

3. Neil is training for a marathon. He has heard that eating lots of pasta, bread, and rice will help his performance. Would this diet be of any benefit to Neil in his quest to run the marathon?

4. Naomi swallows a multivitamin tablet every morning and an antioxidant tablet containing beta-carotene, vitamin C, and vitamin E with her dinner every night. What are the functions of antioxidants in the body? What happens to the antioxidants if any exceed her daily requirements?

ANSWERS TO FIGURE QUESTIONS

20.1 The blue cup is a reminder to include three daily servings of dairy such as milk, yogurt, and cheese.

20.2 The formation of digestive enzymes in the pancreas is part of anabolism, so anabolism would predominate.

20.3 Complete catabolism of one molecule of glucose yields 36–38 ATP.

20.4 Liver cells can carry out gluconeogenesis.

20.5 Liver cells form ketone bodies.

20.6 Exercise, the sympathetic nervous system, hormones (epinephrine, norepinephrine, thyroid hormones, testosterone, human growth hormone), elevated body temperature, and ingestion of food are factors that increase metabolic rate.

CHAPTER 21
THE URINARY SYSTEM

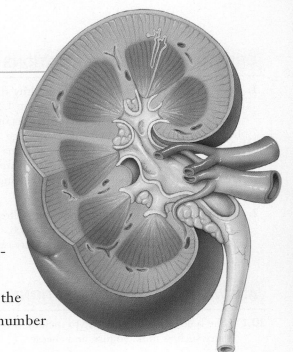

As body cells carry out their metabolic functions, they consume oxygen and nutrients and produce substances, such as carbon dioxide, that have no useful functions and need to be eliminated from the body. While the respiratory system rids the body of carbon dioxide, the urinary system disposes of most other unneeded substances. As you will learn in this chapter, however, the urinary system is not merely concerned with waste disposal; it carries out a number of other important functions as well.

Did you know?   Urine is eliminated from the body through the urethra. The external urethral opening allows contact between the urinary system and the big, wide world. Bacteria can enter the system through this opening. While in most cases the immune system disables harmful pathogens, some people, especially women, are prone to recurrent urinary tract infections. In severe cases, infection can spread up through the urinary system, into the urinary bladder and kidneys. Urinary tract infections can usually be cured and prevented by a combination of antibiotics and good hygiene.

LOOKING BACK TO MOVE AHEAD...

Transport Across the Plasma Membrane (Section 3.3)

Simple Cuboidal Epithelium (Section 4.2)

Transitional Epithelium (Section 4.2)

Actions of Antidiuretic Hormone (ADH) (Section 13.3)

Vitamin D, Calcitriol, and Calcium Homeostasis (Section 13.5)

Renin–Angiotensin–Aldosterone Pathway (Section 13.7)

Filtration and Reabsorption in Capillaries (Section 16.1)

Blood Colloid Osmotic Pressure (Section 16.1)

FOCUS ON WELLNESS
Infection Prevention for Recurrent UTIs

21.1 OVERVIEW OF THE URINARY SYSTEM

OBJECTIVE • **List the components of the urinary system and their general functions.**

The ***urinary system*** consists of two kidneys, two ureters, one urinary bladder, and one urethra (Figure 21.1). After the kidneys filter blood, they return most of the water and many of the solutes to the bloodstream. The remaining water and solutes constitute ***urine***, which passes through the ureters and is stored in the urinary bladder until it is expelled from the body through the urethra. ***Nephrology*** (nef-ROL-ō-jē; *nephro-* = kidney; *-logy* = study of) is the scientific study of the anatomy, physiology, and disorders of the kidneys. The branch of medicine that deals with the male and female urinary systems and the male reproductive system is ***urology*** (ū-ROL-ō-jē; *uro-* = urine). A physician who specializes in this branch of medicine is called a ***urologist*** (ū-ROL-ō-jist).

The kidneys do the major work of the urinary system. The other parts of the system are primarily passageways and temporary storage areas. Functions of the kidneys include the following:

- **Regulation of ion levels in the blood.** The kidneys help regulate the blood levels of several ions, most importantly sodium ions (Na^+), potassium ions (K^+), calcium ions (Ca^{2+}), chloride ions (Cl^-), and phosphate ions (HPO_4^{2-}).

- **Regulation of blood volume and blood pressure.** The kidneys adjust the volume of blood in the body by returning water to the blood or eliminating it in the urine. They help regulate blood pressure by secreting the enzyme renin, which activates the renin–angiotensin–aldosterone pathway (see Figure 13.14), by adjusting blood flow into and out of the kidneys, and by adjusting blood volume.

- **Regulation of blood pH.** The kidneys regulate the concentration of H^+ in the blood by excreting a variable amount of H^+ in the urine. They also conserve blood bicarbonate ions (HCO_3^-), an important buffer of H^+. Both activities help regulate blood pH.

- **Production of hormones.** The kidneys produce two hormones. *Calcitriol*, the active form of vitamin D, helps

Figure 21.1 Organs of the female urinary system in relation to surrounding structures.

🔑 Urine formed by the kidneys passes first into the ureters, then to the urinary bladder for storage, and finally through the urethra for elimination from the body.

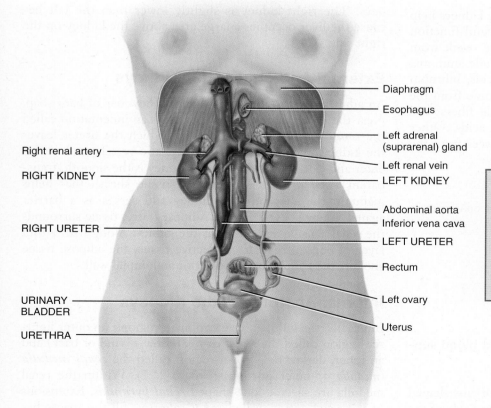

Right renal artery

RIGHT KIDNEY

RIGHT URETER

URINARY BLADDER

URETHRA

Diaphragm

Esophagus

Left adrenal (suprarenal) gland

Left renal vein

LEFT KIDNEY

Abdominal aorta

Inferior vena cava

LEFT URETER

Rectum

Left ovary

Uterus

Functions of the Urinary System

1. The kidneys regulate blood volume and composition, help regulate blood pressure and pH, produce two hormones, and excrete wastes.
2. The ureters transport urine from the kidneys to the urinary bladder.
3. The urinary bladder stores urine and expels it into the urethra.
4. The urethra discharges urine from the body.

(a) Anterior view of urinary system

FIGURE 21.1 CONTINUES

Figure 21.1 CONTINUED

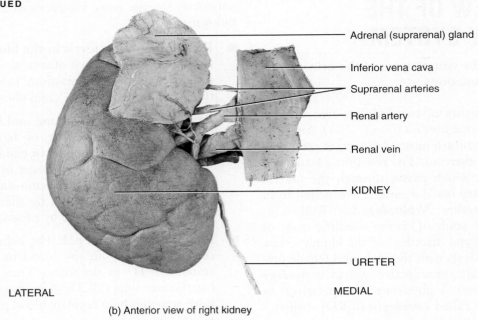

Adrenal (suprarenal) gland

Inferior vena cava

Suprarenal arteries

Renal artery

Renal vein

KIDNEY

URETER

LATERAL   MEDIAL

(b) Anterior view of right kidney

 Which organ of the urinary system does most of the work to form urine?

regulate calcium homeostasis (see Figure 13.10), and *erythropoietin* stimulates production of red blood cells (see Figure 14.4).

■ **Excretion of wastes.** By forming urine, the kidneys help excrete *wastes*—substances that have no useful function in the body. Some wastes excreted in urine result from metabolic reactions in the body. These include ammonia and urea from the breakdown of amino acids; bilirubin from the breakdown of hemoglobin; creatinine from the breakdown of creatine phosphate in muscle fibers; and uric acid from the breakdown of nucleic acids. Other wastes excreted in urine are foreign substances from the diet, such as drugs and environmental toxins.

✓ **CHECKPOINT**

1. What are wastes, and how do the kidneys take part in their removal from the body?

21.2 STRUCTURE OF THE KIDNEYS

OBJECTIVE • **Describe the structure and blood supply of the kidneys.**

The *kidneys* (KID-nēz) are a pair of reddish organs shaped like kidney beans (Figure 21.2). They lie on either side of the vertebral column between the peritoneum and the back wall

of the abdominal cavity at the level of the 12th thoracic and first three lumbar vertebrae. The 11th and 12th pairs of ribs provide some protection for the superior parts of the kidneys. The right kidney is slightly lower than the left because the liver occupies a large area above the kidney on the right side.

External Anatomy of the Kidneys

An adult kidney is about the size of a new bar of bath soap. Near the center of the medial border is an indentation called the *renal hilum* (HĪ-lum), through which the ureter leaves the kidney, and blood vessels, lymphatic vessels, and nerves enter and exit. Surrounding each kidney is the smooth, transparent *renal capsule*, a connective tissue sheath that helps maintain the shape of the kidney and serves as a barrier against trauma (Figure 21.2). Adipose (fatty) tissue surrounds the renal capsule and cushions the kidney. Along with a thin layer of dense irregular connective tissue, the adipose tissue anchors the kidney to the posterior abdominal wall.

Internal Anatomy of the Kidneys

Internally, the kidneys have two main regions: an outer light-red region called the *renal cortex* (*cortex* = rind or back) and an inner, darker red-brown region called the *renal medulla* (*medulla* = inner portion) (Figure 21.2). Within the renal medulla are several cone-shaped *renal pyramids*. Extensions of the renal cortex, called *renal columns*, fill the spaces between renal pyramids.

Figure 21.2 **Structure of the kidney.**

A renal capsule covers the kidney; internally, the two main regions are the renal cortex and the renal medulla.

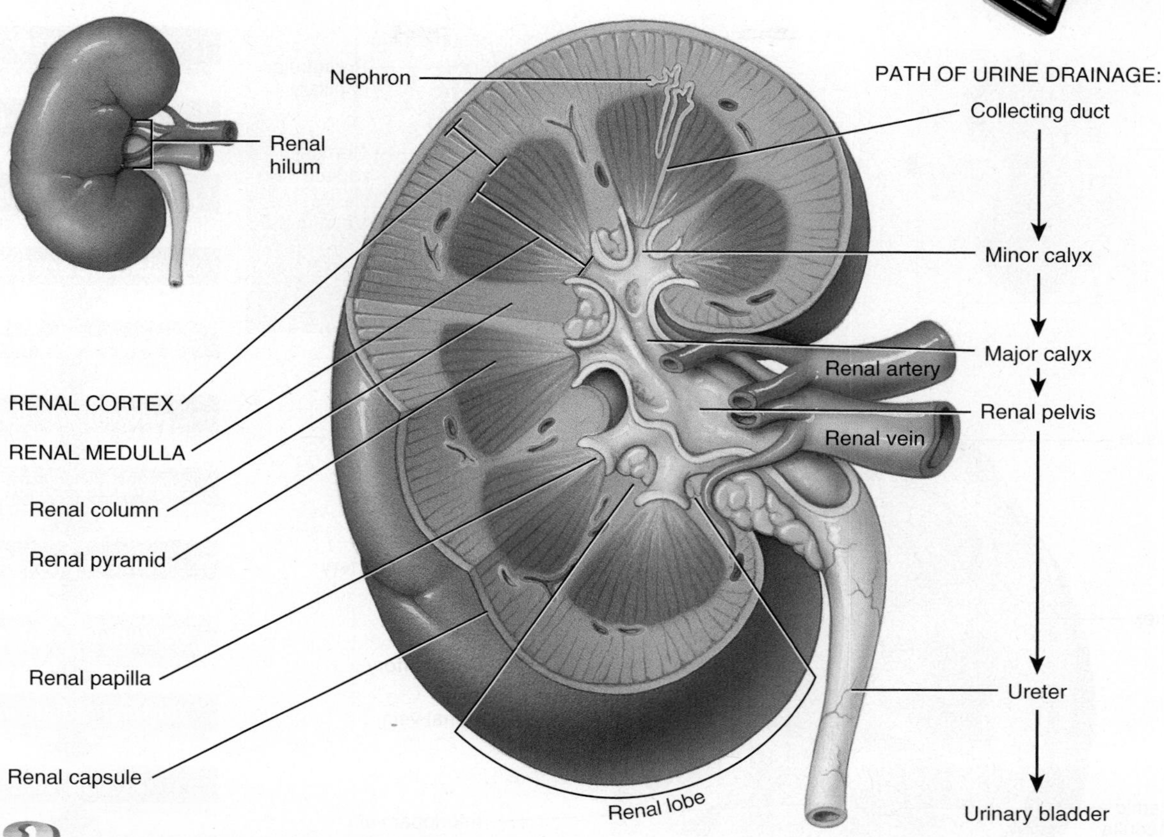

Nephron

Renal hilum

RENAL CORTEX

RENAL MEDULLA

Renal column

Renal pyramid

Renal papilla

Renal capsule

Renal lobe

PATH OF URINE DRAINAGE:

Collecting duct

Minor calyx

Major calyx

Renal artery

Renal vein

Renal pelvis

Ureter

Urinary bladder

? Where in the kidney are the renal pyramids located?

Urine formed in the kidney drains into a large, funnel-shaped cavity called the ***renal pelvis*** (*pelv-* = basin). The rim of the renal pelvis contains cuplike structures called ***major*** and ***minor calyces*** (KĀL-i-sēz = cups; singular is *calyx*). Urine flows from several ducts within the kidney into a minor calyx and from there through a major calyx into the renal pelvis, which connects to a ureter. Water and solutes in the fluid that drains into the renal pelvis remain in the urine and are *excreted* (eliminated from the body).

Renal Blood Supply

About 20–25 percent of the resting cardiac output—1200 milliliters of blood per minute—flows into the kidneys through the right and left ***renal arteries*** (Figure 21.3). Within each kidney, the renal artery divides into smaller and smaller vessels (*segmental, interlobar, arcuate* (AR-kū-āt), *cortical radiate*

(KOR-ti-kal RĀ-dē-āt) *arteries*) that eventually deliver blood to the ***afferent arterioles*** (*af-* = toward; *-ferre* = to carry). Each afferent arteriole divides into a tangled capillary network called a ***glomerulus*** (glō-MER-ū-lus = little ball; plural is *glomeruli*).

The capillaries of the glomerulus reunite to form an ***efferent arteriole*** (*ef-* = out). Upon leaving the glomerulus, each efferent arteriole divides to form a network of capillaries around the kidney tubules (described shortly). These ***peritubular capillaries*** (*peri-* = around) eventually reunite to form *peritubular veins*, which merge into *cortical radiate, arcuate*, and *interlobar veins*. Ultimately, all of these smaller veins drain into the ***renal vein***.

Nephrons

The functional units of the kidney are called the ***nephrons***

Figure 21.3 **Blood supply of the right kidney.**

The renal arteries deliver about 25 percent of the resting cardiac output to the kidneys.

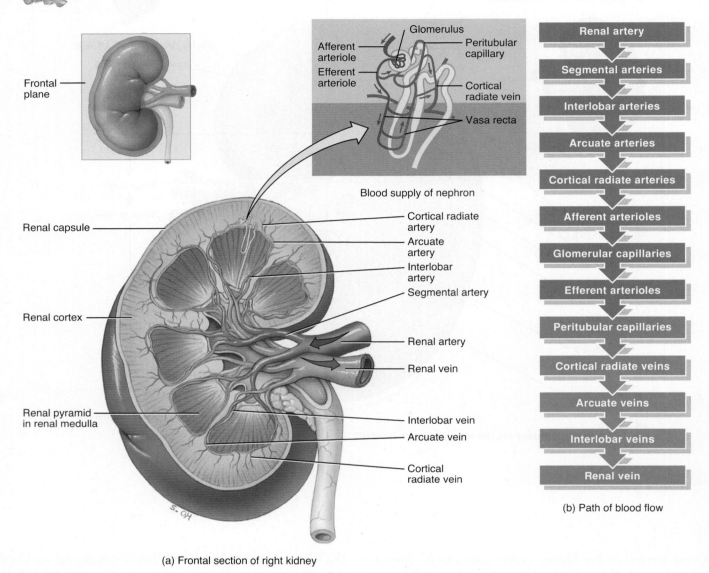

Blood supply of nephron

(a) Frontal section of right kidney

(b) Path of blood flow

How much blood enters the renal arteries each minute?

(NEF-ronz), numbering about a million in each kidney (Figure 21.4). A nephron consists of two parts: a **renal corpuscle** (KOR-pus-ul = tiny body), where blood plasma is filtered, and a **renal tubule** into which the filtered fluid, called **glomerular filtrate**, passes. Closely associated with a nephron is its blood supply. As the fluid moves through the renal tubules, wastes and excess substances are added, and useful materials are returned to the blood in the peritubular capillaries.

The two parts that make up a renal corpuscle are the **glomerulus** and the **glomerular (Bowman's) capsule**, a double-walled cup of epithelial cells that surrounds the glomerular capillaries. Glomerular filtrate first enters the glomerular capsule and then passes into the renal tubule. In the order that fluid passes through them, the three main sections of the renal tubule are the **proximal convoluted tubule**, the **nephron loop**, and the **distal convoluted tubule**. *Proximal* denotes the part of the tubule attached to the glomerular capsule, and *distal* denotes the part that is farther away. *Convoluted* means the tubule is tightly coiled rather than straight. The renal corpuscle and both convoluted tubules lie within the renal cortex; the nephron loop extends into the renal medulla. The first part of the nephron loop begins at the point where the proximal convoluted tubule takes its final turn downward. It begins in the renal cortex and extends downward into the renal medulla and is called the **descending limb of the nephron loop** (Figure 21.4). It then makes a hairpin turn and returns to the renal cortex

Figure 21.4 The parts of one type of nephron (cortical nephron), collecting duct, and associated blood vessels.
Most nephrons are cortical nephrons; their renal corpuscles lie in the outer renal cortex and their short loops of Henle are mostly in the renal cortex.

🔑 **Nephrons are the functional units of the kidneys.**

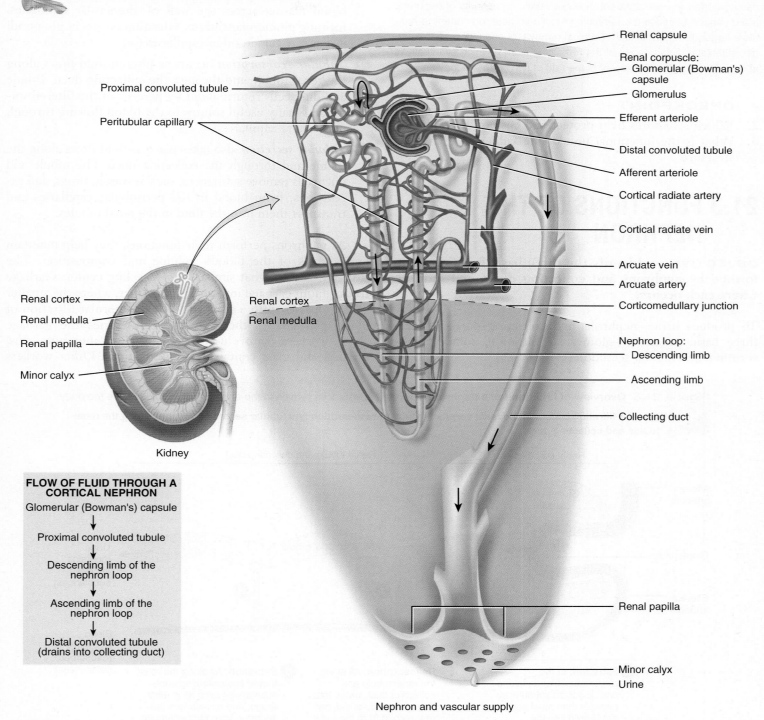

FLOW OF FLUID THROUGH A CORTICAL NEPHRON

Glomerular (Bowman's) capsule
↓
Proximal convoluted tubule
↓
Descending limb of the nephron loop
↓
Ascending limb of the nephron loop
↓
Distal convoluted tubule (drains into collecting duct)

Kidney

Nephron and vascular supply

❓ **A water molecule has just entered the proximal convoluted tubule of a nephron. Which parts of the nephron will it travel through (in order) to reach the renal pelvis in a drop of urine?**

where it terminates at the distal convoluted tubule and is known as the *ascending limb of the nephron loop*. The distal convoluted tubules of several nephrons empty into a common *collecting duct*.

✓ **CHECKPOINT**

2. Which structures help protect and cushion the kidneys?
3. What is the functional unit of the kidney? Describe its structure.

21.3 FUNCTIONS OF THE NEPHRON

OBJECTIVE • **Identify the three basic functions performed by nephrons and collecting ducts and indicate where each occurs.**

To produce urine, nephrons and collecting ducts perform three basic processes—glomerular filtration, tubular reabsorption, and tubular secretion (Figure 21.5):

1 Filtration is the forcing of fluids and dissolved substances smaller than a certain size through a membrane by pressure. *Glomerular filtration* is the first step of urine production: Blood pressure forces water and most solutes in blood plasma across the wall of glomerular capillaries, forming glomerular filtrate. Filtration occurs in glomeruli just as it occurs in other capillaries (see Figure 16.3).

2 *Tubular reabsorption* occurs as filtered fluid flows along the renal tubule and through the collecting duct: Tubule and duct cells return about 99 percent of the filtered water and many useful solutes to the blood flowing through peritubular capillaries.

3 *Tubular secretion* also takes place as fluid flows along the tubule and through the collecting duct: The tubule and duct cells remove substances, such as wastes, drugs, and excess ions, from blood in the peritubular capillaries and transport them into the fluid in the renal tubules.

As nephrons perform their functions, they help maintain homeostasis of the blood's volume and composition. The situation is somewhat similar to a recycling center: Garbage trucks dump garbage into an input hopper, where the smaller garbage passes onto a conveyor belt (glomerular filtration of blood plasma). As the conveyor belt carries the garbage along, workers remove useful items, such as aluminum cans, plastics, and glass containers (reabsorption). Other workers

Figure 21.5 Overview of functions of a nephron. Excreted substances remain in the urine and eventually leave the body.

Glomerular filtration occurs in the renal corpuscle; tubular reabsorption and tubular secretion occur all along the renal tubule and collecting duct.

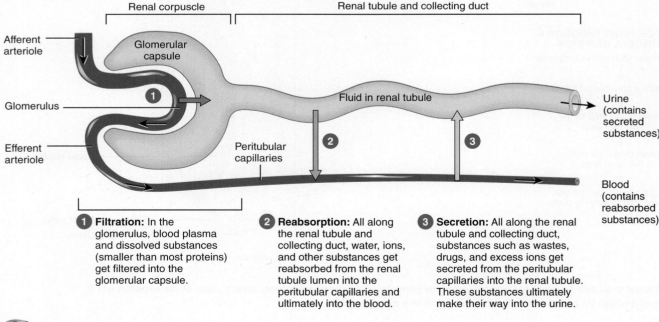

1 Filtration: In the glomerulus, blood plasma and dissolved substances (smaller than most proteins) get filtered into the glomerular capsule.

2 Reabsorption: All along the renal tubule and collecting duct, water, ions, and other substances get reabsorbed from the renal tubule lumen into the peritubular capillaries and ultimately into the blood.

3 Secretion: All along the renal tubule and collecting duct, substances such as wastes, drugs, and excess ions get secreted from the peritubular capillaries into the renal tubule. These substances ultimately make their way into the urine.

 When the renal tubules secrete the drug penicillin, is the drug being added to or removed from the blood?

place additional garbage and larger items onto the conveyor belt (secretion). At the end of the belt, all remaining garbage falls into a truck for transport to the landfill (excretion of wastes in urine).

Table 21.1 compares the substances that are filtered, reabsorbed, and secreted in urine per day in an adult male. Although the values shown are typical, they vary considerably according to diet. The following sections describe each of the three steps that contribute to urine formation in more detail.

CLINICAL CONNECTION | Kidney Transplant

A **kidney transplant** is the transfer of a kidney from a donor to a recipient whose kidneys no longer function. In the procedure, the donor kidney is placed in the pelvis of the recipient through an abdominal incision. The renal artery and vein of the transplanted kidney are attached to a nearby artery or vein in the pelvis of the recipient and the ureter of the transplanted kidney is then attached to the urinary bladder. During a kidney transplant, the patient receives only one donor kidney, since only one kidney is needed to maintain sufficient renal function. The nonfunctioning diseased kidneys are usually left in place. As with all organ transplants, kidney transplant recipients must be ever vigilant for signs of infection or organ rejection. The transplant recipient will take immunosuppressive drugs for the rest of his or her life to avoid rejection of the "foreign" organ. •

Glomerular Filtration

Two layers of cells compose the capsule that surrounds the glomerular capillaries (Figure 21.6). Think of the renal corpuscle as a fist (the glomerular capillaries) pushed into a limp balloon (the glomerular capsule) until the fist is covered by two layers of the balloon with a space, the *capsular space*, in between. The cells that make up the inner wall of the glomerular capsule, called *podocytes*, adhere closely to the endothelial cells of the glomerulus. Together, the podocytes and glomerular endothelium form a *filtration membrane* that permits the passage of water and solutes from the blood into the capsular space. Blood cells and most plasma proteins remain in the blood because they are too large to pass through the filtration membrane. Simple squamous epithelial cells form the outer layer of the glomerular capsule.

Net Filtration Pressure

The pressure that causes filtration is the blood pressure in the glomerular capillaries. Two other pressures oppose glomerular filtration: (1) blood colloid osmotic pressure (see Chapter 16) and (2) glomerular capsule pressure (due to fluid already in the capsular space and renal tubule). When either of these pressures increases, glomerular filtration decreases. Normally, blood pressure is greater than the two opposing pressures, producing a *net filtration pressure* of about 10 mm Hg. Net filtration pressure forces a large volume of fluid into the capsular space, about 150 liters daily in females and 180 liters daily in males. Net filtration pressure can be summarized as follows:

Net filtration pressure = glomerular capillary blood pressure − (blood colloidal osmotic pressure + glomerular capsule pressure)

Because the efferent arteriole is smaller in diameter than the afferent arteriole, it helps raise the blood pressure in the glomerular capillaries. When blood pressure increases or decreases slightly, changes in the diameters of the afferent and efferent arterioles can actually keep net filtration pressure steady to maintain normal glomerular filtration. Constriction of the afferent arteriole decreases blood flow into the glomerulus, which decreases net filtration pressure. Constriction of the efferent arteriole slows outflow of blood and increases net filtration pressure.

TABLE 21.1

Substances Filtered, Reabsorbed, and Excreted in Urine per Day

SUBSTANCE	FILTERED* (ENTERS RENAL TUBULE)	REABSORBED (RETURNED TO BLOOD)	SECRETED IN URINE
Water	180 liters	178–179 liters	1–2 liters
Chloride ions (Cl^-)	640 g	633.7 g	6.3 g
Sodium ions (Na^+)	579 g	575 g	4 g
Bicarbonate ions (HCO_3^-)	275 g	274.97 g	0.03 g
Glucose	162 g	162 g	0
Urea	54 g	24 g	30 g[†]
Potassium ions (K^+)	29.6 g	29.6 g	2.0 g[‡]
Uric acid	8.5 g	7.7 g	0.8 g
Creatinine	1.6 g	0	1.6 g

*Assuming glomerular filtration is 180 liters per day.
[†]In addition to being filtered and reabsorbed, urea is secreted.
[‡]After virtually all filtered K^+ is reabsorbed in the convoluted tubules and loop of Henle, a variable amount of K^+ is secreted in the collecting duct.

Figure 21.6 Glomerular filtration, the first step in urine formation.

 Glomerular filtrate (red arrows) passes into the capsular space and then into the proximal convoluted tubule.

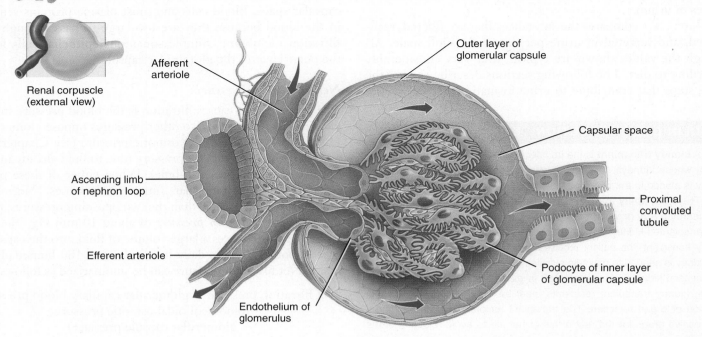

Renal corpuscle (external view)

Afferent arteriole

Outer layer of glomerular capsule

Capsular space

Ascending limb of nephron loop

Proximal convoluted tubule

Efferent arteriole

Podocyte of inner layer of glomerular capsule

Endothelium of glomerulus

Renal corpuscle (internal view)

? Which cells make up the filtration membrane in the renal corpuscle?

CLINICAL CONNECTION | Oliguria and Anuria

Conditions that greatly reduce blood pressure, for instance severe hemorrhage, may cause glomerular blood pressure to fall so low that net filtration pressure drops despite constriction of efferent arterioles. Then, glomerular filtration slows, or even stops entirely. The result is **oliguria** (*olig-* = scanty; *-uria* = urine production), a daily urine output between 50 and 250 mL, or **anuria**, a daily urine output of less than 50 mL. Obstructions, such as a kidney stone that blocks a ureter or an enlarged prostate that blocks the urethra in a male, can also decrease net filtration pressure and thereby reduce urine output. •

Glomerular Filtration Rate

The amount of filtrate that forms in both kidneys every minute is called the **glomerular filtration rate (GFR)**. In adults, the GFR is about 105 mL/min in females and 125 mL/min in males. It is very important for the kidneys to maintain a constant GFR. If the GFR is too high, needed substances pass so quickly through the renal tubules that they are unable to be reabsorbed and pass out of the body as part of urine. On the other hand, if the GFR is too low, nearly all the filtrate is reabsorbed, and waste products are not adequately excreted.

Atrial natriuretic peptide (ANP) is a hormone that promotes loss of sodium ions and water in the urine in part because it increases glomerular filtration rate. Cells in the

atria of the heart secrete more ANP if the heart is stretched more, as occurs when blood volume increases. ANP then acts on the kidneys to increase loss of sodium ions and water in urine, which reduces the blood volume back to normal.

Like most blood vessels of the body, those of the kidneys are supplied by sympathetic neurons of the autonomic nervous system. When these neurons are active, they cause vasoconstriction. At rest, sympathetic stimulation is low and the afferent and efferent arterioles are relatively dilated. With greater sympathetic stimulation, as occurs during exercise or hemorrhage, the afferent arterioles are constricted more than the efferent arterioles. As a result, blood flow into glomerular capillaries is greatly decreased, net filtration pressure decreases, and GFR drops. These changes reduce urine output, which helps conserve blood volume and permits greater blood flow to other body tissues.

Tubular Reabsorption

Tubular reabsorption—returning most of the filtered water and many of the filtered solutes to the blood—is the second basic function of the nephrons and collecting ducts. The filtered fluid becomes *tubular fluid* once it enters the proximal convoluted tubule. Due to reabsorption and secretion, the composition of tubular fluid changes as it flows along the nephron tubule and through a collecting duct. Typically, about 99 percent of the filtered water is reabsorbed. Only 1 percent of the water in glomerular filtrate actually leaves the body in *urine*, the fluid that drains into the renal pelvis.

Epithelial cells all along the renal tubules and collecting ducts carry out tubular reabsorption (Figure 21.7). Some solutes are passively reabsorbed by diffusion; others are reab-sorbed by active transport. Proximal convoluted tubule cells make the largest contribution, reabsorbing 65 percent of the filtered water, 100 percent of the filtered glucose and amino acids, and large quantities of various ions such as sodium (Na^+), potassium (K^+), chloride (Cl^-), bicarbonate (HCO_3^-), calcium (Ca^{2+}), and magnesium (Mg^{2+}). Reabsorption of solutes also promotes reabsorption of water in the following way. The

Figure 21.7 Filtration, reabsorption, and secretion in the nephrons and collecting ducts. Percentages refer to the amounts initially filtered at the glomerulus.

Filtration occurs in the renal corpuscle; reabsorption occurs all along the renal tubule and collecting ducts.

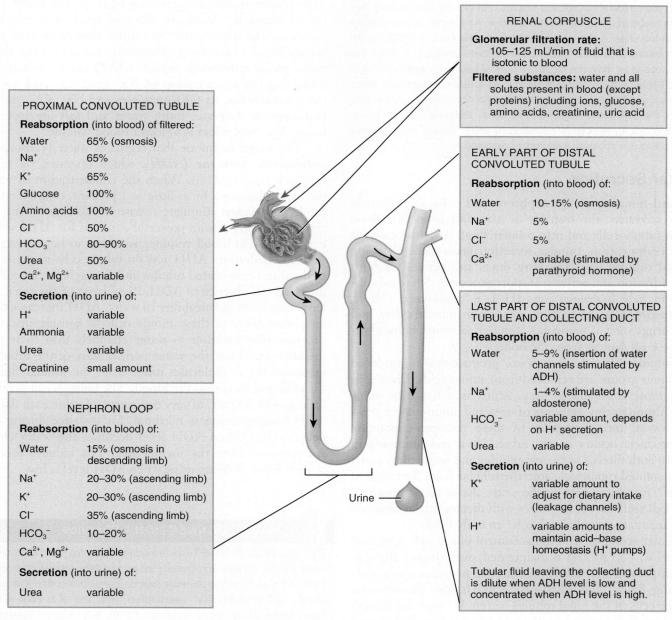

PROXIMAL CONVOLUTED TUBULE

Reabsorption (into blood) of filtered:

Water	65% (osmosis)
Na^+	65%
K^+	65%
Glucose	100%
Amino acids	100%
Cl^-	50%
HCO_3^-	80–90%
Urea	50%
Ca^{2+}, Mg^{2+}	variable

Secretion (into urine) of:

H^+	variable
Ammonia	variable
Urea	variable
Creatinine	small amount

NEPHRON LOOP

Reabsorption (into blood) of:

Water	15% (osmosis in descending limb)
Na^+	20–30% (ascending limb)
K^+	20–30% (ascending limb)
Cl^-	35% (ascending limb)
HCO_3^-	10–20%
Ca^{2+}, Mg^{2+}	variable

Secretion (into urine) of:

Urea	variable

RENAL CORPUSCLE

Glomerular filtration rate:
105–125 mL/min of fluid that is isotonic to blood

Filtered substances: water and all solutes present in blood (except proteins) including ions, glucose, amino acids, creatinine, uric acid

EARLY PART OF DISTAL CONVOLUTED TUBULE

Reabsorption (into blood) of:

Water	10–15% (osmosis)
Na^+	5%
Cl^-	5%
Ca^{2+}	variable (stimulated by parathyroid hormone)

LAST PART OF DISTAL CONVOLUTED TUBULE AND COLLECTING DUCT

Reabsorption (into blood) of:

Water	5–9% (insertion of water channels stimulated by ADH)
Na^+	1–4% (stimulated by aldosterone)
HCO_3^-	variable amount, depends on H^+ secretion
Urea	variable

Secretion (into urine) of:

K^+	variable amount to adjust for dietary intake (leakage channels)
H^+	variable amounts to maintain acid–base homeostasis (H^+ pumps)

Tubular fluid leaving the collecting duct is dilute when ADH level is low and concentrated when ADH level is high.

Urine

 In which segments of the nephrons and collecting ducts does secretion occur?

movement of solutes into peritubular capillaries decreases the solute concentration of the tubular fluid but increases the solute concentration in the peritubular capillaries. As a result, water moves by osmosis into peritubular capillaries. Cells located distal to the proximal convoluted tubule fine-tune reabsorption to maintain homeostatic balances of water and selected ions. To appreciate the huge extent of tubular reabsorption, look back at Table 21.1 and compare the amounts of substances that are filtered, reabsorbed, and excreted in urine.

CLINICAL CONNECTION | Glucosuria and Polyuria

When the blood concentration of glucose rises above normal, transporters in the proximal convoluted tubules may not be able to work fast enough to reabsorb all of the filtered glucose. As a result, some glucose remains in the urine, a condition called **glucosuria** (gloo′-kō-SOO-rē-a). The most common cause of glucosuria is diabetes mellitus, in which the blood glucose level may rise far above normal because insulin activity is deficient. Because "water follows solutes" as tubular reabsorption takes place, any condition that reduces reabsorption of filtered solutes also increases the amount of water lost in urine. **Polyuria** (pol′-ē-Ū-rē-a; *poly-* = too much), excessive excretion of urine, usually accompanies glucosuria and is a common symptom of diabetes. •

Tubular Secretion

The third function of the nephrons and collecting ducts is *tubular secretion*, the transfer of materials from the blood through tubule cells and into tubular fluid. As is the case for tubular reabsorption, tubular secretion takes place all along the renal tubules and collecting ducts and occurs via both passive diffusion and active transport processes. Secreted substances include hydrogen ions (H$^+$), K$^+$, ammonia (NH$_3$), urea, creatinine (a waste from creatine in muscle cells), and certain drugs such as penicillin. Tubular secretion helps eliminate these substances from the body.

Ammonia is a poisonous waste product that is produced when amino groups are removed from amino acids. Liver cells convert most ammonia to urea, which is a less-toxic compound. Although tiny amounts of urea and ammonia are present in sweat, most excretion of these nitrogen-containing waste products occurs in the urine. Urea and ammonia in blood are both filtered at the glomerulus and secreted by proximal convoluted tubule cells into the tubular fluid. Secretion of excess K$^+$ for elimination in the urine also is very important. Tubule cell secretion of K$^+$ varies with dietary intake of potassium to maintain a stable level of K$^+$ in body fluids.

Tubular secretion also helps control blood pH. A normal blood pH of 7.35 to 7.45 is maintained, even though the typical high-protein diet in North America provides more acid-producing foods than alkali-producing foods. To eliminate acids, the cells of the renal tubules secrete H$^+$ into the tubular fluid, which helps maintain the pH of blood in the normal range. Due to H$^+$ secretion, urine is typically acidic (has a pH below 7).

Hormonal Regulation of Nephron Functions

Hormones affect the extent of Na$^+$, Cl$^-$, Ca^{2+}, and water reabsorption as well as K$^+$ secretion by the renal tubules. The most important hormonal regulators of ion reabsorption and secretion are *angiotensin II* and *aldosterone*. In the proximal convoluted tubules, angiotensin II enhances reabsorption of Na$^+$ and Cl$^-$. Angiotensin II also stimulates the adrenal cortex to release aldosterone, a hormone that in turn stimulates the tubule cells in the last part of the distal convoluted tubules and throughout the collecting ducts to reabsorb more Na$^+$ and Cl$^-$ and secrete more K$^+$. When more Na$^+$ and Cl$^-$ are reabsorbed, then more water is also reabsorbed by osmosis. Aldosterone-stimulated secretion of K$^+$ is the major regulator of blood K$^+$ level. An elevated level of K$^+$ in plasma causes serious disturbances in cardiac rhythm or even cardiac arrest. Besides increasing glomerular filtration rate, the hormone *atrial natriuretic peptide (ANP)* plays a minor role in inhibiting the reabsorption of Na$^+$ (and Cl$^-$ and water) by the renal tubules. As GFR increases and Na$^+$, Cl$^-$, and water reabsorption decrease, more water and salt are lost in the urine. The final effect is to lower blood volume.

The major hormone that regulates water reabsorption is *antidiuretic hormone (ADH)*, which operates via negative feedback (Figure 21.8). When the concentration of water in the blood decreases by as little as 1 percent, osmoreceptors in the hypothalamus stimulate release of ADH from the posterior pituitary. A second powerful stimulus for ADH secretion is a decrease in blood volume, as occurs in hemorrhaging or severe dehydration. ADH acts on tubule cells in the last part of the distal convoluted tubules and throughout the collecting ducts. In the absence of ADH, these parts of the renal tubule have a very low permeability to water. ADH increases the water permeability of these tubule cells by causing insertion of proteins that function as water channels into their plasma membranes. When the water permeability of the tubule cells increases, water molecules move from the tubular fluid into the cells and then into the blood. The kidneys can produce as little as 400–500 mL of very concentrated urine each day when ADH concentration is maximal, for instance during severe dehydration. When ADH level declines, the water channels are removed from the membranes. The kidneys produce a large volume of dilute urine when ADH level is low.

CLINICAL CONNECTION | Diuretics

Diuretics (dī′-ū-RET-iks) are substances that slow reabsorption of water by the kidneys and thereby cause *diuresis*, an elevated urine flow rate. Naturally occurring diuretics include *caffeine* in coffee, tea, and cola sodas, which inhibits Na$^+$ reabsorption, and *alcohol* in beer, wine, and mixed drinks, which inhibits secretion of ADH. In a condition known as *diabetes insipidus*, ADH secretion is inadequate or the ADH receptors are faulty, and a person may excrete up to 20 liters of very dilute urine daily. •

Figure 21.8 Negative feedback regulation of water reabsorption by ADH.

When ADH level is high, the kidneys reabsorb more water.

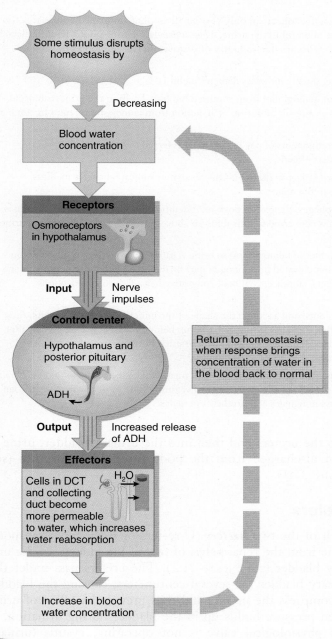

Would blood ADH level be higher or lower than normal in a person who has just completed a 5-km run without drinking any water?

Although the hormones mentioned thus far involve regulation of water loss as urine, the kidney tubules also respond to a hormone that regulates ionic composition. For example, a lower-than-normal level of Ca^{2+} in the blood stimulates the parathyroid glands to release **parathyroid hormone (PTH)**.

PTH in turn stimulates cells in the early distal convoluted tubules to reabsorb more Ca^{2+} into the blood. PTH also inhibits HPO_4^{2-} (phosphate) reabsorption in proximal convoluted tubules, thereby promoting phosphate excretion.

Components of Urine

An analysis of the volume and physical, chemical, and microscopic properties of urine, called a ***urinalysis***, tells us much about the state of the body. Table 21.2 summarizes the principal physical characteristics of urine.

The volume of urine eliminated per day in a normal adult is 1 to 2 liters (about 1 to 2 quarts). Water accounts for about 95 percent of the total volume of urine. In addition to urea, creatinine, potassium, and ammonia, typical solutes normally present in urine include uric acid as well as sodium, chloride, magnesium, sulfate, phosphate, and calcium ions.

If disease alters body metabolism or kidney function, traces of substances not normally present may appear in the urine, or normal constituents may appear in abnormal amounts. Table 21.3 lists several abnormal constituents in urine that may be detected as part of a urinalysis.

TABLE 21.2	
Physical Characteristics of Normal Urine	
CHARACTERISTIC	**DESCRIPTION**
Volume	One to two liters (about 1 to 2 quarts) in 24 hours but varies considerably.
Color	Yellow or amber, but varies with urine concentration and diet. Color is due to urochrome (pigment produced from breakdown of bile) and urobilin (from breakdown of hemoglobin). Concentrated urine is darker in color. Diet (reddish urine from beets), medications, and certain diseases affect color. Kidney stones may produce blood in urine.
Turbidity	Transparent when freshly voided, but becomes turbid (cloudy) after a while.
Odor	Mildly aromatic but becomes ammonia-like after a time. Some people inherit the ability to form methylmercaptan from digested asparagus, which gives urine a characteristic odor.
pH	Ranges between pH 4.6 and 8.0; average 6.0; varies considerably with diet. High-protein diets increase acidity; vegetarian diets increase alkalinity.
Specific gravity	Specific gravity (density) is the ratio of the weight of a volume of a substance to the weight of an equal volume of distilled water. Urine specific gravity ranges from 1.001 to 1.035. The higher the concentration of solutes, the higher the specific gravity.

TABLE 21.3

Summary of Abnormal Constituents in Urine

ABNORMAL CONSTITUENT	COMMENTS
Albumin	A normal constituent of blood plasma that usually appears in only very small amounts in urine because it is too large to be filtered. The presence of excessive albumin in the urine, *albuminuria* (al′-bū-mi-NOO-rē-a), indicates an increase in the permeability of filtering membranes due to injury or disease, increased blood pressure, or damage to kidney cells.
Glucose	*Glucosuria*, the presence of glucose in the urine, usually indicates diabetes mellitus.
Red blood cells (erythrocytes)	*Hematuria* (hēm-a-TOO-rē-a), the presence of hemoglobin from ruptured red blood cells in the urine, can occur with acute inflammation of the urinary organs as a result of disease or irritation from kidney stones, tumors, trauma, and kidney disease.
White blood cells (leukocytes)	The presence of white blood cells and other components of pus in the urine, referred to as *pyuria* (pī-Ū-rē-a), indicates infection in the kidneys or other urinary organs.
Ketone bodies	High levels of ketone bodies in the urine, called *ketonuria* (kē-tō-NOO-rē-a), may indicate diabetes mellitus, anorexia, starvation, or too little carbohydrate in the diet.
Bilirubin	When red blood cells are destroyed by macrophages, the globin portion of hemoglobin is split off and the heme is converted to biliverdin. Most of the biliverdin is converted to bilirubin. An above-normal level of bilirubin in urine is called *bilirubinuria* (bil′-ē-roo-bi-NOO-rē-a).
Urobilinogen	The presence of urobilinogen (breakdown product of hemoglobin) in urine is called *urobilinogenuria* (ū′-rō-bi-lin′-ō-jē-NOO-rē-a). Trace amounts are normal, but elevated urobilinogen may be due to hemolytic or pernicious anemia, infectious hepatitis, obstruction of bile ducts, jaundice, cirrhosis, congestive heart failure, or infectious mononucleosis.
Casts	*Casts* are tiny masses of material that have hardened and assumed the shape of the lumen of a tubule in which they formed. They are flushed out of the tubule when glomerular filtrate builds up behind them. Casts are named after the cells or substances that compose them or based on their appearance. For example, there are white blood cell casts, red blood cell casts, and epithelial cell casts (cells from renal tubules).
Microbes	The number and type of bacteria vary with specific infections in the urinary tract. One of the most common is *E. coli*. The most common fungus to appear in urine is *Candida albicans*, a cause of vaginitis. The most frequent protozoan seen is *Trichomonas vaginalis*, a cause of vaginitis in females and urethritis in males.

✓ CHECKPOINT

4. How does blood pressure promote filtration of blood in the kidneys?
5. What solutes are reabsorbed and secreted as fluid moves along the renal tubules?
6. How do angiotensin II, aldosterone, and antidiuretic hormone regulate tubular reabsorption and secretion?
7. What are the characteristics of normal urine?

21.4 TRANSPORTATION, STORAGE, AND ELIMINATION OF URINE

OBJECTIVE • **Describe the structure and functions of the ureters, urinary bladder, and urethra.**

As you learned earlier in the chapter, urine produced by the nephrons drains into the minor calyces, which join to become major calyces that unite to form the renal pelvis (see Figure 21.2). From the renal pelvis, urine drains first into the ureters and then into the urinary bladder; urine is then discharged from the body through the urethra (see Figure 21.1).

Ureters

Each of the two **ureters** (Ū-re-ters or ū-RĒ-ters) transports urine from the renal pelvis of one of the kidneys to the urinary bladder (see Figure 21.1). The ureters pass under the urinary bladder for several centimeters, causing the bladder to compress the ureters and thus prevent backflow of urine when pressure builds up in the bladder during urination. If this physiological valve is not operating, cystitis (urinary bladder inflammation) may develop into a kidney infection.

The wall of the ureter consists of three layers. The inner layer is the mucosa, containing *transitional epithelium* (see Table 4.1I) with an underlying layer of areolar connective tissue. Transitional epithelium is able to stretch—a marked advantage for any organ that must accommodate a variable volume of fluid. Mucus secreted by the goblet cells of the mucosa prevents the cells from coming in contact with urine, the solute concentration and pH of which may differ drastically from the cytosol of cells that form the wall of the ureters. The middle layer consists of smooth muscle. Urine is transported from the renal pelvis to the urinary bladder primarily by peristaltic contractions of this smooth muscle, but the fluid pressure of the urine and gravity may also contribute. The outer layer consists of areolar connective tissue containing blood vessels, lymphatic vessels, and nerves.

Urinary Bladder

The *urinary bladder* is a hollow muscular organ situated in the pelvic cavity behind the pubic symphysis (Figure 21.9). In males, it is directly in front of the rectum (see Figure 23.1). In females, it is in front of the vagina and below the uterus. Folds of the peritoneum hold the urinary bladder in position. The shape of the urinary bladder depends on how much urine

it contains. When empty, it looks like a deflated balloon. It becomes spherical when slightly stretched and, as urine volume increases, becomes pear-shaped and rises into the abdominal cavity. Urinary bladder capacity averages 700–800 mL. It is smaller in females because the uterus occupies the space just superior to the urinary bladder. Toward the base of the urinary bladder, the ureters drain into the urinary bladder via the *ureteral openings*. Like the ureters, the mucosa of the urinary bladder contains transitional epithelium, which permits stretching. The mucosa also contains folds called *rugae*, which also permit expansion of the urinary bladder. The muscular layer of the urinary bladder wall consists of three layers of smooth muscle called the *detrusor muscle* (de-TROO-ser = to push down). The peritoneum, which covers the superior surface of the urinary bladder, forms a serous outer coat; the rest of the urinary bladder has a fibrous outer covering.

Figure 21.9 Ureters, urinary bladder, and urethra (female).

🔑 **Urine is stored in the urinary bladder until it is expelled by micturition.**

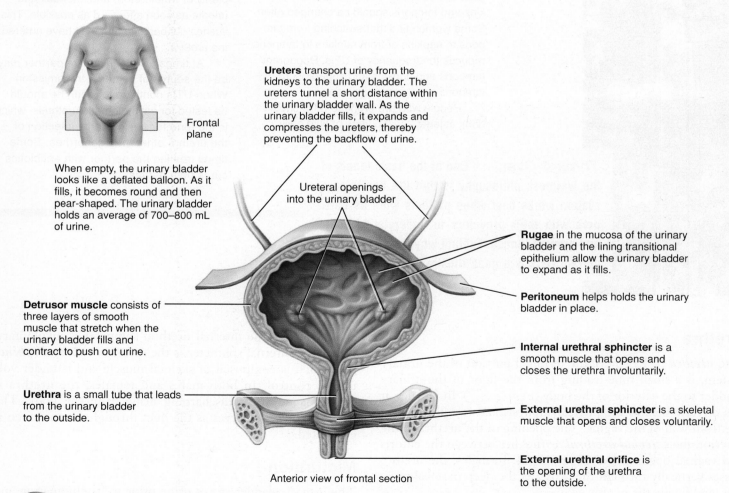

Ureters transport urine from the kidneys to the urinary bladder. The ureters tunnel a short distance within the urinary bladder wall. As the urinary bladder fills, it expands and compresses the ureters, thereby preventing the backflow of urine.

Frontal plane

When empty, the urinary bladder looks like a deflated balloon. As it fills, it becomes round and then pear-shaped. The urinary bladder holds an average of 700–800 mL of urine.

Ureteral openings into the urinary bladder

Rugae in the mucosa of the urinary bladder and the lining transitional epithelium allow the urinary bladder to expand as it fills.

Peritoneum helps holds the urinary bladder in place.

Detrusor muscle consists of three layers of smooth muscle that stretch when the urinary bladder fills and contract to push out urine.

Internal urethral sphincter is a smooth muscle that opens and closes the urethra involuntarily.

Urethra is a small tube that leads from the urinary bladder to the outside.

External urethral sphincter is a skeletal muscle that opens and closes voluntarily.

External urethral orifice is the opening of the urethra to the outside.

Anterior view of frontal section

❓ **What is a lack of voluntary control over micturition called?**

Infection Prevention for Recurrent UTIs

Urinary tract infections (UTIs) are the most common bacterial infections and the second most common illness (after colds) among women. About 10–15 percent of women develop UTIs several times a month. Men get UTIs, too, but much less frequently. The female's shorter urethra allows bacteria to enter the urinary bladder more easily. In addition, the urethral and anal openings are closer in females. Most first-time UTIs are caused by *Escherichia coli (E. coli),* bacteria that have migrated to the urethra from the anal area. *E. coli* bacteria are necessary for proper digestion and are welcome in the intestinal tract, but they cause much pain and suffering if they infect the urinary system.

Infection Prevention

Personal hygiene is the first line of prevention. Care must be taken to avoid transporting bacteria from the anal area to the urethra. Girls should be taught to wipe from front to back and to wash hands thoroughly after using the toilet. When bathing, women and girls should wash from front to back as well.

Menstrual blood provides an excellent growth medium for bacteria. Sanitary napkins and tampons should be changed often. Some women find that switching from tampons to napkins or from napkins to tampons reduces the frequency of UTIs. Deodorant tampons and napkins and superabsorbent tampons can increase irritation.

People who are prone to UTIs should drink at least 2 to 2.5 liters of fluid daily.

Drinking cranberry and blueberry juice may help to decrease bacterial growth in the urinary bladder. Voiding frequently, every 2 to 3 hours, helps prevent recurrent UTIs because it expels bacteria and eliminates the urine needed for their growth. Some women find that alcohol and caffeine consumption irritate the urethra and increase discomfort.

Panties with a cotton crotch allow moisture to escape, making a less hospitable environment for bacteria. Thongs can irritate the urethra and trap moisture in this area.

Partners in Health

Sexual intercourse is frequently associated with the onset of UTIs in women. Women who find that sex brings on UTIs learn to develop and teach their partners stringent personal hygiene. Women should drink plenty of water before and after sex and urinate as soon afterward as possible. This flushes out bacteria that may have entered the urethra.

At times a woman's male partner may be the source of bacterial transmission. When UTIs continue to recur, he should be tested for asymptomatic urethritis, which is the term for any bacterial infection of the urethra other than gonorrhea. Sometimes treating the partner with antibiotics cures both parties.

Think It Over . . . **One of the basic tenets of the wellness philosophy is that the health-care system works best when patients work as partners with their providers to understand, treat, and prevent illness. Explain why treatment of recurrent UTIs is a good illustration of this belief.**

Urethra

The *urethra* (ū-RĒ-thra), the terminal portion of the urinary system, is a small tube leading from the floor of the urinary bladder to the exterior of the body (Figure 21.9). In females, it lies directly behind the pubic symphysis and is embedded in the front wall of the vagina. The opening of the urethra to the exterior, the *external urethral orifice*, lies between the clitoris and vaginal opening (see Figure 23.6). In males, the urethra passes vertically through the prostate, the deep perineal muscles, and finally the penis (see Figure 23.1).

Around the opening to the urethra is an *internal urethral sphincter* composed of smooth muscle. The opening and closing of the internal urethral sphincter is involuntary. Below the internal sphincter is the *external urethral sphincter*, which is composed of skeletal muscle and is under voluntary control. In both males and females, the urethra is the passageway for discharging urine from the body. The male urethra also serves as the duct through which semen is ejaculated.

Micturition

The urinary bladder stores urine prior to its elimination and then expels urine into the urethra by an act called *micturition* (mik′-too-RI-shun = to urinate), commonly known as *urination*. Micturition requires a combination of involuntary and voluntary muscle contractions. When the volume of urine in the urinary bladder exceeds 200 to 400 mL, pressure within the bladder increases considerably, and stretch receptors in its wall transmit nerve impulses into the spinal cord. These impulses propagate to the lower part of the spinal cord and trigger a reflex called the *micturition reflex*. In this reflex, parasympathetic impulses from the spinal cord cause *contraction* of the detrusor muscle and *relaxation* of the internal urethral sphincter muscle. Simultaneously, the spinal cord inhibits somatic motor neurons, causing relaxation of skeletal muscle in the external urethral sphincter. Upon contraction of the urinary bladder wall and relaxation of the sphincters, urination takes place. Urinary bladder filling causes a sensation of fullness that initiates a conscious desire to urinate before the micturition reflex actually occurs. Although emptying of the urinary bladder is a reflex, in early childhood we learn to initiate it and stop it voluntarily. Through learned control of the external urethral sphincter muscle and certain muscles of the pelvic floor, the cerebral cortex can initiate micturition or delay it for a limited period of time.

CLINICAL CONNECTION | Urinary Incontinence

A lack of voluntary control over micturition is termed **urinary incontinence**. Under about 2–3 years of age, urinary incontinence is normal because neurons to the external urethral sphincter muscle are not completely developed. Infants void whenever the urinary bladder is sufficiently distended to trigger the reflex. In *stress incontinence*, the most common type of urinary incontinence, physical stresses that increase abdominal pressure, such as coughing, sneezing, laughing, exercising, straining, lifting heavy objects, pregnancy, or simply walking, cause leakage of urine from the urinary bladder. Smokers have twice the risk of developing urinary incontinence as nonsmokers. •

✓ CHECKPOINT

8. What forces help propel urine from the renal pelvis to the urinary bladder?

9. What is micturition? How does the micturition reflex occur?
10. How does the location of the urethra compare in males and females?

21.5 AGING AND THE URINARY SYSTEM

OBJECTIVE • **Describe the effects of aging on the urinary system.**

With aging, the kidneys shrink in size, have a decreased blood flow, and filter less blood. The mass of the two kidneys decreases from an average of 260 g in 20-year-olds to less than 200 g by age 80. Likewise, renal blood flow and filtration rate decline by 50 percent between ages 40 and 70. Kidney diseases that become more common with age include acute and chronic kidney inflammations and renal calculi (kidney stones). Because the sensation of thirst diminishes with age, older individuals also are susceptible to dehydration. Urinary tract infections are more common among the elderly, as are polyuria, nocturia (excessive urination at night), increased frequency of urination, dysuria (painful urination), urinary retention or incontinence, and hematuria (blood in the urine).

✓ CHECKPOINT

11. Why are older individuals more susceptible to dehydration?

• • •

To appreciate the many ways that the urinary system contributes to homeostasis of other body systems, examine Focus on Homeostasis: The Urinary System. Next, in Chapter 22, we will see how the kidneys and lungs contribute to maintenance of homeostasis of body fluid volume, ion levels in body fluids, and acid–base balance.

Focus on Homeostasis

BODY SYSTEM	CONTRIBUTION OF THE URINARY SYSTEM

For all body systems
The kidneys regulate the volume, composition, and pH of body fluids by removing wastes and excess substances from blood and excreting them in the urine. The ureters transport urine from the kidneys to the urinary bladder, which stores urine until it is eliminated through the urethra.

Integumentary system
The kidneys and skin both contribute to the synthesis of calcitriol, the active form of vitamin D.

Skeletal system
The kidneys help adjust the levels of blood calcium and phosphates needed for building bone extracellular matrix.

Muscular system
The kidneys help adjust the level of blood calcium, needed for contraction of muscles.

Nervous system
The kidneys perform gluconeogenesis (synthesis of glucose from certain amino acids and lactic acid), thereby providing glucose for ATP production in neurons, especially during fasting or starvation.

Endocrine system
The kidneys participate in the synthesis of calcitriol, the active form of vitamin D, and release erythropoietin, the hormone that stimulates the production of red blood cells.

Cardiovascular system
By increasing or decreasing reabsorption of water filtered from the blood, the kidneys help adjust the blood volume and blood pressure. Renin released by the kidneys raises blood pressure. Some bilirubin (from hemoglobin breakdown) is converted to a yellow pigment (urobilin), which is excreted in the urine.

Lymphatic system and immunity
By increasing or decreasing the reabsorption of water filtered from blood, the kidneys help adjust the volume of interstitial fluid and lymph. Urination flushes microbes out of the urethra.

Respiratory system
The kidneys and lungs cooperate in adjusting the pH of body fluids.

Digestive system
The kidneys help synthesize calcitriol, the active form of vitamin D, which is needed for absorption of dietary calcium.

Reproductive systems
In males, the portion of the urethra that extends through the prostate and penis is a passageway for semen as well as urine.

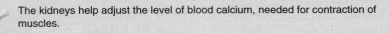

THE URINARY SYSTEM

COMMON DISORDERS

Glomerulonephritis

Glomerulonephritis (glo-mer′-ū-lō-ne-FRĪ-tis) is an inflammation of the glomeruli of the kidney. One of the most common causes is an allergic reaction to the toxins produced by streptococcal bacteria that have recently infected another part of the body, especially the throat. Because inflamed and swollen glomeruli allow blood cells and plasma proteins to enter the filtrate, the urine contains many red blood cells (hematuria) and large amounts of protein (proteinuria).

Renal Failure

Renal failure is a decrease or cessation of glomerular filtration. In *acute renal failure (ARF)* the kidneys abruptly stop working entirely (or almost entirely). The main feature of ARF is the suppression of urine flow, leading to oliguria or anuria. Causes include low blood volume (for example, due to hemorrhage); decreased cardiac output; damaged renal tubules; kidney stones; or reactions to the dyes used to visualize blood vessels in angiograms, nonsteroidal anti-inflammatory drugs, and some antibiotic drugs.

Chronic renal failure (CRF) refers to a progressive and usually irreversible decline in glomerular filtration rate (GFR). CRF may result from chronic glomerulonephritis, pyelonephritis, polycystic kidney disease, or traumatic loss of kidney tissue. The final stage of CRF is called *end-stage renal failure* and occurs when about 90 percent of the nephrons have been lost. At this stage, GFR diminishes to 10–15 percent of normal, oliguria is present, and blood levels of nitrogen-containing wastes and creatinine are high. People with end-stage renal failure require dialysis therapy and are possible candidates for a kidney transplant operation.

Polycystic Kidney Disease

Polycystic kidney disease (PKD) is one of the most common inherited disorders. In PKD, the kidney tubules become riddled with hundreds or thousands of cysts (fluid-filled cavities). In addition, inappropriate apoptosis (programmed cell death) of cells in noncystic tubules leads to progressive impairment of renal function and eventually to end-stage renal failure.

People with PKD also may have cysts and apoptosis in the liver, pancreas, spleen, and gonads; increased risk of cerebral aneurysms; heart valve defects; and diverticuli in the colon. Typically, symptoms are not noticed until adulthood, when patients may have back pain, urinary tract infections, blood in the urine, hypertension, and large abdominal masses. Using drugs to restore normal blood pressure, restricting protein and salt in the diet, and controlling urinary tract infections may slow progression to renal failure.

MEDICAL TERMINOLOGY AND CONDITIONS

Dialysis (dī-AL-i-sis; *dialyo* = to separate)   The separation of large solutes from smaller ones by diffusion through a selectively permeable membrane. It is used to cleanse a person's blood artificially when the kidneys are so impaired by disease or injury that they are unable to function adequately. One method of dialysis is *hemodialysis* (hē-mō-dī-AL-i-sis; *hemo-* = blood), which filters the patient's blood directly by removing wastes and excess electrolytes and fluid and then returning the cleansed blood to the patient. Blood removed from the body is delivered to a *hemodialyzer* (artificial kidney). Inside the hemodialyzer, blood flows through a *dialysis membrane*, which contains pores large enough to permit the diffusion of small solutes. A special solution, called the *dialysate* (dī-AL-i-sāt), is pumped into the hemodialyzer so that it surrounds the dialysis membrane. The dialysate is specially formulated to maintain diffusion gradients that remove wastes from the blood (for example, urea, creatinine, uric acid, excess phosphate, potassium, and sulfate ions) and add needed substances (for example, glucose and bicarbonate ions) to it. As a rule, most people on hemodialysis require about 6–12 hours a week, typically divided into three sessions.

Dysuria (dis-Ū-rē-a; *dys* = painful; *uria* = urine)   Painful urination.

Enuresis (en′-ū-RĒ-sis; = to void urine)   Involuntary voiding of urine after the age at which voluntary control has typically been attained.

Intravenous pyelogram (IVP) (in′-tra-VĒ-nus PĪ-el-ō-gram′; *intra-* = within; *veno-* = vein; *pyelo-* = pelvis of kidney; *-gram* = record)   Radiograph (x-ray film) of the kidneys after venous injection of a dye.

Kidney stones   Insoluble stones occasionally formed from solidification of the crystals of urine salts. Can be caused by ingestion of excessive mineral salts, insufficient water intake, abnormally alkaline or acidic urine, or overactive parathyroid glands. Usually form in the renal pelvis. Often cause intense pain. Also termed *renal calculi*.

Nocturnal enuresis (nok-TUR-nal en′-ū-RĒ-sis)   Discharge of urine during sleep, resulting in bed-wetting; occurs in about 15 percent of 5-year-old children and generally resolves spontaneously, afflicting only about 1 percent of adults. Possible causes include smaller-than-normal urinary bladder capacity, failure to awaken in response to a full urinary bladder, and above-normal production of urine at night. Also termed *nocturia*.

Urinary retention   A failure to completely or normally void urine; may be due to an obstruction in the urethra or neck of the urinary bladder, to nervous contraction of the urethra, or to lack of urge to urinate. In men, an enlarged prostate may constrict the urethra and cause urinary retention. If urinary retention is prolonged, a catheter (slender rubber drainage tube) must be placed into the urethra to drain the urine.

CHAPTER REVIEW AND RESOURCE SUMMARY

REVIEW	RESOURCES

21.1 Overview of the Urinary System

1. The organs of the urinary system include the kidneys, ureters, urinary bladder, and urethra.
2. After the kidneys filter blood and return most of the water and many solutes to the blood, the remaining water and solutes constitute **urine**.
3. The kidneys regulate blood ionic composition, blood volume, blood pressure, and blood pH.
4. The kidneys also release calcitriol and erythropoietin and excrete wastes and foreign substances.

Anatomy Overview—
    The Urinary System
Animation—Regulation of pH

21.2 Structure of the Kidneys

1. The **kidneys** lie on either side of the vertebral column between the peritoneum and the back wall of the abdominal cavity.
2. Each kidney is enclosed in a renal capsule, which is surrounded by adipose tissue.
3. Internally, the kidneys consist of a **renal cortex, renal medulla, renal pyramids, renal columns, major** and **minor calyces**, and a **renal pelvis**.
4. Blood enters the kidney through the **renal artery** and leaves through the **renal vein**.
5. The **nephron** is the functional unit of the kidney. A nephron consists of a **renal corpuscle** (**glomerulus** and **glomerular [Bowman's] capsule**) and a **renal tubule** (**proximal convoluted tubule, descending limb of the nephron loop, ascending limb of the nephron loop**, and **distal convoluted tubule**). Each nephron also has its own blood supply. The distal convoluted tubules of several nephrons empty into a common **collecting duct**.

Anatomy Overview—
    Kidney Overview

Figure 21.2 Structure of
the kidney

Exercise—Assemble the
    Urinary Tract
Exercise—Magic Renal Ride
Exercise—Paint the Nephron

21.3 Functions of the Nephron

1. Nephrons perform three basic tasks: **glomerular filtration, tubular reabsorption**, and **tubular secretion**.
2. Together, the podocytes and glomerular endothelium form a leaky filtration membrane that permits the passage of water and solutes from the blood into the **capsular space**. Blood cells and most plasma proteins remain in the blood because they are too large to pass through the filtration membrane. The pressure that causes filtration is the blood pressure in the glomerular capillaries.
3. Table 21.1 describes the substances that are filtered, reabsorbed, and excreted in urine on a daily basis.
4. The amount of filtrate that forms in both kidneys every minute is the **glomerular filtration rate (GFR)**. **Atrial natriuretic peptide (ANP)** increases GFR; sympathetic stimulation decreases GFR.
5. Epithelial cells all along the renal tubules and collecting ducts carry out tubular reabsorption and tubular secretion. Tubular reabsorption retains substances needed by the body, including water, glucose, amino acids, and ions such as sodium (Na^+), potassium (K^+), chloride (Cl^-), bicarbonate (HCO_3^-), calcium (Ca^{2+}), and magnesium (Mg^{2+}).
6. **Angiotensin II** enhances reabsorption of Na^+ and Cl^-. Angiotensin II also stimulates the adrenal cortex to release **aldosterone**, which stimulates the collecting ducts to reabsorb more Na^+ and Cl^- and secrete more K^+. **Atrial natriuretic peptide (ANP)** inhibits reabsorption of Na^+ (and Cl^- and water) by the renal tubules, which reduces blood volume.
7. Most water is reabsorbed by osmosis together with reabsorbed solutes, mainly in the proximal convoluted tubule. Reabsorption of the remaining water is regulated by **antidiuretic hormone (ADH)** in the last part of the distal convoluted tubule and collecting duct.
8. Tubular secretion discharges chemicals not needed by the body into the urine. Included are excess ions, nitrogenous wastes, hormones, and certain drugs. The kidneys help maintain blood pH by secreting H^+. Tubular secretion also helps maintain proper levels of K^+ in the blood.

Anatomy Overview—Overview
    of Nephron
Animation—Renal Filtration
Animation—Renal Reabsorption
    and Secretion
Animation—Hormonal Control
    of Blood Volume and
    Pressure
Animation—Water Homeostasis
Animation—The Urinary
    System: Water and Fluid Flow
Exercise—Pick the Urinary
    Process
Exercise—Renal Regulation
Exercise—Renal Trip

REVIEW	RESOURCES

REVIEW

9. Table 21.2 describes the physical characteristics of urine that are evaluated in a **urinalysis**: color, odor, turbidity, pH, and specific gravity. Chemically, normal urine contains about 95 percent water and 5 percent solutes.

10. Table 21.3 lists the abnormal constituents that can be diagnosed through urinalysis, including albumin, glucose, red blood cells, white blood cells, ketone bodies, bilirubin, urobilinogen, casts, and microbes.

21.4 Transportation, Storage, and Elimination of Urine

1. The **ureters** transport urine from the renal pelves of the right and left kidneys to the urinary bladder and consist of a mucosa, muscularis, and adventitia.

2. The **urinary bladder** is posterior to the pubic symphysis. Its function is to store urine prior to **micturition**.

3. The mucosa of the urinary bladder contains stretchy transitional epithelium. The muscular layer of the wall consists of three layers of smooth muscle together referred to as the **detrusor muscle**.

4. The **urethra** is a tube leading from the floor of the urinary bladder to the exterior. Its function is to discharge urine from the body.

5. The **micturition reflex** discharges urine from the urinary bladder by means of parasympathetic impulses that cause contraction of the detrusor muscle and relaxation of the internal urethral sphincter muscle, and by inhibition of somatic motor neurons to the external urethral sphincter.

Anatomy Overview—Ureters, Urinary Bladder, and Urethra

21.5 Aging and the Urinary System

1. With aging, the kidneys shrink in size, have lowered blood flow, and filter less blood.

2. Common problems related to aging include urinary tract infections, increased frequency of urination, urinary retention or incontinence, and renal calculi (kidney stones).

Homeostatic Imbalance—The Case of the Thirsty Woman
Homeostatic Imbalance—The Case of the Man with the Swollen Kidneys
Homeostatic Imbalance—The Case of the Girl with Relentless Pelvic Pain
Homeostatic Imbalance—The Case of the Consistently Full Bladder

SELF-QUIZ

1. Which of the following is NOT a function of the urinary system?
   a. regulation of blood volume and composition
   b. stimulation of red blood cell production
   c. regulation of body temperature
   d. regulation of blood pressure
   e. regulation of blood pH

2. Which of the following structures is located in the renal cortex?
   a. the renal pyramid   **b.** the renal column
   c. the major calyx   **d.** the minor calyx
   e. the renal corpuscle

3. Match the following:
   _____ **a.** enhances reabsorption of Na^+ and secretion of K^+
   _____ **b.** inhibits Na^+ and H_2O reabsorption
   _____ **c.** increases H_2O reabsorption in the distal convoluted tubule
   _____ **d.** stimulates Ca^{2+} reabsorption and inhibits phosphate reabsorption
   _____ **e.** decreases H_2O reabsorption
   _____ **f.** enhances Na^+ and Cl^- reabsorption in proximal convoluted tubule

   A. antidiuretic hormone (ADH)
   B. angiotensin II
   C. atrial natriuretic peptide (ANP)
   D. diuretics
   E. parathyroid hormone
   F. aldosterone

4. The major openings located in the base of the bladder are the
   a. renal artery, renal vein, and urethra.
   b. renal artery, renal vein, and ureter.
   c. ureter, urethra, and collecting tubes.
   d. urethra and two ureters.
   e. external urethral sphincter and two ureters.

5. A protective barrier that helps prevent kidney trauma is the
   a. renal pelvis.   **b.** renal corpuscle.
   c. renal capsule.   **d.** detrusor muscle.
   e. renal hilum.

6. Place the following structures in the correct order for the flow of urine:
   1. renal tubules   2. minor calyx   3. renal pelvis
   4. major calyx   5. collecting ducts   6. ureters
      a. 1, 2, 4, 3, 6, 5   b. 5, 1, 4, 2, 3, 6   c. 5, 1, 2, 4, 3, 6
      d. 3, 5, 1, 2, 4, 6   e. 1, 5, 2, 4, 3, 6

7. The functional unit of the kidney where urine is produced is the
   a. nephron.   b. pyramid.   c. pelvis.
   d. glomerulus.   e. calyx.

8. What causes filtration of plasma across the filtration membrane?
   a. a full urinary bladder
   b. control by the nervous system
   c. water retention
   d. the pressure of the blood in the glomerular capillaries
   e. the pressure of urine in the capsular space

9. Glomerular filtration rate (GFR) is the
   a. rate of urinary bladder filling.
   b. amount of filtrate formed in both kidneys each minute.
   c. amount of filtrate reabsorbed at the collecting ducts.
   d. amount of blood delivered to the kidneys each minute.
   e. amount of urine formed per hour.

10. Which of the following is normally secreted into the tubular fluid from the blood?
    a. hydrogen ions (H^+)   b. amino acids   c. glucose
    d. water   e. white blood cells

11. In the nephron, tubular fluid that is reabsorbed from the renal tubules enters the
    a. glomerulus.   b. peritubular capillaries.
    c. efferent arteriole.   d. afferent arteriole.   e. renal artery.

12. Place the following structures in the correct order as they are involved in the formation of urine in the nephrons.
    1. distal convoluted tubule
    2. renal corpuscle
    3. descending limb of nephron loop
    4. proximal convoluted tubule
    5. collecting duct
    6. ascending limb of nephron loop
       a. 4, 1, 6, 3, 2, 5   b. 2, 6, 3, 1, 5, 4   c. 2, 4, 3, 6, 5, 1
       d. 2, 4, 3, 6, 1, 5   e. 5, 1, 4, 3, 6, 2

13. Blood is carried out of the glomerulus by the
    a. renal artery.   b. afferent arteriole.
    c. peritubular venule.   d. segmental artery.
    e. efferent arteriole.

14. Which of the following increases glomerular filtration rate (GFR)?
    a. atrial natriuretic peptide (ANP)
    b. constriction of the afferent arterioles

c. increased sympathetic stimulation to the afferent arterioles
    d. ADH   e. angiotensin II

15. Which of the following statements concerning tubular reabsorption is NOT true?
    a. Most reabsorption occurs in the proximal convoluted tubules.
    b. Reabsorption restores all of the filtered glucose and amino acids.
    c. Tubular reabsorption is a primary controller of blood pH.
    d. The reabsorption of water in the proximal convoluted tubules depends on reabsorption of solutes.
    e. Tubular reabsorption allows the body to retain most filtered nutrients.

16. The micturition reflex
    a. is under the control of hormones.
    b. is activated by low pressure in the urinary bladder.
    c. depends on contraction of the internal urethral sphincter muscle.
    d. is an involuntary reflex over which normal adults have voluntary control.
    e. is also known as incontinence.

17. Which of the following is NOT normally present in glomerular filtrate?
    a. blood cells
    b. glucose
    c. nitrogenous wastes such as urea
    d. amino acids
    e. water

18. Urine formation requires which of the following?
    a. glomerular filtration and tubular secretion only
    b. glomerular filtration and tubular reabsorption only
    c. glomerular filtration, tubular reabsorption, and tubular secretion
    d. tubular reabsorption, tubular filtration, and tubular secretion
    e. tubular secretion and tubular reabsorption only

19. The transport of urine from the renal pelvis into the urinary bladder is the function of the
    a. urethra.   b. efferent arteriole.
    c. afferent arteriole.   d. renal pyramids.
    e. ureters.

20. Which of the following characteristics do the urinary bladder and ureters have in common?
    a. ability to store urine
    b. mucosa layer with transitional epithelium
    c. three layers of smooth muscle
    d. sphincters to control urine flow
    e. peritoneal covering

CRITICAL THINKING APPLICATIONS

1. Yesterday, you attended a large, outdoor party where beer was the only beverage available. You remember having to urinate many, many times yesterday, and today you're very thirsty. What hormone is affected by alcohol, and how does this affect your kidney function?

2. Ellie is an "above average" 1-year-old whose parents would like her to be the first toilet-trained child in preschool. However, in this case at least, Ellie is average for her age and remains incontinent. Should her parents be concerned by this lack of success?

3. Thomas is a healthy, VERY active 4-year old. He doesn't like to take the time to go to the bathroom because, as he says, "I might miss something." His parents are worried that Thomas's kidneys may stop working when his urinary bladder is full. Should they be concerned?

4. Valerie is irritable today because for the second time this month she is experiencing frequent and urgent urination, dysuria, and a slight fever. Her doctor confirms what she suspects is wrong and prescribes antibiotics. Describe her condition, why it is recurring, and how it is prevented.

ANSWERS TO FIGURE QUESTIONS

21.1 By forming urine, the kidneys do the major work of the urinary system.

21.2 The renal pyramids are located in the renal medulla.

21.3 About 1200 mL of blood enters the kidneys each minute.

21.4 The water molecule will travel from the proximal convoluted tubule → descending limb of the nephron loop → ascending limb of the nephron loop → distal convoluted tubule → collecting duct → minor calyx → major calyx → renal pelvis.

21.5 Secreted penicillin is being removed from the blood.

21.6 Podocytes and the glomerular endothelium make up the filtration membrane.

21.7 Secretion occurs in the proximal convoluted tubule, the loop of Henle, the last part of the distal convoluted tubule, and the collecting duct.

21.8 The blood level of ADH would be higher than normal after a 5-km run, due to loss of body water in sweat.

21.9 A lack of voluntary control over micturition is termed incontinence.

CHAPTER 22

FLUID, ELECTROLYTE, AND ACID–BASE BALANCE

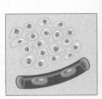

In Chapter 21 you learned how the kidneys form urine. One important function of the kidneys is to help maintain fluid balance in the body. The water and dissolved solutes in the body constitute the *body fluids*. Regulatory mechanisms involving the kidneys and other organs normally maintain homeostasis of the body fluids. Malfunction in any or all of them may seriously endanger the functioning of organs throughout the body. In this chapter, we will explore the mechanisms that regulate the volume and distribution of body fluids and examine the factors that determine the concentrations of solutes and the pH of body fluids.

Did you know?   If the level of calcium in the blood gets too low, this important electrolyte is drawn from the bones. Over time, this can lead to low bone density. Many people think calcium supplements when they think osteoporosis. But simply popping calcium pills does not necessarily mean the calcium will find its way into the bones. In addition, high levels of calcium supplements may have unwanted side effects. Calcium does its best work as part of a healthful diet. While adequate calcium, preferably from food, is important, other dietary factors are equally important. Vitamins D and K help the calcium get into the bones, and other minerals, including magnesium, help build strong bones. Consuming too much sodium leads to calcium losses, as do diets that are too low in fruits and vegetables.

FOCUS ON WELLNESS
Diet, Acid–Base Balance, and Osteoporosis

LOOKING BACK TO MOVE AHEAD...

Acids, Bases, and pH (Section 2.2)

Intracellular and Extracellular Fluid (Section 3.3)

Osmosis (Section 3.3)

Antidiuretic Hormone (ADH) (Section 13.3)

Hormonal Regulation of Calcium in Body Fluids (Section 13.5)

Renin–Angiotensin–Aldosterone Pathway (Section 13.7)

Control of Breathing Rate and Depth (Section 18.5)

Ions Reabsorbed and Secreted in the Kidneys (Section 21.3)

Negative Feedback Regulation of ADH Secretion (Section 21.3)

22.1 FLUID COMPARTMENTS AND FLUID BALANCE

OBJECTIVES • **Compare the locations of intracellular fluid (ICF) and extracellular fluid (ECF), and describe the various fluid compartments of the body.**

• **Describe the sources of water and solute gain and loss, and explain how each is regulated.**

In lean adults, body fluids make up between 55 percent and 60 percent of total body mass (Figure 22.1). Fluids are present in two main "compartments"—inside cells and outside cells. About two-thirds of body fluid is **intracellular fluid (ICF)** (*intra-* = within) or **cytosol**, the fluid within cells. The other third, called **extracellular fluid (ECF)** (*extra-* = outside), is outside cells and includes all other body fluids. About 80 percent of the ECF is **interstitial fluid** (*inter-* = between), which occupies the spaces between tissue cells, and about 20 percent of the ECF is **blood plasma**, the liquid portion of the blood. Other extracellular fluids that are grouped with interstitial fluid include lymph in lymphatic vessels; cerebrospinal fluid in the nervous system; synovial fluid in joints; aqueous humor and vitreous body in the eyes; endolymph and perilymph in the ears; and pleural, pericardial, and peritoneal fluids between serous membranes of the lungs, heart, and abdominal organs.

Two "barriers" separate intracellular fluid, interstitial fluid, and blood plasma.

1. The *plasma membrane* of each cell separates intracellular fluid from the surrounding interstitial fluid. You learned in Section 3.2 that the plasma membrane is a selectively permeable barrier: It allows some substances to cross but blocks the movement of other substances. In addition, active transport pumps work continuously to maintain different concentrations of certain ions in the cytosol and interstitial fluid.

2. *Blood vessel walls* separate the interstitial fluid from blood plasma. Only in capillaries, the smallest blood vessels, are the walls thin enough and leaky enough to permit the exchange of water and solutes between blood plasma and interstitial fluid.

The body is in **fluid balance** when the required amounts of water and solutes are present and are correctly proportioned among the various compartments. Water is by far the largest single component of the body, making up 45–75 percent of total body mass, depending on age and gender.

Figure 22.1 Body fluid compartments.

🔑 **In lean adults, fluids make up 55–60 percent of body mass.**

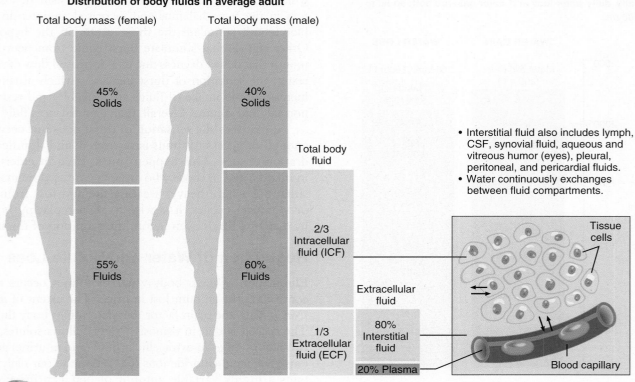

Distribution of body fluids in average adult

Total body mass (female)
- 45% Solids
- 55% Fluids

Total body mass (male)
- 40% Solids
- 60% Fluids

Total body fluid
- 2/3 Intracellular fluid (ICF)
- 1/3 Extracellular fluid (ECF)

Extracellular fluid
- 80% Interstitial fluid
- 20% Plasma

- Interstitial fluid also includes lymph, CSF, synovial fluid, aqueous and vitreous humor (eyes), pleural, peritoneal, and pericardial fluids.
- Water continuously exchanges between fluid compartments.

Tissue cells

Blood capillary

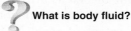
What is body fluid?

The processes of filtration, reabsorption, diffusion, and osmosis provide for the continual exchange of water and solutes among body fluid compartments (Figure 22.1b). Yet, the volume of fluid in each compartment remains remarkably stable. Because osmosis is the primary means of water movement between intracellular fluid and interstitial fluid, the concentration of solutes in these fluids determines the *direction* of water movement. Most solutes in body fluids are *electrolytes*, inorganic compounds that break apart into ions when dissolved in water. They are the main contributors to the osmotic movement of water. Fluid balance depends primarily on electrolyte balance so the two are closely interrelated. Because intake of water and electrolytes rarely occurs in exactly the same proportions as their presence in body fluids, the ability of the kidneys to excrete excess water by producing dilute urine, or to excrete excess electrolytes by producing concentrated urine, is of utmost importance in the maintenance of homeostasis.

Sources of Body Water Gain and Loss

The body can gain water by ingestion and by metabolic reactions (Figure 22.2). The main sources of body water are ingested liquids (about 1600 mL) and moist foods (about 700 mL)

Figure 22.2 Water balance: Sources of daily water gain and loss under normal conditions. Numbers are average volumes for adults.

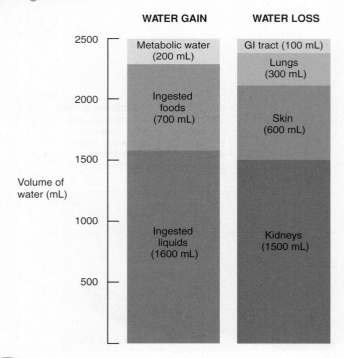

Normally, daily water loss and water gain are both equal to 2500 mL.

How would a diuretic drug affect a person's water balance?

absorbed from the gastrointestinal (GI) tract, which total about 2300 mL/day. The other source of water is *metabolic water* that is produced in the body during chemical reactions. Most of it is produced during aerobic cellular respiration (see Figure 20.3) and to a smaller extent during dehydration synthesis reactions (see Figure 2.8). Metabolic water gain accounts for about 200 mL/day. Thus, daily water gain totals about 2500 mL.

Normally, body fluid volume remains constant because water loss equals water gain. Water loss occurs in four ways (Figure 22.2). Each day the kidneys excrete about 1500 mL in urine, about 600 mL evaporates from the skin surface, the lungs exhale about 300 mL as water vapor, and the gastrointestinal tract eliminates about 100 mL in feces. In women of reproductive age, additional water is lost in menstrual flow. On average, daily water loss totals about 2500 mL. The amount of water lost by a given route can vary considerably over time. For example, water may literally pour from the skin in the form of sweat during strenuous exertion. In other cases, water may be lost in vomit or diarrhea during a GI tract infection.

Regulation of Body Water Gain

An area in the hypothalamus known as the *thirst center* governs the urge to drink. When water loss is greater than water gain, *dehydration*—a decrease in volume and an increase in osmotic pressure of body fluids—stimulates thirst (Figure 22.3). When body mass decreases by 2 percent due to fluid loss, mild dehydration exists. A decrease in blood volume causes blood pressure to fall. This change stimulates the kidneys to release renin, which promotes the formation of angiotensin II. Osmoreceptors in the hypothalamus and increased angiotensin II in the blood both stimulate the thirst center in the hypothalamus. Other signals that stimulate thirst come from neurons in the mouth that detect dryness due to a decreased flow of saliva. As a result, the sensation of thirst increases, which usually leads to increased fluid intake (if fluids are available) and restoration of normal fluid volume. Overall, fluid gain balances fluid loss.

Sometimes the sensation of thirst does not occur quickly enough or access to fluids is restricted, and significant dehydration ensues. This happens most often in elderly people, in infants, and in those who are in a confused mental state. In situations where heavy sweating or fluid loss from diarrhea or vomiting occurs, it is wise to start replacing body fluids by drinking fluids even before the sensation of thirst appears.

Regulation of Water and Solute Loss

Elimination of *excess* body water or solutes occurs mainly by controlling the amount lost in urine. The extent of *urinary salt (NaCl) loss* is the main factor that determines body fluid *volume*. The reason is that in osmosis "water follows solutes," and the two main solutes in extracellular fluid (and in urine) are sodium ions (Na^+) and chloride ions (Cl^-). Because our daily diet contains a highly variable amount of NaCl, urinary excretion of Na^+ and Cl^- must also vary to maintain homeostasis. Two hormones regulate the extent of renal Na^+ and Cl^-

Figure 22.3 Pathways through which dehydration stimulates thirst.

Dehydration occurs when water loss is greater than water gain.

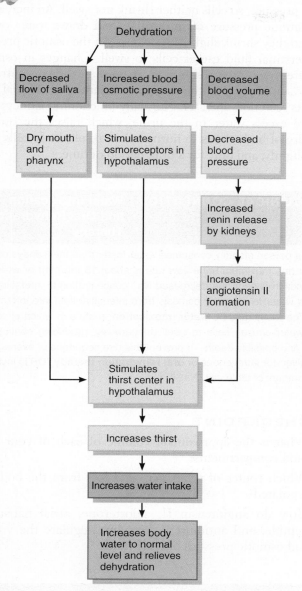

? Does regulation of these pathways occur via negative or positive feedback? Why?

Figure 22.4 Hormonal regulation of renal Na⁺ and Cl⁻ reabsorption.

The two main hormones that regulate renal Na^+ and Cl^- reabsorption (and thus the amount lost in the urine) are aldosterone and atrial natriuretic peptide.

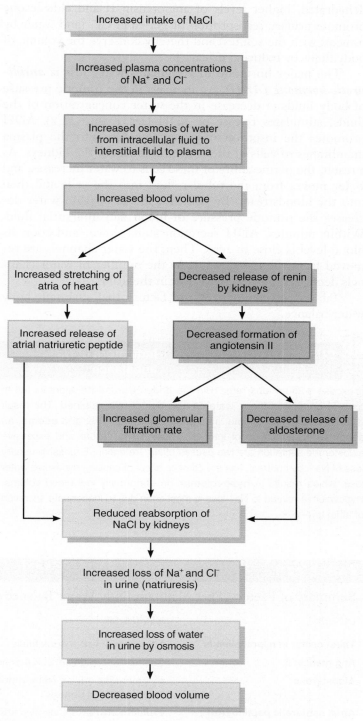

 How does excessive aldosterone secretion cause edema?

reabsorption (and thus how much is lost in the urine): *atrial natriuretic peptide (ANP)* and *aldosterone*.

Figure 22.4 depicts the sequence of changes that occur after a salty meal. The resulting increase in blood volume stretches the atria of the heart and promotes the release of atrial natriuretic peptide. ANP promotes *natriuresis* (*natrium* = sodium), elevated urinary loss of Na⁺ (and Cl⁻) and water, which decreases blood volume. The initial increase in blood volume also slows release of renin from the kidneys. As renin level declines, less angiotensin II is formed. Less angiotensin II leads to less aldosterone, which in turn causes the kidney tubules to reabsorb less Na^+ and Cl^-. More filtered Na^+ and Cl^- thus remain in the tubular fluid to be excreted in the urine. The osmotic consequence of excreting more Na^+ and Cl^- is loss of more water in urine, which decreases blood volume and blood pressure. By contrast, when someone becomes dehydrated, higher levels of angiotensin II and aldosterone promote urinary reabsorption of Na^+ and Cl^- (and water by osmosis with the solutes) and thereby conserve the volume of body fluids by reducing urinary loss.

The major hormone that regulates water loss is ***antidiuretic hormone (ADH)***. An increase in the osmotic pressure of body fluids (a decrease in the water concentration of the fluids) stimulates release of ADH (see Figure 21.8). ADH promotes the insertion of water channels into the plasma membranes of cells in the collecting ducts of the kidneys. As a result, the permeability of these cells to water increases, and water moves from the tubular fluid into the cells and then into the bloodstream. By contrast, intake of plain water decreases the osmotic pressure of blood and interstitial fluid. Within minutes, ADH secretion shuts down, and soon its blood level is close to zero. Then, the water channels are removed from the membranes. As the number of water channels decreases, more water is lost in the urine.

Table 22.1 summarizes the factors that maintain body water balance.

CLINICAL CONNECTION | Indicators of Na^+ Imbalance

If excess sodium ions remain in the body because the kidneys fail to excrete enough of them, water is also osmotically retained. The result is increased blood volume, increased blood pressure, and **edema**, an abnormal accumulation of interstitial fluid. Renal failure and excessive aldosterone secretion are two causes of Na^+ retention. Excessive urinary loss of Na^+, by contrast, has the osmotic effect of causing excessive water loss, which results in **hypovolemia**, an abnormally low blood volume. Hypovolemia related to Na^+ loss is most often due to inadequate secretion of aldosterone. •

Movement of Water Between Fluid Compartments

Intracellular and interstitial fluids normally have the same osmotic pressure, so cells neither shrink nor swell. An increase in the osmotic pressure of interstitial fluid draws water out of cells, so they shrink slightly. A decrease in the osmotic pressure of interstitial fluid causes cells to swell. Changes in osmotic pressure most often result from changes in the concentration of Na^+. A decrease in the osmotic pressure of interstitial fluid inhibits secretion of ADH. Normally functioning kidneys then excrete excess water in the urine, which raises the osmotic pressure of body fluids to normal. As a result, body cells swell only slightly, and only for a brief period of time.

CLINICAL CONNECTION | Water Intoxication and Oral Rehydration Therapy

When a person steadily consumes water faster than the kidneys can excrete it (the maximum urine flow rate is about 15 mL/min) or when kidney function is poor, the decreased Na^+ concentration of interstitial fluid causes water to move by osmosis from interstitial fluid into intracellular fluid. The result may be **water intoxication**, a state in which excessive body water causes cells to swell dangerously, producing convulsions, coma, and possibly death. To prevent this dire sequence of events, solutions given for intravenous or **oral rehydration therapy (ORT)** include a small amount of table salt (NaCl). •

✓ CHECKPOINT

1. What is the approximate volume of each of your body fluid compartments?

2. Which routes of water gain and loss from the body are regulated?

3. How do angiotensin II, aldosterone, atrial natriuretic peptide, and antidiuretic hormone regulate the volume and osmotic pressure of body fluids?

TABLE 22.1

Summary of Factors That Maintain Body Water Balance

FACTOR	MECHANISM	EFFECT
Thirst center in hypothalamus	Stimulates desire to drink fluids	Water gain if thirst is quenched
Angiotensin II	Stimulates secretion of aldosterone	Reduces loss of water in urine
Aldosterone	By promoting urinary reabsorption of Na^+ and Cl^-, increases water reabsorption via osmosis	Reduces loss of water in urine
Atrial natriuretic peptide (ANP)	Promotes natriuresis, elevated urinary excretion of Na^+ (and Cl^-), accompanied by water	Increases loss of water in urine
Antidiuretic hormone (ADH)	Promotes insertion of water-channel proteins into the plasma membranes of cells in the collecting ducts of the kidneys; as a result, water permeability of these cells increases and more water is reabsorbed	Reduces loss of water in urine

22.2 ELECTROLYTES IN BODY FLUIDS

OBJECTIVES • **Compare the electrolyte composition of the three major fluid compartments: plasma, interstitial fluid, and intracellular fluid.**

• **Discuss the functions of sodium, chloride, potassium, and calcium ions, and explain how their concentrations are regulated.**

The ions formed when electrolytes break apart serve four general functions in the body:

1. Because they are largely confined to particular fluid compartments and are more numerous than nonelectrolytes, certain ions *control the osmosis of water between fluid compartments.*

2. Certain ions *help maintain the acid–base balance* required for normal cellular activities.

3. Ions *carry electrical current*, which allows production of action potentials.

4. Several ions *serve as cofactors* needed for optimal activity of enzymes.

Figure 22.5 compares the concentrations of the main electrolytes and protein anions in extracellular fluid (blood plasma and interstitial fluid) and intracellular fluid. The chief difference between the two extracellular fluids is that blood plasma contains many protein anions, but interstitial fluid has very few. Because normal capillary membranes are virtually impermeable to proteins, only a few plasma proteins leak out of blood vessels into the interstitial fluid. This difference in protein concentration is largely responsible for the blood colloid osmotic pressure, the difference in osmotic pressure between blood plasma and interstitial fluid. The other components of the two extracellular fluids are similar.

The electrolyte content of intracellular fluid differs considerably from that of extracellular fluid. Sodium ions (Na^+) are the most abundant extracellular ions, representing about 90 percent of extracellular cations. Na^+ plays a pivotal role in fluid and electrolyte balance because it accounts for almost half of the osmotic pressure of extracellular fluid. Na^+ is necessary for the generation and conduction of action potentials in neurons and muscle fibers. As you learned earlier in this chapter, the Na^+ level in the blood is controlled by aldosterone, antidiuretic hormone, and atrial natriuretic peptide.

Chloride ions (Cl^-) are the most prevalent anions in extracellular fluid. Because most plasma membranes contain many

Figure 22.5 Electrolyte and protein anion concentrations in blood plasma, interstitial fluid, and intracellular fluid. The height of each column represents the milliequivalents per liter (mEq/liter), the total number of cations or anions (positive or negative electrical charges) in a given volume of solution.

The electrolytes present in extracellular fluids are different from those present in intracellular fluid.

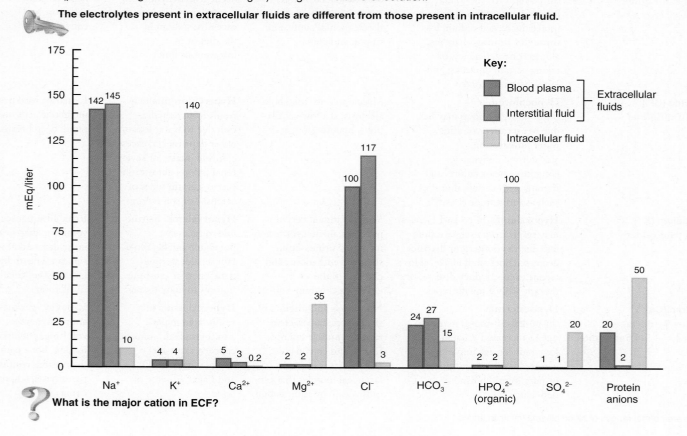

? What is the major cation in ECF?

Cl⁻ leakage channels, Cl⁻ moves easily between the extracellular and intracellular compartments. For this reason, Cl⁻ can help balance the level of anions in different fluid compartments. As mentioned earlier, processes that increase or decrease renal reabsorption of sodium ions also affect reabsorption of chloride ions. The negatively charged Cl⁻ follows the positively charged Na⁺ due to the electrical attraction of oppositely charged particles.

Potassium ions (K⁺), the most abundant cations in intracellular fluid, play a key role in establishing the resting membrane potential and in the repolarization phase of action potentials in neurons and muscle fibers. When K⁺ moves into or out of cells, it often is exchanged for H⁺ and thereby helps regulate the pH of body fluids. The level of K⁺ in blood plasma is controlled mainly by aldosterone. When blood plasma K⁺ is high, more aldosterone is secreted into the blood. Aldosterone then stimulates the renal collecting ducts to secrete more K⁺ and excess K⁺ is lost in the urine. Conversely, when blood plasma K⁺ is low, aldosterone secretion decreases and less K⁺ is excreted in urine.

About 98 percent of the calcium in adults is in the skeleton and teeth, where it is combined with phosphates to form mineral salts. In body fluids, calcium is mainly an extracellular cation (Ca^{2+}). Besides contributing to the hardness of bones and teeth, Ca^{2+} plays important roles in blood clotting, neurotransmitter release, maintenance of muscle tone, and excitability of nervous and muscle tissue.

The two main regulators of Ca^{2+} level in blood plasma are parathyroid hormone (PTH) and calcitriol, the form of vitamin D that acts as a hormone (see Figure 13.10). A low plasma Ca^{2+} level promotes release of more PTH, which increases bone *resorption* by stimulating osteoclasts in bone tissue to release Ca^{2+} (and phosphate) from mineral salts of bone matrix. PTH also enhances *reabsorption* of Ca^{2+} from glomerular filtrate back into blood and increases production of calcitriol (which in turn increases Ca^{2+} *absorption* from the gastrointestinal tract).

Table 22.2 describes the imbalances that result from the deficiency or excess of several electrolytes.

TABLE 22.2

Blood Electrolyte Imbalances

| ELECTROLYTE* | DEFICIENCY | | EXCESS | |
	NAME AND CAUSES	SIGNS AND SYMPTOMS	NAME AND CAUSES	SIGNS AND SYMPTOMS
Sodium (Na⁺) 136–148 mEq/liter	**Hyponatremia** (hī-pō-na-TRĒ-mē-a) may be due to decreased sodium intake; increased sodium loss through vomiting, diarrhea, aldosterone deficiency, or taking certain diuretics; and excessive water intake	Muscular weakness; dizziness, headache, and hypotension; tachycardia and shock; mental confusion, stupor, and coma	**Hypernatremia** may occur with dehydration, water deprivation, or excessive sodium in the diet or in intravenous fluids	Intense thirst, hypertension, edema, agitation, and convulsions
Chloride (Cl⁻) 95–105 mEq/liter	**Hypochloremia** (hī-pō-klō-RĒ-mē-a) may be due to excessive vomiting, water intoxication, aldosterone deficiency, congestive heart failure, and therapy with certain diuretics such as furosemide (Lasix®)	Muscle spasms, metabolic alkalosis, shallow ventilations, hypotension, and tetany	**Hyperchloremia** may result from dehydration due to water loss or water deprivation, excessive chloride intake, or severe renal failure, aldosterone excess, certain types of acidosis, or some drugs	Lethargy, weakness, metabolic acidosis, and rapid, deep breathing
Potassium (K⁺) 3.5–5.0 mEq/liter	**Hypokalemia** (hī-pō-ka-LĒ-mē-a) may result from excessive fluid loss due to vomiting or diarrhea, decreased potassium intake, aldosterone excess, kidney disease, or therapy with some diuretics	Muscle fatigue, flaccid paralysis, mental confusion, increased urine output, shallow ventilations, and changes in the electrocardiogram	**Hyperkalemia** may be due to excessive potassium intake, renal failure, aldosterone deficiency, or crushing injuries to body tissues	Irritability, nausea, vomiting, diarrhea, muscular weakness; can cause death by inducing ventricular fibrillation
Calcium (Ca²⁺) Total = 9–10.5 mg/dL; ionized = 4.5–5.5 mEq/liter	**Hypocalcemia** (hī-pō-kal-SĒ-mē-a) may be due to increased calcium loss, reduced calcium intake, elevated levels of phosphate, or parathyroid hormone deficiency	Numbness and tingling of the fingers; hyperactive reflexes, muscle cramps, tetany, and convulsions; bone fractures; spasms of laryngeal muscles that can cause death by asphyxiation	**Hypercalcemia** may result from hyperparathyroidism, some cancers, excessive intake of vitamin D, and Paget's disease of bone	Lethargy, weakness, anorexia, nausea, vomiting, polyuria, itching, bone pain, depression, confusion, paresthesia, stupor, and coma

*Values are normal ranges of blood plasma levels in adults.

People who are at risk for **fluid and electrolyte imbalances** include those who depend on others for fluid and food, such as infants, the elderly, and the hospitalized. Also at risk are individuals undergoing medical treatment that involves intravenous infusions, drainages or suctions, and urinary catheters. People who receive diuretics experience excessive fluid losses and require increased fluid intake; those who experience fluid retention and have fluid restrictions are also at risk. Finally at risk are postoperative individuals, severe burn or trauma cases, individuals with chronic diseases (congestive heart failure, diabetes, chronic obstructive lung disease, and cancer), people in confinement, and individuals with altered levels of consciousness who may be unable to communicate needs or respond to thirst. •

✓ CHECKPOINT

4. What are the functions of electrolytes in the body?

22.3 ACID–BASE BALANCE

OBJECTIVES • **Compare the roles of buffers, exhalation of carbon dioxide, and kidney excretion of H^+ in maintaining the pH of body fluids.**

- **Define acid–base imbalances, describe their effects on the body, and explain how they are treated.**

From our discussion thus far, it should be clear that various ions play different roles in helping to maintain homeostasis. A major homeostatic challenge is keeping the H^+ level (pH) of body fluids in the appropriate range. This task—the maintenance of acid–base balance—is of critical importance because the three-dimensional shape of all body proteins, which enables them to perform specific functions, is very sensitive to the most minor changes in pH. When the diet contains a large amount of protein, as is typical in North America, cellular metabolism produces more acids than bases and thus tends to acidify the blood.

In a healthy person, the pH of systemic arterial blood remains between 7.35 and 7.45. The removal of H^+ from body fluids and its subsequent elimination from the body depend on three major mechanisms: buffer systems, exhalation of carbon dioxide, and kidney excretion of H^+ into the urine.

The Actions of Buffer Systems

Buffers are substances that act quickly to temporarily bind H^+, removing the highly reactive, excess H^+ from solution but not from the body. They may also release H^+ into solution if the H^+ concentration is too low. Buffers prevent rapid, drastic changes in the pH of a body fluid by converting strong acids and bases into weak acids and bases. Strong acids release H^+ more readily than weak acids and thus contribute more free hydrogen ions. Similarly, strong bases raise pH more than weak ones. The principal buffer systems of the body fluids are the protein buffer system, the carbonic acid–bicarbonate buffer system, and the phosphate buffer system.

Protein Buffer System

Many proteins can act as buffers. Altogether, proteins in body fluids comprise the ***protein buffer system***, which is the most abundant buffer in intracellular fluid and plasma. Hemoglobin is an especially good buffer within red blood cells, and albumin is the main protein buffer in blood plasma. Recall that proteins are composed of amino acids, organic molecules that contain at least one carboxyl group ($-COOH$) and at least one amino group ($-NH_2$); these groups are the functional components of the protein buffer system. The carboxyl group releases H^+ when pH rises. The H^+ is then able to react with any excess OH^- in the solution to form water. The amino group combines with H^+, forming an $-NH_3^+$ group, when pH falls. Thus, proteins can buffer both acids and bases.

Carbonic Acid–Bicarbonate Buffer System

The ***carbonic acid–bicarbonate buffer system*** is based on the *bicarbonate ion* (HCO_3^-), which can act as a weak base, and *carbonic acid* (H_2CO_3), which can act as a weak acid. HCO_3^- is a significant anion in both intracellular and extracellular fluids (see Figure 22.5). Because the kidneys reabsorb filtered HCO_3^-, this important buffer is not lost in the urine. If there is an excess of H^+, the HCO_3^- can function as a weak base and remove the excess H^+ as follows:

$$\underset{\text{Hydrogen ion}}{H^+} + \underset{\substack{\text{Bicarbonate ion} \\ \text{(weak base)}}}{HCO_3^-} \longrightarrow \underset{\text{Carbonic acid}}{H_2CO_3}$$

Conversely, if there is a shortage of H^+, the H_2CO_3 can function as a weak acid and provide H^+ as follows:

$$\underset{\substack{\text{Carbonic acid} \\ \text{(weak acid)}}}{H_2CO_3} \longrightarrow \underset{\text{Hydrogen ion}}{H^+} + \underset{\text{Bicarbonate ion}}{HCO_3^-}$$

Phosphate Buffer System

The ***phosphate buffer system*** acts via a mechanism similar to the carbonic acid–bicarbonate buffer system. The components of the phosphate buffer system are the ions *dihydrogen phosphate* ($H_2PO_4^-$) and *monohydrogen phosphate* (HPO_4^{2-}). Recall that phosphates are major anions in intracellular fluid and minor ones in extracellular fluids (see Figure 22.5). The dihydrogen phosphate ion acts as a weak acid and is capable of buffering strong bases, such as OH^-, as follows:

$$\underset{\substack{\text{Hydroxide ion} \\ \text{(strong base)}}}{OH^-} + \underset{\substack{\text{Dihydrogen} \\ \text{phosphate} \\ \text{(weak acid)}}}{H_2PO_4^-} \longrightarrow \underset{\text{Water}}{H_2O} + \underset{\substack{\text{Monohydrogen} \\ \text{phosphate} \\ \text{(weak base)}}}{HPO_4^{2-}}$$

Diet, Acid–Base Balance, and Osteoporosis

One of the important electrolytes found in the blood is calcium. The body maintains blood calcium concentration within a fairly narrow range, since calcium is essential for many physiological functions. An adequate calcium intake is important for bone health, because if blood calcium falls too low, then the body moves this important mineral from the bone tissue into the bloodstream. Small calcium losses over the years may result in low bone density in midlife and old age. But other dietary factors affect bone metabolism and mineral density as well.

Excess Dietary Sodium = Calcium Loss

The kidneys treat calcium ions like sodium ions. When they sense that blood sodium levels are too high, the kidneys begin to excrete sodium into the urine, to lower blood levels. Unfortunately for the bones, whenever the kidneys go into sodium excretion mode, they excrete calcium as well. Blood calcium levels fall, and the bones may later be called upon for a donation. A diet high in sodium is associated with low bone density in later life.

Acid–Base Balance

The composition of the diet affects acid–base balance in the body. Interestingly, it is not whether the foods themselves are acidic or basic, but whether the processes of digestion

lead to excess hydrogen ions (H^+). Grains and meats generate H^+, while fruits and vegetables are broken down into bicarbonate ions (HCO_3^-) when they are metabolized, so they can soak up some of the excess H^+.

The kidneys become less efficient at regulating acid–base balance as they age. Researchers have found that middle-aged and older adults can develop a chronic, mild acidosis, which seems to be related to bone mineral loss. Older adults who consume low levels of fruits and vegetables have higher markers of bone turnover. Bone tissue has receptors for H^+, so bones sense blood pH levels. It is unclear exactly why bone responds by losing minerals when blood pH gets too low (H^+ concentration gets too high), but researchers suggest that bone may participate in buffering processes.

Should people try to consume more fruits and vegetables? Bone researchers suggest replacing a couple of grain servings each day with fruits and vegetables to reduce H^+ production. Because older adults are often protein-deficient, and protein is needed to produce the collagen that holds the bones together, older adults are cautioned not to decrease their consumption of protein-rich foods.

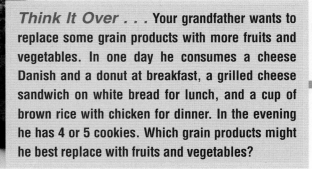

Think It Over . . . **Your grandfather wants to replace some grain products with more fruits and vegetables. In one day he consumes a cheese Danish and a donut at breakfast, a grilled cheese sandwich on white bread for lunch, and a cup of brown rice with chicken for dinner. In the evening he has 4 or 5 cookies. Which grain products might he best replace with fruits and vegetables?**

The monohydrogen phosphate ion, in contrast, acts as a weak base and is capable of buffering the H^+ released by a strong acid such as hydrochloric acid (HCl):

$$H^+ \quad + \quad HPO_4^{2-} \quad \longrightarrow \quad H_2PO_4^-$$

Hydrogen ion (strong acid)   Monohydrogen phosphate (weak base)   Dihydrogen phosphate (weak acid)

Because the concentration of phosphates is highest in intracellular fluid, the phosphate buffer system is an important regulator of pH in the cytosol. It also acts to a smaller degree in extracellular fluids, and it buffers acids in urine.

Exhalation of Carbon Dioxide

Breathing plays an important role in maintaining the pH of body fluids. An increase in the carbon dioxide (CO_2) concentration in body fluids increases H^+ concentration and

thus lowers the pH (makes body fluids more acidic). Conversely, a decrease in the CO_2 concentration of body fluids raises the pH (makes body fluids more alkaline). These chemical interactions are illustrated by the following reversible reactions.

$$CO_2 + H_2O \rightleftharpoons H_2CO_3 \rightleftharpoons H^+ + HCO_3^-$$

Carbon dioxide   Water   Carbonic acid   Hydrogen ion   Bicarbonate ion

Changes in the rate and depth of breathing can alter the pH of body fluids within a couple of minutes. With increased ventilation, more CO_2 is exhaled, the reaction goes from right to left, H^+ concentration falls, and blood pH rises. If ventilation is slower than normal, less carbon dioxide is exhaled, and the blood pH falls.

The pH of body fluids and the rate and depth of breathing interact via a negative feedback loop (Figure 22.6). When

Figure 22.6 Negative feedback regulation of blood pH by the respiratory system.

Exhalation of carbon dioxide lowers the H$^+$ concentration of blood.

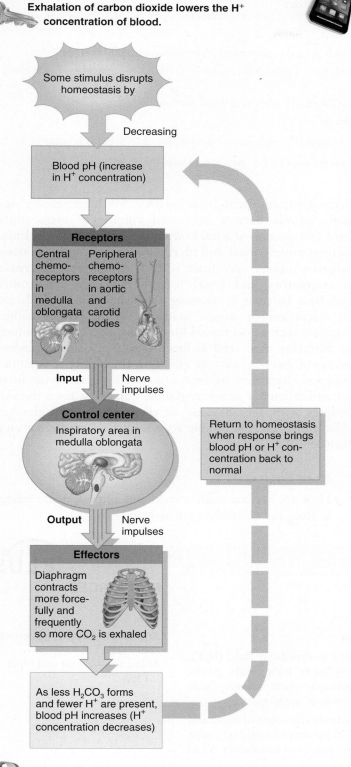

If you hold your breath for 30 seconds, what is likely to happen to your blood pH?

the blood acidity increases, the decrease in pH (increase in concentration of H$^+$) is detected by chemoreceptors in the medulla oblongata and in the aortic and carotid bodies, both of which stimulate the inspiratory area in the medulla oblongata. As a result, the diaphragm and other respiratory muscles contract more forcefully and frequently, so more CO$_2$ is exhaled, driving the reaction to the left. As less H$_2$CO$_3$ forms and fewer H$^+$ are present, blood pH increases. When the response brings blood pH (H$^+$ concentration) back to normal, there is a return to acid–base homeostasis.

By contrast, if the pH of the blood increases, the respiratory center is inhibited and the rate and depth of breathing decreases. Then, CO$_2$ accumulates in the blood and its H$^+$ concentration increases. This respiratory mechanism is powerful, but it can regulate the concentration of only one acid—carbonic acid.

Kidney Excretion of H$^+$

The slowest mechanism for removal of acids is also the only way to eliminate most acids that form in the body: Cells of the renal tubules secrete H$^+$, which then is excreted in urine. Also, because the kidneys synthesize new HCO$_3^-$ and reabsorb filtered HCO$_3^-$, this important buffer is not lost in the urine. Because of the contributions of the kidneys to acid–base balance, it's not surprising that renal failure can quickly cause death.

Table 22.3 summarizes the mechanisms that maintain pH of body fluids.

Acid–Base Imbalances

Acidosis is a condition in which arterial blood pH is below 7.35. The principal physiological effect of acidosis is depression of the central nervous system through depression of synaptic transmission. If the systemic arterial blood pH falls below 7, depression of the nervous system is so severe that the individual becomes disoriented, then becomes comatose, and may die.

In *alkalosis*, arterial blood pH is higher than 7.45. A major physiological effect of alkalosis is overexcitability in both the central nervous system and peripheral nerves. Neurons conduct impulses repetitively, even when not stimulated; the results are nervousness, muscle spasms, and even convulsions and death.

A change in blood pH that leads to acidosis or alkalosis may be countered by *compensation*, the physiological response to an acid–base imbalance that acts to normalize arterial blood pH. Compensation may be either *complete*, if pH indeed is brought within the normal range, or *partial*, if systemic arterial blood pH is still lower than 7.35 or higher than 7.45. If a person has altered blood pH due to metabolic causes, hyperventilation or hypoventilation can help bring blood pH back toward the normal range; this form of compensation, termed *respiratory compensation*, occurs within minutes and reaches its maximum within hours. If, however, a person has altered blood pH due to respiratory causes, then *renal compensation*—changes in secretion of H$^+$ and reabsorption of HCO$_3^-$ by the kidney tubules—can help reverse

TABLE 22.3

Mechanisms That Maintain pH of Body Fluids

MECHANISM	COMMENTS
Buffer Systems	Convert strong acids and bases into weak acids and bases, preventing drastic changes in body fluid pH
Proteins	The most abundant buffers in body cells and blood. Hemoglobin is a buffer in the cytosol of red blood cells; albumin is a buffer in blood plasma
Carbonic acid–bicarbonate	Important regulators of blood pH; the most abundant buffers in extracellular fluid
Phosphates	Important buffers in intracellular fluid and in urine
Exhalation of CO_2	With increased exhalation of CO_2, pH rises (fewer H^+); with decreased exhalation of CO_2, pH falls (more H^+)
Kidneys	Kidney tubules secrete H^+ into the urine and reabsorb HCO_3^- so it is not lost in the urine

the change. Renal compensation may begin in minutes, but it takes days to reach maximum effectiveness.

✓ CHECKPOINT

5. How do proteins, bicarbonate ions, and phosphate ions help maintain the pH of body fluids?

6. What are the major physiological effects of acidosis and alkalosis?

22.4 AGING AND FLUID, ELECTROLYTE, AND ACID–BASE BALANCE

OBJECTIVE • **Describe the changes in fluid, electrolyte, and acid–base balance that may occur with aging.**

By comparison with children and younger adults, older adults often have an impaired ability to maintain fluid, electrolyte, and acid–base balance. With increasing age, many people have a decreased volume of intracellular fluid and decreased total body potassium due to declining skeletal muscle mass and increasing mass of adipose tissue (which contains very little water). Age-related decreases in respiratory and renal functioning may compromise acid–base balance by slowing the exhalation of CO_2 and the excretion of excess acids in urine. Other kidney changes, such as decreased blood flow, decreased glomerular filtration rate, and reduced sensitivity to antidiuretic hormone, have an adverse effect on the ability to maintain fluid and electrolyte balance. Due to a decrease in the number and efficiency of sweat glands, water loss from the skin declines with age. Because of these age-related changes, older adults are susceptible to several fluid and electrolyte disorders.

✓ CHECKPOINT

7. How are skeletal muscle mass and adipose tissue related to fluid and electrolyte imbalance?

CHAPTER REVIEW AND RESOURCE SUMMARY

REVIEW

22.1 Fluid Compartments and Fluid Balance

1. The water and dissolved solutes in the body constitute the **body fluids**.

2. About two-thirds of the body's fluid is located within cells and is called **intracellular fluid (ICF)**. The other one-third, called **extracellular fluid (ECF)**, includes all other body fluids. About 80 percent of the ECF is **interstitial fluid**, which occupies the microscopic spaces between tissue cells, and about 20 percent of the ECF is **blood plasma**, the liquid portion of the blood.

3. **Fluid balance** means that the various body compartments contain the normal amount of water and solutes. Water is the largest single component in the body, about 45–70 percent of total body mass in lean adults. An **electrolyte** is an inorganic substance that dissociates into ions in solution. Fluid balance and electrolyte balance are interrelated.

4. Daily water gain and loss are each about 2500 mL. Sources of water gain are ingested liquids and foods and water produced by metabolic reactions (**metabolic water**). Water is lost from the body through urination, evaporation from the skin surface, exhalation of water vapor, and defecation. In women, menstrual flow is an additional route for loss of body water.

RESOURCES

Anatomy Overview—Overview of Fluids
Animation—Water and Fluid Flow

REVIEW	RESOURCES

REVIEW

5. The main way to regulate body water gain is by adjusting the volume of water intake. The **thirst center** in the hypothalamus governs the urge to drink.

6. **Aldosterone** reduces urinary loss of Na^+ and Cl^- and thereby increases the volume of body fluids. **Atrial natriuretic peptide (ANP)** promotes natriuresis, elevated excretion of Na^+ (and Cl^-) and water, which decreases blood volume.

7. Table 22.1 summarizes the factors that maintain water balance.

22.2 Electrolytes in Body Fluids

1. Electrolytes control the osmosis of water between fluid compartments, help maintain acid–base balance, carry electrical current, and act as enzyme cofactors.

2. Sodium ions (Na^+) are the most abundant extracellular ions. They are involved in action potentials, muscle contraction, and fluid and electrolyte balance. Na^+ level is controlled by aldosterone, antidiuretic hormone, and atrial natriuretic peptide.

3. Chloride ions (Cl^-) are the major extracellular anions. They play a role in regulating osmotic pressure and forming HCl in gastric juice. Cl^- level is controlled by processes that increase or decrease kidney reabsorption of Na^+.

4. Potassium ions (K^+) are the most abundant cations in intracellular fluid. They play a key role in establishing the resting membrane potential in neurons and muscle fibers, and contribute to regulation of pH. K^+ level is controlled by aldosterone.

5. Calcium is the most abundant mineral in the body. Calcium salts are structural components of bones and teeth. Ca^{2+}, which are principally extracellular cations, function in blood clotting, neurotransmitter release, and contraction of muscle. Ca^{2+} level is controlled mainly by parathyroid hormone and calcitriol.

6. Table 22.2 describes the imbalances that result from deficiency or excess of important body electrolytes.

RESOURCES (22.2)

Anatomy Overview—
Electrolytes
Exercise—Calcium
Homeostasis
Concepts and Connections—
Blood Calcium Regulation

22.3 Acid–Base Balance

1. The normal pH of systemic arterial blood is 7.35 to 7.45. Homeostasis of pH is maintained by buffer systems (**protein buffer system**, **carbonic acid–bicarbonate buffer system**, **phosphate buffer system**), by exhalation of carbon dioxide, and by kidney excretion of H^+ and reabsorption of HCO_3^-. Table 22.3 summarizes the mechanisms that maintain pH of body fluids.

2. **Acidosis** is a systemic arterial blood pH below 7.35; its principal effect is depression of the central nervous system (CNS). **Alkalosis** is a systemic arterial blood pH above 7.45; its principal effect is overexcitability of the CNS.

RESOURCES (22.3)

Animation—Acids and Bases

Figure 22.6 Negative feedback regulation of blood pH by the respiratory system

Exercise—Destination: Acid/Base Balance

22.4 Aging and Fluid, Electrolyte, and Acid–Base Balance

1. With increasing age, there is decreased intracellular fluid volume and decreased potassium due to declining skeletal muscle mass.

2. Decreased kidney function adversely affects fluid and electrolyte balance.

RESOURCES (22.4)

Homeostatic Imbalance—The Case of the Thirsty Woman

SELF-QUIZ

1. Normally, most of the body's water is lost through
   a. the gastrointestinal tract.   b. cellular respiration.
   c. exhalation by the lungs.   d. excretion of urine.
   e. evaporation from the skin.

2. Substances that dissociate into ions when dissolved in body fluids are
   a. neurotransmitters.   b. enzymes.   c. nonelectrolytes.
   d. hormones.   e. electrolytes.

3. Which of the following statements about sodium is NOT true?

   a. Sodium ions are the most abundant intracellular ions.

   b. Sodium is necessary for generating action potentials in neurons.

   c. Excess sodium ions can cause edema.

   d. Sodium levels are regulated by the kidneys.

   e. Aldosterone helps regulate the concentration of sodium in the blood.

4. Parathyroid hormone (PTH) controls blood levels of

   a. magnesium.   b. sodium.   c. calcium.

   d. potassium.   e. chloride.

5. Fluid movement between intracellular fluid and extracellular fluid depends primarily on the concentration of which ion in extracellular fluid?

   a. sodium   b. potassium   c. calcium

   d. phosphate   e. magnesium

6. The slowest but most effective way to remove acids from the body is by

   a. attaching the acids to hemoglobin.

   b. utilizing the phosphate buffer system.

   c. increasing ventilation.

   d. secreting H^+ into the kidney tubules.

   e. producing large amounts of bases.

7. Which of the following statements concerning acid–base balance in the body is NOT true?

   a. An increase in respiration rate increases pH of body fluids.

   b. Normal pH of extracellular fluid is 7.35 to 7.45.

   c. Buffers are an important mechanism in the maintenance of pH balance.

   d. A blood pH of 7.2 is called alkalosis.

   e. High levels of CO_2 in body fluids can result in a drop in pH.

8. In the carbonic acid–bicarbonate buffer system, the bicarbonate ion acts as a

   a. weak base.   b. strong base.   c. strong acid.

   d. nonelectrolyte.   e. weak acid.

9. The most abundant buffer in intracellular fluid and plasma is the _____ buffer system.

   a. respiratory   b. carbonic acid   c. protein

   d. bicarbonate   e. phosphate

10. Most (80 percent) of the extracellular fluid is part of the body's

    a. interstitial fluid.   b. lymph.   c. cerebrospinal fluid.

    d. plasma.   e. synovial fluid.

11. Which hormone stimulates the kidneys to secrete more K^+?

    a. atrial natriuretic peptide   b. angiotensin

    c. aldosterone   d. antidiuretic hormone

    e. parathyroid hormone

12. Most of the body's water comes from

    a. cellular respiration.   b. adipose tissue.

    c. urine production.   d. water intoxication.

    e. ingested liquids and foods.

13. The thirst center can be activated by all of the following EXCEPT

    a. angiotensin II.

    b. an increase in blood volume.

    c. a decrease in flow from salivary glands.

    d. a decrease in blood pressure.

    e. an increase in blood osmotic pressure.

14. The center for thirst is located in the

    a. kidneys.   b. adrenal cortex.   c. hypothalamus.

    d. cerebral cortex.   e. liver.

15. Which of the following is NOT one of the functions of electrolytes in the body?

    a. control of fluid movement between the extracellular and intracellular compartments

    b. regulation of pH

    c. enzyme cofactor

    d. energy source

    e. carrier of electric current

16. Aldosterone is secreted in response to

    a. increased blood pressure.   b. decreased blood volume.

    c. increased calcium levels.   d. increased sodium levels.

    e. increased water levels.

17. What is the importance of buffer systems in the body?

    a. They help maintain the calcium and phosphate balances of bone.

    b. They control the body's water balance.

    c. They prevent drastic changes in the body's pH.

    d. They help regulate blood volume.

    e. They are responsible for the operation of the body's sodium pump.

18. Match the following:

    ____ a. extracellular cation; structural component of bones and teeth

    ____ b. most abundant anion in extracellular fluid

    ____ c. most abundant extracellular cation; needed for generation and conduction of action potentials

    ____ d. most abundant cation in intracellular fluid; involved in nerve and muscle homeostasis

    A. calcium
    B. chloride
    C. potassium
    D. sodium

CRITICAL THINKING APPLICATIONS

1. Stefan was eating his lunch at a fast food restaurant. He had a large order of fries and a double cheeseburger with ketchup (a very high sodium content lunch). Next, Stefan bought a large bottled water and drank the entire bottle. How will his body respond to this lunch?

2. One-year-old Josh had a busy morning at the swimming pool. Today's lesson included lots of underwater exercises in blowing bubbles. After the lesson, Josh seemed disoriented and then suffered a convulsion. The emergency room nurse thinks the swim class has something to do with Josh's problem. What is wrong with Josh?

3. Many years of heavy smoking have taken a toll on Christina's lungs. Emphysema makes it so difficult to breathe that Christina can't walk through a shopping center without frequently sitting to rest and "catch her breath." Describe what is occurring with Christina's acid–base balance as related to her emphysema.

4. Theo was 15 minutes late for A&P class. While searching for his pen, he thought he heard the instructor say something about the heart affecting water balance but he had thought it was the other way around. Theo just decided to ignore the whole thing. Explain the relationship of the heart to fluid balance to Theo.

ANSWERS TO FIGURE QUESTIONS

22.1 The term body fluid refers to body water and its dissolved substances.

22.2 A diuretic drug increases urine flow rate, thereby increasing loss of fluid from the body and decreasing the volume of body fluids.

22.3 Negative feedback is in operation because the result (an increase in fluid intake) is opposite to the initiating stimulus (dehydration).

22.4 Elevated aldosterone promotes abnormally high renal reabsorption of NaCl and water, which expands blood volume and increases blood pressure. Increased blood pressure causes more fluid to filter out of capillaries and accumulate in the interstitial fluid, a condition called edema.

22.5 The major cation in ECF is Na^+.

22.6 Breath holding causes blood pH to decrease slightly as CO_2 and H^+ accumulate.

CHAPTER 23
THE REPRODUCTIVE SYSTEMS

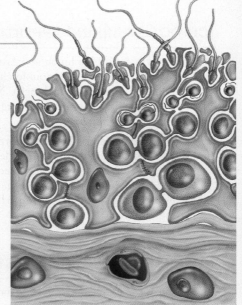

S*exual reproduction* is the process by which organisms produce offspring by making germ cells called ***gametes*** (GAM-ēts = spouses). After ***fertilization***, when the male gamete (sperm cell) unites with the female gamete (secondary oocyte), the resulting cell contains one set of chromosomes from each parent. The organs that make up the male and female reproductive systems can be grouped by function. The ***gonads***—testes in males and ovaries in females—produce gametes and secrete sex hormones. Various ***ducts*** then store and transport the gametes, and ***accessory sex glands*** produce substances that protect the gametes and facilitate their movement. Finally, ***supporting structures***, such as the penis and the uterus, assist the delivery and joining of gametes and, in females, the growth of the embryo and fetus during pregnancy.

   Gynecology (gī′-ne-KOL-ō-jē; *gyneco-* = woman; *-logy* = study of) is the specialized branch of medicine concerned with the diagnosis and treatment of diseases of the female reproductive system. As noted in Chapter 21, ***urology*** (ū-ROL-ō-jē) is the study of the urinary system. Urologists also diagnose and treat diseases and disorders of the male reproductive system. The branch of medicine that deals with male disorders, especially infertility and sexual dysfunction, is called ***andrology*** (an-DROL-ō-jē; *andro-* = masculine).

Did you know?   *Sexually transmitted diseases (STDs) can lead to long-term negative health consequences, including cancers, infertility, blindness, and even death. STDs such as infection with human immunodeficiency virus (HIV) can devastate a population, as STDs tend to strike young people in the prime of life. They can also be transmitted to infants and children in utero or during birth. When young adults are lost to STDs, a country loses its work force, and children their parents. The high rates of disability and death from STDs worldwide are especially tragic, since STDs are totally preventable (with abstinence), and when sexual intercourse does occur, risk can be greatly reduced with safer sex practices.*

FOCUS ON WELLNESS
Sexually Transmitted Diseases—
Why Are Females More Vulnerable?

LOOKING BACK TO MOVE AHEAD...

Somatic Cell Division (Section 3.7)

Sympathetic and Parasympathetic Divisions of the Autonomic Nervous System (Section 11.1)

Hormones of the Hypothalamus and Pituitary Gland (Section 13.3)

23.1 MALE REPRODUCTIVE SYSTEM

OBJECTIVES • **Describe the location, structure, and functions of the organs of the male reproductive system.**

• **Describe how sperm cells are produced.**

• **Explain the roles of hormones in regulating male reproductive functions.**

The organs of the ***male reproductive system*** are the testes; a system of ducts (epididymis, ductus deferens, ejaculatory ducts, and urethra); accessory sex glands (seminal vesicles, prostate, and bulbourethral glands); and several supporting structures, including the scrotum and the penis (Figure 23.1). The testes produce sperm and secrete hormones. Sperm are transported and stored, helped to mature, and conveyed to the exterior by a system of ducts. Semen contains sperm plus the secretions provided by the accessory sex glands.

Scrotum

The ***scrotum*** (SKRŌ-tum = bag) is a pouch that supports the testes; it consists of loose skin, superficial fascia, and smooth muscle (Figure 23.1). Internally, a septum divides the scrotum into two sacs, each containing a single testis.

The production and survival of sperm is optimal at a temperature that is about 2–3°C below normal body temperature. This lowered body temperature is maintained within the scrotum because it is outside the pelvic cavity. On exposure to cold, skeletal muscles contract to elevate the testes, moving them closer to the pelvic cavity, where they can absorb body

Figure 23.1 **Male organs of reproduction and surrounding structures.**

Reproductive organs are adapted to produce new individuals and pass genetic material from one generation to the next.

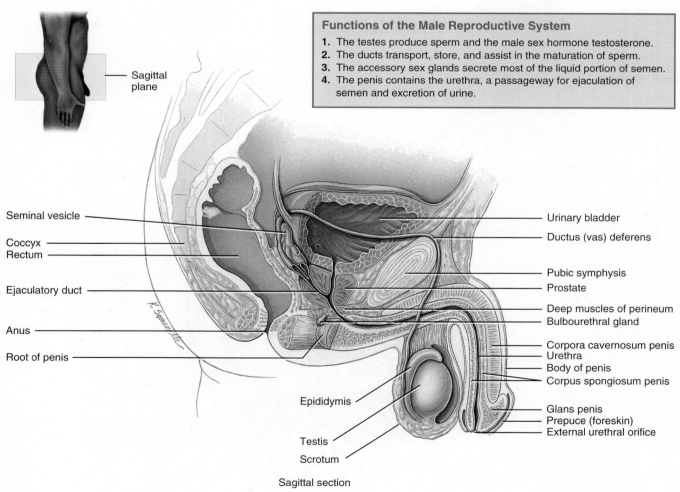

Functions of the Male Reproductive System

1. The testes produce sperm and the male sex hormone testosterone.
2. The ducts transport, store, and assist in the maturation of sperm.
3. The accessory sex glands secrete most of the liquid portion of semen.
4. The penis contains the urethra, a passageway for ejaculation of semen and excretion of urine.

Sagittal plane

Seminal vesicle

Coccyx

Rectum

Ejaculatory duct

Anus

Root of penis

Epididymis

Testis

Scrotum

Urinary bladder

Ductus (vas) deferens

Pubic symphysis

Prostate

Deep muscles of perineum

Bulbourethral gland

Corpora cavernosum penis

Urethra

Body of penis

Corpus spongiosum penis

Glans penis

Prepuce (foreskin)

External urethral orifice

Sagittal section

 Among the male organs of reproduction, how is the penis classified functionally?

heat. Exposure to warmth causes relaxation of the skeletal muscles and descent of the testes, increasing the surface area exposed to the air, so that the testes can give off excess heat to their surroundings.

Testes

The **testes** (TES-tēz; singular is *testis*; Figure 23.2), or *testicles*, are paired oval glands that develop on the embryo's posterior

Figure 23.2 Anatomy and histology of the testes. (a) Spermatogenesis occurs in the seminiferous tubules. (b) The stages of spermatogenesis. Arrows indicate the progression from least mature to most mature spermatogenic cells. The (*n*) and (2*n*) refer to haploid and diploid chromosome number, to be described shortly.

🔑 **The male gonads are the testes, which produce haploid sperm.**

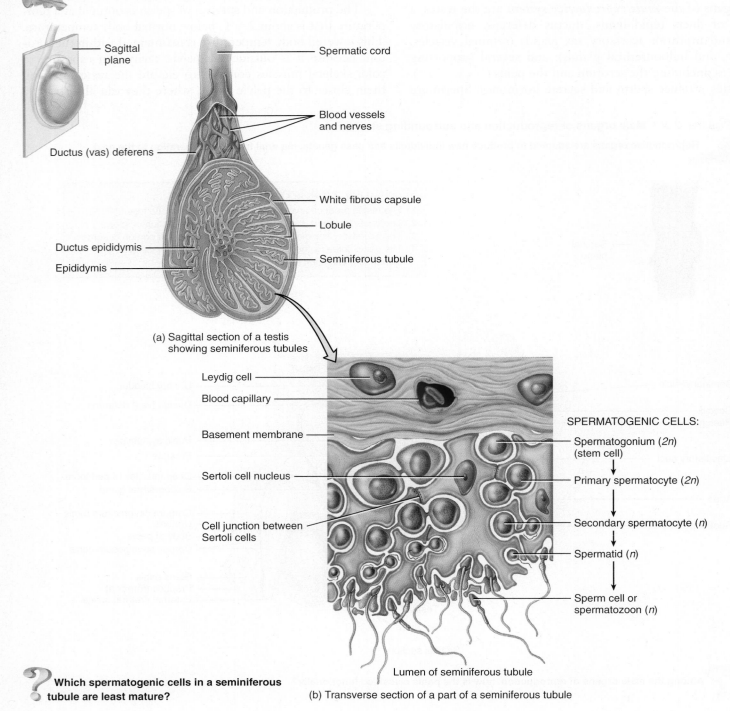

Sagittal plane

Spermatic cord

Blood vessels and nerves

Ductus (vas) deferens

White fibrous capsule

Lobule

Ductus epididymis

Seminiferous tubule

Epididymis

(a) Sagittal section of a testis showing seminiferous tubules

Leydig cell

Blood capillary

Basement membrane

SPERMATOGENIC CELLS:

Spermatogonium (2*n*) (stem cell)

Primary spermatocyte (2*n*)

Sertoli cell nucleus

Secondary spermatocyte (*n*)

Cell junction between Sertoli cells

Spermatid (*n*)

Sperm cell or spermatozoon (*n*)

Lumen of seminiferous tubule

? **Which spermatogenic cells in a seminiferous tubule are least mature?**

(b) Transverse section of a part of a seminiferous tubule

abdominal wall and usually begin their descent into the scrotum in the seventh month of fetal development.

The testes are covered by a dense *white fibrous capsule* that extends inward and divides each testis into internal compartments called *lobules* (Figure 23.2a). Each of the 200 to 300 lobules contains one to three tightly coiled *seminiferous tubules* (semin- = seed; fer- = to carry) that produce sperm by a process called spermatogenesis (described shortly).

Seminiferous tubules are lined with sperm-forming cells called *spermatogenic cells* (Figure 23.2b). Positioned against the basement membrane, toward the outside of the tubules, are the *spermatogonia* (sper-ma′-tō-GŌ-nē-a; -gonia = offspring), the stem cell precursors. Toward the lumen of the tubule are layers of cells in order of advancing maturity: primary spermatocytes, secondary spermatocytes, spermatids, and sperm cells. After a *sperm cell* or *spermatozoon* (sper′-ma-tō-ZŌ-on; -zoon = life) has formed, it is released into the lumen of the seminiferous tubule.

Located between the developing sperm cells in the seminiferous tubules, large *Sertoli cells* (ser-TŌ-lē) support, protect, and nourish spermatogenic cells; phagocytize degenerating spermatogenic cells; secrete fluid for sperm transport; and release the hormone inhibin, which helps regulate sperm production. Between the seminiferous tubules are clusters of *Leydig cells*. These cells secrete the hormone *testosterone*, the most important androgen. An *androgen* (AN-drō-jen) is a hormone that promotes the development of masculine characteristics. Testosterone also promotes a man's libido (sex drive).

CLINICAL CONNECTION | Cryptorchidism

The condition in which the testes do not descend into the scrotum is called **cryptorchidism** (krip-TOR-ki-dizm; *crypt-* = hidden; *orchid* = testis). It occurs in about 3 percent of full-term infants and about 30 percent of premature infants. Untreated bilateral cryptorchidism causes sterility due to the higher temperature of the pelvic cavity. The chance of testicular cancer is 30 to 50 times greater in cryptorchid testes, possibly due to abnormal division of germ cells caused by the higher temperature of the pelvic cavity. The testes of about 80 percent of boys with cryptorchidism will descend spontaneously during the first year of life. When the testes remain undescended, the condition can be corrected surgically, ideally before 18 months of age. •

Spermatogenesis

The process by which the seminiferous tubules of the testes produce sperm is called *spermatogenesis* (sper′-ma-tō-JEN-e-sis). It consists of three stages: meiosis I, meiosis II, and spermiogenesis. We begin with meiosis.

OVERVIEW OF MEIOSIS   As you learned in Chapter 3, most body cells (somatic cells), such as brain cells, stomach cells,

kidney cells, and so forth, contain 23 pairs of chromosomes, or a total of 46 chromosomes. One member of each pair is inherited from each parent. The two chromosomes that make up each pair are called *homologous chromosomes* (hō-MOL-ō-gus; *homo-* = same); they contain similar genes arranged in the same (or almost the same) order. Because somatic cells contain two sets of chromosomes, they are termed *diploid cells* (DIP-loyd; *dipl-* = double; *-oid* = form), symbolized as *2n*. Gametes differ from somatic cells because they contain a single set of 23 chromosomes, symbolized as *n*; they are thus said to be *haploid* (HAP-loyd; *hapl-* = single).

In sexual reproduction, an organism results from the fusion of two different gametes, one produced by each parent. If each gamete had the same number of chromosomes as somatic cells, then the number of chromosomes would double each time fertilization occurred. Instead, gametes receive a single set of chromosomes by means of a special type of reproductive cell division called *meiosis* (mī-Ō-sis; *mei-* = lessening; *-osis* = condition of). Meiosis occurs in two successive stages: *meiosis I* and *meiosis II*. First, we will examine how meiosis occurs during spermatogenesis. Later in the chapter, we will follow the steps of meiosis during oogenesis, the production of female gametes.

STAGES OF SPERMATOGENESIS   Spermatogenesis begins during puberty and continues throughout life. The time from onset of cell division in a *spermatogonium* until sperm are released into the lumen of a seminiferous tubule is 65 to 75 days. The spermatogonia contain the diploid number of chromosomes (46). After a spermatogonium undergoes mitosis, one cell stays near the basement membrane as a spermatogonium, so stem cells remain for future mitosis (Figure 23.3a). The other cell differentiates into a *primary spermatocyte* (sper-MA-tō-sīt′). Like spermatogonia, primary spermatocytes are diploid.

Unlike mitosis, which is complete after a single round, meiosis occurs in two successive stages: *meiosis I* and *meiosis II*. During the interphase that precedes meiosis I, the chromosomes of the diploid cell start to replicate. As a result of replication, each chromosome consists of two sister (genetically identical) chromatids, which are attached at their centromeres. This replication of chromosomes is similar to the one that precedes mitosis in somatic cell division.

MEIOSIS I   Meiosis I, which begins once chromosomal replication is complete, consists of four phases: prophase I, metaphase I, anaphase I, and telophase I (Figure 23.3b). In prophase I the chromosomes shorten and thicken, the nuclear envelope and nucleoli disappear, and the mitotic spindle forms. Two events that are not seen in mitotic prophase occur during prophase I of meiosis (Figure 23.3c). First, the two sister chromatids of each pair of homologous chromosomes pair off, an event called *synapsis* (sin-AP-sis).

The resulting four chromatids form a structure called a *tetrad*. Second, parts of the chromatids of two homologous chromosomes may be exchanged with one another. Such an exchange between parts of nonsister (genetically different) chromatids is called *crossing-over.* This process, among others, permits an exchange of genes between chromatids of homologous chromosomes. Due to crossing-over, the resulting cells are genetically unlike each other and genetically unlike the starting cell that produced them. Crossing-over results in *genetic recombination*—that is, the formation of new combinations of genes—and accounts for part of the great genetic variation among humans and other organisms that form gametes via meiosis.

In metaphase I, the tetrads formed by the homologous pairs of chromosomes line up along the metaphase plate of the cell, with homologous chromosomes side by side (Figure 23.3b). During anaphase I, the members of each homologous pair of chromosomes separate as they are pulled to opposite poles of the cell by the microtubules attached to the centromeres. The paired chromatids, held by a centromere, remain together. (Recall that during mitotic anaphase, the centromeres split and the sister chromatids separate.) Telophase I and cytokinesis of meiosis are similar to telophase and cytokinesis of mitosis. The net effect of meiosis I is that each resulting cell (*secondary spermatocyte*) contains the haploid number of chromosomes because it contains only one member of each pair of the homologous chromosomes present in the starting cell.

MEIOSIS II   The second stage of meiosis, meiosis II, also consists of four phases: prophase II, metaphase II, anaphase II, and telophase II (Figure 23.3b). These phases are similar to those that occur during mitosis; the centromeres split, and the sister chromatids separate and move toward opposite poles of the cell.

In summary, meiosis I begins with a diploid starting cell and ends with two cells, each with the haploid number of chromosomes. During meiosis II, each of the two haploid cells formed during meiosis I divides; the net result is four haploid gametes that are genetically different from the original diploid starting cell. The haploid cells formed from meiosis II are called *spermatids*.

SPERMIOGENESIS   In the final stage of spermatogenesis, called *spermiogenesis* (sper′-mē-ō-JEN-e-sis), each haploid spermatid develops into a single *sperm cell* (see Figure 23.2b).

Sperm

Sperm are produced at the rate of about 300 million per day. Once ejaculated, most do not survive more than 48 hours in the female reproductive tract. The major parts of a sperm cell

Figure 23.3 Spermatogenesis and meiosis. The designation 2*n* means diploid (46 chromosomes); *n* means haploid (23 chromosomes). Compare meiosis to mitosis, which is shown in Figure 3.21.

🔑 **Spermiogenesis is the process whereby spermatids mature into sperm.**

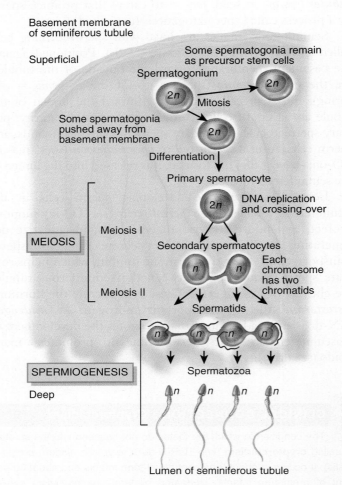

(a) Spermatogenesis

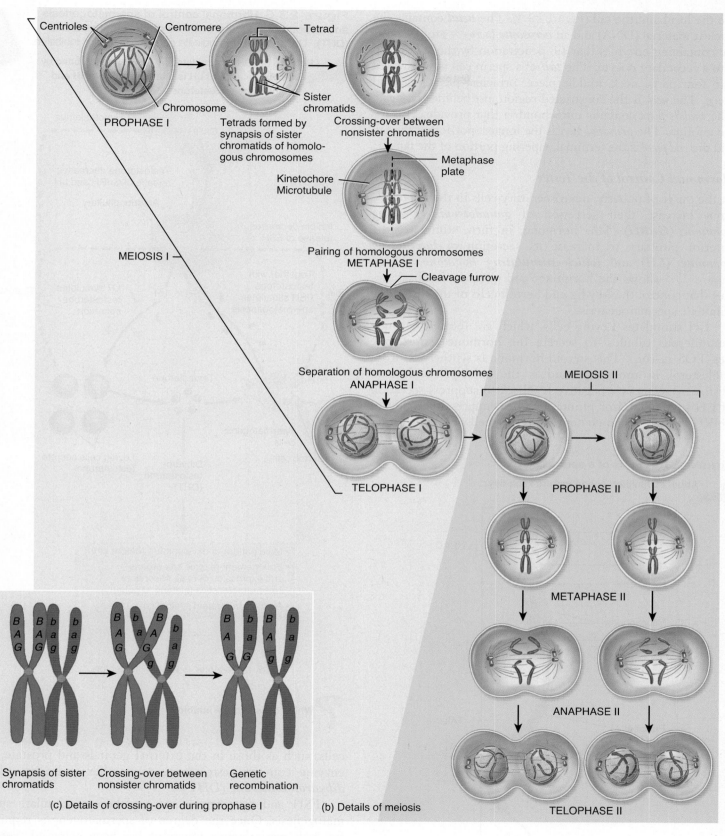

Centrioles · Centromere · Tetrad

PROPHASE I

Chromosome

Sister chromatids

Tetrads formed by synapsis of sister chromatids of homologous chromosomes

Crossing-over between nonsister chromatids

MEIOSIS I

Kinetochore Microtubule

Metaphase plate

Pairing of homologous chromosomes
METAPHASE I

Cleavage furrow

Separation of homologous chromosomes
ANAPHASE I

MEIOSIS II

TELOPHASE I

PROPHASE II

METAPHASE II

ANAPHASE II

TELOPHASE II

(b) Details of meiosis

Synapsis of sister chromatids

Crossing-over between nonsister chromatids

Genetic recombination

(c) Details of crossing-over during prophase I

What is the significance of crossing-over?

are the head and the tail (Figure 23.4). The *head* contains the nuclear material (DNA) and an *acrosome* (*acro-* = atop), a vesicle containing enzymes that aid penetration by the sperm cell into a secondary oocyte. The *tail* of a sperm cell is subdivided into four parts: neck, middle piece, principal piece, and end piece. The *neck* is the constricted region just behind the head. The *middle piece* contains mitochondria that provide ATP for locomotion. The *principal piece* is the longest portion of the tail and the *end piece* is the terminal, tapering portion of the tail.

Hormonal Control of the Testes

At the onset of puberty, neurosecretory cells in the hypothalamus increase their secretion of *gonadotropin-releasing hormone* (*GnRH*). This hormone, in turn, stimulates the anterior pituitary to increase its secretion of *luteinizing hormone (LH)* and *follicle-stimulating hormone (FSH)*. Figure 23.5 shows the hormones and negative feedback cycles that control the Leydig and Sertoli cells of the testes and stimulate spermatogenesis.

LH stimulates Leydig cells, which are located between seminiferous tubules, to secrete the hormone *testosterone* (tes-TOS-te-rōn). This steroid hormone is synthesized from cholesterol in the testes and is the principal androgen. Testosterone acts via negative feedback to suppress secretion of LH by the anterior pituitary and to suppress secretion of GnRH by hypothalamic neurosecretory cells. In some target

Figure 23.4 Parts of a sperm cell.

About 300 million sperm mature each day.

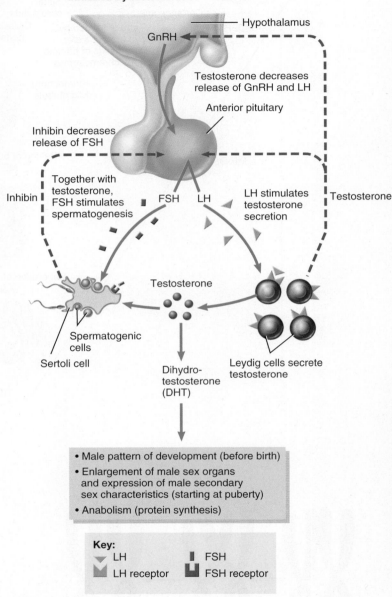

Figure 23.5 Hormonal control of spermatogenesis and actions of testosterone and dihydrotestosterone (DHT). Dashed red lines indicate negative feedback inhibition.

Release of FSH is stimulated by GnRH and inhibited by inhibin; release of LH is stimulated by GnRH and inhibited by testosterone.

? Which cells secrete inhibin?

? What is the function of the sperm middle piece?

cells, such as those in the external genitals and prostate, an enzyme converts testosterone to another androgen called *dihydrotestosterone (DHT)*.

FSH and testosterone act together to stimulate spermatogenesis. Once the degree of spermatogenesis required for male reproductive functions has been achieved, Sertoli cells release *inhibin*, a hormone named for its inhibition of FSH secretion by the anterior pituitary (Figure 23.5). Inhibin thus inhibits the secretion of hormones needed for spermatogenesis. If spermatogenesis is proceeding too slowly, less in-

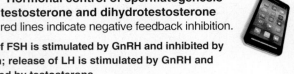

hibin is released, which permits more FSH secretion and an increased rate of spermatogenesis.

Testosterone and dihydrotestosterone both bind to the same androgen receptors, producing several effects:

- **Prenatal development.** Before birth, testosterone stimulates the male pattern of development of reproductive system ducts and the descent of the testes. DHT, by contrast, stimulates development of the external genitals. Testosterone also is converted in the brain to estrogens (feminizing hormones), which may play a role in the development of certain regions of the brain in males.

- **Development of male sexual characteristics.** At puberty, testosterone and DHT bring about development and enlargement of the male sex organs and the development of masculine secondary sexual characteristics. These include muscular and skeletal growth that results in wide shoulders and narrow hips; pubic, axillary, facial, and chest hair (within hereditary limits); thickening of the skin; increased sebaceous (oil) gland secretion; and enlargement of the larynx and consequent deepening of the voice.

- **Development of sexual function.** Androgens contribute to male sexual behavior and spermatogenesis and to sex drive (libido) in both males and females. Recall that the adrenal cortex is the main source of androgens in females.

- **Stimulation of anabolism.** Androgens are anabolic hormones; that is, they stimulate protein synthesis. This effect is obvious in the heavier muscle and bone mass of most men as compared to women.

✓ **CHECKPOINT**

1. How does the scrotum protect the testes?
2. What are the principal events of spermatogenesis and where do they occur?
3. What are the roles of FSH, LH, testosterone, and inhibin in the male reproductive system? How is secretion of these hormones controlled?

Ducts

Following spermatogenesis, pressure generated by the continual release of sperm and fluid secreted by Sertoli cells propels sperm and fluid through the seminiferous tubules and into the epididymis (see Figure 23.2a).

Epididymis

The **epididymis** (ep′-i-DID-i-mis; *epi-* = above or over; *-didymis* = testis; plural is *epididymides*) is a comma-shaped organ that lies along the posterior border of the testis (see Figures 23.1 and 23.2a). Each epididymis consists mostly of the tightly coiled **ductus epididymis**. Functionally, the ductus epididymis is the site of *sperm maturation*, the process by which sperm acquire motility and the ability to fertilize a secondary oocyte. This occurs over a period of 10 to 14 days. The ductus epididymis also stores sperm and helps propel them during sexual arousal by peristaltic contraction of its smooth muscle into the ductus (vas) deferens. Sperm may remain in storage in the ductus epididymis for several months. Any stored sperm that are not ejaculated by that time are eventually phagocytized and reabsorbed.

Ductus Deferens

At the end of the epididymis, the ductus epididymis becomes less convoluted, and its diameter increases. Beyond the epididymis, the duct is termed the **ductus deferens** or **vas deferens** (VAS DEF-er-enz; *vas* = vessel; *de-* = away; see Figure 23.2a). The ductus deferens ascends along the posterior border of the epididymis and penetrates the inguinal canal, a passageway in the front abdominal wall. Then, it enters the pelvic cavity, where it loops over the side and down the posterior surface of the urinary bladder (see Figure 23.1). The ductus deferens has a heavy coat of three layers of muscle. Functionally, the ductus deferens stores sperm, which can remain viable here for up to several months. The ductus deferens also conveys sperm from the epididymis toward the urethra during sexual arousal by peristaltic contractions of the muscular coat.

Accompanying the ductus deferens as it ascends in the scrotum are blood vessels, autonomic nerves, and lymphatic vessels that together make up the **spermatic cord**, a supporting structure of the male reproductive system (see Figure 23.2a).

Ejaculatory Ducts

The **ejaculatory ducts** (e-JAK-ū-la-tō′-rē; *ejacul-* = to expel) (see Figure 23.1) are formed by the union of the duct from the ductus deferens and the seminal vesicles (to be described shortly). The short ejaculatory ducts carry sperm into the urethra.

Urethra

The **urethra** is the terminal duct of the male reproductive system, serving as a passageway for both sperm and urine. In the male, the urethra passes through the prostate, deep perineal muscles, and penis (see Figure 23.1). The opening of the urethra to the exterior is called the **external urethral orifice**.

Accessory Sex Glands

The ducts of the male reproductive system store and transport sperm cells, but the **accessory sex glands** secrete most of the liquid portion of semen.

The paired **seminal vesicles** (VES-i-kuls) are pouchlike structures, lying posterior to the base of the urinary bladder and anterior to the rectum (see Figure 23.1). They secrete an alkaline, viscous fluid that contains fructose, prostaglandins, and clotting proteins (unlike those found in blood). The alkaline nature of the fluid helps to neutralize the acidic environment of the male urethra and female reproductive tract that otherwise would inactivate and kill sperm. The fructose is used for ATP production by sperm. Prostaglandins contribute to sperm motility and viability and may also stimulate muscular contraction within the female reproductive tract. Clotting proteins help semen coagulate after ejaculation. Fluid secreted by the seminal vesicles normally constitutes about 60 percent of the volume of semen.

The *prostate* (PROS-tāt) is a single, doughnut-shaped gland about the size of a golf ball (see Figure 23.1). It is inferior to the urinary bladder and surrounds the upper portion of the urethra. The prostate slowly increases in size from birth to puberty, and then it expands rapidly. The size attained by age 30 remains stable until about age 45, when further enlargement may occur. The prostate secretes a milky, slightly acidic fluid (pH about 6.5) that contains (1) *citric acid*, which can be used by sperm for ATP production via the Krebs cycle (see Section 20.2); (2) acid phosphatase (the function of which is unknown); and (3) several protein-digesting enzymes, such as *prostate-specific antigen (PSA)*. Prostatic secretions make up about 25 percent of the volume of semen.

The paired *bulbourethral glands* (bul'-bō-ū-RĒ-thral) are about the size of peas. They are located inferior to the prostate on either side of the urethra (see Figure 23.1). During sexual arousal, the bulbourethral glands secrete an alkaline substance into the urethra that protects the passing sperm by neutralizing acids from urine in the urethra. At the same time, they secrete mucus that lubricates the end of the penis and the lining of the urethra, thereby decreasing the number of sperm damaged during ejaculation.

Semen

Semen (= seed) is a mixture of sperm and the secretions of the seminal vesicles, prostate, and bulbourethral glands. The volume of semen in a typical ejaculation is 2.5 to 5 milliliters, with 50 to 150 million sperm per milliliter. When the number falls below 20 million per milliliter, the male is likely to be infertile. A very large number of sperm is required for fertilization because only a tiny fraction ever reaches the secondary oocyte.

Despite the slight acidity of prostatic fluid, semen has a slightly alkaline pH of 7.2 to 7.7 due to the higher pH and larger volume of fluid from the seminal vesicles. The prostatic secretion gives semen a milky appearance, and fluids from the seminal vesicles and bulbourethral glands give it a sticky consistency. Semen also contains an antibiotic that can destroy certain bacteria. The antibiotic may help control the abundance of naturally occurring bacteria in the semen and in the lower female reproductive tract. The presence of blood in semen is called *hemospermia* (hē-mō-SPER-mē-a; *hemo-* = blood; *-sperma* = seed). In most cases, it is caused by inflammation of the blood vessels lining the seminal vesicles; it is usually treated with antibiotics.

Penis

The *penis* contains the urethra and is a passageway for the ejaculation of semen and the excretion of urine. It is cylindrical in shape and consists of a root, a body, and the glans penis (see Figure 23.1). The *root of the penis* is the attached portion (proximal portion). The *body of the penis* is composed of three cylindrical masses of tissue. The two dorsolateral masses are called the *corpora cavernosa penis* (*corpora* = main bodies; *cavernosa* = hollow). The smaller midventral mass, the *corpus spongiosum penis*, contains the urethra. All three masses are enclosed by fascia (a sheet of fibrous connective tissue) and skin and consist of erectile tissue permeated by blood sinuses.

The distal end of the corpus spongiosum penis is a slightly enlarged region called the *glans penis*. In the glans penis is the opening of the urethra (the *external urethral orifice*) to the exterior. Covering the glans in an uncircumcised penis is the loosely fitting *prepuce* (PRĒ-poos), or *foreskin*.

CLINICAL CONNECTION | Circumcision

Circumcision (= to cut around) is a surgical procedure in which part of the entire prepuce is removed. It is usually performed just after delivery, 3 to 4 days after birth, or on the eighth day as part of a Jewish religious rite. Although most health-care professionals find no medical justification for circumcision, some feel that it has benefits, such as a lower risk of urinary tract infections, protection against penile cancer, and possibly a lower risk for sexually transmitted diseases. Indeed, studies in several African villages have found lower rates of HIV infection among circumcised men. •

Most of the time, the penis is flaccid (limp) because its arteries are vasoconstricted, which limits blood flow. The first visible sign of sexual excitement is *erection*, the enlargement and stiffening of the penis. Parasympathetic impulses cause release of neurotransmitters and local hormones, including the gas nitric oxide, which relaxes vascular smooth muscle in the penile arteries. The arteries supplying the penis dilate, and large quantities of blood enter the blood sinuses. Expansion of these spaces compresses the veins draining the penis, so blood outflow is slowed.

Ejaculation (ē-jak-ū-LĀ-shun; *ejectus-* = to throw out), the powerful release of semen from the urethra to the exterior, is a sympathetic reflex coordinated by the lumbar portion of the spinal cord. As part of the reflex, the smooth muscle sphincter at the base of the urinary bladder closes. Thus, urine is not expelled during ejaculation, and semen does not enter the urinary bladder. Even before ejaculation occurs, peristaltic contractions in the ductus deferens, seminal vesicles, ejaculatory ducts, and prostate propel semen into the penile portion of the urethra. Typically, this leads to *emission* (ē-MISH-un), the discharge of a small volume of semen before ejaculation. Emission may also occur during sleep (nocturnal emission). The penis returns to its flaccid state when the arteries constrict, and pressure on the veins is relieved.

23.2 FEMALE REPRODUCTIVE SYSTEM

OBJECTIVES • **Describe the location, structure, and functions of the organs of the female reproductive system.**

• **Explain how oocytes are produced.**

The organs of the *female reproductive system* (Figure 23.6) include the ovaries; the uterine (fallopian) tubes, or oviducts; the uterus; the vagina; and external organs, which are collectively called the vulva, or pudendum. The mammary glands also are considered part of the female reproductive system.

✓ **CHECKPOINT**

4. Trace the course of sperm through the system of ducts from the seminiferous tubules through the urethra.

5. What is semen? What is its function?

Figure 23.6 Female organs of reproduction and surrounding structures.

The female organs of reproduction include the ovaries, uterine (fallopian) tubes, uterus, vagina, vulva, and mammary glands.

Functions of the Female Reproductive System

1. The ovaries produce secondary oocytes and hormones, including estrogens, progesterone, inhibin, and relaxin.
2. The uterine tubes transport a secondary oocyte to the uterus, and normally are the sites where fertilization occurs.
3. The uterus is the site of implantation of a fertilized ovum, development of the fetus during pregnancy, and labor.
4. The vagina receives the penis during sexual intercourse and is a passageway for childbirth.
5. The mammary glands synthesize, secrete, and eject milk for nourishment of the newborn.

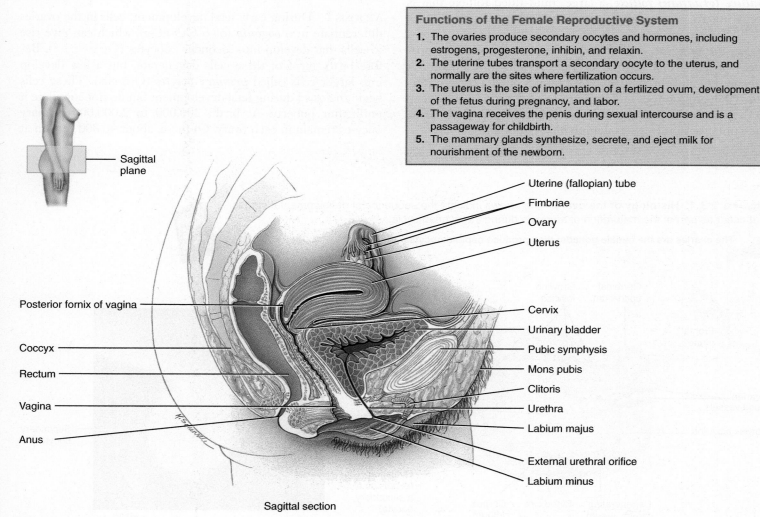

Sagittal section

What term refers to the external genitals of the female?

Ovaries

The *ovaries* (= egg receptacles) are paired organs that produce secondary oocytes (cells that develop into mature ova, or eggs, following fertilization) and hormones, such as progesterone and estrogens (the female sex hormones), inhibin, and relaxin. The ovaries arise from the same embryonic tissue as the testes, and they are the size and shape of unshelled almonds. One ovary lies on each side of the pelvic cavity, held in place by ligaments. Figure 23.7a shows the histology of an ovary.

The *germinal epithelium* is a layer of simple epithelium (low cuboidal or squamous) that covers the surface of the ovary. Deep to the germinal epithelium is the *ovarian cortex*, a region of dense connective tissue that contains ovarian follicles. Each *ovarian follicle* (*folliculus* = little bag) consists of an *oocyte* and a variable number of surrounding cells that nourish the developing oocyte and begin to secrete estrogens as the follicle grows larger. The follicle enlarges until it is a *mature (graafian) follicle*, a large, fluid-filled follicle that is preparing to rupture and expel a secondary oocyte (Figure 23.7b). The remnants of an ovulated follicle develop into a *corpus luteum* (= yellow body). The corpus luteum produces progesterone, estrogens, relaxin, and inhibin until it degenerates and turns into fibrous tissue called a *corpus albicans* (= white body). The *ovarian medulla* is a region deep to the ovarian cortex that consists of loose connective tissue and contains blood vessels, lymphatic vessels, and nerves.

Oogenesis

Formation of gametes in the ovaries is termed *oogenesis* (ō′-ō-JEN-e-sis; *oo-* = egg). Unlike spermatogenesis, which begins in males at puberty, oogenesis begins in females before they are even born. Also, males produce new sperm throughout life, while females have all the eggs they will ever have by birth. Oogenesis occurs in essentially the same manner as spermatogenesis (see Figure 23.3). It involves meiosis and maturation.

MEIOSIS I   During early fetal development, cells in the ovaries differentiate into *oogonia* (ō′-ō-GŌ-nē-a), which can give rise to cells that develop into secondary oocytes (Figure 23.8). Before birth, most of these cells degenerate, but a few develop into larger cells called *primary oocytes* (Ō-ō-sīts). These cells begin meiosis I during fetal development but do not complete it until after puberty. At birth, 200,000 to 2,000,000 primary oocytes remain in each ovary. Of these, about 40,000 remain at

Figure 23.7 Histology of the ovary. The arrows indicate the sequence of developmental stages that occur as part of the maturation of an ovum during the ovarian cycle.

 The ovaries are the female gonads; they produce haploid oocytes.

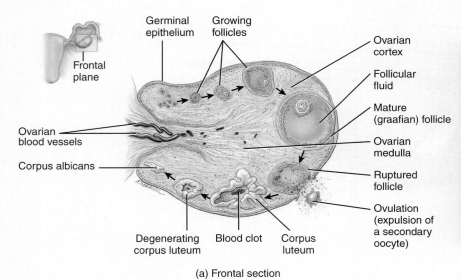

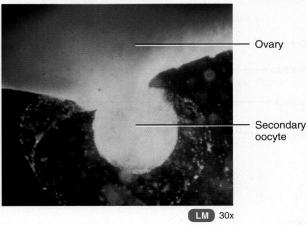

(a) Frontal section

(b) Ovulation of a secondary oocyte

LM 30x

What structures in the ovary contain endocrine tissue, and what hormones do they secrete?

Figure 23.8 Oogenesis. Diploid cells (2*n*) have 46 chromosomes; haploid cells (*n*) have 23 chromosomes.

🗝 **In an oocyte, meiosis II is completed only if fertilization occurs.**

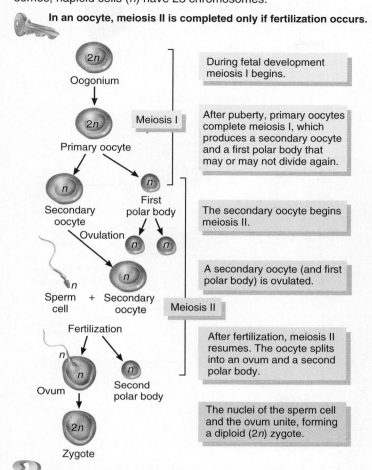

During fetal development meiosis I begins.

After puberty, primary oocytes complete meiosis I, which produces a secondary oocyte and a first polar body that may or may not divide again.

The secondary oocyte begins meiosis II.

A secondary oocyte (and first polar body) is ovulated.

After fertilization, meiosis II resumes. The oocyte splits into an ovum and a second polar body.

The nuclei of the sperm cell and the ovum unite, forming a diploid (2*n*) zygote.

❓ **How does the age of a primary oocyte in a female compare with the age of a primary spermatocyte in a male?**

puberty, but only 400 go on to mature and ovulate during a woman's reproductive lifetime. The remainder degenerate.

After puberty, hormones secreted by the anterior pituitary stimulate the resumption of oogenesis each month. Meiosis I resumes in several primary oocytes, although in each cycle only one follicle typically reaches the maturity needed for ovulation. The diploid primary oocyte completes meiosis I, resulting in two haploid cells of unequal size, both with 23 chromosomes (*n*) of two chromatids each. The smaller cell produced by meiosis I, called the *first polar body*, is essentially a packet of discarded nuclear material; the larger cell, known as the **secondary oocyte**, receives most of the cytoplasm. Once a secondary oocyte is formed, it begins meiosis II and then stops. The follicle in which these events are taking place—the mature (graafian) follicle—soon ruptures and releases its secondary oocyte, a process known as **ovulation**.

Meiosis II At ovulation, usually a single secondary oocyte (with the first polar body) is expelled into the pelvic cavity and swept into the uterine (fallopian) tube. If a sperm penetrates the secondary oocyte (fertilization), meiosis II resumes. The secondary oocyte splits into two haploid (*n*) cells of unequal size. The larger cell is the **ovum**, or mature egg; the smaller one is

the *second polar body*. The nuclei of the sperm cell and the ovum then unite, forming a diploid (2*n*) **zygote**. The first polar body may also undergo another division to produce two polar bodies. If it does, the primary oocyte ultimately gives rise to a single haploid (*n*) ovum and three haploid (*n*) polar bodies. Thus, each primary oocyte gives rise to a single gamete (secondary oocyte, which becomes an ovum after fertilization); in contrast, each primary spermatocyte produces four gametes (sperm).

Uterine Tubes

Females have two **uterine (fallopian) tubes** that extend laterally from the uterus and transport the secondary oocytes from the ovaries to the uterus (Figure 23.9). The open, funnel-shaped end of each tube, the **infundibulum**, lies close to the ovary but is open to the pelvic cavity. It ends in a fringe of fingerlike projections called **fimbriae** (FIM-brē-ē = fringe). From the infundibulum, the uterine tubes extend medially, attaching to the upper and outer corners of the uterus.

After ovulation, local currents produced by movements of the fimbriae, which surround the surface of the mature follicle just before ovulation occurs, sweep the secondary oocyte into the uterine tube. The oocyte is then moved along the tube by cilia in the tube's mucous lining and peristaltic contractions of its smooth muscle layer.

The usual site for fertilization of a secondary oocyte by a sperm cell is in the uterine tube. Fertilization may occur any time up to about 24 hours after ovulation. The fertilized ovum (zygote) descends into the uterus within seven days. Unfertilized secondary oocytes disintegrate.

Uterus

The **uterus** (*womb*) serves as part of the pathway for sperm deposited in the vagina to reach the uterine tubes. It is also the site of implantation of a fertilized ovum, development of the fetus during pregnancy, and labor. During reproductive cycles when implantation does not occur, the uterus is the source of menstrual flow. The uterus is situated between the urinary bladder and the rectum and is shaped like an inverted pear.

Parts of the uterus include the dome-shaped portion superior to the uterine tubes called the **fundus**, the tapering central portion called the **body**, and the narrow portion opening into the vagina called the **cervix**. The interior of the body of the uterus is called the **uterine cavity** (Figure 23.9).

Histologically, the uterus consists of three layers of tissue: perimetrium, myometrium, and endometrium (Figure 23.9). The outer layer—the **perimetrium** (per′-i-MĒ-trē-um; *peri-* = around; *-metrium* = uterus) or serosa—is part of the visceral peritoneum; it is composed of simple squamous epithelium and areolar connective tissue.

The middle muscular layer of the uterus, the **myometrium** (*myo-* = muscle), consists of smooth muscle and forms the bulk of the uterine wall. During childbirth, coordinated contractions of uterine muscles help expel the fetus.

The innermost part of the uterine wall, the **endometrium** (*endo* = within), is a mucous membrane. It nourishes a growing

Figure 23.9 Uterus and associated structures. In the left side of the drawing, the uterine tube and uterus have been sectioned to show internal structures.

🔑 **The uterus is the site of menstruation, implantation of a fertilized ovum, development of a fetus, and labor.**

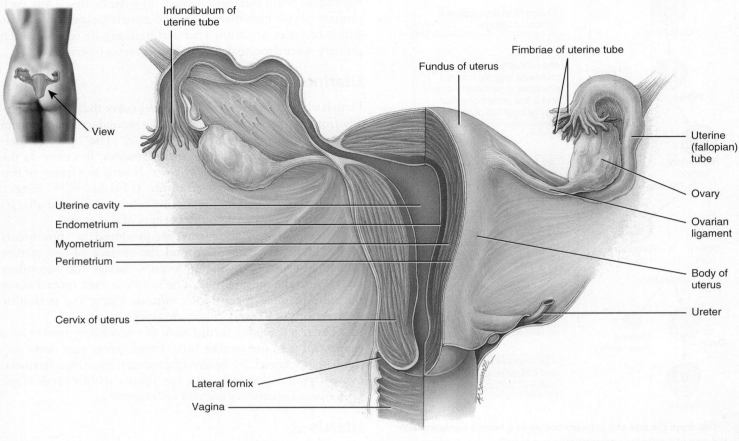

Infundibulum of uterine tube

Fimbriae of uterine tube

Fundus of uterus

View

Uterine (fallopian) tube

Ovary

Uterine cavity

Endometrium

Myometrium

Perimetrium

Ovarian ligament

Body of uterus

Ureter

Cervix of uterus

Lateral fornix

Vagina

Posterior view

? **Which part of the uterine lining rebuilds after each menstruation?**

fetus or is shed each month during menstruation if fertilization does not occur. The endometrium contains many *endometrial glands* whose secretions nourish sperm and the zygote.

⚕ CLINICAL CONNECTION | Hysterectomy

Hysterectomy (hiss-ter-EK-tō-mē; *hyster-* = uterus), the surgical removal of the uterus, is the most common gynecological operation. It may be indicated in conditions such as fibroids (noncancerous tumors composed of muscular and fibrous tissue); endometriosis; pelvic inflammatory disease; recurrent ovarian cysts; excessive uterine bleeding; and cancer of the cervix, uterus, or ovaries. In a *partial (subtotal) hysterectomy,* the body of the uterus is removed but the cervix is left in place. A *complete hysterectomy* is the removal of both the body and cervix of the uterus. A *radical hysterectomy* includes removal of the body and cervix of the uterus, uterine tubes, possibly the ovaries, the superior portion of the vagina, pelvic lymph nodes, and supporting structures, such as ligaments. A hysterectomy can be performed either through an incision in the abdominal wall or through the vagina. •

Vagina

The **vagina** (va-JĪ-na = sheath) is a tubular canal that extends from the exterior of the body to the uterine cervix (Figure 23.9). It is the receptacle for the penis during sexual intercourse, the outlet for menstrual flow, and the passageway for childbirth. The vagina is situated between the urinary bladder and the rectum. A recess, called the **fornix** (= arch or vault), surrounds the cervix. When properly inserted, a contraceptive diaphragm rests on the fornix, covering the cervix.

The mucosa of the vagina contains large stores of glycogen, the decomposition of which produces organic acids. The resulting acidic environment retards microbial growth, but it also is harmful to sperm. Alkaline components of semen, mainly from the seminal vesicles, neutralize the acidity of the vagina and increase viability of sperm. The muscular layer is composed of smooth muscle that can stretch to receive the penis during intercourse and allow for childbirth. There may be a thin fold of mucous membrane called the **hymen**

(= membrane) partially covering the *vaginal orifice*, the vaginal opening (see Figure 23.10).

Perineum and Vulva

The *perineum* (per'-i-NĒ-um) is the diamond-shaped area between the thighs and buttocks of both males and females that contains the external genitals and anus (Figure 23.10).

The term *vulva* (VUL-va = to wrap around), or *pudendum* (pū-DEN-dum), refers to the external genitals of the female (Figure 23.10). The *mons pubis* (MONZ PŪ-bis; *mons* = mountain) is an elevation of adipose tissue covered by coarse pubic hair, which cushions the pubic symphysis. From the mons pubis, two longitudinal folds of skin, the *labia majora* (LĀ-bē-a ma-JŌ-ra; *labia* = lips; *majora* = larger), extend down and back (singular is *labium majus*). In females the labia majora develop from the same embryonic tissue that the scrotum develops from in males. The labia majora contain adipose tissue and sebaceous (oil) and sudoriferous

(sweat) glands. Like the mons pubis, they are covered by pubic hair. Medial to the labia majora are two folds of skin called the *labia minora* (mi-NŌ-ra = smaller; singular is *labium minus*). The labia minora do not contain pubic hair or fat and have few sudoriferous (sweat) glands; they do, however, contain numerous sebaceous (oil) glands.

The *clitoris* (KLIT-o-ris) is a small, cylindrical mass of erectile tissue and nerves. It is located at the anterior junction of the labia minora. A layer of skin called the *prepuce* (PRĒ-poos), also known as the *foreskin*, is formed at a point where the labia minora unite and cover the body of the clitoris. The exposed portion of the clitoris is the *glans*. Like the penis, the clitoris is capable of enlargement upon sexual stimulation.

The region between the labia minora is called the *vestibule*. In the vestibule are the hymen (if present); *vaginal orifice*, the opening of the vagina to the exterior; *external urethral orifice*, the opening of the urethra to the exterior; and on either side of the external urethral orifice, the openings of the ducts of the *paraurethral glands*. These glands in the wall of

Figure 23.10 Components of the vulva.

Like the penis, the clitoris is capable of erection upon sexual stimulation.

Mons pubis

Prepuce of clitoris
Clitoris

Labia majora (spread)

Labia minora (spread exposing vestibule)

External urethral orifice

Vaginal orifice (dilated)

Hymen

Anus

Inferior view

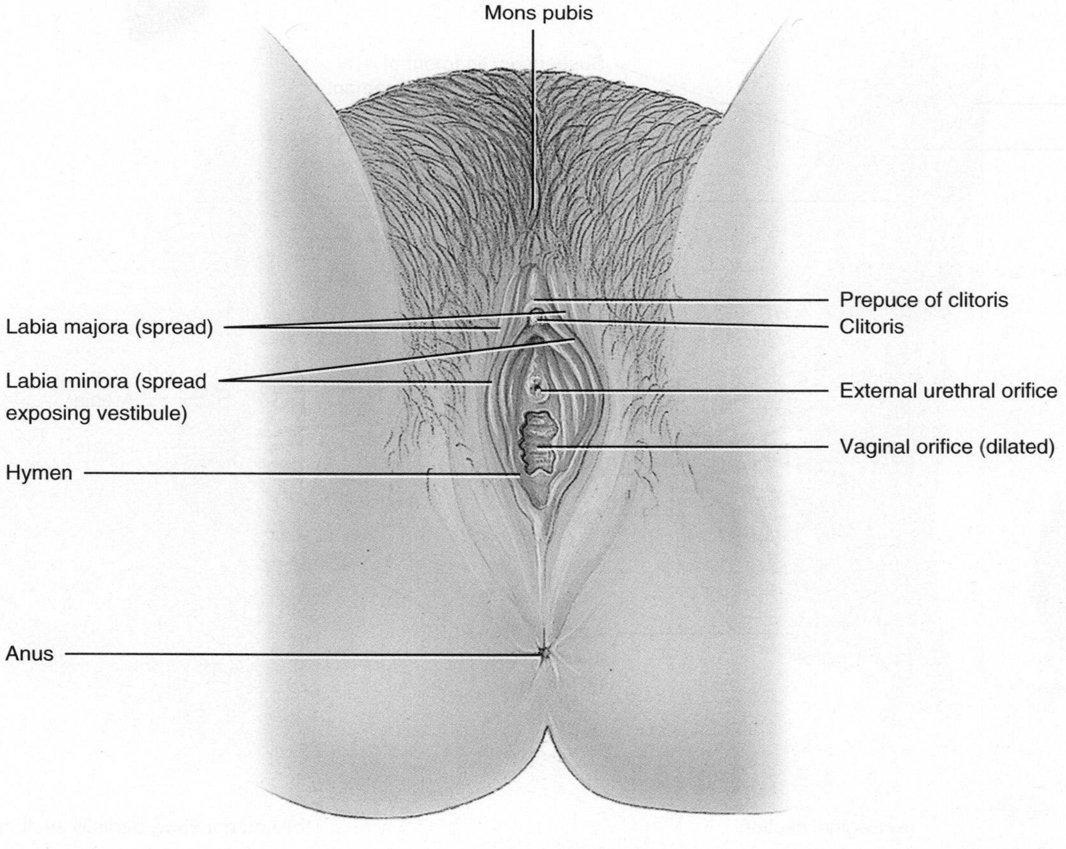

What surface structures are anterior to the vaginal opening?

the urethra secrete mucus. The male's prostate develops from the same embryonic tissue as the female's paraurethral glands. On either side of the vaginal orifice itself are the ***greater vestibular glands***, which produce a small quantity of mucus during sexual arousal and intercourse that adds to cervical mucus and provides lubrication. In males, the bulbourethral glands are equivalent structures.

CLINICAL CONNECTION | Episiotomy

In specific maternal–fetal situations that require a quick delivery, an **episiotomy** (e-piz-ē-OT-ō-mē; *episi-* = vulva or pubic region; *-otomy* = incision) may be performed. In the procedure, a perineal cut is made with surgical scissors. The cut may be made along the midline or at an angle of approximately 45 degrees to the midline. In effect, a straight, more easily sutured cut is substituted for the jagged tear that would otherwise be caused by passage of the fetus. The incision is closed in layers with sutures that are absorbed within a few weeks, so that the busy new mom does not have to worry about making time to have them removed. •

Mammary Glands

The ***mammary glands*** (*mamma* = breast), located in the breasts, are modified sudoriferous (sweat) glands that produce milk. The breasts lie over the pectoralis major and serratus anterior muscles and are attached to them by a layer of connective tissue (Figure 23.11). Each breast has one pigmented projection, the ***nipple***, with a series of closely spaced openings of ducts where milk emerges. The circular pigmented area of skin surrounding the nipple is called the ***areola*** (a-RĒ-ō-la = small space). This region appears rough because it contains modified sebaceous (oil) glands. Internally, each mammary gland consists of 15 to 20 ***lobes*** arranged radially and separated by adipose tissue and strands of connective tissue called ***suspensory ligaments of the breast (Cooper's ligaments)***, which support the breast. In each lobe are smaller ***lobules***, in which milk-secreting glands called ***alveoli*** (= small cavities) are found. When milk is being produced, it passes from the alveoli into a series of tubules that drain toward the nipple.

Figure 23.11 Mammary glands.

The mammary glands function in the synthesis, secretion, and ejection of milk (lactation).

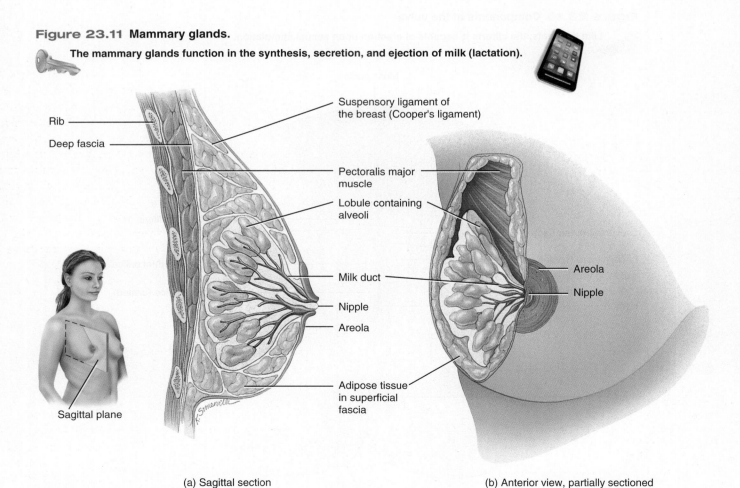

Rib

Deep fascia

Sagittal plane

Suspensory ligament of the breast (Cooper's ligament)

Pectoralis major muscle

Lobule containing alveoli

Milk duct

Nipple

Areola

Adipose tissue in superficial fascia

Areola

Nipple

(a) Sagittal section

(b) Anterior view, partially sectioned

? **What hormone regulates the ejection of milk from the mammary glands?**

Breast augmentation (awg-men-TĀ-shun = enlargement), technically called **augmentation mammaplasty** (mam-a-PLAS-tē), is a surgical procedure to increase breast size and shape. It may be done to enhance breast size for females who feel that their breasts are too small, to restore breast volume due to weight loss or following pregnancy, to improve the shape of breasts that are sagging, and to improve breast appearance following surgery, trauma, or congenital abnormalities. The most commonly used implants are filled with either a saline solution or silicone gel. The incision for the implant is made under the breast, around the areola, in the armpit, or in the navel. Then a pocket is made to place the implant either directly behind the breast tissue or beneath the pectoralis major muscle.

Breast reduction or **reduction mammaplasty** is a surgical procedure that involves decreasing breast size by removing fat, skin, and glandular tissue. This procedure is done because of chronic back, neck, and shoulder pain; poor posture; circulation or breathing problems; a skin rash under the breasts; restricted levels of activity; self-esteem problems; deep grooves in the shoulders from bra strap pressure; and difficulty wearing or fitting into certain bras and clothing. The most common procedure involves an incision around the areola, down the breast toward the crease between the breast and abdomen, and then along the crease. The surgeon removes excess tissue through the incision. In most cases, the nipple and areola remain attached to the breast. However, if the breasts are extremely large, the nipple and areola may have to be reattached at a higher position. •

At birth, the mammary glands are undeveloped and appear as slight elevations on the chest. With the onset of puberty, under the influence of estrogens and progesterone, the female breasts begin to develop. The duct system matures and fat is deposited, which increases breast size. The areola and nipple also enlarge and become more darkly pigmented.

The functions of the mammary glands are the synthesis, secretion, and ejection of milk; these functions, called *lactation*, are associated with pregnancy and childbirth. Milk production is stimulated largely by the hormone prolactin from the anterior pituitary, with contributions from progesterone and estrogens. The ejection of milk is stimulated by oxytocin, which is released from the posterior pituitary in response to the sucking of an infant on the mother's nipple (suckling).

The breasts of females are highly susceptible to cysts and tumors. In **fibrocystic disease**, the most common cause of breast lumps in females, one or more cysts (fluid-filled sacs) and thickening of alveoli develop. The condition, which occurs mainly in females between the ages of 30 and 50, is probably due to a relative excess of estrogens or a deficiency of progesterone in the postovulatory phase of the reproductive cycle (discussed shortly). Fibrocystic disease usually causes one or both breasts to become lumpy, swollen, and tender a week or so before menstruation begins. •

✓ CHECKPOINT

6. Describe the principal events of oogenesis.

7. Where are the uterine tubes located? What is their function?

8. Describe the histology of the uterus.

9. How does the histology of the vagina contribute to its function?

10. Describe the structure and support of the mammary glands.

23.3 THE FEMALE REPRODUCTIVE CYCLE

OBJECTIVE • **Describe the major events of the ovarian and uterine cycles.**

During their reproductive years, nonpregnant females normally exhibit cyclical changes in the ovaries and uterus. Each cycle takes about a month and involves both oogenesis and preparation of the uterus to receive a fertilized ovum. Hormones secreted by the hypothalamus, anterior pituitary, and ovaries control the main events. You have already learned about the *ovarian cycle*, the series of events in the ovaries that occur during and after the maturation of an oocyte. Steroid hormones released by the ovaries control the *uterine (menstrual) cycle*, a concurrent series of changes in the endometrium of the uterus to prepare it for the arrival of a fertilized ovum that will develop there until birth. If fertilization does not occur, the levels of ovarian hormones decrease, which causes part of the endometrium to slough off. The general term *female reproductive cycle* encompasses the ovarian and uterine cycles, the hormonal changes that regulate them, and the related cyclical changes in the breasts and cervix.

Hormonal Regulation of the Female Reproductive Cycle

Gonadotropin-releasing hormone (GnRH) secreted by the hypothalamus controls the ovarian and uterine cycles (Figure 23.12). GnRH stimulates the release of *follicle-stimulating hormone (FSH)* and *luteinizing hormone (LH)* from the anterior pituitary. FSH, in turn, initiates follicular growth and the secretion of estrogens by the growing follicles. LH stimulates the further development of ovarian follicles and their full secretion of estrogens. At midcycle, LH triggers ovulation and then promotes formation of the corpus luteum, the reason for the name luteinizing hormone. Stimulated by LH, the corpus luteum produces and secretes estrogens, progesterone, relaxin, and inhibin.

Figure 23.12 The female reproductive cycle. The length of the female reproductive cycle typically is 24 to 36 days; the preovulatory phase is more variable in length than the other phases. (a) Events in the ovarian and uterine cycles and the release of anterior pituitary hormones are correlated with the sequence of the cycle's four phases. In the cycle shown, fertilization and implantation have not occurred. (b) Relative concentrations of anterior pituitary hormones (FSH and LH) and ovarian hormones (estrogens and progesterone) during the phases of a normal female reproductive cycle.

Estrogens are secreted by the dominant follicle before ovulation; after ovulation, both progesterone and estrogens are secreted by the corpus luteum.

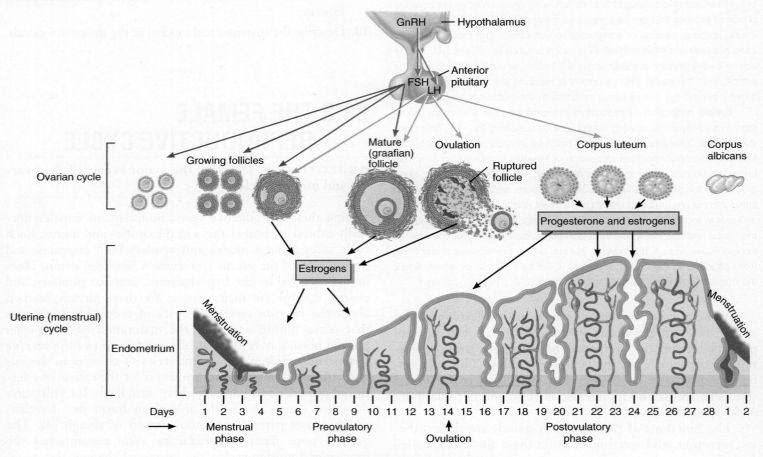

(a) Hormonal regulation of changes in the ovary and uterus

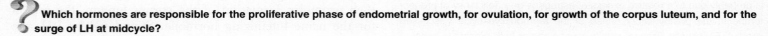

(b) Changes in concentration of anterior pituitary and ovarian hormones

Which hormones are responsible for the proliferative phase of endometrial growth, for ovulation, for growth of the corpus luteum, and for the surge of LH at midcycle?

Estrogens secreted by ovarian follicles have several important functions throughout the body:

- Estrogens promote the development and maintenance of female reproductive structures, feminine secondary sex characteristics, and the mammary glands. The secondary sex characteristics include distribution of adipose tissue in the breasts, abdomen, mons pubis, and hips; a broad pelvis; and the pattern of hair growth on the head and body.

- Estrogens stimulate protein synthesis, acting together with insulinlike growth factors, insulin, and thyroid hormones.

- Estrogens lower blood cholesterol level, which is probably the reason that women under age 50 have a much lower risk of coronary artery disease than do men of comparable age.

Progesterone, secreted mainly by cells of the corpus luteum, acts together with estrogens to prepare and then maintain the endometrium for implantation of a fertilized ovum and to prepare the mammary glands for milk secretion.

A small quantity of *relaxin*, produced by the corpus luteum during each monthly cycle, relaxes the uterus by inhibiting contractions of the myometrium. Presumably, implantation of a fertilized ovum occurs more readily in a "quiet" uterus. During pregnancy, the placenta produces much more relaxin, and it continues to relax uterine smooth muscle. At the end of pregnancy, relaxin also increases the flexibility of the pubic symphysis and helps dilate the uterine cervix, both of which ease delivery of the baby.

Inhibin is secreted by growing follicles and by the corpus luteum after ovulation. It inhibits secretion of FSH and, to a lesser extent, LH.

Phases of the Female Reproductive Cycle

The duration of the female reproductive cycle varies from 24 to 36 days. For this discussion we assume a duration of 28 days and divide it into four phases: the menstrual phase, the preovulatory phase, ovulation, and the postovulatory phase (Figure 23.12). Because they occur at the same time, the events of the ovarian cycle (events in the ovaries) and menstrual cycle (events in the uterus) will be discussed together.

Menstrual Phase

The *menstrual phase* (MEN-stroo-al), also called *menstruation* (men′-stroo-ā-shun) or *menses* (= month), lasts for roughly the first five days of the cycle. (By convention, the first day of menstruation marks the first day of a new cycle.)

EVENTS IN THE OVARIES   During the menstrual phase, several ovarian follicles grow and enlarge.

EVENTS IN THE UTERUS   Menstrual flow from the uterus consists of 50 to 150 mL of blood and tissue cells from the endometrium. This discharge occurs because the declining level of ovarian hormones (progesterone and estrogens) causes the uterine arteries to constrict. As a result, the cells they supply become oxygen-deprived and start to die. Eventually, part of the endometrium sloughs off. The menstrual flow passes from the uterine cavity to the cervix and through the vagina to the exterior.

Preovulatory Phase

The *preovulatory phase* is the time between the end of menstruation and ovulation. The preovulatory phase of the cycle accounts for most of the variation in cycle length. In a 28-day cycle, it lasts from days 6 to 13.

EVENTS IN THE OVARIES   Under the influence of FSH, several follicles continue to grow and begin to secrete estrogens and inhibin. By about day 6, a single follicle in one of the two ovaries has outgrown all the others to become the *dominant follicle*. Estrogens and inhibin secreted by the dominant follicle decrease the secretion of FSH (Figure 23.12b, see days 8 to 11), which causes other, less well-developed follicles to stop growing and die.

The one dominant follicle becomes the *mature (graafian) follicle*. The mature follicle continues to enlarge until it is ready for ovulation, forming a blisterlike bulge on the surface of the ovary. During maturation, the follicle continues to increase its production of estrogens under the influence of an increasing level of LH.

With reference to the ovarian cycle, the menstrual phase and preovulatory phase together are termed the *follicular phase* (fō-LIK-ū-lar) because ovarian follicles are growing and developing.

EVENTS IN THE UTERUS   Estrogens liberated into the blood by growing ovarian follicles stimulate the repair of the endometrium. As the endometrium thickens, the short, straight endometrial glands develop, and the arterioles coil and lengthen.

Ovulation

Ovulation, the rupture of the mature (graafian) follicle and the release of the secondary oocyte into the pelvic cavity, usually occurs on day 14 in a 28-day cycle.

The high levels of estrogens during the last part of the preovulatory phase exert a *positive feedback* effect on both LH and GnRH. A high level of estrogens stimulates the hypothalamus to release more gonadotropin-releasing hormone (GnRH) and the anterior pituitary to produce more LH. GnRH then promotes the release of even more LH. The resulting surge of LH (Figure 23.12b) brings about rupture of the mature (graafian) follicle and expulsion of a secondary oocyte. An over-the-counter home test that detects the LH surge associated with ovulation can be used to predict ovulation a day in advance.

Postovulatory Phase

The *postovulatory phase* of the female reproductive cycle is the time between ovulation and onset of the next menstruation.

This phase is the most constant in duration and lasts for 14 days, from days 15 to 28 in a 28-day cycle.

EVENTS IN ONE OVARY After ovulation, the mature follicle collapses. Stimulated by LH, the remaining follicular cells enlarge and form the corpus luteum, which secretes progesterone, estrogens, relaxin, and inhibin. With reference to the ovarian cycle, this phase is also called the *luteal phase*.

Subsequent events depend on whether the oocyte is fertilized. If the oocyte is not fertilized, the corpus luteum lasts

for only two weeks, after which its secretory activity declines, and it degenerates into a corpus albicans (Figure 23.12a). As the levels of progesterone, estrogens, and inhibin decrease, release of GnRH, FSH, and LH rises due to loss of negative feedback suppression by the ovarian hormones. Then, follicular growth resumes and a new ovarian cycle begins.

If the secondary oocyte is fertilized and begins to divide, the corpus luteum persists past its normal two-week lifespan. It is "rescued" from degeneration by *human chorionic gonadotropin*

Figure 23.13 Summary of hormonal interactions in the ovarian and menstrual cycles.

Hormones from the anterior pituitary regulate ovarian function, and hormones from the ovaries regulate the changes in the endometrial lining of the uterus.

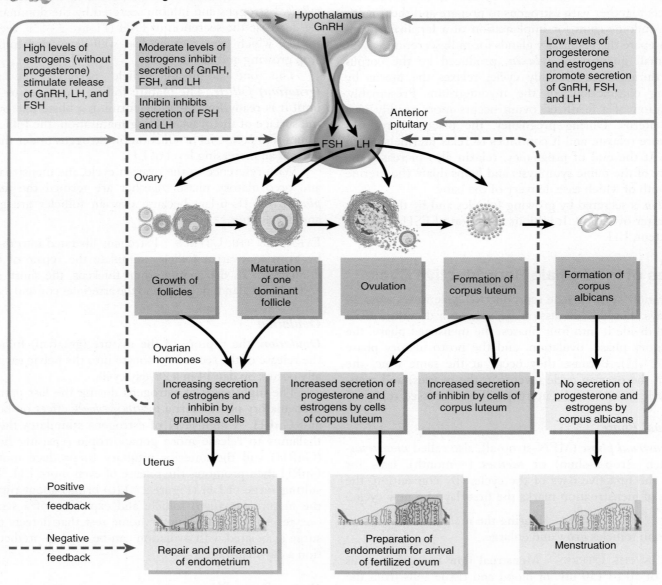

When declining levels of estrogens and progesterone stimulate secretion of GnRH, is this a positive or negative feedback effect? Why?

(hCG) (kō-rē-ON-ik), a hormone produced by the embryo beginning about eight days after fertilization. Like LH, hCG stimulates the secretory activity of the corpus luteum. The presence of hCG in maternal blood or urine is an indicator of pregnancy, and hCG is the hormone detected by home pregnancy tests.

EVENTS IN THE UTERUS   Progesterone and estrogens produced by the corpus luteum promote growth of the endometrial glands, which begin to secrete glycogen, and vascularization and thickening of the endometrium. These preparatory changes peak about one week after ovulation, at the time a fertilized ovum might arrive at the uterus.

Figure 23.13 summarizes the hormonal interactions and cyclical changes in the ovaries and uterus during the ovarian and menstrual cycles.

✓ CHECKPOINT

11. Describe the function of each of the following hormones in the uterine and ovarian cycles: GnRH, FSH, LH, estrogens, progesterone, and inhibin.

12. Briefly outline the major events and hormonal changes of each phase of the uterine cycle, and correlate them with the events of the ovarian cycle.

13. Prepare a labeled diagram of the major hormonal changes that occur during the uterine and ovarian cycles.

23.4 BIRTH CONTROL METHODS AND ABORTION

OBJECTIVE • **Explain the differences among the various types of birth control methods and compare their effectiveness.**

Birth control refers to restricting the number of children by various methods designed to control fertility and prevent conception. No single, ideal method of birth control exists. The only method of preventing pregnancy that is 100 percent reliable is **complete abstinence**, the avoidance of sexual intercourse. Several other methods are available; each has its advantages and disadvantages. These include surgical sterilization, hormonal methods, intrauterine devices, spermicides, barrier methods, and periodic abstinence. Table 23.1 provides the failure rates for various methods of birth control. Although it is not a form of birth control, in this section we will also discuss abortion, the premature expulsion of the products of conception from the uterus.

Birth Control Methods

Surgical Sterilization

Sterilization is a procedure that renders an individual incapable of further reproduction. The principal method for ster-

TABLE 23.1

Failure Rates for Several Birth Control Methods

METHOD	FAILURE RATES* (%)	
	PERFECT USE†	TYPICAL USE
Complete abstinence	0	0
Surgical sterilization		
Vasectomy	0.10	0.15
Tubal ligation	0.5	0.5
Non-incisional sterilization (Essure®)	0.2	0.2
Hormonal methods		
Oral contraceptives		
Combined pill	0.3	1–2
Seasonale®	0.3	1–2
Minipill	0.5	2
Non-oral contraceptives		
Contraceptive skin patch	0.1	1–2
Vaginal contraceptive ring	0.1	1–2
Emergency contraception	25	25
Hormone injections	0.3	1–2
Intrauterine devices (Copper T 380A®)	0.6	0.8
Spermicides (alone)	15	29
Barrier methods		
Male condom	2	15
Vaginal pouch	5	21
Diaphragm (with spermicide)	6	16
Cervical cap (with spermicide)	9	16
Periodic abstinence		
Rhythm	9	25
Sympto-thermal	2	20
No method	85	85

*Defined as percentage of women having an unintended pregnancy during the first year of use.
†Failure rate when the method is used correctly and consistently.

ilization of males is a **vasectomy** (va-SEK-tō-mē; *-ectomy* = cut out), in which a portion of each ductus deferens is removed. In order to gain access to the ductus deferens, an incision is made with a scalpel (conventional procedure) or a puncture is made with special forceps (non-scalpel vasectomy). Next the ducts are located and cut, each is tied (ligated) in two places with stitches, and the portion between the ties is removed. Although sperm production continues in the testes, sperm can no longer reach the exterior. The sperm degenerate and are destroyed by phagocytosis. Because the blood vessels are not cut, testosterone levels in the blood remain normal, so vasectomy has no effect on sexual desire or performance. If done correctly, it is close to 100 percent effective. The procedure can be reversed, but the chance of regaining fertility is only 30–40 percent. Sterilization in females most often is achieved by performing a ***tubal ligation*** (lī-GĀ-shun), in which both uterine tubes are tied closed and then cut. This can be achieved in a few different ways. "Clips" or "clamps" can be placed on the uterine tubes, the tubes can be tied and/or cut, and sometimes they are cauterized. In any case, the result is that the secondary oocyte cannot pass through the uterine tubes, and sperm cannot reach the oocyte.

Non-incisional Sterilization

Essure® is a non-incisional irreversible procedure that is an alternative to tubal ligation. In the Essure procedure, a soft micro-insert coil made of polyester fibers and metals (nickel–titanium and stainless steel) is inserted with a catheter into the vagina, through the uterus, and into each uterine tube. Over a three-month period, the insert stimulates tissue growth (scar tissue) in and around itself, blocking the uterine tubes. As with tubal ligation, the secondary oocyte cannot pass through the uterine tubes, and sperm cannot reach the oocyte. Unlike tubal ligation, Essure® does not require general anesthesia.

Hormonal Methods

Aside from complete abstinence or surgical sterilization, hormonal methods are the most effective means of birth control. ***Oral contraceptives*** (the pill) contain hormones designed to prevent pregnancy. Some, called *combined oral contraceptives (COCs)*, contain both progestin (hormone with actions similar to progesterone) and estrogens. The primary action of COCs is to inhibit ovulation by suppressing the gonadotropins FSH and LH. The low levels of FSH and LH usually prevent the development of a dominant follicle in the ovary. As a result, levels of estrogens do not rise, the midcycle LH surge does not occur, and ovulation does not take place. Even if ovulation does occur, as it does in some cases, COCs may also block implantation in the uterus and inhibit the transport of ova and sperm in the uterine tubes.

Progestins thicken cervical mucus and make it more difficult for sperm to enter the uterus. *Progestin-only pills* thicken cervical mucus and may block implantation in the uterus, but they do not consistently inhibit ovulation.

Among the noncontraceptive benefits of oral contraceptives are regulation of the length of the menstrual cycle and decreased menstrual flow (and therefore decreased risk of anemia). The pill also provides protection against endometrial and ovarian cancers and reduces the risk of endometriosis. However, oral contraceptives may not be advised for women with a history of blood clotting disorders, cerebral blood vessel damage, migraine headaches, hypertension, liver malfunction, or heart disease. Women who take the pill and smoke face far higher odds of having a heart attack or stroke than do nonsmoking pill users. Smokers should quit smoking or use an alternative method of birth control.

Following are several variations of *oral* hormonal methods of contraception:

- **Combined pill.** Contains both progestin and estrogens and is typically taken once a day for three weeks to prevent pregnancy and regulate the menstrual cycle. The pills taken during the fourth week are inactive (do not contain hormones) and permit menstruation to occur.

- **Seasonale**®. Contains both progestin and estrogens and is taken once a day in three-month cycles of 12 weeks of hormone-containing pills followed by 1 week of inactive pills. Menstruation occurs during the thirteenth week.

- **Minipill.** Contains progestin only and is taken every day of the month.

Non-oral hormonal methods of contraception are also available. Among these are the following:

- **Contraceptive skin patch (Ortho Evra®).** Contains both progestin and estrogens delivered in a skin patch placed on the skin (upper outer arm, back, lower abdomen, or buttocks) once a week for three weeks. After 1 week, the patch is removed from one location and then a new one is placed elsewhere. During the fourth week no patch is used.

- **Vaginal contraceptive ring (NuvaRing®).** A flexible doughnut-shaped ring about 5 cm (2 in.) in diameter that contains estrogens and progesterone and is inserted by the female herself into the vagina. It is left in the vagina for three weeks to prevent conception and then removed for one week to permit menstruation.

- **Emergency contraception (EC) (morning after pill).** Consists of progestin and estrogens or progestin alone to prevent pregnancy following unprotected sexual intercourse. The relatively high levels of progestin and estrogens in EC pills provide inhibition of FSH and LH secretion. Loss of the stimulating effects of these gonadotropic hormones causes the ovaries to cease secretion of their own estrogens and progesterone. In turn, declining levels of estrogens and progesterone induce shedding of the uterine lining, thereby blocking implantation. One pill is taken as soon as possible but within 72 hours of unprotected sexual intercourse. The second pill must be taken 12 hours after the first. The pills work in the same way as regular birth control pills.

■ **Hormone injections (Depo-provera®).** An injectable progestin given intramuscularly by a health-care practitioner once every three months.

Intrauterine Devices

An **intrauterine device (IUD)** is a small object made of plastic, copper, or stainless steel that is inserted by a health-care professional into the cavity of the uterus. IUDs prevent fertilization from taking place by blocking sperm from entering the uterine tubes. The IUD most commonly used in the United States today is the Copper T 380A®, which is approved for up to 10 years of use and has long-term effectiveness comparable to that of tubal ligation. Some women cannot use IUDs because of expulsion, bleeding, or discomfort.

Spermicides

Various foams, creams, jellies, suppositories, and douches that contain sperm-killing agents, or **spermicides** (SPER-mi-sīds), make the vagina and cervix unfavorable for sperm survival and are available without prescription. They are placed in the vagina before sexual intercourse. The most widely used spermicide is *nonoxynol-9*, which kills sperm by disrupting their plasma membranes. A spermicide is more effective when used with a barrier method such as a male condom, vaginal pouch, diaphragm, or cervical cap.

Barrier Methods

Barrier methods use a physical barrier and are designed to prevent sperm from gaining access to the uterine cavity and uterine tubes. In addition to preventing pregnancy, certain barrier methods (male condom and vaginal pouch) may also provide some protection against sexually transmitted diseases (STDs) such as AIDS. In contrast, oral contraceptives and IUDs confer no such protection. Among the barrier methods are the male condom, vaginal pouch, diaphragm, and cervical cap.

A **male condom** is a nonporous, latex covering placed over the penis that prevents deposition of sperm in the female reproductive tract. A **vaginal pouch**, sometimes called a **female condom**, is designed to prevent sperm from entering the uterus. It is made of two flexible rings connected by a polyurethane sheath. One ring lies inside the sheath and is inserted to fit over the cervix; the other ring remains outside the vagina and covers the female external genitals. A **diaphragm** is a rubber, dome-shaped structure that fits over the cervix and is used in conjunction with a spermicide. It can be inserted by the female up to six hours before intercourse. The diaphragm stops most sperm from passing into the cervix and the spermicide kills most sperm that do get by. Although diaphragm use does decrease the risk of some STDs, it does not fully protect against HIV infection because the vagina is still exposed. A **cervical cap** resembles a diaphragm but is smaller and more rigid. It fits snugly over the cervix and must be fitted by a health-care professional. Spermicides should be used with the cervical cap.

Periodic Abstinence

A couple can use their knowledge of the physiological changes that occur during the female reproductive cycle to decide either to abstain from intercourse on those days when pregnancy is a likely result, or to plan intercourse on those days if they wish to conceive a child. In females with normal and regular menstrual cycles, these physiological events help to predict the day on which ovulation is likely to occur.

The first physiologically based method, developed in the 1930s, is known as the **rhythm method.** It involves abstaining from sexual activity on the days that ovulation is likely to occur in each reproductive cycle. During this time (3 days before ovulation, the day of ovulation, and 3 days after ovulation) the couple abstains from intercourse. The effectiveness of the rhythm method for birth control is poor in many women due to the irregularity of the female reproductive cycle.

Another system is the **sympto-thermal method,** in which couples are instructed to know and understand certain signs of fertility. The signs of ovulation include increased basal body temperature; the production of abundant clear, stretchy cervical mucus; and pain associated with ovulation (*mittelschmerz*). If a couple abstains from sexual intercourse when the signs of ovulation are present and for three days afterward, the chance of pregnancy is decreased. A big problem with this method is that fertilization is very likely if intercourse occurs one or two days *before* ovulation.

Abortion

Abortion refers to the premature expulsion of the products of conception from the uterus, usually before the twentieth week of pregnancy. An abortion may be **spontaneous** (naturally occurring; also called a *miscarriage*) or **induced** (intentionally performed).

There are several types of induced abortions. One involves **mifepristone** (MIF-pris-tōn), called **miniprex** in the United States and **RU 486** in Europe. It is a hormone approved only for pregnancies of nine weeks or less when taken with misoprostol (a prostaglandin). Mifepristone is an antiprogestin; it blocks the action of progesterone by binding to and blocking progesterone receptors. Progesterone prepares the uterine endometrium for implantation and then maintains the uterine lining after implantation. If the level of progesterone falls during pregnancy or if the action of the hormone is blocked, menstruation occurs, and the embryo sloughs off along with the uterine lining. Within 12 hours after taking mifepristone, the endometrium starts to degenerate, and within 72 hours it begins to slough off. Misoprostol

Sexually Transmitted Diseases—Why Are Females More Vulnerable?

Sexual intercourse provides the perfect storm for the spread of sexually transmitted diseases (STDs). The pathogens responsible for these infections dwell in bodily fluids, including the saliva, blood, and mucus. In men, pathogens are also found in the semen, and in women, cervical and vaginal mucus. While both males and females can easily acquire STDs, females have an elevated risk. Women and girls are more likely to acquire sexually transmitted diseases from sexual intercourse than boys and men for several reasons.

The Act of Intercourse

Both women and men can acquire an STD from sexual intercourse. If a female has an STD, the male is exposed during intercourse to her cervical secretions, but exposure time is usually somewhat limited, as the penis is in and out, so to speak. On the other hand, if the male is infected, the pathogens in his semen may remain in the female's vaginal canal for several hours, allowing more time for infection transmission.

Anatomy and Physiology

Unless the penis has open sores, the skin on the penis provides a helpful barrier to infection. On the other hand, the tissue on the cervix is more vulnerable to pathogenic invaders, such as HIV. The younger the female, the more susceptible she is to STDs. This is because her cervix contains a higher proportion of columnar cells to squamous cells (see Chapter 4). Columnar cells are more susceptible to infection. As girls and women mature, the cervix develops a higher proportion of squamous cells.

Gender Roles and Inequality

Not only are women biologically more vulnerable to STDs, but in many relationships and cultures they lack the ability to make their male partners use condoms. Even under the best of circumstances, sex often proceeds without adequate protection. Add to this a lack of access to condoms and conditions of abuse and/or intoxication, and the likelihood of condom use during sexual intercourse approaches zero.

Cultural Norms and Beliefs

Some cultures around the world have especially dangerous practices and beliefs. For example, in many cultures men are encouraged to have sex with multiple partners, including sex workers. This increases their risk of acquiring STDs that they then pass around to all of their partners.

Especially damaging are practices that target very young females. Sexual intercourse with very young girls is more likely to cause tearing of the vaginal tissue, making the girls even more vulnerable to STDs.

Think It Over . . . **Condom availability alone does not ensure safer sexual practices. How can male and female partners work together to be sure condoms are used during sexual intercourse?**

---

stimulates uterine contractions and is given after mifepristone to aid in expulsion of the endometrium.

Another type of induced abortion is called *vacuum aspiration* (suction) and can be performed up to the sixteenth week of pregnancy. A small, flexible tube attached to a vacuum source is inserted into the uterus through the vagina. The embryo or fetus, placenta, and lining of the uterus are then removed by suction. For pregnancies between 13 and 16 weeks, a technique called *dilation and evacuation* is commonly used. After the cervix is dilated, suction and forceps are used to remove the fetus, placenta, and uterine lining. From the sixteenth to twenty-fourth week, a *late-stage abortion* may be employed using surgical methods similar to dilation and evacuation or through nonsurgical methods using a saline solution or medications to induce abortion. Labor may be induced by using vaginal suppositories, intravenous infusion, or injections into the amniotic fluid through the uterus.

✓ CHECKPOINT

14. How do oral contraceptives reduce the likelihood of pregnancy?

15. Why do some methods of birth control protect against sexually transmitted diseases, but others do not?

23.5 AGING AND THE REPRODUCTIVE SYSTEMS

OBJECTIVE • **Describe the effects of aging on the reproductive systems.**

During the first decade of life, the reproductive system is in a juvenile state. At about age 10, hormone-directed changes start to occur in both sexes. *Puberty* (PŪ-ber-tē = a ripe age) is the period when secondary sexual characteristics begin to develop and the potential for sexual reproduction is reached. Onset of puberty is marked by bursts of LH and FSH secretion, each triggered by a burst of GnRH. The stimuli that cause the GnRH bursts are still unclear, but a role for the hormone leptin is starting to unfold. Just before puberty, leptin levels rise in proportion to adipose tissue mass. Leptin may signal the hypothalamus that long-term energy stores (triglycerides in adipose tissue) are adequate for reproductive functions to begin.

In females, the reproductive cycle normally occurs once each month from *menarche* (me-NAR-kē), the first menses, to *menopause*, the permanent cessation of menses. Thus, the female reproductive system has a time-limited span of fertility between menarche and menopause. Between the ages of 40 and 50 the pool of remaining ovarian follicles becomes exhausted. As a result, the ovaries become less responsive to hormonal stimulation. The production of estrogens declines, despite copious secretion of FSH and LH by the anterior pituitary. Many women experience hot flashes and heavy sweating, which coincide with bursts of GnRH release. Other symptoms of menopause are headache, hair loss, muscular pains, vaginal dryness, insomnia, depression, weight gain, and mood swings. Some atrophy of the ovaries, uterine tubes, uterus, vagina, external genitalia, and breasts occurs in postmenopausal women. Due to loss of estrogens, most women also experience a decline in bone mineral density after menopause. Sexual desire (libido) does not show a parallel decline; it may be maintained by adrenal androgens. The risk of having uterine cancer peaks at about 65 years of age, but cervical cancer is more common in younger women.

In males, declining reproductive function is much more subtle than in females. Healthy men often retain reproductive capacity into their eighties or nineties. At about age 55 a decline in testosterone synthesis leads to reduced muscle strength, fewer viable sperm, and decreased sexual desire. However, abundant sperm may be present even in old age.

Enlargement of the prostate to two to four times its normal size occurs in approximately one-third of all males over age 60. This condition, called *benign prostatic hyperplasia (BPH)*, is characterized by frequent urination, bed-wetting, hesitancy in urination, decreased force of urinary stream, postvoiding dribbling, and a sensation of incomplete emptying.

✓ **CHECKPOINT**

16. What changes occur in males and females at puberty?

17. What do the terms menarche and menopause mean?

• • •

To appreciate the many ways that the reproductive systems contribute to homeostasis of other body systems, examine Focus on Homeostasis: The Reproductive Systems. Next, in Chapter 24, you will explore the major events that occur during pregnancy and you will discover how genetics (inheritance) plays a role in the development of a child.

BODY SYSTEM	CONTRIBUTION OF THE REPRODUCTIVE SYSTEMS

For all body systems
The male and female reproductive systems produce gametes (oocytes and sperm) that unite to form embryos and fetuses, which contain cells that divide and differentiate to form all of the organ systems of the body.

Integumentary system
Androgens promote the growth of body hair. Estrogens stimulate the deposition of fat in the breasts, abdomen, and hips. Mammary glands produce milk. Skin stretches during pregnancy as the fetus enlarges.

Skeletal system
Androgens and estrogens stimulate the growth and maintenance of bones of the skeletal system.

Muscular system
Androgens stimulate the growth of skeletal muscles.

Nervous system
Androgens influence libido (sex drive). Estrogens may play a role in the development of certain regions of the brain in males.

Endocrine system
Testosterone and estrogens exert feedback effects on the hypothalamus and anterior pituitary gland.

Cardiovascular system
Estrogens lower blood cholesterol level and may reduce the risk of coronary artery disease in women under age 50.

Lymphatic system and immunity
The presence of an antibiotic-like chemical in semen and the acidic pH of vaginal fluid provide innate immunity against microbes in the reproductive tract.

Respiratory system
Sexual arousal increases the rate and depth of breathing.

Digestive system
The presence of the fetus during pregnancy crowds the digestive organs, which leads to heartburn and constipation.

Urinary system
In males, the portion of the urethra that extends through the prostate and penis is a passageway for urine as well as semen.

THE REPRODUCTIVE SYSTEMS

626

COMMON DISORDERS

Reproductive System Disorders in Males

Testicular Cancer

Testicular cancer is the most common cancer in males between the ages of 20 and 35. More than 95 percent of testicular cancers arise from spermatogenic cells within the seminiferous tubules. An early sign of testicular cancer is a mass in the testis, often associated with a sensation of testicular heaviness or a dull ache in the lower abdomen; pain usually does not occur. To increase the chance for early detection of a testicular cancer, all males should perform regular self-examinations of the testes. The examination should be done starting in the teen years and once each month thereafter. After a warm bath or shower (when the scrotal skin is loose and relaxed) each testis should be examined as follows. The testis is grasped and gently rolled between the index finger and thumb, feeling for lumps, swellings, hardness, or other changes. If a lump or other change is detected, a physician should be consulted as soon as possible.

Prostate Disorders

Because the prostate surrounds part of the urethra, any prostatic infection, enlargement, or tumor can obstruct the flow of urine. Acute and chronic infections of the prostate are common in adult males, often in association with inflammation of the urethra. In *acute prostatitis*, the prostate becomes swollen and tender. *Chronic prostatitis* is one of the most common chronic infections in men of the middle and later years; on examination, the prostate feels enlarged, soft, and very tender, and its surface outline is irregular.

Prostate cancer is the leading cause of death from cancer in men in the United States. A blood test can measure the level of prostate-specific antigen (PSA) in the blood. The amount of PSA, which is produced only by prostate epithelial cells, increases with enlargement of the prostate and may indicate infection, benign enlargement, or prostate cancer. Males over the age of 40 should have an annual examination of the prostate. In a *digital rectal exam*, a physician palpates the prostate through the rectum with the fingers (digits). Many physicians also recommend an annual PSA test for males over 50. Treatment for prostate cancer may involve surgery, radiation, hormonal therapy, and chemotherapy. Because many prostate cancers grow very slowly, some urologists recommend "watchful waiting" before treating small tumors in men over age 70.

Reproductive System Disorders in Females

Premenstrual Syndrome

Premenstrual syndrome (PMS) is a cyclical disorder of severe physical and emotional distress. It appears during the postovulatory phase of the female reproductive cycle and dramatically disappears when menstruation begins. The signs and symptoms are highly variable from one woman to another. They may include edema, weight gain, breast swelling and tenderness, abdominal distension, backache, joint pain, constipation, skin eruptions, fatigue and lethargy, greater need for sleep, depression or anxiety, irritability, mood swings, headache, poor coordination and clumsiness, and cravings for sweet or salty foods. The cause of PMS is unknown. For some women, getting regular exercise; avoiding caffeine, salt, and alcohol; and eating a diet that is high in complex carbohydrates and lean proteins can bring considerable relief.

Endometriosis

Endometriosis (en′-dō-mē′-trē-Ō-sis; *endo-* = within; *metri-* = uterus; *-osis* = condition of or disease) is characterized by the growth of endometrial tissue outside the uterus. The tissue enters the pelvic cavity via the open uterine tubes and may be found in any of several sites—on the ovaries, the outer surface of the uterus, the sigmoid colon, pelvic and abdominal lymph nodes, the cervix, the abdominal wall, the kidneys, and the urinary bladder. Endometrial tissue responds to hormonal fluctuations, whether it is inside or outside the uterus, by first proliferating and then breaking down and bleeding. When this occurs outside the uterus, it can cause inflammation, pain, scarring, and infertility. Symptoms include premenstrual pain or unusually severe menstrual pain.

Breast Cancer

One in eight women in the United States faces the prospect of *breast cancer*, the second-leading cause of female deaths from cancer. Early detection by breast self-examination and mammograms is the best way to increase the chance of survival.

The most effective technique for detecting tumors less than 1 cm (0.4 in.) in diameter is *mammography* (mam-OG-ra-fē; *-graphy* = to record), a type of radiography using very sensitive x-ray film. The image of the breast, called a *mammogram*, is best obtained by compressing the breasts, one at a time, using flat plates. A supplementary procedure for evaluating breast abnormalities is *ultrasonography*. Although ultrasonography cannot detect tumors smaller than 1 cm in diameter, it can be used to determine whether a lump is a benign, fluid-filled cyst or a solid (and therefore possibly malignant) tumor.

Among the factors that increase the risk of developing breast cancer are (1) a family history of breast cancer, especially in a mother or sister; (2) never having borne a child or having a first child after age 35; (3) previous cancer in one breast; (4) exposure to ionizing radiation, such as x-rays; (5) excessive alcohol intake; and (6) cigarette smoking.

The American Cancer Society recommends the following steps to help diagnose breast cancer as early as possible:

- All women over 20 should develop the habit of monthly breast self-examination.

- A physician should examine the breasts every three years when a woman is between the ages of 20 and 40, and every year after age 40.

- A mammogram should be taken in women between the ages of 35 and 39, to be used later for comparison (baseline mammogram).

- Women with no symptoms should have a mammogram every year after age 40.

- Women of any age with a history of breast cancer, a strong family history of the disease, or other risk factors should consult a physician to determine a schedule for mammography.

In November 2009, the United States Preventive Services Task Force (USPSTF) issued a series of recommendations relative to breast cancer screening for females at normal risk for breast cancer, that is, for females who have no signs or symptoms of breast cancer and who are not at a higher risk for breast cancer (for example, no family history). These recommendations are as follows:

- Women aged 50–74 should have a mammogram every 2 years.
- Women over 75 should not have mammograms.
- Breast self-examination is not required.

Treatment for breast cancer may involve hormone therapy, chemotherapy, radiation therapy, *lumpectomy* (removal of the tumor and the immediate surrounding tissue), a modified or radical mastectomy, or a combination of these approaches. A *radical mastectomy* (*mast-* = breast) involves removal of the affected breast along with the underlying pectoral muscles and the axillary lymph nodes. (Lymph nodes are removed because the spread of cancerous cells usually occurs through lymphatic or blood vessels.) Radiation treatment and chemotherapy may follow the surgery to ensure the destruction of any stray cancer cells. In some cases of metastatic (spreading) breast cancer, Herceptin®, a monoclonal antibody drug that targets an antigen on the surface of breast cancer cells, can cause regression of the tumors and retard progression of the disease. Finally, two promising drugs for breast cancer *prevention* are now on the market—tamoxifen (Nolvadex®) and raloxifene (Evista®).

Ovarian Cancer

Ovarian cancer is the sixth most common form of cancer in females, but the leading cause of death from all gynecological malignancies (excluding breast cancer) because it is difficult to detect before it metastasizes (spreads) beyond the ovaries. Risk factors associated with ovarian cancer include age (usually over age 50); race (whites are at highest risk); family history of ovarian cancer; more than 40 years of active ovulation; *nulliparity* (no pregnancies) or first pregnancy after age 30; a high-fat, low-fiber, vitamin A–deficient diet; and prolonged exposure to asbestos and talc. Early ovarian cancer may have no symptoms or mild ones such as abdominal discomfort, heartburn, nausea, loss of appetite, bloating, and flatulence. Later-stage signs and symptoms include an enlarged abdomen, abdominal and/or pelvic pain, persistent gastrointestinal disturbances, urinary complications, menstrual irregularities, and heavy menstrual bleeding.

Cervical Cancer

Cervical cancer, cancer of the uterine cervix, starts with *cervical dysplasia* (dis-PLĀ-zha), a change in the shape, growth, and number of cervical cells. The cells may either return to normal or progress to cancer. In most cases cervical cancer may be detected in its earliest stages by a Pap smear. Some evidence links cervical cancer to the virus that causes genital warts (human papilloma virus). Increased risk is associated with a large number of sexual partners, first intercourse at a young age, and smoking cigarettes.

Vulvovaginal Candidiasis

Candida albicans is a yeastlike fungus that commonly grows on mucous membranes of the gastrointestinal and genitourinary tracts. The organism is responsible for *vulvovaginal candidiasis* (vul′-vō-

VAJ-i-nal can′-di-DĪ-a-sis), the most common form of vaginitis (vaj′-i-NĪ-tis), inflammation of the vagina. Candidiasis, commonly referred to as a yeast infection, is characterized by severe itching; a thick, yellow, cheesy discharge; a yeasty odor; and pain. The disorder, experienced at least once by about 75% of females, is usually a result of proliferation of the fungus following antibiotic therapy for another condition. Predisposing conditions include the use of oral contraceptives or cortisone-like medications, pregnancy, and diabetes.

Sexually Transmitted Diseases

A *sexually transmitted disease (STD)* is one that is spread by sexual contact. AIDS and hepatitis B, which are sexually transmitted diseases that also may be contracted in other ways, are discussed in Chapters 17 and 19, respectively.

Chlamydia

Chlamydia (kla-MID-ē-a) is a sexually transmitted disease caused by the bacterium *Chlamydia trachomatis* (*chlamy-* = cloak). This unusual bacterium cannot reproduce outside body cells; it "cloaks" itself inside cells, where it divides. At present, chlamydia is the most prevalent sexually transmitted disease in the United States. In most cases the initial infection is asymptomatic and thus difficult to recognize clinically. In males, urethritis is the principal result, causing a clear discharge and burning, frequent, and painful urination. Without treatment, the epididymides may also become inflamed, leading to male sterility. In 70% of females with chlamydia, symptoms are absent, but chlamydia is the leading cause of pelvic inflammatory disease. The uterine tubes may also become inflamed, which increases the risk of female infertility due to the formation of scar tissue in the tubes.

Gonorrhea

Gonorrhea (gon′-ō-RĒ-a) or "*the clap*" is caused by the bacterium *Neisseria gonorrhoeae*. Discharges from infected mucus membranes are the source of transmission of the bacteria either during sexual contact or during the passage of a newborn through the birth canal. Males usually experience urethritis with profuse pus drainage and painful urination. In females, infection typically occurs in the vagina, often with a discharge of pus. In females, the infection and consequent inflammation can proceed from the vagina into the uterus, uterine tubes, and pelvic cavity. Thousands of women are made infertile by gonorrhea every year as a result of scar tissue formation that closes the uterine tubes. Transmission of bacteria in the birth canal to the eyes of a newborn can result in blindness.

Syphilis

Syphilis, caused by the bacterium *Treponema pallidum*, is transmitted through sexual contact or exchange of blood, or through the placenta to a fetus. The disease progresses through several stages. During the *primary stage*, the chief sign is a painless open sore, called a *chancre* (SHANG-ker), at the point of contact. The chancre heals within 1 to 5 weeks. From 6 to 24 weeks later, signs and symptoms such as a skin rash, fever, and aches in the joints and muscles usher in the *secondary stage*, which is systemic—the infection spreads to all major body systems. When signs of organ degeneration appear, the disease is said to be in the *tertiary stage*. If the nervous system is involved, the tertiary stage is called *neurosyphilis*. As motor areas become extensively damaged, victims may be unable to control urine and bowel movements; eventually they may become bedridden, unable even to feed themselves. Damage to the cerebral cortex produces memory loss and personality changes that range from irritability to hallucinations.

Genital Herpes

Genital herpes is caused by type 2 herpes simplex virus (HSV-2), producing painful blisters on the prepuce, glans penis, and penile shaft in males, and on the vulva or sometimes high up in the vagina in females. The blisters disappear and reappear in most patients, but the virus itself remains in the body; there is no cure. A related virus, type 1 herpes simplex virus (HSV-1), which is not an STD,

causes cold sores on the mouth and lips. Infected individuals typically experience recurrences of symptoms several times a year.

Genital Warts

Warts are an infectious disease caused by viruses. *Human papillomavirus (HPV)* causes **genital warts**, which can be transmitted sexually. Nearly 1 million people a year develop genital warts in the United States. Patients with a history of genital warts may be at increased risk for cancers of the cervix, vagina, anus, vulva, and penis. There is no cure for genital warts. A vaccine (Gardasil®) against certain types of HPV that cause cervical cancer and genital warts is available and recommended for 11- and 12-year-old girls.

MEDICAL TERMINOLOGY AND CONDITIONS

Amenorrhea (ā-men′-ō-RĒ-a; *a-* = without; *men-* = month; *-rrhea* = a flow) The absence of menstruation; it may be caused by a hormone imbalance, obesity, extreme weight loss, or very low body fat, as may occur during rigorous athletic training.

Castration (kas-TRĀ-shun = to prune) Removal, inactivation, or destruction of the gonads; commonly used in reference to removal of the testes only.

Colposcopy (kol-POS-kō-pē; *colpo-* = vagina; *-scopy* = to view) Visual inspection of the vagina and cervix of the uterus using a culposcope, an instrument that has a magnifying lens (between 5× and 50×) and a light. The procedure generally takes place after an unusual Pap smear.

Dysmenorrhea (dis-men′-ō-RĒ-a; *dys-* = difficult or painful) Painful menstruation; the term is usually reserved to describe menstrual symptoms that are severe enough to prevent a woman from functioning normally for one or more days each month. Some cases are caused by uterine tumors, ovarian cysts, pelvic inflammatory disease, or intrauterine devices.

Dyspareunia (dis-pa-ROO-nē-a; *dys-* = difficult; *para* = beside; *-enue* = bed) Pain during intercourse. It may occur in the genital area or in the pelvic cavity, and may be due to inadequate lubrication, inflammation, infection, an improperly fitting diaphragm or cervical cap, endometriosis, pelvic inflammatory disease, pelvic tumors, or weakened uterine ligaments.

Endocervical curettage (kū′-re-TAHZH; *curette* = scraper) A procedure in which the cervix is dilated and the endometrium of the uterus is scraped with a spoon-shaped instrument called a curette; commonly called a D and C (dilation and curettage).

Fibroids (FĪ-broyds; *fibro-* = fiber; *-eidos* = resemblance) Noncancerous tumors in the myometrium of the uterus composed of muscular and fibrous tissue. Their growth appears to be related

to high levels of estrogens. They do not occur before puberty and usually stop growing after menopause. Symptoms include abnormal menstrual bleeding, and pain or pressure in the pelvic area.

Menorrhagia (men-ō-RA-jē-a; *meno-* = menstruation; *-rhage* = to burst forth) Excessively prolonged or profuse menstrual period. May be due to a disturbance in hormonal regulation of the menstrual cycle, pelvic infection, medications (anticoagulants), fibroids, endometriosis, or intrauterine devices.

Oophorectomy (ō′-of-ō-REK-tō-mē; *oophor-* = bearing eggs) Removal of the ovaries.

Ovarian cyst The most common form of ovarian tumor, in which a fluid-filled follicle or corpus luteum persists and continues growing.

Papanicolaou test (pa′-pa-ni′-kō-LĀ-oo), or **Pap smear** A test to detect uterine cancer in which a few cells from the cervix and the part of the vagina surrounding the cervix are removed with a swab and examined microscopically. Malignant cells have a characteristic appearance that allows diagnosis even before symptoms occur.

Pelvic inflammatory disease (PID) A collective term for any extensive bacterial infection of the pelvic organs, especially the uterus, uterine tubes, or ovaries, which is characterized by pelvic soreness, lower back pain, abdominal pain, and urethritis. Often the early symptoms of PID occur just after menstruation. As infection spreads and cases advance, fever may develop, along with painful abscesses of the reproductive organs.

Salpingectomy (sal′-pin-JEK-tō-mē; *salpingo* = tube) Removal of a uterine (fallopian) tube.

Smegma (SMEG-ma) The secretion, consisting principally of sloughed off epithelial cells, found chiefly around the external genitals and especially under the foreskin of the male.

CHAPTER REVIEW AND RESOURCE SUMMARY

REVIEW	RESOURCES

Introduction

1. **Sexual reproduction** is the process of producing offspring by the union of **gametes** (oocytes and sperm).
2. The organs of reproduction are grouped as **gonads** (produce gametes), **ducts** (transport and store gametes), **accessory sex glands** (produce materials that support gametes), and **supporting structures**.

| REVIEW | RESOURCES |

23.1 Male Reproductive System

1. The **male reproductive system** includes the testes, epididymis, ductus (vas) deferens, ejaculatory ducts, urethra, seminal vesicles, prostate, bulbourethral (Cowper's) glands, scrotum, and penis.
2. The **scrotum** is a sac that supports and regulates the temperature of the testes. The male gonads include the **testes**, oval-shaped organs in the scrotum that contain the **seminiferous tubules**, in which **sperm cells** develop; **Sertoli cells**, which nourish sperm cells and produce the hormone inhibin; and **Leydig cells**, which produce the sex hormone testosterone.
3. **Spermatogenesis** occurs in the testes and consists of **meiosis I**, **meiosis II**, and **spermiogenesis**. It results in the formation of four haploid sperm cells from a **primary spermatocyte**.
4. Mature sperm consist of a **head** and a **tail**. Their function is to fertilize a secondary oocyte.
5. At puberty, **gonadotropin-releasing hormone (GnRH)** stimulates anterior pituitary secretion of **luteinizing hormone (LH)** and **follicle-stimulating hormone (FSH)**. LH stimulates Leydig cells to produce testosterone. FSH and testosterone initiate spermatogenesis.
6. **Testosterone** controls the growth, development, and maintenance of sex organs; stimulates bone growth, protein anabolism, and sperm maturation; and stimulates development of male secondary sex characteristics. **Inhibin** is produced by Sertoli cells; its inhibition of FSH helps regulate the rate of spermatogenesis.
7. Sperm are transported out of the testes into an adjacent organ, the **epididymis**, where their motility increases. The **ductus (vas) deferens** stores sperm and propels them toward the **urethra** during **ejaculation**. The **ejaculatory ducts** are formed by the union of the ducts from the seminal vesicles and vas deferens, and they eject sperm into the urethra. The male urethra passes through the prostate, deep perineal muscles, and penis.
8. The **seminal vesicles** secrete an alkaline, viscous fluid that constitutes about 60% of the volume of semen and contributes to sperm viability. The **prostate** secretes a slightly acidic fluid that constitutes about 25% of the volume of semen and contributes to sperm motility. The **bulbourethral glands** secrete mucus for lubrication and an alkaline substance that neutralizes acid.
9. **Semen** is a mixture of sperm and seminal fluid; it provides the fluid in which sperm are transported, supplies nutrients, and neutralizes the acidity of the male urethra and the vagina.
10. The **penis** consists of three parts: the **root of the penis**, the **body of the penis**, and the glans penis. It functions to introduce sperm into the vagina. Expansion of its blood sinuses under the influence of sexual excitation is called **erection**.

Anatomy Overview—The Reproductive System: Structures and Hormones

Figure 23.2 Anatomy and histology of the testes

Figure 23.5 Hormonal control of spermatogenesis and actions of testosterone and dihydrotestosterone (DHT)

23.2 Female Reproductive System

1. The female organs of reproduction include the ovaries, uterine (fallopian) tubes, uterus, vagina, and vulva. The mammary glands are also considered part of the reproductive system.
2. The female gonads are the **ovaries**, located in the upper pelvic cavity on either side of the uterus. Ovaries produce secondary oocytes; discharge secondary oocytes (the process of ovulation); and secrete estrogens, progesterone, relaxin, and inhibin.
3. **Oogenesis** (production of haploid secondary oocytes) begins in the ovaries. The oogenesis sequence includes meiosis I and meiosis II. Meiosis II is completed only after an ovulated secondary oocyte is fertilized by a sperm cell.
4. The **uterine (fallopian) tube**, which transports a secondary oocyte from an ovary to the uterus, is the normal site of fertilization.
5. The **uterus** is an organ the size and shape of an inverted pear that functions in menstruation, implantation of a fertilized ovum, development of a fetus during pregnancy, and labor. It also is part of the pathway for sperm to reach a uterine tube to fertilize a secondary oocyte. The innermost layer of the uterine wall is the **endometrium**, which undergoes marked changes during the menstrual cycle.
6. The **vagina** is a passageway for the menstrual flow, the receptacle for the penis during sexual intercourse, and the lower portion of the birth canal. The smooth muscle of the vaginal wall makes it capable of considerable stretching.
7. The **vulva**, a collective term for the external genitals of the female, consists of the **mons pubis**, **labia majora**, **labia minora**, **clitoris**, **vestibule**, **vaginal** and **urethral orifices**, **paraurethral glands**, and **greater vestibular glands**.

Anatomy Overview—Reproductive System Histology
Animation—Spermatogenesis
Animation—Hormonal Control of Male Reproductive Function

Figure 23.7 Histology of the ovary

Figure 23.9 Uterus and associated structures

Figure 23.11 Mammary glands

Exercise—Match the Male Hormones
Exercise—Spermatogenesis Selections
Concepts and Connections—Regulation of Male Reproduction

REVIEW	RESOURCES

8. The **mammary glands** of the female breasts are modified sweat glands located over the pectoralis major muscles. Their function is to secrete and eject milk (**lactation**). Mammary gland development depends on estrogens and progesterone. Milk production is stimulated by prolactin, estrogens, and progesterone; milk ejection is stimulated by oxytocin.

23.3 Female Reproductive Cycle

1. The **female reproductive cycle** includes the ovarian and **uterine** (menstrual) cycles. The function of the **ovarian cycle** is development of a secondary oocyte; that of the **uterine cycle** is preparation of the endometrium each month to receive a fertilized egg.

2. The ovarian and **uterine** cycles are controlled by GnRH from the hypothalamus, which stimulates the release of FSH and LH by the anterior pituitary. FSH stimulates development of follicles and initiates secretion of estrogens by the follicles. LH stimulates further development of the follicles, secretion of estrogens by follicular cells, ovulation, formation of the **corpus luteum**, and the secretion of progesterone and estrogens by the corpus luteum.

3. **Estrogens** stimulate the growth, development, and maintenance of female reproductive structures; the development of secondary sex characteristics; and protein synthesis.

4. **Progesterone** works together with estrogens to prepare the endometrium for implantation and the mammary glands for milk synthesis.

5. **Relaxin** increases the flexibility of the pubic symphysis and helps dilate the uterine cervix to ease delivery of a baby.

6. During the **menstrual phase**, part of the endometrium is shed, discharging blood and tissue cells.

7. During the **preovulatory phase**, a group of follicles in the ovaries begins to undergo maturation. One follicle outgrows the others and becomes dominant while the others die. At the same time, endometrial repair occurs in the uterus. Estrogens are the dominant ovarian hormones during the preovulatory phase.

8. **Ovulation** is the rupture of the dominant **mature (graafian) follicle** and the release of a secondary oocyte into the pelvic cavity. It is brought about by a surge of LH.

9. During the **postovulatory phase**, both progesterone and estrogens are secreted in large quantity by the corpus luteum of the ovary, and the uterine endometrium thickens in readiness for implantation.

10. If fertilization and implantation do not occur, the corpus luteum degenerates, and the resulting low levels of progesterone and estrogens allow discharge of the endometrium (menstruation) followed by the initiation of another reproductive cycle. If fertilization and implantation occur, the corpus luteum is maintained by **human chorionic gonadotropin (hCG)**.

Animation—Oogenesis

Figure 23.12 The female reproductive cycle

Exercise—Organize Oogenesis Concepts and Connections— Regulation of Female Reproduction

23.4 Birth Control Methods and Abortion

1. Birth control methods include surgical **sterilization (vasectomy, tubal ligation)**, non-incisional sterilization, hormonal methods, **intrauterine devices**, **spermicides**, **barrier methods** (**male condom, vaginal pouch, diaphragm, cervical cap**), and periodic abstinence. Table 23.1 provides failure rates of the various methods of birth control. Abstinence is the only foolproof method of birth control.

2. **Oral contraceptives** of the combination type contain estrogens and progestins in concentrations that decrease the secretion of FSH and LH and thereby inhibit development of ovarian follicles and ovulation.

3. An **abortion** is the premature expulsion from the uterus of the products of conception; it may be **spontaneous** or **induced**. **RU 486 (miniprex)** can induce abortion by blocking the action of progesterone.

23.5 Aging and the Reproductive Systems

1. **Puberty** is the period of time when secondary sex characteristics begin to develop and the potential for sexual reproduction arises. In older females, levels of progesterone and estrogens decrease, resulting in changes in menstruation and then **menopause**.

2. In older males, decreased levels of testosterone are associated with decreased muscle strength, waning sexual desire, and fewer viable sperm; prostate disorders are common.

BBC Video—Chlamydia

SELF-QUIZ

1. The role of FSH in the male reproductive system is to
   a. convert testosterone into dihydrotestosterone.
   b. stimulate spermatogenesis.
   c. inhibit GnRH.
   d. stimulate the secretion of testosterone.
   e. influence the development of male secondary sexual characteristics.

2. Match the following:
   ____ a. cells that support, protect, and nourish developing spermatogonia
   ____ b. contain developing oocytes
   ____ c. immature sperm cells
   ____ d. cells that secrete testosterone
   ____ e. secretes progesterone and estrogens

   A. corpus luteum
   B. Leydig cells
   C. Sertoli cells
   D. follicles
   E. spermatogonia

3. Secretions from the seminal vesicles
   a. appear milky.
   b. inhibit testosterone release.
   c. help lubricate the end of the penis.
   d. contribute to semen's acidity.
   e. affect semen clotting.

4. Which of the following is true?
   a. Meiosis is the process by which somatic (body) cells divide.
   b. The haploid chromosome number is symbolized by 2n.
   c. Meiosis II results in diploid spermatocytes and oocytes.
   d. Gametes contain the haploid chromosome number.
   e. Gametes contain 46 chromosomes in their nuclei.

5. The uterus is the site of all of the following except
   a. menstrual fluids.   b. implantation of a fertilized ovum.
   c. ovulation.   d. labor.   e. development of the fetus.

6. Menstruation is triggered by a
   a. rapid rise in luteinizing hormone (LH).
   b. rapid fall in luteinizing hormone (LH).
   c. drop in estrogens and progesterone.
   d. rise in estrogens and progesterone.
   e. rise in inhibin.

7. An inflammation of the seminiferous tubules would interfere with the ability to
   a. secrete testosterone.   b. produce sperm.
   c. secrete LH.   d. make semen alkaline.
   e. regulate the temperature in the scrotum.

8. What causes erection of the penis?
   a. sympathetic nerve impulses
   b. relaxation of smooth muscle in penile arteries
   c. dilation of penile veins
   d. peristaltic contractions
   e. all of the above

9. Which of the following is NOT a function of semen?
   a. transport sperm
   b. lubricate the reproductive tract
   c. provide an acidic environment needed for fertilization
   d. provide nourishment for sperm
   e. produce antibiotics to destroy some bacteria

10. Prior to ejaculation, sperm are stored in the
    a. spermatic cord.   b. scrotum.   c. seminal vesicles.
    d. ejaculatory ducts.   e. ductus (vas) deferens and epididymis.

11. Place the following in the correct order for the passage of sperm from the testes to the outside of the body.
    1. urethra   2. ductus (vas) deferens
    3. seminiferous tubules   4. ejaculatory duct
    5. external urethral orifice   6. epididymis
       a. 6, 3, 2, 4, 1, 5   b. 3, 2, 6, 4, 1, 5   c. 3, 6, 2, 4, 1, 5
       d. 3, 6, 2, 4, 5, 1   e. 2, 4, 6, 1, 3, 5

12. In males, the gland that surrounds the urethra at the base of the urinary bladder is the
    a. glans penis.   b. prostate.   c. seminal vesicle.
    d. bulbourethral gland.   e. greater vestibular gland.

13. An oocyte is moved toward the uterus by
    a. peristaltic contractions of the uterine (fallopian) tubes.
    b. contraction of the uterus.   c. gravity.
    d. swimming.   e. flagella.

14. Fertilization normally occurs in the
    a. vagina.   b. cervix.   c. uterus.
    d. ovary.   e. uterine tube.

15. In the female reproductive system, lubricating mucus is produced by the
    a. vulva.   b. hymen.   c. mons pubis.
    d. greater vestibular glands.   e. labia minora.

16. Ovarian follicles mature during
    a. the menstrual phase.   b. ovulation.
    c. the postovulatory phase.   d. the preovulatory phase.
    e. the luteal phase.

17. Match the following:
    ____ a. area between thighs and buttocks
    ____ b. contains modified sebaceous glands
    ____ c. opening of the uterus to the vagina
    ____ d. opening of the vagina to the body's exterior
    ____ e. external genitals of the female
    ____ f. erectile tissue
    ____ g. milk-secreting glands

    A. vaginal orifice
    B. vulva
    C. cervix
    D. alveoli
    E. clitoris
    F. perineum
    G. areola

18. The portion of the uterus responsible for its contraction is the
    a. fundus.   **b.** infundibulum.   **c.** endometrium.
    d. myometrium.   **e.** perineum.

19. Match the following:

  ____ **a.** released by the hypothalamus to regulate the ovarian cycle

  ____ **b.** stimulates the initial secretion of estrogens by growing follicles

  ____ **c.** stimulates ovulation

  ____ **d.** stimulate growth, development, and maintenance of the female reproductive system

  ____ **e.** works with estrogens to prepare the uterus for implantation of a fertilized ovum

  ____ **f.** assists with labor by helping to dilate the cervix and increase flexibility of the pubic symphysis

  ____ **g.** inhibits release of FSH by the anterior pituitary

A. luteinizing hormone (LH)
B. gonadotropin-releasing hormone (GnRH)
C. relaxin
D. progesterone
E. inhibin
F. follicle-stimulating hormone (FSH)
G. estrogens

20. Birth control pills are a combination of ovarian hormones that prevent pregnancy by
    a. neutralizing the pH of the vagina.
    b. inhibiting motility of the sperm.
    c. causing early ovulation, before the follicle is mature.
    d. preventing sperm from entering the uterus.
    e. inhibiting the secretion of LH and FSH from the pituitary gland.

CRITICAL THINKING APPLICATIONS

1. Yearning to have a child, thirty-two-year-old Maria has no trouble getting pregnant but has trouble maintaining a pregnancy. She miscarries early in her pregnancies. What vital hormone may be insufficient and contributing to miscarriage?

2. Darren has promised his wife that he will get a vasectomy after the birth of their next child. He is a little concerned, however, about the possible effects on his virility. What would you tell Darren about the procedure?

3. Andrew and his wife have been trying unsuccessfully to become pregnant. The fertility clinic suggested that the problem may have something to do with Andrew's habits of wearing very close-fitting briefs during the day and taking a long nightly soak in a hot bath. What effect could this have on fertility?

4. Peter has just been diagnosed with an enlarged prostate (benign prostatic hyperplasia). What are the symptoms of this condition? What is the effect on the semen of removal of the prostate?

ANSWERS TO FIGURE QUESTIONS

23.1 Functionally, the penis is considered a supporting structure.

23.2 Spermatogonia (stem cells) are the least mature.

23.3 Crossing-over permits the formation of new combinations of genes from maternal and paternal chromosomes.

23.4 The middle piece contains mitochondria, which produce ATP that provides energy for locomotion of sperm.

23.5 The Sertoli cells secrete inhibin.

23.6 The female external genitals are collectively referred to as the vulva or pudendum.

23.7 Ovarian follicles secrete estrogens, and the corpus luteum secretes estrogens, progesterone, relaxin, and inhibin.

23.8 Primary oocytes are present in the ovary at birth, so they are as old as the woman is. In males, primary spermatocytes are continually being formed from spermatogonia and thus are only a few days old.

23.9 The endometrium is rebuilt after each menstruation.

23.10 The mons pubis, clitoris, prepuce, and external urethral orifice are anterior to the vaginal opening.

23.11 Oxytocin regulates milk ejection from the mammary glands.

23.12 The hormones responsible for the proliferative phase of endometrial growth are estrogens; for ovulation, LH; for growth of the corpus luteum, LH; and for the midcycle surge of LH, estrogens.

23.13 This is negative feedback because the response is opposite to the stimulus. Decreasing levels of estrogens and progesterone stimulate release of GnRH, which, in turn, increases production and release of estrogens.

CHAPTER 24
DEVELOPMENT AND INHERITANCE

Once sperm and a secondary oocyte have developed through meiosis and maturation, and the sperm have been deposited in the vagina, pregnancy can occur. *Pregnancy* is a sequence of events that begins with fertilization, proceeds to implantation, embryonic development, and fetal development, and normally ends with birth about 38 weeks later, or 40 weeks after the last menstrual period.

Developmental biology is the study of the extraordinary sequence of events from the fertilization of a secondary oocyte to the formation of an adult organism. From fertilization through the eighth week of development, the developing human is called an *embryo* (*em-* = into; *-bryo* = grow), and this is the *embryonic period*. *Embryology* (em-brē-OL-ō-jē) is the study of development from the fertilized egg through the eighth week. The *fetal period* begins at week nine and continues until birth. During this time, the developing human is called a *fetus* (FĒ-tus = offspring).

Obstetrics (ob-STET-riks; *obstetrix* = midwife) is the branch of medicine that deals with the management of pregnancy, labor, and the *neonatal period*, the first 28 days after birth. *Prenatal development* (prē-NĀ-tal; *pre-* = before; *natal* = birth) is the time from fertilization to birth and includes both the embryonic and fetal periods.

In this chapter, we focus on the developmental sequence from fertilization through implantation, embryonic and fetal development, labor, and birth. We will also consider the concept of inheritance.

Did you know? *Human milk provides perfect nutrition for human infants. It also supplies important digestive enzymes and hormones that promote healthy development. Breast milk contains important substances that help babies fight infections. These include secretory immunoglobulin A (IgA) antibodies, formed by the mother in response to infectious agents in the mother's (and baby's) environment. These antibodies summon an immune response without harming helpful flora in the gastrointestinal (GI) tract or causing inflammation, a process that can harm the baby more than the infection. Human milk also contains mucins, oligosaccharides (sugar chains), and glycoproteins (carbohydrate–protein compounds) that bind to microbes and prevent them from infecting the baby's GI tract.*

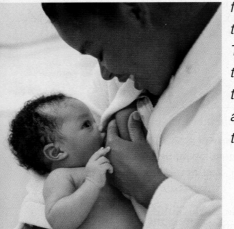

FOCUS ON WELLNESS

Breast Milk—Mother Nature's Approach to Infection Prevention

LOOKING BACK TO MOVE AHEAD...

Somatic Cell Division (Section 3.7)

Testes and Ovaries (Sections 23.1 and 23.2)

Uterine Tubes and Uterus (Section 23.2)

Estrogens and Progesterone (Section 23.3)

Positive Feedback System (Section 1.4)

Mammary Glands (Section 23.2)

Oxytocin (Section 13.3)

Prolactin (Section 13.3)

24.1 EMBRYONIC PERIOD

OBJECTIVE • Explain the major developmental events that occur during the embryonic period.

First Week of Development

The first week of development is characterized by several significant events including fertilization, cleavage of the zygote, blastocyst formation, and implantation.

Fertilization

During ***fertilization*** (fer-til-i-ZĀ-shun; *fertil-* = fruitful), the genetic material from a haploid sperm cell and a haploid secondary oocyte merges into a single diploid nucleus (Figure 24.1). Of approximately 300 million sperm introduced into the vagina, fewer than 2 million reach the cervix of the uterus and only about 200 reach the secondary oocyte. Fertilization normally occurs in the uterine (fallopian) tube within 12 to 24 hours after ovulation. Sperm can remain viable for about 48 hours after deposition in the vagina, although a secondary oocyte is viable for only about 24 hours after ovulation. Thus, pregnancy is *most likely* to occur if intercourse takes place during a 3-day "window"—from 2 days before ovulation to 1 day after ovulation.

Sperm swim from the vagina into the cervical canal propelled by the whiplike movements of their tails (flagella). The

Figure 24.1 Fertilization. A sperm cell is shown penetrating the corona radiata and zona pellucida around a secondary oocyte.

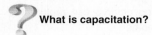 **During fertilization, genetic material from a sperm cell and a secondary oocyte merge to form a single diploid nucleus.**

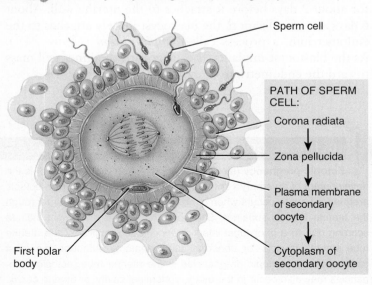

Sperm cell

PATH OF SPERM CELL:

Corona radiata
↓
Zona pellucida
↓
Plasma membrane of secondary oocyte
↓
Cytoplasm of secondary oocyte

First polar body

Sperm cell penetrating secondary oocyte

? What is capacitation?

passage of sperm through the rest of the uterus and then into the uterine tube results mainly from contractions of the walls of these organs. Prostaglandins in semen are believed to stimulate uterine motility at the time of intercourse and to aid in the movement of sperm through the uterus and into the uterine tube. Sperm that reach the vicinity of the oocyte within minutes after ejaculation *are not capable* of fertilizing it until about seven hours later. During this time in the female reproductive tract, mostly in the uterine tube, sperm undergo ***capacitation*** (ka-pas'-i-TĀ-shun; *capacit-* = capable of), a series of functional changes that cause the sperm's tail to beat even more vigorously and prepare its plasma membrane to fuse with the oocyte's plasma membrane.

For fertilization to occur, a sperm cell first must penetrate the ***corona radiata*** (kō-RŌ-na = crown; rā-dē-A-ta = to shine), the cells that surround the secondary oocyte, and then the ***zona pellucida*** (ZŌ-na = zone; pe-LOO-si-da = allowing passage of light), the clear glycoprotein layer between the corona radiata and the oocyte's plasma membrane (Figure 24.1). One of the glycoproteins in the zona pellucida acts as a sperm receptor. Its binding to specific membrane proteins in the sperm head triggers the release of enzymes from the acrosome. The acrosomal enzymes digest a path through the zona pellucida as the lashing sperm tail pushes the sperm cell onward. Although many sperm bind to the zona pellucida and release their enzymes, only the first sperm cell to penetrate the entire zona pellucida and reach the oocyte's plasma membrane fuses with the oocyte. The fusion of a sperm with a secondary oocyte sets in motion events that block fertilization by more than one sperm cell.

Once a sperm cell enters a secondary oocyte, the oocyte first must complete meiosis II. It divides into a larger ovum (mature egg) and a smaller second polar body that fragments and disintegrates (see Figure 23.8). The nucleus in the head of the sperm and the nucleus of the fertilized ovum fuse, producing a single diploid nucleus that contains 23 chromosomes from each cell. Thus, the fusion of the haploid (*n*) cells restores the diploid number (*2n*) of 46 chromosomes. The fertilized ovum now is called a ***zygote*** (ZĪ-gōt; *zygon* = yolk).

CLINICAL CONNECTION | Dizygotic and Monozygotic Twins

Dizygotic (fraternal) twins are produced from the independent release of two secondary oocytes and the subsequent fertilization of each by different sperm. They are the same age and in the uterus at the same time, but they are genetically as dissimilar as are any other siblings. Dizygotic twins may or may not be the same sex. Because **monozygotic (identical) twins** develop from a single fertilized ovum, they contain exactly the same genetic material and are always the same sex. Monozygotic twins arise from separation of the developing zygote into two embryos, which occurs within 8 days after fertilization 99 percent of the time. Separations that occur later than 8 days are likely to produce **conjoined twins**, a situation in which the twins are joined together and share some body structures. •

Early Embryonic Development

After fertilization, rapid mitotic cell divisions of the zygote called *cleavage* (KLĒV-ij) take place (Figure 24.2). The first division of the zygote begins about 24 hours after fertilization and is completed about 6 hours later. Each succeeding division takes slightly less time. By the second day after fertilization, the second cleavage is completed and there are four cells (Figure 24.2b). By the end of the third day, there are 16 cells. The progressively smaller cells produced by cleavage are called ***blastomeres*** (BLAS-tō-mērz; *blasto-* = germ or sprout; *-meres* = parts). Successive cleavages eventually produce a solid sphere of cells called the ***morula*** (MOR-ū-la = mulberry). The morula is still surrounded by the zona pellucida and is about the same size as the original zygote (Figure 24.2c).

By the end of the fourth day, the number of cells in the morula increases as it continues to move through the uterine tube toward the uterine cavity. When the morula enters the uterine cavity on day 4 or 5, a glycogen-rich secretion from the glands of the endometrium of the uterus penetrates the morula, collects between the blastomeres, and reorganizes them around a large fluid-filled cavity called the ***blastocyst cavity*** (BLAS-tō-sist; *blasto-* = germ or sprout; *-cyst* = bag) (Figure 24.2e). With the formation of this cavity, the developing mass is then called the ***blastocyst***. Though it now has hundreds of cells, the blastocyst is still about the same size as the original zygote. Further rearrangement of the blastomeres results in the formation of two distinct structures: the inner cell mass and trophoblast (Figure 24.2e). The ***inner cell mass*** is located internally and eventually develops into the embryo. The ***trophoblast*** (TRŌF-ō-blast; *tropho-* = develop or nourish) is an outer superficial layer of cells that forms the wall of the blastocyst. It will ultimately develop into the fetal portion of the placenta, the site of exchange of nutrients and wastes between the mother and fetus.

The blastocyst remains free within the uterine cavity for about 2 days before it attaches to the uterine wall. About 6 days after fertilization, the blastocyst loosely attaches to the endometrium, a process called ***implantation*** (Figure 24.3). As the blastocyst implants, it orients with the inner cell mass toward the endometrium (Figure 24.3).

Figure 24.2 Cleavage and the formation of the morula and blastocyst.

Cleavage refers to the early, rapid mitotic divisions of a zygote.

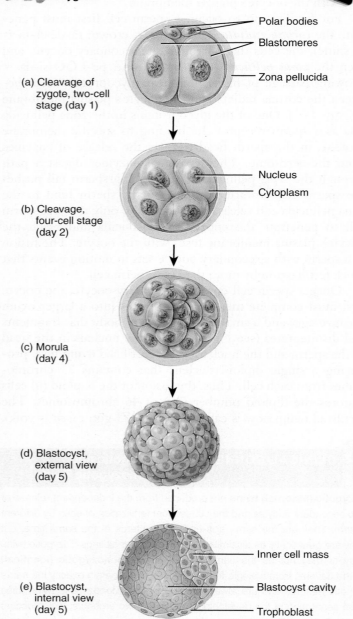

(a) Cleavage of zygote, two-cell stage (day 1)
- Polar bodies
- Blastomeres
- Zona pellucida

(b) Cleavage, four-cell stage (day 2)
- Nucleus
- Cytoplasm

(c) Morula (day 4)

(d) Blastocyst, external view (day 5)

(e) Blastocyst, internal view (day 5)
- Inner cell mass
- Blastocyst cavity
- Trophoblast

? **What is the histological difference between a morula and a blastocyst?**

CLINICAL CONNECTION | Ectopic Pregnancy

Ectopic pregnancy (ek-TOP-ik; *ec-* = out of; *-topic* = place) is the development of an embryo or fetus outside the uterine cavity. An ectopic pregnancy usually occurs when movement of the fertilized ovum through the uterine tube is impaired. Situations that impair movement include scarring due to a prior tubal infection, decreased motility of the uterine tube smooth muscle, or abnormal tubal anatomy. Although the most common site of ectopic pregnancies is the uterine tube, ectopic pregnancies may also occur in the ovary, abdominal cavity, or uterine cervix. The signs and symptoms of ectopic pregnancy include one or two missed menstrual cycles followed by bleeding and acute abdominal and pelvic pain. Unless removed, the developing embryo can rupture the uterine tube, often resulting in death of the mother. •

Stem cells are unspecialized cells (cells without a particular function) that have the ability to divide for long periods and develop into specialized cells. Based on their potential, stem cells are classified into three types:

1. *Totipotent stem cells* (tō-TIP-ō-tent; *totus-* = whole; *-potentia* = power) have the potential to form all cells of an entire organism. An example is a zygote (fertilized ovum).

2. *Pluripotent stem cells* (ploo-RIP-ō-tent; *plur-* = several) have the potential to develop into many (but not all) different types of cells of an organism. Examples are inner cell mass cells.

3. *Multipotent stem cells* (mul-TIP-ō-tent) have the potential to develop into a few different types of cells of an organism. Examples are myeloid and lymphoid stem cells that develop into blood cells.

Pluripotent stem cells currently used in research are derived from (1) extra embryos that were destined to be used for infertility treatments but were not needed, and (2) nonliving fetuses terminated during the first trimester of pregnancy. Because pluripotent stem cells give rise to almost all cell types in the body, they are extremely important in research and health care. For example, they might be used to generate cells and tissues for transplantation to treat conditions such as cancer, Parkinson's and Alzheimer's disease, spinal cord injury, diabetes, heart disease, stroke, burns, birth defects, osteoarthritis, and rheumatoid arthritis.

On October 13, 2001, researchers reported cloning of the first human embryo to grow cells to treat human diseases. **Therapeutic cloning** is envisioned as a procedure in which the genetic material of a patient with a particular disease is used to create pluripotent stem cells to treat the disease. Using the principles of therapeutic cloning, scientists hope to make an embryo clone of a patient, remove the pluripotent stem cells from the embryo, and then use them to grow tissues to treat particular diseases and disorders.

Scientists are also investigating the potential clinical applications of using *adult stem cells*, stem cells that remain in the body throughout adulthood. Studies have suggested that stem cells in human adult red bone marrow have the ability to differentiate into cells of the liver, kidney, heart, lung, skeletal muscle, skin, and organs of the gastrointestinal tract. In theory, adult stem cells from red bone marrow could be harvested from a patient and then used to repair other tissues and organs in that patient's body without having to use stem cells from embryos. •

Figure 24.3 Relationship of a blastocyst to the endometrium of the uterus at the time of implantation.

🔑 Implantation, the attachment of a blastocyst to the endometrium, occurs about 6 days after fertilization.

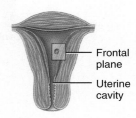

Frontal plane

Uterine cavity

Frontal section through uterus

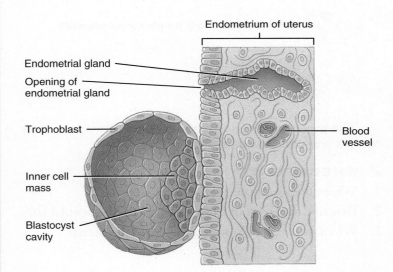

Endometrium of uterus

Endometrial gland

Opening of endometrial gland

Trophoblast

Inner cell mass

Blastocyst cavity

Blood vessel

Frontal section through endometrium of uterus and blastocyst, about 6 days after fertilization

❓ **How does the blastocyst merge with and burrow into the endometrium?**

Figure 24.4 Summary of events associated with the first week of development.

Fertilization usually occurs in the uterine tube.

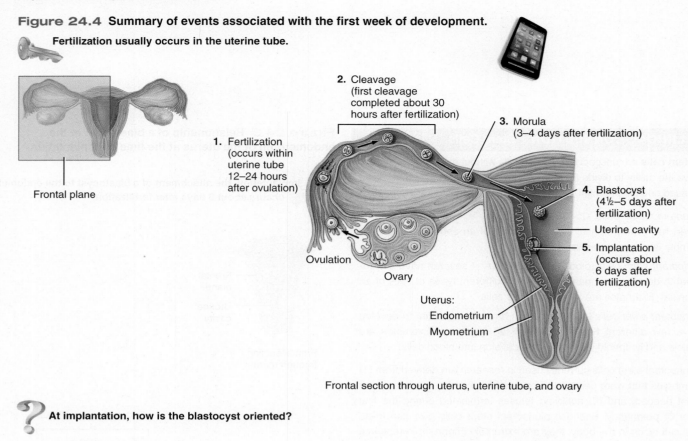

Frontal section through uterus, uterine tube, and ovary

At implantation, how is the blastocyst oriented?

The major events associated with the first week of development are summarized in Figure 24.4.

✓ **CHECKPOINT**

1. Where does fertilization normally occur?
2. Describe the layers of a blastocyst and their eventual fates.
3. When, where, and how does implantation occur?

Second Week of Development

About 8 days after fertilization, the trophoblast develops into two layers: a *syncytiotrophoblast* (sin-sīt′-ē-ō-TRŌF-ō-blast) and a *cytotrophoblast* (sī-tō-TRŌF-ō-blast) (Figure 24.5a). The two layers of trophoblast become part of the chorion (one of the fetal membranes) as they undergo further growth (see Figure 24.8 inset). During implantation, the syncytiotrophoblast secretes enzymes that enable the blastocyst to penetrate the uterine lining. Another secretion of the trophoblast is human chorionic gonadotropin (hCG), a hormone that sustains secretion of progesterone and estrogens by the corpus luteum. These hormones maintain the uterine lining in a secretory state and thereby prevent menstruation. About the

ninth week of pregnancy the placenta is fully developed and produces the progesterone and estrogens that continue to sustain the pregnancy.

Cells of the inner cell mass also differentiate into two layers around 8 days after fertilization: a *hypoblast (primitive endoderm)* and *epiblast (primitive ectoderm)* (Figure 24.5b). Cells of the hypoblast and epiblast together form a flat disc referred to as the *bilaminar embryonic disc* (bī-LAM-in-ar = two-layered). In addition, a small cavity appears within the epiblast and eventually enlarges to form the *amniotic cavity* (am-nē-OT-ik; *amnio-* = lamb).

As the amniotic cavity enlarges, a thin protective membrane called the *amnion* (AM-nē-on) develops from the epiblast (Figure 24.5a). With growth of the embryo, the amnion eventually surrounds the entire embryo (see Figure 24.8 inset), creating the amniotic cavity that becomes filled with *amniotic fluid*. Amniotic fluid serves as a shock absorber for the fetus, helps regulate fetal body temperature, helps prevent drying out, and prevents adhesions between the skin of the fetus and surrounding tissues. Because embryonic cells are normally sloughed off into amniotic fluid, they can be examined in a procedure called *amniocentesis* (see Medical Terminology section).

Figure 24.5 Principal events of the second week of development.

About 8 days after fertilization, the trophoblast develops into a syncytiotrophoblast and a cytotrophoblast; the inner cell mass develops into a hypoblast and epiblast (bilaminar embryonic disc).

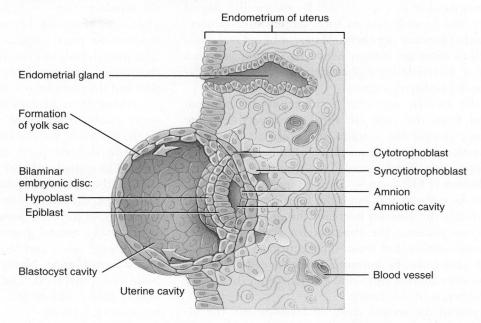

(a) Frontal section through endometrium of uterus showing blastocyst, about 8 days after fertilization

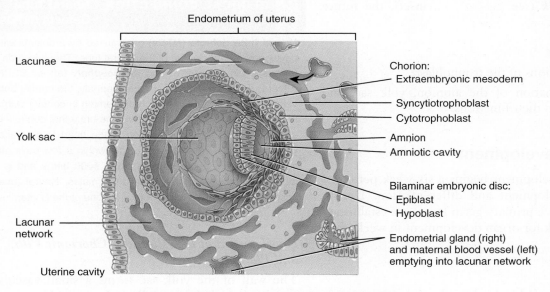

(b) Frontal section through endometrium of uterus showing blastocyst, about 12 days after fertilization

How is the bilaminar embryonic disc connected to the trophoblast?

Also on the eighth day after fertilization, cells of the hypoblast migrate and cover the inner surface of the blastocyst wall (Figure 24.5a), forming the wall of the *yolk sac*, formerly called the blastocyst cavity (Figure 24.5b). The yolk sac has several important functions in humans. It supplies nutrients to the embryo during the second and third weeks of development, is the source of blood cells from the third through sixth weeks, contains the first cells (primordial germ cells) that will eventually migrate into the developing gonads, and forms part of the gut (gastrointestinal tract). Finally, the yolk sac functions as a shock absorber and helps prevent drying out of the embryo.

On the ninth day after fertilization, the blastocyst becomes completely embedded in the endometrium and small spaces called *lacunae* (la-KOO-nē = little lakes) develop within the trophoblast (Figure 24.5b). By the twelfth day of development, the lacunae fuse to form larger, interconnecting spaces called *lacunar networks*. Maternal blood and glandular secretions enter the lacunar networks, serving as both a rich source of materials for embryonic nutrition and a disposal site for the embryo's wastes.

About the twelfth day after fertilization, mesodermal cells derived from the yolk sac form a connective tissue (mesenchyme) around the amnion and yolk sac called the *extraembryonic mesoderm* (Figure 24.5b). The extraembryonic mesoderm and the two layers of the trophoblast together form the *chorion* (KOR-ē-on = membrane) (Figure 24.5b). It surrounds the embryo and, later, the fetus (see Figure 24.8 inset). Eventually the chorion becomes the principal embryonic part of the placenta, the structure for exchange of materials between mother and fetus. The chorion protects the embryo and fetus from the immune responses of the mother and also produces human chorionic gonadotropin (hCG), an important hormone of pregnancy.

By the end of the second week of development, the bilaminar embryonic disc becomes connected to the trophoblast by a band of extraembryonic mesoderm called the *connecting (body) stalk* (see Figure 24.6 inset), the future umbilical cord.

✓ CHECKPOINT

4. What are the functions of the trophoblast?

5. Describe the formation of the amnion, yolk sac, and chorion and explain their functions.

Third Week of Development

The third week of development begins a six-week period of rapid embryonic development and differentiation. During the third week, the three primary germ layers are established and lay the groundwork for organ development in weeks four through eight.

Gastrulation

The first major event of the third week of development is called *gastrulation* (gas'-troo-LĀ-shun) (Figure 24.6). In this process, the bilaminar (two-layered) embryonic disc transforms into a trilaminar (three-layered) embryonic disc consisting of three primary germ layers, the ectoderm, mesoderm, and endoderm. The *primary germ layers* are the major embryonic tissues from which the various tissues and organs of the body develop.

As part of gastrulation, cells of the epiblast move inward and detach from it (Figure 24.6b). Some of the cells push out other cells of the hypoblast, forming the *endoderm* (endo- =

inside; -derm = skin). Other cells remain between the epiblast and newly formed endoderm to form the *mesoderm* (meso- = middle). Cells remaining in the epiblast then form the *ectoderm* (ecto- = outside). As the embryo develops, the endoderm ultimately becomes the epithelial lining of the gastrointestinal tract, respiratory tract, and several other organs. The mesoderm gives rise to muscle, bone, and other connective tissues. The ectoderm develops into the epidermis of the skin and the nervous system.

About 22 to 24 days after fertilization, mesodermal cells form a solid cylinder of cells called the *notochord* (nō-tō-KORD; noto- = back; -chord = cord). It stimulates mesodermal cells to form parts of the backbone and intervertebral discs. The notochord also stimulates ectodermal cells over it to form the *neural plate* (see Figure 24.9a). By the end of the third week, the lateral edges of the neural plate become more elevated and form the *neural fold*. The depressed midregion is called the *neural groove*. Generally, the neural folds approach each other and fuse, thus converting the neural plate into a *neural tube*. Neural tube cells then develop into the brain and spinal cord. The process by which the neural plate, neural folds, and neural tube form is called *neurulation* (noor-oo-LĀ-shun).

CLINICAL CONNECTION | Neural Tube Defects (NTDs)

Neural tube defects (NTDs) are caused by problems with the normal development and closure of the neural tube. These include *spina bifida* (discussed in Chapter 6) and *anencephaly* (an'-en-SEPH-a-lē; an- = without; encephal = brain). In anencephaly, the cranial bones fail to develop and certain parts of the brain remain in contact with amniotic fluid and degenerate. Usually, the part of the brain that controls vital functions such as breathing and regulation of the heart is also affected. Infants with anencephaly are stillborn or die within a few days after birth. The condition occurs about once in every 1000 births and is 2 to 4 times more common in female infants than males. Neural tube defects are associated with low levels of folic acid, one of the B vitamins. •

Development of the Allantois, Chorionic Villi, and Placenta

The wall of the yolk sac forms a small vascularized outpouching called the *allantois* (a-LAN-tō-is; allant- = sausage) (see Figure 24.8 inset). In most other mammals, the allantois is used for gas exchange and waste removal. Because of the role of the human placenta in these activities, the allantois is not a prominent structure in humans. Nevertheless, it does function in early formation of blood and blood vessels and it is associated with the development of the urinary bladder.

By the end of the second week of development, *chorionic villi* (ko-rē-ON-ik VIL-ī) begin to develop. These fingerlike projections consist of chorion (syncytiotrophoblast surrounded

Figure 24.6 Gastrulation.

Gastrulation involves the rearrangement and migration of cells from the epiblast.

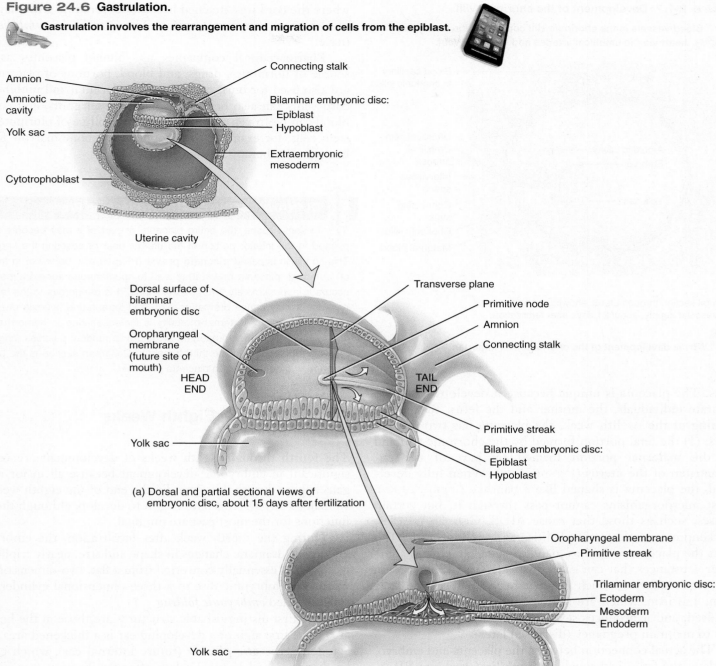

(a) Dorsal and partial sectional views of embryonic disc, about 15 days after fertilization

(b) Transverse section of trilaminar embryonic disc, about 16 days after fertilization

? **What is the significance of gastrulation?**

by cytotrophoblast) and contain fetal blood vessels (Figure 24.7). By the end of the third week, blood capillaries that develop in the chorionic villi connect to the embryonic heart by way of the umbilical arteries and umbilical vein. As a result, maternal and fetal blood vessels are in close proximity. Note, however, that maternal and fetal blood vessels do not join, and the blood they carry *does not normally mix*. Instead, oxygen and nutrients in the mother's blood diffuse across the cell membranes into the capillaries of the chorionic villi. Waste products such as carbon dioxide diffuse in the opposite direction.

The *placenta* (pla-SEN-ta = flat cake) is the site of the exchange of nutrients and wastes between the mother and

Figure 24.7 Development of the chorionic villi.

 Blood vessels in the chorionic villi connect to the embryonic heart via the umbilical arteries and umbilical vein.

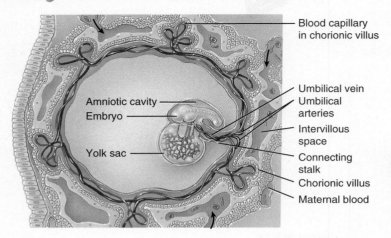

Blood capillary in chorionic villus

Amniotic cavity
Embryo
Yolk sac

Umbilical vein
Umbilical arteries
Intervillous space
Connecting stalk
Chorionic villus
Maternal blood

Frontal section through uterus showing an embryo and its vascular supply, about 21 days after fertilization

? Why is development of the chorionic villi important?

fetus. The placenta is unique because it develops from two separate individuals, the mother and the fetus. By the beginning of the twelfth week, the placenta has two distinct parts: (1) the fetal portion formed by the chorionic villi and (2) the maternal portion formed by part of the endometrium of the uterus (Figure 24.8a). When fully developed, the placenta is shaped like a pancake (Figure 24.8b). Most microorganisms cannot pass through it, but certain viruses, such as those that cause AIDS, German measles, chickenpox, measles, encephalitis, and poliomyelitis, can cross the placenta as well as many drugs, alcohol, and some other substances that can cause birth defects. The placenta also stores nutrients such as carbohydrates, proteins, calcium, and iron, which are released into fetal circulation as required, and it produces several hormones that are necessary to maintain pregnancy (discussed later).

The actual connection between the placenta and embryo, and later the fetus, is through the *umbilical cord* (um-BIL-i-kul = navel), which develops from the connecting stalk. The umbilical cord consists of two umbilical arteries that carry deoxygenated fetal blood to the placenta, one umbilical vein that carries oxygenated maternal blood into the fetus, and supporting mucous connective tissue. A layer of amnion surrounds the entire umbilical cord and gives it a shiny appearance (Figure 24.8a).

After the birth of the baby, the placenta detaches from the uterus and is therefore termed the *afterbirth*. At this time, the umbilical cord is tied off and then severed, leaving the baby on its own. The small portion (about an inch) of the cord that remains attached to the infant begins to wither and falls off, usually within 12 to 15 days after birth. The area

where the cord was attached becomes covered by a thin layer of skin, and scar tissue forms. The scar is the *umbilicus* (navel).

Pharmaceutical companies use human placentas as a source of hormones, drugs, and blood; portions of placentas are also used for burn coverage. The placental and umbilical cord veins can also be used in blood vessel grafts, and cord blood can be frozen to provide a future source of pluripotent stem cells, for example, to repopulate red bone marrow following radiotherapy for cancer.

CLINICAL CONNECTION | Placenta Previa

In some cases, the entire placenta or part of it may become implanted in the inferior portion of the uterus, near or covering the cervix. This condition is called **placenta previa** (PRĒ-vē-a = before or in front of). Although placenta previa may lead to spontaneous abortion, it also occurs in approximately 1 in 250 live births. It is dangerous to the fetus because it may cause premature birth and intrauterine hypoxia due to maternal bleeding. Maternal mortality is increased due to hemorrhage and infection. The most important symptom is sudden, painless, bright-red vaginal bleeding in the third trimester. Cesarean section is the preferred method of delivery in placenta previa. •

Fourth Through Eighth Weeks of Development

The fourth through eighth weeks of development are very significant in embryonic development because all major organs appear during this time. By the end of the eighth week, all major body systems have begun to develop, although their functions for the most part are minimal.

During the fourth week after fertilization, the embryo undergoes dramatic changes in shape and size, nearly tripling its size. It is essentially converted from a flat, two-dimensional trilaminar embryonic disc to a three-dimensional cylinder, a process called *embryonic folding*.

The first distinguishable structures are those in the head area. The first sign of a developing ear is a thickened area of ectoderm, the *otic placode* (future internal ear), which can be distinguished about 22 days after fertilization (see Figure 24.9d). The eyes also begin their development about 22 days after fertilization. This is evidenced by a thickened area of ectoderm called the *lens placode* (see Figure 24.9c).

By the middle of the fourth week, the upper limbs begin their development as outgrowths of mesoderm covered by ectoderm called *upper limb buds* (see Figure 24.9c, d). By the end of the fourth week, the *lower limb buds* develop. The heart also forms a distinct projection on the ventral surface of the embryo called the *heart prominence* (see Figure 24.9c). A *tail* is also a distinguishing feature of an embryo at the end of the fourth week (see Figure 24.9c).

During the fifth week of development, there is very rapid development of the brain, so growth of the head is consider-

Figure 24.8 Placenta and umbilical cord.

The placenta is formed by the chorionic villi of the embryo and part of the endometrium of the mother.

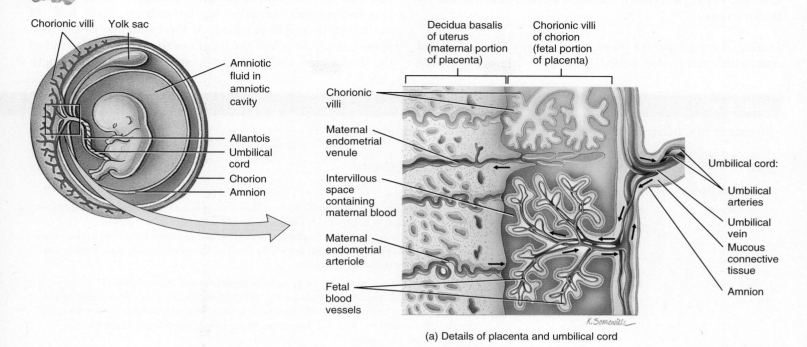

(a) Details of placenta and umbilical cord

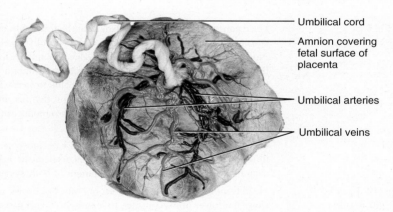

(b) Fetal surface of placenta

What is the function of the placenta?

able. By the end of the sixth week, the head grows even larger relative to the trunk, and the limbs show substantial development. In addition, the neck and trunk begin to straighten, and the heart is now four-chambered. By the seventh week, the various regions of the limbs become distinct and the beginnings of digits appear (see Figure 24.9e). At the start of the eighth week, the final week of the embryonic period, the digits of the hands are short and webbed and the tail is still visible, but shorter. In addition, the eyes are open and the auricles of the ears are visible. By the end of the eighth week, all regions of the limbs are apparent and the digits are distinct and no longer webbed. Also, the eyelids come together and may fuse, the tail disappears, and the external genitals

begin to differentiate. The embryo now has clearly human characteristics.

✓ **CHECKPOINT**

6. How do the three primary germ layers form? Why are they important?

7. Describe how neurulation occurs. Why is it significant?

8. How does the placenta form and what is its function?

9. Why are the second through fourth weeks of development so crucial?

10. What changes occur in the limbs during the second half of the embryonic period?

24.2 FETAL PERIOD

OBJECTIVE • **Define the fetal period and outline its major events.**

During the fetal period, tissues and organs that developed during the embryonic period grow and differentiate. Very few new structures appear during the fetal period, but the rate of body growth is remarkable, especially during the second half of intrauterine life. For example, during the last two-and-one-half months of intrauterine life, half of the full-term weight is added. At the beginning of the fetal period, the head is half the length of the body. By the end of the fetal period, the head size is only one-quarter the length

TABLE 24.1

Summary of Changes During Embryonic and Fetal Development

TIME	APPROXIMATE SIZE AND WEIGHT	REPRESENTATIVE CHANGES
Embryonic Period		
1–4 weeks	0.6 cm (3/16 in.)	Primary germ layers and notochord develop. Neurulation occurs. Brain development begins. Blood vessel formation begins and blood forms in yolk sac, allantois, and chorion. Heart forms and begins to beat. Chorionic villi develop and placental formation begins. The embryo folds. The primitive gut and limb buds develop. Eyes and ears begin to develop, tail forms, and body systems begin to form.
5–8 weeks	3 cm (1.25 in.) 1 g (1/30 oz)	Brain development continues. Limbs become more distinct and digits appear. Heart becomes four-chambered. Eyes are far apart and eyelids are fused. Nose develops and is flat. Face is more humanlike. Ossification begins. Blood cells start to form in liver. External genitals begin to differentiate. Tail disappears. Major blood vessels form. Many internal organs continue to develop.
Fetal Period		
9–12 weeks	7.5 cm (3 in.) 30 g (1 oz)	Head constitutes about half the length of the fetal body, and fetal length nearly doubles. Brain continues to enlarge. Face is broad, with eyes fully developed, closed, and widely separated. Nose develops a bridge. External ears develop and are low set. Ossification continues. Upper limbs almost reach final relative length but lower limbs are not quite as well developed. Heartbeat can be detected. Gender is distinguishable from external genitals. Urine secreted by fetus is added to amniotic fluid. Red bone marrow, thymus, and spleen participate in blood cell formation. Fetus begins to move, but its movements cannot be felt yet by the mother. Body systems continue to develop.
13–16 weeks	18 cm (6.5–7 in.) 100 g (4 oz)	Head is relatively smaller than rest of body. Eyes move medially to their final positions, and ears move to their final positions on the sides of the head. Lower limbs lengthen. Fetus appears even more humanlike. Rapid development of body systems occurs.
17–20 weeks	25–30 cm (10–12 in.) 200–450 g (0.5–1 lb)	Head is more proportionate to rest of body. Eyebrows and head hair are visible. Growth slows but lower limbs continue to lengthen. Vernix caseosa (fatty secretions of sebaceous glands and dead epithelial cells) and lanugo (delicate fetal hair) cover fetus. Brown fat forms and is the site of heat production. Fetal movements are commonly felt by mother (quickening).
21–25 weeks	27–35 cm (11–14 in.) 550–800 g (1.25–1.5 lb)	Head becomes even more proportionate to rest of body. Weight gain is substantial, and skin is pink and wrinkled. By 24 weeks, lung cells begin to produce surfactant.
26–29 weeks	32–42 cm (13–17 in.) 1110–1350 g (2.5–3 lb)	Head and body are more proportionate and eyes are open. Toenails are visible. Body fat is 3.5% of total body mass and additional subcutaneous fat smoothes out some wrinkles. Testes begin to descend toward scrotum at 28 to 32 weeks. Red bone marrow is major site of blood cell production. Many fetuses born prematurely during this period survive if given intensive care because lungs can provide adequate ventilation and central nervous system is developed sufficiently to control breathing and body temperature.
30–34 weeks	41–45 cm (16.5–18 in.) 2000–2300 g (4.5–5 lb)	Skin is pink and smooth. Fetus assumes upside-down position. Pupillary reflex is present by 30 weeks. Body fat is 8 percent of total body mass. Fetuses 33 weeks and older usually survive if born prematurely.
35–38 weeks	50 cm (20 in.) 3200–3400 g (7–7.5 lb)	By 38 weeks, circumference of fetal abdomen is greater than that of head. Skin is usually bluish-pink, and growth slows as birth approaches. Body fat is 16% of total body mass. Testes are usually in scrotum in full-term male infants. Even after birth, an infant is not completely developed; an additional year is required, especially for complete development of the nervous system.

of the body. During the same period, the fetal limbs also increase in size from one-eighth to one-half the fetal length. The fetus is also less vulnerable to the damaging effects of drugs, radiation, and microbes than it was as an embryo.

A summary of the major developmental events of the embryonic and fetal periods is presented in Table 24.1 and illustrated in Figure 24.9.

✓ **CHECKPOINT**

11. What are the general developmental trends during the fetal period?

12. Using Table 24.1 as a guide, select any one body structure in weeks 9 through 12 and trace its development through the remainder of the fetal period.

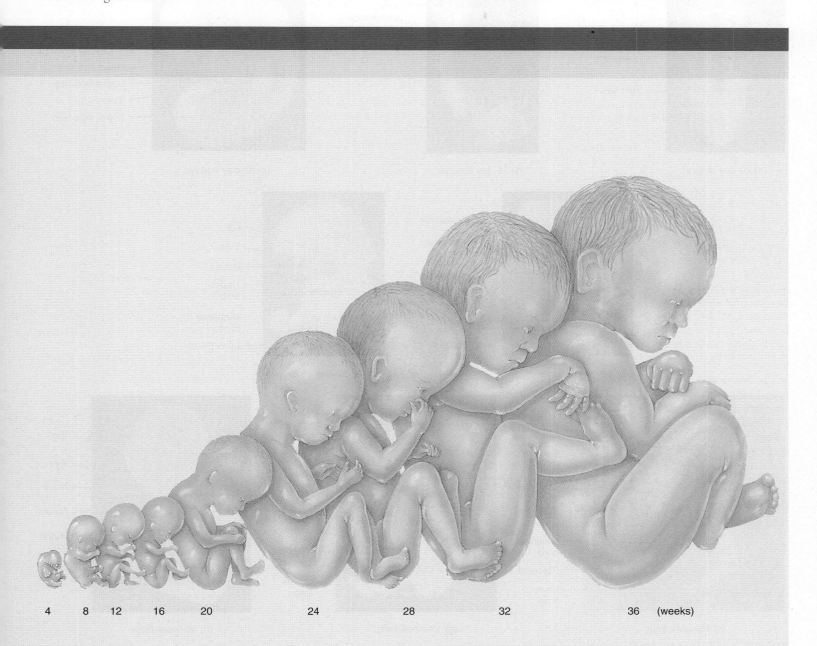

4   8   12   16   20   24   28   32   36   (weeks)

Figure 24.9 Summary of representative developmental events of the embryonic and fetal periods.
The embryos and fetuses are not shown at their actual sizes.

Development during the fetal period is mostly concerned with the growth and differentiation of tissues and organs formed during the embryonic period.

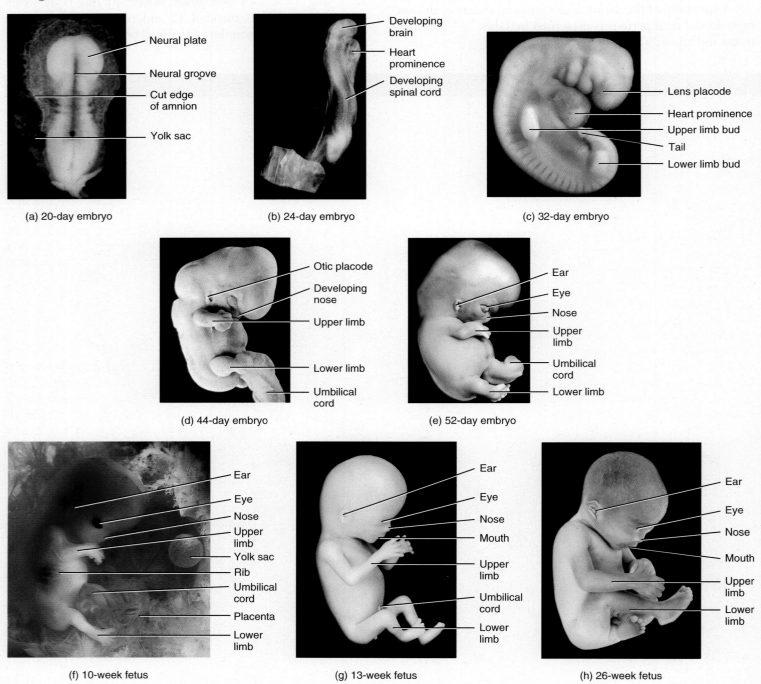

(a) 20-day embryo

(b) 24-day embryo

(c) 32-day embryo

(d) 44-day embryo

(e) 52-day embryo

(f) 10-week fetus

(g) 13-week fetus

(h) 26-week fetus

How does mid-fetal weight compare to end-fetal weight?

24.3 MATERNAL CHANGES DURING PREGNANCY

OBJECTIVES • **Describe the sources and functions of the hormones secreted during pregnancy.**

• **Describe the hormonal, anatomical, and physiological changes in the mother during pregnancy.**

Hormones of Pregnancy

During the first three to four months of pregnancy, the corpus luteum in the ovary continues to secrete progesterone and estrogens, which maintain the lining of the uterus during pregnancy and prepare the mammary glands to secrete milk. The amounts secreted by the corpus luteum, however, are only slightly more than those produced after ovulation in a normal menstrual cycle. From the third month through the remainder of the pregnancy, the placenta itself provides the high levels of progesterone and estrogens required. The chorion secretes *human chorionic gonadotropin (hCG)* into the blood. In turn, hCG stimulates the corpus luteum to continue production of progesterone and estrogens—an activity required to prevent menstruation and for the continued attachment of the embryo and fetus to the lining of the uterus. By the eighth day after fertilization, hCG can be detected in the blood and urine of a pregnant woman. Peak secretion of hCG occurs at about the ninth week of pregnancy. During the fourth and fifth months the hCG level decreases sharply and then levels off until childbirth.

The chorion begins to secrete estrogens after the first three to four weeks of pregnancy and progesterone by the sixth week. These hormones are secreted in increasing quantities until the time of birth. From the third month to the ninth month, the placenta supplies the levels of progesterone and estrogens needed to maintain the pregnancy. A high level of progesterone ensures that the uterine myometrium is relaxed and that the cervix is tightly closed. After delivery, estrogens and progesterone in the blood decrease to normal levels.

Relaxin, a hormone produced first by the corpus luteum of the ovary and later by the placenta, increases the flexibility of the pubic symphysis and ligaments of the sacroiliac and sacrococcygeal joints and helps dilate the uterine cervix during labor. Both of these actions ease delivery of the baby.

A third hormone produced by the chorion of the placenta is *human placental lactogen (hPL)*. The rate of secretion of hPL increases in proportion to placental mass, reaching maximum levels after 32 weeks and remaining relatively constant after that. It is thought to help prepare the mammary glands for lactation, enhance maternal growth by increasing protein synthesis, and regulate certain aspects of metabolism in the mother and fetus.

The hormone most recently found to be produced by the placenta is *corticotropin-releasing hormone (CRH)*, which in nonpregnant people is secreted only by the hypothalamus. CRH is now thought to be part of the "clock" that establishes the timing of birth. Women who have higher levels of CRH earlier in pregnancy are more likely to deliver prematurely; those who have low levels are more likely to deliver after the due date. CRH from the placenta has a second important effect: It increases secretion of cortisol, which is needed for maturation of the fetal lungs and the production of surfactant (see Section 18.1).

CLINICAL CONNECTION | Early Pregnancy Tests

Early pregnancy tests detect the tiny amounts of human chorionic gonadotropin (hCG) in the urine that begin to be excreted about 8 days after fertilization. The test kits can detect pregnancy as early as the first day of a missed menstrual period—that is, at about 14 days after fertilization. Chemicals in the kits produce a color change if a reaction occurs between hCG in the urine and hCG antibodies included in the kit. •

Changes During Pregnancy

By about the end of the third month of pregnancy, the uterus occupies most of the pelvic cavity. As the fetus continues to grow, the uterus extends higher into the abdominal cavity. Toward the end of a full-term pregnancy, the uterus fills almost the entire abdominal cavity, reaching almost to the xiphoid process of the sternum. It pushes the maternal intestines, liver, and stomach superiorly, elevates the diaphragm, and widens the thoracic cavity.

Changes in the skin during pregnancy are more apparent in some women than in others. Included are increased pigmentation around the eyes and cheekbones in a masklike pattern, in the areolae of the breasts, and in the lower abdomen. Striae (stretch marks) over the abdomen can occur as the uterus enlarges, and hair loss increases. Pregnancy-induced physiological changes include weight gain due to the fetus, amniotic fluid, the placenta, uterine enlargement, and increased total body water; increased storage of proteins, triglycerides, and minerals; marked breast enlargement in preparation for lactation; and lower back pain due to lordosis.

Several changes occur in the maternal cardiovascular system. Stroke volume increases by about 30 percent and cardiac output rises by 20–30 percent due to increased maternal blood flow to the placenta and increased metabolism. Heart rate increases 10–15 percent and blood volume increases 30–50 percent, mostly during the second half of pregnancy. These increases are necessary to meet the additional demands of the fetus for nutrients and oxygen.

Pulmonary function is also altered during pregnancy to meet the added oxygen demands of the fetus. Tidal volume can increase by 30–40 percent, expiratory reserve volume can be reduced by up to 40 percent, minute ventilation (the total volume of air inhaled and exhaled each minute) can increase by up to 40 percent, and total body oxygen consumption can increase by about 10–20 percent. Dyspnea (difficult breathing) also occurs as the expanding uterus pushes on the diaphragm.

With regard to the gastrointestinal tract, pregnant women experience an increase in appetite. Pressure on the stomach may force the stomach contents superiorly into the esophagus, resulting in heartburn. A general decrease in GI tract motility can cause constipation, delay gastric emptying time, and produce nausea, vomiting, and heartburn. Pressure on the urinary bladder by the enlarging uterus can produce urinary symptoms, such as increased frequency and urgency of urination, and stress incontinence.

Changes in the reproductive system include edema and increased blood flow to the vagina. The uterus increases from its nonpregnant mass of 60–80 g to 900–1200 g at term because of increased numbers of muscle fibers in the myometrium in early pregnancy and enlargement of muscle fibers during the second and third trimesters.

✓ CHECKPOINT

13. List the hormones involved in pregnancy, and describe the functions of each.

14. What structural and functional changes occur in the mother during pregnancy?

24.4 EXERCISE AND PREGNANCY

OBJECTIVE • **Explain the effects of pregnancy on exercise and of exercise on pregnancy.**

Only a few changes in early pregnancy affect exercise. A pregnant woman may tire more easily than usual, or morning sickness (nausea and sometimes vomiting) may interfere with regular exercise. As the pregnancy progresses, weight is gained and posture changes, so more energy is needed to perform activities, and certain maneuvers (sudden stopping, changes in direction, rapid movements) are more difficult to execute. In addition, certain joints, especially the pubic symphysis, become less stable in response to the increased level of the hormone relaxin. As compensation, many mothers-to-be walk with widely spread legs and a shuffling motion.

Although blood shifts from viscera (including the uterus) to the muscles and skin during exercise, there is no evidence of inadequate blood flow to the placenta. The heat generated during exercise may cause dehydration and further increase body temperature. During early pregnancy especially, excessive exercise and heat buildup should be avoided because elevated body temperature has been implicated in neural tube defects. Exercise has no known effect on lactation, provided a woman remains hydrated and wears a bra that provides good support. Overall, moderate physical activity does not endanger the fetus of a healthy woman who has a normal pregnancy.

Among the benefits of exercise to the mother during pregnancy are a greater sense of well-being and fewer minor complaints.

✓ CHECKPOINT

15. How do changes during early and late pregnancy affect the ability to exercise?

24.5 LABOR AND DELIVERY

OBJECTIVE • **Explain the events associated with the three stages of labor.**

Labor is the process by which the fetus is expelled from the uterus through the vagina. *Parturition* (par′-toor-ISH-un; *parturit-* = childbirth) also means giving birth.

Progesterone inhibits uterine contractions. Toward the end of pregnancy, the levels of estrogens in the mother's blood rise sharply, producing changes that overcome the inhibiting effects of progesterone. Estrogens also stimulate the placenta to release prostaglandins. Prostaglandins induce production of enzymes that digest collagen fibers in the cervix, causing it to soften. High levels of estrogens cause uterine muscle fibers to display receptors for oxytocin, the hormone that stimulates uterine contractions. Relaxin assists by increasing the flexibility of the pubic symphysis and helping dilate the uterine cervix.

The control of labor contractions occurs via a positive feedback cycle. Uterine contractions force the baby's head or body into the uterine cervix, which stretches the cervix. This stimulates stretch receptors in the cervix to send nerve impulses to the hypothalamus, causing it to release oxytocin. Oxytocin stimulates more forceful uterine contractions, which stretches the cervix more, and promotes secretion of more oxytocin. The positive feedback system is broken with the birth of the infant, which decreases stretching of the cervix.

Uterine contractions occur in waves (quite similar to peristaltic waves) that start at the top of the uterus and move downward, eventually expelling the fetus. *True labor* begins when uterine contractions occur at regular intervals, usually producing pain. As the interval between contractions shortens, the contractions intensify. Another symptom of true labor in some women is localization of pain in the back that is intensified by walking. The reliable indicator of true labor is dilation of the cervix and the "show," a discharge of a blood-containing mucus that appears in the cervical canal during labor. In *false labor*, pain is felt in the abdomen at irregular intervals, but it does not intensify and walking does not alter it significantly. There is no "show" and no cervical dilation.

True labor can be divided into three stages:

1. **Stage of dilation.** The time from the onset of labor to the complete dilation of the cervix is the *stage of dilation*. This stage, which typically lasts 6–12 hours, features regular contractions of the uterus, usually a rupturing of the amniotic sac, and complete dilation (to 10 cm) of the cervix. If the amniotic sac does not rupture spontaneously, it is ruptured intentionally.

2. **Stage of expulsion.** The time (10 minutes to several hours) from complete cervical dilation to delivery of the baby is the *stage of expulsion*.

3. **Placental stage.** The time (5–30 minutes or more) after delivery until the placenta or "afterbirth" is expelled by powerful uterine contractions is the *placental stage*. These contractions also constrict blood vessels that were torn during delivery, thereby reducing the likelihood of hemorrhage.

As a rule, labor lasts longer with first babies, typically about 14 hours. For women who have previously given birth, the average duration of labor is about 8 hours—although the time varies enormously among births.

Delivery of a physiologically immature baby carries certain risks. A *premature infant* or "preemie" is generally considered a baby who weighs less than 2500 g (5.5 lb) at birth. Poor prenatal care, drug abuse, history of a previous premature delivery, and mother's age below 16 or above 35 increase the chance of premature delivery. The body of a premature infant is not yet ready to sustain some critical functions, and thus its survival is uncertain without medical intervention. The major problem after delivery of an infant under 36 weeks of gestation is respiratory distress syndrome (RDS) of the newborn due to insufficient surfactant. RDS can be eased by use of artificial surfactant and a ventilator that delivers oxygen until the lungs can operate on their own.

About 7 percent of pregnant women do not deliver by two weeks after their due date. Such infants are called *post-term babies* or *post-date babies*. They carry an increased risk of brain damage to the fetus, and even fetal death, due to inadequate supplies of oxygen and nutrients from an aging placenta. Post-term deliveries may be facilitated by inducing labor, initiated by administration of oxytocin (Pitocin®), or by surgical delivery (cesarean section).

Following the delivery of the baby and placenta is a six-week period during which the maternal reproductive organs and physiology return to the prepregnancy state. This period is called the *puerperium* (pū'-er-PER-ē-um).

CLINICAL CONNECTION | Dystocia and Cesarean Section

Dystocia (dis-TŌ-sē-a; *dys-* = painful or difficult; *toc-* = birth), or difficult labor, may result either from an abnormal position (presentation) of the fetus or a birth canal of inadequate size to permit vaginal delivery. In a **breech presentation**, for example, the fetal buttocks or lower limbs, rather than the head, enter the birth canal first; this occurs most often in premature births. If fetal or maternal distress prevents a vaginal birth, the baby may be delivered surgically through an abdominal incision. A low, horizontal cut is made through the abdominal wall and lower portion of the uterus, through which the baby and placenta are removed. Even though it is popularly associated with the birth of Julius Caesar, the true reason this procedure is termed a **cesarean section (C-section)** is because it was described in Roman Law, *lex cesarea*, about 600 years before Julius Caesar was born. Even a history of multiple C-sections need not exclude a pregnant woman from attempting a vaginal delivery. •

✓ CHECKPOINT

16. What hormonal changes induce labor?

17. What happens during the stage of dilation, the stage of expulsion, and the placental stage of true labor?

24.6 LACTATION

OBJECTIVE • Discuss the hormonal control of lactation.

Lactation (lak'-TĀ-shun; *lact-* = milk) is the production and ejection of milk from the mammary glands. A principal hormone in promoting milk production is *prolactin (PRL)*, which is secreted from the anterior pituitary gland. Even though prolactin levels increase as the pregnancy progresses, no milk production occurs because progesterone inhibits the effects of prolactin. After delivery, the levels of progesterone and estrogens in the mother's blood decrease, and the inhibition is removed. The principal stimulus in maintaining prolactin production during lactation is the sucking action of the infant. Suckling initiates nerve impulses from stretch receptors in the nipples to the hypothalamus, and more prolactin is released by the anterior pituitary.

Oxytocin causes the release of milk into the mammary ducts. Milk formed by the glandular cells of the breasts is stored until the baby begins active suckling. Stimulation of touch receptors in the nipple initiates sensory nerve impulses that are relayed to the hypothalamus. In response, secretion of oxytocin from the posterior pituitary increases.

Breast Milk—Mother Nature's Approach to Infection Prevention

Physicians in both industrialized and developing countries have long observed that breast-fed babies contract fewer infections than do babies who are fed formula. This difference is due, in part, to a number of ingredients in breast milk that enhance an infant's ability to fight disease, including antibodies and other immune cells.

Keeping the Tract Intact When Under Attack

Several substances in breast milk enhance immunity in the baby's gastrointestinal (GI) tract. One family of these is the secretory immunoglobulin A (IgA) antibodies. When the baby's mother encounters pathogens, she manufactures antibodies specific to each one. The antibodies pass into her breast milk and escape breakdown in the baby's GI tract because they are protected by the so-called secretory component. Once in the baby's GI tract, the antibodies bind with the targeted infectious agents and prevent them from passing through the lining of the GI tract. This protection is especially important in the earliest days of life, because the infant does not begin to make his or her own secretory IgA until several weeks or months after birth.

The secretory IgA antibodies disable pathogens without harming helpful GI tract flora or causing inflammation. This is important because, although inflammation helps fight infection, sometimes the process overwhelms the GI tract. An infant may suffer more from the inflammatory process than the infection itself when inflammation destroys healthy tissue.

The large quantities of the immune system molecule interleukin-10 found in breast milk also help inhibit inflammation. And a substance called fibronectin enhances the phagocytic activity of macrophages, inhibits inflammation, and helps repair tissues damaged by inflammation.

Several other molecules in breast milk help disable harmful microbes. Mucins, certain oligosaccharides (sugar chains), and glycoproteins (carbohydrate–protein compounds) bind to microbes and prevent them from gaining a foothold on the lining of the GI tract. Many of breast milk's immune cells, including T lymphocytes and macrophages, attack invading microbes directly.

Breast milk compounds help in other ways as well. Some decrease the supply of nutrients such as iron and vitamin B_{12} needed by harmful bacteria to survive. A substance called bifidus factor promotes the growth of helpful gut flora, which help crowd out pathogens. Retinoic acids, a group of vitamin A precursors, reduce the ability of viruses to replicate. And some of the hormones and growth factors present in breast milk stimulate the baby's GI tract to mature more quickly, making it less vulnerable to dangerous invaders.

Think It Over . . . **Why do you think that preventing infection through breast feeding is preferable to giving babies antibiotics?**

Oxytocin stimulates contraction of smooth-muscle-like cells surrounding the glandular cells and ducts. The resulting compression moves the milk from the alveoli of the mammary glands into the mammary ducts, where it can be suckled.

During late pregnancy and the first few days after birth, the mammary glands secrete a cloudy fluid called *colostrum* (kō-LOS-trum). Although it is not as nutritious as milk—it contains less lactose and virtually no fat—colostrum serves adequately until the appearance of true milk on about the fourth day. Colostrum and maternal milk contain important antibodies that protect the infant during the first few months of life.

Lactation often blocks ovarian cycles for the first few months following delivery, if the frequency of sucking is about 8–10 times a day. This effect is inconsistent, however, and ovulation commonly precedes the first menstrual period after delivery of a baby. As a result, the mother can never be certain she is not fertile. Breast feeding is therefore not a very reliable birth control measure.

A primary benefit of breast feeding is nutritional: Human milk is a sterile solution that contains amounts of fatty acids, lactose, amino acids, minerals, vitamins, and water that are ideal for the baby's digestion, brain development, and growth. Breast feeding also benefits infants in other ways, as indicated in the Focus on Wellness feature.

Years before oxytocin was discovered, it was common practice in midwifery to let a firstborn twin nurse at the mother's breast to speed the birth of the second child. Now we know why this practice is helpful—it stimulates the release of oxytocin. Even after a single birth, nursing promotes expulsion of the placenta (afterbirth) and helps the uterus return to its normal size. Synthetic oxytocin (Pitocin®) is often given to induce labor or to increase uterine tone and control hemorrhage just after parturition.

✓ CHECKPOINT

18. Which hormones contribute to lactation? What is the function of each?

24.7 INHERITANCE

OBJECTIVE • **Define inheritance, and explain the inheritance of dominant, recessive, and sex-linked traits.**

As previously indicated, the genetic material of a father and a mother unite when a sperm cell fuses with a secondary oocyte to form a zygote. Children resemble their parents because they inherit traits passed down from both parents. We now examine some of the principles involved in that process, called inheritance.

Inheritance is the passage of hereditary traits from one generation to the next. It is the process by which you acquired your characteristics from your parents and may transmit some of your traits to your children. The branch of biology that deals with inheritance is called *genetics* (je-NET-iks). The area of health care that offers advice on genetic problems (or potential problems) is called *genetic counseling*.

Genotype and Phenotype

The nuclei of all human cells except gametes contain 23 pairs of chromosomes—the diploid number (*2n*). One chromosome in each pair came from the mother, and the other came from the father. Each *homolog*—one of the two chromosomes that make up a pair—contains genes that control the same traits. If a chromosome contains a gene for body hair, for example, its homolog will also contain a gene for body hair in the same position on the chromosome. Such alternative forms of a gene that code for the same trait and are at the same location on homologous chromosomes are called *alleles* (ah-LĒLZ). For example, one allele of a body hair gene might code for coarse hair, and another allele for fine hair. A *mutation* (mū-TĀ-shun; *muta-* = change) is a permanent heritable change in an allele that produces a different variant of the same trait.

The relationship of genes to heredity is illustrated by examining the alleles involved in a disorder called *phenylketonuria* (*PKU*). People with PKU lack phenylalanine hydroxylase, an enzyme that converts the amino acid phenylalanine into tyrosine, another amino acid. If infants with PKU eat foods containing phenylalanine, high levels of

phenylalanine build up in the blood. The result is severe brain damage and mental retardation. The allele that codes for phenylalanine hydroxylase is symbolized as *P*; the mutated allele that fails to produce a functional enzyme is symbolized as *p*. The chart in Figure 24.10, which shows the possible combinations of gametes from two parents who each have one *P* and one *p* allele, is called a *Punnett square*. In constructing a Punnett square, the possible paternal alleles in sperm are written at the left side and the possible maternal alleles in ova (or secondary oocytes) are written at the top. The four spaces on the chart show how the alleles can combine in zygotes formed by the union of these sperm and ova to produce the three different genetic makeups, or *genotypes* (JĒ-nō-tīps): *PP*, *Pp*, or *pp*. Notice from the Punnett square that 25 percent of the offspring will have the *PP* genotype, 50 percent will have the *Pp* genotype, and 25 percent will have the *pp* genotype. People who inherit *PP* or *Pp* genotypes do not have PKU; those with a *pp* genotype suffer from the disorder. Although people with a *Pp* genotype have one PKU allele (*p*), the allele that codes for the normal trait (*P*) is more dominant. An allele that dominates or masks the presence of

Figure 24.10 Inheritance of phenylketonuria (PKU).

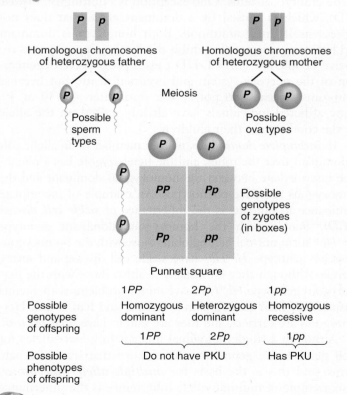

Genotype refers to genetic makeup; phenotype refers to the physical or outward expression of a gene.

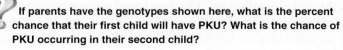

If parents have the genotypes shown here, what is the percent chance that their first child will have PKU? What is the chance of PKU occurring in their second child?

another allele and is fully expressed (*P* in this example) is said to be a ***dominant allele***, and the trait expressed is called a dominant trait. The allele whose presence is completely masked (*p* in this example) is said to be a ***recessive allele***, and the trait it controls is called a recessive trait.

By tradition, the symbols for genes are written in italics, with dominant alleles written in capital letters and recessive alleles in lowercase letters. A person with the same alleles on homologous chromosomes (for example, *PP* or *pp*) is said to be ***homozygous*** for the trait. *PP* is homozygous dominant, and *pp* is homozygous recessive. An individual with different alleles on homologous chromosomes (for example, *Pp*) is said to be ***heterozygous*** for the trait.

Phenotype (FĒ-nō-tīp; *pheno-* = showing) refers to how the genetic makeup is expressed in the body; it is the physical or outward expression of a gene. A person with *Pp* (a heterozygote) has a different *genotype* from a person with *PP* (a homozygote), but both have the same *phenotype*—normal production of phenylalanine hydroxylase. Heterozygous individuals who carry a recessive gene but do not express it (*Pp*) can pass the gene on to their offspring. Such individuals are called *carriers* of the recessive gene.

Alleles that code for normal traits do not always dominate over those that code for abnormal ones, but dominant alleles for severe disorders usually are lethal and cause death of the embryo or fetus. One exception is Huntington disease (HD), which is caused by a dominant allele that does not express itself until adulthood. Both homozygous dominant and heterozygous people exhibit the disease; homozygous recessive people are normal. HD causes progressive degeneration of the nervous system and eventual death, but because symptoms typically do not appear until after age 30 or 40, many afflicted individuals have already passed on the allele for the condition to their children.

In ***incomplete dominance***, neither member of an allelic pair is dominant over the other, and the heterozygote has a phenotype intermediate between the homozygous dominant and the homozygous recessive phenotypes. An example of incomplete dominance in humans is the inheritance of ***sickle cell disease (SCD)***. People with the homozygous dominant genotype *Hb^A Hb^A* form normal hemoglobin; those with the homozygous recessive genotype *Hb^S Hb^S* have sickle cell disease and severe anemia. Although they are usually healthy, those with the heterozygous genotype *Hb^A Hb^S* have minor problems with anemia because half their hemoglobin is normal and half is not. Heterozygotes are carriers, and they are said to have *sickle cell trait*.

Although a single individual inherits only two alleles for each gene, some genes may have more than two alternate forms, and this is the basis for ***multiple-allele inheritance***. One example of multiple-allele inheritance is the inheritance of the ABO blood group. The four blood types (phenotypes) of the ABO group—A, B, AB, and O—result from the inheritance of six combinations of three different alleles of a single gene called the *I* gene: (1) allele *I^A* produces the A antigen, (2) allele

I^B produces the B antigen, and (3) allele *i* produces neither A nor B antigen. Each person inherits two *I*-gene alleles, one from each parent, that give rise to the various phenotypes. The six possible genotypes produce four blood types, as follows:

Genotype	Blood Type (Phenotype)
I^A I^A or *I^A i*	A
I^B I^B or *I^B i*	B
I^A I^B	AB
ii	O

Notice that both *I^A* and *I^B* are inherited as dominant traits, and *i* is inherited as a recessive trait. An individual with type AB blood has characteristics of both type A and type B red blood cells.

Autosomes and Sex Chromosomes

When viewed under a microscope, the 46 human chromosomes in a normal somatic cell can be identified by their size, shape, and staining pattern to be members of 23 different pairs of chromosomes. In 22 of the pairs, the homologous chromosomes look alike and have the same appearance in both males and females; these 22 pairs are called ***autosomes***. The two members of the 23rd pair are termed the ***sex chromosomes***; they look different in males and females (Figure 24.11a). In females, the pair consists of two chromosomes called X chromosomes. One X chromosome is also present in males, but its mate is a much smaller chromosome called a Y chromosome.

When a spermatocyte undergoes meiosis to reduce its chromosome number, it gives rise to two sperm that contain an X chromosome and two sperm that contain a Y chromosome. Oocytes have no Y chromosomes and produce only X-containing gametes. If the secondary oocyte is fertilized by an X-bearing sperm, the offspring normally is female (XX). Fertilization by a Y-bearing sperm produces a male (XY). Thus, an individual's sex is determined by the father's chromosomes (Figure 24.11b). The prime male-determining gene is one called ***SRY (sex-determining region of the Y chromosome)***. *SRY* acts as a switch to turn on the male pattern of development. Only if the *SRY* gene is present and functional in a fertilized ovum will the fetus develop testes and differentiate into a male; in the absence of *SRY*, the fetus will develop ovaries and differentiate into a female.

The sex chromosomes also are responsible for the transmission of several nonsexual traits. Many of the genes for these traits are present on X chromosomes but are absent from Y chromosomes. This feature produces a pattern of heredity, termed ***sex-linked inheritance***, that is different from the patterns already described.

One example of sex-linked inheritance is ***red–green color blindness***, the most common type of color blindness. This condition is characterized by a deficiency in either red- or green-sensitive cones, so red and green are seen as the same

Figure 24.11 Inheritance of gender (sex). In (a) the sex chromosomes, pair 23, are indicated in the colored box.

Gender is determined at the time of fertilization by the sex chromosome of the sperm cell.

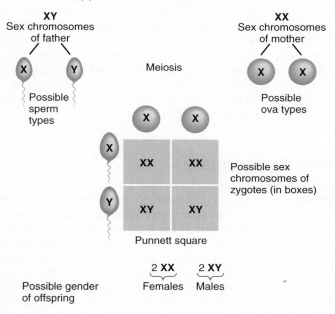

(a) Normal human male chromosomes

(b) Sex determination

? What are chromosomes other than sex chromosomes called?

color (either red or green, depending on which cone is present). The gene for red–green color blindness is a recessive one designated *c*. Normal color vision, designated *C*, dominates. The *C/c* genes are located only on the X chromosome, and thus the ability to see colors depends entirely on the X chromosomes. The possible combinations are as follows:

Genotype	Phenotype
$X^C X^C$	Normal female
$X^C X^c$	Normal female (but a carrier of the recessive gene)
$X^c X^c$	Red–green color-blind female
$X^C Y$	Normal male
$X^c Y$	Red–green color-blind male

Only females who have two X^c genes are red–green color blind. This rare situation can result only from the mating of a color-blind male and a color-blind or carrier female. (In $X^C X^c$ females the trait is masked by the normal, dominant gene.) Because males do not have a second X chromosome that could mask the trait, all males with an X^c gene will be red–green color blind. Figure 24.12 illustrates the inheritance of red–green color blindness in the offspring of a normal male and a carrier female. Traits inherited in the manner just described are called *sex-linked traits*. The most common type of *hemophilia*—a condition in which the blood fails to clot or clots very slowly after an injury—is also a sex-linked trait.

✓ CHECKPOINT

19. What do the terms genotype, phenotype, dominant, recessive, homozygous, and heterozygous mean?
20. Define incomplete dominance and give an example.
21. What is multiple-allele inheritance? Give an example.
22. How is the development of gender determined?
23. Define and provide an example of sex-linked inheritance.

Figure 24.12 An example of the inheritance of red–green color blindness.

Red–green color blindness is an example of a sex-linked trait.

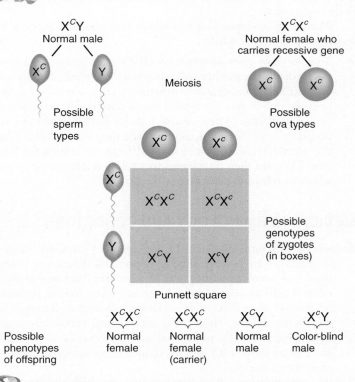

? What is the genotype of a red–green color-blind female?

COMMON DISORDERS

Infertility

Female infertility, or the inability to conceive, occurs in about 10% of all women of reproductive age in the United States. Female infertility may be caused by ovarian disease, obstruction of the uterine tubes, or conditions in which the uterus is not adequately prepared to receive a fertilized ovum. *Male infertility (sterility)* is an inability to fertilize a secondary oocyte; it does not imply erectile dysfunction (impotence). Male fertility requires production of adequate quantities of viable, normal sperm by the testes, unobstructed transport of sperm through the ducts, and satisfactory deposition of sperm in the vagina. The seminiferous tubules of the testes are sensitive to many factors—x-rays, infections, toxins, malnutrition, and higher-than-normal scrotal temperatures—that may cause degenerative changes and produce male sterility.

To begin and maintain a normal reproductive cycle, a female must have a minimum amount of body fat. Even a moderate deficiency of fat—10 percent to 15 percent below normal weight for height—may delay the onset of menstruation, inhibit ovulation during the reproductive cycle, or cause amenorrhea (cessation of menstruation). Both dieting and intensive exercise may reduce body fat below the minimum amount and lead to infertility that is reversible, if weight gain or reduction of intensive exercise or both occurs. Studies of very obese women indicate that they, like very lean ones, experience problems with amenorrhea and infertility. Males also experience reproductive problems in response to undernutrition and weight loss. For example, they produce less prostatic fluid and reduced numbers of sperm having decreased motility.

Many fertility-expanding techniques now exist for assisting infertile couples to have a baby.

■ To achieve *in vitro fertilization (IVF)*—fertilization in a laboratory dish—the mother-to-be is given follicle-stimulating hormone (FSH) soon after menstruation, so that several secondary oocytes, rather than the typical single oocyte, will be produced (superovulation). When several follicles have reached the appropriate size, a small incision is made near the umbilicus, and the secondary oocytes are aspirated from the stimulated follicles. They are then transferred to a solution containing sperm, where fertilization takes place.

■ *Intracytoplasmic sperm injection (ICSI)*, the injection of a sperm or spermatid into an oocyte's cytoplasm, has been used when infertility is due to impairments in sperm motility or to the failure of spermatids to develop into spermatozoa. When the zygote achieved by IVF or ICSI reaches the 8-cell or 16-cell stage, it is introduced into the uterus for implantation and subsequent growth.

■ In *embryo transfer*, a man's semen is used to artificially inseminate a fertile secondary oocyte donor. After fertilization in the donor's uterine tube, the morula or blastocyst is transferred from the donor to the infertile woman, who then carries it (and subsequently the fetus) to term.

■ In *gamete intrafallopian transfer (GIFT)* the goal is to mimic the normal process of conception by uniting sperm and secondary oocyte in the prospective mother's uterine tubes. It is an attempt to bypass conditions in the female reproductive tract that might prevent fertilization, such as high acidity or inappropriate mucus. In this procedure, a woman is given FSH and LH to stimulate the production of several secondary oocytes, which are aspirated from the mature follicles, mixed outside the body with a solution containing sperm, and then immediately inserted into the uterine tubes.

Down Syndrome

Down syndrome is a disorder that most often results during meiosis when an extra chromosome 21 passes to one of the gametes. Most of the time the extra chromosome comes from the mother, a not-too-surprising finding given that all her oocytes began meiosis when she herself was a fetus. They may have been exposed to chromosome-damaging chemicals and radiation for years. (Sperm, by contrast, usually are less than 10 weeks old at the time they fertilize a secondary oocyte.) The chance of conceiving a baby with this syndrome, which is less than 1 in 3000 for women under age 30, increases to 1 in 300 in the 35 to 39 age group, and to 1 in 9 at age 48.

Down syndrome is characterized by mental retardation; retarded physical development (short stature and stubby fingers); distinctive facial structures (large tongue, flat profile, broad skull, slanting eyes, and round head); and malformations of the heart, ears, hands, and feet. Sexual maturity is rarely attained.

MEDICAL TERMINOLOGY AND CONDITIONS

Amniocentesis (am′-nē-ō-sen-TĒ-sis; *amnio-* = amnion; *-centesis* = puncture to remove fluid) A prenatal diagnostic procedure that involves withdrawal of amniotic fluid and analysis of fetal cells and dissolved substances to test for the presence of genetic disorders such as Down syndrome, hemophilia, Tay-Sachs disease, sickle cell disease, and certain muscular dystrophies. Usually performed at 14–18 weeks of gestation and carries about a 0.5 percent chance of spontaneous abortion after the procedure.

Breech presentation A malpresentation in which the fetal buttocks or lower limbs present into the maternal pelvis; the most common cause is prematurity.

Chorionic villi sampling (CVS) (kō-rē-ON-ik VIL-i) A prenatal diagnostic procedure that involves removal of chorionic villi tissue to examine it for the same genetic disorder as amniocentesis. May be performed as early as eight weeks gestation; results are available in a few days. Causes about a 1–2 percent chance of spontaneous abortion after the procedure.

Conceptus (kon-SEP-tus) Includes all structures that develop from a zygote: embryo plus the embryonic part of the placenta and associated membranes (chorion, amnion, yolk sac, and allantois).

Emesis gravidarum (EM-e-sis gra-VID-ar-um; *emeo* = to vomit; *gravida* = a pregnant woman) Episodes of nausea and possibly vomiting that are most likely to occur in the morning during the early stages of pregnancy; also called *morning sickness*. Its cause is unknown, but the high levels of human chorionic gonadotropin (hCG) secreted by the placenta, and of progesterone secreted by the ovaries, have been implicated. In some women the severity of these symptoms requires hospitalization for intravenous feeding.

Fertilization age Two weeks less than the gestational age, since a secondary oocyte is not fertilized until about two weeks after the last normal menstrual period (LNMP).

Fetal alcohol syndrome (FAS) A specific pattern of fetal malformation due to intrauterine exposure to alcohol. FAS is one of the most common causes of mental retardation and the most common preventable cause of birth defects in the United States. The symptoms of FAS may include slow growth before and after birth, characteristic facial features (short palpebral fissures, a thin upper lip, and sunken nasal bridge), defective heart and other organs, malformed limbs, genital abnormalities, and central nervous system damage. Behavioral problems, such as hyperactivity, extreme nervousness, reduced ability to concentrate, and an inability to appreciate cause-and-effect relationships, are common.

Fetal surgery A surgical procedure performed on a fetus; in some cases the uterus is opened and the fetus is operated on directly. Fetal surgery has been used to repair diaphragmatic hernias and remove lesions in the lungs.

Fetal ultrasonography (ul'-tra-son-OG-ra-fē) A prenatal diagnostic procedure that uses ultrasound to confirm pregnancy, identify multiple pregnancies, determine fetal age, evaluate fetal viability and growth, determine fetal position, identify fetal–maternal abnormalities, and assist in procedures such as amniocentesis.

Gestational age (jes-TĀ-shun-al; *gestatus* = to bear) The age of an embryo or fetus calculated from the presumed first day of the last normal menstrual period (LNMP).

Lethal gene (LĒ-thal jēn; *lethum* = death) A gene that, when expressed, results in death either in the embryonic state or shortly after birth.

Metafemale syndrome A sex chromosome disorder characterized by at least three X chromosomes (XXX) that occurs about once in every 700 births. These females have underdeveloped genital organs and limited fertility. Generally, they are mentally retarded.

Preeclampsia (prē'-e-KLAMP-sē-a) A syndrome of pregnancy characterized by sudden hypertension, large amounts of protein in urine, and generalized edema; possibly related to an autoimmune or allergic reaction to the presence of a fetus. When the condition is also associated with convulsions and coma, it is referred to as *eclampsia*.

Puerperal fever (pū-ER-per-al; *puer* = child) A maternal infectious disease of childbirth, also called puerperal sepsis and childbed fever. The disease, which results from an infection originating in the birth canal, affects the endometrium. It may spread to other pelvic structures and lead to septicemia.

Teratogen (TER-a-tō-jen; *terato-* = monster; *-gen* = creating) Any agent or influence that causes developmental defects in the embryo. Examples include alcohol, pesticides, industrial chemicals, antibiotics, thalidomide, LSD, and cocaine.

CHAPTER REVIEW AND RESOURCE SUMMARY

REVIEW

24.1 Embryonic Period

1. Pregnancy is a sequence of events that begins with fertilization and proceeds to implantation, embryonic development, and fetal development. It normally ends in birth.

2. During **fertilization** a sperm cell penetrates a secondary oocyte and their nuclei unite. Penetration of the zona pellucida is facilitated by enzymes in the sperm's acrosome. The resulting cell is a **zygote**. Normally, only one sperm cell fertilizes a secondary oocyte.

3. Early rapid cell division of a zygote is called **cleavage**, and the cells produced by cleavage are called **blastomeres**. The solid sphere of cells produced by cleavage is a **morula**.

4. The morula develops into a **blastocyst**, a hollow ball of cells differentiated into a **trophoblast** and an **inner cell mass**. The attachment of a blastocyst to the endometrium is termed **implantation**.

5. The trophoblast develops into the **syncytiotrophoblast** and **cytotrophoblast**. The inner cell mass differentiates into **hypoblast** and **epiblast**, the **bilaminar** (two-layered) **embryonic disc**. The **amnion** is a thin protective membrane that develops from the cytotrophoblast.

6. The hypoblast forms the **yolk sac**, which transfers nutrients to the embryo, forms blood cells, produces primordial germ cells, and forms part of the gut. Blood and secretions enter **lacunar networks** to supply nutrition to and remove wastes from the embryo. The **extraembryonic mesoderm** and trophoblast form the **chorion**, the principal embryonic part of the placenta.

7. The third week of development is characterized by **gastrulation**, the conversion of the bilaminar disc into a trilaminar (three-layered) embryo consisting of **ectoderm**, **mesoderm**, and **endoderm**. The three primary germ layers form all tissues and organs of the developing organism. The process by which the **neural plate**, **neural folds**, and **neural tube** form is called **neurulation**. The brain and spinal cord develop from the neural tube.

RESOURCES

Anatomy Overview—Developmental Stages
Animation—Fertilization and Development

Figure 24.2 Cleavage and the formation of the morula and blastocyst

Figure 24.4 Summary of events associated with the first week of development

Figure 24.6 Gastrulation

REVIEW	RESOURCES

8. **Chorionic villi**, projections of the chorion, connect to the embryonic heart so that maternal and fetal blood vessels are brought into close proximity. Thus, nutrients and wastes are exchanged between maternal and fetal blood.

9. The **placenta** is the site of exchange of nutrients and wastes between the mother and fetus. The placenta also functions as a protective barrier, stores nutrients, and produces several hormones to maintain pregnancy. The actual connection between the placenta and embryo (and later the fetus) is the **umbilical cord**.

10. The formation of body organs and systems occurs during the fourth week of development. By the end of the fourth week, **upper** and **lower limb buds** develop and by the end of the eighth week the embryo has clearly human features.

24.2 Fetal Period

1. The fetal period is primarily concerned with the growth and differentiation of tissues and organs that developed during the embryonic period.

2. The rate of body growth is remarkable, especially during the ninth and sixteenth weeks.

3. The principal changes associated with embryonic and fetal growth are summarized in Table 24.1.

Anatomy Overview— Developmental Stages

24.3 Maternal Changes During Pregnancy

1. Pregnancy is maintained by **human chorionic gonadotropin (hCG)**, estrogens, and progesterone.

2. **Relaxin** increases flexibility of the pubic symphysis and helps dilate the uterine cervix near the end of pregnancy.

3. **Human placental lactogen (hPL)** contributes to breast development, protein anabolism, and catabolism of glucose and fatty acids.

4. **Corticotropin-releasing hormone (CRH)**, produced by the placenta, is thought to establish the timing of birth, and stimulates the secretion of cortisol by the fetal adrenal gland.

5. During pregnancy, several anatomical and physiological changes occur in the mother.

Animation—Hormonal Regulation of Pregnancy and Childbirth

24.4 Exercise and Pregnancy

1. During pregnancy, some joints become less stable, and certain maneuvers are more difficult to execute.

2. Moderate physical activity does not endanger the fetus in a normal pregnancy.

24.5 Labor and Delivery

1. **Labor** is the process by which the fetus is expelled from the uterus through the vagina to the outside. **True labor** involves dilation of the cervix, expulsion of the fetus, and delivery of the placenta.

2. Oxytocin stimulates uterine contractions.

Animation—Positive Feedback Control of Labor

24.6 Lactation

1. **Lactation** refers to the production and ejection of milk by the mammary glands.

2. Milk production is influenced by **prolactin (PRL)**, estrogens, and progesterone.

3. Milk ejection is stimulated by oxytocin.

4. A few of the many benefits of breast feeding include ideal nutrition for the infant, protection from disease, and decreased likelihood of developing allergies.

Animation—Regulation of Lactation

24.7 Inheritance

1. **Inheritance** is the passage of hereditary traits from one generation to the next.

2. The genetic makeup of an organism is called its **genotype**; the traits expressed are called its **phenotype**.

Animation—The Cell Cycle and Division Processes

REVIEW

3. **Dominant alleles** control a particular trait; expression of **recessive alleles** is masked by dominant alleles.

4. In **incomplete dominance**, neither member of an allelic pair dominates; phenotypically, the heterozygote is intermediate between the homozygous dominant and the homozygous recessive. An example is sickle cell disease.

5. In **multiple-allele inheritance**, genes have more than two alternative forms. An example is the inheritance of ABO blood groups.

6. Each somatic cell has 46 chromosomes—22 pairs of **autosomes** and 1 pair of **sex chromosomes**.

7. In females, the sex chromosomes are two X chromosomes; in males, they are one X chromosome and a much smaller Y chromosome, which normally includes the prime male-determining gene, called *SRY*.

8. If the *SRY* gene is present and functional in a fertilized ovum, the fetus will develop testes and differentiate into a male. In the absence of *SRY*, the fetus will develop ovaries and differentiate into a female.

9. **Red–green color blindness** and **hemophilia** result from recessive genes located on the X chromosome. They are **sex-linked traits** that occur primarily in males because of the absence of any counterbalancing dominant genes on the Y chromosome.

RESOURCES

Biology Basics—Mendelian Inheritance
Biology Basics—Meiosis

SELF-QUIZ

1. The change in a sperm that allows it to fertilize an egg is known as
   a. vasocongestion.   b. capacitation.   c. cleavage.
   d. gestation.   e. meiosis.

2. Match the following:
   ____ a. early division of the zygote that increases the cell number but not size
   ____ b. solid mass of cells three to four days following fertilization
   ____ c. a hollow ball of cells found in the uterine cavity about five days after fertilization
   ____ d. portion of the blastocyst that develops into the embryo
   ____ e. portion of the blastocyst that forms the fetal portion of the placenta
   ____ f. results from fertilization of the sperm and egg

   A. blastocyst
   B. trophoblast
   C. morula
   D. inner cell mass
   E. zygote
   F. cleavage

3. The principal hormone in promoting milk production is
   a. relaxin.   b. prolactin.   c. oxytocin.
   d. cortisol. e. human chorionic gonadotropin (hCG).

4. The placental hormone that appears to affect the timing of birth is
   a. cortisol.   b. human placental lactogen (hPL).
   c. relaxin.   d. corticotropin-releasing hormone (CRH).
   e. human chorionic gonadotropin (hCG).

5. Most of the major body systems in the fetus begin development
   a. during the first week.   b. during the 16th week.
   c. during the fourth through eighth weeks.
   d. while cleavage occurs.   e. as part of gastrulation.

6. Fingerlike projections containing fetal blood vessels
   a. are chorionic villi.
   b. become umbilical veins.
   c. are part of the notochord.
   d. develop into morulas.
   e. emerge from the zona pellucida.

7. The hormone that causes a home pregnancy test to show a positive result for pregnancy is
   a. follicle-stimulating hormone (FSH).
   b. progesterone.
   c. human placental lactogen (hPL).
   d. luteinizing hormone (LH).
   e. human chorionic gonadotropin (hCG).

8. The transformation of a two-layered embryonic disc into a three-layered embryonic disc is called
   a. gastrulation.   b. neurulation.   c. fertilization.
   d. implantation.   e. cleavage.

9. Match the following:
   ____ a. becomes part of the placenta
   ____ b. becomes muscle and bone
   ____ c. develops into the epidermis and nervous system
   ____ d. is the source of blood cells from the third through the sixth week of development
   ____ e. becomes the epithelial lining of the respiratory and gastrointestinal tracts
   ____ f. membrane that surrounds the embryo creating a fluid-filled cavity

   A. yolk sac
   B. chorion
   C. amnion
   D. endoderm
   E. ectoderm
   F. mesoderm

10. The period of time from conception of the zygote to delivery of the fetus is called
    a. fertilization.   b. parturition.   c. gestation.
    d. implantation.   e. gastrulation.

11. Homologous chromosomes
    a. contain genes that control the same trait.
    b. contain genes that control different traits.
    c. are inherited only from the mother.
    d. are inherited only from the father.
    e. contain all identical alleles.

12. Sex-linked traits are carried on the
    a. autosomes.   b. X and Y chromosomes.
    c. X chromosomes only.   d. Y chromosomes only.
    e. Y chromosomes in males and X chromosomes in females.

13. A person who is homozygous for a dominant trait on an autosome would have the genotype
    a. Aa.   b. AA.   c. aa.   d. X^AX^A.   e. X^aX^a.

14. The genotype of a normal female is
    a. XY and 44 autosomes.   b. 46 autosomes.
    c. 46 X chromosomes.   d. XX and 44 autosomes.
    e. XX and 46 autosomes.

15. Which of the following is NOT a change that occurs in a female during pregnancy?
    a. increased cardiac output
    b. decreased pulmonary expiratory reserve volume
    c. decreased gastrointestinal tract motility
    d. increased frequency and urgency of urination
    e. decreased production of estrogens

16. The afterbirth is expelled from the uterus during the _____ stage of labor.
    a. parturition   b. placental   c. dilation
    d. expulsion   e. puerperium

CRITICAL THINKING APPLICATIONS

1. Your friend posts an update on Facebook to announce the birth of their twins, a girl and a boy. Another friend comments "Congratulations! Are they identical?" Without even seeing the twins, what can you tell her?

2. Matt was studying genetics at school and one day came home upset. He told his older sister, "We were doing our family tree and when I filled in our traits, I discovered that Mum and Dad can't be my real parents 'cause the traits don't match!" It turns out that Mum and Dad can roll their tongues but Matt can't. Could Mum and Dad still be Matt's parents?

3. Jenny is concerned about the health of her unborn baby. Due to Jenny's medical history, her doctor wants to test for the presence of a genetic disorder. Jenny is afraid that this will hurt the baby. The physician reassures her that the procedure will not touch the baby even though it will get a sample of fetal tissue. How is this possible?

4. Melanie's baby is due in three weeks. She is debating about whether to breast-feed or bottle-feed her baby. She works as a child care assistant at a creche and has observed that breast-fed infants don't seem to get as sick as often as those who are bottle-fed infant formula. She knows you take anatomy and physiology classes and wants you to explain to her the benefits of breast feeding.

ANSWERS TO FIGURE QUESTIONS

24.1 Capacitation refers to the functional changes in sperm after they have been deposited in the female reproductive tract that enable them to fertilize a secondary oocyte.

24.2 A morula is a solid ball of cells; a blastocyst consists of a rim of cells (trophoblast) surrounding a cavity (blastocyst cavity) and an inner cell mass.

24.3 The blastocyst secretes digestive enzymes that eat away the endometrial lining at the site of implantation.

24.4 At implantation, the blastocyst is oriented so that the inner cell mass is closest to the endometrium.

24.5 The bilaminar embryonic disc is attached to the trophoblast by the connecting stalk.

24.6 Gastrulation converts a bilaminar embryonic disc into a trilaminar embryonic disc.

24.7 Chorionic villi help to bring the fetal and maternal blood vessels close to each other.

24.8 The placenta functions in the exchange of materials between the fetus and mother, serves as a protective barrier against many microbes, and stores nutrients.

24.9 During this time, fetal weight doubles.

24.10 The odds that a child will have PKU are the same for each child—25%.

24.11 The chromosomes that are not sex chromosomes are called autosomes.

24.12 A red–green color-blind female has an X^cX^c genotype.

ANSWERS

ANSWERS TO SELF-QUIZ QUESTIONS

Chapter 1 **1.** d **2.** b **3a.** G **b.** B **c.** E **d.** C **e.** F **f.** A **g.** D **4a.** Nervous **b.** Brain, spinal cord, nerves, sense organs **c.** Lymphatic **d.** returns proteins and fluid to blood; sites of B and T cell maturation and proliferation to protect against disease; carries lipids from digestive system to blood **e.** Respiratory **f.** Lungs, pharynx, larynx, trachea, bronchial tubes **g.** Testes, ovaries, vagina, uterine tubes, uterus, penis **h.** Reproduces the organism and releases hormones **5.** d **6a.** E **b.** C **c.** F **d.** A **e.** B **f.** D **7.** b & c **8.** e **9a.** C **b.** D **c.** A **d.** B **10.** a **11.** a **12.** d **13.** c **14.** e **15.** d **16.** c **17.** b **18.** a **19.** e **20a.** D **b.** A **c.** H **d.** F **e.** G **f.** E **g.** C **h.** B

Chapter 2 **1.** d **2.** c **3.** a **4.** e **5.** d **6.** b **7.** b **8.** c **9.** a **10.** d **11.** e **12.** d **13.** c **14a.** R, D **b.** D **c.** D **d.** R **e.** R **f.** D **g.** D **h.** R **i.** R, D **j.** R, D **15a.** I **b.** O **c.** I **d.** O **e.** O **f.** I **g.** O **16.** d **17.** e **18.** a **19.** carbon, hydrogen, oxygen, nitrogen **20a.** D **b.** C **c.** A **d.** E **e.** B

Chapter 3 **1.** b **2.** e **3.** a **4.** a **5.** e **6.** d **7.** b **8.** d **9.** a **10a.** B **b.** F **c.** G **d.** H **e.** C **f.** E **g.** D **h.** A **11.** e **12.** c **13.** e **14.** d **15.** b **16.** b **17a.** C **b.** E **c.** F **d.** B **e.** D **f.** A **18.** e **19.** b **20.** c

Chapter 4 **1.** c **2.** d **3.** e **4.** d **5.** a **6a.** C **b.** G **c.** E **d.** D **e.** H **f.** F **g.** B **h.** A **7.** a **8.** e **9.** c **10.** b **11.** a **12.** c **13.** c **14.** d **15.** e **16.** d **17.** a **18.** c **19.** a **20.** b

Chapter 5 **1.** b **2.** d **3.** a **4.** e **5a.** C **b.** D **c.** B **d.** E **e.** A **6.** a **7.** c **8.** c **9.** a **10.** a **11.** b **12.** a **13.** c **14.** b **15.** d **16.** e **17.** b **18a.** D **b.** F **c.** E **d.** G **e.** H **f.** A **g.** C **h.** B **i.** I **19.** a **20.** b

Chapter 6 **1a.** C **b.** E **c.** D **d.** A **e.** B **2.** d **3.** a **4a.** E **b.** D **c.** A **d.** C **e.** B **5.** b **6.** e **7.** a **8.** a **9.** b **10.** d **11.** b **12a.** C **b.** D **c.** B **d.** A **13.** e **14.** b **15.** e **16a.** AX **b.** AP **c.** AP **d.** AX **e.** AP **f.** AP **g.** AP **h.** AX **i.** AP **j.** AP **k.** AX **l.** AP **m.** AX **n.** AP **o.** AX **p.** AP **q.** AX **r.** AX **s.** AP **t.** AX **u.** AX **v.** AP **w.** AX **x.** AX **y.** AX **z.** AP **aa.** AP **bb.** AP **cc.** AX **dd.** AX

Chapter 7 **1.** d **2.** a **3.** e **4.** b **5.** d **6.** e **7.** b **8a.** D **b.** A **c.** E **d.** C **e.** B **9.** a **10.** d **11.** c **12.** a **13.** b **14.** e **15a.** A **b.** G **c.** B **d.** C **e.** D **f.** E **g.** F **h.** J **i.** I **j.** H

Chapter 8 **1.** c **2.** e **3.** d **4a.** C **b.** D **c.** A **d.** E **e.** B **5.** b **6.** a **7.** a **8a.** SM, CA **b.** SK **c.** SK, CA **d.** CA **e.** SK **f.** SK **g.** SM **h.** SM **i.** SK, some SM (multiunit) **j.** CA, some SM (visceral) **9.** b **10.** b **11.** e **12.** b **13.** d **14.** e **15.** c **16.** a **17.** b **18a.** C **b.** F **c.** D **d.** B **e.** G **f.** H **g.** A **h.** E **19a.** E **b.** D **c.** G **d.** H **e.** P **f.** M **g.** O **h.** F **i.** J **j.** C **k.** K **l.** A **m.** L **n.** I **o.** N **p.** B **20a.** C **b.** E **c.** B **d.** D **e.** F **f.** A

Chapter 9 **1.** e **2a.** C **b.** B **c.** A **d.** A **e.** C **f.** B **3.** b **4.** d **5.** a **6.** a **7.** c **8.** a **9.** c **10.** e **11.** b **12.** d **13.** b **14.** d **15a.** D **b.** E **c.** F **d.** C **e.** A **f.** B **16a.** I **b.** A **c.** D **d.** O **e.** E **f.** P **g.** H **h.** J **i.** G **j.** C **k.** B **l.** L **m.** N **n.** K **o.** F **p.** M

Chapter 10 **1.** e **2.** a **3.** b **4.** c **5.** d **6.** b **7.** c **8.** a **9.** c **10.** a **11.** b **12.** e **13.** e **14.** a **15.** b **16a.** C **b.** D **c.** A **d.** B **17.** b **18.** d **19.** a, b, d **20a.** G **b.** H **c.** K **d.** J **e.** F **f.** I **g.** A **h.** D **i.** C **j.** B **k.** E

Chapter 11 **1.** b **2.** a **3.** c **4.** e **5.** d **6.** a **7.** d **8.** b **9.** d **10.** e **11.** b **12.** d **13.** b **14a.** C **b.** F **c.** E **d.** B **e.** A **f.** D **15a.** S **b.** P **c.** P **d.** S **e.** S **f.** P **g.** S **h.** S **i.** S

Chapter 12 **1.** b **2.** b **3.** a **4.** d **5.** e **6.** c **7a.** J **b.** E **c.** F **d.** I **e.** C **f.** G **g.** H **h.** A **i.** B **j.** D **8.** b **9.** d **10.** c **11.** a **12.** e **13.** e **14.** c **15a.** D **b.** F **c.** G **d.** B **e.** C **f.** A **g.** E **16.** a **17.** e **18.** c **19.** d **20.** a

Chapter 13 **1.** d **2.** e **3.** c **4.** a **5.** d **6.** b **7.** e **8.** a **9.** b **10.** c **11.** c **12.** a **13.** b **14.** e **15.** b **16.** d **17a.** C **b.** F **c.** G **d.** A **e.** D **f.** E **g.** B **18.** b **19a.** B **b.** C **c.** E **d.** A **e.** H **f.** D **g.** I **h.** G **i.** F **20a.** R **b.** F **c.** F **d.** R **e.** F **f.** E **g.** R **h.** E **i.** F

Chapter 14 **1.** c **2a.** C **b.** F **c.** A **d.** D **e.** E **f.** B **3.** c **4a.** F **b.** C **c.** G **d.** A **e.** D **f.** H **g.** B **h.** E **5.** e **6.** d **7.** b **8.** b **9.** d **10.** a **11.** c **12.** a **13a.** E **b.** D **c.** C **d.** A **e.** B **14.** e **15.** b **16.** a **17.** e **18.** a **19.** b **20.** c

Chapter 15 **1a.** D **b.** J **c.** H **d.** I **e.** C **f.** B **g.** A **h.** E **i.** K **j.** F **k.** G **2.** e **3.** e **4.** e **5.** c **6.** c **7.** a **8.** c **9.** b **10.** d **11.** d **12.** a **13.** a **14.** a **15.** d **16.** d **17.** b **18.** d **19a.** A **b.** A **c.** A **d.** B **e.** A **f.** B **g.** B **h.** A **i.** B **j.** A **k.** B **l.** A **20a.** B **b.** G **c.** A **d.** F **e.** D **f.** C **g.** E

Chapter 16 **1.** b **2.** a **3.** b **4.** c **5.** d **6.** e **7a.** E **b.** D **c.** A **d.** G **e.** C **f.** F **g.** B **8.** d **9.** c **10.** b **11.** c **12a.** H **b.** D **c.** E **d.** F **e.** A **f.** B **g.** C **h.** I **i.** G **13a.** I **b.** D **c.** D **d.** I **e.** I **f.** I **g.** I **h.** D **i.** D **j.** I **14.** e **15.** b **16.** a **17.** d

Chapter 17 **1.** c **2.** d **3.** e **4.** c **5.** a **6.** b **7.** b **8.** a **9.** d **10.** e **11.** b **12.** c **13.** d **14a.** E **b.** B **c.** D **d.** A **e.** C **15.** c **16.** b **17.** e **18.** b **19.** e **20.** a

Chapter 18 **1.** d **2.** c **3.** b **4.** a **5.** c **6.** a **7.** a **8.** d **9a.** B **b.** D **c.** A **d.** C **e.** E **10.** a **11a.** F **b.** D **c.** H **d.** A **e.** G **f.** C **g.** E **h.** B **12.** b **13.** c **14.** a **15.** d **16.** b **17.** d **18.** c **19.** e **20.** c **21a.** D **b.** A **c.** F **d.** C **e.** B **f.** G **g.** E

Chapter 19 **1.** e **2.** b **3.** b **4.** c **5.** d **6.** a **7.** b **8a.** D **b.** G **c.** A **d.** H **e.** B **f.** C **g.** E **h.** F **9.** e **10.** e **11.** a **12.** a **13a.** D **b.** G **c.** B **d.** F **e.** E **f.** H **g.** A **h.** I **i.** K **j.** J **k.** C **14.** b **15.** d **16.** a **17.** e **18.** c **19.** e **20.** a

Chapter 20 **1.** d **2.** c **3.** a **4a.** C **b.** A **c.** D **d.** B **5.** c **6.** a **7.** b **8.** e **9.** d **10.** a **11.** d **12.** a **13.** e **14.** b **15.** c **16.** c **17.** d **18.** a **19.** d **20a.** A **b.** B **c.** A **d.** C **e.** D **f.** C & D **g.** A **h.** B

Chapter 21 **1.** c **2.** e **3a.** F **b.** C **c.** A **d.** E **e.** D **f.** B **4.** d **5.** c **6.** e **7.** a **8.** d **9.** b **10.** a **11.** b **12.** d **13.** e **14.** a **15.** c **16.** d **17.** a **18.** c **19.** e **20.** b

Chapter 22 **1.** d **2.** e **3.** a **4.** c **5.** a **6.** d **7.** d **8.** a **9.** c **10.** a **11.** c **12.** e **13.** b **14.** c **15.** d **16.** b **17.** c **18a.** A **b.** B **c.** D **d.** C

Chapter 23 1. b 2a. C b. D c. E d. B e. A 3. e 4. d 5. c
6. c 7. b 8. b 9. c 10. e 11. c 12. b 13. a 14. e 15. d
16. d 17a. F b. G c. C d. A e. B f. E g. D 18. d 19a. B
b. F c. A d. G e. D f. C g. E 20. e

Chapter 24 1. b 2a. F b. C c. A d. D e. B f. E 3. b 4. d
5. c 6. a 7. e 8. a 9a. B b. F c. E d. A e. D f. C 10. c
11. a 12. c 13. b 14. d 15. e 16. b

ANSWERS TO CRITICAL THINKING APPLICATIONS

Chapter 1

1. In the anatomical position, the arm would hang along the lateral side of the trunk with the palm facing forward.

2. The minimal level of organization at which life occurs is the cellular level. To ensure the organism is alive, you would need to observe life processes such as metabolism, responsiveness, movement, growth, differentiation, and reproduction.

3. Anatomically speaking, Guy's answer does not make sense. Caudal means inferior or away from the head, dorsal means the back, and sural is the calf of the leg. The groin area is located on the anterior of the trunk, near the top of the leg.

4. The midsagittal plane divides the body into equal right and left sides. The transverse (cross-sectional or horizontal) plane divides the top from the bottom of the body.

Chapter 2

1. The protein in the milk is denatured by the acid in the lemon juice. Denaturation changes the characteristic shape of the milk protein so that it is no longer soluble.

2. Cooking the heart-healthy salmon in the liquid corn oil would be preferable to the corn oil margarine. Pure corn oil contains polyunsaturated fats which are considered to decrease the risk of heart disease. However, in order to manufacture the corn oil margarine, the liquid corn oil is hydrogenated, which creates very unhealthy *trans* fatty acids.

3. Alfie doesn't understand pH. Each increase of one whole number on the pH scale represents a 10-fold decrease in H^+ concentration. The pH 3.5 mixture is 10 times less acidic (or 10 times more alkaline) than the pH 2.5 mixture.

4. Simply adding water to table sugar does not cause it to break apart into monosaccharides. The water acts as a solvent, dissolving the sucrose, and forming a sucrose-water solution. To complete the breakdown of table sugar to glucose and fructose would require the presence of the enzyme sucrase.

Chapter 3

1. Lysosomes contain digestive enzymes that digest bone tissue and release the stored calcium.

2. Seawater is hypertonic to the body. It contains a higher concentration of solutes (NaCl) than the body's cells. Drinking seawater would cause crenation of cells.

3. Mucin (protein) is synthesized on the ribosomes attached to the rough ER. The mucin will travel from the rough ER to a transport vesicle to the Golgi complex. The protein will be modified to a glycoprotein while being transported through the Golgi's cisterns (again using transfer vesicles). The mucin will be packaged in a secretory vesicle and moved to the plasma membrane where it will be released by exocytosis.

4. Steve has several risk factors that could be contributing to his "premature" aging. High levels of stress can shorten the protective telomeres on the ends of chromosomes, contributing to aging and eventual death of cells. Glucose cross-links that form between proteins can contribute to loss of elasticity, which ages tissues. His immune system may be malfunctioning, producing autoimmune responses that can also affect the aging process.

Chapter 4

1. The pins are stuck through the keratinized, stratified squamous layer of the skin. There is no bleeding because epithelium is avascular.

2. Collagen is the most abundant protein in the body. Bone, cartilage, and dense connective tissue contain abundant collagen. Collagen is found in tendons, ligaments, and bones.

3. The uterine tube is lined with ciliated simple columnar epithelium. The cilia (hair) help move the ovum.

4. Alex's "broken cartilage" involved the hyaline cartilage in his tibia, in an area where bones grow in length. Because cartilage doesn't have a blood supply, the cartilage healed slowly and delayed growth in his right leg. Because his uninjured left leg grew normally, the difference in length between his right and left legs resulted in his limp.

Chapter 5

1. The hair shaft is fused, dead keratinized cells. At the base of the hair, the hair bulb contains the living hair matrix where cell division occurs. Hair root plexuses (nervous tissue) surround each hair follicle.

2. The epidermal layer is repaired by cell division of keratinocytes starting at the deepest layer of the epidermis, the stratum basale. The keratinocytes move toward the surface through the stratum spinosum, stratum granulosum, and stratum corneum. The cells become fully keratinized, flat, and dead as they travel to the surface. The process takes about 2–4 weeks.

3. The extensibility and elasticity of the skin is due primarily to the presence of collagen and elastic fibers in the deep region of the dermis. The white streaks, or striae, result from small tears in the dermis as a result of overstretching of the skin with pregnancy.

4. Frank's blackheads are caused by oil (sebum) accumulation in the sebaceous glands. Sebum secretion often increases after puberty. The color is due to melanin and oxidized oil.

Chapter 6

1. Iain fractured his tibia and fibula, the styloid process of the radius, and the scaphoid.

2. You can compare the sizes of long bones, especially the articular ends, which tend to be larger and thicker in males. Bumps, lines, tuberosities, and ridges for muscle attachments are also typically more pronounced in males. There should be differences in the pelvic bone structure. The female pelvis should be wider and shallower and have more space in the pelvic inlet and outlet for childbirth.

3. Due to her age and gender, your great-grandmother probably has osteoporosis. Bone loss is due to increased calcium loss and decreased production of growth hormone and estrogen. Shrinkage of the vertebrae results in hunched back and loss of height.

4. Keith fractured the top (under the kneecap) of the larger of the two leg bones. The body requires calcium, phosphorus, magnesium, vitamins A, C, and D, hGH and other hormones, and protein for bone matrix in order for Keith's injury to heal.

Chapter 7

1. Flex the knees, flex the hip (on the side with knee up), hyperextend the neck, flex fingers, flex and extend at elbow and shoulder, and depress mandible.

2. The hip joint is a diarthrotic, synovial joint of the ball-and-socket type formed by the head of the femur fitting into the acetabulum of the hip bone. The movements are extension/flexion, abduction/adduction, rotation, and circumduction.

3. The ACL is the anterior cruciate ligament. It connects the tibia to the femur, running posteriorly and laterally. The ACL works along with other internal and external ligaments to stabilize the knee joint.

4. Uncle Bob is likely suffering from osteoarthritis—the most common degenerative joint disease in the elderly. Although Bob can't buy "legs," he could discuss with her physician the pros and cons of arthroplasty—the replacement of damaged joints with artificial joints.

Chapter 8

1. The skeletal muscles will not contract without receiving a signal from the neurotransmitter acetylcholine. Since release of ACh is blocked, skeletal muscles will not function.

2. Sarah used the orbicularis oris (puckering), frontalis (eyebrows), zygomaticus (cheeks), and buccinator (cheeks).

3. Keith's leg muscles atrophied from loss of myofibrils due to lack of use of the muscles. Keith will need to exercise in order to increase muscle size by building up myofibrils, mitochondria, and sarcoplasmic reticulum.

4. The sprinters have higher proportions of fast glycolytic muscle fibers, which are large in diameter as compared to the marathon runners, who would have a higher proportion of fast oxidative-glycolytic fibers with smaller diameters. Sprinters would have training regimens that would tend to encourage muscle hypertrophy, while the marathon runners' training would encourage greater respiratory and cardiovascular changes but not an increase in muscle mass.

Chapter 9

1. Smelling coffee and hearing an alarm are somatic sensory, stretching and yawning are somatic motor, salivating is autonomic (parasympathetic) motor.

2. Acetylcholine is the excitatory neurotransmitter released by motor neurons to initiate muscle contraction. The drug given to James blocks the receptor sites on the muscle fibers so ACh cannot attach and stimulate the muscle cells.

3. Neuropeptides such as endorphins are found in the brain. They are related to feelings of pleasure and are natural painkillers.

4. A myelin sheath increases the speed of nerve impulse conduction (propagation). Because myelination is not complete in infants, their responses are slower and less coordinated than in older children.

Chapter 10

1. The brachial plexus runs through the axillary region and supplies the arm and hands. Keith's weight has put pressure on the brachial plexus and interrupted nerve impulse transmission.

2. Alexa suffered a head injury from her car accident that has caused a blockage of CSF circulation in the subarachnoid space near the occipital lobe, where visual function is centered. The accumulating CSF will be drained with a shunt to prevent permanent and possibly life-threatening brain damage.

3. The primary motor area in the left precentral gyrus and Broca's area in the left frontal lobe were damaged by the stroke. Both of these areas are in the cerebral cortex.

4. A receptor in Jack's foot detected the pain from the pin. The impulse traveled along a sensory nerve to the spinal cord, which passed the signal along to a motor neuron. The motor neuron(s) stimulated the muscles in Jack's leg muscles to contract, resulting in a withdrawal reflex.

Chapter 11

1. The parasympathetic division of the ANS directs rest-and-digest activities. The organs of the digestive system will have increased activity to digest food, absorb nutrients, and defecate wastes. In the relaxed condition, the body will also exhibit slower heart rate and airway constriction.

2. The anxiety of giving the presentation has activated your sympathetic nervous system. Even though you are not in physical danger, your body has initiated the fight-or-flight response, resulting in the symptoms you experienced. The release of epinephrine and norepinephrine from the adrenal medullae ensures that the effects don't go away any time soon!

3. The goose bumps are a sympathetic nervous system response. The cell bodies of sympathetic preganglionic neurons are in the thoracolumbar (T1–L2) segments of the spinal cord; their axons exit in the anterior roots of spinal nerves and extend out to a sympathetic ganglion. From there, postganglionic neurons extend to hair follicle smooth muscles (arrector pili), which produce goose bumps when they contract.

4. Dual innervation refers to the innervation of most organs by both the sympathetic and parasympathetic divisions of the autonomic nervous system, not to the presence of more than one head.

Chapter 12

1. There are several kinds of receptors that are activated by Kate's actions, including: warm thermoreceptors from the bath; free nerve endings from tickling; corpuscles of touch from the kiss on the lips; hair root plexuses from stroking his arms.

2. The saccule and utricle in the vestibular apparatus of the internal ear respond to head position and maintain static equilibrium. The otolithic membrane of the maculae moves in response to head movement, stimulating hair cells and triggering impulses that travel along cranial nerve VIII.

3. The cornea is responsible for about 75% of the total refraction of light rays entering the eye. Changing the shape of the cornea by shaving off surface layers will alter refraction of light and change the focus of the image on the retina, hopefully improving visual acuity.

4. The optometrist used eye drops to temporarily paralyze the muscles of the iris during the examination. The radial muscles are contracted in the paralyzed state, resulting in dilation of Louise's pupils. Light sensitivity occurs because the circular muscles are also paralyzed and cannot contract to constrict the pupil in response to the bright light.

Chapter 13

1. Sam has Type 1 diabetes mellitus due to the destruction of the pancreatic beta cells. He must have injections of insulin to metabolize glucose. His aunt has Type 2 diabetes mellitus. She still produces insulin but her body cells have decreased sensitivity to the hormone.

2. As aging occurs, there is a decreased release of many hormones, including hGH. Lack of hGH is one cause of muscle loss with aging. A decrease in testosterone levels with age could also affect muscle loss.

3. Melatonin is released by the pineal gland during darkness and sleep. Melatonin helps set the biological clock, controlled by the hypothalamus, which sets sleep patterns. SAD may be caused by excess melatonin. Bright light inhibits melatonin secretion and is a treatment for SAD.

4. Dehydration will stimulate the release of ADH from the posterior pituitary. ADH will increase water retention by the kidneys, decrease sweating, and constrict arterioles, which will raise BP. Epinephrine and norepinephrine will be released by the adrenal medullae in response to stress.

Chapter 14

1. Bilirubin is a pigment formed from the breakdown of heme from the hemoglobin of old RBCs phagocytized by the liver. If the bile ducts do not transport the bilirubin in bile away from the liver, the bilirubin will build up in blood and other tissues, causing a yellow color in the skin and eyes called jaundice.

2. Person 1 has blood type A, person 2 has blood type AB, and person 3 has blood type O.

3. A bluish-purple color in the nail beds is seen in cyanosis, which is caused by prolonged lack of oxygen (hypoxia).

4. Pluripotent stem cells can transform into myeloid stem cells and lymphoid stem cells and, from these, can produce all of the formed elements of blood: RBCs, platelets, and WBCs (monocytes, eosinophils, neutrophils, basophils, and lymphocytes).

Chapter 15

1. A pacemaker sends electrical impulses into the right side of the heart that stimulate contraction of the heart muscle. A pacemaker is used in conditions in which the heart rhythm is irregular to take over the function of the SA node.

2. The sudden appearance of the car activated the sympathetic division of the nervous system. The sympathetic signals from the cardiovascular (CV) center in the medulla oblongata travel down the spinal cord to the cardiac accelerator nerves, which release norepinephrine that increases heart rate and forcefulness of contraction.

3. SV = CO ÷ HR (beats/min). Assuming an average CO of 5250 mL/min at rest, SV = 5250 mL/min ÷ 40 beats/min = 131.25 mL. Rearranging the equation, CO = SV × HR (beats/min). With exercise, Samantha's CO = 131.25 mL × 60 beats/min = 7875 mL/min.

4. Tricia should view her husband's news as good. HDLs (high-density lipoproteins) are associated with removal of excess cholesterol from the body, so are considered the "good cholesterol." LDLs (low-density lipoproteins) are associated with promoting atherosclerotic plaques and are considered as "bad cholesterol." Patients want HDLs to be high and LDLs to be low.

Chapter 16

1. One effect of adrenaline is to cause vasoconstriction of arterioles. If these vessels were constricted temporarily, this would reduce the blood flow locally at the site of dental work and reduce bleeding.

2. Varicose veins are caused by weak venous valves that allow the backflow of blood. With aging, venous walls may lose their elasticity and become stretched and distended with blood. Arteries rarely become distended because they have thicker inner and middle layers than veins and do not contain valves.

3. The exchange of nutrients, wastes, oxygen, and carbon dioxide occurs through the placenta, which joins the maternal and fetal circulatory systems. The umbilical cord contains vessels that transfer materials between the mother and fetus. The umbilical vein transfers nutrients and oxygen from the placenta to the fetus. Some of the blood from the umbilical vein enters the liver, while most flows into the ductus venosus and then into the inferior vena cava. The deoxygenated and waste-filled blood from the fetus returns to the placenta via two umbilical arteries, vessels that form from branches of the fetal internal iliac arteries.

4. Philip cut an artery. Blood flows from arteries in rapid spurts due to the high pressure generated by ventricular contraction.

Chapter 17

1. The right axillary lymph nodes were probably removed because lymph flowing in lymphatics away from the tumor (breast lump) was filtered by the axillary lymph nodes. Cancerous cells from the tumor may be carried in the lymph to the axillary nodes and spread the cancer by metastasis.

2. The spleen contains lymphocytes (B cells, T cells, plasma cells), granular leukocytes, and macrophages, all of which are involved in combating pathogens. The spleen also stores platelets, destroys defective and worn out red blood cells and platelets, and produces blood cells (during fetal life). Although it is preferable to maintain the spleen, the functions of the spleen can be performed by other body organs, such as the red bone marrow and liver.

3. Initial tetanus immunization provided artificially acquired active immunity. The booster dose is needed to maintain immunity against the tetanus bacteria and toxin.

4. Without a blood supply, antibodies and T cells do not have easy access to the cornea. Therefore no immune response occurs to reject the transplanted foreign cornea.

Chapter 18

1. Holding the breath will cause blood levels of CO_2 and H^+ to increase and O_2 to decrease. These changes will strongly stimulate the inspiratory area, which will send impulses to resume breathing, whether Toby is still conscious or not.

2. Exercise normally induces the sympathetic nervous system to send signals to dilate the bronchioles, which increases air flow and oxygen supply. Asthma causes constriction of the bronchioles, making inhalation more difficult and reducing the airflow.

3. Zoe has a viral infection—the common cold (coryza). Laryngitis, an inflammation of the larynx, is a common complication from a cold. Inflammation of the true vocal cords can result in loss of voice.

4. As you swallowed, the epiglottis did not close completely, allowing the liquid to enter the larynx. The irritation caused by the liquid resulted in temporary apnea, followed by the coughing reflex to expel the liquid.

Chapter 19

1. You should brush your teeth to remove food residue and bacteria. When bacteria metabolize the sugars left on teeth after a meal, they produce acid that can demineralize enamel. Dental caries (cavities) can result.

2. Pam and Jane are both wrong. Lactose intolerance causes symptoms of cramping, diarrhea, and flatulence from excess gas in the large intestine.

3. The pyloric sphincter is located at this junction. The spider traveled from the mouth through the oro- and laryngopharynx, then the esophagus, and finally entered the stomach. Gastroscopy of the stomach with an endoscope would reveal mucus-coated epithelium, gastric pits, and rugae.

4. Jane's heartburn is due to an inadequate closure of the lower esophageal sphincter, which is causing her stomach acids to back up into the esophagus and result in a burning sensation. Diets that stimulate gastric acid secretion (such as tomatoes, chocolate, and coffee) as well as alcohol (which can cause the sphincter to relax) will worsen the condition. If Jane has antacids in her medicine cabinet, she could consume them to temporarily neutralize the acids. If the condition is persistent, she will need to modify her eating habits and may want to discuss prescription medications with her physician.

Chapter 20

1. *See* Figure 20.1.

2. Body temperature will increase due to radiation from the sun and hot surrounding sand, and possibly conduction from lying on the hot sand. Heat will be lost to the water by conduction and convection.

3. The diet Neil is considering is typically known as "carbohydrate loading." There is some evidence that eating a diet high in carbohydrates 2–3 days before an endurance athletic event will encourage high levels of glycogen stores in the muscles. During the event, the glycogen is catabolized into glucose, which is subsequently used for ATP production.

4. Antioxidants protect against free radical damage to cell membranes, DNA, and blood vessel walls. Vitamin C is water-soluble, so excess quantities will be excreted in the urine. Vitamins A (from the beta-carotene) and E are lipid-soluble and might accumulate to toxic levels in tissues such as the liver.

Chapter 21

1. Alcohol inhibits the secretion of antidiuretic hormone (ADH). ADH is secreted when the hypothalamus detects a decrease in the amount of water in the blood. ADH makes the collecting ducts and distal portion of the distal convoluted tubules more permeable to water so that it can be reabsorbed.

2. Incontinence (lack of involuntary control over micturition) is normal in children of Ellie's age. Neurons to the external urethral sphincter are not completely developed until after about 2 years of age. The desire to control micturition voluntarily must also be present and initiated by the cerebral cortex.

3. No. Glomerular filtration is mainly driven by blood pressure and opposed by glomerular capsular pressure, not pressure from urine in the bladder. Under normal physiological conditions, urine remains in the bladder and does not back up into the kidneys.

4. From the symptoms described, Valerie has a UTI (urinary tract infection). Most UTIs can be cured by antibiotics. Females are more prone to recurrent UTIs due to the shortness of the female's urethra and its proximity to the anal area. Bacteria can easily enter and colonize the female's urinary tract. Prevention includes proper hygiene when wiping and during intercourse, changing sanitary tampons and napkins regularly, consuming large amounts of fluid daily, and urinating frequently to help wash out bacteria.

Chapter 22

1. The high sodium concentration will stimulate secretion of ANP to reduce Na^+ concentration in the blood and to reduce blood volume.

2. Josh has water intoxication. He had excessive intake of water during his swim class, causing his body fluids to become too dilute or hypotonic. Water moved into his cells by osmosis, swelling the cells and resulting in convulsions.

3. Christina's emphysema results in an inability to fully exhale the carbon dioxide her body produces. As CO_2 levels rise, the CO_2 reacts with water to form carbonic acid, which then dissociates into H^+ and the bicarbonate ion (HCO_3^-). The increase in H^+ produces respiratory acidosis. Her kidneys attempt to compensate by secreting more H^+ and reabsorbing more HCO_3^-.

4. An increase in blood volume stretches the atria, causing the release of atrial natriuretic peptide (ANP). ANP promotes excretion of Na^+ (natriuresis) in the urine. Increased loss of water goes along with loss of Na^+ and reduces blood volume.

Chapter 23

1. Maria may have insufficient levels of progesterone, which is secreted by the corpus luteum and which prepares and maintains the uterus for pregnancy. Low levels of progesterone may contribute to miscarriage.

2. Vasectomy cuts the ductus deferens so that sperm cannot be transported out of the body. The function of the testes is not affected. The Leydig cells secrete the hormone testosterone, which maintains male sex characteristics and sex drive. A vasectomy will not affect production of the hormone or its transport to the rest of the body via the blood.

3. Sperm production is optimal at a temperature slightly below normal body temperature. The higher temperature that results from Julio wearing very close-fitting briefs and soaking in the hot tub inhibits sperm production and survival and therefore fertility.

4. Symptoms of BPH include frequent urination, nocturia, hesitancy, decreased force, and incomplete emptying. The prostatic secretions supply a milky appearance to the semen, nutrients for the sperm, and clotting enzymes such as PSA (prostate specific antigen), which liquefy semen. Without the prostatic contribution, the semen volume would decrease about 25%.

Chapter 24

1. The friend's twins are fraternal or dizygotic. If the cells resulting from cleavage of a single fertilized ovum split into two separate groups and continue developing into two babies, then identical or monozygotic twins result. Since identical twins came from the same original fertilized ovum, they must contain the same genetic information and be the same gender.

2. Yes. Tongue rolling is a dominant trait. Matt's parents are both heterozygous for the tongue rolling gene (Tt). The dominant (T) gene determines their tongue rolling ability but they each passed on the recessive (t) gene to Matt. Matt is homozygous recessive (tt) and can't roll his tongue.

3. The physician may obtain a sample of fetal tissue from the amniotic fluid, which contains sloughed off fetal cells, or from the chorionic villi, which is fetal placental tissue. Amniocentesis and chorionic villi sampling do not take samples from the actual baby (fetus).

4. Breast-fed infants obtain certain secretory antibodies (IgAs) from their mothers through the breast milk. These antibodies protect the infants from many pathogens through the first few months of life until they can begin to produce their own antibodies. Other important materials passed from the mother's milk to the infant that help prevent inflammation and illness include interleukin-10, mucins, glycoproteins, T lymphocytes, and macrophages. Compounds that inhibit growth of harmful bacteria and promote growth of beneficial bacteria are also present in breast milk. In addition, breast milk contains the proper amounts of nutrients, hormones, and growth factors required for the infant's development. Nursing stimulates the release of oxytocin, which helps return the uterus back to pre-pregnancy condition.

GLOSSARY

Pronunciation Key

1. The most strongly accented syllable appears in capital letters, for example, bilateral (bī-LAT-er-al) and diagnosis (dī-ag-NŌ-sis).

2. If there is a secondary accent, it is noted by a prime ('), for example, constitution (kon'-sti-TOO-shun) and physiology (fiz'-ē-OL-ō-jē). Any additional secondary accents are also noted by a prime, for example, decarboxylation (dē'-kar-bok'-si-LĀ-shun).

3. Vowels marked by a line above the letter are pronounced with the long sound, as in the following common words:

ā as in *māk*
ē as in *bē*
ī as in *īvy*

ō as in *pōle*
ū as in *cūte*

4. Vowels not marked by a line above the letter are pronounced with the short sound, as in the following words:

a as in *above* or *at*
e as in *bet*
i as in *sip*
o as in *not*
u as in *bud*

5. Other vowel sounds are indicated as follows:

oy as in *oil*
oo as in *root*

6. Consonant sounds are pronounced as in the following words:

b as in *bat*
ch as in *chair*
d as in *dog*
f as in *father*
g as in *get*
h as in *hat*
j as in *jump*
k as in *can*
ks as in *tax*
kw as in *quit*
l as in *let*
m as in *mother*
n as in *no*

p as in *pick*
r as in *rib*
s as in *so*
t as in *tea*
v as in *very*
w as in *welcome*
z as in *zero*
zh as in *lesion*

A

Abdomen (ab-DŌ-men *or* AB-dō-men) The area between the diaphragm and pelvis.

Abdominal (ab-DŌM-i-nal) **cavity** Superior portion of the abdominopelvic cavity that contains the stomach, spleen, liver, gallbladder, pancreas, kidneys, small intestine, and part of the large intestine.

Abdominal thrust maneuver A first-aid procedure for choking. Employs a quick, upward thrust against the diaphragm that forces air out of the lungs with sufficient force to eject any lodged material.

Abdominopelvic (ab-dom'-i-nō-PEL-vik) **cavity** Inferior to the diaphragm and subdivided into a superior abdominal cavity and an inferior pelvic cavity.

Abduction (ab-DUK-shun) Movement away from the midline of the body.

Abortion (a-BOR-shun) The premature loss **(spontaneous)** or removal **(induced)** of the embryo or nonviable fetus; miscarriage due to a failure in the normal process of developing or maturing.

Abscess (AB-ses) A localized collection of pus and liquefied tissue in a cavity.

Absorption (ab-SORP-shun) Intake of fluids or other substances by cells of the skin or mucous membranes; the passage of digested foods from the gastrointestinal tract into blood or lymph.

Accessory duct A duct of the pancreas that empties into the duodenum about 2.5 cm (1 in.) superior to the hepatopancreatic ampulla. Also called the **duct of Santorini** (san'-tō-RĒ-nē).

Acetabulum (as'-e-TAB-ū-lum) The rounded cavity on the external surface of the hip bone that receives the head of the femur.

Acetylcholine (as'-ē-til-KŌ-lēn) **(ACh)** A neurotransmitter liberated by many peripheral nervous system neurons and some central nervous system neurons. It is excitatory at neuromuscular junctions but inhibitory at some other synapses (for example, it slows heart rate).

Achalasia (ak'-a-LĀ-zē-a) A condition, caused by malfunction of the myenteric plexus, in which the lower esophageal sphincter fails to relax normally as food approaches. A whole meal may become lodged in the esophagus and enter the stomach very slowly. Distension of the esophagus results in chest pain that is often confused with pain originating from the heart.

Achilles tendon *See* **Calcaneal tendon.**

Acini (AS-i-nē) Groups of cells in the pancreas that secrete digestive enzymes. *Singular* is **acinus** (AS-i-nus).

Acoustic (a-KOOS-tik) Pertaining to sound or the sense of hearing.

Acquired immunodeficiency syndrome (AIDS) A fatal disease caused by the human immunodeficiency virus (HIV). Characterized by a positive HIV-antibody test, low helper T cell count, and certain indicator diseases (for example, Kaposi's sarcoma, pneumocystis carinii pneumonia, tuberculosis, fungal diseases). Other symptoms include fever or night sweats, coughing, sore throat, fatigue, body aches, weight loss, and enlarged lymph nodes.

Acrosome (AK-rō-sōm) A lysosomelike organelle in the head of a sperm cell containing enzymes that facilitate the penetration of a sperm cell into a secondary oocyte.

Actin (AK-tin) A contractile protein that is part of thin filaments in muscle fibers.

Action potential (AP) An electrical signal that propagates along the membrane of a neuron or muscle fiber (cell); a rapid change in membrane potential that involves a depolarization followed by a repolarization. Also called a **nerve action potential** or **nerve impulse** as it relates to a neuron, and a **muscle action potential** as it relates to a muscle fiber.

Activation (ak'-ti-VĀ-shun) **energy** The minimum amount of energy required for a chemical reaction to occur.

Active transport The movement of substances across cell membranes against a concentration gradient, requiring the expenditure of cellular energy (ATP).

Acute (a-KŪT) Having rapid onset, severe symptoms, and a short course; not chronic.

Adaptation (ad'-ap-TĀ-shun) The adjustment of the pupil of the eye to changes in light intensity. The property by which a sensory neuron relays a decreased frequency of action potentials from a receptor, even though the strength of the stimulus remains constant; the decrease in perception of a sensation over time while the stimulus is still present.

Adduction (ad-DUK-shun) Movement toward the midline of the body.

Adenoids (AD-e-noyds) The pharyngeal tonsils.

Adenosine triphosphate (a-DEN-ō-sēn trī-FOS-fāt) **(ATP)** The main energy currency in living cells; used to transfer the chemical energy needed for metabolic reactions. ATP consists of the purine base *adenine* and the five-carbon sugar *ribose*, to which are added, in linear array, three *phosphate* groups.

Adhesion (ad-HĒ-zhun) Abnormal joining of parts to each other.

Adipocyte (AD-i-pō-sīt) Fat cell, derived from a fibroblast. Also called **fat cell** or **adipose cell.**

Adipose (AD-i-pōz) **tissue** Tissue composed of adipocytes specialized for triglyceride storage and present in the form of soft pads between various organs for support, protection, and insulation.

Adrenal cortex (a-DRĒ-nal KOR-teks) The outer portion of an adrenal gland, divided into three zones; the zona glomerulosa secretes mineralocorticoids, the zona fasciculata secretes glucocorticoids, and the zona reticularis secretes androgens.

Adrenal glands Two glands located superior to each kidney. Also called the **suprarenal** (soo′-pra-RĒ-nal) **glands.**

Adrenal medulla (me-DUL-a) The inner part of an adrenal gland, consisting of cells that secrete epinephrine, norepinephrine, and a small amount of dopamine in response to stimulation by sympathetic preganglionic neurons.

Adrenergic (ad′-ren-ER-jik) **neuron** A neuron that releases epinephrine (adrenaline) or norepinephrine (noradrenaline) as its neurotransmitter.

Adrenocorticotropic (ad-rē′-nō-kor-ti-kō-TRŌP-ik) **hormone (ACTH)** A hormone produced by the anterior pituitary that influences the production and secretion of certain hormones of the adrenal cortex. Also called **corticotropin** (kor′-ti-kō-TRŌ-pin).

Adventitia (ad-ven-TISH-a) The outermost covering of a structure or organ.

Aerobic (air-Ō-bik) Requiring molecular oxygen.

Aerobic (ār-Ō-bik) **cellular respiration** The production of ATP (36 molecules) from the complete oxidation of pyruvic acid in mitochondria. Carbon dioxide, water, and heat are also produced.

Afferent arteriole (AF-er-ent ar-TĒ-rē-ōl) A blood vessel of a kidney that divides into the capillary network called a glomerulus; there is one afferent arteriole for each glomerulus.

Agglutination (a-gloo-ti-NĀ-shun) Clumping of microorganisms or blood cells, typically due to an antigen–antibody reaction.

Aggregated lymphatic follicles Clusters of lymph nodules that are most numerous in the ileum. Also called **Peyer's** (PĪ-erz) **patches.**

Albinism (AL-bin-izm) Abnormal, nonpathological, partial, or total absence of pigment in skin, hair, and eyes.

Aldosterone (al-DOS-ter-ōn) A mineralocorticoid produced by the adrenal cortex that promotes sodium and water reabsorption by the kidneys and potassium excretion in urine.

All-or-none principle If a stimulus depolarizes a neuron to threshold, the neuron fires at its maximum voltage (all); if threshold is not reached, the neuron does not fire at all (none). Given above threshold, stronger stimuli do not produce stronger action potentials.

Allantois (a-LAN-tō-is) A small, vascularized outpouching of the yolk sac that serves as an early site for blood formation and development of the urinary bladder.

Alleles (a-LĒLZ) Alternate forms of a single gene that control the same inherited trait (such as type A blood) and are located at the same position on homologous chromosomes.

Allergen (AL-er-jen) An antigen that evokes a hypersensitivity reaction.

Alopecia (al′-ō-PĒ-shē-a) The partial or complete lack of hair as a result of factors such as genetics, aging, endocrine disorders, chemotherapy, and skin diseases.

Alpha (AL-fa) **cell** A type of cell in the pancreatic islets (islets of Langerhans) that secretes the hormone glucagon. Also termed an **A cell.**

Alpha (α) receptor A type of receptor for norepinephrine and epinephrine; present on visceral effectors innervated by sympathetic postganglionic neurons.

Alveolar duct Branch of a respiratory bronchiole around which alveoli and alveolar sacs are arranged.

Alveolar macrophage (MAK-rō-fāj) Highly phagocytic cell found in the alveolar walls of the lungs. Also called a **dust cell.**

Alveolar sac A cluster of alveoli that share a common opening.

Alveolus (al-VĒ-ō-lus) A small hollow or cavity; an air sac in the lungs; milk-secreting portion of a mammary gland. *Plural* is **alveoli** (al-VĒ-ol-ī).

Alzheimer (ALTZ-hī-mer) **disease (AD)** Disabling neurological disorder characterized by dysfunction and death of specific cerebral neurons, resulting in widespread intellectual impairment, personality changes, and fluctuations in alertness.

Amenorrhea (ā-men-ō-RĒ-a) Absence of menstruation.

Amnesia (am-NĒ-zē-a) A lack or loss of memory.

Amnion (AM-nē-on) A thin, protective fetal membrane that develops from the epiblast; holds the fetus suspended in amniotic fluid. Also called the "**bag of waters.**"

Amniotic (am′-nē-OT-ik) **fluid** Fluid in the amniotic cavity, the space between the developing embryo (or fetus) and amnion; the fluid is initially produced as a filtrate from maternal blood and later includes fetal urine. It functions as a shock absorber, helps regulate fetal body temperature, and helps prevent desiccation.

Amphiarthrosis (am′-fē-ar-THRŌ-sis) A slightly movable joint, in which the articulating bony surfaces are separated by fibrous connective tissue or fibrocartilage to which both are attached; types are syndesmosis and symphysis.

Ampulla (am-PUL-la) A saclike dilation of a canal or duct.

Ampulla of Vater *See* **Hepatopancreatic ampulla.**

Anabolism (a-NAB-ō-lizm) Synthetic, energy-requiring reactions whereby small molecules are built up into larger ones.

Anaerobic (an-ar-Ō-bik) Not requiring oxygen.

Anal (Ā-nal) **canal** The last 2 or 3 cm (1 in.) of the rectum; opens to the exterior through the anus.

Anal column A longitudinal fold in the mucous membrane of the anal canal that contains a network of arteries and veins.

Anal triangle The subdivision of the female or male perineum that contains the anus.

Analgesia (an-al-JĒ-zē-a) Pain relief; absence of the sensation of pain.

Anaphase (AN-a-fāz) The third stage of mitosis in which the chromatids that have separated at the centromeres move to opposite poles of the cell.

Anaphylaxis (an′-a-fi-LAK-sis) A hypersensitivity (allergic) reaction in which IgE antibodies attach to mast cells and basophils, causing them to produce mediators of anaphylaxis (histamine, leukotrienes, kinins, and prostaglandins) that bring about increased blood permeability, increased smooth muscle contraction, and increased mucus production. Examples are hay fever, hives, and anaphylactic shock.

Anastomosis (a-nas′-tō-MŌ-sis) An end-to-end union or joining of blood vessels, lymphatic vessels, or nerves.

Anatomic dead space Spaces of the nose, pharynx, larynx, trachea, bronchi, and bronchioles totaling about 150 mL of the 500 mL in a quiet breath (tidal volume); air in the anatomic dead space does not reach the alveoli to participate in gas exchange.

Anatomical (an′-a-TOM-i-kal) **position** A position of the body universally used in anatomical descriptions in which the body is erect, the head is level, the eyes face forward, the upper limbs are at the sides, the palms face forward, and the feet are flat on the floor.

Anatomy (a-NAT-ō-mē) The structure or study of the structure of the body and the relationship of its parts to each other.

Androgens (AN-drō-jenz) Masculinizing sex hormones produced by the testes in males and the adrenal cortex in both sexes; responsible for libido (sexual desire); the two main androgens are testosterone and dihydrotestosterone.

Anemia (a-NĒ-mē-a) Condition of the blood in which the number of functional red blood cells or their hemoglobin content is below normal.

Anesthesia (an′-es-THĒ-zē-a) A total or partial loss of feeling or sensation; may be general or local.

Aneurysm (AN-ū-rizm) A saclike enlargement of a blood vessel caused by a weakening of its wall.

Angina pectoris (an-JĪ-na *or* AN-ji-na PEK-tō-ris) A pain in the chest related to reduced coronary circulation due to coronary artery disease (CAD) or spasms of vascular smooth muscle in coronary arteries.

Angiogenesis (an′-jē-ō-JEN-e-sis) The formation of blood vessels in the extraembryonic mesoderm of the yolk sac, connecting stalk, and chorion at the beginning of the third week of development.

Ankylosis (ang′-ki-LŌ-sis) Severe or complete loss of movement at a joint as the result of a disease process.

Antagonist (an-TAG-ō-nist) A muscle that has an action opposite that of the prime mover (agonist) and yields to the movement of the prime mover.

Antagonistic (an-tag-ō-NIST-ik) **effect** A hormonal interaction in which the effect of one hormone on a target cell is opposed by another hormone. For example, calcitonin (CT) lowers blood calcium level, whereas parathyroid hormone (PTH) raises it.

Anterior (an-TĒR-ē-or) Nearer to or at the front of the body. Equivalent to **ventral** in bipeds.

Anterior pituitary Anterior lobe of the pituitary gland. Also called the **adenohypophysis** (ad′-e-nō-hī-POF-i-sis).

Anterior root The structure composed of axons of motor (efferent) neurons that emerges from the anterior aspect of the spinal cord and extends laterally to join a posterior root, forming a spinal nerve. Also called a **ventral root.**

Anterolateral (an′-ter-ō-LAT-er-al) **pathway** Sensory pathway that conveys information related to pain, temperature, tickle, and itch. Also called **spinothalamic pathway.**

Antibody (AN-ti-bod′-ē) **(Ab)** A protein produced by plasma cells in response to a specific antigen; the antibody combines with that antigen to neutralize, inhibit, or destroy it. Also called an **immunoglobulin** (im-ū-nō-GLOB-ū-lin) or **Ig.**

Anticoagulant (an-tī-cō-AG-ū-lant) A substance that can delay, suppress, or prevent the clotting of blood.

Antidiuretic (an′-ti-dī-ū-RET-ik) Substance that inhibits urine formation.

Antidiuretic hormone (ADH) Hormone produced by neurosecretory cells in the paraventricular and supraoptic nuclei of the hypothalamus that stimulates water reabsorption from kidney tubule cells into the blood and vasoconstriction of arterioles. Also called **vasopressin** (vāz-ō-PRES-in).

Antigen (AN-ti-jen) **(Ag)** A substance that has immunogenicity (the ability to provoke an immune response) and reactivity (the ability to react with the antibodies or cells that result from the immune response); contraction of *anti*body *gen*erator. Also termed a **complete antigen.**

Antigen-presenting cell (APC) Special class of migratory cell that processes and presents antigens to T cells during an immune response; APCs include macrophages, B cells, and dendritic cells, which are present in the skin, mucous membranes, and lymph nodes.

Antioxidant A substance that inactivates oxygen-derived free radicals. Examples are selenium, zinc, beta carotene, and vitamins C and E.

Antrum (AN-trum) Any nearly closed cavity or chamber, especially one within a bone, such as a sinus.

Anuria (an-Ū-rē-a) Absence of urine formation or daily urine output of less than 50 mL.

Anus (Ā-nus) The distal end and outlet of the rectum.

Aorta (ā-OR-ta) The main systemic trunk of the arterial system of the body that emerges from the left ventricle.

Aortic (ā-OR-tik) **body** Cluster of chemoreceptors on or near the arch of the aorta that respond to changes in blood levels of oxygen, carbon dioxide, and hydrogen ions (H^+).

Aortic reflex A reflex that helps maintain normal systemic blood pressure; initiated by baroreceptors in the wall of the ascending aorta and arch of the aorta. Nerve impulses from aortic baroreceptors reach the cardiovascular center via sensory axons of the vagus (X) nerves.

Apex (Ā-peks) The pointed end of a conical structure, such as the apex of the heart.

Aphasia (a-FĀ-zē-a) Loss of ability to express oneself properly through speech or loss of verbal comprehension.

Apnea (AP-nē-a) Temporary cessation of breathing.

Apneustic (ap-NOO-stik) **area** A part of the respiratory center in the pons that sends stimulatory nerve impulses to the inspiratory area that activate and prolong inhalation and inhibit exhalation.

Apocrine (AP-ō-krin) **gland** A type of gland in which the secretory products gather at the free end of the secreting cell and are pinched off, along with some of the cytoplasm, to become the secretion, as in mammary glands.

Aponeurosis (ap′-ō-noo-RŌ-sis) A sheetlike tendon joining one muscle with another or with bone.

Apoptosis (ap′-ōp-TŌ-sis *or* ap-ō-TŌ-sis) Programmed cell death; a normal type of cell death that removes unneeded cells during embryological development, regulates the number of cells in tissues, and eliminates many potentially dangerous cells such as cancer cells. During apoptosis, the DNA fragments, the nucleus condenses, mitochondria cease to function, and the cytoplasm shrinks, but the plasma membrane remains intact. Phagocytes engulf and digest the apoptotic cells, and an inflammatory response does not occur.

Appositional (a-pō-ZISH-o-nal) **growth** Growth due to surface deposition of material, as in the growth in diameter of cartilage and bone. Also called **exogenous** (eks-OJ-e-nus) **growth.**

Aqueous humor (A-kwē-us HŪ-mor) The watery fluid, similar in composition to cerebrospinal fluid, that fills the anterior cavity of the eye.

Arachnoid (a-RAK-noyd) **mater** The middle of the three meninges (coverings) of the brain and spinal cord. Also termed the **arachnoid.**

Arachnoid villus (VIL-us) Berrylike tuft of the arachnoid mater that protrudes into the superior sagittal sinus and through which cerebrospinal fluid is reabsorbed into the bloodstream.

Arbor vitae (AR-bor VĪ-tē) The white matter tracts of the cerebellum, which have a treelike appearance when seen in midsagittal section.

Arch of the aorta The most superior portion of the aorta, lying between the ascending and descending segments of the aorta.

Areola (a-RĒ-ō-la) Any tiny space in a tissue. The pigmented ring around the nipple of the breast.

Arm The part of the upper limb from the shoulder to the elbow.

Arousal (a-ROW-zal) Awakening from sleep, a response due to stimulation of the reticular activating system (RAS).

Arrector pili (a-REK-tor PĪ-lē) Smooth muscles attached to hairs; contraction pulls the hairs into a vertical position, resulting in "goose bumps."

Arrhythmia (a-RITH-mē-a) An irregular heart rhythm. Also called a **dysrhythmia.**

Arteriole (ar-TĒ-rē-ōl) A small, almost microscopic, artery that delivers blood to a capillary.

Arteriosclerosis (ar-tē-rē-ō-skle-RŌ-sis) Group of diseases characterized by thickening of the walls of arteries and loss of elasticity.

Artery (AR-ter-ē) A blood vessel that carries blood away from the heart.

Arthritis (ar-THRĪ-tis) Inflammation of a joint.

Arthrology (ar-THROL-ō-jē) The study or description of joints.

Arthroplasty (AR-thrō-plas′-tē) Surgical replacement of joints, for example, the hip and knee joints.

Arthroscopy (ar-THROS-kō-pē) A procedure for examining the interior of a joint, usually the knee, by inserting an arthroscope into a small incision; used to determine extent of damage, remove torn cartilage, repair cruciate ligaments, and obtain samples for analysis.

Arthrosis (ar-THRŌ-sis) A joint or articulation.

Articular (ar-TIK-ū-lar) **capsule** Sleevelike structure around a synovial joint composed of a fibrous capsule and a synovial membrane.

Articular cartilage (KAR-ti-lij) Hyaline cartilage attached to articular bone surfaces.

Articular disc Fibrocartilage pad between articular surfaces of bones of some synovial joints. Also called a **meniscus** (men-IS-kus).

Articulation (ar-tik-ū-LĀ-shun) A joint; a point of contact between bones, cartilage and bones, or teeth and bones.

Arytenoid (ar′-i-TĒ-noyd) **cartilages** A pair of small, pyramidal cartilages of the larynx that attach to the vocal folds and intrinsic pharyngeal muscles and can move the vocal folds.

Ascending colon (KŌ-lon) The part of the large intestine that passes superiorly from the cecum to the inferior border of the liver, where it bends at the right colic (hepatic) flexure to become the transverse colon.

Ascites (a-SĪ-tēz) Abnormal accumulation of serous fluid in the peritoneal cavity.

Association areas Large cortical regions on the lateral surfaces of the occipital, parietal, and temporal lobes and on the frontal lobes anterior to the motor areas connected by many motor and sensory axons to other parts of the cortex. The association areas are concerned with motor patterns, memory, concepts of word-hearing and word-seeing, reasoning, will, judgment, and personality traits.

Asthma (AZ-ma) Usually allergic reaction characterized by smooth muscle spasms in bronchi resulting in wheezing and difficult breathing. Also called **bronchial asthma.**

Astigmatism (a-STIG-ma-tizm) An irregularity of the lens or cornea of the eye causing the image to be out of focus and producing faulty vision.

Astrocyte (AS-trō-sīt) A neuroglial cell having a star shape that participates in brain development and the metabolism of neurotransmitters, helps form the blood–brain barrier, helps maintain the proper balance of K^+ for generation of nerve impulses, and provides a link between neurons and blood vessels.

Ataxia (a-TAK-sē-a) A lack of muscular coordination, lack of precision.

Atherosclerosis (ath-er-ō-skle-RŌ-sis) A progressive disease characterized by the formation in the walls of large and medium-sized arteries of lesions called atherosclerotic plaques.

Atherosclerotic (ath′-er-ō-skle-RO-tik) **plaque** (PLAK) A lesion that results from accumulated cholesterol and smooth muscle fibers (cells) of the tunica media of an artery; may become obstructive.

Atom Unit of matter that makes up a chemical element; consists of a nucleus (containing positively charged protons and uncharged neutrons) and negatively charged electrons that orbit the nucleus.

Atresia (a-TRĒ-zē-a) Degeneration and reabsorption of an ovarian follicle before it fully matures and ruptures; abnormal closure of a passage, or absence of a normal body opening.

Atrial fibrillation (Ā-trē-al fib-ri-LĀ-shun) **(AF)** Asynchronous contraction of cardiac muscle fibers in the atria that results in the cessation of atrial pumping.

Atrial natriuretic (nā′-trē-ū-RET-ik) **peptide (ANP)** Peptide hormone, produced by the atria of the heart in response to stretching, that inhibits aldosterone production and thus lowers blood pressure; causes natriuresis, increased urinary excretion of sodium.

Atrioventricular (AV) (ā′-trē-ō-ven-TRIK-ū-lar) **bundle** The part of the conduction system of the heart that begins at the atrioventricular (AV) node, passes through the cardiac skeleton separating the atria and the ventricles, then extends a short distance down the interventricular septum before splitting into right and left bundle branches. Also called the **bundle of His** (HISS).

Atrioventricular (AV) node The part of the conduction system of the heart made up of a compact mass of conducting cells located in the septum between the two atria.

Atrioventricular (AV) valve A heart valve made up of membranous flaps or cusps that allows blood to flow in one direction only, from an atrium into a ventricle.

Atrium (Ā-trē-um) A superior chamber of the heart.

Atrophy (AT-rō-fē) Wasting away or decrease in size of a part, due to a failure, abnormality of nutrition, or lack of use.

Auditory ossicle (AW-di-tō-rē OS-si-kul) One of the three small bones of the middle ear called the malleus, incus, and stapes.

Auditory tube The tube that connects the middle ear with the nose and nasopharynx region of the throat. Also called the **eustachian** (ū-STĀ-kē-an *or* ū-STĀ-shun) **tube** or **pharyngotympanic tube.**

Auscultation (aws-kul-TĀ-shun) Examination by listening to sounds in the body.

Autoimmunity An immunological response against a person's own tissues.

Autolysis (aw-TOL-i-sis) Self-destruction of cells by their own lysosomal digestive enzymes after death or in a pathological process.

Autonomic ganglion (aw′-tō-NOM-ik GANG-lē-on) A cluster of cell bodies of sympathetic or parasympathetic neurons located outside the central nervous system.

Autonomic nervous system (ANS) Visceral sensory (afferent) and visceral motor (efferent) neurons. Autonomic motor neurons, both sympathetic and parasympathetic, conduct nerve impulses from the central nervous system to smooth muscle, cardiac muscle, and glands. So named because this part of the nervous system was thought to be self-governing or spontaneous.

Autonomic plexus (PLEK-sus) A network of sympathetic and parasympathetic axons; examples are the cardiac, celiac, and pelvic plexuses, which are located in the thorax, abdomen, and pelvis, respectively.

Autophagy (aw-TOF-a-jē) Process by which worn-out organelles are digested within lysosomes.

Autopsy (AW-top-sē) The examination of the body after death.

Autorhythmic (aw′-tō-RITH-mik) **cells** Cardiac or smooth muscle fibers that are self-excitable (generate impulses without an external stimulus); act as the heart's pacemaker and conduct the pacing impulse through the conduction system of the heart; self-excitable neurons in the central nervous system, as in the inspiratory area of the brain stem.

Autosome (AW-tō-sōm) Any chromosome other than the X and Y chromosomes (sex chromosomes).

Axilla (ak-SIL-a) The small hollow beneath the arm where it joins the body at the shoulders. Also called the **armpit.**

Axon (AK-son) The usually single, long process of a nerve cell that propagates a nerve impulse toward the axon terminals.

Axon terminal Terminal branch of an axon where synaptic vesicles undergo exocytosis to release neurotransmitter molecules. Also called **telodendria** (tel′-o-DEN-drea).

B

B cell A lymphocyte that can develop into a clone of antibody-producing plasma cells or memory cells when properly stimulated by a specific antigen.

Babinski (ba-BIN-skē) **sign** Extension of the great toe, with or without fanning of the other toes, in response to stimulation of the outer margin of the sole; normal up to 18 months of age and indicative of damage to descending motor pathways such as the corticospinal tracts after that.

Back The posterior part of the body; the dorsum.

Ball-and-socket joint A synovial joint in which the rounded surface of one bone moves within a cup-shaped depression or socket of another bone, as in the shoulder or hip joint. Also called a **spheroid** (SFĒ-royd) **joint.**

Baroreceptor (bar′-ō-re-SEP-tor) Neuron capable of responding to changes in blood or air or fluid pressure. Also called a **pressoreceptor** or **stretch receptor.**

Basal nuclei Paired clusters of gray matter deep in each cerebral hemisphere including the globus pallidus, putamen, and caudate nucleus. Together, the caudate nucleus and putamen are known as the corpus striatum. Nearby structures that are functionally linked to the basal nuclei are the substantia nigra of the midbrain and the subthalamic nuclei of the diencephalon.

Basement membrane Thin, extracellular layer between epithelium and connective tissue consisting of a basal lamina and a reticular lamina.

Basilar (BĀS-i-lar) **membrane** A membrane in the cochlea of the internal ear that separates the cochlear duct from the scala tympani and on which the spiral organ (organ of Corti) rests.

Basophil (BĀ-sō-fil) A type of white blood cell characterized by a pale nucleus and large granules that stain blue-purple with basic dyes.

Belly The abdomen. The gaster or prominent, fleshy part of a skeletal muscle.

Beta (BĀ-ta) **cell** A type of cell in the pancreatic islets (islets of Langerhans) in the pancreas that secretes the hormone insulin. Also called a **B cell.**

Beta (β) receptor A type of adrenergic receptor for epinephrine and norepinephrine; found on visceral effectors innervated by sympathetic postganglionic neurons.

Bicuspid (bī-KUS-pid) **valve** Atrioventricular (AV) valve on the left side of the heart. Also called the **mitral valve.**

Bilateral (bī-LAT-er-al) Pertaining to two sides of the body.

Bile (BĪL) A secretion of the liver consisting of water, bile salts, bile pigments, cholesterol, lecithin, and several ions; it emulsifies lipids prior to their digestion.

Bilirubin (bil-ē-ROO-bin) An orange pigment that is one of the end products of hemoglobin breakdown in the hepatocytes and is excreted as a waste material in bile.

Biopsy (BĪ-op-sē) The removal of a sample of living tissue to help diagnose a disorder, for example, cancer.

Blastocyst (BLAS-tō-sist) In the development of an embryo, a hollow ball of cells that consists of a blastocele (the internal cavity), trophoblast (outer cells), and embryoblast (inner cell mass).

Blastocyst (BLAS-tō-sist) **cavity** The fluid-filled cavity within the blastocyst. Also called the **blastocele.**

Blastomere (BLAS-tō-mēr) One of the cells resulting from the cleavage of a fertilized ovum.

Blastula (BLAS-tyū-la) An early stage in the development of a zygote.

Blind spot Area in the retina at the end of the optic (II) nerve in which there are no photoreceptors. Also called **optic disc.**

Blood The fluid that circulates through the heart, arteries, capillaries, and veins and that constitutes the chief means of transport within the body.

Blood–brain barrier (BBB) A barrier consisting of specialized brain capillaries and astrocytes that prevents the passage of materials from the blood to the cerebrospinal fluid and brain.

Blood clot A gel that consists of the formed elements of blood trapped in a network of insoluble protein fibers.

Blood island Isolated mass of mesoderm derived from angioblasts and from which blood vessels develop.

Blood pressure (BP) Force exerted by blood against the walls of blood vessels due to contraction of the heart and influenced by the elasticity of the vessel walls; clinically, a measure of the pressure in arteries during ventricular systole and ventricular diastole.

Blood reservoir (REZ-er-vwar) Systemic veins and venules that contain large amounts of blood that can be moved quickly to parts of the body requiring the blood.

Blood–testis barrier (BTB) A barrier formed by Sertoli cells that prevents an immune response against antigens produced by spermatogenic cells by isolating the cells from the blood.

Body cavity A space within the body that contains various internal organs.

Bolus (BŌ-lus) A soft, rounded mass, usually food, that is swallowed.

Bone remodeling Replacement of old bone by new bone tissue.

Bony labyrinth (LAB-i-rinth) A series of cavities within the petrous portion of the temporal bone forming the vestibule, cochlea, and semicircular canals of the inner ear.

Bowman's capsule See **Glomerular capsule.**

Brachial plexus (BRĀ-kē-al PLEK-sus) A network of nerve axons of the ventral rami of spinal nerves C5, C6, C7, C8, and T1. The nerves that emerge from the brachial plexus supply the upper limb.

Bradycardia (brād′-i-KAR-dē-a) A slow resting heart or pulse rate (under 50 beats per minute).

Brain The part of the central nervous system contained within the cranial cavity.

Brain stem The portion of the brain immediately superior to the spinal cord, made up of the medulla oblongata, pons, and midbrain.

Brain waves Electrical signals that can be recorded from the skin of the head due to electrical activity of brain neurons.

Broad ligament A double fold of parietal peritoneum attaching the uterus to the side of the pelvic cavity.

Broca's (BRŌ-kaz) **speech area** Motor area of the brain in the frontal lobe that translates thoughts into speech. Also called the **motor speech area.**

Bronchi (BRON-kī) Branches of the respiratory passageway including primary bronchi (the two divisions of the trachea), secondary or lobar bronchi (divisions of the primary bronchi that are distributed to the lobes of the lung), and tertiary or segmental bronchi (divisions of the secondary bronchi that are distributed to bronchopulmonary segments of the lung). *Singular* is **bronchus.**

Bronchial (BRON-kē-al) **tree** The trachea, bronchi, and their branching structures up to and including the terminal bronchioles.

Bronchiole (BRONG-kē-ōl) Branch of a tertiary bronchus further dividing into terminal bronchioles (distributed to lobules of the lung), which divide into respiratory bronchioles (distributed to alveolar sacs).

Bronchitis (brong-KĪ-tis) Inflammation of the mucous membrane of the bronchial tree; characterized by hypertrophy and hyperplasia of seromucous glands and goblet cells that line the bronchi, which results in a productive cough.

Bronchopulmonary (brong′-kō-PUL-mō-ner-ē) **segment** One of the smaller divisions of a lobe of a lung supplied by its own branches of a bronchus.

Brunner's gland See **Duodenal gland.**

Buccal (BUK-al) Pertaining to the cheek or mouth.

Buffer system A weak acid and the salt of that acid (that functions as a weak base). Buffers prevent drastic changes in pH by converting strong acids and bases to weak acids and bases.

Bulb of penis Expanded portion of the base of the corpus spongiosum penis.

Bulbourethral (bul′-bō-ū-RĒ-thral) **gland** One of a pair of glands located inferior to the prostate on either side of the urethra that secretes an alkaline fluid into the cavernous urethra. Also called a **Cowper's** (KOW-perz) **gland.**

Bulimia (boo-LIM-ē-a *or* boo-LĒ-mē-a) A disorder characterized by overeating at least twice a week followed by purging by self-induced vomiting, strict dieting or fasting, vigorous exercise, or use of laxatives or diuretics. Also called **binge–purge syndrome.**

Bulk-phase endocytosis A process by which most body cells can ingest membrane-surrounded droplets of interstitial fluid.

Bundle branch One of the two branches of the atrioventricular (AV) bundle made up of specialized muscle fibers (cells) that transmit electrical impulses to the ventricles.

Bundle of His See **Atrioventricular (AV) bundle.**

Burn Tissue damage caused by excessive heat, electricity, radioactivity, or corrosive chemicals that denature (break down) proteins in the skin.

Bursa (BUR-sa) A sac or pouch of synovial fluid located at friction points, especially about joints.

Bursitis (bur-SĪ-tis) Inflammation of a bursa.

Buttocks (BUT-oks) The two fleshy masses on the posterior aspect of the inferior trunk, formed by the gluteal muscles.

C

Calcaneal (kal-KĀ-nē-al) **tendon** The tendon of the soleus, gastrocnemius, and plantaris muscles at the back of the heel. Also called the **Achilles** (a-KIL-ēz) **tendon.**

Calcification (kal′-si-fi-KĀ-shun) Deposition of mineral salts, primarily hydroxyapatite, in a framework formed by collagen fibers in which the tissue hardens. Also called **mineralization** (min′-e-ral-i-ZĀ-shun).

Calcitonin (kal-si-TŌ-nin) **(CT)** A hormone produced by the parafollicular cells of the thyroid gland that can lower the amount of blood calcium and phosphates by inhibiting bone resorption (breakdown of bone extracellular matrix) and by accelerating uptake of calcium and phosphates into bone matrix.

Calculus (KAL-kū-lus) A stone, or insoluble mass of crystallized salts or other material, formed within the body, as in the gallbladder, kidney, or urinary bladder.

Callus (KAL-lus) A growth of new bone tissue in and around a fractured area, ultimately replaced by mature bone. An acquired, localized thickening.

Calyx (KĀL-iks) Any cuplike division of the kidney pelvis. *Plural* is **calyces** (KĀ-li-sēz).

Canal (ka-NAL) A narrow tube, channel, or passageway.

Canaliculus (kan′-a-LIK-ū-lus) A small channel or canal, as in bones, where they connect lacunae. *Plural* is **canaliculi** (kan′-a-LIK-ū-lī).

Canal of Schlemm See **Scleral venous sinus.**

Cancer A group of diseases characterized by uncontrolled or abnormal cell division.

Capacitation (ka-pas′-i-TĀ-shun) The functional changes that sperm undergo in the female reproductive tract that allow them to fertilize a secondary oocyte.

Capillary (KAP-i-lar′-ē) A microscopic blood vessel located between an arteriole and venule through which materials are exchanged between blood and interstitial fluid.

Carbohydrate Organic compound consisting of carbon, hydrogen, and oxygen; the ratio of hydrogen to oxygen atoms is usually 2:1. Examples include sugars, glycogen, starches, and glucose.

Carcinogen (kar-SIN-ō-jen) A chemical substance or radiation that causes cancer.

Cardiac (KAR-dē-ak) **arrest** Cessation of an effective heartbeat in which the heart is completely stopped or in ventricular fibrillation.

Cardiac cycle A complete heartbeat consisting of systole (contraction) and diastole (relaxation) of both atria plus systole and diastole of both ventricles.

Cardiac muscle Striated muscle fibers (cells) that form the wall of the heart; stimulated by an intrinsic conduction system and autonomic motor neurons.

Cardiac notch An angular notch in the anterior border of the left lung into which part of the heart fits.

Cardiac output (CO) The volume of blood ejected from the left ventricle (or the right ventricle) into the aorta (or pulmonary trunk) each minute.

Cardinal ligament A ligament of the uterus, extending laterally from the cervix and vagina as a continuation of the broad ligament.

Cardiogenic area (kar-dē-ō-JEN-ik) A group of mesodermal cells in the head end of an embryo that gives rise to the heart.

Cardiology (kar-dē-OL-ō-jē) The study of the heart and diseases associated with it.

Cardiovascular (kar-dē-ō-VAS-kū-lar) **(CV) center** Groups of neurons scattered within the medulla oblongata that regulate heart rate, force of contraction, and blood vessel diameter.

Cardiovascular physiology Study of the functions of the heart and blood vessels.

Cardiovascular system System that consists of blood, the heart, and blood vessels.

Carotene (KAR-ō-tēn) Antioxidant precursor of vitamin A, which is needed for synthesis of photopigments; yellow-orange pigment present in the stratum corneum of the epidermis. Accounts for the yellowish coloration of skin. Also termed **beta carotene.**

Carotid (ka-ROT-id) **body** Cluster of chemoreceptors on or near the carotid sinus that respond to changes in blood levels of oxygen, carbon dioxide, and hydrogen ions.

Carotid sinus A dilated region of the internal carotid artery just superior to where it branches from the common carotid artery; it contains baroreceptors that monitor blood pressure.

Carpal bones The eight bones of the wrist. Also called **carpals.**

Carpus (KAR-pus) A collective term for the eight bones of the wrist.

Cartilage (KAR-ti-lij) A type of connective tissue consisting of chondrocytes in lacunae embedded in a dense network of collagen and elastic fibers and an extracellular matrix of chondroitin sulfate.

Cartilaginous (kar′-ti-LAJ-i-nus) **joint** A joint without a synovial (joint) cavity where the articulating bones are held tightly together by cartilage, allowing little or no movement.

Catabolism (ka-TAB-ō-lizm) Chemical reactions that break down complex organic compounds into simple ones, with the net release of energy.

Cataract (KAT-a-rakt) Loss of transparency of the lens of the eye or its capsule or both.

Cauda equina (KAW-da ē-KWĪ-na) A tail-like array of roots of spinal nerves at the inferior end of the spinal cord.

Caudal (KAW-dal) Pertaining to any tail-like structure; inferior in position.

Cecum (SĒ-kum) A blind pouch at the proximal end of the large intestine that attaches to the ileum.

Celiac plexus (SĒ-lē-ak PLEK-sus) A large mass of autonomic ganglia and axons located at the level of the superior part of the first lumbar vertebra. Also called the **solar plexus.**

Cell The basic structural and functional unit of all organisms; the smallest structure capable of performing all of the activities vital to life.

Cell biology The study of cellular structure and function. Also called **cytology.**

Cell cycle Growth and division of a single cell into two identical cells; consists of interphase and cell division.

Cell division Process by which a cell reproduces itself that consists of a nuclear division (mitosis) and a cytoplasmic division (cytokinesis); types include somatic and reproductive cell division.

Cell junction Point of contact between plasma membranes of tissue cells.

Cellular respiration The oxidation of glucose to produce ATP that involves glycolysis, acetyl coenzyme A formation, the Krebs cycle, and the electron transport chain.

Cementum (se-MEN-tum) Calcified tissue covering the root of a tooth.

Central canal A microscopic tube running the length of the spinal cord in the gray commissure. A circular channel running longitudinally in the center of an osteon (haversian system) of mature compact bone, containing blood and lymphatic vessels and nerves. Also called a **haversian** (ha-VER-shun) **canal.**

Central fovea (FŌ-vē-a) A depression in the center of the macula lutea of the retina, containing cones only and lacking blood vessels; the area of highest visual acuity (sharpness of vision).

Central nervous system (CNS) That portion of the nervous system that consists of the brain and spinal cord.

Centrioles (SEN-trē-ōlz) Paired, cylindrical structures of a centrosome, each consisting of a ring of microtubules and arranged at right angles to each other.

Centromere (SEN-trō-mēr) The constricted portion of a chromosome where the two chromatids are joined; serves as the point of attachment for the microtubules that pull chromatids during anaphase of cell division.

Centrosome (SEN-trō-sōm) A dense network of small protein fibers near the nucleus of a cell, containing a pair of centrioles and pericentriolar material.

Cephalic (se-FAL-ik) Pertaining to the head; superior in position.

Cerebellar peduncle (ser-e-BEL-ar pe-DUNG-kul) A bundle of nerve axons connecting the cerebellum with the brain stem.

Cerebellum (ser′-e-BEL-um) The part of the brain lying posterior to the medulla oblongata and pons; governs balance and coordinates skilled movements.

Cerebral aqueduct (SER-ē-bral AK-we-dukt) A channel through the midbrain connecting the third and fourth ventricles and containing cerebrospinal fluid. Also termed the **aqueduct of the midbrain.**

Cerebral arterial circle A ring of arteries forming an anastomosis at the base of the brain between the internal carotid and basilar arteries and arteries supplying the cerebral cortex. Also called the **circle of Willis.**

Cerebral cortex The surface of the cerebral hemispheres, 2–4 mm thick, consisting of gray matter; arranged in six layers of neuronal cell bodies in most areas.

Cerebral peduncle (pe-DUNG-kul or PĒ-dung-kul) One of a pair of nerve axon bundles located on the anterior surface of the midbrain, conducting nerve impulses between the pons and the cerebral hemispheres.

Cerebrospinal (se-rē′-brō-SPĪ-nal) **fluid (CSF)** A fluid produced by ependymal cells that cover choroid plexuses in the ventricles of the brain; the fluid circulates in the ventricles, the central canal, and the subarachnoid space around the brain and spinal cord.

Cerebrovascular (se-rē′-brō-VAS-kū-lar) **accident (CVA)** Destruction of brain tissue (infarction) resulting from obstruction or rupture of blood vessels that supply the brain. Also called a **stroke** or **brain attack.**

Cerebrum (se-RĒ-brum or SER-e-brum) The two hemispheres of the forebrain (derived from the telencephalon), making up the largest part of the brain.

Cerumen (se-ROO-men) Waxlike secretion produced by ceruminous glands in the external auditory meatus (ear canal). Also termed **earwax.**

Ceruminous (se-RŪ-mi-nus) **gland** A modified sudoriferous (sweat) gland in the external auditory meatus that secretes cerumen (ear wax).

Cervical ganglion (SER-vi-kul GANG-glē-on) A cluster of cell bodies of postganglionic sympathetic neurons located in the neck, near the vertebral column.

Cervical plexus (PLEK-sus) A network formed by nerve axons from the ventral rami of the first four cervical nerves and receiving gray rami communicantes from the superior cervical ganglion.

Cervix (SER-viks) Neck; any constricted portion of an organ, such as the inferior cylindrical part of the uterus.

Chemical reaction The formation of new chemical bonds or the breaking of old chemical bonds between atoms.

Chemistry (KEM-is-trē) The science of the structure and interactions of matter.

Chemoreceptor (kē′-mō-rē-SEP-tor) Sensory receptor that detects the presence of a specific chemical.

Chiasm (KĪ-azm) A crossing; especially the crossing of axons in the optic (II) nerve.

Chief cell The secreting cell of a gastric gland that produces pepsinogen, the precursor of the enzyme pepsin, and the enzyme gastric lipase. Also called a **zymogenic** (zī′-mō-JEN-ik) **cell.** Cell in the parathyroid glands that secretes parathyroid hormone (PTH). Also called a **principal cell.**

Cholecystectomy (kō′-lē-sis-TEK-tō-mē) Surgical removal of the gallbladder.

Cholecystitis (kō′-lē-sis-TĪ-tis) Inflammation of the gallbladder.

Cholesterol (kō-LES-te-rol) Classified as a lipid, the most abundant steroid in animal tissues; located in cell membranes and used for the synthesis of steroid hormones and bile salts.

Cholinergic (kō′-lin-ER-jik) **neuron** A neuron that liberates acetylcholine as its neurotransmitter.

Chondrocyte (KON-drō-sīt) Cell of mature cartilage.

Chondroitin (kon-DROY-tin) **sulfate** An amorphous extracellular matrix material found outside connective tissue cells.

Chordae tendineae (KOR-dē TEN-di-nē-ē) Tendonlike, fibrous cords that connect atrioventricular valves of the heart with papillary muscles.

Chorion (KŌ-rē-on) The most superficial fetal membrane that becomes the principal embryonic portion of the placenta; serves a protective and nutritive function.

Chorionic villi (kō-rē-ON-ik VIL-li) Fingerlike projections of the chorion that grow into the decidua basalis of the endometrium and contain fetal blood vessels.

Chorionic villi sampling (CVS) The removal of a sample of chorionic villus tissue by means of a catheter to analyze the tissue for prenatal genetic defects.

Choroid (KŌ-royd) One of the vascular coats of the eyeball.

Choroid plexus (PLEK-sus) A network of capillaries located in the roof of each of the four ventricles of the brain; ependymal cells around choroid plexuses produce cerebrospinal fluid.

Chromaffin (KRŌ-maf-in) **cell** Cell that has an affinity for chrome salts, due in part to the presence of the precursors of the neurotransmitter epinephrine; found, among other places, in the adrenal medulla.

Chromatid (KRŌ-ma-tid) One of a pair of identical connected nucleoprotein strands that are joined at the centromere and separate during cell division, each becoming a chromosome of one of the two daughter cells.

Chromatin (KRŌ-ma-tin) The threadlike mass of genetic material, consisting of DNA and histone proteins, that is present in the nucleus of a nondividing or interphase cell.

Chromatolysis (krō′-ma-TOL-i-sis) The breakdown of Nissl bodies into finely granular masses in the cell body of a neuron whose axon has been damaged.

Chromosome (KRŌ-mō-sōm) One of the small, threadlike structures in the nucleus of a cell, normally 46 in a human diploid cell, that bears the genetic material; composed of DNA and proteins (histones) that form a delicate chromatin thread during interphase; becomes packaged into compact rodlike structures that are visible under the light microscope during cell division.

Chronic (KRON-ik) Long term or frequently recurring; applied to a disease that is not acute.

Chronic obstructive pulmonary disease (COPD) A disease, such as bronchitis or emphysema, in which there is some degree of obstruction of airways and consequent increase in airway resistance.

Chyle (KĪL) The milky-appearing fluid found in the lacteals of the small intestine after absorption of lipids in food.

Chyme (KĪM) The semifluid mixture of partly digested food and digestive secretions found in the stomach and small intestine during digestion of a meal.

Ciliary (SIL-ē-ar′-ē) **body** One of the three parts of the vascular tunic of the eyeball, the others being the choroid and the iris; includes the ciliary muscle and the ciliary processes.

Ciliary ganglion (GANG-glē-on) A very small parasympathetic ganglion whose preganglionic axons come from the oculomotor (III) nerve and whose postganglionic axons carry nerve impulses to the ciliary muscle and the sphincter muscle of the iris.

Cilium (SIL-ē-um) A hair or hairlike process projecting from a cell that may be used to move the entire cell or to move substances along the surface of the cell. *Plural* is **cilia.**

Circadian (ser-KĀ-dē-an) **rhythm** The pattern of biological activity on a 24-hour cycle, such as the sleep–wake cycle.

Circle of Willis *See* **Cerebral arterial circle.**

Circular folds Permanent, deep, transverse folds in the mucosa and submucosa of the small intestine that increase the surface area for absorption. Also called **plicae circulares** (PLĪ-kē SER-kū-lar-ēs).

Circulation time The time required for a drop of blood to pass from the right atrium, through pulmonary circulation, back to the left atrium, through systemic circulation down to the foot, and back again to the right atrium.

Circumduction (ser-kum-DUK-shun) A movement at a synovial joint in which the distal end of a bone moves in a circle while the proximal end remains relatively stable.

Cirrhosis (si-RŌ-sis) A liver disorder in which the parenchymal cells are destroyed and replaced by connective tissue.

Cisterna chyli (sis-TER-na KĪ-lē) The origin of the thoracic duct.

Cleavage (KLĒV-ij) The rapid mitotic divisions following the fertilization of a secondary oocyte, resulting in an increased number of progressively smaller cells, called blastomeres.

Clitoris (KLI-to-ris) An erectile organ of the female, located at the anterior junction of the labia minora, that is homologous to the male penis.

Clone (KLŌN) A population of identical cells.

Coarctation (kō′-ark-TĀ-shun) **of the aorta** A congenital heart defect in which a segment of the aorta is too narrow. As a result, the flow of oxygenated blood to the body is reduced, the left ventricle is forced to pump harder, and high blood pressure develops.

Coccyx (KOK-siks) The fused bones at the inferior end of the vertebral column.

Cochlea (KOK-lē-a) A winding, cone-shaped tube forming a portion of the inner ear and containing the spiral organ (organ of Corti).

Cochlear duct The membranous cochlea consisting of a spirally arranged tube enclosed in the bony cochlea and lying along its outer wall. Also called the **scala media** (SCA-la MĒ-dē-a).

Collagen (KOL-a-jen) A protein that is the main organic constituent of connective tissue.

Collateral circulation The alternate route taken by blood through an anastomosis.

Colliculus (ko-LIK-ū-lus) A small elevation.

Colon The portion of the large intestine consisting of ascending, transverse, descending, and sigmoid portions.

Colony-stimulating factor (CSF) One of a group of molecules that stimulates development of white blood cells. Examples are macrophage CSF and granulocyte CSF.

Colostrum (kō-LOS-trum) A thin, cloudy fluid secreted by the mammary glands a few days prior to or after delivery before true milk is produced.

Column (KOL-um) Group of white matter tracts in the spinal cord.

Common bile duct A tube formed by the union of the common hepatic duct and the cystic duct that empties bile into the duodenum at the hepatopancreatic ampulla (ampulla of Vater).

Compact (dense) bone tissue Bone tissue that contains few spaces between osteons (haversian systems); forms the external portion of all bones and the bulk of the diaphysis (shaft) of long bones; is found immediately deep to the periosteum and external to spongy bone.

Concha (KON-ka) A scroll-like bone found in the skull. *Plural* is **conchae** (KON-kē).

Concussion (kon-KUSH-un) Traumatic injury to the brain that produces no visible bruising but may result in abrupt, temporary loss of consciousness.

Conduction system A group of autorhythmic cardiac muscle fibers that generates and distributes electrical impulses to stimulate coordinated contraction of the heart chambers; includes the sinoatrial (SA) node, the atrioventricular (AV) node, the atrioventricular (AV) bundle, the right and left bundle branches, and the Purkinje fibers.

Condyloid (KON-di-loyd) **joint** A synovial joint structured so that an oval-shaped condyle of one bone fits into an elliptical cavity of another bone, permitting side-to-side and back-and-forth movements, such as the joint at the wrist between the radius and carpals. Also called an **ellipsoidal** (ē-lip-SOYD-al) **joint.**

Cone (KŌN) The type of photoreceptor in the retina that is specialized for highly acute color vision in bright light.

Congenital (kon-JEN-i-tal) Present at the time of birth.

Conjunctiva (kon′-junk-TĪ-va) The delicate membrane covering the eyeball and lining the eyes.

Connective tissue One of the most abundant of the four basic tissue types in the body, performing the functions of binding and supporting; consists of relatively few cells in a generous extracellular matrix (the ground substance and fibers between the cells).

Consciousness (KON-shus-nes) A state of wakefulness in which an individual is fully alert, aware, and oriented, partly as a result of feedback between the cerebral cortex and reticular activating system.

Continuous conduction (kon-DUK-shun) Propagation of an action potential (nerve impulse) in a step-by-step depolarization of each adjacent area of an axon membrane.

Contraception (kon′-tra-SEP-shun) The prevention of fertilization or impregnation without destroying fertility.

Contractility (kon′-trak-TIL-i-tē) The ability of cells or parts of cells to actively generate force to undergo shortening for movements. Muscle fibers (cells) exhibit a high degree of contractility.

Contralateral (KON-tra-lat-er-al) On the opposite side; affecting the opposite side of the body.

Conus medullaris (KŌ-nus med-ū-LAR-is) The tapered portion of the spinal cord inferior to the lumbar enlargement.

Convergence (con-VER-jens) A synaptic arrangement in which the synaptic end bulbs of several presynaptic neurons terminate on one postsynaptic neuron. The medial movement of the two eyeballs so that both are directed toward a near object being viewed in order to produce a single image.

Cornea (KOR-nē-a) The nonvascular, transparent fibrous coat through which the iris of the eye can be seen.

Corona (kō-RŌ-na) Margin of the glans penis.

Corona radiata The innermost layer of granulosa cells that is firmly attached to the zona pellucida around a secondary oocyte.

Coronary artery disease (CAD) A condition such as atherosclerosis that causes narrowing of coronary arteries so that blood flow to the heart is reduced. The result is **coronary heart disease (CHD),** in which the heart muscle receives

inadequate blood flow due to an interruption of its blood supply.

Coronary circulation The pathway followed by the blood from the ascending aorta through the blood vessels supplying the heart and returning to the right atrium. Also called **cardiac circulation.**

Coronary sinus (SĪ-nus) A wide venous channel on the posterior surface of the heart that collects the blood from the coronary circulation and returns it to the right atrium.

Corpus albicans (KOR-pus AL-bi-kanz) A white fibrous patch in the ovary that forms after the corpus luteum regresses.

Corpus callosum (kal-LŌ-sum) The great commissure of the brain between the cerebral hemispheres.

Corpuscle of touch *See* **Meissner corpuscle.**

Corpus luteum (LOO-tē-um) A yellowish body in the ovary formed when a follicle has discharged its secondary oocyte; secretes estrogens, progesterone, relaxin, and inhibin.

Corpus striatum (strī-Ā-tum) An area in the interior of each cerebral hemisphere composed of the caudate and putamen of the basal ganglia and white matter of the internal capsule, arranged in a striated manner.

Cortex (KOR-teks) An outer layer of an organ. The convoluted layer of gray matter covering each cerebral hemisphere.

Costal (KOS-tal) Pertaining to a rib.

Cowper's gland *See* **Bulbourethral gland.**

Cramp A spasmodic, usually painful contraction of a muscle.

Cranial (KRĀ-nē-al) **cavity** A body cavity formed by the cranial bones and containing the brain.

Cranial nerve One of 12 pairs of nerves that leave the brain; pass through foramina in the skull; and supply sensory and motor neurons to the head, neck, part of the trunk, and viscera of the thorax and abdomen. Each is designated by a Roman numeral and a name.

Craniosacral (krā-nē-ō-SĀK-ral) **outflow** The axons of parasympathetic preganglionic neurons, which have their cell bodies located in nuclei in the brain stem and in the lateral gray matter of the sacral portion of the spinal cord.

Cranium (KRĀ-nē-um) The skeleton of the skull that protects the brain and the organs of sight, hearing, and balance; includes the frontal, parietal, temporal, occipital, sphenoid, and ethmoid bones.

Crista (KRIS-ta) A crest or ridged structure. A small elevation in the ampulla of each semicircular duct that contains receptors for dynamic equilibrium. *Plural* is **cristae.**

Crossing-over The exchange of a portion of one chromatid with another during meiosis. It permits an exchange of genes among chromatids and is one factor that results in genetic variation of progeny.

Crus (KRUS) **of penis** Separated, tapered portion of the corpora cavernosa penis. *Plural* is **crura** (KROO-ra).

Crypt of Lieberkühn *See* **Intestinal gland.**

Cryptorchidism (krip-TOR-ki-dizm) The condition of undescended testes.

Cuneate (KŪ-nē-āt) **nucleus** A group of neurons in the inferior part of the medulla oblongata in which axons of the cuneate fasciculus terminate.

Cupula (KU-pū-la) A mass of gelatinous material covering the hair cells of a crista; a sensory receptor in the ampulla of a semicircular canal stimulated when the head moves.

Cushing's syndrome Condition caused by a hypersecretion of glucocorticoids characterized by spindly legs, "moon face," "buffalo hump," pendulous abdomen, flushed facial skin, poor wound healing, hyperglycemia, osteoporosis, hypertension, and increased susceptibility to disease.

Cutaneous (kū-TĀ-nē-us) Pertaining to the skin.

Cyanosis (sī-a-NŌ-sis) A blue or dark purple discoloration, most easily seen in nail beds and mucous membranes, that results from an increased concentration of deoxygenated (reduced) hemoglobin (more than 5 g/dL).

Cyst (SIST) A sac with a distinct connective tissue wall, containing a fluid or other material.

Cystic (SIS-tik) **duct** The duct that carries bile from the gallbladder to the common bile duct.

Cystitis (sis-TĪ-tis) Inflammation of the urinary bladder.

Cytokinesis (sī′-tō-ki-NĒ-sis) Distribution of the cytoplasm into two separate cells during cell division; coordinated with nuclear division (mitosis).

Cytolysis (sī-TOL-i-sis) The rupture of living cells in which the contents leak out.

Cytoplasm (SĪ-tō-plasm) Cytosol plus all organelles except the nucleus.

Cytoskeleton Complex internal structure of cytoplasm consisting of microfilaments, microtubules, and intermediate filaments.

Cytosol (SĪ-tō-sol) Semifluid portion of cytoplasm in which organelles and inclusions are suspended and solutes are dissolved. Also called **intracellular fluid.**

D

Dartos (DAR-tōs) The contractile smooth muscular tissue deep to the skin of the scrotum.

Decidua (dē-SID-ū-a) That portion of the endometrium of the uterus (all but the deepest layer) that is modified during pregnancy and shed after childbirth.

Deciduous (dē-SID-ū-us) Falling off or being shed seasonally or at a particular stage of development. In the body, referring to the first set of teeth.

Decussation (dē′-ku-SĀ-shun) A crossing-over to the opposite (contralateral) side; an example is the crossing of 90% of the axons in the large motor tracts to opposite sides in the medullary pyramids.

Decussation (dē′-ku-SĀ-shun) **of pyramids** The crossing of most axons (90%) in the left pyramid of the medulla to the right side and the crossing of most axons (90%) in the right pyramid to the left side.

Deep Away from the surface of the body or an organ.

Deep abdominal inguinal (IN-gwi-nal) **ring** A slitlike opening in the aponeurosis of the transversus abdominis muscle that represents the origin of the inguinal canal.

Deep vein thrombosis (DVT) The presence of a thrombus in a vein, usually a deep vein of the lower limbs.

Defecation (def-e-KĀ-shun) The discharge of feces from the rectum.

Deglutition (dē-gloo-TISH-un) The act of swallowing.

Dehydration (dē-hī-DRĀ-shun) Excessive loss of water from the body or its parts.

Delta cell A cell in the pancreatic islets (islets of Langerhans) that secretes somatostatin. Also termed a **D cell.**

Demineralization (dē-min′-er-al-i-ZĀ-shun) Loss of calcium and phosphorus from bones.

Dendrite (DEN-drīt) A neuronal process that carries electrical signals, usually graded potentials, toward the cell body.

Dendritic (den-DRIT-ik) **cell** One type of antigen-presenting cell with long branchlike projections that commonly is present in mucosal linings such as the vagina, in the skin (Langerhans cells in the epidermis), and in lymph nodes (follicular dendritic cells).

Dental caries (KA-rēz) Gradual demineralization of the enamel and dentin of a tooth that may invade the pulp and alveolar bone. Also called **tooth decay.**

Denticulate (den-TIK-ū-lāt) Finely toothed or serrated; characterized by a series of small, pointed projections.

Dentin (DEN-tin) The bony tissues of a tooth enclosing the pulp cavity.

Dentition (den-TI-shun) The eruption of teeth. The number, shape, and arrangement of teeth.

Deoxyribonucleic (dē-ok′-sē-rī-bō-nū-KLĒ-ik) **acid (DNA)** A nucleic acid constructed of nucleotides consisting of one of four bases (adenine, cytosine, guanine, or thymine), deoxyribose, and a phosphate group; encoded in the nucleotides is genetic information.

Depression (de-PRESH-un) Movement in which a part of the body moves inferiorly.

Dermal papilla (pa-PIL-a) Fingerlike projection of the papillary region of the dermis that may contain blood capillaries or corpuscles of touch (Meissner corpuscles).

Dermatology (der′-ma-TOL-ō-jē) The medical specialty dealing with diseases of the skin.

Dermatome (DER-ma-tōm) The cutaneous area developed from one embryonic spinal cord segment and receiving most of its sensory innervation from one spinal nerve. An instrument for incising the skin or cutting thin transplants of skin.

Dermis (DER-mis) A layer of dense irregular connective tissue lying deep to the epidermis.

Descending colon (KŌ-lon) The part of the large intestine descending from the left colic (splenic) flexure to the level of the left iliac crest.

Detrusor (de-TROO-ser) **muscle** Smooth muscle that forms the wall of the urinary bladder.

Developmental biology The study of development from the fertilized egg to the adult form.

Deviated nasal septum A nasal septum that does not run along the midline of the nasal cavity. It deviates (bends) to one side.

Diabetes mellitus (dī-a-BĒ-tēz MEL-i-tus) An endocrine disorder caused by an inability to produce or use insulin. It is characterized by the three "polys": polyuria (excessive urine production), polydipsia (excessive thirst), and polyphagia (excess eating).

Diagnosis (dī′-ag-NŌ-sis) Distinguishing one disease from another or determining the nature of a disease from signs and symptoms by inspection, palpation, laboratory tests, and other means.

Dialysis (dī-AL-i-sis) The removal of waste products from blood by diffusion through a selectively permeable membrane.

Diaphragm (DĪ-a-fram) Any partition that separates one area from another, especially the dome-shaped skeletal muscle between the thoracic and abdominal cavities. Also a dome-shaped device that is placed over the cervix, usually with a spermicide, to prevent conception.

Diaphysis (dī-AF-i-sis) The shaft of a long bone.

Diarrhea (dī-a-RĒ-a) Frequent defecation of liquid feces caused by increased motility of the intestines.

Diarthrosis (dī-ar-THRŌ-sis) A freely movable joint; types are gliding, hinge, pivot, condyloid, saddle, and ball-and-socket.

Diastole (dī-AS-tō-lē) In the cardiac cycle, the phase of relaxation or dilation of the heart muscle, especially of the ventricles.

Diastolic (dī-as-TOL-ik) **blood pressure (DBP)** The force exerted by blood on arterial walls during ventricular relaxation; the lowest blood pressure measured in the large arteries, normally about 70 mmHg in a young adult.

Diencephalon (DĪ-en-sef′-a-lon) A part of the brain consisting of the thalamus, hypothalamus, and epithalamus.

Differentiation (dif′-er-en-shē-Ā-shun) Development of a cell from an unspecialized to a specialized one.

Diffusion (di-FŪ-zhun) A passive process in which there is a net or greater movement of molecules or ions from a region of high concentration to a region of low concentration until equilibrium is reached.

Digestion (dī-JES-chun) The mechanical and chemical breakdown of food to simple molecules that can be absorbed and used by body cells.

Digestive system A system that consists of the gastrointestinal tract (mouth, pharynx, esophagus, stomach, small intestine, and large intestine) and accessory digestive organs (teeth, tongue, salivary glands, liver, gallbladder, and pancreas). Its function is to break down foods into small molecules that can be used by body cells.

Dilate (DĪ-lāt) To expand or swell.

Diploid (DIP-loyd) **cell** Having the number of chromosomes characteristically found in the somatic cells of an organism; having two haploid sets of chromosomes, one each from the mother and father. Symbolized $2n$.

Direct motor pathways Collections of upper motor neurons with cell bodies in the motor cortex that project axons into the spinal cord, where they synapse with lower motor neurons or interneurons in the anterior horns. Also called the **pyramidal pathways.**

Disease Any change from a state of health. Any illness characterized by a recognizable set of signs and symptoms.

Dislocation (dis′-lō-KĀ-shun) Displacement of a bone from a joint with tearing of ligaments, tendons, and articular capsules. Also called **luxation** (luks-Ā-shun).

Dissect (di-SEKT) To separate tissues and parts of a cadaver or an organ for anatomical study.

Distal (DIS-tal) Farther from the attachment of a limb to the trunk; farther from the point of origin or attachment.

Diuretic (dī-ū-RET-ik) A chemical that increases urine volume by decreasing reabsorption of water, usually by inhibiting sodium reabsorption.

Divergence (dī-VER-jens) A synaptic arrangement in which the synaptic end bulbs of one presynaptic neuron terminate on several postsynaptic neurons.

Diverticulum (dī′-ver-TIK-ū-lum) A sac or pouch in the wall of a canal or organ, especially in the colon.

Dorsal ramus (RĀ-mus) A branch of a spinal nerve containing motor and sensory axons supplying the muscles, skin, and bones of the posterior part of the head, neck, and trunk.

Dorsiflexion (dor-si-FLEK-shun) Bending the foot in the direction of the dorsum (upper surface).

Down-regulation Phenomenon in which there is a decrease in the number of receptors in response to an excess of a hormone or neurotransmitter.

Dual innervation The concept by which most organs of the body receive impulses from sympathetic and parasympathetic neurons.

Duct of Santorini *See* **Accessory duct.**

Duct of Wirsung *See* **Pancreatic duct.**

Ductus arteriosus (DUK-tus ar-tē-rē-O-sus) A small vessel connecting the pulmonary trunk with the aorta; found only in the fetus.

Ductus (vas) deferens (DEF-er-ens) The duct that carries sperm from the epididymis to the ejaculatory duct. Also called the **seminal duct.**

Ductus epididymis (ep′-i-DID-i-mis) A tightly coiled tube inside the epididymis, distinguished into a head, body, and tail, in which sperm undergo maturation.

Ductus venosus (ve-NŌ-sus) A small vessel in the fetus that helps the circulation bypass the liver.

Duodenal (doo-ō-DĒ-nal) **gland** Gland in the submucosa of the duodenum that secretes an alkaline mucus to protect the lining of the small intestine from the action of enzymes and to help neutralize the acid in chyme. Also called a **Brunner's** (BRUN-erz) **gland.**

Duodenal papilla (pa-PIL-a) An elevation on the duodenal mucosa that receives the hepatopancreatic ampulla (ampulla of Vater).

Duodenum (doo′-ō-DĒ-num *or* doo-OD-e-num) The first 25 cm (10 in.) of the small intestine, which connects the stomach and the ileum.

Dura mater (DOO-ra MĀ-ter) The outermost of the three meninges (coverings) of the brain and spinal cord.

Dynamic equilibrium (ē-kwi-LIB-rē-um) The maintenance of body position, mainly the head, in response to sudden movements such as rotation.

Dysmenorrhea (dis′-men-ō-RĒ-a) Painful menstruation.

Dysplasia (dis-PLĀ-zē-a) Change in the size, shape, and organization of cells due to chronic irritation or inflammation; may either revert to normal if stress is removed or progress to neoplasia.

Dyspnea (DISP-nē-a) Shortness of breath; painful or labored breathing.

E

Ectoderm The primary germ layer that gives rise to the nervous system and the epidermis of skin and its derivatives.

Ectopic (ek-TOP-ik) Out of the normal location, as in ectopic pregnancy.

Edema (e-DĒ-ma) An abnormal accumulation of interstitial fluid.

Effector (e-FEK-tor) An organ of the body, either a muscle or a gland, that is innervated by somatic or autonomic motor neurons.

Efferent arteriole (EF-er-ent ar-TĒ-rē-ōl) A vessel of the renal vascular system that carries blood from a glomerulus to a peritubular capillary.

Efferent (EF-er-ent) **ducts** A series of coiled tubes that transport sperm from the rete testis to the epididymis.

Ejaculation (ē-jak-ū-LĀ-shun) The reflex ejection or expulsion of semen from the penis.

Ejaculatory (ē-JAK-ū-la-tō-rē) **duct** A tube that transports sperm from the ductus (vas) deferens to the prostatic urethra.

Elasticity (e-las-TIS-i-tē) The ability of tissue to return to its original shape after contraction or extension.

Electrocardiogram (e-lek′-trō-KAR-dē-ō-gram) (**ECG** or **EKG**) A recording of the electrical changes that accompany the cardiac cycle that can be detected at the surface of the body; may be resting, stress, or ambulatory.

Elevation (el-e-VĀ-shun) Movement in which a part of the body moves superiorly.

Embolus (EM-bō-lus) A blood clot, bubble of air or fat from broken bones, mass of bacteria, or other debris or foreign material transported by the blood.

Embryo (EM-brē-ō) The young of any organism in an early stage of development; in humans, the developing organism from fertilization to the end of the eighth week of development.

Embryoblast (EM-brē-ō-blast) A region of cells of a blastocyst that differentiates into the three primary germ layers—ectoderm, mesoderm, and endoderm—from which all tissues and organs develop; also called an **inner cell mass.**

Embryology (em′-brē-OL-ō-jē) The study of development from the fertilized egg to the end of the eighth week of development.

Emesis (EM-e-sis) Vomiting.

Emigration (em′-i-GRĀ-shun) Process whereby white blood cells (WBCs) leave the bloodstream by rolling along the endothelium, sticking to it, and squeezing between the endothelial cells. Adhesion molecules help WBCs stick to the endothelium. Also known as **migration** or **extravasation.**

Emission (ē-MISH-un) Propulsion of sperm into the urethra due to peristaltic contractions of the ducts of the testes, epididymides, and ductus (vas) deferens as a result of sympathetic stimulation.

Emphysema (em-fi-SĒ-ma) A lung disorder in which alveolar walls disintegrate, producing abnormally large air spaces and loss of elasticity in the lungs; typically caused by exposure to cigarette smoke.

Emulsification (e-mul-si-fi-KĀ-shun) The dispersion of large lipid globules into smaller, uniformly distributed particles in the presence of bile.

Enamel (e-NAM-el) The hard, white substance covering the crown of a tooth.

Endocardium (en-dō-KAR-dē-um) The layer of the heart wall, composed of endothelium and smooth muscle, that lines the inside of the heart and covers the valves and tendons that hold the valves open.

Endochondral (en′-dō-KON-dral) **ossification** Bone formation within hyaline cartilage that develops from mesenchyme.

Endocrine (EN-dō-krin) **gland** A gland that secretes hormones into interstitial fluid and then the blood; a ductless gland.

Endocrine system (EN-dō-krin) All endocrine glands and hormone-secreting cells.

Endocrinology (en′-dō-kri-NOL-ō-jē) The science concerned with the structure and functions of endocrine glands and the diagnosis and treatment of disorders of the endocrine system.

Endocytosis (en′-dō-sī-TŌ-sis) The uptake into a cell of large molecules and particles in which a segment of plasma membrane surrounds the substance, encloses it, and brings it in; includes phagocytosis, pinocytosis, and receptor-mediated endocytosis.

Endoderm (EN-dō-derm) A primary germ layer of the developing embryo; gives rise to the gastrointestinal tract, urinary bladder, urethra, and respiratory tract.

Endodontics (en′-dō-DON-tiks) The branch of dentistry concerned with the prevention, diagnosis, and treatment of diseases that affect the pulp, root, periodontal ligament, and alveolar bone.

Endolymph (EN-dō-limf′) The fluid within the membranous labyrinth of the internal ear.

Endometriosis (en′-dō-me′-trē-Ō-sis) The growth of endometrial tissue outside the uterus.

Endometrium (en′-dō-MĒ-trē-um) The mucous membrane lining the uterus.

Endomysium (en′-dō-MĪZ-ē-um) Invagination of the perimysium separating each individual muscle fiber (cell).

Endoneurium (en′-dō-NOO-rē-um) Connective tissue wrapping around individual nerve axons.

Endoplasmic reticulum (en′-dō-PLAS-mik re-TIK-ū-lum) (**ER**) A network of channels running through the cytoplasm of a cell that serves in intracellular transportation, support, storage, synthesis, and packaging of molecules. Portions of ER where ribosomes are attached to the outer surface are called **rough ER;** portions that have no ribosomes are called **smooth ER.**

End organ of Ruffini *See* **Type II cutaneous mechanoreceptor.**

Endosteum (end-OS-tē-um) The membrane that lines the medullary (marrow) cavity of bones, consisting of osteogenic cells and scattered osteoclasts.

Endothelium (en′-dō-THĒ-lē-um) The layer of simple squamous epithelium that lines the cavities of the heart, blood vessels, and lymphatic vessels.

Enteric (en-TER-ik) **nervous system** A portion of the autonomic nervous system within the wall of the gastrointestinal tract, pancreas, and gallbladder. Its sensory neurons monitor tension in the intestinal wall and assess the composition of intestinal contents; its motor neurons exert control over the motility and secretions of the gastrointestinal tract.

Enteroendocrine (en-ter-ō-EN-dō-krin) **cell** A cell of the mucosa of the gastrointestinal tract that secretes a hormone that governs function of the GI tract; hormones secreted include gastrin, cholecystokinin, glucose-dependent insulinotropic peptide (GIP), and secretin.

Enzyme (EN-zīm) A substance that accelerates chemical reactions; an organic catalyst, usually a protein.

Eosinophil (ē-ō-SIN-ō-fil) A type of white blood cell characterized by granules that stain red or pink with acid dyes.

Ependymal (ep-EN-de-mal) **cells** Neuroglial cells that cover choroid plexuses and produce cerebrospinal fluid (CSF); they also line the ventricles of the brain and probably assist in the circulation of CSF.

Epicardium (ep′-i-KAR-dē-um) The thin outer layer of the heart wall, composed of serous tissue and mesothelium. Also called the **visceral pericardium.**

Epidemiology (ep′-i-dē-mē-OL-ō-jē) Study of the occurrence and transmission of diseases and disorders in human populations.

Epidermis (ep′-i-DERM-is) The superficial, thinner layer of skin, composed of keratinized stratified squamous epithelium.

Epididymis (ep′-i-DID-i-mis) A comma-shaped organ that lies along the posterior border of the testis and contains the ductus epididymis, in which sperm undergo maturation. *Plural* is **epididymides** (ep′-i-di-DIM-i-dēz).

Epidural (ep′-i-DOO-ral) **space** A space between the spinal dura mater and the vertebral canal, containing areolar connective tissue and a plexus of veins.

Epiglottis (ep′-i-GLOT-is) A large, leaf-shaped piece of cartilage lying on top of the larynx, attached to the thyroid cartilage; its unattached portion is free to move up and down to cover the glottis (vocal folds and rima glottidis) during swallowing.

Epimysium (ep-i-MĪZ-ē-um) Fibrous connective tissue around muscles.

Epinephrine (ep-ē-NEF-rin) Hormone secreted by the adrenal medulla that produces actions similar to those that result from sympathetic stimulation. Also called **adrenaline** (a-DREN-a-lin).

Epineurium (ep′-i-NOO-rē-um) The superficial connective tissue covering around an entire nerve.

Epiphyseal (ep′-i-FIZ-ē-al) **line** The remnant of the epiphyseal plate in the metaphysis of a long bone.

Epiphyseal plate The hyaline cartilage plate in the metaphysis of a long bone; site of lengthwise growth of long bones.

Epiphysis (e-PIF-i-sis) The end of a long bone, usually larger in diameter than the shaft (diaphysis).

Epiphysis cerebri (se-RĒ-brē) Pineal gland.

Episiotomy (e-piz′-ē-OT-ō-mē) A cut made with surgical scissors to avoid tearing of the perineum at the end of the second stage of labor.

Epistaxis (ep′-i-STAK-sis) Loss of blood from the nose due to trauma, infection, allergy, neoplasm, and bleeding disorders. Also called **nosebleed.**

Epithalamus (ep′-i-THAL-a-mus) Part of the diencephalon superior and posterior to the thalamus, comprising the pineal gland and associated structures.

Epithelial (ep-i-THĒ-lē-al) **tissue** The tissue that forms the innermost and outermost surfaces of body structures and forms glands.

Eponychium (ep′-o-NIK-ē-um) Narrow band of stratum corneum at the proximal border of a nail that extends from the margin of the nail wall. Also called the **cuticle.**

Equilibrium (ē-kwi-LIB-rē-um) The state of being balanced.

Erectile dysfunction (ED) Failure to maintain an erection long enough for sexual intercourse. Previously known as **impotence** (IM-pō-tens).

Erection (ē-REK-shun) The enlarged and stiff state of the penis or clitoris resulting from the engorgement of the spongy erectile tissue with blood.

Eructation (e-ruk′-TĀ-shun) The forceful expulsion of gas from the stomach. Also called **belching.**

Erythema (er-e-THĒ-ma) Skin redness usually caused by dilation of the capillaries.

Erythrocyte (e-RITH-rō-sīt) A mature red blood cell.

Erythropoietin (e-rith′-rō-POY-e-tin) **(EPO)** A hormone released by the juxtaglomerular cells of the kidneys that stimulates red blood cell production.

Esophagus (e-SOF-a-gus) The hollow muscular tube that connects the pharynx and the stomach.

Estrogens (ES-tro-jenz) Feminizing sex hormones produced by the ovaries; govern development of oocytes, maintenance of female reproductive structures, and appearance of secondary sex characteristics; also affect fluid and electrolyte balance, and protein anabolism. Examples are β-estradiol, estrone, and estriol.

Eupnea (ŪP-nē-a) Normal quiet breathing.

Eustachian tube *See* **Auditory tube.**

Eversion (ē-VER-zhun) The movement of the sole laterally at the ankle joint or of an atrioventricular valve into an atrium during ventricular contraction.

Excitability (ek-sīt′-a-BIL-i-tē) The ability of muscle fibers to receive and respond to stimuli; the ability of neurons to respond to stimuli and generate nerve impulses.

Excretion (eks-KRĒ-shun) The process of eliminating waste products from the body; also the products excreted.

Exercise physiology Study of the changes in cell and organ function due to muscular activity.

Exhalation (eks-ha-LĀ-shun) Breathing out; expelling air from the lungs into the atmosphere. Also called **expiration.**

Exocrine (EK-sō-krin) **gland** A gland that secretes its products into ducts that carry the secretions into body cavities, into the lumen of an organ, or to the outer surface of the body.

Exocytosis (ek-sō-sī-TŌ-sis) A process in which membrane-enclosed secretory vesicles form inside the cell, fuse with the plasma membrane, and release their contents into the interstitial fluid; achieves secretion of materials from a cell.

Extensibility (ek-sten′-si-BIL-i-tē) The ability of muscle tissue to stretch when it is pulled.

Extension (eks-TEN-shun) An increase in the angle between two bones; restoring a body part to its anatomical position after flexion.

External Located on or near the surface.

External auditory (AW-di-tōr-ē) **canal** or **meatus** (mē-Ā-tus) A curved tube in the temporal bone that leads to the middle ear.

External ear The **outer ear,** consisting of the pinna, external auditory canal, and tympanic membrane (eardrum).

External nares (NĀ-rez) The openings into the nasal cavity on the exterior of the body. Also called the **nostrils.**

External respiration The exchange of respiratory gases between the lungs and blood. Also called **pulmonary respiration** or **pulmonary gas exchange.**

Exteroceptor (EKS-ter-ō-sep′-tor) A sensory receptor adapted for the reception of stimuli from outside the body.

Extracellular fluid (ECF) Fluid outside body cells, such as interstitial fluid and plasma.

Extracellular matrix (MĀ-triks) The ground substance and fibers between cells in a connective tissue.

Eyebrow The hairy ridge superior to the eye.

F

F cell A cell in the pancreatic islets (islets of Langerhans) that secretes pancreatic polypeptide.

Face The anterior aspect of the head.

Falciform ligament (FAL-si-form LIG-a-ment) A sheet of parietal peritoneum between the two principal lobes of the liver. The ligamentum teres, or remnant of the umbilical vein, lies within its fold.

Fallopian tube *See* **Uterine tube.**

Falx cerebelli (FALKS′ ser-e-BEL-lī) A small triangular process of the dura mater attached to the occipital bone in the posterior cranial fossa and projecting inward between the two cerebellar hemispheres.

Falx cerebri (FALKS SER-e-brē) A fold of the dura mater extending deep into the longitudinal fissure between the two cerebral hemispheres.

Fascia (FASH-ē-a) Large connective tissue sheet that wraps around groups of muscles.

Fascicle (FAS-i-kul) A small bundle or cluster, especially of nerve or muscle fibers (cells). Also called a **fasciculus** (fa-SIK-ū-lus). *Plural* is **fasciculi** (fa-SIK-yoo-lī).

Fasciculation (fa-sik-ū-LĀ-shun) Abnormal, spontaneous twitch of all skeletal muscle fibers in one motor unit that is visible at the skin surface; not associated with movement of the affected muscle; present in progressive diseases of motor neurons, for example, poliomyelitis.

Fat A triglyceride that is a solid at room temperature.

Fatty acid A simple lipid that consists of a carboxyl group and a hydrocarbon chain; used to synthesize triglyceride and phospholipids.

Fauces (FAW-sēs) The opening from the mouth into the pharynx.

Feces (FĒ-sēz) Material discharged from the rectum and made up of bacteria, excretions, and food residue. Also called **stool.**

Feedback system (loop) A cycle of events in which the status of a body condition is monitored, evaluated, changed, remonitored, reevaluated, and so on.

Female reproductive cycle General term for the ovarian and uterine cycles, the hormonal changes that accompany them, and cyclic changes in the breasts and cervix; includes changes in the endometrium of a nonpregnant female that prepares the lining of the uterus to receive a fertilized ovum. Less correctly termed the **menstrual cycle.**

Fertilization (fer-til-i-ZĀ-shun) Penetration of a secondary oocyte by a sperm cell, meiotic division of a secondary oocyte to form an ovum, and subsequent union of the nuclei of the gametes.

Fetal circulation The cardiovascular system of the fetus, including the placenta and special blood vessels involved in the exchange of materials between fetus and mother.

Fetus (FĒ-tus) In humans, the developing organism *in utero* from the beginning of the third month to birth.

Fever An elevation in body temperature above the normal temperature of 37°C (98.6°F) due to a resetting of the hypothalamic thermostat.

Fibroblast (FĪ-brō-blast) A large, flat cell that secretes most of the extracellular matrix of areolar and dense connective tissues.

Fibrosis The process by which fibroblasts synthesize collagen fibers and other extracellular matrix materials that aggregate to form scar tissue.

Fibrous (FĪ-brus) **joint** A joint that allows little or no movement, such as a suture or a syndesmosis.

Fibrous tunic (TOO-nik) The superficial coat of the eyeball, made up of the posterior sclera and the anterior cornea.

Fight-or-flight response The effects produced on stimulation of the sympathetic division of the autonomic nervous system.

Filiform papilla (FIL-i-form pa-PIL-a) One of the conical projections that are distributed in parallel rows over the anterior two-thirds of the tongue and lack taste buds.

Filtration (fil-TRĀ-shun) The flow of a liquid through a filter (or membrane that acts like a filter) due to a hydrostatic pressure; occurs in capillaries due to blood pressure.

Filum terminale (FĪ-lum ter-mi-NAL-ē) Nonnervous fibrous tissue of the spinal cord that extends inferiorly from the conus medullaris to the coccyx.

Fimbriae (FIM-brē-ē) Fingerlike structures, especially the lateral ends of the uterine (fallopian) tubes.

Fissure (FISH-ur) A groove, fold, or slit that may be normal or abnormal.

Fixator A muscle that stabilizes the origin of the prime mover so that the prime mover can act more efficiently.

Fixed macrophage (MAK-rō-fāj) Stationary phagocytic cell found in the liver, lungs, brain, spleen, lymph nodes, subcutaneous tissue, and red bone marrow. Also called a **tissue macrophage** or **histiocyte** (HIS-tē-ō-sīt).

Flaccid (FLAK-sid) Relaxed, flabby, or soft; lacking muscle tone.

Flagellum (fla-JEL-um) A hairlike, motile process on the extremity of a bacterium, protozoan, or sperm cell. *Plural* is **flagella** (fla-JEL-a).

Flatus (FLĀ-tus) Gas in the stomach or intestines; commonly used to denote expulsion of gas through the anus.

Flexion (FLEK-shun) Movement in which there is a decrease in the angle between two bones.

Follicle (FOL-i-kul) A small secretory sac or cavity; the group of cells that contains a developing oocyte in the ovaries.

Follicle-stimulating hormone (FSH) Hormone secreted by the anterior pituitary; it initiates development of ova and stimulates the ovaries to secrete estrogens in females, and initiates sperm production in males.

Fontanel (fon-ta-NEL) A mesenchyme-filled space where bone formation is not yet complete, especially between the cranial bones of an infant's skull.

Foot The terminal part of the lower limb, from the ankle to the toes.

Foramen (fō-RĀ-men) A passage or opening; a communication between two cavities of an organ, or a hole in a bone for passage of vessels or nerves. *Plural* is **foramina** (fō-RAM-i-na).

Foramen ovale (fō-RĀ-men ō-VAL-ē) An opening in the fetal heart in the septum between the right and left atria. A hole in the greater wing of the sphenoid bone that transmits the mandibular branch of the trigeminal (V) nerve.

Forearm (FOR-arm) The part of the upper limb between the elbow and the wrist.

Fornix (FOR-niks) An arch or fold; a tract in the brain made up of association fibers, connecting the hippocampus with the mammillary bodies; a recess around the cervix of the uterus where it protrudes into the vagina.

Fossa (FOS-a) A furrow or shallow depression.

Fourth ventricle (VEN-tri-kul) A cavity filled with cerebrospinal fluid within the brain lying between the cerebellum and the medulla oblongata and pons.

Fracture (FRAK-choor) Any break in a bone.

Free radical An atom or group of atoms with an unpaired electron in the outermost shell. It is unstable, highly reactive, and destroys nearby molecules.

Frontal plane A plane at a right angle to a midsagittal plane that divides the body or organs into anterior and posterior portions. Also called a **coronal** (kō-RŌ-nal) **plane.**

Fundus (FUN-dus) The part of a hollow organ farthest from the opening.

Fungiform papilla (FUN-ji-form pa-PIL-a) A mushroomlike elevation on the upper surface of the tongue appearing as a red dot; most contain taste buds.

Furuncle (FŪ-rung-kul) A boil; painful nodule caused by bacterial infection and inflammation of a hair follicle or sebaceous (oil) gland.

G

Gallbladder A small pouch, located inferior to the liver, that stores bile and empties by means of the cystic duct.

Gallstone A solid mass, usually containing cholesterol, in the gallbladder or a bile-containing duct; formed anywhere between bile canaliculi in the liver and the hepatopancreatic ampulla

(ampulla of Vater), where bile enters the duodenum. Also called a **biliary calculus.**

Gamete (GAM-ēt) A male or female reproductive cell; a sperm cell or secondary oocyte.

Ganglion (GANG-glē-on) Usually, a group of neuronal cell bodies lying outside the central nervous system (CNS). *Plural* is **ganglia** (GANG-glē-a).

Gastric (GAS-trik) **glands** Glands in the mucosa of the stomach composed of cells that empty their secretions into narrow channels called gastric pits. Types of cells are chief cells (secrete pepsinogen), parietal cells (secrete hydrochloric acid and intrinsic factor), surface mucous and mucous neck cells (secrete mucus), and G cells (secrete gastrin).

Gastroenterology (gas'-trō-en'-ter-OL-ō-jē) The medical specialty that deals with the structure, function, diagnosis, and treatment of diseases of the stomach and intestines.

Gastrointestinal (gas-trō-in-TES-ti-nal) **(GI) tract** A continuous tube running through the ventral body cavity extending from the mouth to the anus. Also called the **alimentary** (al'-i-MEN-tar-ē) **canal.**

Gastrulation (gas-troo-LĀ-shun) The migration of groups of cells from the epiblast that transform a bilaminar embryonic disc into a trilaminar embryonic disc with three primary germ layers; transformation of the blastula into the gastrula.

Gene (JĒN) Biological unit of heredity; a segment of DNA located in a definite position on a particular chromosome; a sequence of DNA that codes for a particular mRNA, rRNA, or tRNA.

Genetic engineering The manufacture and manipulation of genetic material.

Genetics The study of genes and heredity.

Genome (JĒ-nōm) The complete set of genes of an organism.

Genotype (JĒ-nō-tīp) The genetic makeup of an individual; the combination of alleles present at one or more chromosomal locations, as distinguished from the appearance, or phenotype, that results from those alleles.

Geriatrics (jer'-ē-AT-riks) The branch of medicine devoted to the medical problems and care of elderly persons.

Gestation (jes-TĀ-shun) The period of development from fertilization to birth.

Gingivae (jin-JI-vē) Gums. They cover the alveolar processes of the mandible and maxilla and extend slightly into each socket.

Gland Specialized epithelial cell or cells that secrete substances; may be exocrine or endocrine.

Glans penis (glanz PĒ-nis) The slightly enlarged region at the distal end of the penis.

Glaucoma (glaw-KŌ-ma) An eye disorder in which there is increased intraocular pressure due to an excess of aqueous humor.

Glomerular (glō-MER-ū-lar) **capsule** A double-walled globe at the proximal end of a nephron that encloses the glomerular capillaries. Also called **Bowman's** (BŌ-manz) **capsule.**

Glomerular filtrate (glō-MER-ū-lar FIL-trāt) The fluid produced when blood is filtered by the filtration membrane in the glomeruli of the kidneys.

Glomerular filtration The first step in urine formation in which substances in blood pass through the filtration membrane and the filtrate enters the proximal convoluted tubule of a nephron.

Glomerular filtration rate (GFR) The amount of filtrate formed in all renal corpuscles per minute. It averages 125 mL/min in males and 105 mL/min in females.

Glomerulus (glō-MER-ū-lus) A rounded mass of nerves or blood vessels, especially the microscopic tuft of capillaries that is surrounded by the glomerular (Bowman's) capsule of each kidney tubule. *Plural* is **glomeruli** (glō-MER-ū-li).

Glottis (GLOT-is) The vocal folds (true vocal cords) in the larynx plus the space between them (rima glottidis).

Glucagon (GLOO-ka-gon) A hormone produced by the alpha cells of the pancreatic islets (islets of Langerhans) that increases blood glucose level.

Glucocorticoids (gloo′-kō-KOR-ti-koyds) Hormones secreted by the cortex of the adrenal gland, especially cortisol, that influence glucose metabolism.

Glucose (GLOO-kōs) A hexose (six-carbon sugar), $C_6H_{12}O_6$, that is a major energy source for the production of ATP by body cells.

Glucosuria (gloo′-kō-SOO-rē-a) The presence of glucose in the urine; may be temporary or pathological. Also called **glycosuria.**

Glycogen (GLĪ-kō-jen) A highly branched polymer of glucose containing thousands of subunits; functions as a compact store of glucose molecules in liver and muscle fibers (cells).

Goblet cell A goblet-shaped unicellular gland that secretes mucus; present in epithelium of the airways and intestines.

Goiter (GOY-ter) An enlarged thyroid gland.

Golgi (GOL-jē) **complex** An organelle in the cytoplasm of cells consisting of four to six flattened sacs (cisternae), stacked on one another, with expanded areas at their ends; functions in processing, sorting, packaging, and delivering proteins and lipids to the plasma membrane, lysosomes, and secretory vesicles.

Golgi tendon organ *See* **Tendon organ.**

Gomphosis (gom-FŌ-sis) A fibrous joint in which a cone-shaped peg fits into a socket.

Gonad (GŌ-nad) A gland that produces gametes and hormones; the ovary in the female and the testis in the male.

Gonadotropic hormone Anterior pituitary hormone that affects the gonads.

Gout (GOWT) Hereditary condition associated with excessive uric acid in the blood; the acid crystallizes and deposits in joints, kidneys, and soft tissue.

Graafian follicle *See* **Mature follicle.**

Gracile (GRAS-īl) **nucleus** A group of nerve cells in the inferior part of the medulla oblongata in which axons of the gracile fasciculus terminate.

Gray commissure (KOM-mi-shur) A narrow strip of gray matter connecting the two lateral gray masses within the spinal cord.

Gray matter Areas in the central nervous system and ganglia containing neuronal cell bodies, dendrites, unmyelinated axons, axon terminals, and neuroglia; Nissl bodies impart a gray color and there is little or no myelin in gray matter.

Gray ramus communicans (RĀ-mus kō-MŪ-ni-kans) A short nerve containing axons of sympathetic postganglionic neurons; the cell bodies of the neurons are in a sympathetic chain ganglion, and the unmyelinated axons extend via the gray ramus to a spinal nerve and then to the periphery to supply smooth muscle in blood vessels, arrector pili muscles, and sweat glands. *Plural* is **rami communicantes** (RĀ-mē kō-mū-ni-KAN-tēz).

Greater omentum (ō-MEN-tum) A large fold in the serosa of the stomach that hangs down like an apron anterior to the intestines.

Greater vestibular (ves-TIB-ū-lar) **glands** A pair of glands on either side of the vaginal orifice that open by a duct into the space between the hymen and the labia minora. Also called **Bartholin's** (BAR-to-linz) **glands.**

Groin (GROYN) The depression between the thigh and the trunk; the inguinal region.

Gross anatomy The branch of anatomy that deals with structures that can be studied without using a microscope. Also called **macroscopic anatomy.**

Growth An increase in size due to an increase in (1) the number of cells, (2) the size of existing cells as internal components increase in size, or (3) the size of intercellular substances.

Gustation (gus-TĀ-shun). The sense of taste.

Gustatory (GUS-ta-tō′-rē) Pertaining to taste.

Gynecology (gī′-ne-KOL-ō-jē) The branch of medicine dealing with the study and treatment of disorders of the female reproductive system.

Gynecomastia (gīn′-e-kō-MAS-tē-a) Excessive growth (benign) of the male mammary glands due to secretion of estrogens by an adrenal gland tumor (feminizing adenoma).

Gyrus (JI-rus) One of the folds of the cerebral cortex of the brain. *Plural* is **gyri** (JĪ-rī). Also called a **convolution.**

H

Hair A threadlike structure produced by hair follicles that develops in the dermis. Also called a **pilus** (PĪ-lus).

Hair follicle (FOL-li-kul) Structure composed of epithelium and surrounding the root of a hair from which hair develops.

Hair root plexus (PLEK-sus) A network of dendrites arranged around the root of a hair as free or naked nerve endings that are stimulated when a hair shaft is moved.

Hand The terminal portion of an upper limb, including the carpals, metacarpals, and phalanges.

Haploid (HAP-loyd) **cell** Having half the number of chromosomes characteristically found in the somatic cells of an organism; characteristic of mature gametes. Symbolized *n*.

Hard palate (PAL-at) The anterior portion of the roof of the mouth, formed by the maxillae and palatine bones and lined by mucous membrane.

Haustra (HAWS-tra) A series of pouches that characterize the colon; caused by tonic contractions of the teniae coli. *Singular* is **haustrum.**

Haversian canal *See* **Central canal.**

Haversian system *See* **Osteon.**

Head The superior part of a human, cephalic to the neck. The superior or proximal part of a structure.

Hearing The ability to perceive sound.

Heart A hollow muscular organ lying slightly to the left of the midline of the chest that pumps the blood through the cardiovascular system.

Heart block An arrhythmia (dysrhythmia) of the heart in which the atria and ventricles contract independently because of a blocking of electrical impulses through the heart at some point in the conduction system.

Heart murmur (MER-mer) An abnormal sound that consists of a flow noise that is heard before, between, or after the normal heart sounds, or that may mask normal heart sounds.

Hemangioblast (hē-MAN-jē-ō-blast) A precursor mesodermal cell that develops into blood and blood vessels.

Hematocrit (he-MAT-ō-krit) **(Hct)** The percentage of blood made up of red blood cells. Usually measured by centrifuging a blood sample in a graduated tube and then reading the volume of red blood cells and dividing it by the total volume of blood in the sample.

Hematology (hēm-a-TOL-ō-jē) The study of blood.

Hematoma (hē′-ma-TŌ-ma) A tumor or swelling filled with blood.

Hemiplegia (hem-i-PLĒ-jē-a) Paralysis of the upper limb, trunk, and lower limb on one side of the body.

Hemodialysis (hē-mō-dī-AL-i-sis) Direct filtration of blood by removing wastes and excess electrolytes and fluid and then returning the cleansed blood.

Hemodynamics (hē-mō-dī-NAM-iks) The forces involved in circulating blood throughout the body.

Hemoglobin (hē′-mō-GLŌ-bin) **(Hb)** A substance in red blood cells consisting of the protein globin and the iron-containing red pigment heme that transports most of the oxygen and some carbon dioxide in blood.

Hemolysis (hē-MOL-i-sis) The escape of hemoglobin from the interior of a red blood cell into the surrounding medium; results from disruption of the cell membrane by toxins or drugs, freezing or thawing, or hypotonic solutions.

Hemolytic disease of the newborn (HDN) A hemolytic anemia of a newborn child that results from the destruction of the infant's erythrocytes (red blood cells) by antibodies produced by the mother; usually the antibodies are due to an Rh blood type incompatibility. Also called **erythroblastosis fetalis** (e-rith′-rō-blas-TŌ-sis fe-TAL-is).

Hemophilia (hē′-mō-FIL-ē-a) A hereditary blood disorder in which there is a deficient production of certain factors involved in blood clotting, resulting in excessive bleeding into joints, deep tissues, and elsewhere.

Hemopoiesis (hēm-ō-poy-Ē-sis) Blood cell production, which occurs in red bone marrow after birth. Also called **hematopoiesis** (hem′-a-tō-poy-Ē-sis).

Hemorrhage (HEM-o-rij) Bleeding; the escape of blood from blood vessels, especially when the loss is profuse.

Hemorrhoids (HEM-ō-royds) Dilated or varicosed blood vessels (usually veins) in the anal region. Also called **piles.**

Hepatic (he-PAT-ik) Refers to the liver.

Hepatic duct A duct that receives bile from the bile capillaries. Small hepatic ducts merge to form the larger right and left hepatic ducts that unite to leave the liver as the common hepatic duct.

Hepatic portal circulation The flow of blood from the gastrointestinal organs to the liver before returning to the heart.

Hepatocyte (he-PAT-ō-cyte) A liver cell.

Hepatopancreatic (hep′-a-tō-pan′-krē-A-tik) **ampulla** A small, raised area in the duodenum where the combined common bile duct and main pancreatic duct empty into the duodenum. Also called the **ampulla of Vater** (FAH-ter).

Hernia (HER-nē-a) The protrusion or projection of an organ or part of an organ through a membrane or cavity wall, usually the abdominal cavity.

Herniated (HER-nē-ā′-ted) **disc** A rupture of an intervertebral disc so that the nucleus pulposus protrudes into the vertebral cavity. Also called a **slipped disc.**

Hiatus (hī-Ā-tus) An opening; a foramen.

Hilum (HĪ-lum) An area, depression, or pit where blood vessels and nerves enter or leave an organ. Also called a **hilus.**

Hinge joint A synovial joint in which a convex surface of one bone fits into a concave surface of another bone, such as the elbow, knee, ankle, and interphalangeal joints. Also called a **ginglymus** (JIN-gli-mus) **joint.**

Hirsutism (HER-soo-tizm) An excessive growth of hair in females and children, with a distribution similar to that in adult males, due to the conversion of vellus hairs into large terminal hairs in response to higher-than-normal levels of androgens.

Histamine (HISS-ta-mēn) Substance found in many cells, especially mast cells, basophils, and

platelets, that is released when the cells are injured; results in vasodilation, increased permeability of blood vessels, and constriction of bronchioles.

Histology (his′-TOL-ō-jē) Microscopic study of the structure of tissues.

Holocrine (HŌ-lō-krin) **gland** A type of gland in which entire secretory cells, along with their accumulated secretions, make up the secretory product of the gland, as in the sebaceous (oil) glands.

Homeostasis (hō′-mē-ō-STĀ-sis) The condition in which the body's internal environment remains relatively constant within physiological limits.

Homologous (hō-MOL-ō-gus) **chromosomes** Two chromosomes that belong to a pair. Also called **homologs.**

Hormone (HOR-mōn) A secretion of endocrine cells that alters the physiological activity of target cells of the body.

Horn An area of gray matter (anterior, lateral, or posterior) in the spinal cord.

Human chorionic gonadotropin (kō-rē-ON-ik gō-nad-ō-TRŌ-pin) **(hCG)** A hormone produced by the developing placenta that maintains the corpus luteum.

Human chorionic somatomammotropin (sō-mat-ō-mam-ō-TRŌ-pin) **(hCS)** Hormone produced by the chorion of the placenta that stimulates breast tissue for lactation, enhances body growth, and regulates metabolism. Also called **human placental lactogen (hPL).**

Human growth hormone (hGH) Hormone secreted by the anterior pituitary that stimulates growth of body tissues, especially skeletal and muscular tissues. Also known as **somatotropin** (sō′-ma-tō-TRŌ-pin) and **somatotropic hormone (STH).**

Hyaluronic (hī′-a-loo-RON-ik) **acid** A viscous, amorphous extracellular material that binds cells together, lubricates joints, and maintains the shape of the eyeballs.

Hymen (HĪ-men) A thin fold of vascularized mucous membrane at the vaginal orifice.

Hyperextension (hī′-per-ek-STEN-shun) Continuation of extension beyond the anatomical position, as in bending the head backward.

Hyperplasia (hī-per-PLĀ-zē-a) An abnormal increase in the number of normal cells in a tissue or organ, increasing its size.

Hypersecretion (hī′-per-se-KRĒ-shun) Overactivity of glands resulting in excessive secretion.

Hypersensitivity (hī′-per-sen-si-TI-vi-tē) Overreaction to an allergen that results in pathological changes in tissues. Also called **allergy.**

Hypertension (hī′-per-TEN-shun) High blood pressure.

Hyperthermia (hī′-per-THERM-ē-a) An elevated body temperature.

Hypertonia (hī′-per-TŌ-nē-a) Increased muscle tone that is expressed as spasticity or rigidity.

Hypertonic (hī′-per-TON-ik) Solution that causes cells to shrink due to loss of water by osmosis.

Hypertrophy (hī-PER-trō-fē) An excessive enlargement or overgrowth of tissue without cell division.

Hyperventilation (hī′-per-ven-ti-LĀ-shun) A rate of inhalation and exhalation higher than that required to maintain a normal partial pressure of carbon dioxide in the blood.

Hyponychium (hī′-pō-NIK-ē-um) Free edge of the fingernail.

Hypophyseal fossa (hī′-pō-FIZ-ē-al FOS-a) A depression on the superior surface of the sphenoid bone that houses the pituitary gland.

Hypophyseal (hī′-pō-FIZ-ē-al) **pouch** An outgrowth of ectoderm from the roof of the mouth from which the anterior pituitary develops. Also called **Rathke's pouch.**

Hypophysis (hī-POF-i-sis) Pituitary gland.

Hyposecretion (hī′-pō-se-KRĒ-shun) Underactivity of glands resulting in diminished secretion.

Hypothalamohypophyseal (hī′-pō-thal′-a-mō-hī-pō-FIZ-ē-al) **tract** A bundle of axons containing secretory vesicles filled with oxytocin or antidiuretic hormone that extend from the hypothalamus to the posterior pituitary.

Hypothalamus (hī′-pō-THAL-a-mus) A portion of the diencephalon, lying beneath the thalamus and forming the floor and part of the wall of the third ventricle.

Hypothermia (hī′-pō-THER-mē-a) Lowering of body temperature below 35°C (95°F); in surgical procedures, it refers to deliberate cooling of the body to slow down metabolism and reduce oxygen needs of tissues.

Hypotonia (hī′-pō-TŌ-nē-a) Decreased or lost muscle tone in which muscles appear flaccid.

Hypotonic (hī′-pō-TON-ik) Solution that causes cells to swell and perhaps rupture due to gain of water by osmosis.

Hypoventilation (hī-pō-ven-ti-LĀ-shun) A rate of inhalation and exhalation lower than that required to maintain a normal partial pressure of carbon dioxide in plasma.

Hypoxia (hī-POKS-ē-a) Lack of adequate oxygen at the tissue level.

Hysterectomy (hiss-te-REK-tō-mē) The surgical removal of the uterus.

I

Ileocecal (il-ē-ō-SĒ-kal) **sphincter** A fold of mucous membrane that guards the opening from the ileum into the large intestine. Also called the **ileocecal valve.**

Ileum (IL-ē-um) The terminal part of the small intestine.

Immunity (i-MŪ-ni-tē) The state of being resistant to injury, particularly by poisons, foreign proteins, and invading pathogens.

Immunoglobulin (im-ū-nō-GLOB-ū-lin) **(Ig)** An antibody synthesized by plasma cells derived from B lymphocytes in response to the introduction of an antigen. Immunoglobulins are divided into five kinds (IgG, IgM, IgA, IgD, IgE).

Immunology (im′-ū-NOL-ō-jē) The study of the responses of the body when challenged by antigens.

Imperforate (im′-PER-fō-rāt) Abnormally closed.

Implantation (im′-plan-TĀ-shun) The insertion of a tissue or a part into the body. The attachment of the blastocyst to the stratum basalis of the endometrium about 6 days after fertilization.

Incontinence (in-KON-ti-nens) Inability to retain urine, semen, or feces through loss of sphincter control.

Indirect motor pathways Motor tracts that convey information from the brain down the spinal cord for automatic movements, coordination of body movements with visual stimuli, skeletal muscle tone and posture, and balance. Also known as **extrapyramidal pathways.**

Induction (in-DUK-shun) The process by which one tissue (inducting tissue) stimulates the development of an adjacent unspecialized tissue (responding tissue) into a specialized one.

Infarction (in-FARK-shun) A localized area of necrotic tissue, produced by inadequate oxygenation of the tissue.

Infection (in-FEK-shun) Invasion and multiplication of microorganisms in body tissues, which may be inapparent or characterized by cellular injury.

Inferior (in-FĒR-ē-or) Away from the head or toward the lower part of a structure. Also called **caudal** (KAW-dal).

Inferior vena cava (VĒ-na KĀ-va) **(IVC)** Large vein that collects blood from parts of the body inferior to the heart and returns it to the right atrium.

Infertility Inability to conceive or to cause conception. Also called **sterility** in males.

Inflammation (in′-fla-MĀ-shun) Localized, protective response to tissue injury designed to destroy, dilute, or wall off the infecting agent or injured tissue; characterized by redness, pain, heat, swelling, and sometimes loss of function.

Infundibulum (in-fun-DIB-ū-lum) The stalklike structure that attaches the pituitary gland to the hypothalamus of the brain. The funnel-shaped, open, distal end of the uterine (fallopian) tube.

Ingestion (in-JES-chun) The taking in of food, liquids, or drugs, by mouth.

Inguinal (IN-gwin-al) Pertaining to the groin.

Inguinal canal An oblique passageway in the anterior abdominal wall just superior and parallel to the medial half of the inguinal ligament that transmits the spermatic cord and ilioinguinal nerve in the male and round ligament of the uterus and ilioinguinal nerve in the female.

Inhalation (in-ha-LĀ-shun) The act of drawing air into the lungs. Also termed **inspiration.**

Inheritance The acquisition of body traits by transmission of genetic information from parents to offspring.

Inhibin A hormone secreted by the gonads that inhibits release of follicle-stimulating hormone (FSH) by the anterior pituitary.

Inhibiting hormone Hormone secreted by the hypothalamus that can suppress secretion of hormones by the anterior pituitary.

Insertion (in-SER-shun) The attachment of a muscle tendon to a movable bone or the end opposite the origin.

Insula (IN-soo-la) A triangular area of the cerebral cortex that lies deep within the lateral cerebral fissure, under the parietal, frontal, and temporal lobes.

Insulin (IN-soo-lin) A hormone produced by the beta cells of a pancreatic islet (islet of Langerhans) that decreases the blood glucose level.

Integrins (IN-te-grinz) A family of transmembrane glycoproteins in plasma membranes that function in cell adhesion; they are present in hemidesmosomes, which anchor cells to a basement membrane, and they mediate adhesion of neutrophils to endothelial cells during emigration.

Integumentary (in-teg-ū-MEN-tar-ē) Relating to the skin.

Integumentary (in-teg-ū-MEN-tar-ē) **system** A system composed of organs such as the skin, hair, oil and sweat glands, nails, and sensory receptors.

Intercalated (in-TER-ka-lāt-ed) **disc** An irregular transverse thickening of sarcolemma that contains desmosomes, which hold cardiac muscle fibers (cells) together, and gap junctions, which aid in conduction of muscle action potentials from one fiber to the next.

Intercostal (in′-ter-KOS-tal) **nerve** A nerve supplying a muscle located between the ribs. Also called **thoracic nerve.**

Intermediate (in′-ter-MĒ-dē-at) Between two structures, one of which is medial and one of which is lateral.

Intermediate filament Protein filament, ranging from 8 to 12 nm in diameter, that may provide structural reinforcement, hold organelles in place, and give shape to a cell.

Internal Away from the surface of the body.

Internal capsule A large tract of projection fibers lateral to the thalamus that is the major connection between the cerebral cortex and the brain stem and spinal cord; contains axons of sensory neurons carrying auditory, visual, and somatic sensory signals to the cerebral cortex plus axons of motor neurons descending from the cerebral cortex to the thalamus, subthalamus, brain stem, and spinal cord.

Internal ear The inner ear or labyrinth, lying inside the temporal bone, containing the organs of hearing and balance.

Internal nares (NĀ-rez) The two openings posterior to the nasal cavities opening into the nasopharynx. Also called the **choanae** (kō-Ā-nē).

Internal respiration The exchange of respiratory gases between blood and body cells. Also called **tissue respiration** or **systemic gas exchange.**

Interneurons (in′-ter-NOO-ronz) Neurons whose axons extend only for a short distance and contact nearby neurons in the brain, spinal cord, or a ganglion; they comprise the vast majority of neurons in the body. Also called **association neurons.**

Interoceptor (IN-ter-ō-sep′-tor) Sensory receptor located in blood vessels and viscera that provides information about the body's internal environment. Also called **visceroceptor.**

Interphase (IN-ter-fāz) The period of the cell cycle between cell divisions, consisting of the G_0 phase; G_1 (gap or growth) phase, when the cell is engaged in growth, metabolism, and production of substances required for division; S (synthesis) phase, during which chromosomes are replicated; and G_2 phase.

Interstitial cell of Leydig *See* **Interstitial endocrinocyte.**

Interstitial (in′-ter-STISH-al) **endocrinocyte** A cell that is located in the connective tissue between seminiferous tubules in a mature testis that secretes testosterone. Also called an **interstitial cell of Leydig** (LĪ-dig).

Interstitial (in′-ter-STISH-al) **fluid** The portion of extracellular fluid that fills the microscopic spaces between the cells of tissues; the internal environment of the body. Also called **intercellular** or **tissue fluid.**

Interstitial growth Growth from within, as in the growth of cartilage. Also called **endogenous** (en-DOJ-e-nus) **growth.**

Interventricular (in′-ter-ven-TRIK-ū-lar) **foramen** A narrow, oval opening through which the lateral ventricles of the brain communicate with the third ventricle. Also called the **foramen of Monro.**

Intervertebral (in′-ter-VER-te-bral) **disc** A pad of fibrocartilage located between the bodies of two vertebrae.

Intestinal gland A gland that opens onto the surface of the intestinal mucosa and secretes digestive enzymes. Also called a **crypt of Lieberkühn** (LĒ-ber-kūn).

Intracellular (in′-tra-SEL-yū-lar) **fluid (ICF)** Fluid located within cells.

Intrafusal (in′-tra-FŪ-sal) **fibers** Three to ten specialized muscle fibers (cells), partially enclosed in a spindle-shaped connective tissue capsule, that make up a muscle spindle.

Intramembranous (in′-tra-MEM-bra-nus) **ossification** Bone formation within mesenchyme arranged in sheetlike layers that resemble membranes.

Intramuscular injection An injection that penetrates the skin and subcutaneous layer to enter a skeletal muscle. Common sites are the deltoid, gluteus medius, and vastus lateralis muscles.

Intraocular (in′-tra-OK-ū-lar) **pressure (IOP)** Pressure in the eyeball, produced mainly by aqueous humor.

Intrinsic factor (IF) A glycoprotein, synthesized and secreted by the parietal cells of the gastric mucosa, that facilitates vitamin B_{12} absorption in the small intestine.

Invagination (in-vaj′-i-NĀ-shun) The pushing of the wall of a cavity into the cavity itself.

Inversion (in-VER-zhun) The movement of the sole medially at the ankle joint.

In vitro (VĒ-trō) Literally, in glass; outside the living body and in an artificial environment such as a laboratory test tube.

Ipsilateral (ip-si-LAT-er-al) On the same side, affecting the same side of the body.

Iris The colored portion of the vascular tunic of the eyeball seen through the cornea that contains circular and radial smooth muscle; the hole in the center of the iris is the pupil.

Irritable bowel syndrome (IBS) Disease of the entire gastrointestinal tract in which a person reacts to stress by developing symptoms (such as cramping and abdominal pain) associated with alternating patterns of diarrhea and constipation. Excessive amounts of mucus may appear in feces, and other symptoms include flatulence, nausea, and loss of appetite. Also known as **irritable colon** or **spastic colitis.**

Ischemia (is-KĒ-mē-a) A lack of sufficient blood to a body part due to obstruction or constriction of a blood vessel.

Islet of Langerhans *See* **Pancreatic islet.**

Isotonic (ī′-sō-TON-ik) Having equal tension or tone. A solution having the same concentration of impermeable solutes as cytosol.

Isthmus (IS-mus) A narrow strip of tissue or narrow passage connecting two larger parts.

J

Jaundice (JON-dis) A condition characterized by yellowness of the skin, the white of the eyes, mucous membranes, and body fluids because of a buildup of bilirubin.

Jejunum (je-JOO-num) The middle part of the small intestine.

Joint A point of contact between two bones, between bone and cartilage, or between bone and teeth. Also called an **articulation** or **arthrosis**.

Joint kinesthetic (kin′-es-THET-ik) **receptor** A proprioceptive receptor located in a joint, stimulated by joint movement.

Juxtaglomerular (juks-ta-glō-MER-ū-lar) **apparatus (JGA)** Consists of the macula densa (cells of the distal convoluted tubule adjacent to the afferent and efferent arteriole) and juxtaglomerular cells (modified cells of the afferent and sometimes efferent arteriole); secretes renin when blood pressure starts to fall.

K

Keratin (KER-a-tin) An insoluble protein found in the hair, nails, and other keratinized tissues of the epidermis.

Keratinocyte (ker-a-TIN-ō-sīt) The most numerous of the epidermal cells; produces keratin.

Kidney (KID-nē) One of the paired reddish organs located in the lumbar region that regulates the composition, volume, and pressure of blood and produces urine.

Kidney stone A solid mass, usually consisting of calcium oxalate, uric acid, or calcium phosphate crystals, that may form in any portion of the urinary tract. Also called a **renal calculus.**

Kinesiology (ki-nē-sē′-OL-ō-jē) The study of the movement of body parts.

Kinesthesia (kin′-es-THĒ-zē-a) The perception of the extent and direction of movement of body parts; this sense is possible due to nerve impulses generated by proprioceptors.

Kinetochore (ki-NET-ō-kor) Protein complex attached to the outside of a centromere to which kinetochore microtubules attach.

Kupffer's cell *See* **Stellate reticuloendothelial cell.**

Kyphosis (kī-FŌ-sis) An exaggeration of the thoracic curve of the vertebral column, resulting in a "round-shouldered" appearance. Also called **hunchback.**

L

Labial frenulum (LĀ-bē-al FREN-ū-lum) A medial fold of mucous membrane between the inner surface of the lip and the gums.

Labia majora (LĀ-bē-a ma-JŌ-ra) Two longitudinal folds of skin extending downward and backward from the mons pubis of the female.

Labia minora (min-OR-a) Two small folds of mucous membrane lying medial to the labia majora of the female.

Labium (LĀ-bē-um) A lip. A liplike structure. *Plural* is **labia** (LA-bē-a).

Labor The process of giving birth in which a fetus is expelled from the uterus through the vagina.

Labyrinth (LAB-i-rinth) Intricate communicating passageway, especially in the internal ear.

Lacrimal canal A duct, one on each eyelid, beginning at the punctum at the medial margin of an eyelid and conveying tears medially into the nasolacrimal sac.

Lacrimal gland Secretory cells, located at the superior anterolateral portion of each orbit, that secrete tears into excretory ducts that open onto the surface of the conjunctiva.

Lacrimal sac The superior expanded portion of the nasolacrimal duct that receives the tears from a lacrimal canal.

Lactation (lak-TĀ-shun) The secretion and ejection of milk by the mammary glands.

Lacteal (LAK-tē-al) One of many lymphatic vessels in villi of the intestines that absorb triglycerides and other lipids from digested food.

Lacuna (la-KOO-na) A small, hollow space, such as that found in bones in which the osteocytes lie. *Plural* is **lacunae** (la-KOO-nē).

Lambdoid (LAM-doyd) **suture** The joint in the skull between the parietal bones and the occipital bone; sometimes contains sutural (Wormian) bones.

Lamellae (la-MEL-ē) Concentric rings of hard, calcified extracellular matrix found in compact bone.

Lamellated corpuscle *See* **Pacinian corpuscle.**

Lamina (LAM-i-na) A thin, flat layer or membrane, as the flattened part of either side of the arch of a vertebra. *Plural* is **laminae** (LAM-i-nē).

Lamina propria (PRŌ-prē-a) The connective tissue layer of a mucosa.

Langerhans (LANG-er-hans) **cell** Epidermal dendritic cell that functions as an antigen-presenting cell (APC) during an immune response.

Lanugo (la-NOO-gō) Fine downy hairs that cover the fetus.

Large intestine The portion of the gastrointestinal tract extending from the ileum of the small intestine to the anus, divided structurally into the cecum, colon, rectum, and anal canal.

Laryngopharynx (la-rin′-gō-FAIR-inks) The inferior portion of the pharynx, extending downward from the level of the hyoid bone that divides posteriorly into the esophagus and anteriorly into the larynx. Also called the **hypopharynx.**

Larynx (LAIR-inks) The **voice box,** a short passageway that connects the pharynx with the trachea.

Lateral (LAT-er-al) Farther from the midline of the body or a structure.

Lateral ventricle (VEN-tri-kul) A cavity within a cerebral hemisphere that communicates with the lateral ventricle in the other cerebral hemisphere and with the third ventricle by way of the interventricular foramen.

Leg The part of the lower limb between the knee and the ankle.

Lens A transparent organ constructed of proteins (crystallins) lying posterior to the pupil and iris of the eyeball and anterior to the vitreous body.

Lesion (LĒ-zhun) Any localized, abnormal change in a body tissue.

Lesser omentum (ō-MEN-tum) A fold of the peritoneum that extends from the liver to the lesser curvature of the stomach and the first part of the duodenum.

Lesser vestibular (ves-TIB-ū-lar) **gland** One of the paired mucus-secreting glands with ducts that open on either side of the urethral orifice in the vestibule of the female.

Leukemia (loo-KĒ-mē-a) A malignant disease of the blood-forming tissues characterized by either uncontrolled production and accumulation of immature leukocytes in which many cells fail to reach maturity (acute) or an accumulation of mature leukocytes in the blood because they do not die at the end of their normal life span (chronic).

Leukocyte (LOO-kō-sīt) A white blood cell.

Leydig (LĪ-dig) **cell** A type of cell that secretes testosterone; located in the connective tissue between seminiferous tubules in a mature testis. Also known as **interstitial cell of Leydig** or **interstitial endocrinocyte.**

Ligament (LIG-a-ment) Dense regular connective tissue that attaches bone to bone.

Ligand (LĪ-gand) A chemical substance that binds to a specific receptor.

Limbic system A part of the forebrain, sometimes termed the visceral brain, concerned with various aspects of emotion and behavior; includes the limbic lobe, dentate gyrus, amygdala, septal nuclei, mammillary bodies, anterior thalamic nucleus, olfactory bulbs, and bundles of myelinated axons.

Lingual frenulum (LIN-gwal FREN-ū-lum) A fold of mucous membrane that connects the tongue to the floor of the mouth.

Lipase An enzyme that splits fatty acids from triglycerides and phospholipids.

Lipid (LIP-id) An organic compound composed of carbon, hydrogen, and oxygen that is usually insoluble in water, but soluble in alcohol, ether, and chloroform; examples include triglycerides (fats and oils), phospholipids, steroids, and eicosanoids.

Lipid bilayer Arrangement of phospholipid, glycolipid, and cholesterol molecules in two parallel sheets in which the hydrophilic "heads" face outward and the hydrophobic "tails" face inward; found in cellular membranes.

Lipoprotein (lip′-ō-PRŌ-tēn) One of several types of particles containing lipids (cholesterol and triglycerides) and proteins that make it water-soluble for transport in the blood; high levels of **low-density lipoproteins (LDLs)** are associated with increased risk of atherosclerosis, whereas high levels of **high-density lipoproteins (HDLs)** are associated with decreased risk of atherosclerosis.

Liver Large organ under the diaphragm that occupies most of the right hypochondriac region and part of the epigastric region. Functionally, it produces bile and synthesizes most plasma proteins; interconverts nutrients; detoxifies substances; stores glycogen, iron, and vitamins; carries on phagocytosis of worn-out blood cells and bacteria; and helps synthesize the active form of vitamin D.

Long-term potentiation (po-ten′-she-Ā-shun) **(LTP)** Prolonged, enhanced synaptic transmission that occurs at certain synapses within the hippocampus of the brain; believed to underlie some aspects of memory.

Lordosis (lor-DŌ-sis) An exaggeration of the lumbar curve of the vertebral column. Also called **hollow back.**

Lower limb The appendage attached at the pelvic (hip) girdle, consisting of the thigh, knee, leg, ankle, foot, and toes. Also called the **lower extremity** or **lower appendage.**

Lumbar (LUM-bar) Region of the back and side between the ribs and pelvis; loin.

Lumbar plexus (PLEK-sus) A network formed by the anterior (ventral) branches of spinal nerves L1 through L4.

Lumen (LOO-men) The space within an artery, vein, intestine, renal tubule, or other tubular structure.

Lungs Main organs of respiration that lie on either side of the heart in the thoracic cavity.

Lunula (LOO-noo-la) The moon-shaped white area at the base of a nail.

Luteinizing (LOO-tē-in′-īz-ing) **hormone (LH)** A hormone secreted by the anterior pituitary that stimulates ovulation, stimulates progesterone secretion by the corpus luteum, and readies the mammary glands for milk secretion in females; stimulates testosterone secretion by the testes in males.

Lymph (LIMF) Fluid confined in lymphatic vessels and flowing through the lymphatic system until it is returned to the blood.

Lymph node An oval or bean-shaped structure located along lymphatic vessels.

Lymphatic (lim-FAT-ik) **capillary** Closed-ended microscopic lymphatic vessel that begins in spaces between cells and converges with other lymphatic capillaries to form lymphatic vessels.

Lymphatic system (lim-FAT-ik) A system consisting of a fluid called lymph, vessels called lymphatics that transport lymph, a number of organs containing lymphatic tissue (lymphocytes within a filtering tissue), and red bone marrow.

Lymphatic tissue A specialized form of reticular tissue that contains large numbers of lymphocytes.

Lymphatic vessel A large vessel that collects lymph from lymphatic capillaries and converges with other lymphatic vessels to form the thoracic and right lymphatic ducts.

Lymphocyte (LIM-fō-sīt) A type of white blood cell that helps carry out cell-mediated and antibody-mediated immune responses; found in blood and in lymphatic tissues.

Lysosome (LĪ-sō-sōm) An organelle in the cytoplasm of a cell, enclosed by a single membrane and containing powerful digestive enzymes.

Lysozyme (LĪ-sō-zīm) A bactericidal enzyme found in tears, saliva, and perspiration.

M

Macrophage (MAK-rō-fāj) Phagocytic cell derived from a monocyte; may be fixed or wandering.

Macula (MAK-ū-la) A discolored spot or a colored area. A small, thickened region on the wall of the utricle and saccule that contains receptors for static equilibrium.

Macula lutea (LOO-tē-a) The yellow spot in the center of the retina.

Major histocompatibility (MHC) antigens Surface proteins on white blood cells and other nucleated cells that are unique for each person (except for identical siblings); used to type tissues and help prevent rejection of transplanted tissues. Also known as **human leukocyte antigens (HLA).**

Malignant (ma-LIG-nant) Referring to diseases that tend to become worse and cause death, especially the invasion and spreading of cancer.

Mammary (MAM-ar-ē) **gland** Modified sudoriferous (sweat) gland of the female that produces milk for the nourishment of the young.

Mammillary (MAM-i-ler-ē) **bodies** Two small rounded bodies on the inferior aspect of the hypothalamus that are involved in reflexes related to the sense of smell.

Marrow (MAR-ō) Soft, spongelike material in the cavities of bone. *Red bone marrow* produces blood cells; *yellow bone marrow* contains adipose tissue that stores triglycerides.

Mast cell A cell found in areolar connective tissue that releases histamine, a dilator of small blood vessels, during inflammation.

Mastication (mas′-ti-KĀ-shun) Chewing.

Mature follicle A large, fluid-filled follicle containing a secondary oocyte and surrounding granulosa cells that secrete estrogens. Also called a **graafian** (GRAF-ē-an) **follicle.**

Meatus (mē-Ā-tus) A passage or opening, especially the external portion of a canal.

Mechanoreceptor (me-KAN-ō-rē-sep-tor) Sensory receptor that detects mechanical deformation of the receptor itself or adjacent cells; stimuli so detected include those related to touch, pressure, vibration, proprioception, hearing, equilibrium, and blood pressure.

Medial (MĒ-dē-al) Nearer the midline of the body or a structure.

Medial lemniscus (lem-NIS-kus) A white matter tract that originates in the gracile and cuneate nuclei of the medulla oblongata and extends to the thalamus on the same side; sensory axons in this tract conduct nerve impulses for the sensations of proprioception, fine touch, vibration, hearing, and equilibrium.

Median aperture (AP-er-choor) One of the three openings in the roof of the fourth ventricle through which cerebrospinal fluid enters the subarachnoid space of the brain and cord. Also called the **foramen of Magendie.**

Median plane A vertical plane dividing the body into right and left halves. Situated in the middle.

Mediastinum (mē′-dē-as-TĪ-num) The anatomical region on the thoracic cavity between the pleurae of the lungs that extends from the sternum to the vertebral column and from the first rib to the diaphragm.

Medulla (me-DOOL-la) An inner layer of an organ, such as the medulla of the kidneys.

Medulla oblongata (me-DOOL-la ob′-long-GA-ta) The most inferior part of the brain stem. Also termed the **medulla.**

Medullary (MED-ū-lar′-ē) **cavity** The space within the diaphysis of a bone that contains yellow bone marrow. Also called the **marrow cavity.**

Medullary rhythmicity (rith-MIS-i-tē) **area** The neurons of the respiratory center in the medulla oblongata that control the basic rhythm of respiration.

Meibomian gland *See* **Tarsal gland.**

Meiosis (mī-Ō-sis) A type of cell division that occurs during production of gametes, involving two successive nuclear divisions that result in cells with the haploid *(n)* number of chromosomes.

Meissner (MĪS-ner) **corpuscle** A sensory receptor for touch; found in dermal papillae, especially in the palms and soles. Also called a **corpuscle of touch.**

Melanin (MEL-a-nin) A dark black, brown, or yellow pigment found in some parts of the body such as the skin, hair, and pigmented layer of the retina.

Melanocyte (MEL-a-nō-sīt′) A pigmented cell, located between or beneath cells of the deepest layer of the epidermis, that synthesizes melanin.

Melanocyte-stimulating hormone (MSH) A hormone secreted by the anterior pituitary that stimulates the dispersion of melanin granules in melanocytes in amphibians; continued administration produces darkening of skin in humans.

Melatonin (mel-a-TŌN-in) A hormone secreted by the pineal gland that helps set the timing of the body's biological clock.

Membrane A thin, flexible sheet of tissue composed of an epithelial layer and an underlying connective tissue layer, as in an epithelial membrane, or of areolar connective tissue only, as in a synovial membrane.

Membranous labyrinth (mem-BRA-nus LAB-i-rinth) The part of the labyrinth of the internal ear that is located inside the bony labyrinth and separated from it by the perilymph; made up of the semicircular ducts, the saccule and utricle, and the cochlear duct.

Memory The ability to recall thoughts; commonly classifed as short-term (activated) and long-term.

Menarche (me-NAR-kē) The first menses (menstrual flow) and beginning of ovarian and uterine cycles.

Meninges (me-NIN-jēz) Three membranes covering the brain and spinal cord, called the dura mater, arachnoid mater, and pia mater. *Singular* is **meninx** (MEN-inks).

Menopause (MEN-ō-pawz) The termination of the menstrual cycles.

Menstruation (men′-stroo-Ā-shun) Periodic discharge of blood, tissue fluid, mucus, and epithelial cells that usually lasts for 5 days; caused by a sudden reduction in estrogens and progesterone. Also called the **menstrual phase** or **menses.**

Merkel (MER-kel) **cell** *See* **Tactile cell.**

Merkel disc *See* **Tactile disc.**

Merocrine (MER-ō-krin) **gland** Gland made up of secretory cells that remain intact throughout the process of formation and discharge of the secretory product, as in the salivary and pancreatic glands.

Mesenchyme (MEZ-en-kīm) An embryonic connective tissue from which all other connective tissues arise.

Mesentery (MEZ-en-ter′-ē) A fold of peritoneum attaching the small intestine to the posterior abdominal wall.

Mesocolon (mez′-ō-KŌ-lon) A fold of peritoneum attaching the colon to the posterior abdominal wall.

Mesoderm The middle primary germ layer that gives rise to connective tissues, blood and blood vessels, and muscles.

Mesothelium (mez′-ō-THĒ-lē-um) The layer of simple squamous epithelium that lines serous membranes.

Mesovarium (mez′-ō-VAR-ē-um) A short fold of peritoneum that attaches an ovary to the broad ligament of the uterus.

Metabolism (me-TAB-ō-lizm) All of the biochemical reactions that occur within an organism, including the synthetic (anabolic) reactions and decomposition (catabolic) reactions.

Metacarpus (met′-a-KAR-pus) A collective term for the five bones that make up the palm.

Metaphase (MET-a-fāz) The second stage of mitosis, in which chromatid pairs line up on the metaphase plate of the cell.

Metaphysis (me-TAF-i-sis) Region of a long bone between the diaphysis and epiphysis that contains the epiphyseal plate in a growing bone.

Metarteriole (met′-ar-TĒ-rē-ōl) A blood vessel that emerges from an arteriole, traverses a capillary network, and empties into a venule.

Metastasis (me-TAS-ta-sis) The spread of cancer to surrounding tissues (local) or to other body sites (distant).

Metatarsus (met′-a-TAR-sus) A collective term for the five bones located in the foot between the tarsals and the phalanges.

Microfilament (mī-krō-FIL-a-ment) Rodlike protein filament about 6 nm in diameter; constitutes contractile units in muscle fibers (cells) and provides support, shape, and movement in nonmuscle cells.

Microglia (mī-KROG-lē-a) Neuroglial cells that carry on phagocytosis.

Microtubule (mī-krō-TOO-būl) Cylindrical protein filament, from 18 to 30 nm in diameter, consisting of the protein tubulin; provides support, structure, and transportation.

Microvilli (mī-krō-VIL-ī) Microscopic, finger-like projections of the plasma membranes of cells that increase surface area for absorption, especially in the small intestine and proximal convoluted tubules of the kidneys.

Micturition (mik′-choo-RISH-un) The act of expelling urine from the urinary bladder. Also called **urination** (ū-ri-NĀ-shun).

Midbrain The part of the brain between the pons and the diencephalon. Also called the **mesencephalon** (mes′-en-SEF-a-lon).

Middle ear A small, epithelial-lined cavity hollowed out of the temporal bone, separated from the external ear by the eardrum and from the internal ear by a thin bony partition containing the oval and round windows; extending across the middle ear are the three auditory ossicles. Also called the **tympanic** (tim-PAN-ik) **cavity.**

Midline An imaginary vertical line that divides the body into equal left and right sides.

Midsagittal plane A vertical plane through the midline of the body that divides the body or organs into *equal* right and left sides. Also called a **median plane.**

Mineralocorticoids (min′-er-al-ō-KOR-ti-koyds) A group of hormones of the adrenal cortex that help regulate sodium and potassium balance.

Mitochondrion (mī-tō-KON-drē-on) A double-membraned organelle that plays a central role in the production of ATP; known as the "powerhouse" of the cell. *Plural* is **mitochondria.**

Mitosis (mī-TŌ-sis) The orderly division of the nucleus of a cell that ensures that each new nucleus has the same number and kind of chromo-

somes as the original nucleus. The process includes the replication of chromosomes and the distribution of the two sets of chromosomes into two separate and equal nuclei.

Mitotic spindle Collective term for a football-shaped assembly of microtubules (nonkinetochore, kinetochore, and aster) that is responsible for the movement of chromosomes during cell division.

Modality (mō-DAL-i-tē) Any of the specific sensory entities, such as vision, smell, taste, or touch.

Modiolus (mō-DĪ-ō′-lus) The central pillar or column of the cochlea.

Molecule (mol′-e-KŪL) A substance composed of two or more atoms chemically combined.

Monocyte (MON-ō-sīt′) The largest type of white blood cell, characterized by agranular cytoplasm.

Monounsaturated fat A fatty acid that contains one double covalent bond between its carbon atoms; it is not completely saturated with hydrogen atoms. Plentiful in triglycerides of olive and peanut oils.

Mons pubis (MONZ PŪ-bis) The rounded, fatty prominence over the pubic symphysis, covered by coarse pubic hair.

Morula (MOR-ū-la) A solid sphere of cells produced by successive cleavages of a fertilized ovum about 4 days after fertilization.

Motor area The region of the cerebral cortex that governs muscular movement, particularly the precentral gyrus of the frontal lobe.

Motor end plate Region of the sarcolemma of a muscle fiber (cell) that includes acetylcholine (ACh) receptors, which bind ACh released by synaptic end bulbs of somatic motor neurons.

Motor neurons (NOO-ronz) Neurons that conduct impulses from the brain toward the spinal cord or out of the brain and spinal cord into cranial or spinal nerves to effectors that may be either muscles or glands. Also called **efferent neurons.**

Motor unit A motor neuron together with the muscle fibers (cells) it stimulates.

Mucosa associated lymphatic tissue (MALT) Lymphatic nodules scattered throughout the lamina propria (connective tissue) of mucous membranes lining the gastrointestinal tract, respiratory airways, urinary tract, and reproductive tract.

Mucous (MŪ-kus) **cell** A unicellular gland that secretes mucus. Two types are mucous neck cells and surface mucous cells in the stomach.

Mucous membrane A membrane that lines a body cavity that opens to the exterior. Also called the **mucosa** (mū-KŌ-sa).

Mucus The thick fluid secretion of goblet cells, mucous cells, mucous glands, and mucous membranes.

Muscarinic (mus′-ka-RIN-ik) **receptor** Receptor for the neurotransmitter acetylcholine found on all effectors innervated by parasympathetic postganglionic axons and on sweat glands innervated by cholinergic sympathetic postganglionic axons; so named because muscarine activates these receptors but does not activate nicotinic receptors for acetylcholine.

Muscle An organ composed of one of three types of muscle tissue (skeletal, cardiac, or smooth), specialized for contraction to produce voluntary or involuntary movement of parts of the body.

Muscle action potential A stimulating impulse that propagates along the sarcolemma and transverse tubules; in skeletal muscle, it is generated by acetylcholine, which increases the permeability of the sarcolemma to cations, especially sodium ions (Na^+).

Muscle fatigue (fa-TĒG) Inability of a muscle to maintain its strength of contraction or tension; may be related to insufficient oxygen, depletion of glycogen, and/or lactic acid buildup.

Muscle spindle An encapsulated proprioceptor in a skeletal muscle, consisting of specialized intrafusal muscle fibers and nerve endings; stimulated by changes in length or tension of muscle fibers.

Muscle strain Tearing of skeletal muscle fibers or tendon. Also called a **muscle pull** or **muscle tear.**

Muscle tone A sustained, partial contraction of portions of a skeletal or smooth muscle in response to activation of stretch receptors or a baseline level of action potentials in the innervating motor neurons.

Muscular dystrophies (DIS-trō-fēz) Inherited muscle-destroying diseases, characterized by degeneration of muscle fibers (cells), which causes progressive atrophy of the skeletal muscle.

Muscularis (MUS-kū-la′-ris) A muscular layer (coat or tunic) of an organ.

Muscularis mucosae (mū-KŌ-sē) A thin layer of smooth muscle fibers that underlie the lamina propria of the mucosa of the gastrointestinal tract.

Muscular system Usually refers to the approximately 100 voluntary muscles of the body that are composed of skeletal muscle tissue.

Muscular tissue A tissue specialized to produce motion in response to muscle action potentials by its qualities of contractility, extensibility, elasticity, and excitability; types include skeletal, cardiac, and smooth.

Musculoskeletal (mus′-kyū-lō-SKEL-e-tal) **system** An integrated body system consisting of bones, joints, and muscles.

Mutation (mū-TĀ-shun) Any change in the sequence of bases in a DNA molecule resulting in a permanent alteration in some inheritable trait.

Myasthenia (mī-as-THĒ-nē-a) **gravis** Weakness and fatigue of skeletal muscles caused by antibodies directed against acetylcholine receptors.

Myelin (MĪ-e-lin) **sheath** Multilayered lipid and protein covering, formed by Schwann cells and oligodendrocytes, around axons of many peripheral and central nervous system neurons.

Myenteric (mī-en-TER-ik) **plexus** A network of autonomic axons and postganglionic cell bodies located in the muscularis of the gastrointestinal tract. Also called the **plexus of Auerbach** (OW-er-bak).

Myocardial infarction (mī′-ō-KAR-dē-al in-FARK-shun) **(MI)** Gross necrosis of myocardial tissue due to interrupted blood supply. Also called a **heart attack.**

Myocardium (mī′-ō-KAR-dē-um) The middle layer of the heart wall, made up of cardiac muscle tissue, lying between the epicardium and the endocardium and constituting the bulk of the heart.

Myofibril (mī-ō-FĪ-bril) A threadlike structure, extending longitudinally through a muscle fiber (cell), consisting mainly of thick filaments (myosin) and thin filaments (actin, troponin, and tropomyosin).

Myoglobin (mī-ō-GLŌB-in) The oxygen-binding, iron-containing protein present in the sarcoplasm of muscle fibers (cells); contributes the red color to muscle.

Myogram (MĪ-ō-gram) The record or tracing produced by a myograph, an apparatus that measures and records the force of muscular contractions.

Myology (mī-OL-ō-jē) The study of muscles.

Myometrium (mī′-ō-MĒ-trē-um) The smooth muscle layer of the uterus.

Myopathy (mī-OP-a-thē) Any abnormal condition or disease of muscle tissue.

Myopia (mī-Ō-pē-a) Defect in vision in which objects can be seen distinctly only when very close to the eyes; nearsightedness.

Myosin (MĪ-ō-sin) The contractile protein that makes up the thick filaments of muscle fibers.

Myotome (MĪ-ō-tōm) A group of muscles innervated by the motor neurons of a single spinal segment. In an embryo, the portion of a somite that develops into some skeletal muscles.

N

Nail A hard plate, composed largely of keratin, that develops from the epidermis of the skin to form a protective covering on the dorsal surface of the distal phalanges of the fingers and toes.

Nail matrix (MĀ-triks) The part of the nail beneath the body and root from which the nail is produced.

Nasal (NĀ-zal) **cavity** A mucosa-lined cavity on either side of the nasal septum that opens onto the face at the external nares and into the nasopharynx at the internal nares.

Nasal septum (SEP-tum) A vertical partition composed of bone (perpendicular plate of ethmoid and vomer) and cartilage, covered with a mucous membrane, separating the nasal cavity into left and right sides.

Nasolacrimal (nā′-zō-LAK-ri-mal) **duct** A canal that transports the lacrimal secretion (tears) from the nasolacrimal sac into the nose.

Nasopharynx (nā′-zō-FAR-inks) The superior portion of the pharynx, lying posterior to the nose and extending inferiorly to the soft palate.

Neck The part of the body connecting the head and the trunk. A constricted portion of an organ, such as the neck of the femur or uterus.

Necrosis (ne-KRŌ-sis) A pathological type of cell death that results from disease, injury, or lack of blood supply in which many adjacent cells swell, burst, and spill their contents into the interstitial fluid, triggering an inflammatory response.

Negative feedback system A feedback cycle that reverses a change in a controlled condition.

Neonatal (nē-ō-NĀ-tal) Pertaining to the first 4 weeks after birth.

Neoplasm (NĒ-ō-plazm) A new growth that may be benign or malignant.

Nephron (NEF-ron) The functional unit of the kidney.

Nerve A cordlike bundle of neuronal axons and/or dendrites and associated connective tissue coursing together outside the central nervous system.

Nerve fiber General term for any process (axon or dendrite) projecting from the cell body of a neuron.

Nerve impulse A wave of depolarization and repolarization that self-propagates along the plasma membrane of a neuron; also called a **nerve action potential.**

Nervous system A network of billions of neurons and even more neuroglia that is organized into two main divisions, central nervous system (brain and spinal cord) and peripheral nervous system (nerves, ganglia, enteric plexuses, and sensory receptors outside the central nervous system).

Nervous tissue Tissue containing neurons that initiate and conduct nerve impulses to coordinate homeostasis, and neuroglia that provide support and nourishment to neurons.

Neuralgia (noo-RAL-jē-a) Attacks of pain along the entire course or branch of a peripheral sensory nerve.

Neural plate A thickening of ectoderm, induced by the notochord, that forms early in the third week of development and represents the beginning of the development of the nervous system.

Neural tube defect (NTD) A developmental abnormality in which the neural tube does not close properly. Examples are spina bifida and anencephaly.

Neuritis (noo-RĪ-tis) Inflammation of one or more nerves.

Neurofibral node *See* **Node of Ranvier.**

Neuroglia (noo-RŌG-lē-a) Cells of the nervous system that perform various supportive functions. The neuroglia of the central nervous system are the astrocytes, oligodendrocytes, microglia, and ependymal cells; neuroglia of the peripheral nervous system include Schwann cells and satellite cells. Also called **glial** (GLĒ-al) **cells.**

Neurohypophyseal (noo′-rō-hī′-pō-FIZ-ē-al) **bud** An outgrowth of ectoderm located on the floor of the hypothalamus that gives rise to the posterior pituitary.

Neurolemma (noo-rō-LEM-ma) The peripheral, nucleated cytoplasmic layer of the Schwann cell. Also called **sheath of Schwann** (SCHWON).

Neurology (noo-ROL-ō-jē) The study of the normal functioning and disorders of the nervous system.

Neuromuscular (noo-rō-MUS-kū-lar) **junction (NMJ)** A synapse between the axon terminals of a motor neuron and the sarcolemma of a muscle fiber (cell).

Neuron (NOO-ron) A nerve cell, consisting of a cell body, dendrites, and an axon.

Neurophysiology (NOOR-ō-fiz-ē-ol′-ō-jē) Study of the functional properties of nerves.

Neurosecretory (noo-rō-SĒK-re-tō-rē) **cell** A neuron that secretes a hypothalamic releasing hormone or inhibiting hormone into blood capillaries of the hypothalmus; a neuron that secretes oxytocin or antidiuretic hormone into blood capillaries of the posterior pituitary.

Neurotransmitter (noo′-rō-trans′-MIT-er) One of a variety of molecules within axon terminals that are released into the synaptic cleft in response to a nerve impulse and that change the membrane potential of the postsynaptic neuron.

Neurulation (noor-oo-LĀ-shun) The process by which the neural plate, neural folds, and neural tube develop.

Neutrophil (NOO-trō-fil) A type of white blood cell characterized by granules that stain pale lilac with a combination of acidic and basic dyes.

Nicotinic (nik′-ō-TIN-ik) **receptor** Receptor for the neurotransmitter acetylcholine found on both sympathetic and parasympathetic postganglionic neurons and on skeletal muscle in the motor end plate; so named because nicotine activates these receptors but does not activate muscarinic receptors for acetylcholine.

Nipple A pigmented, wrinkled projection on the surface of the breast that is the location of the openings of the lactiferous ducts for milk release.

Nociceptor (nō′-sē-SEP-tor) A free (naked) nerve ending that detects painful stimuli.

Node of Ranvier (RON-vē-ā) A space along a myelinated axon between the individual Schwann cells that form the myelin sheath and the neurolemma. Also called a **neurofibral node.**

Norepinephrine (nor′-ep-ē-NEF-rin) **(NE)** A hormone secreted by the adrenal medulla that produces actions similar to those that result from sympathetic stimulation. Also called **noradrenaline** (nor-a-DREN-a-lin).

Notochord (NŌ-tō-cord) A flexible rod of mesodermal tissue that lies where the future vertebral column will develop and plays a role in induction.

Nucleic (noo-KLĒ-ik) **acid** An organic compound that is a long polymer of nucleotides, with each nucleotide containing a pentose sugar, a phosphate group, and one of four possible nitrogenous bases (adenine, cytosine, guanine, and thymine or uracil).

Nucleolus (noo′-KLĒ-ō-lus) Spherical body within a cell nucleus composed of protein, DNA, and RNA that is the site of the assembly of small and large ribosomal subunits. *Plural* is **nucleoli.**

Nucleosome (NOO-klē-ō-sōm) Structural subunit of a chromosome consisting of histones and DNA.

Nucleus (NOO-klē-us) A spherical or oval organelle of a cell that contains the hereditary factors of the cell, called genes. A cluster of unmyelinated nerve cell bodies in the central nervous system. The central part of an atom made up of protons and neutrons.

Nucleus pulposus (pul-PŌ-sus) A soft, pulpy, highly elastic substance in the center of an intervertebral disc; a remnant of the notochord.

Nutrient A chemical substance in food that provides energy, forms new body components, or assists in various body functions.

O

Obesity (ō-BĒS-i-tē) Body weight more than 20% above a desirable standard due to excessive accumulation of fat.

Oblique (ō-BLĒK) **plane** A plane that passes through the body or an organ at an angle between the transverse plane and either the midsagittal, parasagittal, or frontal plane.

Obstetrics (ob-STET-riks) The specialized branch of medicine that deals with pregnancy, labor, and the period of time immediately after delivery (about 6 weeks).

Olfaction (ōl-FAK-shun) The sense of smell.

Olfactory (ōl-FAK-tō-rē) Pertaining to smell.

Olfactory bulb A mass of gray matter containing cell bodies of neurons that form synapses with neurons of the olfactory (I) nerve, lying inferior to the frontal lobe of the cerebrum on either side of the crista galli of the ethmoid bone.

Olfactory receptor A bipolar neuron with its cell body lying between supporting cells located in the mucous membrane lining the superior portion of each nasal cavity; transduces odors into neural signals.

Olfactory tract A bundle of axons that extends from the olfactory bulb posteriorly to olfactory regions of the cerebral cortex.

Oligodendrocyte (OL-i-gō-den′-drō-sīt) A neuroglial cell that supports neurons and produces a myelin sheath around axons of neurons of the central nervous system.

Oliguria (ol′-i-GŪ-rē-a) Daily urinary output usually less than 250 mL.

Olive A prominent oval mass on each lateral surface of the superior part of the medulla oblongata.

Oncogene (ON-kō-jēn) Cancer-causing gene; it derives from a normal gene, termed a proto-oncogene, that encodes proteins involved in cell growth or cell regulation but has the ability to transform a normal cell into a cancerous cell when it is mutated or inappropriately activated. One example is *p53.*

Oncology (on-KOL-ō-jē) The study of tumors.

Oogenesis (ō-ō-JEN-e-sis) Formation and development of female gametes (oocytes).

Oophorectomy (ō′-of-ō-REK-tō-me) Surgical removal of the ovaries.

Ophthalmic (of-THAL-mik) Pertaining to the eye.

Ophthalmologist (of′-thal-MOL-ō-jist) A physician who specializes in the diagnosis and treatment of eye disorders using drugs, surgery, and corrective lenses.

Ophthalmology (of-thal-MOL-ō-jē) The study of the structure, function, and diseases of the eye.

Optic (OP-tik) Refers to the eye, vision, or properties of light.

Optic chiasm (kī-AZM) A crossing point of the two branches of the optic (II) nerve, anterior to the pituitary gland. Also called **optic chiasma.**

Optic disc A small area of the retina containing openings through which the axons of the ganglion cells emerge as the optic (II) nerve. Also called the **blind spot.**

Optic tract A bundle of axons that carry nerve impulses from the retina of the eye between the optic chiasm and the thalamus.

Ora serrata (Ō-ra ser-RĀ-ta) The irregular margin of the retina lying internal and slightly posterior to the junction of the choroid and ciliary body.

Orbit (OR-bit) The bony, pyramidal-shaped cavity of the skull that holds the eyeball.

Organ A structure composed of two or more different kinds of tissues with a specific function and usually a recognizable shape.

Organelle (or-ga-NEL) A permanent structure within a cell with characteristic morphology that is specialized to serve a specific function in cellular activities.

Organism (OR-ga-nizm) A total living form; one individual.

Organogenesis (or′-ga-nō-JEN-e-sis) The formation of body organs and systems. By the end of the eighth week of development, all major body systems have begun to develop.

Orifice (OR-i-fis) Any aperture or opening.

Origin (OR-i-jin) The attachment of a muscle tendon to a stationary bone or the end opposite the insertion.

Oropharynx (or′-ō-FAR-inks) The intermediate portion of the pharynx, lying posterior to the mouth and extending from the soft palate to the hyoid bone.

Orthopedics (or′-thō-PĒ-diks) The branch of medicine that deals with the preservation and restoration of the skeletal system, articulations, and associated structures.

Osmoreceptor (oz′-mō-rē-CEP-tor) Receptor in the hypothalamus that is sensitive to changes in blood osmolarity and, in response to high osmolarity (low water concentration), stimulates synthesis and release of antidiuretic hormone (ADH).

Osmosis (oz-MŌ-sis) The net movement of water molecules through a selectively permeable membrane from an area of higher water concentration to an area of lower water concentration until equilibrium is reached.

Osseous (OS-ē-us) Bony.

Ossicle (OS-si-kul) One of the small bones of the middle ear (malleus, incus, stapes).

Ossification (os′-i-fi-KĀ-shun) Formation of bone. Also called **osteogenesis.**

Ossification (os′-i-fi-KĀ-shun) **center** An area in the cartilage model of a future bone where the cartilage cells hypertrophy, secrete enzymes that calcify their extracellular matrix, and die, and the area they occupied is invaded by osteoblasts that then lay down bone.

Osteoblast (OS-tē-ō-blast′) Cell formed from an osteogenic cell that participates in bone formation by secreting some organic components and inorganic salts.

Osteoclast (OS-tē-ō-klast′) A large, multinuclear cell that resorbs (destroys) bone matrix.

Osteocyte (OS-tē-ō-sīt′) A mature bone cell that maintains the daily activities of bone tissue.

Osteogenic (os′-tē-ō-JEN-ik) **cell** Stem cell derived from mesenchyme that has mitotic potential and the ability to differentiate into an osteoblast.

Osteogenic layer The inner layer of the periosteum that contains cells responsible for forming new bone during growth and repair.

Osteology (os-tē-OL-ō-jē) The study of bones.

Osteon (OS-tē-on) The basic unit of structure in adult compact bone, consisting of a central (haversian) canal with its concentrically arranged lamellae, lacunae, osteocytes, and canaliculi. Also called a **haversian** (ha-VER-shun) **system.**

Osteoporosis (os′-tē-ō-pō-RŌ-sis) Age-related disorder characterized by decreased bone mass and increased susceptibility to fractures, often as a result of decreased levels of estrogens.

Otic (Ō-tik) Pertaining to the ear.

Otolith (Ō-tō-lith) A particle of calcium carbonate embedded in the otolithic membrane that functions in maintaining static equilibrium.

Otolithic (ō-tō-LITH-ik) **membrane** Thick, gelatinous, glycoprotein layer located directly over hair cells of the macula in the saccule and utricle of the internal ear.

Otorhinolaryngology (ō-tō-rī′-nō-lar-in-GOL-ō-jē) The branch of medicine that deals with the diagnosis and treatment of diseases of the ears, nose, and throat.

Oval window A small, membrane-covered opening between the middle ear and inner ear into which the footplate of the stapes fits.

Ovarian (ō-VAR-ē-an) **cycle** A monthly series of events in the ovary associated with the maturation of a secondary oocyte.

Ovarian follicle (FOL-i-kul) A general name for oocytes (immature ova) in any stage of develop-

ment, along with their surrounding epithelial cells.

Ovarian ligament (LIG-a-ment) A rounded cord of connective tissue that attaches the ovary to the uterus.

Ovary (Ō-var-ē) Female gonad that produces oocytes and the estrogen, progesterone, inhibin, and relaxin hormones.

Ovulation (ov′-ū-LĀ-shun) The rupture of a mature ovarian (graafian) follicle with discharge of a secondary oocyte into the pelvic cavity.

Ovum (Ō-vum) The female reproductive or germ cell; an egg cell; arises through completion of meiosis in a secondary oocyte after penetration by a sperm.

Oxyhemoglobin (ok′-sē-HĒ-mō-glō-bin) **(Hb–O₂)** Hemoglobin combined with oxygen.

Oxytocin (ok′-sē-TŌ-sin) **(OT)** A hormone secreted by neurosecretory cells in the paraventricular and supraoptic nuclei of the hypothalamus that stimulates contraction of smooth muscle in the pregnant uterus and myoepithelial cells around the ducts of mammary glands.

P

P wave The deflection wave of an electrocardiogram that signifies atrial depolarization.

Pacinian corpuscle (pa-SIN-ē-an) Oval-shaped pressure receptor located in the dermis or subcutaneous tissue and consisting of concentric layers of a connective tissue wrapped around the dendrites of a sensory neuron. Also called a **lamellated corpuscle.**

Palate (PAL-at) The horizontal structure separating the oral and the nasal cavities; the roof of the mouth.

Palpate (PAL-pāt) To examine by touch; to feel.

Pancreas (PAN-krē-as) A soft, oblong organ lying along the greater curvature of the stomach and connected by a duct to the duodenum. It is both an exocrine gland (secreting pancreatic juice) and an endocrine gland (secreting insulin, glucagon, somatostatin, and pancreatic polypeptide).

Pancreatic (pan′-krē-AT-ik) **duct** A single large tube that unites with the common bile duct from the liver and gallbladder and drains pancreatic juice into the duodenum at the hepatopancreatic ampulla (ampulla of Vater). Also called the **duct of Wirsung** (VĒR-sung).

Pancreatic islet (Ī-let) Cluster of endocrine gland cells in the pancreas that secretes insulin, glucagon, somatostatin, and pancreatic polypeptide. Also called an **islet of Langerhans** (LAHNG-er-hanz).

Papanicolaou (pa-pa-NI-kō-lō) **test** A cytological staining test for the detection and diagnosis of premalignant and malignant conditions of the female genital tract. Cells scraped from the epithelium of the cervix of the uterus are examined microscopically. Also called a **Pap test** or **Pap smear.**

Papilla (pa-PIL-a) A small nipple-shaped projection or elevation.

Paralysis (pa-RAL-a-sis) Loss or impairment of motor function due to a lesion of nervous or muscular origin.

Paranasal sinus (par′-a-NĀ-zal SĪ-nus) A mucus-lined air cavity in a skull bone that communicates with the nasal cavity. Paranasal sinuses are located in the frontal, maxillary, ethmoid, and sphenoid bones.

Paraplegia (par-a-PLĒ-jē-a) Paralysis of both lower limbs.

Parasagittal plane (par-a-SAJ-i-tal) A vertical plane that does not pass through the midline and that divides the body or organs into *unequal* left and right portions.

Parasympathetic (par′-a-sim-pa-THET-ik) **division** One of the two subdivisions of the autonomic nervous system, having cell bodies of preganglionic neurons in nuclei in the brain stem and in the lateral gray horn of the sacral portion of the spinal cord; primarily concerned with activities that conserve and restore body energy.

Parathyroid (par′-a-THĪ-royd) **gland** One of usually four small endocrine glands embedded in the posterior surfaces of the lateral lobes of the thyroid gland.

Parathyroid hormone (PTH) A hormone secreted by the chief (principal) cells of the parathyroid glands that increases blood calcium level and decreases blood phosphate level. Also called **parathormone.**

Paraurethral (par′-a-ū-RĒ-thral) **gland** Gland embedded in the wall of the urethra whose duct opens on either side of the urethral orifice and secretes mucus. Also called **Skene's** (SKĒNZ) **gland.**

Parenchyma (pa-RENG-ki-ma) The functional parts of any organ, as opposed to tissue that forms its stroma or framework.

Parietal (pa-RĪ-e-tal) Pertaining to or forming the outer wall of a body cavity.

Parietal cell A type of secretory cell in gastric glands that produces hydrochloric acid and intrinsic factor. Also called an **oxyntic cell.**

Parietal pleura (PLOO-ra) The outer layer of the serous pleural membrane that encloses and protects the lungs; the layer that is attached to the wall of the pleural cavity.

Parkinson disease (PD) Progressive degeneration of the basal nuclei and substantia nigra of the cerebrum resulting in decreased production of dopamine (DA) that leads to tremor, slowing of voluntary movements, and muscle weakness.

Parotid (pa-ROT-id) **gland** One of the paired salivary glands located inferior and anterior to the ears and connected to the oral cavity via a duct (Stensen's) that opens into the inside of the cheek opposite the maxillary (upper) second molar tooth.

Pars intermedia A small avascular zone between the anterior and posterior pituitary glands.

Parturition (par-toor-ISH-un) Act of giving birth to young; childbirth, delivery.

Patent (PĀ-tent) **ductus arteriosus (PDA)** A congenital heart defect in which the ductus arteriosus remains open. As a result, aortic blood flows into the lower-pressure pulmonary trunk, increasing pulmonary trunk pressure and overworking both ventricles.

Pathogen (PATH-ō-jen) A disease-producing microbe.

Pathological (path′-ō-LOJ-i-kal) **anatomy** The study of structural changes caused by disease.

Pathophysiology (PATH-ō-fez-ē-ol-ō-jē) Study of functional changes associated with disease and aging.

Pectinate (PEK-ti-nāt) **muscles** Projecting muscle bundles of the anterior atrial walls and the lining of the auricles.

Pectoral (PEK-tō-ral) Pertaining to the chest or breast.

Pedicel (PED-i-sel) Footlike structure, as on podocytes of a glomerulus.

Pelvic (PEL-vik) **cavity** Inferior portion of the abdominopelvic cavity that contains the urinary bladder, sigmoid colon, rectum, and internal female and male reproductive structures.

Pelvic splanchnic (PEL-vik SPLANGK-nik) **nerves** Consist of preganglionic parasympathetic axons from the levels of S2, S3, and S4 that supply the urinary bladder, reproductive organs, and the descending and sigmoid colon and rectum.

Pelvis The basinlike structure formed by the two hip bones, the sacrum, and the coccyx. The expanded, proximal portion of the ureter, lying within the kidney and into which the major calyces open.

Penis (PĒ-nis) The organ of urination and copulation in males; used to deposit semen into the female vagina.

Pepsin Protein-digesting enzyme secreted by chief cells of the stomach in the inactive form pepsinogen, which is converted to active pepsin by hydrochloric acid.

Peptic ulcer An ulcer that develops in areas of the gastrointestinal tract exposed to hydrochloric acid; classified as a gastric ulcer if in the lesser curvature of the stomach and as a duodenal ulcer if in the first part of the duodenum.

Percussion (pur-KUSH-un) The act of striking (percussing) an underlying part of the body with short, sharp taps as an aid in diagnosing the part by the quality of the sound produced.

Perforating canal A minute passageway by means of which blood vessels and nerves from the periosteum penetrate into compact bone. Also called **Volkmann's** (FŌLK-mans) **canal.**

Pericardial (per′-i-KAR-dē-al) **cavity** Small potential space between the visceral and parietal layers of the serous pericardium that contains pericardial fluid.

Pericardium (per′-i-KAR-dē-um) A loose-fitting membrane that encloses the heart, consisting of a superficial fibrous layer and a deep serous layer.

Perichondrium (per′-i-KON-drē-um) A covering of dense irregular connective tissue that surrounds the surface of most cartilage.

Perilymph (PER-i-limf) The fluid contained between the bony and membranous labyrinths of the inner ear.

Perimetrium (per′-i-MĒ-trē-um) The serosa of the uterus.

Perimysium (per-i-MĪZ-ē-um) Invagination of the epimysium that divides muscles into bundles.

Perineum (per′-i-NĒ-um) The pelvic floor; the space between the anus and the scrotum in the male and between the anus and the vulva in the female.

Perineurium (per′-i-NOO-rē-um) Connective tissue wrapping around fascicles in a nerve.

Periodontal (per′-ē-ō-DON-tal) **disease** A collective term for conditions characterized by degeneration of gingivae, alveolar bone, periodontal ligament, and cementum.

Periodontal ligament The periosteum lining the alveoli (sockets) for the teeth in the alveolar processes of the mandible and maxillae.

Periosteum (per′-ē-OS-tē-um) The covering of a bone that consists of connective tissue, osteogenic cells, and osteoblasts; is essential for bone growth, repair, and nutrition.

Peripheral (pe-RIF-er-al) Located on the outer part or a surface of the body.

Peripheral nervous system (PNS) The part of the nervous system that lies outside the central nervous system, consisting of nerves and ganglia.

Peristalsis (per′-i-STAL-sis) Successive muscular contractions along the wall of a hollow muscular structure.

Peritoneum (per′-i-tō-NĒ-um) The largest serous membrane of the body that lines the abdominal cavity and covers the viscera within it.

Peritonitis (per′-i-tō-NĪ-tis) Inflammation of the peritoneum.

Peroxisome (pe-ROKS-i-sōm) Organelle similar in structure to a lysosome that contains enzymes that use molecular oxygen to oxidize various organic compounds; such reactions produce hydrogen peroxide; abundant in liver cells.

Perspiration Sweat; produced by sudoriferous (sweat) glands and containing water, salts, urea, uric acid, amino acids, ammonia, sugar, lactic acid, and ascorbic acid. Helps maintain body temperature and eliminate wastes.

Peyer's patches *See* **Aggregated lymphatic follicles.**

pH A measure of the concentration of hydrogen ions (H^+) in a solution. The **pH scale** extends from 0 to 14, with a value of 7 expressing neutrality, values lower than 7 expressing increasing acidity, and values higher than 7 expressing increasing alkalinity.

Phagocytosis (fag′-ō-sī-TŌ-sis) The process by which phagocytes ingest and destroy microbes, cell debris, and other foreign matter.

Phalanx (FĀ-lanks) The bone of a finger or toe. *Plural* is **phalanges** (fa-LAN-jēz).

Pharmacology (far′-ma-KOL-ō-jē) The science of the effects and uses of drugs in the treatment of disease.

Pharynx (FAR-inks) The throat; a tube that starts at the internal nares and runs partway down the neck, where it opens into the esophagus posteriorly and the larynx anteriorly.

Phenotype (FĒ-nō-tīp) The observable expression of genotype; physical characteristics of an organism determined by genetic makeup and influenced by interaction between genes and internal and external environmental factors.

Phlebitis (fle-BĪ-tis) Inflammation of a vein, usually in a lower limb.

Photopigment A substance that can absorb light and undergo structural changes that can lead to the development of a receptor potential. An example is rhodopsin. In the eye, also called **visual pigment.**

Photoreceptor Receptor that detects light shining on the retina of the eye.

Physiology (fiz′-ē-OL-ō-jē) Science that deals with the functions of an organism or its parts.

Pia mater (PĪ-a MĀ-ter *or* PĒ-a MA-ter) The innermost of the three meninges (coverings) of the brain and spinal cord.

Pineal (PĪN-ē-al) **gland** A cone-shaped gland located in the roof of the third ventricle that secretes melatonin. Also called the **epiphysis cerebri** (ē-PIF-i-sis se-RĒ-brē).

Pinealocyte (pin-ē-AL-ō-sīt) Secretory cell of the pineal gland that releases melatonin.

Pinna (PIN-na) The projecting part of the external ear composed of elastic cartilage and covered by skin and shaped like the flared end of a trumpet. Also called the **auricle** (OR-i-kul).

Pituicyte (pi-TOO-i-sīt) Supporting cell of the posterior pituitary.

Pituitary (pi-TOO-i-tār-ē) **gland** A small endocrine gland occupying the hypophyseal fossa of the sphenoid bone and attached to the hypothalamus by the infundibulum. Also called the **hypophysis** (hī-POF-i-sis).

Pivot joint A synovial joint in which a rounded, pointed, or conical surface of one bone articulates with a ring formed partly by another bone and partly by a ligament, as in the joint between the atlas and axis and between the proximal ends of the radius and ulna. Also called a **trochoid** (TRŌ-koyd) **joint.**

Placenta (pla-SEN-ta) The special structure through which the exchange of materials between fetal and maternal circulations occurs. Also called the **afterbirth.**

Plane joint A synovial joint having articulating surfaces that are usually flat, permitting only side-to-side and back-and-forth movements, as between carpal bones, tarsal bones, and the scapula and clavicle. Also called an **arthrodial** (ar-THRŌ-dē-al) **joint.**

Plantar flexion (PLAN-tar FLEK-shun) Bending the foot in the direction of the plantar surface (sole).

Plaque (PLAK) A layer of dense proteins on the inside of a plasma membrane in adherens junctions and desmosomes. A mass of bacterial cells, dextran (polysaccharide), and other debris that adheres to teeth (dental plaque). *See also* **Athero-sclerotic plaque.**

Plasma (PLAZ-ma) The extracellular fluid found in blood vessels; blood minus the formed elements.

Plasma cell Cell that develops from a B cell (lymphocyte) and produces antibodies.

Plasma (cell) membrane Outer, limiting membrane that separates the cell's internal parts from extracellular fluid or the external environment.

Platelet (PLĀT-let) A fragment of cytoplasm enclosed in a cell membrane and lacking a nucleus; found in the circulating blood; plays a role in hemostasis. Also called a **thrombocyte** (THROM-bō-sīt).

Platelet plug Aggregation of platelets (thrombocytes) at a site where a blood vessel is damaged that helps stop or slow blood loss.

Pleura (PLOO-ra) The serous membrane that covers the lungs and lines the walls of the chest and the diaphragm.

Pleural cavity Small potential space between the visceral and parietal pleurae.

Plexus (PLEK-sus) A network of nerves, veins, or lymphatic vessels.

Plexus of Auerbach *See* **Myenteric plexus.**

Plexus of Meissner *See* **Submucosal plexus.**

Pluripotent (ploo-RI-pō-tent) **stem cell** Immature stem cell in red bone marrow that gives rise to precursors of all of the different mature blood cells.

Pneumotaxic (noo-mō-TAK-sik) **area** A part of the respiratory center in the pons that continually sends inhibitory nerve impulses to the inspiratory area, limiting inhalation and facilitating exhalation.

Polycythemia (pol′-ē-sī-THĒ-mē-a) Disorder characterized by an above-normal hematocrit (above 55%) in which hypertension, thrombosis, and hemorrhage can occur.

Polyunsaturated fat A fatty acid that contains more than one double covalent bond between its carbon atoms; abundant in triglycerides of corn oil, safflower oil, and cottonseed oil.

Polyuria (pol′-ē-Ū-rē-a) An excessive production of urine.

Pons (PONZ) The part of the brain stem that forms a "bridge" between the medulla oblongata and the midbrain, anterior to the cerebellum.

Portal system The circulation of blood from one capillary network into another through a vein.

Positive feedback system A feedback cycle that strengthens or reinforces a change in a controlled condition.

Postcentral gyrus Gyrus of cerebral cortex located immediately posterior to the central sulcus; contains the primary somatosensory area.

Posterior (pos-TER-ē-or) Nearer to or at the back of the body. Equivalent to **dorsal** in bipeds.

Posterior column–medial lemniscus pathway Sensory pathway that carries information related to proprioception, fine touch, two-point discrimination, pressure, and vibration. First-order neurons project from the spinal cord to the ipsilat-eral medulla in the posterior columns (gracile fasciculus and cuneate fasciculus). Second-order neurons project from the medulla to the contralateral thalamus in the medial lemniscus. Third-order neurons project from the thalamus to the somatosensory cortex (postcentral gyrus) on the same side.

Posterior pituitary Posterior lobe of the pituitary gland. Also called the **neurohypophysis** (noo-rō-hī-POF-i-sis).

Posterior root The structure composed of sensory axons lying between a spinal nerve and the dorsolateral aspect of the spinal cord. Also called the **dorsal (sensory) root.**

Posterior root ganglion (GANG-glē-on) A group of cell bodies of sensory neurons and their supporting cells located along the posterior root of a spinal nerve. Also called a **dorsal (sensory) root ganglion.**

Postganglionic neuron (pōst′-gang-lē-ON-ik NOO-ron) The second autonomic motor neuron in an autonomic pathway, having its cell body and dendrites located in an autonomic ganglion and its unmyelinated axon ending at cardiac muscle, smooth muscle, or a gland.

Postsynaptic (pōst-sin-AP-tik) **neuron** The nerve cell that is activated by the release of a neurotransmitter from another neuron and carries nerve impulses away from the synapse.

Pouch of Douglas *See* **Rectouterine pouch.**

Precapillary sphincter (SFINGK-ter) The distalmost muscle fiber (cell) at the metarteriole–capillary junction that regulates blood flow into capillaries.

Precentral gyrus Gyrus of cerebral cortex located immediately anterior to the central sulcus; contains the primary motor area.

Preganglionic (pre′-gang-lē-ON-ik) **neuron** The first autonomic motor neuron in an autonomic pathway, with its cell body and dendrites in the brain or spinal cord and its myelinated axon ending at an autonomic ganglion, where it synapses with a postganglionic neuron.

Pregnancy Sequence of events that normally includes fertilization, implantation, embryonic growth, and fetal growth, and terminates in birth.

Premenstrual syndrome (PMS) Severe physical and emotional stress ocurring late in the postovulatory phase of the menstrual cycle and sometimes overlapping with menstruation.

Prepuce (PRĒ-poos) The loose-fitting skin covering the glans of the penis and clitoris. Also called the **foreskin.**

Presbyopia (prez-bē-Ō-pē-a) A loss of elasticity of the lens of the eye due to advancing age with resulting inability to focus clearly on near objects.

Presynaptic (pre-sin-AP-tik) **neuron** A neuron that propagates nerve impulses toward a synapse.

Prevertebral ganglion (pre-VER-te-bral GANG-glē-on) A cluster of cell bodies of postganglionic sympathetic neurons anterior to the spinal column and close to large abdominal arteries. Also called a **collateral ganglion.**

Primary germ layer One of three layers of embryonic tissue, called ectoderm, mesoderm, and endoderm, that give rise to all tissues and organs of the body.

Primary motor area A region of the cerebral cortex in the precentral gyrus of the frontal lobe of the cerebrum that controls specific muscles or groups of muscles.

Primary somatosensory area A region of the cerebral cortex posterior to the central sulcus in the postcentral gyrus of the parietal lobe of the cerebrum that localizes exactly the points of the body where somatic sensations originate.

Prime mover The muscle directly responsible for producing a desired motion. Also called an **agonist** (AG-ō-nist).

Primitive gut Embryonic structure formed from the dorsal part of the yolk sac that gives rise to most of the gastrointestinal tract.

Primordial (prī-MOR-dē-al) Existing first; especially primordial egg cells in the ovary.

Principal cell Cell type in the distal convoluted tubules and collecting ducts of the kidneys that is stimulated by aldosterone and antidiuretic hormone.

Proctology (prok-TOL-ō-jē) The branch of medicine concerned with the rectum and its disorders.

Progeny (PROJ-e-nē) Offspring or descendants.

Progesterone (prō-JES-te-rōn) A female sex hormone produced by the ovaries that helps prepare the endometrium of the uterus for implantation of a fertilized ovum and the mammary glands for milk secretion.

Prognosis (prog-NŌ-sis) A forecast of the probable results of a disorder; the outlook for recovery.

Prolactin (prō-LAK-tin) **(PRL)** A hormone secreted by the anterior pituitary that initiates and maintains milk secretion by the mammary glands.

Prolapse (PRŌ-laps) A dropping or falling down of an organ, especially the uterus or rectum.

Proliferation (prō-lif′-er-Ā-shun) Rapid and repeated reproduction of new parts, especially cells.

Pronation (prō-NĀ-shun) A movement of the forearm in which the palm is turned posteriorly.

Prophase (PRŌ-fāz) The first stage of mitosis during which chromatid pairs are formed and aggregate around the metaphase plate of the cell.

Proprioception (prō-prē-ō-SEP-shun) The perception of the position of body parts, especially the limbs, independent of vision; this sense is possible due to nerve impulses generated by proprioceptors.

Proprioceptor (PRŌ-prē-ō-sep′-tor) A receptor located in muscles, tendons, joints, or the internal ear (muscle spindles, tendon organs, joint kinesthetic receptors, and hair cells of the vestibular apparatus) that provides information about body position and movements.

Prostaglandin (pros′-ta-GLAN-din) **(PG)** A membrane-associated lipid; released in small quantities and acts as a local hormone.

Prostate (PROS-tāt) A doughnut-shaped gland inferior to the urinary bladder that surrounds the superior portion of the male urethra and secretes a slightly acidic solution that contributes to sperm motility and viability.

Proteasome (PRŌ-tē-a-sōm) Tiny cellular organelle in cytosol and nucleus containing proteases that destroy unneeded, damaged, or faulty proteins.

Protein An organic compound consisting of carbon, hydrogen, oxygen, nitrogen, and sometimes sulfur and phosphorus; synthesized on ribosomes and made up of amino acids linked by peptide bonds.

Prothrombin (prō-THROM-bin) An inactive blood-clotting factor synthesized by the liver, released into the blood, and converted to active thrombin in the process of blood clotting by the activated enzyme prothrombinase.

Proto-oncogene (prō′-tō-ON-kō-jēn) Gene responsible for some aspect of normal growth and development; it may transform into an oncogene, a gene capable of causing cancer.

Protraction (prō-TRAK-shun) The movement of the mandible or shoulder girdle forward on a plane parallel with the ground.

Proximal (PROK-si-mal) Nearer the attachment of a limb to the trunk; nearer to the point of origin or attachment.

Pseudopod (SOO-dō-pod) Temporary protrusion of the leading edge of a migrating cell; cellular projection that surrounds a particle undergoing phagocytosis.

Pterygopalatine ganglion (ter′-i-gō-PAL-a-tīn GANG-glē-on) A cluster of cell bodies of parasympathetic postganglionic neurons ending at the lacrimal and nasal glands.

Ptosis (TŌ-sis) Drooping, as of the eyelid or the kidney.

Puberty (PŪ-ber-tē) The time of life during which the secondary sex characteristics begin to appear and the capability for sexual reproduction is possible; usually occurs between the ages of 10 and 17.

Pubic symphysis A slightly movable cartilaginous joint between the anterior surfaces of the hip bones.

Puerperium (pū-er-PER-ē-um) The period immediately after childbirth, usually 4–6 weeks.

Pulmonary (PUL-mo-ner′-ē) Concerning or affected by the lungs.

Pulmonary circulation The flow of deoxygenated blood from the right ventricle to the lungs and the return of oxygenated blood from the lungs to the left atrium.

Pulmonary edema (e-DĒ-ma) An abnormal accumulation of interstitial fluid in the tissue spaces and alveoli of the lungs due to increased pulmonary capillary permeability or increased pulmonary capillary pressure.

Pulmonary embolism (EM-bō-lizm) **(PE)** The presence of a blood clot or a foreign substance in a pulmonary arterial blood vessel that obstructs circulation to lung tissue.

Pulmonary ventilation The inflow (inhalation) and outflow (exhalation) of air between the atmosphere and the lungs. Also called **breathing.**

Pulp cavity A cavity within the crown and neck of a tooth, which is filled with pulp, a connective tissue containing blood vessels, nerves, and lymphatic vessels.

Pulse (PULS) The rhythmic expansion and elastic recoil of a systemic artery after each contraction of the left ventricle.

Pupil The hole in the center of the iris, the area through which light enters the posterior cavity of the eyeball.

Purkinje (pur-KIN-jē) **fiber** Muscle fiber (cell) in the ventricular tissue of the heart specialized for conducting an action potential to the myocardium; part of the conduction system of the heart.

Pus The liquid product of inflammation containing leukocytes or their remains and debris of dead cells.

Pyloric (pī-LOR-ik) **sphincter** A thickened ring of smooth muscle through which the pylorus of the stomach communicates with the duodenum. Also called the **pyloric valve.**

Pyorrhea (pī-ō-RĒ-a) A discharge or flow of pus, especially in the alveoli (sockets) and the tissues of the gums.

Pyramid (PIR-a-mid) A pointed or cone-shaped structure. One of two roughly triangular structures on the anterior aspect of the medulla oblongata composed of the largest motor tracts that run from the cerebral cortex to the spinal cord. A triangular structure in the renal medulla.

Pyramidal (pi-RAM-i-dal) **tracts (pathways)** See **Direct motor pathways.**

Q

QRS complex The deflection waves of an electrocardiogram that represent onset of ventricular depolarization.

Quadrant (KWOD-rant) One of four parts.

Quadriplegia (kwod′-ri-PLĒ-jē-a) Paralysis of four limbs: two upper and two lower.

R

Radiographic (rā′-dē-ō-GRAF-ik) **anatomy** Diagnostic branch of anatomy that includes the use of x-rays.

Rami communicantes (RĀ-mē kō-mū-ni-KAN-tēz) Branches of a spinal nerve. *Singular* is **ramus communicans** (RĀ-mus kō-MŪ-ni-kans).

Rathke's pouch See **Hypophyseal pouch.**

Receptor A specialized cell or a distal portion of a neuron that responds to a specific sensory modality, such as touch, pressure, cold, light, or sound, and converts it to an electrical signal (generator or receptor potential). A specific molecule or cluster of molecules that recognizes and binds a particular ligand.

Receptor-mediated endocytosis A highly selective process whereby cells take up specific ligands, which usually are large molecules or particles, by enveloping them within a sac of plasma membrane. Ligands are eventually broken down by enzymes in lysosomes.

Recombinant DNA Synthetic DNA, formed by joining a fragment of DNA from one source to a portion of DNA from another.

Rectouterine (rek-tō-Ū-ter-in) **pouch** A pocket formed by the parietal peritoneum as it moves posteriorly from the surface of the uterus and is reflected onto the rectum; the most inferior point in the pelvic cavity. Also called the **pouch of Douglas.**

Rectum (REK-tum) The last 20 cm (8 in.) of the gastrointestinal tract, from the sigmoid colon to the anus.

Recumbent (re-KUM-bent) Lying down.

Red bone marrow A highly vascularized connective tissue located in microscopic spaces between trabeculae of spongy bone tissue.

Red nucleus A cluster of cell bodies in the midbrain, occupying a large part of the tectum from which axons extend into the rubroreticular and rubrospinal tracts.

Red pulp That portion of the spleen that consists of venous sinuses filled with blood and thin plates of splenic tissue called splenic (Billroth's) cords.

Referred pain Pain that is felt at a site remote from the place of origin.

Reflex Fast response to a change (stimulus) in the internal or external environment that attempts to restore homeostasis.

Reflex arc The most basic conduction pathway through the nervous system, connecting a receptor and an effector and consisting of a receptor, a sensory neuron, an integrating center in the central nervous system, a motor neuron, and an effector. Also called **reflex circuit.**

Regional anatomy The division of anatomy dealing with a specific region of the body, such as the head, neck, chest, or abdomen.

Regurgitation (rē-gur′-ji-TĀ-shun) Return of solids or fluids to the mouth from the stomach; backward flow of blood through incompletely closed heart valves.

Relaxin (RLX) A female hormone produced by the ovaries and placenta that increases flexibility of the pubic symphysis and helps dilate the uterine cervix to ease delivery of a baby.

Releasing hormone Hormone secreted by the hypothalamus that can stimulate secretion of hormones of the anterior pituitary.

Renal (RĒ-nal) Pertaining to the kidneys.

Renal corpuscle (KOR-pus-l) A glomerular (Bowman's) capsule and its enclosed glomerulus.

Renal pelvis A cavity in the center of the kidney formed by the expanded, proximal portion of the ureter, lying within the kidney, and into which the major calyces open.

Renal physiology Study of the functions of the kidneys.

Renal pyramid A triangular structure in the renal medulla containing the straight segments of renal tubules and the vasa recta.

Reproduction (rē-prō-DUK-shun) The formation of new cells for growth, repair, or replacement; the production of a new individual.

Reproductive cell division Type of cell division in which gametes (sperm and oocytes) are produced; consists of meiosis and cytokinesis.

Respiration (res-pi-RĀ-shun) Overall exchange of gases between the atmosphere, blood, and body cells consisting of pulmonary ventilation, external respiration, and internal respiration.

Respiratory center Neurons in the pons and medulla oblongata of the brain stem that regulate the rate and depth of pulmonary ventilation.

Respiratory (RES-pir-a-tōr-ē) **physiology** Study of the functions of the air passageways and lungs.

Respiratory (RES-pi-ra-tōr-ē) **system** System composed of the nose, pharynx, larynx, trachea, bronchi, and lungs that obtains oxygen for body cells and eliminates carbon dioxide from them.

Retention (rē-TEN-shun) A failure to void urine due to obstruction, nervous contraction of the urethra, or absence of sensation of desire to urinate.

Rete (RĒ-tē) **testis** The network of ducts in the testes.

Reticular (re-TIK-ū-lar) **activating system (RAS)** A portion of the reticular formation that has many ascending connections with the cerebral cortex; when this area of the brain stem is active, nerve impulses pass to the thalamus and widespread areas of the cerebral cortex, resulting in generalized alertness or arousal from sleep.

Reticular formation A network of small groups of neuronal cell bodies scattered among bundles of axons (mixed gray and white matter) beginning in the medulla oblongata and extending superiorly through the central part of the brain stem.

Reticulocyte (re-TIK-ū-lō-sīt) An immature red blood cell.

Reticulum (re-TIK-ū-lum) A network.

Retina (RET-i-na) The deep coat of the posterior portion of the eyeball consisting of nervous tissue (where the process of vision begins) and a pigmented layer of epithelial cells that contact the choroid.

Retinaculum (ret-i-NAK-ū-lum) A thickening of fascia that holds structures in place, for example, the superior and inferior retinacula of the ankle.

Retraction (rē-TRAK-shun) The movement of a protracted part of the body posteriorly on a plane parallel to the ground, as in pulling the lower jaw back in line with the upper jaw.

Retroperitoneal (re′-trō-per-i-tō-NĒ-al) External to the peritoneal lining of the abdominal cavity.

Rh factor An inherited antigen on the surface of red blood cells in Rh+ individuals; not present in Rh− individuals.

Rhinology (rī-NOL-ō-jē) The study of the nose and its disorders.

Ribonucleic (rī-bō-noo-KLĒ-ik) **acid (RNA)** A single-stranded nucleic acid made up of nucleotides, each consisting of a nitrogenous base (adenine, cytosine, guanine, or uracil), ribose, and a phosphate group; major types are messenger RNA (mRNA), transfer RNA (tRNA), and ribosomal RNA (rRNA), each of which has a specific role during protein synthesis.

Ribosome (RĪ-bō-sōm) A cellular structure in the cytoplasm of cells, composed of a small subunit and a large subunit that contain ribosomal RNA and ribosomal proteins; the site of protein synthesis.

Right lymphatic (lim-FAT-ik) **duct** A vessel of the lymphatic system that drains lymph from the upper right side of the body and empties it into the right subclavian vein.

Rigidity (ri-JID-i-tē) Hypertonia characterized by increased muscle tone, but reflexes are not affected.

Rigor mortis State of partial contraction of muscles after death due to lack of ATP; myosin heads (cross-bridges) remain attached to actin, thus preventing relaxation.

Rod One of two types of photoreceptors in the retina of the eye; specialized for vision in dim light.

Root canal A narrow extension of the pulp cavity lying within the root of a tooth.

Root of penis Attached portion of penis that consists of the bulb and crura.

Rotation (rō-TĀ-shun) Moving a bone around its own axis, with no other movement.

Rotator cuff Refers to the tendons of four deep shoulder muscles (subscapularis, supraspinatus, infraspinatus, and teres minor) that form a complete circle around the shoulder; they strengthen and stabilize the shoulder joint.

Round ligament (LIG-a-ment) A band of fibrous connective tissue enclosed between the folds of the broad ligament of the uterus, emerging from the uterus just inferior to the uterine tube, extending laterally along the pelvic wall and through the deep inguinal ring to end in the labia majora.

Round window A small opening between the middle and internal ear, directly inferior to the oval window, covered by the secondary tympanic membrane.

Ruffini corpuscle A sensory receptor embedded deeply in the dermis and deeper tissues that detects the stretching of the skin.

Rugae (ROO-gē) Large folds in the mucosa of an empty hollow organ, such as the stomach or vagina.

S

Saccule (SAK-ūl) The inferior and smaller of the two chambers in the membranous labyrinth inside the vestibule of the internal ear containing a receptor organ for static equilibrium.

Sacral plexus (SĀ-kral PLEK-sus) A network formed by the ventral branches of spinal nerves L4 through S3.

Sacral promontory (PROM-on-tor′-ē) The superior surface of the body of the first sacral vertebra that projects anteriorly into the pelvic cavity; a line from the sacral promontory to the superior border of the pubic symphysis divides the abdominal and pelvic cavities.

Saddle joint A synovial joint in which the articular surface of one bone is saddle-shaped and the articular surface of the other bone is shaped like the legs of the rider sitting in the saddle, as in the joint between the trapezium and the metacarpal of the thumb.

Sagittal (SAJ-i-tal) **plane** A plane that divides the body or organs into left and right portions. Such a plane may be **midsagittal (median),** in which the divisions are equal, or **parasagittal,** in which the divisions are unequal.

Saliva (sa-LĪ-va) A clear, alkaline, somewhat viscous secretion produced mostly by the three pairs of salivary glands; contains various salts, mucin, lysozyme, salivary amylase, and lingual lipase (produced by glands in the tongue).

Salivary amylase (SAL-i-ver-ē AM-i-lās) An enzyme in saliva that initiates the chemical breakdown of starch.

Salivary gland One of three pairs of glands that lie external to the mouth and pour their secretory product (saliva) into ducts that empty into the oral cavity; the parotid, submandibular, and sublingual glands.

Sarcolemma (sar′-kō-LEM-ma) The cell membrane of a muscle fiber (cell), especially of a skeletal muscle fiber.

Sarcomere (SAR-kō-mēr) A contractile unit in a striated muscle fiber (cell) extending from one Z disc to the next Z disc.

Sarcoplasm (SAR-kō-plazm) The cytoplasm of a muscle fiber (cell).

Sarcoplasmic reticulum (sar′-kō-PLAZ-mik re-TIK-ū-lum) **(SR)** A network of saccules and tubes surrounding myofibrils of a muscle fiber (cell), comparable to endoplasmic reticulum; functions to reabsorb calcium ions during relaxation and to release them to cause contraction.

Satellite (SAT-i-līt) **cell** Flat neuroglial cells that surround cell bodies of peripheral nervous system ganglia to provide structural support and regulate the exchange of material between a neuronal cell body and interstitial fluid.

Saturated fat A fatty acid that contains only single bonds (no double bonds) between its carbon atoms; all carbon atoms are bonded to the maximum number of hydrogen atoms; prevalent in triglycerides of animal products such as meat, milk, milk products, and eggs.

Scala tympani (SKA-la TIM-pan-ē) The inferior spiral-shaped channel of the bony cochlea, filled with perilymph.

Scala vestibuli (ves-TIB-ū-lē) The superior spiral-shaped channel of the bony cochlea, filled with perilymph.

Schwann (SCHVON or SCHWON) **cell** A neuroglial cell of the peripheral nervous system that forms the myelin sheath and neurolemma around a nerve axon by wrapping around the axon in a jellyroll fashion.

Sciatica (sī-AT-i-ka) Inflammation and pain along the sciatic nerve; felt along the posterior aspect of the thigh extending down the inside of the leg.

Sclera (SKLE-ra) The white coat of fibrous tissue that forms the superficial protective covering over the eyeball except in the most anterior portion; the posterior portion of the fibrous tunic.

Scleral venous sinus A circular venous sinus located at the junction of the sclera and the cornea through which aqueous humor drains from the anterior chamber of the eyeball into the blood. Also called the **canal of Schlemm** (SHLEM).

Sclerosis (skle-RŌ-sis) A hardening with loss of elasticity of tissues.

Scoliosis (skō-lē-Ō-sis) An abnormal lateral curvature from the normal vertical line of the backbone.

Scrotum (SKRŌ-tum) A skin-covered pouch that contains the testes and their accessory structures.

Sebaceous (se-BĀ-shus) **gland** An exocrine gland in the dermis of the skin, almost always associated with a hair follicle, that secretes sebum. Also called an **oil gland.**

Sebum (SĒ-bum) Secretion of sebaceous (oil) glands.

Secondary sex characteristic A characteristic of the male or female body that develops at puberty under the influence of sex hormones but is not directly involved in sexual reproduction; examples are distribution of body hair, voice pitch, body shape, and muscle development.

Secretion (se-KRĒ-shun) Production and release from a cell or a gland of a physiologically active substance.

Selective permeability (per′-mē-a-BIL-i-tē) The property of a membrane by which it permits the passage of certain substances but restricts the passage of others.

Semen (SĒ-men) A fluid discharged at ejaculation by a male that consists of a mixture of sperm and the secretions of the seminiferous tubules, seminal vesicles, prostate, and bulbourethral (Cowper's) glands.

Semicircular canals Three bony channels (anterior, posterior, lateral), filled with perilymph, in which lie the membranous semicircular canals filled with endolymph. They contain receptors for equilibrium.

Semicircular ducts The membranous semicircular canals filled with endolymph and floating in the perilymph of the bony semicircular canals; they contain cristae that are concerned with dynamic equilibrium.

Semilunar (sem′-ē-LOO-nar) **(SL) valve** A valve between the aorta or the pulmonary trunk and a ventricle of the heart.

Seminal vesicle (SEM-i-nal VES-i-kul) One of a pair of convoluted, pouchlike structures, lying posterior and inferior to the urinary bladder and anterior to the rectum, that secrete a component of semen into the ejaculatory ducts. Also termed **seminal gland.**

Seminiferous tubule (sem′-i-NI-fer-us TOO-būl) A tightly coiled duct, located in the testis, where sperm are produced.

Sensation A state of awareness of external or internal conditions of the body.

Sensory area A region of the cerebral cortex concerned with the interpretation of sensory impulses.

Sensory neurons (NOO-ronz) Neurons that carry sensory information from cranial and spinal nerves into the brain and spinal cord or from a lower to a higher level in the spinal cord and brain. Also called **afferent neurons.**

Septal defect An opening in the atrial septum (atrial septal defect) because the foramen ovale fails to close, or the ventricular septum (ventricular septal defect) due to incomplete development of the ventricular septum.

Septum (SEP-tum) A wall dividing two cavities.

Serous (SĒR-us) **membrane** A membrane that lines a body cavity that does not open to the exterior. The external layer of an organ formed by a serous membrane. The membrane that lines the pleural, pericardial, and peritoneal cavities. Also called a **serosa** (se-RŌ-sa).

Sertoli (ser-TŌ-lē) **cell** A supporting cell in the seminiferous tubules that secretes fluid for supplying nutrients to sperm and the hormone inhibin, removes excess cytoplasm from spermatogenic cells, and mediates the effects of FSH and testosterone on spermatogenesis. Also called a **sustentacular** (sus′-ten-TAK-ū-lar) **cell.**

Serum Blood plasma minus its clotting proteins.

Sesamoid (SES-a-moyd) **bones** Small bones usually found in tendons.

Sex chromosomes The twenty-third pair of chromosomes, designated X and Y, which determines the genetic sex of an individual; in males, the pair is XY; in females, XX.

Sexual intercourse The insertion of the erect penis of a male into the vagina of a female. Also called **coitus** (KŌ-i-tus).

Sheath of Schwann See **Neurolemma.**

Shock Failure of the cardiovascular system to deliver adequate amounts of oxygen and nutrients to meet the metabolic needs of the body due to inadequate cardiac output. It is characterized by hypotension; clammy, cool, and pale skin; sweating; reduced urine formation; altered mental state; acidosis; tachycardia; weak, rapid pulse; and thirst. Types include hypovolemic, cardiogenic, vascular, and obstructive.

Shoulder joint A synovial joint where the humerus articulates with the scapula.

Sigmoid colon (SIG-moyd KŌ-lon) The S-shaped part of the large intestine that begins at the level of the left iliac crest, projects medially, and terminates at the rectum at about the level of the third sacral vertebra.

Sign Any objective evidence of disease that can be observed or measured, such as a lesion, swelling, or fever.

Sinoatrial (si-nō-Ā-trē-al) **(SA) node** A small mass of cardiac muscle fibers (cells) located in the right atrium inferior to the opening of the superior vena cava that spontaneously depolarize and generate a cardiac action potential about 100 times per minute. Also called the natural **pacemaker.**

Sinus (SĪ-nus) A hollow in a bone (paranasal sinus) or other tissue; a channel for blood (vascular sinus); any cavity having a narrow opening.

Sinusoid (SĪ-nū-soyd) A large, thin-walled, and leaky type of capillary, having large intercellular clefts that may allow proteins and blood cells to pass from a tissue into the bloodstream; present in the liver, spleen, anterior pituitary, parathyroid glands, and red bone marrow.

Skeletal muscle An organ specialized for contraction, composed of striated muscle fibers (cells), supported by connective tissue, attached to a bone by a tendon or an aponeurosis, and stimulated by somatic motor neurons.

Skeletal system Framework of bones and their associated cartilages, ligaments, and tendons.

Skene's gland See **Paraurethral gland.**

Skin The external covering of the body that consists of a superficial, thinner epidermis (epithelial tissue) and a deep, thicker dermis (connective tissue) that is anchored to the subcutaneous layer. Also called **cutaneous membrane.**

Skin graft The transfer of a patch of healthy skin taken from a donor site to cover a wound.

Skull The skeleton of the head consisting of the cranial and facial bones.

Sleep A state of partial unconsciousness from which a person can be aroused; associated with a low level of activity in the reticular activating system.

Sliding filament mechanism A model that describes muscle contraction in which thin filaments slide past thick ones so that the filaments overlap, causing shortening of a sarcomere, and thus shortening of muscle fibers and alternately shortening of the entire muscle.

Small intestine A long tube of the gastrointestinal tract that begins at the pyloric sphincter of the stomach, coils through the central and inferior part of the abdominal cavity, and ends at the large intestine; divided into three segments: duodenum, jejunum, and ileum.

Smooth muscle A tissue specialized for contraction, composed of smooth muscle fibers (cells), located in the walls of hollow internal organs, and innervated by autonomic motor neurons.

Sodium–potassium ATPase An active transport pump located in the plasma membrane that transports sodium ions out of the cell and potassium ions into the cell at the expense of cellular ATP. It functions to keep the ionic concentrations of these ions at physiological levels. Also called the **sodium–potassium pump.**

Soft palate (PAL-at) The posterior portion of the roof of the mouth, extending from the palatine bones to the uvula. It is a muscular partition lined with mucous membrane.

Somatic (sō-MAT-ik) **cell division** Type of cell division in which a single starting cell duplicates itself to produce two identical cells; consists of mitosis and cytokinesis.

Somatic motor pathway Pathway that carries information from the cerebral cortex, basal nuclei, and cerebellum that stimulates contraction of skeletal muscles.

Somatic nervous system (SNS) The portion of the peripheral nervous system consisting of somatic sensory (afferent) neurons and somatic motor (efferent) neurons.

Somatic sensory pathway Pathway that carries information from somatic sensory receptor to the primary somatosensory area in the cerebral cortex and cerebellum.

Somite (SŌ-mīt) Block of mesodermal cells in a developing embryo that is distinguished into a myotome (which forms most of the skeletal muscles), dermatome (which forms connective tissues), and sclerotome (which forms the vertebrae).

Spasm (SPAZM) A sudden, involuntary contraction of large groups of muscles.

Spasticity (spas-TIS-i-tē) Hypertonia characterized by increased muscle tone, increased tendon reflexes, and pathological reflexes (Babinski sign).

Spermatic (sper-MAT-ik) **cord** A supporting structure of the male reproductive system, extending from a testis to the deep inguinal ring, that includes the ductus (vas) deferens, arteries, veins, lymphatic vessels, nerves, cremaster muscle, and connective tissue.

Spermatogenesis (sper'-ma-tō-JEN-e-sis) The formation and development of sperm in the seminiferous tubules of the testes.

Sperm cell A mature male gamete. Also termed **spermatozoon** (sper'-ma-tō-ZŌ-on).

Spermiogenesis (sper'-mē-ō-JEN-e-sis) The maturation of spermatids into sperm.

Sphincter (SFINGK-ter) A circular muscle that constricts an opening.

Sphincter of Oddi *See* **Sphincter of the hepatopancreatic ampulla.**

Sphincter of the hepatopancreatic ampulla A circular muscle at the opening of the common bile and main pancreatic ducts in the duodenum. Also called the **sphincter of Oddi** (OD-ē).

Spinal (SPĪ-nal) **cord** A mass of nerve tissue located in the vertebral canal from which 31 pairs of spinal nerves originate.

Spinal nerve One of the 31 pairs of nerves that originate on the spinal cord from posterior and anterior roots.

Spinal shock A period from several days to several weeks following transection of the spinal cord that is characterized by the abolition of all reflex activity.

Spinothalamic (spī-nō-tha-LAM-ik) **tract** Sensory (ascending) tract that conveys information up the spinal cord to the thalamus for sensations of pain, temperature, itch, and tickle.

Spinous (SPĪ-nus) **process** A sharp or thornlike process or projection. Also called a **spine**. A sharp ridge running diagonally across the posterior surface of the scapula.

Spiral organ The organ of hearing, consisting of supporting cells and hair cells that rest on the basilar membrane and extend into the endolymph of the cochlear duct. Also called the **organ of Corti** (KOR-tē).

Splanchnic (SPLANK-nik) Pertaining to the viscera.

Spleen (SPLĒN) Large mass of lymphatic tissue between the fundus of the stomach and the diaphragm that functions in formation of blood cells during early fetal development, phagocytosis of ruptured blood cells, and proliferation of B cells during immune responses.

Spongy (cancellous) bone tissue Bone tissue that consists of an irregular latticework of thin plates of bone called trabeculae; spaces between trabeculae of some bones are filled with red bone marrow; found inside short, flat, and irregular bones and in the epiphyses (ends) of long bones.

Sprain Forcible wrenching or twisting of a joint with partial rupture or other injury to its attachments without dislocation.

Squamous (SKWĀ-mus) Flat or scalelike.

Starvation (star-VĀ-shun) The loss of energy stores in the form of glycogen, triglycerides, and proteins due to inadequate intake of nutrients or inability to digest, absorb, or metabolize ingested nutrients.

Static equilibrium (ē-kwi-LIB-rē-um) The maintenance of posture in response to changes in the orientation of the body, mainly the head, relative to the ground.

Stellate reticuloendothelial (STEL-āt re-tik'-ū-lō-en'-dō-THĒ-lē-al) **cell** Phagocytic cell bordering a sinusoid of the liver. Also called a **Kupffer** (KOOP-fer) **cell.**

Stem cell An unspecialized cell that has the ability to divide for indefinite periods and give rise to a specialized cell.

Stenosis (sten-Ō-sis) An abnormal narrowing or constriction of a duct or opening.

Stereocilia (ste'-rē-ō-SIL-ē-a) Groups of extremely long, slender, nonmotile microvilli projecting from epithelial cells lining the epididymis.

Sterile (STE-ril) Free from any living microorganisms. Unable to conceive or produce offspring.

Sterilization (ster'-i-li-ZĀ-shun) Elimination of all living microorganisms. Any procedure that renders an individual incapable of reproduction (for example, castration, vasectomy, hysterectomy, or oophorectomy).

Stimulus Any stress that changes a controlled condition; any change in the internal or external environment that excites a sensory receptor, a neuron, or a muscle fiber.

Stomach The J-shaped enlargement of the gastrointestinal tract directly inferior to the diaphragm in the epigastric, umbilical, and left hypochondriac regions of the abdomen, between the esophagus and small intestine.

Straight tubule (TOO-būl) A duct in a testis leading from a convoluted seminiferous tubule to the rete testis.

Stratum (STRĀ-tum) A layer.

Stratum basalis (ba-SAL-is) The layer of the endometrium next to the myometrium that is maintained during menstruation and gestation and produces a new stratum functionalis following menstruation or parturition.

Stratum functionalis (funk'-shun-AL-is) The layer of the endometrium next to the uterine cavity that is shed during menstruation and that forms the maternal portion of the placenta during gestation.

Stretch receptor Receptor in the walls of blood vessels, airways, or organs that monitors the amount of stretching. Also termed **baroreceptor.**

Striae (STRĪ-ē) Internal scarring due to overstretching of the skin in which collagen fibers and blood vessels in the dermis are damaged. Also called **stretch marks.**

Stroma (STRŌ-ma) The tissue that forms the ground substance, foundation, or framework of an organ, as opposed to its functional parts (parenchyma).

Subarachnoid (sub'-a-RAK-noyd) **space** A space between the arachnoid mater and the pia mater that surrounds the brain and spinal cord and through which cerebrospinal fluid circulates.

Subcutaneous (sub'-kū-TĀ-nē-us) Beneath the skin. Also called **hypodermic** (hī-pō-DER-mik).

Subcutaneous (subQ) layer A continuous sheet of areolar connective tissue and adipose tissue between the dermis of the skin and the deep fascia of the muscles. Also called the **hypodermis.**

Subdural (sub-DOO-ral) **space** A space between the dura mater and the arachnoid mater of the brain and spinal cord that contains a small amount of fluid.

Sublingual (sub-LING-gwal) **gland** One of a pair of salivary glands situated in the floor of the mouth deep to the mucous membrane and to the side of the lingual frenulum, with a duct (Rivinus') that opens into the floor of the mouth.

Submandibular (sub'-man-DIB-ū-lar) **gland** One of a pair of salivary glands found inferior to the base of the tongue deep to the mucous membrane in the posterior part of the floor of the mouth, posterior to the sublingual glands, with a duct (Wharton's) situated to the side of the lingual frenulum. Also called the **submaxillary** (sub'-MAK-si-ler-ē) **gland.**

Submucosa (sub-mū-KŌ-sa) A layer of connective tissue located deep to a mucous membrane, as in the gastrointestinal tract or the urinary bladder; the submucosa connects the mucosa to the muscularis layer.

Submucosal plexus A network of autonomic nerve fibers located in the superficial part of the submucous layer of the small intestine. Also called the **plexus of Meissner** (MĪZ-ner).

Substrate A molecule on which an enzyme acts.

Subthalamus (sub-THAL-a-mus) Part of the diencephalon inferior to the thalamus; the substantia nigra and red nucleus extend from the midbrain into the subthalamus.

Sudoriferous (soo′-dor-IF-er-us) **gland** An apocrine or eccrine exocrine gland in the dermis or subcutaneous layer that produces perspiration. Also called a **sweat gland.**

Sulcus (SUL-kus) A groove or depression between parts, especially between the convolutions of the brain. *Plural* is **sulci** (SUL-sī).

Superficial (soo′-per-FISH-al) Located on or near the surface of the body or an organ. Also called **external.**

Superficial subcutaneous inguinal (IN-gwi-nal) **ring** A triangular opening in the aponeurosis of the external oblique muscle that represents the termination of the inguinal canal.

Superior (soo-PĒR-ē-or) Toward the head or upper part of a structure. Also called **cephalic** or **cranial.**

Superior vena cava (VĒ-na KĀ-va) **(SVC)** Large vein that collects blood from parts of the body superior to the heart and returns it to the right atrium.

Supination (soo-pi-NĀ-shun) A movement of the forearm in which the palm is turned anteriorly.

Surface anatomy The study of the structures that can be identified from the outside of the body.

Surfactant (sur-FAK-tant) Complex mixture of phospholipids and lipoproteins, produced by type II alveolar (septal) cells in the lungs, that decreases surface tension.

Suspensory ligament (sus-PEN-so-rē LIG-a-ment) A fold of peritoneum extending laterally from the surface of the ovary to the pelvic wall.

Sustentacular cell *See* **Sertoli cell.**

Sutural (SOO-chur-al) **bone** A small bone located within a suture between certain cranial bones. Also called **Wormian** (WER-mē-an) **bone.**

Suture (SOO-chur) An immovable fibrous joint that joins skull bones.

Sympathetic (sim′-pa-THET-ik) **division** One of the two subdivisions of the autonomic nervous system, having cell bodies of preganglionic neurons in the lateral gray columns of the thoracic segment and the first two or three lumbar segments of the spinal cord; primarily concerned with processes involving the expenditure of energy.

Sympathetic trunk ganglion (GANG-glē-on) A cluster of cell bodies of sympathetic postganglionic neurons lateral to the vertebral column, close to the body of a vertebra. These ganglia extend inferiorly through the neck, thorax, and abdomen to the coccyx on both sides of the vertebral column and are connected to one another to form a chain on each side of the vertebral column. Also called **sympathetic chain ganglia, vertebral chain ganglia,** or **paravertebral ganglia.**

Symphysis (SIM-fi-sis) A line of union. A slightly movable cartilaginous joint such as the pubic symphysis.

Symptom (SIMP-tum) A subjective change in body function not apparent to an observer, such as pain or nausea, that indicates the presence of a disease or disorder of the body.

Synapse (SIN-aps) The functional junction between two neurons or between a neuron and an effector, such as a muscle or gland; may be electrical or chemical.

Synapsis (sin-AP-sis) The pairing of homologous chromosomes during prophase I of meiosis.

Synaptic (sin-AP-tik) **cleft** The narrow gap at a chemical synapse that separates the axon terminal of one neuron from another neuron or muscle fiber (cell) and across which a neurotransmitter diffuses to affect the postsynaptic cell.

Synaptic end bulb Expanded distal end of an axon terminal that contains synaptic vesicles. Also called a **synaptic knob.**

Synaptic vesicle Membrane-enclosed sac in a synaptic end bulb that stores neurotransmitters.

Synarthrosis (sin′-ar-THRŌ-sis) An immovable joint such as a suture, gomphosis, or synchondrosis.

Synchondrosis (sin′-kon-DRŌ-sis) A cartilaginous joint in which the connecting material is hyaline cartilage.

Syndesmosis (sin′-dez-MŌ-sis) A slightly movable joint in which articulating bones are united by fibrous connective tissue.

Synergist (SIN-er-jist) A muscle that assists the prime mover by reducing undesired action or unnecessary movement.

Synergistic (syn-er-JIS-tik) **effect** A hormonal interaction in which the effects of two or more hormones acting together is greater or more extensive than the sum of each hormone acting alone.

Synostosis (sin′-os-TŌ-sis) A joint in which the dense fibrous connective tissue that unites bones at a suture has been replaced by bone, resulting in a complete fusion across the suture line.

Synovial (si-NŌ-vē-al) **cavity** The space between the articulating bones of a synovial joint, filled with synovial fluid. Also called a **joint cavity.**

Synovial fluid Secretion of synovial membranes that lubricates joints and nourishes articular cartilage.

Synovial joint A fully movable or diarthrotic joint in which a synovial (joint) cavity is present between the two articulating bones.

Synovial membrane The deeper of the two layers of the articular capsule of a synovial joint, composed of areolar connective tissue that secretes synovial fluid into the synovial (joint) cavity.

System An association of organs that have a common function.

Systemic (sis-TEM-ik) Affecting the whole body; generalized.

Systemic anatomy The anatomical study of particular systems of the body, such as the skeletal, muscular, nervous, cardiovascular, or urinary systems.

Systemic circulation The routes through which oxygenated blood flows from the left ventricle through the aorta to all the organs of the body and deoxygenated blood returns to the right atrium.

Systole (SIS-tō-lē) In the cardiac cycle, the phase of contraction of the heart muscle, especially of the ventricles.

Systolic (sis-TOL-ik) **blood pressure (SBP)** The force exerted by blood on arterial walls during ventricular contraction; the highest pressure measured in the large arteries, about 110 mmHg under normal conditions for a young adult.

T

T cell A lymphocyte that becomes immunocompetent in the thymus and can differentiate into a helper T cell or a cytotoxic T cell, both of which function in cell-mediated immunity.

T wave The deflection wave of an electrocardiogram that represents ventricular repolarization.

Tachycardia (tak′-i-KAR-dē-a) An abnormally rapid resting heartbeat or pulse rate (over 100 beats per minute).

Tactile (TAK-tīl) Pertaining to the sense of touch.

Tactile cell Type of cell in the epidermis of hairless skin that makes contact with a tactile disc, which functions in touch. Also called **Merkel** (MER-kel) **cell.**

Tactile disc Soucer-shaped free nerve endings that make contact with tactile cells in the epidermis and function as touch receptors. Also called **Merkel disc.**

Target cell A cell whose activity is affected by a particular hormone.

Tarsal bones The seven bones of the ankle. Also called **tarsals.**

Tarsal gland Sebaceous (oil) gland that opens on the edge of each eyelid. Also called a **Meibomian** (mī-BŌ-mē-an) **gland.**

Tarsal plate A thin, elongated sheet of connective tissue, one in each eyelid, giving the eyelid form and support. The aponeurosis of the levator palpebrae superioris is attached to the tarsal plate of the superior eyelid.

Tarsus (TAR-sus) A collective term for the seven bones of the ankle.

Tectorial (tek-TŌ-rē-al) **membrane** A gelatinous membrane projecting over and in contact with the hair cells of the spiral organ (organ of Corti) in the cochlear duct.

Teeth (TĒTH) Accessory structures of digestion, composed of calcified connective tissue and embedded in bony sockets of the mandible and maxilla, that cut, shred, crush, and grind food. Also called **dentes** (DEN-tēz).

Telophase (TEL-ō-fāz) The final stage of mitosis.

Tendon (TEN-don) A white fibrous cord of dense regular connective tissue that attaches muscle to bone.

Tendon organ A proprioceptive receptor, sensitive to changes in muscle tension and force of contraction, found chiefly near the junctions of tendons and muscles. Also called a **Golgi** (GOL-jē) **tendon organ.**

Tendon reflex A polysynaptic, ipsilateral reflex that protects tendons and their associated muscles from damage that might be brought about by excessive tension. The receptors involved are called tendon organs (Golgi tendon organs).

Teniae coli (TĒ-nē-ē KŌ-lī) The three flat bands of thickened, longitudinal smooth muscle running the length of the large intestine, except in the rectum. *Singular* is **tenia coli.**

Tentorium cerebelli (ten-TŌ-rē-um ser′-e-BEL-ī) A transverse shelf of dura mater that forms a partition between the occipital lobe of the cerebral hemispheres and the cerebellum and that covers the cerebellum.

Teratogen (TER-a-tō-jen) Any agent or factor that causes physical defects in a developing embryo.

Terminal ganglion (TER-min-al GANG-glē-on) A cluster of cell bodies of parasympathetic postganglionic neurons either lying very close to the visceral effectors or located within the walls of the visceral effectors supplied by the postganglionic neurons. Also called **intramural ganglion.**

Testis (TES-tis) Male gonad that produces sperm and the hormones testosterone and inhibin. Also called a **testicle.**

Testosterone (tes-TOS-te-rōn) A male sex hormone (androgen) secreted by interstitial (Leydig) cells of a mature testis; needed for development of sperm; together with a second androgen termed **dihydrotestosterone** (dī-hī-drō-tes-TOS-ter-ōn) **(DHT),** controls the growth and development of male reproductive organs, secondary sex characteristics, and body growth.

Tetralogy of Fallot (tet-RAL-ō-jē of fal-Ō) A combination of four congenital heart defects: (1) constricted pulmonary semilunar valve, (2) interventricular septal opening, (3) emergence of the aorta from both ventricles instead of from the left only, and (4) enlarged right ventricle.

Thalamus (THAL-a-mus) A large, oval structure located bilaterally on either side of the third ventricle, consisting of two masses of gray matter organized into nuclei; main relay center for sensory impulses ascending to the cerebral cortex.

Thermoreceptor (THER-mō-rē-sep-tor) Sensory receptor that detects changes in temperature.

Thermoregulation Homeostatic regulation of body temperature through sweating and adjustment of blood flow in the dermis.

Thigh The portion of the lower limb between the hip and the knee.

Third ventricle (VEN-tri-kul) A slitlike cavity between the right and left halves of the thalamus and between the lateral ventricles of the brain.

Thoracic (thor-AS-ik) **cavity** Cavity superior to the diaphragm that contains two pleural cavities, the mediastinum, and the pericardial cavity.

Thoracic duct A lymphatic vessel that begins as a dilation called the cisterna chyli, receives lymph from the left side of the head, neck, and chest, left arm, and the entire body below the ribs, and empties into the junction between the internal jugular and left subclavian veins. Also called the **left lymphatic** (lim-FAT-ik) **duct.**

Thoracolumbar (thōr′-a-kō-LUM-bar) **outflow** The axons of sympathetic preganglionic neurons, which have their cell bodies in the lateral gray columns of the thoracic segments and first two or three lumbar segments of the spinal cord.

Thorax (THŌ-raks) The chest.

Thrombosis (THROM-BŌ-sis) The formation of a clot in an unbroken blood vessel, usually a vein.

Thrombus (THROM-bus) A stationary clot formed in an unbroken blood vessel, usually a vein.

Thymus (THĪ-mus) A bilobed organ, located in the superior mediastinum posterior to the sternum and between the lungs, in which T cells develop immunocompetence.

Thyroid cartilage (THĪ-royd KAR-ti-lij) The largest single cartilage of the larynx, consisting of two fused plates that form the anterior wall of the larynx. Also called **Adam's apple.**

Thyroid follicle (FOL-i-kul) Spherical sac that forms the parenchyma of the thyroid gland and consists of follicular cells that produce thyroxine (T_4) and triiodothyronine (T_3).

Thyroid gland An endocrine gland with right and left lateral lobes on either side of the trachea connected by an isthmus; located anterior to the trachea just inferior to the cricoid cartilage; secretes thyroxine (T_4), triiodothyronine (T_3), and calcitonin.

Thyroid-stimulating hormone (TSH) A hormone secreted by the anterior pituitary that stimulates the synthesis and secretion of thyroxine (T_4) and triiodothyronine (T_3). Also called **thyrotropin** (THĪ-rō-TRŌ-pin).

Thyroxine (thī-ROK-sēn) **(T_4)** A hormone secreted by the thyroid gland that regulates metabolism, growth and development, and the activity of the nervous system. Also called **tetraiodothyronine** (tet-ra-ī-ō-dō-THĪ-rō-nēn).

Tic Spasmodic, involuntary twitching of muscles that are normally under voluntary control.

Tissue A group of similar cells and their intercellular substance joined together to perform a specific function.

Tissue rejection Phenomenon by which the body recognizes the protein (HLA antigens) in transplanted tissues or organs as foreign and produces antibodies against them.

Tongue A large skeletal muscle covered by a mucous membrane located on the floor of the oral cavity.

Tonsil (TON-sil) An aggregation of large lymphatic nodules embedded in the mucous membrane of the throat.

Topical (TOP-i-kal) Applied to the surface rather than ingested or injected.

Torn cartilage A tearing of an articular disc (meniscus) in the knee.

Trabecula (tra-BEK-ū-la) Irregular latticework of thin plates of spongy bone tissue. Fibrous cord of connective tissue serving as supporting fiber by forming a septum extending into an organ from its wall or capsule. *Plural* is **trabeculae** (tra-BEK-ū-lē).

Trabeculae carneae (KAR-nē-ē) Ridges and folds of the myocardium in the ventricles.

Trachea (TRĀ-kē-a) Tubular air passageway extending from the larynx to the fifth thoracic vertebra. Also called the **windpipe.**

Tract A bundle of nerve axons in the central nervous system.

Transplantation (tranz-plan-TĀ-shun) The transfer of living cells, tissues, or organs from a donor to a recipient or from one part of the body to another in order to restore a lost function.

Transverse colon (trans-VERS KŌ-lon) The portion of the large intestine extending across the abdomen from the right colic (hepatic) flexure to the left colic (splenic) flexure.

Transverse fissure (FISH-er) The deep cleft that separates the cerebrum from the cerebellum.

Transverse plane A plane that divides the body or organs into superior and inferior portions. Also called a **cross-sectional** or **horizontal plane.**

Transverse tubules (TOO-būls) **(T tubules)** Small, cylindrical invaginations of the sarcolemma of striated muscle fibers (cells) that conduct muscle action potentials toward the center of the muscle fiber.

Tremor (TREM-or) Rhythmic, involuntary, purposeless contraction of opposing muscle groups.

Triad (TRĪ-ad) A complex of three units in a muscle fiber composed of a transverse tubule and the sarcoplasmic reticulum terminal cisterns on both sides of it.

Tricuspid (trī-KUS-pid) **valve** Atrioventricular (AV) valve on the right side of the heart.

Triglyceride (trī-GLI-ser-īd) A lipid formed from one molecule of glycerol and three molecules of fatty acids that may be either solid (fats) or liquid (oils) at room temperature; the body's most highly concentrated source of chemical potential energy. Found mainly within adipocytes. Also called a **neutral fat** or a **triacylglycerol.**

Trigone (TRĪ-gōn) A triangular region at the base of the urinary bladder.

Triiodothyronine (trī-ī-ō-dō-THĪ-rō-nēn) **(T_3)** A hormone produced by the thyroid gland that regulates metabolism, growth and development, and the activity of the nervous system.

Trophoblast (TRŌF-ō-blast) The superficial covering of cells of the blastocyst.

Tropic (TRŌ-pik) **hormone** A hormone whose target is another endocrine gland.

Trunk The part of the body to which the upper and lower limbs are attached.

Tubal ligation (lī-GĀ-shun) A sterilization procedure in which the uterine (fallopian) tubes are tied and cut.

Tubular reabsorption The movement of filtrate from renal tubules back into blood in response to the body's specific needs.

Tubular secretion The movement of substances in blood into renal tubular fluid in response to the body's specific needs.

Tumor-suppressor gene A gene coding for a protein that normally inhibits cell division; loss or alteration of a tumor suppressor gene called *p53* is the most common genetic change in a wide variety of cancer cells.

Tunica albuginea (TOO-ni-ka al′-bū-JIN-ē-a) A dense white fibrous capsule covering a testis or deep to the surface of an ovary.

Tunica externa (eks-TER-na) The superficial coat of an artery or vein, composed mostly of elastic and collagen fibers. Also called the **adventitia.**

Tunica interna (in-TER-na) The deep coat of an artery or vein, consisting of a lining of endothelium, basement membrane, and internal elastic lamina. Also called the **tunica intima** (IN-ti-ma).

Tunica media (MĒ-dē-a) The intermediate coat of an artery or vein, composed of smooth muscle and elastic fibers.

Tympanic antrum (tim-PAN-ik AN-trum) An air space in the middle ear that leads into the mastoid air cells or sinus.

Tympanic (tim-PAN-ik) **membrane** A thin, semitransparent partition of fibrous connective tissue between the external auditory meatus and the middle ear. Also called the **eardrum.**

U

Umbilical (um-BIL-i-kul) **cord** The long, rope-like structure containing the umbilical arteries and vein that connect the fetus to the placenta.

Umbilicus (um-BIL-i-kus *or* um-bi-LĪ-kus) A small scar on the abdomen that marks the former attachment of the umbilical cord to the fetus. Also called the **navel.**

Upper limb The appendage attached at the shoulder girdle, consisting of the arm, forearm, wrist, hand, and fingers. Also called **upper extremity** or **upper appendage.**

Uremia (ū-RĒ-mē-a) Accumulation of toxic levels of urea and other nitrogenous waste products in the blood, usually resulting from severe kidney malfunction.

Ureter (Ū-rē-ter) One of two tubes that connect the kidney with the urinary bladder.

Urethra (ū-RĒ-thra) The duct from the urinary bladder to the exterior of the body that conveys urine in females and urine and semen in males.

Urinalysis (ū-ri-NAL-i-sis) An analysis of the volume and physical, chemical, and microscopic properties of urine.

Urinary (Ū-ri-ner-ē) **bladder** A hollow, muscular organ situated in the pelvic cavity posterior to the pubic symphysis; receives urine via two ureters and stores urine until it is excreted through the urethra.

Urinary system A system that consists of the kidneys, ureters, urinary bladder, and urethra. The system regulates the ionic composition, pH, volume, pressure, and osmolarity of blood.

Urine The fluid produced by the kidneys that contains wastes and excess materials; excreted from the body through the urethra.

Urogenital (ū′-rō-JEN-i-tal) **triangle** The region of the pelvic floor inferior to the pubic symphysis, bounded by the pubic symphysis and the ischial tuberosities, and containing the external genitalia.

Urology (ū-ROL-ō-jē) The specialized branch of medicine that deals with the structure, function, and diseases of the male and female urinary systems and the male reproductive system.

Uterine cycle A series of changes in the endometrium of a nonpregnant female that prepares the lining of the uterus to receive a fertilized ovum. Also called **menstrual cycle.**

Uterine (Ū-ter-in) **tube** Duct that transports ova from the ovary to the uterus. Also called the **fallopian** (fal-LŌ-pē-an) **tube** or **oviduct.**

Uterosacral ligament (ū-ter-ō-SĀ-kral LIG-a-ment) A fibrous band of tissue extending from the cervix of the uterus laterally to the sacrum.

Uterus (Ū-te-rus) The hollow, muscular organ in females that is the site of menstruation, implantation, development of the fetus, and labor. Also called the **womb.**

Utricle (Ū-tri-kul) The larger of the two divisions of the membranous labyrinth located inside the vestibule of the inner ear, containing a receptor organ for static equilibrium.

Uvea (Ū-vē-a) The three structures that together make up the vascular tunic of the eye.

Uvula (Ū-vū-la) A soft, fleshy mass, especially the U-shaped pendant part, descending from the soft palate.

V

Vagina (va-JĪ-na) A muscular, tubular organ that leads from the uterus to the vestibule, situated between the urinary bladder and the rectum of the female.

Vallate papilla (VAL-āt pa-PIL-a) One of the circular projections that is arranged in an inverted V-shaped row at the back of the tongue; the largest of the elevations on the upper surface of the tongue containing taste buds. Also called **circumvallate papilla.**

Varicocele (VAR-i-kō-sēl) A twisted vein; especially, the accumulation of blood in the veins of the spermatic cord.

Varicose (VAR-i-kōs) Pertaining to an unnatural swelling, as in the case of a varicose vein.

Vas A vessel or duct.

Vasa recta (VĀ-sa REK-ta) Extensions of the efferent arteriole of a juxtamedullary nephron that run alongside the nephron loop (loop of Henle) in the medullary region of the kidney.

Vasa vasorum (va-SŌ-rum) Blood vessels that supply nutrients to the larger arteries and veins.

Vascular (VAS-kū-lar) Pertaining to or containing many blood vessels.

Vascular (venous) sinus A vein with a thin endothelial wall that lacks a tunica media and externa and is supported by surrounding tissue.

Vascular spasm Contraction of the smooth muscle in the wall of a damaged blood vessel to prevent blood loss.

Vascular tunic (TOO-nik) The middle layer of the eyeball, composed of the choroid, ciliary body, and iris. Also called the **uvea** (Ū-vē-a).

Vasectomy (va-SEK-tō-mē) A means of sterilization of males in which a portion of each ductus (vas) deferens is removed.

Vasoconstriction (vāz-ō-kon-STRIK-shun) A decrease in the size of the lumen of a blood vessel caused by contraction of the smooth muscle in the wall of the vessel.

Vasodilation (vāz′-ō-dī-LĀ-shun) An increase in the size of the lumen of a blood vessel caused by relaxation of the smooth muscle in the wall of the vessel.

Vein A blood vessel that conveys blood from tissues back to the heart.

Vena cava (VĒ-na KĀ-va) One of two large veins that open into the right atrium, returning to the heart all of the deoxygenated blood from the systemic circulation except from the coronary circulation.

Ventral (VEN-tral) Pertaining to the anterior or front side of the body; opposite of dorsal.

Ventral ramus (RĀ-mus) The anterior branch of a spinal nerve, containing sensory and motor fibers to the muscles and skin of the anterior surface of the head, neck, trunk, and the limbs.

Ventricle (VEN-tri-kul) A cavity in the brain filled with cerebrospinal fluid. An inferior chamber of the heart.

Ventricular fibrillation (ven-TRIK-ū-lar fib-ri-LĀ-shun) **(VF** or **V-fib)** Asynchronous ventricular contractions; unless reversed by defibrillation, results in heart failure.

Venule (VEN-ūl) A small vein that collects blood from capillaries and delivers it to a vein.

Vermiform appendix (VER-mi-form a-PEN-diks) A twisted, coiled tube attached to the cecum.

Vermis (VER-mis) The central constricted area of the cerebellum that separates the two cerebellar hemispheres.

Vertebrae (VER-te-brē) Bones that make up the vertebral column.

Vertebral (VER-te-bral) **canal** A cavity within the vertebral column formed by the vertebral foramina of all the vertebrae and containing the spinal cord. Also called the **spinal canal.**

Vertebral column The 26 vertebrae of an adult and 33 vertebrae of a child; encloses and protects the spinal cord and serves as a point of attachment for the ribs and back muscles. Also called the **backbone, spine,** or **spinal column.**

Vesicle (VES-i-kul) A small bladder or sac containing liquid.

Vesicouterine (ves′-ik-ō-Ū-ter-in) **pouch** A shallow pouch formed by the reflection of the peritoneum from the anterior surface of the uterus, at the junction of the cervix and the body, to the posterior surface of the urinary bladder.

Vestibular (ves-TIB-ū-lar) **apparatus** Collective term for the organs of equilibrium, which include the saccule, utricle, and semicircular ducts.

Vestibular membrane The membrane that separates the cochlear duct from the scala vestibuli.

Vestibule (VES-ti-būl) A small space or cavity at the beginning of a canal, especially the inner ear, larynx, mouth, nose, and vagina.

Villus (VIL-lus) A projection of the intestinal mucosal cells containing connective tissue, blood vessels, and a lymphatic vessel; functions in the absorption of the end products of digestion. *Plural is* **villi** (VIL-ī).

Viscera (VIS-er-a) The organs inside the ventral body cavity. *Singular is* **viscus** (VIS-kus).

Visceral (VIS-er-al) Pertaining to the organs or to the covering of an organ.

Visceral effectors (e-FEK-torz) Organs of the ventral body cavity that respond to neural stimulation, including cardiac muscle, smooth muscle, and glands.

Vision The act of seeing.

Vitamin An organic molecule necessary in trace amounts that acts as a catalyst in normal metabolic processes in the body.

Vitreous (VIT-rē-us) **body** A soft, jellylike substance that fills the vitreous chamber of the eyeball, lying between the lens and the retina.

Vocal folds Pair of mucous membrane folds below the ventricular folds that function in voice production. Also called **true vocal cords.**

Volkmann's canal *See* **Perforating canal.**

Vulva (VUL-va) Collective designation for the external genitalia of the female. Also called the **pudendum** (poo-DEN-dum).

W

Wallerian (wal-LE-rē-an) **degeneration** Degeneration of the portion of the axon and myelin sheath of a neuron distal to the site of injury.

Wandering macrophage (MAK-rō-fāj) Phagocytic cell that develops from a monocyte, leaves the blood, and migrates to infected tissues.

White matter Aggregations or bundles of myelinated and unmyelinated axons located in the brain and spinal cord.

White pulp The regions of the spleen composed of lymphatic tissue, mostly B lymphocytes.

White ramus communicans (RĀ-mus kō-MŪ-ni-kans) The portion of a preganglionic sympathetic axon that branches from the anterior ramus of a spinal nerve to enter the nearest sympathetic trunk ganglion.

Wormian bone *See* **Sutural bone.**

X

Xiphoid (ZĪ-foyd) Sword-shaped.

Xiphoid (ZĪ-foyd) **process** The inferior portion of the sternum **(breastbone).**

Y

Yolk sac An extraembryonic membrane composed of the exocoelomic membrane and hypoblast. It transfers nutrients to the embryo, is a source of blood cells, contains primordial germ cells that migrate into the gonads to form primitive germ cells, forms part of the gut, and helps prevent desiccation of the embryo.

Z

Zona fasciculata (ZŌ-na fa-sik′-ū-LA-ta) The middle zone of the adrenal cortex consisting of cells arranged in long, straight cords that secrete glucocorticoid hormones, mainly cortisol.

Zona glomerulosa (glo-mer′-ū-LŌ-sa) The outer zone of the adrenal cortex, directly under the connective tissue covering, consisting of cells arranged in arched loops or round balls that secrete mineralocorticoid hormones, mainly aldosterone.

Zona pellucida (pe-LOO-si-da) Clear glycoprotein layer between a secondary oocyte and the surrounding granulosa cells of the corona radiata.

Zona reticularis (ret-ik′-ū-LAR-is) The inner zone of the adrenal cortex, consisting of cords of branching cells that secrete sex hormones, chiefly androgens.

Zygote (ZĪ-gōt) The single cell resulting from the union of male and female gametes; the fertilized ovum.

CREDITS

ART CREDITS

Chapter 1 Figure 1.1, 1.9–1.11: Kevin Somerville/Imagineering. 1.2, 1.3, 1.7: Imagineering. 1.4: Molly Borman. 1.5: Kevin Somerville. 1.6, 1.8, Table 1.1: DNA Illustrations.

Chapter 2 Figure 2.1–2.16, Table 2.1: Imagineering.

Chapter 3 Figure 3.1, 3.2, 3.12, 3.14–3.17: Tomo Narashima. 3.4–3.11, 3.13, 3.18–3.22: Imagineering.

Chapter 4 Figure 4.1–4.4: Kevin Somerville/Imagineering. Tables 4.1–4.5: Imagineering.

Chapter 5 Figure 5.1, 5.3, 5.4: Kevin Somerville. 5.2, 5.6, 5.7: Imagineering.

Chapter 6 Figure 6.1, 6.4, 6.6–6.10, 6.13–6.27: John Gibb. 6.2a, 6.5: Imagineering. 6.2b, 6.3: Kevin Somerville. 6.11, 6.12: John Gibb/Imagineering.

Chapter 7 Figure 7.1, 7.2, 7.10: John Gibb. 7.11, 7.12: John Gibb/Imagineering.

Chapter 8 Figure 8.1, 8.2a: Kevin Somerville. 8.2b, 8.3, 8.5–8.11, Table 8.1: Imagineering. 8.4: Kevin Somerville/Imagineering. 8.12–8.24: John Gibb.

Chapter 9 Figure 9.1a, Table 9.1: Kevin Somerville/Imagineering. 9.2: Kevin Somerville. 9.3–9.7: Imagineering.

Chapter 10 Figure 10.1, 10.2, 10.4, 10.6–10.9, 10.11–10.13: Kevin Somerville. 10.3, 10.10: Kevin Somerville/Imagineering. 10.5: Leonard Dank/Imagineering. 10.14, 10.15: Imagineering.

Chapter 11 Table 11.1: Kevin Somerville/Imagineering. 11.2, 11.3: Imagineering.

Chapter 12 Figure 12.1: Kevin Somerville. 12.2: Imagineering. 12.3, 12.6, 12.12–12.14: Tomo Narashima. 12.4: Molly Borman. 12.5: Sharon Ellis. 12.7, 12.8, 12.9–12.11, Tables 12.2 and 12.3: Imagineering. 12.15, 12.16: Tomo Narashima/Sharon Ellis.

Chapter 13 Figure 13.1: Kevin Somerville. 13.2, 13.3, 13.6, 13.8, 13.10, 13.12, 13.14: Imagineering. 13.4, 13.5, 13.7, 13.9, 13.13: Lynn O'Kelley. 13.11: Lynn O'Kelley/Imagineering.

Chapter 14 Figure 14.1–14.6, Table 14.2: Imagineering.

Chapter 15 Figure 15.1–15.4, 15.6: Kevin Somerville. 15.5: Kevin Somerville/Imagineering. 15.7–15.9: Imagineering.

Chapter 16 Figure 16.1, 16.4, 16.8, 16.9, 16.12, 16.17: Kevin Somerville. 16.2, 16.3, 16.5, 16.7: Imagineering. 16.16: Kevin Somerville/Imagineering. 16.10, 16.11, 16.13–16.15: John Gibb. 16.16: Kevin Somerville/Imagineering.

Chapter 17 Figure 17.1, 17.4: Kevin Somerville/Imagineering. 17.2, 17.3, 17.5–17.13, Table 17.2: Imagineering.

Chapter 18 Figure 18.1, 18.5, 18.6: Kevin Somerville. 18.2–18.4: Molly Borman. 18.7: John Gibb. 18.8–18.13: Imagineering.

Chapter 19 Figure 19.1, 19.2, 19.8, 19.10–19.12, 19.15: Kevin Somerville. 19.3, 19.13, 19.14: Imagineering. 19.4–19.6: Nadine Sokol. 19.7, 19.9: Steve Oh.

Chapter 20 Figure 20.2–20.6: Imagineering.

Chapter 21 Figure 21.1, 21.6: Kevin Somerville. 21.2, 21.3, 21.9: Steve Oh/Imagineering. 21.4, 21.7, 21.8: Imagineering.

Chapter 22 Figure 22.1–22.6: Imagineering.

Chapter 23 Figure 23.1, 23.4, 23.6, 23.7, 23.9– 23.11: Kevin Somerville. 23.2: Kevin Somerville/Imagineering. 23.3, 23.5, 23.8, 23.12, 23.13: Imagineering.

Chapter 24 Figure 24.1–24.8, Table 24.1: Kevin Somerville. 24.10–24.12: Imagineering.

Orientation Diagrams and Focus on Homeostasis icons: Imagineering

PHOTO CREDITS

Chapter 1 Focus on Wellness: Masterfile. Figure 1.1 (top): Kenneth Eward/BioGrafx/Photo Researchers, Inc. Figure 1.1 (center): Rubberball Productions/Getty Images. Figure 1.7a, 1.7b, 1.7c: Dissection Shawn Miller; Photograph Mark Nielsen.

Chapter 2 Focus on Wellness: Don Farrall/Getty Images, Inc.

Chapter 3 Focus on Wellness: Masterfile. Figure 3.3: Andy Washnik. Figure 3.8b: David Phillips/Photo Researchers, Inc. Figure 3.21: Courtesy Michael Ross, University of Florida.

Chapter 4 Focus on Wellness: SPL/Photo Researchers, Inc. Table 4.1a, 4.1b, 4.1c, 4.1d, 4.1e, 4.1f, 4.1g, 4.1h, 4.1i, 4.2a, 4.2b, 4.3a, 4.3b, 4.3c, 4.4a, 4.4b, 4.4c, 4.5a, 4.5b, 4.5c: Mark Nielsen.

Chapter 5 Focus on Wellness: Masterfile. Figure 5.5a: Alain Dex/Photo Researchers, Inc. Figure 5.5b: Biophoto Associates/Photo Researchers, Inc. Figure 5.6a: Sheila Terry/Science Photo Library/Photo Researchers, Inc. Figure 5.6b, 5.6c: St. Stephen's Hospital/Science Photo Library/Photo Researchers, Inc.

Chapter 6 Focus on Wellness: G Flume/Getty Images, Inc. Figure 6.1: Mark Nielsen. Figure 6.2b: John Burbidge/Photo Researchers, Inc. Figure 6.4: Scott Camazine/Photo Researchers, Inc. Figure 6.28a, 6.28b: P. Motta/Photo Researchers, Inc Page 155: ISM/Phototake.

Chapter 7 Focus on Wellness: AFP/Stringer/Getty Images, Inc. Figure 7.1c: Dissection Shawn Miller; Photograph Mark Nielsen. Figure 7.3b, 7.4: Mark Nielsen. Figure 7.5, 7.6, 7.7, 7.8, 7.9: John Wilson White.

Chapter 8 Focus on Wellness: Moritz Steiger/Riser/Getty Images, Inc. Figure 8.2: Courtesy Michael Ross, University of Florida. Figure 8.11: Mark Nielsen.

Chapter 9 Focus on Wellness: Nicole S. Young/Getty Images, Inc. Figure 9.2b: Mark Nielsen.

Chapter 10 Focus on Wellness: Masterfile. Figure 10.6b: Dissection Shawn Miller, Photograph Mark Nielsen. Table 10.1: Dissection Shawn Miller, Photograph Mark Nielsen.

Chapter 11 Focus on Wellness: Masterfile.

Chapter 12 Focus on Wellness: Cultura/Masterfile. Figure 12.6: Geirge Diebold/Getty Images, Inc.

Chapter 13 Focus on Wellness: Bill Freeman/PhotoEdit. Figure 13.4: Dissection Shawn Miller, Photograph Mark Nielsen. Figure 13.7b: Mark Nielsen. Figure 13.7c: Dissection Shawn Miller, Photograph Mark Nielsen. Figure 13.11c: Courtesy Michael Ross, University of Florida. Figure 13.13c: Dissection Shawn Miller, Photograph Mark Nielsen. Figure 13.13d: Mark Nielsen. Figure 13.15a: From New England Journal of Medicine, February 18, 1999, vol. 340, No. 7, page 524. Photo provided courtesy of Robert Gagel, Department of Internal Medicine, University of Texas M.D. Anderson Cancer Center, Houston Texas. Reproduced with permission. Figure 13.15b: Lester Bergman/The Bergman Collection. Figure 13.15c: Dr. M.A. Ansary/Photo Researchers, Inc. Figure 13.15d: ISM/Phototake. Figure 13.15e: Biophoto Associates/Photo Researchers, Inc.

Chapter 14 Focus on Wellness: Masterfile. Figure 14.2b (top left): Courtesy Michael Ross, University of Florida. Figure 14.2b (top center): Courtesy Michael Ross, University of Florida. Figure 14.2b (top right): Courtesy Michael Ross, University of Florida. Figure 14.2b (center): Mark Nielsen. Figure 14.2b (bottom left): Courtesy Michael Ross, University of Florida. Figure 14.2b (bottom center): Courtesy Michael Ross, University of Florida. Figure 14.5: Dennis Kunkel Microscopy, Inc. /Phototake. Figure 14.7: Jean Claude Revy/Phototake.

Chapter 15 Focus on Wellness: altrendo images/Stockbyte/Getty Images, Inc. Figure 15.2c: Mark Nielsen. Figure 15.10a: Chuck Brown/Photo Researchers, Inc. Figure 15.10b: Carolina Biological Supply Company/Phototake.

Chapter 16 Focus on Wellness: Dougal Walters/Digital Vision/Getty Images, Inc. Figure 16.2: Courtesy Michael Ross, University of Florida.

Chapter 17 Focus on Wellness: Peter Dazeley/Getty Images, Inc.

Chapter 18 Focus on Wellness: David De Lossy/Getty Images. Figure 18.5: Biophoto Associates/Photo Researchers, Inc.

Chapter 19 Focus on Wellness: Donna Day/Getty Images. Figure 19.1b, 19.7b, 19.11b: Dissection Shawn Miller, Photograph Mark Nielsen.

Chapter 20 Focus on Wellness: JGI/Jamie Grill/Getty Images, Inc.

Chapter 21 Focus on Wellness: Jacqui Hurst/©Corbis. Figure 21.1b: Dissection Shawn Miller, Photograph Mark Nielsen

Chapter 22 Focus on Wellness: PhotoDisc, Inc./Getty Images.

Chapter 23 Focus on Wellness: Thomas Rodriguez/Masterfile. Figure 23.7b: Claude Edelmann/Photo Researchers, Inc.

Chapter 24 Focus on Wellness: Camille Tokerud/Photographer s Choice/Getty Images, Inc. Figure 24.8b: Preparation and photography Mark Nielsen. Figure 24.9a: Photo provided courtesy of Kohei Shiota, Congenital Anomaly Research Center, Kyoto University, Graduate School of Medicine. Figure 24.9b, 24.9c, 24.9d, 24.9e: Courtesy National Museum of Health and Medicine, Armed Forces Institute of Pathology. Figure 24.9f: Photo by Lennart Nilsson/Scanpix. Figure 24.9g, 24.9h: Photo provided courtesy of Kohei Shiota, Congenital Anomaly Research Center, Kyoto University, Graduate School of Medicine.

INDEX

f following a page number refers to an illustration or photo; *t* indicates a table; and *e* indicates an exhibit.

COMBINING FORMS, WORD ROOTS, PREFIXES, AND SUFFIXES

Many of the terms used in anatomy and phsiology are compound words; that is, they are made up of word roots and one or more prefixes or suffixes. For example, *leukocyte* is formed from the word roots *leuk-* meaning "white", a connecting vowel (o), and *cyte* meaning "cell." Thus, a leukocyte is a white blood cell. The following list includes some of the most commonly used combining forms, word roots, prefixes, ad suffixes used in the study of anatomy and physiology. Each entry includes a usage example. Learning the meanings of these fundamental word parts will help you remember terms that, at first glance, may seem long or complicated.

COMBINING FORMS AND WORD ROOTS

Acous-, Acu- hearing Acoustics.
Acr- extremity Acromegaly.
Aden- gland Adenoma.
Alg-, Algia- pain Neuralgia.
Angi- vessel Angiocardiography.
Anthr- joint Arthropathy.
Aut-, Auto- self Autolysis.
Audit- hearing Auditory canal.

Bio- life, living Biopsy.
Blast- germ, bud Blastula.
Blephar- eyelid Blepharitis.
Brachi- arm Brachial plexus.
Bronch- trachea, windpipe Bronchoscopy.
Bucc- cheek Buccal.

Capit- head Decapitate.
Carcin- cancer Carcinogenic.
Cardi-, Cardia-, Cardio- heart Cardiogram.
Cephal- head Hydrocephalus.
Cerebro- brain Cerebrospinal fluid.
Chole- bile, gall Cholecystogram.
Chondr-, cartilage Chondrocyte.
Cor-, Coron- heart Coronary.
Cost- rib Costal.
Crani- skull Craniotomy.
Cut- skin Subcutaneous.
Cyst- sac, bladder Cystoscope.

Derma-, Dermato- skin Dermatosis.
Dura- hard Dura mater.

Enter- intestine Enteritis.
Erythr- red Erythrocyte.

Gastr- stomach Gastrointestinal.
Gloss- tongue Hypoglossal.
Glyco- sugar Glycogen.
Gyn-, Gynec- female, woman Gynecology.

Hem-, Hemat- blood Hematoma.
Hepar-, Hepat- liver Hepatitis.
Hist-, Histio- tissue Histology.
Hydr- water Dehydration.
Hyster- uterus Hysterectomy.

Ischi- hip, hip joint Ischium.

Kines- motion Kinesiology.

Labi- lip Labial.
Lacri- tears Lacrimal glands.
Laparo- loin, flank, abdomen Laparoscopy.
Leuko- white Leukocyte.
Lingu- tongue Sublingual glands.
Lip- fat Lipid.
Lumb- lower back, loin Lumbar.

Macul- spot, blotch Macula.
Malign- bad, harmful Malignant.
Mamm-, Mast- breast Mammography, Mastitis.
Meningo- membrane Meningitis.
Myel- marrow, spinal cord Myeloblast.
My-, Myo- muscle Myocardium.

Necro- corpse, dead Necrosis.
Nephro- kidney Nephron.
Neuro- nerve Neurotransmitter.

Ocul- eye Binocular.
Odont- tooth Orthodontic.
Onco- mass, tumor Oncology.
Oo- egg Oocyte.
Opthalm- eye Ophthalmology.
Or- mouth Oral.
Osm- odor, sense of small Anosmia.
Os-, Osseo-, Osteo- bone Osteocyte.
Ot- ear Otitus media.

Palpebr- eyelid Palpebra.
Patho- disease Pathogen.
Pelv- basin Renal pelvis.
Phag- to eat Phagocytosis.
Phleb- vein Phlebitis.
Phren- diaphragm Phrenic.
Pilo- hair Depilatory.
Pneumo- lung, air Pneumothorax.
Pod- foot Podocyte.
Procto- anus, rectum Proctology.
Pulmon- lung Pulmonary.

Ren- kidneys Renal artery.
Rhin- nose Rhinitis.

Scler-, Sclero- hard Atherosclerosis.
Sep-, Spetic- toxic condition due to micoorganisms Septicemia.
Soma-, Somato- body Somatotropin.
Sten- narrow Stenosis.
Stasis-, Stat- stand still Homeostasis.

Tegument- skin, covering Integumentary.
Therm- heat Thermogenesis.
Thromb- clot, lump Thrombus.

Vas- vessel, duct Vasoconstriction.

Zyg- joined Zygote.

PREFIXES

A-, An- without, lack of, deficient Anesthesia.
Ab- away from, from Abnormal.
Ad-, Af- to, toward Adduction, Afferent neuron.
Alb- white Albino.
Alveol- cavity, socket Alveolus.
Andro- male, masculine Androgen.
Ante- before Antebrachial vein.
Anti- against Anticoagulant.

Bas- base, foundation Basal ganglia.
Bi- two, double Biceps.
Brady- slow Bradycardia.

Cata- down, lower, under Catabolism.
Circum- around Circumduction.
Cirrh- yellow Cirrhosis of the liver.
Co-, Con-, Com with, together Congenital.
Contra- against, opposite Contraception.
Crypt- hidden, concealed Cryptorchidism.
Cyano- blue Cyanosis.

De- down, from Deciduous.
Demi-, hemi- half Hemiplegia.
Di-, Diplo- two Diploid.
Dis- separation, apart, away from Dissection.
Dys- painful, difficult Dyspnea.

E-, Ec-, Ef- out from, out of Efferent neuron.
Ecto-, Exo- outside Ectopic pregnancy.
Em-, En- in, on Emmetropia.
End-, Endo- within, inside Endocardium.
Epi- upon, on, above Epidermis.
Eu- good, easy, normal Eupnea.
Ex-, Exo- outside, beyond Exocrine gland.
Extra- outside, beyond, in addition to Extracellular fluid.

Fore- before, in front of Forehead.

Gen- originate, produce, form Genitalia.
Gingiv- gum Gingivitis.

Hemi- half Hemiplegia.
Heter-, Hetero- other, different Heterozygous.
Homeo-, Homo- unchanging, the same, steady Homeostasis.
Hyper- over, above, excessive Hyperglycemia.
Hypo- under, beneath, deficient Hypothalamus.

In-, Im- in, inside, not Incontinent.
Infra- beneath Infraorbital.
Inter- among, between Intercostal.
Intra- within, inside Intracellular fluid.
Ipsi- same Ipsilateral.
Iso- equal, like Isotonic.

Juxta- near to Juxtaglomerular apparatus.

Later- side Lateral.

Macro- large, great Macrophage.
Mal- bad, abnormal Malnutrition.
Medi-, Meso- middle Medial.
Mega-, Megalo- great, large Magakaryocyte.
Melan- black Melanin.
Meta- after, beyond Metacarpus.
Micro- small Microfilament.
Mono- one Monounsaturated fat.

Neo- new Neonatal.

Oligo- small, few Oliguria.
Ortho- straight, normal Orthopedics.

Para- near, beyond, beside Paranasal sinus.
Peri- around Pericardium.
Poly- much, many, too much Polycythemia.
Post- after, beyond Postnatal.
Pre-, Pro- before, in front of Presynaptic.
Pseudo- false Pseudostratified.

Retro- backward, behind Retroperitoneal.

Semi- half Semicircular canals.
Sub- under, beneath, below Submucosa.
Super- above, beyond Superficial.
Supra- above, over Suprarenal.
Sym-, Syn- with, together Symphysis.

Tachy- rapid Tachycardia.
Trans- across, through, beyond Transudation.
Tri- three Trigone.

SUFFIXES

-able capable of, having ability to Viable.
-ac, -al pertaining to Cardiac.
-algia painful condition Myalgia.
-an, -ian pertaining to Circadian.
-ant having the characteristic of Malignant.
-ary connected with Ciliary.
-asis, -asia, -esis, -osis condition or state of Hemostasis.

-asthenia weakness Myasthenia.
-ation process, action, condition Inhalation.
-centesis puncture, usually for drainage Amniocentesis.
-cid, -cide, -cis, cut, kill destroy Spermicide.
-ectomize, -ectomy excision of, removal of Thyroidectomy.
-emia condition of blood Anemia.
-esthesia sensation, feeling Anesthesia.

-fer carry Efferent arteriole.

-gen agent that produces or originates Pathogen.
-genic producing Pyogenic.
-gram record Electrocardiogram.
-graph instrument for recording Electroencephalograph.

-ia state, condition Hypermetropia.
-ician person associated with Pediatrician.
-ics art of, science of Optics.
-ism condition, state Rheumatism.
-itis inflammation Neuritis.

-logy the study or science of Physiology.
-lysis dissolution, loosening, destruction Hemolysis.

-malacia softening Osteomalacia.
-megaly enlarged Cardiomegaly.
-mers, -meres parts Polymers.

-oma tumor Fibroma.
-osis condition, disease Necrosis.
-ostomy create an opening Colostomy.
-otomy surgical incision Tracheotomy.

-pathy disease Myopathy.
-penia deficiency Thrombocytopenia
-philic to like, have an affinity for Hydrophilic.
-phobe, -phobia fear of, aversion to Photophobia.
-plasia, -plasty forming, molding Rhinoplasty.
-pnea breath Apnea.
-poiesis making Hemopoiesis.
-ptosis falling, sagging Blepharoptosis.

-rrhage bursting forth, abnormal discharge Hemorrhage.
-rrhea flow, discharge Diarrhea.

-scope instrument for viewing Bronchoscope.
-stomy creation of a mouth or artificial opening Tracheostomy.

-tomy cutting into, incision into Laparotomy.
-tripsy crushing Lithotripsy.
-trophy relating to nutrition or growth Atrophy.

-uria urine Polyuria.